CHEMISTRY and PHARMACOLOGY of NATURALLY OCCURRING BIOACTIVE COMPOUNDS

CHEMISTRY and PHARMACOLOGY of NATURALLY OCCURRING BIOACTIVE COMPOUNDS

Edited by
Goutam Brahmachari

CRC Press
Taylor & Francis Group
Boca Raton London New York

CRC Press is an imprint of the
Taylor & Francis Group, an informa business

CRC Press
Taylor & Francis Group
6000 Broken Sound Parkway NW, Suite 300
Boca Raton, FL 33487-2742

Version Date: 20121203

International Standard Book Number-13: 978-1-4398-9168-1 (Hardback)

Visit the Taylor & Francis Web site at
http://www.taylorandfrancis.com

and the CRC Press Web site at
http://www.crcpress.com

Dedication

To all those who are working globally with bioactive natural products for the cause of human welfare

Contents

Forewords

Natural Products as Drugs—The Road Ahead

For millennia, natural products, mostly plants, were the only source of drugs. Some of the original knowledge has undoubtedly been lost. We know very little about the ancient use of plants in North and South America, most of Europe, Africa, and Australia. But some of the advanced ancient civilizations have left us amazing pharmacopeias. Many of the plants described in them have yet to be explored. The Assyrians (about the second millennium BC to the sixth century BC) have left us hundreds of tablets and plant lists, which were used by Campbell Thomson (1949) to compile a detailed Assyrian herbal. The Chinese classic medical pharmacopeia Ben Ts'ao, originally compiled around the first century AD and finalized in the sixteenth century AD, details thousands of prescriptions (Needham, 1978). The plant drugs used in Ancient Egypt (summarized by Von Deines and Grapow, 1959) are mentioned in many of the papyri. The outstanding herbal by Dioscorides (died circa AD 90) was one of the most influential drug books over 18 centuries. It summarizes in detail the existing knowledge on plant drugs in the Middle East and was widely copied by medieval herbalists (Dioscorides, 1934). Apparently, the well-established herbal use in India was first compiled and published by Da Orta (1563)—a Portuguese-Jewish physician who was on the run from the Inquisition. One can find in these herbals most of the medicinal plants that have been evaluated over the last century and from which many of the modern drugs have been developed—be they either the molecules as found in the plant or their synthetic derivatives. And we continue to find new natural medicinal agents in them. Artemisinin, the very valuable antimalarial drug from a Chinese plant, is just one of them.

The development of methods for the isolation and structure elucidation of natural products during the late nineteenth and over the twentieth century slowly decreased traditional herbal use. Many naturally occurring bioactive compounds and/or their derivatives have become drugs of central importance and represent a high percentage of the drugs used today. Antibiotics, hormones, and statins are well-known examples. Today, treatments that involve the use of plants or plant extracts are part of "alternative medicine." However, still a large part of the population in third world countries relies on herbal products and, surprisingly, in the rest of the world their use is coming back.

Several decades ago, an isolated natural product with proven therapeutic activity could be introduced in the clinic shortly after its isolation. Thus, in the 1920s insulin became a drug within months after its identification. Various naturally occurring steroids became drugs within a few years after their discovery in the 1930s and 1940s. The picture has changed. For example, anandamide, a lipid brain constituent, with a wide spectrum of therapeutic properties as shown by animal trials, has never been administered to a human being nearly 20 years after its discovery (Devane et al., 1992). Due to lack of patent protection of most identified molecules of natural origin, few pharmaceutical companies can afford to spend hundreds of millions on the toxicology and clinical evaluations needed to get approval. Many useful natural products are lost for medicine.

Treatments with herbal drugs have positive as well as negative aspects. The positive side is associated with the accumulated knowledge over centuries, or even millennia, of their therapeutic action. Our knowledge on the activities of these drugs comes from their human use in diseases, many of which do not have good animal models, and hence modern drugs are difficult to develop. In some plants, or mixtures of plants, the therapeutic effect may be due to synergism or an "entourage effect" of several constituents. Again, such plants are mostly lost to medicine. The use of herbal products as drugs also has its negative aspects. In most countries, the content of the active constituents of plant drugs sold to the public is unknown. The secondary metabolites, which are the active constituents of plant drugs, are notorious in their variability, which depends on the soil, the climate, the plant pathogens, etc. Hence, a patient can never be sure of the level of the active drug consumed.

Few herbal products have been investigated for their chronic effects. While some medicinal plants or extracts are used on an acute basis and hence their long-term effects are presumably mostly negligible, many are consumed on a daily basis for long periods of time and may lead to various pathologies. This is not the case with most drugs, of either natural or synthetic origin, introduced over the last few decades, whose effects are followed for years after market introduction by the companies that have developed them.

Some of the "modern" trends of herbal plant use are not based on traditional use but on premises, which are unacceptable. Thus, one of the basic rules of homeopathy, a nineteenth-century-invented method of medicinal plant use, is that dilution of a drug solution leads to better results. When extreme dilution actually leads to the disappearance of the active molecule in the formulation, the answer by homeopathic dispensers is that "water retains drug memory"—a statement of questionable value.

In the United States, the Food and Drug Administration (FDA) has to approve all new drugs sold to the public. This approval is based on thorough evaluations of pharmacological and clinical data. An FDA approval is generally accepted in most countries. However, U.S. law does not allow the FDA to use its criteria for "alternative medicine"—hence in the United States, as well as in many other countries, regulations as regards herbal drugs are lax.

How do we balance the positive with the negative aspects? There are some obvious steps that have to be taken to close the gap between plant drugs and extracts and approved drugs based on detailed pharmacological and clinical investigations:

1. Every batch of a herbal drug on the market should indicate the concentrations of the active molecule and any possible side effects.
2. Drug regulatory bodies should be given wider responsibility over herbal plant drugs.
3. Therapeutic claims should be based on scientific data.

But the best way to apply the knowledge gathered over millennia is to evaluate and develop it by modern methods—chemical, pharmacological, and clinical—and thus help to introduce novel, single molecule drugs based both on ancient practice and on contemporary science.

This volume, edited by Dr. Brahmachari, presents not only an extensive spectrum of chemical and pharmacological aspects on bioactive natural products and their derivatives, written by leading researchers in the field, but also a significant amount of what might best be covered under the rubric of alternative medicine or even functional foodstuffs. The book contains excellent chapters on the chemistry and the methodologies of production of natural products that could be drug molecules in their own right. Besides, there are chapters by well-known experts on not only classical fermentation processes using microorganisms but also on methodologies involved in plant tissue culture and biosynthetic pathway modulations directed toward the enhancement of natural product production. Natural products clearly are an important source for future therapeutic options for an array of acute and chronic diseases, but more research is needed on efficacy, standardization, toxicity, and long-term effects. This book, by providing state-of-the-art evidence on bioactive natural products, will obviously help to fill this gap. I am very happy to recommend this volume to academics, researchers, and students interested in understanding the chemistry and pharmacology of bioactive natural products and their enormous role in the modern drug discovery processes.

Raphael Mechoulam
Hebrew University, Jerusalem, Israel

REFERENCES

Campbell Thomson, R., *A Dictionary of Assyrian Botany*, The British Academy, London, 1949.
Da Orta, G., *Coloquios dos Simples e Drogas e Cousas Medicinais da India*, Goa, India, 1563. Reproduced in 1872 by The Academia das Cientias de Lisboa, Portugal.
Devane, W. et al., Isolation and structure of a brain constituent that binds to the cannabinoid receptor. *Science* 258, 1946–1949 (1992).
Dioscorides, *The Greek Herbal*, Gunther, R.T. Ed., Oxford University Press, Oxford, 1934.
Needham, J., *The Shorter Science and Civilization in China*, vol. 1, Cambridge University Press, 1978
Von Deines, H. and Grapow, H., *Grundriss der Medicine der Alten Agypter, Vl. Worterbuch der Agyptischen Drogennamen*, Academie Verlage, Berlin, Germany,1959.

Mankind has been utilizing natural products, mostly from plants, for ages as traditional medicines, fragrances, spices, and colorants. Naturally occurring bioactive compounds are still the major source of lead molecules in modern drug discovery and development. In the last decade, natural products have gained a renewed interest in the pharmaceutical industry, and as a consequence a new drive toward the study of bioactive compounds, both directly derived from natural sources or derived/inspired from those obtained from natural products, has been initiated worldwide. It is also interesting to recall that human bodies and health are composed of natural products and supported by everyday intake of natural products, in spite of remarkable renovations occurring in the individual natural medicinal products and the ways of utilizing them.

The content of this book, well organized by Dr. Brahmachari, comprises wide areas in the natural resources, their production, bioactive compounds produced by them, and new facts that could lead to novel methods of creating new medicines. Many readers will find various buried jewels based on their own specialties. Summarized contents of all chapters are very precisely documented in the overview in Chapter 1, where the readers will easily find how to reach the chapters of most to interest them. I am very happy to recommend this book without any reservation to graduate and PhD students of medicinal chemistry as well as to scientists and professionals working in the domain of bioactive natural products to explore their potential as prospective molecules of medicinal interests.

Takuo Okuda
Okayama University, Japan

Preface

This single volume entitled *Chemistry and Pharmacology of Naturally Occurring Bioactive Compounds* is an endeavor to focus on the recent cutting-edge research advances in the field of bioactive natural products and their significant contributions in the domain of discovery and development of new medicinal agents. This book consists of a total of 22 chapters contributed by eminent researchers from several countries in response to my personal invitation. I am most grateful to the contributors for their generous and timely response in spite of their busy and tight schedules with academics, research, and other responsibilities.

Natural products have played a crucial role in modern drug development and still constitute a prolific source of novel lead compounds, or pharmacophores, for ongoing drug discovery programs. Natural products present in the plant and animal kingdom offer a huge diversity of chemical structures, which are the result of biosynthetic processes that have been modulated over the millennia through genetic effects, and hence the search for bioactive molecules from nature (plants, animals, microflora) continues to play an important role in fashioning new medicinal entities. With the advent of modern techniques, particularly the rapid improvements in spectroscopic as well as accompanying advances in high-throughput screening techniques, it has become possible to have an enormous repository of bioactive natural compounds, thus opening up exciting new opportunities in the field of new drug development to the pharmaceutical industry. Medicinal chemistry of such bioactive compounds encompasses a vast area that includes their isolation and characterization from natural sources, structure modification for optimization of their activity and other physical properties, and also total and semisynthesis for a thorough scrutiny of structure–activity relationships. It has been well documented that natural products played a crucial role in modern drug development, especially for antibacterial and antitumor agents; however, their uses in the treatment of other epidemics such as AIDS and cardiovascular, cancerous, neurodegradative, infective, and metabolic diseases have also been extensively explored. The need for leads to solve such health problems threatening the world population makes all natural sources important for the search of novel molecules, diversified and unique structural architectures of which inspired scientists to pursue new chemical entities with completely different structures from known drugs. Researchers around the globe are deeply engaged in exploring the detailed chemistry and pharmacology of such potent and efficacious naturally occurring bioactive compounds.

This book, which comprises a variety of 22 chapters written by active researchers and leading experts working in the field of chemistry of biologically active natural products, brings together an overview of current discoveries and trends in this remarkable field. Chapter 1 presents an overview of the book and summarizes the contents of the other chapters so as to offer glimpses of the subject matter covered to the readers before they go in for a detailed study. Chapters 2 through 22 are devoted to exploring the ongoing chemical and pharmacological advances in naturally occurring organic compounds and describe their spectral and x-ray properties, chemical transformations, and structure–activity relationships, including mode of action, toxicology, pharmacokinetics, and metabolism of certain bioactive molecules of medicinal interest.

This timely volume encourages interdisciplinary work among chemists, pharmacologists, biologists, botanists, and agronomists with an interest in bioactive natural products. It is also an outstanding source of information with regard to the industrial application of natural products for medicinal purposes. The broad interdisciplinary approach dealt with in this book would surely make the work much more interesting for scientists deeply engaged in the research and/or use of bioactive natural products. It will serve not only as a valuable resource for researchers in their own fields to predict promising leads for developing pharmaceuticals to treat various ailments and disease manifestations but also to motivate young scientists to the dynamic field of bioactive natural products research.

Representation of facts and their discussions in each chapter are exhaustive, authoritative, and deeply informative; hence, the book would serve as a key reference for recent developments in the frontier research on bioactive natural products and would also be of much utility to scientists working in this area. I would like to express my sincere thanks once again to all the contributors for the excellent reviews on the chemistry and pharmacology of bioactive natural products. It is their participation that makes my effort to organize such a book possible. Their masterly accounts will surely provide the readers with a strong awareness of current cutting-edge research approaches being followed in some of the promising fields of biologically active natural products.

I would like to express my sincere thanks and deep sense of gratitude to Professor Raphael Mechoulam, Institute for Drug Research, Medical Faculty, Hebrew University, Israel, and Professor Takuo Okuda, Okayama University, Japan, for their keen interests in the manuscript and for writing forewords to the book.

I would also like to express my deep sense of appreciation to all of the editorial and publishing staff members associated with Taylor & Francis Group/CRC Press, United States, for their keen interest in publishing the works as well as their all-round help so as to ensure that the highest standards of publication have been maintained in bringing out this book.

Goutam Brahmachari
Chemistry Department, Visva-Bharati University
Santiniketan, India

Editor

Dr. Goutam Brahmachari received his high school degree in scientific studies in 1986 at Barala R. D. Sen High School under the West Bengal Council of Higher Secondary Education (WBCHSE). He then moved to Visva-Bharati (a central university founded by Rabindranath Tagore at Santiniketan, West Bengal, India) to study chemistry at the undergraduate level. After graduating from this university in 1990, Dr. Brahmachari completed his master's in 1992 with specialization in organic chemistry and thereafter received his PhD in 1997 in natural products chemistry from the same university. He was appointed as assistant professor of organic chemistry at Visva-Bharati University, Department of Chemistry, in 1998. In 2008, he became an associate professor of organic chemistry in the same faculty. At present, he is responsible for teaching courses in organic chemistry, natural products chemistry, and physical methods in organic chemistry. Several students have received their PhDs under his supervision during this period, and a dozen of research fellows are presently working with him both in the fields of natural products and synthetic organic chemistry. Dr. Brahmachari also serves as a member of the Indian Association for the Cultivation of Science (IACS), Kolkata, and as an editor in chief for *Signpost Open Access Journal of Organic and Biomolecular Chemistry*.

Dr. Brahmachari's research interests include (1) isolation, structural determination, and/or detailed NMR study of new natural products from medicinal plants; (2) evaluation of biological activities and pharmacological potential of such phytochemicals; (3) semisynthetic studies with natural products; and (4) synthetic organic chemistry. With more than 15 years of teaching experience, he has produced so far about 70 publications, including original research papers, review articles, and invited book chapters in edited books in the field of natural products and organic synthesis from internationally reputed presses. Dr. Brahmachari has authored/edited a number of text and reference books that include *Organic Name Reactions: A Unified Approach* (Narosa Publishing House, New Delhi; copublished by Alpha Science International, Oxford, 2006), *Chemistry of Natural Products: Recent Trends and Developments* (Research Signpost, 2006), *Organic Chemistry Through Solved Problems* (Narosa Publishing House, New Delhi; copublished by Alpha Science International, Oxford, 2007), *Natural Products: Chemistry, Biochemistry and Pharmacology* (Alpha Science International, Oxford, 2009), *Handbook of Pharmaceutical Natural Products*—two-volume set (Wiley-VCH Verlag GmbH & Co. KGaA, Weinheim, Germany, 2010), and *Bioactive Natural Products: Opportunities and Challenges in Medicinal Chemistry* (World Scientific Publishing Co. Pte. Ltd., Singapore, 2011). He is regularly consulted as a referee by *RSC Advances*; *Tetrahedron Letters*; *Journal of Molecular Catalysis A: Chemical*; *Spectroscopy Letters*; *Journal of Heterocyclic Chemistry*; *Phosphorus, Sulfur, and Silicon and the Related Elements*; *Archives der Pharmazie*; *Current Organic Chemistry*; *Natural Product Communications*; *Journal of Antimicrobial Chemotherapy*; *Mini Reviews in Medicinal Chemistry*; *Phytochemistry Reviews*; *Pharmaceutical Biology*; *Indian Journal of Chemistry: Sec B*, *Journal of the Indian Chemical Society*; *Annals of the Brazilian Academy of Sciences*; *Journal of Medicinal Plants Research*; *International Journal of Green Pharmacy*; *Journal of Essential Oil Bearing Plants*, and some other international journals and financial commissions.

Dr. Brahmachari enjoys songs of Rabindranath Tagore and finds interests in literature as well!

Contributors

Bimal K. Banik
Department of Chemistry
The University of Texas—Pan American
Edinburg, Texas

Vagner A. Benedito
Plant and Soil Sciences Division
Genetics and Developmental Biology Program
West Virginia University
Morgantown, West Virginia

Ira Bhatnagar
Marine Biochemistry Laboratory
Department of Chemistry
Pukyong National University
Busan, Korea

and

Laboratory of Infectious Diseases
Centre for Cellular and Molecular Biology
Hyderabad, India

Goutam Brahmachari
Laboratory of Natural Products and Organic Synthesis
Department of Chemistry
Visva-Bharati University
West Bengal, India

Philip C. Calder
Faculty of Medicine
Human Development and Health Academic Unit
Southampton General Hospital
University of Southampton
Southampton, United Kingdom

Devdutt Chaturvedi
Laboratory of Medicinal Chemistry
Amity Institute of Pharmacy
Amity University Lucknow Campus
Uttar Pradesh, India

Ana M. Damas
Instituto de Biologia Molecular e Celular
and
Instituto de Ciências Biomédicas Abel Salazar
Universidade do Porto
Porto, Portugal

Danilo Davyt
Facultad de Química
Cátedra de Química Farmacéutica
Universidad de la República
Montevideo, Uruguay

Arnold L. Demain
Research Institute for Scientists Emeriti
Drew University
Madison, New Jersey

Zelalem Y. Desta
Department of Chemistry
University of Botswana
Gaborone, Botswana

Anna Dzierżęga-Lęcznar
Faculty of Pharmacy
Department of Instrumental Analysis
Medical University of Silesia in Katowice
Sosnowiec, Poland

Francesco Epifano
Dipartimento di Farmacia
Università "G. D'Annunzio" of Chieti-Pescara
Chieti Scalo, Italy

Raquel O. Faria
Grupo de Estudos em Bioquímica de Plantas
Departamento de Botânica
Instituto de Ciências Biológicas
Universidade Federal de Minas Gerais
Belo Horizonte, Brazil

Ângelo de Fátima
Grupo de Estudos em Química Orgânica e Biológica
Departamento de Química
Instituto de Ciências Exatas
Universidade Federal de Minas Gerais
Minas Gerais, Brazil

Luís Gales
Instituto de Biologia Molecular e Celular
and
Instituto de Ciências Biomédicas Abel Salazar
Universidade do Porto
Porto, Portugal

Salvatore Genovese
Dipartimento di Farmacia
Università "G. D'Annunzio" of Chieti-Pescara
Chieti Scalo, Italy

Paolo Girardi
Mental Health and Sensory Organs
Department of Neurosciences
Sant'Andrea Hospital
Sapienza University of Rome
Rome, Italy

Francisco V. González
Facultad de Farmacia
Departamento de Biología Vegetal
Universidad de La Laguna
Grupo de Biología Vegetal Aplicada
España, Spain

Maria Y. González-Padrón
Facultad de Farmacia
Departamento de Biología Vegetal
Universidad de La Laguna
Grupo de Biología Vegetal Aplicada
España, Spain

Vincent Gullo
Research Institute for Scientists Emeriti
Drew University
Madison, New Jersey

Vivek K. Gupta
Post-Graduate Department of Physics and Electronics
University of Jammu
Jammu Tawi, India

S.W.A. Himaya
Marine Biochemistry Laboratory
Department of Chemistry
Pukyong National University
Busan, Republic of Korea

Se-Kwon Kim
Marine Biochemistry Laboratory
Department of Chemistry
and
Marine Bioprocess Research Center
Pukyong National University
Busan, Republic of Korea

Vladimir V. Kouznetsov
Laboratorio de Química Orgánica y Biomolecular
Escuela de Química
Universidad Industrial de Santander
Bucaramanga, Colombia

Dhananjay Kumar
Department of Chemistry
Centre of Advanced Study
Banaras Hindu University
Varanasi, India

Slawomir Kurkiewicz
Faculty of Pharmacy
Department of Instrumental Analysis
Medical University of Silesia in Katowice
Sosnowiec, Poland

Divya Kushwaha
Department of Chemistry
Centre of Advanced Study
Banaras Hindu University
Varanasi, India

Debomoy K. Lahiri
Department of Psychiatry
Institute of Psychiatric Research
and
Department of Medical and Molecular Genetics
Indiana University School of Medicine
Indianapolis, Indiana

Juan C. Luis
Facultad de Farmacia
Departamento de Biología Vegetal
Universidad de La Laguna
Grupo de Biología Vegetal Aplicada
España, Spain

Diego R. Merchan Arenas
Laboratorio de Química Orgánica y Biomolecular
Escuela de Química
Universidad Industrial de Santander
Bucaramanga, Colombia

Bhuwan B. Mishra
Department of Chemistry
Centre of Advanced Study
Banaras Hindu University
Varanasi, India

Luzia V. Modolo
Grupo de Estudos em Bioquímica de Plantas
Departamento de Botânica
Instituto de Ciências Biológicas
Universidade Federal de Minas Gerais
Belo Horizonte, Brazil

Dandara R. Muniz
Grupo de Estudos em Bioquímica de Plantas
Departamento de Botânica
Instituto de Ciências Biológicas
Universidade Federal de Minas Gerais
Belo Horizonte, Brazil

Kanakapura K. Namitha
Fruit and Vegetable Technology Department
Central Food Technological Research Institute
Council of Scientific and Industrial Research
Mysore, India

Pradeep S. Negi
Fruit and Vegetable Technology Department
Central Food Technological Research Institute
Council of Scientific and Industrial Research
Mysore, India

Raquel M. Pérez
Facultad de Farmacia
Departamento de Biología Vegetal
Universidad de La Laguna
Grupo de Biología Vegetal Aplicada
España, Spain

Marina P. Polovinka
Laboratory of Natural and Bioactive Compounds
The Novosibirsk Institute of Organic Chemistry SB RAS
Novosibirsk, Russia

Maurizio Pompili
Mental Health and Sensory Organs
Department of Neurosciences
Sant'Andrea Hospital
Sapienza University of Rome
Rome, Italy

Balmiki Ray
Department of Psychiatry
Institute of Psychiatric Research
Indiana University School of Medicine
Indianapolis, Indiana

Fernando A. Rojas Ruiz
Laboratorio de Química Orgánica y Biomolecular
Escuela de Química
Universidad Industrial de Santander
Bucaramanga, Colombia

Nariman F. Salakhutdinov
Laboratory of Natural and Bioactive Compounds
The Novosibirsk Institute of Organic Chemistry SB RAS
Novosibirsk, Russia

Sumit Sarkar
Division of Neurotoxicology
National Center for Toxicological Research/the Food and Drug Administration
Jefferson, Arkansas

Larry Schmued
Division of Neurotoxicology
National Center for Toxicological Research/the Food and Drug Administration
Jefferson, Arkansas

Gianluca Serafini
Mental Health and Sensory Organs
Department of Neurosciences
Sant'Andrea Hospital
"Sapienza" University of Rome
Rome, Italy

Gloria Serra
Facultad de Química
Cátedra de Química Farmacéutica
Universidad de la República
Montevideo, Uruguay

Jay Sharma
Department of Pharmacology
Celprogen Inc.
San Pedro, California

Girija S. Singh
Chemistry Department
University of Botswana
Gaborone, Botswana

Krystyna Stępień
Faculty of Pharmacy
Department of Instrumental Analysis
Medical University of Silesia in Katowice
Sosnowiec, Poland

Irena Tam
Faculty of Pharmacy
Department of Instrumental Analysis
Medical University of Silesia in Katowice
Sosnowiec, Poland

Vinod K. Tiwari
Department of Chemistry
Centre of Advanced Study
Banaras Hindu University
Varanasi, India

Leonor Y. Vargas Méndez
Laboratorio de Química Orgánica y Biomolecular
Escuela de Química
Universidad Industrial de Santander
Bucaramanga, Colombia

Ignacio F. Viera
Facultad de Farmacia
Departamento de Biología Vegetal
Universidad de La Laguna
Grupo de Biología Vegetal Aplicada
España, Spain

1 Chemistry and Pharmacology of Naturally Occurring Bioactive Compounds
An Overview

Goutam Brahmachari

CONTENTS

1.1 INTRODUCTION

This book entitled *Chemistry and Pharmacology of Naturally Occurring Bioactive Compounds* is an endeavor to present cutting-edge research in the chemistry of bioactive natural products and helps the reader understand how natural product research continues to make significant contributions in the discovery and development of new medicinal entities. The reference is meant for phytochemists, synthetic chemists, combinatorial chemists, as well as other practitioners and advanced students in related fields. The book comprising 21 technical chapters highlights chemistry and pharmaceutical potential of natural products in modern drug discovery processes, and covers the synthesis and semisynthesis of potentially bioactive natural products. It also features chemical advances in naturally occurring organic compounds describing their chemical transformations and structure–activity relationships.

This introductory chapter (Chapter 1) presents an overview of the book and summarizes the contents and subject matter of each chapter so as to offer certain glimpses of the coverage of discussion to the readers before they go for detailed study.

1.2 AN OVERVIEW OF THE BOOK

The present book contains a total of 21 technical chapters—Chapters 2 through 22; this section summarizes the contents and subject matter of each of these chapters.

1.2.1 Chapter 2

In Chapter 2, Tiwari and his group describe the impact of solid-supported cyclization–elimination strategies toward the development of natural product inspired molecules in drug discovery research. During the past few years, there has been an increasing demand for natural product inspired drug-like molecules and their libraries in pharmaceutical industries worldwide, and to meet with this ever-increasing demand, combinatorial chemistry has emerged as one of the most potential methodologies developed to replace the conventional sequential approach in constructing new drug candidates within a reasonable time frame, thereby cutting down preclinical development costs significantly leading to a massive change in the fundamental approach to drug discovery program. In this present review, the authors have highlighted the practically high-yielding and high-purity approach for the solid phase combinatorial synthesis of diverse pharmacologically active natural product inspired molecules through cyclo-release strategy which has become a practical method mostly for the preparation of small cyclic molecules using a solid-phase synthesis on polymeric support. The cyclo-release strategy is gaining widespread popularity over the last twenty years because of its useful feature which minimizes chemical and tethering implications by releasing intact, desired target molecules in the final step of a reaction. The simultaneous cyclization and release of a desired product from solid support with adequate purity without adding an extra time-consuming step is the major advantage of this method. It has recently emerged as an efficient strategy to detect and to evaluate affinity between the library products and a target molecule as well. The authors have demonstrated the implication and usefulness of the solid-supported cyclization–elimination strategy by describing synthetic routes of a handful of pharmaceutically promising natural product inspired molecules including epothilone A, fumiquinazoline alkaloid, (*S*)-zearalenone, tetramic acids, 1,4-benzodiazepine-2,5-diones, benzodiazepines, diketopiperazines, 1,3-disubstitued quinazolinediones, 4-hydroxyquinolin-2(1*H*)-ones, 4-hydroxy-3-(2′-pyridyl)coumarin, dihydropyrimidones, heterosteroids, imidazoquinazolinones, and many more. The present discussion in Chapter 2 by Tiwari and his group would thus be much helpful to the readers at large and must boost the ongoing research in this direction.

1.2.2 Chapter 3

Chapter 3 by Banik is devoted to the synthesis and biological activities of novel β-lactam compounds; β-lactam nucleus is necessary for the biological activity of a large number of antibiotics. β-Lactams are four-membered ring compounds, occasionally fused with side chains, unsaturated groups, heteroatoms, and cyclic ring systems; these are widely used as useful therapeutic agents against various ailments for many years. Hence, significant attention has been paid by chemists round the globe to undertake systematic works on novel β-lactam synthesis, based either on new or established methodologies, or on the chemical manipulation of preexisting groups preserving at the four-membered ring. The present chapter deeply overviews some of the most significant contributions on the synthesis of the β-lactam derivatives reported since 2000 and also includes the biological and pharmacological applications of some important 2-azetidinone compounds.

1.2.3 Chapter 4

Singh and Desta have contributed to the applications of isatin (indole-1*H*-2,3-dione) chemistry in organic synthesis and medicinal chemistry in Chapter 4 covering the period 2000–2010. Isatin, a naturally occurring bioactive molecule, bears a five-membered cyclic amide (γ-lactam) ring fused to benzene nucleus. The γ-lactam ring also possesses a ketone functionality, and such structural features of isatin molecule eventually have drawn keen attention of the organic chemists at large as an appealing compound of synthetic interests. The phenyl ring in isatin undergoes electrophilic aromatic substitution mainly at C-5, and the amide nitrogen is highly reactive and can be substituted with alkyl, acyl, and aryl groups. The authors have discussed all these reactions along with reactivity of the keto-carbonyl group with various carbon and nitrogen nucleophiles with ample examples. Applications of isatin as a building block in the synthesis of the various kinds of heterocyclic compounds, including spiro-fused cyclic compounds, quinolines, indoles, oxindoles, and spiro-fused β-lactams, have also been described in the present review. It has been demonstrated that oxidation, reduction, and ring-opening reactions of isatin and its derivatives offer attractive routes to diverse types of heterocyclic compounds as well. During the synthesis of such organic compounds, isatins are used both as electrophilic and nucleophilic components; the authors have summed up numerous reported reactions of these kinds that include *N*-alkylation, *N*-acylation, *N*-arylation, and electrophilic aromatic substitutions, etc. *N*-Alkylation reaction of isatins finds application in attaching isatins to other heterocyclic moieties; using isatin as an electrophilic reagent, reactions of ketone group with carbon, oxygen, and nitrogen-centered nucleophiles are discussed herein. These reactions include the Baylis–Hillman reaction, aldol reaction, Henry reaction, reactions forming isatin oximes, hydrazones, semicarbazones, thiosemicarbazones and imines, and many more. The authors have taken care of all these significant reactions of isatin and its derivatives with a huge number of illustrative schemes (more than 80). This illuminating review on isatin chemistry would obviously enrich the readers with these synthetically and biologically useful isatin molecules, as well as would attract the attention of researchers in the area of organic synthesis and medicinal chemistry.

1.2.4 Chapter 5

In Chapter 5, Chatuvedi has highlighted the role of incorporation of carbamate functionality into the various kinds of structurally diverse, biologically active anticancer natural/semisynthetic/synthetic molecules delineating a comprehensive literature since 2005. Organic carbamates have clearly been demonstrated to be extremely useful and stable compounds, exhibiting unique applications in medicinal chemistry, mainly as drugs and prodrugs. Organic carbamates have frequently been used as synthons for the synthesis of various kinds of structurally diverse synthetic intermediates, which find broad applications in drug discovery synthesis. In recent years, it has been realized by various researchers that the introduction of a carbamate functionality into various biologically active synthetic/natural/semisynthetic molecules significantly increases their biological activities, and many of these derivatives either have been approved as drugs or are in various phases of clinical trials. The author has demonstrated the applicability of introducing the carbamate functionality into ample of biologically relevant natural products such as fumagillin, rhazinilam, geldenamycin, vitamin D_3, podophyllotoxin, butelin, butelinic acid, taxol, staurosporine, mitomycin C, and several synthetically modified carbamates analogs in regard to their comparative anticancer potentials. A comprehensive discussion of this important functional-group class would serve the purpose of a source of valuable information to the organic/medicinal chemists working along the same line of the relevant subject.

1.2.5 Chapter 6

Chapter 6 deals with rational design of several new *N*- or *O*-heterocyclic molecules of biological significance from phenolic constituents of some tropical plants by Kouznetsov and his group. The authors have presented a comprehensive discussion in three distinguished sections such as: (i) propenyl (allyl) C_6-C_3 phenolic compounds as activated substituted alkenes in [4 + 2] and [3 + 2] cycloaddition processes; (ii) formyl substituted C_6-C_1 phenolic compounds (piperonal, (iso)vanillins) as aromatic aldehydes in [4 + 2] cycloaddition process (multicomponent imino Diels–Alder reaction) and multicomponent condensation; (iii) bio-screening and in silico calculated physicochemical properties of the molecules obtained. Each thematic area has been supplemented with a wide range of illustrations. Chemical diversity of functionalities for plant products allows one in generating a novel structural and skeletal diversity, and hence, the present review would help in inspiring readers into further discoveries and innovations in chemical transformations of plant products as inexpensive, available, and renewable reagents.

1.2.6 Chapter 7

Davyt and Serra have reviewed the isolation, synthesis, and biological activity of two promising azole marine products, namely largazole and neopeltolide. The unique structural features of marine chemical entities have already invoked tremendous interests among the chemists and pharmacologists in using them as models or scaffolds for synthesis of promising analogs and lead candidates in drug discovery programs.

1.2.7 Chapter 8

In Chapter 8, Calder has described the impact and functions of omega-3 (ω-3) polyunsaturated fatty acids like eicosapentaenoic acid (EPA) and docosahexaenoic acid (DHA) in humans. Following their incorporation into cells, EPA and DHA can influence the physical nature of cell membranes and membrane protein-mediated responses, lipid mediator generation, cell signaling, and gene expression in many different cell types. As a result of their effects on cell and tissue physiology, very long chain ω-3 fatty acids play a role in achieving optimal health and in protection against disease. Calder has focused on all these areas in the present review, highlighting the importance of these fatty acids in human diet.

1.2.8 Chapter 9

Chapter 9 deals with the structure and biological activity of natural melanin pigments by Stępień and her group. The bioactivity of melanins, the heterogeneous natural polymeric pigments, results from their unique physicochemical properties strongly associated with the pigment structure. In their review, the authors have focused on the methods of structural investigations of the biopolymers, in particular pyrolysis coupled with gas chromatography and mass spectrometry; in addition, a variety of biological activities including photoprotective, antioxidant, radioprotective, and immunomodulatory activities of synthetic, microbial, plant, and human melanins are discussed in detail, emphasizing the possibility of their practical use in fields related to human health. In recent years, the development of nanotechnology is offering new opportunities to use melanin bioactivity; for instance, melanin-coated nanospheres are considered as a novel approach to the protection of bone marrow in cancer radiotherapy. The present discussion by Stępień and her group has incorporated all these significant areas of interest.

1.2.9 Chapter 10

Epifano and Genovese have contributed to the recent acquisitions on naturally occurring oxyprenylated secondary plant metabolites in Chapter 10. Oxyprenylated natural products (isopentenyloxy-, geranyloxy-, and farnesyloxy derivatives) represent a family of secondary plant metabolites exhibiting promising biological activities that include anticancer, anti-inflammatory, antimicrobial, neuroprotective, and antifungal effects. The authors have presented a detailed survey on the oxyprenylated natural compounds reported during the last 5 years, highlighting their phytochemical and pharmacological aspects.

1.2.10 CHAPTER 11

In Chapter 11, Lahiri and his group have described the role of curcumin in ameliorating neuroinflammation and neurodegeneration associated with Alzheimer's disease (AD) which is a progressive neurodegenerative disorder and the fifth leading cause of death in the United States, and the most common cause of adult-onset dementia. AD is a multifactorial disorder with many pathological sequels; the contributors have also offered an insight in the pathology of AD in this chapter. It has been shown that curcumin can emerge as a potential therapeutic/preventive agent in the treatment of several neurodegenerative disorders including AD through suppressing the activation of nuclear factor kappa beta (NFκB), defibrilling Aβ plaques, preserving neurons, upregulating neurotrophic factors, and facilitating neurogenesis. Hence, curcumin and its newer formulations can open a new horizon in the treatment of AD.

1.2.11 CHAPTER 12

Nitric oxide (NO) production inhibitory potential of a wide range of plant secondary metabolites including terpenoids, phenolics, alkaloids, flavonoids, and their glycosides has been thoroughly reviewed by Polovinka and Salakhutdinov in Chapter 12. Nitric oxide, one of the most important biological mediators, participates in many physiological and pathophysiological processes. A discussion on structure–activity relationship (SAR) has also been taken into account by the authors. The present review includes data on NO- production inhibitory plant secondary metabolites covering the literature since 2000, and the resumé hence offers huge information to the readers in this particular area of interest.

1.2.12 CHAPTER 13

Chapter 13 is dedicated to the x-ray structural behavior of some significant bioactive steroids and their chemistry in the crystal packing and related matters by Gupta. Crystal and molecular structure determinations have been reported for single crystals of eleven steroids including one sterol (Z-guggulsterone); five withanolides (withaferin A, withanone, α,7α:24α,25α-diepoxy-5α,12α,dihydroxy-1-oxo-20*S*,22*R*-witha-2-enolide, (20*R*,22*R*)-6α,7α-epoxy-5α,27-dihydroxy-1-oxowitha-2,24-dienolide and 6α,7α-epoxy-5α,17α,27-trihydroxy-1-oxo-22*R*-witha-2,24-dienolide); three pregnanes (16-dehydropregnenolone acetate, 3β-acetoxy-17α-hydroxy-16α-methylallopregnan-20-one and 3β-hydroxy-16α-methylpregn-5-en-20-one); one steroid sapogenin (25*R*-spirost-5-en-3β-acetate); and one androstane (3β-hydroxy-17-oximinoandrost-5-ene monohydrate). A general inference on the systematic crystallographic analysis of such group of compounds has been drawn.

1.2.13 CHAPTER 14

Gales and Damas have presented a detailed discussion on the available crystal structures of biologically potent xanthones and their derivatives in Chapter 13 so as to contribute to a better understanding of their multiple biological activities at the molecular level. A variety of such compounds have been considered for the present discussion that include hydroxylated and methoxylated xanthones, glycosidic xanthones, prenylated and related xanthones, xanthones containing a *bis*-dihydrofuran ring system, halogenated xanthones, xanthones containing a crown ether, and xanthones forming metal complexes. This valuable information would find immense applications in understanding the activity of such group of chemical entities that will eventually offer impetus to design new and more potent xanthone derivatives in the ongoing drug discovery program.

1.2.14 Chapter 15

Chapter 15 by Brahmachari is devoted to the anticancer potential of gambogic acid, a "caged prenylated xanthone"—the major chemical constituent of gamboge, the dried resin of *Garcinia hanburyi* Hook.f (Clusiaceae). The bioactive natural xanthone has been established as a potent anticancer agent, and the outcomes of such investigations are very much encouraging. This chapter overviews not only the exhaustive studies of the anticancer potential of gambogic acid but also offers an insight into the mode of action, semisynthesis and the structure–activity relationship (SAR), pharmacokinetics, toxicology, metabolism, and bioavailability of this drug molecule. The potential role of gambogic acid in exhibiting anticancer efficacies has created a stir among the scientific community at large to undertake extensive research for exploring the possibility of its prospective use as a "lead molecule" in the ongoing drug discovery process against cancerous diseases. Semisynthetic derivatives of the compound also showed better activity profiles and bioavailability in certain situations. It is expected that interest in this molecule grows even more, and further in-depth research on this molecule is strongly recommended to explore its potential in becoming a promising anticancer drug in the near future.

1.2.15 Chapter 16

Serafini and his group have presented a thorough discussion on neuroplasticity as a new approach to the pathophysiology of depression and the role of antidepressants in inhibiting brain stress-induced changes in Chapter 16. The authors have thoroughly reviewed the current literature on the pathophysiological mechanisms underlying depression, stress-related disorders, and antidepressant treatment, and commented that the neuroplasticity hypothesis may explain the therapeutic and prophylactic action of antidepressant drugs and thus represents a new innovative approach to the pathophysiology of depression and stress-related disorders. The concept of neuroplasticity can help a lot in understanding the complex pathophysiology underlying depressive illness as well. The present chapter would surely enlighten the readers at large providing impressive and valuable information in the disease area concerned.

1.2.16 Chapter 17

Chapter 17 deals with the statins—the fermentation products for cholesterol control in humans. The statins occupy quite significant position in the present-day drug market worldwide, and the sales of statins in U.S. pharmaceutical business alone is estimated to be ~$30 billion per annum, thus representing the great significance of these natural product derived compounds. In this chapter, Gullo and Demain have presented an up-to-date overview on the subject delineating their discovery and developments; improvements in microbial production, synthesis, and semisynthesis; and various pharmaceutical applications including their established hypocholesterolemic potential.

1.2.17 Chapter 18

Chapter 18 is aimed to provide an insight into the molecular aspects of fungal bioactive polyketides by Bhatnagar and Kim. Polyketides are produced by a non-ribosomal biosynthesis process involving polyketide synthetases as multi-enzymatic templates. Polyketide synthetases possess multiple modules that bind, activate, and condense each specific amino acid to form polyketide product; the number, organization, and the order of these modules in the polyketide synthetase reflect the size, complexity, and the sequence of the synthesized polyketide. Both terrestrial and marine fungi have been known to be prolific sources of dozens of secondary metabolites that find useful applications as antibiotic, antitumor, immunosuppressive, hypocholesterolaemic, antimigraine, and antiparasitic agents. Novel active secondary metabolites, especially polyketides, have also been discovered with pharmacological activities both in terrestrial and marine fungi and some of them have already progressed to clinical trials. With advancement in the molecular biological arena, a closer look at the molecular production of these pharmacologically important entities has become possible. The authors have described all such molecular aspects of fungal polyketide biosynthesis in the present chapter which is indeed helpful to identify and produce related polyketides of pharmaceutical importance.

1.2.18 Chapter 19

In Chapter 19, Himaya and Kim have presented a detailed discussion on marine microalgal metabolites that have recently emerged as a potential source of pharmaceutical products. Microalgae are still an untapped resource of potential pharmaceutical compounds; however, the awareness and knowledge about this resource is dramatically increasing. Structural novelty and a wide array of selective molecular targets, marine microalgal derived metabolites such as carotenoids, fatty acids, active peptides, natural compounds, and polysaccharides have already evoked potent pharmaceutical interest. Identification of pharmaceutical agents from marine algae is advantageous not only because of this wide spectrum of active compounds but also due to environmental considerations; production of "green natural metabolites" which can be reproduced is gaining more attention over scarce marine resources. In this regard, marine microalgae have an added advantage as the primary owners of most active marine ingredients which can be readily cultured for the continuous production of biologically active metabolites in larger amounts. In the present chapter, the authors have presented an overview of the natural metabolites isolated from microalgae as potential candidates for the pharmaceutical industry with a special focus on the pharmacological potential of marine microalgae-derived carotenoids, fatty acids, peptides, and polysaccharides. Besides, the economic feasibility and future perspectives are also mentioned.

1.2.19 Chapter 20

Chapter 20 is devoted to rosmarinic acid, a bioactive plant-derived phenolic compound; Luis and his group have reviewed this significant phytochemical in regard to its biology, pharmacology, and *in vitro* plant cell culture approximation. Biotechnological approaches to the production of this secondary metabolite have been discussed in detail. Significant progress has already been made in understanding rosmarinic acid production dynamics in plant cell cultures on a number of levels. An integrated approach will most likely be necessary in successful engineering efforts to reliably increase the production. The present chapter provides a detailed literature on this subject.

1.2.20 Chapter 21

Chapter 21 deals with biosynthetic approaches for enhancing antioxidants in plants by Namitha and Negi. Plant secondary metabolites can act as promising antioxidants, and these natural products include mainly, carotenoids, flavonoids, tocopherol, and ascorbic acid. Besides the antioxidant activity, these phytochemicals are also involved in various other functions in plants such as maintaining membrane fluidity, photoprotection during photosynthesis, signaling between plant and microbes, act as antimicrobial agents, and help in pollination and seed dispersion. As far as human health and nutrition are concerned, such bioactive natural products find immense pharmaceutical applications such as reducing risk of certain cancers, cardiovascular diseases, aging, osteoporosis, menopausal symptoms, and enhancing immune response. Animals do not synthesize these compounds *de novo*, hence rely on a diet rich in fruits, vegetables, nuts, grains, and oilseeds as their source. Due to their enormous beneficial effect, the current focus in research is to increase these phytonutrients in crop plants; several attempts have been made to successfully bioengineer the crop plants to enhance these secondary metabolites—the present chapter is greatly devoted to encompass their structure, distribution, functions, biosynthesis, and the genetic/metabolic engineering approaches involved in enhancing these antioxidant compounds in various plants.

1.2.21 Chapter 22

Modolo and her group have contributed a detailed discussion on plants as biofactories of pharmaceuticals and nutraceuticals in Chapter 22. Plant cells follow complex pathways to synthesize such significant secondary metabolites. The authors have produced a state-of-the-art review of plant research focusing on the production of health-promoting substances; the discussion features the major points: (i) specific transcription factors and gene promoters for rational genetic and metabolic engineering, (ii) elicitation techniques to boost secondary metabolism, (iii) hairy root cultivation system as a powerful tool for producing valuable substances, (iv) plant tissue and cell culture systems as platforms for metabolic engineering, and (v) next-generation technologies for the discovery of pharma(nutra)ceuticals-related genes. Besides, notable examples of *in vitro* conditions and field-grown crops used to produce valuable substances to attend the market demand are also mentioned herein.

1.3 CONCLUDING REMARKS

This introductory chapter summarizes each technical chapter of the book for which representation of facts and their discussions are exhaustive, authoritative, and deeply informative. The readers will find interest in each of the chapters which practically cover a wide area of bioactive natural product research. The reference encourages interdisciplinary work among chemists, pharmacologists, biologists, botanists, and agronomists with an interest in bioactive natural products. Hence, the present book would surely serve as a key reference for recent developments in the frontier research on bioactive natural products, and also would find much utility for scientists working in this area.

2 Impact of Solid-Supported Cyclization–Elimination Strategies toward the Development of Natural Product Inspired Molecules in Drug Discovery Research*

Dhananjay Kumar, Bhuwan B. Mishra, Divya Kushwaha, and Vinod K. Tiwari

CONTENTS

* This chapter is dedicated to Dr. R. P. Tripathi, Scientist, F at Central Drug Research Institute, Lucknow, India on the occasion of his 57th birthday.

2.1 INTRODUCTION

Natural products (NPs) originated from the phenomenon of biodiversity in which the interactions among organisms and their environment formulate the diverse complex chemical entities within the organisms are usually responsible for enhancing their survival and competitiveness (Mishra and Tiwari 2011). Diverse chemical scaffolds of NPs offer promising templates for combinatorial chemistry since being evolutionarily selected for their ability to display chemical information in three-dimensional space. The therapeutic areas of infectious diseases and oncology have already benefited a lot from these numerous drug classes that are able to interact with many specific targets within the cell, and indeed for many years, they have been the central molecules in the drug discovery and development processes. Libraries built around such scaffolds thus have the potential for both lead discovery (against targets unrelated to the original activity of the natural product) and lead optimization (analogs with improved properties over the natural product) (Hajduk et al. 2011).

In recent years, the increasing demands for natural product inspired drug-like molecules or their libraries have stipulated the development of reaction sequences and linking strategies that allow complex and diverse targets to be constructed efficiently and reliably (Samiulla et al. 2005). Combinatorial chemistry in this regard is among the most important novel methodologies developed to replace the sequential approach with the most effective parallel technique, where this powerful methodology has been found to be helpful to the pharmaceutical companies in developing new drug candidates within a reasonable time frame (Merrifield 1963), thereby, cutting down preclinical development costs significantly, leading to a massive change in the fundamental approach to drug discovery program (Atrash and Bradley 1997; Tripathi et al. 2012). However, construction of conformationally restricted cyclic bioactive molecules is still a challenge for the chemists. Nowadays, one most attractive and versatile approach is the cyclo-elimination strategy which has become a practical method mostly for the preparation of small cyclic molecules using a solid-phase synthesis on polymeric support (Thompson and Ellman 1996).

The cyclo-elimination strategy can minimize the chemical and tethering implications by releasing the intact desired target molecule in the final step of a reaction. The simultaneous cyclization and release of the product from the solid support with adequate purity without adding an extra time-consuming step is the major advantage of this method. It has recently emerged as an efficient strategy to detect and to evaluate affinity between the library products and a target molecule (Eifler-Lima et al. 2010; Mishra et al. 2012). The present review highlights the practical high-yielding with high-purity approach for the solid-phase combinatorial synthesis of natural products inspired diverse pharmacologically active carbo/heterocyclic skeletons using cyclo-release strategy.

2.2 APPLICATION OF CYCLO-RELEASE STRATEGY TOWARD THE DEVELOPMENT OF NATURAL PRODUCT INSPIRED MOLECULES IN DRUG RESEARCH

Nature is a true inventor of combinatorial chemistry that generates molecules with all possible structures which are highly sophisticated and unique in their own (Verdine 1996). A good number of evolutionarily shaped molecules are historically proven important and effective therapeutic agents. Collections of natural products exhibit physicochemical property profiles that are favorable compared to those of drugs and complementary to those provided by synthetic compounds generated from combinatorial chemistry (Feher and Schmidt 2003; Park and Kurth 2000). Despite these advantages, the classical processes to identify discrete new chemical entities from natural product sources are too inefficient to have survived in many of the current drug discovery programs at pharmaceutical companies; however, interest in natural products and their analogs as considerable sources of pharmaceutical agents has shown recent resilience and has regained importance (Newman et al. 2003). To capitalize more efficiently on the effective features of naturally occurring substances, solid-phase cyclo-elimination strategy may serve as a competent platform for natural product based library production for lead discovery. This approach combines the attractive biological and physicochemical properties of natural product scaffolds, provided by eons of natural selection, with the chemical diversity available from parallel synthetic methods. A good number of promising and biologically relevant compounds are screened under the present discussion so as to underline the significance of cyclo-release strategy for their effective syntheses.

2.2.1 Epothilone A

Epothilone A, a 16-membered ring macrolide, originally isolated from the bacterium *Sorangium cellulosum*, is a new kind of microtubule function inhibitor with an IC_{50} of 4.4 nM for tubulin polymerization and cytotoxicity (Hofle et al. 1996). Epothilone A significantly prevents cancer cells from dividing by interfering with tubulin, and has better efficacy and milder adverse effects than taxanes (Muhlradt and Sasse 1997).

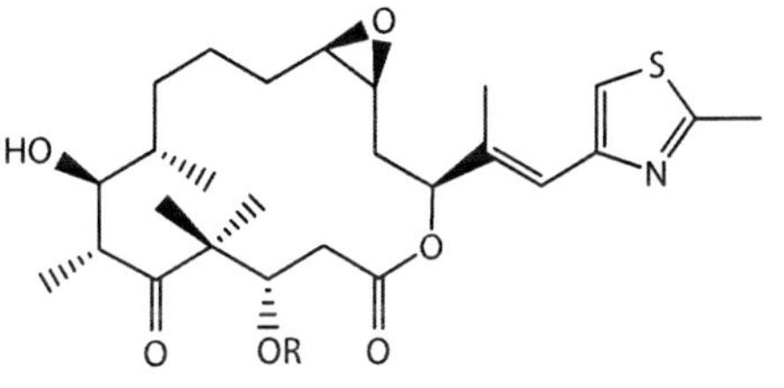

Epothilone A

The synthesis of epothilone A under the solid-phase condition utilizes olefin metathesis reaction (Nicolaou et al. 1997) and involves the preparation of solid-phase phosphonium salt using Merrifield-tethered 1,4-butanediol. Treatment with base in the next step affords an ylide that delivers the TBS-protected olefin on reaction with aldehyde. Desilylation followed by Swern oxidation of the resulting alcohol affords the solid-phase aldehyde which on aldol condensation with the delta-keto acid gives a mixture of diastereomers (1:1). The DCC-mediated alcohol coupling yields corresponding solid-phase ester which on macrocyclization using ring closing metathesis (RCM) affords a mixture of four diastereomers. Separation of the targeted molecule by HPLC or preparative layer silica-gel chromatography following desilylation and epoxidation affords naturally occurring epothilone A (Scheme 2.1).

2.2.2 Fumiquinazoline Alkaloid

Building up and maintaining a high-quality natural product inspired library requires a skill set that can be achieved easily by using solid-phase cyclo-elimination strategies. For example, total synthesis of cytotoxic fumiquinazoline alkaloid isolated from marine *Aspergillus clavatus* was investigated by Wang and Ganesan under solid-phase conditions (Wang and Ganesan 2000). Methodology involves the loading of commercially available Wang resin with Fmoc-L-Trp **1** followed by deprotection and coupling with anthranilic acid in the presence of an activating agent EDC (Scheme 2.1). The acylation of aniline **2** using Fmoc-Gly-Cl affords linear tripeptide **3** which on subsequent dehydrative cyclization in the presence of triphenylphosphine (10 equivalents) yields oxazine **4**.

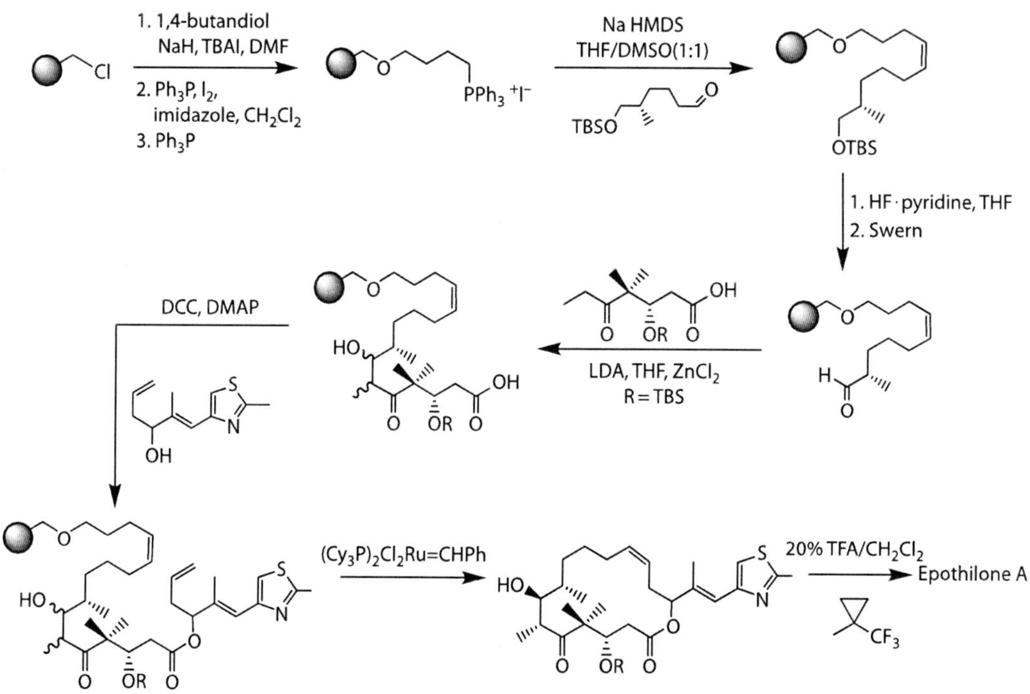

SCHEME 2.1 Synthesis of epothilone A through cyclo-release strategy.

Deprotection of the Fmoc group and rearrangement of oxazine **4** affords amidine carboxamide **5**. After washing, the resin on heating in acetonitrile affords fumiquinazoline alkaloid **6** through cyclative release strategy (Scheme 2.2).

2.2.3 Freidinger Lactam

Peptidomimetics are crucial in the drug development process. In the early 1980s, Freidinger proposed the concept of protected lactam-bridged dipeptides, a milestone in the design of conformationally constrained peptides (Perdih and Kikelj 2006). These types of compounds, now widely known as Freidinger lactams, have been of interest to many medicinal and peptide chemists, and can be synthesized easily under solid-phase conditions via ring closing metathesis (RCM) (Piscopio et al. 1997, 1998, 1999). Methodology involves the synthesis of 2,4-dinitrobenzenesulfonamide resin by the reaction of cinnamyl alcohol resin with phenylalanine methyl ester-2,4-dinitrobenzenesulfonamide under Mitsunobu condition. Sulfonamide cleavage in the next step followed by acylation with racemic Boc-allylglycine affords a resin-bound diene which on cyclo-elimination via RCM in the final step provides the Freidinger lactam as a 1:1 mixture of diastereomers (Scheme 2.3).

2.2.4 Macrocycles

Nicolaou et al. have reported the use of Stille coupling for a solid-phase cyclo-elimination process, where in the solid-phase polystyrene(di-*n*-butyltin)hydride (PBTH) on addition across the acetylenic bond of 3-butynol affords an *E*:*Z*-mixture of alcohols (1:1) which on treatment with iodine yield vinyl iodide(s). Coupling with glutaric anhydride in the next step delivers a resin-bound acid which on DCC coupling with either vinyl iodide alcohol or 2-iodobenzyl alcohol affords corresponding esters. Subsequent treatment with [Pd(PPh_3)$_4$] resulted in macrocyclization with concomitant cyclo-elimination from the support (Scheme 2.4).

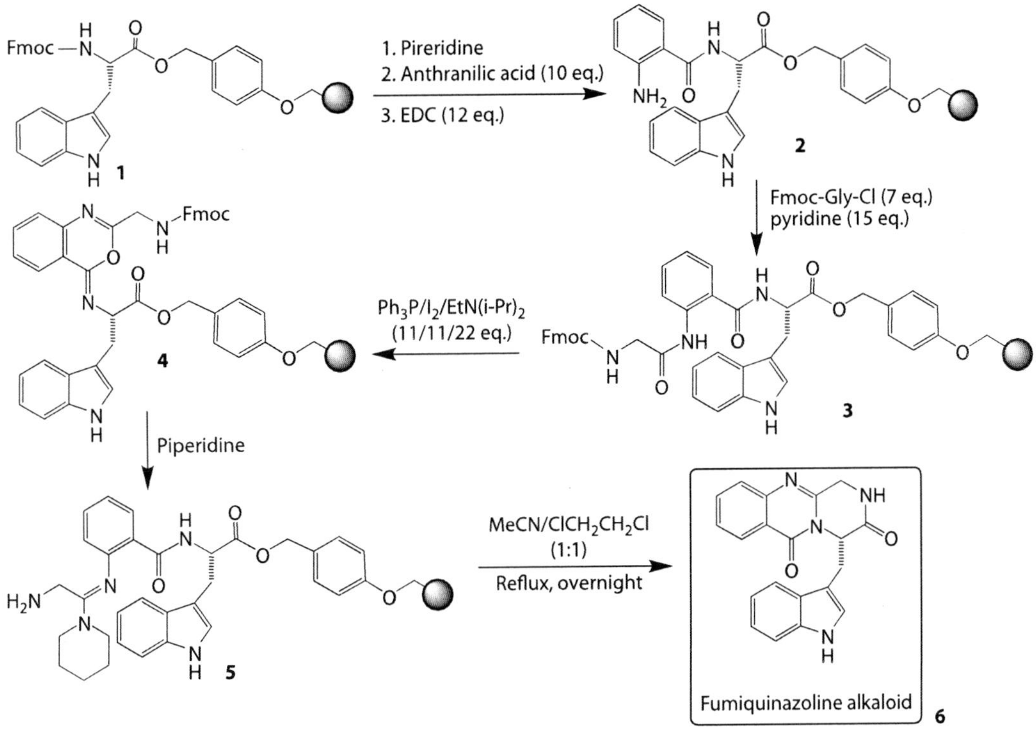

SCHEME 2.2 Synthesis of fumiquinazoline alkaloid through cyclative release strategy.

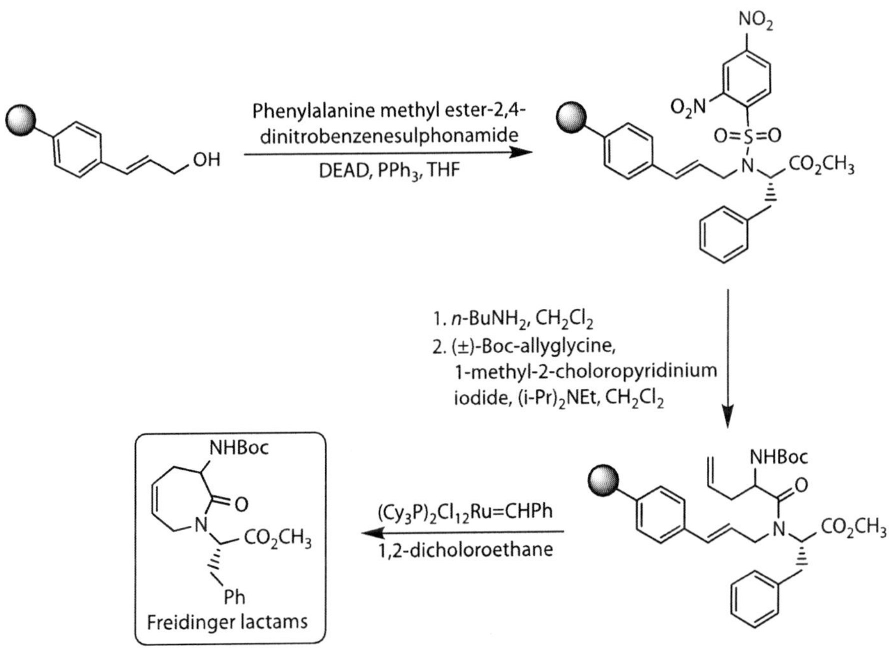

SCHEME 2.3 Second-generation solid-phase approach to Freidinger lactum using cyclative release via ring closing metathesis.

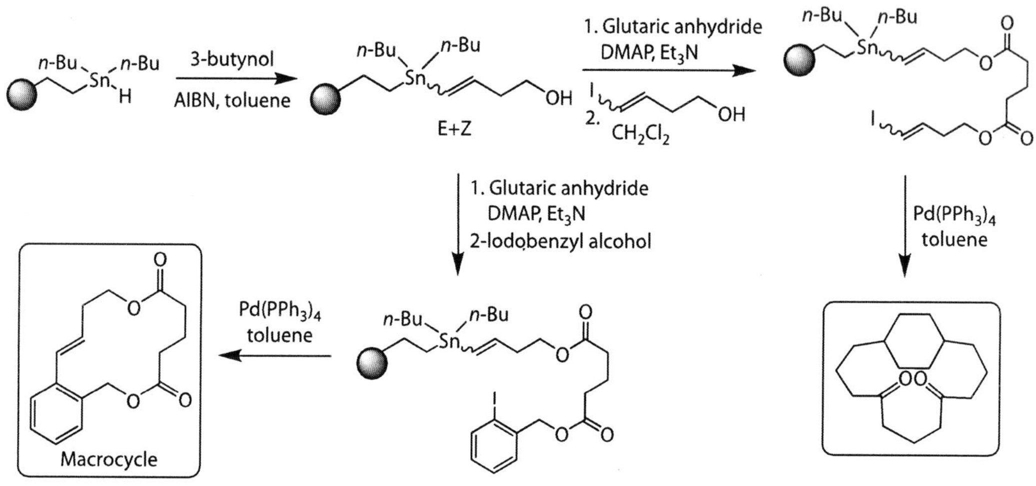

SCHEME 2.4 Stille coupling for the synthesis of macrocycles through cyclo-release strategy.

2.2.5 (*S*)-Zearalenone

(*S*)-Zearalenone, a naturally occurring 14-membered orsellinic acid-type macrolide, exhibits anabolic, uterotropic, and antibacterial activity (Mirocha et al. 1978; Olsen 1985; Stob et al. 1962).

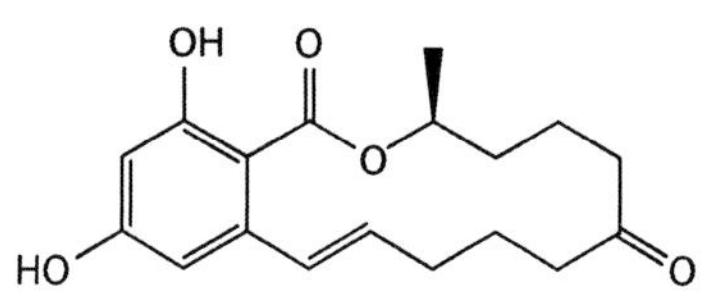

(*S*)-Zearalenone

This targeted 14-membered orsellinic acid-type macrolide can be synthesized under solid-phase condition via Stille coupling. For a synthesis of (*S*)-zearalenone, PBTH on reaction with Weinreb amide affords a vinyltin conjugate ($E{:}Z = 1{:}1$). A solid-phase (*E*)-vinyl intermediate is obtained by the reaction of solid-phase tin chloride (PBTC) with vinyllithium reagent following the deprotection and oxidation reactions (Nicolaou et al. 1998b). Addition of Grignard reagent to the Weinreb amide or aldehyde followed by Corey–Kim oxidation affords the corresponding ketone. Desilylation and subsequent coupling with an iodobenzoic acid gives the Stille precursor, which on subsequent treatment with [$Pd(PPh_3)_4$] results in cyclo-elimination by delivering (*S*)-zearalenone after acid-induced deprotection (Scheme 2.5).

2.2.6 Tetramic Acids

The cyclo-elimination strategy finds tremendous application in the synthesis of a wide range of bioactive cyclic amides. Tetramic acids are an important class of naturally occurring molecules exhibiting various significant biological activities including antibiotic, antiviral, antifungal, cytotoxic, and enzyme inhibitory activities against bacterial DNA-directed RNA polymerases (Karwowski et al. 1992; Rinehart and Borders 1963). These nitrogen heterocycles with pyrrolidine-2,4-dione moieties are key structural motifs in many natural products of terrestrial and marine origin (Brodyck 1995; Lang et al. 2006; Wolf et al. 1999). 3-Acyl-2,4-pyrrolidinediones or 3-aceyl tetramic acids can easily be synthesized under mild solid-phase condition (Raillard et al. 1999; Romoff et al. 1998; Weber et al. 1998) by loading of Merrifield resin with amino acid

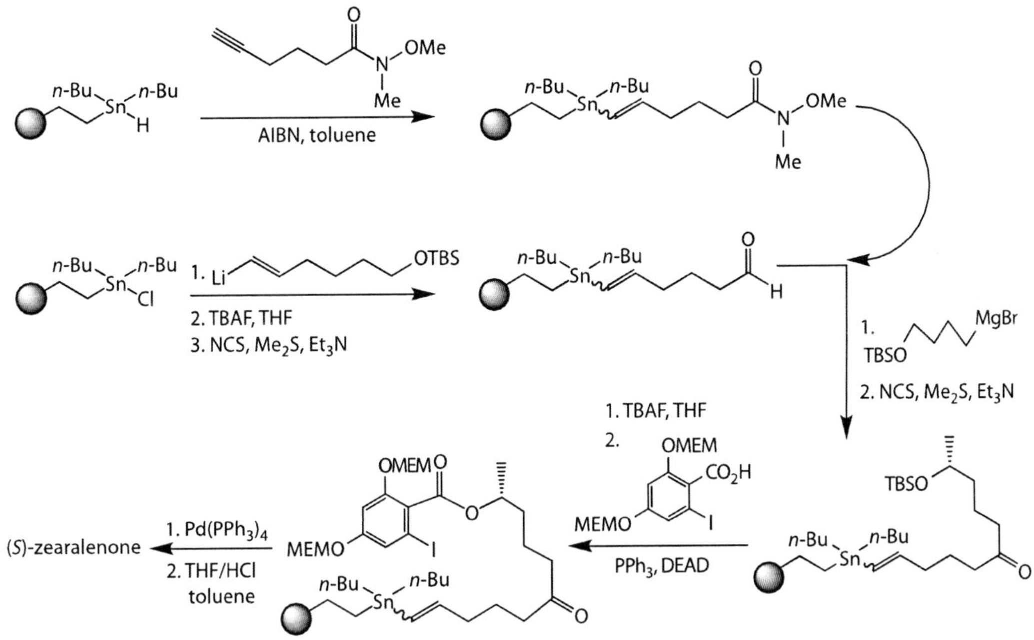

SCHEME 2.5 Stille coupling for the synthesis of (*S*)-zearalenone.

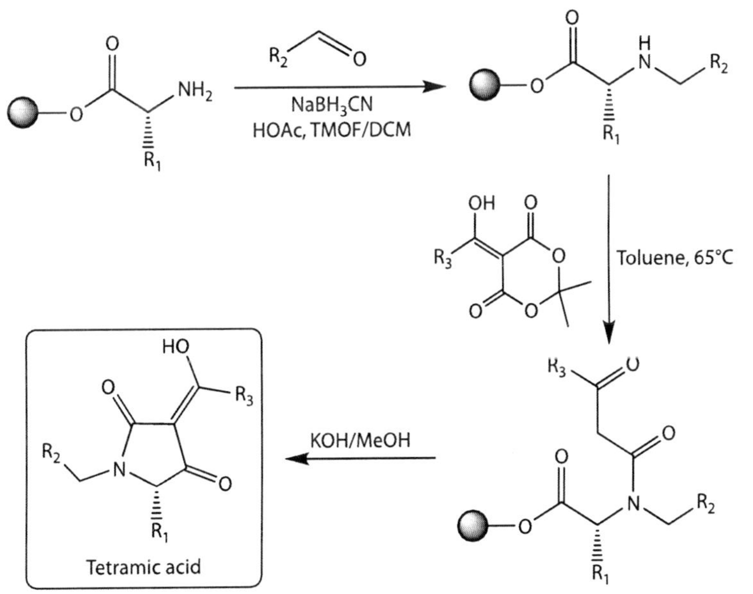

SCHEME 2.6 Synthesis of tetramic acid through cyclo-release strategy.

ester and subsequent alkylation with an aldehyde to afford solid-phase secondary amine. The secondary amino group on acylation with a β-ketoester equivalent (i.e., Meldrum's acid derivatives) gives the β-keto amide-bound resin which under Beckmann intramolecular *C*-alkylation delivers a library of 3-acyl tetramic acids (Scheme 2.6).

Another solid-phase route to tetramic acids using cyclo-elimination Claisen-type condensation involves the loading of Wang resin with an Fmoc-protected amino acid which on deprotection followed by acylation affords corresponding amide. A unidirectional Claisen-like condensation in the

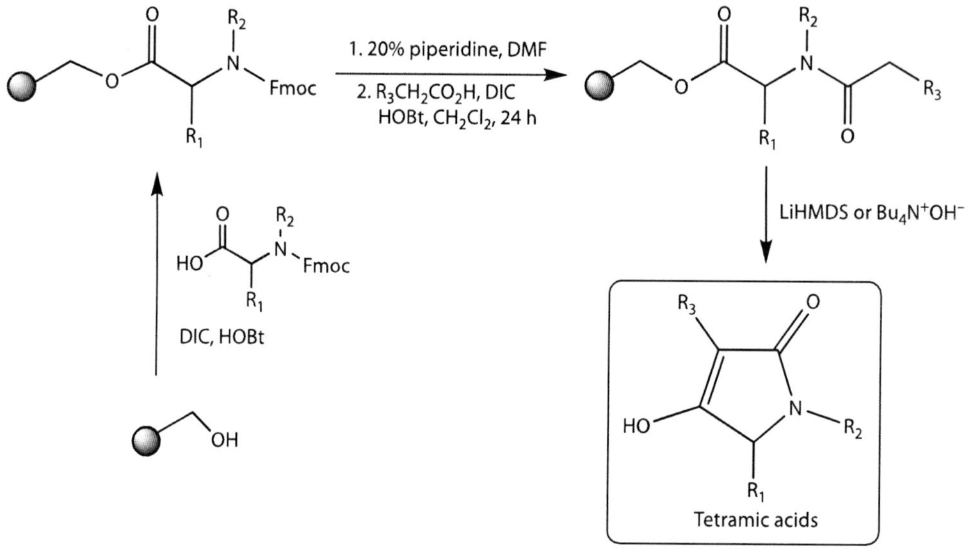

SCHEME 2.7 Synthesis of tetramic acid through cyclo-elimination Claisen-type condensation.

presence of tetrabutylammonium hydroxide results in cyclo-elimination of the targeted tetramic acid (Kulkarni and Ganesan 1997, 1998). The base employed can be effectively scavenged by using Amberlyst A-15 resin (Scheme 2.7).

A three-step solid-phase protocol for the synthesis of 1,3,5-trisubstituted tetramic acids includes reductive alkylation of a solid-phase α-amino acid with aldehydes (Mattews and Rivero 1998). Acylation of the resulting secondary amine with either malonic acids or aryl acetic acids affords acyl tertiary amides which on base-promoted cyclo-elimination give tetramic acids (Scheme 2.8).

2.2.7 Muscone and Its Derivatives

Muscone, obtained from musk (a glandular secretion) of the male muskdeer *Moschus moschiferus*, has been used in perfumery and medicine for thousands of years (Fujimoto et al. 2002). Muscone naturally occurs as (−)-enantiomer consisting of a 15-membered ring ketone with one methyl substituent in the 3-position (Kamat et al. 2000). A cyclo-elimination based solid-phase protocol provides an easy access to a muscone library via an intramolecular ketophosphonate-aldehyde reaction (Nicolaou et al. 1998a). The methodology involves the reaction of Merrifield resin with 1,4-butanediol followed by the addition of $CH_3P(O)(OCH_3)Cl$. Treatment of this resin with *n*-BuLi followed by the reaction with protected ω-hydroxy methyl ester affords a resin-bound ketophosphonate. Desilylation followed by a DCC-mediated condensation with various protected ω-hydroxy carboxylic acids delivers the corresponding esters. Desilylation and oxidation furnish precursor aldehydes which on treatment with 18-crown-6 in toluene release the macrocyclic products (Scheme 2.9).

Radio frequency-encoded MICROKANS have been also used in the synthesis of a muscone library. Claisen-type coupling of the starting solid-phase methylphosphonate with ω-olefinic esters followed by cross olefin metathesis with ω-olefinic alcohols affords the solid-phase ω-hydroxyl esters as an *E*:*Z*-mixture. Dess–Martin oxidation followed by treatment with crown ether releases the macrocyclic enones which on subsequent solution phase cuprate addition followed by hydrogenation afford *dl*-muscone library (Scheme 2.10).

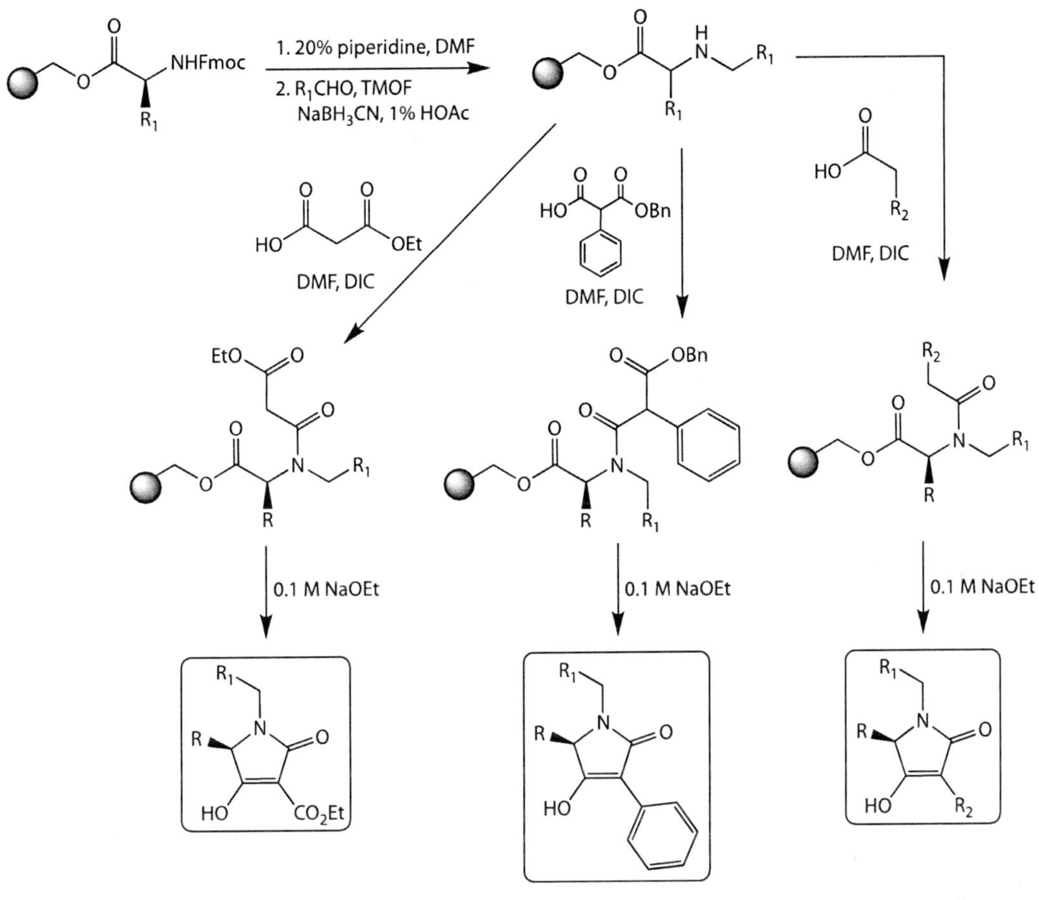

SCHEME 2.8 Synthesis of 1,3,5-trisubstituted tetramic acids through cyclo-elimination protocol.

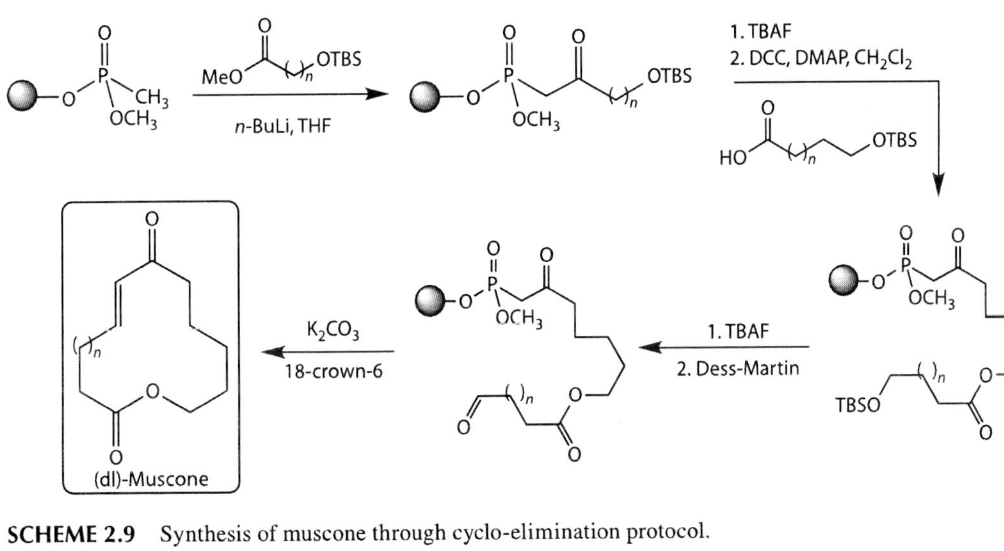

SCHEME 2.9 Synthesis of muscone through cyclo-elimination protocol.

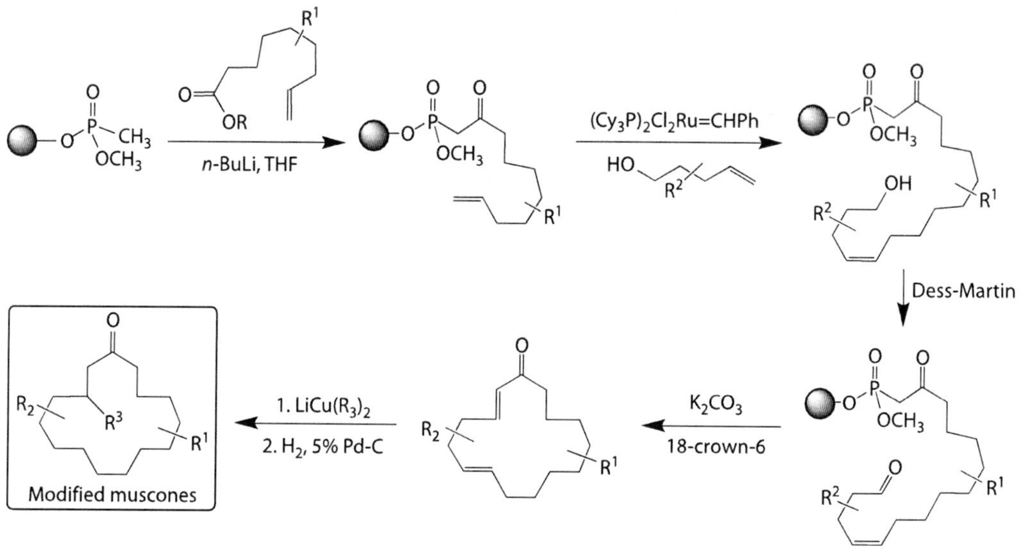

SCHEME 2.10 Synthesis of muscone derivatives through cyclo-elimination strategy.

2.2.8 1,4-Benzodiazepine-2,5-Diones

1,4-Benzodiazepine-2,5-diones, the small molecular templates acting as opiate receptor antagonist, anticonvulsant, glycoprotein mimic and cholecystokinin receptor antagonist (Joseph et al. 2008), have been synthesized using solid-phase technology. The methodology involves the deprotection and coupling of Fmoc amino acid-bound Wang resin with Fmoc-protected *o*-anthranilic acid or *o*-nitrobenzoic acid (Mayer et al. 1996). Fmoc deprotection or reduction of the nitro group gives amido-anthranalate intermediate which on treatment with sodium *tert*-butoxide releases the pharmacophore in high yield (Scheme 2.11).

2.2.9 Benzodiazepine

Benzodiazepine, a psychoactive drug whose core chemical structure arising out of the fusion of a benzene ring and a diazepine ring, enhances the effect of the neurotransmitter gamma-aminobutyricacid (GABA) resulting in sedative, hypnotic (sleep inducing), anxiolytic (anti-anxiety), anticonvulsant, muscle relaxant, and amnesic action (Lader 2008; Olkkola and Ahonen 2008). For the synthesis of benzodiazepine, DeWitt and coworkers have employed cyclo-elimination strategy in which the *trans*-amidation reaction between amino acid resin and 2-aminobenzophenone imines affords the support-bound imine (DeWitt et al. 1993). Further heating in TFA results in cyclo-elimination (Scheme 2.12).

2.2.10 Diketopiperazines

Diketopiperazines (DKPs), a class of cyclic organic compounds resulting from peptide bond between two aminoacids to form a lactam, are quite common in nature and have a wide variety of biological potentials including antitumor, antiviral, antifungal, and antibacterial activities (Houston et al. 2004; Kowalski and Lipton 1996; Kwon et al. 2000; Martins and Carvalho 2007; Nicholson et al. 2006; Sinha et al. 2004). Synthesis of DKP under solid-phase condition involves the preparation of an ester-bound resin by esterification of either Tentagel S-OH (Rapp Polymere) or PAM (Novabiochem) resin with an amino acid in the presence of acyl fluorides generated *in situ* from *N*-protected amino acids by treatment with 1,3-dimethyl-2-fluoropyridinium

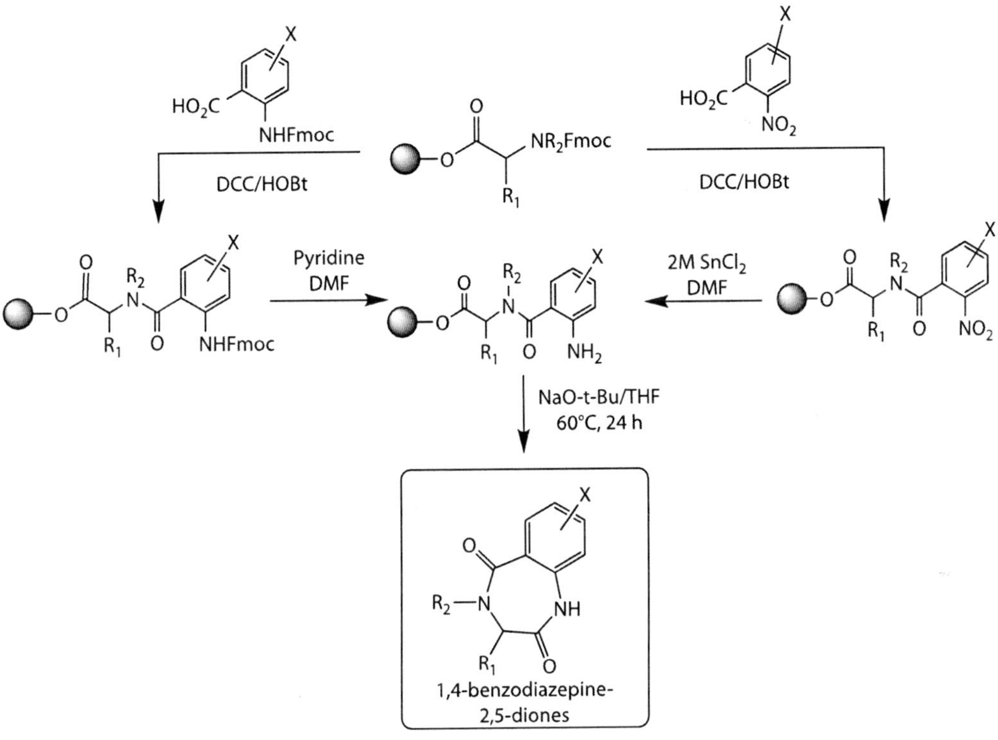

SCHEME 2.11 Synthesis of 1,4-benzodiazepine-2,5-diones through cyclo-elimination strategy.

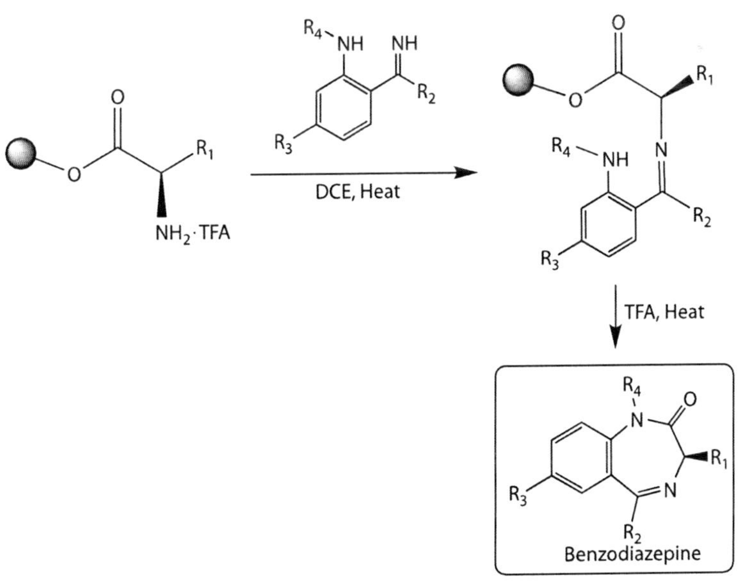

SCHEME 2.12 Synthesis of benzodiazepine through cyclo-elimination strategy.

4-toluenesulfonate (DMFP) and diisoproplethylamine (DIPEA) (Szardenings et al. 1997). Deprotection followed by reductive alkylation with different aldehydes gives the secondary amine. *N*-acylation with Boc-protected amino acids delivers the peptide-bound resin via a double-coupling method which on deprotection with TFA followed by cyclization under acidic or basic conditions affords pure DKPs (Scheme 2.13).

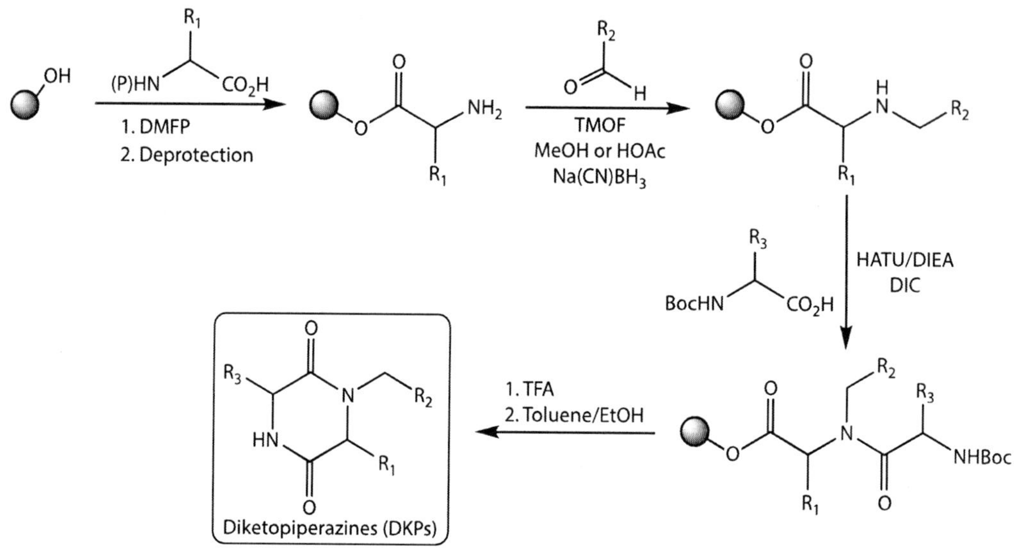

SCHEME 2.13 Synthesis of diketopiperazines through cyclo-elimination strategy.

2.2.11 Demethoxyfumitremorgin C

Demethoxyfumitremorgin C, a fungal inhibitor of mammalian cell cycle progression at the G_2/M transition (Wang et al. 2000) first isolated from *Aspergillus fumigatus*, contains both a tetrahydro-β-carboline and a diketopiperazine. Analogs of this M-phase inhibitor of the mammalian cell cycle have been synthesized using a solid-phase cyclo-elimination protocol, wherein Wang resin-bound Fmoc-L-tryptophan after deprotection reacts with senecioaldehyde in TMOF (Hartog et al. 1996; Loevezijn et al. 1998; Wang and Ganesan 1999). Treatment of imine with Fmoc-proline acid chloride results in a Pictet–Spengler reaction (Scheme 2.14). Subsequent Fmoc deprotection delivers demethoxyfumitremorgin C and its *trans*-epimer (*cis:trans* = 53:47)

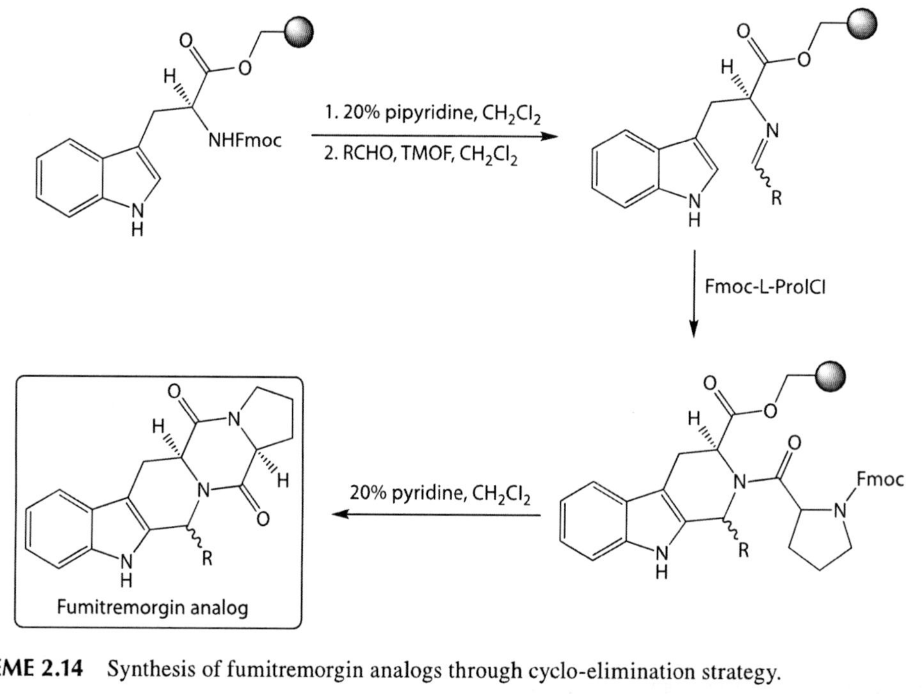

SCHEME 2.14 Synthesis of fumitremorgin analogs through cyclo-elimination strategy.

2.2.12 1,3-Disubstituted Quinazolinediones

The quinazolinedione moiety is an important scaffold embedded in a variety of natural alkaloids and many biologically active molecules including serotonergic, dopaminergic, and adrenergic receptor ligands and inhibitors of aldose reductase, lipoxygenase, cyclooxygenase, collagenase, and carbonic anhydrase (Dreyer and Brenner 1980; Michael 2007; Rivero et al. 2004; Vogtle and Marzinzik 2004). A simple, reliable, and efficient solid-phase route to 1,3-disubstituted quinazolinediones involves the treatment of chloroform-functionalized polystyrene with a differently substituted anthranilic acids in the presence of base to afford carbamate-linked resin. This resin is doubly coupled with a diverse range of primary amines in the presence of PyBOP to deliver anthranilamide (Smith et al. 1996). Heating these anthranilamides at 125°C releases the 1,3-disubstituted quinazolinediones (Scheme 2.15).

2.2.13 4-Hydroxyquinolin-2(1*H*)-One

Likewise, therapeutically significant 4-hydroxyquinolin-2(1*H*)-ones have been synthesized under solid-phase condition using cyclo-elimination protocol, wherein cyanoacetate-functionalized Wang resin reacts with isatoic anhydride and Et_3N in anhydrous DMF to afford a *C*-acylated intermediate (Sim et al. 1998). Heating (80°C) this α-cyano-β-keto ester results in intramolecular *trans*-amination releasing the 4-hydroxyquinolin-2(1*H*)-one (Scheme 2.16).

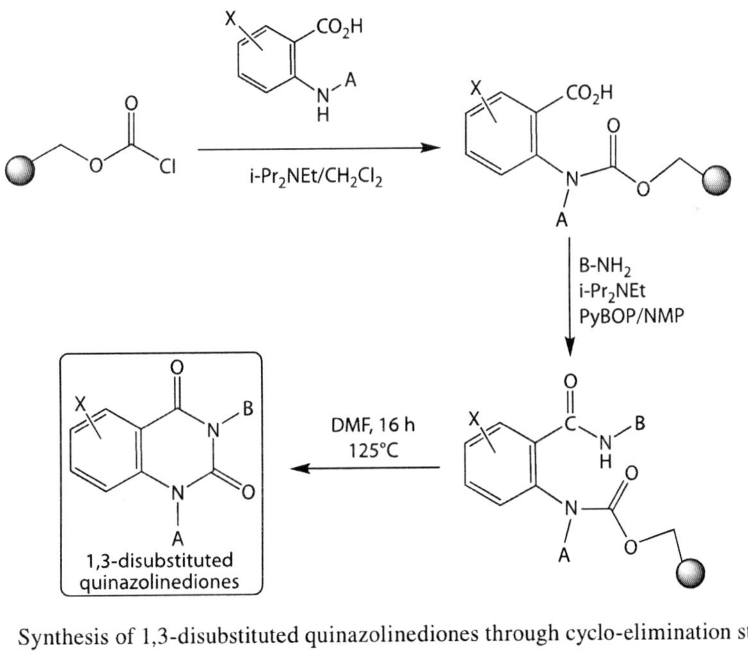

SCHEME 2.15 Synthesis of 1,3-disubstituted quinazolinediones through cyclo-elimination strategy.

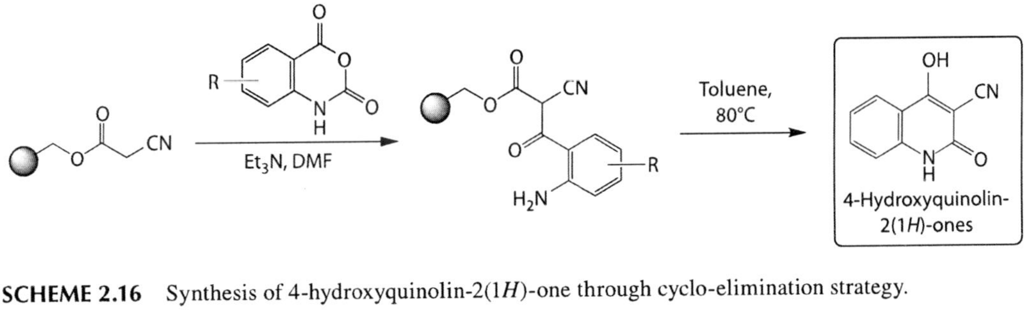

SCHEME 2.16 Synthesis of 4-hydroxyquinolin-2(1*H*)-one through cyclo-elimination strategy.

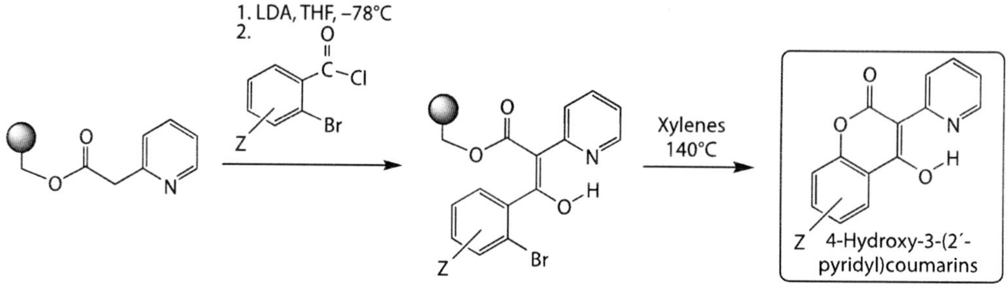

SCHEME 2.17 Synthesis of 4-hydroxy-3-(2′-pyridyl)coumarin through cyclo-elimination strategy.

2.2.14 4-Hydroxy-3-(2′-Pyridyl)Coumarin

The 1-benzopyran-2-one moiety, a structural core of coumarins, is often found in more complex natural products and is frequently associated with numerous biological activities including anticancer, antifungal, anti-HIV, and anticlotting. Kurth et al. have synthesized a library of 4-hydroxy-3-(2′-pyridyl)coumarins using the cyclo-elimination protocol (Liu et al. 2006). The strategy begins with the attachment of 2-pyridylacetic acid to Merrifield resin through an ester bond using Cs_2CO_3 in DMF in catalytic amount of KI. Treatment of the substrate with LDA in THF at –78°C and the resulting lithium enolate on reaction with various 2-bromobenzoyl chlorides give *C*-acylated products which on key intramolecular *ipso*-substitution step in xylenes at 140°C afford 4-hydroxy-3-(2′-pyridyl)coumarins (Scheme 2.17).

2.2.15 *C*-Nucleosides

The *C*-nucleosides are a unique class of nucleosides in which the glycosidic chain is connected to the pendant heterocyclic base by a C–C bond instead of the C–N bond of the natural nucleosides. As a result, they are resistant to the chemical and the enzymatic hydrolytic cleavage of the glycosidic bond (Cai et al. 2004). The *C*-nucleosides having C–C linkage between the aglycon and the sugar moiety are known mainly for their anticancer, antiviral, and antileukemic activities (Burchenal et al. 1976; Robins et al. 1982; Schaeffer et al. 1978). Tripathi Research Group at CDRI utilizes cyclo-release strategy to achieve glycosyl ureas in rigid form using Wang resin. The methodology involves the loading of combinatorial scaffold glycosyl amino acid on Siber amide resin in the presence of DIC/HOBT/TbTU as the coupling agent. Reductive amination and reaction with different isocynates followed by the removal of polymer support using 2% TFA in CH_2Cl_2 afford glycosyl ureas in flexible form in good yield. A combinatorial library of glycosyl ureas in rigid form have been reported via loading of scaffold glycosyl amino acid with Wang resin in the presence of an appropriate coupling agent, followed by different sets of reactions through the cyclo-release strategy in high yield and with excellent purity (Scheme 2.18) (Mishra et al. 2003).

These *C*-nucleosides have been screened for anti-filarial activities and few of them have shown significant activity. For detailed biological screening, Tripathi and colleagues have successfully developed an efficient and versatile method for introducing dihydropyrimidinone skeleton on protected glucofuranose as well galactopyranose derivatives by reacting glycosyl ureas with DBU as catalyst and TBAB as co-catalyst in 4 A°MS (Tewari et al. 2002).

2.2.16 *O*-Glycosylation

The *O*-glycosylation typically performed under Lewis acidic conditions is an essential reaction in the synthesis of oligosaccharides. A majority of the linkers developed for solid-phase peptide synthesis can seldom be applied directly to oligosaccharide synthesis due to limited stability under

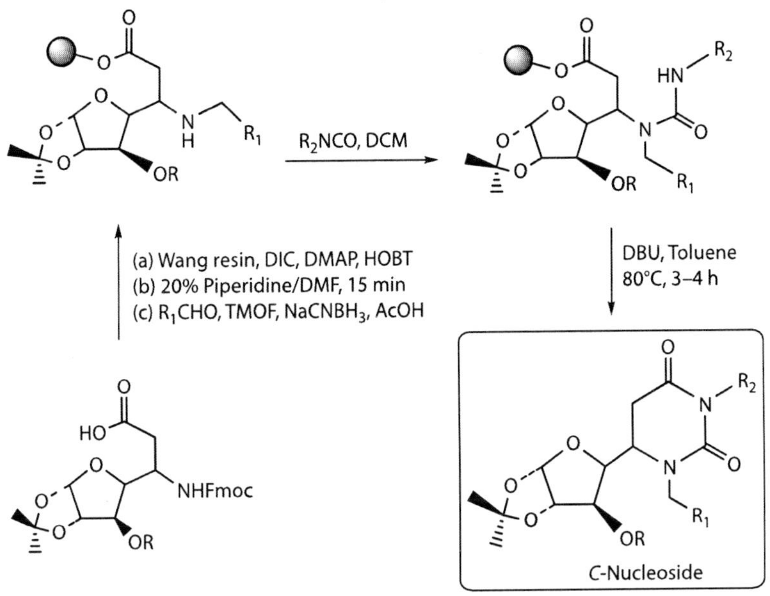

SCHEME 2.18 Synthesis of *C*-nucleoside through cyclo-elimination strategy.

glycosylation conditions (Ito and Manabe 1998). Large numbers of modified linkers have been investigated so far in order to maximize their suitability to solid-phase oligosaccharide synthesis (James 1999). Ito et al. have reported one such linker that not just tolerates the glycosylation conditions but also can be cleaved easily under mild conditions. Synthesis of the linker commences with reaction of 4-hydroxy-3-nitrobenzaldehyde and methyl bromoacetate or *tert*-butyl bromoacetate in the presence of K_2CO_3 in CH_3CN. Subsequent reduction by $NaBH_4$ affords alcohols which after protection as TBS ether and hydrolysis under alkaline conditions afford the free acid. Coupling of free acid with poly(ethylene-glycol) methyl ether (PEG) using 1-ethyl-3-(3-dimethylaminopropyl) carbodiimide·HCl (WSCDI) and DMAP followed by desilylation using 5% aq. HF in CH_3CN or $BF_3 \cdot OEt_2$ affords the linker-attached PEG **1**. Schmidt glycosylation of linker-attached PEG **1** using imidate **2** in the presence of TMSOTf as promoter affords resin-bound *O*-glycoside **3** which on

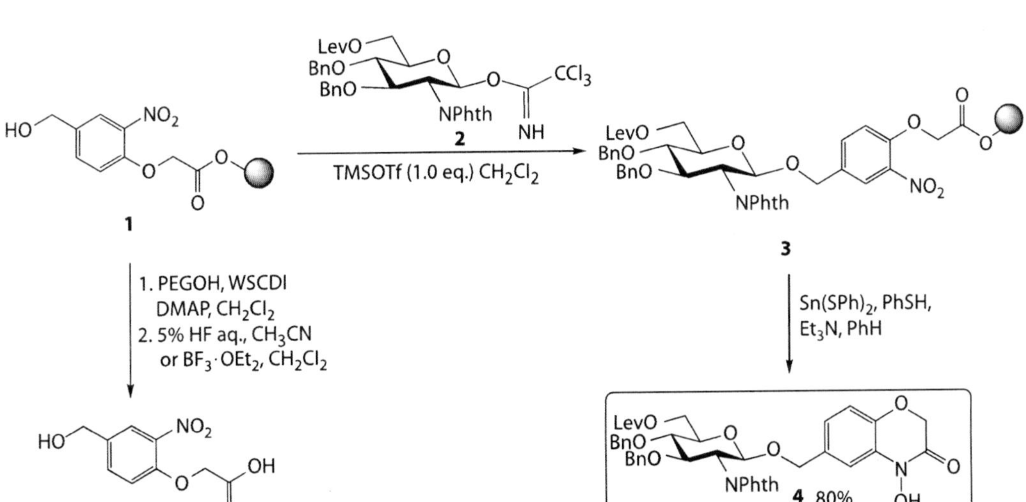

SCHEME 2.19 *O*-Glycosylation through glucose-bound linker.

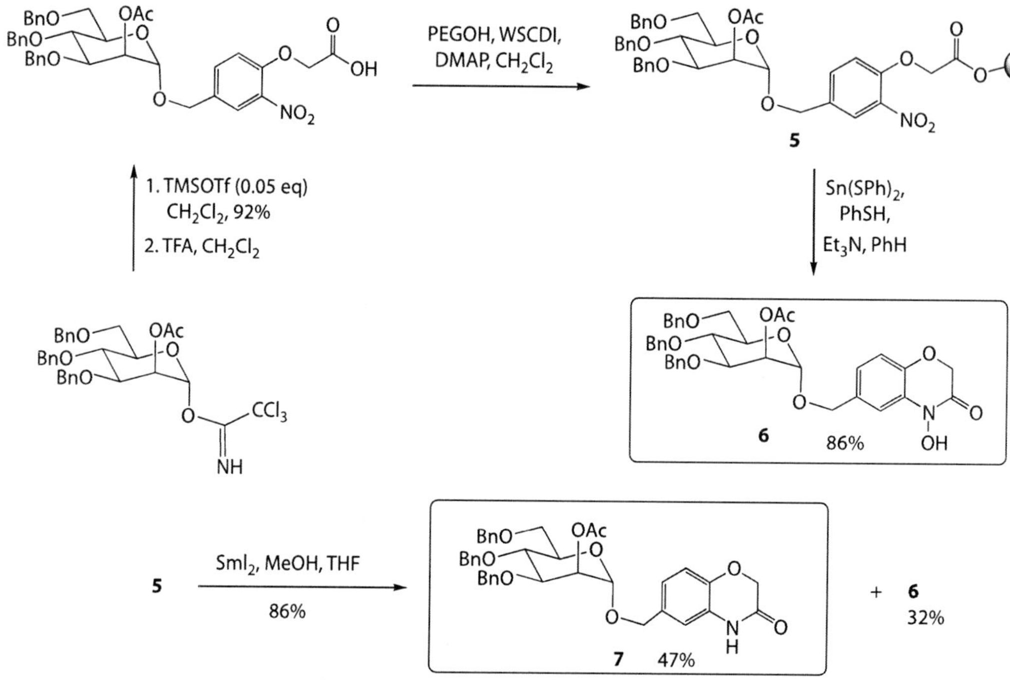

SCHEME 2.20 *O*-Glycosylation through mannose-bound linker.

treatment with $Sn(SPh)_2$-$PhSHEt_3N_6$ affords cyclo-released product **4** in 80% yield, without affecting phthalimide and levulinate groups (Scheme 2.19) (Manabe et al. 2000).

Likewise, mannose-bound, linker-attached PEG **5** under cyclo-release conditions affords hydroxamic acid **6** in 86% yield. The cleavage of sugar from polymer support by using SmI_2, a more powerful reducing reagent, affords lactam **7** along with hydroxamic acid **6**, respectively, in 47% and 32% yield (Scheme 2.20).

2.2.17 Dihydropyrimidones

Dihydropyrimidones (DHPMs) represent a heterocyclic system of remarkable pharmacological efficiency. More recently, appropriately functionalized DHPMs have emerged as potent calcium channel blockers, antihypertensive agents, α_{1a}-adrenergic antagonists, mitotic kinesin Eg5 inhibitors, and neuropeptide Y (NPY) antagonists (Atwal et al. 1991; Falsone and Kappe 2001; Kappe 2000). Kappe et al. have used solid-phase cyclo-elimination protocol for the synthesis of a variety of interesting bicyclic scaffolds such as furo-[3,4-*d*]-pyrimidines, pyrrolo-[3,4-*d*]-pyrimidines, and pyrimido-[4,5-*d*]-pyridazines. The methodology involves the synthesis of a key support-bound ester by rapid, microwave-assisted acetoacetylation of the hydroxymethylpolystyrene resin with methyl 4-chloroacetoacetate in 1,2-dichlorobenzene to facilitate transesterification involving a highly reactive *R*-oxoketene intermediate. The immobilized 4-chloroacetoacetate precursor on subsequent three-component Biginelli-type condensation with aromatic aldehydes and urea in the presence of dioxane/con.HCl (catalyst) affords 6-chloromethyl-functionalized resin-bound dihydropyrimidones which on microwave (MW) flash heating in sealed vessels at 150°C using DMF as solvent release the corresponding furo-[3,4-*d*]-pyrimidines with high purity. Further reaction of resin-bound chloromethyl precursor with primary amines and hydrazines under MW condition results in the release of pyrrolo-[3,4-*d*]-pyrimidines and pyridazino-[4,5-*d*]-pyrimidines, respectively (Scheme 2.21) (Perez et al. 2002).

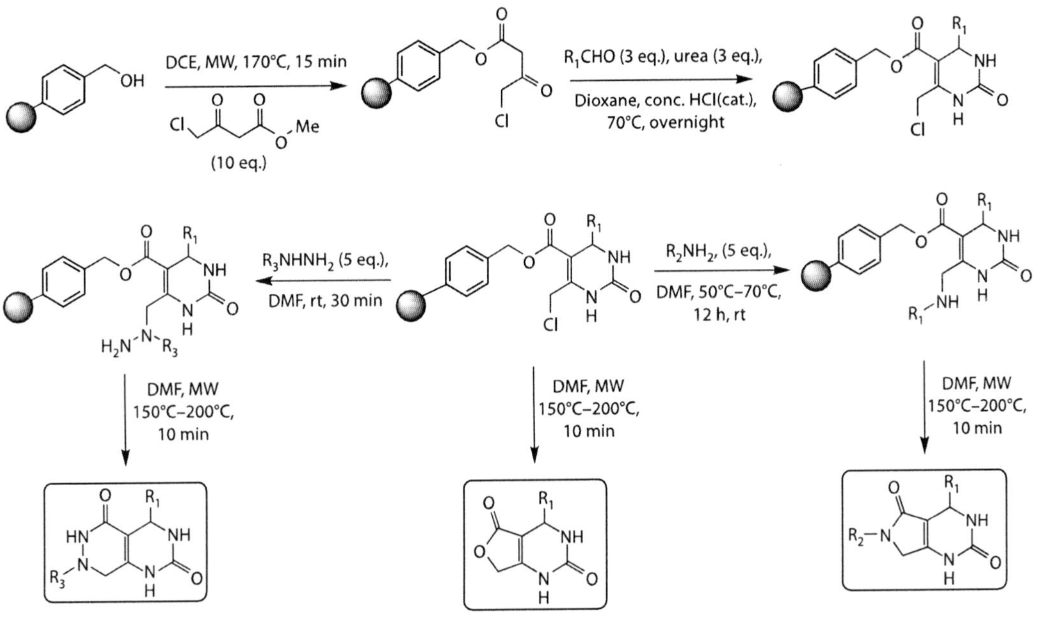

SCHEME 2.21 Synthesis of bicyclic dihydropyrimidones through cyclo-elimination strategy.

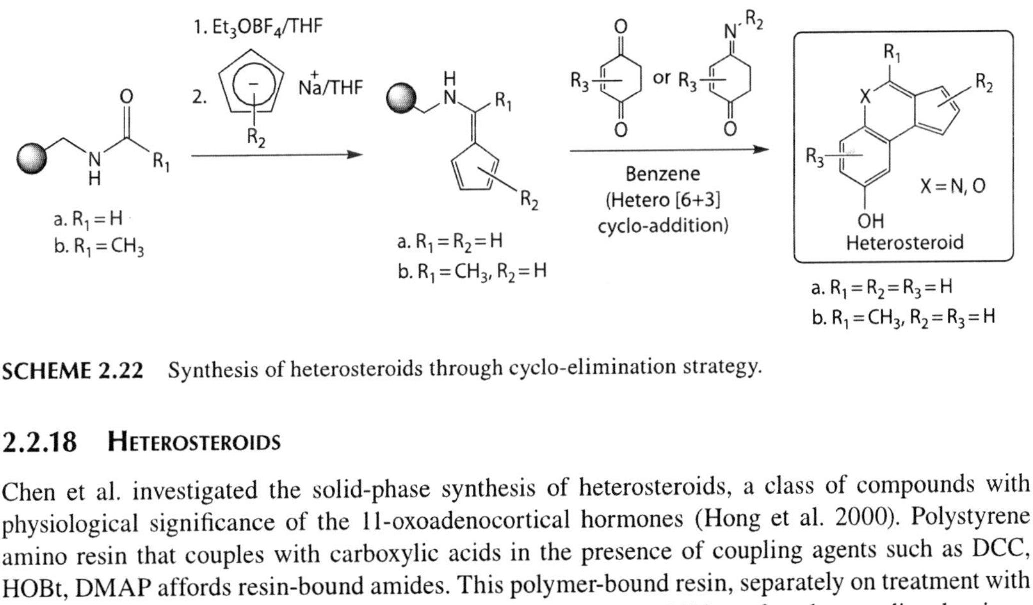

SCHEME 2.22 Synthesis of heterosteroids through cyclo-elimination strategy.

2.2.18 Heterosteroids

Chen et al. investigated the solid-phase synthesis of heterosteroids, a class of compounds with physiological significance of the 11-oxoadenocortical hormones (Hong et al. 2000). Polystyrene amino resin that couples with carboxylic acids in the presence of coupling agents such as DCC, HOBt, DMAP affords resin-bound amides. This polymer-bound resin, separately on treatment with Et_3OBF_4 in THF at 0°C for 1 h followed by the subsequent addition of cyclopentadienyl anions, affords resin-bound fulvene precursors which on reaction with benzoquinones in C_6H_6 undergo hetero [6+3] cyclo-addition to afford a library of heterosteroids (Scheme 2.22).

2.2.19 Imidazoquinazolinones

Solid-phase cyclo-elimination strategy through intramolecular nucleophilic substitution is an attractive route for the preparation of a variety of medicinally important five-, six-, and seven-membered heterocyclic compounds in high purity. Kundu et al. have synthesized imidazoquinazolinones by starting with the coupling of substituted 2-nitro benzoic acid and Fmoc anthranilic

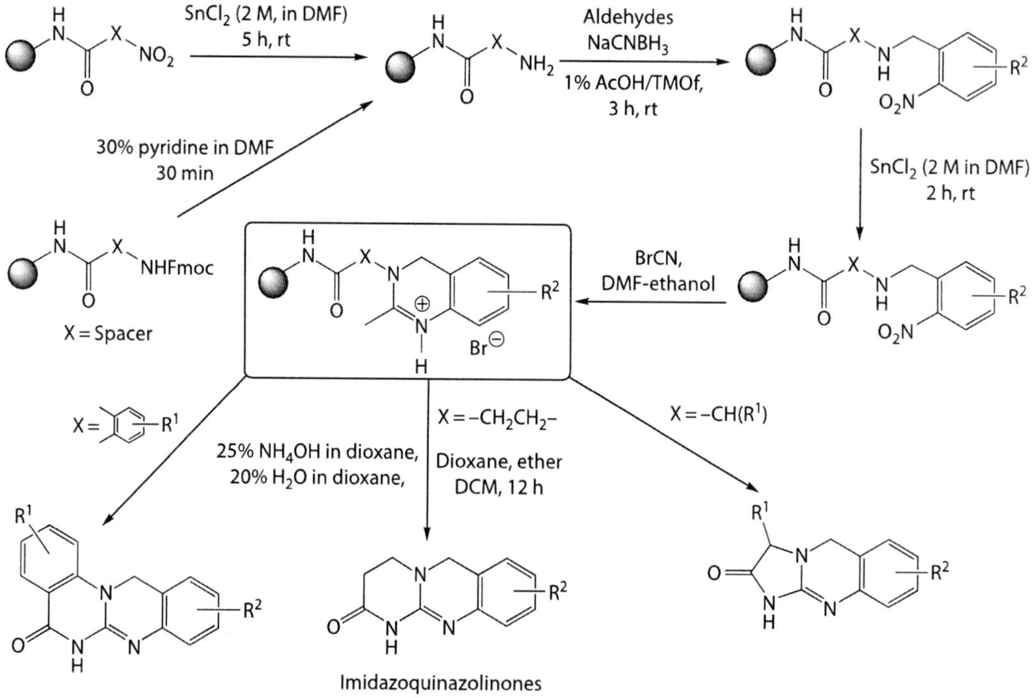

SCHEME 2.23 Synthesis of imidazoquinazolinones through cyclo-elimination strategy.

acid to the Rink Amide AM resin using DIC/HOBt method duly monitored by a negative Kaiser test for complete loading. Reduction of the *o*-nitro group with $SnCl_2 \cdot 2H_2O$ affords an amine (Grover et al. 2005; Kesarwani et al. 2005). The amine on further treatment with *o*-nitro benzaldehyde followed by reduction with $SnCl_2 \cdot 2H_2O$ affords the precursor amine. Subsequent cyclization using cyanogen bromide gives immobilized 2-aminoquinazoline which on treatment with ammonium hydroxide results in imidazoquinazolinone (Scheme 2.23).

2.3 CONCLUSION

Natural products have long been an excellent source of pharmaceutical drug discovery and development. Solid-phase cyclo-elimination strategy not just offers the opportunity of synthesizing natural products via novel routes, which may be extremely difficult using traditional solution phase methods, but also tenders the possibility for rapidly synthesizing drug-like molecules without tedious and time-consuming purification. Solid-supported cyclization–elimination strategies can effectively be utilized for the synthesis of numerous heterocyclic structures such as indolines, tetrahydroquinolines, hydrobenzofuranes, and many more that occur frequently in natural products offering a high degree of structural diversity, and have proven to be broadly useful as therapeutic agents. The results of such studies have enriched pharmaceutical research both in the area of new lead discovery and in gathering more information on structure–activity relationship (SAR), thereby strengthening the field of combinatorial chemistry. Cyclo-elimination strategies have been found to be effective in the construction of libraries because of the high purity of the final product released from the support backbone. Furthermore, the simultaneous cyclization and release of product makes the strategy more efficient than simple traceless cleavage. Thus, solid-phase synthesis of various natural product inspired heterocycles that have been reported to date illustrates several different approaches to the challenge of preparing libraries of bioactive products and allows the synthesis of many novel chemical structures.

ACKNOWLEDGMENTS

The authors thank the Council of Scientific and Industrial Research, New Delhi, for the funding. VKT thanks Dr. B. Kundu, senior scientist at Central Drug Research Institute, Lucknow, for his support during CDRI-JNU course work (2001).

ABBREVIATIONS

DBU	1,8-diazabicyclo[5.4.0]undec-7-ene
DCC	dicyclohexylcarbodiimide
DHPMs	dihydropyrimidones
DIC	diisopropylcarbodiimide
DIPEA	diisoproplethylamine
DKPs	diketopiperazines
DMAP	4-dimethylaminopyridine
DMFP	1,3-dimethyl-2-fluoropyridinium 4-toluenesulfonate
DNA	deoxyribonucleic acid
EDC	1-ethyl-3-(3-dimethylaminopropyl) carbodiimide
Fmoc	fluorenylmethyloxycarbonyl
GABA	gamma-aminobutyric acid
HOBt	hydroxybenzotriazole
HPLC	high-performance liquid chromatography
IC	inhibitory concentration
LDA	lithium diisopropylamide
MW	microwave
NPs	natural products
NPY	neuropeptide Y
PBTC	polystyrene(di-*n*-butyltin)chloride
PBTH	polystyrene(di-*n*-butyltin)hydride
PEG	poly(ethylene-glycol)
PyBOP	benzotriazol-1-yl-oxytripyrrolidinophosphonium hexafluorophosphate
RCM	ring closing metathesis
RNA	ribonucleic acid
TBAB	tetra butyl ammonium bromide
TBS	*tert*-butyldimethylsilyl
TMSOTf	trimethylsilyl trifluoromethanesulfonate
WSCDI	1-ethyl-3-(3-dimethylaminopropyl)carbodiimide

REFERENCES

Atrash B. and M. Bradley. 1997. A *pH* cleavable linker for zone dimension assays and single bead solution screens in combinatorial chemistry. *J. Chem. Soc. Chem. Commun.* 13:97–98.

Atwal K.S., B.N. Swanson, S.E. Unger et al. 1991. Dihydropyrimidine calcium channel blockers. 3′,3-Carbamoyl-4-aryl-1,2,3,4-tetrahydro-6-methyl-5-pyrimidine-carboxylic acid esters as orally effective antihypertensive agents. *J. Med. Chem.* 34:806–811.

Brodyck J.L.R. 1995. Naturally occurring tetramic acids: Structure, isolation, and synthesis. *Chem. Rev.* 95:1981–2001.

Burchenal J.H., K. Ciovacco, K. Kalaher et al. 1976. Antileukemic effects of pseudoisocytidine, a new synthetic pyrimidine *C*-nucleoside. *Cancer Res.* 36:1520–1523.

Cai D.M., M.J. Li, D.L. Li et al. 2004. Synthesis of C-nucleoside analogues: 2-[2-(Hydroxymethyl)-1, 3-dioxolan-5-yl]1, 3-thiazole-4-carboxamide and 2-[2-(Mercaptometh-yl)-1, 3-dioxolan-5-yl] 1,3-thiazole-4-carboxamide. *Chin. Chem. Lett.* 15:163–166.

DeWitt S.H., J.S. Kiely, C.J. Stankovic et al. 1993. Diversomers, an approach to nonpeptidek nonoligomeric chemical diversity. *Proc. Natl. Acad. Sci. USA* 90:6909–6913.
Dreyer D.L. and R.C. Brenner. 1980. Alkaloids of some Mexican *Zanthoxylum* species. *Photochemistry* 19:935–939.
Eifler-Lima V.L., C.S. Graebin, F.T. Uchoa et al. 2010. Highlights in the solid-phase organic synthesis of natural products and analogues. *J. Braz. Chem. Soc.* 21:1401–1423.
Falsone F.S. and C.O. Kappe. 2001. The Biginelli dihydropyrimidone synthesis using polyphosphate ester as a mild and efficient cyclocondensation/dehydration reagent. *ARKIVOC* (ii):122–134.
Feher M. and J.M. Schmidt. 2003. Property distributions: Differences between drugs, natural products, and molecules from combinatorial chemistry. *J. Chem. Inf. Comput. Sci.* 43:218–227.
Fujimoto S., K. Yoshikawa, M. Itoh et al. 2002. Synthesis of (*R*)- and (*S*)-muscone. *Biosci. Biotechnol. Biochem.* 66:1389–1392.
Grover R.K., A.P. Kesarwani, G.K. Srivastava et al. 2005. Base catalyzed intramolecular transamidation of 2-aminoquinazoline derivatives on solid phase, *Tetrahedron* 61:5011–5018.
Hajduk P.J., W.R.J.D. Galloway and D.R. Spring. 2011. Drug discovery: A question of library design. *Nature* 470:42–43.
Hofle G., N. Bedorf, H. Steinmertz et al. 1996. Epothilone A and B-novel 16-membered macrolides with cytotoxic activity: Isolation, crystal structure, and conformation in solution. *Angew. Chem.* 35:1567–1569.
Hong B.C., Z.Y. Chen and W.H. Chen. 2000. Traceless solid-phase synthesis of heterosteroid framework. *Org. Lett.* 2:2647–2649.
Houston D.R., B. Synstad, V.G.H. Eijsink et al. 2004. Structure-based exploration of cyclic dipeptide chitinase inhibitors. *J. Med. Chem.* 47:5713–5720.
Ito Y. and S. Manabe. 1998. Solid-phase oligosaccharide synthesis and related technologies. *Curr. Opin. Chem. Biol.* 2:701–708.
James I.W. 1999. Linkers for solid phase organic synthesis. *Tetrahedron* 55:4855–4946.
Joseph C.G., K.R. Wilson, M.S. Wood et al. 2008. The 1,4-benzodiazepine-2,5-dione small molecule template results in melanocortin receptor agonists with nanomolar potencies. *J. Med. Chem.* 51:1423–1431.
Kamat V.P., H. Hagiwara, T. Katsumi et al. 2000. Ring closing metathesis directed synthesis of (*R*)-(−)-Muscone from (+)-Citronellal. *Tetrahedron* 56:4397–4403.
Kappe C.O. 2000. Biologically active dihydripyrimidones of the bignelli-type, a literature survey. *Eur. J. Med. Chem.* 35:1043–1052.
Karwowski J.P., M. Jackson, R.J. Theriault et al. 1992. Tirandalydigin, a novel tetramic acid of the tirandamycin-streptolydigin type I taxonomy of the producing organism, fermentation and biological activity. *J. Antibiot.* 45:1125–1132.
Kesarwani A.P., R.K. Grover, R. Roy et al. 2005. Solid-phase synthesis of imidazoquinazolinone derivatives with three-point diversity. *Tetrahedron* 61:629–635.
Kowalski J. and M.A. Lipton. 1996. Solid-phase synthesis of diketopiperazine catalyst containing the unnatural amino acid (*S*)-norarginine. *Tetrahedron Lett.* 37:5839–5840.
Kulkarni B.A. and A. Ganesan. 1997. Ion-exchange resins for combinatorial synthesis: 2.4-pyrrilidinediones by Dickmann condensation. *Angew. Chem. Int. Ed.* 36:2454–2455.
Kulkarni B.A. and A. Ganesan. 1998. Solid-phase synthesis of tetramic acids. *Tetrahedron Lett.* 39:4369–4372.
Kwon O.S., S.H. Park, B.-S. Yun et al. 2000. Cyclo(dehydroala-L-Leu), an a-glucosidase inhibitor from *Penicillium* sp. F70614. *J. Antibiot.* 53:954–958.
Lader M. 2008. Effectiveness of benzodiazepines: Do they work or not? *Expert Rev. Neurother.* 8:1189–1191.
Lang G., A.L.J. Cole, J.W. Blunt et al. 2006. An unusual oxalylated tetramic acid from the New Zealand Basidiomycete *Chamonixia pachydermis*. *J. Nat. Prod.* 69:151–153.
Liu Y., A.D. Mills, and M.J. Kurth. 2006. Solid phase synthesis of 3-(5-arylpyridin-2-yl)-4-hydroxycoumarins. *Tetrahedron Lett.* 47:1985–1988.
Loevezijn A., J.H. Maarseveen, and K. Stegman. 1998. Solid-phase synthesis of fumitremorgin, verruculogen, and tryprostatin analogs based on a cyclization/cleavage strategy. *Tetrahedron Lett.* 39:4737–4740.
van Maarseveen J.H., J.A.J. den Hartog, V. Engelen et al. 1996. Solid phase ring closing metathesis: Cyclization/cleavage approach towards a seven membered cyclolefin. *Tetrahedron Lett.* 37:8249–8252.
Manabe S., Y. Nakahara and Y. Ito. 2000. Novel nitro Wang type linker for polymer support oligosaccharide synthesis; Polymer supported acceptor. *Synlett* 1241–1244.
Martins M.B. and I. Carvalho. 2007. Diketopiperazines: Biological activity and synthesis. *Tetrahedron* 63:9923–9932.
Mattews J. and R.A. Rivero. 1998. Solid-phase synthesis of substituted tetramic acids. *J. Org. Chem.* 63:4808–4810.
Mayer J.P., J. Zhang, K. Bjergarde et al. 1996. Solid-phase synthesis of 1,4-benzodiazepine-2,5-diones. *Tetrahedron Lett.* 37:8081–8084.

Merrifield R.B. 1963. Solid phase peptide synthesis I: The synthesis of a tetrapeptide. *J. Am. Chem. Soc.* 85:2149–2154.

Michael J.P. 2007. Quinoline, quinazoline and acridone alkaloids. *Nat. Prod. Rep.* 24:223–246.

Mirocha C. J., S.V. Pathre, J. Behrens et al. 1978. Uterotropic activity of *cis* and *trans* isomers of zearalenone and zearalenol. *Appl. Environ. Microbiol.* 35:986–987.

Mishra R.C., N. Tewari, K. Arora et al. 2003. DBU-Assisted cyclorelease elimination: Combinatorial synthesis and γ-glutamyl cysteine synthetase and glutathione-S-transeferase modulatory effect of C-nucleoside analogs. *Comb. Chem. High Throughput Screen.* 6:37–51.

Mishra B.B. and V.K. Tiwari. 2011. Natural products: An evolving role in future drug discovery. *Eur. J. Med. Chem.* 46:4769–4807.

Mishra B.B., D. Kumar, A. Mishra, P.P. Mohapatra and V.K. Tiwari. 2012. Cyclorelease strategy in solid-phase combinatorial synthesis of heterocyclic skeletons. *Adv. Heterocycl. Chem.* 107:41–99.

Muhlradt P.F. and F. Sasse. 1997. Epothilone B stabilizes microtubuli of macrophages like taxol without showing taxol-like endotoxin activity. *Cancer Res.* 57:3344–3346.

Newman D.J., G.M. Cragg and K.M. Snader. 2003. Natural products as sources of new drugs over the period 1981–2002. *J. Nat. Prod.* 66:1022–1037.

Nicholson B., G.K. Lloyd, B.R. Miller et al. 2006. NPI-2358 is a tubulin-depolymerizing agent: In-vitro evidence for activity as a tumor vascular-disrupting agent. *Anti-Cancer Drugs* 17:25–31.

Nicolaou K.C., J. Pastor, N. Winssinger et al. 1998a. Solid-phase synthesis of macrocycles by an intramolecular ketophosphonate reaction. Synthesis of a (*dl*)-Muscone library. *J. Am. Chem. Soc.* 120:5132–5133.

Nicolaou K.C., N. Winssinger, J. Pastor et al. 1997. Synthesis of epithiolones A and B in solid and solution phase. *Nature* 387:268–272.

Nicolaou K.C., N. Winssinger, J. Pastor et al. 1998b. Solid-phase synthesis of macrocyclic system by cyclo-releases strategy: Application of the Stille coupling to a synthesis of (*S*) zearalenone. *Angew. Chem. Int. Ed.* 37:2534–2537.

Olkkola K.T. and J. Ahonen. 2008. Midazolam and other benzodiazepines. *Handb. Exp. Pharmacol.* 182:335–360.

Olsen M. 1985. Inducing effect of testosterone on the hepatic reduction of zearalenone in the female prepubertal rat. *Mycotoxin Res.*1:51–56.

Park K.-H. and M.J. Kurth. 2000. Cyclo-elimination release strategies applied to solid-phase organic synthesis in drug discovery. *Drugs Fut.* 25:1265–1294.

Perdih A. and D. Kikelj. 2006. The application of Freidinger lactams and their analogs in the design of conformationally constrained peptidomimetics. *Curr. Med. Chem.* 13:1525–1556.

Perez R., T. Beryozkina, O.I. Zbruyev et al. 2002. Traceless solid-phase synthesis of bicyclic dihydropyrimidones using multidirectional cyclization cleavage. *J. Comb. Chem.* 4:501–510.

Piscopio A.D., J.F. Miller and K. Koch. 1997. Solid phase heterocyclic synthesis via ring closing metathesis: Traceless linking and cyclative cleavage through a carbon-carbon double bond. *Tetrahedron Lett.* 38:7143–7146.

Piscopio A.D., J.F. Miller and K. Koch. 1998. A second generation solid phase approach to Freidinger lactum: Application to Fukuyama's amine synthesis and cyclative release via ring closing metathesis. *Tetrahedron Lett.* 39:2667–2670.

Piscopio A.D., J.F. Miller and K. Koch. 1999. Ring closing metathesis in organic synthesis: Evolution of a high speed, solid phase method for the preparation of 8-turn mimics. *Tetrahedron* 55:8189–8198.

Raillard S.P., G. Ji, A.D. Mann et al. 1999. Fast scale-up using solid-phase chemistry. *Org. Proc. Res. Dev.* 3:177–183.

Rinehart K.L. and D. Borders. 1963. Streptolydigin. II. Ydiginic Acid. *J. Am. Chem. Soc.* 85:4037–4038.

Rivero I.A., K. Espinoza and R. Somanathan. 2004. Synthesis of quinazoline-2,4-dione alkaloids and analogues from Mexican *Zanthoxylum* species. *Molecules* 9:609–616.

Robins R.K., P.C. Srivastava, V.L. Narayanan et al. 1982. 2-β-D-ribofuranosylthiazole-4-carboxamide, a novel potential antitumor agent for lung tumors and metastases. *J. Med. Chem.* 25:107–108.

Romoff T.T., L. Ma, Y. Wang et al. 1998. Solid-phase synthesis of 3-scyl-2,4-pyrrolidiones (3-acyl tetramic acids) via mild cyclative cleavage. *Synlett* 1341–1342.

Samiulla D.S., V.V. Vaidyanathan, P.C. Arun et al. 2005. Rational selection of structurally diverse natural product scaffolds with favorable ADME properties for drug discovery. *Mol. Divers.* 9:131–139.

Schaeffer H.J., L. Beauchamp, P. de-Miranda et al. 1978. 9-(2-Hydroxyethoxymethyl)-guanine activity against viruses of the herpes group. *Nature* 272:583–585.

Sim M.M., C.L. Lee and A. Ganesan. 1998. Solid-phase combinatorial synthesis of 4-hydroxyquinolin-2(1H)-ones. *Tetrahedron Lett.* 39:6399–6402.

Sinha S., R. Srivastava, E. De Clercq et al. 2004. Synthesis and antiviral properties of arabino and ribonucleosides of 1,3-dideazaadenine, 4-nitro-1,3-dideazaadenine and diketopiperazine. *Nucleos. Nucleot. Nucleic Acids* 23:1815–1824.

Smith A.L., C.G. Thomson and P.D. Leeson. 1996. An efficient solid phase synthetic route to 1,3-disubstituted 2,4(1*H*,3*H*)-quinazolinediones suitable for combinatorial synthesis. *Bioorg. Med. Chem. Lett.* 6:1483–1486.

Stob M., R.S. Baldwin, J. Tuite et al. 1962. Isolation of an anabolic, uterotrophic compound from corn infected with *Gibberella zeae*. *Nature* 196:1318.

Szardenings A.K., T.S. Burkoth, H.H. Lu et al. 1997. A simple procedure for the solid-phase synthesis of diketopiperazine and diketomorpholine derivatives. *Tetrahedron* 53:6573–6593.

Tewari N., R.C. Mishra, V.K. Tiwari and R.P. Tripathi. 2002. DBU catalyzed cyclative amidation reaction: A convenient synthesis of c-nucleoside analogs. *Synlett* 11:1779–1782.

Thompson L.A. and J.A. Ellman. 1996. Synthesis and applications of small molecule libraries. *Chem. Rev.* 96:555–600.

Tripathi R.P., R.C. Mishra and V.K. Tiwari. 2012. Solid-phase combinatorial synthesis of carbohydrate-containing ureas with four point diversity. *Trends Carbohydr. Chem.* 4(3):28–44.

Verdine G.L. 1996. The combinatorial chemistry of nature. *Nature* 384:11–13.

Vogtle M.M. and A.L. Marzinzik. 2004. Synthetic approaches towards quinazolines, quinazolinones and quinazolinediones on solid phase. *Mol. Informat.* 23:440–459.

Wang H. and A. Ganesan. 1999. The *N*-acylimium Pictet-Spengler condensation as a multicomponent combinatorial reaction on solid and its application to the synthesis of demethoxyfumitremorgin C analogues. *Org. Lett.* 1:1647–1649.

Wang H. and A. Ganesan. 2000. Total synthesis of the fumiquinazoline alkaloids: Solid-phase studies. *J. Comb. Chem.* 2:186–194.

Wang H., T. Usui, H. Osada et al. 2000. Synthesis and evaluation of tryprostatin B and demethoxyfumitremorgin C analogues. *J. Med. Chem.* 43:1577–1585.

Weber L., P. Iaiza, G. Bbringer et al. 1998. Solid-phase synthesis of 3-acetyltetramic acids. *Synlett* 1156–1158.

Wolf D., F.J. Schmitz, F. Qiu et al. 1999. Aurantoside C, a new tetramic acid glycoside from the sponge *Homophymia conferta*. *J. Nat. Prod.* 62:170–172.

3 Synthesis and Biological Studies of Novel β-Lactams*

Bimal K. Banik

CONTENTS

3.1 INTRODUCTION

β-lactam nucleus is necessary for the biological activity of a large number of antibiotics. These are four-membered ring compounds and occasionally fused with side chains, unsaturated groups, heteroatoms, and cyclic ring systems. After the discovery of β-lactam antibiotics as useful active agents, past research has witnessed a remarkable growth in the field of β-lactams chemistry (Bose et al. 2000a,b, Singh 2004). The need for challenging β-lactam antibiotics as well as effective β-lactamase inhibitors has prompted synthetic organic and medicinal chemists to synthesize new functionalized 2-azetidionones. Besides their clinical use as antibiotics, these compounds have also been used in the preparation of various herterocyclics of biological importance (Bose et al. 2000a,b Manhas et al. 2000a,b). The use of β-lactams as therapeutic agents for lowering cholesterol has been demonstrated (Burnett et al. 1994, Clader et al. 1996). Studies on the human leukocyte elastase inhibitory mechanisms of this class of compounds have also been conducted (Finke et al. 1995). As a result, significant attention has been paid by chemists to continue working on novel β-lactam synthesis, based either on new or established methodologies or on the chemical manipulation of preexisting groups preserving at the four-membered ring.

We present herein a description of some of the most significant contributions on the preparation of the β-lactam structures published in various journals since 2000. The structural modification of functional groups linked to the nitrogen N-1, the C-3, or the C-4 carbon atoms is examined. Moreover, a concise survey of the literature on the biological and pharmacological applications of the 2-azetidinone compounds is also included.

* This chapter is dedicated to Professor LeMaster for his constant support, encouragement, and interest in our work.

3.2 BIOLOGICAL ACTIVITIES

In this section, a synthetic survey of the significant biological and pharmacological applications of β-lactam derivatives is discussed. This has been described in a book edited by Banik (Banik 2010) and particularly in a similar contribution by Troisi et al. (2010).

3.2.1 Antibacterial Activity: Inhibitors of β-Lactamases

The emergence of pathogenic microorganisms resistant to many antibiotics is a crucial clinical challenge (Cohn 1992, Neu 1992, Davies 1994). The most useful mechanism for resistance to β-lactams is the ability of bacteria to generate β-lactamases (Yang et al. 1999, Kuzin et al. 2001). These enzymes hydrolyze the β-lactam ring, thus inactivating the antibiotics. Studies of amino acid sequence homology have identified four classes of β-lactamase: A, B, C, and D. However, classes A and C are currently the most common in human disease. A successful treatment in overcoming the undesired action of these enzymes has been the coadministration of β-lactamase inhibitors with common β-lactam antibiotics. However, this method has been compromised by the discovery of new variants of β-lactamases (Belaaouaj et al. 1994, Brun et al. 1994, Sirot et al. 1994). On this basis, the development of novel β-lactam inhibitors to withstand inactivation by the diversity of β-lactamases has been considered an ongoing research agenda. Several monocyclic β-lactams have also been proved to have antibacterial activity with unique mechanisms of action.

3.2.2 Inhibitors of Various Enzymes

Leukocyte elastase (LE) is expressed by polymorphonuclear (PMN) leukocytes, mainly nutrophils, which acts to kill engulfed pathogens through an intracellular process (Sternlicht and Werb 1999). Because LE has the ability to degrade some proteins of the extracellular matrix (ECM), for example, elastin, fibronectin, and collagens, excess of LE activity has been found in a number of pathological situations that lead to the impairment of ECM organization (rheumatoid arthritis, emphysema, cystic fibrosis, and tumor progression). LE also stimulates the proenzymatic components of matrix metalloproteinase (MMP)-9 (Sternlicht and Werb 1999) released by the PMN leukocytes and key to their extravasation (Delclaux 1996, Esparza 2004). A number of β-lactams, widely used as antimicrobials, have been shown to act as inhibitors of these serine enzymes. Inhibitors of LE and HLE have similarities in their structures. Many of them are built on the cephem system and are bicyclic compounds (clavams and cephalosporins).

Many other representatives of the class of β-lactams have also been found to effectively inhibit proteases. 2-Azetidinones have also been discovered as novel inhibitors of thrombin, a serine protease involved in both venous and arterial thrombotic episodes. Analogous compounds have also been found to display inhibition toward tryptase.

3.2.3 Azetidin-2-ones as Vasopressin V1a Antagonists

The neurohypophysical hormones vasopressin and oxytocin exhibit many physiological properties. They bind to specific membrane receptors belonging to the G protein-coupled receptor superfamily (Jard et al. 1998). Three pharmacologically active vasopressin receptor subtypes and one oxytocin receptor have been known (Jard et al. 1998). Vasopressin acts in the cardiovascular system. However, it also works in the central nervous system (CNS). Arginine vasopressin expresses a neurochemical signal in the brain. Vasopressin, through the vasopressin 1A receptor (V1a) in particular, can stimulate aggressive behavior of animal and human. It has been demonstrated that specific β-lactam can also work as the essential scaffold of several antagonists directed to the vasopressin V1a receptor.

3.2.4 Hypocholesterolemic and Antihyperglycemic Activity

Atherosclerotic coronary artery disease (CAD) is one of the major causes of death. Reducing dietary fat and cholesterol is considered the best therapy. But use of effective pharmacological compounds has led to an increased use of drug therapy to control cholesterol. Serum cholesterol can be minimized by targeting endogenous cholesterol biosynthesis, promoting hepatic cholesterol clearance from the plasma, and inhibiting the absorption of dietary and biliary cholesterol. 2-Azetidinones have been studied as important inhibitors of cholesterol absorption. It is interesting to note that monocyclic β-lactams have been reported for antidiabetic activity, because they can control hypercholesterolemia.

3.2.5 Anticancer Activity

Recently discovered anticancer monocyclic and bicyclic β-lactam systems (Veinberg et al. 2000) indicate that azetidin-2-one pharmacophore is of inexhaustible pharmacological potential on account of the specific ability of its numerous derivatives to inhibit not only bacterial enzymes but also mammalian serin and cyctein proteases (Finke et al. 1993). As a measure of cytotoxicity, several compounds have been assayed against nine human cancer cell lines (Banik et al. 2003 and subsequent papers in this series). A clan of new β-lactam antibiotics based on *N*-methylthio-substituted 2-azetidinones also demonstrated apoptosis-inducing properties against human solid tumor cell lines such as breast, prostate, and head-and-neck (Smith 2002).

3.2.6 Antiviral Activity

Human cytomegalovirus (HCMV) is a crucial member of the herpes virus family. Most infections are asymptomatic. Severe manifestations of HCMV can also be seen in individuals whose immune system has been weakened by diseases (cancers and AIDS) (Holwerda 1997, Field 1999, Waxman and Darke 2000). Due to its critical function in capsid assembly and viral maturation, studies on HMCV serine protease has become an attractive target for the clinical development of anti-HMCV drugs (Deziel and Malenfant 1998).

3.3 SYNTHESIS OF IMPORTANT β-LACTAMS

The results of the experiments carried out with *N*-(silyl) imine **2** and TIPS-ketene **1** (Bacchi et al. 1998) for the preparation of β-lactam were noteworthy. Under certain conditions, [4 + 2] cycloaddition of this imine with TAS-vinylketene gave products in perfect yield; however, in some instances, a good yield of the [2 + 2] cycloadduct **3** was obtained. This was performed by a reaction without solvent at higher temperature. The operation of unique and different mechanisms for the two types of cycloadditions—stepwise [2 + 2] for **1** and concerted [4 + 2] for the much more facile reaction of vinylketene—explained this serious observation. In contrast, both reactions proceeded by stepwise pathways where the determining rate 6π electrocyclic closure in the [4 + 2] process occurs far sooner than the 4π electrocyclization involved in the [2 + 2] cycloaddition reaction of **1** (Scheme 3.1).

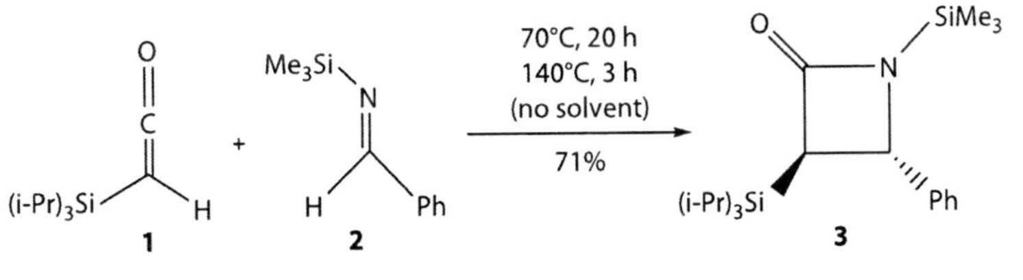

SCHEME 3.1 [2 + 2]-Cycloaddition between *N*-(silyl) imine **2** and TIPS-ketene **1**.

Ketene-imine cycloaddition is one of the best ways of synthesizing the framework of interest (Staudinger 1907). Bose et al. demonstrated that penam derivatives formed with the desired stereochemistry if Δ^2-thiazolines were reduced with ketenes and generated *in situ* from base and acid chlorides (Bose et al. 1974). However, the compounds were not optically active and the reported yields were a low 11% when Δ^2-thiazolines such as **5** were used. To overcome the problem, an attempt to increase the yield in the cycloaddition step via a selenium-Δ^2-thiazoline was reported afterward. Most importantly, the yields in the cycloadditions reached up to 92%. Four steps, with a total yield of <20%, were necessary to synthesize the selenium-containing thiazoline. A reductive demethylselenation step had to be performed after the cycloaddition. Additionally, acid chlorides were not suitable for the assembly of acyl ketenes **7**. Other reports (Yamamoto et al. 1987) suggested Meldrum's acid derivatives (e.g., **4**) as acyl ketene components. The simplicity with which ketenes were created from derivative **4** was exceptional since a suitable method tolerant of different functional groups was required to facilitate the synthesis of optically active β-lactams. There were no previous reports on cycloadditions between ketenes generated *in situ* from Meldrum's acid derivatives and Δ^2-thiazolines. They reacted under anhydrous and acidic conditions to give the preferred optically active β-lactams with complete stereoselectivity (Scheme 3.2). The Δ^2-thiazoline **5** was prepared in two steps. It was prepared from commercially available L-cysteine methyl ester hydrochloride (Almqvist 1998) with an overall yield of >70%. Additionally, the derivatives **4** were easily synthesized from Meldrum's acid and acid chlorides in >80% yields (Yamamoto et al. 1987). No decomposition was detected in all derivatives prepared while stored in the freezer. A series of penam derivatives were created with aryl-, *n*-hexyl-, and cyclohexyl substituents. Excellent yields of the corresponding β-lactams **6a**, **6b**, **6e**, and **6f** were accomplished (72%–93%). However, the yields were somewhat lower when a methylene group was present between the carbonyl group and the aryl moiety. Compounds **6c** and **6d** were obtained in 62% and 65% yields, respectively (Scheme 3.2).

The regio and diastereoselectivity of the carbon–carbon bond formation were investigated through the indium-mediated reaction between the 2,3-azetidinedione (+)-**8** and propargyl bromide in aqueous tetrahydrofuran at room temperature. The product was obtained with total diastereoselectivity of the 3-substituted 3-hydroxy-β-lactam. The observed regioselectivity however was very poor (42:58) in favor of the allenic product. The regiochemical preference was reversed on the indium-promoted reaction simply by changing the solvent. A saturated aqueous solution of NH_4Cl in Tetrahydrofuran (THF) was used instead of aqueous tetrahydrofuran, with the expected alcohols (+)-**9** and (+)-**10** being obtained as a mixture of regioisomers in a ratio of 71:29. This initial result encouraged to find a better reagent for this transformation. The optically pure homopropargyl alcohol (+)-**9** as single regio

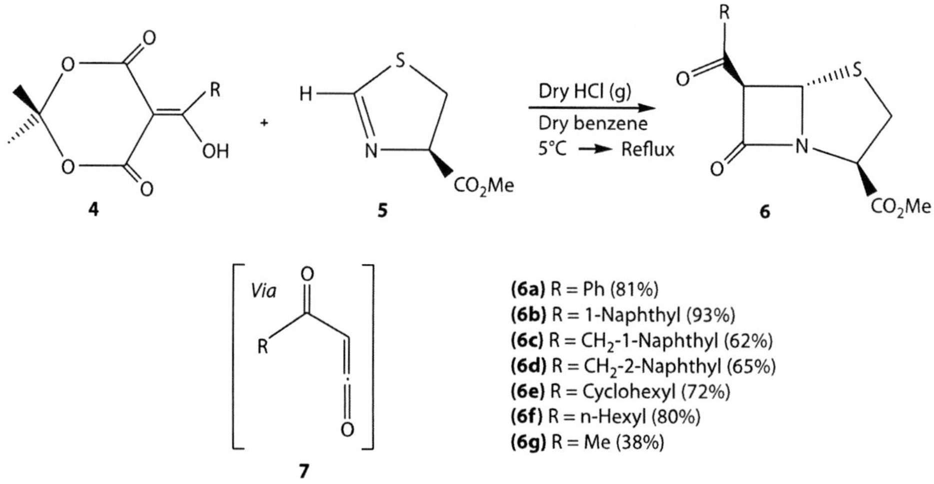

SCHEME 3.2 Synthesis of β-lactams **6a–g**.

and diastereoisomer in a reasonable 70% yield was obtained when the above reaction was mediated by zinc and conducted in a saturated aqueous solution of NH_4Cl in THF at 0°C. Nonetheless, the yield could not be improved when the zinc-mediated coupling of the 2,3-azetidinedione (+)-**8** and propargyl bromide in anhydrous THF in the presence of solid NH_4Cl was carried out. There was no observed reaction in anhydrous THF when the NH_4Cl was suppressed. The change of the solvent from tetrahydrofuran/NH_4Cl (aqueous saturated) to methanol/NH_4Cl (aqueous saturated) resulted in loss of regioselectivity. Similarly, when propargylmagnesium bromide was added to the dione (+)-**8**, the homopropargylic alcohol was prepared as a major product. This product contained 15% of the homoallenic alcohol. The tin-mediated reaction between ketone (+)-**8** and propargylbromide in aqueous tetrahydrofuran resulted in lack of coupling. On the other hand, when the same experiment was performed in a saturated aqueous solution of NH_4Cl in THF, the homoallenyl alcohol was formed as major product. Allenylation with propargyl bromide in anhydrous THF using the copper (II)/tin(II) as the promoter lowered the regioselectivity (34:66) (Yi et al. 1998). Efforts to promote the copper (II)/tin(II)-mediated propargylation in Dimethylformamide (DMF) were also demonstrated to be sluggish. Comparable results were obtained in the metal-mediated Barbier reactions of different *N*-substituted azetidine-2,3-diones **8**, with propargyl bromide (Scheme 3.3).

The next focus was to identify an allenylation method that proceeds in a regio and diastereoselective fashion. Metal-mediated reactions of 2,3-azetidinediones **11** with propargyl bromides with an aliphatic or an aromatic substituent at the terminal position afforded the homoallenyl alcohol **13** (Scheme 3.4). This result was in contrast to the metal-promoted reaction of propargyl bromide (Isaac and Chan 1995, Yi et al. 1998). Due to structural differences in the organometallic reagents, the difference in behavior in different products was attributed. A metallotropic rearrangement between the propargylmetal and allenylmetal species was speculated. This suggests that both intermediates were able to react with the 2,3-azetidinediones **11**, through a six-membered transition state, leading to homoallenyl or homopropargylic alcohols (Scheme 3.5) (Isaac and Chan 1995, Yi et al. 1998).

The ester enolate–imine condensation was tested in the liquid phase on model compound **16**. This reaction afforded β-lactam **17** in 71% yield. Relative configuration of **17** was determined by NOE experiments (Scheme 3.6).

Optically pure substrates starting from (–)-(1*R*)-tricarbonyl[*N*-(2-fluorobenzylidene)-4-methoxyaniline] chromium **18** were reported (Scheme 3.7). Imine **18** was obtained in nearly quantitative yield.

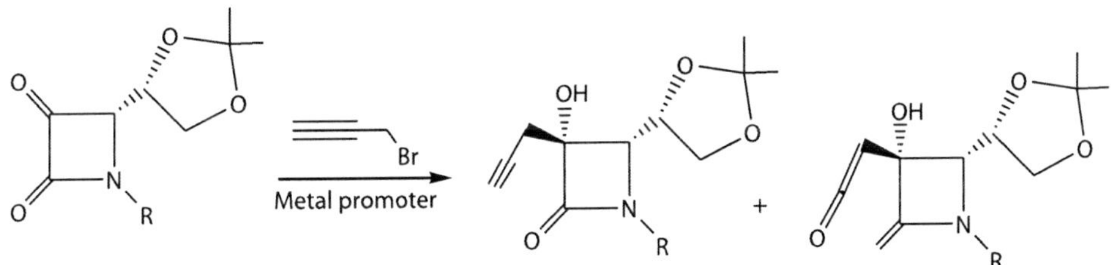

SCHEME 3.3 Metal-mediated Barbier reactions of different *N*-substituted azetidine-2,3-diones **8** with propargyl bromide.

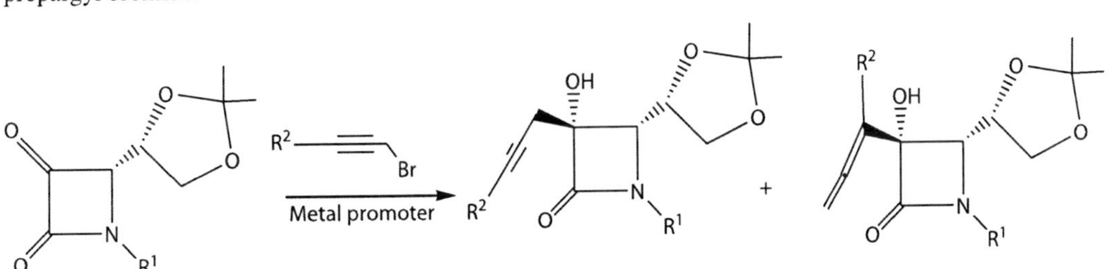

SCHEME 3.4 Formation of homoallenyl alcohol **13** through metal-mediated Barbier reaction.

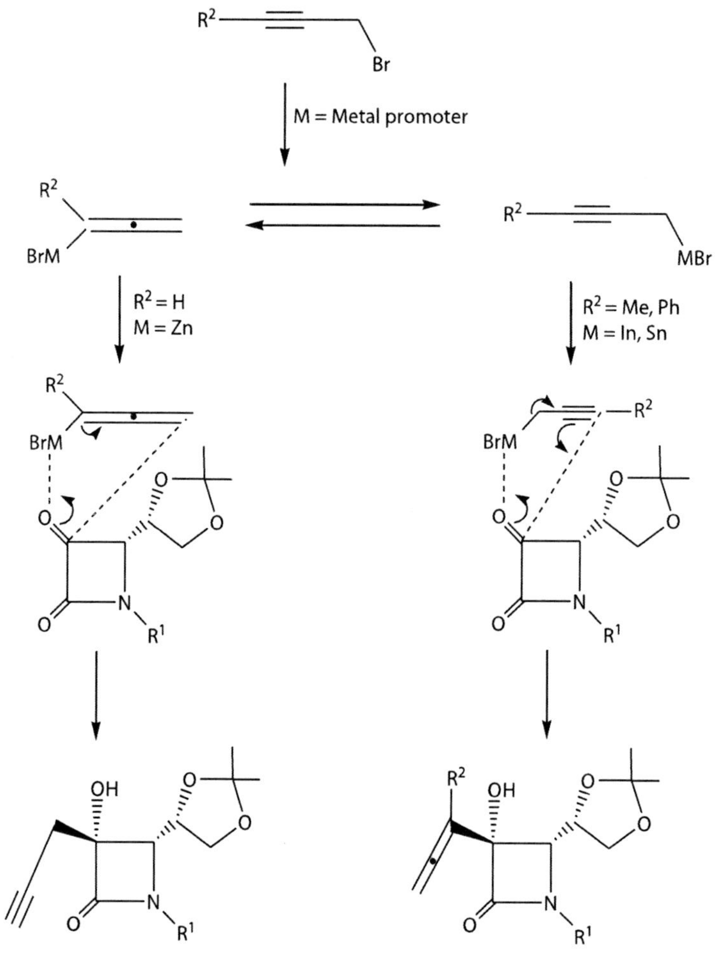

14 **15**

SCHEME 3.5 Plausible metallotropic rearrangement between the propargylmetal and allenylmetal leading to the formation of homoallenyl or homopropargylic alcohols.

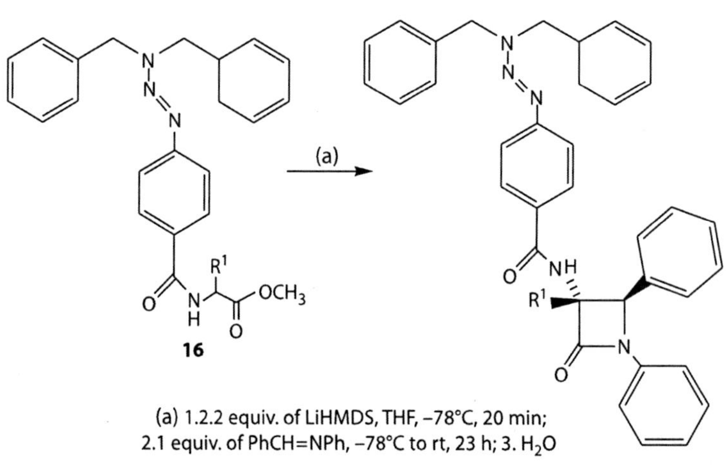

SCHEME 3.6 Synthesis of β-lactam **17**.

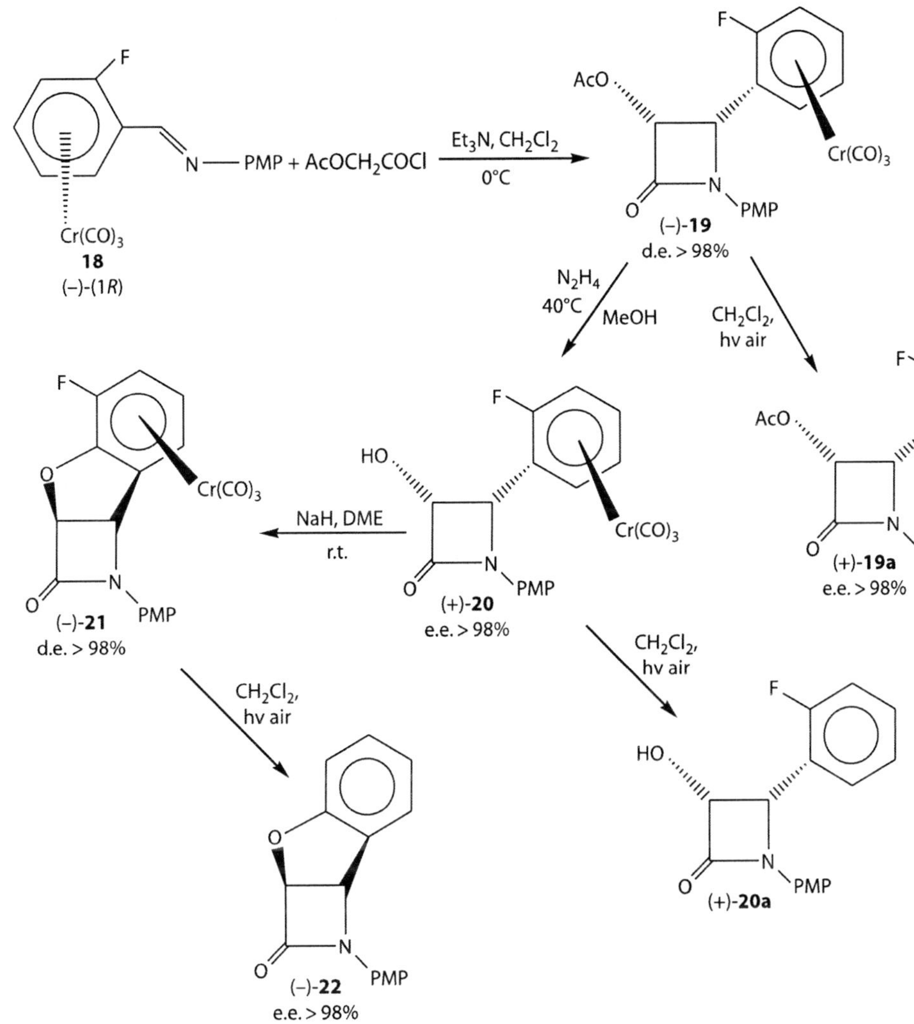

SCHEME 3.7 Optically pure β-lactams starting from (−)-(1*R*)-tricarbonyl[*N*-(2-fluorobenzylidene)-4-methoxyaniline]chromium **18**.

This was obtained from the corresponding (−)-(1*R*) benzaldehyde complex (Baldoli et al. 1991). *Cis*-β-lactam **19** was obtained in 94% yield as a single diastereoisomer on [2 + 2] cycloaddition of imine (±)-**1** and acetoxyacetyl chloride at 0°C in the presence of Et_3N in CH_2Cl_2. The corresponding *cis* 3-hydroxy β-lactam **20** was then prepared by treatment of **19** with hydrazine in methanol in 85% yield was prepared, intermediate for the tricyclic structure. The intramolecular displacement of the fluorine atom in **20** was performed by treatment with NaH at room temperature. This reaction produced product **21** as a single diastereoisomer in 50% yield. Spectroscopic and analytical data were consistent with the tricyclic structure. An exposure of **21** to air and sunlight in CH_2Cl_2 solution gave uncomplexed **22**. A single crystal of racemic complex **21** by x-ray confirmed the (Burla et al. 1989) tricyclic structure. The $Cr(CO)_3$ group lies on the opposite side of the C-3 and C-4 hydrogens of the β-lactam ring.

Synthesis of β-lactams **25a–d** at moderate temperature in good yields and excellent stereoselectivity (Scheme 3.8, Table 3.1) was achieved using ketene compounds derived from carbohydrate and different acyclic imines. When triethylamine and the imine were subsequently added at 0°C to a solution of the acid **23** and **24** in dichloromethane, best results were found. For all compounds, a ratio of >99:1 was found of the two possible diastereomeric *cis* products. No *trans*-configured product could be observed. The absolute configuration was established by x-ray structure analysis of

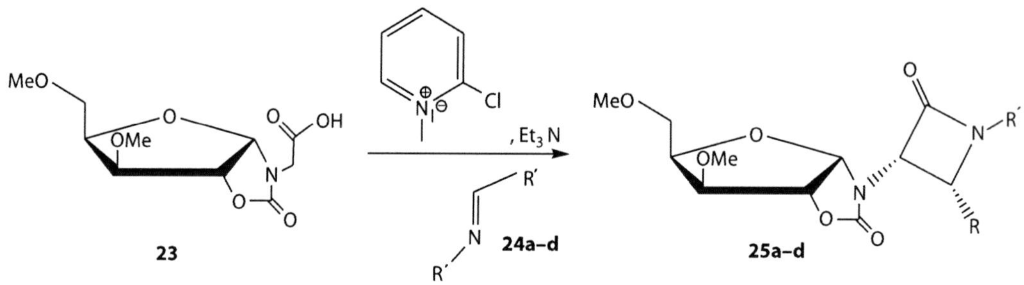

SCHEME 3.8 Synthesis of β-lactams **25a–d**.

TABLE 3.1
Preparation of Optically Active β-Lactams Starting from Furanose Derivative

Compound	*R*	R′	Yield [%][a]	drb
25a	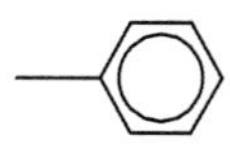	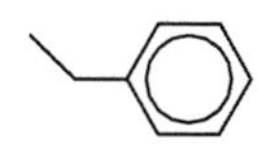	67	>99:1
25b	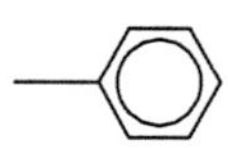	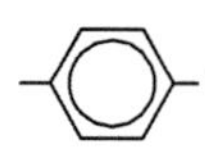	69	>99:1
25c	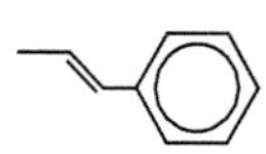	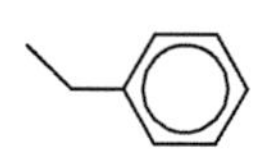	71	>99:1
25d	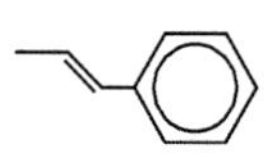	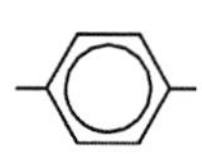	58	>99:1

[a] Isolated yield.

25d (Sheldrick 1990). Since there are small variations in the imine structures, an identical (3*S*,4*R*)-configuration for **25a–c** was expected.

Introduction of chiral groups onto the enamidic structure was achieved by the usual synthetic procedure (Speckamp and Hiemstra 1985). Enamides **26** were reacted with $Mn(OAc)_3 \cdot 2H_2O$ in glacial acetic acid (Scheme 3.9). As expected, *trans* azetidinones **29** were formed. With regard to product stereochemistry, enamides (obtained from (*R*)-(+)- and (*S*)-(−)-phenylethylamine, respectively) gave no diastereoselective reaction. Enamides obtained from cyclohexyl and naphthylamine gave better stereochemical results. Consequently, different enamide **26** were prepared, and their reactions with Mn(III) were investigated. A considerable level of diastereoselection (ca. 80:20 dr) was observed when the enamide chiral center was linked to a secondary or tertiary carbon as R^2. Disappointingly, the structure of the prevalent oily diastereoisomers could not be determined by either x-ray diffraction data or normal spectral analysis. However, semiempirical molecular calculations suggested that, when a stereoselection is detected, the prevailing compound is the diastereoisomer **29**″.

Experiments were performed using achiral hydrazones **30a** and **30b** as well as *in situ*-generated benzyloxyketene as model compounds. These reactions in toluene led to the configuration of desired cycloadducts **32a** and **32b** in 84% and 67% yields, respectively. Differences in nucleophilicity determined the product distribution (Lassaletta et al. 1996, Fernandez et al. 1998). Additionally a strong influence of the solvent was recognized. Use of chloroform, CH_2Cl_2, THF, or Et_2O resulted in the formation of complex mixtures. 1-benzyloxy-3-(pyrrolidin-1-ylimino)-propan-2-one and its dimethyl analog (1,2 adducts to the ketene) were major products. Formaldehyde SAMP hydrazone **30** (Enders et al. 1996) reacted with

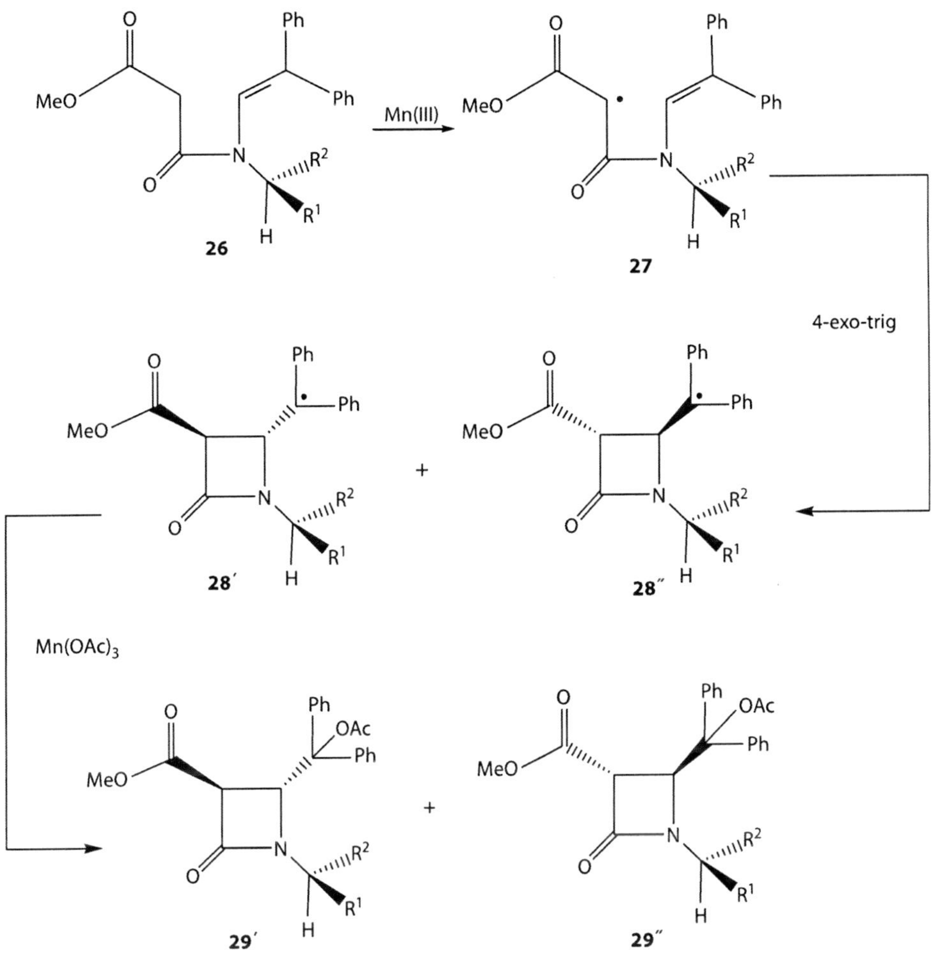

SCHEME 3.9 Introduction of chiral groups onto enamidic structure.

ketene **31** in a comparable way. The low inductions observed, however, prompted investigation of the behavior of available hydrazones **30** (Fernandez et al. 1998) which have quaternary residues. The higher steric hindrance in these reagents resulted in an increase in activity in the cycloaddition method (d.r. values of 81:19 and 84:16). In some cases, both diastereomers (*R*, *S*)—and (*S*, *S*)-**32e** could be separated by flash chromatography. Using this method afforded a satisfactory 80% yield of the major diastereomer (*R*, *S*)-**32** in a single step (Scheme 3.10). Furthermore, the behavior of the C_2-symmetric reagent **30** was investigated. This reagent is of considerable importance because of several reasons. A better stereoselectivity could be expected due to the frequently observed benefits for C_2-symmetric auxiliaries (Whitesell 1989). The incidence of two neighboring groups at both sides (C-2 and C-5) of the C–N double bond and the rigidity present in the 1,3 dioxane rings should impede the planar conformation associated with the opposing nucleophilic reactivity of the azomethine carbon (Pareja et al. 1999). The parent hydrazine was prepared from D-mannitol in bulk quantities in a few steps (Defoin et al. 1991).

Cycloaddition of hydrazones **30** to various α-amino-ketenes **33** and (*R*)-**34** was also investigated. The best results were obtained by using suitable α-amino acid derivatives and 2-chloro-*N*-methyl pyridinium iodine in this case for the ketene formation (Amin et al. 1979). With these conditions, adduct **35** was obtained in 93% yield as a 82:18 mixture of diastereomers. As expected, a "matched" double induction experiment using (*R*)-**34** (Evans and Sjogren 1985) provided the corresponding adduct **36** as a single diastereoisomer. Moreover, compound **35**, which has the opposite S configuration at C-3, was also obtained in good yield, as a single diastereoisomer from the reaction of **30** with **33** (Scheme 3.10).

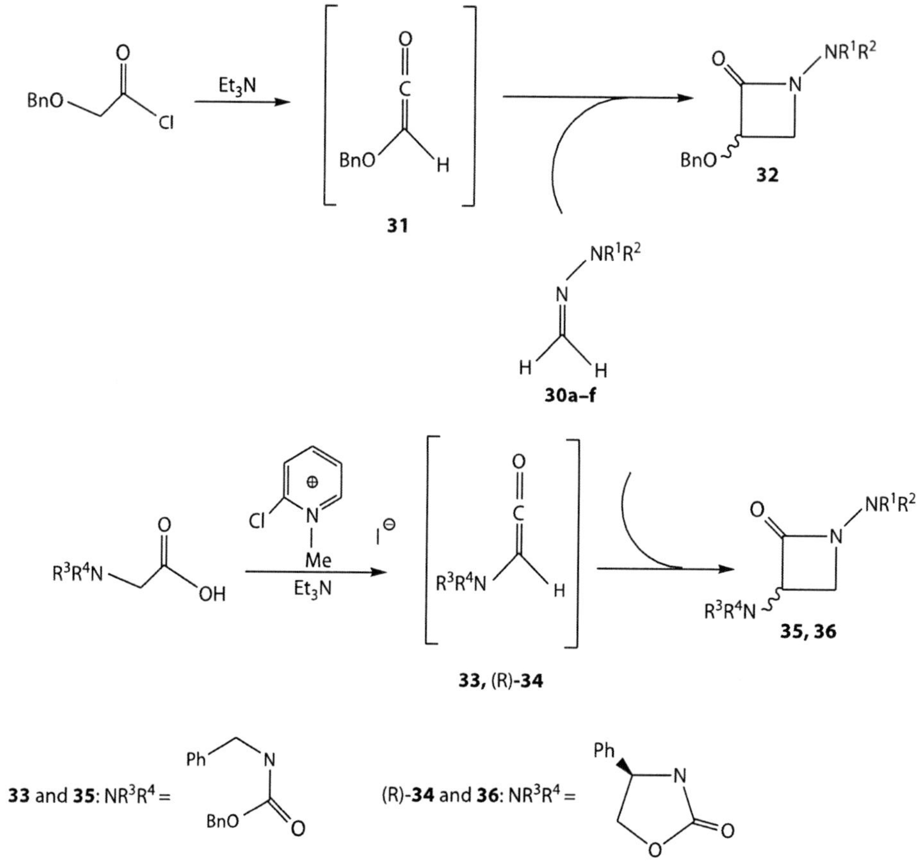

SCHEME 3.10 Diastereosynthesis of β-lactams **25a–d**.

α-Acetoxy-β-lactams **(37)** were hydrolyzed to alcohols **(38)** in excellent yields. This was done under very mild conditions (Brieva et al. 1993). Subsequent oxidations were performed by treatment with DMSO in the presence of phosphorus pentoxide (Palomo et al. 1994). This was conducted to give α-keto-β-lactams **(39)** in good yields as shown in Scheme 3.11.

Aldimines would be anticipated to react with ynolates **42** to yield β-lactam enolates. Barrett et al. (Adlington et al. 1981) and Murai et al. (Kai et al. 1996) reported that phenyl-substituted and silyl-substituted ynolates were added to aldimines bearing an electron-withdrawing substituent to provide β-lactams (2:1 adducts) at −60°C and α,β-unsaturated amides, respectively. It was also reported that lithium ynolates added to *N*-sulfonyl imines afforded β-lactam **43** (1:1 adducts) at −78°C (Scheme 3.12) (Shindo et al. 1998). The cycloaddition of the lithium ynolate with the *N*-2-methoxyethyl imine was attempted; however, the reaction proceeded very slowly. The *N*-2-methoxyphenyl (OMP) aldimine as a substrate was then investigated. It was expected that the 2-methoxy group may act as a

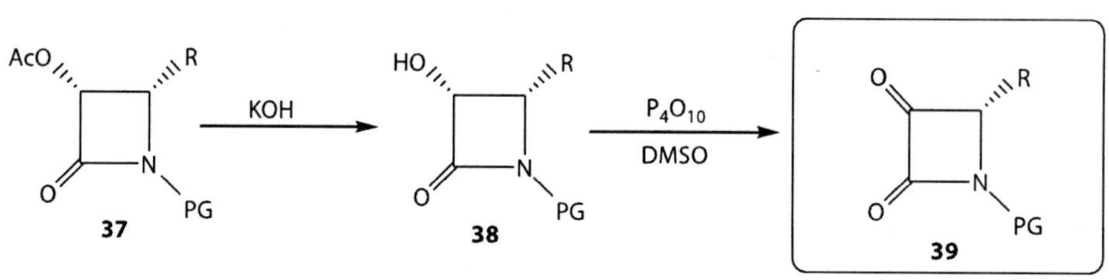

SCHEME 3.11 Formation of α-keto-β-lactams **39**.

SCHEME 3.12 Formation of β-lactam **43** via addition of lithium ynolate to *N*-sulfonyl imine.

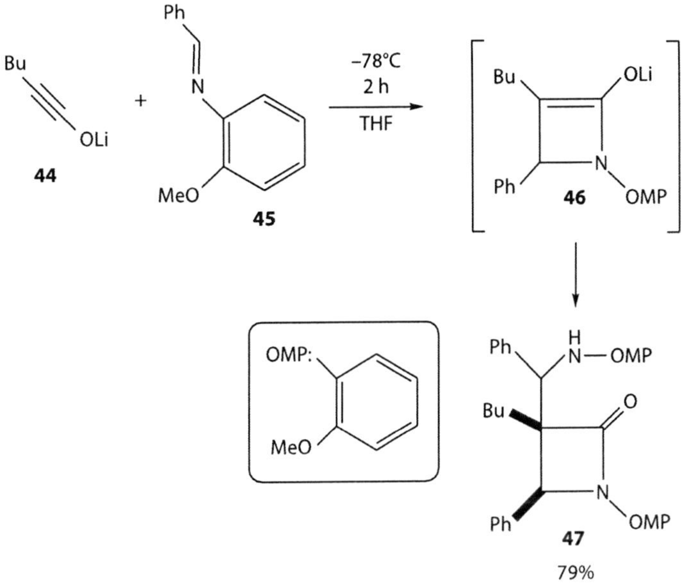

SCHEME 3.13 Formation of β-lactam **47**.

coordination site. The imine **45** was added to a THF solution of the lithium ynolate **44**, prepared from the dibromo ester and *tert*-BuLi, at –78°C. Notably, the imine disappeared in 2 h at –78°C. After workup, followed by isolation of the major products, the β-lactam **47** was produced in 79% yield as a single isomer (Scheme 3.13).

A two-step sequence oxidative hydrolysis of compound **48** was performed without isolation of the nitrone intermediate **49** (Dirat et al. 1998). β-aminoaldehyde **50** and ketol **51** (Berranger and Langlois 1995) were isolated after aqueous acidic hydrolysis, in 64% yield. The aldehyde group in compound **50** was next oxidized into β-urethane acid derivative **52** (Dalcanale and Montanari 1986). Excess of palladium on charcoal was necessary for the deprotection of the benzyloxycarbonyl group. Cyclization to β-lactam derivative **53** was completed with dicyclohexylcarbodiimide in acetonitrile (Dalcanale and Montanari 1986). Furthermore, compound **53** has been identified with β-lactam **55**, a synthetic precursor of carpetimycin A (**1**) (Limori et al. 1983). As a result, treatment of **53** with trimethylsilyl chloride produced a mixture of protected β-lactams **54a** and **54b**. Saponification of the ester group in **54a** was possible to prepare compound **55**. This proved to be identical in all respects with the compound described by Ohno (Limori et al. 1983) (Scheme 3.14).

The trisubstituted amidines **56** on cycloaddition reaction with ketenes derived from acid chloride (**57a,b**), in the presence of triethylamine, afforded exclusively *trans*-β-lactams (**58**) in very good yields (Scheme 3.15). Surprisingly, these β-lactams (**58**) were found to be very stable since they also did not undergo any decomposition even after storing for several months at room temperature.

Effect of a variety of imines **59** with ethyl bromoacetate in the presence of indium metal using anhydrous tetrahydrofuran as the solvent created the β-lactams **60** (Scheme 3.16). No reaction was observed with *tert*-butyl bromoacetate and the imine was recovered unchanged from this in 65% and 48% yields, respectively. However, a small amount of unsaturated ester (8%) was also formed

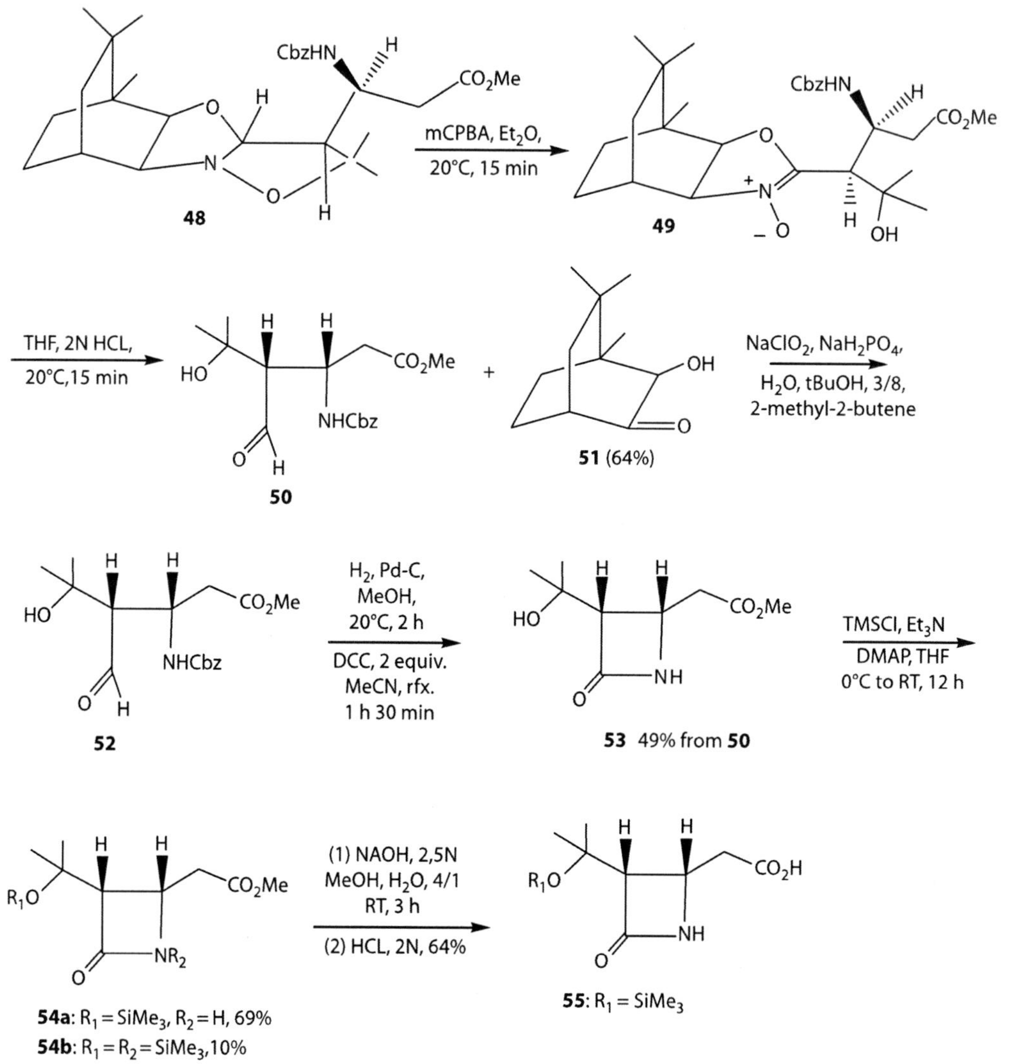

SCHEME 3.14 Formation of β-lactam **55**.

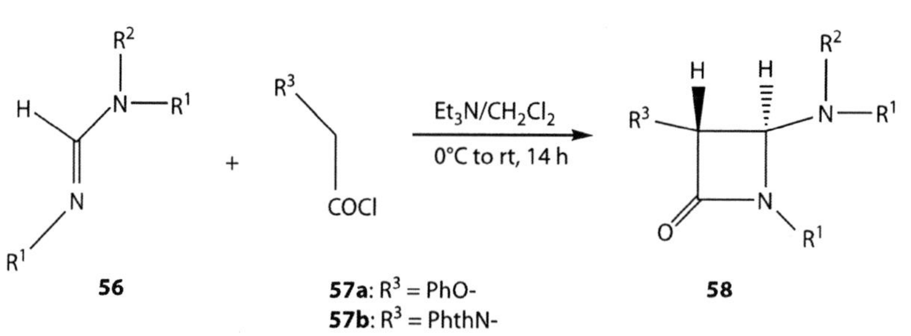

SCHEME 3.15 Formation of *trans*-β-lactams (**58**).

from the reaction of **61a** and ethyl iodoacetate. The formation of the unsaturated ester could be explained by the decomposition of the imine **61a** in the presence of the iodoester, Reformatsky-type addition to the resulting benzaldehyde and a subsequent elimination reaction (Table 3.2).

In order to estimate the enantioselectivity of the intramolecular alkylation reaction, 2-azetidinone-4-carboxylic acids obtained after hydrolysis of the corresponding alkyl esters was altered into the

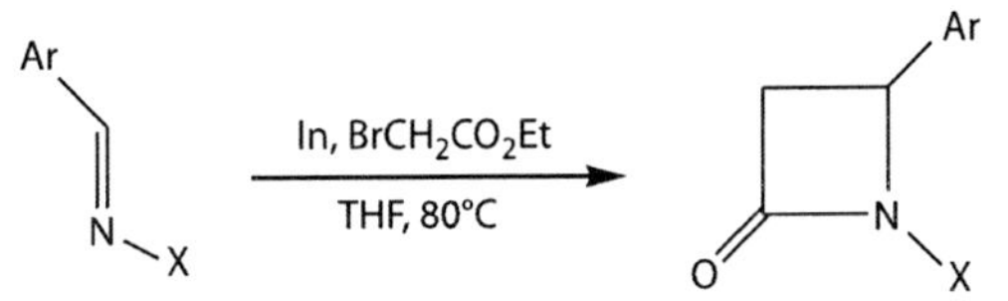

59 60

SCHEME 3.16 Formation of β-lactam **60**.

TABLE 3.2
Preparation of 3-Unsubstituted β-Lactams

Entry	Imine	β-Lactam [Yield (%)[b]]
1	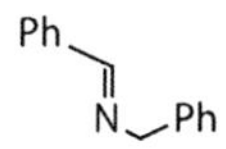 **61a**	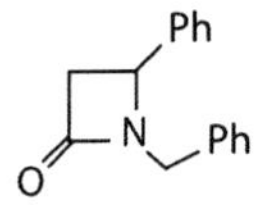 **62a**
2	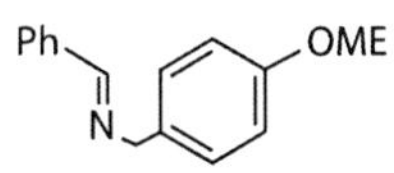 **61b**	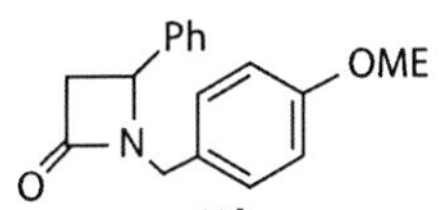 **62b**
3	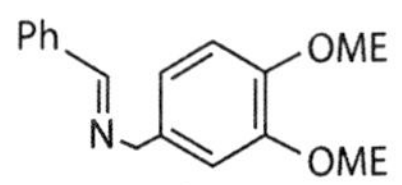 **61c**	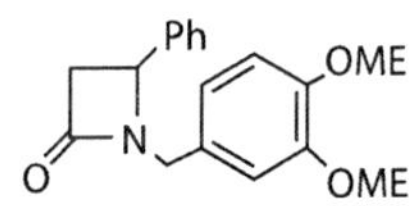 **62c**
4	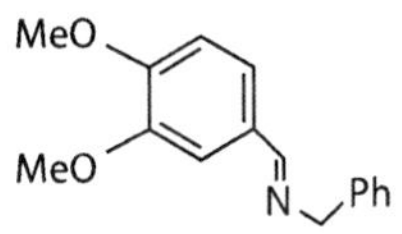 **61d**	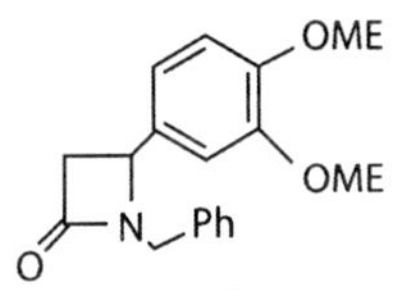 **62d**

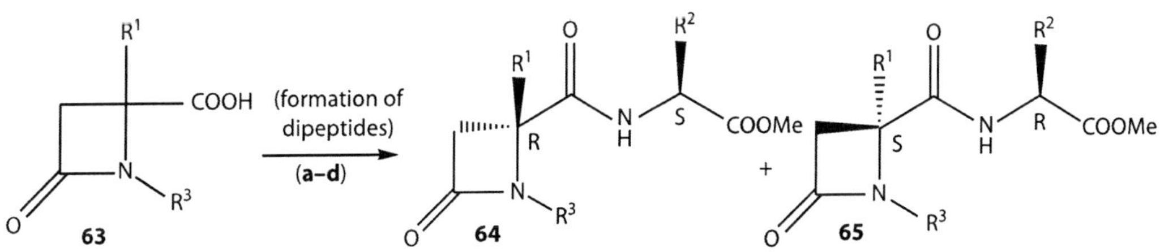

a: $R^1=CH_2Ph$; $R^2=CH_3$; $R^3=Bzl$
b: $R^1=CH_2Ph$; $R^2=CH_3$; $R^3=Pmb$
c: $R^1=CH_3$; $R^2=CH_2Ph$; $R^3=Bzl$
d: $R^1=CH_2CH(CH_3)_2$; $R^2=CH_2Ph$; $R^3=Pmb$

SCHEME 3.17 Preparation of dipeptide derivatives of β-lactams **64** and **65**.

dipeptide derivatives **64** and **65** (Scheme 3.17). The stereoselectivity was much higher when CH_3CN was used as solvent than when DMF was used. When the formation of the starting 4-benzyl-2-azetidinone was induced by Cs_2CO_3, compared to NaH, the diastereomeric excess was increased by 22%, indicating the importance of the countercation on the observed stereoselectivity. With regard to the effects of the R_1 substituent, Leu-derived dipeptides were obtained in a 71:29 ratio (42% d.e.).

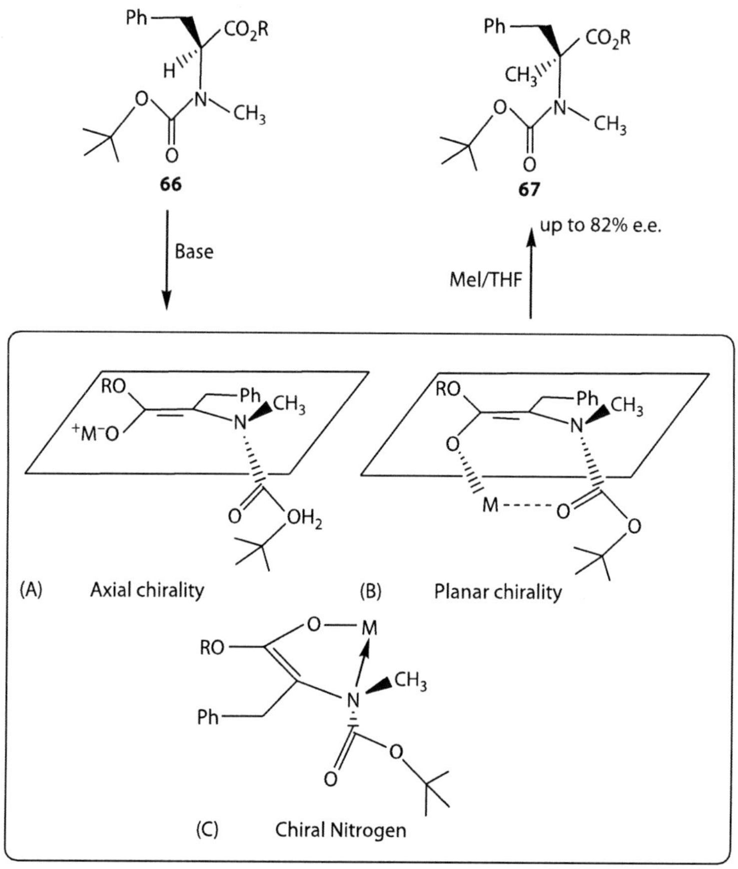

SCHEME 3.18 Transition state model for the conversion of **66** to **67**.

In contrast, the reaction of **63** with H-L-Phe-OMe led to the corresponding dipeptides **64** and **65** in a ratio of 50:50. It has been described that *N*-methyl-*N*-Boc-(*S*)-phenylalanine alkyl ester **66** can be converted into enantiomerically enriched *R*-methyl derivatives **67** via direct alkylation without employing any external chiral source. The transition state model of this reaction is shown in Scheme 3.18 (Kawabata et al. 1994).

The attempted cyclization of compound **71** obtained from **68** following conditions originally reported by Miller (Miller et al. 1980) failed to produce appreciable β-lactam products. Consequently, the amounts of DEAD and PPh_3 were increased. The reaction was attempted in different solvents at room and high temperature or with heating. TCT-AliR test was positive in each case. Alternatively, FT-IR of the beads showed the incidence of a weak signal around 1770 cm^{-1} demonstrating that the β-lactam was formed. Freshly distilled DEAD (Pansare et al. 1991) in THF promoted cyclization to **73** with high conversion. In a similar way, compound **72** produced **74**. At this point, two alternatives were feasible for product removal from the resin. N–O bond was then cleaved to give β-lactams **75–78**) cleave the *N*-trityl bond to give 1-hydroxy-β-lactams **77** and **78**. The first method was accomplished using a reductive cleavage with SmI_2 (Myers et al. 2000). The products were isolated from the solution after passage through a silica gel column as well as hydrolytic workup. Products **75** and **76** were achieved in good yields (45% and 52% calculated from the loading of the original trityl-hydroxylamine resin **69**). Conversely, acidic cleavage with 5% TFA in CH_2Cl_2 for 3 h and aqueous workup gave compounds **77** and **78** in moderate yields (35%) (Cainelli et al. 1997) (Scheme 3.19).

The ^1H-NMR spectrum of the racemic complex **79** shows H(4) and H(6) at δ 4.64 and 5.47 with a coupling constant of 2.4 Hz and the two nonequivalent CH_2 protons. Imines **80a**, **80b**, **80c**, and **80d** were prepared as yellow crystals by condensation of **79** with 4-fluoroaniline, benzylamine,

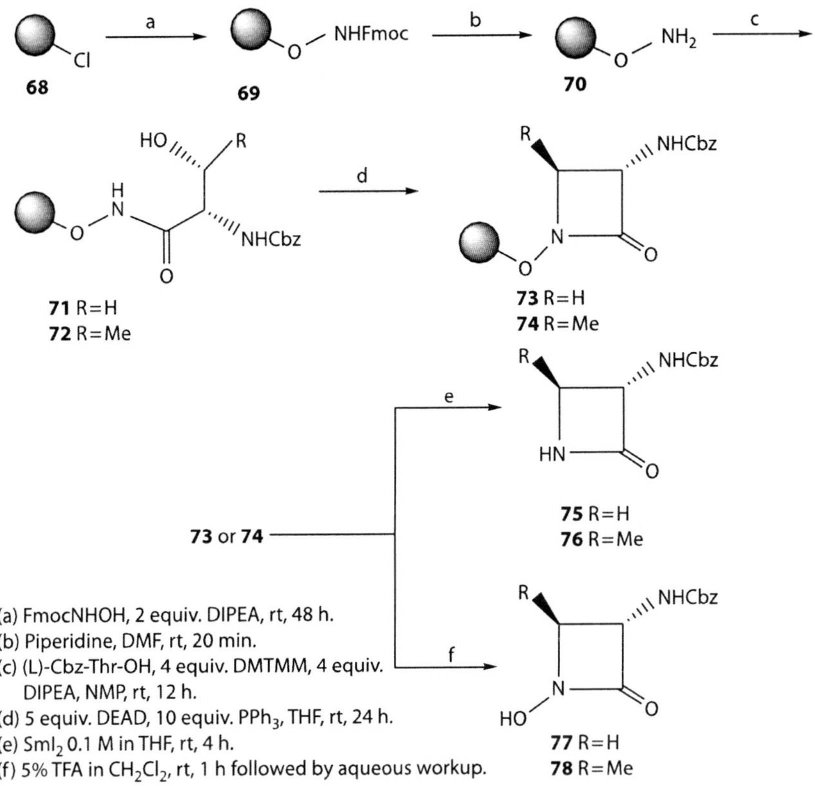

SCHEME 3.19 Preparation of β-lactams **75–78**.

aniline, and phenyl hydrazine in good yields (Scheme 3.20). The cycloaddition reactions were examined between the reactive ketenes generated *in situ* from acetoxyacetyl chloride and phenylacetyl chloride. Treatment of the imine complexes **80a–c** with acetoxyacetyl chloride and triethylamine in dichloromethane as solvent at 0°C led to the *cis*-β-lactam complexes **81a–c**, respectively. Inspection of the ^{1}H-NMR spectral data of their crude mixtures confirmed that only one stereoisomer had been formed. The *cis* disposition of the vicinal methine protons at C(3′) and C(4′) in each azetidin-2′-one ring of the complexes was confirmed easily on the basis of their coupling constants ($J_{3',4'} \approx 5.4$ Hz). These complexes were separated as single *cis*-diastereomers following column chromatography and recrystallization. The above method was extended to imine **80d** and the ^{1}H-NMR spectrum of the crude mixture. This did not show the formation of the corresponding β-lactam complex. Following distillation, complex **82** was isolated as an orange powder in 50% yield and its ^{1}H-NMR spectrum showed the presence of the CH_3COOCH_2CO group and two *meta* protons in the complexed arene ring and the two nonequivalent CH_2 proton singlets of the dioxolane ring. The ^{13}C-NMR spectrum also confirmed the presence of the CH_3COOCH_2CO group and DEPT. This lack of cycloaddition [2 + 2] could be explained when considering the presence of the NHPh group attached to the nitrogen of the imine in **80d**. Probably, the nucleophilicity of the nitrogen lone pair of the imine complex was diminished by the electron-withdrawing effect of the NHPh group attached to it (March 1985). As a result, this did not allow the cycloaddition reaction to the ketene. But, abstraction of the acidic NH proton by triethylamine occurred, which in turn allowed the reaction with the acid chloride to proceed, and this route was designed to access complex **82**. Phenylacetyl chloride was then used to produce *in situ* the corresponding ketene, and the imine complex **80a** was allowed to react. Regardless of several reactions, there was no trace of the corresponding β-lactam complexes detected. It was not easy to predict how phenyl and aryl substituents would behave because its conformation may vary. Due to the interactions between the ketene substituents and the substituents

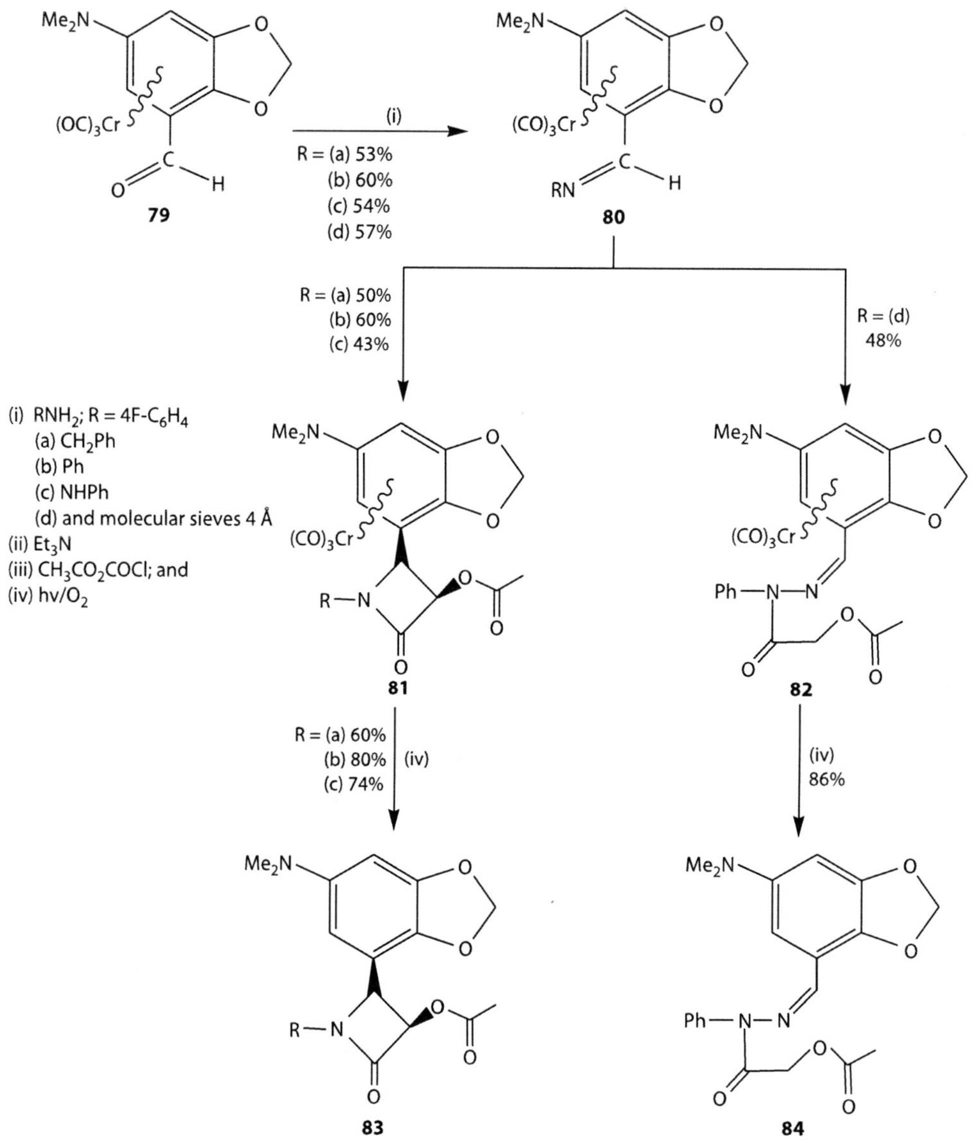

SCHEME 3.20 Preparation of β-lactams **83**.

of the imine complexes, different results were obtained. Complexes **81a–c**, and **82** were subjected to decomplexation reactions by exposing their ethereal solutions to air and sunlight. This was continued until the yellow color of the solutions disappeared. Following purification, the decomplexed *cis*-β-lactams **83a–c**, and compound **84** were isolated (Scheme 3.20).

The preparation of monobactam like 1,4-bisaryl β-lactams **88** was investigated. The condensation reaction of **85** was carried out with numerous bisarylimines. These all gave very good results concerning loading of the lactam resins **86** and purity of the cleaved diazonium salts **87**. These diazonium salts were stable at room temperature and were analyzed by ^1H NMR spectroscopy in d_4-MeOH. The β-lactams **88**, obtained from diazonium salts **87**, were easily separated from yellow by-products by dissolving in EtOAc/pentane (2:3) and eluting through silica gel. Many substituted β-lactams **87a** were prepared in high purity (84%–98%). Good-to-excellent diastereomeric excess (50% to ≥96%) and reasonable-to-good yields (26%–71%) were also seen. However, this reaction was challenging; challenging cases were R^1 = H and phenyl, where either no product was formed or

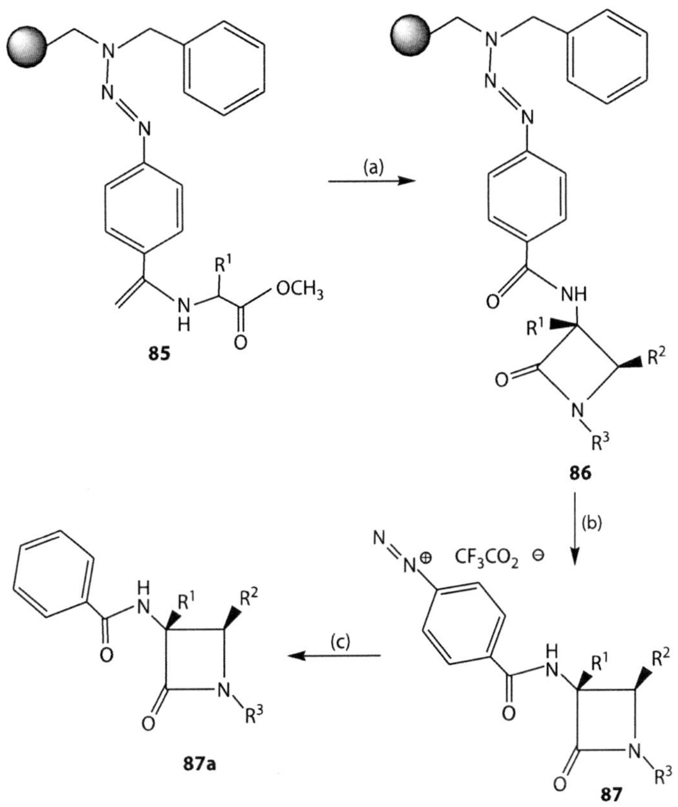

Reagents and conditions: (a) (1) p-Aminobenzoic acid (5 equiv.), BF_3Et_2O (10 equiv.), t-BuONO (10 equiv.), THF, −10°C, 1 h. (2) 6, pyridine/DMF (1:1), rt, 1 h. (b) 7, amino acid methyl esterase HCl (3 equiv.), 2-chloro-1-methylpyridinium iodide (2 equiv.), NEt_3 (20 equiv.), CH_2Cl_2, rt, 12 h. (c) 5% TFA in CH_2Cl_2.

SCHEME 3.21 Preparation of *trans*-β-lactams **88f,h,j,k**.

a product was formed in low yield and purity. This supported that the nucleophilicity of the R-group is of higher importance than steric hindrance. Optimum results were obtained with tertiary esters. In the case of phenylic substitution, that reduced the nucleophilicity of the R-position, the corresponding β-lactam was achieved, in low yield. Very good diastereomeric excesses were obtained from all β-lactams. No change was observed in diastereomeric purity by comparing the ^{1}H NMR spectra of diazonium salts **87** and lactam **87a**. NOE experiments confirmed the *trans* configuration of β-lactams **88f**, **88h**, **88j**, and **88k** (Scheme 3.21).

Synthetically attractive conversion of *R*-oxoamides (**88**) into β-lactams (**89**) was achieved (Scheme 3.22) (Akermark et al. 1969). Consequently, the carboxylic acid-containing R-oxoamide **88a** (Scheffer and Wang 2001) was treated with a variety of optically pure amines. Irradiation of the crystalline salts was performed under nitrogen on 5 mg samples squeezed in between Pyrex microscope slides. L-Prolinamide proved to be the best chiral auxiliary. It also produced optically pure photoproduct **89c** at nearly 100% conversion. Interestingly, exceedingly low e.e. was observed in the case of the 1-phenylethylamine salt. Photolysis of the salts in methanol gave negligible enantiomeric excesses, for example, 5% e.e. in the case of the L-prolinamide salt that dictates the importance of the crystalline state to these results.

To aid the formation of the β-lactam, the Staudinger reaction was chosen (Ruhland et al. 1996) using Wang resin as a solid support. Compound **91** was used as starting material and then deprotected by treatment with 30% piperidine in DMF to obtain resin **92** (Scheme 3.23). The solid-supported

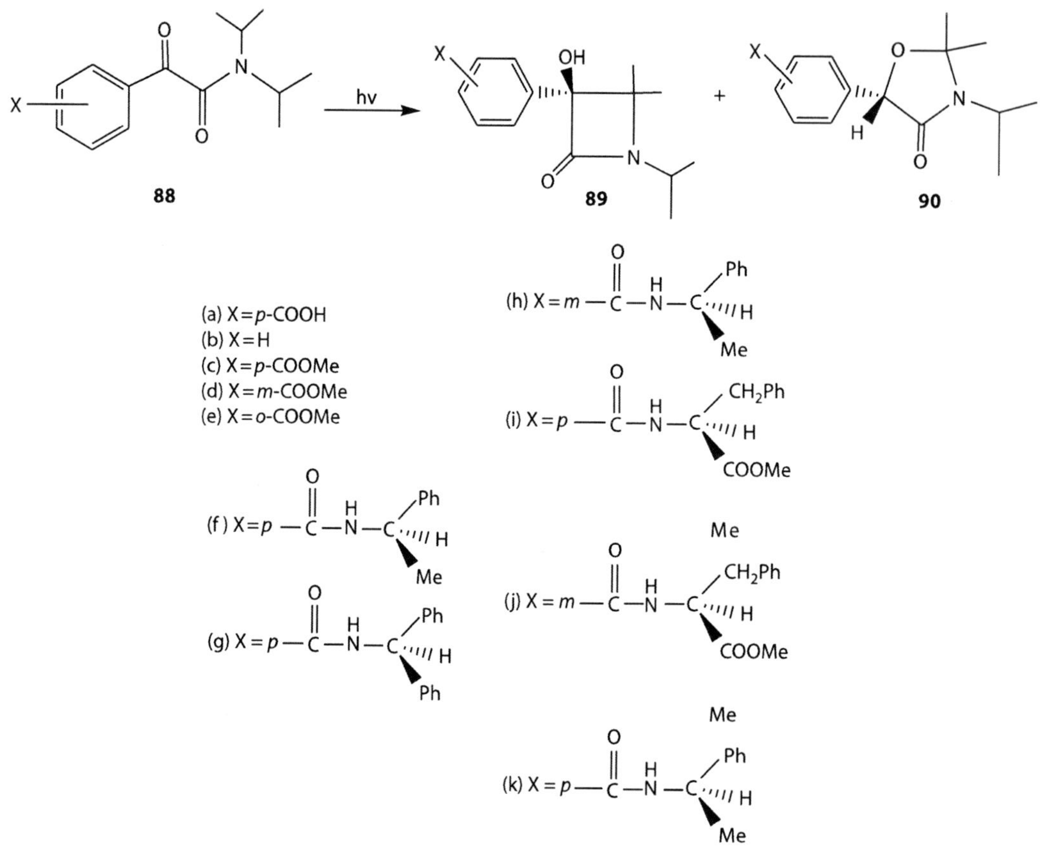

SCHEME 3.22 Conversion of *R*-oxoamides (**88**) into β-lactams (**89**).

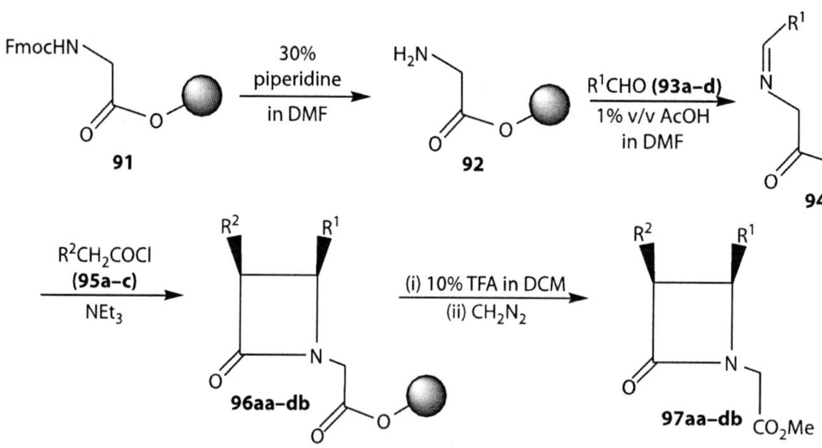

SCHEME 3.23 Solid-supported synthesis of β-lactams **97**.

amine **92** was condensed with 3,4-dimethoxybenzaldehyde **93a** (R_1 = 3,4-dimethoxyphenyl) in DMF containing 1% acetic acid based on the method developed by Boyd (Boyd et al. 1996) and the aldimine **94a** was produced. Cycloaddition with the ketene achieved *in situ* from the phenoxyacetyl chloride **95a** (R_2 = PhO) and Et_3N provided the preferred resin-bound β-lactam **96aa**. This reaction sequence was followed using ^{13}C gel-phase NMR and FT-IR spectroscopy. Resin-bound β-lactam **96aa** was then cleaved. Whereas treatment with aluminum chloride **96a** gave 40% yield, the use of

10% trifluoroacetic acid in dichloromethane was found to be the effective method for the cleavage. Given that it afforded the β-lactam **97aa** as its methyl ester in 78% isolated yield. Subsequently, an asymmetric version of this solid-phase synthesis was investigated in order to develop a useful procedure for the generation of chiral intermediates for the synthesis of biologically interested β-lactams. The asymmetric Staudinger reaction was carried out on the solid support between **95d** and the different resin-bound aldimines **94a–e** and in the presence of triethylamine.

Starting from commercially available (+)-3-carene **98**, the cycloaddition of CSI produced the enantiomeric β-lactam **99** in a regio and stereoselective reaction. The CSI addition was possible at room temperature in a respectable yield (76%). There are numerous methods in the literature for the ring opening of azetidinones (Palomo et al. 2001). Refluxing **99** with ethanol-containing HCl to obtain the corresponding amino ester (Szakonyi et al. 2000) was attempted. With aqueous HCl, hydrolysis of the β-lactam of **99** to the amino acid was also attempted (Furet et al. 1999). None of these applied methods resulted in the expected compounds. It appeared that under highly acidic conditions, the strongly constrained carene ring system decomposes. Treatment of the β-lactam **99** with di-*tert*-butyl dicarbonate formed *N*-Boc β-lactam **100**. Interestingly, this β-lactam was readily opened under mild conditions. Homochiral *N*-Boc β-amino acid **103** in excellent yield (98%) was produced when reaction of **100** with aqueous LiOH in THF was preformed (Furet et al. 1999). The *N*-Boc amino ester was prepared in a different way. This was done by ring opening of *N*-Boc β-lactam **100** in the presence of a catalytic amount of sodium methoxide in dry methanol. Alternatively, the esterification of *N*-Boc β-amino acid **103** with diazomethane in dry diethyl ether produced **102** in excellent yield. *N*-Boc amino ester **102** was reduced by lithium aluminum hydride to *N*-methyl amino alcohol **101**. This was performed after the deprotection of **102** to the corresponding amino alcohol **104**. β-Amino ester **105** was converted to β-amino acid **106** by refluxing in a dioxane:water = 1:1 mixture (Tamagnan et al. 1996) (Scheme 3.24).

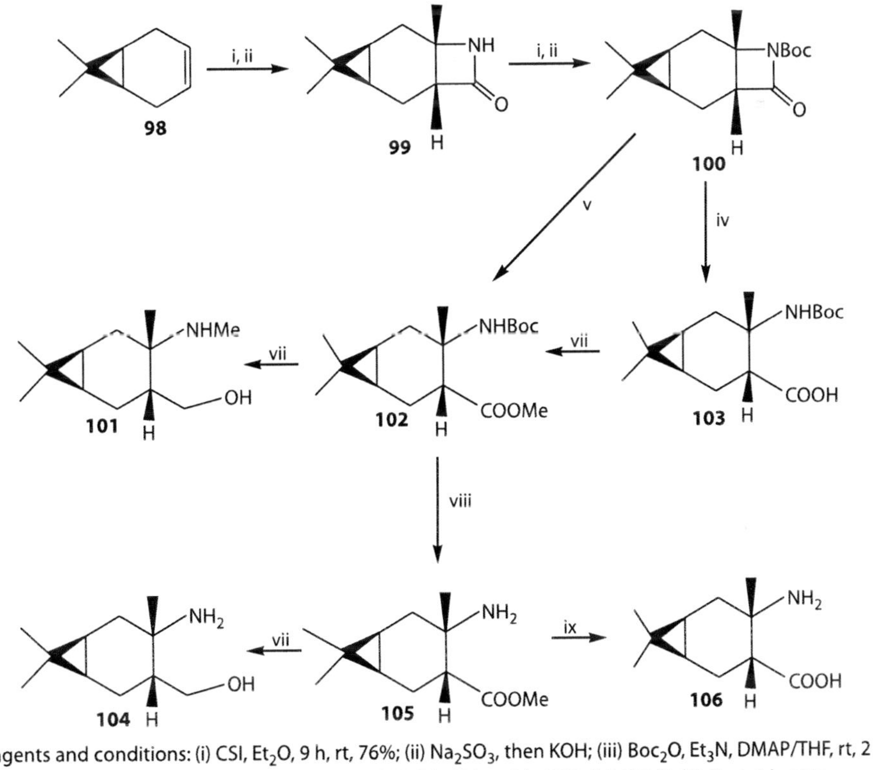

Reagents and conditions: (i) CSI, Et_2O, 9 h, rt, 76%; (ii) Na_2SO_3, then KOH; (iii) Boc_2O, Et_3N, DMAP/THF, rt, 2 h, 82%; (iv) $LiOH/H_2O$, THF, rt, 7 h, 94%; (v) cat. NaOMe/MeOH, rt, 2 h, 89%; (vi) CH_2N_2/Et_2O, rt, 2 h, 98%; (vii) $LiAlH_4$/THF, rt, 2 h, **101**: 95%, **104**: 85%; (viii) TFA/CH_2Cl_2, rt, 2 h, 96%; (ix) dioxane/H_2O, reflux, 48 h, 88%.

SCHEME 3.24 Reactions of (+)-Carene **98**.

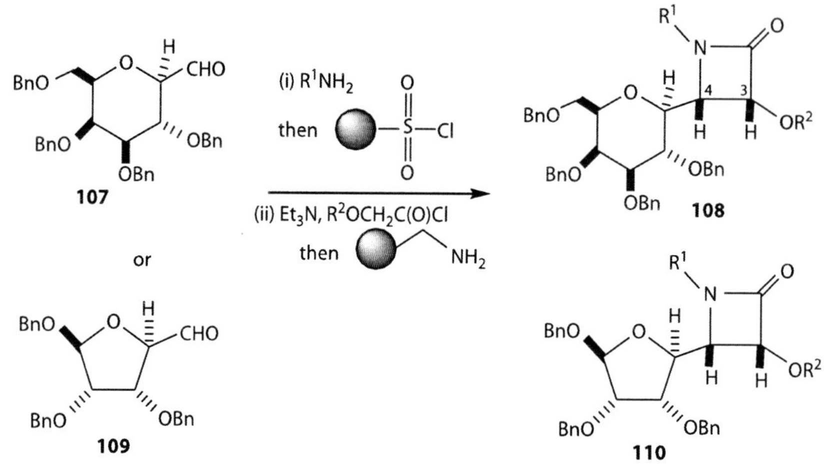

SCHEME 3.25 Conversion of β-*C*-galactosyl formaldehyde **107** or *C*-ribosyl formaldehyde **109** to the corresponding β-lactams.

A chiral *C*-glycosylimine was generated in CH_2Cl_2. This was done by mixing the β-C-galactosyl formaldehyde **107** (Dondoni and Scherrmann 1994) with an excess of p-methoxybenzylamine (R^1 = PMB) (Scheme 3.25). By treatment with resin-supported sulfonyl chloride, the unreacted alkylamine was easily removed. The resulting heterogeneous mixture was treated with acetoxy acetyl chloride (R^2 = Ac) and triethylamine to produce the corresponding acetoxy ketene (Taggi et al. 2002). After a period of time, the reaction mixture was treated with nucleophilic aminomethylated resin. Simple workup of the resulting suspension and solvent evaporation afforded a mixture of 4-(C-galactosyl)-β-lactams (3*R*, 4*S*)-3a and (3*S*, 4*R*)-3a **108**. The conservation of the β-linkage at the anomeric carbon of the sugar group and the *cis*-relationship of the C-3 and C-4 protons of the β-lactam ring were established by ^{1}HNMR analysis. Complete configuration of the latter carbon atom was appointed by chemical correlation of (3*R*, 4*S*)-3a. Following the optimized cyclocondensation protocol, the C-ribosyl formaldehyde **109** (Dondoni and Scherrmann 1994) was also effectively transformed into the corresponding **110**-(C-ribosyl)-β-lactam (3*R*, 4*S*)-**110a** (Scheme 3.25).

D-Phenylalanine ethyl ester **111** was reacted with cinnamaldehyde to afford the chiral imine phenylalanine (*N*-cinnamylidene) ethyl ester **112** in 85% yield (Hakimelahi and Jarrahpour 1989). Reaction of phthalimidoacetyl chloride with chiral Schiff base **112** and triethylamine produced a single stereoisomer β-Lactam **113** in 80% yield (Mukerjee and Singh 1978). ^{1}H NMR was used to confirm the *cis*-configuration of **113**. Ozonolysis of β-lactam **113** at −78°C provided formyl β-lactam **114** in 90%. The formyl β-lactam was then reduced to the 4-hydroxymethyl β-lactam **115**. This was achieved in 60% yield by treatment with lithium tri(*tert*-butoxy)aluminum hydride in dry THF. β-Lactam **116** was obtained through mesylation of **115** with methanesulfonyl chloride and triethylamine at −78°C in dichloromethane. The mesyl β-lactam **116** was treated with 1,5-diazabicyclo[5.4.0] undec-5-ene (DBU) in 9:1 THF–DMF at reflux temperature for 4 h. **117** was obtained as a yellow solid in 20% yield (Scheme 3.26).

Staudinger-like [2 + 2] cycloaddition of chiral, aliphatic *N*, *N*-dialkylhydrazones **120** to *R*-aminoketenes **119** for the synthesis of 3-amino-4-alkylazetidin-2-ones **121** and derivatives was investigated. Hydrazones **120** containing *C*2-symmetric (2*R*, 5*R*)-2,5-dimethylpyrrolidine was chosen as the auxiliary. The excellent stereocontrol and high reactivity observed in cycloadditions to benzyloxyketene was very helpful (Fernandez et al. 2002). The *N*-benzyloxycarbonyl-*N*-benzylglycine **118** as the source of aminoketene **119** and 2-chloro-*N*-methyl pyridinium iodide was used as an activating agent (Scheme 3.27). This selection was based on the previous experience with 4-unsubstituted derivatives (Fernandez et al. 2000). Cycloadduct **121** in low yields (25%–50%) resulted from experiments carried out under certain conditions (Et_3N, toluene, heat). However, reaction mixture analysis indicated the formation of a single stereoisomer. Coincidentally, a screening of different reaction

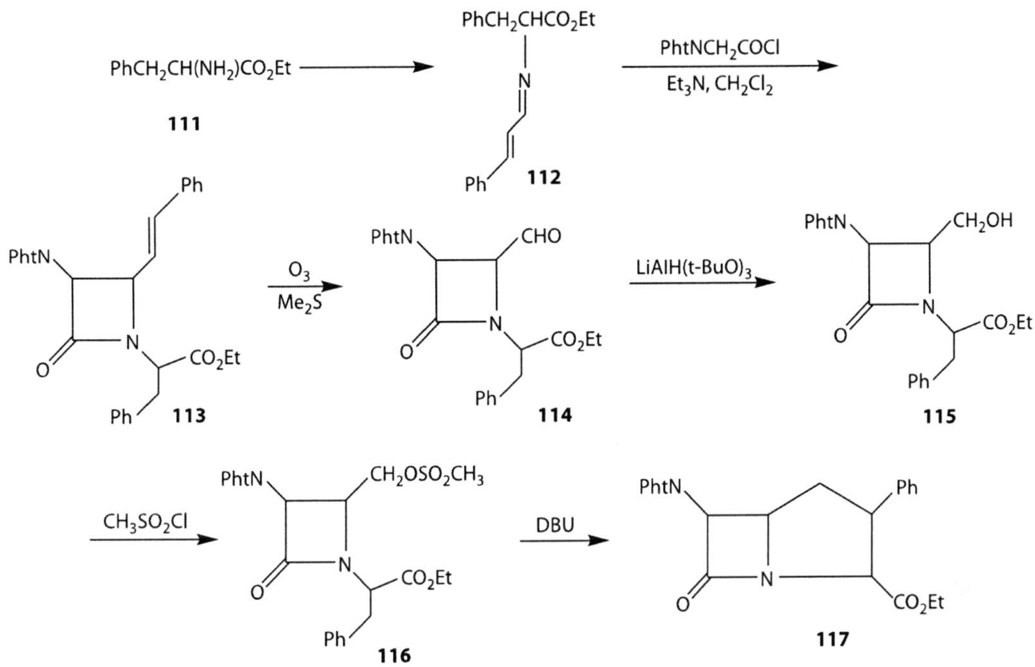

SCHEME 3.26 Preparation of β-lactam **117**.

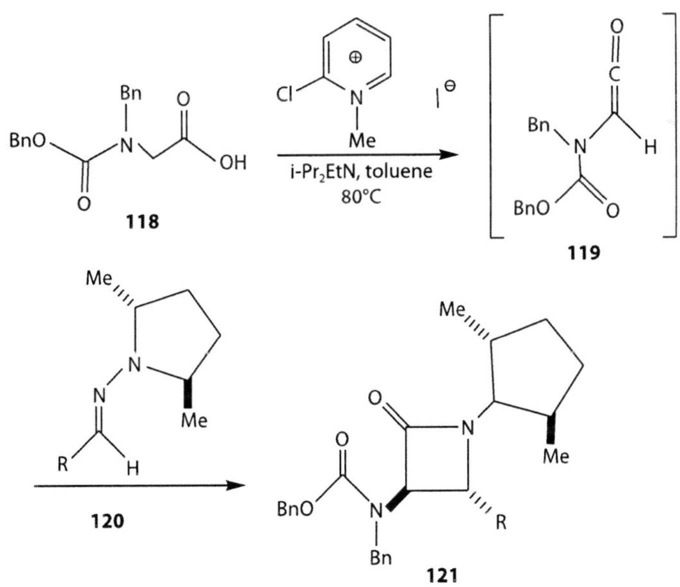

SCHEME 3.27 Preparation of β-lactams **121**.

conditions revealed the key importance of the base used for ketene generation. Interestingly, replacement of Et_3N by hindered $(i\text{-Pr})_2EtN$ resulted in a significant increase in results. This led to the isolation of the corresponding products **121** in moderate-to-good yields.

The Staudinger [2 + 2] cycloaddition of chiral carbohydrate Schiff base **125** with phthalimidoacetyl chloride produced the sugar-based monocyclic β-lactam **127** as a single isomer. Treatment of protected β-lactam **127** with methylhydrazine provided free amino β-lactam. Starting material for these studies D-(+)-Galactose **122** was selected. It was converted into

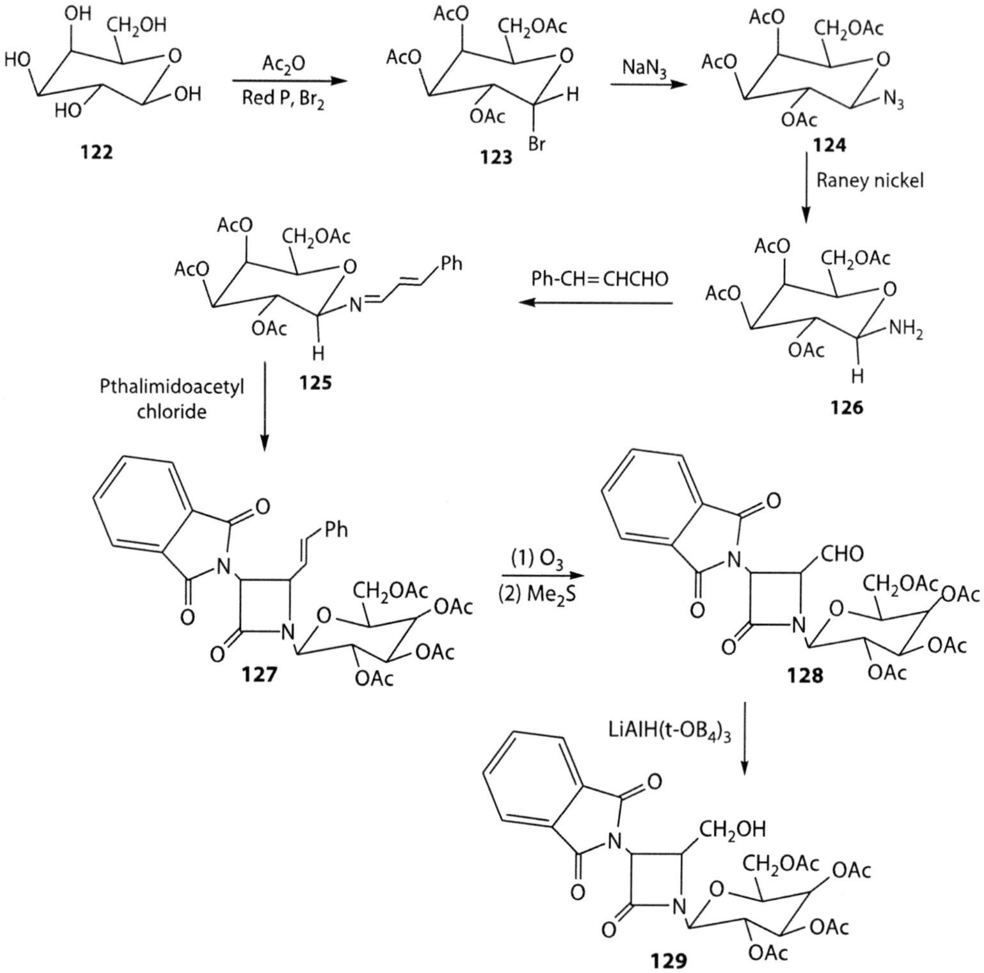

SCHEME 3.28 Preparation of β-lactam **129**.

2,3,4,6-Tetra-*O*-acetyl-α-D-galactopyranosyl bromide **123**. The α-anomer was formed which is thermodynamically more stable. The bromo group in the acylgalactosyl compound is reactive and therefore may be easily displaced by an azido group. Accordingly, 2,3,4,6-tetra-*O*-acetyl-β-D-galactopyranosyl azide **124** as a white crystalline substance in 70% yield (Scheme 3.28) was obtained. The replacement involved inversion of configuration at the anomeric site; thus, the α-galactopyranosyl halide **123** yields a β-galactosyl azide **124**. However, the pyranose ring structure was retained in the presence of the acyl groups in the acylgalactosyl halide. 2,3,4,6-Tetra-*O*-acetyl-β-D-galactopyranosyl amine **126** was obtained in 90% yield when heterogeneous reduction of the azide group of **124** with Raney Nickel in ethyl acetate under reflux for 2.5 h was performed. The imine β-D-galactopyranosyl amino-(*N*-cinnamylidene)-2,3,4,6-tetra-*O*-acetate **125** was prepared in quantitative yield. The formation of E-isomer as the predominant product was observed with β-D-galactopyranosyl imine **125**. Compound **125** was treated with phthalimidoacetyl chloride in the presence of triethylamine to give 1-[(2,3,4,6-Tetra-*O*-acetyl-β-D-galactopyranosyl)-3-phthalimido-4-styryl]azetidin-2-one **127** in 75% yield. 1-[(2,3,4,6-Tetra-*O*-acetyl-β-D-galactopyranosyl-3-phthalimido-4-formyl]azetidin-2-one **128** in 90% yield was obtained when the ozonolysis of the styryl group was performed. Reduction of **127** using $LiA_lH(t\text{-}OBlu)_3$ in dry THF at 0°C yielded 1-[(2,3,4,6-tetra-*O*-acetyl-β-D-galactopyranosyl)-3-phthalimido-4-hydroxymethyl]-azetidin-2-one **129** in 60% yield (Scheme 3.28).

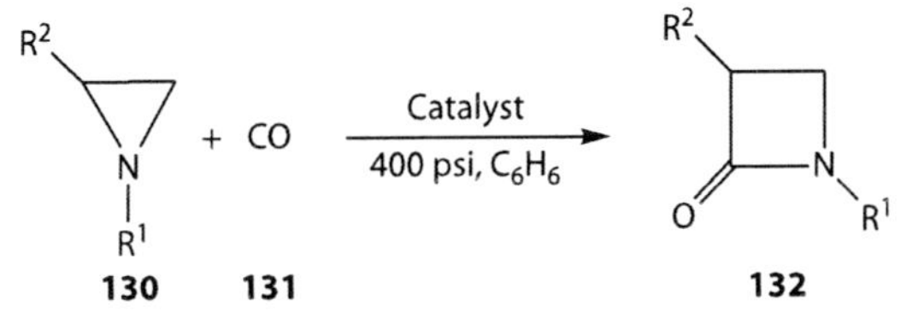

SCHEME 3.29 Expansion of aziridine ring to β-lactam using carbon monoxide.

Aziridine ring was expanded to β-lactam using carbon monoxide (Scheme 3.29). The carbonylation reaction was simple in workup and execution. After the complete reaction, the catalyst was recovered by simple filtration, and washing with benzene for the subsequent cycles. For the carbonylative ring expansion of 1-(1-adamantyl)-2-phenylaziridine, 1-*tert*butyl-2-(biphenyl-4-yl)aziridine, 1-(1-adamantyl)-2-(biphenyl-4-yl)aziridine, and 1-*tert*-butyl-2-(4-bromophenyl) aziridine, similar results were observed.

The Staudinger reaction between a ketene derived from optically active acid chloride **133a** and *N*-benzyl-*N*-[(1*E*)-phenylmethylene] amine was conducted in dichloromethane in the presence of Et_3N (Scheme 3.30). Although four diastereomers **134a** through **138a** are possible, analysis of the crude reaction mixture by TLC indicated two major products. Importantly, HPLC showed a 99:1 ratio of two peaks. Flash chromatography purification of the crude reaction afforded two diastereomerically pure β-lactams **134a** in 51% and **137a** in 20% isolated yields (single enantiomers). These two, respectively, exhibited identical retention times in HPLC. The Cbz group of **134a** was removed to give pyrrolidine **136a** as a single diastereomer as evidenced by its simplified ^{1}H NMR. Additionally, the structure of **134a** was x-ray crystal structure. To afford the "proline-derived" Staudinger product **140**, olefin **139** was deprotected and hydrogenated (Donohoe and House 2002). The structure of **137a** was shown to be as indicated by converting it to proline-derived product **142** via the elimination of methanesulfonic acid, by hydrogenation of the resultant double bond of **141**, and removal of the Cbz-protecting group. ^{1}H NMR of **142** was identical to **140**.

Being derived from two different compounds (or diastereomers) **134a** and **137a**, compounds **140** and **142** are therefore enantiomeric to each other. Since **140** and **142** are enantiomeric in nature, and given the fact that **142** was derived from **137a**, β-lactam **137a** should have the structure as shown (Scheme 3.30).

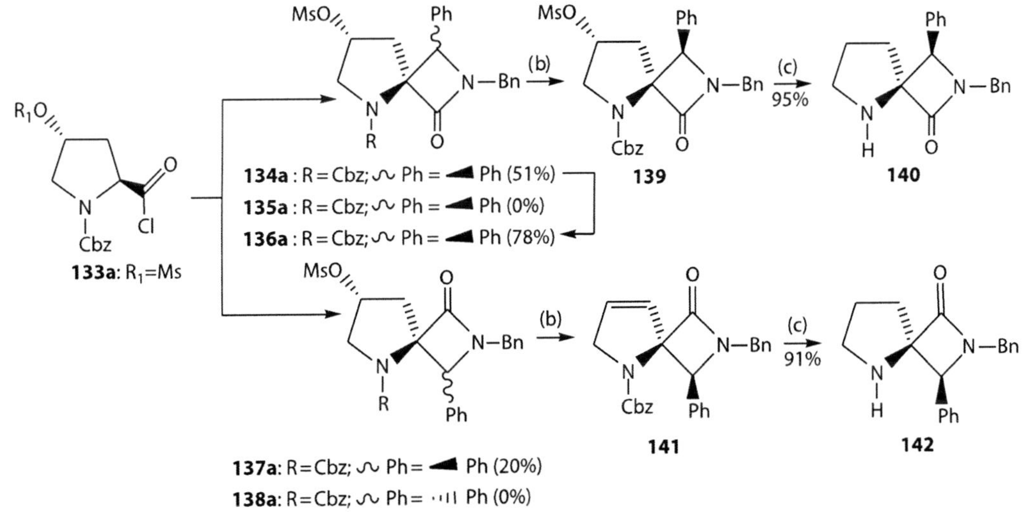

SCHEME 3.30 Involvement of Staudinger reaction in synthesizing β-lactams.

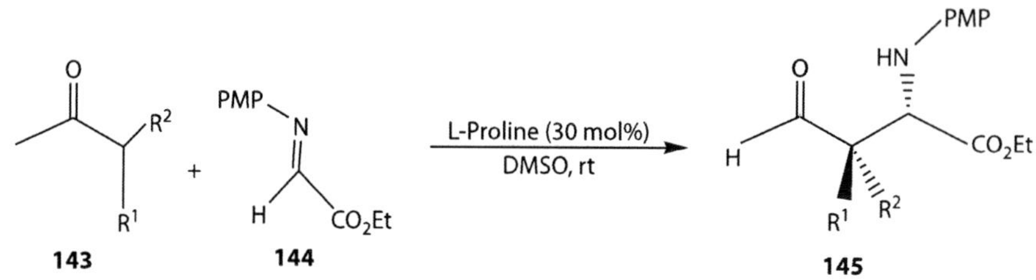

SCHEME 3.31 Mannich type reaction.

The Mannich reaction was studied using R,R-disubstituted aldehyde **145** as nucleophiles or donors in this reaction (Scheme 3.31). The quaternary β-formyl R-amino acid derivatives were obtained with excellent yields. Alkyl-, aryl-, benzyl-, and heteroaromatic-substituted aldehydes were used to make functionalized amino acids. Reactions with aryl-substituted aldehydes are faster than those of benzyl-substituted ones. The Mannich product **146** was also synthesized in DMF and NMP with good yield (77%). Compound **146** was converted to β-Lactam **147** very easily (Scheme 3.32).

Spiro-linked β-lactam-dihydropyridines **149a**, **150a**, and **151b** were prepared (Fraenkel et al. 1972). Ozonolysis of **149a** gave a single diastereoisomer of the monocyclic β-lactam **148**. Hydrogenation of **149a** and **149b** yielded saturated 2,7-diazaspiro[3.5]nonan-2-ones **150a** and **150b** (Scheme 3.33).

A sequence of heterocycle-fused β-lactams under thermal conditions was synthesized (Xu et al. 2001). This was done in order to further investigate the stereochemical process and to prepare bicyclic β-lactams. These β-lactams were prepared with different stereosubstitutions. A Staudinger reaction of R-diazoketone **152** with imines **153a–c**, respectively, was performed. These reactions were done under microwave and photoirradiation conditions. **153c** for *N*-alkyl and sterically hindered

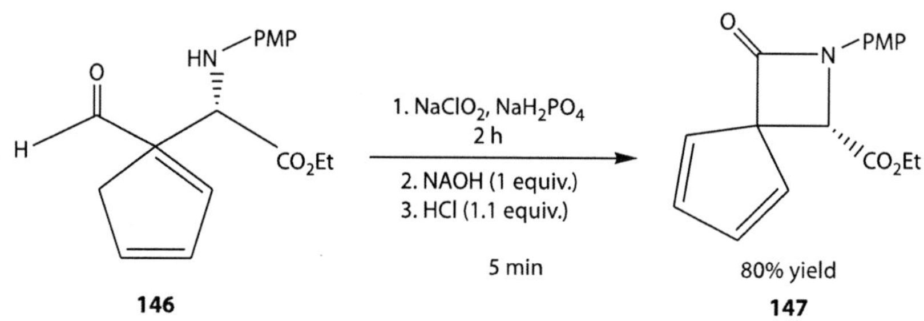

SCHEME 3.32 Conversion of Compound **146** to β-lactam **147**.

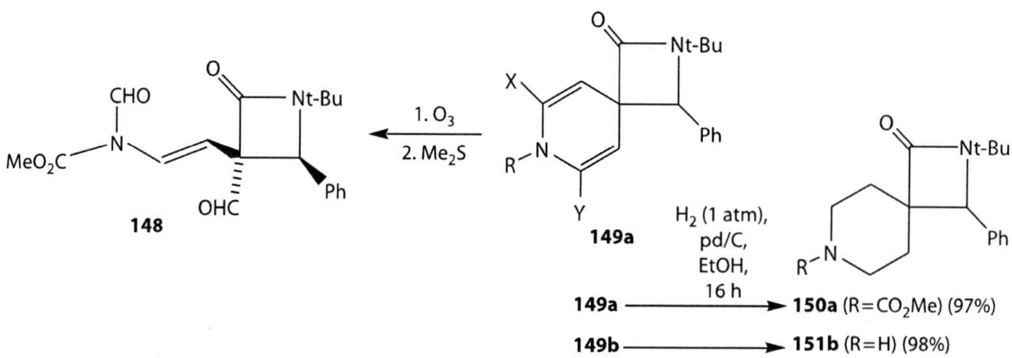

SCHEME 3.33 Preparation of spiro-β-lactams.

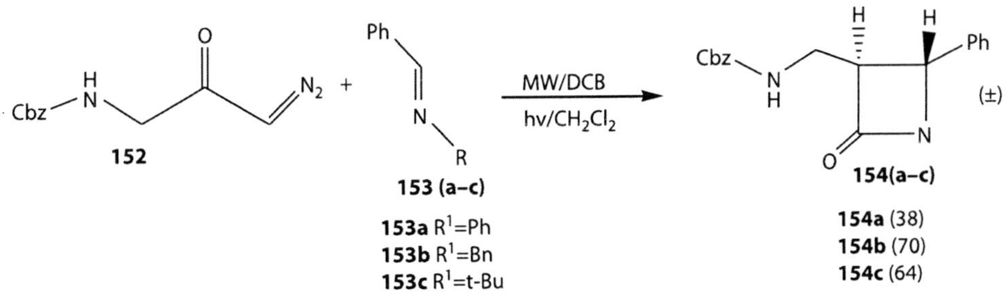

SCHEME 3.34 Microwave-assisted synthesis of *trans*-β-lactams.

imine and *trans*-β-lactams **154a–c** were obtained exclusively under both conditions (Scheme 3.34) with a series of acyclic imines **153a–c (153a** for *N*-aryl imine; **153b** for *N*-benzyl, nonconjugated imine). This is in agreement with the results of Podlech et al. (Podlech and Linder 1997, Linder and Podlech 2001) (Scheme 3.34).

N-p-chlorobenzyl-*N-tert*-butyl-R-ethoxycarbonyl-R-diazoacetamide **155a** treatment with $[RuCl_2(p\text{cymene})]_2$ (0.5 mol%) in toluene at 70°C under an argon atmosphere afforded *N-tert*-butyl-*cis*-1-ethoxycarbonyl-2-*p*-chlorophenyl-β-lactam **155b**. This β-lactam was produced in quantitative yield. ^{1}H NMR analysis detected the absence of *trans*-β-lactam in the crude mixture but the presence of *cis* β-lactam. Slow addition of *R*-diazo compounds and inert atmosphere were necessary for ruthenium-catalyzed carbenoid transformations. However, it was found that the $[RuCl_2$-(*p*-cymene)] 2-catalyzed intramolecular carbenoid C-H insertion reaction could also be completed without using the slow addition procedure or an inert atmosphere. For example, by heating a mixture of **155a** and $[RuCl_2(p\text{-cymene})]_2$ at 70°C in open atmosphere provided *cis*-β-lactam in quantitative yield (Table 3.3, entry 1). ^{1}H NMR analysis detected no diazo-coupling products. Other ruthenium complexes, such as [Ru^{II}(TTP)(CO)] [H_2TPP) *meso*-tetrakis(*p*-tolyl)porphyrin], [Ru^{II}(salen)-$(PPh_3)_2$] (salen) *N,N′*-bis(2,4-dibromosalicyclidene)-1,2-cyclohexanediamine)], Ru^{II} (6,6′-Cl_2 bpy)$_2$(H_2O)$_2$] $(CF_3SO_3)_2$ (6,6′-Cl2-bpy) 6,6′-dichloro-2,2′-bipyridine), [Ru^{II}$(PPh_3)_2$-Cl_2], and [Ru(COD)Cl_2]*n* (COD) 1,8-cyclooctadiene), were unsuccessful in causing catalytic cyclization of **155a**. But $[Cp^*RuCl_2]_2$ (Cp*) pentamethylcyclopentadienyl) was determined to catalyze cyclization of **155a**. This method produced *cis*-lactam **155b** exclusively in 96% yield within 2 h. However, using an open atmosphere, 62% yield at 80% substrate conversion to *cis* β-lactam after 5 h of reaction was observed. Other solvents such as toluene, $CHCl_3$, CH_2Cl_2, acetone, EtOAc, and THF were employed for the cyclization of **155a** with >95% yields and complete *cis* selectivity was accomplished. The reaction depended on the nature of the solvents since no substrate conversion was observed within 3 h when DMF, CH_3CN, and MeOH were used. The results of the scope of the $[RuCl_2(p\text{-cymene})]$2-catalyzed intramolecular carbenoid C-H insertion were explored. Similar to **155a**, other *N-para*-Y-substituted benzyl-*N-tert*-butyl *R*-diazoacetamides [Y) H (**155a**), OMe (**157a**)] were converted to the corresponding *cis*-β-lactams. This was done under the Ru-catalyzed conditions **(entries 2 and 3)**. The catalytic reaction of R-diazoketone **158a** was established to give *trans*-lactam **158b** solely in quantitative yield **(entry 4)**. The Ru-mediated carbenoid insertion was performed to the methane (tertiary) C-H bond to produce β-lactam **159b** in 89% isolated yield **(entry 5)**. No γ-lactam due to insertion at the primary C-H bond was detected by ^{1}H NMR analysis with **1e** as substrate. These reactivity preferences (i.e., tertiary C-H > primary C-H bonds) were related to the systems with $[Rh_2(CH_3CO_2)_4]$ as catalyst (Padwa et al. 1993).

Using triphenylphosphine (Ph_3P) and diethyl azodicarboxylate (DEAD) in 72% yield (Arnold et al. 1985), conversion of **160** into the corresponding β-lactone **161** was achieved (Scheme 3.35). No reaction occurred when the β-lactone was combined with **162**. Pre-activation of **162** by reaction with trimethyl aluminum produced a dimethylaluminum–hydrazide complex. This method was helpful to the reaction at the acyl group of β-lactone **161**. Consequently, the desired azapeptide **164**

TABLE 3.3
Preparation of 3,4-Disubstituted β-Lactams

Entry	Substrate	Product	% Yield
1	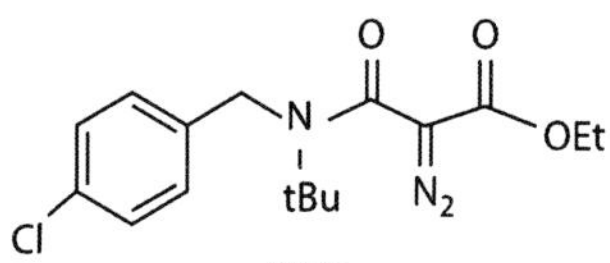 **155a**	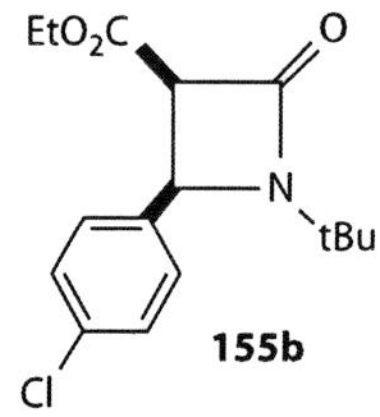 **155b**	99
2	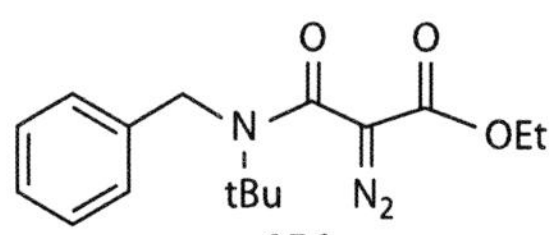 **156a**	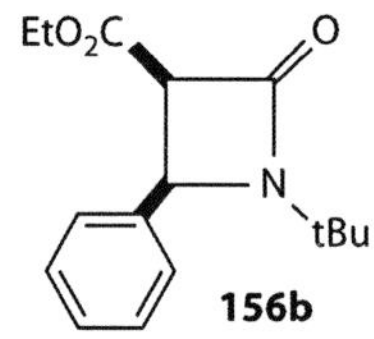 **156b**	99
3	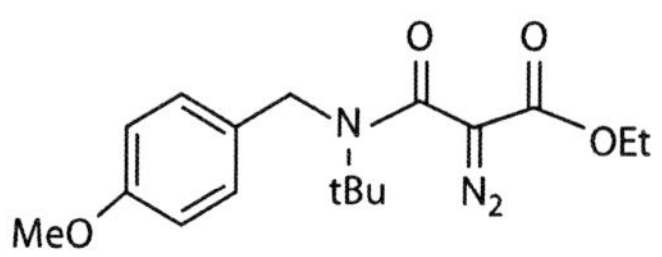 **157a**	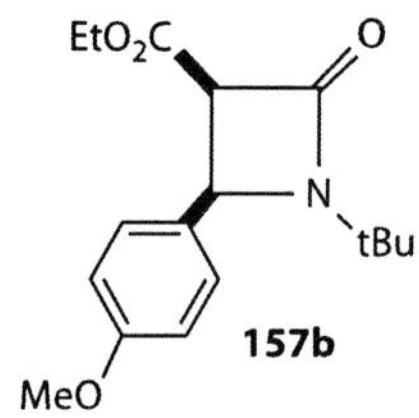 **157b**	99
4	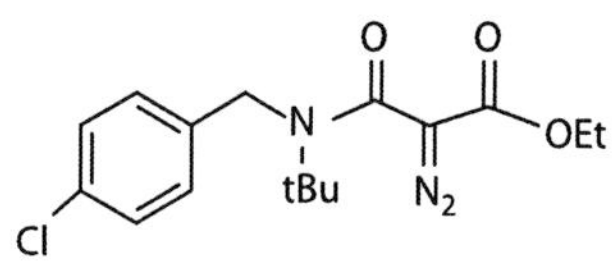 **158a**	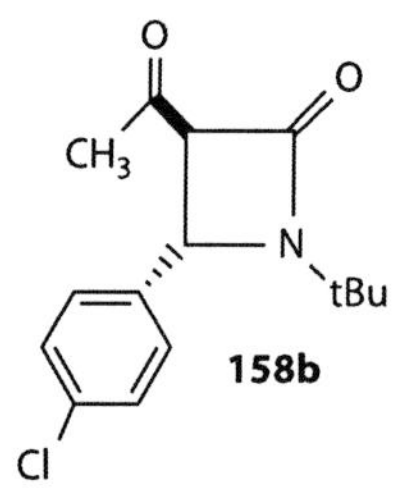 **158b**	99
5	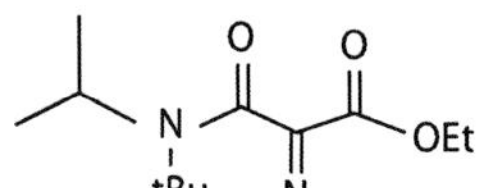 **159a**	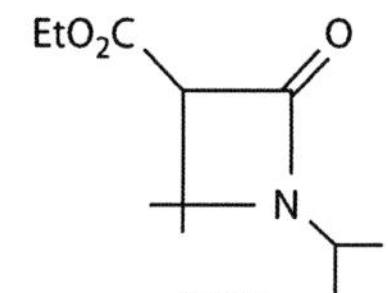 **159b**	99

was formed. The azapeptide was then subjected to Mitsunobu conditions (Ph_3P, DEAD) to afford the C-3 benzyl β-lactam azapeptidomimetic **163** in high yield.

The efficiency of cross-metathesis on the solid phase was documented. Its application to the generation of biologically interesting 3-(aryl)alkenyl-β-lactams was reported. The resin-bound 3-vinyl-β-lactam **169** (Scheme 3.36) was the key. The resin-bound aniline **166** was deprotected. This was converted to the aldimine **168** by condensation with aldehyde **167**. Next, synthesis of the β-lactam ring was performed by a solid-supported Staudinger reaction using Mukaiyama's reagent as an acid-activating agent (Delpiccolo et al. 2003). As a result, reaction between crotonic acid and the corresponding imine **168** effectively gave the supported 3-vinyl-β-lactam **169**. Formation of **169** was confirmed by FT-IR and gel-phase ^{13}C NMR.

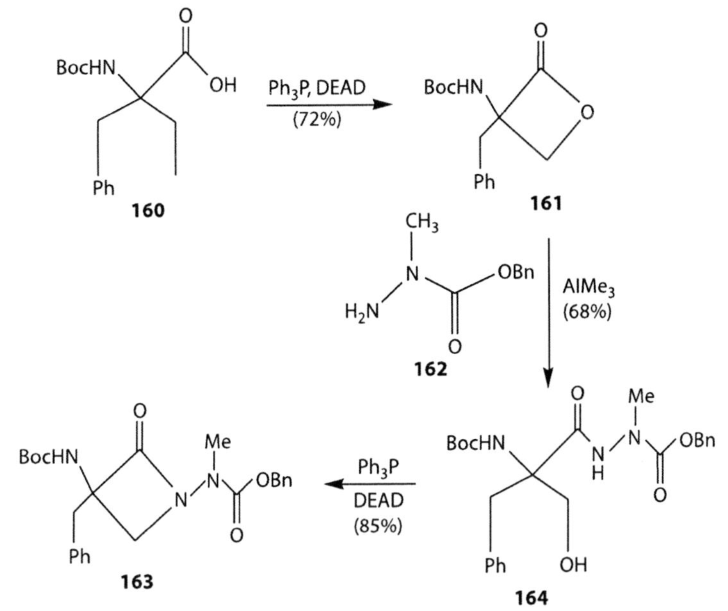

SCHEME 3.35 Use of Mitsunobu conditions in synthesizing β-lactam **163**.

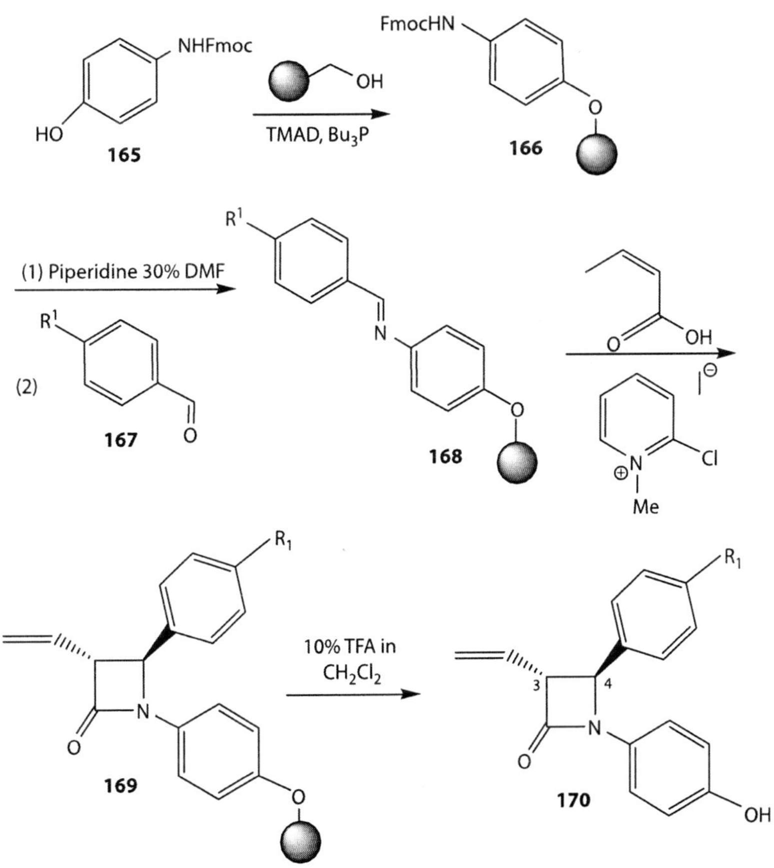

SCHEME 3.36 Application of cross-metathesis on the solid phase in synthesizing β-lactam.

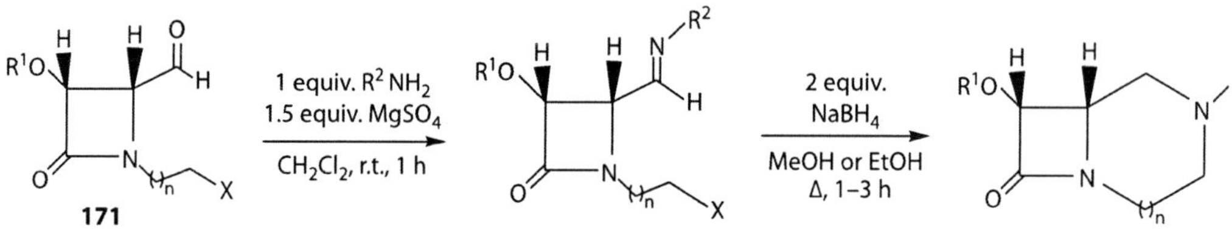

SCHEME 3.37 Synthesis of optically active bicyclic β-lactams **174** and **175**.

Conditions for releasing the 3-vinyl-β-lactam from the support were explored. Treatment of the resin with 10% trifluoroacetic acid in CH_2Cl_2 for 1 h at room temperature was found to be a very efficient procedure for the cleavage. It afforded the 3-vinyl-β-lactam **170** in 32% overall isolated yield. FT-IR of the cleaved resin confirmed quantitative release of the β-lactam.

The optically active bicyclic β-lactams **174** and **175** were synthesized in good yields in a one-step procedure (Scheme 3.37). Conversion of 4-formyl-1-(w-haloalkyl)-β-lactams **171** into the corresponding 4-imidoyl-β-lactams **172** and **173** upon condensation with different primary amines followed by reduction of the latter azetidin-2-ones **172** and **173** with $NaBH_4$ in refluxing methanol or ethanol produced the product. The corresponding 4-aminomethyl-2-azetidinones were formed as minor constituents (<20%). This method is new and very elegant in organic synthesis. In literature, related piperazine annulated β-lactams have been prepared either by intramolecular dipolar cycloaddition of 4-vinyl-2-azetidinones (Murthy and Hassner 1987) or by cyclization of Boc-protected 4-aminomethyl-1-(2-hydroxyethyl)-β-lactams in a stepwise approach, which involved mesylation, *N*-deprotection, base-induced ring closure, and *N*-protection (Palomo et al. 1997). The method presented, therefore, contained an efficient alternative for these approaches. Sterically hindering substituents at the imidoyl nitrogen (R^2) *t*Bu, *i*Pr lowered the yields of the products. Due to the steric hindrance, probably ring closure proceeded slower. The 1,4-diazabicyclo [4.2.0]octan-8-ones **174** and 1,5-diazabicyclo [5.2.0]-nonan-9-ones **175** were considered as novel bicyclic β-lactam skeletons. This was also considered as bicyclic piperazine and 1,4-diazepane derivatives. β-Lactams **174** is very important due to their structural resemblance to the unsaturated isodethiaazacephems. These act as potent antibacterial agents (Hwu et al. 1998). Piperazines are used as antifungals, antidepressants, antivirals, and serotonin receptor (5-HT) antagonists/agonists. Carbon-monosubstituted piperazines have worked as farnesyl transferase inhibitors and neurokinin-1 antagonists (Berkheij et al. 2005). Because of their value in psychotherapy (e.g., diazepam, the active compound in Valium), these types of compounds have received a lot of attention from the scientific community.

Novel optically active β-Lactams were prepared by cycloaddition reaction. The reaction between **176a** and an excess of phthalimidoacetyl chloride was performed in the presence of a base. An excellent yield of β-lactam **177a** was obtained as a single diastereoisomer (Scheme 3.38). Good yields of chloro- and 2,5 dimethoxyphenyl β-lactams **177** and **177** were obtained in toluene while maintaining complete diastereoselectivity. Benzyloxyacetyl chloride led to an 82:12 mixture of diastereoisomeric β-lactams (80%) along with 14% of **178**. A decrease in the reaction temperature favored the synthesis of **177** as a single diastereomer. The behavior of 5,6-dihydropyrazin-2(1H)-one **176b**, with the imine flanked by an aromatic group (R^1 = 1-naphthyl), was examined in order to extend the scope of this procedure. The results showed that **176b** underwent a highly diastereoselective Staudinger reaction. Preparation of 3-alkyl/vinyl β-lactams and 3-alkyl/vinyl β-lactams were not successful under thermodynamic and kinetically controlled conditions (Scheme 3.38).

A mixture of *cis*- and *trans*-β-lactams were formed when an excess (3 equivalents) of methoxy- or benzyloxyacetyl chlorides in dichloromethane was added to a solution of imine **179** containing diisopropylethylamine (3.5 equivalents) (Coantic et al. 2007). Under conditions previously described, the same reaction with acetoxyacetyl chloride produced *trans*-β-lactam **179** in good yield (Scheme 3.39).

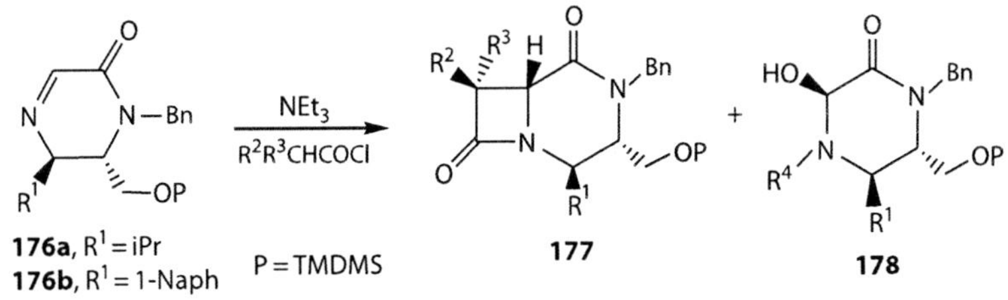

SCHEME 3.38 Preparation of optically active β-lactam **177**.

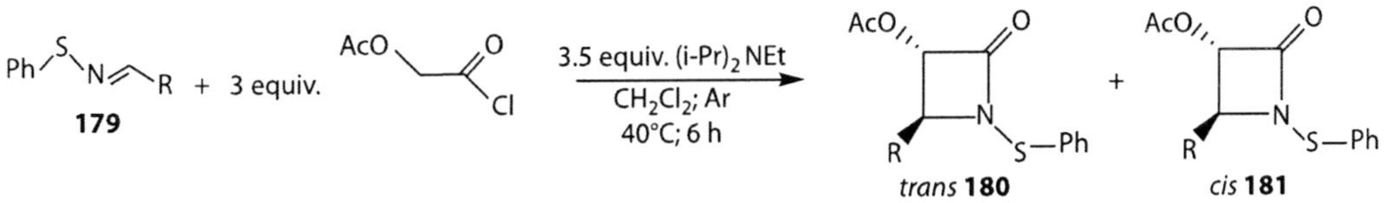

SCHEME 3.39 Formation of the mixture of *trans*-β-lactams **180** and **181**.

The selective formation of *trans*-β-lactam with acetoxyketene is noteworthy. This observation is in sharp contrast to the poor selectivity observed with alkoxyketenes. A possible *cis–trans* isomerization of the 3,4 disubstituted-β-lactams, under the reaction conditions, was ruled out. The formation of pure *trans* **183** in 79% yield resulted from the reaction of *N*-phenylsulfenylimine of pyrrole-2-carboxaldehyde **182** with of acetoxyacetyl chloride (Scheme 3.40).

The carbonyl insertion into aziridines to afford β-lactams catalyzed by nickel (Chamchaang and Pinhas 1990), rhodium (Calet et al. 1989), or palladium (Tanner and Somafai 1993) complexes proceeds with retention of configuration at the carbon atoms of the aziridine ring. As a result of this process, *cis*-aziridines produced *cis*-β-lactams, whereas *trans*-β-lactams were obtained from *trans*-aziridines. Further experimental works demonstrated that different cobalt complexes also carbonylate aziridines to generate β-lactams. However, these processes proceeded with inversion of configuration at the site of attack (Piotti and Alper 1996, Davoli et al. 1999). It had been postulated that when the $Co_2(CO)_8$ or NaCo-$(CO)_4$ complexes were used as catalysts (Piotti and Alper 1996), the reaction proceeded via an S_N2-like mechanism. The putative active species, the $[Co(CO)_4]$- ion, opened the aziridine ring by attacking the least-substituted carbon atom, with inversion of its configuration. Subsequently, an external CO was inserted into the C (attacked)–Co bond of this intermediate with retention of configuration to afford another open intermediate. In turn, this method finally underwent a ring closure to render the β-lactam and regenerated the catalyst. It had also been suggested that path II mechanistic route differed from the above-mentioned one in the second step—the CO insertion when the cationic component of the $[Co(CO)_4]$- anion was a Lewis acid (Mahadevan et al. 2002). This followed a stepwise process. First, one of the carbonyl ligands of $Co(CO)_4$- inserted into the C(attacked)–Co bond, and second, an extra molecular CO added to the cobalt atom to regenerate the catalyst following this suggestion. This mechanism was

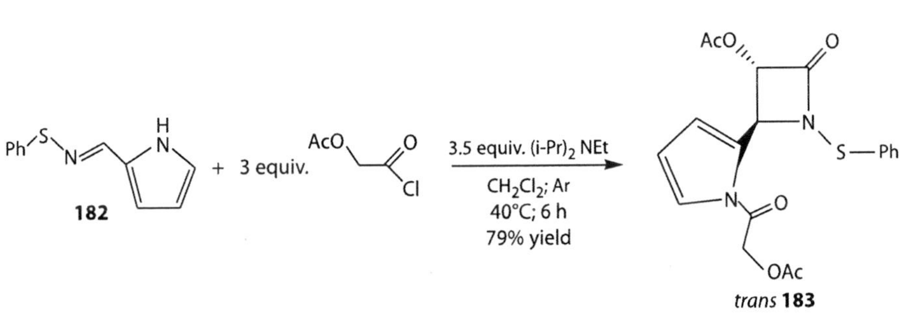

SCHEME 3.40 Formation of pure *trans*-β-lactam **183**.

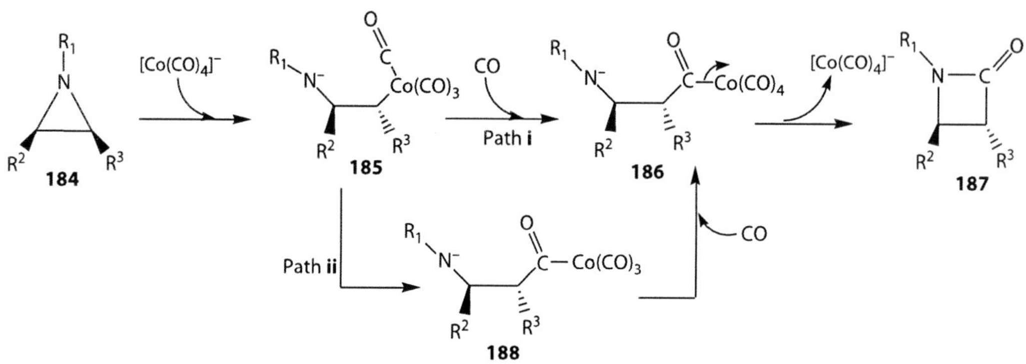

SCHEME 3.41 Metal-catalyzed carbonyl insertion into aziridines to afford β-lactams.

investigated in a study of a related reaction by density functional theory (DFT) calculations (Molnar et al. 2003) (Scheme 3.41).

Reformatsky reaction of the imine **189** with ethyl bromodifluoroacetate **190** and ethyl bromoacetate **191** was performed in order to determine the influence of the gem-difluoro moiety on the β lactam/β-aminoester ratio. Previous studies showed that β-lactams and β-aminoesters were formed in ratios depending upon temperature and time of the reaction (Dardoize et al. 1972, Ross et al. 2004). Results obtained are presented in Scheme 3.42.

Novel C-4-disubstituted optically active β-lactams were prepared from chiral acid. Intermediates **199a** and **199b** were prepared by acylation with racemic 2-chloropropionyl chloride (**198ab**), followed by separation of the diastereoisomers using Pmb-L-Phe-OMe (**196a**) as a starting amino acid derivative. The absolute configuration of *N*-chloropropionyl Phe derivatives was assigned. Synthesis of isomer **199a** was done by coupling **196a** with enantiomerically pure 2(*S*)-chloropropionic acid (**197a**) in the presence of BOP. This coupling reaction evolved with lower yield than the acylation with the acyl chloride. Some unwanted racemization of the 2(*S*)-chloropropionic acid was also observed. The racemic 2-chloropropionyl chloride (**198**) was also used for the synthesis of diastereoisomeric intermediates **199c** and **199d** from H-Pmb-D-Phe-OMe (**196b**). A unique 3,4-*cis* β-lactam, indicating the high degree of diastereoselectivity in this reaction (Scheme 3.43), resulting from the base-promoted cyclization of each diastereoisomer of **199** was observed. Furthermore, 2*S*-intermediates afforded the same 3*S*,4*S* 2-azetidinone, **201a** (64%, e.e. > 98%). The 3*R*,4*R* β-lactam **201b** (66%), the enantiomer of **201a**, resulted from the same cyclization reaction with derivatives **199b** and **199d**, both having a 2*R* configuration. These results suggested the enantiocontrol of this reaction with the construction of the quaternary stereogenic center completely directed by the configuration of the 2-chloropropionyl group. A comparable 2,3-*cis* selectivity was observed in a intramolecular alkylation leading to azetidine-derived amino acids (Sivaprakasam et al. 2006).

Triphenylphosphazene derived from (4-methyloxyphenyl)-azide **202** was used for the preparation of β-lactam (Scheme 3.44). Azide **202** reacted with triphenylphosphine in 1,2-dichloroethane

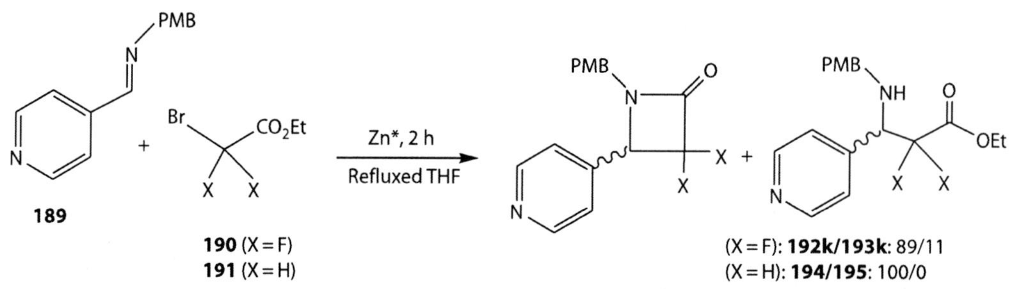

SCHEME 3.42 Involvement of Reformatsky reaction in synthesizing β-lactams.

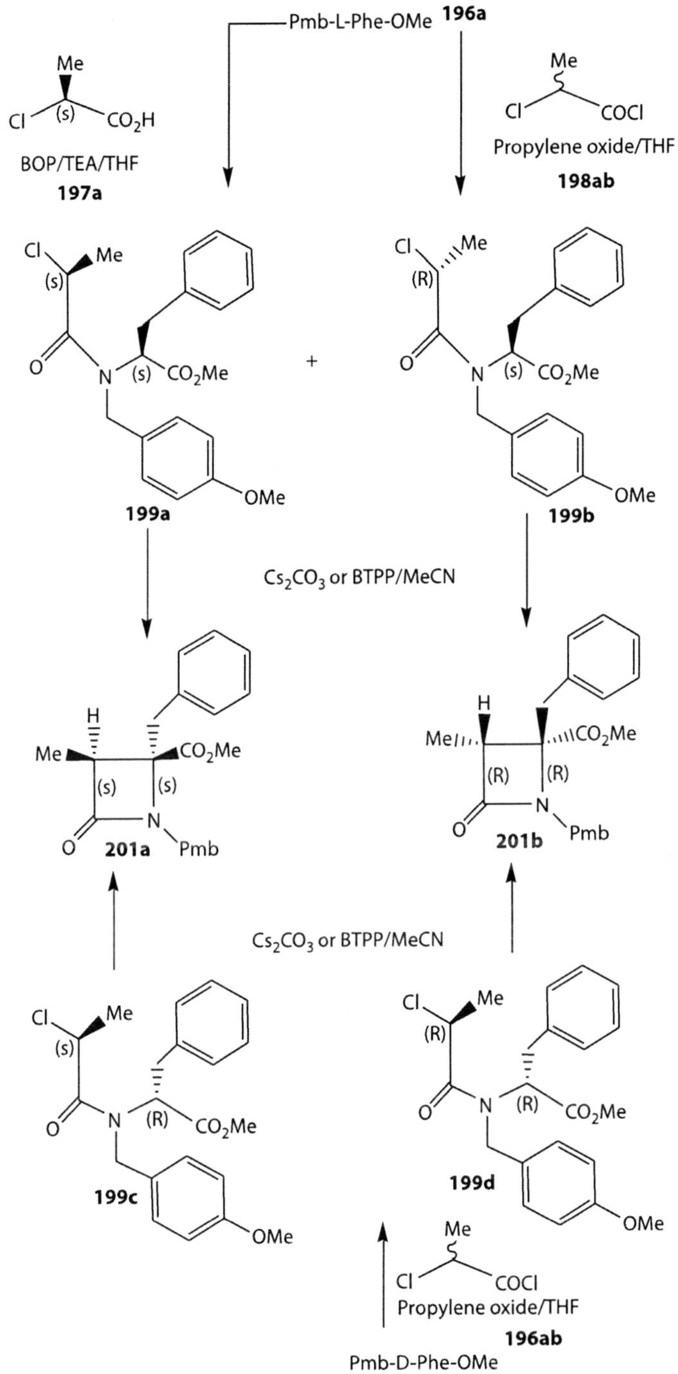

SCHEME 3.43 Synthesis of novel C-4-disubstituted optically active β-lactams from chiral acids.

(DCE) to form triphenylphosphazene **203** via the Staudinger–Meyer reaction, triethylamine at −5°C, and then at room temperature to afford 4-phenoxymethylene-β-lactam **206** in a one-pot procedure. It is believed that this cascade process involves an aza-Wittig reaction of triphenylphosphazenes **203** with ketene. The reaction required longer reaction time (5 h) and gave **206** in good yield (72%). High temperature could shorten the first-step reaction time and increase the yield, although the formation

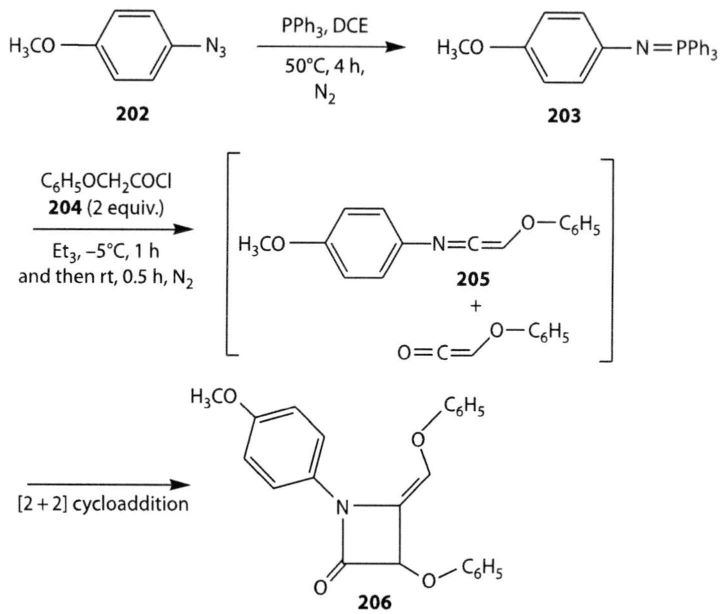

SCHEME 3.44 Synthesis of β-lactams from triphenylphosphazenes.

of **203** could take place at room temperature. β-lactam **206** in 76% yield under the optimum reaction conditions was formed in the first step (50°C, 4 h). A variety of aryl azides **202** and aryloxyacetyl chlorides **204** were investigated using the optimized reaction conditions.

A powerful new MCR based upon the highly strained 2-methyleneaziridine ring system for the synthesis method of β-lactam has been developed (Hayes et al. 2000). The reaction involves ring opening of methyleneaziridine **207** at C-3 using a Grignard reagent under Cu(I) catalysis and of the resultant metalloenamine with electrophile (R^2X). By combining this approach to ketimines with a Staudinger [2π + 2π] cycloaddition, it was possible to develop a flexible approach to 1,3,4,4-tetrasubstituted β-lactams (Scheme 3.45). There were several crucial features in this sequence. Three new intermolecular C–C bonds via a "one-pot" process and four points of chemical diversity were generated with this four-component reaction (4-CR). Furthermore, because it created one quaternary center (Denissova and Barriault 2003) as well as one tertiary center, there may be synthetic value of the products (He and Bode 2008). **207** and three methyleneaziridines were prepared and used in this study. 1-(4-Methoxybenzyl)-2-methyleneaziridine **207a** was prepared in 87% yield.

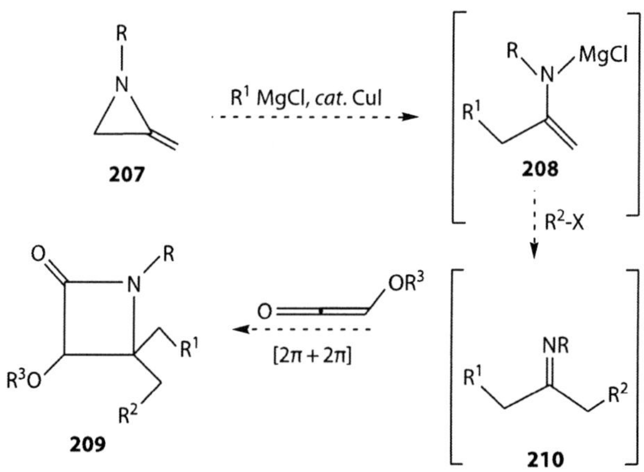

SCHEME 3.45 Synthesis of 1,3,4,4-tetrasubstituted β-lactams.

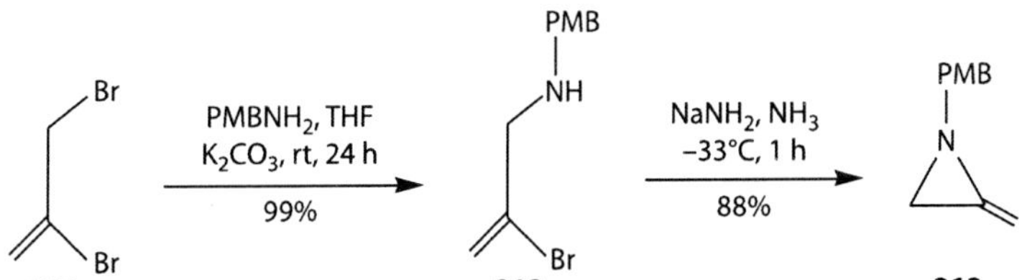

211 212 213

SCHEME 3.46 Sodium amide-induced ring closure.

This was obtained from 2,3-dibromopropene and 4-methoxybenzylamine using a sodium amide-induced ring closure (Scheme 3.46) (Shiers et al. 2004). Aziridine **207a** was selected as *N*-deprotection of the PMB group. The resultant β-lactam was straightforward (vide infra) (Maruyama et al. 1985). A similar manner was followed for the preparation of 1-benzyl-2-methyleneaziridine **207b** and 1-cyclohexyl-2-methyleneaziridine **207c**. This method finally produced C-4 distinguished β-lactam **209** and **214** (Scheme 3.47).

Reactions of chiral ynamides **215** (Scheme 3.48) (Zificsak et al. 2001, Oppenheimer et al. 2007) with nitrones provided a direct synthesis of chiral R-amino-β-lactams **218**. This reaction was found to be highly stereoselective and provided a single optically active β-lactam.

In continuation of our research in this area, we reported stereocontrolled synthesis of novel anticancer β-lactams starting from imines, with pendent polyaromatic substituents (Becker and Banik 1998,

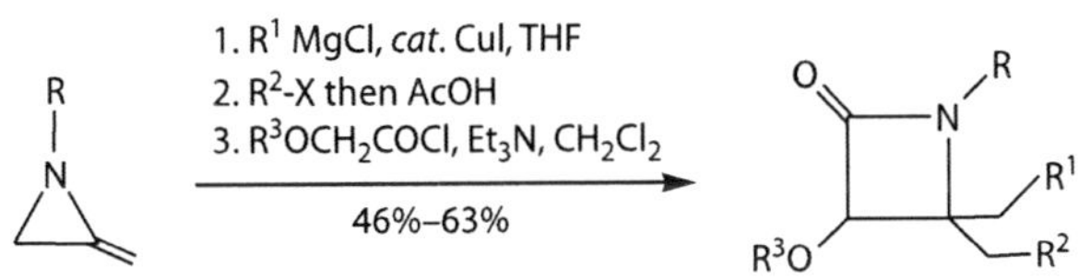

207a (R = PMB)
207b (R = Bn)
207c (R = c-Hex)

214

SCHEME 3.47 Conversion of aziridines to β-lactams.

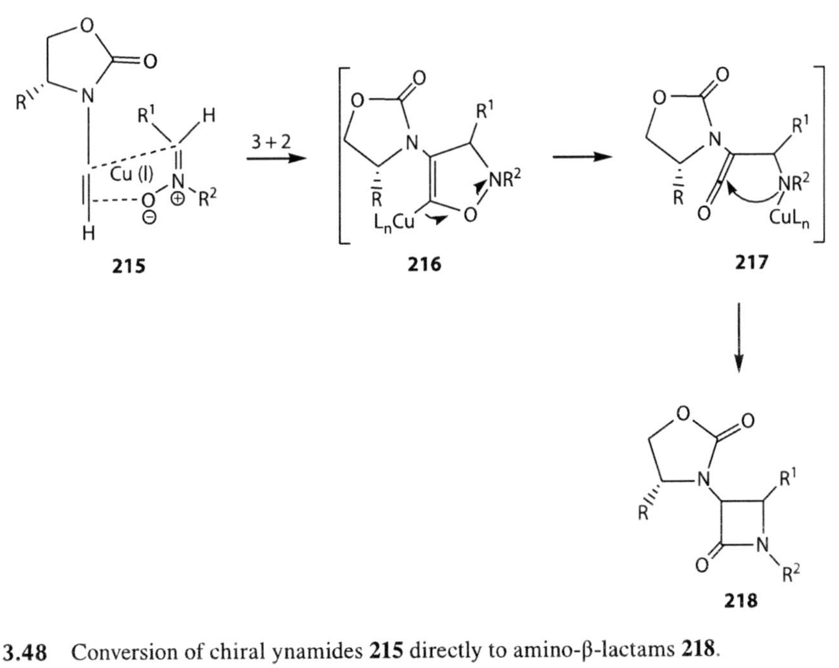

215 216 217 218

SCHEME 3.48 Conversion of chiral ynamides **215** directly to amino-β-lactams **218**.

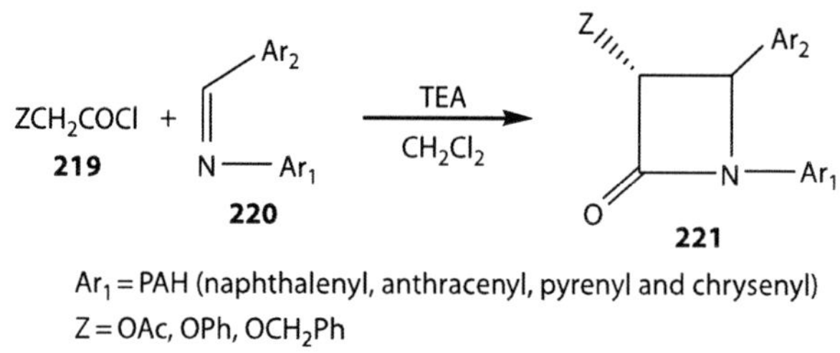

SCHEME 3.49 Microwave-assisted synthesis of *trans*-β-lactams.

Banik et al. 2003). Although the chemistry of β-lactam as antibiotics is very rich, studies of these agents as anticancer agents have been investigated very poorly (Finke et al. 1995, Mascaretti et al. 1995, Suffness 1995, Clader et al. 1996, Ruhland et al. 1996, Buynak et al. 1997, Bonneau et al. 1999, Kidwai et al. 1999, Ghatak et al. 2000, Taggi et al. 2000, Buynak 2004). Despite notable developments in the development of new anticancer agents, new compounds with less toxicity are required. Based on our and others work, we realize that novel, less toxic β-lactams can be identified as anticancer agents that may have increased activity against cancer cells. We have published some of our results previously. During the course of study on β-lactams, we anticipated that conformationally constrained analogs of our open chain diamides may increase activity against cancer cell lines (Lin et al. 1996).

Microwave irradiation of **219** and **220** under identical conditions afforded the *trans* product **221** as the only isomers (Scheme 3.49) (Bose et al. 1991, Georg and Ravikumar 1992, Alcaide and Vincente-Rodriguez 1999, Perreux and Loupy 2001).

Several of our β-lactams were tested using nine human cancer cell lines with cisplatin.

The structure–activity study revealed β-lactams containing naphthalene and anthracene pyrene derivatives demonstrated no activity against any of these cancer cell lines. The *trans* acetoxy phenanthrene and chrysene derivatives demonstrated reasonable activity. Phenoxy and phthalimido β-lactams were inactive. It is clear that the minimal structural requirement of the aromatic system for cytotoxicity is at least three aromatic rings in an angular configuration. This is confirmed by the fact that only phenanthrene and chrysene derivatives demonstrated cytotoxicity against the tumor cell lines. Interestingly, the presence of the acetoxy group proved to be very crucial for anticancer activity.

The anticancer activity of our racemic β-lactams has prompted us to devise a method for the preparation of the optically active analogs. Optically active isomer of a racemic compound may demonstrate better and much selective biological activity. Our goal was to use the glycosides (α- and β-) as the ketene component. We predicted that the absolute stereochemistry of the anomeric center in the ketene component of the carbohydrate would be the most important. Reaction of the activated acid **222** with polyaromatic imine **223** in the presence of triethylamine produced a mixture of diastereomeric O-glycosides of trans β-lactams **239** and **226** in the ratio of 45:55. The diasteromers **225** and **226** were separated through column chromatography. Acid-mediated reaction was used to cleave the anomeric bond and this resulted in the *trans*-hydroxyl β-lactams (+)-**227** and (−)-**228** in excellent yield. The hydroxy compounds were converted to the acetates (+)-**229** and (−)-**230** (Scheme 3.50).

Cycloaddition of the acid **231** with imine **223** was performed using imine in the presence of *N*-methyl-2-chloropyridiniumiodide and triethylamine. NMR analyses of the crude reaction mixture showed the presence of two diasteromeric *trans* β-lactams **232** and **233** in 60:40 ratios. The diasteromeric *O*-glycosides after separation were treated with mild aqueous acid to the hydroxy compounds **234** and **235** and the resulting alcohols were converted to acetates. The absolute stereochemistry of the *trans* acetoxy-β-lactam **236** and **237** was confirmed by a comparison with known *trans* β-lactam as described earlier (Scheme 3.51). The mechanism of formation of β-lactam via the Staudinger reaction had been advanced (Bose et al. 1991, Georg and Ravikumar 1992, Alcaide and Vincente-Rodriguez 1999, Perreux and Loupy 2001).

The cell growth inhibition data confirmed that of the optically active β-lactams **229** and **236** is extremely active. The results of the mutagenicity assay indicated that these β-lactams demonstrated

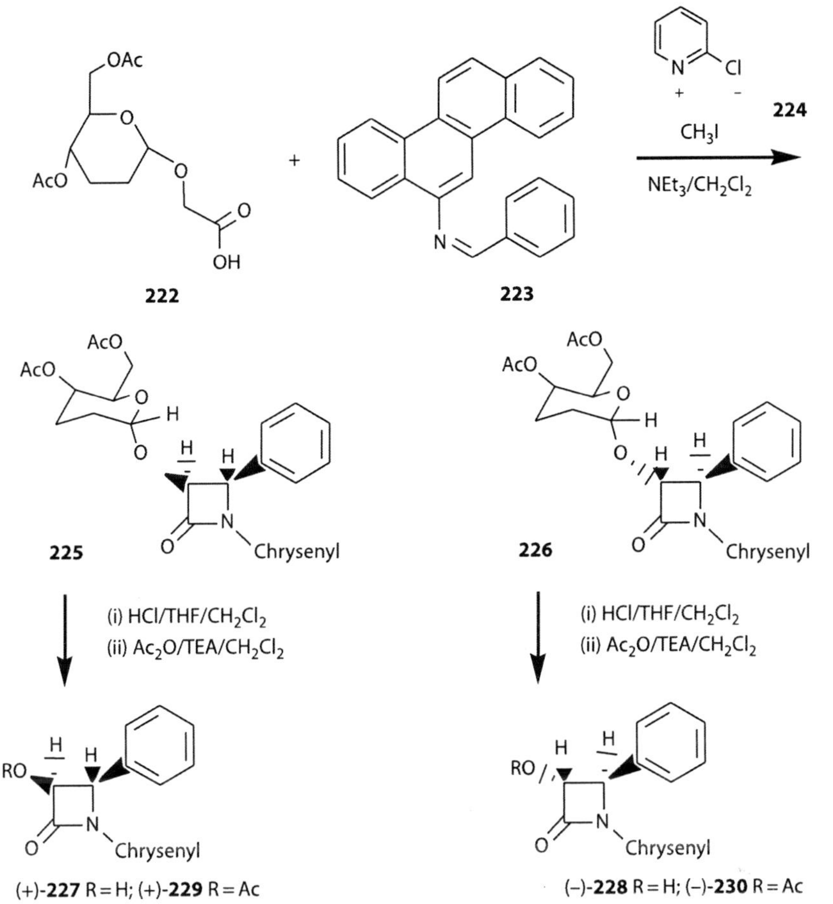

SCHEME 3.50 Synthesis of *trans*-hydroxyl β-lactams.

a negative response with these tester strains at any concentration in either the presence or absence of Aroclor-induced rat liver S9, confirming that neither of these compounds demonstrate any mutagenicity.

Necessary kits to identify the interaction of organic compounds with topoisomerases were purchased from TOPOGEN, Inc. These kits contained reagents required for the detection of topoisomerase I and II with DNA markers for the detection of enzymatic action and standard topoisomerase inhibitors. The activity of the active β-lactam was determined using these systems. However, no inhibition of either topoisomerase I or topoisomerase II was detected in the HL-60 cancer cell lines using much higher concentration that were required to inhibit the growth of these cell lines *in vitro*. There was no evidence that the active β-lactams showed their cytotoxic activity in sensitive cell lines through interaction with DNA or DNA-associated enzyme systems.

A few other outstanding researchers also performed significant studies on anticancer β-lactams. For example, β-lactams **238** to **241** (Scheme 3.52) induced DNA damage, inhibited DNA replication, and activated the apoptotic death program in human leukemic Jurkat T cells, in a time and concentration-dependent manner. Importantly, β-lactam **238** also inhibited proliferation and induced apoptosis in other human solid tumor cell lines. It was believed that induction of apoptosis by **238** is associated with the activation of p38 mitogen-activated protein (MAP) kinase, release of mitochondrial cytochrome c, and activation of the caspases.

They had also found other two β-lactam analogs **242** and **243**, both containing a branched-chain system at C_3 of the ring, exhibited potent apoptosis-inducing activity (Scheme 3.53).

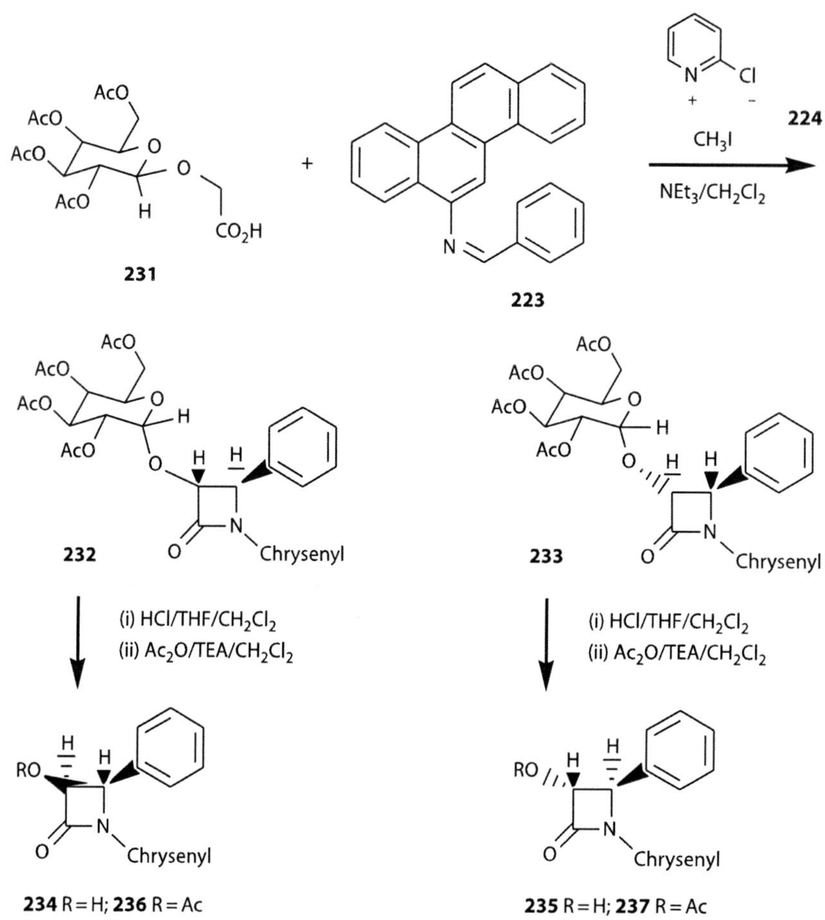

SCHEME 3.51 Synthesis of *trans*-acetoxyl β-lactams.

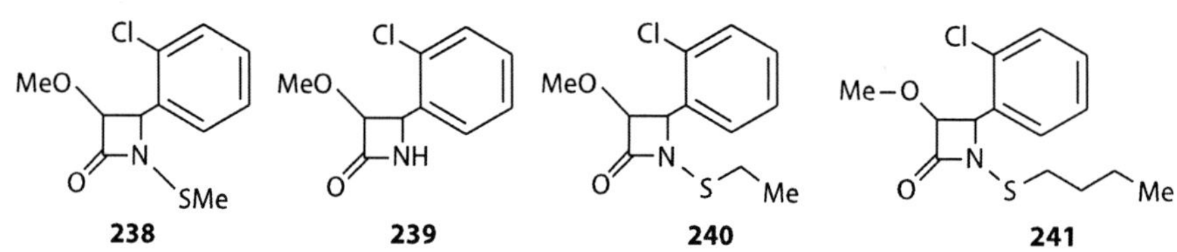

SCHEME 3.52 Anticancer β-lactams.

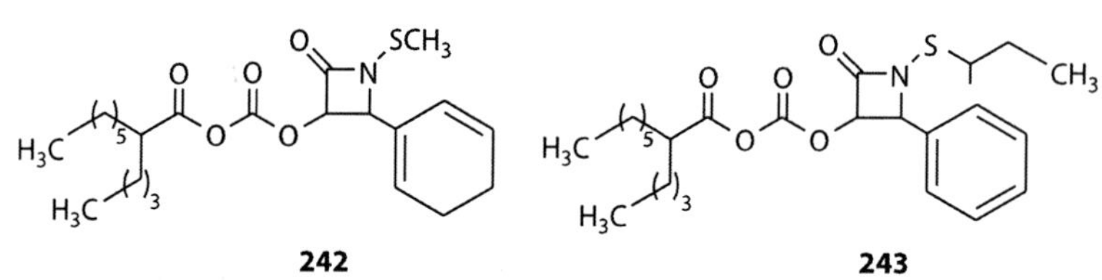

SCHEME 3.53 Apoptosis-inducing anticancer β-lactams.

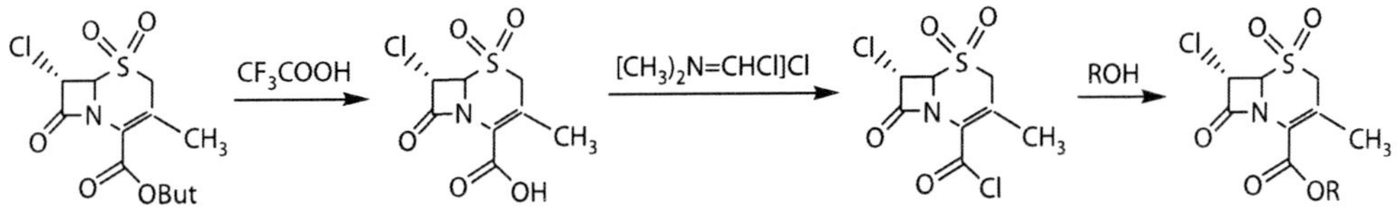

244

SCHEME 3.54 Synthesis of 7α-chloro-3-methyl-1,1-dioxoceph-3-EM-4-carboxylic acid esters.

Veinberg et al. reported synthesis and anticancer properties of 7α-chloro-3-methyl-1,1-Dioxoceph-3-EM-4-carboxylic acid esters **244** (Scheme 3.54).

Ruf et al. synthesized and tested β-lactams at a concentration of <20 μM in an *in vitro* screening method with the following human cancer cell lines: 5637 (urinary bladder carcinoma), RT-4 (urinary bladder carcinoma), A-427 (lung carcinoma), and LCLC-103H (large cell lung carcinoma).

Meegan et al. reported the synthesis of a few of β-lactams and these were evaluated via *in vitro* assays which determined their antiproliferative activity in MCF-7 and MDA-MB-231 breast cancer cell lines. Most of the compounds showed low cytotoxicity. Cytotoxicity values considerably below that obtained for tamoxifen (13.4%, 10 μM) were observed.

3.4 CONCLUDING REMARKS

In this chapter, most significant developments on the synthesis of β-lactams and their biological properties have been addressed. This chapter will help chemists to design and develop new β-lactams that will provide potential medicinal activities. It is our expectation that novel β-lactams will be synthesized and tested on different types of medical disorders.

ACKNOWLEDGMENT

We gratefully acknowledge the financial support for this research project from the National Institutes of Health (2S06M008038-39).

REFERENCES

Adlington, R. M., A. G. M. Barett, P. Quayle, A. Walker, and M. J. Betts. 1981. Novel syntheses of 3-methyleneazetidin-2-one derivatives and related systems. *J Chem Soc Chem Commun* 9:404–405.

Akermark, B., N. G. Johanson, and B. Sjoberg. 1969. Synthesis of strained heterocyclic rings. I. 2-hydroxy-β-lactams and 4-oxazolidinones by photocyclization of 2-oxo amides. *Tetrahedron Lett* 5:371–372.

Alcaide, B. and A. Vicente-Rodriguez. 1999. A convenient trans-stereoselective synthesis of phenanthridine derived 2-azetidinones using the Staudinger ketene-imine cycloaddition. *Tetrahedron Lett* 40:2005–2006.

Almqvist, F., D. Guillaume, S. J. Hultgren, and G. R. Marshall. 1998. Efficient regioselective synthesis of enantiomerically pure 4-hydroxymethyl-Δ^2-thiazoline. *Tetrahedron Lett* 39:2293–2294.

Amin, S. G., R. D. Glazer, and M. S. Manhas. 1979. Heterocyclic compounds. Part X. A convenient method for the synthesis of β-lactams via 1-methyl-2-halopyridinium salts. *Synthesis* 1979:210–213.

Arnold, L. D., T. H. Kalantar, and J. C. Vederas. 1985. Conversion of serine to stereochemically pure β-substituted α-amino acids via β-lactones. *J Am Chem Soc* 107:7105–7109.

Bacchi, S., A. Bongini, M. Panunzio, and M. Willa. 1998. Highly syn-diastereoselective synthesis of NH-3-benzyloxy-4-aryl-azetidin-2-ones via a two-step Staudinger reaction. *Synlett* 8:843–844.

Baldoli, C., P. Del Buttero, S. Maiorana, G. Jaouen, and S. Top. 1991. Microbial resolution of organometallic planar chirality. Enantioselective reduction of ortho and meta substituted tricarbonylchromium benzaldehydes by bakers' yeast. *J Organomet Chem* 413:125–135.

Banik, B. K. 2010. Topics in heterocyclic chemistry. In: *Heterocyclic Scaffolds I*, eds. Maes, B. U. W. and Banik, B. K. pp. 349–373. Berlin, Germany: Springer Verlag.

Becker, F. F. and B. K. Banik. 1998. Polycyclic aromatic compounds as anticancer agents: Synthesis and biological evaluation of some chrysene derivatives. *Bioorg Med Chem Lett* 8:2877–2880.
Belaaouaj, A., C. Lapoumeroulie, M. M. Canica, G. Vedel, P. Nevot, R. Krishnamoorthy, and G. Paul. 1994. Nucleotide sequences of the genes coding for the TEM-like β-lactamases IRT-1 and IRT-2 (formerly called TRI-1 and TRI-2). *FEMS Microbiol Lett* 120:75–80.
Berkheij, M., L. van der Sluis, C. Sewing, D. J. den Boer, J. W. Terpstra, H. Hiemstra, W. I. Iwema Bakker et al. 2005. Synthesis of 2-substituted piperazines via direct α-lithiation. *Tetrahedron Lett* 46:2369–2371.
Berranger, T. and Y. Langlois. 1995. [2 + 3] Cycloadditions of enantiomerically pure oxazoline N-oxides: An alternative to the asymmetric aldol condensation. *J Org Chem* 60:1720–1726.
Bonneau, P. R., F. Hasani, C. Plouffe, E. Malenfant, S. R. Laplante, I. Guse, W. W. Ogilvie et al. 1999. Inhibition of human cytomegalovirus protease by monocyclic β-lactam derivatives: Kinetic characterization using a fluorescent probe. *J Am Chem Soc* 121:2965–2973.
Bose, A. K., M. S. Manhas, B. K. Banik, and V. Srirajan. 2000a. β-Lactams: Cyclic amides of distinction. In: *The Amide Linkage: Selected Structural Aspects in Chemistry, Biochemistry, and Material Science*, eds. Greenberg, A., Breneman, C. M., and Liebman, J. F., p. 157. New York: Wiley-Interscience.
Bose, A. K., M. S. Manhas, J. S. Chib, H. P. S. Chawla, and B. Dayal. 1974. β-lactams XXXVI monocyclic *cis* β-lactams *via* penams and cephams. *J Org Chem* 39:2877–2884.
Bose, A. K., M. S. Manhas, M. Ghosh, M. Shah, V. S. Raju, S. S. Bari, S. N. Newaz et al. 1991. Microwave-induced organic reaction enhancement chemistry. 2. Simplified techniques. *J Org Chem* 56:6968–6970.
Bose, A. K., B. K. Banik, C. Mathur, D. R. Wagle, and M. S. Manhas. 2000b. Polyhydroxy amino acid derivatives via β-lactams using enantiospecific approaches and microwave techniques. *Tetrahedron* 56:5603–5619.
Boyd, E. A., W. C. Chan, and V. M. Lohn, Jr. 1996. Multiple solid phase synthesis of (RS)-1-aminophosphinic acids. *Tetrahedron Lett* 37:1647–1650.
Brieva, R., J. Z. Crich, and C. J. Sih. 1993. Chemoenzymic synthesis of the C-13 side chain of taxol: Optically active 3-hydroxy-4-phenyl β-lactam derivatives. *J Org Chem* 58:1068–1075.
Brun, T., J. Peduzzi, M. M. Canica, G. Paul, P. Nevot, M. Barthelemy, and R. Labia. 1994. Characterization and amino acid sequence of IRT-4, a novel TEM-type enzyme with a decreased susceptibility to β-lactamase inhibitors. *FEMS Microbiol Lett* 120:111–118.
Burla, M. C., M. Camalli, G. Cascarano, G. Giacovazzo, G. Polidori, R. Spagna, and D. Viterbo. 1989. SIR88-a direct-methods program for the automatic solution of crystal structures. *J Appl Crystallogr* 22:389–393.
Burnett, D. A., M. A. Caplen, H. R. Davis, R. E. Burrier, and J. W. Clader, 1994. 2-Azetidinones as inhibitors of cholesterol absorption. *J Med Chem* 37:1733–1736.
Buynak, J. 2004. The discovery and development of modified penicillin- and cephalosporin-derived β-lactamase inhibitors. *Curr Med Chem* 11:1951–1964.
Buynak, J. D., A. S. Rao, G. P. Fod, C. Carver, G. Adam, B. Geng, B. Bachmann et al. 1997. 7-alkylidenecephalosporin esters as inhibitors of human leukocyte elastase. *J Med Chem* 40:3423–3433.
Cainelli, G., D. Giacomini, P. Galletti, and M. DaCol. 1997. Penicillin G acylase mediated synthesis of the enantiopure (S)-3-amino-azetidin-2-one. *Tetrahedron Asymmetry* 8:3231–3235.
Calet, S., F. Urso, and H. Alper. 1989. Enantiospecific and stereospecific rhodium(I)-catalyzed carbonylation and ring expansion of aziridines. Asymmetric synthesis of β-lactams and the kinetic resolution of aziridines. *J Am Chem Soc* 111:931–934.
Chamchaang, W. and A. R. Pinhas. 1990. The conversion of an aziridine to a β-lactam. *J Org Chem* 55:2943–2950.
Clader, J. W., D. A. Burnett, M. A. Caplen, M. S. Domalski, S. Dugar, W. Vaccaro, R. Sher et al. 1996. 2-Azetidinone cholesterol absorption inhibitors: Structure–activity relationships on the heterocyclic nucleus. *J Med Chem* 39:3684–3693.
Coantic, S., D. Mouysset, S. Mignani, M. Tabart, and L. Stella. 2007. The use of N-sulfenylimines in the β-lactam synthon method: Staudinger reaction, oxidation of the cycloadducts and ring opening of β-lactams. *Tetrahedron* 63:3205–3216.
Cohn, M. L. 1992. Epidemiology of drug resistance: Implications for a post antimicrobial era. *Science* 257:1050–1055.
Dalcanale, E. and F. Montanari. 1986. Selective oxidation of aldehydes to carboxylic acids with sodium chlorite-hydrogen peroxide. *J Org Chem* 51:567–569.
Dardoize, F., J. L. Moreau, and M. Gaudemar. 1972. Reformatskii reaction on schiff bases. I. Preparation of β-aminoesters. *Bull Soc Chim Fr* 10:3841–3846.
Davies, J. 1994. Inactivation of antibiotics and the dissemination of resistance genes. *Science* 264:375–382.
Davoli, P., I. Moretti, F. Prati, and H. Alper. 1999. Carbonylation of silylated hydroxymethyl aziridines to β-lactams. *J Org Chem* 64:518–521.
Defoin, A., A. Brouillard-Poichet, and J. Streith. 1991. Asymmetric diels-alder cycloadditions with C_2-symmetrical chiral carbamoylnitroso dienophiles. *Helv Chim Acta* 74:103–109.

Delclaux, C., C. Delacourt, M. P. D'Ortho, V. Boyer, C. Lafuma, and A. Harf. 1996. Role of gelatinase B and elastase in human polymorphonuclear neutrophil migration across basement membrane. *Am J Respir Cell Mol Biol* 14:288–295.

Delpiccolo, C. M. L., M. A. Fraga, and E. G. Mata. 2003. An efficient, stereoselective solid-phase synthesis of β-lactams using mukaiyama's salt for the Staudinger reaction. *J Comb Chem* 5:208–210.

Denissova, I. and L. Barriault. 2003. Stereoselective formation of quaternary carbon centers and related functions. *Tetrahedron* 59:10105–10146.

Deziel, R. and E. Malenfant. 1998. Inhibition of human cytomegalovirus protease no with monocyclic β-lactams. *Bioorg Med Chem Lett* 8:1437–1442.

Dirat, O., C. Kouklovsky, and Y. Langlois. 1998. Oxazoline N-oxide-mediated [2 + 3] cycloadditions: Application to a total synthesis of the hypocholesterolemic agent 1233A. *J Org Chem* 63:6634–6642.

Dondoni, A. and M. C. Scherrmann. 1994. Thiazole-based synthesis of formyl C-glycosides. *J Org Chem* 59:6404–6412.

Donohoe, T. J. and D. House. 2002. Ammonia-free partial reduction of aromatic compounds using lithium di-tert-butylbiphenyl (LiDBB). *J Org Chem* 67:5015–5018.

Enders, D., R. Syrig, G. Raabe, C. Fernandez, J. M. Gasch, J. M. Lassaletta, and J. M. Llera. 1996. Formaldehyde SAMP-hydrazone as a neutral chiral formyl anion and cyanide equivalent: Asymmetric Michael additions to nitroalkenes. *Synthesis* 1:48–52.

Esparza, J., M. Kruse, J. Lee, M. Michaud, and J. A. Madri. 2004. MMP-2 null mice exhibit an early onset and severe experimental autoimmune encephalomyelitis due to an increase in MMP-9 expression and activity. *FASEB J* 18:1682–1691.

Evans, D. A. and E. B. Sjogren. 1985. The asymmetric synthesis of β-lactam antibiotics. I. Application of chiral oxazolidones in the Staudinger reaction. *Tetrahedron Lett* 26:3783–3786.

Fernandez, R., A. Ferrete, J. M. Lassaletta, J. M. Llera, and E. Martin-Zamora. 2002. N,N-dialkylhydrazones as the imine component in the Staudinger-like [2 + 2] cycloaddition to benzyloxyketene. *Angew Chem Int Ed* 41:831–833.

Fernandez, R., A. Ferrete, J. M. Lassaletta, J. M. Llera, and A. Monge. 2000. Enantioselective synthesis of 4-unsubstituted 3-alkoxy- and 3-aminoazetidin-2-ones from formaldehyde N,N-dialkylhydrazones. *Angew Chem Int Ed* 39:2893–2897.

Fernandez, R., E. Martin, C. Pareja, J. Vasquez, E. Diez, A. Monge, and J. M. Lassaletta. 1998. Synthese von enantiomerenreinen α-Alkoxy-α-trifluormethyl aldehyde nund-carbonsäuren aus trifluormethyl ketonen. *Angew Chem* 110:3598–3600.

Field, A. K. 1999. Human cytomegalovirus: Challenges opportunities and new drug development. *Antiviral Chem Chemother* 10:219–232.

Finke, P. E., M. E. Dahlgren, H. Weston, A. L. Maycock, and J. B. Doherty. 1993. Inhibition of human leukocyte elastase. 5. Inhibition by 6-alkyl substituted penem benzyl esters. *Bioorg Med Chem Lett* 3:2277–2282.

Finke, P. E., S. K. Shah, D. S. Fletcher, B. M. Ashe, K. A. Brause, G. O. Chandler, P. S. Dellea et al. 1995. Orally active β-lactam inhibitors of human leukocyte elastase. 3. Stereospecific synthesis and structure-activity relationships for 3,3-dialkylazetidin-2-ones. *J Med Chem* 38:2449–2462.

Fraenkel, G., C. C. Ho, Y. Liang, and S. Yu. 1972. Generation of a stable spiro dihydroaromatic anion. *J Am Chem Soc* 94:4732–4734.

Furet, P., C. Garcia-Eccheveria, B. Gay, J. Schoepfer, M. Zeller, and J. Rahuel. 1999. Structure-based design, synthesis, and x-ray crystallography of a high-affinity antagonist of the Grb2-SH2 domain containing an asparagine mimetic. *J Med Chem* 42:2358–2363.

Georg, G. I. and V. T. Ravikumar. 1992. Stereocontrolled Ketene-Imine cycloaddition reactions. In: *The Organic Chemistry of β-Lactams*, ed. Georg, G. I. New York: VCH Publishers.

Ghatak, A., F. F. Becker, and B. K. Banik. 2000. Indium-mediated facile synthesis of 3-unsubstituted ferrocenyl β-lactams. *Heterocycles* 53:2769–2772.

Hakimelahi, G. H. and A. A. Jarrahpour. 1989. Synthesis of ethyl cis-2-[(diethoxyphosphoryl)methyl]-7-oxo-3-phenyl-6-phthalimido-1-azabicyclo[3.2.0]hept-3-ene-2-carboxylate and methyl cis-2-bromo-3-methyl-8-oxo-7-phthalimido-4-oxa-1-azabicyclo[4.2.0]octane-2-carboxylate. *Helv Chim Acta* 72:1501–1505.

Hayes, J. F., M. Shipman, and H. Twin. 2000. Generation of metalloenamines by carbon–carbon bond formation: Ring opening reactions of 2-methyleneaziridines with organometallic reagents. *Chem Commun* 18:1791–1792.

He, M. and J. W. Bode. 2008. Enantioselective, NHC-catalyzed bicyclo-β-lactam formation via direct annulations of enals and unsaturated N-sulfonyl ketimines. *J Am Chem Soc* 130:418–419.

Holwerda, B. C. 1997. Herpes virus proteases: Targets for novel antiviral drugs. *Antiviral Res* 35:1–21.

Hwu, J. R., S. C. Tsay, and S. Hakimelahi. 1998. Syntheses of new isodethiaazacephems as potent antibacterial agents. *J Med Chem* 41:4681–4685.

Isaac, M. B. and T. H. Chan. 1995. Indium-mediated coupling of aldehydes with prop-2-ynyl bromides in aqueous media. *J Chem Soc Chem Commun* 10:1003–1004.

Jard, S., J. Elands, A. Schmidt, and C. Barberis, 1998. Vasopressin and oxytocin receptors: An overview. In: *Progress in Endocrinology*, eds. Imura, H. and Shizume, K., p. 1183. Amsterdam, the Netherlands: Elsevier.

Kai, H., K. Iwamoto, N. Chantani, and S. Murai. 1996. Ynolates from the reaction of lithiosilyldiazomethane with carbon monoxide. New ketenylation reactions. *J Am Chem Soc* 118:7634–7635.

Kawabata, T., T. Wirth, K. Yahiro, H. Suzuki, and K. Fuji. 1994. Direct asymmetric α-alkylation of phenylalanine derivatives using no external chiral sources. *J Am Chem Soc* 116:10809–10810.

Kidwai, M., P. Sapra, and K. R. Bhushan. 1999. Synthetic strategies and medicinal properties of β-lactams. *Curr Med Chem* 6:195–215.

Lassaletta, J. M., R. Fernandez, E. Martin-Zamora, and C. Pareja. 1996. Stereospecific addition of formaldehyde dialkyl hydrazones to sugar aldehydes. Synthesis of cyanohydrins and α-hydroxyaldehydes. *Tetrahedron Lett* 37:5787.

Limori, T., Y. Takahashi, T. Izawa, S. Kobayashi, and M. Ohno. 1983. Stereocontrolled synthesis of a *cis*-carbapenem antibiotic (−)-carpetimycin A. *J Am Chem Soc* 105:1659–1600.

Lin, T. H., T. S. Rogers, D. L. Hill, L. Simpson-Herren, D. R. Farnell, D. M. Kochhar, M. Alam et al. 1996. Murine toxicology and pharmacology of UAB-8, a conformationally constrained analog of retinoic acid. *Toxicol Appl Pharmacol* 139:310–316.

Linder, M. R. and J. Podlech. 2001. Synthesis of β-lactams from diazoketones and imines: The use of microwave irradiation. *Org Lett* 3:1849–1851.

Mahadevan, V., Y. D. Y. L. Getzler, and G. W. Coates. 2002. [Lewis acid] + [$Co(CO)_4$]- complexes: A versatile class of catalysts for carbonylative ring expansion of epoxides and aziridines. *Angew Chem Int Ed* 41:2781–2784.

Manhas, M. S., B. K. Banik, A. Mathur, J. E. Vincent, and A. K. Bose. 2000. Vinyl-β-lactams as efficient synthons. Eco-friendly approaches via microwave assisted reactions. *Tetrahedron* 56:5587–5601.

March, J. 1985. *Advances in Organic Chemistry*, vols. 218–236, p. 33. New York: Wiley.

Maruyama, H., M. Shiozaki, S. Oida, and T. Hiraoka. 1985. Cyclization reaction of N-propargyl epoxyamide to acetylenic 2-azetidinone, a precursor to thienamycin and related carbapenems. *Tetrahedron Lett* 26:4521–4522.

Mascaretti, O. A., C. E. Boschetti, G. O. Danelon, E. G. Mata, and O. Roveri. 1995. β-Lactam compounds. Inhibitors of transpeptidases, β-lactamases and elastases: A review. *Curr Med Chem* 1:441–470.

Miller, M. J., P. G. Mattingly, M. A. Morrison, and J. F. Kerwin, Jr. 1980. Synthesis of β-lactams from substituted hydroxamic acids. *J Am Chem Soc* 102:7026–7032.

Molnar, F., G. Luinstra, M. Allmendinger, and B. Rieger. 2003. Multisite catalysis: A mechanistic study of β-lactone synthesis from epoxides and CO-insights into a difficult case of homogeneous catalysis. *Chem Eur J* 9:1273–1280.

Mukerjee, A. K. and A. K. Singh. 1978. β-lactams: Retrospect and prospect. *Tetrahedron* 34: 1731–1767.

Murthy, K. S. K. and A. Hassner. 1987. Fused β-lactams via intramolecular dipolar cycloaddition. *Tetrahedron Lett* 28:97–100.

Myers, R. M., S. P. Langston, S. P. Conway, and C. Abell. 2000. Reductive cleavage of N-O bonds using samarium(II) iodide in a traceless release strategy for solid-phase synthesis. *Org Lett* 2:1349–1352.

Neu, H. C. 1992. The crisis in antibiotic resistance. *Science* 257:1064–1073.

Nukaga, P. M., Y. Nukaga, A. Hujer, R. A. Bomono, and J. R. Knox. 2001. Inhibition of the SHV-1 β-lactamase by sulfones: Crystallographic observation of two reaction intermediates with tazobactam. *Biochemistry* 40:1861–1866.

Oppenheimer, J., W. L. Johnson, M. R. Tracey, R. P. Hsung, P. Y. Yao, R. Liu, and K. Zhao. 2007. A rhodium (I)-catalyzed demethylation-cyclization of o-anisole-substituted ynamides in the synthesis of chiral 2-amido benzofurans. *Org Lett* 9:2361–2364.

Padwa, A., D. J. Austin, A. T. Price, M. A. Semones, M. P. Doyle, M. N. Protopopova, W. R. Winchester et al. 1993. Ligand effects on dirhodium(II) carbene reactivities. Highly effective switching between competitive carbenoid transformations. *J Am Chem Soc* 115:8669–8680.

Palomo, C., Aizpurua, J., M. Ganboa, I. Carrequx, F. Cuevas, C. Maneiro, E., and Ontoria J. M. 1994. New synthesis of alpha-amino acid N-carboxy anhydrides through Baeyer-Villiger oxidation of α-keto β-lactams. *J Org Chem* 59:3123–3130.

Palomo, C., J. M. Aizpurua, I. Ganboa, and M. Oiarbide. 2001. β-lactams as versatile intermediates in α- and β-amino acid synthesis. *Synlett* 12:1813–1826.

Palomo, C., I. Ganboa, C. Cuevas, C. Boschetti, and A. Linden. 1997. A concise synthesis of piperazine-2-carboxylic acids via β-lactam-derived α-amino acid N-carboxy anhydrides. *Tetrahedron Lett* 38:4643.

Pansare, S. V., G. Huyer, L. D. Arnold, and J. C. Vederas. 1991. Synthesis of N-protected α-amino acids from N-(benzyloxycarbonyl)-L-serine via its β-lactone: N^{α}-(benzyloxycarbonyl)-β-(pyrazol-1-yl)-L-aline. *Org Synth* 70:1–6.

Pareja, C., E. Martin-Zamora, R. Fernandez, and J. M. Lassaletta. 1999. Stereoselective synthesis of trifluoromethylated compounds: Nucleophilic addition of formaldehyde N,N-dialkylhydrazones to trifluoromethyl ketones. *J Org Chem* 64:8846.

Perreux, L. and A. Loupy. 2001. A tentative rationalization of microwave effects in organic synthesis according to the reaction medium, and mechanistic considerations. *Tetrahedron* 57:9199–9223.

Piotti, M. E. and H. Alper. 1996. Inversion of stereochemistry in the $Co_2(CO)_8$-catalyzed carbonylation of aziridines to β-lactams. The first synthesis of highly strained trans-bicyclic β-lactams. *J Am Chem Soc* 118:111–116.

Podlech, J. and M. R. Linder. 1997. Cycloadditions of ketenes generated in the Wolff rearrangement. Stereoselective synthesis of aminoalkyl-substituted β-lactams from α-amino acids. *J Org Chem* 62:5873–5883.

Ross, N. A., R. R. McGregor, and R. A. Bartsch. 2004. Synthesis of β-lactams and β-aminoesters via high intensity ultrasound-promoted Reformatsky reactions. *Tetrahedron* 60:2035–2041.

Ruhland, B., A. Bhandari, E. M. Gordon, and M. A. Gallop. 1996. Solid-supported combinatorial synthesis of structurally diverse β-lactams. *J Am Chem Soc* 118:253–254.

Scheffer, J. R. and K. Wang. 2001. Enantioselective photochemical synthesis of a β-lactam via the solid state ionic chiral auxiliary method. *Synthesis* 8:1253–1257.

Sheldrick, G. M. 1990. Phase annealing in SHELX-90: Direct methods for larger structures. *Acta Crystallogr Sect A* 46:467–473.

Shiers, J. J., M. Shipman, J. F. Hayes, and A. M. Z. Slawin. 2004. Rare example of nucleophilic substitution at vinylic carbon with inversion: Mechanism of methyleneaziridine formation by sodium amide induced ring closure revisited. *J Am Chem Soc* 126:6868–6869.

Shindo, M. S., Oya, Y. Sato, and K. Shishido. 1998. Cycloaddition of lithium ynolate to imines: Synthesis of 3,4-disubstituted β-lactams. *Heterocycles* 49:113–116.

Singh, G. S. 2004. β-Lactams in the new millenium. Part I: Monobactams and carbapenems. *Mini Rev Med Chem* 4:69–92.

Sirot, D., C. Chanal, C. Henquell, R. Labia, J. Sirot, and R. J. Cluzel. 1994. Clinical isolates of *Escherichia coli* producing multiple TEM mutants resistant to β-lactamase inhibitors. *Antimicrob Chemother* 33:1117–1126.

Sivaprakasam, M., F. Couty, G. Evano, B. Srinivas, R. Sridhar, and K. Rama Rao. 2006. Stereocontrolled synthesis of 3-substituted azetidinic amino acids. *Synlett* 5:781–785.

Smith, D. M., A. Kazi, L. Smith, T. E. Long, B. Heldreth, E. Turos, and Q. P. Dou. 2002. A novel β-lactam antibiotic activates tumor cell apoptotic program by inducing DNA damage. *Mol Pharmacol* 61:1348–1358.

Speckamp, W. N. and H. Hiemstra. 1985. Intramolecular reactions of N-acyliminium intermediates. *Tetrahedron* 41:4367–4416.

Staudinger, H. 1907. Ketenes. 1. Diphenylketene. *Liebigs Ann Chem* 356:51–123.

Sternlicht, M. D. and Z. Werb. 1999. Neutrophil elastase and cathepsin G. In: *Extracellular Matrix, Anchor and Adhesion Proteins*, eds. Kreis, T. and Vale, R., p. 543. Oxford, U.K.: Oxford University Press.

Suffness, M. 1995. *Taxol Science and Applications*. Boca Raton, FL: CRC Press.

Szakonyi, Z., T. Martinek, A. Hetenyi, and F. Fulop. 2000. Synthesis and transformations of enantiomeric 1,2-disubstituted monoterpene derivatives. *Tetrahedron Asymmetry* 11:4571–4579.

Taggi, A. E., A. M. Hafez, H. Wack, B. Young, W. J. Drury III, and T. Lectka. 2000. Catalytic, asymmetric synthesis of β-lactams. *J Am Chem Soc* 122:7831–7832.

Taggi, A. E., A. M. Hafez, H. Wack, B. Young, D. Ferraris, and T. Lectka. 2002. The development of the first catalyzed reaction of ketenes and imines: Catalytic, asymmetric synthesis of β-lactams. *J Am Chem Soc* 124:6626–6635.

Tamagnan, G., Y. Gao, R. M. Baldwin, S. S. Zoghbi, and J. L. Neumeyer. 1996. Synthesis of β-CIT-BAT, a potential technetium-99m imaging ligand for dopamine transporter. *Tetrahedron Lett* 37:4353–4356.

Tanner, D. and P. Somafai. 1993. Palladium-catalyzed transformation of a chiral vinylaziridine to a β-lactam. An enantioselective route to the carbapenem (+)-PS-5. *Bioorg Med Chem Lett* 3:2415–2418.

Troisi, L., C. Granito, and E. Pindinelli. 2010. Novel and recent synthesis and applications of β-lactams. *Top Het Chem* 22:101–209.

Veinberg, G., M. Vorona, I. Shestakova, I. Kanepe, O. Zharkova, R. Mezapuke, I. Turovskis et al. 2000. Synthesis and antitumor activity of selected 7-alkylidene substituted cephems. *Bioorg Med Chem* 8:1033–1040.

Waxman, L. and P. L. Darke. 2000. The herpes virus proteases as targets for antiviral chemotherapy. *Antiviral Chem Chemother* 11:1–22.

Whitesell, J. K. 1989. C_2 symmetry and asymmetric induction. *Chem Rev* 89:1581–1590.

Xu, J. X., G. Zuo, and W. L. Chan. 2001. Reactions of 2,3-dihydro-1H-1,5-benzodiazepines and chloroacetyl chlorides: Synthesis of 2a,3,4,5-tetrahydroazeto[1,2-a][1,5]benzodiazepin-1(2H)-ones. *Heteroatom Chem* 12:636–640.

Yamamoto, Y., Y. Watanabe, S., and Ohnishi. 1987. 1, 3-Oxazines and related compounds. XIII. Reaction of acyl meldrum's acids with schiff bases giving 2, 3-disubstituted 5-acyl–3, 4, 5, 6-tetrahydro-2*H*-1, 3-oxazine-4, 6-diones and 2, 3, 6-trisubstituted 2, 3-dihydro-1, 3-oxazin-4-ones. *Chem Pharm Bull* 35:1860–1870.

Yang, Y., B. A. Rasmussen, and D. M. Shales. 1999. Class A β-lactamases-enzyme-inhibitor interactions and resistance. *Pharmacol Ther* 83:141–151.

Yi, X. H., Y. Meng, X. G. Hua, and C. J. Li. 1998. Regio and diastereoselective allenylation of aldehydes in aqueous media: Total synthesis of (+)-goniofufurone. *J Org Chem* 63:7472–7480.

Zificsak, C. A., J. A. Mulder, R. P. Hsung, C. Rameshkumar, and L. L. Wei. 2001. Recent advances in the chemistry of ynamines and ynamides. *Tetrahedron* 57:7575–7606.

4 Applications of Isatin Chemistry in Organic Synthesis and Medicinal Chemistry

Girija S. Singh and Zelalem Y. Desta

CONTENTS

4.1 INTRODUCTION

Isatin (1*H*-indole-2,3-dione, indoline-2,3-dione) **1** (Figure 4.1) is a structurally simple natural product found in the plants of genus *Isatis* and in *Couropita guianancis aubl* (Bergman et al. 1988, Silva et al. 2001). It has also been found as a metabolic derivative of adrenaline in humans (Chiyanzu et al. 2003, Almeida et al. 2010). Isatin, possessing an indole motif with a ketone and a γ-lactam moiety fused to the benzene ring, has drawn considerable interest to the researchers in the field of organic synthesis and medicinal chemistry. Isatin undergoes electrophilic aromatic substitutions at positions C-5 and C-7 of its benzene ring (Silva et al. 2001). *N*-Alkylations/arylations/acylations, nucleophilic additions at the carbonyl group, chemoselective reductions, oxidations, and ring expansion are also reported. The diverse reactivity of isatin has made it a valuable building block for the synthesis of various other heterocyclic frameworks such as quinolines, indoles, oxindole, and β-lactams, etc. The chemistry of isatin has been reviewed in the past by Sumpter (1944), Popp (1975), Mesropyan and Avetisyan (2009), and Silva et al. (2001). Recent literature shows resurgence of interest in the chemistry and bioactivity of isatin and its derivatives leading to improvement in several already known reactions and synthesis of many isatin derivatives with different types of biological activity (Pandeya et al. 2005). Isatin derivatives having antitubercular activity have been reviewed recently (Aboul-Fadl and Bin-Jubair 2010). This chapter is based on the chemistry of isatin reported from 2000 to 2010. Some examples from early 2011 are also included.

4.2 SYNTHESIS OF ISATINS

Although the review of synthetic methodologies is not the main objective of this review, it would be useful by way of introduction to give a brief idea about the synthesis of isatins. The classical methods for the synthesis of isatins are Sandmeyer's method (Scheme 4.1), the Stolle procedure (Scheme 4.2), and Gassman procedure (Scheme 4.3), all using aniline as substrate.

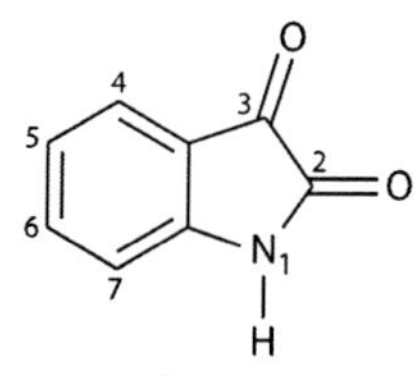

1

FIGURE 4.1 Structure of isatin.

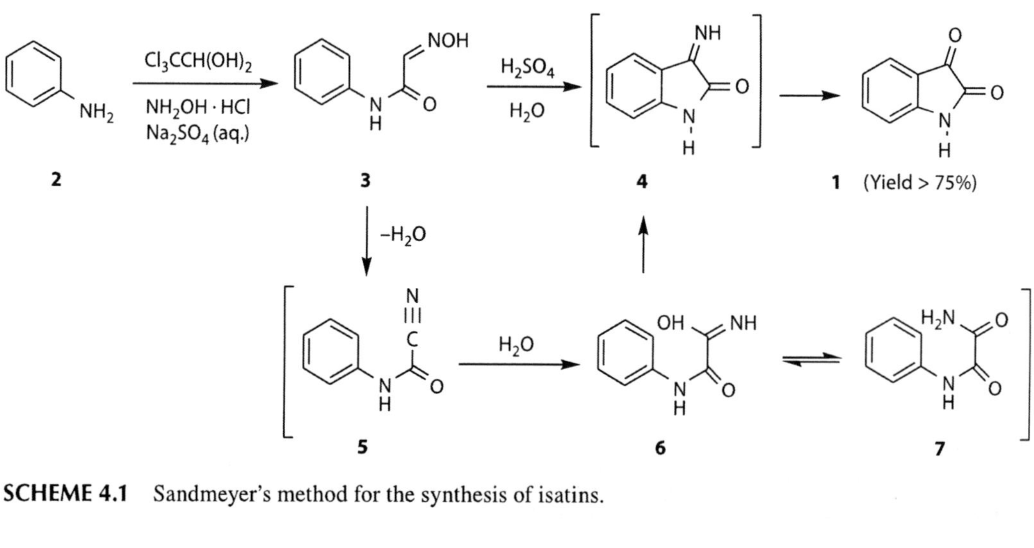

SCHEME 4.1 Sandmeyer's method for the synthesis of isatins.

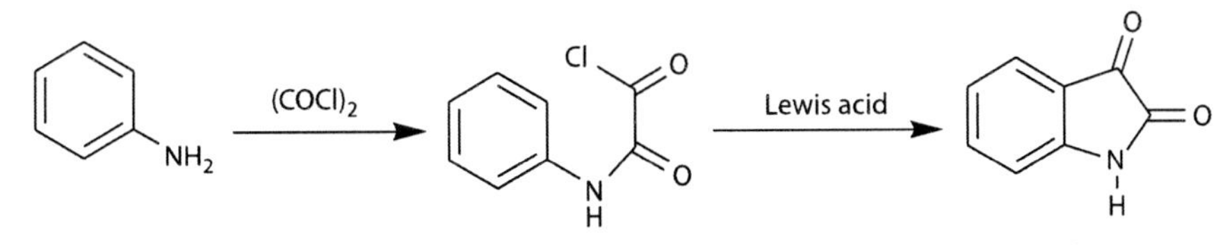

SCHEME 4.2 Stolle's method for the synthesis of isatins.

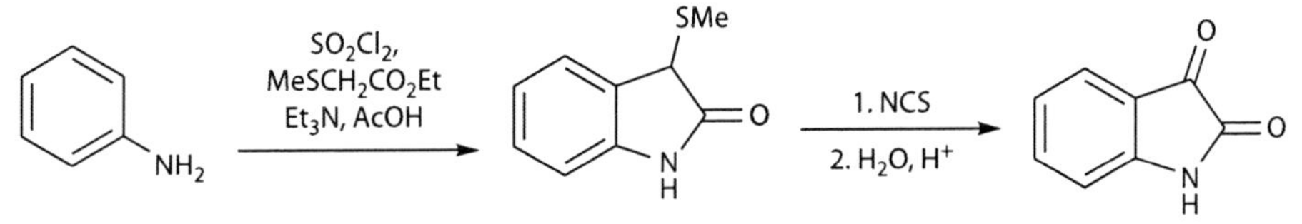

SCHEME 4.3 Gassman's method for the synthesis of isatins.

4.2.1 Sandmeyer's Method

A three-component reaction of aniline **2**, hydroxylamine hydrochloride, and 2,2,2-trichloroethane-1,1-diol (chloral hydrate) affords isonitrosoacetanilide **3**, which on treatment with sulfuric acid leads to the formation of isatin **1** (Scheme 4.1) (Sandmeyer 1919). This method is applicable to anilines bearing both electron-donating and electron-withdrawing groups. The mechanism of this reaction has been subject of much discussion. Sandmeyer explained the formation of isatin through the imine **4**. Later, the mechanism was elaborated by Piozzi and Favini who proposed the formation of imine **4** from **3** via compounds **5** and **6**. The latter compound was also proposed to be in equilibrium with compound **7** (Silva et al. 2011a). Recently, the mechanism has been investigated by microreactor-electrospray ionization mass spectrometry (Silva et al. 2011b).

4.2.2 Stolle's Method

In the Stolle method, aniline and its derivatives react with excess oxalyl chloride to give *N*-chlorooxalylanilide which in the presence of Lewis acids ($AlCl_3$, $BF_3 \cdot Et_2O$, or $TiCl_4$) cyclizes to isatin (Scheme 4.2). This reaction, however, is not applicable to substrates containing electron-withdrawing groups (Kurkin et al. 2011). Kurkin and coworkers have reported 30%–60% yields of isatins using the Stolle method. Ma and coworkers, however, reported a very poor yield of 5% by this method (Ma et al. 2003).

4.2.3 Gassman's Method

This methodology constitutes the formation of 3-methylthio-2-oxindole from aniline and oxidation of C-3 methine carbon in it with *N*-chlorosuccinimide followed by hydrolysis of the chlorinated intermediate (Scheme 4.3) (Gassman et al. 1977). The reaction is compatible with anilines having strongly electron-withdrawing and electron-donating groups.

Efforts are still on to improve these three methodologies for the synthesis of isatins and several variations are reported in literature. Besides these three, some other interesting methodologies have also been reported in recent literature. Selected examples are discussed in the succeeding paragraphs.

4.2.4 Some Recent Syntheses of Isatins

Palladium-catalyzed *N*-heteroannulations of 1-(2-bromoalkyn-1-yl)-2-nitrobenzenes using carbon monoxide as the ultimate reducing agent have emerged as a viable method for the synthesis of a variety of indoles. 1-(2-Bromoethynyl)-2-nitrobenzene **10** reacts with carbon monoxide in the

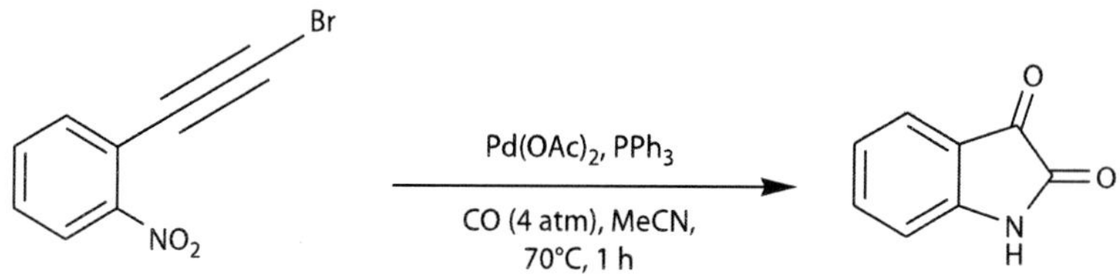

10 **1** (Yield = 35%)

SCHEME 4.4 Synthesis of isatin by palladium-catalyzed *N*-heteroannulation of 1-(2-bromoalkyn-1-yl)-2-nitrobenzene.

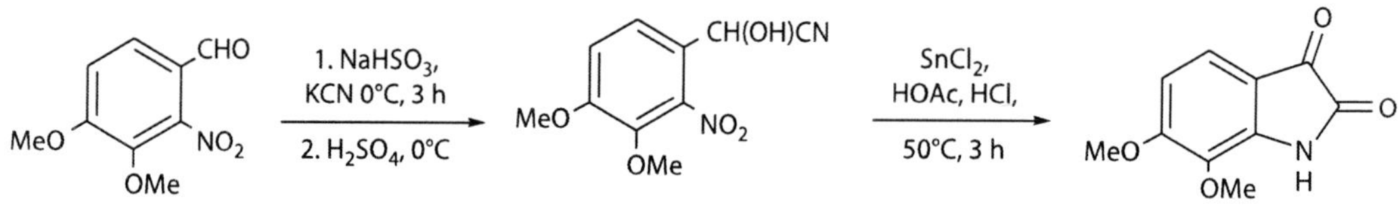

11 **12** (Yield = 85.5%) **13** (Yield = 83.6%)

SCHEME 4.5 Synthesis of 6,7-dimethoxyisatin.

presence of a catalytic amount of palladium diacetate and triphenylphosphine to yield isatin **1**. The starting material in this reaction was completely consumed within 1 h at 70°C (Scheme 4.4) (Soderberg et al. 2009).

A method for the synthesis of 6,7-dimethoxyisatin **13** involves the formation of the cyanohydrins **12** from 2-nitroveratraldehye **11**. The reduction of nitro group in 2-nitro-3,4-dimethoxymandelonitrile **12** followed by cyclization affords the product (Scheme 4.5) (Ma et al. 2003).

A one-pot procedure for the synthesis of isatins starting from anilines **2** and **14** by using oxalyl chloride as acylating agent and H-β zeolite as a reusable catalyst under heterogeneous conditions provides a simple and efficient method (Scheme 4.6) (Raj et al. 2010). Recently, a one-pot procedure for the synthesis of isatins based on the oxidation of indoles with hypervalent iodine as an oxidant and indium(III) chloride as a catalyst has been reported (Scheme 4.7) (Yadav et al. 2007).

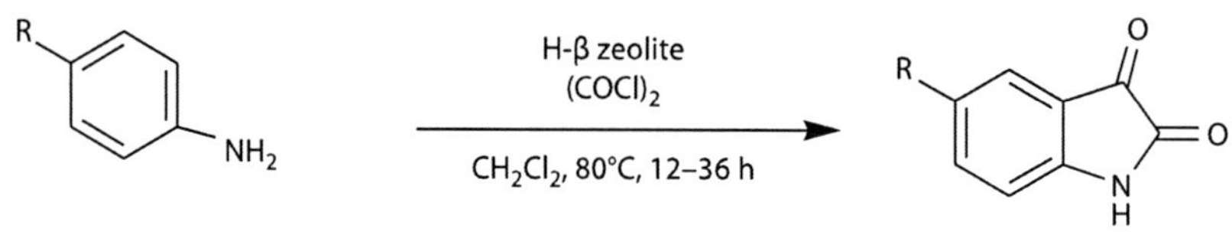

1 and **2**: R = H
14 and **15**: R = Me, Cl, F, OMe, NO_2, *i*-Pr, CO_2Me

1
15 (Yield = 48%–79%)

SCHEME 4.6 Synthesis of isatins by H-β zeolite-catalyzed reactions of anilines with oxalyl chloride.

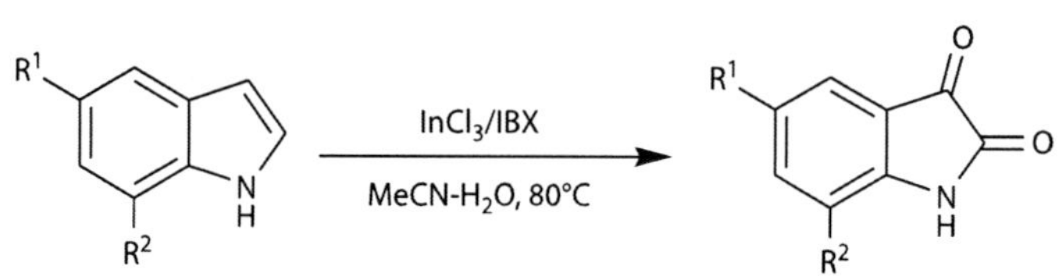

16 **17**

SCHEME 4.7 Synthesis of isatins by indium(III)-catalyzed oxidation of indoles with hypervalent iodine.

4.3 REACTIVITY OF ISATINS

Isatin is a highly reactive molecule and has been exploited in organic synthesis both as an electrophile and a nucleophile. As a nucleophile it undergoes electrophilic substitution on the aromatic ring, substitution at γ-lactam nitrogen. The most common reactions of isatin as an electrophile are nucleophilic additions to the ketone group. Besides this, isatin undergoes oxidation, reduction, and ring expansion reactions forming different types of heterocyclic compounds, and also ring opening products. The literature is arranged according to the reactivity of a particular group such as phenyl ring, ring nitrogen, and ketone group followed by oxidation, reduction, ring opening, and synthesis of isatin-based spiro-fused heterocycles.

4.3.1 Reactivity of Phenyl Ring

Isatin is known to undergo electrophilic aromatic substitution either at C-5 or C-7. Halogenation of isatin at C-5 has been achieved by reacting isatin **1** with acidic trichloroisocyanuric acid (TCCA) in ethyl acetate and sodium bicarbonate affording 5-chloroisatin **35** (Scheme 4.8) (Bhardwaj et al. 2010). TCCA in sulfuric acid is also reported to form 5-chloroisatin and 5,7-dichloroisatin (Mendonca et al. 2005). TCCA is used as a source of electrophilic chlorine. Acidic media strongly promoted the formation of superelectrophilic species where TCCA is either polyprotonated or protosolvated, causing more efficient "Cl^+" transfer to isatin due to the charge–charge repulsion. The reaction of isatin bearing an electron-donating group such as methyl group at C-5 with TCCA in different molar ratios in sulfuric acid led to the formation of a mixture of chlorinated products (Scheme 4.9) (Silva et al. 2011a). However, when sulfuric acid was replaced by acetic acid, in order to get milder reaction conditions, *N*-chlorinated isatin derivative was formed in different yields depending upon the reaction conditions. A maximum yield of 82% was obtained when 5-methylisatin **15** reacted with TCCA in 1:2 molar ratios at 25°C for 1 h.

A palladium-catalyzed ring metathesis by intramolecular aryl–aryl coupling in **20** has led to the synthesis of novel isatins tethered to eight-membered ring **21** (Scheme 4.10) (Lee et al. 2010).

4.3.2 Reactivity of Amide Nitrogen

4.3.2.1 *N*-Alkylation

N-Alkylations of isatins have been achieved either by direct synthesis from *N*-alkylanilines as shown in the Gassman procedure or by *N*-alkylation of isatin (Silva et al. 2001). The simplest *N*-alkylated isatin, *N*-methylisatin **22**, is obtained by treating isatin **1** with dimethylsulfate in dil. aqueous sodium hydroxide (Scheme 4.11) (Silva et al 2001). This method has been employed recently by Luntha (2009) and Bhardwaj et al. (2010) in the synthesis of various bioactive isatin derivatives affording the product in quantitative yield.

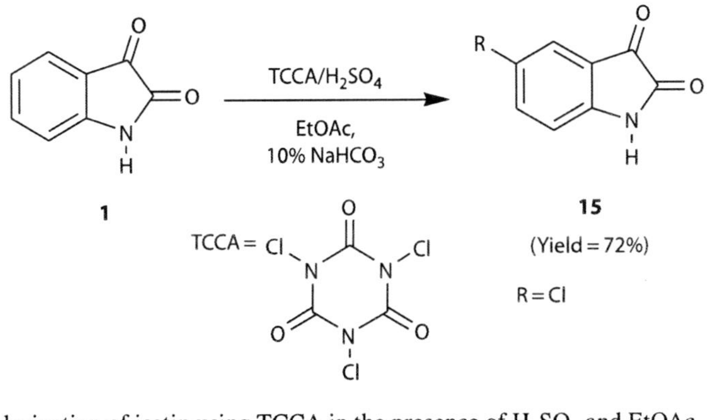

SCHEME 4.8 Chlorination of isatin using TCCA in the presence of H_2SO_4 and EtOAc.

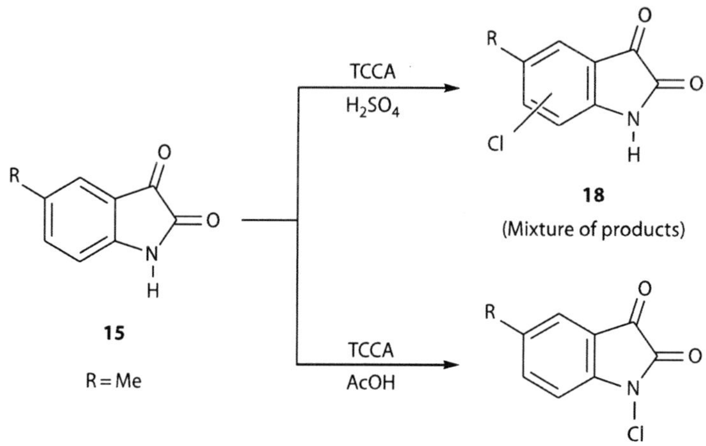

15
R = Me

18
(Mixture of products)

19 (Yield = 46%–82%)

SCHEME 4.9 Chlorination of isatin using TCCA in the presence of H_2SO_4, and H_2SO_4 and AcOH.

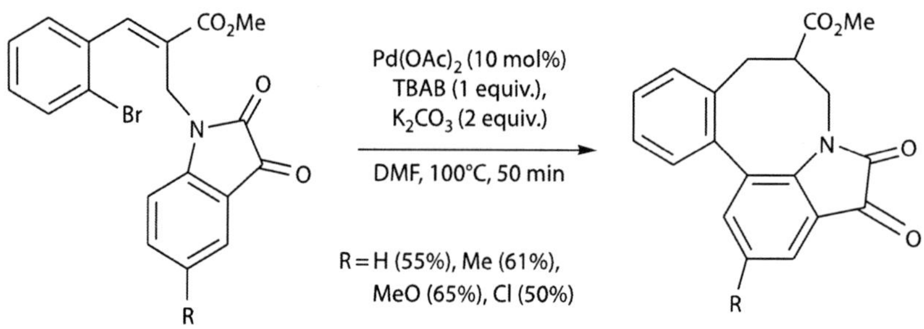

20

21 (Yield = 50%–65%)

SCHEME 4.10 A palladium-catalyzed ring-metathesis in *N*-substituted isatin.

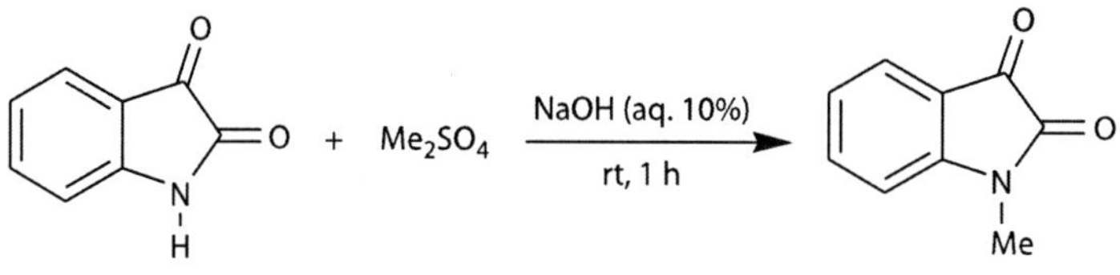

1

22

SCHEME 4.11 *N*-Methylation of isatin using dimethylsulfate.

Several other *N*-alkylated isatins **23** have been obtained by reacting isatin **1** or 5-bromoisatin **15** with alkyl halides using 1.5 equiv. of potassium carbonate in DMF at 80°C (Scheme 4.12) (Aboul-Fadl et al. 2010). The use of magnesium carbonate in refluxing acetone is also reported in *N*-alkylation of isatin-forming products **24** (Scheme 4.13) (Garden et al. 1998, Rekhter 1999).

The reaction of isatin with sodium or calcium hydride in toluene leads to the formation of corresponding isatides **25** that reacts with α-halogenated ketones **26** at the NH group and produces *N*-alkylated isatin **27** (Scheme 4.14) (Rekhter 2005). Azizian et al. and Schmidt et al. have reported facile *N*-alkylation of isatins under microwave irradiation (Azizian et al. 2003, Schmidt et al. 2008).

4.3.2.2 *N*-Acylation

N-Acyl derivatives of isatin are obtained by acylation of isatin with carboxylic acids anhydrides in the presence of perchloric acid (Mesropyan and Avetisyan 2009). The formation of *N*-acetyl-substituted

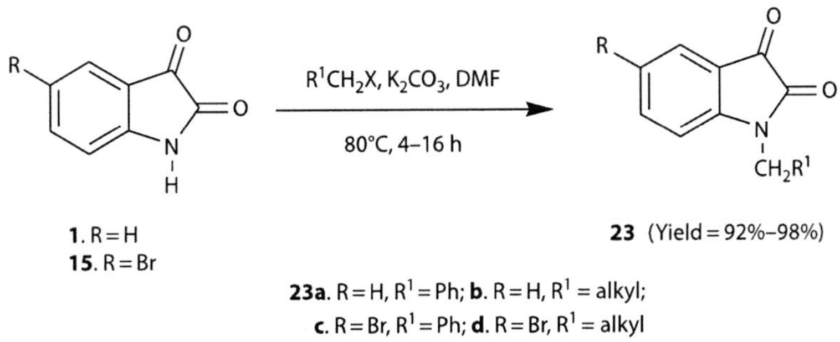

SCHEME 4.12 *N*-Alkylation of isatin and 5-bromoisatin.

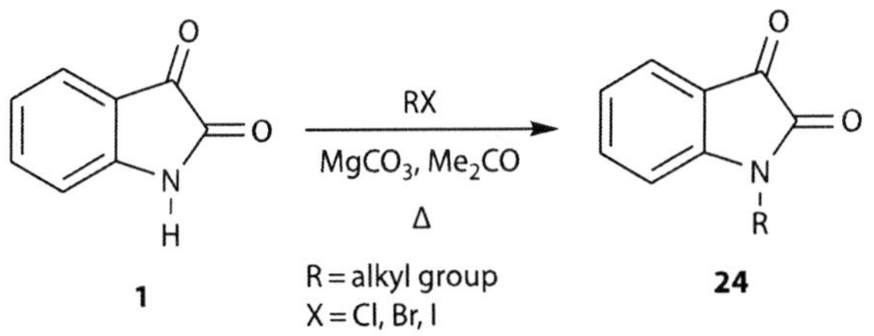

SCHEME 4.13 *N*-Alkylation of isatin.

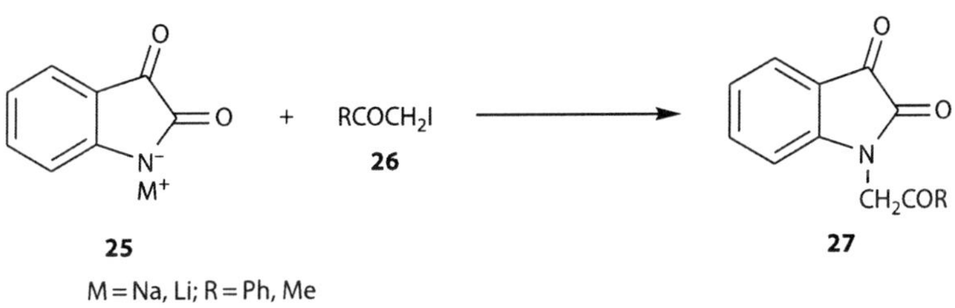

SCHEME 4.14 *N*-Acylation of isatins in the presence of metal hydrides.

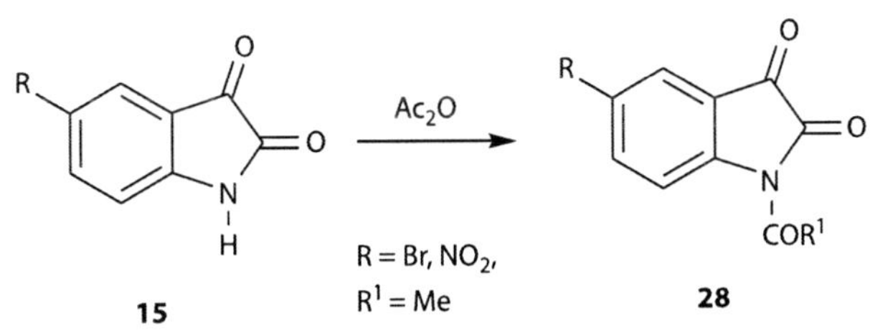

SCHEME 4.15 *N*-Acylation of isatins using acetic anhydride.

isatins **28** has been reported by treatment of isatins **15** with acid anhydride (Scheme 4.15) (Boechat et al. 2008, Smitha et al. 2008). Recently, Lesogo and Singh have synthesized *N*-(diphenylacyl)isatin **30** by reacting isatin **1** with diphenylketene, generated *in situ* by thermal decomposition of 2-diazo-1,2-diphenylethanone **29** (Scheme 4.16) (Masutlha and Singh 2012).

4.3.2.3 *N*-Arylation

N-Phenylisatin **31** has been synthesized by reacting isatin **1** with chlorobenzene **14** in triethylamine (Scheme 4.17) (Bhragual et al. 2010). *N*-Arylisatins were obtained earlier from isatin by reaction with aryl bromides in the presence of cupric oxide (Silva et al. 2001).

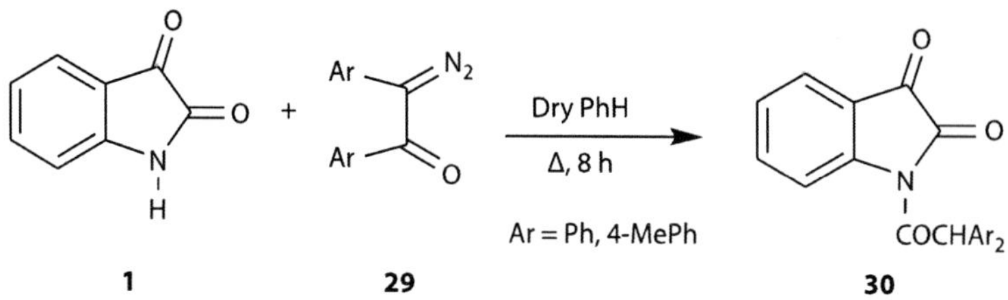

SCHEME 4.16 *N*-Acylation of isatins using 2-diazo-1,2-diarylethanones.

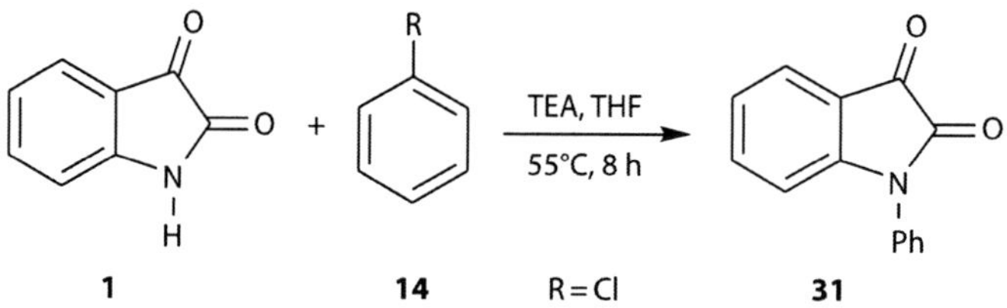

SCHEME 4.17 *N*-Arylation of isatin.

4.3.2.4 Reactivity of Amide Nitrogen: Applications in Heterocycle Synthesis

The reactivity of isatin ring nitrogen has been exploited for attaching different types of heterocyclic moieties to the isatin ring. For example, *N*-alkylation of isatin **1** with various chloromethylquinolines (**32a–e**) in the presence of potassium *tert*-butoxide (KO^{t}Bu) in tetrahydrofuran (THF) at 70°C to afford corresponding *N*-alkylated derivatives (**33a–e**) is reported (Scheme 4.18) (Roopan et al. 2010). Unfortunately, the authors have not mentioned about the yield of the products.

Isatin **1** undergoes reaction with *tert*-butylbromoacetate **34** in the presence of potassium carbonate to afford the corresponding ester **35**, which on treatment with hydroxylamine in the presence of *p*-toluenesulfonic acid in methanol gives 3-oxime **36**. Transformation of the carboxylic ester group in oxime **36** by treating it with oxalyl chloride affords compound **37**. The reaction of the latter compound with *N*-substituted *o*-phenylenediamine **38** followed by cyclization introduces a benzimidazolomethyl group onto the nitrogen atom of isatin ring affording product **39** (Scheme 4.19) (Sin et al. 2009). The reaction has been carried out with differently substituted anilines.

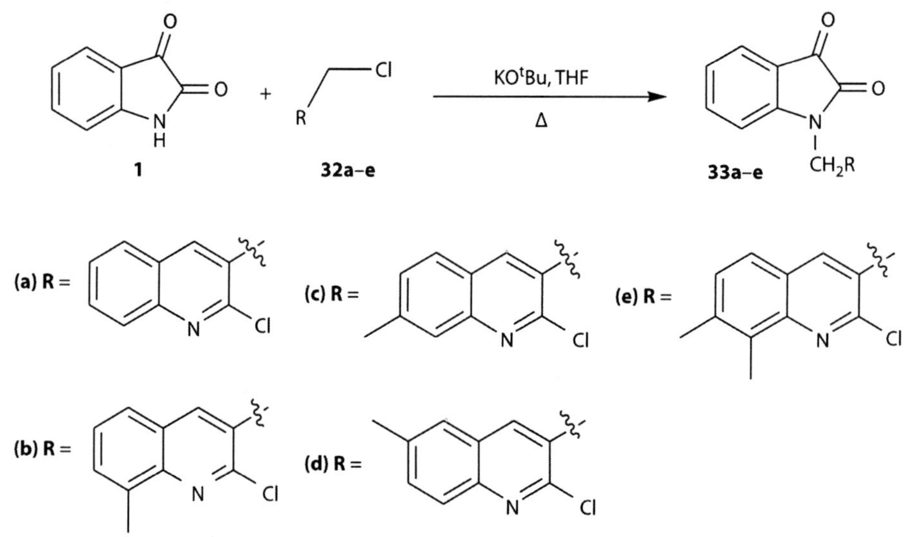

SCHEME 4.18 *N*-Alkylation of isatins by alkyl chlorides containing heterocyclic moieties.

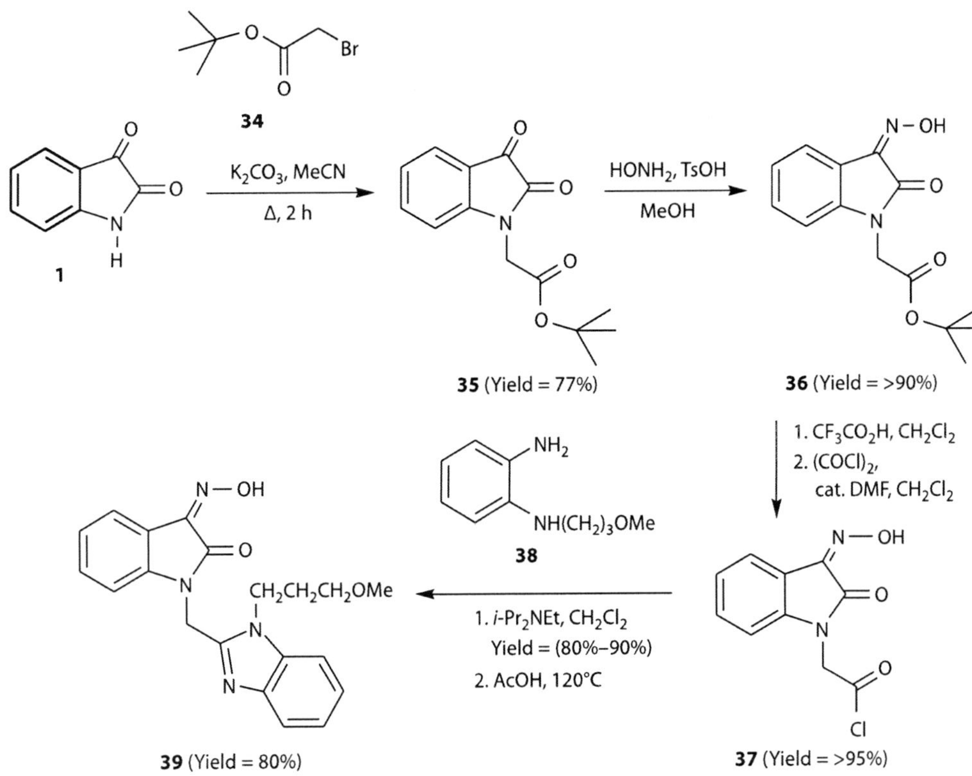

SCHEME 4.19 A multi-step synthesis of 1-substituted isatin-3-oxime.

Also, various *O*-substituted oximes have been synthesized. Many compounds exhibited antiviral activity in the BALB/c mouse model of RSV infection following oral dosing.

A [2 + 3]-cycloaddition of propargyl-substituted 3-hydroxy-3-(2-hydroxynaphthalene-1-yl)-1-prop-2-ynyl-1,3-dihydroindol-2-one **40**, synthesized by the Friedel–Crafts alkylation of 2-naphthol by isatin, with benzyl azide **41** under $CuSO_4$ catalysis furnished 1,4-disubstituted[1,2,3]-triazole **42** regioselectively in one pot in 82% yield (Scheme 4.20) (Ramachary et al. 2007). The cycloaddition of another isatin **43** bearing *N*-propargylic substituent with phenylazide **44** in the presence of $CuSO_4$ affords *N*-(1-phenyl-1,2,3-triazol-4-yl)methylisatin derivative **45** (Scheme 4.21) (Jiang and Hansen 2011). This compound has been observed as an inhibitor of caspase-3 (IC_{50} = 21 nM).

The imines **46**, obtained from the reaction of isatins **1** and **15** with amines **14**, are alkylated at ring nitrogen using ethylchloroacetate forming *N*-1-substituted imines **47**. Treatment of these imines with hydrazine transforms the ester group into hydrazide, affording compounds **48** which

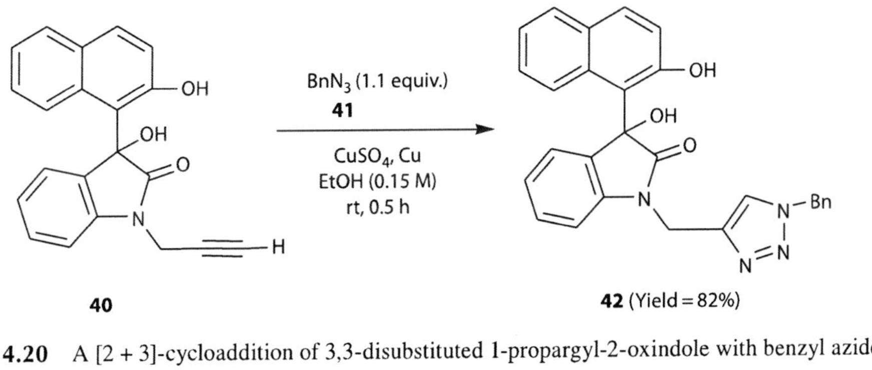

SCHEME 4.20 A [2 + 3]-cycloaddition of 3,3-disubstituted 1-propargyl-2-oxindole with benzyl azide.

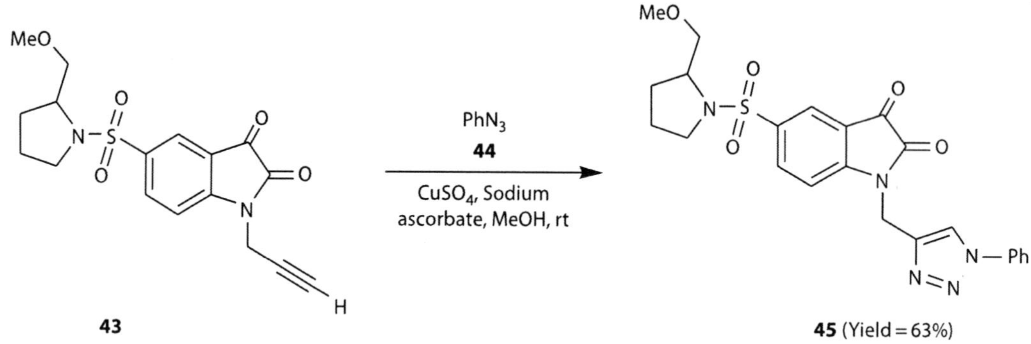

SCHEME 4.21 A [2 + 3]-cycloaddition of 5-substituted 1-propargylisatin with phenyl azide.

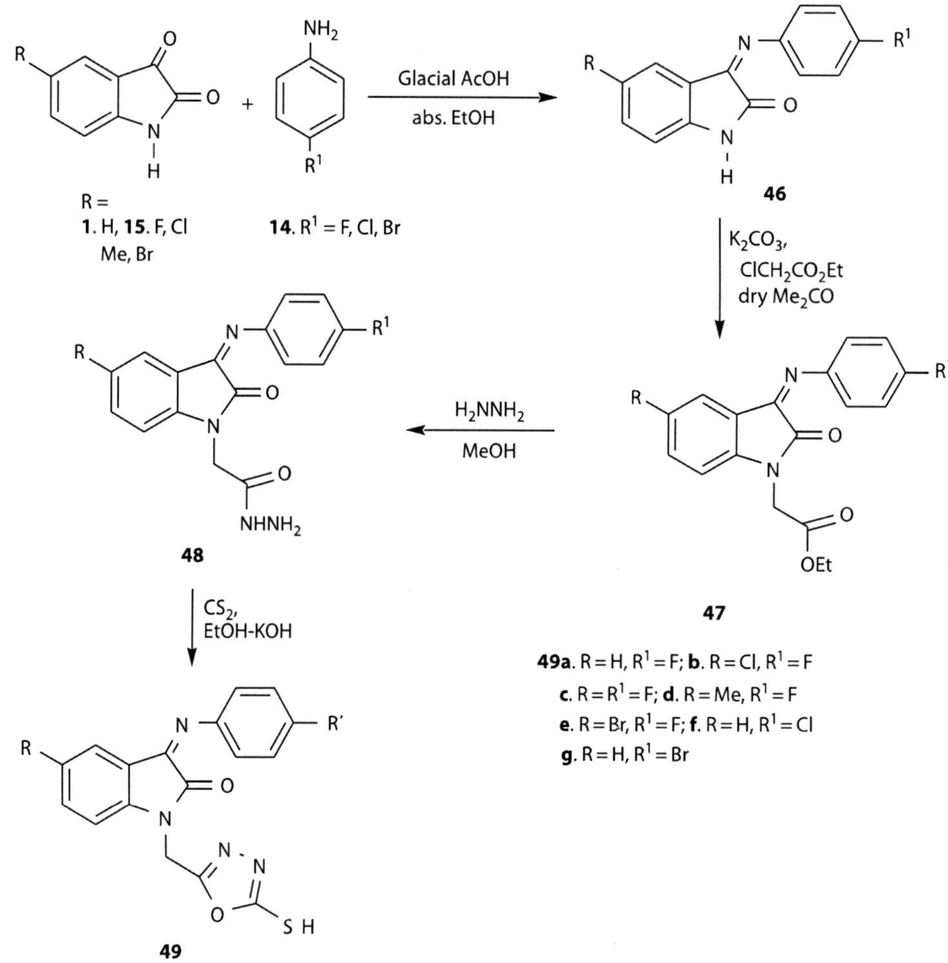

SCHEME 4.22 A multi-step synthesis of 3-iminoisatins with a (2-mercapto-1,3,4-oxadiazol-5-yl)methyl group.

on treatment with CS_2 in ethanolic KOH leads to the formation *N*-(2-mercapto-1,3,4-oxidiazol-5-yl) methylisatin imines **49** (Scheme 4.22) (Bari et al. 2008).

Thiolactone-isatin hybrids **52** and a tetracyclic side-product **51** were obtained by reacting *N*-(bromoalkyl)isatins **23** with potassium salt of thiolactone **50** in *N*,*N*-dimethylformamide at 60°C in low-to-moderate yields (Scheme 4.23) (Hans et al. 2010). The product **52** was a major product in

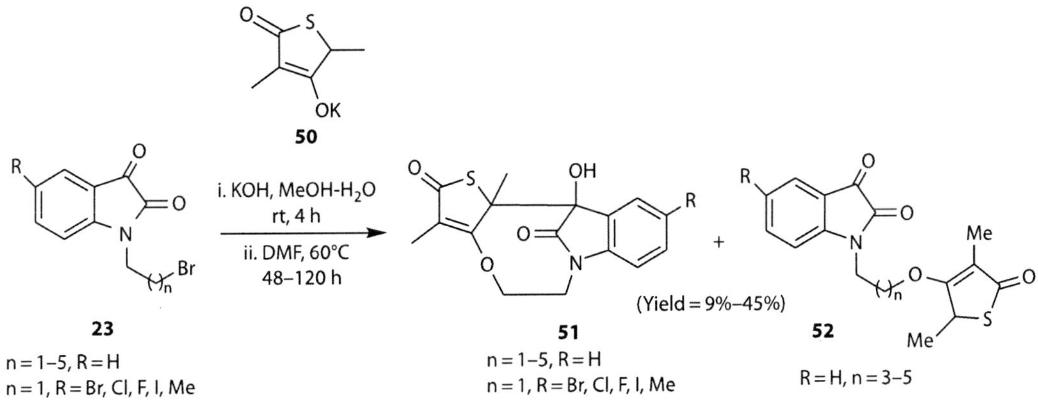

SCHEME 4.23 Synthesis of thiolactone-isatin hybrids.

all the cases except in case of *N*-(1-bromopropyl)isatin where the product ratio was 22:78 (**52**:**51**) in the total yield of 45%. These compounds have been investigated for their antimalarial and antitubercular activity (Hans et al. 2011).

4.3.3 Reactivity of Ketone Group

4.3.3.1 Allylation of Isatin and Its Derivatives

Palladium-catalyzed enantioselective asymmetric allylation of *N*-methylisatin **22** with allyl alcohol in the presence of triethylborane provides an efficient route to generate useful 3-allyl-3-hydroxy-2-oxindole product **53** (Scheme 4.24) (Qiao et al. 2009). The reaction is applicable to differently substituted isatins and allyl alcohols and the products were obtained in excellent yields (74%–99%) but moderate enantioselectivity (56%–71%).

Alkylation of *N*-methylisatin hydrazones **54** is reported in aqueous media promoted by indium. Treatment of a THF/NH_4Cl (aqueous saturated) solution of substrate with allyl bromide in the presence of indium afforded oxindoles **55** and **56**, respectively (Scheme 4.25) (Alcaide et al. 2010).

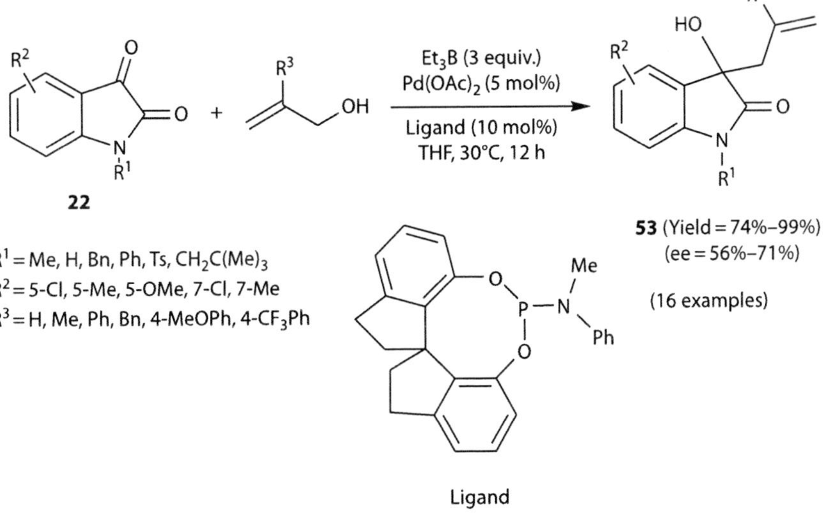

SCHEME 4.24 Palladium-catalyzed asymmetric allylation of isatins.

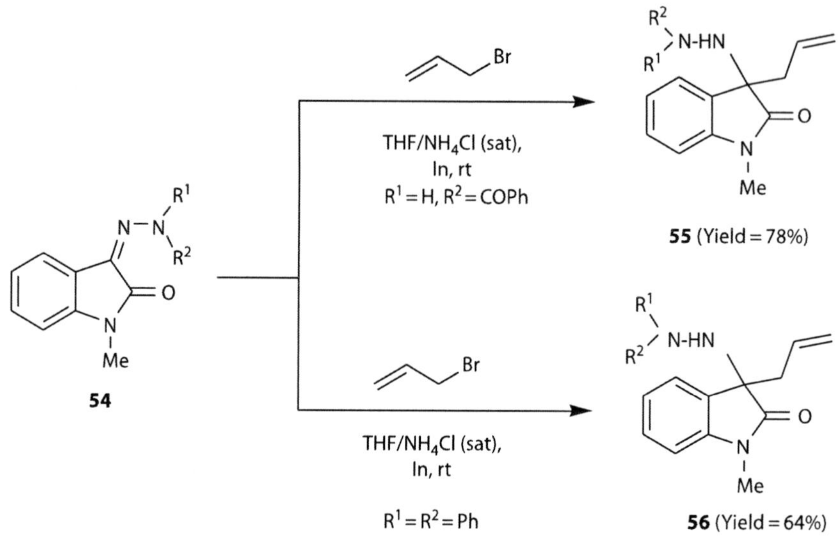

SCHEME 4.25 Allylation of isatin imines.

4.3.3.2 Baylis–Hillman Reaction

The Baylis–Hillman reaction, a carbon–carbon bond forming reaction, which basically involves a reaction between an aldehyde or ketone and an activated alkene in the presence of a tertiary base, affords highly functionalized products (Shanmugam et al. 2006, Shanmugam and Vaithyanathan 2008, Singh and Batra 2008). Highly functionalized Baylis–Hillman adducts have been used as substrates in stereoselective synthesis of a variety of highly functionalized compounds and in natural product synthesis (Basavaiah et al. 2003, Shanmugam et al. 2006). Isatin and its derivatives are used as electrophilic components for the Baylis–Hillman reaction due to the presence of a reactive keto-carbonyl group. Isatin and its alkyl, aryl, and acyl derivatives react with alkenes **57** having electron-withdrawing groups in the presence of DABCO to yield compound **58** (Scheme 4.26) (Tables 4.1 and 4.2) (Chung et al. 2002, Garden and Skakle 2002).

Isatin and its *N*-methyl and *N*-benzyl derivatives also undergo Baylis–Hillman coupling with chromene derivatives **59** in methanolic trimethylamine and lead to the formation of the corresponding adducts **60** (Scheme 4.27) (Basavaiah and Rao 2003).

N-Methylisatin **22** serves as an electrophile in the Morita–Baylis–Hillman reaction and reacts with acrolein **61** and with methyl vinyl ketone **62** in the presence of 10 mol% of phosphinothiourea **63** as a catalyst to afford compounds **64** and **65**, respectively, in moderate to good yields but poor enantioselectivity (Scheme 4.28) (Wang and Wu 2011).

The reaction of *N*-benzylisatin **23a** with methyl vinyl ketone **62** in the presence of catalyst TQO as an efficient catalyst in dichloromethane has led to the formation of a mixture of

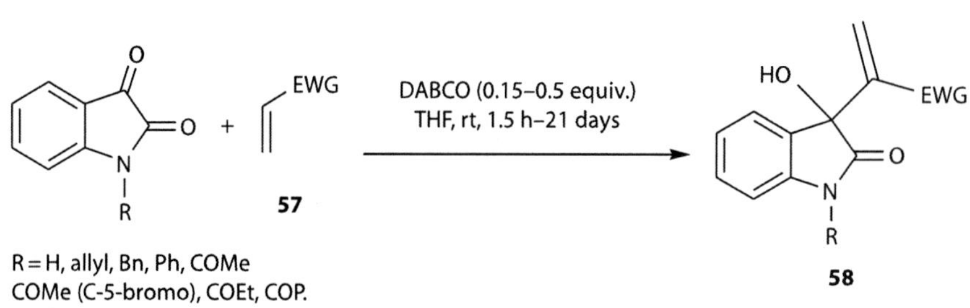

SCHEME 4.26 The Baylis-Hillman reaction of isatins.

TABLE 4.1
Synthesis of Baylis–Hillman Adducts of Isatin (1) and Their Alkyl or Aryl Derivatives

Reagents/Conditions	Product	Compd No.	Yield (%)
Methyl acrylate DABCO (0.15 equiv.) THF, rt, 5 days	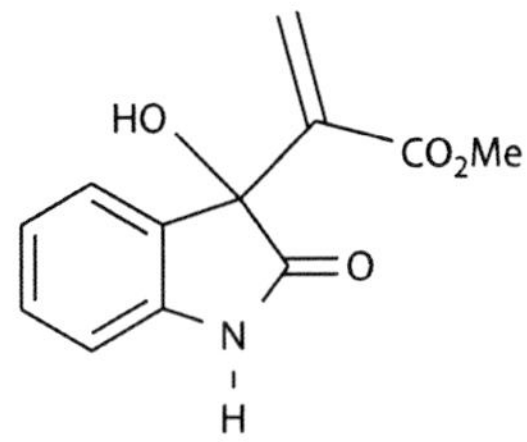	**58a**	63
Ethyl acrylate DABCO (0.15 equiv.) THF, rt, 8 days	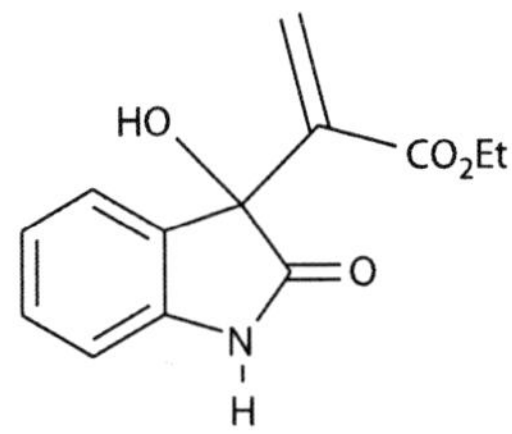	**58b**	71
Acrylonitrile DABCO (0.15 equiv.) THF, rt, 4 days	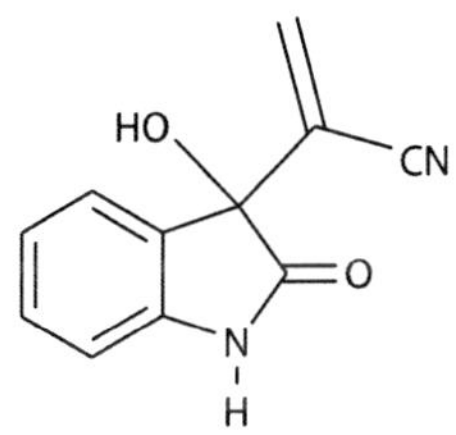	**58c**	69
Methyl acrylate DABCO (0.15 equiv.), rt, 21 days	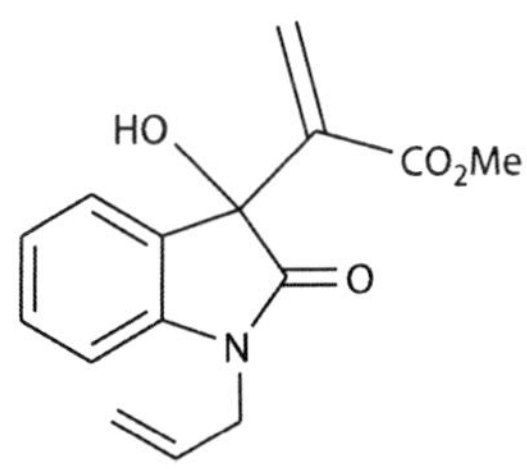	**58d**	81
Acrylonitrile DABCO (0.15 equiv.) THF, rt, 8 days	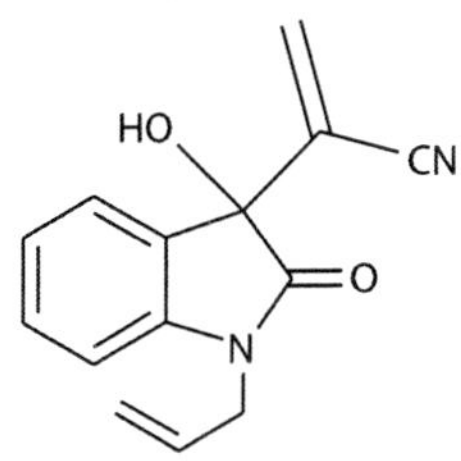	**58e**	81
Methyl acrylate DABCO (0.15 equiv.) THF, rt, 5 days	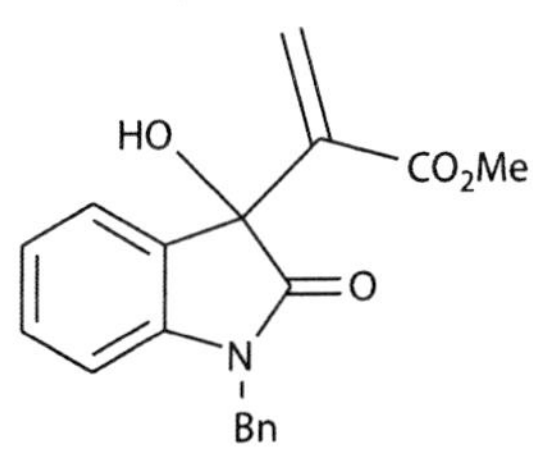	**58f**	50

(continued)

TABLE 4.1 (continued)
Synthesis of Baylis–Hillman Adducts of Isatin (1) and Their Alkyl or Aryl Derivatives

Reagents/Conditions	Product	Compd No.	Yield (%)
Acrylonitrile DABCO (0.15 equiv) THF, rt, 20 h	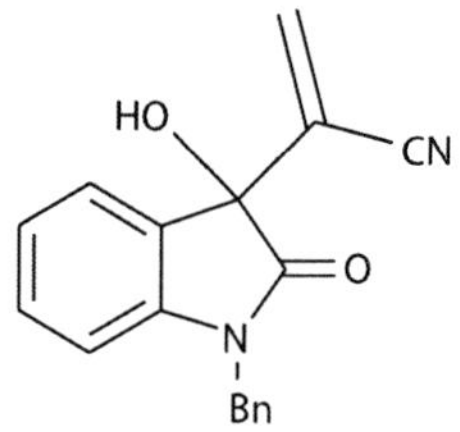	**58g**	94
DABCO (0.15 equiv.) THF, rt, 9 days	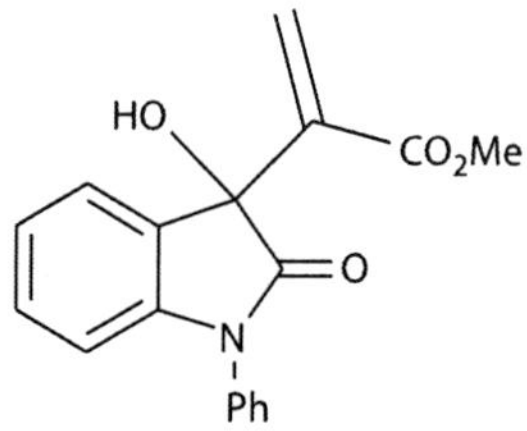	**58h**	87
Acrylonitrile DABCO (0.15 equiv.) THF, rt, 25 h	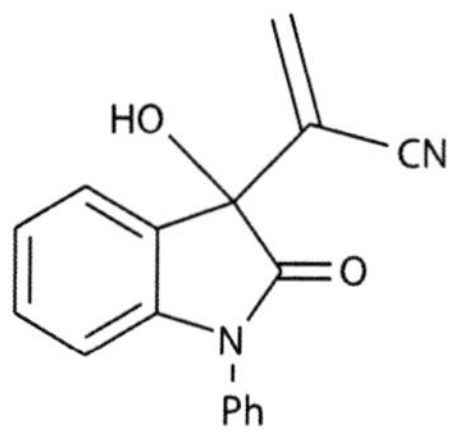	**58i**	79

compounds **67a** and **67b**, resulting from 1:1 and 1:2 molar reaction, respectively, of isatin **23a** with ketone **62** (Scheme 4.29) (Guan et al. 2010).

4.3.3.3 Aldol Reaction

The aldol reaction is a well-known carbon–carbon bond forming reaction. The 3-substituted 3-hydroxy-2-oxindoles resulting from the aldol reactions of isatins are important synthetic intermediates for a variety of biologically active alkaloids (Aikawa et al. 2011, Liu et al. 2011). An electro-aldol reaction of isatin and its *N*-methyl, *N*-benzyl, *N*-acyl, and *N*-chloro derivatives with cyclic 1,3-diketones **68** in alcohol in an undivided cell results in the formation of substituted 2-(3-hydroxy-2-oxo-2,3-dihydro-1*H*-indole-3-yl)cyclohxane-1,3-diones **69** in 70%–85% yields (Scheme 4.30) (Elinson et al. 2010).

There are many examples of enantioselective organocatalytic aldol reaction of isatins with inactivated carbonyl compounds (Chen et al. 2010a, Aikawa et al. 2011, Allu et al. 2011, Peng et al. 2011). A representative example is the reaction of *N*-methylisatin **22** with diphenylphosphite **70** catalyzed by commercially available cinchona alkaloid **71** resulting in an enantioselective phospho-aldol addition forming adduct **72** (Scheme 4.31) (Peng et al. 2011).

Isatin and its derivatives also undergo a nitro-aldol reaction known as the Henry reaction with nitromethane in the presence of diethylamine as a catalyst to form 3-hydroxy-3-nitromethyloxindole

TABLE 4.2
Synthesis of Baylis–Hillman Adducts of *N*-Acylisatins

Reagents/Conditions	Product	Compd No.	Yield (%)
Methyl acrylate DABCO (0.2 equiv.) DMF, rt, 120 min	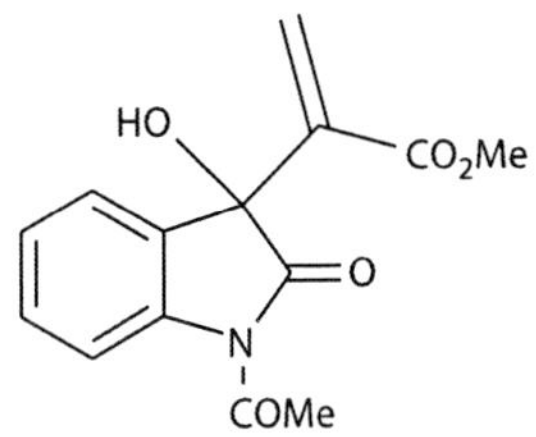	**58j**	58
Acrylonitrile DABCO (0.2 equiv.) DMF, rt, 120 min	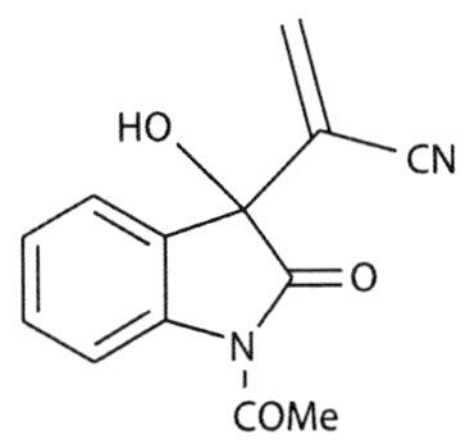	**58k**	70
Methyl acrylate DABCO (0.2 equiv.) DMF, rt, 100 min	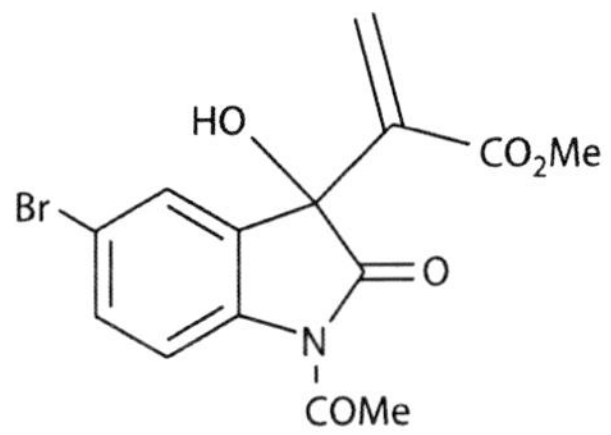	**58l**	59
Acrylonitrile DABCO (0.2 equiv.) DMF, rt, 90 min	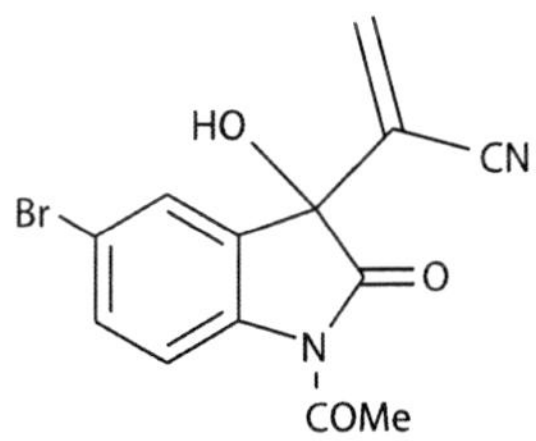	**58m**	50
Methyl acrylate DABCO (0.2 equiv.) DMF, rt, 180 min	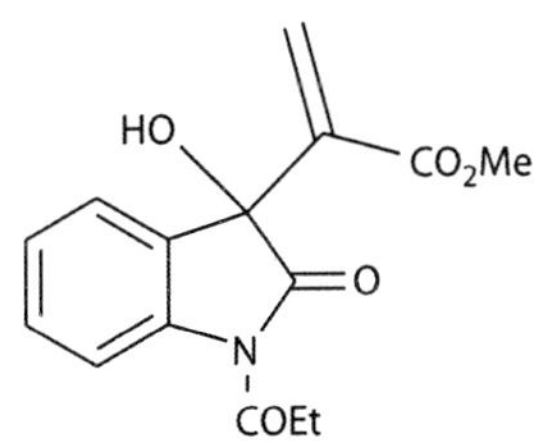	**58n**	55
Acrylonitrile DABCO (0.2 equiv.) DMF, rt, 100 min	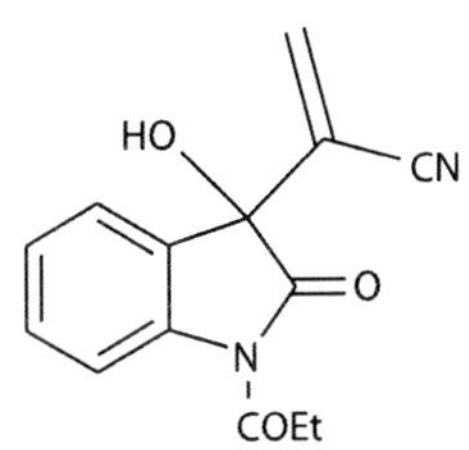	**58o**	84

(continued)

TABLE 4.2 (continued)
Synthesis of Baylis–Hillman Adducts of *N*-Acylisatins

Reagents/Conditions	Product	Compd No.	Yields (%)
Acrylonitrile DABCO (0.2 equiv.) DMF, rt, 90 min	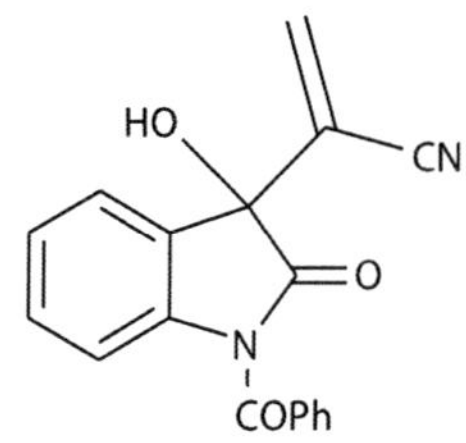	**58p**	52

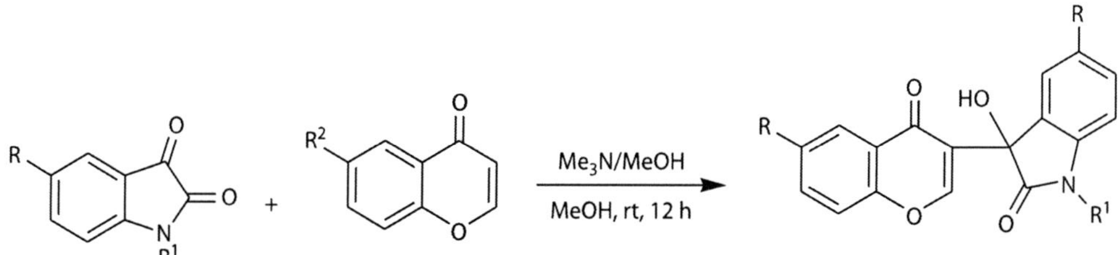

59 R = H, Me

60a–g (Yield = 78%–85%)

60a. $R = R^1 = R = H$, **60b**. $R = R^1 = Me$, $R = NO_2$
60c. $R^2 = H$, $R^1 = Bn$, $R = NO_2$, **60d**. $R^2 = H$, $R^1 = Me$, $R = NO_2$
60e. $R^2 = Me$, $R^1 = Bn$, $R = NO_2$, **60f**. $R^2 = H$, $R^1 = Me$, $R = H$
60g. $R^2 = H$, $R^1 = Bn$, $R = H$

SCHEME 4.27 The Baylis-Hillman reaction of isatins with chromene derivatives.

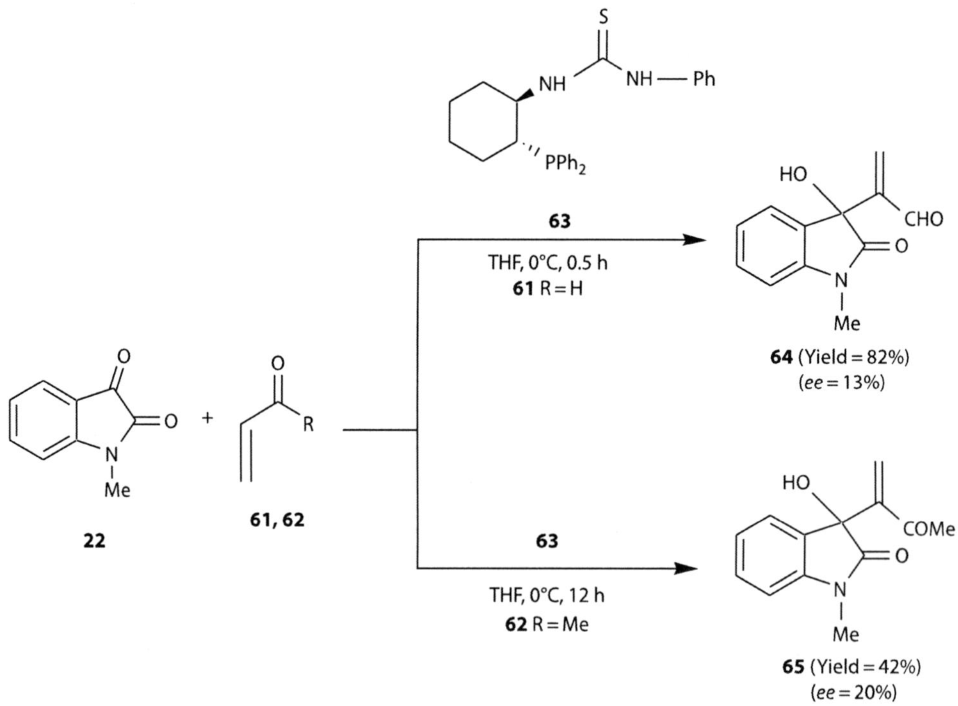

SCHEME 4.28 The Baylis-Hillman reaction of isatins with acrolein and methyl vinyl ketone in the presence of phosphinothiourea.

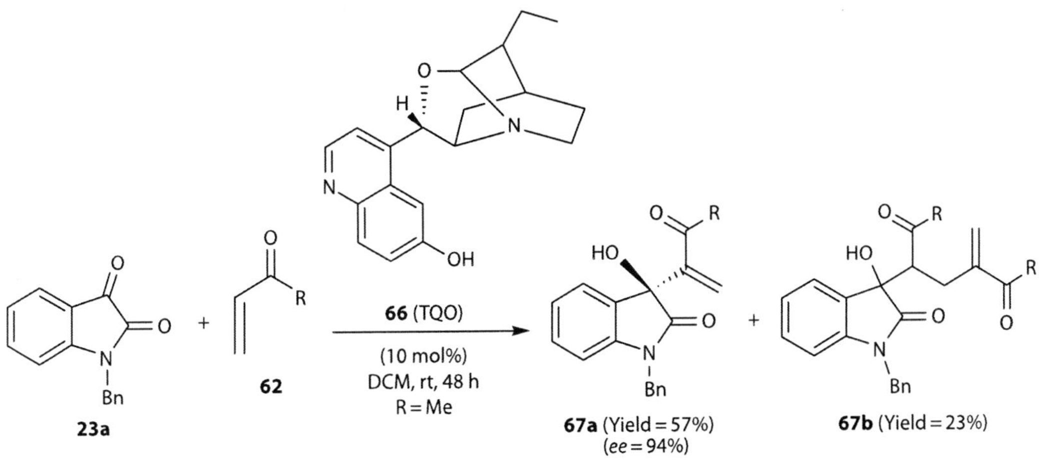

SCHEME 4.29 The Morita-Baylis-Hillman reaction of *N*-benzylisatin with methyl vinyl ketone in the presence of catalyst TQO.

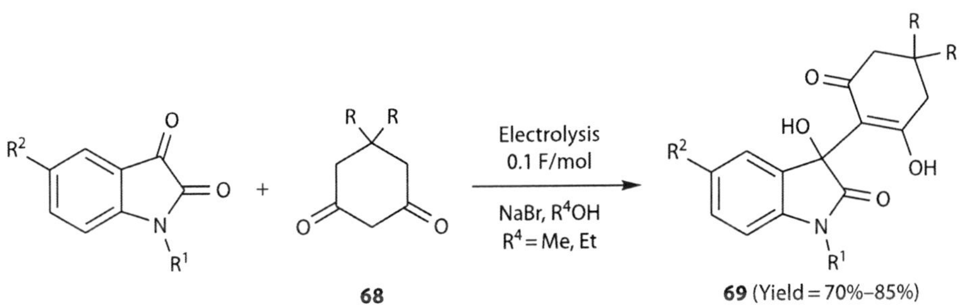

69a. $R^1 = R^2 = R = H$; **69b**. $R^1 = Me$, $R^2 = H$, $R = Me$;
69c. $R^1 = Bn$, $R^2 = H$, $R = Me$; **69d**. $R^1 = Ac$, $R^2 = H$, $R = Me$;
69e. $R^1 = H$, $R^2 = Me$, $R = Me$; **69f**. $R^1 = H$, $R^2 = Cl$, $R = Me$;
69g. $R^1 = R^2 = H$, $R = Me$; **69h**. $R^1 = Me$, $R^2 = H$, $R = H$;
69i. $R^1 = H$, $R^2 = Cl$, $R = H$

SCHEME 4.30 Electrochemical aldol reaction of isatins with cyclohexane-1,3-diones.

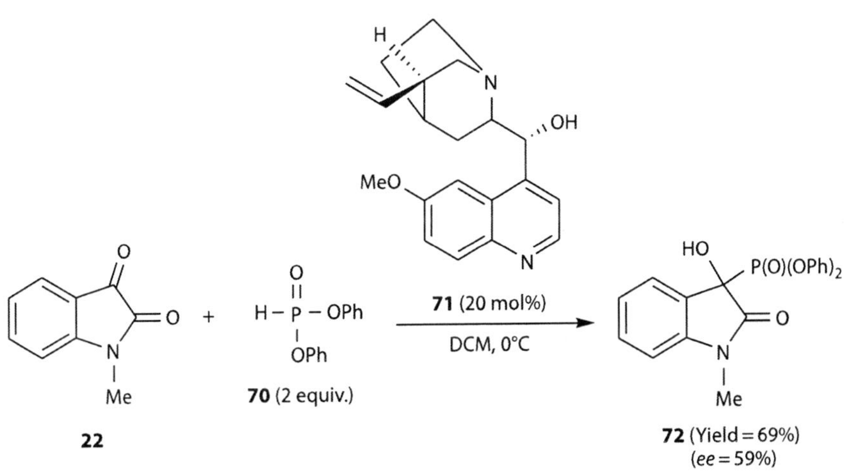

SCHEME 4.31 Enantioselective organocatalytic phospho-aldol reaction of *N*-methylisatin.

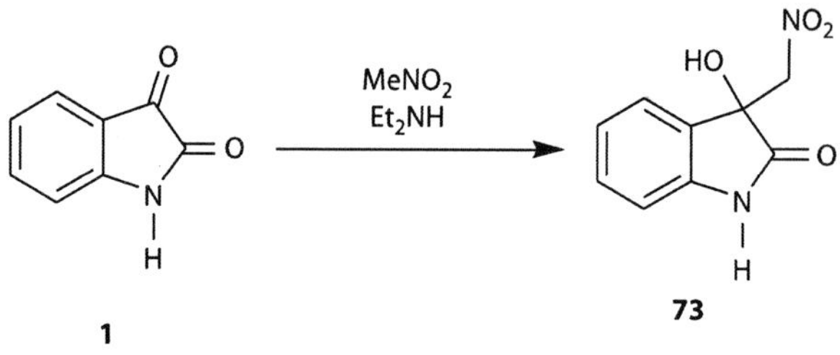

SCHEME 4.32 The Henry reaction of isatin with nitromethane (nitro-aldol reaction).

73 (Scheme 4.32) (Chen et al. 2010b). The nitromethyl adduct is a valuable building block for the total synthesis of natural products and their analogues because the nitro functionality can easily be transformed into a variety of functional groups, such as amine, ketone, nitrile oxide, carboxylic acid, hydrogen, and so on (Liu et al. 2011). Furthermore, asymmetric Henry reaction of isatins offers direct entry to the chiral 3-substituted 3-hydroxyoxindole.

Direct catalytic asymmetric aldol reaction of ketones with isatin **1** using L-proline derived bis-amide organocatalysts **74** represents a general approach to 3-alkyl-3-hydroxyoxindoles **75** and **76** with a quaternary stereocenter. The products are obtained in excellent yields (up to 99%) (Scheme 4.33) (Chen et al. 2007). Isatin **1** also reacts with acetaldehyde in the presence of organocatalyst 4-hydroxydiaryl prolinol **77** to give aldol adduct **78** in almost quantitative yield with good enantioselectivity (Scheme 4.34) (Chen et al. 2010).

An efficient vinylogous Mukaiyama aldol reaction of various *N*-alkylisatins with 2-(trimethylsilyloxy)furan **79** is described in the presence of lanthanum(III) triflates (5 mol%) (Meshram et al. 2011). In this way, the reaction of isatin **1** proceeds rapidly and affords the corresponding diastereomeric 3-hydroxy-(5-oxo-2,5-dihydrofuran-2-yl)indolin-2-ones **80** and **81** in high yields with good diastereoselectivities (Scheme 4.35).

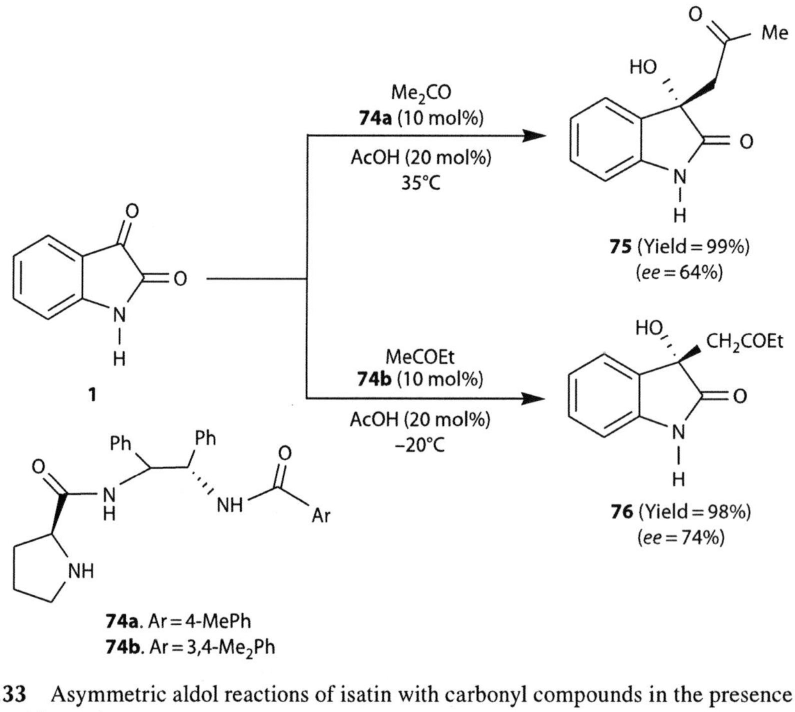

SCHEME 4.33 Asymmetric aldol reactions of isatin with carbonyl compounds in the presence of L-proline derived bis-amide catalysts.

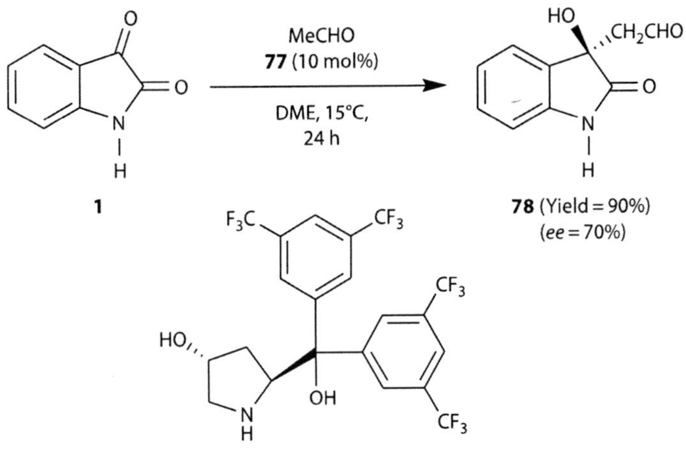

SCHEME 4.34 Asymmetric aldol reactions of isatin with acetaldehyde catalyzed by a 4-hydroxydiarylprolinol catalyst.

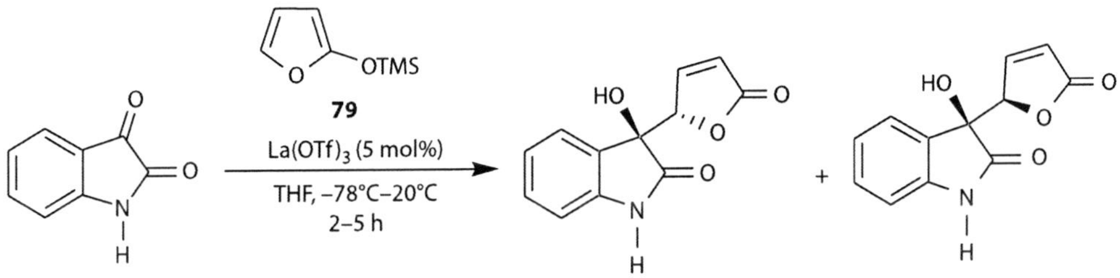

SCHEME 4.35 Mukaiyama's aldol reaction of isatin with 2-(trimethylsilyloxy)furan in the presence of lanthanum(III) triflate.

4.3.3.4 Isatin-3-Oximes

Isatins react with hydroxylamines to form isatin oximes. Isatin oximes are compounds of biological interest (Rad et al. 2010) and their synthesis, chemistry, and biological activity has also been reviewed (Abele et al. 2003). Liu and coworkers have reported the ^{15}N NMR studies on isatin oximes (Liu et al. 2010). In a classical method to synthesize isatin-3-oximes, isatins are reacted with hydroxylamine hydrochloride in 10% aqueous NaOH, NaOH-EtOH, NaOAc, *n*-PrOH-H_2O, NaOAc-dioxane, Na_2CO_3-H_2O, Na_2CO_3-EtOH, and H_2O. Pinto and coworkers have developed an efficient methodology for the preparation of isatin 3-oximes under Lewis or Bronsted acid catalysis in different imidazolium-based ionic liquid solvent (Pinto et al. 2008). The reaction of compounds **82** with hydroxylamine afforded the corresponding 3-oxime **83** (Scheme 4.36). This approach was proved to be useful even with the phenyl ring bearing an electron-withdrawing group, except when the substituent was chlorine.

4.3.3.5 Isatin-3-Semicarbazones/Thiosemicarbazones

The reaction of isatins with semicarbazides and thiosemicarbazides are known to form the corresponding 3-semicarbazones and 3-thiosemicarbazones, respectively. Thiosemicarbazones constitute a class of compounds which have been found to display numerous biological activities (Pervez et al. 2007) such as antitumor, antibacterial, antiviral, and antimalarial activities (Konstantinovic

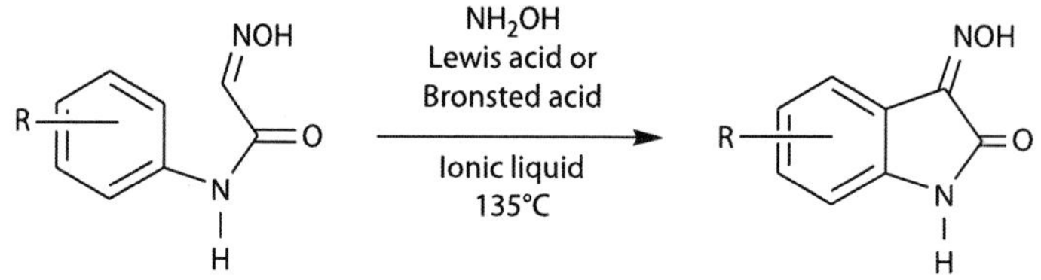

82

83

R = 7-CF_3, 7-Cl, 7-I, 7-OMe, 6-OMe, 5-OMe (Yield = 12%–63%)

SCHEME 4.36 Synthesis of isatin-3-oximes.

et al. 2008). Their activity is thought to be due to their ability to make chelation with traces of metal ions present in biological systems (Konstantinovic et al. 2007). The 3-thiosemicarbazones of isatin have been of interest since 1-methylisatin-3-thiosemicarbazone was found to be active in the treatment of small pox (Rai et al. 2005, Vasta et al. 2005).

Pandeya et al. have reported the synthesis and anticonvulsant activity of 3-semicarbazones of isatins (Pandeya and Raja 2002). The reaction of isatins with thiosemicarbazide **84** in ethanol provided the corresponding isatin-3-thiosemicarbazones **85** (Scheme 4.37) (Chiyanzu et al. 2003). Many isatin-3-(*N*-aryl)thiosemicarbazones **87** have been also synthesized by the reaction of isatin with *N*-arylthiosemicarbazides **86** (Scheme 4.38) (Kang et al. 2011). These thiosemicarbazones have been observed as potent herpes simplex virus inhibitors. The synthesis of 3-*N,N*-diethylthiosemicarbazone **89** has been carried out by reaction of isatin **1** with *N,N*-diethylthiosemicarbazide **88** (Scheme 4.39). The Mannich reaction of the compound **89** affords the 1-substituted thiosemicarbazone **90** (Scheme 4.39) (Bal et al. 2005).

Another approach for the synthesis of isatin-3-(*N*-phenyl)thiosemicarbazone **87** involves the condensation of isatin C-3 carbonyl with methylhydrazinecarbodithioate **91** forming compound **92** which reacts with aniline to afford the final product **87** (Scheme 4.40) (Pervez et al. 2007).

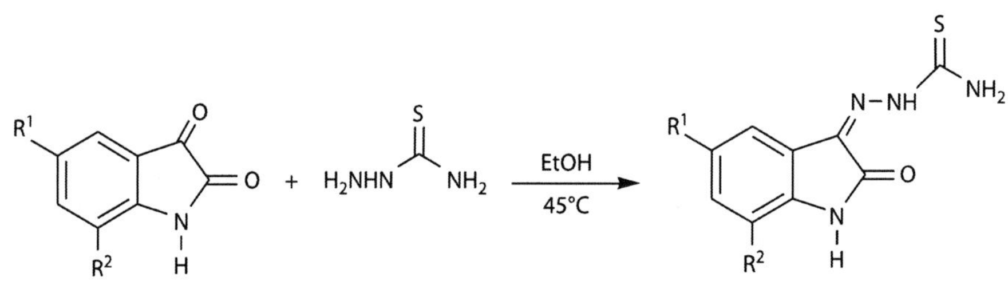

1, 15, 16 **84** **85** (Yield = 40%–98%)

85. $R^1 = R^2 = H$; $R^1 = Me$, $R^2 = H$; $R^1 = F$, $R^2 = H$; $R^1 = Cl$, $R^2 = H$;
$R^1 = Br$, $R^2 = H$; $R^1 = I$, $R^2 = H$; $R^1 = NO_2$, $R^2 = H$; $R^1 = Me$, $R^2 = Me$;
$R^1 = Cl$, $R^2 = Me$

SCHEME 4.37 Synthesis of isatin-3-thiosemicarbazones from isatins.

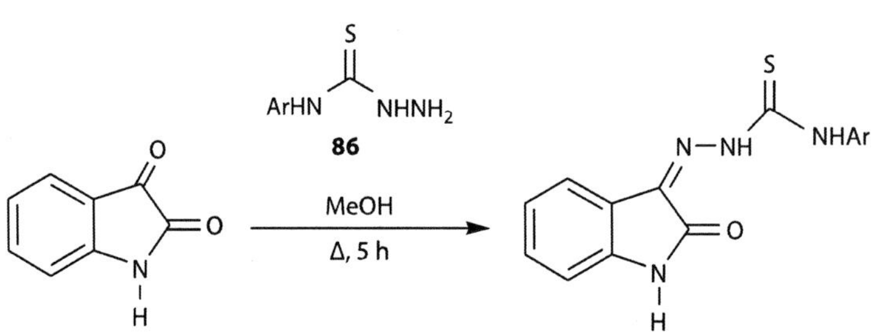

1 **87**

SCHEME 4.38 Synthesis of 3-*N*-arylthiosemicarbazones from isatin.

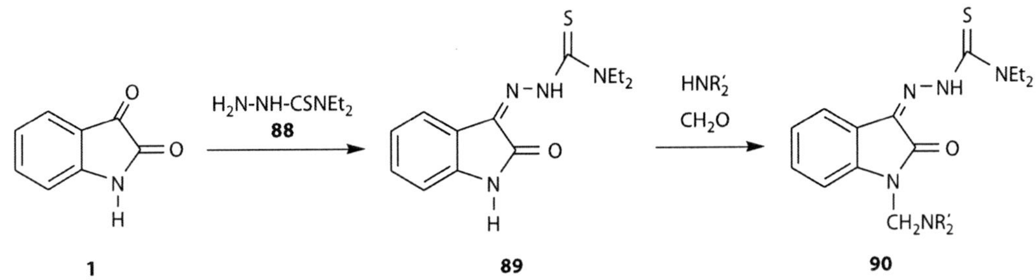

SCHEME 4.39 Formation and Mannich reaction of isatin-3-*N*,*N*-diethylthiosemicarbazones.

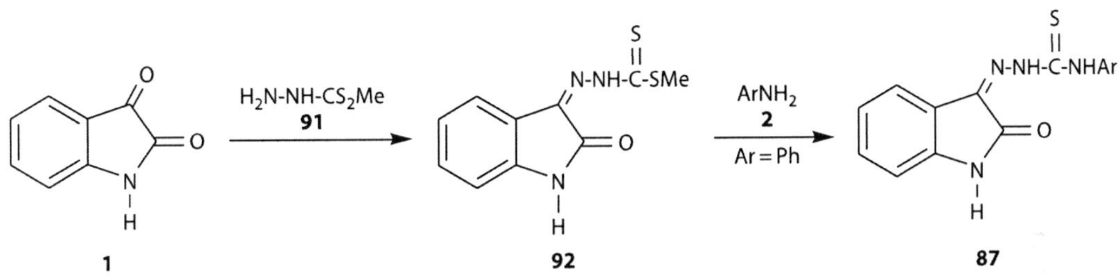

SCHEME 4.40 Synthesis of 3-*N*-phenylthiosemicarbazones from isatin using methylhydrazine carbodothioate.

4.3.3.6 Isatin-3-Hydrazones

Isatin hydrazones are reported to have anticonvulsant activity (Sridhar et al. 2002). Isatin **1** reacts with hydrazine hydrate in methanol to give the corresponding 3-hydrazone **93** (Scheme 4.41) (Sridhar and Ramesh 2001, Srinivas et al. 2010). The reaction of isatin **1** with hydrazines **94a–d** in acidified ethanol affords isatin-3-(*N*-acyl/aroyl)hydrazones **95a–d** (Scheme 4.42) (Adibi et al. 2010). Somogyi has reported the cyclization of isatin-3-(*N*-acyl)hydrazones forming spiro-fused 1,3,4-oxadiazolines (Somogyi 2001). The reaction of isatin **1** and 3-(3,5-di-*tert*-butyl-4-hydroxyphenyl)propionohydrazide **96** yields isatin-3-(*N*-acyl)hydrazone **97** (Scheme 4.43) (Nugumanova et al. 2009).

4.3.3.7 Isatin-3-Imines

Isatin imines are well known as building blocks in organic synthesis. The formation of isatin-3-imines from isatin and some alkyl amines were reported by Piccirilli and Popp by refluxing

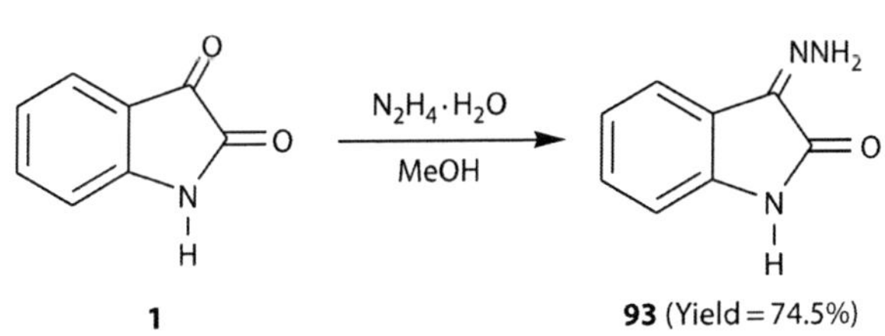

SCHEME 4.41 Preparation of isatin-3-hydrazone from isatin.

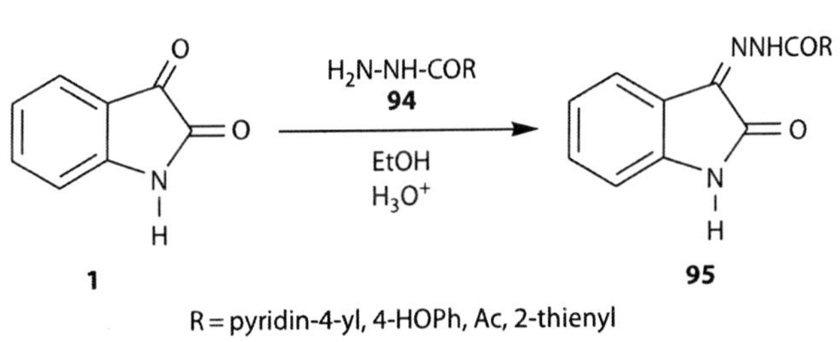

SCHEME 4.42 Preparation of isatin-3-(*N*-acyl)hydrazones from isatin.

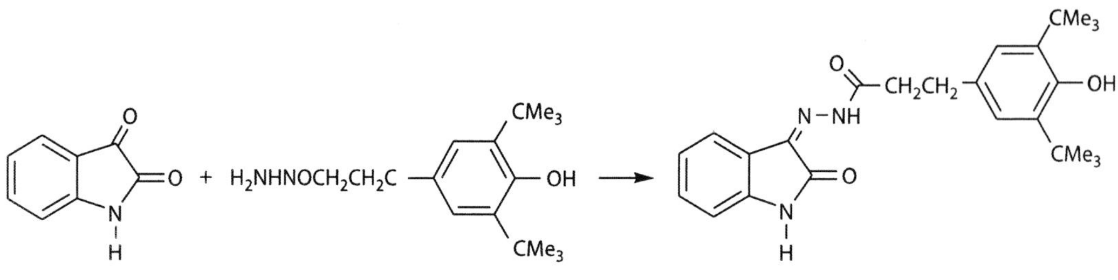

1 96 97 (Yield = 60.8%)

SCHEME 4.43 Preparation of isatin-3-(*N*-acyl)hydrazone.

the substrates in ethanol (Piccirilli and Popp 1973). Many isatin imines have been synthesized since then and evaluated for different kinds of biological activities (Singh et al. 1993a,b). Bari et al. has used glacial acetic acid in ethanol for the synthesis of isatin imines from the reaction of isatin and aromatic amines (see Scheme 4.22) (Bari et al. 2008), and Sharma and coworkers have also utilized this protocol in the synthesis of imines having anticonvulsant activity from isatin and 2-aminobenzothiazoles (Sharma et al. 2009). Recently, synthesis of some antileishmanial isatin imines has been reported in aqueous medium (Khan et al. 2008). Our group has also observed antileishmanial activity in isatin imines (Al-Kahraman et al. 2011). The reaction of isatin **1** or 5-chloroisatin **15** with 5-amino-8-hydroxy quinoline **98** has been reported recently to form imines **99** which undergo the Mannich reaction to afford the *N*-alkylated imines **100** (Scheme 4.44) (Chhajed and Padwal 2010).

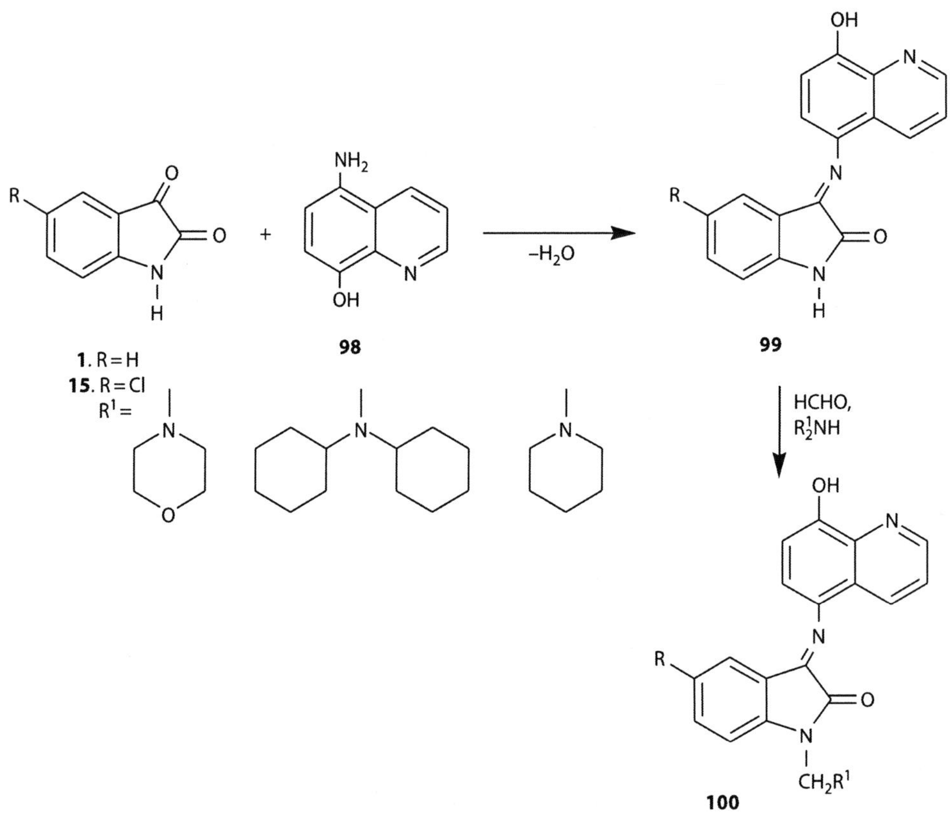

SCHEME 4.44 Preparation of 3-iminoisatins and their Mannich reaction.

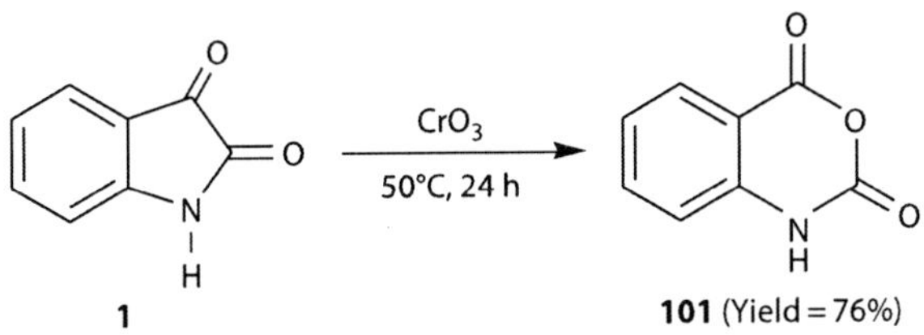

SCHEME 4.45 Oxidation of isatin to isatoic anhydride using chromic anhydride.

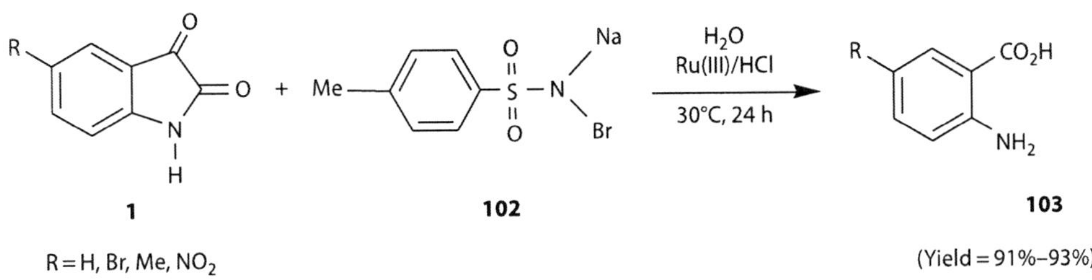

SCHEME 4.46 Oxidation of isatins to anthranilic acids by *N*-bromo-*p*-toluenesulfonamide using Ru(III) catalyst.

4.3.4 Oxidation Reactions of Isatins: Synthesis of Isatoic Anhydride and Anthranilic Acids

Oxidation of isatin **1** using either hydrogen peroxide or chromic anhydride yields isatoic anhydride **101** (Scheme 4.45). In the oxidation of isatin to isatoic anhydride, the oxidizing agent selected should be able to introduce an oxygen atom between the two adjacent carbonyl groups without substantial decomposition of the ring system (Deligeorgiev et al. 2007).

A mixture of aqueous hydrogen peroxide in acetic or formic acid in the presence of catalytic amount of sulfuric acid is also used for the oxidation of isatin to isatoic anhydride. Another economic and environment-friendly procedure for the oxidation of isatins is through the use of the urea–hydrogen peroxide complex (percarbamide, $H_2NCONH_2 \cdot H_2O_2$) (Deligeorgiev et al. 2007).

Isatin **1** is oxidized to anthranilic acids **103** with the *N*-bromo-*p*-tolunesulfonamide **102** or bromamine-T as an oxidant and ruthenium(III) chloride as a catalyst in acidic medium (Scheme 4.46) (Jagadeesh et al. 2008).

4.3.5 Reduction Reactions of Isatins: Synthesis of Indoles and 2-Oxindoles

The reduction reactions of isatin under different conditions lead to the formation of indoles and oxindoles. The reduction of isatin with lithium aluminum hydride in pyridine affords indoles in moderate yields. The use of THF as a solvent under an inert atmosphere, however, gives better yields. Chemoselective alkylation of isatin at C-3 or N-1 accompanied by reduction using metal hydrides leads to the formation of 1- or 3-alkylindoles **104** or **105** (Scheme 4.47) (Silva et al. 2001). The reduction of 5,6-dibromoisatin **106** using a solution of BH_3 in THF furnishes 5,6-dibromoindole **107** in 68% yield (Scheme 4.48) (Mollica et al. 2011). *N*-(3-Chloropropyl)-5-nitroisatin **23** undergoes reduction on treatment with $NaBH_4$ in the presence of $ZrCl_4$ to afford *N*-(3-chloropropyl)-5-nitroindole **108** (Scheme 4.49) (Torisawa et al. 2001).

It is worth mentioning that indoles are well-known alkaloids that occur frequently in plants and other natural resources, and several thousands of alkaloids with indole moiety are known to be having important pharmacological activities (Jaishree et al. 2009). There are numerous reports in literature on new synthetic methodologies to construct indole framework (Labo and Prabhakar 2009).

The partial reduction of the γ-lactam ring in isatin leads to the formation of oxindoles. Catalytic reduction of isatin **1** via 3-hydroxyoxindole **109** offers an easy method for the synthesis of oxindole

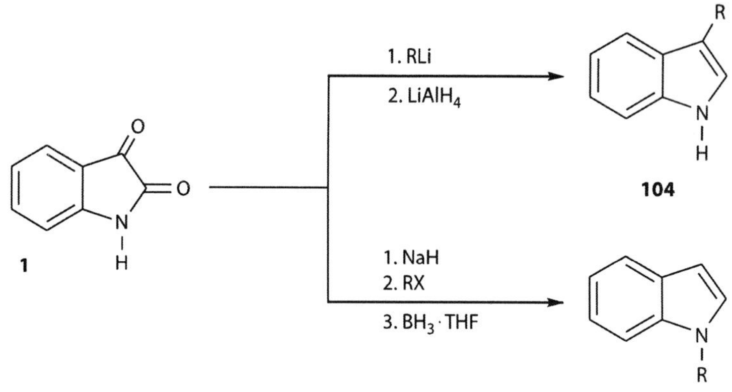

1 104 105

SCHEME 4.47 Reduction of isatin to indole derivatives.

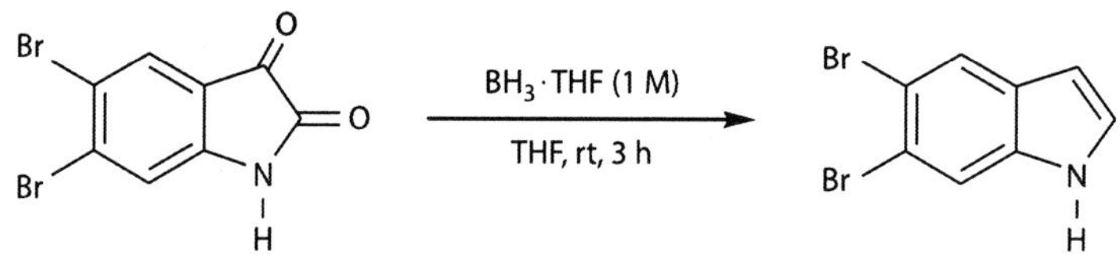

106 107

SCHEME 4.48 Reduction of 5,6-dibromoisatin to 5,6-dibromoindole.

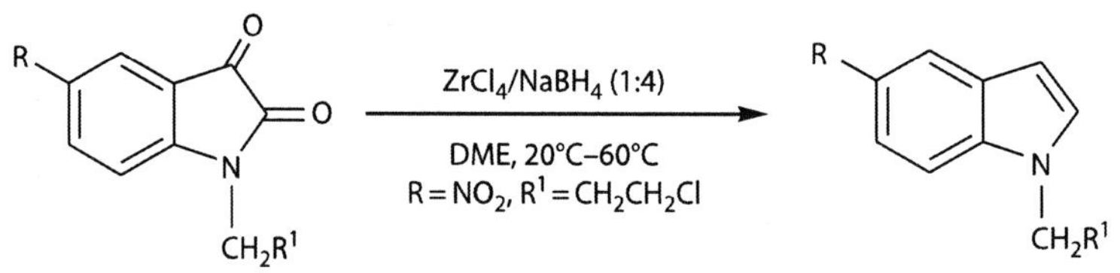

23 108

SCHEME 4.49 Reduction of *N*-alkylisatins to *N*-alkylindole derivatives.

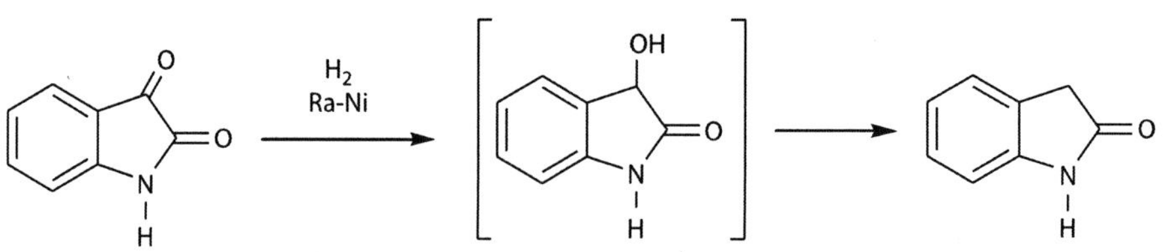

1 109 110

SCHEME 4.50 Catalytic reduction of isatin to 2-oxindole.

110 (Scheme 4.50) (Volk and Simig 2003, Porcs-Makkay et al. 2004). The asymmetric hydrogenations of 5-methylisatin **15** over modified Pt/Al_2O_3 and cinchonidine (CD) **111** catalysts are reported to yield the corresponding 3-hydroxyoxindole **112** at the low concentration of cinchonidine and bisoxindole **113** at either higher concentration of cinchonidine or cinchonidine in the presence of 100 equiv. of TFA (Scheme 4.51) (Sonderegger et al. 2004). Oxindoles are obtained from isatin by other methods as well, which will be discussed separately in the succeeding section on oxindole synthesis.

A tetracyclic 3-hydroxyoxindole **115** has been synthesized from the reaction of isatin **1** with the Baylis–Hillman adduct **114** (Scheme 4.52) (Lee et al. 2010). The *o*-bromophenyl ring in *N*-alkylated

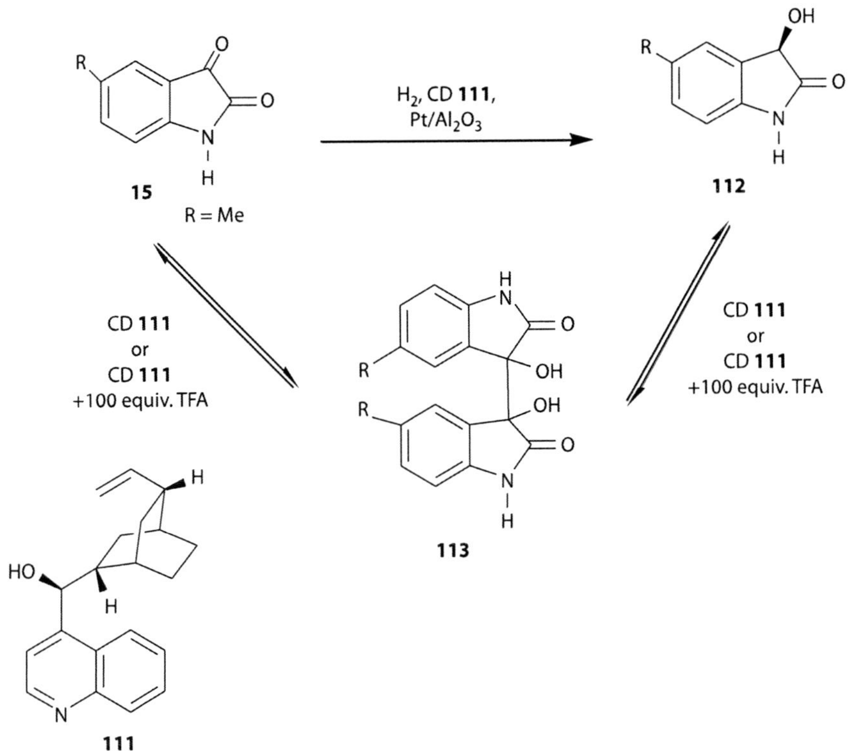

SCHEME 4.51 Catalytic asymmetric reduction of 5-methylisatin to 3-hydroxy-5-methyl-2-oxindole.

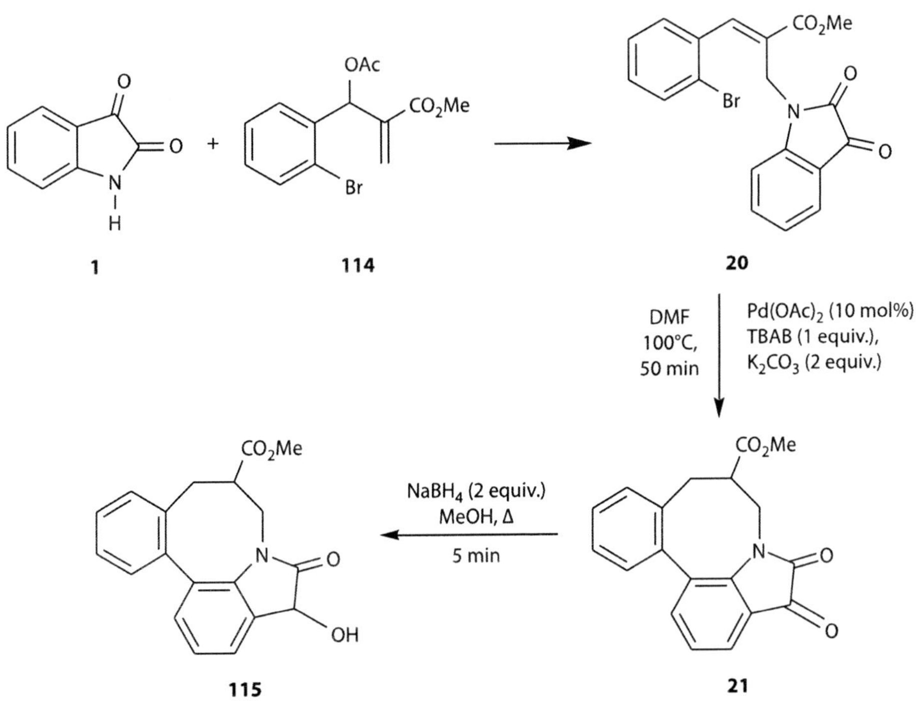

SCHEME 4.52 Synthesis of 3-hydroxy-2-oxindole with a fused ring system.

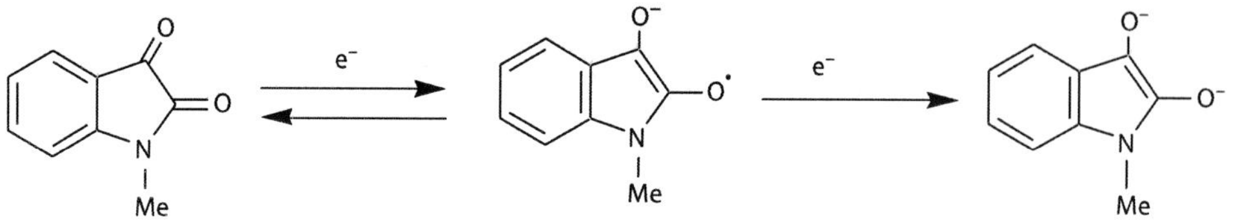

22

116

Radical anion form

117

Dianion form

SCHEME 4.53 Mechanism of reduction of isatin by voltametric studies.

product **20** undergoes a palladium-catalyzed intramolecular nucleophilic aromatic substitution by the phenyl ring of isatin forming a tetracyclic product **21**. The chemoselective reduction of C-3 carbonyl group by sodium borohydride affords the tetracyclic 3-hydroxyoxindole **115**.

Voltametric studies of the reduction of substituted isatins in the aprotic solvent DMF and direct analogy with benzo and naphthoquinones indicate that the reduction of the *O*-quinoid-like carbonyl group of *N*-methylisatin **22** in aprotic medium occurs via successive one-electron transfers forming radical anion **116** and dianion **117** (Scheme 4.53) (Yeagley et al. 2011).

It is thus evident that the reduction of isatins serves as a good methodology for the synthesis of indoles and oxindoles. There are, however, other methods as well for the synthesis of oxindoles using isatins as substrates. It is, therefore, pertinent to have a separate section on the synthesis of oxindoles from isatins. The succeeding sections, thus, give a glance of other methodologies for the synthesis of oxindoles from isatins besides straightforward reduction of the latter.

4.4 SYNTHESIS OF 2-OXINDOLES FROM ISATINS

Oxindoles are well known among different isatin derivatives and are of potential biological interest as antibacterials, kinase inhibitors (Messaoudi et al. 2004), progesterone receptor antagonists (Fensome et al. 2002), CDK2 inhibitors (Dermatakis et al. 2003), PDE4 inhibitors (Hulme et al. 1998), anti-HIV agents (Kumari et al. 2011), and antitumor agents (Girgis 2009). Oxindoles constitute a common structural motif in various natural products and biologically active compounds such as alkaloids (Shintani et al. 2006, Chauhan and Chimni 2010, Trost and Zhang 2011). Substituted 3-hydroxy-2-oxindoles are also important core structures found in many natural products and pharmaceutical compounds (Hanhan et al. 2010). Although the reduction of isatins constitutes the principal methodology for the synthesis of oxindoles, there are several other methods for conversion of isatins into oxindoles, for example, *N*-Benzylation of isatin **1** followed by treatment with hydrazine at 140°C affords *N*-benzyloxindole **118** (Scheme 4.54) (Trost and Zhang 2011).

Reaction of *N*-acetyl isatins **28** with diethylaminosulfurtrifluoride (DAST) in dichloromethane at room temperature leads to the formation of *N*-acetyl-3,3-difluoro-2-oxinoles **119** in 65%–94% yields (Scheme 4.55) (Boechat et al. 2008).

Ultrasonic irradiation of isatin **1** with two molar equivalents of 2*H*-indene-1,3-dione **120** in the presence of *p*-toluenesulfonic acid in ethanol at 40°C afforded 2,2′-(2-oxindoline-3,3-diyl)bis(2*H*-indene-1,3-dione **121** (Scheme 4.56) (Ghahremanzadeh et al. 2011).

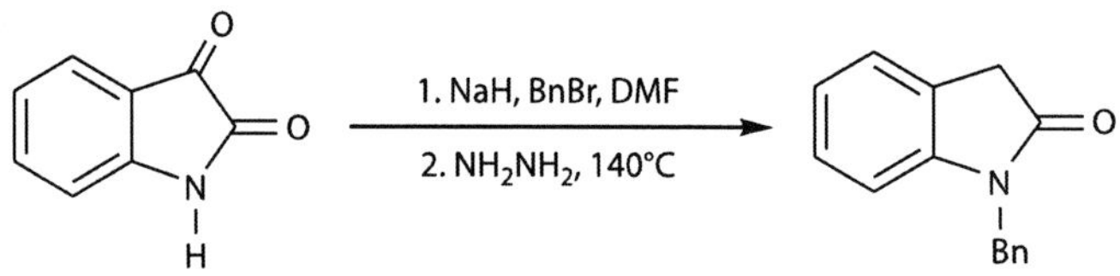

1

118 (Yield=76%)

SCHEME 4.54 Transformation of isatin to *N*-benzyl-2-oxindole.

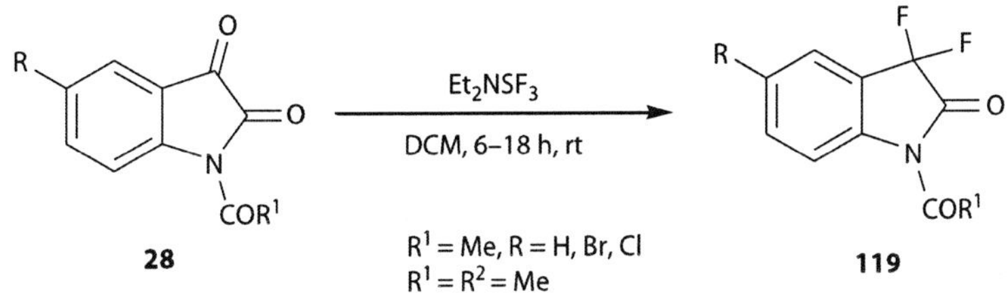

28 119

SCHEME 4.55 Transformation of *N*-acylisatins to *N*-acyl-3,3-difluoro-2-oxindoles.

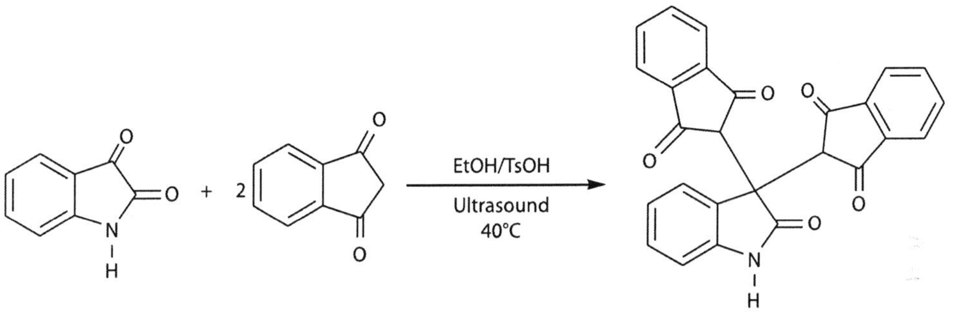

1 120 121

SCHEME 4.56 Reaction of isatin with 2*H*-indene-1,3-dione by ultrasonic irradiation.

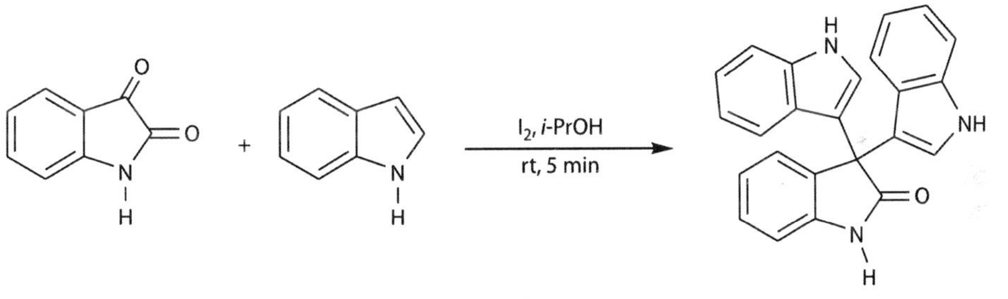

1 122 123 (Yield = 98%)

SCHEME 4.57 Reaction of isatin with indole forming a 3,3-disubstituted 2-oxindole.

The reaction of isatin **1** with indole **122** in the presence of molecular iodine in isopropanol for 5 min yielded the 3,3-bis(indol-3-yl)2-oxindole **123** in 98% yield (Scheme 4.57) (Paira et al. 2009). This reaction has also been carried out in the presence of $RuCl_3.H_2O$ as a catalyst (Messaoudi et al. 2004).

The rhodium-catalyzed addition of arylboronic acids to isatins affords 3-aryl-3-hydroxyindoles. For example, the reaction of isatins **1** and 2 equiv. of phenyl boronic acid **124** in the presence of a catalyst, generated *in situ* from 3 mol% of $[(C_2H_4)_2Rh(acac)]$ and 7 mol% of $P(OPh)_3$, leads to the formation of 3-hydroxy-3-phenyl-2-oxindole **125** in quantitative yield (Scheme 4.58) (Toullec et al. 2006).

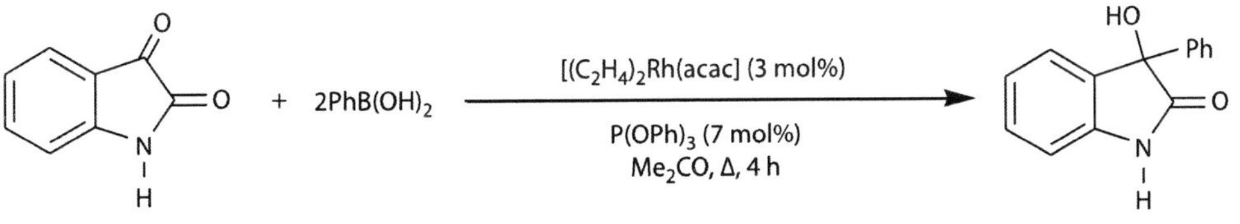

1 124 125

SCHEME 4.58 Transformation of isatin to 3-hydroxy-3-phenyl-2-oxindole using phenyl boronic acid.

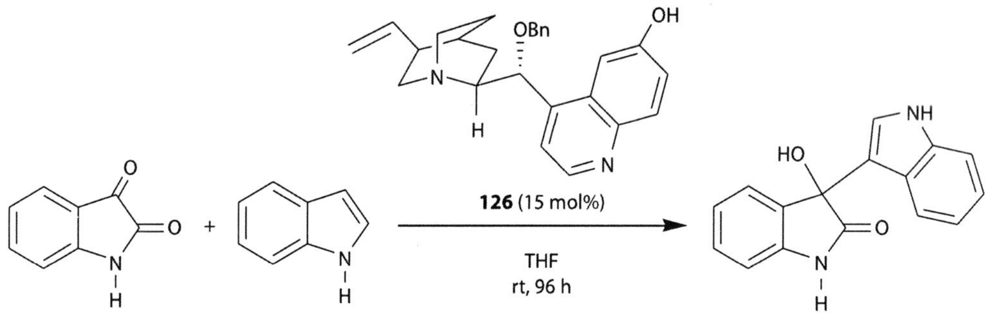

1 (10 mmol) 122 (15 mmol) 127

SCHEME 4.59 Reaction of isatin with indole in the presence of a modified cinchona alkaloid forming 3-hydroxy-3-(indol-3-yl)-2-oxindole.

The nucleophilic addition of indole **122** to the C-3 of the isatin **1** using a bifunctional modified cinchona alkaloid catalyst **126** transforms isatin **1** to 3-hydroxy-3-(indol-3-yl)-2-oxindole **127** (Scheme 4.59) (Hanhan et al. 2010).

The examples described in this section constitute synthesis of oxindoles either unsubstituted or substituted at C-3 by another atom(s) or group(s) such as fluorine, hydroxyl group, phenyl group, or a heteroaryl group(s). However, there are a large number of reports in literature on the synthesis of oxindoles in which C-3 of the oxindoles is spiro-fused to different types of rings. The next section is thus devoted to the synthesis of spiro-fused oxindoles from isatin and its derivatives.

4.5 SYNTHESIS OF SPIRO-FUSED 2-OXINDOLES

The compounds with spiro-fused cyclic frameworks are attractive synthetic targets owing to their broad application in the area of medicinal chemistry, their therapeutic value, and because they are core structures in many natural products. The application of isatins in synthesis of such compounds have been reviewed recently (Singh and Desta 2012). Since a detail treatment of the topic is beyond the scope of this chapter; representative examples of applications of isatin and its derivatives in the synthesis of spiro-oxindoles are described in this section.

The condensation of isatins with 1,3-diamines and with 1,2-diamines leads to the formation of spiro-fused 2-oxindoles. For example, the condensation of isatins **1**, **15**, and **22** with 2-aminobenzylamine **128** in methanol has been reported to produce the 2-oxindoles **129** spiro-fused to tetrahydroquinazoline (Scheme 4.60) (Bergman et al. 2003). This reaction in refluxing acetic acid, however, afforded the quinolinone derivatives as a major product together with traces of **129**. The reaction of isatin **1** and an *o*-diamine 3,4-diaminofurazane **130** affords the product **131** having 2-oxindole ring spiro-fused to 1,2,5-oxadiazoloimidazolidine (Scheme 4.61) (Gurevich et al. 2010).

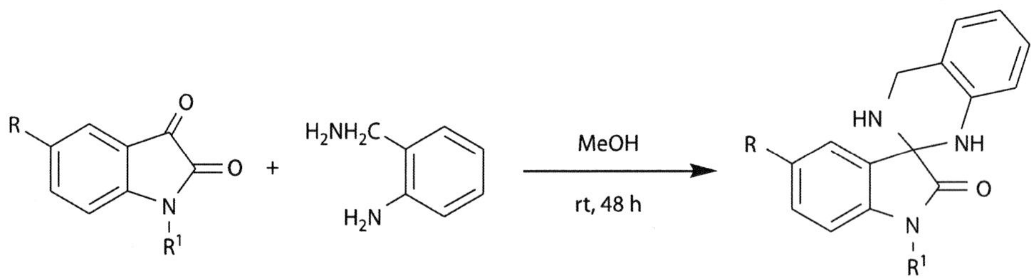

1, 15, 22 128 129

R = R^1 = H; R = Me, R^1 = H; R = H, R^1 = Me (Yield = 65%–68%)

SCHEME 4.60 Synthesis of spiro-tetrahydroquinazoline-oxindoles by the reaction of isatins with 2-aminobenzylamine.

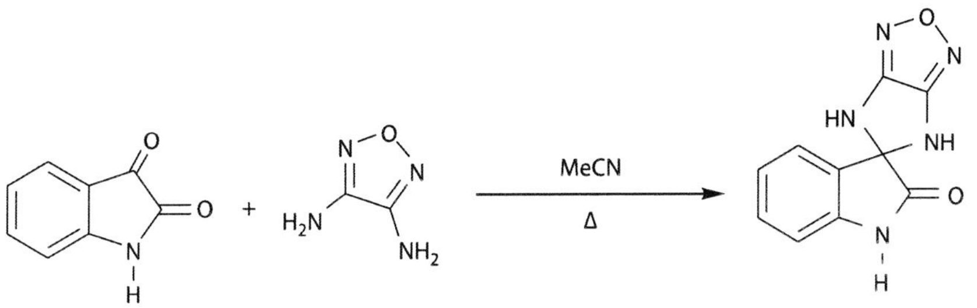

1 130 131 (Yield = 77%)

SCHEME 4.61 Synthesis of spiro-imidazole-oxindole by the reaction of isatin with 3,4-diaminofurazane.

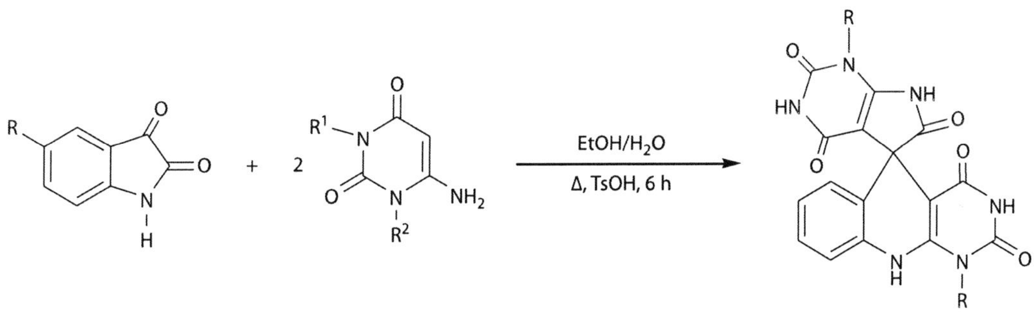

1, 15 132 133 (Yield = 78%–90%)

133. $R = R^1 = R^2 = H$; $R = Br, R^1 = H, R^2 = H$; $R = NO_2, R^1 = H, R^2 = H$;
$R = H, R^1 = Me, R^2 = Me$; $R = NO_2, R^1 = Me, R^2 = Me$; $R = Br, R^1 = Me, R^2 = Me$;
$R = H, R^1 = H, R^2 = Me$; $R = NO_2, R^1 = H, R^2 = Me$; $R = Br, R^1 = H, R^2 = Me$

SCHEME 4.62 A 1:2 molar reaction of isatins with 6-amino-1-methyluracil forming spiro-oxindole derivatives.

A 1:2 molar reaction of isatins **1**, **15**, and 6-amino-1-methyluracil **132** in the presence of catalytic amount of *p*-toluenesulfonic acid afforded 1,1′-dimethyl-1*H*-spiro[pyrimido[4,5-b]quinoline-5,5′-pyrrolo[2,3-d]pyrimidine]-2,2′,4,4′,6′(1′*H*,3*H*,3′*H*,7′*H*,10*H*)-pentaones **133** in good yields (Scheme 4.62) (Dabiri et al. 2008). These products have shown good antibacterial activity against some Gram-positive and Gram-negative bacteria (Ghahremanzadeh et al. 2008).

A 2:1 molar reaction of 2-hydroxynaphthalene-1,4-dione **134** and isatin **1** in the presence of a catalytic amount of *p*-toluenesulfonic acid in aqueous medium furnishes spiro[dibenzo[b,i]-xanthene-13,3′-indoline]-2′,5,7,12,14-pentaone **135** in 80% yield (Scheme 4.63) (Bazgir et al. 2008). A series of such compounds in good yields (75%–82%) have been synthesized from the reactions of *N*-methylisatin, *N*-benzylisatin, 5-bromoisatin, 5-nitroisatin, and *N*-bromo-5-methylisatin with compound **134**.

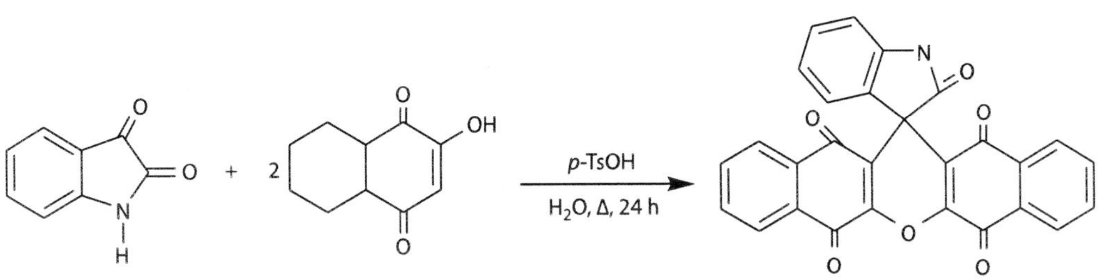

1 134 135 (Yield = 80%)

SCHEME 4.63 A 1:2 molar reaction of isatin with 2-hydroxynaphthalene-1,4-dione in the presence of *p*-TsOH forming spiro-oxindole derivative.

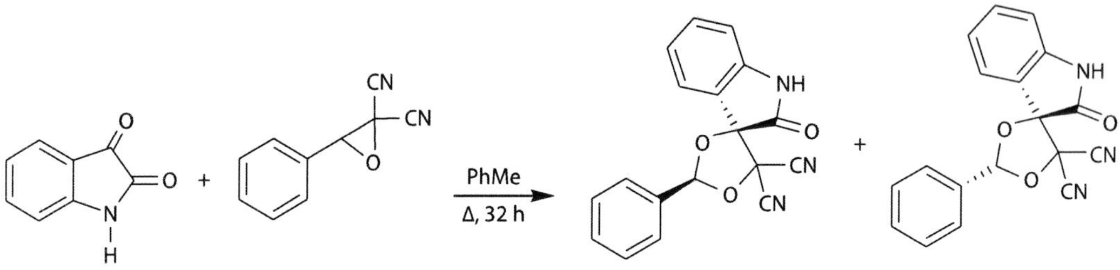

1 136 **137** (Yield = 42%) (**137**:**138** = 57:43) **138**

SCHEME 4.64 A [2 + 3]-cycloaddition reaction of isatin with carbonyl ylide generated *in situ* from 3-phenyloxirane-2,2-carbonitrile to give spiro-dioxolane-oxindoles.

A [2 + 3]-cycloaddition reaction of carbonyl ylides, generated from epoxide **136** and ketone group of isatin **1**, afforded spiro-fused heterocyclic compounds **137** and **138** in which 1,3-dioxolane ring was spiro-fused to C-3 of 2-oxindole (Scheme 4.64) (Bentabed-Ababsa et al. 2008). The reaction has also been carried out with *N*-methylisatin and 5-chloroisatin using epoxides having 4-methoxyphenyl group and 4-chlorophenyl group affording products in moderate yields (30%–73%).

When equimolar amounts of methyl benzoylpyruvate **139** and isatin **1** are gently heated in the presence of *N,N,N′,N′*-tetramethylguanidine in dioxane, the guanidinium salt of spiro-fused compound **140** was formed. An acidic hydrolysis of the salt liberated the 3′-benzoyl-4′-hydroxyspiro[indole-3,2′-furan]2,5′(1*H*)-dione **141** and the tetramethylguanidinium hydrochloride **142** (Scheme 4.65) (Gein et al. 2010).

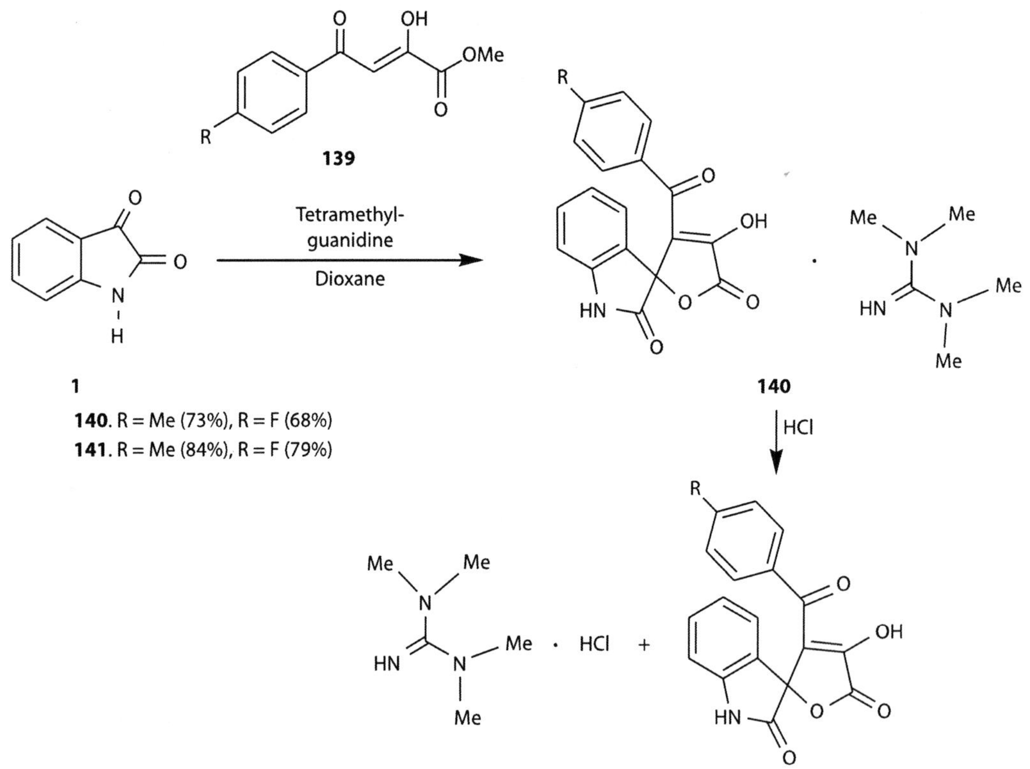

1 140

140. R = Me (73%), R = F (68%)
141. R = Me (84%), R = F (79%)

142 141

SCHEME 4.65 Reaction of isatin with methyl benzoyl pyruvate in the presence of *N,N,N′,N′*-tetramethylguanidine followed by acidic hydrolysis of the salt.

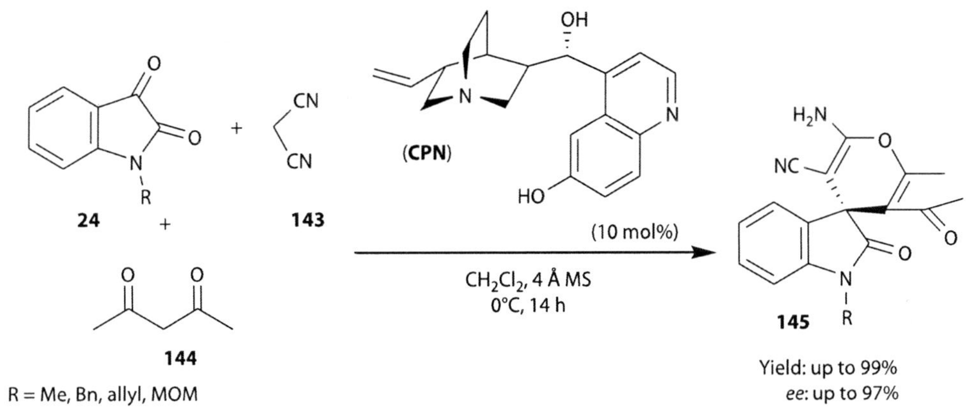

SCHEME 4.66 An enantioselective synthesis of spiro-pyran-oxindole by a three-component reaction of isatin using cupreine as a catalyst.

The first enantioselective two- and three-component reactions via a domino Knoevenagel/Michael/cyclization sequence with cupreine (CPN) as catalyst have been developed (Chen et al. 2010c). Optically active spiro[4*H*-pyran-3,3′-oxindoles] **145** were obtained in excellent yields (up to 99%) with good-to-excellent enantioselectivity (up to 97%) in the reaction of isatin **1**, malononitrile **143**, and pentane-2,4-dione **144** in the presence of cupreine (CPN) as a catalyst (Scheme 4.66). Similar products have been synthesized by an electrocatalytic reaction of isatin, 1,3-diketones, and malononitrile under neutral and mild conditions in undivided cell in alcoholic solvent in the presence of sodium bromide as an electrolyte (Elinson et al. 2007).

A [2 + 3]-cycloaddition of azomethine ylides, generated *in situ* from the reaction of isatins with amino acids such as sarcosine and L-proline, serves as a common strategy to synthesize spiro-oxindoles. A three-component domino reaction of isatin **1**, *N*-methylpiperidin-4-one **146**, and *N*-methylglycine **147** has led to the formation of bis-spiropyrrolidine **148** in moderate yield (Scheme 4.67) (Kumar et al. 2010).

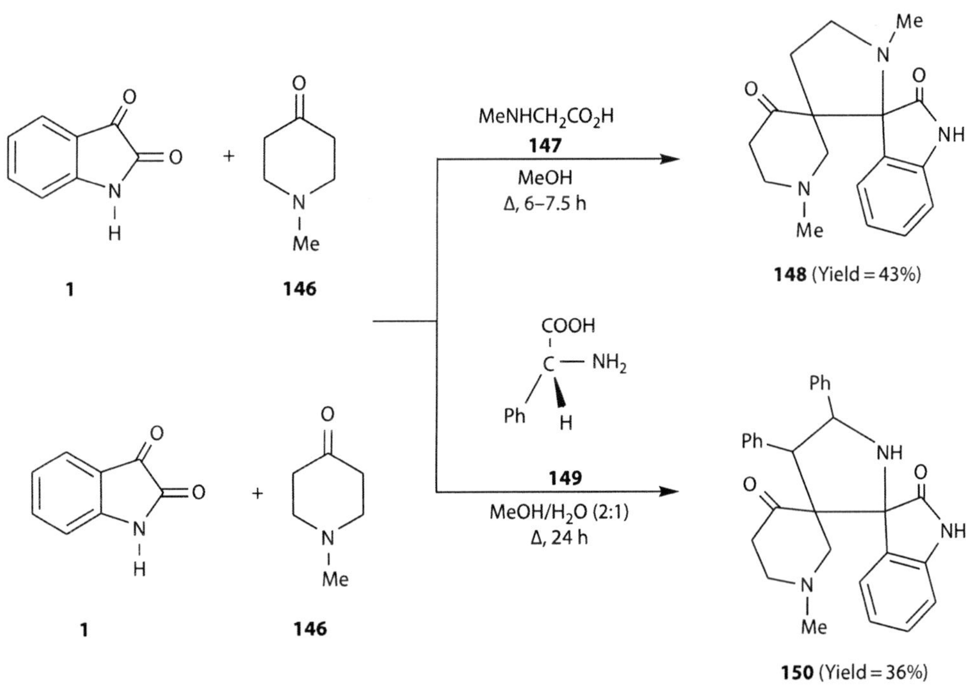

SCHEME 4.67 Three-component reactions of isatin forming spiro-heterocyclic 2-oxindoles.

A similar reaction using 2-phenylglycine **149** afforded another bis-spiropyrrolidine **150**. Other dipolarophiles used in this methodology are 3-acetyl-2*H*-chromen-2-ones (Ghandi et al. 2010) and (*E*)-3-aryl-1-(thiophen-2-yl)prop-2-en-1-ones (Thangamani 2010).

The reaction of isatin **1** with carbohydrazide in glacial acetic acid afforded hydrazone **160**, which underwent oxidative cyclization to yield 2-oxindole **151** spiro-fused to 1,3,4-oxadiazoline. Hydrazinolysis of hydrazide chain in product **152** afforded spiro-fused 2-oxindoles **153** (Scheme 4.68) (Islam and Mohsin 2007).

A 3-spirocyclopentene-2-oxindole **155** has been synthesized by reacting the bromoallyl derivative of 1-methylisatin **154** with methyl acrylate **57** in the presence of Ph_3P and K_2CO_3 in toluene (Scheme 4.69), whereas 3-spiropyrazole-2-oxindole **158** has been synthesized by reacting the bromoallyl derivative **156** of 1-methylisatin with diethylazadicarboxylate (DEAD) **157** in the presence of Me_2S/K_2CO_3 in acetonitrile at room temperature (Scheme 4.70) (Selvakumar et al. 2010).

Aldol-addition of acetophenones **159** to isatin **1** affords 3-hydroxy-3-phenacyloxindoles **160** which undergoes dehydration forming 3-phenacylidene-2-indolinones **161** in quantitative yields. The reactions of compounds **161** with phenylthiourea, phenylhydrazine, and hydrazine resulted in the formation of 2-oxindoles **162–164**, spiro-fused to pyrimidine-2-thione, *N*-phenylpyrazoline, and pyrazoline, respectively (Scheme 4.71) (Ibrahim et al. 2010). Kusanur et al. have reported the formation of (Kusanur et al. 2004) 2-oxindoles spiro-fused to benzodiazepine by the reaction of isatin–coumarin aldol adduct with *o*-phenylenediamines.

The isatin imine **165** reacts with mercaptoacetic acid to give 2-oxindole **166** spiro-fused to 3-phenylthiazolidinone (Scheme 4.72) (Mashelkar and Rane 2005). The reactivity of the nitrogen atom in the 2-oxindole ring of this compound has been further explored to synthesize new products.

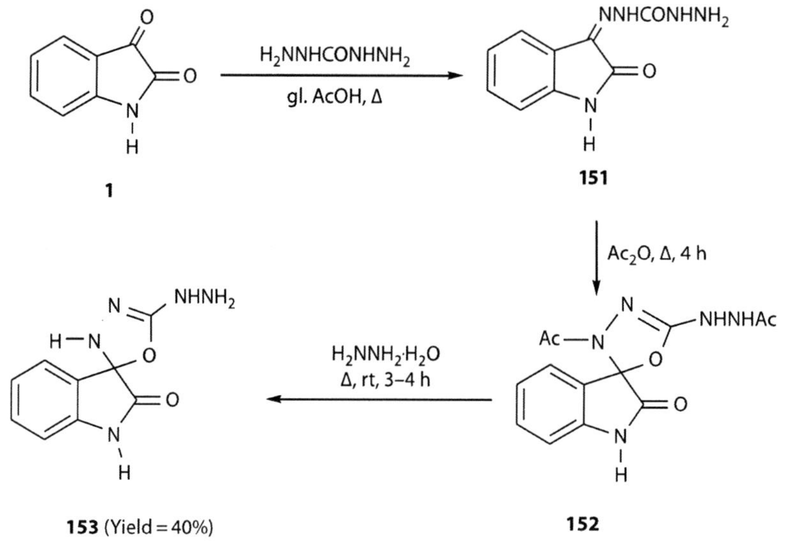

SCHEME 4.68 A multi-step synthesis of spiro-oxadiazoline-oxindole derivative from isatin.

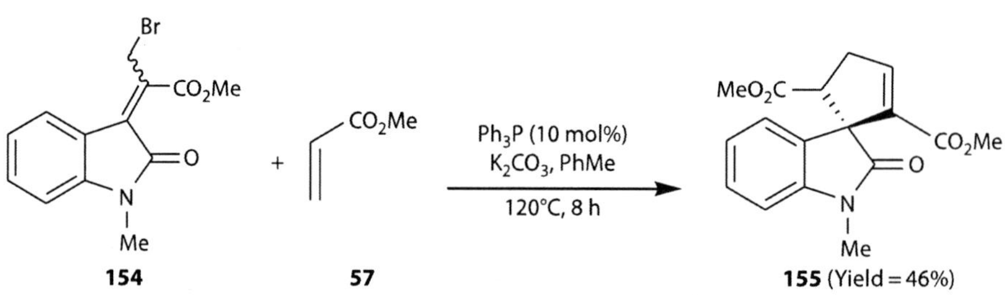

SCHEME 4.69 Synthesis of a spiro-cyclopentene-oxindole derivative from the reaction of bromoallyl derivative of *N*-methylisatin with methyl acrylate in the presence of triphenylphosphine.

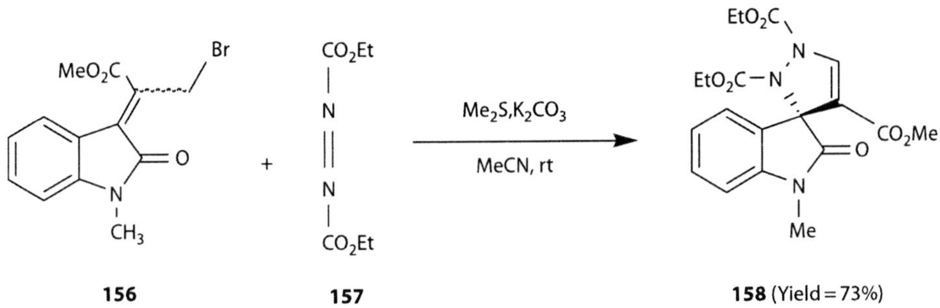

SCHEME 4.70 Synthesis of a spiro-pyrazole-oxindole derivative from the reaction of bromoallyl derivative of *N*-methylisatin with DEAD.

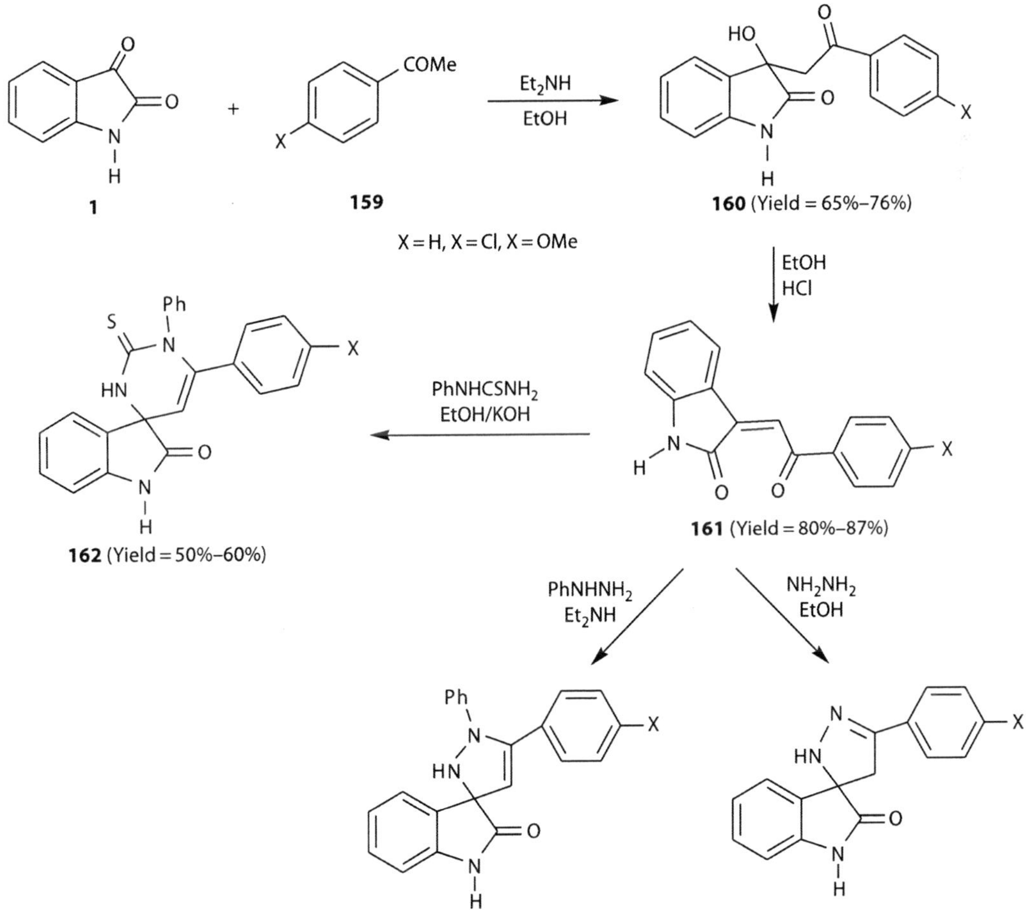

SCHEME 4.71 Synthesis of spiro-heterocyclic 2-oxindoles from aldol adducts of isatin.

The reaction of compound **166** with ethyl chloroacetate in the presence of NaH/DMF yielded the compound **167** which underwent hydrazinolysis to afford the compound **168**. This product on further condensation with azalactone **169** furnished spirocyclic compound **170**.

The reactions of isatin imines with diphenylketene, di-*p*-tolylketene, and di-*p*-anisylketene, generated from the corresponding 2-diazo-1,2-diarylethanone, have been reported to form 2-azetidinones spiro-fused to oxindoles (Singh et al. 1997, Singh and Mmolotsi 2006). Recently, the reactions

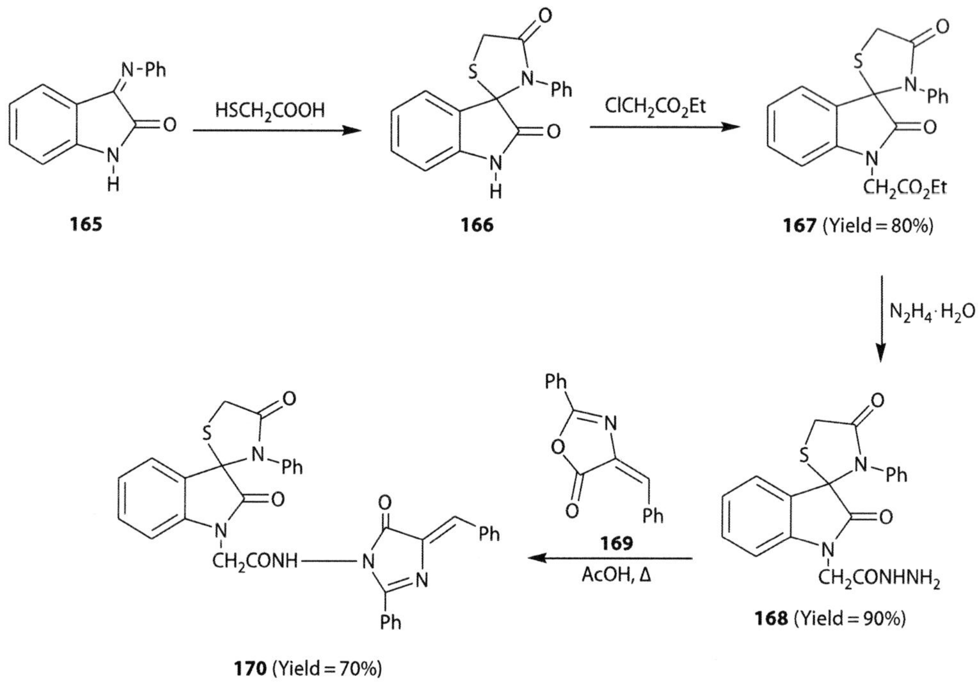

SCHEME 4.72 Synthesis and transformation of spiro-thiazolidinone-oxindole from the reaction of 3-*N*-phenyliminoisatin with mercaptoacetic acid.

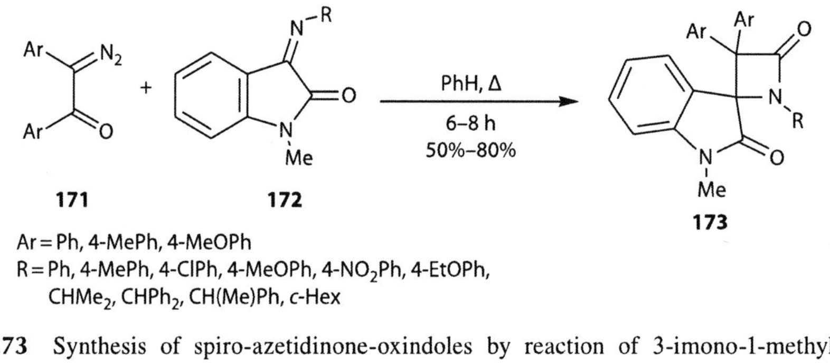

SCHEME 4.73 Synthesis of spiro-azetidinone-oxindoles by reaction of 3-imono-1-methylisatins with 2-diazo-1,2-diarylethanones.

of diarylketenes, obtained from 2-diazo-1,2-diarylethanones **171**, with 3-alkylimino-*N*-methylindolin-2-ones **172**, have been reported to yield spiro-fused 2-azetidinones **173** in good yields (Scheme 4.73), but with poor-to-moderate antibacterial activity (Singh and Luntha 2009). The synthesis of isatin-derived mono- and bis-spiroazetidinones using the Staudinger reaction has been reported by Jarrahpour and Khalili (2007). 3-Arylimino-1-methyl-2-indolinones have also been reacted with dichloroketene to afford the corresponding spiro-fused 2-azetidinones (Azizian et al. 2000).

A $TiCl_4$-catalyzed coupling of 2-acetyl-6-methyl-2,3-dihydro-4*H*-pyran **174** with isatin **1** results into a tandem C–C and C–O bonds formation offering a simple methodology for the stereoselective synthesis of [(1-acetyl-5-methyl-6,8-dioxabicyclo(3.2.1)octane)-7-spiro-3′-(indoline-2′-one)] **175** (Scheme 4.74) (Basavaiah et al. 2005). The reaction has been carried out with isatin, *N*-alkylisatins, *N*-phenylisatin, 5-nitroisatin, and some 1,5-disubstituted isatins in the presence of 20 mol% of catalyst in dichloromethane at room temperature for 6 h to afford the products in 44%–74% yields.

The reaction of isatins with arsonium salts **176** in the presence of K_2CO_3 constitutes a one-pot approach for highly stereoselective synthesis of 2-oxindoles **177** spiro-fused at its C-3 to a

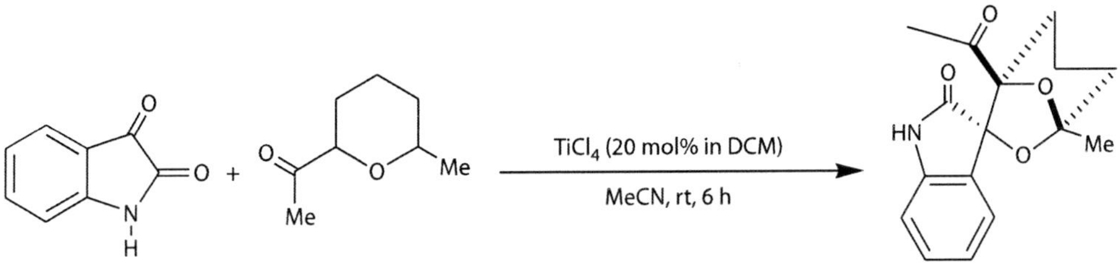

1 **174** **175** (Yield = 74%)

SCHEME 4.74 A $TiCl_4$-catalyzed coupling of isatin with 2-acetyl-6-methyl-2,3-dihydro-4*H*-pyran.

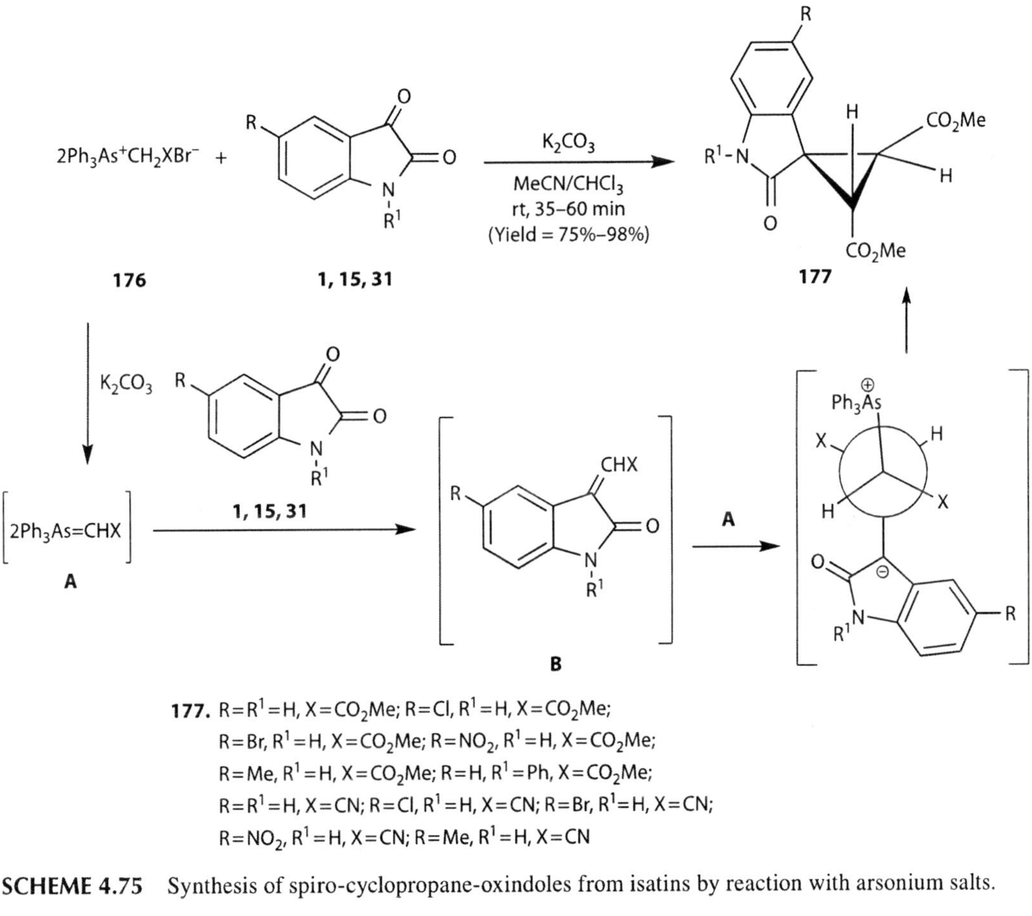

177. R = R[1] = H, X = CO_2Me; R = Cl, R^1 = H, X = CO_2Me;
R = Br, R^1 = H, X = CO_2Me; R = NO_2, R^1 = H, X = CO_2Me;
R = Me, R^1 = H, X = CO_2Me; R = H, R^1 = Ph, X = CO_2Me;
R = R^1 = H, X = CN; R = Cl, R^1 = H, X = CN; R = Br, R^1 = H, X = CN;
R = NO_2, R^1 = H, X = CN; R = Me, R^1 = H, X = CN

SCHEME 4.75 Synthesis of spiro-cyclopropane-oxindoles from isatins by reaction with arsonium salts.

cyclopropane ring (Scheme 4.75) (Yu et al. 2010). According to the proposed mechanism, first the Wittig reaction of isatin **1** and arsonium ylide **A**, derived from arsonium salt **176** with potassium carbonate as base, generates 3-alkylideneisatins **B**. The ylide **A** then adds across the exocyclic carbon–carbon double bond in **B** to form the product **177**.

4.6 RING EXPANSION OF ISATINS TO QUINOLINES

Quinolines constitute an important group of heterocyclic compounds which have been found to possess useful biological activities such as antimalarial, antibacterial, anti-asthmatic, antihypertensive, and anti-inflammatory. In addition, quinolines are valuable synthons for the preparation of nano- and mesostructures with enhanced electronic and photonic functions (Yavari et al. 2010). Quinoline is also known

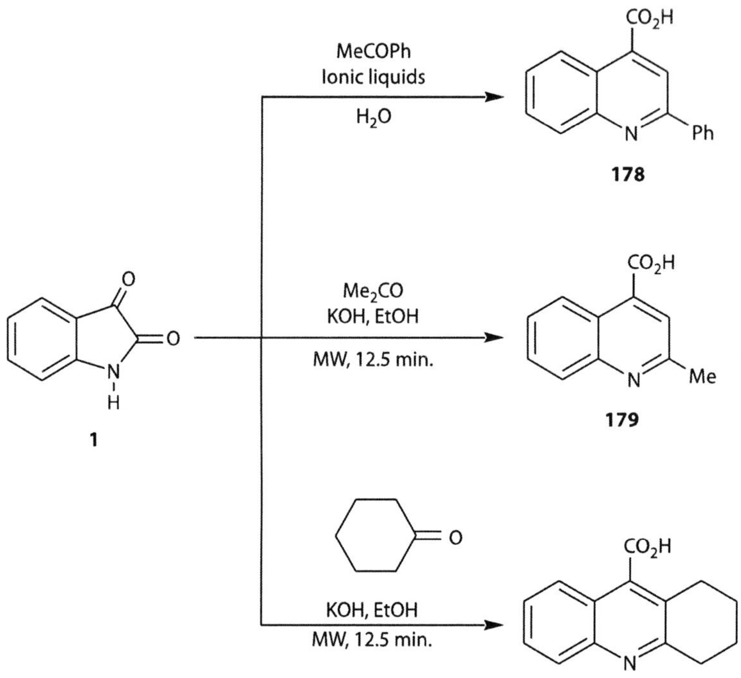

SCHEME 4.76 Reaction of isatin with ketones forming quinoline derivatives.

as 1-azanaphthalene, 1-benzazine, or benzo[b]pyridine (Khan et al. 2009). Many synthetic routes, including those from ring expansion of isatins, are well documented for the formation of quinolines. Isatins undergo ring expansions on reaction with ketones, activated alkynes, and active methylene compounds. For example, the reaction of isatin **1** with acetophenone in the presence of ionic liquids as catalyst affords 2-phenylquinoline-4-carboxylic acid **178** (Scheme 4.76) (Kowsari and Mallakmohammadi 2011). A microwave-assisted synthesis using acetone and cyclohexanone in ethanolic KOH leads to the formation of 2-methylquinoline-4-carboxylic acid **179** and **180**, respectively (Sayed et al. 2005).

The reaction of isatin **1** with dialkylacetylenedicarboxylates **181** in the presence of sodium *O*-alkylcarbonodithionates **182** at room temperature produced trialkylquinoline-2,3,4-tricarboxylates **183** in good yields (Scheme 4.77) (Yavari et al. 2010).

A microwave-assisted reaction of isatin **1** with malonic acid provided quinoline-2-hydroxy-4-carboxylic acid **184** (Scheme 4.78) (Madapa et al. 2008). The reaction of 5-morpholinosulfonylisatin **185** with ethylacetoacetate in the presence of an alkali afforded quinoline-3,4-dicarboxylic acid **186** (Scheme 4.79) (Madapa et al. 2008).

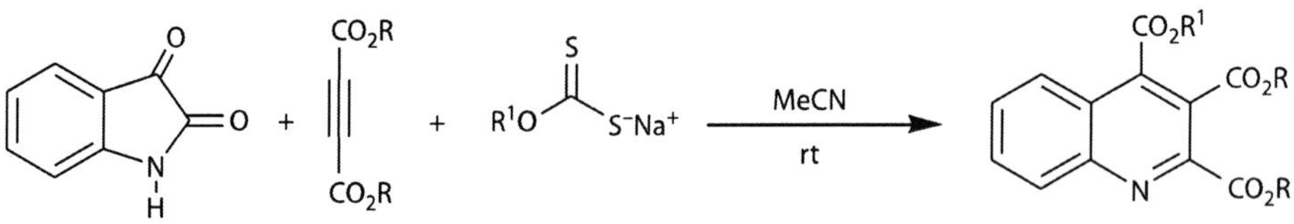

1

181
a. R = Me
b. R = Et

182
a. R^1 = Me
b. R^1 = Et

183 (Yield = 82%–92%)
a. R = R^1 = Me
b. R = Me, R^1 = Et
c. R = Et, R^1 = Me

SCHEME 4.77 Reaction of isatin with dialkylacetylenedicarboxylates in the presence of sodium *O*-alkylcarbondithioate forming quinoline derivatives.

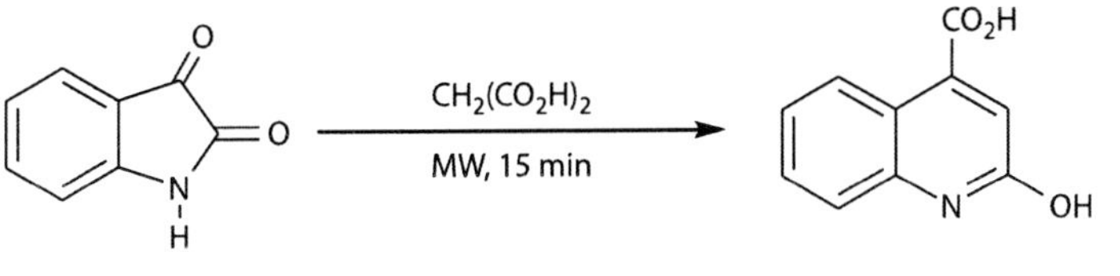

1 184

SCHEME 4.78 Microwave-assisted ring-expansion of isatin to 2-hydroxyquinoline-4-carboxylic acid.

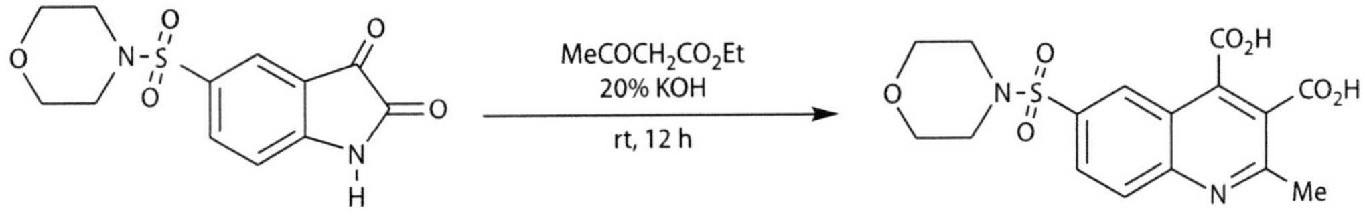

185 186

SCHEME 4.79 Ring-expansion of 5-(morpholinosulfonyl)isatin to a quinoline derivative.

4.7 SUMMARY AND CONCLUSIONS

Isatin, possessing an indole nucleus, is a natural product. It has a ketone group present in the γ-lactam moiety which is condensed to a phenyl ring. Isatin and its derivative have shown diverse types of reactivities that have been used for the synthesis of different kinds of heterocyclic organic compounds. During the synthesis of such organic compounds, isatin and its derivatives have been used both as an electrophilic and as a nucleophilic reagent. When isatin has been used as a nucleophilic reagent, the reaction takes place either at position 5 or position 7 of the aromatic part of isatin and NH of the lactam moiety. The reactions reported include *N*-alkylation, *N*-acylation, *N*-arylation, and electrophilic aromatic substitutions, etc. *N*-Alkylation reaction of isatins finds application in attaching isatins to other heterocyclic moieties. Using isatin as an electrophilic reagent, reactions of the ketone group with carbon, oxygen, and nitrogen-centered nucleophiles are reported. These reactions include the Baylis–Hillman reaction, aldol reaction, Henry reaction, reactions forming isatin oximes, hydrazones, semicarbazones, thiosemicarbazones, and imines, etc. Oxidation, reduction, and ring-enlargement reactions of isatins are reported. These reactions lead to the synthesis of heterocyclic compounds such as isatoic anhydride, oxindoles, indole, and quinolines. Isatin and many of its derivatives have been used as a building block for the synthesis of different types of spiro-fused heterocyclic compounds. Looking at these reports and the number of reports appearing in recent literature, it can be inferred that isatins are synthetically and biologically useful molecules and will continue to attract the attention of researchers in the area of organic synthesis and medicinal chemistry.

ABBREVIATIONS

CD	cinchonidine
CPN	cupreine
DABCO	1,4-diazabicyclo[2.2.2]octane
DCM	dichloromethane
DME	1,2-dimethoxyethane
DMF	*N*,*N*-dimethylformamide
i-PrOH	isopropanol
MW	microwave
OTMS	trimethylsilyloxy

TBAB	tetrabutylammonium bromide
TCCA	trichloroisocyanuric acid
TEA	triethylamine
THF	tetrahydrofuran
TQO	4-(3-ethyl-4-oxa-1-azatricyclo[4.4.0.0]dec-5-yl)quinoline-6-ol
TsOH	*p*-toluenesulfonic acid

REFERENCES

Abele, E., R. Abele, O. Dzenitis, and E. Lukevics. 2003. Indole and isatin oximes: Synthesis, reactions, and biological activity. *Chem. Heterocycl. Compd.* 39: 3–35 and references there in.

Aboul-Fadl, T. and F.A.S. Bin-Jubair. 2010. Anti-tubercular activity of isatin derivatives. *Int. J. Res. Pharm. Sci.* 1: 113–126.

Aboul-Fadl, T., F.A.S. Bin-Jubair, and O. Aboul-Wafa. 2010. Schiff bases of indoline-2,3-dione (isatin) derivatives and nalidixic acid carbohydrazide, synthesis, antitubercular activity and pharmacophoric model building. *Eur. J. Med. Chem.* 45: 4578–4586.

Adibi, H., M.M. Khodaei, P. Pakravan, and R. Abiri. 2010. Synthesis, characterization, and *in vitro* antimicrobial evaluation of hydrazone and bishydrazone derivatives of isatin. *Pharm. Chem. J.* 44: 219–227.

Aikawa, K., S. Mimura, Y. Numata, and K. Mikami. 2011. Palladium-catalyzed enantioselective ene and aldol reactions of isatins, keto esters, and diketones: Reliable approach to chiral tertiary alcohols. *Eur. J. Org. Chem.* 62–65.

Alcaide, B., P. Almendros, and C. Aragoncillo. 2010. Indium-promoted allylation reaction of imino-isatins in aqueous media: Synthesis of quaternary 3-aminooxindoles. *Eur. J. Org. Chem.* 2010: 2845–2848.

Al-Kahraman, Y.M.S.A., G.S. Singh, and M. Yasinzai. 2011. Evaluation of *N*-(2-thienyli-dene)amines, *N*-2-hydroxybenzylideneamines, and 3-iminoindolin-2-ones as antileishmanial agents. *Lett. Drug Design. Dis.* 8: 242–249.

Allu, S., N. Molleti, R. Panem, and V.K. Singh. 2011. Enantioselective organocatalytic aldol reaction of unactivated ketones with isatins. *Tetrahedron Lett.* 52: 4080–4083.

Almeida, M.R., G.G. Leitão, B.V. Silva, J.P. Barbosa, and A.C. Pinto. 2010. Counter-current chromatography separation of isatin derivatives using the Sandmeyer methodology. *J. Braz. Chem. Soc.* 21: 764–769.

Azizian, J., H. Fallah-Bagher-Shaidaei, and H. Kefyati. 2003. A facile one-pot method for the preparation of *N*-alkyl isatins under microwave irradiation. *Synth. Commun.* 33: 789–793.

Azizian, J., Y. Sarrafi, and M. Mehrdad. 2000. Synthesis of some new 1-methyl-3′,3′-dichlorospiro[indol-3,4′-azetidine]-2(3H), 2′-diones and bis[1-methyl-3′,3′-dichlorospiro(indole-3,4′-azetidine)-2(3*H*), 2′-diones]. *Indian J. Chem., Sect B* 39: 304–307.

Bal, T.R., B. Anand, P. Yogeeswari, and D. Sriram. 2005. Synthesis and evaluation of anti-HIV activity of isatin-β-thiosemicarbazone derivatives. *Bioorg. Med. Chem. Lett.* 15: 4451–4455.

Bari, S.B., A.O. Agrawal, and U.K. Patil. 2008. Synthesis and pharmacological evaluation of some novel isatin derivatives for antimicrobial activity. *J. Sci. Islam. Rep. Iran.* 19: 217–221.

Basavaiah, D. and A.J. Rao. 2003. 1-Benzopyran-4(4*H*)-ones as novel activated alkenes in the Baylis–Hillman reaction: A simple and facile synthesis of indolizine-fused-chromones. *Tetrahedron Lett.* 44: 4365–4368.

Basavaiah, D., J.S. Rao, R.J. Reddy, and A.J. Rao. 2005. $TiCl_4$ catalyzed tandem construction of C–C and C–O bonds: A simple and one-pot atom-economical stereoselective synthesis of spiro-oxindoles. *J. Chem. Soc. Chem. Commun.* 2621–2623.

Basavaiah, D., A.J. Rao, and T. Satyanarayana. 2003. Recent advances in the Baylis-Hillman reaction and applications. *Chem. Rev.* 103: 811–891.

Bazgir, A., Z.N. Tisseh, and P. Mirzaei. 2008. An efficient synthesis of spiro[dibenzo[b,i] xanthene-13,3′-indoline]-pentaones and 5*H*-dibenzo[b,i]xanthene-tetraones. *Tetrahedron Lett.* 49: 5165–5168.

Bentabed-Ababsa, G., A. Derdour, T. Roisnel, J.A. S′aez, L.R. Domingo, and F. Mongin. 2008. Polar [3 + 2] cycloaddition of ketones with electrophilically activated carbonyl ylides. Synthesis of spirocyclic dioxolane indolinones. *Org. Biomol. Chem.* 6: 3144–3157.

Bergman, J., R. Engqvist, C. Stalhandsk, and H. Wallberg. 2003. Studies of the reactions between indole-2,3-diones (isatins) and 2-aminobenzylamine. *Tetrahedron* 59: 1033–1048.

Bergman, J., J.O. Lindstrom, and U. Tilstam. 1988. The structure and properties of some indolic constituents in *Couroupita guianansis aubl. Tetrahedron* 41: 2879–2881.

Bhardwaj, S., L. Kumar, R. Verma, and U.K. Singh. 2010. Synthesis, characterization and antimicrobial activity of Schiff bases of isatin and isatin derivatives. *J. Pharm. Res.* 3: 2983–2985.

Bhragual, D.D., N. Kumar, and S. Drabu. 2010. Synthesis and pharmacological evaluation of some substituted imidazoles. *J. Chem. Pharm. Res.* 2: 345–349.

Boechat, N., W.V. Kover, M.M. Bastos et al. 2008. *N*-Acyl-3,3-difluoro-2-oxoindoles as versatile intermediates for the preparation of different 2,2-difluorophenylacetic derivatives. *J. Braz. Chem. Soc.* 19: 445–457.

Chauhan, P. and S. Chimni. 2010. Asymmetric addition of indoles to isatins catalysed by bifunctional modified cinchona alkaloid catalysts. *Chem. Eur. J.* 16: 7709–7713.

Chen, W.-B., X.-L. Du, L.-F. Cun, X.M. Zhang, and W.C. Yuan. 2010a. Highly enantioselective aldol reaction of acetaldehyde and isatins only with 4-hydroxydiarylprolinol as catalyst: Concise stereoselective synthesis of (*R*)-convolutamydines B and *E*,(-)-donaxaridine and (*R*)-chimonamidine. *Tetrahedron* 66: 1441–1446.

Chen, J.-R, X.-P. Liu, X.-Y. Zhu, L. Li, Y.-F. Qiao, J.-M. Zhang, and W.-J. Xiao. 2007. Organocatalytic asymmetric aldol reaction of ketones with isatins: Straightforward stereoselective synthesis of 3-alkyl-3-hydroxyindolin-2-ones. *Tetrahedron* 63: 10437–10444.

Chen, G., Y. Tang, Q.-Z. Zhang, Y. Wu, and S.-Z. Mu. 2010b. Rapid eco-friendly synthesis and structure of 3-hydroxy-3-nitromethyl-1,3-dihydro-indol-2-one. *J. Chem. Crystallogr.* 40: 369–372.

Chen, W.-B., Z.-J. Wu, Q.-L. Pei, L.-F. Cun, X.-M. Zhang, and W.C. Yuan. 2010c. Highly enantioselective construction of spiro[4*H*-pyran-3,3′-oxindoles] through a domino Knoevenagel/Michael/cyclization sequence catalyzed by cupreine. *Org. Lett.* 12: 3132–3135.

Chhajed, S.S. and M.S. Padwal. 2010. Antimicrobial evaluation of some novel Schiff and Mannich bases of isatin and its derivatives with quinoline. *Int. J. Chem. Technol. Res.* 2: 209–213.

Chiyanzu, I., E. Hansell, J. Gut, P.J. Rosenthal, J.H. McKerrowb, and K. Chibale. 2003. Synthesis and evaluation of isatins and thiosemicarbazone derivatives against cruzain, falcipain-2 and rhodesain. *Bioorg. Med. Chem. Lett.* 13: 3527–3530.

Chung, Y.M., Y.J. Im, and J.N. Kim. 2002. Baylis-Hillman reaction of isatin derivatives: Isatins as a new entry for the Baylis-Hillman reaction. *Bull. Korean Chem. Soc.* 23: 1651–1654.

Dabiri, M., S.C. Azimi, H.R. Khavasi, and A. Bazgir. 2008. A novel reaction of 6-amino-uracils and isatins. *Tetrahedron* 64: 7307–7311.

Deligeorgiev, T., A. Vasilev, J.J. Vaquero, and J. Alvarez-Bulla. 2007. A green synthesis of isatoic anhydrides from isatins with urea–hydrogen peroxide complex and ultrasound. *Ultrasonics Sonochem.* 14: 497–501.

Dermatakis, A., K.-C. Luk, and W. De Pinto. 2003. Synthesis of potent oxindole CDK2 inhibitors. *Bioorg. Med. Chem.* 11: 1873–1881.

Elinson, M.N., A.I. Ilovaisky, A.S. Dorofeev, V.M. Merkulova, N.O. Stepanov, F.M. Miloserdov, and Y.N. Ogibin et al. 2007. Electrocatalytic multicomponent transformation of cyclic1,3-diketones, isatins, and malononitrile: Facile and convenient way to functionalized spirocyclic (5,6,7,8-tetrahydro-4*H* chromene)-4, 3′-oxindole system. *Tetrahedron* 63: 10543–10548.

Elinson, M.N., V.M. Merkulova, A.I. Ilovaisky, A.O. Chizhov, P.A. Belyakov, F. Barba, and B. Batanero. 2010. Electrochemically induced aldol reaction of cyclic 1,3-diketones with isatins. *Electrochim. Acta* 55: 2129–2133.

Fensome, A., R. Bender, J. Cohen, M.A. Collins, V.A. Mackner, L.L. Miller, and J.W. Ullrich. 2002. New progesterone receptor antagonists: 3,3-disubstituted 5-aryloxindoles. *Bioorg. Med. Chem. Lett.* 12: 3487–3490.

Garden, S.J. and J.M.S. Skakle. 2002. Isatin derivatives are reactive electrophilic components for the Baylis–Hillman reaction. *Tetrahedron Lett.* 43: 1969–1972.

Garden, S.J., J.C. Torres, L.E. da Silva, and A.C. Pinto. 1998. A convenient methodology for the *N*-alkylation of isatin compounds. *Synth. Commun.* 28: 1679–1689.

Gassman, P.G., B.W. Cue Jr., and T.-Y. Juh. 1977. A general method for the synthesis of isatins. *J. Org. Chem.* 42: 1344–1348.

Gein, V.L., E.B. Levandovskaya, and V.N. Vichegjanina. 2010. Synthesis of 3′-aroyl-4′-hydroxyspiro- [indole-3,2′-furan]-2,5′(1*H*)-diones. *Chem. Heterocycl. Compd.* 46: 931–933.

Ghahremanzadeh, R., S.C. Azimi, N. Gholami, and A. Bazgir. 2008. Clean synthesis and antibacterial activities of spiro[pyrimido[4,5-*b*]-quinoline-5,5′-pyrrolo[2,3-*d*]pyrimidine]-pentaones. *Chem. Pharm. Bull.* 56: 1617–1620.

Ghahremanzadeh, R., F. Fereshtehnejad, P. Mirzaei, and A. Bazgir. 2011. Ultrasound- assisted synthesis of 2,2′-(2-oxoindoline-3,3-diyl)bis(1*H*-indene-1,3(2*H*)-dione) derivatives. *Ultrason. Sonochem.* 18: 415–418.

Ghandi, M., A. Taheri, and A. Abbasi. 2010. A facile synthesis of chromeno[3,4-c]spiro- pyrrolidine-oxindoles via 1,3-dipolar cycloadditions. *Tetrahedron* 66: 6744–6748.

Girgis, A.S. 2009. Regioselective synthesis and stereochemical structure of anti-tumor active dispiro[3*H*-indole-3,2′-pyrrolidene 3′,3″-piperidine]-2(1*H*), 4″-diones. *Eur. J. Med. Chem.* 44: 1257–1264.

Guan, X.-Y., Y. Wei, and M. Shi. 2010. Construction of chiral quaternary carbon through Morita–Baylis–Hillman reaction: An enantioselective approach to 3-substituted 3-hydroxyoxindole derivatives. *Chem. Eur. J.* 16: 13617–13621.

Gurevich, P.G., L.F. Sattarova, A.S. Petrovskiy, N.A. Frolova, B.P. Strunin, and R.Z. Musin. 2010. Interaction of spiro-heterocyclic oxindole system with sodium diformylimide. *Chem. Heterocycl. Compd.* 46: 1527–1530.

Hanhan, N.V., A.H. Sahin, T.W. Chang, J.C. Fettinger, and A.K. Franz. 2010. Catalytic asymmetric synthesis of substituted 3-hydroxy-2-oxindoles. *Angew. Chem. Int. Ed.* 49: 744–747.

Hans, R.H., H. Su, and K. Chibale. 2010. Novel tetracyclic structures from the synthesis of thiolactone-isatin hybrids. *Beil. J. Org. Chem.* 6:78.

Hans, R.H., I.J.F. Wiid, P.D.V. Helden, B. Wan, S.G. Franzblau, J. Gut, and P.J. Rosenthal. 2011. Novel thiolactone–isatin hybrids as potential antimalarial and antitubercular agents. *Bioorg. Med. Chem. Lett.* 21: 2055–2058.

Hulme, C., G.B. Poli, F.-C. Huang, J.E. Souness, and S.W. Djuric. 1998. Quaternary substituted PDE4 inhibitors I: The synthesis and *in vitro* evaluation of a novel series of oxindoles. *Bioorg. Med. Chem. Lett.* 8: 175–178.

Ibrahim, M.N., M.F. El-Messmary, and M.G.A. Elarfi. 2010. Synthesis of spiro heterocyclic compounds. *Eur. J. Chem.* 7: 55–58.

Islam, M.R. and M. Mohsin. 2007. Synthesis of isatin, 5-chloroisatin and their Δ2–1,3,4 oxadizoline derivatives for comparative cytotoxicity study on brine shrimp. *Bangladesh J. Pharmacol.* 2: 7–12.

Jagadeesh, R.V., Puttaswamy, N. Vaz, and N.M.M. Gowda. 2008. Ruthenium catalyzed oxidative conversion of isatins to anthranilic acids: Mechanistic study. *AIChE J.* 54: 756–765.

Jaishree, B., K. Manjunatha, M. Girish, S. Adil, and M.G. Purohit. 2009. Synthesis and biological evaluation of some *N*-substituted indoles. *ARKIVOC*. XII: 217–231 and references there in.

Jarrahpour, A. and D. Khalili. 2007. Synthesis of some mono- and bis-spiro-β-lactams of benzylisatin. *Tetrahedron Lett.* 48: 7140–7143.

Jiang, Y. and T.V. Hansen. 2011. Isatin 1,2,3-triazoles as potent inhibitors against caspase-3. *Bioorg. Med. Chem. Lett.* 21: 1626–1629.

Kang, I.-J., L.-W. Wang, T.-A. Hsu, A. Yueh, C.-C. Lee, Y.-C. Lee, and C.-Y. Lee et al. 2011. Isatin-β-thiosemicarbazones as potent herpes simplex virus inhibitors. *Bioorg. Med. Chem. Lett.* 21: 1948–1952.

Khan, K.M., U.R. Mughal, Samreen, S. Parveen, and M.I. Choudhary. 2008. Schiff bases of isatin—Potential anti-leishmanial agents. *Lett. Drug Des. Discovery* 5: 243–249.

Khan, K.M., S. Saied, U.R. Mughal, M. Munawar, M., Samreen, A. Khan, and S. Perveen. 2009. Synthesis, leishmanicidal and enzyme inhibitory activities of quinoline-4-carboxylic acids. *J. Chem. Soc. Pak.* 31: 809–818.

Konstantinovic, S.S., B.C. Radovanović, S.P. Sovilj, and S. Stanojević. 2008. Antimicrobial activity of some isatin-3—Thiosemicarbazone complexes. *J. Serb. Chem. Soc.* 73: 7–13.

Konstantinovic, S.S., B.C. Radovanović, Z.B. Todorovic, and S.B. Ilic. 2007. Spectrophotometric study of Co(II), Ni(II), Cu(II), Zn(II), Pd(II) and Hg(II) complexes with isatin-β-thiosemicarbazone. *J. Serb. Chem. Soc.* 72: 975–981.

Kowsari, E. and M. Mallakmohammadi. 2011. Ultrasound promoted synthesis of quinolines using basic ionic liquids in aqueous media as a green procedure. *Ultrason. Sonochem.* 18: 447–454.

Kumar, R.S., S.M. Rajesh, S. Perumal, D. Banerjee, P. Yogeeswari, and D. Sriram. 2010. Novel three-component domino reactions of ketones, isatin and amino acids: Synthesis and discovery of antimycobacterial activity of highly functionalised novel dispiropyrrolidines. *Eur. J. Med. Chem.* 45: 411–422.

Kumari, G., M. Nutan, M. Modi, S.K. Gupta, and R.K. Singh. 2011. Rhodium(II)-catalyzed stereoselective synthesis, SAR, and anti-HIV activity of novel oxindoles bearing cyclopropane ring. *Eur. J. Med. Chem.* 46: 1181–1188.

Kurkin, A.V., A.A. Bernovskaya, and M.A. Yurovskaya. 2011. Comparative study of the different approaches to the synthesis of isatins with a chiral substituent at the nitrogen atom. *Chem. Heterocycl. Compd.* 46: 1497–1504.

Kusanur, R.A., M. Ghate, and M.V. Kulkarni. 2004. Synthesis of spiro[indolo-1,5-benzo- diazepines] from 3-acetyl coumarins for use as possible anti-anxiety agents. *J. Chem. Sci.* 116: 265–270.

Labo, A.M. and S. Prabhakar. 2009. Recent developments in the synthesis of biologically active indole alkaloids. *J. Heterocycl. Chem.* 39: 429–436.

Lee, H.S., K.H. Kim, Y.M. Kim, and J.N. Kim. 2010. Synthesis of tetracyclic oxindoles from isatin containing Baylis-Hillman adducts *via* Pd-catalyzed aryl-aryl coupling and reduction with NaBH4. *Bull. Korean Chem. Soc.* 31: 1761–1764.

Liu, X., X.S. Huang, N. Sin, B.L. Venables, and V. Roongta. 2010. ^{15}N chemical shifts of a series of isatin oxime ethers and their corresponding nitrone isomers. *Magn. Reson. Chem.* 48: 873–876.

Liu, L., S. Zhang, F. Xue, G. Lou, H. Zhang, S. Ma, and W. Duan et al. 2011. Catalytic enantioselective Henry reactions of isatins: Application in the concise synthesis of (*S*)-(−)-spirobrassinin. *Chem. Eur. J.* 17: 1791–1795.

Luntha, P. 2009. Synthesis, reactivity and antimicrobial activity of β-lactams spiro-fused to *N*-methylindolin-2-ones. MSc dissertation, University of Botswana, Gaborone, Botswana.

Ma, H.M., Z.Z. Liu, and S.Z. Chen. 2003. New approach to synthesis of 6,7-dimethoxyisatin. *Chin. Chem. Lett.* 14: 468–470.

Madapa, S., Z. Tusi, and S. Batra. 2008. Advances in the syntheses of quinoline and quinoline-annulated ring systems. *Curr. Org. Chem.* 12: 1116–1183.

Mashelkar, U.C. and D.M. Rane. 2005. Synthesis of some isatin based novel spiro heterocycles and their biological activity studies. *Indian J. Chem. Sect. B.* 44: 1937–1939.

Masutlha, L.L. and G.S. Singh. 2012. *N*-acylation of isatin using 2-diazo-1,2-diphenylethanone as an acylating agent. *Natl. Acad. Sci. Lett. India* 82: 147–149.

Mendonca, G.F., R.R. Magalhães, M.C.C. de Mattos, and P.M. Esteves. 2005. Trichloroisocyanuric acid in H_2SO_4: An efficient superelectrophilic reagent for chlorination of isatin and benzene derivatives. *J. Braz. Chem. Soc.*16: 695–698.

Meshram, H.M., P. Ramesh, B.C. Reddy, B. Sridhar, and J.S. Yadav. 2011. Diastereoselective Mukaiyama aldol reaction of (*N*-alkyl)isatins with 2-(trimethylsilyloxy)furan by using lanthanum triflate. *Tetrahedron* 67: 3150–3155.

Mesropyan, E.G. and A.A. Avetisyan. 2009. New isatin derivatives. *Zhur. Org. Khim.* 45: 1583–1593.

Messaoudi, S., M. Sancelme, V. Polard-Housset, B. Aboab, P. Moreau, and M. Prudhomme. 2004. Synthesis and biological evaluation of oxindole and benzimidazolinone. *Eur. J. Med. Chem.* 39: 453–458.

Mollica, A., A. Stefanucci, F. Feliciani, G. Lucente, and F. Pinnen. 2011. Synthesis of (*S*)-5,6-dibromo-tryptophan derivatives as building blocks for peptide chemistry. *Tetrahedron Lett.* 52: 2583–2585.

Nugumanova, G.N., R.G. Tagasheva, S.V. Bukharov, D.B. Krivolapov, I.A. Lativinov, V.V. Syakaev, and N.A. Mukmeneva. 2009. Isatin acylhydrazones with sterically hindered phenolic fragments: Synthesis and structures. *Russ. Chem. Bull. Int. Ed.* 58: 1934–1938.

Paira, P., A. Hazra, S. Kumar, R. Paira, K.B. Sahu, S. Naskar, and P. Saha. 2009. Efficient synthesis of 3,3-diheteroaromatic oxindole analogues and their *in vitro* evaluation for spermicidal potential. *Bioorg. Med. Chem. Lett.* 19: 4786–4789.

Pandeya, S.N. and A.S. Raja. 2002. Synthesis of isatin semicarbazones as novel anticonvulsants—Role of hydrogen bonding. *J. Pharm. Pharmaceut. Sci.* 5: 266–271.

Pandeya, S.N., S. Smitha, M. Jyoti, and S.K. Sridhar. 2005. Biological activities of isatin and its derivatives. *Acta Pharm. (Croatia)* 55: 27–46.

Peng, L., L.-L. Wang, J.-F. Bai, L.-N. Jia, Q.-C. Yang, Q.-C. Huang, and X.-Y. Xu. 2011. Highly effective and enantioselective phospho-aldol reaction of diphenyl phosphite with *N*-alkylated isatins catalyzed by quinine. *Tetrahedron Lett.* 52: 1157–1160.

Pervez, H., M.S. Iqbal, M.Y. Tahir, M.I. Choudhary, and K.M. Khan. 2007. Synthesis of some *N*-substituted isatin-3-thiosemicarbazones. *Natl. Prod. Res.* 21: 1178–1186.

Piccirilli, R.M. and F.D. Popp. 1973. The reaction of isatin with cycloalkylamines. *J. Heterocycl. Chem.* 10: 671–673.

Pinto, A.C., A.A.M. Lapis, B.V. da Silva, R.A. Bastos, J. Dupont, and B.A.D. Neto. 2008. Pronounced ionic liquid effect in the synthesis of biologically active isatin-3-oxime derivatives under acid catalysis. *Tetrahedron Lett.* 49: 5639–5641.

Popp, F.D. 1975. The chemistry of isatin. *Adv. Heterocycl. Chem.* 18: 1–58.

Porcs-Makkay, M., B. Volk, R. Kapiller-Dezsofi, T. Mezei, and G. Simig. 2004. New routes to oxindole derivatives. *Monatsh fur Chem.* 135: 697–711.

Qiao, X.-C., S.-F. Zhu, and Q.-L. Zhou. 2009. From allylic alcohols to chiral tertiary homoallylic alcohol: Palladium-catalyzed asymmetric allylation of isatins. *Tetrahedron Asymmetry* 20: 1254–1261.

Rad, M.N.S., A. Khalafi-Nezhad, S. Babamohammadi, and S. Behrouz. 2010. Microwave-assisted three-component synthesis of some novel 1-alkyl-1*H*-indole-2,3-dione 3-(*O*-alkyloxime) derivatives as potential chemotherapeutic agents. *Helv. Chim. Acta* 93: 2454–2466.

Rai, A., S.K. Sengupta, and O.P. Pandey. 2005. Lanthanum (III) and praseodymium (III) complexes with isatin thiosemicarbazones. *Spectrochim. Acta Part A* 61: 2761–2765.

Raj, I.V.P., T.M. Shaikh, and A. Sudalai. 2010. H-β zeolite: An efficient, reusable catalyst for one-pot synthesis of isatins from anilines. *Acta Chim. Slov.* 57: 466–469.

Ramachary, D.B., G.B. Reddy, and R. Mondal. 2007. A new organocatalyst for Friedel–Crafts alkylation of 2-naphthols with isatins: Application of an organo-click strategy for the cascade synthesis of highly functionalized molecules. *Tetrahedron Lett.* 48: 7618–7623.

Rekhter, M.A. 1999. *N*-Alkylation trimethylsilyl derivatives of isatin with halomethyl ketones. *Chem. Heterocycl. Compd.* 35: 1165–1166.

Rekhter, M.A. 2005. Direct *N*-alkylation of isatin by halomethyl ketones. *Chem. Heterocycl. Compd.* 41: 1119–1120.

Roopan, S.M., F.R.N. Khan, and N.T. Selven. 2010. Synthesis of 1-[(2-chloroquinolin-3-yl)methyl]indoline-2,3-dione derivatives as potential antimicrobials. *J. Pharm. Res.* 3: 950–952.

Sandmeyer, T. 1919. Über Isonitrosoacetanilide und deren Kondensation zu Isatinen. *Helv. Chim. Acta* 2: 234–242.

Sayed, E., E. Ashry, E.S. Ramadan, H.A. Hamid, and M. Hagar. 2005. Microwave-assisted synthesis of quinoline derivatives from isatin. *Synth. Commun.* 35: 2243–2250.

Schmidt, M.M., A.M. Reverdito, L. Kremenchuzky, I.A. Perillo, and M.M. Blanco. 2008. Simple and efficient microwave assisted *N*-alkylation of isatin. *Molecules* 13: 831–840.

Selvakumar, K., V. Vaithiyanathan, and P. Shanmugam. 2010. An efficient stereoselective synthesis of 3-spirocyclopentene- and 3-spiropyrazole-2-oxindoles via 1,3-dipolar cycloaddition reaction. *J. Chem. Soc. Chem. Commun.* 46: 2826–2828.

Shanmugam, P. and V. Vaithiyanathan. 2008. Stereoselective synthesis of 3-spiro-α methylene-γ-butyrolactone oxindoles from Morita-Baylis-Hillman adducts of isatin. *Tetrahedron* 64: 3322–3330.

Shanmugam, P., V. Vaithiyanathan, and B. Viswambharan. 2006. Synthesis of functionalized 3-spirocyclopropane-2-indolones from isomerized Baylis–Hillman adducts of isatin. *Tetrahedron* 62: 4342–4348.

Sharma, P.P., S.N. Pandeya, R.K. Roy, Anurag, K. Verma, and S. Gupta. 2009. Synthesis and anticonvulsant activity of some novel isatin Schiff's bases. *Int. J. Chem. Technol. Res.* 1: 758–763.

Shintani, R., M. Inoue, and T. Hayashi. 2006. Rhodium-catalyzed asymmetric addition of aryl- and alkenylboronic acids to isatins. *Angew. Chem. Int. Ed.* 118: 3431–3434.

Silva, B.V., P.M. Esteves, and A.C. Pinto. 2011a. Chlorination of isatins with trichloroisocyanuric acid. *J. Braz. Chem. Soc.* 22: 257–263.

da Silva, J.F.M., S.J. Garden, and A.C. Pinto. 2001. The chemistry of isatins: A review from 1975 to 1999. *J. Braz. Chem. Soc.* 12: 273–324 and references there in.

Silva, B.V., F.A. Violante, A.C. Pinto, and L.S. Santos. 2011b. The mechanism of Sandmeyer's cyclization reaction by electrospray ionization mass spectrometry. *Rapid Commun. Mass Spectrom.* 25: 423–428 and references there in.

Sin, N., B.L. Venables, K.D. Combrink, H.B. Gulgeze, K.-L. Yu, R.L. Civiello, and J. Thuring et al. 2009. Respiratory syncytial virus fusion inhibitors. Part 7: Structure–activity relationships associated with a series of isatin oximes that demonstrate antiviral activity in vivo. *Bioorg. Med. Chem. Lett.* 19: 4857–4862.

Singh, V. and S. Batra. 2008. Advances in the Baylis-Hillman reaction-assisted synthesis of cyclic frameworks. *Tetrahedron* 64: 4511–4574.

Singh, G.S. and P. Luntha. 2009. Synthesis and antimicrobial activity of new 1-alkyl/cyclohexyl-3,3-diaryl-1′-methylspiro[azetidine-2,3′-indoline]-2′,4-diones. *Eur. J. Med. Chem.* 44: 2265–2269.

Singh, G.S. and B.J. Mmolotsi. 2006. Reactions of α-diazoketones with indolinone imines: Synthesis of new 1,3,3-triaryl-1′-methylspiro[azetidine-2,3′-indoline]-2′,4-diones. *J. Heterocycl. Chem.* 43: 1665–1668.

Singh, G.S., N. Siddiqui, and S.N. Pandeya. 1993a. Synthesis and anticonvulsant and anti-inflammatory activities of 3-aryl/alkylimino-1-methylindol-2-ones. *Arch. Pharm. Res.* 15: 272–274.

Singh, G.S., N. Siddiqui, and S.N. Pandeya. 1993b. Synthesis and anticonvulsant activity of 3-aryl/alkyliminoindol-2-ones. *Asian J. Chem.* 4: 788–791.

Singh, G.S., T. Singh, and R. Lakhan. 1997. Synthesis, 13-C NMR and anticonvulsant activity of some new isatin based spiroazetidinones. *Indian J. Chem.* 36B: 951–954.

Singh, G.S. and Z.Y. Desta. 2012. Isatins as privileged molecules in design and synthesis of spiro-fused cyclic frameworks. *Chem. Rev.* 112: 6104–6155. DOI: 10.1021/cr300135y.

Smitha, S., S.N. Pandeya, J.P. Stables, and S. Ganapathy. 2008. Anticonvulsant and sedative hypnotic activities of *N*-acetyl/methyl isatin derivatives. *Sci. Pharm.* 76: 621–636.

Soderberg, B.C.G., S.P. Gorugantula, C.R. Howerton, J.L. Petersen, and S.W. Dantale. 2009. A palladium-catalyzed synthesis of isatins (1*H*-Indole-2,3-diones) from 1-(2-haloethynyl)-2-nitrobenzenes. *Tetrahedron* 65: 7357–7363.

Somogyi, L. 2001. Transformation of isatin 3-acylhydrazones under acetylating conditions: Synthesis and structures elucidation of 1,5′-disubstituted 3′-acetylspiro[oxindole-3,2′-[1,3,4]oxadiazolines]. *Bull. Chem. Soc. Jpn.* 74: 873–881.

Sonderegger, O.J., T. Bürgi, L.K. Limbach, and A. Baiker. 2004. Enantioselective reduction of isatin derivatives over cinchonidine modified Pt/alumina. *J. Mol. Catal. A Chem.* 217: 93–101.

Sridhar, S.K., S.N. Pandeya, J.P. Stables, and A. Ramesh. 2002. Anticonvulsant activity of isatin hydrazones, Schiff and Mannich bases of isatin derivatives. *Eur. J. Pharm. Sci.* 16: 129–132.

Sridhar, S.K. and A. Ramesh. 2001. Synthesis and pharmacological activities of hydrazones, Schiff and Mannich bases of isatin derivatives. *Biol. Pharm. Bull.* 24: 1149–1152.

Srinivas, B., V.R. Priya, G.S. Babu et al. 2010. Synthesis and screening of new isatin derivatives. *Der. Pharm. Chem.* 2: 378–384.

Sumpter, W.C. 1944. The chemistry of isatin. *Chem. Rev.* 34: 393–434.

Thangamani, A. 2010. Regiospecific synthesis and biological evaluation of spiro-oxindolo-pyrrolizidines via [3 + 2] cycloaddition of azomethine ylide. *Eur. J. Med. Chem.* 45: 6120–6126.

Torisawa, Y., T. Nishi, and J.-I. Minamikawa. 2001. An efficient conversion of 5-nitroisatin into 5-nitroindole derivative. *Bioorg. Med. Chem. Lett.* 11: 829–832.

Toullec, P.Y., R.B.C. Jagt, D.G. de Vries, B.L. Feringa, and A.J. Minnaard. 2006. Rhodium-catalyzed addition of arylboronic acids to isatins: An entry to diversity in 3-aryl-3- hydroxyoxindoles. *Org. Lett.* 8: 2715–2718.

Trost, B.M. and Y. Zhang. 2011. Molybdenum-catalyzed asymmetric allylic alkylation of 3-alkyloxindoles: Reaction development and applications. *Chem. Eur. J.* 17: 2916–2922.

Vasta, G., O.P. Pandey, and S.K. Sengupta. 2005. Synthesis, spectroscopic and toxicity studies of titanocene chelates of isatin-3-thiosemicarbazones. *Bioinorg. Chem. Appl.* 3: 151–160.

Volk, B. and G. Simig. 2003. New one-pot synthesis of 3-alkyl- and 3-(ω-hydroxyalkyl) oxindoles from isatins. *Eur. J. Org. Chem.* 2003: 3991–3996.

Wang, C.-C. and X.-Y. Wu. 2011. Catalytic asymmetric synthesis of 3-hydroxyl-2-oxindoles via enantioselective Morita-Baylis-Hillman reaction of isatins. *Tetrahedron* 67: 2974–2978.

Yadav, J.S., V.S. Reddy, C.S. Reddy, and A.D. Krishna. 2007. Indium(III) chloride/2-iodoxybenzoic acid: A novel reagent for the conversion of indoles into isatins. *Synthesis* 693–696.

Yavari, I., S. Seyfi, and Z. Hossaini. 2010. Formation of trialkylquinoline-2,3,4-tricarb- oxylates by reaction of isatin, dialkyl acetylenedicarboxylates, and sodium *O*-alkyl carbonodithioates. *Tetrahedron Lett.* 51: 2193–2194.

Yeagley, A., A.J. Weigand-Heller, D. Hinds et al. 2011. Substituent and solvent dependence of the one-electron reduction of 5-substituted-*N*-methylisatins in aprotic solvents. *J. Electroanal. Chem.* 651: 228–232.

Yu, H., Y. Liu, H. Zhang, J. Chen, H. Deng, M. Shao, and Z. Ren. 2010. One-pot approach for the stereoselective synthesis of spirocyclopropyl oxindoles from isatins and arsonium salts. *Tetrahedron* 66: 2598–2501.

5 Role of Organic Carbamates in Anticancer Drug Design

Devdutt Chaturvedi

CONTENTS

5.1 INTRODUCTION

Organic carbamates are the stable class of compounds derived from the unstable carbamic acid (H_2N–COOH) by the substitution of amino and acid ends through various kinds of structurally diverse alkyl/aryl, aryl–alkyl or substituted alkyl/aryl, and aryl–alkyl groups, and are identified by the presence of the linkage –O–CO-NH– (Adams and Baron 1965; Chaturvedi 2003, 2011, 2012; Chaturvedi and Ray 2007a,b). When the carbamate linkage is present in a cyclic system, this class of compounds is referred to as cyclic carbamates (Ager et al. 1996; Arya and Qin 2000; Johnson and Evans 2000). When the carbamate group is attached to any inorganic atom, either metal or nonmetal, such compounds are referred to as inorganic carbamates (Aoki et al. 2001; Boyle et al. 1992).

The reaction of carbamation of amines has frequently been utilized in the synthesis of organic carbamates which holds unique applications in the field of pharmaceuticals (Asaka et al. 2003; Hutchinson 2003; Ray and Chaturvedi 2004; Ray et al. 2005); agrochemicals (pesticides, herbicides, insecticides, fungicides etc.) (Goto et al. 2006; Ma et al. 2006; The Pesticidal Manual 1994); as intermediates in organic synthesis (Dangerfield et al. 2009; Han et al. 2004; Smith et al. 2005; Wills et al. 2002); for the protection of amino groups in peptide chemistry (Greene and Wuts 2007; Kociensiki 2003); as linkers in combinatorial chemistry (Buchstaller 1998; Mayer et al. 1997); etc. Functionalization of amines as carbamates offers an attractive method for the generation of derivatives, which may have interesting medicinal and biological properties (Alaxander and Cravatt 2005;

Chang et al. 2006; Tully et al. 2006). Organic carbamates have been extensively used as useful synthons for the synthesis of structurally diverse synthetic intermediates/molecules of biological significance (Becker et al. 2007; Chedid et al. 2007; Han and Widenhoefer 2006; Nicolaou and Mathison 2005; Qin et al. 2007). Therefore, considerable interest has been generated in the recent past for the development of efficient and safe methodologies for carbamate esters synthesis. Our group has been engaged since the past several years for the development of efficient and safer protocols for the synthesis of carbamates and related compounds employing diversity of reagents and catalytic systems (Chaturvedi et al. 2008a,b,c, 2009, 2010, 2011).

Organic carbamates have frequently been employed as demandable pharmaceuticals in the forms of drugs and prodrugs (Rahmanthullan et al. 2008; Ray and Chaturvedi 2004). In recent years, several reports have indicated that the carbamate linkage present in between the active pharmacophores of various structurally diverse molecules increases manifold biological activities of semisynthetic/synthetic natural/synthetic molecules (Borrel et al. 2005; Chaturvedi 2003; Takaoka et al. 2004). Furthermore, the role of carbamate linkage has been extensively studied in structurally diverse natural/semisynthetic molecules against various diseases such as anticancer, antibacterial, antifungal, antimalarial, antiviral, anti-HIV, anti-estrogenic, antiprogestational, anti-osteoporosis, anti-inflammatory, antifilarial, antitubercular, antidiabetic, anti-obesity, anticonvulsant, antihelminthes, Alzheimer's disease, CNS and CVS active, etc. (Giannessi et al. 2003; Ishihara et al. 2000; Kuznetsova et al. 2006; Li et al. 2006; Ouellet et al. 1984; Palomo et al. 2004; Reiss and Vagell 2006; Sharma et al. 2007; Wu and Ojima 2004; Wu and Su 2001; Zega 2005). Some of the important biologically active drug molecules bearing carbamates are shown in Figure 5.1.

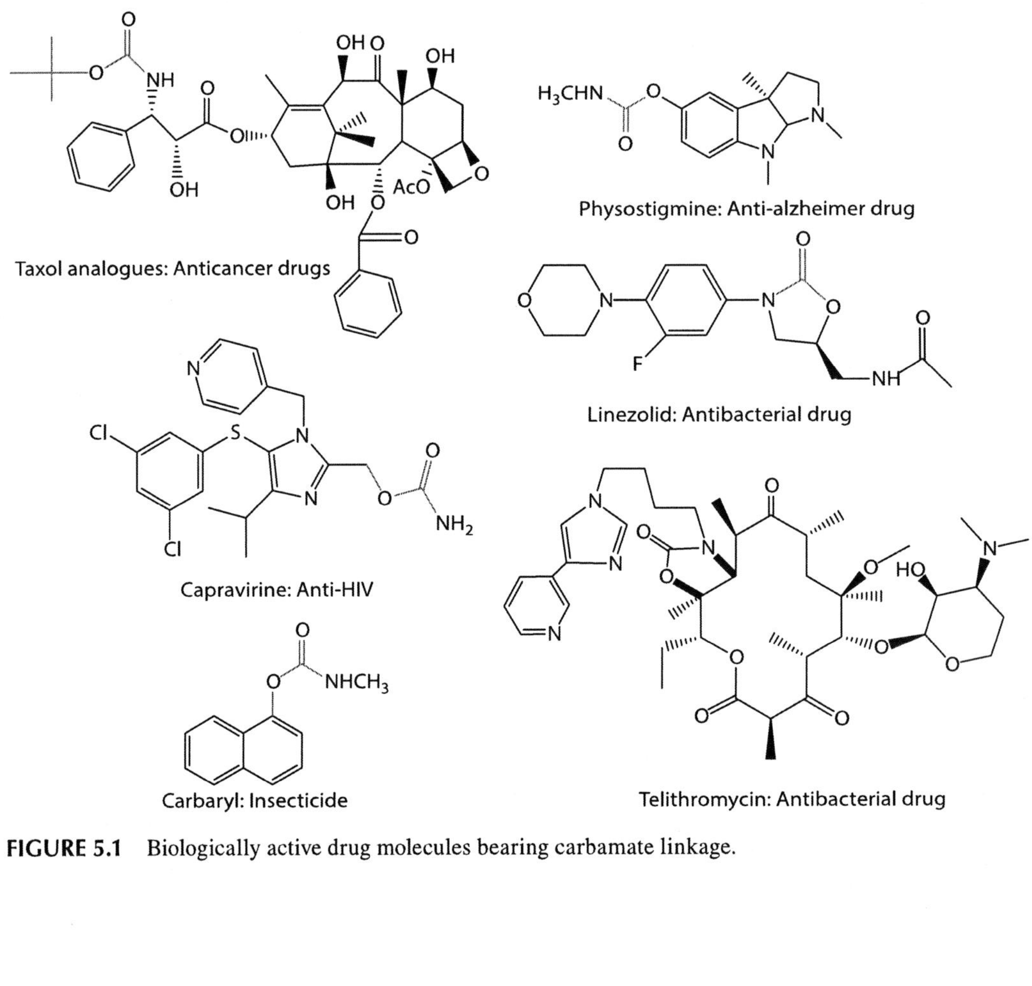

FIGURE 5.1 Biologically active drug molecules bearing carbamate linkage.

In recent years, several workers from different parts of the world have incorporated carbamates in between the active pharmacophores of structurally diverse natural products and found that carbamates play a crucial role in increasing the biological activity of these molecules. Several carbamate derivatives of natural products have emerged as drugs and prodrugs (Chaturvedi et al. 2008a,b,c, 2009, 2010, 2011). It is well reported that manifold anticancer activities of anticancer drug taxol can be increased by incorporating a variety of carbamates at different positions in the taxol molecule; some of these taxol derivatives bearing carbamates are in the various phases of clinical trials (Kuznetsova et al. 2006). Some of the recent molecules in which the extensive role of carbamates have been studied are discodermolide (Shaw et al. 2006), camptothecin (Wang et al. 2006), podophyllotoxin (Liu et al. 2007), mitomycins (Andez 2009), vitamin D_3 (Agoston et al. 2006), geldanamycin (Tian et al. 2004), fumagillin analogues (Fardis et al. 2003), betulinic acid (Santos et al. 2009), amphotericin-B (Sedlak et al. 2008), cephalosporins (Yoshizawa et al. 2004), doxorubicin (Jeffrey et al. 2006), rapamycin (Wagner et al. 2005), anisomycin (Shi et al. 2005), quiniclidine (Mazurov et al. 2005), phytostigmine (Yu et al. 1997), novobiocin (Shen et al. 2004), estradiol (Sharma et al. 2004), cholesterol (Lee et al. 2004), sphingomyelin (Taguchi et al. 2003), vancomycinn (McComas et al. 2003), marphinan (Peng et al. 2007), rifampicin (Combrink et al. 2007), vulmbactin (Bergeron et al. 2009), pregnelone (Slavíková et al. 2009), himbacine (Chackalamannil et al. 2008), iejimalides (Schweitzer et al. 2007), rhazinilam (Dećor et al. 2006), maytansine (Jaracz et al. 2005), calcheamycin (Borrel et al. 2005), combretastanin (Billich et al. 2005), cyclosporin (Jang et al. 2011), duocarmycins (Ashoorzadeh et al. 2011), etc. Besides the above-mentioned molecules, several of other structurally diverse natural/synthetic molecules have also been reported in recent years wherein carbamates play a crucial role in improving the biological activity profile compared to the parent molecules. Some of the important potential carbamates of structurally diverse biologically active anticancer/antibacterial natural products are depicted in Figures 5.2 and 5.3.

Some of the important potential carbamate derivatives of structurally diverse biologically active anticancer (Arico-Muendel et al. 2009; Oves et al. 2006; Tripathi et al. 2008; Xu et al. 2009), antibacterial (Asaka et al. 2003; Takashima 2003; Yan et al. 2010), antimalarial (Bova et al. 2010), antidiabetic (Mizutani et al. 2009), antioxidant (Yekini et al. 2009), anti-inflammatory (Ali et al. 2008), antitubercular (Janin 2007), antiprogestational (Kern et al. 2007), anti-HIV (Chen et al. 2007), anticoagulant (Franciskovich et al. 2005), antiestrogenic (Ohta et al. 2009), CNS-active (Mazurov et al. 2005) molecules are depicted in Figures 5.2 through 5.5, respectively. Several of natural, semisynthetic, synthetic lead molecules bearing carbamate functionality have been discovered in the recent past and are in the various phases of drug development (Asaka et al. 2003; Giannessi et al. 2003; Hutchinson 2003; Ishihara and Goto 2000; Kuznetsova et al. 2006; Li et al. 2006; Ouellet et al. 1984; Palomo et al. 2004; Ray and Chaturvedi 2004; Ray et al. 2005; Reiss and Vagell 2006; Sharma et al. 2007; Wu and Ojima 2004; Wu and Su 2001; Zega 2005).

5.2 CLASSIFICATION OF CARBAMATES

Carbamates can be classified mainly into two groups, namely inorganic and organic. Depending upon the structural variations in the attached moieties, they are further classified as shown in Figure 5.6.

Our group has already reviewed the role of organic carbamates in anticancer drug discovery research (Asaka et al. 2003; Hutchinson 2003; Ray and Chaturvedi 2004; Ray et al. 2005). Hence, the present chapter deals with the recent developments on the role of organic carbamates in a variety of biologically active natural products/semisynthetic molecules in anticancer drug discovery research since 2005 onward.

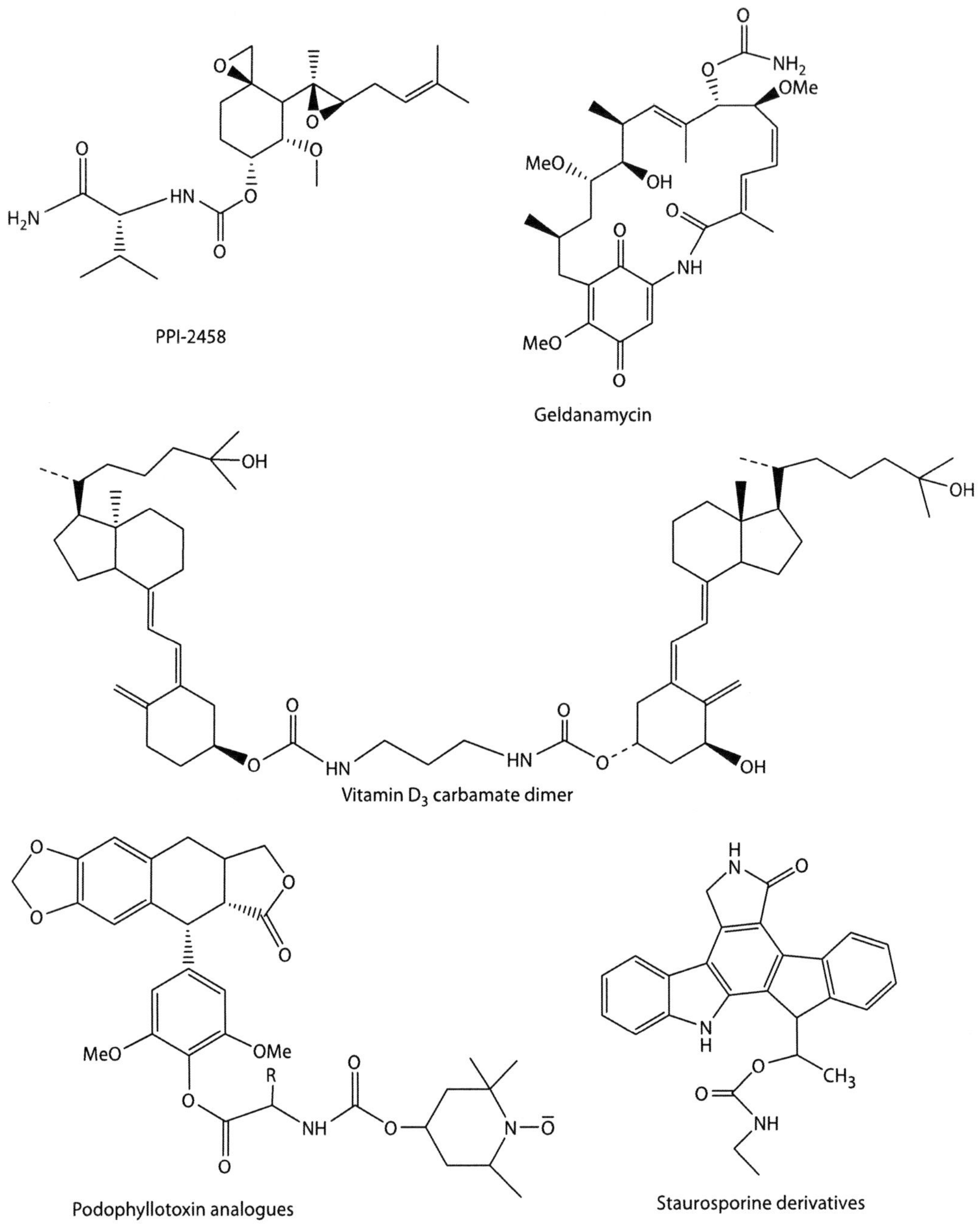

FIGURE 5.2 Potential anticancer carbamates of various natural products.

5.3 CARBAMATES AS ANTICANCER AGENTS

5.3.1 Carbamates of Natural Products

5.3.1.1 Carbamates of Fumagillin

The natural antibiotic fumagillin **1** exerts protective effects against endothelial cell proliferation *in vitro* and tumor-induced angiogenesis *in vivo* by inhibition of methionine aminopeptidase-2 (MetAP-2) as well as tumor growth in mice. Prolonged administration of the drug causes weight loss,

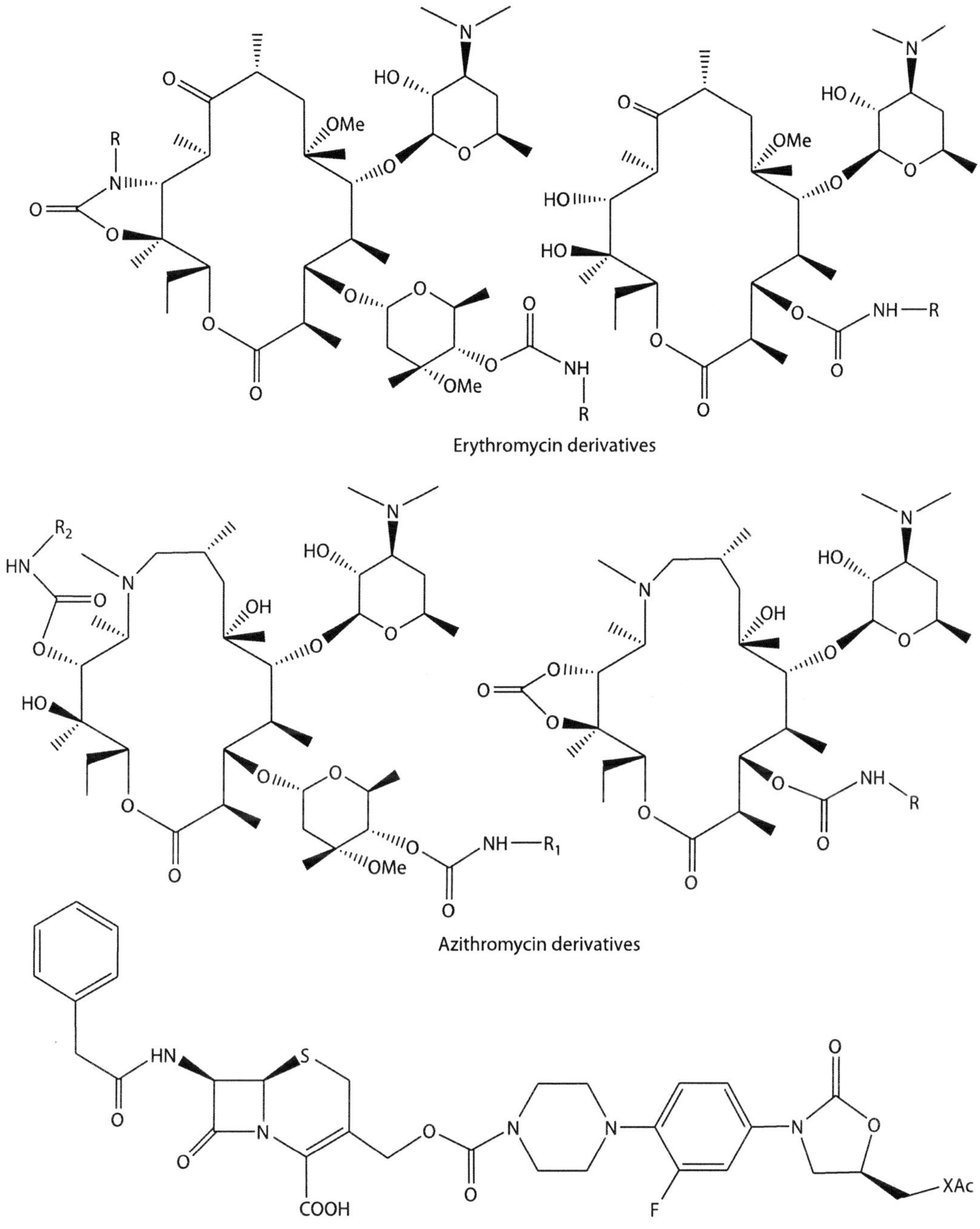

FIGURE 5.3 Potential antibacterial carbamates of various natural products.

and hence the need for structural modification. Replacement of the unsaturated ester chain at C-6 by an *O*-(chloroacetyl)carbamoyl moiety resulted in the potent anticancer compound **2** (TNP-470), which is 50 times more active than fumagillin and devoid of its side effects and was subjected to clinical trials (Fardis et al. 2003). Replacement of C-4 by benzyloxime moiety and C-6 by ethyl piperazinyl carbamate of fumagillin resulted in compound **3**, which exhibited antiangiogenic effect similar to TNP-470 on matrigel plug assay and rat corneal micropocket assay (Pyun et al. 2004). Recently, it has also been observed that replacement of the unsaturated ester chain at C-6 of fumagillin by

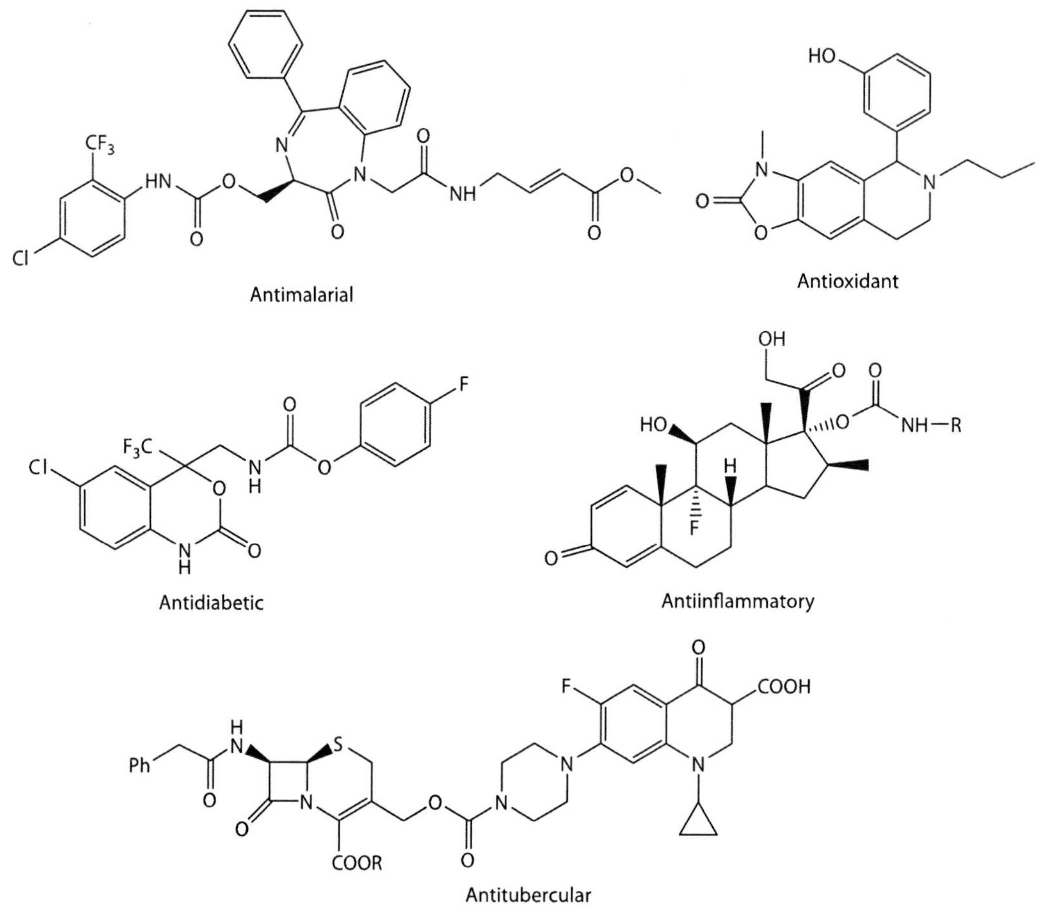

FIGURE 5.4 Biologically potent carbamates of natural/synthetic molecules.

D-valine amide side chain resulted in compound **4** (PPI-2458), demonstrating improved pharmacokinetic profile relative to the earlier clinical candidate TNP-470 and has advanced into phase I clinical studies in non-Hodgkin's lymphoma and solid cancers (Arico-Muendel et al. 2009) (Figure 5.7).

5.3.1.2 B-Ring Carbamate Analogues of (–) Rhazinilam

(–)-Rhazinilam (**5**), a natural compound isolated from *Apocynaceae* plants, whose tetracyclic structure possesses an axially chiral phenyl–pyrrole subunit bridged by a nine-membered lactam ring, was found to have unique antimitotic properties, with *in vitro* inhibition of both microtubule assembly and disassembly and the formation of abnormal tubulin spirals. As a consequence of these tubulin-binding properties, rhazinilam showed *in vitro* cytotoxicity toward various cancer cell lines in a low micromolar range; however, no activity was found *in vivo*. In order to search for more potent compounds, a series of carbamate analogues of B ring was synthesized. Biphenyl-carbamate analogue **6** (Figure 5.8) was found to be the most active analogue of rhazinilam so far, with a twofold activity on tubulin compared to rhazinilam itself—however, its *in vitro* cytotoxicity is still in the low micromolar range, very close to that of the natural compound (Boudoin et al. 2002). The role of the B-ring size of rhazinilam on the activity of carbamates was further demonstrated by synthesizing an 11-membered B-ring carbamate analogue of rhazinilam **7** and it was found that this compound did not show promising activity on tubulin (Décor et al. 2006). A series of phenylpyridine carbamate analogues **8a–c** was synthesized recently but none of the compounds showed promising activity on tubulin (Bonneau et al. 2007) (Figure 5.8).

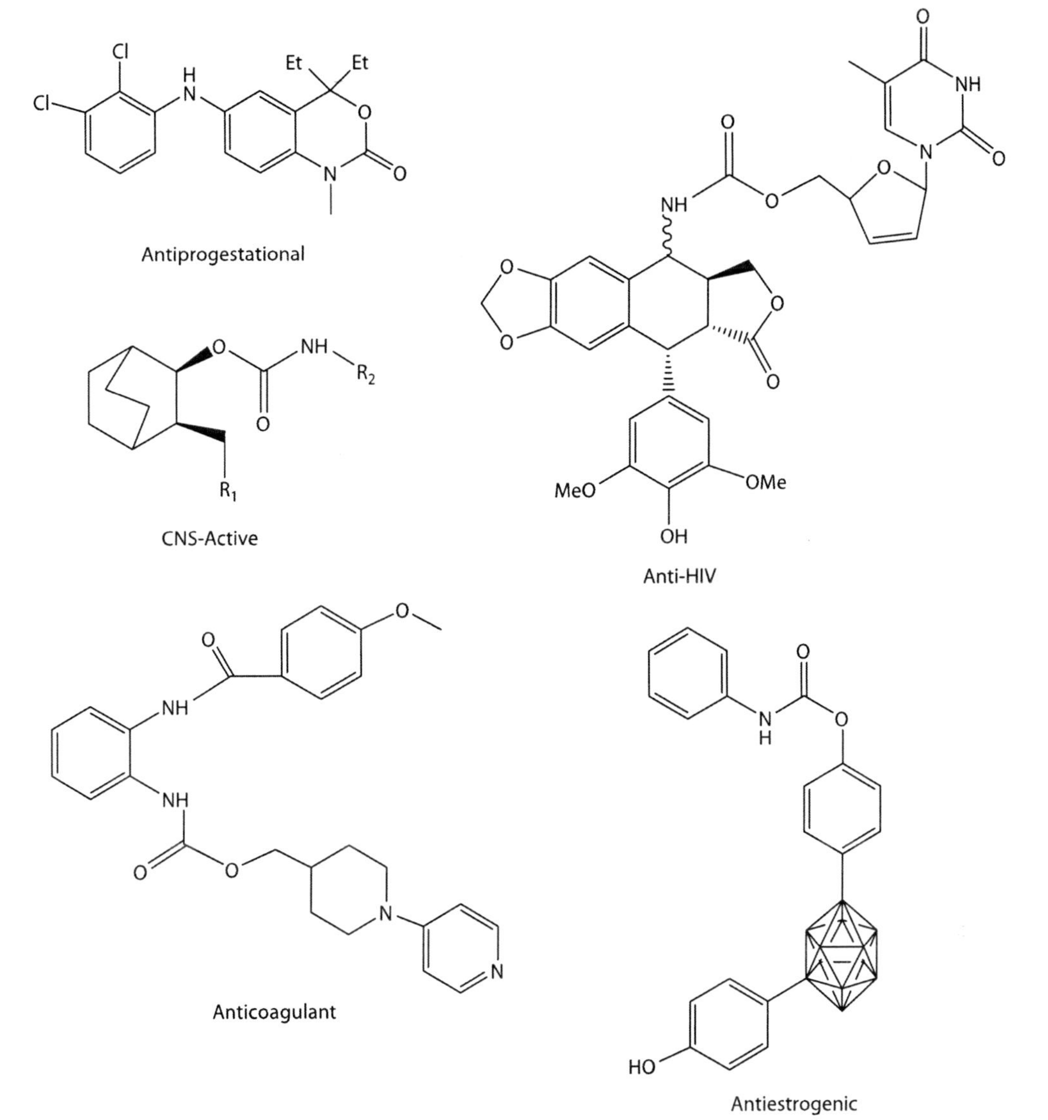

FIGURE 5.5 Biologically potent carbamates of natural/synthetic molecules.

5.3.1.3 Carbamates of Geldanamycin

Geldanamycin (**9**), a naturally occurring benzoquinone ansamycin isolated from culture broth of *Streptomyces hygroscopicus*, binds very tightly to the *N*-terminal ATPase domain of rHsp90 (Kd = 1.2 μM). Because geldanamycin is too chemically and metabolically unstable to become a drug, it has been derivatized, mainly on the 17-position of the quinone moiety, leading to a plethora of 17-alkylaminogeldanamycins including 17-allylamino-17-demethoxygeldanamycin **10** (17-AAG) (Hadden et al. 2006). This latter derivative is a potent inhibitor of Hsp90, and largely as a result of its excellent *in vitro* potency, the National Cancer Institute (NCI) has initiated phase I clinical trials in advanced cancer patients. Although, 17-AAG is a very potent Hsp90 inhibitor, it also suffers from pharmaceutic deficiencies including difficult formulation challenges. Further efforts are being made in order to improve the pharmacokinetic and pharmacodynamic profile of the drug. In this approach, *tert*-butyl carbamates of the 17-amino group of the geldanamycin, that is, **10a** and **10b**, have also been synthesized, but both of the compounds have shown poor activity compared to the

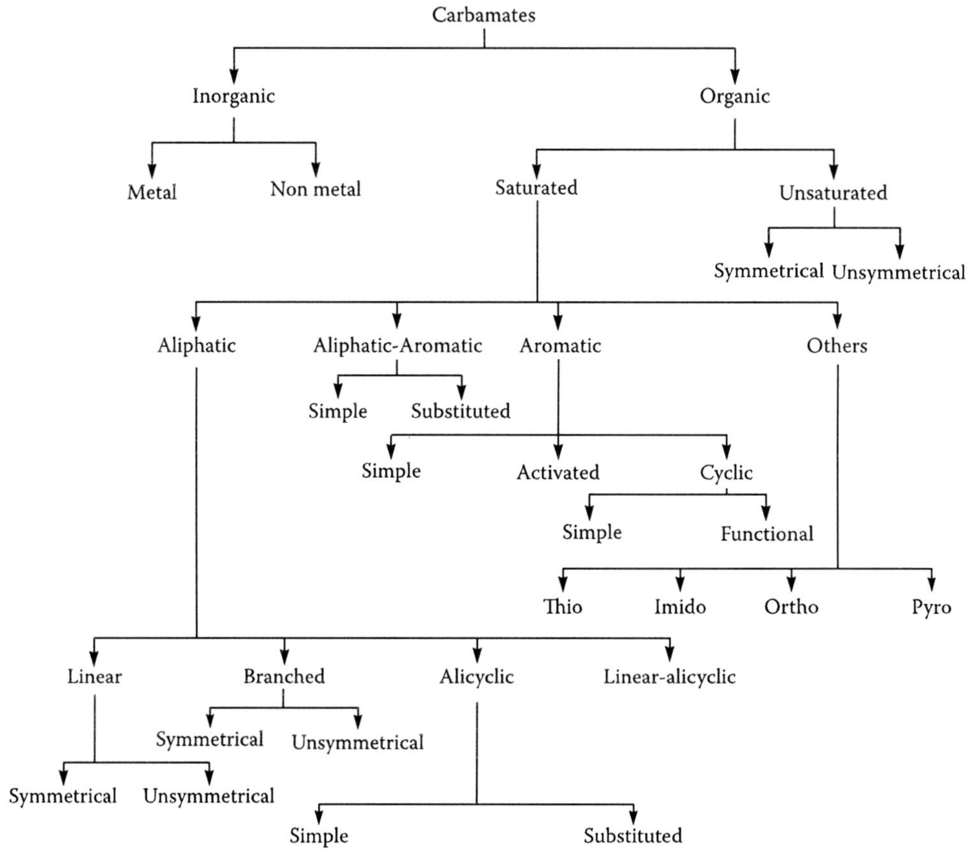

FIGURE 5.6 Classification of carbamates.

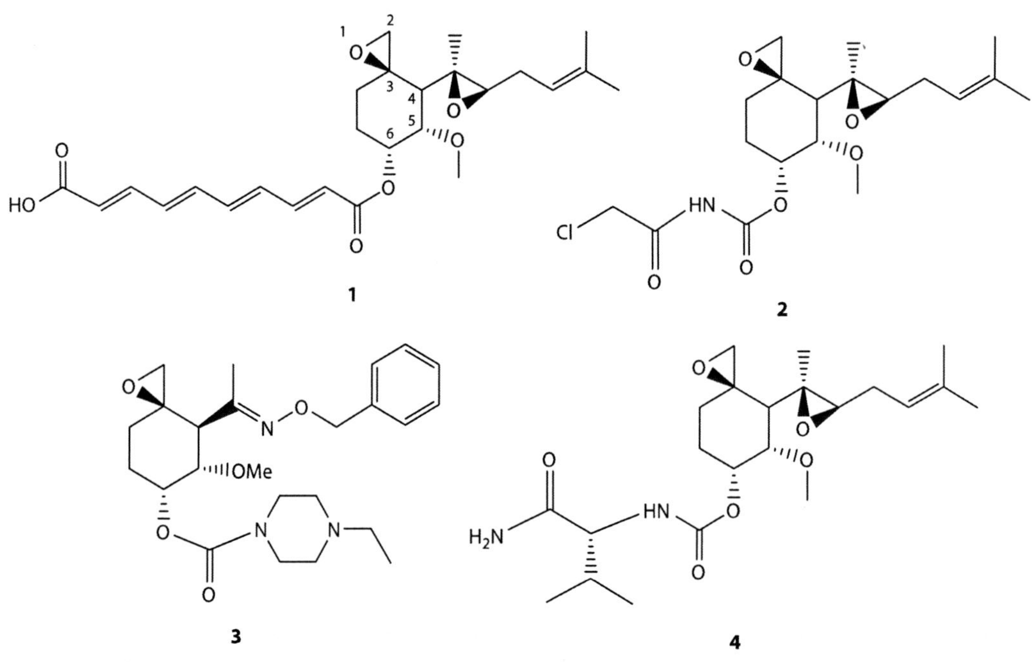

FIGURE 5.7 Carbamates of fumagillin analogues.

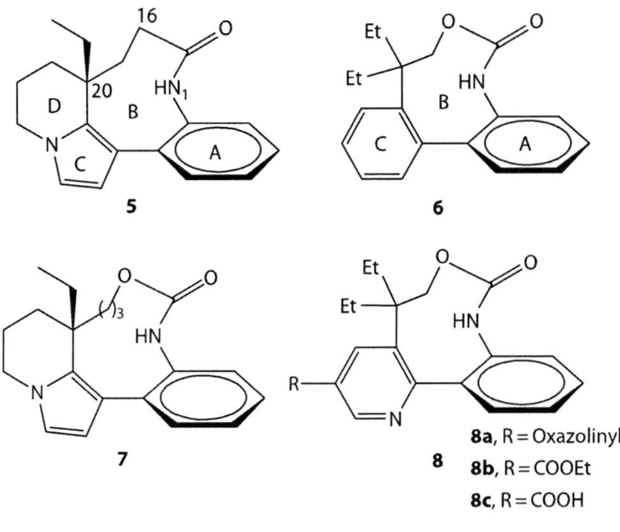

FIGURE 5.8 Carbamates analogues of rhazinilam.

17-amino derivatives (Figure 5.9) (Rastelli et al. 2005). A series of carbamates **11** (**11a–d**) of the 17-amino group of the geldanamycin have been synthesized and it was found that compound **11b** bearing *p*-acetoxybenzyl functionality was the most active due to release of 17-AG in the cell *via* hydrolysis followed by 1,6-elimination of quinine methide moiety (Brazidec et al. 2004).

5.3.1.4 Carbamates of Vitamin D_3

1α,25-Dihydroxyvitamin D_3 [1α,25-$(OH)_2$-D_3, **12**, Figure 5.10], the most active metabolite of vitamin D_3 **13**, plays a major role in many biological processes including calcium–phosphorus homeostasis, cell differentiation and proliferation, and immune reactions. However, the mechanisms for these differential actions have not been clearly defined. In the last two decades, various analogues of 1α,25-$(OH)_2$-D_3 have been developed to improve the biological profile of the natural hormone for a potential therapeutic application. Some of these derivatives have similar or more potent antiproliferative potential, yet reduced hypercalcemic actions, than the natural hormone. An increasing number of synthetic vitamin D derivatives are currently in use as drugs for treatment of various human diseases and new candidates are in human clinical trials (Posner and Kahraman 2003). In recent years, keeping in view the anticancer activity of vitamin D_3 analogues, some of the C-3 carbamate/bis-carbamate derivatives of compound **13** have been synthesized. The most potent analogue to inhibit the cell proliferation of MCF-7 cells or keratinocytes or to stimulate the HL 60 cell differentiation is the analogue 3-*O*-carbamoyl-1α,25$(OH)_2$-D_3 (**14a**); however, this compound is still less potent than the parent compound **13** but has lower calcemic effects *in vivo* (Fernandes et al. 2004). Novel A-ring homodimeric C-3 carbamate analogues **15a–c** have also been synthesized but none of them have shown promising anticancer activity compared to compound **13** (Oves et al. 2006).

5.3.1.5 Carbamates of Podophyllotoxin

Podophyllotoxin (PPT) **16**, one of the well known, naturally occurring aryl tetralin lignans, is extracted as the main component from the root and rhizomes of *Podophyllum* species such as *Podophyllum hexadrum* and *Podophyllum peltatum* whose medicinal properties such as cathartic, antirheumatic, antiviral, and antitumor activity have been well recognized for centuries (Xu et al. 2009). But attempts to use PPT in the treatment of human neoplasia were mostly unsuccessful due to the complicated side effects such as nausea, vomiting, and damage of normal tissues, etc. Therefore, PPT has been used as lead compound for drug design to obtain more potent and less toxic anticancer agents. Extensive structural modifications of podophyllotoxin have been undertaken which culminated in the clinical introduction of some of the most potent PPT analogues. In order

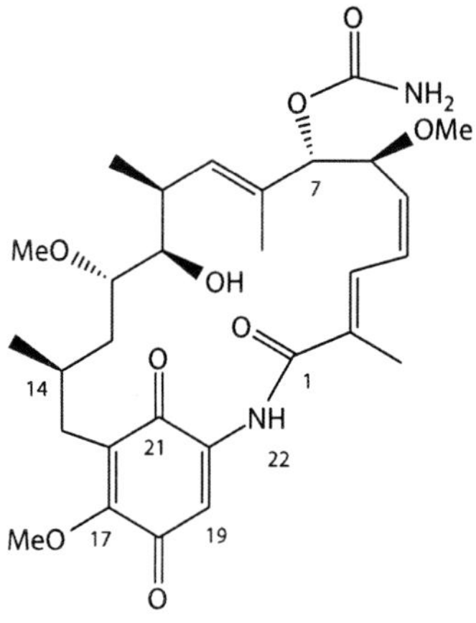

9

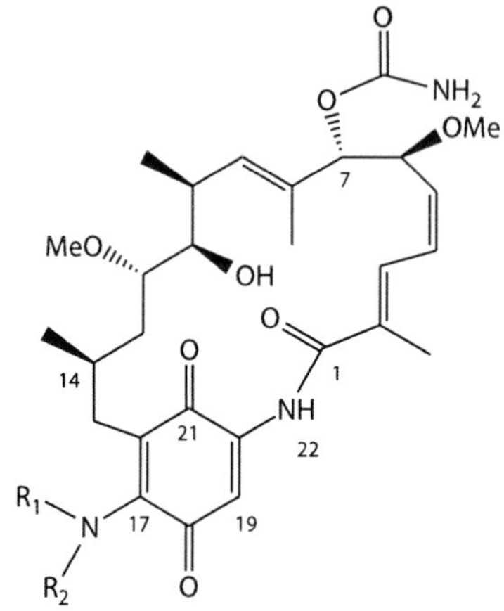

10

10a, R_1 and R_2 =

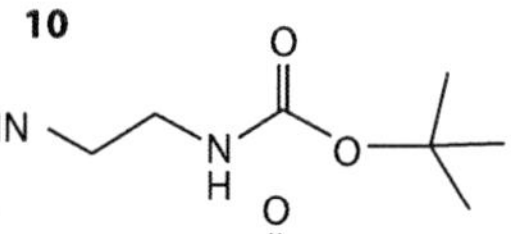

10b, R_1 and R_2 =

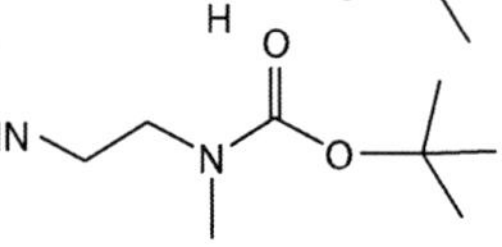

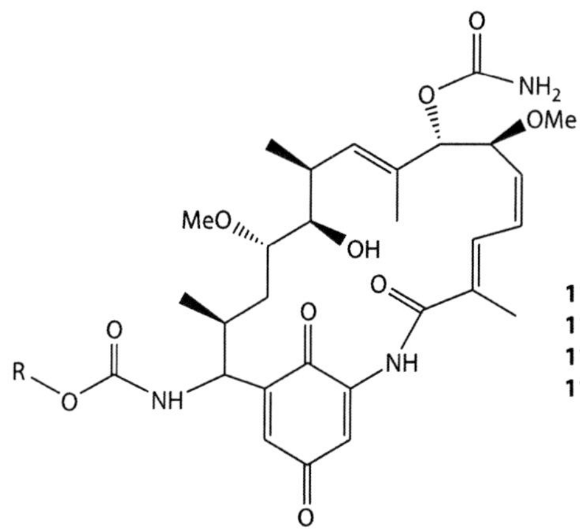

11

11a, R = ethyl
11b, R = p-acetoxybenzyl
11c, R = isobutyl
11d, R = 4-(*N*-morpholino)butyl

FIGURE 5.9 Carbamates of geldanamycin.

search for more potent and less toxic PPT analogues, efforts have been directed to synthesize some of the carbamate derivatives. A series of 4′-demethylepipodophyllotoxin 4-aminoalkylcarbamate **17** analogues and a second series of C-4 carbamate analogues **18** having lactone ring carbon 13–11 (called retrolactone) have been synthesized and evaluated (Figure 5.11). Compound **17a** (IC_{50} = 0.038 μM) and **18a** (IC_{50} = 0.088 μM) displayed potent cytotoxicity against the L1210 cell line (10- to 20-fold higher than the anticancer drug etopside) and proved to be strong topoisomerase II poisons more potent than etopside (IC_{50} = 0.83 μM). From preliminary *in vivo* investigation of both the compounds (**17a** and **18a**) against P388 leukemia and orthotopically grafted human A549 lung carcinoma, it appeared that **17a** and **18a** constitute promising leads for a new class of antitumor agents (Duca et al. 2005b). Furthermore, a simple series of 4β-amino-4′-*O*-desoxypodophyllotoxin **19** (Figure 5.11) have also been synthesized and their effect on human DNA-topoisomerase II and antiproliferative activity was evaluated. Compounds **19a–c**, **19g**, **19j**, and **19k** (IC_{50} = 0.14–18 μM) are topoisomerase II poisons that induce double-stranded breaks in DNA and exhibit increased cytotoxicity compared to etopside (Duca et al. 2005a).

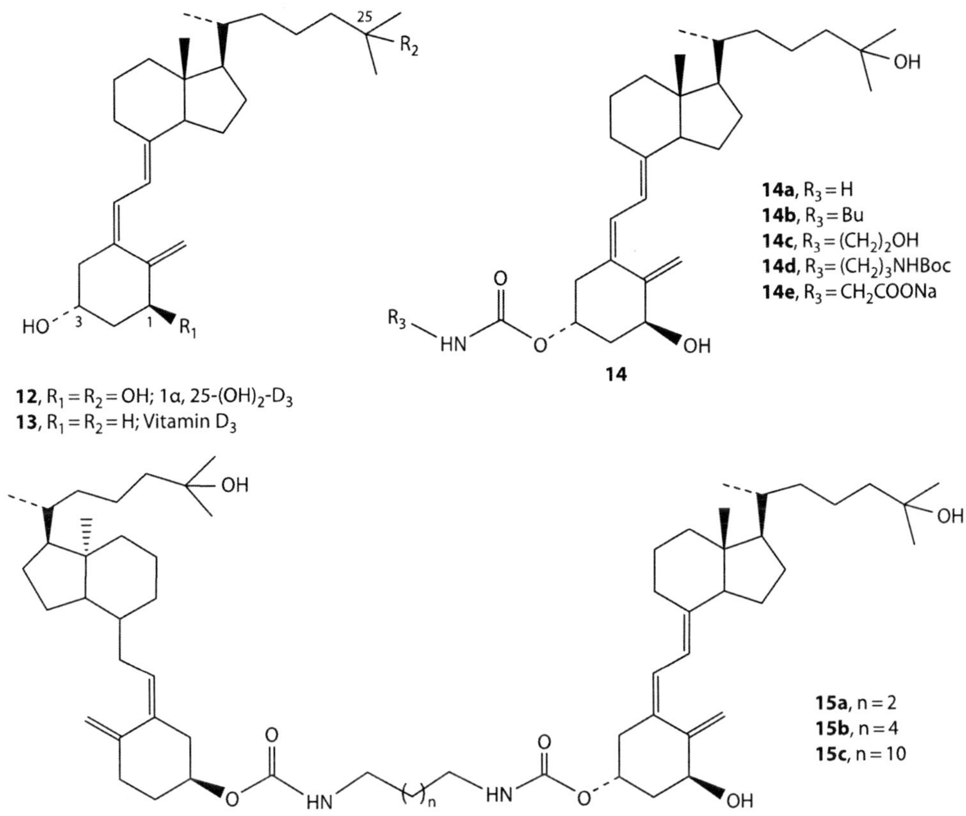

12, $R_1 = R_2 = OH$; 1α, 25-$(OH)_2$-D_3
13, $R_1 = R_2 = H$; Vitamin D_3

14

14a, $R_3 = H$
14b, $R_3 = Bu$
14c, $R_3 = (CH_2)_2OH$
14d, $R_3 = (CH_2)_3NHBoc$
14e, $R_3 = CH_2COONa$

15

FIGURE 5.10 Carbamates of vitamin D_3 analogues.

Recently, a series of spin-labeled derivatives of deoxypodophyllotoxin (DPPT) has been synthesized (Figure 5.12) and their *in vitro* cytotoxic activities against three tumor cell lines (HL-60, RPMI-8226, A-549) were evaluated. The cytotoxic studies indicated that compounds **20a–h** were more potent (IC_{50} = 0.0087–0.11 μM) than the parent drug etopside and DPPT (Zhang et al. 2010).

5.3.1.6 Carbamates of Butelin and Butelinic Acid

Betulin **21** is a main component of birch bark and can be synthetically converted to betulinic acid **22** in a two-step procedure, in high yield (Figure 5.13). Both the compounds **21** and **22** are lupane-type triterpenene, and were reported to display several biological effects including anti-inflammatory, antiviral, antimalarial, and in particular anticancer. Previous reports revealed that compound **22** is a melanoma-specific cytotoxic agent; however, recent evidence has indicated that **22** possesses a broader spectrum of cytotoxic activity against other cancer types (Yogeeswari and Sriram 2005). Moreover, compound **22** has been suggested to induce apoptosis *via* the activation of caspases, regardless of cellular p53 gene status and CD95 activation. This apoptosis-inducing ability, the apparent lack of toxicity on normal cells, and the favorable therapeutic index have made **22** an attractive and a very promising anticancer agent (Mukherjee et al. 2006).

In recent years, in search for more potent and less toxic compounds from butelin **21** and butelinic acid **22**, and keeping in view the importance of carbamates in drug discovery research, researchers have directed their efforts to synthesize their carbamate derivatives (Figure 5.3). A series of C-3 imidazole carbamates have been synthesized and evaluated for *in vitro* cytotoxicity activity

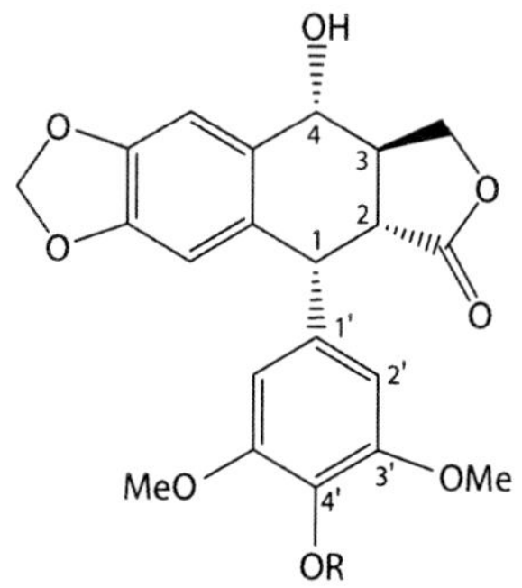

16

R = H, Podophyllotoxin

R = Me, Epipodophyllotoxin

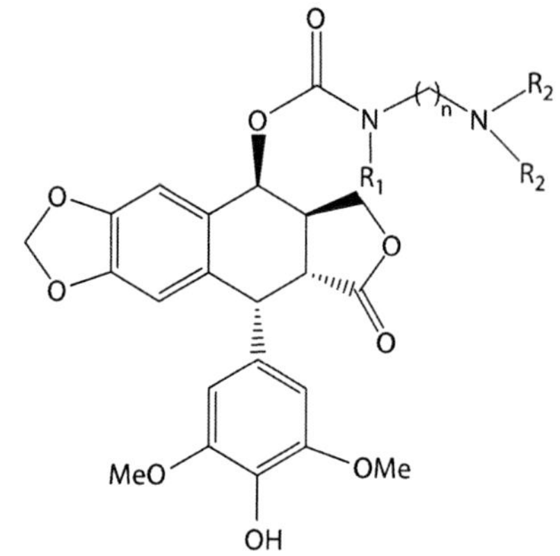

17

17a, $R_1 = H$, $R_2 = Me$, n = 2

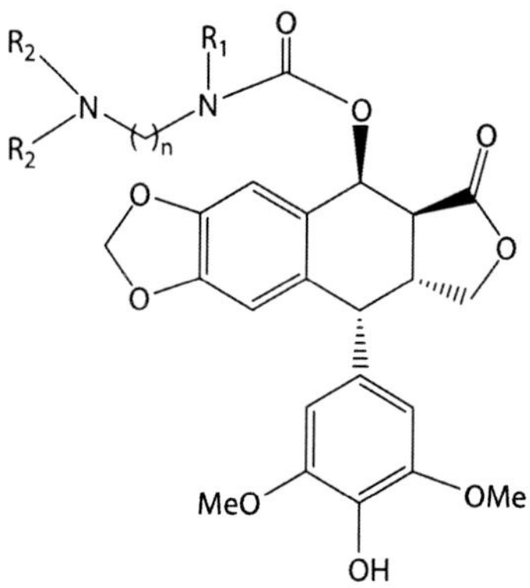

18

18a, $R_1 = H$, $R_2 = CH_3$, n = 2

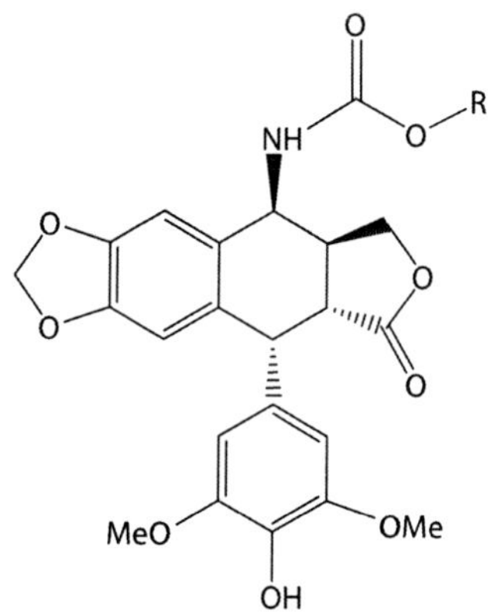

19

19a, $R = CH_3$; **19b**, $R = CH_2CH_3$; **19c**, $R = (CH_2)_3CH_3$,
19d, $R = (CH_2)_3Cl$; **19e**, $R = CH_2CH{=}CH_2$; **19f**, $R = CH_2C{=}CH_2$,
19g, $R = (CH_2)_2OCH_2$; **19h**, R = Benzyl; **19i**, R = 4-F-benzyl

FIGURE 5.11 Carbamates of podophyllotoxin.

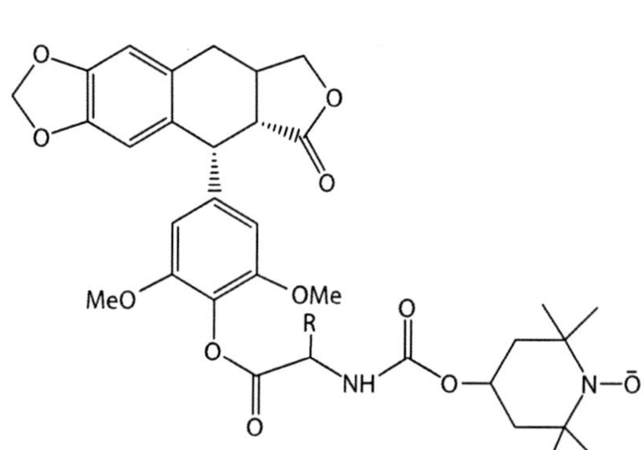

20a–h

20a, R = H;	**20b**, R = Me
20c, $R = CHMe_2$	**20d**, $R = CH_2CHMe_2$
20e, $R = CHMeCH_2Me$,	**20f**, R = Proline
20g, $R = CH_2Ph$,	**20h**, $R = CH_2C_6H_4OH$

FIGURE 5.12 Carbamates of podophyllotoxin analogues.

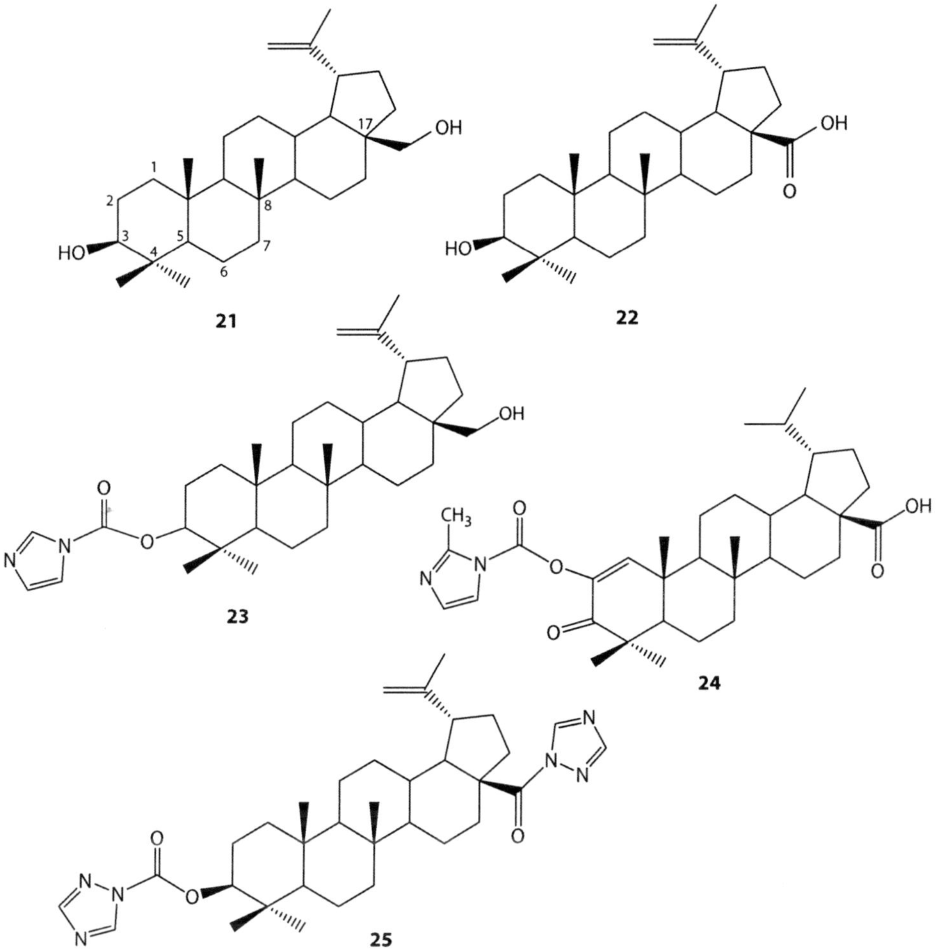

FIGURE 5.13 Carbamates of butelin and butelinic acid.

against human cancer cell lines such as $HepG_2$, Jurkat, and HeLa. Of the whole series synthesized and evaluated, a C-3 imidazole carbamate compound of butelin, that is, **23** (IC_{50} = 2 μM in $HepG_2$ cells), has shown promising activity than butelinic acid **22** (IC_{50} = 36.4 μM) (Santos et al. 2009). Recently, another series of C-2 and C-3 carbamate derivatives along with the *N*-acylheterocyclic ring at C-17 of butelinic acid **22** have been synthesized and evaluated for their *in vitro* cytotoxic activity (Santos et al. 2010). Most of the compounds of this series have shown better cytotoxicity profile than butelinic acid. Two of the imidazole and triazole carbamate compounds **24** (IC_{50} = 1.1 μM) and **25** (IC_{50} = 1.8 μM) were found most promising, being up to 12-fold more potent than betulinic acid **22** (IC_{50} = 21.5 μM) against human PC-3 cell lines.

5.3.1.7 Carbamates of Taxol

Paclitaxel **26** (Taxol, Figure 5.14), a polyoxygenated, naturally occurring diterpene alkaloid, was first isolated by Wani et al. (1971) from the bark of *Taxus brevifolia Nutt.*, and is considered as one of the most important and promising anticancer drug mainly used for the treatment of breast and ovarian cancers. Taxol has also been used for the treatment of skin, lung, head, and neck carcinomas (Fu et al. 2009). In order to search for more potent compounds, several researchers around the world have made significant efforts in the generation of structurally diverse semisynthetic analogues of the

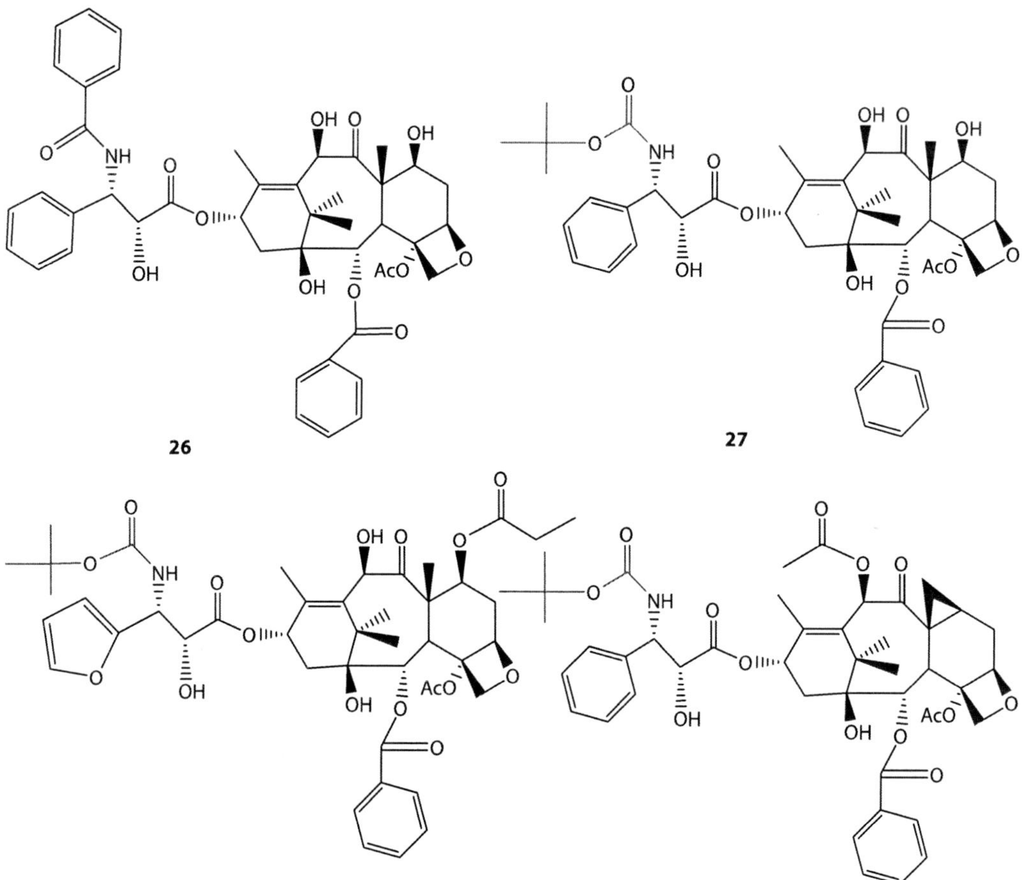

26 27 28 29

FIGURE 5.14 Carbamates analogues of taxol.

taxol. In this connection, various kinds of carbamate analogues of taxol have also been synthesized, such as docetaxel **27**, TL-00139 **28**, and RPR-109881A **29**, etc. (Figure 5.14) (Breen and Walsh 2010). Docetaxel **27**, formed through the replacement of the benzamide group to *tert*- butyl carbamate in taxol, shows a similar spectrum of action compared to taxol but with a fourfold increase in potency and improved water solubility. However, the improved water solubility of docetaxel reduces the complexity and side effects of administration in comparison to taxol; clinical trials to establish the scope of this drug are currently in progress and docetaxel has already been licensed for the treatment of breast and ovarian cancers. Some of the other potential carbamate derivatives of the taxol are **28**, **29**, etc.

5.3.1.8 Carbamates of Staurosporine Derivatives

Staurosporine derivatives are indanocarbazoles alkaloids natural products **30** bearing a sugar moiety. Recently, a series of carbamate derivatives of the staurosporine derivatives **30** have been synthesized using an intermediate obtained from aglycone isostere alcohol **31** where one of the indole nitrogen atom was replaced by carbon and has been sequentially functionalized to generate carbamate compounds and their kinase-inhibiting activity in various cancer cells such as PKC, Trka, FGFR, VEGFR, and IRK CDK_5 (Tripathi et al. 2008) has been tested. One of carbamate compounds **32** (IC_{50} = 8 nM) was evaluated to be the most potent in Trka kinase-inhibiting activity, and majority of compounds have been shown to be 4- to 20-fold more potent than **30** and **31**, respectively (Figure 5.15). Compound **32** has been picked up for further studies.

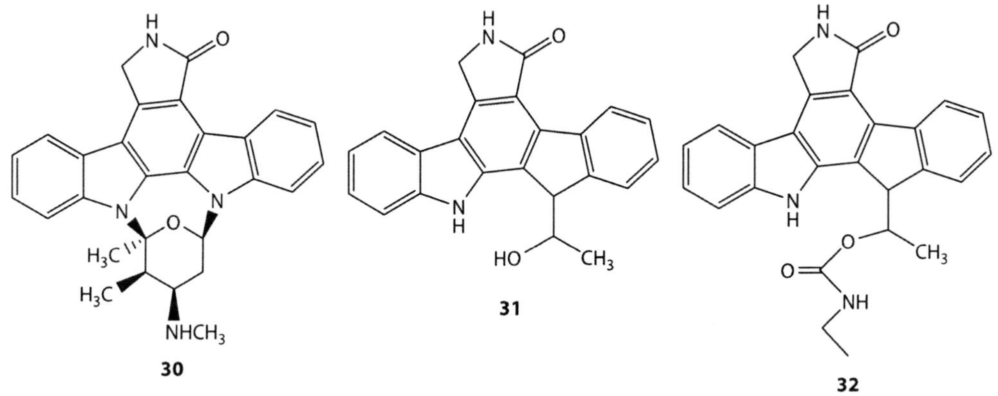

FIGURE 5.15 Carbamates of staurosporine derivatives.

5.3.1.9 Carbamates of Mitomycin C

A series of bifunctional DNA interstrand cross-linking agent bis-(carbamate)-8*H*-3a-azacyclopent[a] indene-1-yl derivatives **34a–k** (Figure 5.16) were synthesized keeping the basis of the structurally similar, known antibiotic anticancer drug mitomycin C (MMC) **33**, and were evaluated for their antitumor activity (Kakadiya et al. 2009). The preliminary antitumor studies reveal that these compounds exhibited potent cytotoxicity *in vitro* and antitumor therapeutic efficacy against human tumor xenografts *in vivo*. Furthermore, these compounds have little or no cross resistance to anticancer drugs, either taxol or vinblastine. Remarkably, complete tumor remission in nude mice bearing human breast carcinoma MX-1 xenograft by **34g**, **34h** was achieved at maximum tolerance dose with relatively low toxicity. In addition, they can induce DNA interstand cross-linking and substantial G2/M-phase arrest in human, non-small lung carcinoma H1299 cells and are promising candidates for preclinical studies.

5.3.2 Carbamates of Synthetic Molecules

In continuation of the search for potential anticancer carbamates, a series of a novel class of pyrrolidinyl acetylenic thieno-[2,3-*d*]-pyrimydines **35** (Figure 5.17) has been identified that are potent and selective inhibitors of both EGFR/ErbB-2 receptor tyrosine kinases (Hubbard et al. 2008). These compounds are found to display a range of enzyme and cellular potency and also to display a varying level of covalent modification of the kinases targets. It was found from the various incorporations on the pyrrolidine ring of compound **35** that incorporations of carbamates have shown

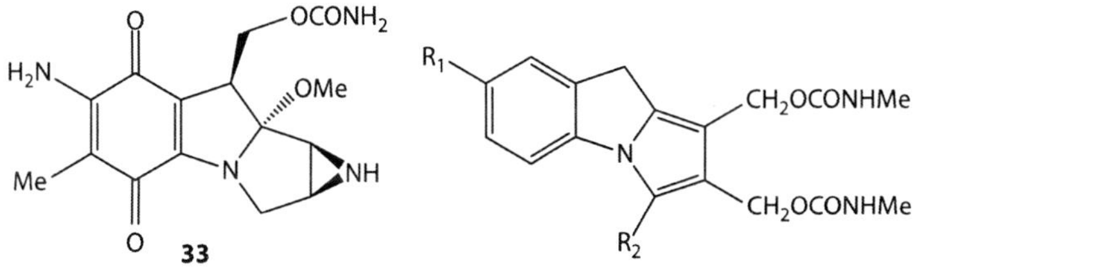

34a, R_1 = H, R_2 = OMe; **34b**, R_1 = H, R_2 = OMe
34c, R_1 = H, R_2 = Ph; **34d**, R_1 = H, R_2 = 4-F-Ph
34e, R_1 = H, R_2 = 4-Cl-Ph **34f**, R_1 = H, R_2 = 3,4-diF-Ph
34g, R_1 = H, R_2 = 4-OMePh **34h**, R_1 = H, R_2 = 2-OMePh
34i, R_1 = H, R_2 = 3,4-MeOPh **34j**, R_1 = H, R_2 = 2,6-OMePh
34k, R_1 = OMe, R_2 = 4-OMePh

FIGURE 5.16 Carbamates of mitomycin C.

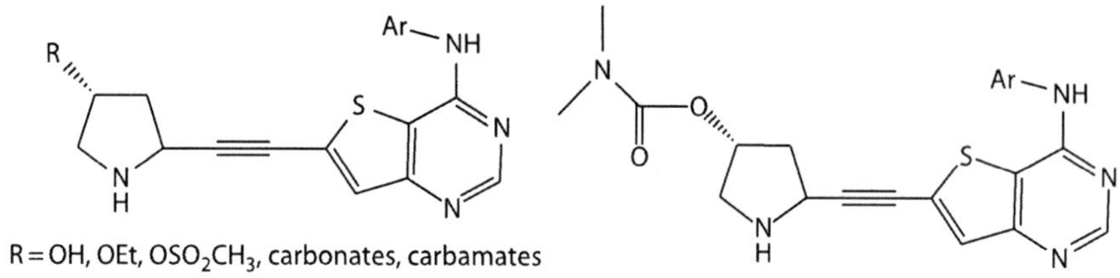

35 36

FIGURE 5.17 Carbamates of pyrrolidinyl acetylenic thieno-[2,3-*d*]-pyrimydines.

potential activity, wherein dimethylaminocarbamate compound **36** was found to be potent in enzymatic and cellular assays and also showed good oral exposure to mouse.

Recently, a series of 4-[1*H*-indazol-5-ylamino]pyrrolo[2,1-*f*] [1,2,4]triazine-6-carbamates **37** have been synthesized (Figure 5.18) by Bristol–Mayers Squibb Research Lab., and were evaluated for their human epidermal growth receptor $(HER)_1/HER_2$ kinase inhibitor activity (Gavai et al. 2009). It was found through structure–activity relationship studies that carbamate compound **38** (BMS-599626) has shown potential activity in HER_1 and HER_2 kinases, which was orally efficacious in human HER_1 (GEO, L2987) or HER_2 (KPL_4, N87)-dependent tumor xenograft models at multiple dose levels. On the basis of its favorable *in vitro* pharmacology, broad spectrum *in vivo* efficacy in multiple tumor models, and a satisfactory pharmacokinetic profile, **38** was utilized in clinical trials.

Recently, two series of phenyl-*N*-mustard 9-anilinoacridine conjugates (**39a–k**, **40a–c**) bearing a AHMA-alkyl carbamate linkers have been synthesized and their *in vitro* studies revealed that these derivatives possess significant cytotoxicity, with IC_{50} in the submicromolar range, in inhibiting human lymphoblastic leukemia (CCRF-CEM), breast carcinoma (MX-1), colon carcinoma (HCT-116), and human non-small cell lung cancer (H1299) cell growth (Kapuria et al. 2009). Structure–activity relationship (SAR) studies show that compounds **39a–k** are more potent than **40a–c** (Figure 5.19). The antitumor therapeutic efficacy studies against human tumor xenografts demonstrated that the newly synthesized carbamate conjugates exhibited potent antitumor efficacy against breast carcinoma MX-1 and colon carcinoma HCT-16 xenografts. Compounds **39a**, **39b**, **39e**, and **39i** were

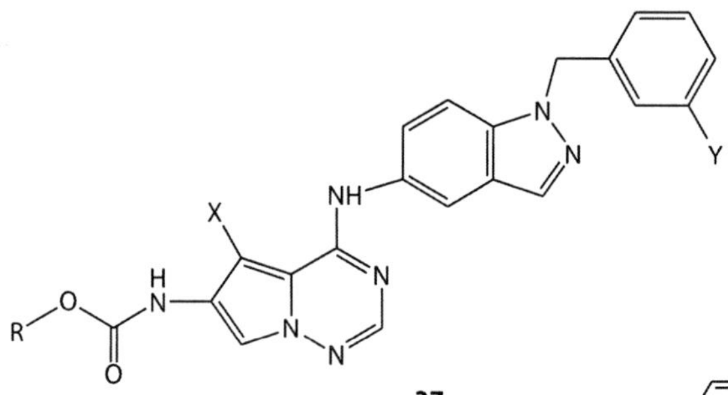

37

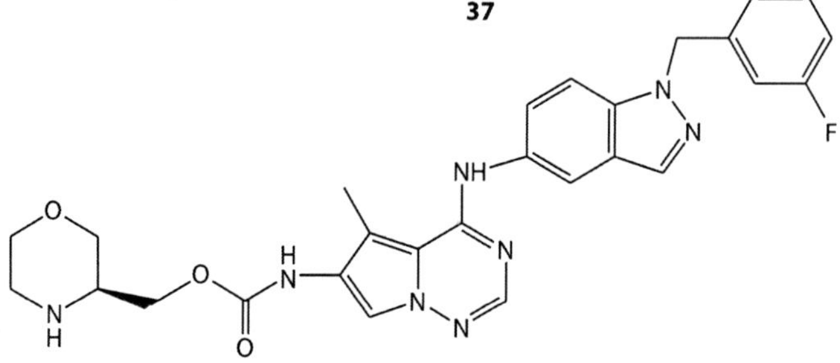

38

FIGURE 5.18 Carbamates of 4-[1*H*-indazol-5-ylamino]pyrrolo[2,1-*f*] [1,2,4]-triazine-6.

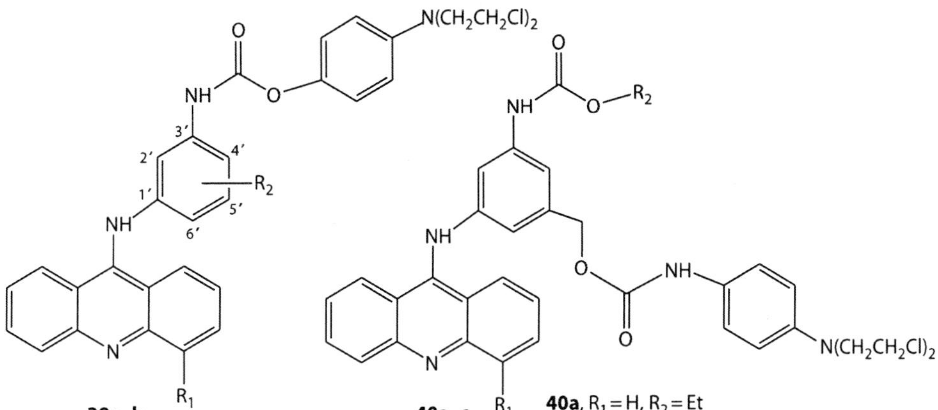

39a–k 40a–c

40a, $R_1 = H$, $R_2 = Et$
40b, $R_1 = H$, $R_2 =$ *iso*-Pro
40c, $R_1 = Me$, $R_2 =$ *tert*-Bu

39a, $R_1 = H$, $R_2 = 5'\text{-}CH_2OH$
39b, $R_1 = Me$, $R_2 = 5'\text{-}CH_2OH$
39c, $R_1 = Me$, $R_2 = 2'\text{-}Me$
39d, $R_1 = H$, $R_2 = 4'\text{-}Me$
39e, $R_1 = Me$, $R_2 = 4'\text{-}Me$
39f, $R_1 = Me$, $R_2 = 6'\text{-}Me$
39g, $R_1 = H$, $R_2 = 4'\text{-}OMe$
39h, $R_1 = Me$, $R_2 = 4'\text{-}OMe$
39i, $R_1 = H$, $R_2 = 5'\text{-}OMe$
39j, $R_1 = H$, $R_2 = 6\text{-}OMe$
39k, $R_1 = Me$, $R_2 = 6\text{-}OMe$

FIGURE 5.19 Carbamates of phenyl-*N*-mustard 9-anilinoacridine conjugates.

selected for evaluating their antitumor activity in nude mice bearing MX-1 and HCT-116 xenografts. Interestingly, no tumor relapse was found in mice treated with **39a**. Interestingly, no tumor relapse was found in mice treated with **39a** over 129 days. This compound is capable of inducing DNA interstand cross-linking in human non-small lung cancer cell H1299 in a dose-dependent manner by modified comet assay and has a long half-life in rat plasma.

Carbamate analogues of 1-*O*-hexadecyl-sn-3-glycerophophocholine compound **41–44** were synthesized and evaluated for their antiproliferative activity against cancer cells derived from a variety of tissues (Figure 5.20) (Byun et al. 2010). Although, all of the compounds are antiproliferative, surprisingly, the carbamates **41** and **42** are more effective against the hormone-independent cell lines DU145 and PC3 than toward other cancer cell lines. This selectivity was not observed with the dicarbamate compounds **43** and **44**. Cell death induced by compound **42** appeared to be mediated by apoptosis, as assessed by caspase activation and loss of mitochondrial membrane potential.

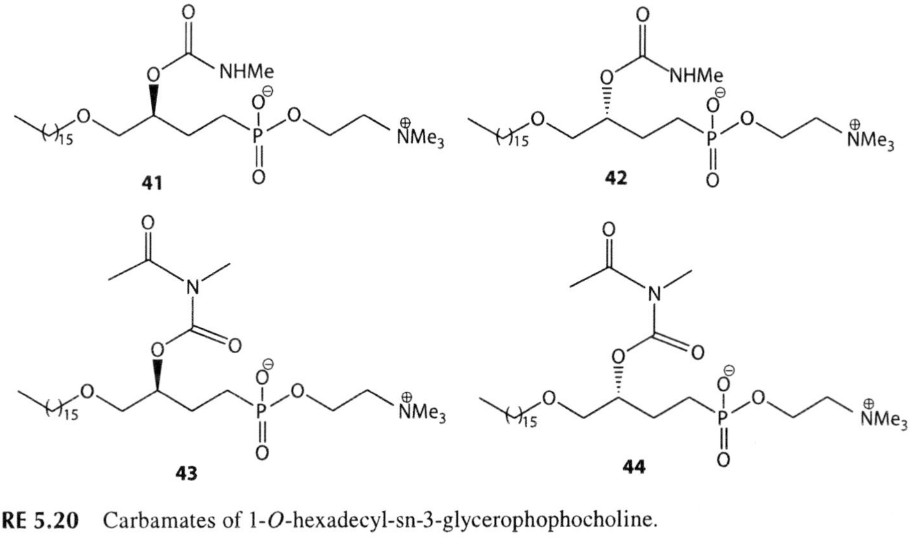

41 42 43 44

FIGURE 5.20 Carbamates of 1-*O*-hexadecyl-sn-3-glycerophophocholine.

The *in vivo* activity of compound **42** was evaluated in a murine prostate cancer xenograft model. Oral and intravenous administration showed that compound **42** is more effective in inhibiting the growth of PC3 tumors in Rag2M mice. Thus, this study shows that carbamate compounds **41–44** are a novel class of prostate cancer selective cytotoxic agents.

5.4 CONCLUSIONS

An effort toward the role of incorporation of carbamate functionality into the various kinds of structurally diverse biologically active anticancer natural/semisynthetic/synthetic molecules has been demonstrated since the year 2005 onward. Among the anticancer natural products, a comprehensive discussion on the incorporation of carbamate functionality into some of the biologically potent natural products such as fumagillin, rhazinilam, geldenamycin, vitamin D_3, podophyllotoxin, butelin, butelinic acid, taxol, staurosporine, mitomycin C, and several synthetically modified carbamate analogues have been discussed. Several of the natural products bearing carbamate residue have shown promising anticancer activity such as compounds **2**, **3**, **4**, **6**, **17a**, **18a**, **19a–c**, **19g**, **19g**, **19k**, **20a–h**, **24**, **25**, **27**, **28**, **29**, **32**, **34g**, **34h**, and are in the various phases of clinical trials. Among synthetic carbamate analogues of anticancer molecules, compounds **36**, **38**, **39a**, **39b**, **39e**, **39i**, **41**, and **42** are in the various phases of the clinical trials. It is greatly anticipated that the present review will serve as a first-line reference to the organic/medicinal chemists working on the same lines of the relevant subject and would inevitably boost the ongoing developments in drug discovery research.

ACKNOWLEDGMENTS

The author is thankful to Director General, Amity University Uttar Pradesh (AUUP), Lucknow Campus, Lucknow, for providing the necessary facilities during the preparation of this chapter and for his constant encouragement for research. The author is also grateful to Prof. G. Brahmachari, Editor of this book, for his invitation and fruitful suggestions. The author is thankful to Suman K. Sen of IIT Kharagpur, India, for providing several references during the preparation of this chapter.

ABBREVIATIONS

17-AAG	17-allylamino-17-demethoxygeldanamycin
BMS	Bristol–Myers Squibb
CNS	central nervous system
CVS	central vascular system
DPPT	deoxypodophyllotoxin
MetAP-2	methionine aminopeptidase-2
MMC	mitomycin C
NCI	National Cancer Institute
PPT	podophyllotoxin
SAR	structure–activity relationship

REFERENCES

Adams, P. and Baron, F. A. 1965. Esters of carbamic acid. *Chem. Rev.* 65: 567–602.

Ager, A. J., Prakash, I., and Schaad, D. R. 1996. 1,2-Amino alcohols and their heterocyclic derivatives as chiral auxiliaries in asymmetric synthesis. *Chem. Rev.* 96: 835–876.

Agoston, E. S., Hatcher, M. A., Kensler, T. W., and Posner, G. H. 2006. Vitamin D_3 analogues as anti-carcinogenic agents. *Anti-Cancer Agents Med. Chem.* 6: 53–71.

Alaxander, J. P. and Cravatt, B. F. 2005. Mechanism of carbamate inactivation of FAAH: Implications for the design of covalent inhibitors and *in-vivo* functional probes for enzymes. *Chem. Biol.* 12: 1179–1187.

Ali, A., Balkovec, J. M., Greenlee, M. et al. 2008. Discovery of betamethasone 17α-carbamates as dissociated glucocorticoid receptor modulators in the rat. *Bioorg. Med. Chem.* 16: 7535–7542.

Andez, J. C. 2009. Mitomycins syntheses: A recent update. *Beilstein J. Org. Chem.* 5(33): 1–36 available online. (mdoi:10.3762/bjoc.5.33).

Aoki, S., Kawatani, H., Goto, T., Kimura, E., and Shiro, M. 2001. A double-functionalized cyclen with carbamoyl and dansyl groups (cyclen = 1,4,7,10-tetraazacyclododecane): A selective fluorescent probe for Y^{3+} and La^{3+}. *J. Am. Chem. Soc.* 123: 1123–1132.

Arico-Muendel, C. C., Benjamin, D. R., Caiazzo, T. M. et al. 2009. Carbamates analogues of fumagillin as potent, targeted inhibitors of methionine aminopeptidase-2. *J. Med. Chem.* 52: 8047–8056.

Arya, P. and Qin, H. 2000. Advances in asymmetric enolate methodology. *Tetrahedron* 56: 917–947.

Asaka, T., Manaka, A., and Sugiyama, H. 2003. Recent development in macrolide antimicrobial research. *Curr. Top Med. Chem.* 3: 961–989.

Ashoorzadeh, A., Atwell, G. J., Pruijn, F. B. et al. 2011. The effect of sulfonate leaving groups on the hypoxia-selective toxicity of nitro-analogs of the duocarmycins. *Bioorg. Med. Chem.* 19: 4851–4860.

Becker, J., Grimme, S., Frohlich, R., and Hoppe, D. 2007. Estimation of the kinetic acidity from substrate conformation—Stereochemical course of the deprotonation of cyclohexenyl carbamates. *Angew. Chem. Int. Ed.* 46: 1645–1649.

Bergeron, R. J., Bharti, N., Singh, S., McManis, J. S., Wiegand, J., and Green, L. G. 2009. Vibriobactin antibodies: A vaccine strategy. *J. Med. Chem.* 52: 3801–3813.

Billich, A., Vyplel, H., Grassberger, M., Schmook, F. P., Steck, A., and Stuetz, A. 2005. Novel cyclosporin derivatives featuring enhanced skin penetration despite increased molecular weight. *Bioorg. Med. Chem.* 13: 3157–3167.

Bonneau, A.-L., Robert, N., Hoarau, C., Boidoin, O., and Marsais, F. 2007. A new synthetic approach to biaryl of rhazinilam type: Application to the synthesis of three novel phenylpyrdine carbamate analogues. *Org. Biomol. Chem.* 5: 175–183.

Borrel, C., Thoret, S., Cachet, X. et al. 2005. New antitubulin derivatives in the combretastatin A_4 series: Synthesis and biological evaluation. *Bioorg. Med. Chem.* 13: 3853–3864.

Boudoin, O., Claveau, F., Thoret, S., Herrbach, A., Guenard, D., and Gueritte, F. 2002. Synthesis and biological evaluation of A-ring biaryl carbamate analogues of rhazinilam. *Bioorg. Med. Chem.* 10: 3395–3400.

Bova, F., Ettari, R., Micale, N. et al. 2010. Constrained peptidomimetics as antiplasmodial falcipain-2 inhibitors. *Bioorg. Med. Chem.* 18: 4928–4938.

Boyle, P. H., Convery, M. A., Davis, A. P., Hosken, G. D., and Murray, B. A. 1992. Deprotonation of nitroalkanes by bicyclic amidine and guanidine bases: Evidence for molecular recognition within a catalytic cycle for C–C bond formation. *J. Chem. Soc. Chem. Commun.* 239–242.

Brazidec, J. Y. L., Kamal, A., Busch, D. et al. 2004. Synthesis and biological evaluation of a new class of geldanamycin derivatives as potent inhibitors of Hsp90. *J. Med. Chem.* 47: 3865–3873.

Breen, E. C. and Walsh, J. J. 2010. Tubulin-targeting agents in hybrid drugs. *Curr. Med. Chem.* 17: 609–639.

Buchstaller, H. P. 1998. Solid phase synthesis of oxazolidinones *via* a novel cyclization/cleavage reaction. *Tetrahedron* 54: 3465–3470.

Byun, H. P., Bittman, R., Samadder, P., and Arthur, G. 2010. Synthesis and antitumor activity of ether glycerolphospholipids bearing a carbamate moiety at the sn-2 position: Selective sensitivity against prostate cancer cell lines. *ChemMedChem* 5: 1045–1052.

Chackalamannil, S., Wang, Y., Greenlee, W. J. et al. 2008. Discovery of a novel, orally active himbacine-based thrombin receptor antagonist (SCH 530348) with potent antiplatelet activity. *J. Med. Chem.* 51: 3061–3064.

Chang, P. A., Wu, Y. J., Li, W., and Leng, X. F. 2006. Effect of carbamate esters on neurite outgrowth in differentiating human SK-N-SH neuroblastoma cells. *Chem. Biol. Interact.* 159: 65–72.

Chaturvedi, D. 2003. Chapter 2: Synthesis of organic carbamates through the various approaches. PhD thesis, Agra University, India.

Chaturvedi, D. 2011. Recent developments on the carbamation of amines. *Curr. Org. Chem.*15: 1593–1524.

Chaturvedi, D. 2012. Perspectives on the synthesis of organic carbamates. *Tetrahedron* 68: 15–45.

Chaturvedi, D., Chaturvedi, A. K., Mishra, N., and Mishra, V. 2008a. An efficient, one-pot, synthesis of trithiocarbonates from the corresponding thiols using the Mitsunobu reagent. *Tetrahedron Lett.* 49: 4886–4888.

Chaturvedi, D., Chaturvedi, A. K., Mishra, N., and Mishra, V. 2009. Basic resin mediated, efficient, one-pot, synthesis of carbazates from the corresponding alkyl halides. *J. Iran. Chem. Soc.* 6: 510–513.

Chaturvedi, A. K., Chaturvedi, D., Mishra, N., and Mishra, V. 2010. A high yielding, one-pot, synthesis of *S,S*-dialkyl dithiocarbonates through the corresponding thiols using Mitsunobu's reagent. *J. Iran. Chem. Soc.* 7: 702–706.

Chaturvedi, A. K., Chaturvedi, D., Mishra, N., and Mishra, V. 2011. An efficient, one-pot, synthesis of carbazates and dithiocarbazates through the corresponding alcohols using Mitsunobu's reagent. *J. Iran. Chem. Soc.* 8: 396–400 and references therein.

Chaturvedi, D., Mishra, N., and Mishra, V. 2008b. An efficient, one-pot, synthesis of *S*-alkyl thiocarbamates from the corresponding thiols using Mitsunobu's reagent. *Synthesis* 355–357.

Chaturvedi, D. and Ray, S. 2007a. Versatile use of carbon dioxide in the synthesis of organic carbamates. *Curr. Org. Chem.* 11: 987–998.

Chaturvedi, D. and Ray, S. 2007b. Various approaches for the synthesis of organic carbamates. *Curr. Org. Synth.* 4: 308–320.

Chaturvedi, D., Ray, S., Srivastava, A. K., and Chander, R. 2008c. ω-(2-Naphthyloxy) amino alkanes as a novel class of anti-hyperglycemic and lipid lowering agents. *Bioorg. Med. Chem.* 16: 2489–2498.

Chedid, R. B., Brummer, M., Wibbeling, B., Fohlich, R., and Hoppe, D. 2007. Stereo- and regiochemical divergence in the substitution of a lithiated alk-1-en-3-yn-2-yl carbamate: Synthesis of highly enantioenriched vinylallenes or alk-3-en-5-yn-1-ols. *Angew. Chem. Int. Ed.* 46: 3131–3134.

Chen, S. W., Wang, Y. H., Jin, Y. et al. 2007. Synthesis and anti-HIV-1 activities of novel podophyllotoxin derivatives. *Bioorg. Med. Chem. Lett.* 17: 2091–2095.

Combrink, K. D., Denton, D. A., Harran, S. et al. 2007. New C_{25} carbamate rifamycin derivatives are resistant to inactivation by ADP-ribosyl transferases. *Bioorg. Med. Chem. Lett.* 17: 522–526.

Dangerfield, E. M., Timmer, M. S. M., and Stocker, B. L. 2009. Total synthesis without protecting groups: Pyrrolidines and cyclic carbamates. *Org. Lett.* 11: 535–538.

Décor, A., Monse, B., Martin, M. T. et al. 2006. Synthesis and biological evaluation of B-ring analogues of (−)-rhazinilam. *Bioorg. Med. Chem.* 14: 2314–2332.

Duca, M., Arimondo, P. B., Leonce, S. et al. 2005a. Novel carbamate derivatives of 4-β-amino-4′-*O*-demethyl-4-desoxypodophyllotoxin as inhibitors of topoisomerase II: Synthesis and biological evaluation. *Org. Biomol. Chem.* 3: 1074–1080.

Duca, M., Guianvarćh, D., Meresse, P. et al. 2005b. Synthesis and biological study of a new series of 4′-demethylepipodophyllotoxin derivatives. *J. Med. Chem.* 48: 593–603.

Fardis, M., Pyun, H. J., Tario, J. et al. 2003. Design, synthesis and evaluation of a series of novel fumagillin analogues. *Bioorg. Med. Chem.* 11: 5051–5058.

Fernandes, V. G., Fernandez, S., Ferrero, M., Gotor, V., Bouillon, R., and Verstuyf, A. 2004. Chemoenzymatic synthesis and biological evaluation of C-3 carbamate analogues of 1α,25-dihydrovitamin D_3. *Bioorg. Med. Chem.* 12: 5443–5451.

Franciskovich, J. B., Masters, J. J., Tinsley, J. M. et al. 2005. Investigation of factor Xa inhibitors containing non-amidine S1 elements. *Bioorg. Med. Chem. Lett.* 15: 4838–4841.

Fu, Y., Li, S., Zu, Y. et al. 2009. Medicinal chemistry of paclitaxel and its analogues. *Curr. Med. Chem.* 16: 3966–3985.

Gavai, A. V., Fink, B. E., Fairfax, D. J. et al. 2009. Discovery and preclinical evaluation of [4-[[1-(3-fluorophenyl)methyl]-1*H*-indazol-5-ylamino]-5-methyl-pyrrolo[2,1-*f*][1,2,4]triazin-6-yl]carbamic acid, (3*S*)-3-morpholinylmethyl ester (BMS-599626), a selective and orally efficacious inhibitor of human epidermal growth factor receptor 1 and 2 kinases. *J. Med. Chem.* 52: 6527–6530.

Giannessi, F., Pessotto, P., Tassoni, E. et al. 2003 Discovery of a long-chain carbamoyl aminocarnitine derivative, a reversible carnitine palmitoyltransferase inhibitor with antiketotic and antidiabetic activity. *J. Med. Chem.* 46: 303–309.

Goto, T., Ito, Y., Yamada, S., Matsumoto, H., Oka, H., and Nagase, H. 2006. The high throughput analysis of *N*-methyl carbamate pesticides in fruits and vegetables by liquid chromatography electrospray ionization tandem mass spectrometry using a short column. *Anal. Chim. Acta* 555: 225–232.

Greene, T. W. and Wuts, P. G. M. 2007. *Protective Group in Organic Synthesis*, 4th edn. John Wiley & Sons, Inc., New York.

Hadden, M. K., Lubbers, D. J., and Blagg, B. S. 2006. Geldanamycin, radicicol, and chimeric inhibitors of the Hsp90 N-terminal ATP-binding site. *Curr. Top. Med. Chem.* 6: 1173–1182.

Han, C., Shen, R., Su, S., and Porco, J. A. 2004. Copper-mediated synthesis of *N*-acyl vinylogous carbamic acids and derivatives: Synthesis of the antibiotic CJ-15,801. *Org. Lett.* 6: 27–30.

Han, X. and Widenhoefer, R. A. 2006. Gold (I) catalyzed intramolecular hydroamination of alkenyl carbamates. *Angew. Chem. Int. Ed.* 45: 1747–1748.

Hubbard, R. D., Dickerson, S. H., Emerson, H. K. et al. 2008. Dual EGFR/ErbB-2 inhibitors from novel pyrrolidinyl-acetylenic thieno [3,2-*d*] pyrimidines. *Bioorg. Med. Chem. Lett.* 18: 5738–5740.

Hutchinson, D. K. 2003. Oxazolidinone antibacterial agents: A critical review. *Curr. Top. Med. Chem.* 3: 1021–1042.

Ishihara, Y., Goto, G., and Miyamoto, M. 2000. Central selective acetylcholinesterase inhibitor with neurotrophic activity: Structure-activity relationships of TAK-147 and related compounds. *Curr. Med. Chem.* 7: 341–354.

Jang, M.-Y., Lin, Y., De, J. S. et al. 2011. Discovery of 7-*N*-piperazinylthiazolo[5,4-d]pyrimidine analogues as a novel class of immunosuppressive agents with in vivo biological activity. *J. Med. Chem.* 54: 655–668.

Janin, Y. L. 2007. Antituberculosis drugs: Ten years of research. *Bioorg. Med. Chem.* 15: 2479–2513.

Jaracz, S., Chen, J., Kuznetsova, L. V., and Ojima, I. 2005. Recent advances in tumor-targeting anticancer drug conjugates. *Bioorg. Med. Chem.* 13: 5043–5054.

Jeffrey, S. C., Nguyen, M. T., Andreyka, J. B., Meyer, D. L., Doronina, S. O., and Senter, P. D. 2006. Dipeptide based highly potent doxorubicin antibody conjugates. *Bioorg. Med. Chem. Lett.* 16: 358–362.

Johnson, J. S. and Evans, D. A. 2000. Chiral *bis*(oxazoline) copper (II) complexes: Versatile catalysts for enantioselective cycloaddition, Aldol, Michael, and carbonyl ene reactions. *Acc. Chem. Res.* 33: 325–335.

Kakadiya, R., Dong, H., Lee, P. C. et al. 2009. Potent antitumor bifunctional DNA alkylating agents, synthesis and biological activities of 3a-aza-cyclopenta[*a*]indenes. *Bioorg. Med. Chem.* 17: 5614–5626.

Kapuria, N., Kapuriya, K., Dong, H. et al. 2009. Novel DNA-directed alkylating agents: Design, synthesis and potent antitumor effect of phenyl *N*-mustard-9-anilinoacridine conjugates *via* a carbamate or carbonate linker. *Bioorg. Med. Chem.* 17: 1264–1275.

Kern, J. C., Terefenko, E. A., Fensome, A. et al. 2007. SAR studies of 6-(arylamino)-4,4-disubstituted-1-methyl-1,4-dihydro-benzo[*d*][1,3]oxazin-2-ones as progesterone receptor antagonists. *Bioorg. Med. Chem. Lett.* 17: 189–192.

Kociensiki, P. J. 2003. *Protective Groups*, 3rd edn. Thieme Verlag, Stuttgart, Germany.

Kuznetsova, L., Chen, J., Sun, L. et al. 2006. Syntheses and evaluation of novel fatty-acid second generation taxoid conjugates as promising anticancer agents. *Bioorg. Med. Chem. Lett.* 16: 974–977.

Lee, Y., Koo, H., Lim, Y. B., Lee, Y., Moa, H., and Park, H. 2004. New cationic lipids for gene transfer with high efficiency and low toxicity: T-shape cholesterol ester derivatives. *Bioorg. Med. Chem. Lett.* 14: 2637–2641.

Li, Q. Y., Zu, Y. G., Shi, R. Z., and Yao, L. P. 2006. Review camptothecin: Current perspectives *Curr. Med. Chem.* 13: 2021–2039.

Liu, Y.-Q., Yang, L., and Tian, X. 2007. Podophyllotoxin: Current perspectives. *Curr. Bioactive Compd.* 3: 37–66.

Ma, J., Lu, N., Qin, W., Xu, R., Wang, Y., and Chen, X. 2006. Differential responses of eight cyanobacterial and green algal species, to carbamate insecticides. *Ecotoxicol. Environ. Safety* 63: 268–274.

Mayer, J. P., Lewis, G. S., Curtius, M. J., and Zhang, J. 1997. Solid phase synthesis of quinazolinones. *Tetrahedron Lett.* 38: 8445–8448.

Mazurov, A., Klucik, J., Miao, L. et al. 2005. 2-(Arylmethyl)-3-substituted quinuclidines as selective 7α-nicotinic receptor ligands. *Bioorg. Med. Chem. Lett.* 15: 2073–2077.

McComas, C., Crowley, B. M., Hwang, I., and Boger, D. L. 2003. Synthesis and evaluation of methyl ether derivatives of the vancomycin, teicoplanin, and ristocetin aglycon methyl esters. *Bioorg. Med. Chem. Lett.* 13: 2933–2936.

Mizutani, T., Ishikawa, S., Nagase, T. et al. 2009. Discovery of novel benzoxazinones as potent and orally active long chain fatty acid elongase 6 inhibitors. *J. Med. Chem.* 52: 7289–7300.

Mukherjee, R., Kumar, V., Srivastava, S. K., Agarwal, S. K., and Burman, A. C. 2006. Betulinic acid derivatives as anticancer agents: Structure-activity relationship. *Anticancer Agents Med. Chem.* 6: 271–279.

Nicolaou, K. C. and Mathison, J. N. 2005. Synthesis of imides, *N*-acyl vinylogous carbamates and ureas, and nitriles by oxidation of amides and amines with Dess-Martin periodinane. *Angew. Chem. Int. Ed.* 44: 5992–5996.

Ohta, K., Ogawa, T., Suzuki, T., Ohta, S., and Endo, Y. 2009. Novel estrogen receptor (ER) modulators: Carbamate and thiocarbamate derivatives with *m*-carborane bisphenol structure. *Bioorg. Med. Chem.* 17: 7958–7962.

Ouellet, R., Rousseau, J., Brasseur, N., Lier, J. E., Doksic, M., and Westera, G. 1984. Synthesis, receptor binding, and target-tissue uptake of carbon-11 labeled carbamate derivatives of estradiol and hexestrol. *J. Med. Chem.* 27: 509–513.

Oves, D., Fernandez, S., Verlinden, L. et al. 2006. Novel A-ring homodimeric C-3-carbamate analogues of 1α,25-dihydroxyvitamin D_3: Synthesis and preliminary biological evaluation. *Bioorg. Med. Chem.* 14: 7512–7519.

Palomo, C., Aizpurua, J. M., Ganboa, I., and Oiarbiode, M. 2004. Asymmetric synthesis of β-lactams through the Staudinger reaction and their use as building blocks of natural and non-natural products. *Curr. Med. Chem.* 11: 1837–1872.

Peng, X., Knapp, B. I., Bidlack, J. M., and Neumeyer, J. N. 2007. High-affinity carbamate analogues of morphinan at opioid receptors. *Bioorg. Med. Chem. Lett.* 17: 1508–1511.

Posner, G. H. and Kahraman, M. 2003. Organic chemistry of vitamin D analogues. *Eur. J. Org. Chem.* 20: 3889–3895.

Pyun, H. J., Fardis, M., Tario, J. et al. 2004. Investigation of novel fumagillin analogues as angiogenesis inhibitors. *Bioorg. Med. Chem. Lett.* 14: 91–94.

Qin, H., Yamagiva, N., Matsunaga, S., and Shibasaki, M. 2007. Bismuth-catalyzed direct substitution of the hydroxy group in alcohols with sulfonamides, carbamates, and carboxamides. *Angew. Chem. Int. Ed.* 46: 409–413.

Rahmanthullan, S. M., Tidwell, R. R., Jones, S. K., Hall, J. E., and Boykin, D. W. 2008. Carbamate prodrugs of *N*-alkylfuramidines. *Eur. J. Med. Chem.* 43: 174–177.

Rastelli, G., Tian, Z. Q., Wang, Z., Myles, D., and Liu, Y. 2005. Structure based design of 7-carbamate analogs of geldanamycin. *Bioorg. Med. Chem. Lett.* 15: 5016–5021.

Ray, S. and Chaturvedi, D. 2004. Application of organic carbamates in drug design. Part 1: Anticancer agents-recent reports. *Drug Future* 29: 343–357.

Ray, S., Pathak, S. R., and Chaturvedi, D. 2005. Organic carbamates in drug development. Part II: Antimicrobial agents-recent reports. *Drug Future* 30: 161–180.

Reiss, A. B. and Vagell, M. E. 2006. PPAR-γ activity in the vessel wall: Anti-atherogenic properties. *Curr. Med. Chem.* 13: 3227–3238.

Santos, R. C., Salvador, J. A. R., Marin, S., and Cascante, M. 2009. Novel semisynthetic derivatives of betulin and betulinic acid with cytotoxic activity. *Bioorg. Med. Chem.* 17: 6241–6250.

Santos, R. C., Salvador, J. A. R., Marin, S., Cascante, M., Moreira, J. N., and Dinis, T. C. P. 2010. Synthesis and structure-activity relationship study of novel cytotoxic carbamate and *N*-acylheterocyclic bearing derivatives of betulin and betulinic acid. *Bioorg. Med. Chem.* 18: 4385–4396.

Schweitzer, D., Zhu, J., Jarori, G. et al. 2007. Synthesis of carbamate derivatives of iejimalides. Retention of normal antiproliferative activity and localization of binding in cancer cells. *Bioorg. Med. Chem.* 15: 3208–3216.

Sedlak, M., Dravina, P., Bilkova, E., Simunek, P., and Buchta, V. 2008. Novel targeting system for antimitotic drugs: β-Glycosidase sensitive Amphotericin-B star (polyethylene glycol) conjugate. *Bioorg. Med. Chem. Lett.* 18: 2952–2956.

Sharma, V., Hupp, C. D., and Tepe, J. J. 2007. Enhancement of chemotherapeutic efficacy by small molecule inhibition of NF-κB and checkpoint kinases. *Curr. Med. Chem.* 14: 1061–1074.

Sharma, U., Marquis, J. C., and Dinaut, A. N. et al. 2004. Design, synthesis, and evaluation of estradiol-linked genotoxicants as anti-cancer agents. *Bioorg. Med. Chem. Lett.* 14: 3829–3833.

Shaw, K. F., Sundermann, K. F., Burlingame, M. A., Zhang, D., Petryka, J., and Myles, D. C. 2006. Syntheses and evaluation of novel fatty-acid second generation taxoid conjugates as promising anticancer agents. *Bioorg. Med. Chem. Lett.* 16: 1961–1964.

Shen, G., Yu, X. M., Blagg, B. S. J. 2004. Syntheses of photolabile novobiocin analogues. *Bioorg. Med. Chem. Lett.* 14: 5903–5906.

Shi, S., Zhu, S., Gerritz, S. W. et al. 2005. Solid-phase synthesis and anti-infective activity of a combinatorial library based on the natural product anisomycin. *Bioorg. Med. Chem. Lett.* 15: 4151–4154.

Slavíková, B., Krištofíková, Z., Chodounská, H. et al. 2009. Allopregnanolone (3α-hydroxy-5α-pregnan-20-one) derivatives with a polar chain in position 16α: Synthesis and activity. *J. Med. Chem.* 52: 2119–2125.

Smith, A. B., Freez, B. S., LaMarche, M. J. et al. 2005. Design, synthesis, and evaluation of carbamate-substituted analogues of (+)-discodermolide. *Org. Lett.* 7: 311–314.

Taguchi, M., Goda, K., Sugimoto, K. et al. 2003. Biological evaluation of sphingomyelin analogues as inhibitors of sphingomyelinase. *Bioorg. Med. Chem. Lett.* 13: 3681–3684.

Takaoka, K., Tatsu, Y., Yumoto, N., Nakajima, T., and Shimamoto, K. 2004. Synthesis of carbamate type caged derivatives of a novel glutamate transporter blocker. *Bioorg. Med. Chem.* 12: 3687–3694.

Takashima, H. 2003. Structural consideration of macrolide antibiotics in relation to the ribosomal interaction and drug design. *Curr. Top. Med. Chem.* 3: 991–999.

The Pesticidal Manual. 1994. *World Compendium*, 10th edn., Thomlin, C. D. S. ed. Crop. Protection Publication, Surrey, U.K.

Tian, Z. Q., Liu, Y., Zhang, D. et al. 2004. Synthesis and biological activities of novel 17-aminogeldenamycin derivatives. *Bioorg. Med. Chem.* 12: 5317–5329.

Tripathi, R., Angeles, T. S., Yang, S. X., and Mallamo, J. P. 2008. TrkA kinase inhibitors from a library of modified and isosteric Staurosporine aglycone. *Bioorg. Med. Chem. Lett.* 18: 3551–3555.

Tully, D. C., Liu, H., Chatterjee, A. K. et al. 2006. Arylaminoethyl carbamates as a novel series of potent and selective cathepsin S inhibitors. *Bioorg. Med. Chem. Lett.* 16: 5107–5111.

Wagner, R., Mollison, K. W., Liu, L. et al. 2005. Rapamycin analogs with reduced systemic exposure. *Bioorg. Med. Chem. Lett.* 15: 5340–5343.

Wang, Y., Li, L., Tian, Z., Jiang, W., and Larrick, J. W. 2006. Synthesis and anticancer activity of CBI-bearing ester and carbamate prodrugs CC-1065 analogues. *Bioorg. Med. Chem.* 14: 7854–7861.

Wani, M., Taylor, H., Wall, M., Coggon, P., and McPhail, A. 1971. Plant antitumor agents. VI. The isolation and structure of taxol, a novel antileukemic and antitumor agent from *Taxus brevifolia*. *J. Am. Chem. Soc.* 93: 2325–2327.

Wills, A. J., Ghosh, Y. K., and Balasubramanian, S. 2002. Synthesis of a polymer-supported oxazolidine aldehyde for asymmetric chemistry. *J. Org. Chem.* 67: 6646–6652.

Wu, X. and Ojima, J. 2004. Tumor specific novel taxoid-monoclonal antibody conjugates. *Curr. Med. Chem.* 11: 429–438.

Wu, Y. J. and Su, W. G. 2001. Recent developments on ketolides and macrolides. *Curr. Med. Chem.* 8: 1727–1758.

Xu, H., Lv, M., and Tian, X. 2009. A review on hemisynthesis, biosynthesis, biological activities, mode of action, and structure-activity relationship of podophyllotoxins: 2003–2007. *Curr. Med. Chem.* 16: 327–349.

Yan, S., Miller, M. J., Wencewicz, T. A., and Mollmann, U. 2010. Syntheses and biological evaluation of new cephalosporin-oxazolidinone conjugates. *Med. Chem. Commun.* 1: 145–148.

Yekini, I., Hammoudi, F., Martin-Nizard, F. et al. 2009. Antioxidant activity of benzoxazolinonic and benzothiazolinonic derivatives in the LDL oxidation model. *Bioorg. Med. Chem.* 17: 7823–7830.

Yogeeswari, P. and Sriram, D. 2005. Betulinic acid and its derivatives: A review on their biological properties. *Curr. Med. Chem.* 12: 657–666.

Yoshizawa, H., Kubota, T., Itani, H., Minami, K., Miwa, H., and Nishitani, H. 2004. New broad-spectrum parenteral cephalosporins exhibiting potent activity against both methicillin-resistant *Staphylococcus aureus* (MRSA) and *Pseudomonas aeruginosa*. Part 3: 7β-[2-(5-Amino-1,2,4-thiadiazol-3-yl)-2-ethoxyiminoacetamido] cephalosporins bearing 4-[3-(aminoalkyl)-ureido]-1-pyridinium at C-3′. *Bioorg. Med. Chem.* 12: 4221–4231.

Yu, Q. S., Pei, X. F., Holloway, H. W., and Greig, N. H. 1997. Total syntheses and anticholinesterase activities of (3a*S*)-*N*(8)-norphysostigmine, (3a*S*)-*N*(8)-norphenserine, their antipodal isomers, and other *N*(8)-substituted analogues. *J. Med. Chem.* 40: 2895–2901.

Zega, A. 2005. Azapeptides as pharmacological agents. *Curr. Med. Chem.* 12: 589–597.

Zhang, Z. W., Zhang, J. Q., Hui, L., Chen, S. W., and Tian, X. 2010. First synthesis and biological evaluation of spin-labeled derivatives of deoxypodophyllotoxin. *Eur. J. Med. Chem.* 45: 1673–1677.

6 Rational Design of New Molecules of Biological Significance from Phenolic Constituents of Some Tropical Plants as Renewable Materials

Vladimir V. Kouznetsov, Diego R. Merchan Arenas, Fernando A. Rojas Ruiz, and Leonor Y. Vargas Méndez

CONTENTS

6.1 INTRODUCTION

The necessity for reducing society's dependence on imported crude oil has directed researchers' attention to the use of vegetable biomass not only as a source of energy but also as fine chemicals. Indeed, some easily isolable biomass components could be used as chemical reagents in the synthesis of novel products with a higher added value, replacing existing chemicals based on petroleum sources. Among these vegetable components, the essential oils of certain tropical aromatic plants are attractive materials to be utilized as chemical agents. Being phenolic compounds in nature with additional functional groups, these compounds appear as attractive renewable precursors in the construction of new and diverse molecules. Moreover, diversified chemical functionalities of such phenolics allow the generation of a variety of products with novel structural and skeletal diversity with a higher added value, for example, pharmacological, biological, and physical properties.

Essential oils (EOs) are highly variable and complex mixtures of constituents that belong almost exclusively to two distinctive groups with different biogenetic origins; these groups correspond to the terpenic and aromatic (C_6-arenes) molecules. The latter are "products of the

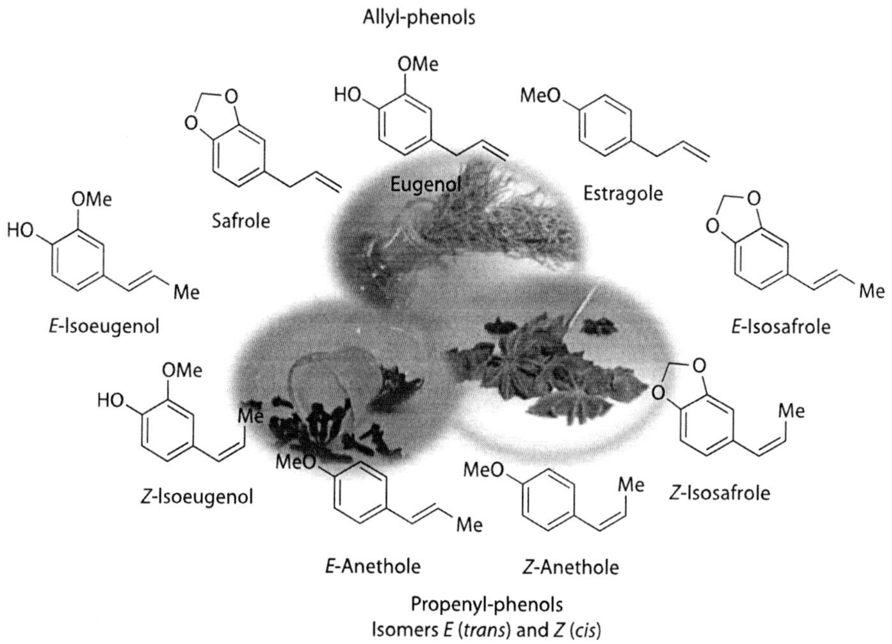

FIGURE 6.1 Allyl and propenyl phenols C_6-C_3 derivatives.

shikimic acid" and less common in EOs. Structurally, the derivatives of this group are divided into two small subgroups: arylpropane C_6-C_3 derivatives and methylbenzene C_6-C_1 derivatives. Mainly, the allyl and propenyl phenols are part of the C_6-C_3 group. Allyl-phenol C_6-C_3 derivatives are characteristic within certain EOs from the Apiaceae family (viz. eugenol, safrole, and estragole). Their analogous derivatives, propenyl phenol molecules, are usually mixtures of *trans* and *cis* isomers. For example, isoeugenols, anetholes, and isosafroles are mixtures of isomers *E* (*trans*) and *Z* (*cis*), whose ratio inclines to the *E*-isomer, and are thermodynamically more stable (Figure 6.1).

Some EOs contain C_6-C_1 compounds as vanillin, isovanillin, anisaldehyde, or piperonal (Figure 6.2). Although the group of aromatic compounds present in EOs is less common than in the terpenic group, their members play an important practical and scientific role. Furthermore, some C_6-C_3 compounds (eugenol, *E*-anethole, and estragole) are major components of the EOs of certain medicinal plants.

The aim of this chapter is to review existing materials on the utilization of the synthetic potential of phenolic constituents (Figure 6.3) extracted from some tropical plants, or even some parts of these tropical plants toward their conversion in new functionalized heterocyclic compounds.

The generation of new libraries of *N*- and *O*-heterocyclic compounds can significantly contribute to the search for promising models for pharmacological studies and to identify potential drugs effective against parasites, fungal pathogens, cancer cells, among others. Given the diverse

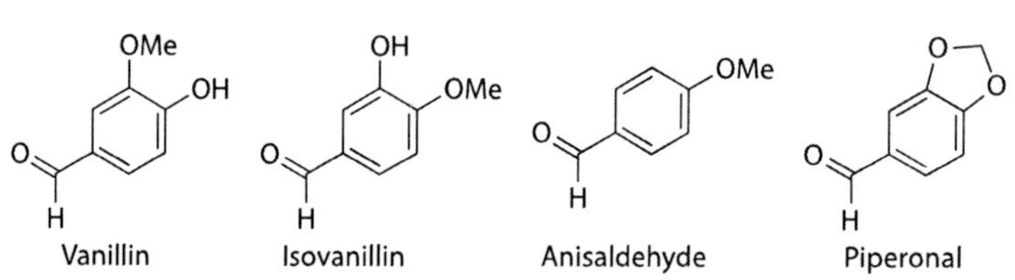

FIGURE 6.2 Formyl-substituted C_6-C_1 compounds.

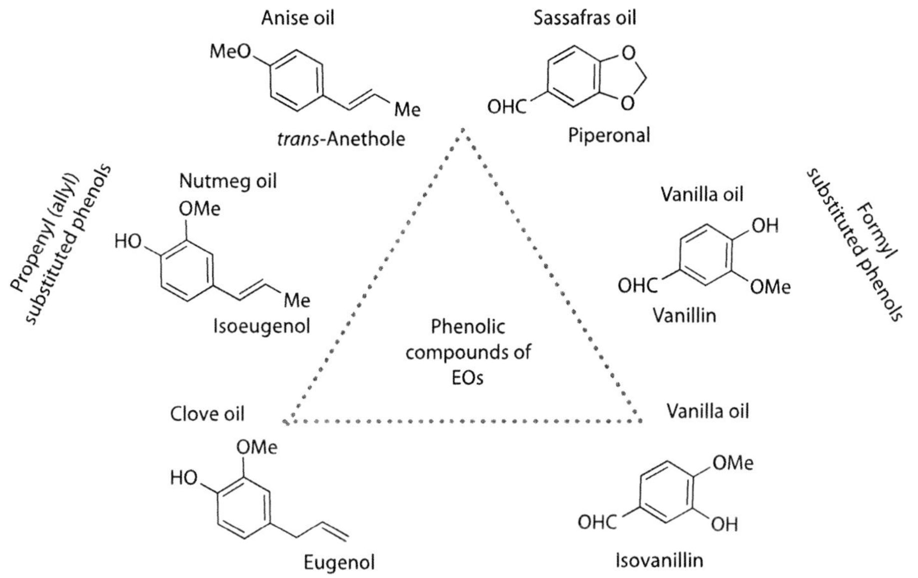

FIGURE 6.3 Utilized phenolic constituents as the synthetic reagents in synthesis of new molecules.

chemical nature of these constituents and prioritizing in the use of raw materials as precursors in organic synthesis and green methodologies, this chapter is divided into three general parts:

1. Propenyl (allyl) C_6-C_3 phenolic compounds as activated substituted alkenes in [4+2] and [3+2] cycloaddition processes
2. Formyl-substituted C_6-C_1 phenolic compounds (piperonal, (iso)vanillins) as aromatic aldehydes in [4+2] cycloaddition process (multicomponent imino Diels–Alder reaction) and multicomponent condensation
3. Bioscreening and in silico calculated physicochemical properties of molecules obtained

6.2 PROPENYL (ALLYL) C_6-C_3 PHENOLIC COMPOUNDS AS ACTIVATED SUBSTITUTED ALKENES IN [4+2] AND [3+2] CYCLOADDITION PROCESSES

Plants' cellular machinery is in charge of the generation of the molecular architecture constituting the basis for science and food, pharmaceutical, agricultural, and cosmetic industries' technology development. The importance of substrates obtained from nature is recognized due to ethnobotany, thanks to the healing properties attained from different plantlets. However, determining the active principle of these blends obtained from vegetables gives place to the performance of organic and analytical chemistry. The extracts and EOs are equipped with a high quantity of compounds (small molecules) derived from cellular metabolism. It has been possible to classify these compounds in numerous natural product families, including alkaloids, polyketides, lipids, polyphenols, carbohydrates, benzofuranoids, tannins, lignans, benzopyranoids, flavonoids, steroids, amino acids and peptides, polypyrroles, terpenoids, and simple aromatic compounds (Bruckingham 2000). The latter compounds are found in high proportion in medicinal and aromatic plants, playing a fundamental protective role against stress as caused to plants; these compounds can also be used for medical, food, fragrance, and flavor purposes (Planta Europa 2011).

Analytical tools have been playing a useful role in the isolation and identification of a great variety of simple aromatic compounds present in EOs (Oprean et al. 2001). Compounds such as propenyl (allyl) phenols have been determined and structurally characterized as C_6-C_3 unities. C_3-Allyl compound such as eugenol has been obtained in 76.8% of the total oil composition, which is extracted from dry clove buds (*Eugenia caryophyllus*) in yields exceeding 12.7% (Jirovetz et al. 2006). Other two phenylpropanes of interest are the estragole, isolated from winter tarragon, *Artemisia dracunculus* L., with a percentage of 60%–75% (De Vincenzi et al. 2000), and safrole, obtained from the trunk wood of *Ocotea pretiosa* or from the root bark or the fruit of tree *Sassafras albidum* (Nutall) Nuss, with a composition of 92.9% and 80.0%, respectively (Hickey 1948). On the other hand, the isomeric isoeugenol, anethole, and isosafrole are also found in the nutmeg *Myristica fragans* in minimal quantities. Star anise fruits (*Illicium verum* Hook. f.) provide *trans*-anethole (9.8% extraction yield and 89%–92% content), while isosafrole is generally obtained by safrole isomerization (Kishore and Kannan 2004). These oils and their major compounds (Figure 5.1) have shown a wide scope of biological activities against fungi (Amiri et al. 2008), protozoans (Santoro et al. 2007), cancer cells (Pisano et al. 2007), platelet aggregation (Tognolini et al. 2007), and stabilization of oxygen reactive species, which cause oxidative stress (Nenadis et al. 2003; Bortolomeazzi et al. 2007; Scherer and Teixeira 2009).

Because of the highlighted importance of phenylpropanes and their role in nature (Solecka 1997), their biogenesis has been studied, identifying the shikimate as the starting product, phenylalanine and tyrosine as intermediate amino acids, the enzymes (PAL and TAL) interacting in the biosynthetic pathway, and posterior lignanic derivatives of C_6-C_3 unities (Naczk and Shahidi 2004; Ferrer et al. 2008). The most important natural derivatives among the phenylpropanoid unities are the lignin, stilbenes, chalcones, the flavonoid structures, tannins, and lignans (Ferrer et al. 2008); the latter compounds have been defined as $(C_6$-$C_3)_2$ dimers of phenylpropanes and are equally present in plants as secondary metabolites. The *Linum album* (Linaceae family), an herbaceous and medicinal plant, biosynthesizes some important lignans as podophyllotoxin, α/β-peltatin, lariciresinol, and matairesinol (Figure 6.4) (Smollny et al. 1998). These compounds also present structures with a privileged topology possessing various biological activities such as anticancer, anti-inflammatory, antimicrobial, anti-rheumatic, antioxidant, and antifungal (Saleem et al. 2005).

Further important lignans are the phenylpropane dimers, which are set up forming an indane-type benzo-carbonylic carbon skeleton (Figure 6.5). Some biological properties common to their monomers (eugenol, anethole, safrole, and asarone) are anxiolytic (Zolyomi et al. 1974), anti-inflammatory (Madrigal et al. 2003), and antioxidant (Kozlova et al. 1968). The asarone and anethole arylindane dimers have been extracted from natural sources such as the psychedelic plant *Accorus calamus* and the sponge *Spheciospongia vesparia* (Bergmann and McAlee 1951).

Due to their natural and biomedical significance, there are a number of reports on the syntheses of these phenol dimers (Kovács 1950; Muller et al. 1954). Although the compounds of structures **5** and **6** (Scheme 6.1) have been reported much earlier, researches on their synthesis continue to be common in

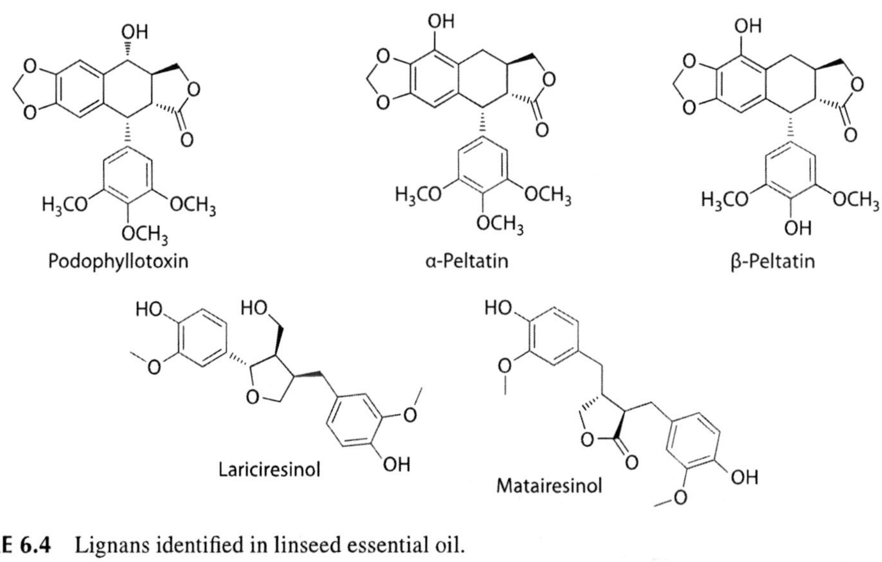

FIGURE 6.4 Lignans identified in linseed essential oil.

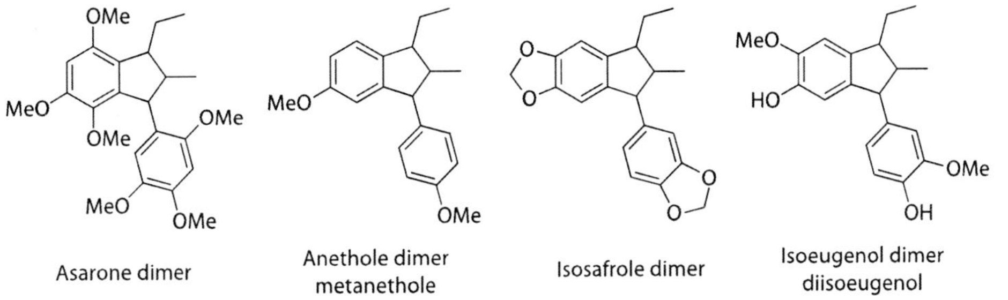

FIGURE 6.5 Arylindane lignans and phenylpropane dimers.

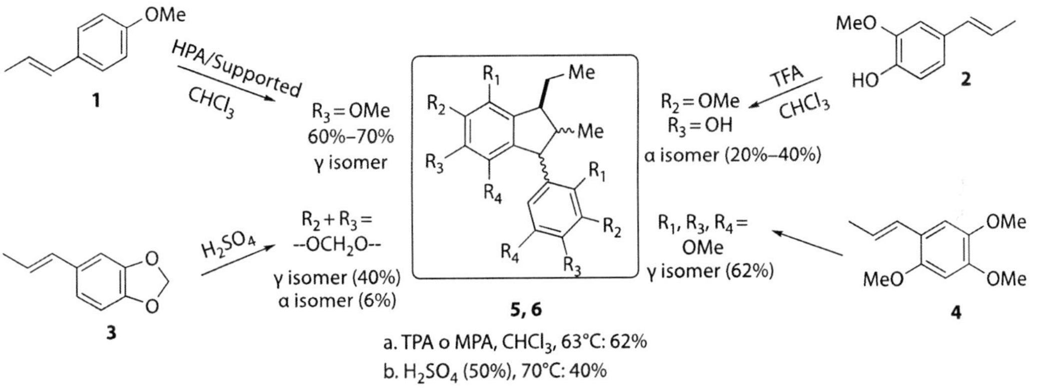

SCHEME 6.1 Different conditions employed in the phenylpropane dimers' synthesis.

recent literature as well as in the reinvestigation of their biological properties. Among different parameters, conventional methods for their syntheses include principally the use of common Brönsted acids as H_2SO_4 (Angle and Arnaiz 1992), TFA (Farhan et al. 1992), Lewis acids such as $FeCl_3$ (Griengl and Foidl 1981), and the well-known heteropolyacids, which promoted the [3+2] cycloaddition reaction of corresponding propenyl C_6-C_3 phenolics **1–4** in yields above 40% (Torviso et al. 2003, 2006) (Scheme 6.1).

There is an ongoing progress in synthetic methodologies for developing such dimeric systems. Emphasizing on the phenylpropanes structural profits, specially the anethole **1** and isoeugenol **2**, Kouznetsov's group has developed the construction of carbocyclic lignanic models, derived from indane as the metanethole **5** and diisoeugenol **6**. In this study, an efficient process, prioritizing on the operational and environmental risks diminution, has been developed. Thus, a solid acid support was elaborated, employing silica gel and $ClSO_3H$. Later, this support was employed as a catalytic agent in MeCN, favoring the formation of arylindanes. The *trans*-isoeugenol and *trans*-anethole were employed from a commercial source without previous purification under different reaction conditions, exploring SiO_2-O-SO_3H in MeCN as the best promotion system for the cycloaddition reaction since the solid support was recovered and reused without any appreciable loss of catalytic capacity (Kouznetsov and Merchan Arenas 2009). For dihydro(1*H*)indene dimers, *r*-1-ethyl-*c*-3-(4-methoxyphenyl)-6-methoxy-*t*-2-methylindane **5** and *r*-1-ethyl-5-hydroxy-*c*-3-(4-hydroxy-5-methoxyphenyl)-6-methoxy-*t*-2-methylindane **6**, it was also possible to define their stereochemistry; hence, the γ-isomers of these dihydro(1*H*)indenes were obtained in good yields (Scheme 6.2).

Studying the bark of *Machilus thunbergii* (Lauraceae family), a small tree, it was possible to determine the presence of another propenyl phenolic derivative, benzofuran neolignans. These benzofuran derivatives with high structural similarity reflect the isoeugenol as the C_6-C_3 unity. These compounds, known as licarin A and (–)-acuminatin, exhibited anticancer activity against the cellular line HL-60 and also potent antioxidant activity (Charlton 1998). In the same way, a benzofuranol structural analogue, known as salvinal, was extracted from *Myristica fragrans* (Kwon et al. 2008)

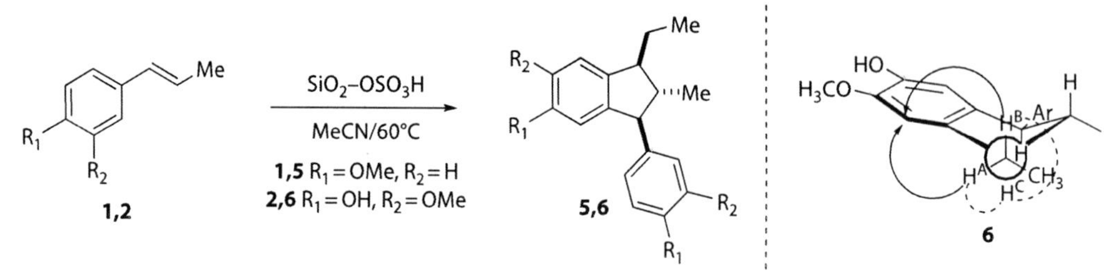

SCHEME 6.2 Stereoselective dimerization process of styrenes **1**,**2** for dihydro(1*H*)indenes **5**,**6** with the γ-configuration.

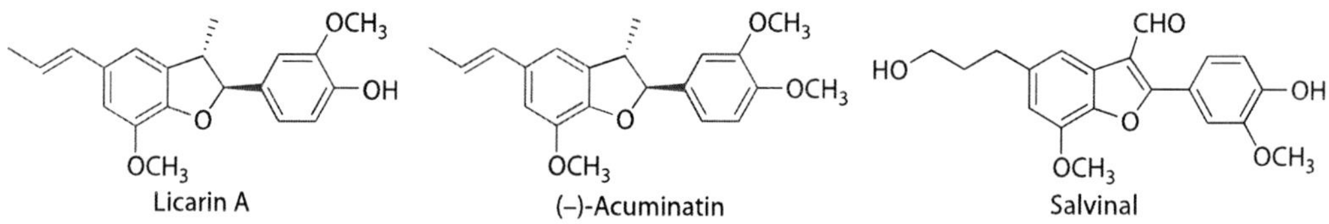

FIGURE 6.6 Isolated neolignans from *Machilus thumbergii* and *Myristica fragrans*.

(Figure 6.6), and its synthesis starting from simpler benzofuranols has already been reported (Wang et al. 2006).

Benzoquinones and certain phenylpropenes have been reported to be used as building blocks for the synthesis of benzofuranols analogues; initially, $Fe(ClO_4)_3$ was used as catalyst. However, in the following years, it was also shown that $BF_3 \cdot OEt_2$, $InCl_3$, and I_2 can act as useful *anti*-selective catalysts to afford diverse substituted *trans*-2,3-dihydrobenzo[*b*]furanols **7** (Wang et al. 1991; Juhász et al. 2000; Ohara et al. 2002; Yadav et al. 2003) (Scheme 6.3, routes a and b).

These conventional methods have been modified by Kouznetsov's group, employing polyethylene glycols, the media of increasing interest (very particularly, polyethyleneglycol with an average mass of 400 g/mol, i.e., PEG 400). The [3+2] cycloaddition using PEG 400 was carried out with benzoquinone and the phenylpropenes, isoeugenol and anethole, in the presence of the Lewis acid, $BF_3 \cdot OEt_2$, to obtain the dihydrobenzo[*b*]furan system (**7e**,**f**) (Scheme 6.3, route b) (Kouznetsov et al. 2008b). Polyethylene glycols are considered to be innocuous media, and are generally employed as vehicles for drug delivery within the organism. For this reason, participation of such a solvent system in the cycloaddition reaction between phenylpropenes and benzoquinones is regarded green and suitable for the construction of *trans*-2,3-dihydrobenzo[*b*]furan-5-ols. As well as the carbocyclic

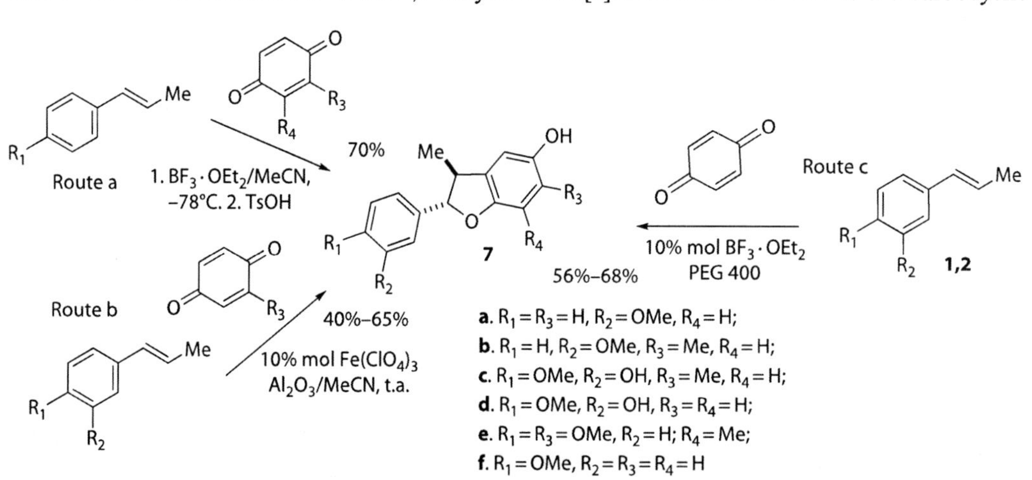

SCHEME 6.3 Synthesis of 2,3-dihydrobenzofuranols **7** from phenylpropenes and benzoquinones.

systems, the dihydrobenzofuranols (i.e., dimers of propenyl phenols) synthesis usually affords *anti*-stereoisomer as the major [3+2] cycloaddition product.

In continuation to this panorama of secondary metabolites of vegetable species, we may turn our attention to a different kind of molecules that differ from phenylpropanes and their derivatives. Antique natural products such as alkaloids still continue to be the leading molecules in organic synthesis, biochemical studies, theoretical studies, and, obviously, in medicinal chemistry as well. There are different kinds of alkaloids depending on their structural patterns. However, among them, the heterocyclic systems based on the quinoline, its reduced analogue 1,2,3,4-tetrahydroquinoline, and the indole skeleton bear a special interest from the historical perspective; these two heterocyclic structures have in common the amino acid, L-tryptophan, from which they are biosynthesized. Quinoline and 1,2,3,4-THQ derivatives are an interesting class of natural compounds; many alkaloids with this core have been well known for a long time. Some important analogues may be exemplified as virantmycin (Omura and Nakagawa 1981), dynemicin A (Konishi et al. 1990), and helquinoline (Asolkara et al. 2004); besides, some quinolines and THQs have also been extracted from vegetal sources such as *Macrorungia longistrobus* (Wuonola and Woodward 1976), *Haplophyllum bucharicum*, and *Euodia roxburghiana* with anti-HIV-1 activity (McCormick et al. 1996). In addition, a number of analogous compounds are also known with broad molecular diversity such as the pyran (*Lyngbya majuscula* Gomont) (Nogle and Gerwick 2003) and 2,3-methylendioxy moiety (*Acanthosyris paulo-alvinii* Barroso) (Chavez et al. 1997; Michael 2001). Nevertheless, only one THQ with phenylpropanoid motif has been isolated from natural sources, the *Galipea officinalis* Hancock, galipeine (Jacquemond-Collet et al. 1999) (Figure 6.7).

On the other hand, a wide variety of THQs and quinolines have been synthesized, generating a great molecular diversity that is very important in regard to medicinal chemistry. These scaffolds are found in many biologically active natural and synthetic products. A large number of reports have shown that these compounds display different activities including antimalarial (Bendale et al. 2007), antioxidant (Nishiyama et al. 2003), and anti-inflammatory (Calhoun et al. 1995; Kouznetsov et al. 2009). Three important biological perspectives with an important relationship are the antioxidant

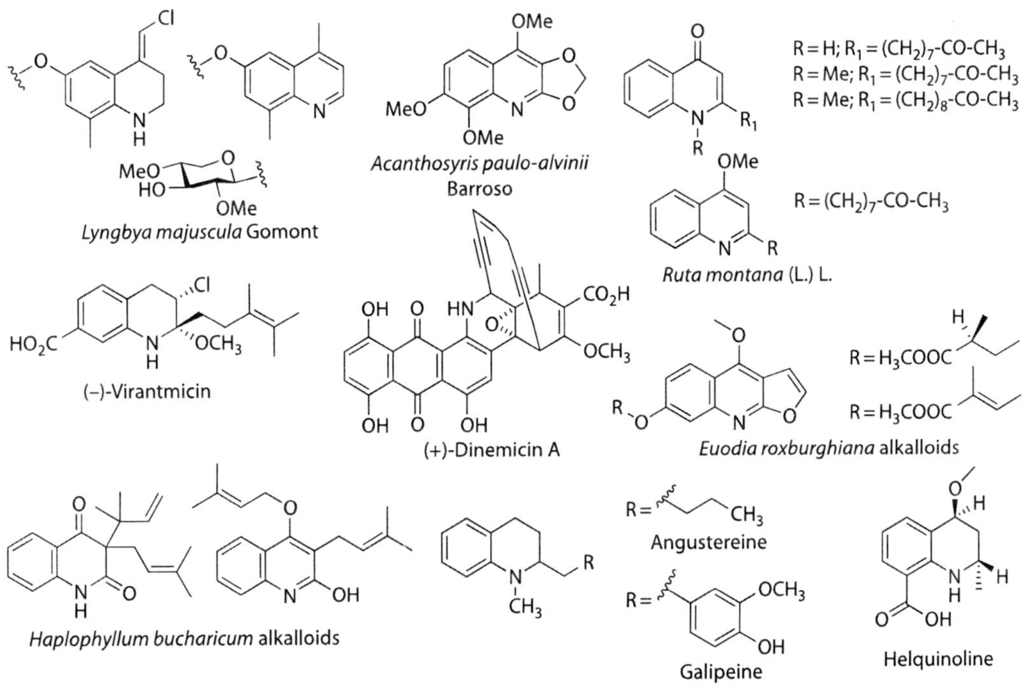

FIGURE 6.7 Natural (tetrahydro)quinolines from microorganism and plants, the base for synthetic research.

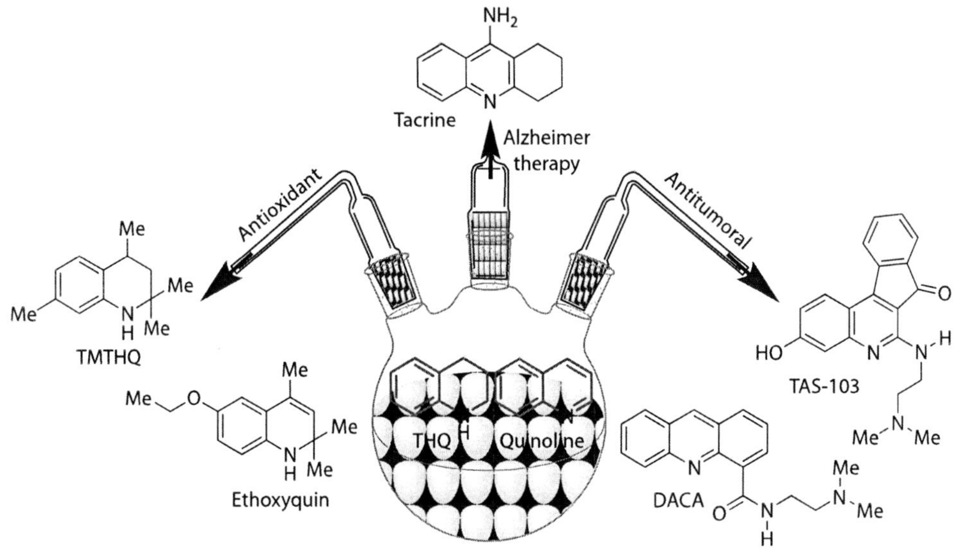

FIGURE 6.8 Synthetic (tetrahydro)quinolines with biological activity as antioxidant, antitumor, and anti-Alzheimer.

activity, the AChE inhibition, and antitumor action. Results of various biological studies indicate that the THQ and quinoline backbones in certain chemical entities play an important role in their beneficial effects against oxidative stress, such as ethoxyquin (1,2-dihydro-6-ethoxy-2,2,4-trimethylquinoline, used as food preservative) (De Koeing 2002) and TMTHQ, a new potential active compound (Blaszczyk and Skolimowski 2006) (Figure 6.8). THQ and quinoline models have also been reported to have a protective effect against Alzheimer's disease. The more classic example is the molecule tacrine (discarded for its toxicity); and some other diverse molecules are also available including several functional groups in the THQ backbone (Gauthier 2001). Examples of two more synthetic lead molecules bearing such scaffold may be cited herein for their potent anticancer activity: the DNA intercalators DACA and TAS-103—both of them exhibit inhibitory activity against topoisomerases I/II and are currently undergoing clinical trials (Ewesuedo et al. 2001) (Figure 6.8).

All such molecules are the fundamental prototypes in our study toward the rational construction of molecular architectures. Numerous methods for the synthesis of THQ and quinoline systems have been described (Kouznetsov et al. 1998; Sridharan et al. 2011); in one occasion, Kouznetsov et al. performed the synthesis of such pharmacophore employing γ–iminopiperidines **9** and homoallylamines **10** as the intermediates. In a recent work, diverse 3′,4′-dihydro-spiro[piperidine-4,2′-(1′*H*) quinolines] **11**,**12** have been synthesized from 1-benzylpiperidin-4-one **8** and anilines. The homoallylic compound **10** was obtained by the Grignard procedure and the final cyclization was carried out with concentrated H_2SO_4 (Vargas Méndez and Kouznetsov 2007) (Scheme 6.4).

The obtained molecules showed high structural similitude with synthetic antioxidants (ethoxyquin and TMTHQ) and the *N*-benzylpiperidine fragment is a moiety from donepezil. Therefore, the antioxidant activity and its capacity to inhibit the AChE were evaluated for these molecules. All compounds showed good Log P and TPSA values ($N + O < 4$; TPSA = 15.27–35.58 Å^2; MW = 216–341) that are required to cross the blood–brain barrier (Clark 2003; Lobell et al. 2003). In this case, all the tested compounds were found to be active as antioxidants and AChE inhibitors; nevertheless, the best AChE inhibitor was the 6-chloro derivative **11e** and the best scavenger of $ABTS^{+\bullet}$ was the 6-methyl derivative **11b** (Kouznetsov et al. 2010) (Figure 6.9).

Despite the effectiveness of Kouznetsov's THQ synthesis coupled with several other methods employed until now, the cycloaddition reactions are the most successful reactions developed for a rapid construction of these scaffolds; similarly, the acid-catalyzed imino DA reaction

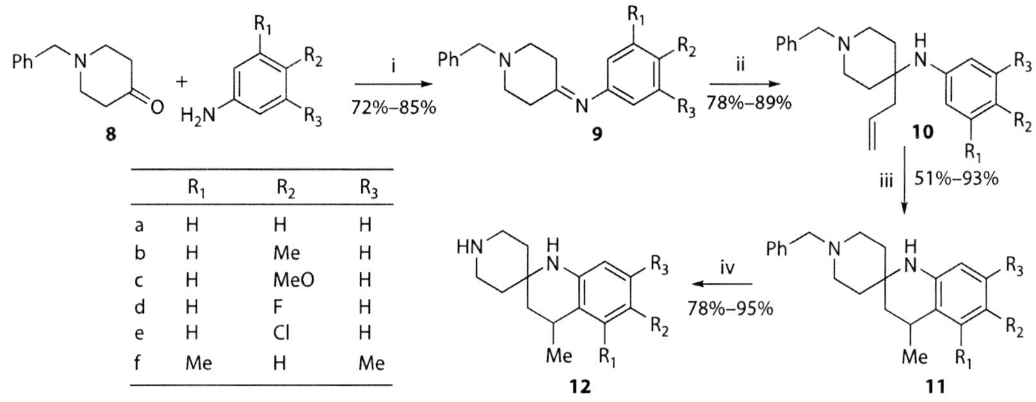

	R_1	R_2	R_3
a	H	H	H
b	H	Me	H
c	H	MeO	H
d	H	F	H
e	H	Cl	H
f	Me	H	Me

Reagents and conditions: (i) Toluene, reflux, 6–8 h, cat. AcOH; (ii) allylmagnesium bromide, Et_2O, 4–5 h and then work-up with NH_4Cl; (iii) 85% H_2SO_4, 80°C, 4–6 h; (iv) $HCOONH_4$/Pd/C/MeOH, reflux, 10 min.

SCHEME 6.4 Kouznetsov' synthetic route to the desired 1-*H*-4′-methyl-3′,4′-dihydrospiro[piperidine-4,2′(1′*H*)quinolines].

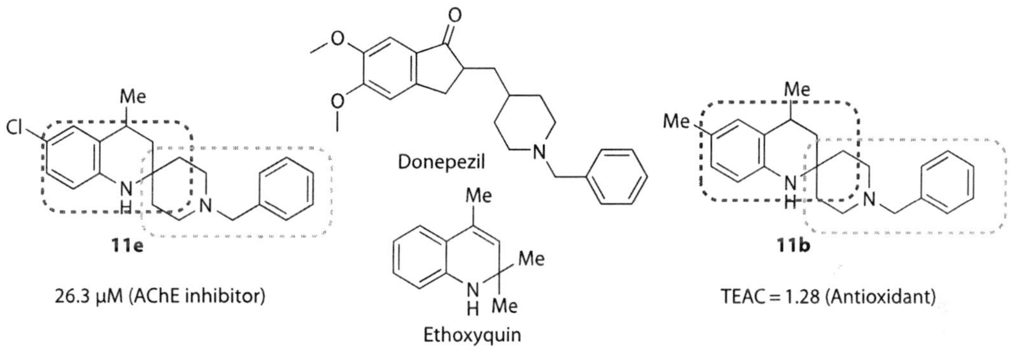

FIGURE 6.9 More active 1-benzyl-4′-methyl-3′,4′-dihydrospiro[piperidine-4,2′(1′*H*)quinolines].

(Povarov reaction) between 2-azadienes (*N*-arylimines) and electron-rich alkenes (mainly, vinyl enol ethers, and vinyl enamides) in its three-component version is one of the most powerful synthetic tools for the production of nitrogen-containing, six-membered heterocyclic compounds (Povarov 1967; Boger and Weinreb 1987; Kouznetsov 2009). Continuing with our studies on the construction of tetrahydroquinoline scaffolds, the three-component reaction of substituted anilines, aryl aldehydes, and electron-rich olefins in the presence of different Lewis acid catalysts offered us the best way to provide molecular diversity to our new natural products analogues. Applying the Diels–Alder methodology, it was possible to prepare efficiently novel hexahydro oxaisoindolo[2,1-*a*] quinoline derivatives **14** from the 2,4-disubstituted 1,2,3,4-tetrahydroquinolines bearing a furan fragment **13**. This new synthetic approach to these molecules was based on the construction of key precursors bearing a furan fragment, which can easily be prepared from readily available materials via two different routes (Kouznetsov et al. 2004) (Scheme 6.5).

Other polycyclic quinoline such as the indenoquinoline derivatives are interesting structures due to their potent antitumoral activity (Byl et al. 1999). The antineoplastic agent TAS-103 that stimulates DNA cleavage by topoisomerases is the most important compound of this family, because it has a broad spectrum of antitumor activity against many human solid tumor xenografts (De Koeing 2002) (Figure 6.8). Available synthetic procedures to obtain indeno[2,1-*c*]quinoline derivatives are based on multistep linear synthesis that are very complex and use poorly accessible starting materials (Anzini 1991). However, using the imino DA reaction between *N*-arylaldimines and indene as

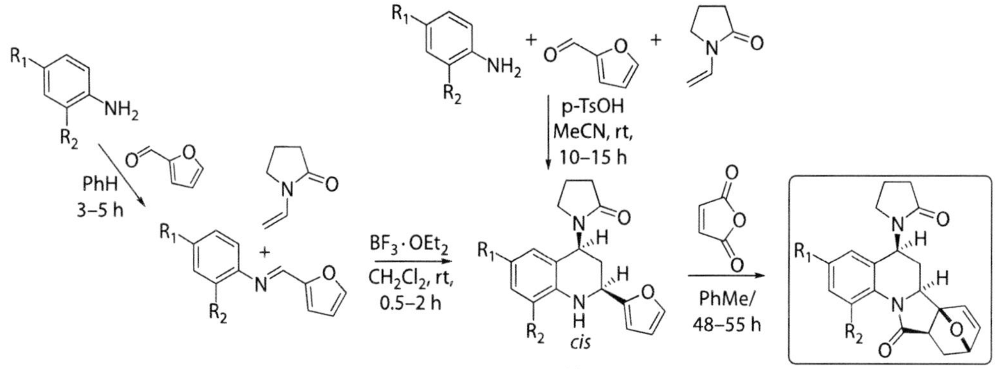

SCHEME 6.5 Synthesis of oxaisoindolo[2,1-*a*]quinoline derivatives via the Diels–Alder reactions.

dienophiles, various groups have reported the synthesis of some tetrahydroindeno[2,1-*c*]quinoline derivatives (Borrione 1989).

Based on these facts and according to the research interests on the preparation of bioactive nitrogen-containing heterocycles (Ochoa Puentes and Kouznetsov 2002), Kouznetsov and coworkers have developed an efficient and general route for the synthesis of new pyridinyl-substituted indeno[2,1-*c*]quinoline derivatives **15**, closer analogues of the potent anticancer TAS-103, from commercially available anilines, α-, β-, or γ-pyridinecarboxyaldehydes and 1*H*-indene, using a three-component imino DA cycloaddition as the key ring-forming step. Subsequent treatment of the 5,6,6a,11b-tetrahydroindeno[2,1-*c*]quinolines **15** with powdered sulfur gives the corresponding indeno[2,1-*c*]quinolones **16** (TAS-103 analogues) (Kouznetsov et al. 2006b, 2009) (Scheme 6.6).

As well as the isoindolo moiety, different structural frameworks can be combined with quinoline by means of the imino DA reactions (Povarov reaction). Using this synthetic methodology, it was possible to explore other commercial unconventional and less reactive alkenes than *N*-vinyl amides and vinyl ethers. Therefore, based on the few reports of the phenylpropanoid derivatives (*trans*-anethole or *trans*-isoeugenol) such as dienophiles in this cycloaddition (Jossang et al. 1991; Fadel et al. 2004), attention was paid to the development of new bioactive tetrahydroquinoline derivatives with anethole or isoeugenol fragments (Juhász et al. 2000; Ding et al. 2005). In the same way, the green reaction parameters such as changing the high values of temperature, long reaction times, hazardous reagents, solvents, and waste generation were taken into consideration. In this order of ideas, synthesis of new THQ by improving the reaction conditions with PEG 400 as a perfect and novel reaction media was carried out. Consequently, the PEG 400 and $BF_3 \cdot OEt_2$ promoted the multicomponent cycloaddition between anilines, benzaldehyde, and *trans*-anethole **1** or *trans*-isoeugenol **2**

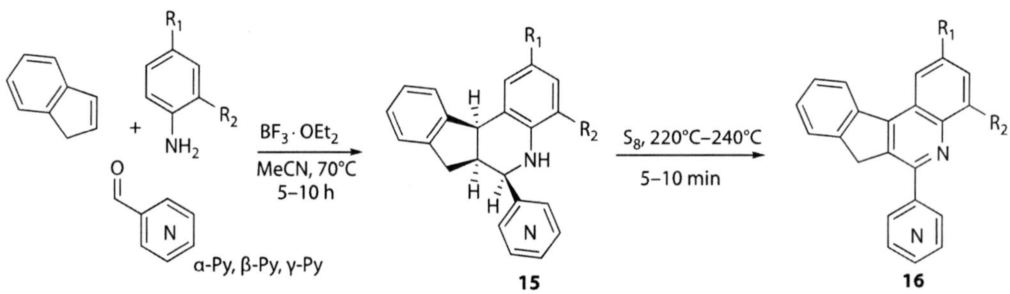

SCHEME 6.6 Pyridinyl substituted 7*H*-indeno[2,1-*c*]quinoline derivatives via three-component imino Diels–Alder reaction.

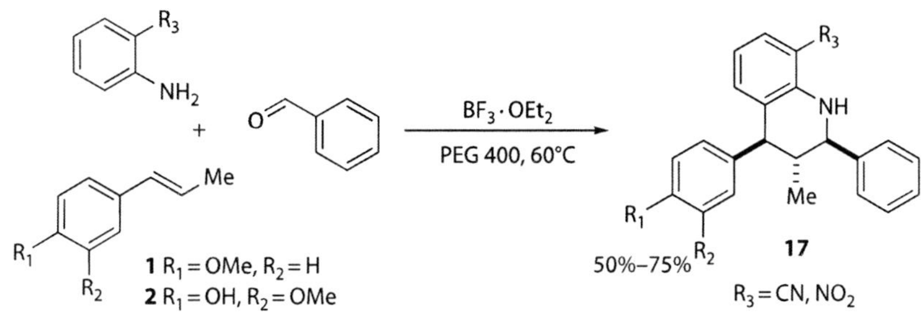

SCHEME 6.7 Synthesis of new tetrahydroquinoline with the phenylpropanoid moiety in PEG 400.

to give new THQs **17** with the phenylpropanoid moiety in excellent yields (Kouznetsov et al. 2008b) (Scheme 6.7).

Exploring new conditions for tetrahydroquinoline preparation with potential natural precursors, we found that $Cu(OTf)_2$ and $Zn(OTf)_2$ used only in 10 mol% allowed the three-component condensation reaction to occur at room temperature, between *trans*-anethole and *trans*-isoeugenol, anilines, and benzaldehyde, contrary to our previous results, in the presence of $BF_3 \cdot OEt_2$ in stoichiometric amounts. When copper(II) triflate was used as a catalyst, GC-MS and HPLC analysis of the crude reaction showed only one peak. ^{1}H NMR analysis confirmed that the prepared tetrahydroquinolines **18** were one unique diastereoisomer, indicating that this is a highly diastereoselective process (Romero Bohórquez et al. 2011) (Scheme 6.8).

Inspired by these results and also to introduce structural diversity coupled with biological activities of the resulting compounds, we have expanded our synthetic studies on new tetrahydroquinoline derivatives unsubstituted at the C-2 position, and to construct this THQ and quinoline topology including the C-4-anisyl or guaiacoyl motifs, the other face of DA reaction was investigated. The cationic imino DA[4$^+$+2]-cycloaddition reactions (Shono et al. 1982; Katrizky et al. 1997) were employed in the synthesis of THQ core with anethole and isoeugenol through a simple protocol with $BF_3 \cdot OEt_2$ as catalyst in MeCN, yielding diverse 3-methyl-4-aryl-1,2,3,4-tetrahydroquinolines, using readily available starting materials. Reaction between the readily available *N*-benzylaniline **19** and the inexpensive formalin (37% formaldehyde in methanol) at 0°C in MeCN afforded smoothly the cationic intermediates **20**; subsequent *in situ* treatment of **20** with either *trans*-anethole **1** or *trans*-isoeugenol **2** in the presence of one equivalent of $BF_3 \cdot OEt_2$ furnished new *N*-benzyl-3-methyl-1,2,3,4-tetrahydroquinolines **21** in moderate-to-good yields (Romero Bohórquez and Kouznetsov 2010). Finally, by using the easily handled system H_2 and Pd/C in a methanol–dichloromethane (3:1) mixture, the desired NH-tetrahydroquinolines **22** were obtained in excellent yields (92%–98%) (Scheme 6.9).

With the excellent results recompiled until now and the interest if our laboratory on the heterocyclic systems, another important ring was introduced in our research. We have evaluated the significance of indole moiety in several natural and synthetic molecules according to the literature. Furthermore, other exclusive molecules like spiro-quinoline derivatives (Katrizky et al. 1996)

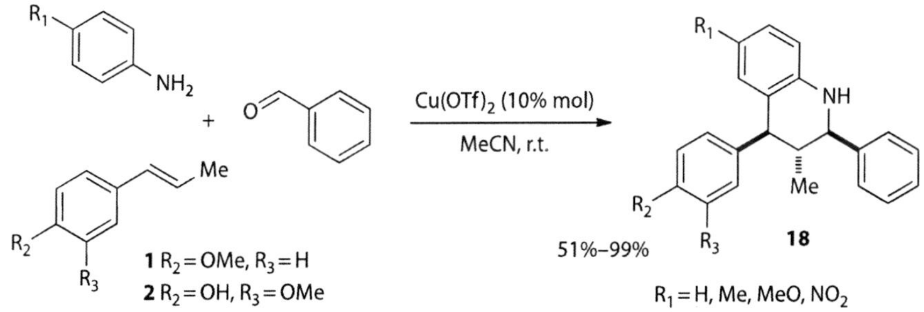

SCHEME 6.8 Synthesis of phenylpropanoid-tetrahydroquinoline core promoted by $Cu(OTf)_2$.

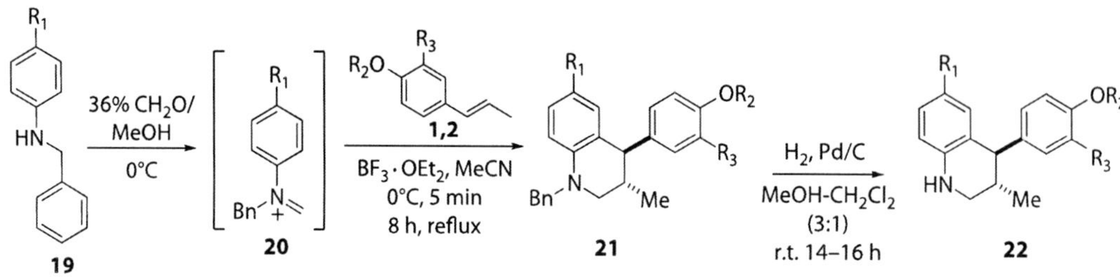

19 20 21 22

SCHEME 6.9 Preparation of new 4-aryl-3-methyl-1,2,3,4-tetrahydroquinolines **21**, **22**.

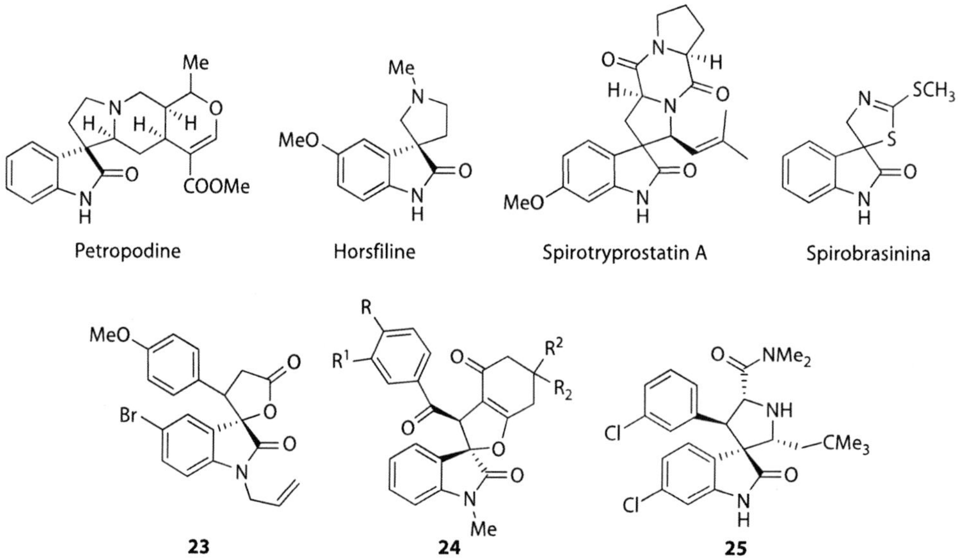

23 24 25

FIGURE 6.10 Important natural and synthetic heterocyclic spiro indolic skeletons.

bearing the indole scaffolds (Figure 6.10) occupy an important position in organic and medicinal chemistry because of their so-called pharmaceutical potentials such as MC4 receptors agonists (Fisher et al. 2005), antipsychotics (Singer et al. 2005), and protein farnesyltransferase inhibitors, an important enzyme for the survival of the pathogenic protozoa *Plasmodium falciparum* (Eastman et al. 2007). Besides, the C-3-spiro-oxindol framework system is the core structure of many natural alkaloids (horsfiline, spirotryprostatin A, pretropodine, etc.) and some reported pharmacological synthetic agents **23–25** (Naczk and Shahidi 2004).

The importance of THQ, quinoline, and indole cores in regard to their biological activities as discussed so far make them quite fascinating for their possible inclusion in the rational design of bioactive molecules. Hence, we investigated the reaction between *trans*-isoeugenol **2** and iminoisatins **26**, derived from isatin and anilines, providing a novel protocol for the preparation of 4′-(4-hydroxy-3-methoxyphenyl)-3′-methyl-3′,4′-dihydro-1′*H*-spiro[indoline-3,2′-quinolin]-2-ones **27** via $BF_3 \cdot OEt_2$-catalyzed imino DA reaction (Povarov reaction) (Kouznetsov et al. 2008a) (Scheme 6.10).

The synthesis of all these alkaloids with the C_6-C_3 scaffolds via the imino DA reaction revealed the interesting reactivity of the commercial products, anethole and *trans*-isoeugenol; these starting materials are not only commercially available but also inexpensive and easily storable. The easy access of *trans*-isoeugenol and anethole could be possible due to their high percentage in natural extracts and essential oils, principally from clove buds (*Eugenia caryophyllus*) and star anise (*Illicium verum* Hook. F.) (Tuan and Ilangantilekete 1997), respectively. Hence, Kouznetsov's group employed the clove buds and anise stars as raw materials to meet their requirements in the cause of green reaction protocols that include the use of innocuous media, solid-catalyst recovery, short reaction times, and less

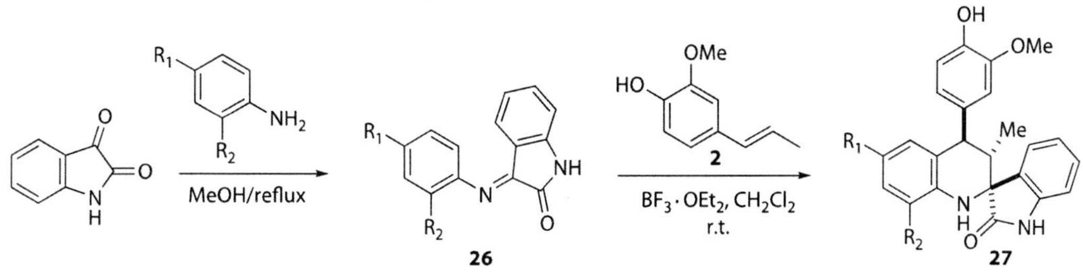

SCHEME 6.10 Synthesis of the dihydrospiro[indoline-3,2′-quinolin]-2-one derivatives from commercial isatin, anilines, and *trans*-isoeugenol.

power consumption (low reaction temperatures). Initially, it was possible to use directly the anise EO as precursor (without previous purification) and the star anise seeds, transforming the complex essential oil mixture into a unique THQ molecule with defined stereochemistry. Firstly, the anise essential oil was extracted by microwave-assisted hydrodistillation technique affording 3% yield from the star anise dry seeds. The oil obtained was characterized by GC-MS, which showed that the *trans*-anethole content was 93% (by weight). This EO, when treated with anilines and benzaldehyde in the presence of $BF_3 \cdot OEt_2$ catalyst, afforded the THQs in good yields. In the case of anise seeds, there was no need for extraction with $scCO_2$ or hydrodistillation because of the use of a supercritical system; on direct treatment with the anilines and the benzaldehyde in the hermetic system, the anise seeds generated the 2,4-diaryl-3-methyl-1,2,3,4-tetrahydroquinolines **28** *in situ* (Kouznetsov et al. 2007) (Scheme 6.11).

As in the anise study, the only minor drawback was the extraction yield on the hydrodistillation process. Therefore, is possible to think on the *trans*-isoeugenol and its role as suitable dienophile for imino DA reaction as well as to look for a natural source of this phenylpropene. However, the abundance of this special molecule in plants is reduced (Suhrez et al. 1993; Nakamura et al. 2006; Setzer et al. 2006). For this reason, the principal form to obtain *trans*-isoeugenol is through its isomerization under basic conditions from isoeugenol, which is found in higher quantities in the clove bud EO (Červeňý et al. 1987). Eco-friendly parameters have been employed for the allyl derivatives' isomerization including basic support solids (10% KOH/Al_2O_3) (Srivastava et al. 2003). In this sense, the natural source of eugenol, that is, aromatic dried flower buds of the tree *Syzygium aromaticum*, was processed. Employing conventional warming in the hydrodistillation procedure, the clove bud essential oil obtained in 11.7% yield indicated the presence of eugenol as the principal component (60.5%). This EO was submitted to an isomerization with the modifications of Jasra's procedure on the basic heterogeneous catalysts (Srivastava et al. 2003). Under these conditions, it was possible to obtain a crude with *cis/trans*-isoeugenol as the principal component. The next step was to take, one more time, a lignan natural model in the synthesis of new aza-lignans. Within the [4+2] cycloaddition parameters, the $BF_3 \cdot OEt_2$-catalyzed reaction of the same alkene

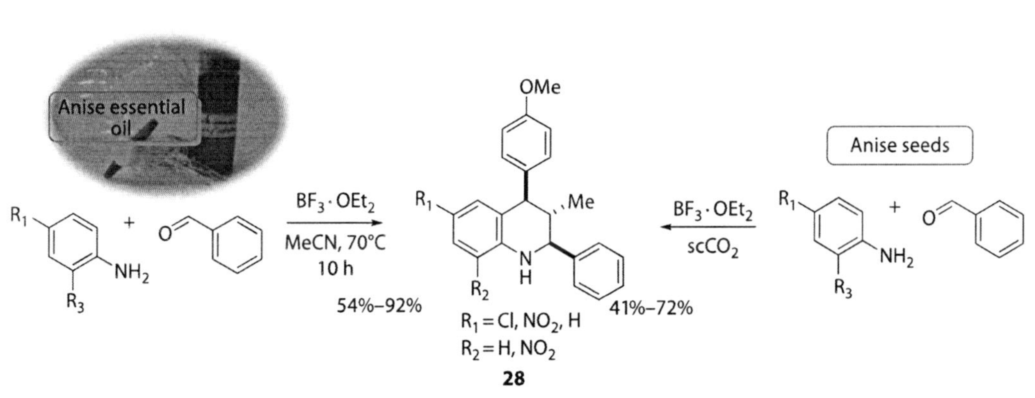

SCHEME 6.11 Synthesis of new C-4 anisyl-substituted THQs with essential oil and seeds of anise.

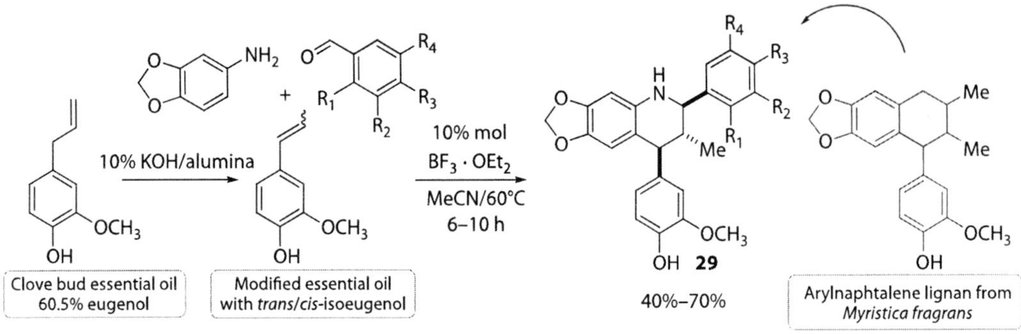

SCHEME 6.12 Synthesis of new heterolignan-like 2,4-diaryl-3-methyl-1,2,3,4-tetrahydroquinolines **29** with clove bud essential oil.

(isoeugenol), 3,4-methylendioxyaniline, and benzaldehydes was carried out to give a new series of lignan-like products **29** in 40%–70% yields (Merchan Arenas et al. 2011) (Scheme 6.12).

Therefore, it appeared that the potential of nature can be utilized by extracting some of its principal compounds (*trans*-anethole and *cis*/*trans*-isoeugenol) by synthetic procedures (Povarov reaction), that conduces to interesting molecular prototypes—the THQ system. Different functional groups have been inserted in the final products, exhibiting potent pharmaceutical properties such as antitumor, antioxidant, and anti-Alzheimer's.

With a new, small molecular library, it was possible to proceed with some biological assays for the synthesized THQ molecules. Firstly, the capacity of the THQ to inhibit the AChE activity employing the modified Ellman's method (Ellman et al. 1961) was compared. Considering cholinergic hypothesis as reference (Francis et al. 1999), biological assays of the THQ series **17** (Scheme 6.7) was performed—the results indicated the 8-cyano-4-(4-hydroxy-3-methoxyphenyl)-3-methyl-2-phenyl-1,2,3,4-tetrahydroquinoline as the best AChE inhibitor with a IC_{50} of 15.35 μM (Figure 6.11), compared with galanthamine (IC_{50} = 0.75 μM), the most popular compound used in Alzheimer therapy (Merchan et al. 2008).

Moreover, the phenylpropanoid-THQ mini-libraries were studied in the growth inhibition of cancer cellular lines due to their structural similitude with other potential anticancer agents such as combretastatin A4 (Srivastava et al. 2005; Tron et al. 2006), indanocine (Leoni et al. 2000), (iso)eugenols, and analogues (Hume et al. 1984; Kozam et al. 1995; Kim et al. 2006; Carrasco et al. 2008), all of which bear the 4-methoxy-3-hydroxyphenyl moiety in common (Figure 6.12).

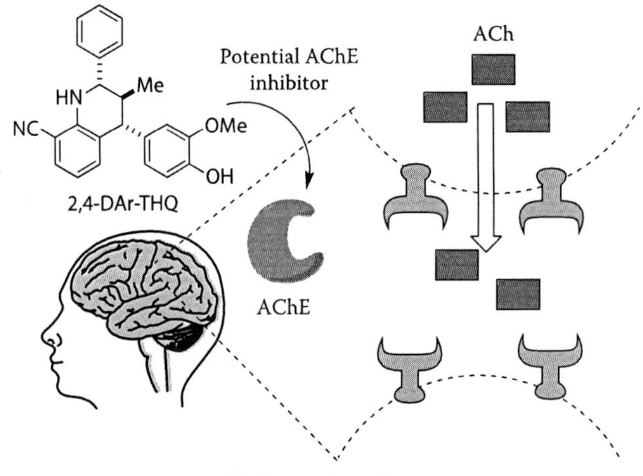

FIGURE 6.11 Cholinergic hypothesis and the THQ-AChE inhibitor.

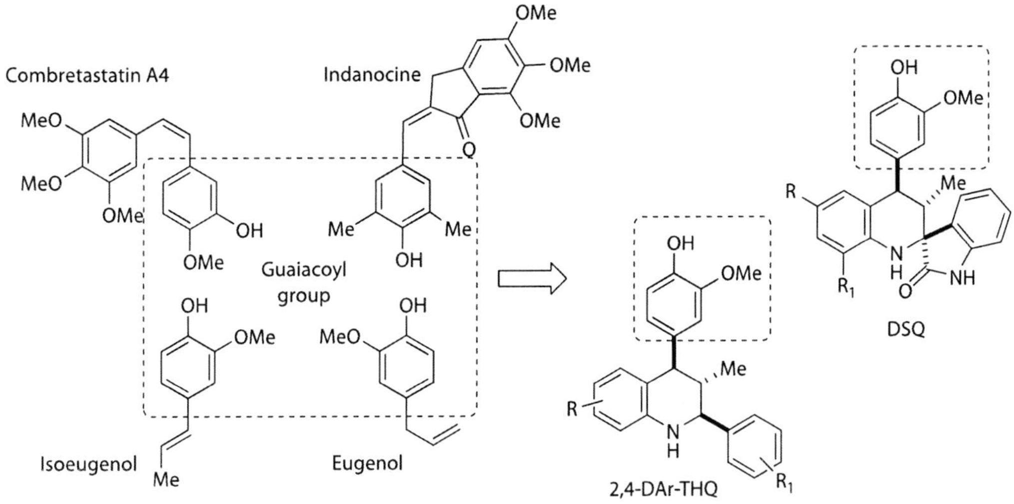

FIGURE 6.12 Logic models anticancer agents with guaiacoyl moiety.

Hence, several polyfunctionalized 2,4-DAr-THQs, comp. **17**, **18** (Schemes 6.7 and 6.8) DSQs, and comp. **27** (Scheme 6.10) were tested for their potential human tumor cell growth inhibitory effect on MCF-7, SKBR-3, PC3, HeLa, and non-tumor cells (primary culture of human dermis fibroblast—control cells).

Interestingly, almost all tetrahydroquinoline compounds from both groups exhibited moderate (IC_{50} = 13.36–15.88 μM) to good (IC_{50} = 7.99–9.48 μM) cytotoxic activity, being more effective on breast carcinoma cell line MCF7 (Kouznetsov et al. 2010a).

Given these excellent results with cytotoxic activity, we continued to explore the antitumor potential of the synthetic 8-substituted THQ derivatives (DM8 and DM12). Thus, the best cancer cell line growth inhibitor, the 8-NO_2-THQ (DM12), was evaluated by its cytotoxic effects in adjuvant therapy combined with two recognized drugs, paclitaxel and gemcitabine. The results showed a great synergism of DM8 (8-NH_2-THQ) and DM12 when mixed with the anticancer drugs, intensifying their cytotoxic activity on both cell lines at concentrations below 1 μg/mL (Figure 6.13).

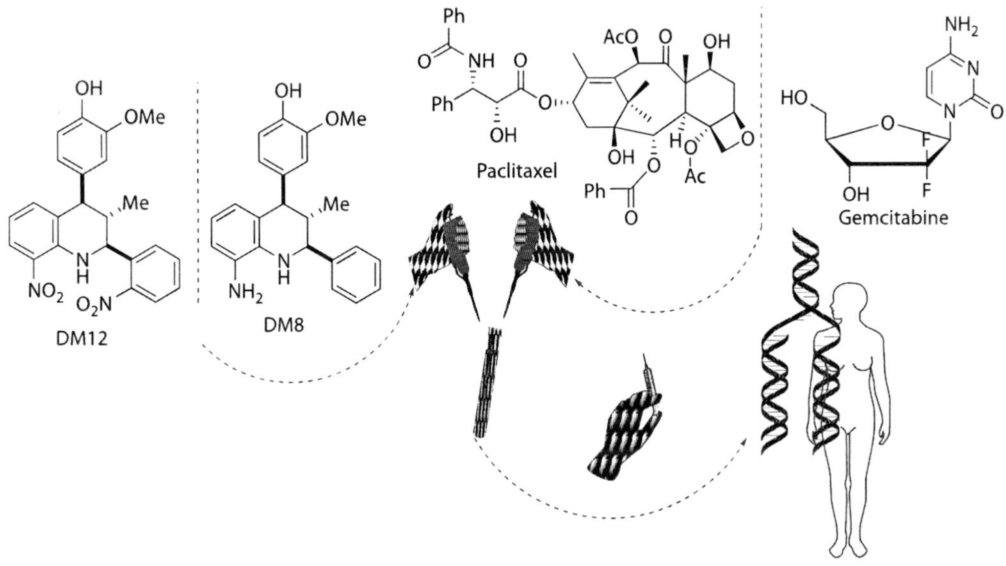

FIGURE 6.13 Cancer therapeutic drug molecules with DM8 and DM12 in adjuvant therapy.

During these studies, the compound DM12 was identified as a new perspective and safe agent for adjuvant therapy (Muñoz et al. 2011).

All these experimental results established the great importance of the rational design of bioactive molecules, keeping in mind the natural models to copy the final product and use the principal components of aromatic plants as precursors in classic synthetic tools. On the other hand, the applicability of these ideas in the chemical industry can contribute to the preparation of new potential drugs throughout green chemistry parameters. Thus, our research group has been deeply engaged in exploring such significant areas with an emphasis on diverse concepts, raw materials, green chemistry, bioactivity, organic synthesis, etc. Moreover, another structural core was included, the C_6-C_1 fragment, while looking for THQ and quinoline molecules with biological action.

6.3 FORMYL-SUBSTITUTED C_6-C_1 PHENOLIC COMPOUNDS (PIPERONAL, (ISO)VANILLINS) AS AROMATIC ALDEHYDES IN [4+2] CYCLOADDITION PROCESS (MULTICOMPONENT IMINO DIELS–ALDER REACTION) AND MULTICOMPONENT CONDENSATION

EOs are made up of various classes of organic compounds such as terpenes, alcohols, esters, phenols, ketones, and aldehydes, possessing a number of biological activities that include antifungal (Battinellia et al. 2006), anti-inflammatory (Borrelli et al. 2002), disinfectant (Johnson et al. 1982), and sedative (Lee and Perez 2003). EOs impart the citrus-like fragrance in plants like *Melissa*, lemongrass, and *Citronella*. Formyl phenolic derivatives, such as vanillin, a major constituent of devil's-claw (*Proboscidea louisianica*) essential oil (Riffle et al. 1990, 1991), are released as root exudates or commonly found in the soil from decomposing plant litter (Whitehead et al. 1982). Furthermore, it has been proved that vanillin is ubiquitous in soil since it constitutes a degradation product of lignin (Sjoblad and Bollag 1981). This phenolic aldehyde is known for its antioxidant and antimicrobial properties and as a food preservative (Fitzgerald et al. 2004; Karathanosa et al. 2007), as well as for other purposes also, such as a cosmetics and drugs constituent (Davidson and Naidu 2000). Likewise, piperonal (heliotropin), another formyl phenol found as a minor natural component of vanilla extract, is a common additive in inexpensive synthetic vanilla flavor and candies (Belay and Poole 1993), and as an anti-inflammatory stabilizer in pharmaceutical and cosmetic products (Stiehm and Baur 2002).

On the other hand, plants from Rutaceae family are studied much due to alkaloids content (Waterman 1975), as alkaloids are pharmacologically important in the development of drug candidates (Salem et al. 2006). Although formyl-substituted phenolic components (piperonal, (iso)vanillins) are not so abundant within the EOs in comparison to the anetholes and eugenols, they can, thus, play an important role in generating new chemical libraries, especially in the construction of alkaloids analogues.

Dubamine, a quinoline alkaloid with 2-(benzo[*d*][1,3]dioxol-5-yl)quinoline structure (Figure 6.14) and the chemical constituent of *Haplophyllum dubium* (Rutaceae), has been reported to exhibit antimicrobial activity (Bessonova and Yunusov 1977). Although a number of strategies are known for the synthesis of this alkaloid employing coupling reactions catalyzed by different metals (Ali et al. 1992), they do not allow the quinoline system functionalization, which is an important factor in the discovery and development of a bioactive molecule to be proposed as a leading compound.

Based on these observations, we decided to design and develop new quinoline or piperidine molecules, analogues of the alkaloid dubamine **31** (Meléndez et al. 2007), as well as some other analogues of girgensohnine **32** (found in *Girgensohnia oppositiflora*, unknown bioactivity) (Figure 6.14) (Kouznetsov et al. 2010). Additionally, within all these new structures, the "natural" piperonaldehyde and hydroxyphenyl or vanillin frameworks were introduced by means of [4+2] cycloaddition reactions as depicted in Scheme 6.13.

Final products obtained here, among other quinoline derivatives prepared by similar procedures, were studied for its antifungal activity (Meléndez Gomez et al. 2008), as they share some structural characteristics with similar antifungal quinolines reported by our group (Urbina et al. 2000;

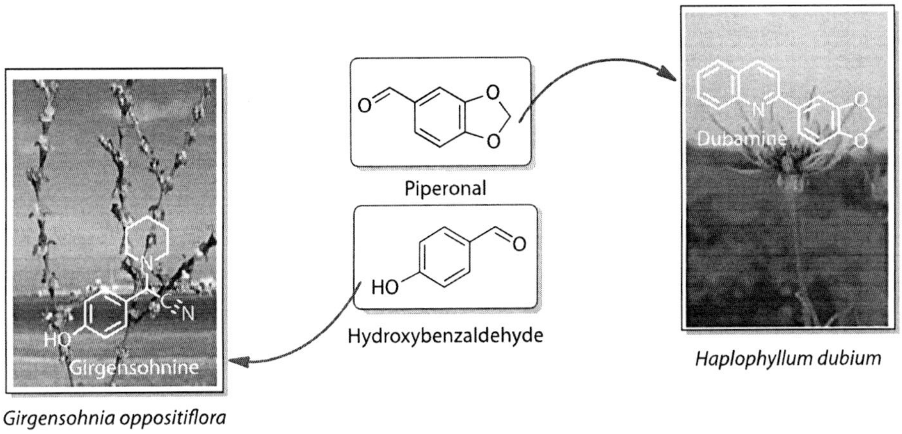

FIGURE 6.14 Structures of alkaloids girgensohnine and dubamine and its natural sources.

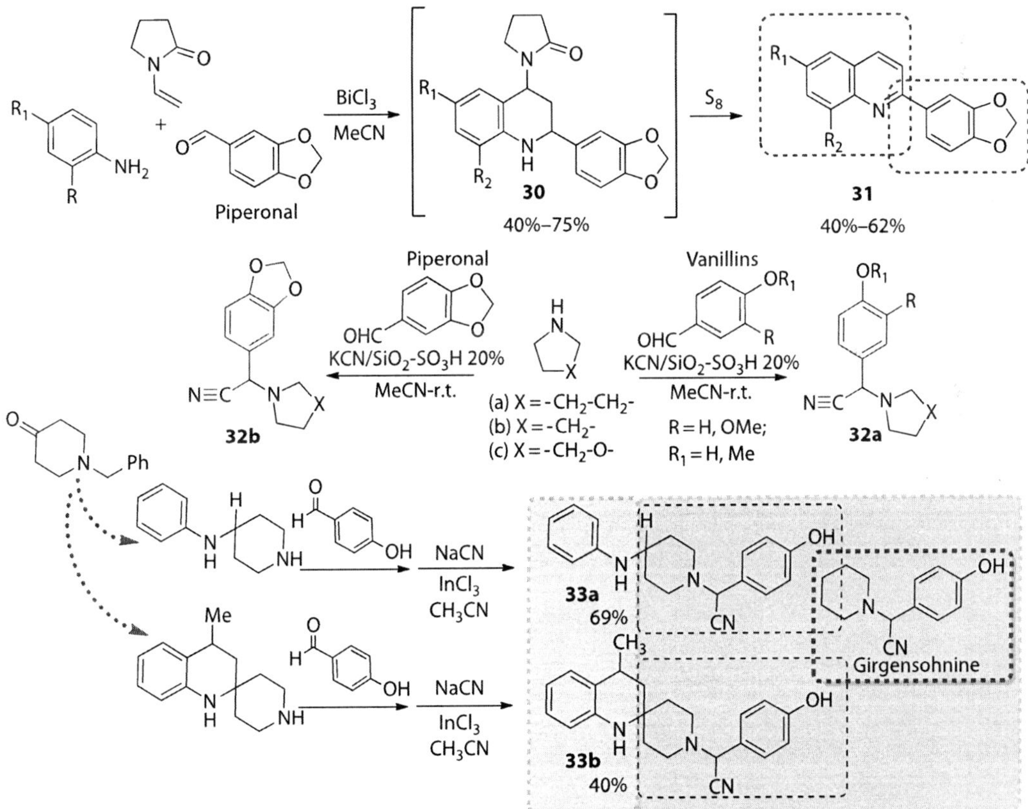

SCHEME 6.13 Synthesis of analogues of alkaloids dubamine **32** and girgensohnine **33** containing the piperonal or vanillin frameworks.

Vargas et al. 2003). Furthermore, in a different work, the antiparasitic activity of intermediates **31** was analyzed alone with other tetrahydroquinolines prepared from different aldehyde precursors (Kouznetsov et al. 2006a, 2007).

As an additional task of our ongoing efforts to introduce natural frameworks in our heterocyclic structures, employing formyl-substituted phenolic precursors, essentially vanillins, and using them in multicomponent cyclo-condensation, it was possible to create a new series of hybrid molecules

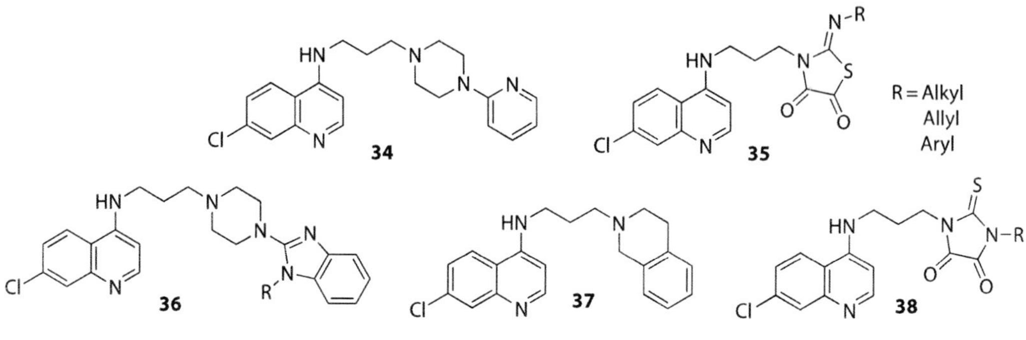

FIGURE 6.15 Hybrids of chloroquine reported in literature.

based on the structural characteristics of diverse drugs as the antimalarial chloroquine. These "double drugs" or molecular hybridization strategies, which utilize new chemical entities with two (or more than two) different *N*-heterocyclic skeletons, are valid and perspective approaches to create new antimalarial agents (Kouznetsov and Gómez-Barrio 2009; Meunier 2008). These strategies have the potential to overcome the drug-resistant parasite problem, and wonderful examples such as trioxaquines or artemisinin-quinine hybrid have been reported (Walsh et al. 2007). Some examples of such hybrid structures are presented in Figure 6.15 (Musonda et al. 2009; Kumar et al. 2010).

Based on these results, we have been motivated to incorporate the (iso)vanillin moiety, thereby, synthesizing new series of aminoquinoline-containing dual inhibitors or "double drugs" that would potentially inhibit hemozoin formation and another target within *P. falciparum*. Amino side functions were introduced in these compounds by interaction of 4,7-dichloroquinoline (4,7-DCQ) with refluxing α,ω-DAA in excess and absence of solvent, affording DAQ. Finally, the target compounds CQT were prepared by one-pot three-component reaction of DAQ, formyl phenols (ArCHO), and α-mercaptoacetic acid in ratios of 1:2.5:2.5, respectively (Scheme 6.14).

Among these compounds, two molecules have shown good pharmacological parameters above the chloroquine efficacy (Rojas Ruiz et al. 2011) that will be described in Section 6.4.

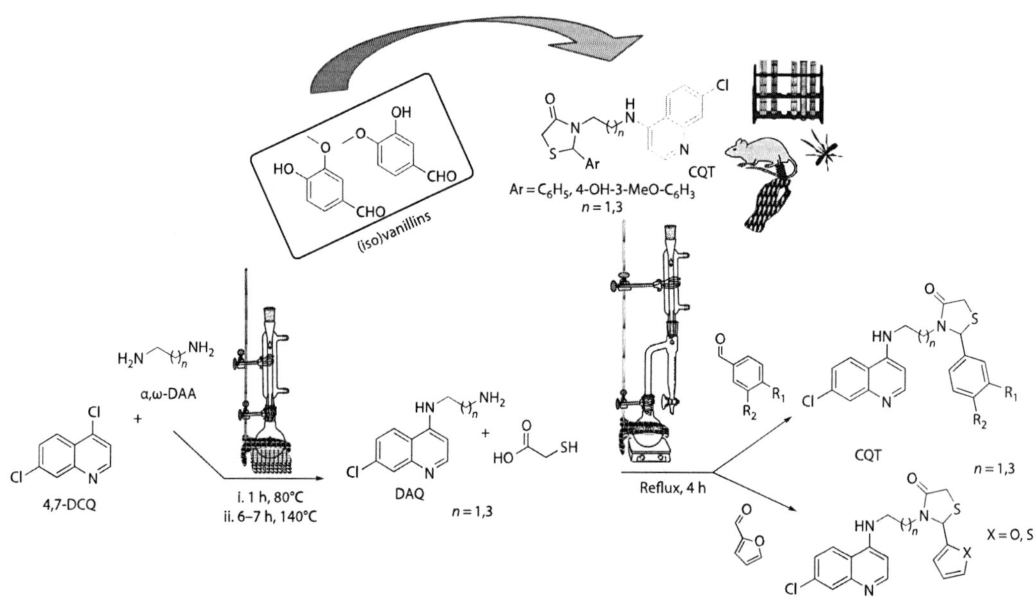

SCHEME 6.14 New antimalarial hybrids based on chloroquine and thiazolidin-4-ones incorporating (iso) vanillin frameworks.

6.4 BIOSCREENING AND *IN SILICO* CALCULATED PHYSICOCHEMICAL PROPERTIES OF MOLECULES OBTAINED

Molecular structure determines every single property of a particular compound. When these properties interact with the physical environment, they generate the physicochemical properties (e.g., solubility). Correspondingly, when these properties interact with proteins, they develop the biochemical properties (metabolism) and at the final point, when these biochemical and physicochemical properties interact with the biological systems as a whole, they generate the pharmacokinetic and toxicity properties (Kerns and Di 2008) (Figure 6.16).

Nowadays, there is thus a considerable discussion about the importance of optimization of organic compounds in regard to ADME/Tox properties, combined with their pharmacological properties (efficacy and selectivity), as directed toward the enhancement of new pharmaceuticals discovery achievements. These properties, commonly termed as drug-like properties after the pivotal work of Lipinski (2000), are an integral element of drug discovery projects and an important area of antiparasitic agents' development, in particular.

Although controlling these properties by means of structural modifications is an organic chemists' task, the previous detailed analysis of the structure of potentially drug-like compounds may help to avoid the synthesis and biological assessment of molecules with a low or null activity. In this sense, we have developed the design and synthesis of new 4-aminoquinoline drug-like molecules through the systematic analysis of their structures and by incorporating some biologically privileged scaffolds employing a new "property-based design" strategy (Rojas and Kouznetsov 2011). Table 6.1 shows how conversely to the chloroquine as reference, the calculations obtained demonstrate that all analyzed compounds contain high bioavailability properties and fulfill all parameters established by Lipinski (molecular weight = 269.73–408.91, log P = 2.66–4.92, nON = 2–6, and nOHNH = 0–4) (Sridharan et al. 2011). TPSA has been shown to be a good descriptor characterizing drug absorption, including intestinal absorption, bioavailability, Caco-2 permeability, and blood–brain barrier penetration (Ertl et al. 2000; Chohan et al. 2010). Prediction results for compounds **39–46** show TPSA values between 63 and 69 Å^2 confirming their drug-relevant properties (Table 6.1).

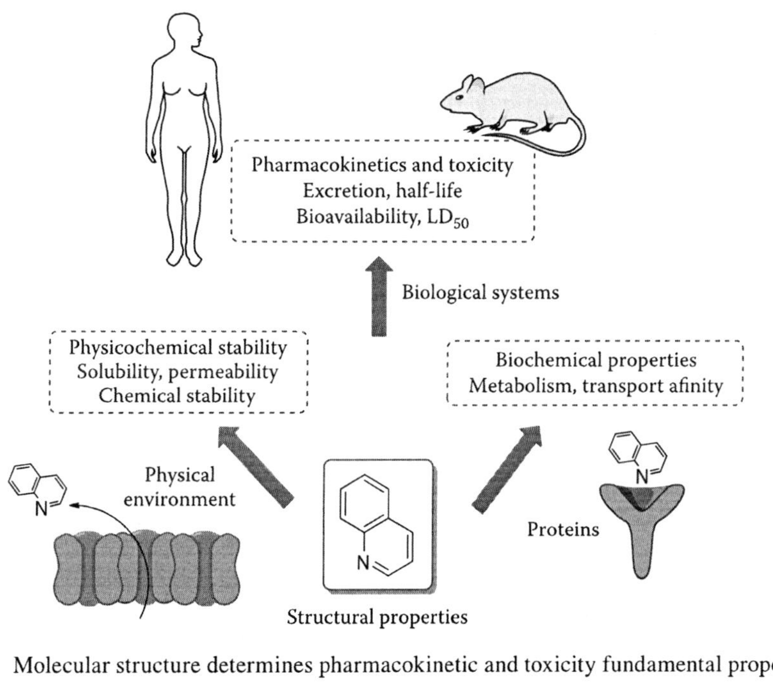

FIGURE 6.16 Molecular structure determines pharmacokinetic and toxicity fundamental properties.

TABLE 6.1
Calculated Lipinski's Rule of Five for the 4-Aminoquinoline Bearing Natural-Like *N*-Heterocyclic Frameworks

Compound			Parameter						
		n	Log P	TPSA[a] (Å²)	MW (g/mol)	*n*ON[b]	*n*OHNH[c]	RBN[d]	Violations
39	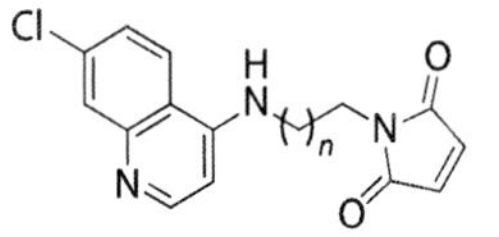	1	2.16	63.99	301.7	5	1	4	0
40		2	2.43	63.99	315.8	5	1	5	0
41	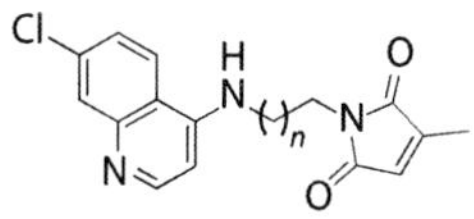	1	2.54	63.99	315.8	5	1	4	0
42		2	2.09	63.99	329.8	5	1	5	0
43	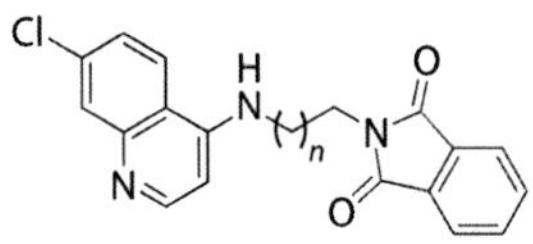	1	2.16	71.53	369.8	6	1	4	0
44		2	2.43	71.53	383.8	6	1	5	0
45	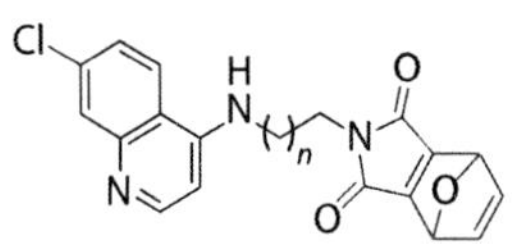	1	3.89	63.99	351.8	5	1	4	0
46		2	4.17	63.99	365.8	5	1	5	0
Chloroquine	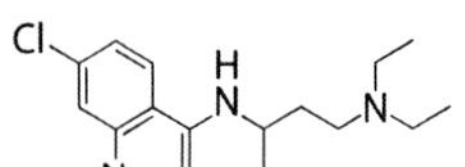 CQ	—	5.00	28.16	319.9	3	1	8	1

[a] Polar surface area (solubility parameter).
[b] Number of hydrogen bond acceptors.
[c] Number of hydrogen bond donors.
[d] Number of routable bonds.

In order to assess the possible pharmacological properties of hybrids **39–46**, a toxicity risk profile evaluation was performed employing the OSIRIS software (Organic Chemistry Portal 2001), as it may point to the presence of some fragments generally responsible for the irritant, mutagenic, tumorigenic, or reproductive effects in these molecules (El-Azab et al. 2010). As shown in Table 6.2, with the exception of compounds **45** and **46**, all desired products represent low or moderate biological risks.

Furthermore, we have used the OSIRIS program to predict the compounds' drug score (Figure 6.17) that clearly indicates the biological potential of the designed 4-aminoquinoline series, demanding for their detailed bioactivity assessments so as to assess their possible use as lead compounds with a low toxicity risk profile. After a similar in silico structural properties analysis of our formyl phenolic skeleton containing quinoline hybrids mentioned earlier, we have successfully achieved the synthesis of two prominent antimalarial agents.

Using chloroquine as 100% of inhibition positive control, it has been evaluated that compounds **47** and **48** (from CQT series, Scheme 6.14) inhibit the growth of *Plasmodium berghei* ANKA in infected mice by 80% and 100%, respectively, at the same dose (Ewesuedo et al. 2001); the results are presented in Table 6.3 (Rojas Ruiz et al. 2011).

TABLE 6.2
Toxicity Risk Profile for Hybrids of 4-Amino-7-Chloroquinoline and Cyclic Imides

Compound	Structure	*n*	Toxicity Risk: Mutagen	Tumor	Irritant	Reproductive	Drug Score
39	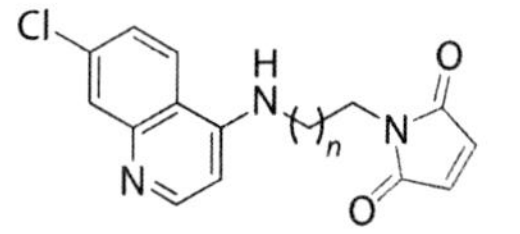	1	moderated risk	nonassociated risk	nonassociated risk	nonassociated risk	0.45
40		2	moderated risk	nonassociated risk	nonassociated risk	nonassociated risk	0.43
41	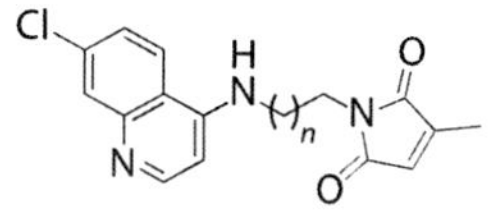	1	moderated risk	nonassociated risk	nonassociated risk	nonassociated risk	0.80
42		2	nonassociated risk	nonassociated risk	nonassociated risk	nonassociated risk	0.77
43	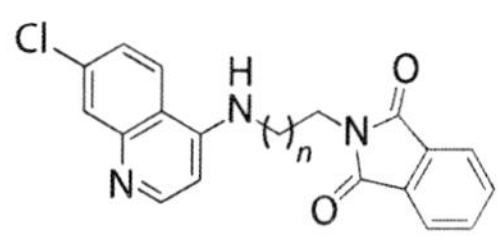	1	moderated risk	nonassociated risk	nonassociated risk	nonassociated risk	0.65
44		2	moderated risk	nonassociated risk	nonassociated risk	nonassociated risk	0.62
45	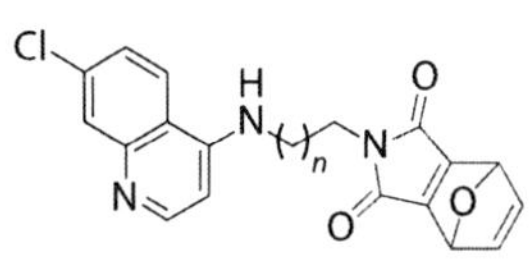	1	moderated risk	nonassociated risk	nonassociated risk	high risk	0.17
46		2	moderated risk	nonassociated risk	nonassociated risk	high risk	0.16
Chloroquine	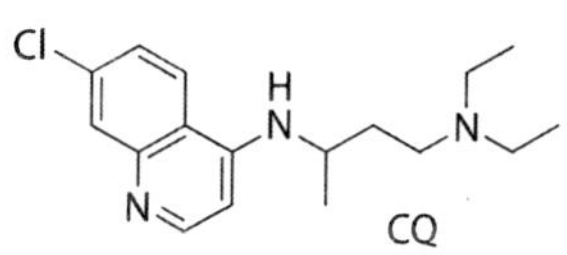	—	nonassociated risk	nonassociated risk	nonassociated risk	nonassociated risk	0.67

Nonassociated risk; moderated risk; high risk.

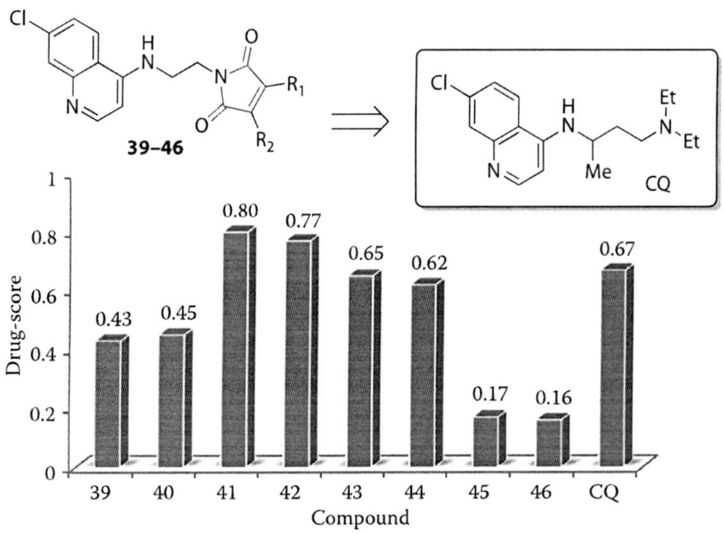

FIGURE 6.17 *Drug-score* for the 4-amino-7-chloroquinolinic hybrids **39–46**.

TABLE 6.3
***In Vivo* Assay (4-Day Suppressive Test) against *P. berghei* ANKA Using 10 mg/kg/Day[a] vs. Distribution Physicochemical Parameters**

Tested Comp.	% Inhibition	Log P^b	Log $D_{5.2}{}^c$	Log $D_{7.4}{}^c$
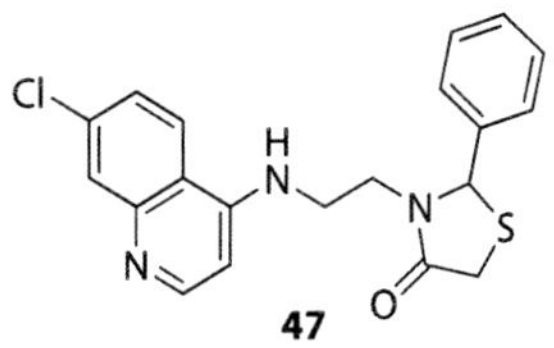 **47**	80	4.46	1.72	3.59
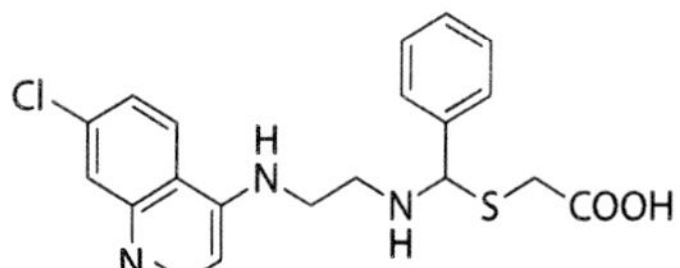 **47 M**	nt[d]	3.81	0.21	0.23
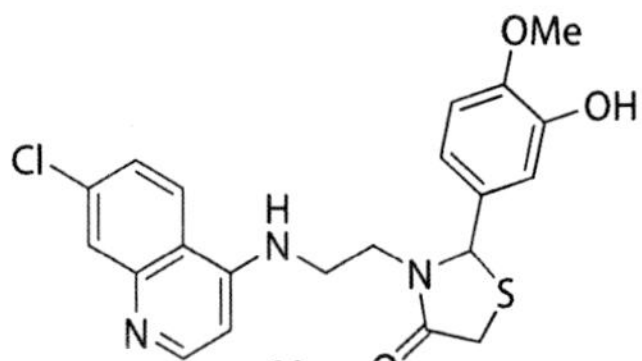 **48**	100	3.33	1.28	3.13
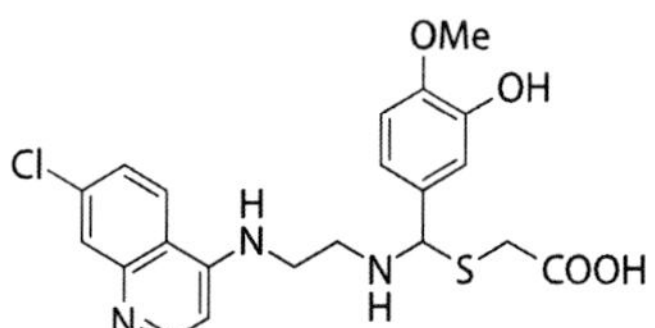 **48 M**	nt[d]	3.15	0.04	0.05
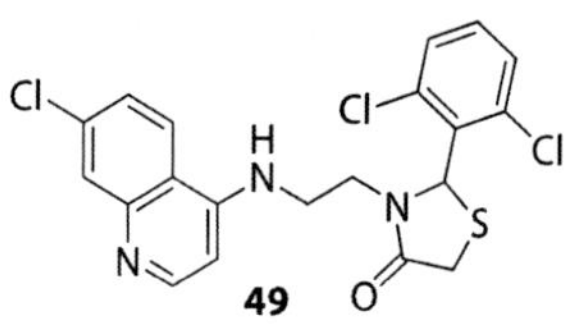 **49**	76[e]	5.25	3.86	4.95
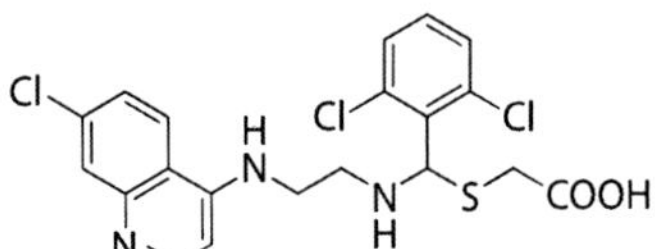 **49 M**	nt[d]	5.07	1.56	1.36
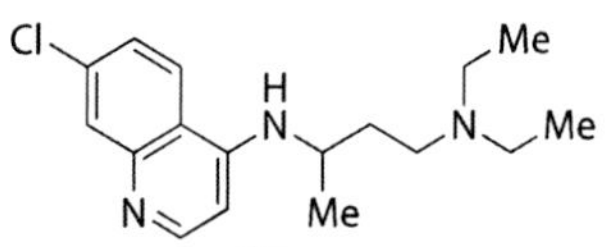 **CQ**	100	3.27	−1.42	1.92

[a] Each group of mice was treated intraperitoneally i.p.
[b] Theoretical values log *P* were calculated using commercially available ACD LAB 6.0 program.
[c] Theoretical values log *D* were calculated using the on line available SPARC V4.5 program.
[d] Not tested *in vivo* assay.
[e] *In vivo* assay (4-day suppressive test) against *P. yoelli* (N-67 strain), data from Solomon et al. 2007.

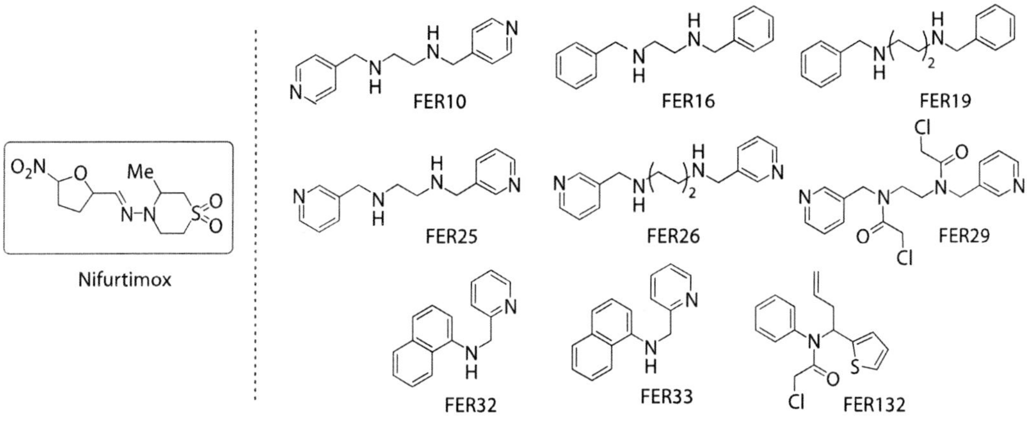

FIGURE 6.18 Structure of nifurtimox and experimentally evaluated compounds.

The parameters analyzed here demonstrated the high GI track absorption (pH 3–7) and lipophilic properties of these hybrids. A good correlation between the calculated distribution coefficients at pH 7.4 and pH 5.2 (log $D_{5.2}$ and log $D_{7.4}$ parameters) (Kerns and Di 2008) and the inhibition percentages for the tested compounds was observed. Moreover, when compared with similar reported hybrids **49** (Solomon et al. 2007), these ADME properties are enhanced by replacing the chlorine atoms over the aryl moiety. Comparing the calculated parameters for the potentially more active metabolites **47M**, **48M**, and **49M** obtained from the metabolic opening of the thiazolidinone ring of hybrids **47**, **48**, and **49** as shown in Table 6.3, it can be confirmed that possible metabolite **48M** has improved lipophilic properties. According to these theoretical values, our most active and lipophilic compound **48** may have an improved absorption and distribution above the chloroquine, as observed in its *in vivo* antimalarial activity, as well as in its *in vitro* activity against the chloroquine-resistant Dd2 *Plasmodium falciparum* strain.

In a different work, a quantitative structure–activity relationship (QSAR) has been undertaken with an intention of classifying and designing new anti-trypanosomal compounds in a rational way by means of non-stochastic and stochastic bond-based quadratic indices. A data set of 440 organic chemicals, 143 with anti-trypanosomal activity and 297 having other clinical uses, were used to develop QSAR models based on linear discriminant analysis (Castillo-Garit et al. 2010). As an experiment of virtual lead generation against epimastigote forms of *Trypanosoma cruzi,* four compounds (FER16, FER32, FER33, and FER132) showed more than 70% of epimastigote inhibition (Anti-epimastigotes percentage,% AE) at a concentration of 100 μg/mL (86.74%, 78.12%, 88.85%, and 72.10%, respectively) and two of these chemicals, FER16 (78.22% of AE) and FER33 (81.31% of AE), also showed good activity at a concentration of 10 μg/mL. At the same concentration, compound FER16 showed a lower value of cytotoxicity (15.44%), and compound FER33 showed a very low value of 1.37% (Figure 6.18).

Taking into account the results obtained here, it is worth mentioning that these three compounds can be optimized in forthcoming works, but we consider that compound FER33 is the best candidate. Even though none of them proved to be more active than nifurtimox, these results constitute a step forward in the search for efficient ways to discover new lead anti-trypanosomal agents.

6.5 CONCLUSION AND FUTURE PROSPECTS

The essential oils of diverse tropical aromatic plants are composed of huge second metabolism products and play an important role in the development of natural sciences. Throughout the history of human civilization, essential oils were used for medicinal, pharmacological, aromatherapy, or culinary purposes.

Nowadays, it is evident that vegetable biomass including essential oils is not only a source of energy and food but also a source of fine chemical reagents. The essential oils are highly variable and complex mixtures of constituents and are also isolable and renewable biomass components.

A great advantage of working with essential oils is that most of them are distilled into a single process offering valuable constituents quickly. Moreover, some aromatic tropical plants grow fast and give a large fraction of essential oil (up to 2%–10%). Thus, if you have a plant that yields the highest possible essential oil, up to 10%, and whose main component (or better, only) content reaches up to 55%, there is great possibility of using this oil as a reagent in chemical synthesis to obtain new products with a higher added value, for example, pharmacological, biological, and physical properties. On this basis, interest in the essential oils as fine chemical reagents will be enhanced in future. Not only aromatic (arenes-C_6) molecules but also other principal constituents of the essential oils (e.g., terpenic compounds) that have interesting structural motifs, can be used in diverse product preparation.

Chemical diversity of both functionalities, phenolics and terpenes, allows the generation of a novel structural and skeletal diversity that are present in numerous natural products. With the spotlight fixed on green synthetic procedures, it would not be surprising to find more utilization of main components of the essential oils in chemical synthesis under green reaction conditions. We hope this review helps to inspire readers into further discoveries and innovations in chemical transformations of the essential oils as inexpensive, available, and renewable reagents.

ABBREVIATIONS

ABTS+•	2,2′-Azino-bis(3-ethylbenzothiazoline-6-sulphonic acid) radical cation
AcOH	acetic acid
AChE	acetylcholinesterase
ADME/Tox	adsorption, distribution, metabolism, excretion, and toxicity
AE	anti-epimastigotes
c	*cis*-configuration
C_6-C_1	methylbenzene
C_6-C_3	arylpropane
Caco-2	human epithelial colorectal adenocarcinoma cells
CQ	chloroquine
CQT	chloroquine thiazolidinone hybrids
DA	Diels–Alder
DAA	diaminoalkenes
DACA	N-[2-(Dimethylamino)ethyl]acridine-4-carboxamide
DAQ	diaminoquinoline
2,4-DAr-THQs	2,4-Diaryl-3-methyl-1,2,3,4-tetrahydroquinolines
DCQ	dichloroquinoline
DNA	deoxyribonucleic acid
DSQs	dihydrospiro[indoline-3,2′-quinolin]-2-ones
EOs	essential oils
GC-MS	gas chromatography-mass spectrometry
HeLa	cervical epithelial carcinoma
HIV	human immunodeficiency virus
HL-60	human promyelocytic leukemia cells
HPA	heteropolyacid
HPLC	high-performance liquid chromatography
IC_{50}	half maximal inhibitory concentration
LD_{50}	median lethal dose
Log *P*	logarithm of the ratio of the concentrations of the un-ionized solute
MC4	melanocortin receptors

MCF-7	breast carcinoma, no overexpression of the HER2/c-erb-2 gene
MeCN	acetonitrile
MeOH	methanol
min	minute
MPA	molybdenum phosphoric acid
MW	molecular weight
NMR	nuclear magnetic resonance
nOHNH	hydrogen bond donors (nitrogen or oxygen atoms with one or more hydrogen atoms)
nON	hydrogen bond acceptors (nitrogen or oxygen atoms)
PAL	phenylalanine ammonia lyase
PC3	prostate carcinoma
PEG	poly ethylene glycol
PhMe	toluene
QSAR	quantitative structure activity relationship
r	reference
r.t.	room temperature
$scCO_2$	supercritical CO_2
SKBR-3	breast carcinoma, overexpresses the HER2/c-erb-2 gene
t	*trans*-configuration
TAL	tyrosine ammonia lyase
TAS-103	6-{[2-(Dimethylamino)ethyl]amino}-3-hydroxy-7*H*-indeno[2,1-*c*]quinolin-7-one dihydrochloride
TEAC	trolox-equivalent antioxidant capacity
TFA	trifluoroacetic acid
THQ	tetrahydroquinoline
TMTHQ	2,3,4,7-tetramethyl-1,2,3,4-tetrahydroquinoline
TPA	tungsten phosphoric acid
TPSA	thermodynamic polar surface area
TsOH	p-toluenesulfonic acid

REFERENCES

Ali, N. M.; Mckillop, A.; Mitchell, M. B.; Rebedo, R., and Wallbank, P. 1992. Palladium-catalysed cross-coupling reactions of arylboronic acids with π-deficient heteroaryl chlorides. *Tetrahedron* 48: 8117–8126.

Amiri, A.; Dugas, R.; Pichot, A., and Bompeix, G. 2008. In vitro and in vivo activity of eugenol oil (*Eugenia caryophylata*) against four important postharvest apple pathogens. *Int. J. Food Mic.* 126: 13–19.

Angle, S. R., and Arnaiz, D. 1992. Formal [3+2] cycloaddition of benzylic cations with alkenes. *J. Org. Chem.* 57: 5937–5947.

Anzini, M.; Cappelli, A., and Vomero, S. 1991. Synthesis of 6-(4-methyl-1-piperazinyl)-7*H*-indeno(2,1-*c*)quinoline derivatives as potential 5-HT receptor ligand. *J. Heterocycl. Chem.* 28: 1809–1812.

Asolkara, R. N.; Schrödera, D.; Heckmannb, R.; Langb, S.; Wagner-Döblerc, I., and Laatsch, H. 2004. Helquinoline, a new tetrahydroquinoline antibiotic from *Janibacter limosus* hel 1. *J. Antibiot.* 57: 17–23.

Battinellia, L.; Danielea, C.; Cristianib, M.; Bisignanob, G.; Saijab, A., and Mazzantia, G. 2006. In vitro antifungal and anti-elastase activity of some aliphatic aldehydes from *Olea europaea* L. fruit. *Phytomedicine* 13: 558–563.

Belay, M. T., and Poole, C. F. 1993. Determination of vanillin and related flavor compounds in natural vanilla extracts and vanilla-flavored foods by thin layer chromatography and automated multiple development. *Chromatographia* 37: 365–373.

Bendale, P.; Olepu, S.; Kumar, S. P.; Buldule, V.; Rivas, K.; Nallan, L.; Smart, B. et al. 2007. Second generation tetrahydroquinoline-based protein farnesyltransferase inhibitors as antimalarials. *J. Med. Chem.* 50: 4585–4605.

Bergmann, W., and McAlee, W. 1951. The isolation of metanethole from the sponge, like the synthetic material, the sponge product *Spheciospongia vesparia*. *J. Am. Chem. Soc.* 73: 4969–4970.

Bessonova, I. A., and Yunusov, S. Yu. 1977. Quinoline alkaloids of *Haplophyllum*. *Chem. Nat. Comp.* 13: 261–277.

Blaszczyk, A., and Skolimowski, J. 2006. Comparative analysis of cytotoxic, genotoxic and antioxidant effects of 2,3,4,7-tetramethyl-1,2,3,4-tetrahydroquinoline and ethoxyquin on human lymphocytes. *Chem. Biol. Interact.* 162: 70–80.

Boger, D. L., and Weinreb, S. M. 1987. *Hetero Diels-Alder Methodology in Organic Synthesis*. San Diego, CA: Academic Press.

Borrelli, F.; Maffia, P.; Pinto, L.; Ianaro, A.; Russo, A.; Capasso, F., and Ialenti, A. 2002. Phytochemical compounds involved in the anti-inflammatory effect of propolis extract. *Fitoterapia* 73: 53–63.

Borrione, E.; Prato, M.; Scorrano, G.; Stivanello, M.; Lucchini, V., and Valle, G. 1989. Diastereofacial selectivity in the cycloaddition of chiral glyoxylate imines to cyclopentadiene and indene: synthesis of optically active tetrahydroquinolines. *J. Chem. Soc. Perkin Trans.* I: 2245–2250.

Bortolomeazzi, R.; Sebastianutto, N.; Toniolo, R., and Pizzariello, A. 2007. Comparative evaluation of the antioxidant capacity of smoke flavouring phenols by crocin bleaching inhibition, DPPH radical scavenging and oxidation potential. *Food Chem.* 100: 1481–1489.

Bruckingham, J. 2000. *Dictionary of Natural Products on CD-ROM*. New York: Chapman and Hall.

Byl, J. A. W.; Fortune, J. M.; Burden, D. A.; Nitiss, J. L.; Utsugi, T.; Yamada, Y., and Osheroff, N. 1999. DNA topoisomerases as targets for the anticancer drug TAS-103: primary cellular target and DNA cleavage enhancement. *Biochemistry* 38: 15573–15579.

Calhoun, W.; Carlson, R. P.; Crossley, R.; Datko, L. J.; Dietrich, S.; Heatherington, K.; Marshall, L. A.; Meade, P. J.; Opalko, A., and Shepherd, R. G. 1995. Synthesis and antiinflammatory activity of certain 5,6,7,8-tetrahydroquinolines and related compounds. *J. Med. Chem.* 38: 1473–1481.

Carrasco, A. H.; Espinoza, C. L.; Cardile, V.; Gallardo, C.; Cardona, W.; Lombardo, L.; Catalán, M. K.; Cuellar, F. M., and Russo, A. 2008. Eugenol and its synthetic analogues inhibit cell growth of human cancer cells (Part I). *J. Braz. Chem. Soc.* 19: 543–548.

Castillo-Garit, J. A.; Vega, M. C.; Rolon, M.; Marrero-Ponce, Y.; Kouznetsov, V. V.; Amado-Torres, D. F.; Gómez-Barrio, A. et al. 2010. Computational discovery of novel trypanosomicidal drug-like chemicals by using bond-based non-stochastic and stochastic quadratic maps and linear discriminant analysis. *Eur. J. Pharm. Sci.* 39: 30–36.

Červený, L.; Krejčiková, A.; Marhoul, A., and Růžička, V. 1987. Isomerization of eugenol to isoeugenol. *React. Kinet. Catal. Lett.* 33: 471–476.

Charlton, J. L. 1998. Antiviral activity of lignans. *J. Nat. Prod.* 61: 1447–1451.

Chavez, J.; Dos Santos, I.; Cru, F.; David, J.; Yang, S. W., and Cordell, G. 1997. A quinoline alkaloid from *Acanthosyris paulo-alvinii*. *Phytochemistry* 46: 967–968.

Chohan, Z. H.; Youssoufi, M. H.; Jarrahpour, A., and Ben Hadda, T. 2010. Inhibition of antiapoptotic BCL-XL, BCL-2, and MCL-1 proteins by small molecule mimetics. *Eur. J. Med. Chem.* 45: 1189–1199.

Clark, D. E. 2003. *In silico* prediction of blood-brain barrier permeation. *Drug Discov. Today* 8: 927–933.

Davidson, P. M., and Naidu, A. S. 2000. Phyto-phenols. In *Natural Food Antimicrobial Systems*, ed. Naidu, A. S., 265–294. London, U.K.: CRC Press.

De Koeing, A. J. 2002. The antioxidant ethoxyquin and its analogues: a review. *Int. J. Food Prop.* 5: 451–461.

DeVincenzi, M.; Silano, M.; Maialetti, F., and Scazzocchio, B. 2000. Constituents of aromatic plants: II. Estragole. *Fitoterapia* 71: 725–729.

Ding, K.; Lu, Y.; Nikolovska-Coleska, Z.; Qiu, S.; Ding, Y.; Gao, W.; Stuckey, J. et al. 2005. Structure-based design of potent non-peptide MDM2 inhibitors. *J. Am. Chem. Soc.* 127: 10130–10131.

Eastman, R. T.; White, J.; Hucke, O.; Yokoyama, K.; Verlinde, C. L. M. J.; Hast, M. A.; Beese, L. S.; Gelb, M. H.; Rathod, P. K., and Van Voorhis, W. C. 2007. Resistance mutations at the lipid substrate binding site of *Plasmodium falciparum* protein farnesyltransferase. *Mol. Biochem. Parasitol.* 152: 66–71.

El-Azab, A. S.; Al-Omar, M. A.; Ala, A. M., and Naglaa, A. A. 2010. Design, synthesis and biological evaluation of novel quinazoline derivatives as potential antitumor agents: molecular docking study. *Eur. J. Med. Chem.* 45: 4188–4198.

Ellman, G.; Courtney, K.; Andres, Jr. V., and Robert, M. F. 1961. A new and rapid colorimetric determination of acetylcholinesterase activity. *Biochem. Pharmacol.* 7: 88–95.

Ertl, P.; Rohde, B., and Selzer, P. 2000. Fast calculation of molecular polar surface area as a sum of fragment-based contributions and its application to the prediction of drug transport properties. *J. Med. Chem.* 43: 3714–3717.

Ewesuedo, R. B.; Iyer, L.; Das, S.; Koenig, A.; Mani, S.; Vogelzang, N. J.; Schilsky, R. L; Brenckman, W., and Ratain, M. J. 2001. Phase 1 clinical and pharmacogenetic of TAS-103 in patients with advanced cancer. *J. Clin. Oncol.* 19: 2084–2090.

Fadel, F.; Lafquih, S.; Soufiaoui, M., and Mazzah, A. 2004. Synthèse de nouveaux dérivés tétrahydroquinoléines et quinoléines via la réaction D'aza-Diels–Alder suivie D'aromatisation. *Tetrahedron Lett.* 45: 5905–5908.

Farhan, A.; Keehn, E.; Philip, M., and Stevenson, R. 1992. Dimerization of isoeugenol, isoeugenyl methyl ether and isoeugenyl acetate. *J. Chem. Res.* 3: 100–101.

Ferrer, J. L.; Austin, M. B.; Stewart, Jr. C., and Noel, J. P. 2008. Structure and function of enzymes involved in the biosynthesis of phenylpropanoids. *Plant Phys. Biochem.* 46: 356–370.

Fisher, M. J.; Backer, R. T.; Husain, S.; Hsiung, H. M.; Mullaney, J. T.; O'Brian, T. P.; Ornstein, P. L.; Rothhaar, R. R.; Zgombick, J. M., and Briner, K. 2005. Privileged structure-based ligands for melanocortin receptors—tetrahydroquinolines, indoles, and aminotetralines. *Bioorg. Med. Chem. Lett.* 15: 4459–4462.

Fitzgerald, D. J.; Stratford, M. J.; Gasson, J.; Ueckert, A., and Narbad, A. 2004. Mode of antimicrobial action of vanillin against *Escherichia coli, Lactobacillus plantarum* and *Listeriainnocua*. *J. Appl. Microbiol.* 97: 104–113.

Francis, P.; Palmer, A. M.; Snape, M., and Wilcock, G. K. 1999.The cholinergic hypothesis of Alzheimer's disease: a review of progress. *J. Neurol. Neurosurg. Psychiat.* 66: 137–147.

Gauthier, S. 2001. Alzheimer's disease: current and future therapeutic perspectives. *Prog. Neuro-Psychopharmacol. Biol. Psychiat.* 25: 73–89.

Griengl, H., and Foidl, G. 1981. Process for making derivatives of dimers of isoeugenol. US Patent 4256764.

Hickey, M. J. 1948. Investigation of the chemical constituents of Brazilian sassafras oil. *J. Org. Chem.* 13: 443–446.

Hume, W. R. 1984. An analysis of the release and the diffusion through dentin of eugenol from zinc oxide-eugenol mixtures. *J. Den. Res.* 63: 1262–1265.

Jacquemond-Collet, I.; Hannedouche, S.; Fabre, N.; Fourasté, I., and Moulis, C. 1999. Two tetrahydroquinoline alkaloids from *Galipea officinalis*. *Phytochemistry* 51: 1167–1169.

Jirovetz, L.; Stoyanova, A.; Buchbauer, G.; Krastanov, A.; Stoilova, I., and Schmidt, E. 2006. Chemical composition and antioxidant properties of clove leaf essential oil. *J. Agric. Food Chem.* 54: 6303–6307.

Johnson, L. L.; Shneider, D. A.; Austin, M. D.; Goodman, F. G.; Bullock, J. M., and DeBruin, J. A. 1982. Two per cent glutaraldehyde: a disinfectant in arthroscopy and arthroscopic surgery. *J. Bone Joint Surg. Am.* 64: 237–239.

Jossang, A.; Jossang, P.; Hadi, H. A.; Sévent, T., and Bodo, B. 1991. Horsfiline, an oxindole alkaloid from Horsfieldia superba. *J. Org. Chem.* 56: 6527–6530.

Juhász, L.; Kürti, L., and Antus, S. 2000. Simple synthesis of benzofuranoid neolignans from *Myristica fragrans*. *J. Nat. Prod.* 63: 866–870.

Karathanosa, V. T.; Mourtzinosa, I.; Yannakopoulou, K., and Andrikopoulosa, N. K. 2007. Study of the solubility, antioxidant activity and structure of inclusion complex of vanillin with β-cyclodextrin. *Food Chem.* 101: 652–658.

Katritzky, A. R.; Nichols, D. A.; Qi, M., and Yang, B. 1997. Lewis acid assisted reactions of n-(*a*-aminoalkyl) benzotriazoles and unactivated alkenes for the facile synthesis of 4-, 2,4-, and 3,4-substituted 1,2,3,4-tetrahydroquinolines. *J. Heterocycl. Chem.* 34: 1259–1262.

Katritzky, A. R.; Rachwal, S., and Rachwal, B. 1996. Recent progress in the synthesis of 1,2,3,4,-tetrahydroquinolines. *Tetrahedron* 52: 15031–15070.

Kerns, E. H. and Di, L. 2008. *Drug-Like Properties: Concepts, Structure Design and Methods*, pp. 6–16. London, U.K.: Elsevier Inc.

Kim, G. C.; Choi, D. S.; Lim, J. S.; Jeong, H. C.; Kim, I. R.; Lee, M. H., and Park, B. S. 2006. Caspases-dependent apoptosis in human melanoma cell by Eugenol. *Korean J. Anat.* 39: 245–253.

Kishore, D., and Kannan, S. 2004. Environmentally benign route for isomerization of safrole-hydrotalcite as solid base catalyst. *J. Mol. Cat. A Chem.* 223: 225–230.

Konishi, M.; Ohkuma, H.; Tsuno, T., and Oki, T. 1990. Crystal and molecular structure of dynemicin A: a Novel 1,5-diyn-3-ene antitumor antibiotic. *J. Am. Chem. Soc.* 112: 3715–3716.

Kouznetsov, V. V. 2009. Recent synthetic developments in a powerful imino Diels–Alder reaction (Povarov reaction): application to the synthesis of *N*-polyheterocycles and related alkaloids. *Tetrahedron* 65: 2721–2750.

Kouznetsov, V. V.; Bello Forero, J. S., and Amado Torres, D. A. 2008a. A simple entry to novel spiro dihydroquinoline-oxindoles using Povarov reaction between 3-*N*-aryliminoisatins and isoeugenol. *Tetrahedron Lett.* 49: 5855–5857.

Kouznetsov, V. V.; Bohórquez Romero, A. R., and Astudillo, S. L. 2009. A convenient procedure for the synthesis of new α-pyridinyl-substituted 7*H*-indeno[2,1-*c*]quinoline derivatives based on a three-component imino Diels–Alder reaction. *Synthesis* 24: 4219–4225.

Kouznetsov, V. V.; Bohórquez Romero, A. R.; Astudillo Saavedra, L., and Fierro Medina, R. 2006a. An efficient synthesis of new C-2 aryl substituted quinolines based on three component imino Diels-Alder reaction. *Mol. Divers.* 10: 29–37.

Kouznetsov, V. V.; Bohórquez Romero, A. R., and Stashenko, E. E. 2007. Three-component imino Diels–Alder reaction with essential oil and seeds of anise: generation of new tetrahydroquinolines. *Tetrahedron Lett.* 48: 8855–8860.

Kouznetsov, V. V., and Gómez-Barrio, A. 2009. Recent developments in the design and synthesis of hybrid molecules based on aminoquinoline ring and their antiplasmodial evaluation. *Eur. J. Med. Chem.* 44: 3091–3113.

Kouznetsov, V. V., and Merchan Arenas, D. R. 2009. First green protocols for the large-scale preparation of γ-diisoeugenol and related dihydro(1*H*)indenes via formal [3+2] cycloaddition reactions. *Tetrahedron Lett.* 50: 1546–1549.

Kouznetsov, V. V.; Merchan Arenas, D. R.; Arvelo, F.; Bello Forero, J. S.; Sojo, F., and Muñoz, A. 2010a. 4-Hydroxy-3-metoxyphenyl substituted 3-methyl-tetrahydroquinoline derivatives obtained through imino Diels-Alder reactions as potential antitumoral agents. *Lett. Drug Des. Discov.* 7: 632–639.

Kouznetsov, V. V.; Merchan Arenas, D. R., and Bohórquez Romero, A. R. 2008b. PEG-400 as green reaction medium for Lewis acid-promoted cycloaddition reactions with isoeugenol and anethole. *Tetrahedron Lett.* 49: 3097–3100.

Kouznetsov, V.; Palma, A.; Ewert, C., and Varlamov, A. 1998. Some aspects of the reduced quinoline chemistry. *J. Heterocycl. Chem.* 35: 761–785.

Kouznetsov, V. V.; Ochoa Puentes, C.; Bohórquez Romero, A. R.; Zacchino, S.A.; Sortino, M.; Gupta, M.; Vázquez, Y.; Bahsas, A., and Amaro-Luis, J. 2006b. A straightforward synthetic approach to antitumoral pyridinyl substituted 7*H*-indeno[2,1-*c*]quinoline derivatives via three-component imino Diels-Alder reaction. *Lett. Org. Chem.* 3: 300–304.

Kouznetsov, V. V.; Vargas Méndez, L. Y.; Leal, S. M.; Mora Cruz, U.; Coronado, C. A.; Meléndez Gómez, C. M.; Romero Bohórquez, A. R., and Escobar Rivero, P. 2007. Target-oriented synthesis of antiparasitic 2-hetaryl substituted quinolines based on imino Diels-Alder reactions. *Lett. Drug Des. Discov.* 4: 293–296.

Kouznetsov, V. V.; Vargas Méndez, L. Y., and Muñoz Acevedo, A. 2010. 3′,4′-Dihydrospiro[piperidine-4,2′-(1′*H*)quinoline] derivatives as new antioxidant agents with acetylcholinesterase inhibitory property. *Lett. Drug Des. Discov.* 7: 710–715.

Kouznetsov, V. V.; Zubkov, F. I.; Mora Cruz, U.; Voskressensky, L. G.; Vargas Mendez, L. Y.; Astudillo, L., and Stashenko, E. E. 2004. An efficient synthesis of hexahydro oxaisoindolo[2,1-*a*]quinoline derivatives via the Diels-Alder reactions. *Lett. Org. Chem.* 1: 37–39.

Kovács, J. 1950. The reaction of propenylphenol ether dibromides with sodium iodide. *J. Org. Chem.* 15: 15–18.

Kozam, G., and Mantell, G. M. 1995. The effect of eugenol on oral mucous membranes. *J. Den. Res.* 57: 954–957.

Kozlova, Z. G.; Ivanova, I. Ya.; Kore, S. A.; Livshits, A. Ya.; Osipova, V. P.; Tsepalov, V. F.; Shlyapintokh, V. Ya., and Livshits, A. G. 1968. Selection of stabilizers for perfumes. US Patent 224741.

Kumar, A.; Srivastava, K.; Kumar, R.; Puri, S. K., and Chauhan, P. M. S. 2010. Synthesis of new 4-aminoquinolines and quinoline-acridine hybrids as antimalarial agents. *Bioorg. Med. Chem.* 15: 7059–7063.

Kwon, H. S.; Kim, M.-J.; Jeong, H. J.; Yang, M. S.; Park, K. H.; Jeong, T. S., and Lee, W. S. 2008. Low-density lipoprotein (LDL)-antioxidant lignans from *Myristica fragrans* seeds. *Bioorg. Med. Chem. Lett.* 18: 194–198.

Lee, C. O., and Perez, C. 2003. Clinical Aromatherapy Part I: an introduction into nursing practice. *Clin. J. Oncol. Nurs.* 37: 595–596.

Leoni, L. M.; Hamel, E.; Genni, D.; Shin, H., and Carrera, C. J. 2000. Indanocine, a microtubule-binding indanone and a selective inducer of apoptosis in multidrug-resistant cancer cells. *J. Natl. Cancer Inst.* 92: 217–224.

Lipinski, C. A. 2000. Drug-like properties and the causes of poor solubility and poor permeability. *J. Pharm. Toxicol. Methods* 44: 235–249.

Lobell, M.; Molmar, L., and Kerserü, G. M. 2003. Recent advances in the prediction of blood-brain partitioning from molecular structures. *J. Pharm. Sci.* 92: 360–370.

Madrigal, B.; Puebla, P.; Peláez, R.; Caballero, E., and Medarde, M. 2003. Naphthalene analogues of lignans. *J. Org. Chem.* 68: 854–864.

McCormick, J. L.; McKee, T. C.; Cardellina, J. H., and Boyd, M. R. 1996. HIV inhibitory natural products. 26.[1] Quinoline alkaloids from *Euodia roxburghiana*. *J. Nat. Prod.* 59: 469–471.

Meléndez, C. M.; Kouznetsov, V., and Astudillo, L. 2007. Síntesis de derivados del alcaloide dubamina vía reacción imino Diels-Alder multi-componente. *Sci. Tech.* 33: 369–372.

Meléndez Gómez, C. M.; Kouznetsov, V. V.; Sortino, A. M.; Álvarez, S. L., and Zacchino, S. A. 2008. In vitro antifungal activity of polyfunctionalized 2-(hetero)arylquinolines prepared through imino Diels–Alder reactions. *Bioorg. Med. Chem.* 16: 7908–7920.

Merchán, D. R.; Vargas, L. Y., and Kouznetsov, V. V. 2008. Nuevos agentes inhibidores de la acetilcolinesterasa con fragmentos estructurales de lignanos. *Salud UIS* 36: 166–168.

Merchan Arenas, D. R.; Rojas Ruiz, F. A., and Kouznetsov, V. V. 2011. Highly diastereoselective synthesis of new heterolignan-like 6,7-methylendioxy-tetrahydroquinolines using the clove bud essential oil as raw material. *Tetrahedron Lett.* 52: 1388–1391.

Meunier, B. 2008. Hybrid molecules with a dual mode of action: dream or reality? *Acc. Chem. Res.* 41: 69–77.

Michael, J. 2001. Quinoline, quinazoline and acridone alkaloids. *Nat. Prod. Rep.* 18: 543–559.

Muller, A.; Toldy, L.; Halmi, G., and Meszaros, M. 1954. Dimeric propenyl phenol ethers. XII. The synthetic stereoisomer of diisohomogenol, diisoeugenol diethyl ether, and metanethole. *J. Org. Chem.* 19: 1533–1547.

Muñoz, A.; Sojo, F.; Merchan Arenas, D. R.; Kouznetsov, V. V., and Arvelo, F. 2011. Cytotoxic effects of new trans 2,4-diaryl-r-3-methyl-1,2,3,4-tetrahydroquinolines and their interaction with antitumoral drugs gemcitabine and paclitaxel on cellular lines of human breast cancer. *Chem. Biol. Interact.* 189: 215–221.

Musonda, C. C.; Whitlock, A.; Witty, M. J.; Burn, R., and Kaiser, M. 2009. Chloroquine-astemizole hybrids with in vitro and in vivo antiplasmodial activity. *Bioorg. Med. Chem.* 19: 481–484.

Naczk, M., and Shahidi, F. 2004. Biosynthesis, classification, and nomenclature of phenolics in food and nutraceuticals. In *Food Phenolics. Sources, Chemistry, Effects, Applications*, ed. M. Naczk, pp. 1–16, Boca Raton, FL: CRC Press.

Nakamura, K.; Matsubara, K.; Watanabe, H.; Kokubun, H.; Ueda, Y.; Oyama-Okubo, N.; Nakayama, M., and Ando, T. 2006. Identification of Petunia hybrida cultivars that diurnally emit floral fragrances. *Sci. Hort.* 108: 61–65.

Nenadis, N.; Zhang, H. Y., and Tsimidou, M. 2003. Structure-antioxidant activity relationship of ferulic acid derivatives: effect of carbon side chain characteristic groups. *J. Agric. Food Chem.* 51: 1874–1879.

Nishiyama, T. T.; Hashiguchi, Y. Y.; Sakata, S. T., and Sakaguchi, T. T. 2003. Antioxidant activity of the fused heterocyclic compounds, 1,2,3,4-tetrahydroquinolines, and related compounds-effect of *ortho*-substituents. *Polym. Degrad. Stab.* 79: 225–230.

Nogle, L., and Gerwick, W. 2003. Diverse secondary metabolites from a Puerto Rican collection of *Lyngbya majuscule*. *J. Nat. Prod.* 66: 217–220.

Ochoa Puentes, C., and Kouznetsov, V. V. 2002. Recent advancements in the homoallylamine chemistry. *J. Heterocycl. Chem.* 39: 595–600.

Ohara, H.; Kiyokane, H., and Itoh, T. 2002. Cycloaddition of styrene derivatives with quinone catalyzed by ferric ion; remarkable acceleration in an ionic liquid solvent system. *Tetrahedron Lett.* 43: 3041–3044.

Omura, S., and Nakagawa, A. 1981. Structure of virantmycin, a novel antiviral antibiotic. *Tetrahedron Lett.* 22: 2199–2202.

Oprean, R.; Oprean, L.; Tamas, M.; Sandulescu, R., and Roman, L. 2001. Essential oils analysis. II. Mass spectra identification of terpene and phenylpropane derivatives. *J. Pharm. Biomed. Anal.* 24: 1163–1168.

Organic Chemistry Portal. 2001. OSIRIS property explorer. http://www.organic-chemistry.org/prog/peo/

Pisano, M.; Pagnan, G.; Loi, M.; Mura, M. E.; Tilocca, M. G.; Palmieri, G.; Fabbri, D. et al. 2007. Antiproliferative and pro-apoptotic activity of eugenol-related biphenyls on malignant melanoma cells. *Mol. Cancer* 6: 1–12.

Planta Europa. 2011. Saving the plants of Europe. http://www.plantaeuropa.org/pe-EPCS-hot issues-MAP.htm

Povarov, L. S. 1967. αβ-Unsaturated ethers and their analogues in reactions of diene synthesis. *Russ. Chem. Rev.* 36: 656–660.

Riffle, M. S.; Waller, G. R., and Murray, D. S. 1991. Composition of essential oil from *Proboscidea louisianica* (Martyniaceae). *Proc. Okla. Acad. Sci.* 71: 35–42.

Riffle, M. S.; Waller, G. R.; Murray, D. S., and Sgaramello, R. P. 1990. Devil's-claw (*Proboscidea louisianica*), Essential oil and its components potential allelochemical agents on cotton and wheat. *J. Chem. Ecol.* 16: 1927–1938.

Rojas, F. A., and Kouznetsov, V. V. 2011. Property-based design and synthesis of new chloroquine hybrids via simple incorporation of 2-imino-thiazolidin-4-one or 1*H*-pyrrol-2,5-dione fragments on the 4-amino-7-chloroquinoline side chain. *J. Braz. Chem. Soc.* 22: 1774–1781.

Rojas Ruiz, F. A.; García-Sánchez, R. N.; Villabona Estupiñan, S.; Gómez-Barrio, A.; Amado Torres, D. F.; Pérez-Solórzano, B. M.; Nogal-Ruiz, J. J.; Martínez-Fernández, A. R., and Kouznetsov, V. V. 2011. Synthesis and antimalarial activity of new heterocyclic hybrids based on chloroquine and thiazolidinone scaffolds. *Bioorg. Med. Chem.* 19: 4562–4573.

Romero Bohórquez, A. R., and Kouznetsov, V. V. 2010. An efficient and short synthesis of 4-aryl-3-methyltetrahydroquinolines from n-benzylanilines and propenylbenzenes through cationic imino Diels-Alder reactions. *Synlett* 6: 970–972.

Romero Bohórquez, A. R.; Kouznetsov, V. V., and Doyle, M. P. 2011. $Cu(OTf)_2$-catalyzed three-component imino Diels-Alder reaction using propenylbenzenes: synthesis of 2,4-diaryl tetrahydroquinoline derivatives. *Lett. Org. Chem.* 8: 5–11.

Saleem, M.; Ja Kim, H.; Saiq Ali, M., and Lee, S. Y. 2005. An update on bioactive plant lignans. *Nat. Prod. Rep.* 22: 696–716.

Salem, M. M., and Werbovetz, K. A. 2006. Natural products from plants as drug candidates and lead compounds against leishmaniasis and trypanosomiasis. *Curr. Med. Chem.* 13: 2571–2598.

Santoro, G. F.; Cardoso, M. G.; Guimarães, L. G.; Mendonça, L. Z., and Soares, M. J. 2007. *Trypanosoma cruzi*: activity of essential oils from *Achillea millefolium* L., *Syzygium aromaticum* L. and *Ocimum basilicum* L. on epimastigotes and trypomastigotes. *Exp. Parasitol.* 116: 283–290.

Scherer, R., and Teixeira, H. 2009. Antioxidant activity index (AAI) by the 2,2-diphenyl-1-picrylhydrazyl method. *Food Chem.* 112: 654–658.

Setzer, W.; Noletto, J. A., and Lawton, R. O. 2006. Chemical composition of the floral essential oil of *Randia matudae* from Monteverde, Costa Rica. *Flav. Fragr. J.* 21: 244–246.

Shono, T.; Matsumura, Y.; Inoue, K.; Ohmizu, H., and Kashimura, S. J. 1982. Electrochemical oxidation of N,N-dimethylanilines. *J. Am. Chem. Soc.* 104: 5753–5756.

Singer, J. M.; Barr, B. M.; Coughenour, L. L.; Gregory, T. F., and Walters, M. A. 2005. 8-Substituted 3,4-dihydroquinolinones as a novel scaffold for atypical antipsychotic activity. *Bioorg. Med. Chem. Lett.* 15: 4560–4563.

Sjoblad, R. D., and Bollag, J. M. 1981. Oxidative coupling of aromatic compounds by enzymes from soil microorganisms. In *Soil Biochemistry*, eds. Paul, E. A., and Ladd, J. N., pp. 114–152. New York: Marcel Dekker, Inc.

Smollny, T.; Wichers, H.; Kalenberg, S.; Shasa, A.; Petersen, M., and Alferman, W. 1998. Accumulation of podophyllotoxin and related lignans in cell suspension cultures of linum album. *Phytochemistry* 48: 975–979.

Solecka, D. 1997. Role of phenylpropanoid compounds in plant responses to different stress factors. *Acta Physiol. Plant* 19: 257–268.

Solomon, V. R.; Haq, W.; Srivastava, K.; Puri, S. K., and Katti, S. B. 2007. Synthesis and antimalarial activity of side chain modified 4-aminoquinoline derivatives. *J. Med. Chem.* 50: 394–398.

Sridharan, V.; Suryavanshi, P., and Menéndez, J. C. 2011. Advances in the chemistry of tetrahydroquinolines. *Chem. Rev.* 111: 7157–7259.

Srivastava, V. K.; Bajaj, H. C., and Jasra, R. V. 2003. Solid base catalysts for isomerization of 1-methoxy-4-(2-propen-1-yl) benzene to 1-methoxy-4-(1-propen-1-yl)benzene. *Catal. Commun.* 4: 543–548.

Srivastava, V.; Negi, A. S.; Kumar, J. K.; Gupta, M. M., and Khanuja, S. P. S. 2005. Plant-based anticancer molecules: a chemical and biological profile of some important leads. *Bioorg. Med. Chem.* 13: 5892–5908.

Stiehm, T., and Baur, M. 2002. Use of piperonal as an anti-inflammatory additive to cosmetics and medicaments. European Patent EP0997137.

Suhrez, M.; Duque, C.; Bicchi, C.; Wintoch, H.; Full, G., and Schreier, P. 1993. Volatile constituents from the peelings of lulo (*Solanurn vestissimum* D.) fruit. *Flav. Fragr. J.* 8: 215–220.

Tognolini, M.; Ballabeni, V.; Bertoni, S.; Bruni, R.; Impicciatore, M., and Barocelli, E. 2007. Protective effect of Foeniculum vulgare essential oil and anethole in an experimental model of thrombosis. *Pharmacol. Res.* 56: 254–260.

Torviso, M.; Alesso, E.; Moltrasio, G.; Vázquez, P.; Pizzio, L.; Cáceres, C., and Blanco, M. 2006. Effect of the support on a new metanethole synthesis heterogeneously catalyzed by Keggin heteropolyacids. *Appl. Catal. A.* 301: 25–31.

Torviso, R.; Lantaño, B.; Erlich, M.; Alesso, E.; Finkielsztein, L.; Moltrasio, G.; Aguirre, J., and Brunet, E. 2003. Synthesis of 1-ethyl-2-methyl-3-arylindanes. Stereochemistry of five-membered ring formation. *Arkivoc.* 10: 283–297.

Tron, G. C.; Pirali, T.; Sorba, G.; Pagliai, F.; Busacca, S., and Genazzani, A. A. 2006. Medicinal chemistry of combretastatin A4: present and future directions. *J. Med. Chem.* 49: 3033–3044.

Tuan, D. Q., and Ilangantilekete, S. G. 1997. Liquid CO_2 extraction of essential oil from star anise fruits (*Illicium verum* H.). *J. Food Eng.* 31: 47–57.

Urbina, J. M.; Cortés, J. C.; Palma, A.; López, S. N.; Zacchino, S. A.; Enriz, D. R.; Ribas, J. C., and Kouznetsov, V. V. 2000. Inhibitors of the fungal cell wall. Synthesis of 4-aryl-4-*N*-arylamino1-butenes and related compounds with inhibitory activities on ß(1–3) glucan and chitin synthases. *Bioorg. Med. Chem.* 8: 691–698.

Vargas, M. L. Y.; Castelli, M. V.; Kouznetsov, V. V.; Urbina González, J. M.; López, S. N.; Sortino, M.; Enriz, R. D.; Ribas, J. C., and Zacchino, S. A. 2003. In vitro antifungal activity of new series of homoallylamines and related compounds with inhibitory properties of the synthesis of fungal cell wall polymers. *Bioorg. Med. Chem.* 11: 1531–1550.

Vargas Méndez, L. Y., and Kouznetsov, V. V. 2007. An efficient synthesis of new 1-*H*-4′ methyl-3′,4′-dihydrospiro[piperidine-4,2′(1′*H*)quinoline] scaffolds. *Tetrahedron Lett.* 48: 2509–2512.

Walsh, J. J.; Coughlan, D.; Heneghan, N.; Gaynor, C., and Bell, A. 2007. A novel artemisinin-quinine hybrid with potent antimalarial activity. *Bioorg. Med. Chem. Lett.* 17: 3599–3602.

Wang, S.; Gates, B. D., and Swenton, J. S. 1991. A convergent route to dihydrobenzofuran neolignans via a formal 1,3-cycloaddition to oxidized phenols. *J. Org. Chem.* 56: 1979–1981.

Wang, E. C.; Wein, Y. S., and Kuo, Y. H. 2006. A concise and efficient synthesis of salvinal from isoeugenol via a phenoxenium ion intermediate. *Tetrahedron Lett.* 47: 9195–9197.

Waterman, P. 1975. Alkaloids of the Rutaceae: their distribution and systematic significance. *Biochem. Syst. Ecol.* 3: 149–180.

Whitehead, D. C., Dibb, H., and Hartley, R. D. 1982. Phenolic compounds in soil as influenced by the growth of different plant species. *J. Appl. Ecol.* 19: 579–588.

Wuonola, M. A., and Woodward, R. B. 1976. Imidazole alkaloids of *Macrorungia longistrobus*. Revised structures and total syntheses. *Tetrahedron* 32: 1085–1095.

Yadav, J. S.; Reddy, B. V. S., and Kondaji, G. 2003. In Cl_3-catalyzed [3+2] cycloaddition reactions: a facile synthesis of trans-dihydrobenzofurans and substituted cyclobutane derivatives. *Synthesis* 4: 1100–1104.

Zolyomi, G.; Banfi, D.; Lang, T., and Korosi, J. 1974. Heterocyclic compounds. III. Synthesis of 14C-labeled 1-(3,4-dimethoxyphenyl)-5-ethyl-7,8-dimethoxy-4-methyl-5*H*-2,3-benzodiazepines. *Chem. Ber.* 107: 3904–3907.

7 Synthesis and Biological Activity of Promising Azole Marine Products

Largazole and Neopeltolide

Danilo Davyt and Gloria Serra

CONTENTS

7.1 INTRODUCTION

The role of natural products in drug discovery has experienced many changes over the past two decades, from a decline in participation by pharmaceutical companies in the 1990s to a renaissance in recent years. In 2007, Newman and Cragg published a comprehensive review analyzing the sources of new drugs, covering the period from 1981 to the middle of 2006 (Newman and Cragg 2007). This analysis demonstrated that more than 60% of new drugs are natural products or natural product–inspired molecules. The potential for finding important new compounds of diverse skeletons in the marine environment is tremendous. More than 10,000 compounds have already been discovered from the marine environment. Examples of approved drugs developed from marine natural products are Ziconotide (Prialt®), Trabectedin (Yondelis®), and Eribulin (Halaven®). Concomitantly at least 20 marine natural products or derivatives are currently being tested in human trials. In this review, synthesis and biological activity of two promising azole marine products, namely, largazole and neopeltolide, will be taken into account.

7.2 SYNTHESIS AND BIOACTIVITY OF LARGAZOLE

Marine cyanobacteria are prolific producers of bioactive secondary metabolites with promising antitumor activities. In 2008, largazole (**1**; Figure 7.1) was isolated by Taori et al. from a sample of a cyanobacterium of the genus *Symploca* collected from Key Largo, Florida Keys (Taori et al. 2008).

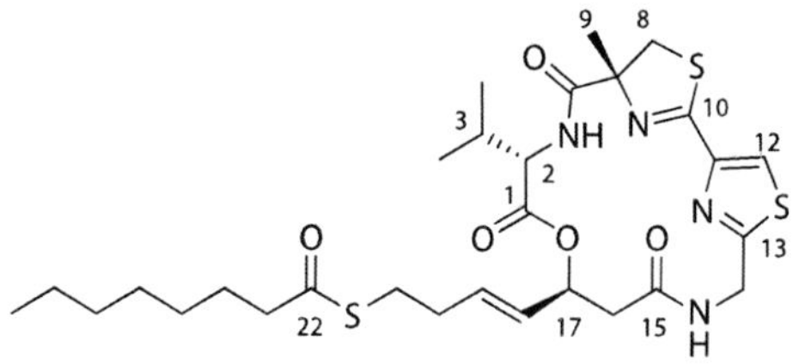

FIGURE 7.1 Structure of largazole.

TABLE 7.1
Growth Inhibitory Activity (GI_{50}) of Natural Product Drugs

Compound	MDA-MB231 GI_{50} (nM)	NMuMG GI_{50} (nM)	U2OS GI_{50} (nM)	NIH3T3 GI_{50} (nM)
1	7.7	122	55	480
Paclitaxel	7.0	5.9	12	6.4
Actinomycin D	0.5	0.3	0.8	0.4
Doxorubicin	310	63	220	47

It possesses a combination of unusual structural features, including a substituted 4-methylthiazoline fused to a thiazole, a thioester moiety that has not been reported in metabolites from cyanobacteria, and the 3-hydroxy-7-mercaptohept-4-enoic acid unit.

Largazole was isolated by a bioassay-guided fractionation and potently inhibited the growth of highly invasive transformed human mammary epithelial cells (MDA-MB-231) (GI_{50} = 7.7 nM). In contrast, nontransformed murine mammary epithelial cells (NMuMG) were less susceptible (GI_{50} = 122 nM). In addition, **1** showed selectivity for transformed fibroblastic osteosarcoma U2OS cells (GI_{50} 55 nM, LC_{50} 94 nM) over nontransformed fibroblasts NIH3T3 (GI_{50} 480 nM, LC_{50} > 8 μM). This remarkable selectivity was not observed for other validated antitumor natural products tested in parallel (Table 7.1). The growth of cancer cell lines derived from colon (HT29) and neuroblastoma (IMR-32) was also strongly inhibited by **1** (GI_{50}/LC_{50} 12 nM/22 nM; 16 nM/22 nM). The results indicated that cancer cells are preferentially targeted by the test compound **1**. As a consequence, several research groups have been embarked in the total synthesis of this potential cancer chemotherapeutic agent.

7.2.1 Synthesis of Largazole and Its Analogs and Evaluation of Their Bioactivity

Four month after the publication of its isolation, the first total synthesis of largazole was reported (Ying et al. 2008a). The synthesis (eight steps, 19% overall yield) involved a macrocyclization reaction for formation of the strained 16-membered depsipeptide core followed by an olefin cross-metathesis reaction for installation of the thioester (Scheme 7.1). The deliberate late stage incorporation of the thioester allowed the access to a series of analogs required to define the biological role of the thioester, the octanoyl group, and the side chain *E*-double bond.

Reagents and conditions: (a) (*R*)-2-methyl cysteine methyl ester hydrochloride, Et_3N, EtOH, 50°C, 72 h; (b) TFA, CH_2Cl_2, 25°C, 1 h; (c) DMAP, CH_2Cl_2, 25°C, 1 h; (d) 2,4,6-trichlorobenzoyl chloride, Et_3N, THF, 0°C, 1 h; then DMAP, 25°C, 10 h; (e) (i) 0.5 N LiOH, THF, H2O, 0°C, 3 h, (ii) TFA, CH_2Cl_2, 25°C, 2 h, (iii) HATU, HOAt, *i*-Pr_2NEt, CH_2Cl_2, 25°C, 24 h; (f) thioacetic acid S-but-3-enyl ester, Grubbs' second-generation catalyst (50 mol%), toluene, reflux, 4 h; (g) 1-triisopropylsilyloxyl-3-butene, Grubbs' second-generation catalyst (30 mol%), toluene, reflux, 3 h; (h) TBAF, THF, 25°C, 1 h; (i) aq. NH_3, CH_3CN, 25°C, 12–18 h.

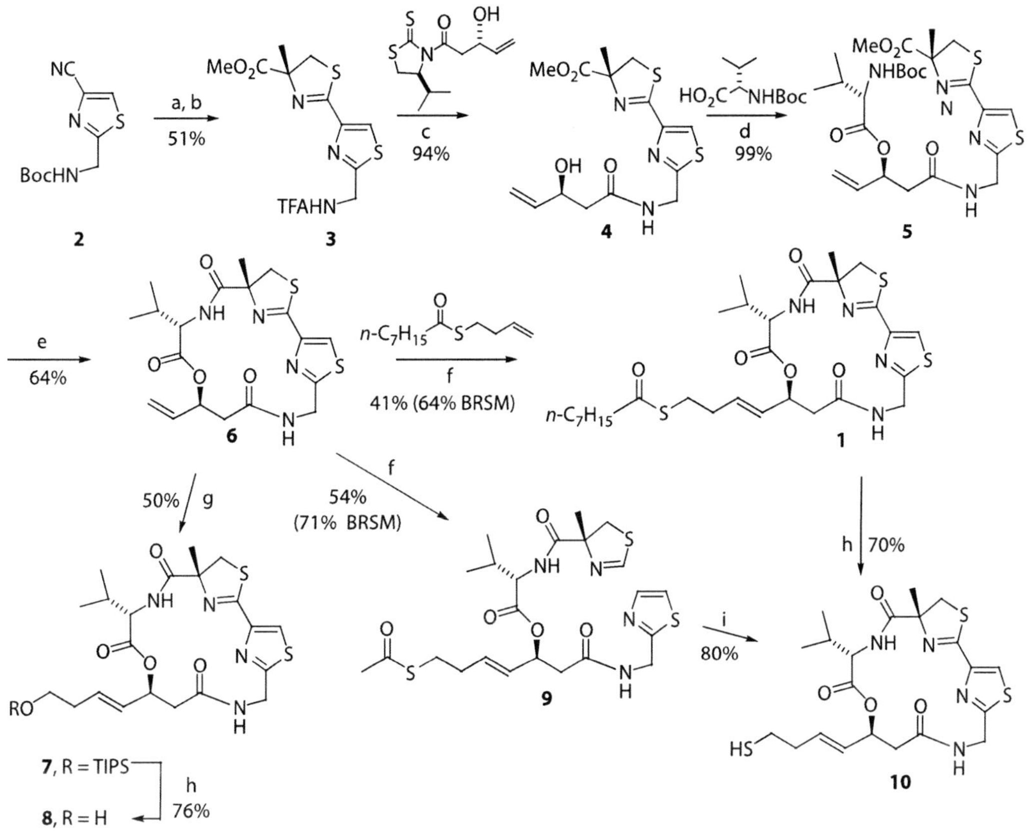

SCHEME 7.1 First total synthesis of largazole (**1**) and its analogues.

The methylthiazolinethiazole (**3**) was obtained by condensation reaction between (*R*)-2-methyl cysteine methyl ester hydrochloride and thiazole **2** followed by deprotection. Coupling reaction to the previously obtained thiazolidinethione using Nagao aldol reaction (Nagao et al. 1986, Hodge and Olivo 2004) provided compound **4**. Yamaguchi esterification of alcohol **4** and *N*-Boc-L-Val rendered the ester **5** in 99% yield. However, DCC-coupling reaction afforded **5** in a lower yield (85%). Attempts to macrocyclization under EDC or FDPP-coupling conditions provided the desired compound **6** in low yield. In contrast, **6** was obtained in high yield (64% for three steps) using HATU-HOAt. Finally, olefin cross-methathesis reaction in the presence of Grubbs' second-generation catalyst provided largazole in 41% (64% BRSM). Starting from compound **6** and **1**, the analogs **7**, **8**, **9** and **10** were prepared.

The authors suggested that thiol **10** could be generated by metabolism of thioester functionality in largazole. They showed the structural similarity between FK228 (Li et al. 1996, Yurek-George et al. 2004, 2007), a histone deacetylase (HDAC) inhibitor, and the thiol **10**, and proposed the hypothesis that largazole inhibits HDACs. Since HDACs I, II and IV are Zn^{+2}-dependent enzymes, they proposed that thiol group of **10** could chelate the Zn^{+2}. HDACs are a family of enzymes found in bacteria, fungi, plants, and animals that remove the acetyl group from the ε-amino groups of lysine residues present within the *N*-terminal extension of the core histones. This has the consequence that the positive charge density on the *N*-termini of the core histones increases, thereby strengthening the interaction with the negatively charged DNA and blocks the access of the transcriptional machinery to the DNA template (Paris et al. 2008). Histone deacetylation is a mechanism that can lead to silencing of tumor suppressor genes, and histone deacetylase inhibitors owe their antitumor action to their ability to reverse some of the aberrant epigenetic states associated with cancer.

TABLE 7.2
IC_{50} and GI_{50} Values for HDACs and Growth Inhibition (nM) by Luesch and Hong's Group

Compound	HCT-116 GI_{50} (nM)	HCT-116-HDAC Cellular Assay (nM)	HeLa Nuclear Extract HDACs (nM)
1	44 ± 10	51 ± 3	37 ± 11
10	38 ± 5	209 ± 15	42 ± 29
9	33 ± 2	50 ± 18	52 ± 27
6	>10,000	>10,000	>10,000
8	>10,000	>10,000	>10,000

To testify the hypothesis that largazole inhibits HDACs, the cellular HDAC activity upon treatment with largazole or its analogues in HCT-116 cells were determined. Largazole treatment for 8 h resulted in a decrease of HDAC activity in a dose–response manner. The IC_{50} for HDAC inhibition closely corresponds with its GI_{50} in this cell line (Table 7.2). This correlation suggested that HDAC is the target responsible for largazole's antiproliferative effect. Largazole (**1**) and the thiol analogue **10** exhibited similar cellular activity against HDACs derived from nuclear HeLa extracts as well as antiproliferative activity. These results substantiated the hypothesis that **10** is the reactive species. In addition, compounds **6** and **8** are inactive since they do not chelate with Zn^{+2}. To establish a preliminary selectivity profile, the authors tested Largazole (**1**) and **10** against recombinant HDAC1 (class I) and HDAC6 (class II). Compound **10** inhibited HDAC1 activity at low nanomolar concentrations and was 150-fold less active against HDAC6. Luesch, Hong and coworkers concluded that largazole is a class I HDAC inhibitor. In addition, structure–activity relationship (SAR) studies revealed that the thiol group is the pharmacophore of the natural product.

Few months later, Luesch and Hong published the synthesis of new analogues of Largazole with linker and macrocycle modifications (Figure 7.2). They reported antiproliferative activity against HCT-116 colon cancer cells and HDAC inhibitory activity using HeLa nuclear extract, Table 7.3 (Ying et al. 2008b).

SARs suggested that the four-atom linker between the macrocycle, an octanoyl group in the side chain and the (*S*)-configuration at the C17 position are critical to HDAC inhibitory effect. In contrast, the valine residue in the macrocycle can be replaced with alanine without significant loss of activity.

During 2008 seven more synthesis of largazole were published. Philips and coworkers reported the synthesis of **1** in eight steps, some SARs, NMR studies and molecular modeling (Nasveschuk et al. 2008). The strategy for the synthesis of largazole involved the introduction of the thioester by cross-metathesis as the final step and preparation of the macrocycle from fragments **16** and **17** (Scheme 7.2).

Readily accessible analogs, such as **6**, **18**, ester **19**, and ketone **20** (Figure 7.3), as well as synthetic largazole, were tested against MDA-MB231 cells and nontransformed human mammary epithelial

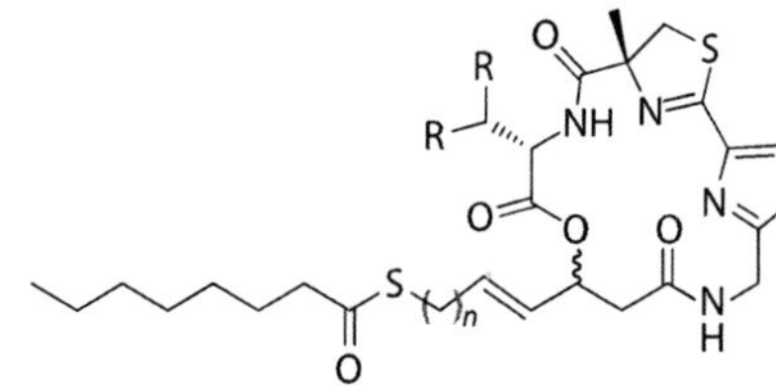

FIGURE 7.2 Largazole analogues reported by Luesch and Hong's group.

TABLE 7.3
Cancer Cell Growth and HDAC Inhibition (GI_{50} and IC_{50} in nM) by Luesch and Hong Group

Compound	HCT 116 Growth Inhibition (GI50) (nM)	HeLa Nuclear Extract HDACs (IC50) (nM)
1	6.8 ± 0.6	32 ±13
11 (*n* = 1, R = Me, 17*S*)	>10,000	>20,000
12 (*n* = 3, R = Me, 17*S*)	620 ±50	7,600 ± 900
13 (*n* = 4, R = Me, 17*S*)	2,500 ± 600	4,100 ± 430
14 (*n* = 2, R = H, 17*S*)	21± 2	72 ±21
15 (*n* = 2, R = Me, 17*R*)	3,900 ± 450	>20,000

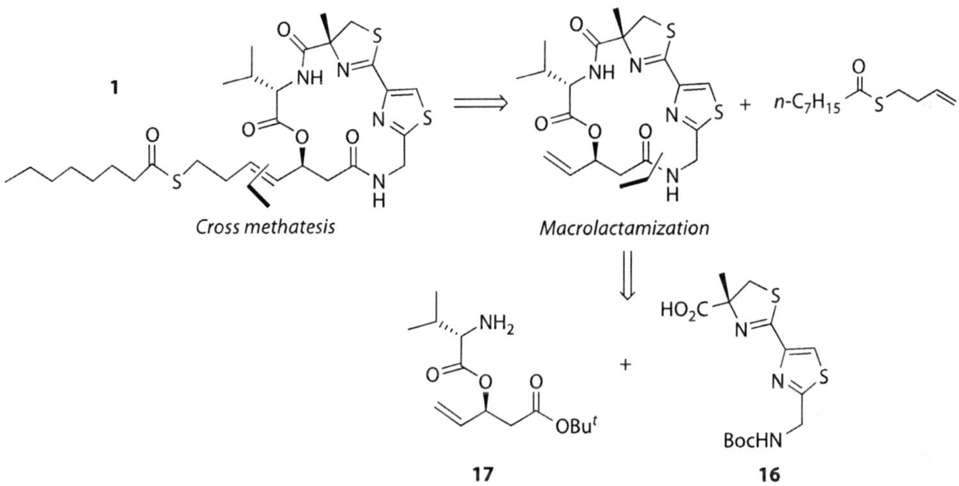

SCHEME 7.2 Retrosynthetic analysis by Phillips and coworkers.

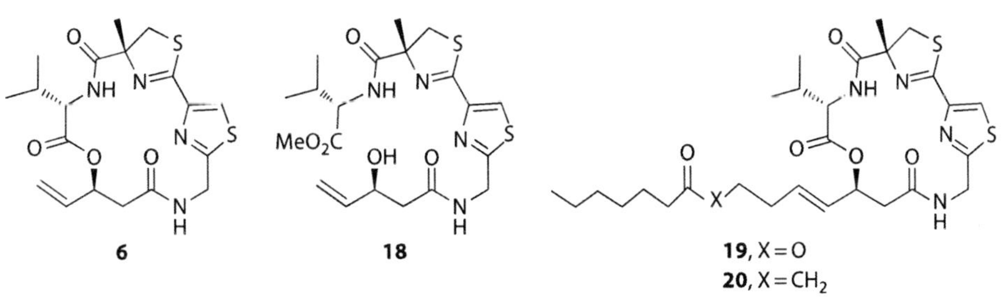

FIGURE 7.3 Largazole analogs obtained by Phillips and coworkers.

cells (HME). GI_{50} value of largazole on MDA-MB231 was higher than what was observed in the Luesch studies and the analogs were inactive (Table 7.4).

NOESY spectra were collected by Phillips group at a variety of mixing times ranging from 150 to 700 ms in order to obtain data about the conformation of largazole. Key transannular and long-range correlations were used for Monte Carlo conformational searching. The structures generated for largazole depicted a relatively rigid and flat macrocycle, with the thiol-ester side chain and Val residue on opposite faces.

Ghosh and Kulkarni completed an enantioselective total synthesis of largazole (**1**) (Ghosh and Kulkarni 2008). Remarkable differences in their synthetic strategy with the previous ones

TABLE 7.4
GI_{50} Values against MDA-MB231 and HME Cells of Largazole and Analogs Synthesized by Phillips and Coworkers

Compound	MDA-MB231 GI_{50} (nM)	HME GI_{50} (nM)
1	71	>600
6	>600	>600
18	>600	>600
19	>600	>600
20	>600	>600

involve preparation of the methylthiazoline by Kelly's procedure (You et al. 2003) and assembles of 3-hydroxy-7-mercaptohept-4-enoic acid derivative to Boc-L-Val. Cramer and coworkers (Seiser et al. 2008) published the synthesis of largazole using a similar disconnection of the molecule to that used by Phillips's group. In this synthetic protocol, the fragment **17** (Scheme 7.2), a β-hydroxy ester, was obtained using Amano lipase PS resolution. The authors optimized the cross-methathesis reaction and informed that *p*-nitro-substituted catalyst developed by Grela and coworkers (Michrowska et al. 2004) shows significantly higher activity and affords largazole in 75% yield with a *trans*/*cis* ratio of 6:1. The antiproliferative activity evaluation of some derivatives against the human epithelial carcinoma cell line A432 and the preadipocyte cell line 3T3L1 was also performed. The intermediate **6**, which contains a terminal double bond, and an analog in which the side chain has been replaced with a C13 alkyl chain, showed no growth-inhibitory activity, even at a concentration of 5 μm. The replacement of the thioester functionality with an ester group also led to a complete loss of activity. The authors have also found that free thiol derivative **10** displayed improved selectivity relative to that of **1** against the wild-type cells (SI(**10**) = 9.5; SI(**1**) = 2.6).

Williams and coworkers also completed an efficient total synthesis of largazole (**1**) in eight linear steps and 37% overall yield, and of its active metabolite, the largazole thiol (**10**), in seven linear steps (Bowers et al. 2008). The obtained **1** and **10**, FK228 and SAHA (Yoshida et al. 1990) have been evaluated for inhibition of HDACs 1, 2, 3, and 6. As presented in Table 7.5, **10** is an extraordinarily potent inhibitor of HDAC1 and HDAC2 (K_i = 70 pM). The parent natural product largazole itself, on the other hand, is a comparatively weak HDAC inhibitor with potency approximating that of the nonselective pharmaceutical product SAHA. There is a significant discrepancy between this result and what was reported by Taori (Taori et al. 2008). Williams et al. emphasized that even a trace contamination of **10** liberated under aqueous assay conditions or by trypsin (presents in the enzyme-coupled reaction) could account for the substantial decrease in enzyme potency observed. Largazole exhibits submicromolar inhibitory effect on melanoma cell proliferation study, and it has

TABLE 7.5
HDAC Inhibitory Activity (K_i nM) of Largazole (1) and Largazole Thiol (10) as Compared to Pharmaceutical HDAC Inhibitors by Williams and Coworkers

Compound	HDAC1 (nM)	HDAC2 (nM)	HDAC3 (nM)	HDAC6 (nM)
1	20	21	48	>1000
10	0.07	0.07	0.17	25
FK228	0.12	0.14	0.28	35
SAHA	10	10	15	9

a superior potency (IC_{50} = 45–315 nM) compared to **10** (IC_{50} = 360–2600 nM). The authors attributed the observed inverse difference in cytotoxicity to the superior cell permeability of the thioester **1** as compared to the thiol **10**.

Ye and coworkers (Ren et al. 2008) have completed the synthesis of largazole in 5.8% overall yield from 3-[(tert-butyldimethylsilyl) oxy] propane and fragment **16,** obtained using Hantzsch reaction and Pattenden procedure (Pattenden et al. 1993). Doi and coworkers also reported the synthesis of largazole using Kelly's method for the formation of the thiazoline and Ishihara's procedure to obtain a bisthiazoline (Numajiri et al. 2008). In 2009, the group of Bradner, Williams, and Wiest (Bowers et al. 2009a) reported the syntheses of an amide isostere of largazole (**21**; Figure 7.4). They prepared the enantiomerically pure β-amino acid analog of the (*S*)-3-hydroxy-7-mercaptohept-4-enoic acid (**22**) from a selectively protected aspartic acid derivative. The largazole amide isostere (**21**) was found to be inactive against HDAC1 (IC_{50} > 3000 nM). The peptide isostere of largazole thiol (**22**) was evaluated to be less potent against HDAC1 by ninefold as compared to the thiol **10**.

In 2009, new analogs of largazole were also prepared and assayed against histone deacetylases (HDACs) 1, 2, 3, and 6 by Williams and coworkers (Figure 7.5 and Table 7.6) (Bowers et al. 2009b). From the biological results for **24**, **25**, and **34**, the authors concluded that there are strict stereochemical, and conformation activity relationship between the natural product and its protein targets. Nevertheless, the single-atom substitutions of the sulfur atoms for oxygen atoms in the oxazoline–oxazole derivative (**26**) provided a compound equipotent to largazole itself. Moreover, a significant increase in potency with pyridine substitution of the thiazole was observed; this compound (**28**) possesses subnanomolar activity against Class I HDACs. They demonstrated that the methyl substituent of the thiazoline ring is nonessential for the dramatic potency of the natural product (see HDAC inhibition for compound **27**).

During 2009, four more reports related to the synthesis of largazole and/or analogs were published (Chen et al. 2009, Seiser and Cramer 2009, Wang and Forsyth 2009, Yan and O'Doherty 2009). In 2010, new analogs of largazole were synthesized by Jiang's group (Zeng et al. 2010). Structure–activity relationship studies suggested that the geometry of the alkene in the side chain is critical. While the largazole's analogs with *trans*-alkene are potent for the antiproliferative effect, those with *cis*-alkene are completely inactive. However, replacement of valine by tyrosine slightly diminishes the potency, but increases selectivity toward human cancer cells over human normal cells more than 100-fold. Besides, De Lera's group reported the synthesis of C-7 modified largazole analogs (**36**, **37**, **38**) and their biological evaluation (Figure 7.6). The compounds showed a potent inhibition of recombinant HDAC1 and HDAC4 with a marked selectivity for HDAC1, for which they exhibit nanomolar potency (Souto et al. 2010).

In 2010, Luesch and Hong's group reported new studies of largazole and analogs (Liu et al. 2010). The screening against the National Cancer Institute's 60 cell lines revealed that largazole is particularly active against several colon cancer cell types. In addition, they tested largazole and some synthetic analogs for HDAC inhibition in human HCT116 colon cancer cells. These studies correlated with grow inhibitory effect. Differential activity of largazole analogs was rationalized by molecular docking to an HDAC1 homology model. The authors investigated the aqueous, plasma,

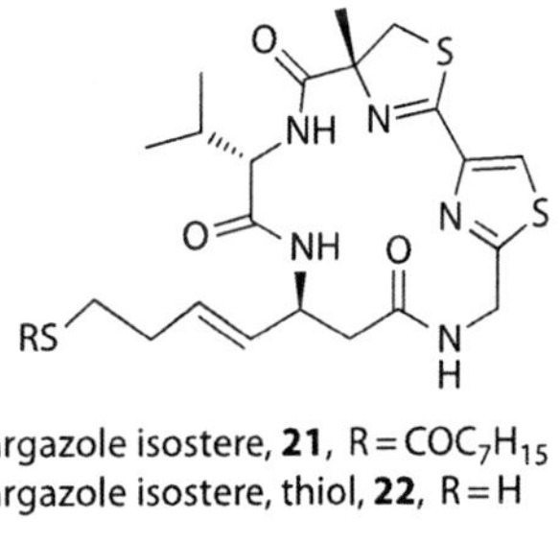

Largazole isostere, **21**, R = COC_7H_{15}
Largazole isostere, thiol, **22**, R = H

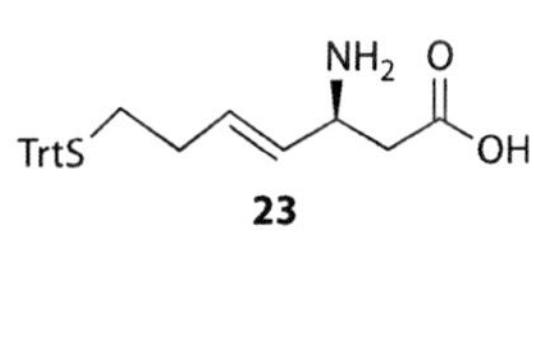

23

FIGURE 7.4 Largazole isostere, thiol isostere, and β-aminoacid by Williams and coworkers.

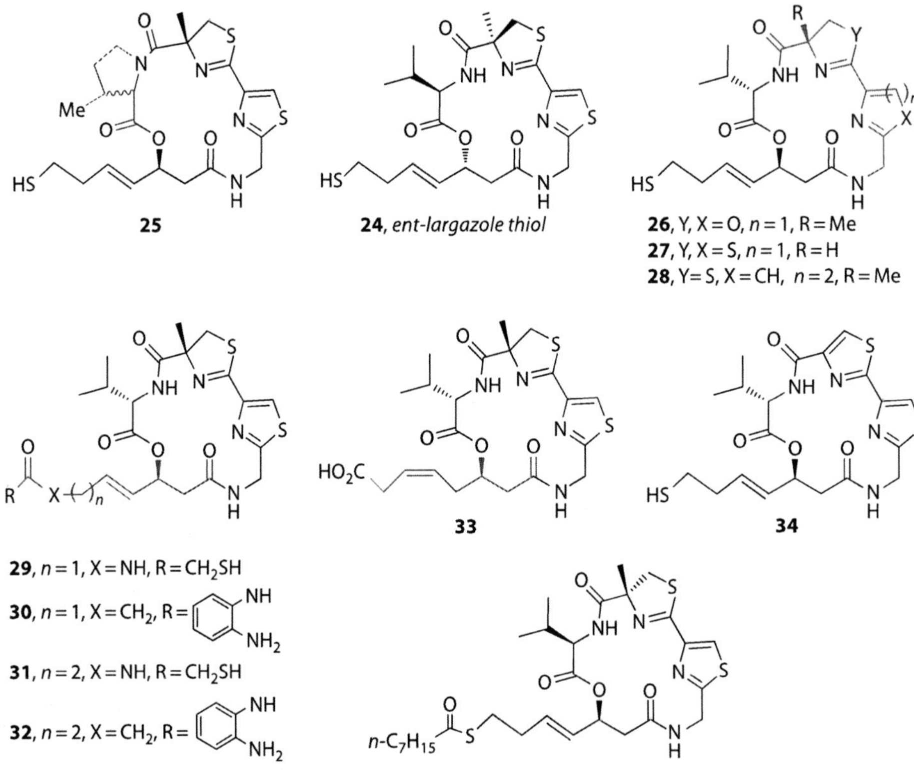

25

24, *ent-largazole thiol*

26, Y, X = O, *n* = 1, R = Me
27, Y, X = S, *n* = 1, R = H
28, Y = S, X = CH, *n* = 2, R = Me

29, *n* = 1, X = NH, R = CH_2SH
30, *n* = 1, X = CH_2, R = 2-aminophenylamino (NH/NH$_2$ aryl group, drawn)
31, *n* = 2, X = NH, R = CH_2SH
32, *n* = 2, X = CH_2, R = 2-aminophenylamino (NH/NH$_2$ aryl group, drawn)

33

34

35, *largazole C-2 epimer*

FIGURE 7.5 Largazole thiol and analogs obtained by Williams's group.

TABLE 7.6
Biochemical Inhibition of Human HDACs by Williams and Coworkers

Compound	HDAC1 (IC_{50}, nM)	HDAC2 (IC_{50}, nM)	HDAC3 (IC_{50}, nM)	HDAC6 (IC_{50}, nM)
10	1.2	3.5	3.4	49
24	1,200	3,100	1,900	2,200
25	110	800	580	13,000
26	0.69	1.7	1.5	45
27	1.9	4.8	3.8	130
28	0.32	0.86	1.1	29
29	670	1,600	960	700
30	270	4,100	4,100	>30,000
31	1,000	1,900	1,500	240
32	230	290	140	>30,000
33	>30,000	>30,000	>30,000	>30,000
34	77	120	85	>30,000
35	30	82	84	680
SAHA	10	26	17	13

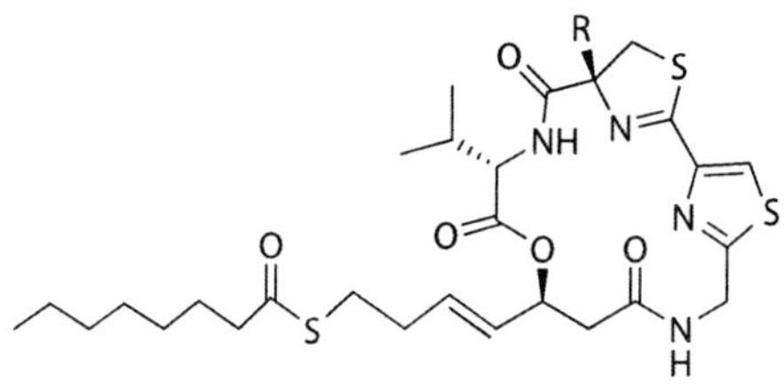

FIGURE 7.6 Largazole analogs developed by de Lera and coworkers.

microsomal, and cellular stability of largazole. It was stable in aqueous solution, but in mouse serum largazole rapidly converted to the largazole thiol. Experiments of cellular stability in the whole-cell protein lysate derived from HCT116 cells confirmed that largazole hydrolyzes and generates the reactive largazole thiol (**10**). This metabolite was found to bind itself with proteins forming an adduct that "reprotect" it. The proteins also act as carriers that release the active largazole thiol at the site of action. The acute toxicity studies in *nu/nu* mice indicated that largazole is well tolerated up to the highest concentration tested (50 mg/kg i.p.). The authors concluded that largazole thiol reaches the tumor and inhibits HDACs showing *in vivo* efficacy.

Since HDACs have been reported to play certain important roles in osteogenesis, Hong and Kim's group investigated *in vitro* and *in vivo* osteogenic activity of largazole; the results indicated that largazole induces the expression of alkaline phosphatase and osteopontin, stimulates bone formation and inhibits bone resorption (Lee et al. 2011). Forsyht and coworkers reported that during the activation of the *C*-terminal valine residue for esterification in a synthetic sequence to obtain largazole, the α-center of valine suffered an unanticipated epimerization to deliver ultimately the C2-epimer **35** (Figure 7.5) along with largazole (**1**) (Wang et al. 2011). The compound **35** displayed more potent activity than **1** in cell viability assays against PC-3 and LNCaP prostate cancer cell line. In 2011, Ganessan and coworkers published the synthesis of a series of largazole analogs (Figure 7.7) and assessed their inhibitory activity against HDAC and growing of MCF7 cells along with pharmacokinetic evaluation of these compounds in terms of metabolic stability (Benelkebir et al. 2011). Largazole thiol was evaluated to be highly potent at a picomolar concentration against HDAC (Table 7.7). Although, analogue **39** was a significantly weaker HDAC inhibitor than largazole thiol, in the growth inhibition assay the two compounds had similar activity. Compounds **40** and **42** are

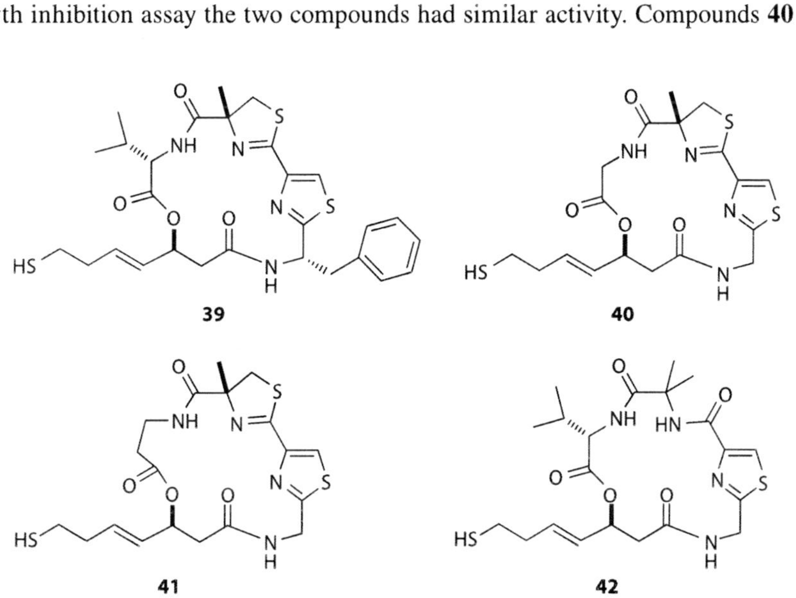

FIGURE 7.7 Largazole analogs developed by Ganessan and coworkers.

TABLE 7.7
HDAC and MCF7 GI_{50} of 1, 10, and Analogues by Ganessan and Coworkers

Compound	HDAC Inhibition (nM)	MCF/GI_{50} (nM)
1	571 ± 29	5 ± 1
10	0.043 ± 0.026	277 ± 130
39	17.2 ± 2.3	377 ± 62
40	0.17 ± 0.05	2,458 ± 1,135
41	3.15 ± 0.35	>10,000
42	0.99 ± 0.07	5,902 ± 1,698

subnanomolar HDAC inhibitor, but **40**, **41**, and **42** are poor in growth inhibition compared to largazole. The authors argued that analogue **39**, the weakest HDAC inhibitor in the series, had the best cell growth due to its increased lipophilicity. Analogues **40–42** are more polar than largazole thiol and this has a deleterious effect on growth inhibition.

They also investigated the stability of largazole and of the analogues in the presence of mouse liver homogenate. Largazole itself was highly unstable with a half-life ≤ 5 min. The metabolism consists of thioester hydrolysis to largazole thiol, which is relatively stable in murine liver homogenate with a half-life of 51 min at 37°C. Analog **40** was similar in stability with a half-life of 32 min. Analogs **41** and **42** had a complex and rapid metabolism with a half-life ≤ 5 min.

7.2.2 X-Ray Crystal Structure of Complexed Largazole Thiol

Christianson and coworkers reported the X-ray crystal structure of HDAC8 complexed with largazole thiol (Cole et al. 2011). They observed that macrocyclic skeleton undergoes minimal conformational changes upon binding to HDAC8 since its conformation is very similar to that of the uncomplexed compound reported by Cramer and coworkers (Seiser et al. 2008) and concluded

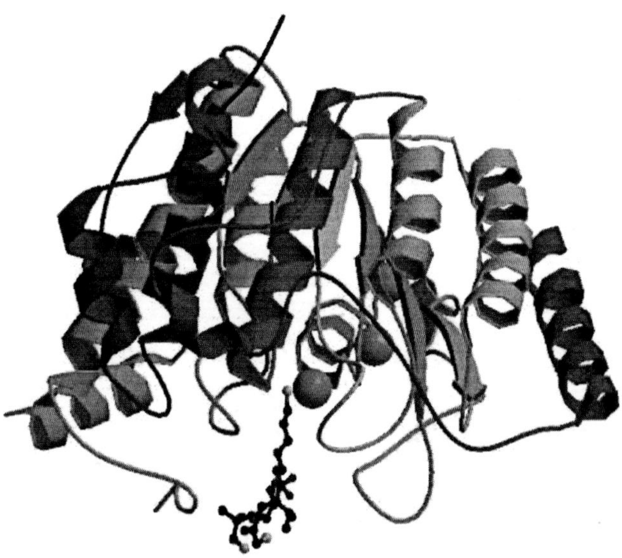

FIGURE 7.8 X-ray structure of largazole thiol binding to HDAC 8; ideal thiolate-zinc coordination geometry. (Image from Protein Data Bank, accession code 3RQD.)

that the thiazoline–thiazole moiety rigidifies the macrocyclic ring with an ideal conformation for binding to HDAC8. In contrast, considerable conformational changes are required by HDAC8 to accommodate the binding of the rigid largazole thiol. The most important observation is the coordination of the thiolate-zinc, which is responsible for the exceptional affinity and biological activity of largazole (Figure 7.8).

The structure of the complex provided the basis for understanding structure–affinity relationships in the previously synthesized largazole derivatives. Structural changes, such as length of the thiol side chain, a *cis* configuration of the side-chain olefin, or an *R* configuration in the macrocycle side-chain linkage, result in significant affinity losses since they would compromise the ideal Zn^{2+} coordination geometry. In contrast, substitution of the methyl group of the 4-methylthiazoline with a hydrogen or the L-valine with L-tyrosine, L-alanine, or glycine have not produced loss of affinity because the methyl group does not contact the protein surface and the L-valine side chain is pointing directly out toward solvent, thereby, not influencing directly the enzyme–inhibitor interface. The thiazole ring is also oriented away from the protein structure toward solvent; so it is possible that this position could tolerate additional substitution without loss of the activity.

7.3 CHEMISTRY AND BIOACTIVITY OF NEOPELTOLIDE

Neopeltolide (**43**; Figure 7.9) is a marine natural product that exhibits potent inhibition of *in vitro* tumor cell proliferation at nanomolar level and also inhibits the growth of the fungal pathogen *Candida albicans*. This compound was isolated from two specimens of a sponge closely related to the genus *Daedalopelta*, collected by manned submersible from the northwest coast of Jamaica. It was isolated and patented by Wright and coworkers in 2004 (Wright et al. 2007a) and then published in 2007 (Wright et al. 2007b). This sponge belongs to the Neopeltidae family and Lithistida order, a prolific order with hundreds of compounds isolated from it, which have been reviewed (Bewley et al. 1998, D'Auria et al. 2002) and many of them have potent bioactivities (Wright 2010). However, as other products isolated from sponges, the biosynthesis of the neopeltolide could be done by an epibiotic microorganism. The chemical structure of neopeltolide has a 2,6-*cis*-tetrahydropyran unit encircled by a 14-membered macrolactone, with a side chain that includes an oxazole and a terminal carbamate.

7.3.1 Isolation, Structure Elucidation, and Related Compounds

The ethanolic extract of 105 g of frozen sponge was partitioned between *n*-butanol and water. The *n*-butanol partition was further fractioned by silica gel chromatography by step gradient of ethyl acetate in heptane and after reversed-phase HPLC 4.2 mg of neopeltolide was obtained as a colorless oil, $[\alpha]_D + 24$. The planar structure of neopeltolide was proposed from the analysis of its 1D and 2D NMR data, including 1D gradient- and sensitivity-enhanced TOCSY (1D-DPFGSE-TOCSY) and 2D

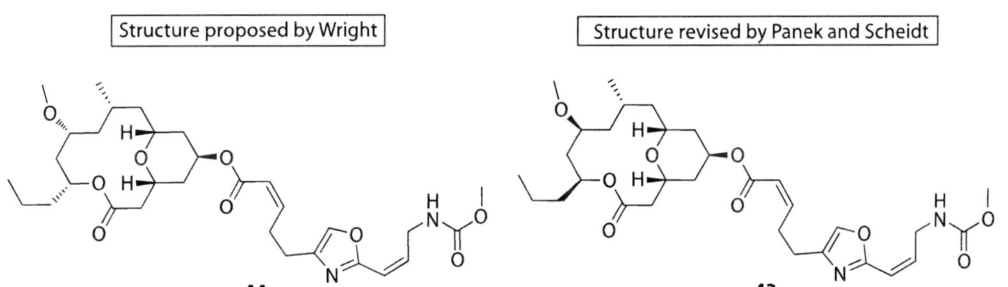

44 **43**

FIGURE 7.9 Structure of neopeltolide proposed by Wright and revised by Panek and Scheidt.

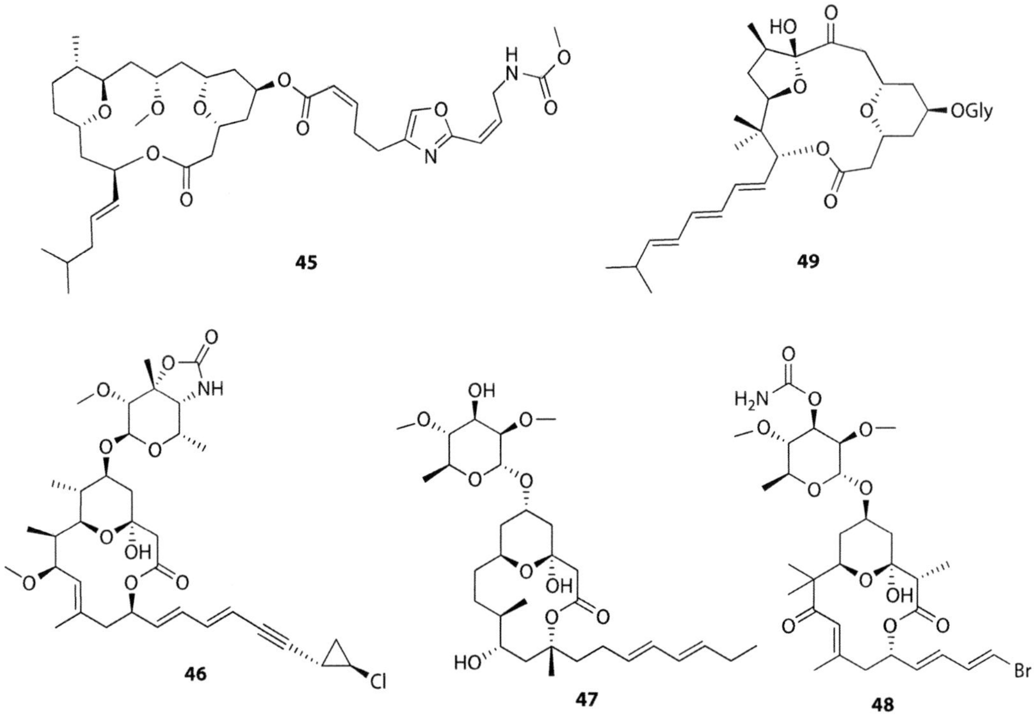

FIGURE 7.10 Marine products related to neopeltolide.

double-quantum filtered COSY (2D-DQF-COSY). Their relative stereochemistry was assigned on the basis of analysis of coupling constants, the 2D-NOESY spectrum, and a series of 1D-DPFGSE-NOE experiments. The investigators reported that the material available of the neopeltolide had not been enough to determine it absolute stereochemistry. The proposed structure (**44**), as shown in the Figure 7.9, was revised (**43**) when Panek and coworkers first (Youngsaye et al. 2007), and few months later Scheidt and coworkers (Custar et al. 2008), reported the total synthesis of neopeltolide.

This compound is structurally related to leucascandrolide A (**45**; Figure 7.10) (D'Ambrosio et al. 1996), a 16-membered macrolide with a tetrahydropyran ring isolated from the calcareous sponge *Leucascandra caveolata* collected along the east coast of New Caledonia. Both compounds have an identical oxazole-bearing side chain. The macrolide portion of neopeltolide shows similarities to the macrolide ring seen in other marine products as callipeltosides (**46**) (Figure 7.10) isolated from the sponge *Callipelta* (Zampella et al. 1996), lyngbouilloside (**47**) from *Lyngbya bouilloni* (Tan et al. 2002), and the aurisides (**48**) isolated from the sea hare *Dolabella auricularia* (Sone et al. 1996). They show 14-membered macrolides with a hemiketal in a C-7 position of the tetrahydropyran ring. Polycavernoside A (**49**) (Figure 7.10) (Yotsu et al. 2004) that presents a 16-membered macrolide and a tetrahydropyran ring as found in neopeltolide is a potent toxin isolated from the red alga *Polycavernosa tsudai*.

7.3.2 Synthesis of Neopeltolide

Since the report of neopeltolide in 2007, 19 papers were published related to its synthesis, 11 total syntheses and 8 formal syntheses. However, only five articles reported studies related to its bioactivity (see Table 7.8).

The first total synthesis of neopeltolide was performed by Panek and coworkers (Youngsaye et al. 2007). They coupled the oxazole side chain by a Still–Gennari olefination between a phosphonate derivative of macrolactone and an aldehyde of the oxazole side chain (see Scheme 7.3).

TABLE 7.8
Articles Published of Neopeltolide

Year	Content	Principal Author	Reference
2007	Neopeltolide bioactivity patent	Amy E. Wright	Wright et al. (2007a)
	Isolation, structure elucidation and bioactivity	Amy E. Wright	Wright et al. (2007b)
	Total synthesis	James S. Panek	Youngsaye et al. (2007)
	Total synthesis	Karl A. Scheidt	Custar et al. (2008)
2008	Total synthesis	Eun Lee	Woo et al. (2008)
	Total synthesis and activity	Sergey A. Kozmin	Ulanovskaya et al. (2008)
	Macrolactone synthesis	Martin E. Maier	Vintonyak and Maier (2008)
	Total synthesis, analogues and bioactivity	Martin E. Maier	Vintonyak et al. (2008)
	Total synthesis	Haruhiko Fuwa and Makoto Sasaki	Fuwa et al. (2008)
	Total synthesis	Ian Paterson	Paterson and Miller (2008)
	Macrolactone synthesis	Richard E. Taylor	Kartika et al. (2008)
2009	Total synthesis, analogues and bioactivity	Karl A. Scheidt	Custar et al. (2009)
	Macrolactone synthesis	Jiyong Hong	Kim et al. (2009)
	Total synthesis	Emmanuel Roulland	Guinchard and Roulland (2009)
	Macrolactone synthesis	Jhillu Singh Yadav	Yadav et al. (2010)
	Total synthesis, analogues and bioactivity	Haruhiko Fuwa and Makoto Sasaki	Fuwa et al. (2009)
2010	Total synthesis and analogues	Paul E. Floreancig	Cui et al. (2010)
	Total synthesis	Haruhiko Fuwa and Makoto Sasaki	Fuwa et al. (2010)
	Macrolactone synthesis	Michael P. Jennings	Martinez-Solorio and Jennings (2010)
	Studies of macrolactone synthesis	Gordon J. Florence	Florence and Cadou (2010)
2011	Macrolactone synthesis	Xuegong She	Yang et al. (2011)

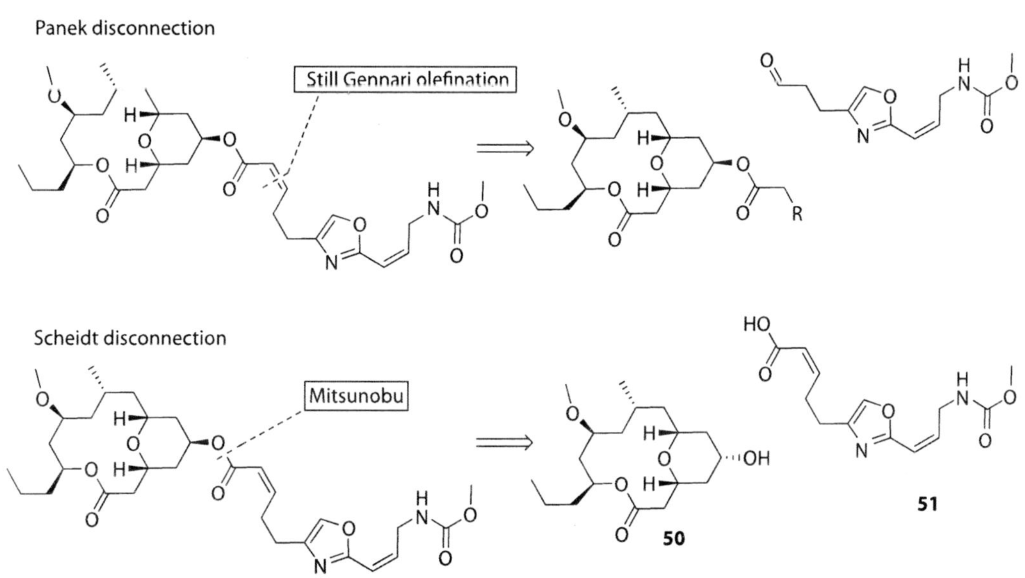

SCHEME 7.3 First disconnection of Panek and Scheidt syntheses.

The total synthesis reported by Scheidt and coworkers (Custar et al. 2008), and the following total syntheses reported until now (see Table 7.8) disconnect the ester side chain to perform a convergent synthesis between the macrolactone and the oxazole side chain by Mitsunobu reaction (see Scheme 7.3). In this case the configuration of macrolactone hydroxyl group is inverted. So, neopeltolide syntheses involve the development of macrolactone and the oxazole side chain syntheses. The methodology to synthesize the side chain has been widely developed in leucascandrolide total syntheses reported previously by Panek (Dakin et al. 2002).

The retrosynthesis of macrolactone (see Scheme 7.4) performed by Panek and coworkers begins with the lactone disconnection. The macrolactone synthesis was performed using Yamaguchi lactonization. This is the most used methodology of the followings reports, with yields between 44% and 87%.

An interesting strategy built the macrolactone and the 2,6-*cis*-tetrahydropyran unit simultaneously. Scheidt and coworkers followed this synthetic strategy using a Lewis acid–catalyzed intramolecular cyclization (Figure 7.11) developed by them in a previous report (Morris et al. 2005). They performed a diastereoselective cyclization between the aldehyde and the β-hydroxy dioxinone group of ester **52** catalyzed by scandium triflate. Then the dioxinone group in **53** was eliminated by heating at 130°C in DMSO-water to yield the tetrahydropyran containing macrolactone **54**. Similar strategy was performed by Lee and coworkers using the Prins reaction (Woo et al. 2008), Figure 7.11. They treated the esters **55** or **56** with triethylsilyl trifluoromethanesulfonate in acetic acid in the presence of trimethylsilyl acetate and then under basic conditions to yield the macrolactone and the 2,6-*cis*-tetrahydropyran. The same synthetic route was performed by Yadav et al. (2010) to build the neopeltolide macrolide, using *S*-citronellol as starting material to build the aldehyde allyl ester **56.**

Floreancig and coworkers (Cui et al. 2010) started their synthesis with the construction of the 12-membered macrolactone **58** using Yamaguchi lactonization. Then they formed the 2,6-*cis* tetrahydropyran by an oxidative cyclization protocol using 2,3-dichloro-5,6-dicyano-1,4-benzoquinone as oxidant (Figure 7.12).

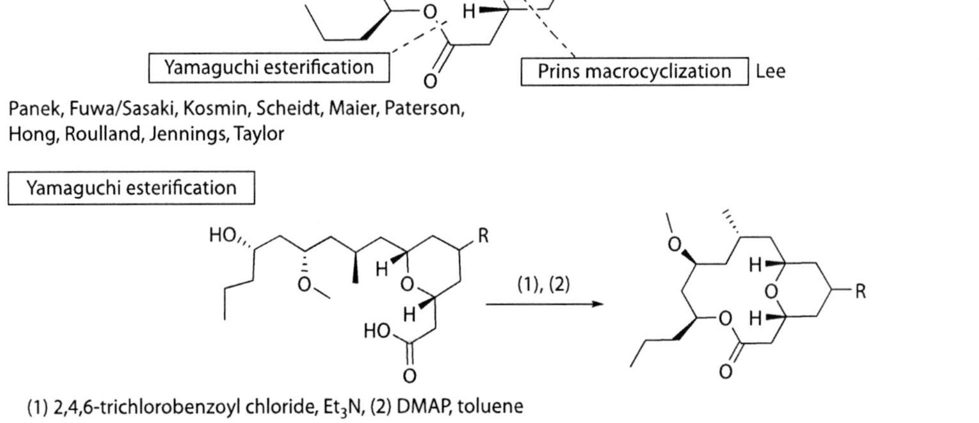

SCHEME 7.4 First disconnection in syntheses of the neopeltolide macrolactone.

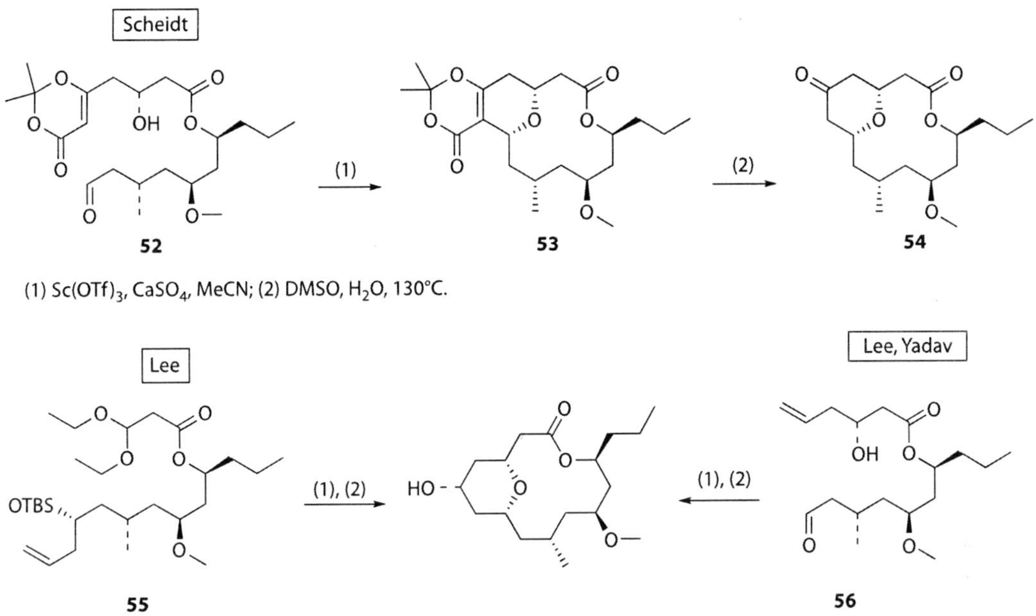

(1) TESOTf (20 equiv), TMSOAc (30 equiv), AcOH (0.01 m), RT, 30 min; (2) K_2CO_3, MeOH.

FIGURE 7.11 Strategies for simultaneous synthesis of tetrahydropyran nuclei and macrolactone by Scheidt, Lee and Yadav.

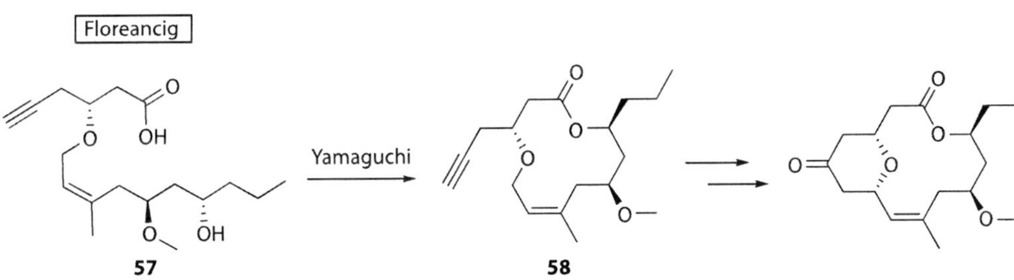

FIGURE 7.12 Synthetic strategy to obtain the macrolactone performed by Floreancig.

Fuwa et al. (2010) performed a second synthesis of the macrolactone, which is shorter and more efficient than their first one, using Grubbs' ring-closing metathesis (RCM) to the macrolactonization of the tetrahydropyran containing ester **59**. The same strategy was performed recently by She and coworkers (Yang et al. 2011). Florence and coworkers intended a similar approach to the macrocyclization by an RCM without the tetrahydropyran being preformed; however, it was unsuccessful and they obtained a cycloheptene **60** (Florence and Cadou 2010) (see Figure 7.13).

In most cases of the neopeltolide syntheses using Yamaguchi macrocyclization approach, the tetrahydropyran core is preformed. In these syntheses several methodologies were used to prepare the tri-substituted 2,6-*cis*tetrahydropyran nuclei as see in Scheme 7.5.

Panek built the tetrahydropyran core using a triflic acid promoted [4 + 2] annulation via oxonium ion between an allylsilane and an aldehyde. Paterson and Miller (2008) performed a similar route using a Jacobsen asymmetric hetero Diels–Alder reaction between an aldehyde and a 2-siloxydiene catalyzed by a chiral tridentate chromium(III) catalyst. Maier and coworkers used an acid-catalyzed Prins reaction to prepare the tetrahydropyran nuclei from the same aldehyde and a homoallylic diol (Vintonyak and Maier 2008, Vintonyak et al. 2008). In a similar way, Kosmin and coworkers (Ulanovskaya et al. 2008) prepared a tetrahydropyran racemic block to build the macrolactone using

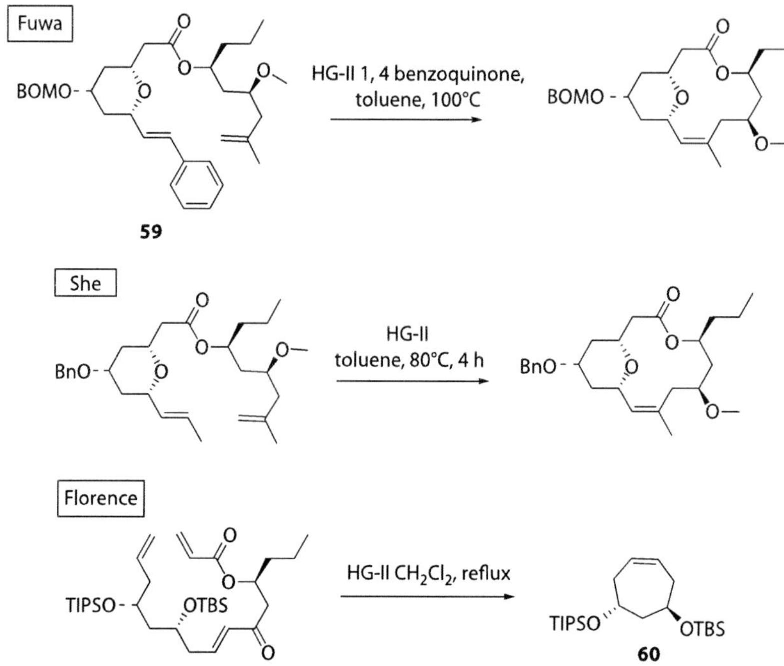

FIGURE 7.13 Different strategies to obtain the macrolactone by ring-closing metathesis.

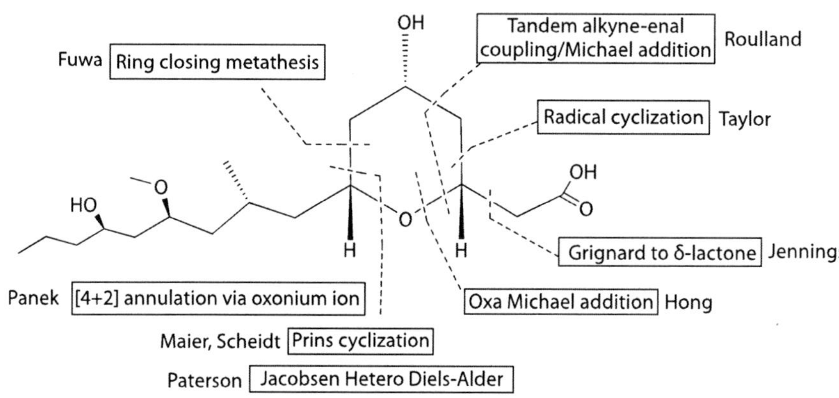

SCHEME 7.5 Tetrahydropyran disconnections in the neopeltolide syntheses.

Prins methodology. Scheidt and coworkers (Custar et al. 2009) proposed a second synthetic plan using their methodology of Lewis acid–catalyzed intramolecular cyclization to build the tetrahydropyran macrocyclic precursor (see Figure 7.14). Also, they reviewed the different syntheses reported using the methodology of Prins to synthesize the neopeltolide macrolide (Crane and Scheidt 2010).

Fuwa et al. (2008, 2009) synthesized the 2,4,6-trisubstituted tetrahydropyran of neopeltolide using a sequence of Suzuki–Miyaura coupling and RCM. Hong and coworkers (Kim et al. 2009) synthesized the tetrahydropyran nuclei by allylic oxidation, oxa-Michael reaction, and other consecutive oxidation with excellent stereoselectivity (d.r. > 20:1) (see Figure 7.15).

Taylor and coworkers (Kartika et al. 2008) performed the tetrahydropyran core via radical cyclization of β-alkoxyacrylate **61** with AIBN and n-Bu_3SnH in refluxing toluene. These conditions afforded **62** in 95% yield as nearly single diastereomer (19:1) (see Figure 7.16). Guinchard and Roulland (2009) performed a stereoselective formation of the tetrahydropyran core by a

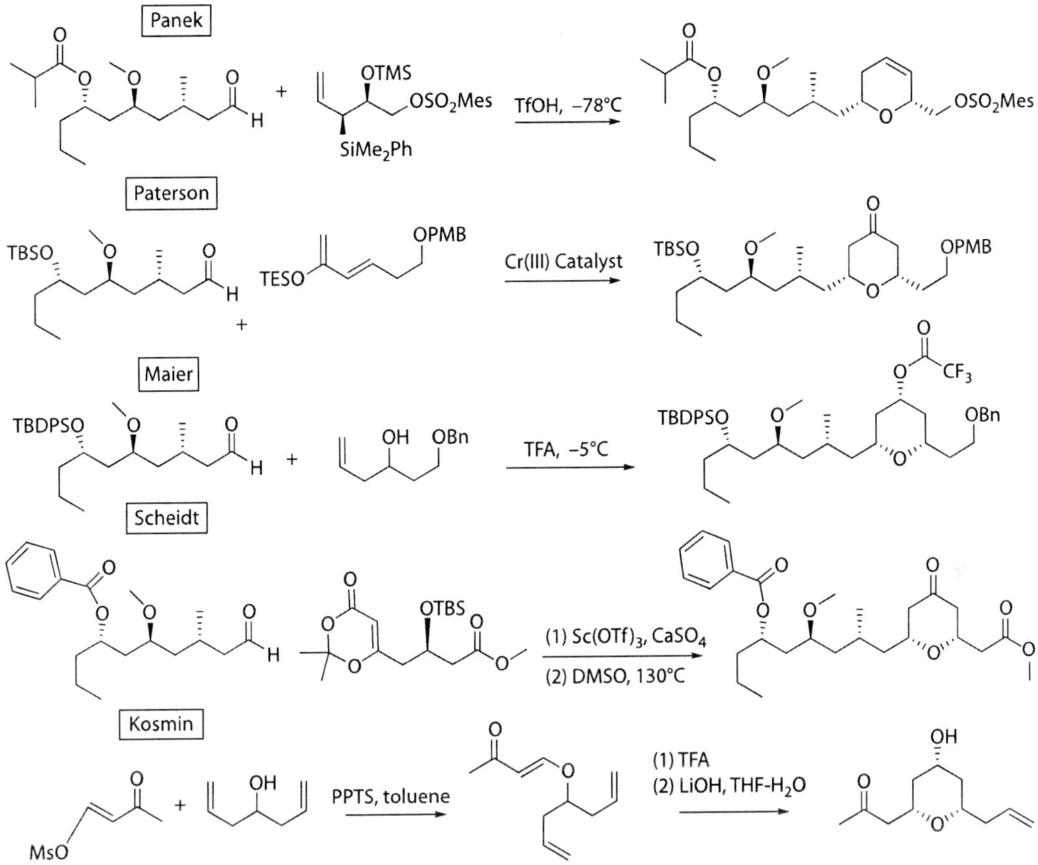

FIGURE 7.14 Different routes to synthesize tetrahydropyran nuclei by Prins reaction.

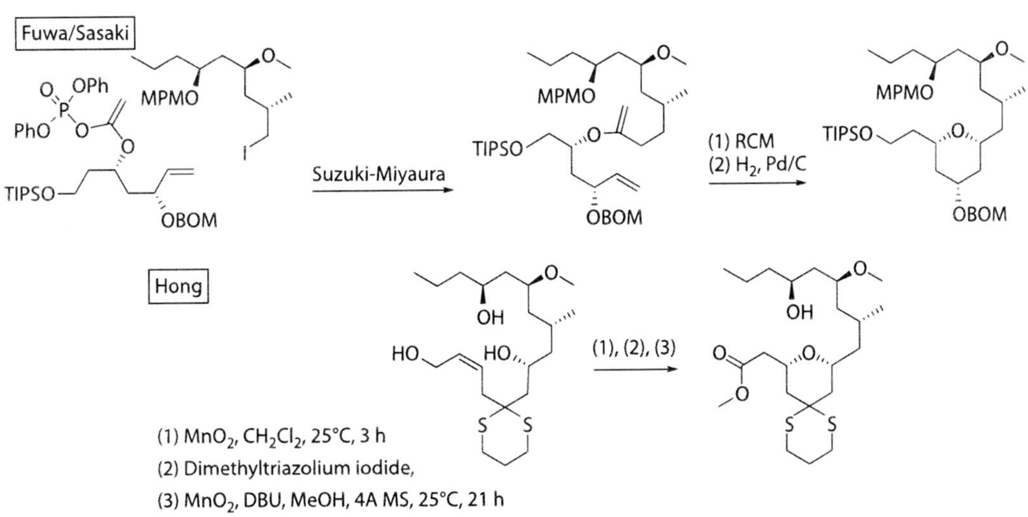

FIGURE 7.15 Syntheses of tetrahydropyran nuclei by Fuwa/ Sasaki and by Hong.

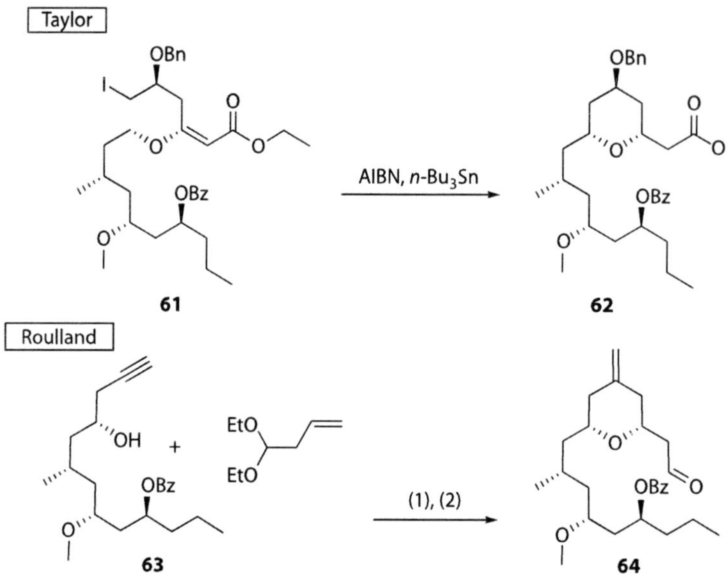

(1) $[CpRu(MeCN)_3]PF_6$, AcOH, acetone
(2) $LiBF_4$, MeCN, H_2O

FIGURE 7.16 Syntheses of tetrahydropyran nuclei by Taylor and Roulland.

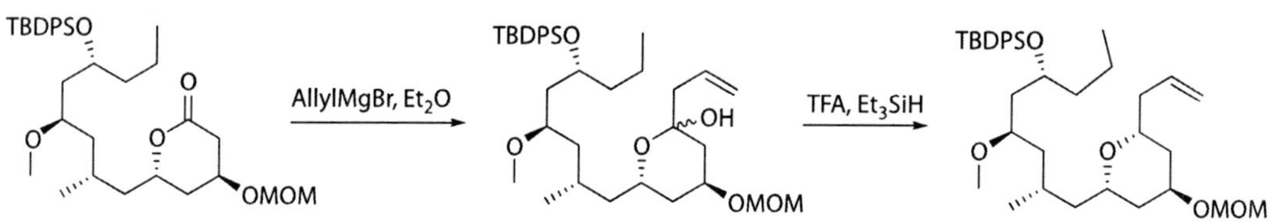

FIGURE 7.17 Synthesis of tetrahydropyran nuclei by Jennings.

[CpRu(MeCN)3]PF6-catalyzed tandem alkyne-enal coupling/Michael addition sequence. This key step allowed assemblage of the alkyne **63** with 3-butenal, an "ene" cross-coupling partner.

Martínez-Solorio and Jennings (2010) reported a singular strategy to the formation of the tetrahydropyran core (Figure 7.17). They performed a stereoselective reduction of an endocyclic oxocarbenium cation mediated by the treatment of hemiketal **66** with a Lewis acid. The intermediate **66** derived from a nucleophilic addition of the allyl Grignard reagent to the δ-lactone **65**.

So far the Maier synthesis of neopeltolide is the most efficient; in the longest linear sequence this synthesis required 18 steps with an overall yield of 18.7%. The other performed syntheses were longer or have not proceeded in higher yields: Panek synthesis in 19 steps, 1.3% of overall yield; Scheidt synthesis in 19 steps, 0.52%; Lee synthesis in 15 steps, 6.7%; Fuwa, Sasaki 2010 synthesis in 12 steps, 14%; Kozmin synthesis in 15 steps, 5.3% (racemic material); Paterson in 18 steps, 5.8%; Roulland in 16 steps, 6.2%; and Floreancig in 14 steps, approximately 2% (see Table 7.8).

7.3.3 Biological Activity

The first report of Wright et al (2007) describes neopeltolide as a potent inhibitor of the *in vitro* proliferation of A-549 human lung adenocarcinoma, NCI-ADR-RES human ovarian sarcoma, and P388 murine leukemia cell lines with IC_{50} values of 1.2, 5.1, and 0.56 nM, respectively. On other cells lines such as PANC-1 of pancreatic cancer and DLD-1 of colorectal adenocarcinoma, neopeltolide showed strong inhibition of cell proliferation (nanomolar), but it showed a cytostatic activity rather than cytotoxic. Cell cycle analysis by flow cytometric methods revealed that neopeltolide causes a block of the cell cycle at G1 at doses of 100 nM in the A549 lung adenocarcinoma cell line. The preliminary investigation about the action mechanism suggests that it does not act via interaction with tubulin or actin.

They also reported that neopeltolide has potent antifungal activity against *Candida albicans*. It showed a growth inhibitory zone of 17 mm when tested at a concentration of 25 μg/disk in the *C. albicans* disk diffusion assay and a minimum inhibitory concentration in liquid culture of 0.625 μg/mL. Kosmin and coworkers (Ulanovskaya et al. 2008) synthesized enough neopeltolide that allowed them to advance into the elucidation of mechanism of action. However, as apparently arises from the reported synthesis, the study was performed with racemic neopeltolide. They confirmed the neopeltolide activity previously reported and also that the structural homolog of neopeltolide, leucascandrolide, has similar potent antiproliferative profiles in mammalian cells and yeast. HCT116 cell line proved to be one of the most sensitive to neopeltolide, which inhibited the growth of these cells by 90% at 1.0 nM. In contrast, PC3 and A549 cell growth was inhibited only by 60% at the same concentration.

Looking for the cellular targets of neopeltolide they followed an approach by genetic studies using an organism with low complexity like yeast. Studies with a wild-type strain of *Sacharomyces cerevisiae* showed increased sensitivity to these compounds over 10,000-fold on agar media containing galactose instead of glucose. Genome-wide yeast deletion screen showed that these products may antagonize a pathway required for ATP biosynthesis in the absence of glucose fermentation, such as mitochondrial oxidative phosphorylation. Their studies on inhibition of oxidative phosphorylation showed that leucascandrolide A and neopeltolide may elicit their potent antiproliferative activity by blocking one or more complexes in the mitochondrial electron-transport chain and so the molecules inhibit mitochondrial ATP synthesis. They have evaluated the activity of the four mitochondrial electron transfer chain complexes in yeast and mammalian cells and found the identification of cytochrome bc1 complex as the principal cellular target. Cytochrome bc1 is a transmembrane protein complex located in the inner mitochondrial membrane. This multi-subunit enzyme is a central component of the mitochondrial respiratory electron transport chain. The function of cytochrome bc1 is to reduce cytochrome c(Fe^{3+}) into cytochrome c(Fe^{2+}) using the membrane localized ubiquinol. They reported that "leucascandrolide A and neopeltolide compare favorably to the most potent inhibitors of cytochrome bc1 complex known today, identifying such compounds as a new class of highly useful biochemical tools for investigation of eukaryotic energy metabolism."

Maier and coworkers (Vintonyak et al. 2008) performed an efficient synthesis that in the longest linear sequence required 18 steps and an overall yield of 18.7%. They also synthesized the 5 and 11 epi-diastereomers and five analogues with side-chain modifications as seen in Figure 7.18. All of these compounds, also the precursor macrolactone and the side chain, were tested for cytotoxicity against L929 mouse fibroblasts and human lung carcinoma A549 line cells, as well as for their inhibitory efficacy on NADH-oxidation in submitochondrial particles of bovine heart. They showed that the macrolactone alone is not sufficient for biological activity but some modifications like the 11 epi and 5 epi neopeltolides keep the cytotoxicity activity (IC_{50} of neopeltolide, 11-epi-, and 5-epi- were 0.16, 0.9, and 5.6 nM against A549, respectively). The modified side-chain analogues showed that the distance from the oxazole ring to the macrolactone is important, with loss of activity with shorter chains. The analogue with an *E* double bond geometry (**70**) was about 20 times less active; however, the analogue with *Z, E* double bonds (**71**) was more active than neopeltolide itself (against L929 IC_{50} of neopeltolide was 250 pM and for the *Z,E* analogue was 160 pM). The correlation of cytotoxicity results and the NADH oxidation assay using submitochondrial particles of bovine heart confirm that the mechanism of action of neopeltolide involves the mitochondrial respiratory chain.

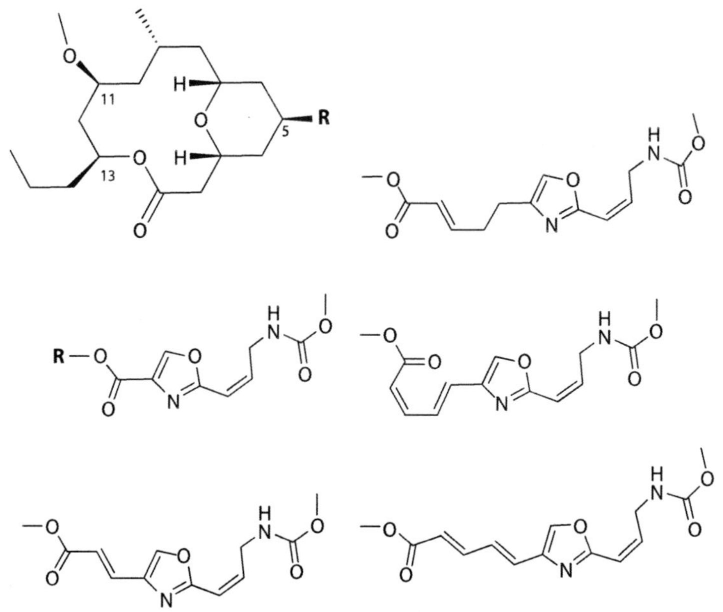

FIGURE 7.18 Neopeltolide analogues synthesized and assayed by Maier and coworkers.

Fuwa et al (2009) assayed synthetic neopeltolide, the originally proposed diastereomer, the analogues 11-demethoxyneopeltolide and 9-demethylneopeltolide, and some advanced precursors against P388 murine leukemia cells. The cytotoxicity results for neopeltolide, the diastereomers, and the advanced precursors like macrolactone or oxazole side chain agreed with those previously reported. The 11-demethoxyneopeltolide analogue showed an activity roughly fourfold lower. However, the 9-demethylneopeltolide is equipotent to or slightly more potent than neopeltolide; this result is significant because the analogue can be prepared more readily than neopeltolide.

Scheidt and coworkers (Custar et al. 2009) performed the biological activity of synthetic neopeltolide and two diastereomers (one with inverse configurations at 5, 11, and 13 carbons and the other corresponding with the structure proposed by Wright). Also, the alcohol and ketone forms of the macrolactone, the side chain, and two analogues of neopeltolide with benzoic or octanoic acid as side chain were assayed. In this report, they confirmed that the revised structure of neopeltolide is the active structure; the diastereomer that corresponds to the structure originally proposed by Wright was approximately 100-fold less active against murine leukemia P388 (0.6 and 42 nM, respectively) and human breast adenocarcinoma MCF-7 cells line (2.2 and 219 nM, respectively). A surprising result was that neopeltolide was minimally active or inactive against the four other cell lines tested: rat adrenal tumor PC12, human cervical carcinoma HeLa, human epidermal carcinoma KB, and human lung carcinoma A549. In the case of the human lung carcinoma A549 line cells they detected no dose-dependent growth inhibition in contrast to the previously results reported as a sensitive line cells. These results showed that neopeltolide exhibits line cells selectivity for its cytotoxic activity. Also the structure–activity relationship studied with synthetic compounds confirmed that both the macrolide and the oxazole side chain bound together play a role in mediating the effect of neopeltolide and neither component on its own is active. The diastereomers assayed were active but less potent than neopeltolide. The benzoyl ester was the unique analogue that kept similar activity to neopeltolide.

Floreancig and coworkers (Cui et al. 2010) reported the synthesis of four analogues: 9-epi-neopeltolide, 8,9-dehydroneopeltolide, 8,9-dihydroxyneopeltolide, and 8,9-epoxyneopeltolide. Despite the initiation of studies having been announced, they have not reported the activity of these compounds yet.

7.4 SUMMARY AND CONCLUSIONS

The recent progress in the synthesis and biology of largazole and analogues has provided significant insight into the structural, stereochemical, functional, and conformational aspects for further investigations. Preclinical studies have shown that largazole and derivatives are promising new drugs that could enter into clinical development. Even though neopeltolide showed exciting activities and efficient syntheses for it were developed, no further studies were reported related to its biological activity since Scheidt's report in 2009 (Custar et al. 2009). Thus, contradictions of neopeltolide activity against some cell lines as well as the extent of its pharmacological potential remains to be answered.

It is the fact that a large number of investigators have quite been inspired by the unique structural features of the chemical entities derived from sea in using them as models or scaffolds for synthesis of promising analogues and lead candidates in drug discovery programs. As a result, a considerable advance in the area of synthetic methodology has already been derived from marine products and will flourish in the near future.

ABBREVIATIONS

1D	one dimensional
2D	two dimensional
4AMS	molecular sieves with pore size of four Angstrom
AIBN	azobisisobutyronitrile
ATP	adenosine triphosphate
Bn	benzyl
Boc	*t*-butoxycarbonyl
BOM	benzyloxymethyl
BRSM	based on recovered starting material
COSY	correlation spectroscopy
DBU	1,8-diazabicyclo[5.4.0]undec-7-ene
DCC	dicyclohexylcarbodiimide
DMAP	4-*N*,*N*-dimethylaminopyridine
DMSO	dimethylsulfoxide
DNA	deoxyribonucleic acid
DPFGSE	double pulsed field gradient spin echo
DQF-COSY	double quantum filtered-correlation spectroscopy
EDC	1-ethyl-3-(dimethylaminopropyl)carbodiimide
EtOH	ethanol
FDPP	pentafluorophenyl diphenylphosphinate
GI_{50}	50% growth inhibition
HATU	*N*-[(dimethylamino)(3*H*-1,2,3triazolo(4,5-*b*)pyridin-3-yloxy)methylene]-*N*-methylmethanaminium hexafluorophosphate
HDAC	histone deacetylase
HOAt	7-aza-1-hydroxybenzotriazole
LC_{50}	50% lethal concentration
L-Val	L-valine
Mes	mesityl or 2,4,6-trimethylphenyl
MOM	methoxymethyl
MPM	*p*-metoxyphenymethyl
NADH	nicotinamide adenine dinucleotide reduced form
NMR	nuclear magnetic resonance
NOESY	nuclear overhauser effect spectroscopy

PMB	*p*-methoxybenzyl
PPTS	pyridinium *p*-toluenesulfonate
RCM	ring closing methatesis
RT	room temperature
SAHA	suberoylanilide hydroxamic acid
TBAF	tetrabutylammonium fluoride
TBDPS	*t*-butyldiphenylsilyl
TBS	*t*-butyldimethylsilyl
TESOTf	triethylsilyl trifluoromethanesulfonate
Tf	trifluoromethanesulfonyl
TFA	trifluoroacetyl or trifluoroacetic acid
THF	tetrahydrofuran
TIPS	triisopropylsilyl
TMS	trimethylsilyl
TMSOAc	trimethylsilyl acetate
TOCSY	total correlation spectroscopy

REFERENCES

Benelkebir H., S. Marie, A. Hayden et al. 2011. Total synthesis of largazole and analogues: HDAC inhibition, antiproliferative activity and metabolic stability. *Bioorg. Med. Chem.* 19:3650–3658.

Bewley C. A. and D. J. Faulkner. 1998. Lithistid sponges: Star performers or hosts to the stars. *Angew. Chem. Int. Ed.* 37:2163–2178.

Bowers A. A., T. J. Greshock, N. West et al. 2009a. Synthesis and conformation-activity relationships of the peptide isosteres of FK228 and largazole. *J. Am. Chem. Soc.* 131:2900–2905.

Bowers A. A., N. West, T. L. Newkirk et al. 2009b. Synthesis and histone deacetylase inhibitory activity of largazole analogs: Alteration of the zinc-binding domain and macrocyclic scaffold. *Org. Lett.* 11:1301–1304.

Bowers A., N. West, J. Taunton et al. 2008. Total synthesis and biological mode of action of largazole: A potent class I histone deacetylase inhibitor. *J. Am. Chem. Soc.* 130:11219–11222.

Chen F., A. H. Gao, J. Li et al. 2009. Synthesis and biological evaluation of C7-demethyl largazole analogues. *ChemMedChem.* 4:1269–1272.

Cole K. E., D. P. Dowling, M. A. Boone et al. 2011. Structural basis of the antiproliferative activity of largazole, a depsipeptide inhibitor of the histone deacetylases. *J. Am. Chem. Soc.* 133:12474–12477.

Crane E. A. and K. A. Scheidt. 2010. Prins-type macrocyclizations as an efficient ring- closing strategy in natural product synthesis. *Angew. Chem. Int. Ed.* 49:8316–8326.

Cui Y., W. Tu and P. E. Floreancig. 2010. Total synthesis of neopeltolide and analogs. *Tetrahedron* 66:4867–4873.

Custar D. W., T. P. Zabawa, J. Hines et al. 2009. Total synthesis and structure-activity investigation of the marine natural product neopeltolide. *J. Am. Chem. Soc.* 131:12406–12414.

Custar D. W., T. P. Zabawa and K. A. Scheidt. 2008. Total synthesis and structural revision of the marine macrolide neopeltolide. *J. Am. Chem. Soc.* 130:804–805.

Dakin, L. A., N. F. Langille and J. S. Panek. 2002. Synthesis of the C1′-C11′ oxazole-containing side chain of leucascandrolide A. Application of a Sonogashira cross-coupling. *J. Org. Chem.* 67:6812–6815.

D'Ambrosio M., A. Guerriero, C. Debitus et al. 1996. Leucascandrolide A, a new type of macrolide. *Helv. Chim. Acta* 79:51–60.

D'Auria M. V., A. Zampella and F. Zollo. 2002. The chemistry of lithistid sponge: A spectacular source of new metabolites. *Stud. Nat. Prod. Chem.* 26:1175–1258.

Florence G. J. and R. F. Cadou. 2010. Studies towards the synthesis of neopeltolide: Synthesis of a ring-closing metathesis macrocyclization precursor. *Tetrahedron Lett.* 51:5761–5763.

Fuwa H., S. Naito, T. Goto et al. 2008. Total synthesis of (+)-neopeltolide. *Angew. Chem. Int. Ed.* 47:4737–4739.

Fuwa H., A. Saito, S. Naito et al. 2009. Total synthesis and biological evaluation of (+)-neopeltolide and its analogues. *Chem. Eur. J.* 15:12807–12818.

Fuwa H., A. Saito and M. Sasaki. 2010. A concise total synthesis of (+)-neopeltolide. *Angew. Chem. Int. Ed.* 49:3041–3044.

Ghosh A. K. and S. Kulkarni. 2008. Enantioselective total synthesis of (+)- largazole, a potent inhibitor of histone deacetylase. *Org. Lett.* 10:3907–3909.

Guinchard X. and E. Roulland. 2009. Total synthesis of the antiproliferative macrolide (+)-neopeltolide. *Org. Lett.* 11:4700–4708.

Hodge M. B. and H. F. Olivo. 2004. Stereoselective aldol additions of titanium enolates of *N*-acetyl-4-isopropyl-thiazolidinethione. *Tetrahedron* 60:9397–9403.

Kartika R., T. R. Gruffi and R. E. Taylor. 2008. Concise enantioselective total synthesis of neopeltolide macrolactone highlighted by ether transfer. *Org. Lett.* 10:5047–5050.

Kim H., Y. Park and J. Hong. 2009. Stereoselective synthesis of 2,6-cis-tetrahydropyrans through a tandem allylic oxidation/oxa-Michael reaction promoted by the gem-disubstituent effect: Synthesis of (+)-neopeltolide macrolactone. *Angew. Chem. Int. Ed.* 48:7577–7581.

Lee S. U., H. B. Kwak, S. H. Pi et al. 2011. In vitro and in vivo osteogenic activity of largazole. *Med. Chem. Lett.* 2:248–251.

Li K. W., J. Wu, W. Xing et al. 1996. Total synthesis of the antitumor depsipeptide FR-901,228 *J. Am. Chem. Soc.* 118:7237–7238.

Liu Y., L. A. Salvador, S. Byeon et al. 2010. Anticolon cancer activity of largazole, a marine-derived tunable histone deacetylase inhibitor. *J. Pharmacol. Exp. Ther.* 335:351–361.

Martinez-Solorio D. and M. P. Jennings. 2010. Formal synthesis of (−)-neopeltolide featuring a highly stereoselective oxocarbenium formation/reduction sequence *J. Org. Chem.* 75:4095–4104.

Michrowska A., R. Bujok, S. Harutyunyan et al. 2004. Nitro-substituted Hoveyda-Grubbs ruthenium carbenes: Enhancement of catalyst activity through electronic activation. *J. Am. Chem. Soc.* 126:9318–9325.

Morris W. J., D. W. Custar and K. A. Scheidt. 2005. Stereoselective synthesis of tetrahydropyran-4-ones from dioxinones catalyzed by scandium(III) triflate. *Org. Lett.* 7:1113–1116.

Nagao Y., Y. Hagiwara, T. Kumagai et al. 1986. New C-4-chiral 1,3-thiazolidine-2-thiones: Excellent chiral auxiliaries for highly diastereo-controlled aldol-type reactions of acetic acid and alpha, beta-unsaturated aldehydes. *J. Org. Chem.* 51:2391–2393.

Nasveschuk C. G., D. Ungermannova, X. Liu et al. 2008. A concise total synthesis of largazole, solution structure, and some preliminary structure activity relationships. *Org. Lett.* 10:3595–3598.

Newman D. J. and G. M. Cragg. 2007. Natural products as sources of new drugs over the last 25 years. *J. Nat. Prod.* 70:461–477.

Numajiri Y., T. Takahashi, M. Takagi et al. 2008. Total synthesis of largazole and its biological evaluation. *Synlett.* 16:2483–2486.

Paris M., M. Porcelloni, M. Binaschi et al. 2008. Histone deacetylase inhibitors: From bench to clinic. *J. Med. Chem.* 51:1505–1529.

Paterson I. and N. A. Miller. 2008. Total synthesis of the marine macrolide (+)-neopeltolide. *Chem. Commun.* 39:4708–4710.

Pattenden G., S. M. Thom and M. F. Jone. 1993. Enantioselective synthesis of 2-alkyl substituted cysteines. *Tetrahedron* 49:2131–2138.

Ren Q., L. Dai, H. Zhang et al. 2008. Total synthesis of largazole. *Synlett.* 15:2379–2383.

Seiser T. and N. Cramer. 2009. Syntheses and biological activity of the HDAC class I inhibitor largazole. *Chimia.* 63:19–22.

Seiser T., F. Kamena and N. Cramer. 2008. Synthesis and biological activity of largazole and derivatives. *Angew. Chem. Int. Ed.* 47:6483–6485.

Sone H., H. Kigoshi and K. Yamada. 1996. Aurisides A and B, cytotoxic macrolide glycosides from the Japanese sea hare *Dolabella auricularia. J. Org. Chem.* 61:8956–8960.

Souto J. A., E. Vaz, I. Lepore et al. 2010. Synthesis and biological characterization of the histone deacetylase inhibitor largazole and C7-modified analogues. *J. Med. Chem.* 53:4654–5467.

Tan L.T., B. L. Márquez and W. H. Gerwick. 2002. Lyngbouilloside, a novel glycosidic macrolide from the marine cyanobacterium *Lyngbya bouillonii. J. Nat. Prod.* 65:925–928.

Taori K., V. J. Paul and H. Luesch. 2008. Structure and activity of largazole, a potent antiproliferative agent from the Floridian marine cyanobacterium *Symploca sp. J. Am. Chem. Soc.* 130:1806–1807.

Ulanovskaya O. A., J. Janjic, M. Suzuki et al. 2008. Synthesis enables identification of the cellular target of leucascandrolide A and neopeltolide. *Nat. Chem. Biol.* 4:418–424.

Vintonyak V. V., B. Kunze, F. Sasse et al. 2008. Total Synthesis and biological activity of neopeltolide and analogues. *Chem. Eur. J.* 14:11132–11140.

Vintonyak V. V. and M. E. Maier. 2008. Formal total synthesis of neopeltolide. *Org. Lett.* 10:1239–1242.

Wang B. and C. J. Forsyth. 2009. Total synthesis of largazole—Devolution of a novel synthetic strategy. *Synthesis* 17:2873–2880.

Wang B., P.-H. Huang, C.-S. Chen et al. 2011. Total syntheses of the histone deacetylase inhibitors largazole and 2-epi-largazole: Application of N-heterocyclic carbene mediated acylations in complex molecule synthesis. *J. Org. Chem.* 76:1140–1150.

Woo S. K., M. S. Kwon and E. Lee. 2008. Total synthesis of (+)-neopeltolide by a Prins macrocyclization. *Angew. Chem. Int. Ed.* 47:3242–3244.

Wright, A. E. 2010. The Lithistida: Important sources of compounds useful in biomedical research. *Curr. Opin. Biotechnol.* 21:801–807.

Wright A. E., J. C. Botelho, E. Guzman et al. 2007b. Neopeltolide, a macrolide from a Lithistid sponge of the family Neopeltidae. *J. Nat. Prod.* 70:412–416.

Wright A. E., S. A. Pomponi and P. J. McCarthy. 2007a. Isolation of biologically active neopeltolide compound. US pat N° 7179828.

Yadav J. S., G. Narayana and G. G. K. S. Kumar. 2010. A concise stereoselective formal total synthesis of the cytotoxic macrolide (+)-neopeltolide via Prins cyclization. *Tetrahedron* 66:480–487.

Yan W. and G. A. O'Doherty. 2009. Total synthesis of (+)- largazole, a histone deacetylase inhibitor. *Chemtracts* 22:50–58.

Yang Z., B. Zhang, G. Zhao et al. 2011. Concise formal synthesis of (+)-neopeltolide. *Org. Lett.* 13:5916–5919.

Ying Y., Y. Liu, S. R. Byeon S. R. et al. 2008b. Synthesis and activity of largazole analogs with linker and macrocycle modification. *Org. Lett.* 10:4021–4024.

Ying Y., K. Taori, H. Kim et al. 2008a. Total synthesis and molecular target of largazole, a histone deacetylase inhibitor. *J. Am. Chem. Soc.* 130:8455–8459.

Yoshida M., M. Kijima, M. Akita et al. 1990. Potent and specific inhibition of mammalian histone deacetylase both in vivo and in vitro by trichostatin A. *J. Biol. Chem.* 265:17174–17179.

Yotsu-Y M., T. Yasumoto, S. Yamada et al. 2004. Identification of Polycavernoside A as the Causative Agent of the Fatal Food Poisoning Resulting from Ingestion of the Red Alga *Gracilaria edulis* in the Philippines. *Chem. Res. Toxicol.* 17:1265–1271.

You S.L., H. Razavi H. and J. W. Kelly. 2003. A biomimetic synthesis of thiazolines using hexaphenyloxodiphosphonium trifluoromethanesulfonate. *Angew. Chem. Int. Ed.* 42:83–85.

Youngsaye W., J. T. Lowe, F. Pohlki et al. 2007. Total synthesis and stereochemical reassignment of (+)-neopeltolide. *Angew. Chem. Int. Ed.* 46:9211–9214.

Yurek-George A., A. R. L. Cecil, A. H. K. Mo et al. 2007. The first biologically active synthetic analogues of FK228, the depsipeptide histone deacetylase inhibitor. *J. Med. Chem.* 50:5720–5726.

Yurek-George A., F. Habens, M. Brimmell et al. 2004. Total synthesis of spiruchostatin A, a potent histone deacetylase inhibitor. *J. Am. Chem. Soc.* 126:1030–1031.

Zampella A., M. V. D'Auria, L. Minale et al. 1996. Callipeltoside A: A cytotoxic aminodeoxy sugar-containing macrolide of a new type from the marine Lithistida sponge *Callipelta sp. J. Am. Chem. Soc.* 118:11085–11088.

Zeng X., B. Yin, Z. Hu et al. 2010. Total synthesis and biological evaluation of largazole and derivatives with promising selectivity for cancer cells. *Org. Lett.* 12:1368–1371.

8 Omega-3 (ω-3) Polyunsaturated Fatty Acids

Philip C. Calder

CONTENTS

8.1 STRUCTURE, NAMING, AND BIOSYNTHESIS OF ω-3 FATTY ACIDS

The term omega-3 (ω-3 or n-3) is a structural descriptor for a family of polyunsaturated fatty acids (PUFAs): ω-3 signifies the position of double bond that is closest to the methyl terminus of the acyl chain. All ω-3 fatty acids have this double bond on carbon three, counting the methyl carbon as carbon one (Figure 8.1). Like other fatty acids, ω-3 fatty acids have systematic and common names (Table 8.1), but they are commonly referred to by a shorthand nomenclature that denotes the number of carbon atoms in the chain, the number of double bonds, and the position of the first double bond relative to the methyl carbon (Table 8.1). The simplest ω-3 fatty acid is α-linolenic acid (18:3ω-3). α-Linolenic acid is synthesized from the ω-6 fatty acid linoleic acid (18:2ω-6) by desaturation, catalyzed by delta-15 desaturase (note that the desaturase enzymes are named according to the first carbon carrying the newly inserted double bond and counting the carboxyl carbon as carbon number one) (Figure 8.2). Animals, including humans, do not possess the delta-15 desaturase enzyme and so cannot synthesize α-linolenic acid. In contrast, plants possess delta-15 desaturase and so are able to synthesize α-linolenic acid.

Although animals cannot synthesize α-linolenic acid, they can metabolize it by further desaturation and elongation; desaturation occurs at carbon atoms below carbon number nine (counting from the carboxyl carbon) and mainly occurs in the liver. α-Linolenic acid can be converted to stearidonic acid (18:4ω-3) by delta-6 desaturase and then stearidonic acid can be elongated to eicosatetraenoic acid (20:4ω-3) (Figure 8.2). This fatty acid can be further desaturated by delta-5 desaturase to yield eicosapentaenoic acid (20:5ω-3; known as EPA) (Figure 8.2). Conversion of α-linolenic acid

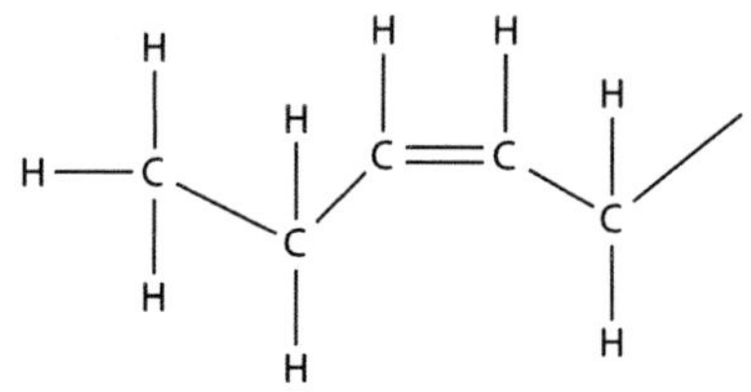

FIGURE 8.1 Generic structure of ω-3 fatty acids.

TABLE 8.1
ω-3 Polyunsaturated Fatty Acid Family

Systematic Name	Common Name	Shorthand Nomenclature
All *cis* 9, 12, 15-octadecatrienoic acid	α-Linolenic acid	18:3ω-3
All *cis* 6, 9, 12, 15-octadecatetraenoic acid	Stearidonic acid	18:4ω-3
All *cis* 8, 11, 14, 17-eicosatetraenoic acid	Eicosatetraenoic acid	20:4ω-3
All *cis* 5, 8, 11, 14, 17-eicosapentaenoic acid	Eicosapentaenoic acid	20:5ω-3
All *cis* 7, 10, 13, 16, 19-docosapentaenoic acid	Docosapentaenoic acid (clupanodonic acid)	22:5ω-3
All *cis* 4, 7, 10, 13, 16, 19-docosahexaenoic acid	Docosahexaenoic acid	22:6ω-3

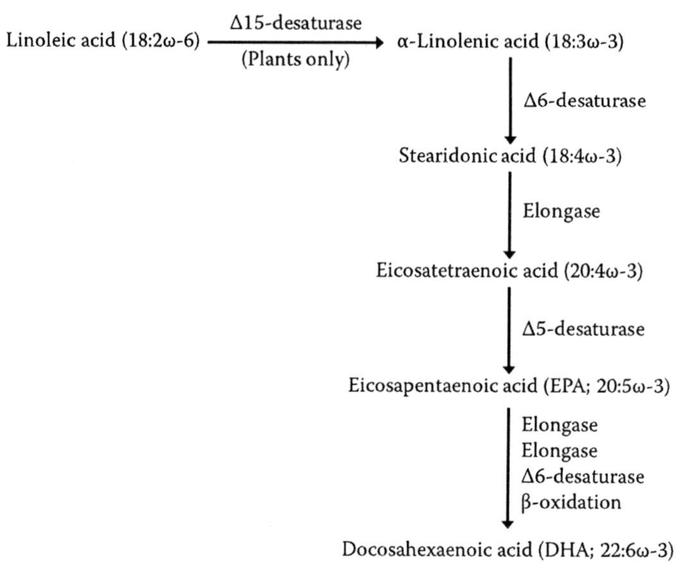

FIGURE 8.2 Pathway of conversion of linoleic acid to α-linolenic acid and of α-linolenic acid to longer chain, more unsaturated ω-3 fatty acids.

to EPA is in competition with the conversion of linoleic acid to arachidonic acid (20:4ω-6) since the same enzymes are used. The delta-6 desaturase reaction is rate limiting in this pathway. Although the preferred substrate for delta-6 desaturase is α-linolenic acid, because linoleic acid is much more prevalent in most human diets than α-linolenic acid, metabolism of ω-6 fatty acids is quantitatively more important. The activities of delta-6 and delta-5 desaturases are regulated by nutritional status, hormones, and by feedback inhibition by end products.

The pathway for conversion of EPA to docosahexaenoic acid (22:6ω-3; known as DHA) involves addition of two carbons to EPA to form docosapentaenoic acid (22:5ω-3; known as DPA), addition of two further carbons to produce 24:5ω-3, desaturation at the delta-6 position to form 24:6ω-3, and translocation of 24:6ω-3 from the endoplasmic reticulum to peroxisomes where two carbons are removed by limited β-oxidation to yield DHA. Short-term studies with isotopically labeled α-linolenic acid and long-term studies using significantly increased intakes of α-linolenic acid have demonstrated that the conversion to EPA, DPA, and DHA is generally poor in humans, with very limited conversion all the way to DHA being observed (Arterburn et al. 2006; Burdge and Calder 2006). EPA and DPA can also be synthesized from DHA by retro-conversion due to limited peroxisomal β-oxidation. In this chapter EPA, DPA, and DHA are referred to as very long chain ω-3 PUFAs; they are sometimes referred to as marine ω-3 PUFAs.

8.2 DIETARY SOURCES AND TYPICAL INTAKES OF ω-3 FATTY ACIDS

8.2.1 α-Linolenic Acid from Plant Sources

Green leaves contain a significant proportion (typically over 50%) of their fatty acids as α-linolenic acid. However, green leaves are not rich sources of fat and so these are not major dietary sources of fatty acids including α-linolenic acid. Several seeds and seed oils and some nuts contain significant amounts of α-linolenic acid. Linseeds (also called flaxseeds) and their oil typically contain 45%–55% of fatty acids as α-linolenic acid, while soybean oil, rapeseed oil, and walnuts typically contain 5%–10% of fatty acids as α-linolenic acid. Corn oil, sunflower oil, and safflower oil are rich in the ω-6 linoleic acid but contain very little α-linolenic acid. Intakes of α-linolenic acid among Western adults are typically 0.5–2 g/d (British Nutrition Foundation 1999; Burdge and Calder 2006). However, the main PUFA in most Western diets is the ω-6 linoleic acid that is typically consumed in 5–20-fold greater amounts than α-linolenic acid (British Nutrition Foundation 1999; Burdge and Calder 2006).

8.2.2 EPA, DPA, and DHA from Seafood

Seafoods are a good source of very long chain ω-3 PUFAs (British Nutrition Foundation 1999). Fish are often classified into lean fish that store lipid in the liver (e.g., cod) or "fatty" ("oily") fish that store lipid in the flesh (e.g., mackerel, herring, salmon, tuna, sardines). Different types of fish contain different amounts of EPA and DHA, and these fatty acids may be present in different ratios (Table 8.2). These characteristics are partly dependent upon the metabolic characteristics of the fish and also upon their diet, water temperature, season, etc. Even so, it seems that a single lean fish meal (e.g., one serving of cod) could provide about 0.2–0.3 g very long chain ω-3 fatty acids, while a single oily fish meal (e.g., one serving of salmon or mackerel) could provide 1.5–3.0 g

TABLE 8.2
Typical Very Long Chain ω-3 Fatty Acid Contents of Fish

	20:5ω-3	22:5ω-3	22:6ω-3	Total Long Chain ω-3 PUFA per Portion
Seafood	g/100 g Food			g
Cod	0.08	0.01	0.16	0.30
Haddock	0.05	0.01	0.10	0.19
Plaice	0.16	0.04	0.10	0.39
Herring	0.51	0.11	0.69	1.56
Mackerel	0.71	0.12	1.10	3.09
Salmon	0.50	0.40	1.30	2.20
Trout	0.23	0.09	0.83	2.65

of these fatty acids (Table 8.2). The latest estimate for fish consumption among adults in the United Kingdom is approximately 100 g lean fish and approximately 50 g oily fish per week (SACN/COT 2004); similar (and in some countries even lower) intakes are expected in other northern and in eastern European, North American, and Australasian countries. Lean fish intake is higher than this in southern European countries and lean and oily fish intake is higher than this in Japan. Average (mean) intakes of very long chain ω-3 fatty acids among adults in the United Kingdom, in other northern and in eastern European, North American, and Australasian countries are usually quoted as approx. 0.15–0.25 g/day (SACN/COT 2004). However, the distribution of intakes is bimodal due to the presence of oily fish consumers and nonconsumers and a fairly recent estimate of very long chain ω-3 fatty acid intake among Australian adults gave a median intake of about 0.03 g/day, compared with a mean intake of about 0.19 mg/day (Meyer et al. 2003). Intakes would be rather higher in those populations, such as the Japanese, who consume oily fish in greater amounts and with greater regularity than seen in Europe, North America, and Australasia.

8.2.3 Fish Oils

The oil obtained from oily fish flesh or lean fish livers (e.g., cod liver) is termed "fish oil" and it is rich in very long chain ω-3 fatty acids. EPA and DHA comprise about 30% of the fatty acids in a typical preparation of fish oil, which means that a one gram fish oil capsule can provide about 0.3 g of EPA plus DHA. However, the amount of ω-3 fatty acids can vary between fish and fish oils and so can the relative proportions of the individual very long chain ω-3 PUFAs (EPA, DPA, and DHA); for example, cod liver oil is richer in EPA than DHA while tuna oil is richer in DHA than EPA. Encapsulated oil preparations that contain ω-3 fatty acids in higher amounts than found in standard fish oils are available ("fish oil concentrates"). In fish oil capsules, the fatty acids are usually present in the form of triacylglycerols, although ω-3 fatty acids are also available in the phospholipid form (e.g., as krill oil) and as ethyl esters (e.g., in the highly concentrated pharmaceutical preparation Omacor also known as Lovazza in North America). Clearly capsules could make a significant contribution to very long chain ω-3 fatty acid intake. Certain algal oils are particularly rich in DHA, which may comprise as much as 45% of total fatty acids. These oils may be useful where provision of DHA, but not EPA, is particularly desired, for example, in infant formulas.

8.3 INCREASED INTAKE OF VERY LONG CHAIN ω-3 FATTY ACIDS ALTERS THE FATTY ACID COMPOSITION OF PLASMA, CELLS, AND TISSUES IN HUMANS

Very long chain ω-3 PUFAs consumed in the diet are handled in the same way as other dietary fatty acids: They are hydrolyzed from triacylglycerols, absorbed into enterocytes, re-esterified back into triacylglycerols, which are assembled into lipoproteins known as chylomicrons, and then released into the lymphatic circulation, later entering the bloodstream. Fatty acids from chylomicron triacylglycerols are targeted toward storage in adipose tissue, with those remaining in the remnant particle being cleared by the liver. Once in the liver fatty acids may be oxidized, metabolized to other fatty acids, or re-secreted into the bloodstream as a component of liver-derived lipoproteins. These lipoproteins form a vehicle for transport of fatty acids between tissues (Figure 8.3). Fatty acids within cell membranes have a number of functional roles (Miles and Calder 1998). Thus, it is possible to identify storage, transport, and functional pools of fatty acids including very long chain ω-3 fatty acids (Figure 8.3).

Different blood plasma lipid pools, cells, and tissues have different and characteristic fatty acid compositions. These compositions are not only influenced by the availability of different fatty acids but also by the metabolic characteristics of the particular pool, cell, or tissue. Modification

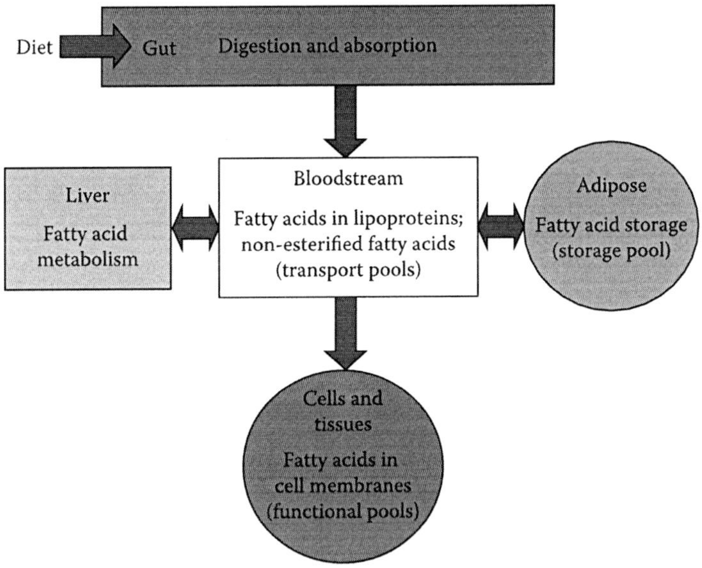

FIGURE 8.3 General scheme of whole body handling of dietary fatty acids showing transport, functional, and storage pools.

of fatty acid profiles has been widely reported after supplementation of the diet with fish oil capsules; such supplementation results in appearance of EPA and DHA in plasma lipids, platelets, erythrocytes (red blood cells), leukocytes (white blood cells), colonic tissue, cardiac tissue, and in many other cell and tissue types. The incorporation of EPA and DHA from fish oil capsules partly displaces ω-6 PUFAs, like arachidonic acid, and occurs in a dose–response fashion. For example, studies using a range of EPA+DHA intakes from 1 to 6 g/day report near linear relationships between EPA and DHA intake and the EPA and DHA contents of plasma phospholipids (Blonk et al. 1990; Harris et al. 1991; Marsen et al. 1992) and of platelet phospholipids (Sanders and Roshanai 1983). In other studies, incorporation of EPA and DHA into blood neutrophils (Healy et al. 2000) and of EPA into plasma phospholipids and blood mononuclear cells (Rees et al. 2006) occurred in a linear dose–response manner (Figure 8.4). In a study combining dose–response and time–course over 12 months in older male subjects, Katan et al. (1997) reported the fatty acid compositions of serum cholesteryl esters, erythrocytes, and adipose tissue. This study confirmed that EPA and DHA are incorporated into circulating lipid pools and into erythrocytes when their intakes are increased. It also demonstrated EPA and DHA incorporation into adipose tissue, a storage pool, when their intakes are increased. However, this study also clearly showed that incorporation into different pools occurs at different rates and to differing extents (i.e., with different efficiencies) and may not be related to intake in a strictly linear fashion, at least over the intakes studied. The study of Katan et al. (1997) showed that near-maximal incorporation of EPA and DHA into serum cholesteryl esters occurs within 30 days of beginning supplementation, whereas maximal incorporation into erythrocytes does not occur until sometime between 56 and 182 days. Yaqoob et al. (2000) reported the time-dependent incorporation of EPA and DHA into blood mononuclear cells; incorporation of both fatty acids was near-maximal after 4 weeks of supplementation (Figure 8.5). Upon cessation of supplementation EPA in mononuclear cells returned to starting levels within 8 weeks, while the cells appeared to retain DHA. The same observations of loss of EPA and selective retention of DHA upon cessation of fish oil supplementation have been made for erthrocytes (Popp-Snijders et al. 1986) and platelets (von Schacky et al. 1985). Thus, a significant body of literature reports that EPA and DHA are incorporated into blood, cell and tissue lipids when their intake is increased.

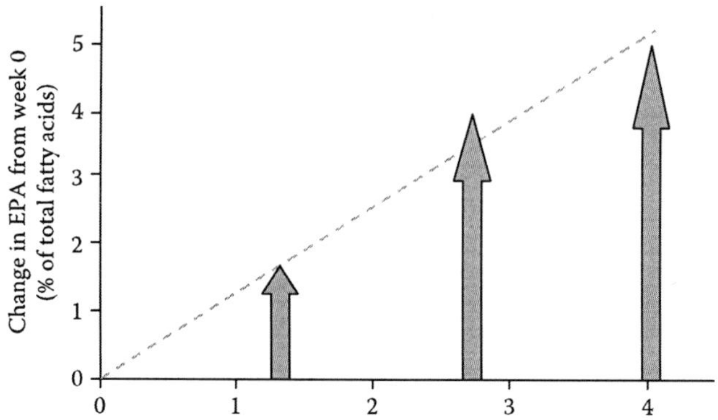

FIGURE 8.4 Dose-dependent incorporation of eicosapentaenoic acid into human plasma phospholipids. Healthy young males supplemented their diet with differing amounts of an EPA-rich oil for a period of 12 weeks. Plasma phospholipids were isolated and their fatty acid composition determined by gas chromatography. Data are expressed as mean change in EPA from week 0 (study entry). (Data from Rees, D. et al., *Am. J. Clin. Nutr.*, 83, 331, 2006.)

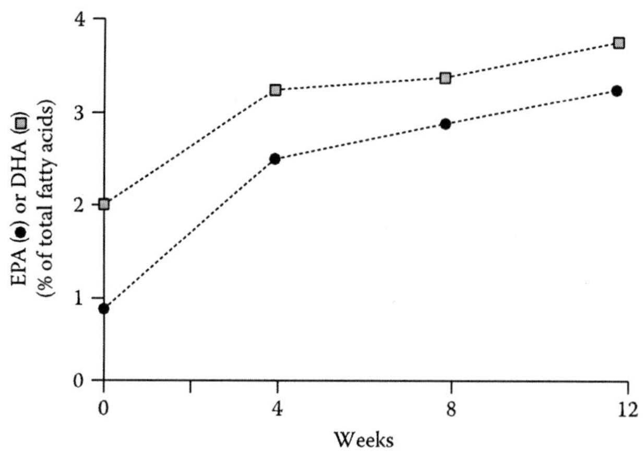

FIGURE 8.5 Time course of changes in eicosapentaenoic and docosahexaenoic acid contents of human blood mononuclear cells in subjects consuming fish oil. Healthy subjects supplemented their diet with fish oil capsules providing 2.1 g EPA plus 1.1 g DHA per day for a period of 12 weeks. Blood mononuclear cell phospholipids were isolated at 0, 4, 8, and 12 weeks and their fatty acid composition determined by gas chromatography. (Data from Yaqoob, P. et al., *Eur. J. Clin. Invest.*, 30, 260, 2000.)

8.4 EXPOSURE TO VERY LONG CHAIN ω-3 FATTY ACIDS CAN MODIFY CELL FUNCTION

Increasing the ω-3 fatty acid content of cells and tissues can modify cell and tissue function through a variety of mechanisms; these are summarized in Figure 8.6.

8.4.1 Alterations in Membrane Structure and Function and Cell Signaling Pathways

Increasing the content of very long chain ω-3 PUFAs in cell membrane phospholipids can lead to modifications of the physical properties of the membrane such as membrane order ("fluidity") and of the structure of rafts (rafts are membrane microdomains with a particular lipid and fatty acid makeup, which play a role as platforms for receptor action and for the initiation of intracellular signaling pathways). In turn, changes in membrane order and in raft structure can influence the activity

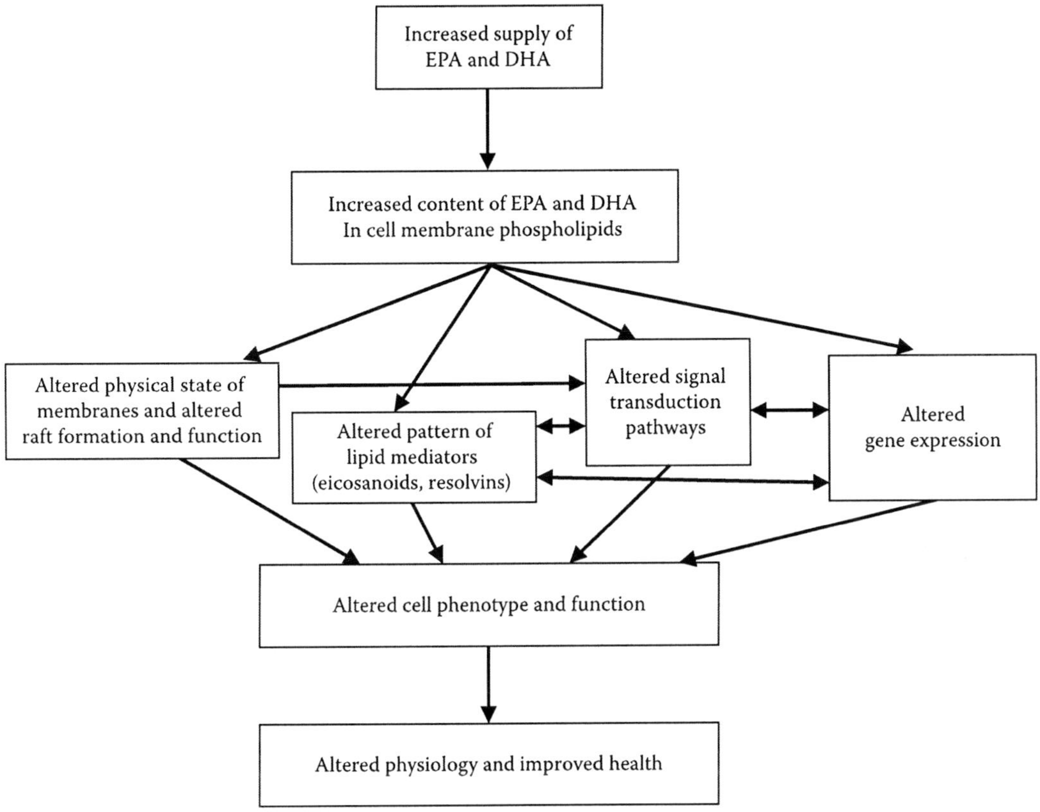

FIGURE 8.6 General scheme of the interacting mechanisms whereby very long chain ω-3 fatty acids influence cell function, physiology, and health.

of membrane proteins including receptors, transporters, ion channels, and signaling enzymes (Miles and Calder 1998; Yaqoob 2009). As a result of these effects, intracellular signal transduction and transcription factor activation may be altered and gene expression modified. Transcription factors reported to be modified by the presence of very long chain ω-3 PUFAs include nuclear factor κB (important in regulating inflammatory processes), peroxisome proliferator activated receptor-α (important in regulating lipid metabolism) and -γ (important in regulating lipid metabolism, adipocyte differentiation, insulin sensitivity, and inflammatory processes), and the sterol regulatory element binding proteins (important in regulating lipid metabolism) (Jump 2002, 2008; Clarke 2004; Lapillonne et al. 2004; Deckelbaum et al. 2006).

8.4.2 Effects on Lipid Mediators

Eicosanoids produced from the ω-6 PUFA arachidonic acid, including various prostaglandins, thromboxanes, and leukotrienes, have well-established physiological roles in regulation of inflammation, immunity, platelet aggregation, smooth muscle contraction, and renal function (Nicolaou and Kafatos 2004). Eicosanoids are produced as a result of the action of cyclooxygenase and lipoxygenase enzymes. Despite their roles in physiology, excess or inappropriate production of eicosanoids is associated with disease processes. For example, cysteinyl-leukotrienes play an important role in asthma. A range of drugs of varying specificity, for example, cyclooxygenase inhibitors, are used clinically to suppress the production of eicosanoids from arachidonic acid. Very long chain ω-3 PUFAs decrease the production of arachidonic–acid derived eicosanoids and so can impact on the actions regulated by those mediators (Calder 2008a). The effect of ω-3 PUFAs on eicosanoid synthesis is through several actions. First, when EPA and DHA are incorporated into cell membranes they displace arachidonic acid, thereby decreasing the availability of the substrate for production of eicosanoids. Second, EPA is able to inhibit the activity of cyclooxygenase toward arachidonic acid.

In addition to decreasing the production of eicosanoids from arachidonic acid, EPA is a substrate for the synthesis of alternative eicosanoids, which are typically less potent than those produced from arachidonic acid (Wada et al. 2007; Calder 2008a). In fact, EPA-derived eicosanoids may antagonize the action of those produced from arachidonic acid (Tull et al. 2009). Relatively recently a new family of lipid mediators, termed resolvins, synthesized from both EPA (E-series resolvins) and DHA (D-series resolvins) have been described. These mediators have been demonstrated in cell culture and animal feeding studies to be potently anti-inflammatory, inflammation resolving, and immunomodulatory (Serhan et al. 2000a,b, 2008). Protectin D1, produced from DHA, appears to have an important role in protecting tissue, including neuronal tissue, from excessive damage in a variety of experimental situations (Serhan et al. 2002).

8.4.3 Receptor-Mediated Effects

Oh et al. (2010) reported that the G-protein coupled receptor GPR120, which is able to bind long chain fatty acids, is highly expressed on adipocytes and on inflammatory macrophages. A GPR120 agonist, GW9508, inhibited inflammatory responsiveness of macrophages, suggesting that GPR120 is involved in anti-inflammatory signaling. EPA and DHA promoted GPR120-mediated gene activation in cultured macrophages, and anti-inflammatory effects of DHA did not occur in GPR120 knockdown cells. Oh et al. (2010) also demonstrated that DHA-induced translocation of the glucose transporter GLUT4 to the surface of cultured adipocytes was abolished by GPR120 knockout, suggesting that GPR120 mediates some of the metabolic actions of DHA. These observations suggest effects of very long chain ω-3 PUFAs that do not involve either their incorporation into cell membrane phospholipids or a modification of lipid mediator production.

8.5 INCREASING INTAKE OF VERY LONG CHAIN ω-3 FATTY ACIDS HAS HEALTH BENEFITS

Through the mechanisms of action outlined earlier and the resulting modifications of cell and tissue function, very long chain ω-3 fatty acids exert physiological actions. These are summarized in Table 8.3 where they are linked to certain health or clinical benefits. A number of risk factors for cardiovascular disease are modified in a beneficial way by increased intake of very long chain ω-3 fatty acids: These include blood pressure (Geleijnse et al. 2002), platelet reactivity and thrombosis (British Nutrition Foundation 1992), plasma triacylglycerol concentrations (Harris 1996), vascular function (Nestel et al. 2002), cardiac arrhythmias (von Schacky 2008), heart rate variability (von Schacky 2008), and inflammation (Calder 2006). As a result, increased very long chain ω-3 fatty acid intake is associated with a reduced risk of cardiovascular morbidity and mortality (Calder 2004). Indeed, supplementation studies with very long chain ω-3 fatty acids have demonstrated reduced mortality (Anonymous 1999; Bucher et al. 2002; Marchioli et al. 2002; Studer et al. 2005; Yokoyama et al. 2007). A number of other, non-cardiovascular, actions of these fatty acids have also been documented (Table 8.3), suggesting that increased intake of these fatty acids could be of benefit in protecting from or treating many conditions. For example, they have been used successfully in rheumatoid arthritis (Calder 2008b) and, in some studies, in inflammatory bowel diseases (Calder 2008c), and may be useful in other inflammatory conditions (Calder 2006). DHA has an important structural role in the eye and brain, and its supply early in life when these tissues are developing is known to be of vital importance in terms of optimizing visual and neurological development (SanGiovanni et al. 2000a,b). Studies have highlighted the potential for very long chain ω-3 fatty acids to contribute to enhanced mental development (Helland et al. 2003) and improved childhood

TABLE 8.3
Summary of the Physiological Roles, Potential Clinical Benefits, and Disease Targets of Very Long Chain ω-3 Fatty Acids

Physiological Role	Potential Clinical Benefit	Target
Regulation of blood pressure	Decreased blood pressure	Hypertension; CVD
Regulation of platelet function	Decreased risk of thrombosis	Thrombosis; CVD
Regulation of blood coagulation	Decreased risk of thrombosis	Thrombosis; CVD
Regulation of plasma triacylglycerol concentrations	Decreased plasma triacylglycerol concentrations	Hypertriglyceridemia; CVD
Regulation of vascular function	Improved vascular reactivity	CVD
Regulation of cardiac rhythm	Decreased risk of cardiac arrhythmias	CVD
Regulation of heart rate	Decreased heart rate; increased heart rate variability	CVD
Regulation of inflammation	Decreased inflammation	Inflammatory diseases; CVD
Regulation of immune function	Improved immune function	Compromised immunity
Regulation of fatty acid and triacylglycerol metabolism	Decreased triacylglycerol synthesis and storage	Weight gain; weight loss; obesity
Regulation of bone turnover	Maintained bone mass	Osteoporosis
Regulation of insulin sensitivity	Improved insulin sensitivity	Type-2 diabetes
Regulation of tumor cell growth	Decreased tumor cell growth and survival	Some cancers
Regulation of visual signaling	Optimized visual signaling	Poor infant visual development (especially pre-term)
Structural component of brain and central nervous system	Optimized brain development—cognitive and learning processes	Poor infant and childhood cognitive processes and learning

Abbreviation used: CVD, cardiovascular disease.

learning and behavior (Richardson 2004) and to reduce the burden of psychiatric illnesses in adults (Freeman et al. 2006), although these remain less certain areas of possible action that require more scientific support. There may also be a role for very long chain ω-3 PUFAs, DHA in particular, in preventing neurodegenerative disease of ageing (Solfrizzi et al. 2010) and the production of protectins, especially protectin D1 (also known as neuroprotectin D1), appears to be crucial for this effect (Lukiw et al. 2005). The effects of very long chain ω-3 PUFAs on health outcomes are likely to be dose-dependent, but clear dose response data have not been identified in most cases.

8.6 HEALTH EFFECTS OF α-LINOLENIC ACID

The discussion so far has centered upon the fish-derived very long chain ω-3 PUFAs for which there is much evidence for human health benefit and an increasing understanding of the multiple mechanisms involved. The major plant ω-3 PUFA, α-linolenic acid, is an essential fatty acid and may have human health benefits either in its own right or by acting as a precursor for synthesis of the longer chain more unsaturated derivatives using the pathway shown in Figure 8.2. These possibilities have been reviewed in some detail elsewhere (Arterburn et al. 2006; Burdge and Calder 2006). Studies in humans using acute ingestion of stable isotopically labeled α-linolenic acid have demonstrated some conversion to EPA and to DPA, but much more limited conversion to DHA, although this may be greater in young adult women than in men (Burdge et al. 2002; Burdge and Wootton 2002), possibly because of upregulation of the delta-6 desaturase by female sex hormones. Little is known about the extent of α-linolenic acid conversion to EPA and DHA in infancy and childhood, in the elderly or during pregnancy and lactation, times when synthesis of very long chain ω-3 PUFAs might be important or desirable. A number of studies have examined the effect of chronic (i.e., weeks to months) consumption of increased amounts of α-linolenic acid. These studies confirm that increasing α-linolenic acid intake increases the EPA (and DPA) content of plasma lipids, platelets, leukocytes, and erythrocytes but that DHA content does not increase (Arterburn et al. 2006; Burdge and Calder 2006); clearly these findings are in agreement with the stable isotope studies. Such studies with α-linolenic acid have demonstrated some effects on cardiovascular risk factors and on inflammatory markers, but where these are reported they are typically weaker than the effects achieved from increasing consumption of EPA+DHA, and may be due to the increased appearance of EPA (Caughey et al. 1996; Zhao et al. 2004).

8.7 CONCLUSIONS

Current intakes of very long chain ω-3 fatty acids EPA and DHA are low in most individuals living in Western countries. A good natural source of these fatty acids is seafood, especially oily fish. Fish oil capsules contain these fatty acids too, with a standard 1 g capsule providing about 0.3 g of EPA plus DHA; more concentrated forms are also available in capsules. Very long chain ω-3 fatty acids are incorporated from capsules into transport (blood lipids), functional (cell and tissue), and storage (adipose) pools in humans. This incorporation is dose-dependent and follows a kinetic pattern that is characteristic for each pool. Incorporation is most rapid into blood lipids, followed by platelets and white cells, followed by erythrocytes. At sufficient levels of incorporation into cells, EPA and DHA influence the physical nature of cell membranes and membrane protein-mediated responses, lipid mediator generation, cell signaling, and gene expression in many different cell types. Through these mechanisms EPA and DHA influence cell and tissue physiology and the way cells and tissues respond to external signals. In most cases the effects seen are compatible with improvements in disease biomarker profiles or in health-related outcomes. An important aspect of this is the requirement for very long chain ω-3 fatty acids, especially DHA, in early growth and development of the brain and visual system, meaning that adequate provision to the fetus and to the newborn infant is essential. As a result of their effects on cell and tissue physiology, very long chain ω-3 fatty acids play a role in achieving optimal health and in protection against disease. Long chain ω-3 fatty acids not only protect against cardiovascular morbidity but also against mortality. In some situations, for example, rheumatoid arthritis, they may be beneficial as therapeutic agents although a high intake is required. The plant ω-3 fatty acid, α-linolenic acid, can be converted to EPA but in humans conversion to DHA appears to be poor. Effects of α-linolenic acid on human health-related outcomes appear to be due to conversion to EPA.

ABBREVIATIONS

DHA	docosahexaenoic acid
DPA	docosapentaenoic acid
EPA	eicosapentaenoic acid
PUFA	polyunsaturated fatty acid

REFERENCES

Anonymous. 1999. Dietary supplementation with n-3 polyunsaturated fatty acids and vitamin E after myocardial infarction: results of the GISSI-Prevenzione trial. *Lancet* 354:447–455.

Arterburn, L.M., Hall, E.B. and Oken, H. 2006. Distribution, interconversion, and dose response of n-3 fatty acids in humans. *Am. J. Clin. Nutr.* 83:1467S–1476S.

Blonk, M.C., Bilo, H.J., Popp-Snijders, C., Mulder, C. and Donker, A.J. 1990. Dose-response effects of fish oil supplementation in healthy volunteers. *Am. J. Clin. Nutr.* 52:120–127.

British Nutrition Foundation. 1992. *Unsaturated Fatty Acids: Nutritional and Physiological Significance.* London, U.K.: Chapman & Hall.

British Nutrition Foundation. 1999. *Briefing Paper: N-3 Fatty Acids and Health*. London, U.K.: British Nutrition Foundation.

Bucher, H.C., Hengstler, P., Schindler, C. and Meier, G. 2002. N-3 polyunsaturated fatty acids in coronary heart disease: a meta-analysis of randomized controlled trials. *Am. J. Med.* 112:298–304.

Burdge, G.C. and Calder, P.C. 2006. Dietary α-linolenic acid and health-related outcomes: a metabolic perspective. *Nutr. Res. Rev.* 19:26–52.

Burdge, G.C., Jones, A.E. and Wootton, S.A. 2002. Eicosapentaenoic and docosapentaenoic acids are the principal products of α-linolenic acid metabolism in young men. *Br. J. Nutr.* 88:355–363.

Burdge, G.C. and Wootton, S.A. 2002. Conversion of α-linolenic acid to eicosapentaenoic, docosapentaenoic and docosahexaenoic acids in young women. *Br. J. Nutr.* 88:411–420.

Calder, P.C. 2004. N-3 fatty acids and cardiovascular disease: evidence explained and mechanisms explored. *Clin. Sci.* 107:1–11.

Calder, P.C. 2006. N-3 polyunsaturated fatty acids, inflammation, and inflammatory diseases. *Am. J. Clin. Nutr.* 83:1505S–1519S.

Calder, P.C. 2008a. The relationship between the fatty acid composition of immune cells and their function. *Prostaglandins Leukot. Essent. Fatty Acids* 79:101–108.

Calder, P.C. 2008b. PUFA, inflammatory processes and rheumatoid arthritis. *Proc. Nutr. Soc.* 67:409–418.

Calder, P.C. 2008c. Polyunsaturated fatty acids, inflammatory processes and inflammatory bowel diseases. *Mol. Nutr. Food Res.* 52:885–897.

Caughey, G.E., Mantzioris, E., Gibson, R.A., Cleland, L.G. and James, M.J. 1996. The effect on human tumor necrosis factor α and interleukin 1β production of diets enriched in n-3 fatty acids from vegetable oil or fish oil. *Am. J. Clin. Nutr.* 63:116–122.

Clarke, S.D. 2004. The multi-dimensional regulation of gene expression by fatty acids: polyunsaturated fats as nutrient sensors. *Curr. Opin. Lipidol.* 15:13–18.

Deckelbaum, R.J., Worgall, T.S. and Seo, T. 2006. N-3 fatty acids and gene expression. *Am. J. Clin. Nutr.* 83:1520S–1525S.

Freeman, M.P., Hibbeln, J.R., Wisner, K.L., Davis, J.M., Mischoulon, D., Peet, M., Keck Jr., P.E. et al. 2006. Omega-3 fatty acids: evidence basis for treatment and future research in psychiatry. *J. Clin. Psychiatry* 67:1954–1967.

Geleijnse, J.M., Giltay, E.J., Grobbee, D.E., Donders, A.R.T. and Kok, F.J. 2002. Blood pressure response to fish oil supplementation: meta-regression analysis of randomized trials. *J. Hypertens.* 20:1493–1499.

Harris, W.S. 1996. N-3 fatty acids and lipoproteins: comparison of results from human and animal studies. *Lipids* 31:243–252.

Harris, W.S., Windsor, S.L. and Dujovne, C.A. 1991. Effects of four doses of n-3 fatty acids given to hyperlipidemic patients for six months. *J. Am. Coll. Nutr.* 10:220–227.

Healy, D.A., Wallace, F.A., Miles, E.A., Calder, P.C. and Newsholme, P. 2000. The effect of low to moderate amounts of dietary fish oil on neutrophil lipid composition and function. *Lipids* 35:763–768.

Helland, I.B., Smith, L., Saarem, K., Saugstad, O.D. and Drevon, C.A. 2003. Maternal supplementation with very-long-chain n-3 fatty acids during pregnancy and lactation augments children's IQ at 4 years of age. *Pediatrics* 111:e39–e44.

Jump, D.B. 2002. Dietary polyunsaturated fatty acids and regulation of gene transcription. *Curr. Opin. Lipidol.* 13:155–164.

Jump, D.B. 2008. N-3 polyunsaturated fatty acid regulation of hepatic gene transcription. *Curr. Opin. Lipidol.* 19:242–247.

Katan, M.B., Deslypere, J.P., van Birgelen, A.P.J.M., Penders, M. and Zegwaars, M. 1997. Kinetics of the incorporation of dietary fatty acids into serum cholesteryl esters, erythrocyte membranes and adipose tissue: an 18 month controlled study. *J. Lipid Res.* 38:2012–2022.

Lapillonne, A., Clarke, S.D. and Heird, W.C. 2004. Polyunsaturated fatty acids and gene expression. *Curr. Opin. Clin. Nutr. Metab. Care* 7:151–156.

Lukiw, W.J., Cui, J.G., Marcheselli, V.L., Bodker, M., Botkjaer, A., Gotlinger, K., Serhan, C.N. and Bazan, N.G. 2005. A role for docosahexaenoic acid-derived neuroprotectin D1 in neural cell survival and Alzheimer disease. *J. Clin. Invest.* 115:2774–2783.

Marchioli, R., Barzi, F., Bomba, E., Chieffo, C., Di Gregorio, D., Di Mascio, R., Franzosi, M.G. et al. 2002. Early protection against sudden death by n-3 polyunsaturated fatty acids after myocardial infarction. Time-course analysis of the results of the Gruppo Italiano per lo Studio della Sopravvivenza nell'Infarto Miocardico (GISSI)-Prevenzione. *Circulation* 105:1897–1903.

Marsen, T.A., Pollok, M., Oette, K. and Baldamus, C.A. 1992. Pharmacokinetics of omega-3 fatty acids during ingestion of fish oil preparations. *Prostaglandins Leukot. Essent. Fatty Acids* 46:191–196.

Meyer, B.J., Mann, N.J., Lewis, J.L., Milligan, G.C., Sinclair, A.J. and Howe, P.R. 2003. Dietary intakes and food sources of omega-6 and omega-3 polyunsaturated fatty acids. *Lipids* 38:391–398.

Miles, E.A. and Calder, P.C. 1998. Modulation of immune function by dietary fatty acids. *Proc. Nutr. Soc.* 57:277–292.

Nestel, P., Shige, H., Pomeroy, S., Cehun, M., Abbey, M. and Raederstorff, D. 2002. The n-3 fatty acids eicosapentaenoic acid and docosahexaenoic acid increase systemic arterial compliance in humans. *Am. J. Clin. Nutr.* 76:326–330.

Nicolaou, A. and Kafatos, G. 2004. *Bioactive Lipids*. Bridgewater, U.K.: The Oily Press.

Oh, D.Y., Talukdar, S., Bae, E.J., Imamura, T., Morinaga, H., Fan, W., Li, P., Lu, W.J., Watkins, S.M. and Olefsky, J.M. 2010. GPR120 is an omega-3 fatty acid receptor mediating potent anti-inflammatory and insulin-sensitizing effects. *Cell* 142:687–698.

Popp-Snijders, C., Schouten, J.A., van Blitterswijk, W.J. and van der Veen, E.A. 1986. Changes in membrane lipid composition of human erythrocytes after dietary supplementation of (n-3) fatty acids: Maintenance of membrane fluidity. *Biochim. Biophys. Acta* 854:31–37.

Rees, D., Miles, E.A., Banerjee, T., Wells, S.J., Roynette, C.E., Wahle, K.W.J.W. and Calder, P.C. 2006. Dose-related effects of eicosapentaenoic acid on innate immune function in healthy humans: a comparison of young and older men. *Am. J. Clin. Nutr.* 83:331–342.

Richardson, A.J. 2004. Clinical trials of fatty acid treatment in ADHD, dyslexia, dyspraxia and the autistic spectrum. *Prostaglandins Leukot. Essent. Fatty Acids* 70:383–390.

Sanders, T.A.B. and Roshanai, F. 1983. The influence of different types of ω-3 polyunsaturated fatty acids on blood lipids and platelet function in healthy volunteers. *Clin. Sci.* 64:91–99.

SanGiovanni, J.P., Berkey, C.S., Dwyer, J.T. and Colditz, G.A. 2000b. Dietary essential fatty acids, long-chain polyunsaturated fatty acids, and visual resolution acuity in healthy fullterm infants: a systematic review. *Early Hum. Dev.* 57:165–188.

SanGiovanni, J.P., Parra-Cabrera, S., Colditz, G.A., Berkey, C.S. and Dwyer, J.T. 2000a. Meta-analysis of dietary essential fatty acids and long-chain polyunsaturated fatty acids as they relate to visual resolution acuity in healthy preterm infants. *Pediatrics* 105:1292–1298.

von Schacky, C. 2008. Omega-3 fatty acids: antiarrhythmic, proarrhythmic or both? *Curr. Opin. Clin. Nutr. Metab. Care* 11:94–99.

von Schacky, C., Fischer, S. and Weber, P.C. 1985. Long term effects of dietary marine ω-3 fatty acids upon plasma and cellular lipids, platelet function, and eicosanoid formation in humans. *J. Clin. Invest.* 76:1626–1631.

Scientific Advisory Committee on Nutrition/Committee on Toxicity. 2004. *Advice on Fish Consumption: Benefits and Risks*. London, U.K.: TSO.

Serhan, C.N., Chiang, N. and van Dyke, T.E. 2008. Resolving inflammation: dual anti-inflammatory and pro-resolution lipid mediators. *Nat. Rev. Immunol.* 8:349–361.

Serhan, C.N., Clish, C.B., Brannon, J., Colgan, S.P., Chiang, N. and Gronert, K. 2000b. Novel functional sets of lipid-derived mediators with antinflammatory actions generated from omega-3 fatty acids via cyclooxygenase 2-nonsteroidal antiinflammatory drugs and transcellular processing. *J. Exp. Med.* 192:1197–1204.

Serhan, C.N., Clish, C.B., Brannon, J., Colgan, S.P., Gronert, K. and Chiang, N. 2000a. Anti-inflammatory lipid signals generated from dietary n-3 fatty acids via cyclooxygenase-2 and transcellular processing: a novel mechanism for NSAID and n-3 PUFA therapeutic actions. *J. Physiol. Pharmacol.* 4:643–654.

Serhan, C.N., Hong, S., Gronert, K., Colgan, S.P., Devchand, P.R., Mirick, G. and Moussignac, R.-L. 2002. Resolvins: a family of bioactive products of omega-3 fatty acid transformation circuits initiated by aspirin treatment that counter pro-inflammation signals. *J. Exp. Med.* 196:1025–1037.

Solfrizzi, V., Frisardi, V., Capurso, C., D'Introno, A., Colacicco, A.M., Vendemiale, G., Capurso, A. and Panza, F. 2010. Dietary fatty acids in dementia and predementia syndromes: epidemiological evidence and possible underlying mechanisms. *Ageing Res. Rev.* 9:184–199.

Studer, M., Briel, M., Leimenstoll, B., Glass, T.R. and Bucher, H.C. 2005. Effect of different anti-lipidemic agents and diets on mortality: a systematic review. *Arch. Int. Med.* 165:725–730.

Tull, S.P., Yates, C.M., Maskrey, B.H., O'Donnell, V.B., Madden, J., Grimble, R.F., Calder, P.C., Nash, G.B. and Rainger, G.E. 2009. Omega-3 Fatty acids and inflammation: novel interactions reveal a new step in neutrophil recruitment. *PLoS Biol.* 7:e1000177.

Wada, M., DeLong, C.J., Hong, Y.H., Rieke, C.J., Song, I., Sidhu, R.S., Yuan, C. et al. 2007. Enzymes and receptors of prostaglandin pathways with arachidonic acid-derived versus eicosapentaenoic acid-derived substrates and products. *J. Biol. Chem.* 282:22254–22266.

Yaqoob, P. 2009. The nutritional significance of lipid rafts. *Annu. Rev. Nutr.* 29:257–282.

Yaqoob, P., Pala, H.S., Cortina-Borja, M., Newsholme, E.A. and Calder, P.C. 2000. Encapsulated fish oil enriched in α-tocopherol alters plasma phospholipid and mononuclear cell fatty acid compositions but not mononuclear cell functions. *Eur. J. Clin. Invest.* 30:260–274.

Yokoyama, M., Origasa, H., Matsuzaki, M., Matsuzawa, Y., Saito, Y., Ishikawa, Y., Oikawa, S. et al. 2007. Effects of eicosapentaenoic acid on major coronary events in hypercholesterolaemic patients (JELIS): a randomised open-label, blinded endpoint analysis. *Lancet* 369:1090–1098.

Zhao, G., Etherton, T.D., Martin, K.R., West, S.G., Gillies, P.J. and Kris-Etherton, P.M. 2004. Dietary alpha-linolenic acid reduces inflammatory and lipid cardiovascular risk factors in hypercholesterolemic men and women. *J. Nutr.* 134:2991–2997.

9 Structure and Biological Activity of Natural Melanin Pigments

Krystyna Stępień, Anna Dzierżęga-Lęcznar, Irena Tam, and Slawomir Kurkiewicz

CONTENTS

9.1 INTRODUCTION

Melanin pigments are heterogeneous biopolymers and are widely distributed in nature. They are produced by animals, plants, and microorganisms, such as pathogenic fungi and bacteria. In humans, melanins are present in the skin, hair, eyes, and in other locations of the body, including the inner ear, and the *substantia nigra* and the *locus coeruleus* of the brain (Marsden 1983; Tolleson 2005). Traditionally, the pigments can be classified into brown to black eumelanins and allomelanins, and yellow or reddish-brown, sulfur-containing pheomelanins (Nicolaus 1968). Allomelanins occur in the plant kingdom, e.g., in certain fungi, and in the seeds of some flowering plants, and are formed by the oxidation of nitrogen-free diphenols, such as catechol, 1,8-dihydroxynaphtalene, and γ-glutaminyl-3,4-dihydroxybenzene (Swan 1974; Wheeler and Bell 1988). Eumelanins are polymers consisting mainly of indole-type units that arise from L-tyrosine or L-DOPA (L-3,4-dihydroxyphenylalanine) oxidation, whereas pheomelanins are derived from the oxidative polymerization of cysteinyl conjugates of DOPA via benzothiazine intermediates (Prota 1992).

The common and obligatory step of both eumelanogenesis and pheomelanogenesis is tyrosinase-catalyzed oxidation of L-tyrosine to L-dopaquinone, which can react with the thiol group of cysteine to produce cysteinyldopa isomers, or can undergo intramolecular cyclization and further oxidation to dopachrome. The rearrangement of dopachrome (with or without decarboxylation) leads to the formation of 5,6-dihydroxyindole (DHI) and 5,6-dihydroxyindole-2-carboxylic acid (DHICA). The two dihydroxyindoles are then oxidized and polymerized to produce eumelanin (Prota et al. 1998; Ito 2003; Ito and Wakamatsu 2006).

Tyrosinase (monophenol, *o*-diphenol: oxygen oxidoreductase, EC 1.14.18.1) is the key regulatory enzyme involved in the biosynthesis of the melanin pigments. This enzyme catalyses the critical rate-limiting hydroxylation of L-tyrosine to L-DOPA and the oxidation of L-DOPA to its *o*-quinone (Olivares and Solano 2009). In plants and lower organisms, the formation of dopaquinone is the only step of melanogenesis that is enzymatically controlled, and the pathway then proceeds spontaneously. In mammalian melanocytes, two tyrosinase-related proteins (TRP-1 and TRP-2) have been shown to regulate melanogenesis, in addition to tyrosinase (Kuzumaki et al. 1993; del Marmol and Beerman 1996). TRP-2, now known as dopachrome tautomerase (DCT, EC 5.3.3.12), catalyses tautomerization of dopachrome to DHICA (Aroca et al. 1990; Tsukamoto et al. 1992). TRP-1 can oxidize DHICA in mice (Jimenez-Cervantez et al. 1994; Kobayashi et al. 1994), but in humans this catalytic function of TRP-1 seems to be lost (Boissy et al. 1998) and DHICA may be further oxidized by tyrosinase (Olivares et al. 2001). It has been suggested that TRP-1 serves as a type of chaperone necessary for the proper processing and trafficking of tyrosinase (Hearing 2005). Melanocytes can synthesize both eumelanin and pheomelanin, and the type of melanin produced depends on the expression and activities of the three melanogenic enzymes as well as on the availability of tyrosine and cysteine (del Marmol and Beerman 1996; Ito and Wakamatsu 2008; Simon et al. 2009).

A major determinant of skin and hair pigmentation is the melanocortin-1 receptor (MC1R), a G protein-coupled receptor that regulates both the quantity and the type of melanin produced (Slominski et al. 2004; Garcia-Borron et al. 2005). MC1R function is controlled by the agonists α-melanocyte-stimulating hormone (α-MSH) and adrenocorticotropic hormone (ACTH), which are produced by the enzymatic cleavage of proopiomelanocortin (POMC) in skin cells, and by the antagonist, *agouti* signaling protein (ASP). Binding of α-MSH or ACTH to MC1R on melanocytes stimulates the expression of the melanogenic cascade and eumelanin synthesis, whereas ASP can reverse those effects and elicit the production of pheomelanin. In mice, over-expression of ASP leads to yellow coat color (Slominski et al. 2004). Polymorphisms within the MC1R gene are largely responsible for the wide range of skin and hair color among different ethnic groups (Garcia-Borron et al. 2005).

Neuromelanin, the dark pigment that deposits with age in the catecholaminergic neurons of the *substantia nigra* (SN) and the *locus coeruleus* (LC) of human brain, derives from the metabolism of catecholamines (dopamine for the SN and norepinephrine for the LC) via the oxidative pathway (Marsden 1983). Neuromelanin synthesis appears to be driven by an excess of cytosolic catecholamines that are not accumulated into synaptic vesicles (Sulzer et al. 2000), but it is not clear whether neuromelanin formation is an autoxidation process or is enzymatically controlled. Neuromelanin is the subject of extensive study, mainly due to its postulated role in brain aging and Parkinson's disease (Double et al. 1999; Zecca et al. 2001).

Melanin biopolymers possess unique optical, electrical, free radical, and redox properties, and the ability to bind metal ions and some organic compounds, including drugs and toxins. The physicochemical properties of melanins determine their bioactivity, although biological functions of melanin pigments are not fully recognized. The present chapter focuses on the photoprotective, antioxidant, and immunomodulatory activities of melanins.

9.2 STRUCTURAL INVESTIGATION OF MELANIN BIOPOLYMERS BY Py-GC/MS

Since the structure of melanin affects strongly its properties, and hence the bioactivity, structural studies of melanin pigments are of great importance. However, such investigations are not easy. The biopolymers are insoluble in most solvents at a wide range of pH and exhibit large heterogeneity in structural features resulting both from the nature of melanin precursor(s) and the presence of some non-melanin components. Moreover, there is no method that allows to split the polymer into individual monomer units, which is essential for accurate qualitative or quantitative analysis. Therefore, in contrast to other biologically important polymers, the structure of melanin pigments is poorly characterized so far. Biosynthetic studies that were carried out *in vitro* using spectroscopic, chromatographic, and mass spectrometric techniques allowed only for the identification of some monomeric and oligomeric intermediates formed oxidatively from a given melanin precursor at very early stages of the pigment synthesis (Nicolaus et al. 1964; Prota et al. 1970; Swan 1974; Allegri et al. 1990; Seraglia et al. 1993; Bertazzo et al. 1995, 1999). The connection mode and the sequence of such structural subunits in melanin polymer, however, remain unknown. Also, very little is known about the way, by which non-melanin components are incorporated into the polymer structure.

The two different strategies are used for structural studies of melanin pigments. The first one consists in the direct analysis of the pigment by the use of various, generally spectroscopic techniques, for which the sample solution is not necessary. The valuable structural information has been acquired using this approach. For instance, on the basis of infrared (IR) and nuclear magnetic resonance (NMR) spectra it was possible to detect lipid, glycidic, and proteinaceous components in human neuromelanin (Zecca et al. 1992, 2000; Aime et al. 1994; Double et al. 2000). Solid state ^{1}H and ^{13}C NMR spectroscopy was also applied for the characterization of non-melanin components of neuromelanin formed in human brain under normal conditions and for the study of the pigment structural changes occurring in course of Parkinson's disease (Aime et al. 2000; Fedorow et al. 2005). Using IR spectroscopy, the interactions between melanins and metal ions were investigated, which allowed for the identification of the metal binding sites in the biopolymer structure (Bridelli et al. 1999; Bilińska 2001). Electron paramagnetic resonance (EPR) spectroscopy provided evidence on the presence of stable organic free radicals in melanin structure as well. Furthermore, it was shown that melanin pigments may be readily distinguishable on the basis of their EPR spectra. Pheomelanins, which contain *N*-centered semiquinoneimine radicals, have a complex EPR spectrum with hyperfine structure, whereas eumelanins are characterized by a single-line spectrum due to the presence of *O*-centered indolesemiquinone radicals (Sealy et al. 1982; Dzierżęga-Lęcznar et al. 1997).

The second strategy used in structural analysis of melanins is based upon the degradative methods, in which structural information about given pigment is inferred from its characteristic decomposition products formed under various conditions. In chemical degradation methods, melanins are treated with strong oxidizing or reducing agents, and the selected degradation products are determined by high-performance liquid chromatography with electrochemical or spectrophotometric detection. Pyrrole di- and tricarboxylic acids, obtained by permanganate or hydrogen peroxide oxidation of melanin in acidic or alkaline medium, are regarded as specific markers of eumelanin-type pigments. Under the same conditions, pheomelanins yield thiazole or benzothiazole carboxylic acids. Other chemical markers of pheomelanin-type pigments, i.e., isomeric aminohydroxyphenylalanines and aminohydroxyphenylethylamines, are formed by the reductive hydrolysis of the pigment in hot hydroiodic acid (Di Donato and Napolitano 2003; Ito and Wakamatsu 2003; Panzella et al. 2006; Greco et al. 2009; Ito et al. 2011). The major disadvantages of chemical degradation methods are usually insufficient yield of the marker products and a high risk of artifacts caused by serious alterations in the pigment framework expected to occur under drastic conditions of the degradation procedures used. Moreover, most of the compounds regarded as chemical markers of eumelanin and pheomelanin have rather poor structural resemblance to the parent monomer units.

In our laboratory, another degradative method is used for structural characterization and differentiation of melanin pigments. The method utilizes pyrolysis in combination with gas chromatography and mass spectrometry (Py-GC/MS). This hyphenated analytical technique is commonly regarded as a valuable tool for compositional analysis and structural studies of many natural and synthetic materials, which, like melanins, are insoluble heteropolymers of high molecular mass (Dworzański and Meuzelaar 2000). Structural analysis of melanin by Py-GC/MS proceeds as follows. Pigment sample is introduced into a pyrolysis device (pyrolyser), where the thermal degradation of the sample in high temperature (usually above 500°C) takes place with the formation of a mixture of fragment molecules termed pyrolysate. The volatile pyrolysis products are transferred by a stream of an inert gas to a gas chromatograph and separated on a capillary column whose outlet is coupled directly to an ion source of a mass spectrometer. As a result, a chromatogram of the pyrolysate (pyrogram) is recorded, and the pyrolysis products are identified on the basis of their mass spectra. If the pyrolysis and GC/MS conditions are properly chosen, the pyrolysate is dominated by the thermal degradation products that retain the characteristic structure of the parent monomer units, from which they arose. Such products are referred to as the pyrolytic markers of the pigment of given type. A fast transfer of the pyrolysate away from a hot pyrolysis zone directly into GC injector port, and appropriate temperature of the sample degradation, prevents the formation of secondary pyrolysis products, thereby discarding the misinterpretation of experimental results. Hence, the risk of possible artifacts is very low. Other advantages of the Py-GC/MS method include high sensitivity, specificity and speed, and a very small amount of the sample, from which structural information can be obtained. Furthermore, a single analysis may provide information about the structure of different parts of a pigment, including its non-melanin components, and no sample pretreatment is usually required for this purpose.

To establish the pyrolytic markers that could be applied for structural investigations of natural melanins, it was necessary to pyrolyze standard pigments of eumelanin and pheomelanin-type, derived oxidatively from various compounds that are thought to be the precursors of naturally occurring pigments. The exemplary pyrograms of eumelanins synthesized in our laboratory are shown in Figure 9.1A and B. Regardless of the eumelanin precursor used, the most characteristic pyrolysis products were identified as pyrrole and indole, and their alkyl derivatives (Dzierżęga-Lęcznar et al. 2002). The compounds are indicative of the monomer units formed during eumelanin synthesis by the cyclization of alanyl or ethylamine side chain of the catechol moiety of the pigment precursor. Another pyrolysis product characteristic of eumelanin-type pigments, especially those synthesized from catecholamines, is an intact catechol (1,2-benzenediol). As shown in Figure 9.1C, the pyrolytic pattern of a pheomelanin-type pigment is completely different, compared with that of any eumelanin. The pyrolysates of pheomelanins are dominated by the heterocyclic compounds that contain a benzene ring fused to a 1,4-thiazine or 1,3-thiazole ring. The same compounds are also substantial constituents of the pyrolysates of eumelanin/pheomelanin copolymers (Figure 9.1D), and the yields of their formation were found to be correlated to the pigment pheomelanin contents (Dzierżęga-Lęcznar et al. 2002). Benzothiazine and benzothiazole derivatives are the thermal degradation products of the pigment monomers formed by the cyclization of cysteinyl side chain of the pigment precursor during its oxidative polymerization. Since the heterocyclic ring closure by the non-sulfur containing side chain of a pheomelanin precursor may also occur, the pyrolysate of pure pheomelanin often contains trace amounts of pyrrole, pyridine, or indole derivatives. There is also a group of products that are always formed during the thermal degradation of melanin pigments, irrespective of their structural type. This group includes benzene, toluene, styrene, phenol, and their alkyl derivatives (Dzierżęga-Lęcznar et al. 2002). However, since their relative contents in eumelanin pyrolysates are much more higher compared with those of pheomelanins, they are numbered among the eumelanin markers. The most probable source of such products is "uncyclized" monomers, i.e., catechol-type structural units, for which side chains have not been converted into the corresponding heterocyclic rings. When "uncyclized" units are incorporated into a pheomelanin structure, the pigment pyrolysate contains additionally substantial amounts of sulfur-containing low molecular weight gases, such as hydrogen sulfide, carbonyl sulfide, and methanethiol. All the earlier discussed pyrolytic markers of melanin pigments are shown in Figure 9.2.

Py-GC/MS method has been successfully applied for the structural studies of natural melanins, isolated from as diverse biological sources as bacteria, soil fungi, insects, bird feathers or human hair, skin or brain tissue (Dworzański 1983; Dworzański and Dębowski 1985; Zecca et al. 1992; Chodurek et al. 1998; Latocha et al. 2000; Dzierżęga-Lęcznar et al. 2004, 2006; Stępień et al. 2009; Gomez-Marin and Sanchez 2010). It was found that the technique allows rapid and efficient differentiation of melanin type and may be used as an alternative to the methods based on chemical degradation in this respect. Figure 9.3 displays some of the pyrograms obtained in our laboratory. Commercially available melanin isolated from ink sacs of a cuttlefish *Sepia officinalis* is commonly recommended as eumelanin standard for comparative structural studies of melanin

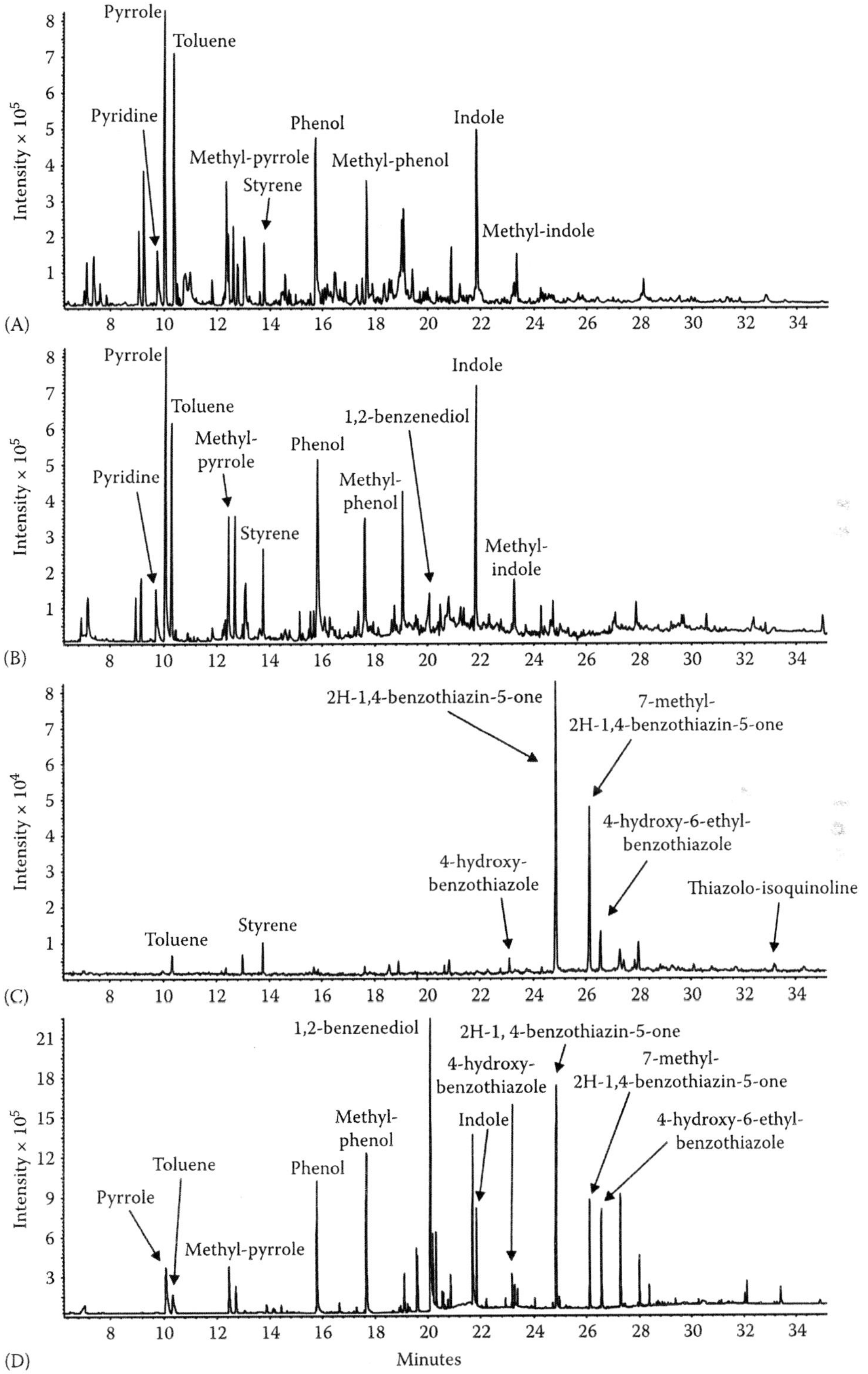

FIGURE 9.1 Reconstructed total ion current chromatograms of the thermal degradation products of synthetic melanin pigments: (A) eumelanin from tyrosine, (B) eumelanin from dopamine, (C) pheomelanin from 5-S-cysteinylDOPA, and (D) mixed-type melanin (eumelanin/pheomelanin copolymer from dopamine/5-S-cysteinyldopamine mixture).

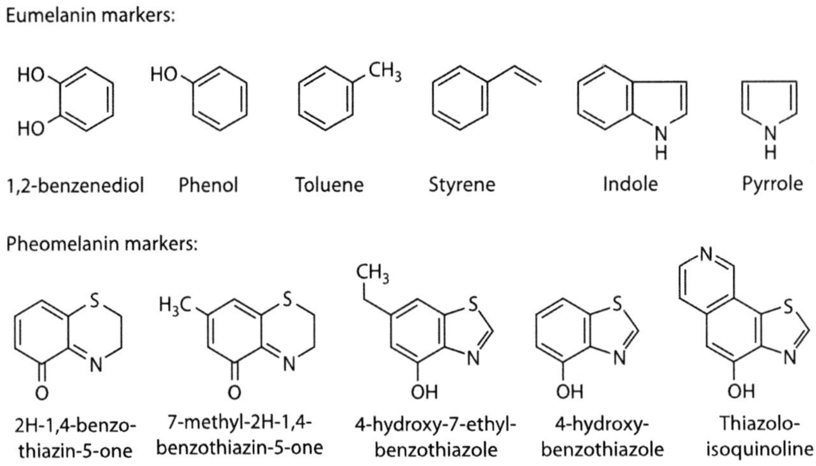

FIGURE 9.2 Pyrolytic markers of melanin pigments.

pigments. Indeed, as shown in Figure 9.3A, its pyrolytic profile is typical for eumelanins with little incorporation of "uncyclized" monomer units. On the basis of the presence of benzothiazole and benzothiazine in the pyrolytic profile shown in Figure 9.3B, the melanin responsible for the body color of the yellow strain of *Drosophila melanogaster* flies was classified as a pheomelanin-containing pigment. The pyrogram shown in Figure 9.3C is dominated by the markers of indole- and catechol-type eumelanin units but contains also some amounts of the derivatives with thiazole ring. Such pyrolytic pattern indicates that the pigment (in this case isolated from the cultured human melanocytes derived from moderately pigmented skin) is eumelanin with little incorporation of pheomelanin-type units. Figure 9.3D displays the pyrogram of neuromelanin obtained post mortem from the human *substantia nigra*. The lack of the pheomelanin markers led us to conclude that no heterocyclic pheomelanin-type units are incorporated into the pigment structure, contrary to the results obtained by chemical degradation. Our Py-GC/MS studies on *substantia nigra* neuromelanin provided also valuable information about non-melanin components of this unique and the most mysterious melanin pigment. The most abundant pyrolysis product of the pigment was identified as limonene. Accordingly, we have concluded that human neuromelanin is tightly associated with an isoprenoid-type compound (Dzierżęga-Lęcznar et al. 2004). Shortly afterward, the compound was identified as dolichol (Fedorow et al. 2005).

The detection of eumelanin or pheomelanin markers in the pyrolysate is necessary to classify the analyzed pigment into the corresponding structural type or to confirm the presence of a pheomelanin component in a mixed-type polymer. It should be noted, however, that both the kind and the relative content of all the pyrolysis products formed is a fingerprint feature of given pigment, and thus may be used to follow its structural changes caused by various factors (Stępień et al. 1989; Chodurek et al. 2003; Dzierżęga-Lęcznar et al. 2003).

Analytical possibilities of Py-GC/MS method with regard to non-melanin constituents of the pigment can be enhanced by conducting pyrolysis in the presence of a derivatizing agent. Such an approach allows determination of more polar and thus thermally instable breakdown products that cannot be detected under conventional conditions. In our experiments, we often use a methanolic solution of tetramethylammonium hydroxide (TMAH) for this purpose. The products generated in the presence of TMAH are a result of various chemical reactions taking place in the pyrolyser, such as pyrolytic bond cleavage, thermally assisted base hydrolysis, and methylation of functional groups (Stępień et al. 2009). Pyrolysis in the presence of TMAH (often termed as thermochemolysis or thermally assisted hydrolysis and methylation) allowed us to confirm the presence of a proteinaceous component in the *substantia nigra* neuromelanin and provide evidence that at least part of the lipid component is chemically bound to the pigment macromolecule

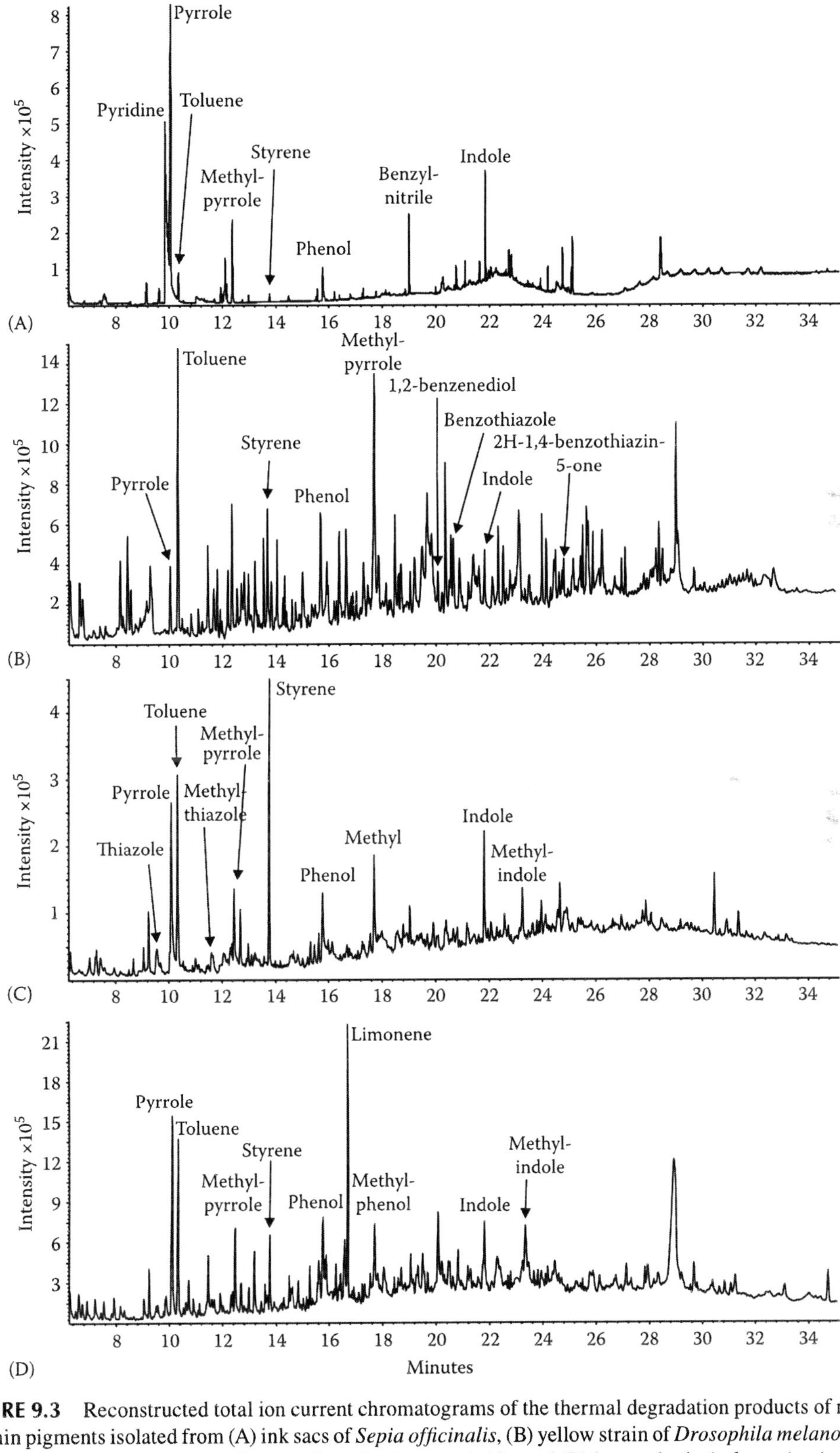

FIGURE 9.3 Reconstructed total ion current chromatograms of the thermal degradation products of natural melanin pigments isolated from (A) ink sacs of *Sepia officinalis*, (B) yellow strain of *Drosophila melanogaster*, (C) cultured human melanocytes from moderately pigmented skin, and (D) human brain (*substantia nigra*).

(Dzierżęga-Lęcznar et al. 2004, 2006). Simultaneously, major constituents of the latter component were identified as saturated and monounsaturated fatty acids with straight chains composed of 14–18 carbon atoms. TMAH thermochemolysis was also found to be useful to assess the purity of melanin pigment isolated from human melanoma for further *in vitro* studies (Chodurek et al. 2008). Adsorption of various tissue impurities on melanin surface is a serious problem that often arises during the pigment isolation from biological material. Furthermore, rich and complex sample matrix may restrict the sensitivity of Py-GC/MS method, especially with respect to the pheomelanin markers. This may lead to the false negative conclusion regarding the presence of a pheomelanin component in a mixed-type pigment with little incorporation of pheomelanin-type units. Very recently, we have developed a method that allows us to overcome this problem and, on the other hand, may be used for accurate quantitation of a pheomelanin content in any natural melanin pigment. The method is based on the analysis of the pyrolytic markers of pheomelanin units with the use of a triple quadrupole tandem mass spectrometer operating in a multiple reaction monitoring (MRM) mode, which offers an extremely specific and sensitive identification of the target molecule by simultaneous measuring of its characteristic precursor and product ion pair(s) (MRM transitions). Using that approach we were able to detect and quantitate the pheomelanin component, which accounted for 0.05% of the total pigment units (Dzierżęga-Lęcznar et al. 2012).

9.3 PHOTOPROTECTIVE AND ANTIOXIDANT FUNCTIONS OF MELANINS

Excessive exposure to solar ultraviolet radiation is an essential etiological factor for skin cancer. UV radiation, directly or indirectly (through the generation of reactive oxygen species, ROS) causes damage to DNA, proteins, and lipids, and this effect eventually induces gene mutation, inflammation, and immunosuppression (Halliday 2005; Ibrahim and Brown 2008). Cutaneous pigmentation afforded by melanocytes is the main photoprotective mechanism in human skin. Epidermal melanocytes, localized at the basal layer of the epidermis, synthesize melanin within melanosomes and transfer melanized melanosomes to adjacent keratinocytes via their elongated dendrites. Melanosomes accumulate in the perinuclear area of keratinocytes as supranuclear "caps" that protect nuclear DNA from impinging UV rays (Brenner and Hearing 2008). It is generally accepted that melanosomal melanin acts as a natural sunscreen that, by absorbing and scattering solar radiation, limits its penetration through the epidermis (Ortonne 2002; Kadekaro et al. 2003). The energy of the absorbed photons is rapidly and efficiently converted into heat within the melanin polymer, and as a result, the risk of potentially damaging photochemical reactions is significantly reduced (Meredith and Sarna 2006). Antioxidant properties of melanins may also play an important role in photoprotection. *In vitro* melanins have the ability to quench excited states of photosensitizing dye molecules and singlet oxygen and scavenge reactive radicals (Meredith and Sarna 2006). They react with the hydroxyl radical and superoxide radical anion and can compete effectively with superoxide dismutase in scavenging of $O_2^{\bullet-}$ (Geremia et al. 1984; Korytowski et al. 1986; Sarna et al. 1986).

Epidemiological and experimental data strongly support the photoprotective role of melanin. Dark skin, which contains numerous large, heavily melanized melanosomes, enriched in eumelanin and distributed individually in keratinocytes, is better protected against UV-induced damage than fair skin, in which smaller, poorly melanized melanosomes tend to form cluster (Hennessy et al. 2005; Brenner and Hearing 2008). Indeed, a distinct correlation between constitutive pigmentation of the skin and the resistance to UV-induced erythema and sunburn is usually observed (Rees 2004), and individuals with high melanin content are less susceptible to skin photoaging (Wlaschek et al. 2001). Furthermore, the incidence of sun-induced skin cancers, including melanoma, is higher in individuals with fair skin and poor ability to tan than in individuals who have dark skin and a good tanning ability (Armstrong and Kricker 2001; Bishop and Bishop 2005).

Several studies have examined a correlation between melanin content and the level of DNA damage after exposure to UV radiation. Comparison of DNA photoproducts yield in UVB-irradiated human melanocytes with a high and a low melanin content revealed that DNA from lightly pigmented melanocytes contained significantly higher numbers of cyclobutane pyrimidine dimers and 6-4 photoproducts than did DNA from heavily pigmented melanocytes (Barker et al. 1995; Smit et al. 2001). Furthermore, increasing melanin content in cultured melanocytes by raising the concentration of tyrosine in the culture medium reduced the formation of DNA photoproduct in response to UV (Smit et al. 2001). Exposure of cultures of melanocytes derived from different skin types resulted in the induction of the highest levels of DNA photoproducts in melanocytes with the least eumelanin content (Hauser et al. 2006). Similarly, an inverse correlation between melanin content and the extent of UVA/UVB-induced DNA damage was demonstrated in the epidermis of individuals with diverse constitutive skin pigmentation (Tadokoro et al. 2003; Del Bino et al. 2006; Yamaguchi et al. 2006). In addition, it was found that pigmentation induced by repeated UV irradiation protected human skin against subsequent DNA damage following UV exposure (Yamaguchi et al. 2008). In cultured melanocytes, melanin was also shown to offer protection against UVA-induced membrane damage (Kwam and Dahle 2003).

The fundamental role of melanin in the protection against harmful effects of solar radiation is not limited to the skin. In the human eye melanin is found in uveal melanocytes located in the choroid and in the stroma of the iris and cilliary body, and in the pigment epithelial cells, especially in the retinal pigment epithelium (RPE). Melanin in RPE is mainly eumelanin, whereas in uveal melanocytes pheomelanin is often present in addition to eumelanin (Prota et al. 1998; Liu et al. 2005a). In melanocytes from eyes with dark-colored irides, the amount of melanin and the ratio of eumelanin to pheomelanin is greater than that from eyes with light-colored irides (Wielgus and Sarna 2005; Wakamatsu et al. 2008). It is believed that iridal melanin protects the ocular cells and tissues from deleterious effect of both UV radiation and visible light acting mainly as a photoscreen, whereas melanin located in the posterior segment of the eye acts as an antioxidant (Hu et al. 2008). The abilities of human RPE melanin and choroidal melanin to scavenge ROS and protect the retina from oxidative damage were documented in a variety of model systems (Hu et al. 2002; Peters et al. 2006; Wang et al. 2006).

Melanin seems to protect the eye against several ocular diseases that can cause blindness, including uveal melanoma and age-related macular degeneration (AMD). Epidemiologic data suggest that the light-colored eye is at higher risk for the occurrence of uveal melanoma (Hu et al. 2005), and meta-analysis (based on 1732 cases) demonstrated that a blue or gray iris is a statistically significant risk factor for the development of uveal melanoma (Weis et al. 2006). Several studies have also revealed an association between light-colored irides and the occurrence or progress of AMD, suggesting that melanin may be protective against AMD development (Friedman et al. 1999; Mitchell et al. 2002; Nicolas et al. 2003).

Melanins may play an important role in the process of lipid peroxidation in biological membranes. It was observed that RPE of pigmented animals was much more resistant to lipid photooxidation than that of albino animals. This finding was attributed to the presence of melanin-containing melanosomes in the pigmented tissue, because the activity of antioxidants such as superoxide dismutase, glutathione peroxidase, and α-tocopherol in the albinos was not lower than that in the pigmented RPE (Sakina et al. 1985). The inhibition action of isolated eye melanosomes on lipid photooxidation was also demonstrated (Ostrovsky et al. 1987). Kvam and Dahle (2003) found an apparent protection from UVA-induced lipid peroxidation and membrane damage in epidermal melanocytes with high melanin content.

Porębska-Budny et al. (1992) examined the effects of synthetic eumelanins derived from various precursors on cardiolipin peroxidation in UV-irradiated liposome membranes and found that the extent of inhibition depends on the type and concentration of melanin polymers. The study demonstrated that the contribution of optical screening effects of melanins to inhibit lipoperoxidation is not higher than 15% for the most active melanin, and that there is a simple correlation between scavenging of superoxide anion radical by melanins and their ability to inhibit cardiolipin peroxidation (Porębska-Budny et al. 1992).

Melanin polymers are thought to be redox systems and electron transfer agents, which are able to interact quite efficiently with oxidizing and reducing radicals, including superoxide anion radical.

The interaction of melanin with such radicals can be explained by the *o*-hydroquinone and *o*-quinone nature of the pigment subunits, which can act as electron donors and acceptors, respectively (Meredith and Sarna 2006). Consistently, reactions of superoxide anion radical with melanins may involve its oxidation to molecular oxygen or its reduction to hydrogen peroxide, and the oxidation/reduction ratio depends on the type of melanin (Korytowski et al. 1986). Induction of transient free radicals in melanins during illumination with UV light seems to be especially important in photoprotection. Photo-induced free radicals of melanin have high reactivity and may effectively participate in various redox reactions, in particular in the reduction of oxygen. On the basis of the effect of catalase and superoxide dismutase, it was proposed that photo-induced radicals of melanin were involved in reduction of oxygen to hydrogen peroxide via a superoxide intermediate (Sarna 1992). The antioxidant efficiency of melanins in lipid photooxidation appears to be related to the levels of intrinsic and photo-induced free radical centers in the melanin polymer, as well as to accessibility of these centers for active species formed during irradiation of lipids (Stępień et al. 1992). It was shown that melanin–copper complexes inhibited UV-induced lecithin peroxidation less effectively than copper-free melanins derived from the same precursors (Stępień et al. 1992).

The ability of melanin to bind metal ions is one of the most characteristic features of the pigment. It has been found that melanins, both *in vivo* and *in vitro*, can accumulate substantial amounts of various metals, including redox-active metal ions (Larson and Tjalve 1978; Fogarty and Tobin 1996; Hong and Simon 2005; Liu et al. 2005b). Neuromelanin localized in the pigmented neurons of the *substantia nigra* has particularly strong chelating ability for iron and is believed to protect neurons against iron-induced oxidative stress. High and low affinity binding sites for iron have been identified on neuromelanin. The former sites can sequester redox active iron, thereby preventing the formation of hydroxyl radicals. In the presence of high iron levels, neuromelanin accumulates iron in low affinity binding sites. Under physiological conditions, the pigment appears to be only partially saturated with iron (Zecca et al. 2002; Double et al. 2003; Zucca et al. 2004, 2006).

Several studies indicate that natural neuromelanin and its synthetic model, dopamine-melanin, may affect the process of lipid peroxidation. DA-melanin has been found to potentiate lipid peroxidation in rat cerebral cortex homogenates after addition of iron (Ben-Shachar et al. 1991). On the other hand, DA-melanin and neuromelanin isolated from human *substantia nigra* have been shown to inhibit Fe(II)—or Fe(II)/ascorbate—initiated lipid peroxidation in lecithin and cardiolipin liposomes, and in methyl linoleate aqueous dispersions (Porębska-Budny et al. 1992; Stępień and Wilczok 1994; Korytowski et al. 1995). Furthermore, it was demonstrated that pure pheomelanin (CysDA-melanin) and mixed type melanin (DA/CysDA-melanin) significantly suppressed oxidation of linoleic acid and liposomal lecithin induced by Fe(II)/ascorbate, although the inhibitory effect of CysDA-melanin was lower than that of DA/CysDA-melanin and DA-melanin (Wilczok et al. 1999).

Several mechanisms have been proposed to explain antioxidative activity of melanins, including scavenging ROS and inhibition of iron-catalyzed free radical decomposition of hydrogen peroxide and lipid hydroperoxides as a result of sequestration of redox-active iron ions by the pigment (Porębska-Budny et al. 1992; Korytowski et al. 1995; Zaręba et al. 1995). We have found that DA-melanin is capable of reducing linoleic acid hydroperoxide to its more stable hydroxyl derivative, both in the absence and in the presence of ferrous ions (Stępień et al. 1998). The ability of DA/CysDA-melanin and CysDA-melanin to reduce the fatty acid hydroperoxide to the corresponding alcohol was also documented (Wilczok et al. 1999). The reductive inactivation of lipid hydroperoxides is known to prevent hydroperoxide-dependent secondary lipid peroxidation. Our results suggest that melanins can act as chain-breaking antioxidants. It has also been reported that DA-melanin is capable of suppressing the yield of hydroxyl radicals generated via Fenton reaction, but after saturation with ferric ions it promotes the formation of hydroxyl radicals by redox activation of the ions (Zaręba et al. 1995). It seems that the amount of iron bound to melanin may determine whether the pigment acts as an antioxidant blocking redox active metal ions or whether it promotes the formation of cytotoxic radicals in the presence of excess iron. Similar mechanism has been proposed to explain protective or cytotoxic activity of human neuromelanin. Saturation of iron-chelating sites of neuromelanin may generate oxidative stress inducing a cascade of events ultimately leading to neuronal death (Zucca et al. 2004, 2006).

Melanin pigments are also able to interact with reactive nitrogen species (RNS). Wang and Casadevall (1994) reported that melanized *Cryptococcus neoformans* cells exposed to sodium nitrite in acidic media (in which nitrite generates nitrogen dioxide radical and other RNS) show significantly higher survival than non-melanized cells, suggesting that melanin could protect cells against nitrogen-derived oxidants. Reszka et al. (1998) demonstrated generation of large amounts of melanin radicals in synthetic DOPA-melanin by lactoperoxidase (LPO)/hydrogen peroxide/nitrite system and proposed the mechanism involving oxidation of nitrite by LPO/H_2O_2 to nitrogen dioxide radical, which reacts with melanin and oxidizes its hydroquinone groups to semiquinones. At the same time $^{\bullet}NO_2$ undergoes reduction back to nitrite. This process appears to be remarkably efficient, indicating a high $^{\bullet}NO_2$ radical scavenging capacity of the melanin (Reszka et al. 1998). Interactions of melanins with $^{\bullet}NO_2$ were confirmed by pulse radiolysis method (Różanowska et al. 1999). Our study has shown the ability of melanins to interact with peroxynitrite, a powerful oxidant and nitrating agent, which is formed by the nearly diffusion-limited reaction between nitric oxide and superoxide radical anion. We have found that DA-melanin markedly inhibited peroxynitrite-mediated nitration of free tyrosine, oxidative loss of tryptophan residues in bovine serum albumin and Ca^{2+}-ATPase inactivation. In the presence of bicarbonate, this inhibitory effect was lower for nitration and insignificant for oxidative protein modifications. These results suggest that DA-melanin can protect against nitrating and oxidizing action of peroxynitrite but is a worse protector against the peroxynitrite-CO_2 adduct (Stępień et al. 2000b). CysDA-melanin and DA/CysDA copolymers also significantly reduced the formation of 3-nitrotyrosine, and this inhibitory effect depends on the type and concentration of melanin polymer. It was found that incorporation of CysDA-derived units into melanin attenuated its protective effect on tyrosine nitration induced by peroxynitrite. In the presence of bicarbonate, the melanins also inhibited 3-nitrotyrosine formation in concentration-dependent manner, although the extent of inhibition was lower than that in the absence of bicarbonate. DA-melanin and CysDA-melanin was shown to inhibit peroxynitrite-induced linoleic acid oxidation, both in the absence and in the presence of bicarbonate (Stępień et al. 2000a). As peroxynitrite is proposed to be a mediator of neurotoxic processes associated with Parkinson's disease, a protective effect of neuromelanin against peroxynitrite may be of physiological importance.

9.4 RADIOPROTECTIVE PROPERTIES OF MELANINS

The observations of the resistance of the melanized fungi to gamma radiation in the highly radioactive environment inside the damaged nuclear reactor in Chernobyl (Mironenko et al. 2000) and cooling pool water in nuclear reactors (Mal'tsev et al. 1996) have drawn attention to potential radioprotective properties of melanin pigments. Dadachova et al. (2007) have found that ionizing radiation changes electron-transfer properties of melanin and enhances the growth of melanized fungi, indicating the capacity of melanin to transduce the energy of ionizing radiation in living cells. Interestingly, the increase of melanin ability to transfer electrons was independent of the energy of the incident photons. It has been shown that radioprotective efficacy of melanins depends on their chemical composition, stable free radical content, and a spherical spatial arrangement (Dadachova et al. 2008; Schweitzer et al. 2009). It has been suggested that the mechanism of the radioprotective action of melanin involves the physical interaction between the pigment and the recoil electrons generated by Compton scattering of incident photons. Controlled dissipation of high-energy recoil electrons by melanin prevents secondary ionizations and the generation of damaging free radical species (Schweitzer et al. 2009).

Recently, it has been postulated that internally administered melanin could protect humans against ionization radiation (Howell et al. 2008). Schweitzer et al. (2010) described the use of melanin for the protection of bone marrow during external beam radiation therapy or radioimmunotherapy of cancer. The silica nanoparticles coated with DOPA- and/or 5-*S*-cysteinyl-DOPA-melanin injected intravenously to melanoma tumor-bearing nude mice reduced the susceptibility of treated animals to the myelotoxic effects of therapeutic radiation. The study has shown that intravenously administered nanoparticles can be used as carriers for delivery of melanin into the bone marrow, where they would protect hematopoietic cells against ionizing radiation and would permit administration of significantly higher and efficient doses of radiation (Schweitzer et al. 2010).

9.5 IMMUNOMODULATORY ACTIVITY OF MELANINS

9.5.1 Microbial Melanins

Certain human pathogenic fungi and bacteria are able to produce melanin pigments, and a large body of evidence indicates a role for melanization in microbial virulence (reviewed by Jacobson 2000; Langfelder et al. 2003; Nosanchuk and Casadevall 2003). The contribution of melanin to virulence has been most extensively studied for *C. neoformans*, a yeast-like fungus that often causes opportunistic infections in immunocompromised individuals. *C. neoformans* expresses laccase (CNLAC1) (Williamson 1994), which can catalyze melanin synthesis from exogenous substrates, such as L-DOPA, catecholamines or homogentisic acid *in vitro* (Wang and Casadevall 1996; Garcia-Rivera et al. 2005; Frases et al. 2007). Melanization of *C. neoformans* during infection in rodents and in human brain tissue has been demonstrated (Nosanchuk et al. 1999, 2000; Rosas et al. 2000). In *C. neoformans*, like in other pathogenic fungi, melanin is deposited in the outer layer of the cell wall (Wang and Casadevall 1996). Analysis of the microstructure of cell wall-associated melanin revealed that the pigment is composed of two to five layers of melanin particles arranged in a concentric manner (Eisenman et al. 2005). Such a localization and arrangement of melanin granules forms a physical barrier, which protects the fungal cell against a variety of lethal insults, including oxidative injury. Indeed, melanized cryptococcal cells were shown to be more resistant to killing by oxidants than non-melanized cells (Nosanchuk and Casadevall 2003). Melanin-deficient mutant strains of *C. neoformans* are less invasive and survive poorly in the spleen, liver, or brain of infected animals, compared to pigmented wild-type strains (Kwon-Chung et al. 1982; Kwon-Chung and Rodes 1986; Salas et al. 1996). Melanization protects cryptococcal cells against killing by macrophages *in vitro* and impedes macrophage phagocytosis of encapsulated *C. neoformans in vitro* and in a murine lung infection model (Wang et al. 1995; Mednick et al. 2005). In addition to reactive oxygen species, phagocytic cells can produce antimicrobial peptides, and melanized *C. neoformans* cells were shown to be less susceptible to the toxic effects of neutrophil defensin and other cationic antimicrobial peptides compared to non-melanized cells (Doering et al. 1999). Melanization of *C. neoformans* cells also reduces their susceptibilities to amphotericin B and caspofungin (van Duin et al. 2002; Ikeda et al. 2003). It has been suggested that the cell wall melanin binds these antifungal drugs, thereby preventing them from reaching their target sites (Ikeda et al. 2003; Nosanchuk and Casadevall 2006).

Further evidence for the contribution of melanin pigments to virulence has been provided by studies of the human fungal pathogens *Aspergillus fumigatus* and *Exophiala* (*Wangiella*) *dermatitidis*. *A. fumigatus* synthesizes melanin from 1,8-dihydroxynaphtalene via the polyketide pathway during its conidial stage of growth (Langfelder et al. 2003). Conidia from an albino mutant strain of *A. fumigatus* were found to be more susceptible to killing by oxidants and by human monocytes *in vitro* and less lethal in a murine infection than conidia from melanized strains (Jahn et al. 1997). Moreover, targeted mutation of the *A. fumigatus alb1* gene, encoding a polyketide synthase, results in a non-pigmented strain with reduced virulence to mice compared to strains with intact expression of the enzyme (Tsai et al. 1998). Elimination of melanin production by *E. dermatitidis* was associated with diminished ability to produce invasive hyphal forms, enhanced susceptibility to neutrophil killing, and reduced virulence in mouse models of infection (Dixon et al. 1992; Schnitzler et al. 1999; Feng et al. 2001). Increased resistance to phagocytosis has been observed for the melanized fungal pathogens *Paracoccidioides brasilliensis* (da Silva et al. 2006) and *Sporothrix schenckii*

(Romero-Martinez et al. 2000) and the melanotic bacteria *Burkholderia cepacia* (Saini et al. 1999) and *Proteus mirabilis* (Agodi et al. 1996). Melanin produced by an epidemic strain of *B. cepacia* has the capacity to scavenge superoxide anion produced by monocytes during oxygen burst (Zughaier et al. 1999). *B. cepacia* is able not only to survive phagocytosis, but also to proliferate within the phagocytes (alveolar macrophages) (Saini et al. 1999).

The data presented above show that melanins contribute to virulence by protecting microbial cells against host defense mechanisms. However, there is also evidence that microbial melanins have immunomodulatory properties and can affect inflammatory and immune responses to infection. Intracerebral infection of mice with an albino strain of *C. neoformans* resulted in minimal tissue damage and triggered production of tumor necrosis factor alpha (TNF-α) and interleukin (IL)-12 and IL-1β, whereas a revertant melanotic strain caused massive CNS tissue damage and inhibited the cytokine response (Barluzzi et al. 2000). Melanin production by *C. neoformans* has been found to be an important determinant of the pathogen ability to induce a pulmonary inflammatory response in mice (Huffnagle et al. 1995; Mednick et al. 2005). Compared to a non-melanized strain of *C. neoformans*, infection with melanized cells resulted in higher levels of IL-4 and monocyte chemoattractant protein-1 and increased the numbers of infiltrating leukocytes early after infection (Mednick et al. 2005). *In vitro* studies have shown that melanin particles ("ghosts") isolated from *C. neoformans* cells and *A. niger* conidia are able to activate the complement cascade via the alternative pathway. Immunofluorescence analysis of lungs from mice injected intratracheally with *C. neoformans*-derived melanin demonstrated deposition of complement C3 fragments onto melanin ghosts, indicating that melanin can activate the complement system *in vivo* (Rosas et al. 2002). The finding suggests a potential mechanism by which melanin could induce an inflammatory response.

Fungal melanins have been shown to be immunogenic. Mice immunized with melanin from *C. neoformans* cells generate specific anti-melanin antibodies that can inhibit fungal growth (Nosanchuk et al. 1998; Rosas et al. 2001). It was also demonstrated that sera from patients with chromoblastomycosis reacted with melanin from the fungus *Fonsecaea pedrosoi,* indicating that anti-melanin antibodies are produced during human infections (Alviano et al. 2004). It has been postulated that fungal melanins belong to the T cell-independent antigens, which can induce an immunological response by binding directly to the immunoglobulin-like receptors on the surface of B lymphocytes (Nosanchuk et al. 1998).

9.5.2 Human Melanins

A large body of evidence suggests that human epidermal melanocytes are an integral part of the skin immune system and can be considered immunocompetent cells. These melanin-producing cells express major histocompatibility complex class II molecules and adhesion molecules, such as vascular cell adhesion molecule-1 (VCAM-1) and intercellular adhesion molecules (ICAM-1 and CD40) (Smit et al. 1993; Lu et al. 2002). Melanocytes can phagocytize microorganisms (Le Poole et al. 1993b) and may be capable of antigen processing and presentation (Le Poole et al. 1993a). In addition, they constitutively produce several cytokines, including the proinflammatory interleukins IL-1 and IL-6 (Mattei et al. 1994; Swope et al. 1994). The melanin pigment itself can act as a physical barrier against microorganisms (Mackintosh 2001), and it can also bind and neutralize bacterial-derived toxins, including the botulinum A (Ishikawa et al. 2000). Furthermore, reactive quinone intermediates and hydrogen peroxide generated during melanin synthesis exert strong antimicrobial activity (Plonka et al. 2009).

Recently, human melanocytes have been shown to express functional toll-like receptors (TLRs), key components of the innate immune response against invading microbial pathogenes (Ahn et al. 2008b; Yu et al. 2009; Jin and Kang 2010). TLRs recognize constituents of microbial cell walls or pathogen-specific nucleic acids, and activate intracellular signaling cascades leading to the induction of inflammatory cytokines and chemokines. In addition to the role in the innate immune response, TLRs are known to link the innate and adaptive immune systems (Doyle and O'Neill 2006; Hari et al. 2010). In melanocytes, TLRs 2–4, 7, and 9 respond to their ligands by activating NF-κB (nuclear factor kappa light chain enhancer of activated B cells) and/or p38 MAPK (mitogen-activated protein kinase) signaling pathway (Ahn et al. 2008a,b; Yu et al. 2009).

Melanocytes could also act as regulators of the skin immune response by producing and releasing several immunosuppressive molecules, including POMC-derived ACTH and α-MSH (Slominski et al. 2000), cortisol, corticosterone, and other steroids (Slominski et al. 1999, 2005). In particular, α-MSH has a powerful antiinflammatory potential and affects various pathways implicated in the regulation of inflammation (Brzoska et al. 2008). Furthermore, intermediates of melanogenesis, especially L-DOPA and/or products of its oxidation, can act as potent immunosuppressors. It has been demonstrated that L-DOPA inhibits lymphocyte proliferation and abolishes production of proinflammatory cytokines by activated lymphocytes (Slominski and Goodman-Snitkoff 1992; Slominski et al. 2009).

Lipopolysaccharide (LPS), a component of the cell wall of Gram-negative bacteria, is a known ligand for TRL4 and elicits a variety of inflammatory responses (Hari et al. 2010). *In vitro* studies have shown that stimulation of epidermal melanocytes with LPS enhances the expression of IL-6, IL-8, and several chemokines (CCL2, CCL3, and CCL5) (Yu et al. 2009) and induces the release of IL-1β and TNF-α from the cells (Tam and Stępień 2011). It was also found that cultured melanocytes are able to express inducible nitric oxide synthase (iNOS) and produce nitric oxide (NO) in response to proinflammatory cytokines and/or LPS (Rocha and Guillo 2001; Fecker et al. 2002). NO is thought to be an important mediator of inflammatory and immune responses in human skin (Bruch-Gerharz et al. 1998). These data suggest that NO, inflammatory cytokines, and chemokines released by melanocytes could affect melanocytes themselves or/and other cells of the epidermis, contributing to a local immune response.

A role of melanin in the induction of inflammatory mediators by epidermal melanocytes has not been established yet. We have compared IL-8 and NO production in LPS-stimulated melanocytes with different melanin contents (Stępień and Tam 2009; Tam and Stępień 2010). Normal human epidermal melanocytes, derived from lightly (HEMn-LP) and darkly pigmented (HEMn-DP) neonatal foreskin, were used in the study. It was found that HEMn-DP released the larger amount of IL-8 than HEMn-LP upon stimulation with LPS (Figure 9.4). In contrast, lightly pigmented melanocytes produced more NO than their heavily pigmented counterparts (Figure 9.5). Moreover, darkly pigmented melanocytes did not respond to low concentrations of LPS. The results suggest immunomodulatory properties of melanin in epidermal melanocytes.

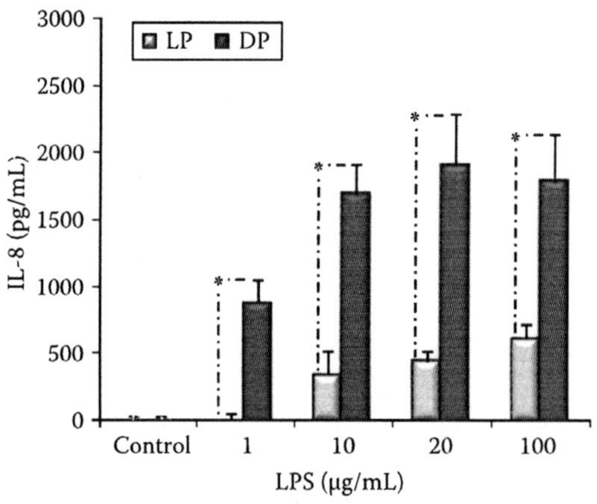

FIGURE 9.4 Effect of LPS on IL-8 secretion by human melanocytes. Human lightly (LP) and darkly (DP) pigmented epidermal melanocytes were incubated for 48 h with LPS. The cytokine secretion levels were assayed in the supernatants by ELISA. Data represent mean values ± SD. *P < 0.05 LP vs. DP.

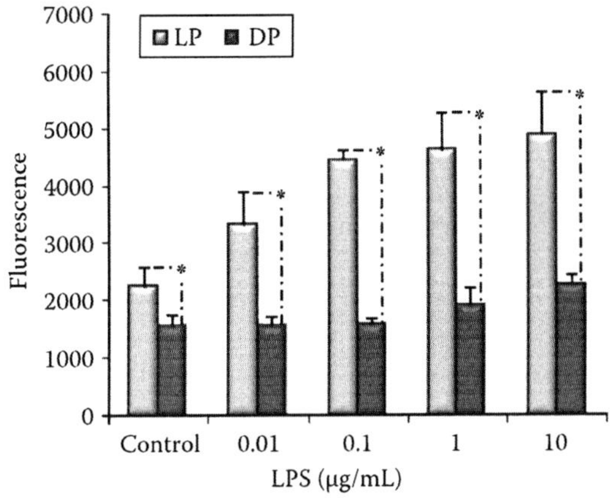

FIGURE 9.5 Induction of nitric oxide synthase (NOS) activity in human melanocytes. Human lightly (LP) and darkly (DP) pigmented epidermal melanocytes were incubated for 24 h with LPS. The intracellular formation of thiazolofluorescein, an indicator of NOS activity, was measured fluorimetrically (λ_{ex} = 485 nm, λ_{em} = 535 nm). Data represent mean values ± SD. *P < 0.05 LP vs. DP.

Recent studies have demonstrated that LPS increases melanin synthesis in human melanocytes and induces pigmentation of cultured skin (Ahn et al. 2008a; Jin and Kang 2010). NO and other mediators of inflammation, such as histamine and eicosanoids, have also been shown to stimulate melanogenesis (Romero-Graillet et al. 1997; Sasaki et al. 2000; Slominski et al. 2004). Furthermore, postinflammatory hyperpigmentation of the skin is frequently observed in clinical practice (Pandya and Guevara 2000; Brajac et al. 2009). These data indicate that epidermal melanocytes can modulate their pigmentation in response to inflammatory conditions. As intermediates of melanogenesis are able to inhibit activity of immune cells (Slominski and Goodman-Snitkoff, 1992; Slominski et al. 2009), the process of melanogenesis could have immunomodulatory functions. Interestingly, inhibition of melanogenesis has been proposed as an adjuvant strategy in the treatment of melanotic melanomas (Slominski et al. 1998, 2009).

Recently, a proinflammatory role for human neuromelanin has been proposed. The study by Wilms et al. (2003) demonstrated that neuromelanin, isolated from the human *substantia nigra*, was able to activate rat microglia *in vitro* with the subsequent release of NO and the proinflammatory cytokines IL-6 and TNF-α. The process involved NF-κB and p38 MAPK activation by neuromelanin. Further studies have shown that neuromelanin particles are phagocytized and degraded by microglia *in vitro* and induce microglial activation and ensuing production of reactive oxygen species, such as superoxide anion and hydrogen peroxide, in addition to proinflammatory factors (Zhang et al. 2011). The ability of extracellular neuromelanin to activate microglia has been confirmed *in vivo*. Human neuromelanin injected into rat *substantia nigra* induced an acute and strong inflammatory microglial activation and degeneration of dopaminergic neurons (Zecca et al. 2008; Zhang et al. 2011). It has been suggested that chronic activation of microglia in the human *substantia nigra* by extracellular neuromelanin released from degenerated dopaminergic neurons plays an important role in the progression of Parkinson's disease (Wilms et al. 2003; Zecca et al. 2006, 2008; Zhang et al. 2011). Insoluble extraneuronal neuromelanin undergoes a slow degradation process, and thus can exist in the extracellular space of Parkinsonian brain in large amounts and for long periods. These findings may be highly relevant to the development of novel therapeutic strategies in Parkinson's disease.

Recent data suggest that extracellular neuromelanin not only causes local inflammation, but may be the initial trigger for an adaptive autoimmune response via activation of dendritic cells. It has been shown that murine dendritic cells treated with neuromelanin from human subjects recognized and effectively phagocytized the pigment. Neuromelanin-activated dendritic cells were able to secrete the proinflammatory cytokines (IL-6 and TNF-α) and trigger T cell proliferation in a mixed lymphocyte reaction, showing that dendritic cell activation was functional to induce a primary T cell response (Oberländer et al. 2011).

9.5.3 Synthetic Melanins

So far, reports on the immunomodulatory properties of synthetic melanins are scarce. Mohagheghpour et al. (2000) examined the cytokine regulatory activity of synthetic melanins produced either by autooxidation of L-DOPA or by tyrosinase-catalyzed oxidation of the dipeptide glycyl-L-tyrosine. They found that both melanins effectively and reversibly suppressed TNF production by LPS-stimulated human peripheral blood monocytes. In addition to TNF, synthetic DOPA-melanin inhibited the release of IL-1β, IL-6, and IL-10 from activated monocytes. The melanin was also able to suppress IL-6 production by IL-1α-stimulated human lung fibroblasts and umbilical vein endothelial cells. In contrast, GM-CSF (granulocyte-macrophage colony-stimulating factor) production by LPS-stimulated monocytes was enhanced by DOPA-melanin treatment (Mohagheghpour et al. 2000). The complementary *in vivo* experiments showed that LPS-stimulated increase in the release of TNF-α was reduced when test animals (BALB/c mice) were injected concomitantly with LPS and DOPA-melanin (Mohagheghpour 2001).

It was reported that synthetic soluble melanins derived from L-tyrosine, L-DOPA, or catecholamines inhibited the replication of human immunodeficiency virus type 1 in human lymphoblastoid cell lines and blocked syncytium formation and cytopathic effects of the virus *in vitro* (Montefiori et al. 1990; Montefiori and Zhou 1991; Sidibe et al. 1996).

Despite the limited data from *in vitro* and *in vivo* studies, the synthetic melanins have been postulated to be used in the treatment and prevention of diseases that are associated with uncontrolled cytokine production, in particular TNF-α, IL-1, and IL-6 (Berliner et al. 1998; Mohagheghpour et al. 2000; Mohagheghpour 2001). It is predicted that the administration of synthetic melanin in an amount sufficient to modulate the immune response might provide a therapeutic benefit to the patients with rheumatoid arthritis, atherosclerosis, wasting syndrome associated with acquired immunodeficiency syndrome (AIDS), Parkinson's disease, psoriasis, and cancer (e.g., myeloma) (Garger and Neidleman 2001; Barcia et al. 2003; Danese et al. 2006).

9.5.4 Plant Melanins

Many botanicals traditionally used to enhance immune functions in humans have been reported to contain melanin or "melanin-like" material. There is some evidence that melanin may be one of the factors responsible for the immunomodulating properties of various plant products. The immunological activity has been reported for the melanin pigments isolated from grape pits (Avramidis et al. 1998), green or black tea leaves (Sava et al. 2001), seeds of black cumin (*Nigella sativa* L.) (El-Obeid et al. 2006a), *Echinacea* species, American ginseng, and alfalfa sprouts (Pugh et al. 2005). Furthermore, potentially immune active "melanin-like" material may also be extracted from lour, black walnut, ginger, shiitake mushroom, and many other botanicals (Pasco et al. 2005; Pugh et al. 2005; Kumar and Deepak 2007).

The activity of extracted melanin varies substantially within botanicals and depends on the average size of melanin particles in the pigment preparation (Pugh et al. 2005). Also the methods used for the extraction and purification of plant melanins may affect biological activity of the pigments. The commonly used procedures for melanin isolation are based on alkaline extraction at high temperature and repeated precipitation with a strong acid. Such a harsh treatment may lead to the degradation of the pigment and hence complete loss of its immunomodulating properties. Moreover, melanin isolated in that way may be contaminated with tissue lipids, carbohydrates, proteins, or nucleic acids.

More gentle method with the use of a weak base at room temperature was applied by Hung et al. (2002) for the isolation of biologically active melanin from tea leaves. The most efficient isolation procedure for plant melanin has been developed by Pasco et al. (2005). The method is based on the pigment extraction with aqueous phenol, and gives immune active melanin of a very high purity.

The ability of plant melanins to modulate the immune response was described for the first time by Avramidis et al. (1998). They demonstrated the inhibitory effect of orally administered melanin isolated from grape pits on adjuvant induced disease (AID) in rats. Melanin treatment led to the normalization of serum levels of proinflammatory cytokines IL-1, IL-6, and TNF-α elevated in AID and to the inhibition of the subpopulation of Th1 lymphocytes responsible for cellular immune response. Unlike grape melanin, most plant melanins can stimulate immune responses. It is believed that orally administered melanin, detected by Peyer's patches of the gut-associated lymphoid tissue (GALT), has the ability to activate the immune effector cells through the TLR-dependent pathway, leading to the production of a number of proinflammatory cytokines, via NF-κB activation (Pugh et al. 2005; El-Obeid et al. 2006b; Öberg et al. 2009). Indeed, Pugh et al. (2005) reported that melanins isolated from *Echinacea*, alfalfa sprouts, and American ginseng were able to activate the cultured monocytes through the TLR2-dependent process. The activated cells substantially increased IL-1β secretion in a dose-dependent manner. Furthermore, *in vivo* experiments showed that oral intake of melanin derived from *Echinacea* species by mice enhanced the production of IL-6 in Peyer's patch cells and IFN-γ in spleen cells. The studies of El-Obeid et al. (2006a,b) support the concept that herbal melanins have a direct modulatory effect on cytokine production. It was shown that melanin extracted from *Nigella sativa* L. induced TNF-α and IL-6 expression, at both mRNA and protein levels, by the human monocytes, monocytic cell lines and peripheral blood mononuclear cells, and IL-8 production by TLR4-transfected cell lines. However, it has been suggested that cytokine induction by the melanin proceeds via TLR4 rather than TLR2-dependent signaling pathway (El-Obeid et al. 2006b). This suggestion has been supported by the recent findings that melanin isolated from *Nigella sativa* L. induces the production of IL-8 and IL-6 via TLR4-dependent activation of the NF-κB signaling pathway (Öberg et al. 2009).

Plant melanins have also been suggested to induce the humoral immunity. Sava et al. (2001) reported that the "melanin-like" material isolated from black tea stimulated B-lymphocytes in mice in a dose-dependent manner. Furthermore, it was found that *Echinacea* melanin administered orally to mice enhanced the release of immunoglobulin A from Peyer's patch cells (Pugh et al. 2005).

In light of the results obtained to date, it can be concluded that melanin pigments extracted from plants possess immunomodulatory activities by stimulating both cell-mediated and humoral immunity. Although it is unlikely that high molecular weight melanin would be absorbed after oral administration, it is possible that it could exert a therapeutic immune enhancing effect by the direct interaction with the mucosal immune system of the gastrointestinal tract (Pugh et al. 2005). Accordingly, plant melanins could be used as dietary pharmaceuticals, and could contribute to future immunotherapies of disorders associated with imbalanced cytokine production (e.g., allergy and autoimmune diseases) and cancer. Obviously, the activity of melanin extracts will be several orders of magnitude more than that of consumed botanicals. Therefore, it is very important to evaluate the biological properties of a particular product very carefully, before any suggestions for its use in a clinical practice is made. Special attention should be paid to microbial purity of plant melanin extracts, since it has been reported that bacterial lipopolysaccharides and lipoproteins may contribute to the immune enhancing activity of some botanicals (Pugh et al. 2008; Tamta et al. 2008).

9.6 CONCLUSIONS

Although melanin pigments are the subject of research for decades, interest in them has not diminished. On the contrary, each year brings new information that allows us to look at these unique biopolymers more broadly than just through the prism of their role in the coloration of living organisms, including humans. As bioactive compounds, melanins may have potential pharmacological importance regarding serious health problems like melanoma or neurodegenerative disorders.

It seems only a matter of time before new, effective therapeutic strategies, based on antioxidant and immunomodulatory properties of melanins, or their ability to bind drugs, are developed. Such therapies could use both endogenous and exogenous pigments, e.g., synthetic or derived from natural plant products or botanicals modified by genetic engineering.

For now, the practical use of melanin pigments is limited mostly to cosmetology. Natural or synthetic melanins or melanin precursors, modified so that they are soluble in aqueous cosmetic buffers at physiological pH and temperature, may be ingredients of face and hand creams, lotions, anti-ageing ointments or foundation make-ups, acting as a screen and antioxidant for the protection against photo-induced skin damages (Herlihy 1985; Pawelek and Platt 1998). Melanin pigments are also recommended as a component of optical lenses for use in sunglasses and other special purpose glasses (Sigimura et al. 2010). The development of nanotechnology offers new opportunities to use melanin bioactivity. For instance, melanin-coated nanospheres are considered as a novel approach to the protection of bone marrow in cancer radiotherapy.

ABBREVIATIONS

α-MSH	α-melanocyte-stimulating hormone
ACTH	adrenocorticotropic hormone
AID	adjuvant-induced disease
AIDS	acquired immunodeficiency syndrome
AMD	age-related macular degeneration
ASP	*agouti* signaling protein
CysDA	cysteinyldopamine
DA	dopamine
DHI	5,6-dihydroxyindole
DHICA	5,6-dihydroxyindole-2-carboxylic acid
EPR	electron paramagnetic resonance
GALT	gut-associated lymphoid tissue
GM-CSF	granulocyte-macrophage colony-stimulating factor
HEMn-DP	human epidermal melanocytes darkly pigmented
HEMn-LP	human epidermal melanocytes lightly pigmented
ICAM-1	intercellular adhesion molecule-1
IL	interleukin
iNOS	inducible nitric oxide synthase
IR	infrared spectroscopy
LC	*locus coeruleus*
L-DOPA	L-3,4-dihydroxyphenylalanine
LPO	lactoperoxidase
LPS	lipopolysaccharide
MAPK	mitogen-activated protein kinase
MC1R	melanocortin-1 receptor
MRM	multiple reaction monitoring
NF-κB	nuclear factor kappa light chain enhancer of activated B cells
NMR	nuclear magnetic resonance
NO	nitric oxide
NOS	nitric oxide synthase
POMC	proopiomelanocortin
Py-GC/MS	pyrolysis coupled with gas chromatography and mass spectrometry
RNS	reactive nitrogen species
ROS	reactive oxygen species
RPE	retinal pigment epithelium

SN	*substantia nigra*
TLR	toll-like receptor
TMAH	tetramethylammonium hydroxide
TNF-α	tumor necrosis factor alpha
TRP	tyrosinase-related protein
UV	ultraviolet
VCAM-1	vascular cell adhesion molecule-1

REFERENCES

Agodi, A., S. Stefani, C. Corsaro, F. Campanile, S. Gribaldo, and G. Sichel. 1996. Study of a melanic pigment of *Proteus mirabilis. Res Microbiol* 147:167–174.

Ahn, J.H., S.H. Jin, and H.Y. Kang. 2008a. LPS induces melanogenesis through p38 MAPK activation in human melanocytes. *Arch Dermatol Res* 300:325–329.

Ahn, J.H., T.J. Park, S.H. Jin, and H.Y. Kang. 2008b. Human melanocytes express functional Toll-like receptor 4. *Exp Dermatol* 17:412–417.

Aime, S., B. Bergamasco, M. Casu et al. 2000. Isolation and ^{13}C-NMR characterization of an insoluble proteinaceous fraction from substantia nigra of Parkinson's disease patients. *Mov Disord* 15:977–981.

Aime, S., M. Fasano, B. Bergamasco, L. Lopiano, and G. Valente. 1994. Evidence for a glycidic-lipidic matrix in human neuromelanin, potentially responsible for the enhanced iron sequestering ability of substantia nigra. *J Neurochem* 62:369–371.

Allegri, G., R. Arban, C. Costa et al. 1990. Fast atom bombardment mass spectrometry in the study of dopamine melanogenesis intermediates. *Pigment Cell Res* 3:181–186.

Alviano, D.S., A.J. Franzen, L.R. Travassos et al. 2004. Melanin from *Fonsecaea pedrosoi* induces production of human antifungal antibodies and enhances the antimicrobial efficacy of phagocytes. *Infect Immun* 72:229–237.

Armstrong, B.K. and A. Kricker. 2001. The epidemiology of UV induced skin cancer. *J Photochem Photobiol B*: 63:8–18.

Aroca, P., J.C. Garcia-Borron, F. Solano, and J.A. Lozano. 1990. Regulation of mammalian melanogenesis. I: Partial purification and characterization of a dopachrome converting factor: Dopachrome tautomerase. *Biochim Biophys Acta* 1035:266–275.

Avramidis, N., A. Kourounakis, L. Hadjipetrou, and V. Senchuk. 1998. Anti-inflammatory and immunomodulating properties of grape melanin. Inhibitory effects on paw edema and adjuvant induced disease. *Arzneimittelforschung* 48:764–771.

Barcia, C., A. Barreiro, M. Poza, and M.-T. Herrero. 2003. Parkinson's disease and inflammatory changes. *Neurotox Res* 5:411–418.

Barker, D., K. Dixon, E. Medrano et al. 1995. Comparison of the response of human melanocytes with different melanin contents to ultraviolet B irradiation. *Cancer Res* 55:4041–4046.

Barluzzi, R., A. Brozzetti, G. Mariucci et al. 2000. Establishment of protective immunity against cerebral cryptoccosis by means of an avirulent, non melanogenic *Cryptococcus neoformans* strain. *J Neuroimmunol* 109:75–86.

Ben-Shachar, D., P. Riederer, and M.B.H. Youdim. 1991. Iron-melanin interaction and lipid peroxidation: Implication for Parkinson's disease. *J Neurochem* 57:1609–1614.

Berliner, D., R. Erwin, and D. McGee. 1998. Therapeutic uses of melanin. U.S. Pat. No. 5817631.

Bertazzo, A., C. Costa, G. Allegri, R. Seraglia, and P. Traldi. 1995. Biosynthesis of melanin from dopamine. An investigation of early oligomerization products. *Rapid Commun Mass Spectrom* 9:634–640.

Bertazzo, A., D. Favretto, C.V. Costa, G. Allegri, and P. Traldi. 1999. Melanogenesis from 5-hydroxytryptamine, 5,6- and 5,7-dihydroxytryptamines. An in vitro study using MALDI-TOF. *Adv Exp Med Biol* 467:779–787.

Bilińska, B. 2001. On the structure of human hair melanins from an infrared spectroscopy analysis of their interactions with Cu^{2+} ions. *Spectrochim Acta Part A* 57:2525–2533.

Bishop, J.A.N. and D.T. Bishop. 2005. The genetics of susceptibility to cutaneous melanoma. *Drugs Today (Barc)* 41:193–203.

Boissy, R.E., C. Sakai, H. Zhao, T. Kobayashi, and V.J. Hearing. 1998. Human tyrosinase-related protein-1 (TRP-1) does not function as a DHICA oxidase in contrast to murine TRP-1. *Exp Dermatol* 7:198–204.

Brajac, I., M. Kaštelan, L. Prpič-Massari, D. Periša, K. Konarek, and D. Malnar. 2009. Melanocyte as a possible key cell in the pathogenesis of *Psoriasis vulgaris*. *Med Hypotheses* 73:254–256.

Brenner, M. and V.J. Hearing. 2008. The protective role of melanin against UV damage in human skin. *Photochem Photobiol* 84:539–549.

Bridelli, M.G., D. Tampellini, and L. Zecca. 1999. The structure of neuromelanin and its iron binding site studied by infrared spectroscopy. *FEBS Lett* 457:18–22.

Bruch-Gerharz, D., T. Ruzicka, and V. Kolb-Bachofen. 1998. Nitric oxide in human skin: current status and future prospects. *J Invest Dermatol* 110:1–7.

Brzoska, T., T. Luger, C. Maaser, C. Abels, and M. Bohm. 2008. "α"-Melanocyte-stimulating hormone and related tripeptides: Biochemistry, antiinflammatory and protective effects in vitro and in vivo, and future perspectives for the treatment of immune-mediated inflammatory diseases. *Endocr Rev* 29:581–602.

Chodurek, E., D. Kuśmierz, A. Dzierżęga-Lęcznar, S. Kurkiewicz, K. Stępień, and Z. Dzierżewicz. 2008. Thermochemolysis as the useful method to assess the purity of melanin isolated from the human *melanoma malignum*. *Acta Pol Pharm Drug Res* 65:531–534.

Chodurek, E., M. Latocha, S. Kurkiewicz, E. Buszman, L. Świątkowska, and T. Wilczok. 1998. Chemical characteristics of melanin from *Cladosporium cladosporioides*. *Bull Pol Acad Sci Biol* 46:51–58.

Chodurek, E., B. Pilawa, A. Dzierżęga-Lęcznar, S. Kurkiewicz, L. Świątkowska, and T. Wilczok. 2003. Effect of Cu^{2+} and Zn^{2+} ions on DOPA-melanin structure as analyzed by pyrolysis-gas chromatography-mass spectrometry and EPR spectroscopy. *J Anal Appl Pyrolysis* 70:43–54.

Da Silva, M.B., A.F. Marques, J.D. Nosanchuk, A. Casadevall, L.R. Travassos, and C.P. Taborda. 2006. Melanin in the dimorphic fungal pathogen *Paracoccidioides brasiliensis*: Effects on phagocytosis, intracellular resistance and drug susceptibility. *Microbes Infect* 8:197–205.

Dadachova, E., R. Bryan, R. Howell et al. 2008. Radioprotective properties of melanin are the function of its chemical composition, free stable radical presence and spatial arrangement. *Pigment Cell Melanoma Res* 21:192–199.

Dadachova, E., R. Bryan, X. Huang et al. 2007. Ionizing radiation changes the electronic properties of melanin and enhances the growth of melanized fungi. *PLoS One* 5:e457.

Dadachova, E. and A. Casadevall. 2008. Ionizing radiation: How fungi cope, adapt, and exploit with the help of melanin. *Curr Opin Microbiol* 11:525–531.

Danese, S., M. Sans, C. de la Motte et al. 2006. Angiogenesis as a novel component of inflammatory bowel disease pathogenesis. *Gastroenterology* 130:2060–2073.

Del Bino, S., J. Sok, E. Bessac, and F. Bernerd. 2006. Relationship between skin response to ultraviolet exposure and skin color type. *Pigment Cell Res* 19:606–614.

Di Donato, P. and A. Napolitano. 2003. 1,4-Benzothiazines as key intermediates in the biosynthesis of red hair pigment pheomelanins. *Pigment Cell Res* 16:532–539.

Dixon, D.M., J. Migliozzi, C.R. Cooper Jr, O. Solis, B. Breslin, and P.J. Szaniszlo. 1992. Melanized and non-melanized multicellular form mutants of *Wangiella dermatitidis* in mice: Mortality and histopathology studies. *Mycoses* 35:17–21.

Doering, T.L., J.D. Nosanchuk, W.K. Roberts, and A. Casadevall. 1999. Melanin as a potential cryptococcal defence against microbicidal proteins. *Med Mycol* 37:175–181.

Double, K.L., M. Gerlach, V. Schunemann et al. 2003. Iron-binding characteristics of neuromelanin of the human substantia nigra. *Biochem Pharmacol* 66:489–494.

Double, K.L., P. Riederer, and M. Gerlach. 1999. The significance of neuromelanin for neurodegeneration in Parkinson's disease. *Drug News Perspect* 12:333–340.

Double, K.L., L. Zecca, P. Costi et al. 2000. Structural characteristics of human substantia nigra neuromelanin and synthetic dopamine melanins. *J Neurochem* 75:2583–2589.

Doyle, S.L. and L.A.J. O'Neill. 2006. Toll-like receptors: From the discovery of NFκB to new insights into transcriptional regulations in innate immunity. *Biochem Pharmacol* 72:1102–1113.

Dworzański, J.P. 1983. Pyrolysis-gas chromatography of natural and synthetic melanins. *J Anal Appl Pyrolysis* 5:69–79.

Dworzański, J.P. and M. Dębowski. 1985. Pyrolysis-gas chromatography of pheomelanins. *J Anal Appl Pyrolysis* 8:463–472.

Dworzański, J.P. and H.L.C. Meuzelaar. 2000. Pyrolysis mass spectrometry, methods. In *Encyclopedia of Spectroscopy and Spectrometry*, eds. J.C Lindon, G.E Tranter, and J.L. Holmes, pp. 1906–1919. San Diego, CA: Academic Press.

Dzierżęga-Lęcznar, A., E. Chodurek, K. Stępień, and T. Wilczok. 2002. Pyrolysis-gas chromatography-mass spectrometry of synthetic neuromelanins. *J Anal Appl Pyrolysis* 62:239–248.

Dzierżęga-Lęcznar, A., S. Kurkiewicz, and K. Stępień. 2012. Detection and quantitation of a pheomelanin component in melanin pigments using pyrolysis-gas chromatography/tandem mass spectrometry system with multiple reaction monitoring mode. *J Mass Spectrom* 47:242–245.

Dzierżęga-Lęcznar, A., S. Kurkiewicz, K. Stępień, E. Chodurek, P. Riederer, and M. Gerlach. 2006. Structural investigations of neuromelanin by pyrolysis-gas chromatography/mass spectrometry. *J Neural Transm* 113:729–734.

Dzierżęga-Lęcznar, A., S. Kurkiewicz, K. Stępień et al. 2004. GC/MS analysis of thermally degraded neuromelanin from the human substantia nigra. *J Am Soc Mass Spectrom* 15:920–926.

Dzierżęga-Lęcznar, A., B. Pilawa, K. Stępień, and T. Wilczok. 1997. EPR studies of synthetic pheo- and mixed type-melanins. *Nukleonika* 42:343–352.

Dzierżęga-Lęcznar, A., K. Stępień, E. Chodurek, S. Kurkiewicz, L. Świątkowska, and T. Wilczok. 2003. Pyrolysis-gas chromatography/mass spectrometry of peroxynitrite-treated melanins. *J Anal Appl Pyrolysis* 70:457–467.

Eisenman, H.C., J.D. Nosanchuk, J.B. Webber, R.J. Emerson, T.A. Camesano, and A. Casadevall. 2005. Microstructure of cell wall-associated melanin in the human pathogenic fungus *Cryptococcus neoformans. Biochemistry* 44:3683–3693.

El-Obeid, A., S. Al-Harbi, N. Al-Jomah, and A. Hassib. 2006a. Herbal melanin modulates tumor necrosis factor alpha (TNF-α), interleukin 6 (IL-6) and vascular endothelial growth factor (VEGF) production. *Phytomedicine* 13:324–333.

El-Obeid, A., A. Hassib, F. Pontén, and B. Westermark. 2006b. Effect of herbal melanin on IL-8: A possible role of Toll-like receptor 4 (TLR). *Biochem Biophys Res Commun* 344:1200–1206.

Fecker, L., J. Eberle, C. Orfanos, and C. Geilen. 2002. Inducible nitric oxide synthase is expressed in normal human melanocytes but not in melanoma cells in response to tumor necrosis factor-alpha, interferon-gamma, and lipopolysaccharide. *J Invest Dermatol* 118:1019–1025.

Fedorow, H., R. Pickford, J.M. Hook et al. 2005. Dolichol is the major lipid component of human substantia nigra neuromelanin. *J Neurochem* 92:990–995.

Feng, B., X. Wang, M. Hauser et al. 2001. Molecular cloning and characterization of WdPKS1, a gene involved in dihydroxynaphthalene melanin biosynthesis and virulence in *Wangiella* (*Exophiala*) *dermatitidis. Infect Immun* 69:1781–1794.

Fogarty, R.V. and J.M. Tobin. 1996. Fungal melanins and their interactions with metals. *Enzyme Microb Technol* 19:311–317.

Frases, S., A. Salazar, E. Dadachova, and A. Casadevall. 2007. *Cryptococcus neoformans* can utilize the bacterial melanin precursor homogentisic acid for fungal melanogenesis. *Appl Environ Microbiol* 73:615–621.

Friedman, D.S., J. Katz, N.M. Bressler, B. Rahmani, and J.M. Tielsch. 1999. Racial differences in the prevalence of age-related macular degeneration: The Baltimore Eye Survey. *Ophthalmology* 106:1049–1055.

Garcia-Borron, J.C., B.L. Sanchez-Laorden, and C. Jimenez-Cervantes. 2005. Melanocortin-1 receptor structure and functional regulation. *Pigment Cell Res* 18:393–410.

Garcia-Rivera, J., H.C. Eisenman, J.D. Nosanchuk et al. 2005. Comparative analysis of *Cryptococcus neoformans* acid-resistant particles generated from pigmented cells grown in different laccase substrates. *Fungal Genet Biol* 42:989–998.

Garger Jr, S. and S. Neidleman. 2001. Melanins with improved ability to inhibit HIV replication. U.S. Pat. No. 6300057 B1.

Geremia, E., C. Corsaro, R. Bonomo, R. Giardinelli, P. Pappalardo, A. Vanella, and G. Sichel. 1984. Eumelanins as free radicals trap and superoxide dismutase activities in Amphibia. *Comp. Biochem Physiol* 79B:67–69.

Gomez-Marin, A.M. and C.I. Sanchez. 2010. Thermal and mass spectroscopic characterization of a sulphur-containing bacterial melanin from *Bacillus subtilis. J Non-Crystal Sol* 356:1576–1580.

Greco, G., K. Wakamatsu, L. Panzella, S. Ito, A. Napolitano, and M. d'Ischia. 2009. Isomeric cysteinyldopas provide a (photo)degradable bulk component and a robust structural element in red human hair pheomelanin. *Pigment Cell Melanoma Res* 22:319–327.

Halliday, G.M. 2005. Inflammation, gene mutation and photoimmunosuppression in response to UVR-induced oxidative damage contributes to photocarcinogenesis. *Mutat Res* 571:107–120.

Hari, A., T.L. Flach, Y. Shi, and P.R. Mydlarski. 2010. Toll-like receptors: role in dermatological disease. *Mediators Inflamm* 2010:1–16.

Hauser, J.E., A.L. Kadekaro, R.J. Kavanagh et al. 2006. Melanin content and MC1R function independently affect UVR-induced DNA damage in cultured human melanocytes. *Pigment Cell Res* 19:303–314.

Hearing, V.J. 2005. Biogenesis of pigment granules: A sensitive way to regulate melanocyte function. *J Dermatol Sci* 37:3–14.

Hennessy, A., C. Oh, B. Diffey, K. Wakamatsu, S. Ito, and J. Rees. 2005. Eumelanin and pheomelanin concentration in human epidermis before and after UVB irradiation. *Pigment Cell Res* 18:220–223.

Herlihy, W. 1985. Skin tanning composition and method. U.S. Pat. No. 4515773.

Hong, L. and J.D. Simon. 2005. Physical and chemical characterization of iris and choroid melanosomes isolated from newborn and mature cows. *Photochem Photobiol* 81:517–523.

Howell, R., A. Schweitzer, A. Casadevall, and E. Dadachova. 2008. Chemosorption of radiometals of interest to nuclear medicine by synthetic melanins. *Nucl Med Biol* 35:353–357.

Hu, D.N., H. Savage, and J.E. Roberts. 2002. Uveal melanocytes, ocular pigment epithelium and Mueller cells in culture: In vitro toxicology. *Int J Toxicol* 21:465–472.

Hu, D.N., J.D. Simon, and T. Sarna. 2008. Role of ocular melanin in ophthalmic physiology and pathology. *Photochem Photobiol* 84:639–644.

Hu, D.N., G.P. Yu, S.A. McCormick, S. Schneider, and P.T. Finger. 2005. Population-based incidence of uveal melanoma in various races and ethnic groups. *Am J Ophthalmol* 140:612–617.

Huffnagle, G.B., G.H. Chen, J.L. Curtis, R.A. McDonald, R.M. Strieter, and G.B. Toews. 1995. Down-regulation of the afferent phase of T cell-mediated pulmonary inflammation and immunity by a high melanin-producing strain of *Cryptococcus neoformans*. *J Immunol* 155:3507–3516.

Hung, Y., V. Sava, C. Juang, T. Yeh, W. Shen, and G. Huang. 2002. Gastrointestinal enhancement of MRI with melanin derived from tea leaves (*Thea sinensis* Linn.). *J Ethnopharmacol* 79:75–79.

Ibrahim, S.F. and M.D. Brown. 2008. Tanning and cutaneous malignancy. *Dermatol Surg* 34:460–474.

Ikeda, R., T. Sugita, E.S. Jacobson, and T. Shinoda. 2003. Effects of melanin upon susceptibility of *Cryptococcus* to antifungals. *Microbiol Immunol* 47:271–277.

Ishikawa, H., Y. Mitsui, T. Yoshitomi et al. 2000. Presynaptic effects of botulinum toxin type A on the neuronally evoked response of albino and pigmented rabbit iris sphincter and dilator muscles. *Jpn J Ophthalmol* 44:106–109.

Ito, S. 2003. IFPCS presidential lecture. A chemist's view of melanogenesis. *Pigment Cell Res* 16:230–236.

Ito, S., Y. Nakanishi, R.K. Valenzuela, M.H. Brilliant, L. Kolbe, and K. Wakamatsu. 2011. Usefulness of alkaline hydrogen peroxide oxidation to analyze eumelanin and pheomelanin in various tissue samples: Application to chemical analysis of human hair melanins. *Pigment Cell Melanoma Res* 24:605–613.

Ito, S. and K. Wakamatsu. 2003. Quantitative analysis of eumelanin and pheomelanin in humans, mice, and other animals: A comparative review. *Pigment Cell Res* 16:523–531.

Ito, S. and K. Wakamatsu. 2006. Chemistry of melanins. In *The Pigmentary System. Physiology and Pathophysiology*, 2nd edn., eds. J.J. Nordlund, R.E. Boissy, V.J. Hearing, R.A. King, W.S. Oetting, and J.P. Ortonne, pp. 282–310. Oxford, U.K.: Blackwell Publishing.

Ito, S. and K. Wakamatsu. 2008. Chemistry of mixed melanogenesis—Pivotal roles of dopaquinone. *Photochem Photobiol* 84:582–592.

Jacobson, E. 2000. Pathogenic roles for fungal melanins. *Clin Microbiol Rev* 13:708–717.

Jahn, B., A. Koch, A. Schmidt et al. 1997. Isolation and characterization of a pigmentless-conidium mutant of *Aspergillus fumigatus* with altered conidial surface and reduced virulence. *Infect Immun* 65:5110–5117.

Jimenez-Cervantes, C., F. Solano, T. Kobayashi et al. 1994. A new enzymatic function in the melanogenic pathway. The 5,6-dihydroxyindole-2-carboxylic acid oxidase activity of tyrosinase-related protein-1 (TRP1). *J Biol Chem* 269:17993–18001.

Jin, S.H. and H.Y. Kang. 2010. Activation of toll-like receptors 1, 2, 4, 5, and 7 on human melanocytes modulate pigmentation. *Ann Dermatol* 22:486–489.

Kadekaro, A.L., R.J. Kavanagh, K. Wakamatsu, S. Ito, M.A. Pipitone, and Z.A. Abdel-Malek. 2003. Cutaneous photobiology. The melanocyte vs. the sun: Who will win the final round? *Pigment Cell Res* 16:434–447.

Kobayashi, T., K. Urabe, A.J. Winder et al. 1994. Tyrosinase-related protein-1 (TRP1) functions as a DHICA oxidase in melanin biosynthesis. *EMBO J* 13:5818–5825.

Korytowski, W., B. Kalyanaraman, I.A. Menon, T. Sarna, and R.C. Sealy. 1986. Reaction of superoxide anions with melanin: Electron spin resonance and spin trapping studies. *Biochim Biophys Acta* 882:145–153.

Korytowski, W., T. Sarna, and M. Zaręba. 1995. Antioxidant action of neuromelanin: The mechanism of inhibitory effect on lipid peroxidation. *Arch Biochem Biophys* 319:142–148.

Kumar, A. and G. Deepak. 2007. Antimicrobial properties of *Osmanthus fragrans* (Lour). *Res J Med Plant* 1:21–24.

Kuzumaki, T., A. Matsuda, K. Wakamatsu, S. Ito, and K. Ishikawa. 1993. Eumelanin biosynthesis is regulated by coordinate expression of tyrosinase and tyrosinase-related protein-1 genes. *Exp Cell Res* 207:33–40.

Kvam, E. and J. Dahle. 2003. Pigmented melanocytes are protected against ultraviolet-A-induced membrane damage. *J Invest Dermatol* 121:564–569.

Kwon-Chung, K.J., I. Polacheck, and T.J. Popkin. 1982. Melanin-lacking mutants of *Cryptococcus neoformans* and their virulence for mice. *J Bacteriol* 150:1414–1421.
Kwon-Chung, K.J. and J.C. Rhodes. 1986. Encapsulation and melanin formation as indicators of virulence in *Cryptococcus neoformans. Infect Immun* 51:218–223.
Langfelder, K., M. Streibel, B. Jahn, G. Haase, and A. Brakhage. 2003. Biosynthesis of fungal melanins and their importance for human pathogenic fungi. *Fungal Genet Biol* 38:143–158.
Larsson, B. and H. Tjälve. 1978. Studies on the melanin affinity of metal ions. *Acta Physiol Scand* 104:479–484.
Latocha, M., E. Chodurek, S. Kurkiewicz, L. Świątkowska, and T. Wilczok. 2000. Pyrolytic GC-MS analysis of melanin from black, gray and yellow strains of *Drosophila melanogaster. J Anal Appl Pyrolysis* 56:89–98.
Le Poole, I.C., T. Mutis, R. Van Der Wijngaard et al. 1993a. A novel antigen-presenting function of melanocytes and its possible relationship to hypopigmentary disorders. *J Immunol* 151:7284–7292.
Le Poole, I.C., R. Van Der Wijngaard, W. Westerhof et al. 1993b. Phagocytosis by normal human melanocytes in vitro. *Exp Cell Res* 205:388–395.
Liu, Y., L. Hong, K. Wakamatsu et al. 2005a. Comparisons of the structural and chemical properties of melanosomes isolated from retinal pigment epithelium, iris and choroid of newborn and mature bovine eyes. *Photochem Photobiol* 81:510–516.
Liu, Y., L. Hong, K. Wakamatsu et al. 2005b. Comparison of structural and chemical properties of black and red human hair melanosomes. *Photochem Photobiol* 81:135–144.
Lu, Y., W. Zhu, C. Tan, G. Yu, and J. Gu. 2002. Melanocytes are potential immunocompetent cells: Evidence from recognition of immunological characteristics of cultured human melanocytes. *Pigment Cell Res* 15:454–460.
Mackintosh, J. 2001. The antimicrobial properties of melanocytes, melanosomes and melanin and the evolution of black skin. *J Theor Biol* 211:101–113.
Mal'tsev, V., A. Saadavi, A. Aiad, O. El'gauli, and M. Shlip. 1996. Microecology of nuclear reactor pool water. *Radiats Biol Radioecol* 36:52–57.
del Marmol, V. and F. Beerman. 1996. Tyrosinase and related proteins in mammalian pigmentation. *FEBS Lett* 381:165–168.
Marsden, C.D. 1983. Neuromelanin and Parkinson's disease. *J Neural Transm Suppl* 19:121–141.
Mattei, S., M. Colombo, C. Melani, A. Silvani, G. Parmiani, and M. Herlyn. 1994. Expression of cytokine/growth factors and their receptors in human melanoma and melanocytes. *Int J Cancer* 56:853–857.
Mednick, A.J., J.D. Nosanchuk, and A. Casadevall. 2005. Melanization of *Cryptococcus neoformans* affects lung inflammatory responses during cryptococcal infection. *Infect Immun* 73:2012–2019.
Meredith, P. and T. Sarna. 2006. The physical and chemical properties of eumelanin. *Pigment Cell Res* 19:572–594.
Mironenko, N., I. Alekhina, N. Zhdanova, and S. Bulat. 2000. Intraspecific variation in gamma-radiation resistance and genomic structure in the filamentous fungus *Alternaria alternata*: A case study of strains inhabiting Charnobyl reactor no 4. *Ecotoxicol Environ Saf* 45:177–187.
Mitchell, P., J.J. Wang, S. Foran, and W. Smith. 2002. Five-year incidence of age-related maculopathy lesions: The Blue Mountains Eye Study. *Ophthalmology* 109:1092–1097.
Mohagheghpour, N. 2001. Mediation of cytokines by melanin. U.S. Pat. No. 6242415 B1.
Mohagheghpour, N., N. Waleh, S.J. Garger, L. Dousman, L.K. Grill, and D. Tuse. 2000. Synthetic melanin suppresses production of proinflammatory cytokines. *Cell Immunol* 199:25–36.
Montefiori, D.C., A. Modliszewski, D.I. Shaff, and J. Zhou. 1990. Inhibition of human immunodeficiency virus type 1 replication and cytopathicity by synthetic soluble catecholamine melanins in vitro. *Biochem Biophys Res Commun* 168:200–205.
Montefiori, D.C. and J.Y. Zhou. 1991. Selective antiviral activity of synthetic soluble l-tyrosine and l-dopa melanins against human immunodeficiency virus in vitro. *Antiviral Res* 15:11–25.
Nicolaus, R.A. 1968. *Melanins*. Paris, France: Herman Press.
Nicolaus, R.A., M. Piatelli, and E. Fattorusso. 1964. The structure of melanins and melanogenesis. IV. On some natural melanins. *Tetrahedron* 20:1163–1172.
Nicolas, C.M., L.D. Robman, G. Tikelis et al. 2003. Iris colour, ethnic origin and progression of age-related macular degeneration. *Clin Exp Ophthalmol* 31:465–469.
Nosanchuk, J.D. and A. Casadevall. 2003. The contribution of melanin to microbial pathogenesis. *Cell Microbiol* 5:203–223.
Nosanchuk, J.D. and A. Casadevall. 2006. Impact of melanin on microbial virulence and clinical resistance to antimicrobial compounds. *Antimicrob Agents Chemother* 50:3519–3528.

Nosanchuk, J.D., A.L. Rosas, and A. Casadevall. 1998. The antibody response to fungal melanin in mice. *J Immunol* 160:6026–6031.

Nosanchuk, J.D., A.L. Rosas, S.C. Lee, and A. Casadevall. 2000. Melanisation of *Cryptococcus neoformans* in human brain tissue. *Lancet* 355:2049–2050.

Nosanchuk, J.D., P. Valadon, M. Feldmesser, and A. Casadevall. 1999. Melanization of *Cryptococcus neoformans* in murine infection. *Mol Cell Biol* 19:745–750.

Öberg, F., A. Haseeb, M. Ahnfelt, F. Pontén, B. Westermark, and A. El-Obeid. 2009. Herbal melanin activates TLR4/NF-κB signaling pathway. *Phytomedicine* 16:477–484.

Oberländer, U., K. Pletinckx, A. Döhler et al. 2011. Neuromelanin is an immune stimulator for dendritic cells in vitro. *BMC Neuroscience* 12:116.

Olivares, C., C. Jimenez-Cervantes, J.A. Lozano, F. Solano, and J.C. Garcia-Borron. 2001. The 5,6-dihydroxyindole-2-carboxylic acid (DHICA) oxidase activity of human tyrosinase. *Biochem J* 354:131–139.

Olivares, C. and F. Solano. 2009. New insights into the active site structure and catalytic mechanism of tyrosinase and its related proteins. *Pigment Cell Melanoma Res* 22:750–760.

Ortonne, J.P. 2002. Photoprotective properties of skin melanin. *Br J Dermatol* 146:7–10.

Ostrovsky, M.A., N.L. Sakina, and A.E. Dontsov. 1987. An antioxidative role of ocular screening pigments. *Vision Res* 27:893–899.

Pandya, A. and I. Guevara. 2000. Disorders of hyperpigmentation. *Dermatol Clin* 18:91–98.

Panzella, L., P. Manini, G. Monfrecola, M. d'Ischia, and A. Napolitano. 2006. An easy-to-run method for routine analysis of eumelanin and pheomelanin in pigmented tissues. *Pigment Cell Res* 20:128–133.

Pasco, D., N. Pugh, I. Khan, and R. Moraes. 2005. Immunostimulatory agents in botanicals. U.S. Pat. No. US 2005/0002962 A1.

Pawelek, J. and J. Platt. 1998. Cosmetic melanin. U.S. Pat. No. 5744125.

Peters, S., T. Lamah, D. Kokkinou, K.U. Bartz-Schmidt, and U. Schraermeyer. 2006. Melanin protects choroidal blood vessels against light toxicity. *Z Naturforsch* 61:427–433.

Plonka, P., T. Passeron, M. Brenner et al. 2009. What are melanocytes *really* doing all day long ...? *Exp Dermatol* 18:799–819.

Porębska-Budny, M., N.L. Sakina, K.B. Stępień, A.E. Dontsov, and T. Wilczok. 1992. Antioxidative activity of synthetic melanins. Cardiolipin liposome model. *Biochim Biophys Acta* 1116:11–16.

Prota, G. 1992. *Melanins and Melanogenesis*. San Diego, CA: Academic Press.

Prota, G., S. Crescenzi, G. Misuraca, and R.A. Nicolaus. 1970. New intermediates in phaeomelanogenesis in vitro. *Experientia* 26:1058–1059.

Prota, G., D.N. Hu, M.R. Vincensi, S.A. McCormick, and A. Napolitano. 1998. Characterization of melanins in human irides and cultured uveal melanocytes from eyes of different colors. *Exp Eye Res* 67:293–299.

Prota, G., M. d'Ischia, and A. Napolitano. 1998. The chemistry of melanins and related metabolites. In *The Pigmentary System. Physiology and Pathophysiology*, 1st edn., eds. J.J. Nordlund, R.E. Boissy, V.J. Hearing, R.A. King, and J.P. Ortonne, pp. 307–332. Oxford, U.K.: Oxford University Press.

Pugh, N., P. Balachandran, H. Lata et al. 2005. Melanin: Dietary mucosal immune modulator from *Echinacea* and other botanical supplements. *Int Immunopharmacol* 5:637–647.

Pugh, N., H. Tamta, P. Balachandran et al. 2008. The majority of in vitro macrophage activation exhibited by extracts of some immune enhancing botanicals is due to bacterial lipoproteins and lipopolysaccharides. *Int Immunopharmacol* 8:1023–1032.

Rees, J.L. 2004. The genetics of sun sensitivity in human. *Am J Hum Genet* 75:739–751.

Reszka, K.J., Z. Matuszak, and C.F. Chignell. 1998. Lactoperoxidase-catalyzed oxidation of melanin by reactive nitrogen species derived from nitrite (NO_2^-): An EPR study. *Free Radic Biol Med* 25:208–216.

Rocha, I. and L. Guillo. 2001. Lipopolysaccharide and cytokines induce nitric oxide synthase and produce nitric oxide in cultured human melanocytes. *Arch Dermatol Res* 293:245–248.

Romero-Graillet, C., E. Aberdam, M. Clement, J. Ortonne, and R. Ballotti.1997. Nitric oxide produced by ultraviolet irradiated keratinocytes stimulates melanogenesis. *J Clin Invest* 99:635–642.

Romero-Martinez, R., M. Wheeler, A. Guerrero-Plata, G. Rico, and H. Torres-Guerrero. 2000. Biosynthesis and functions of melanin in *Sporothrix schenckii*. *Infect Immun* 68:3696–3703.

Rosas, A.L., R.S. MacGill, J.D. Nosanchuk, T.R. Kozel, and A. Casadevall. 2002. Activation of the alternative complement pathway by fungal melanins. *Clin Diagn Lab Immunol* 9:144–148.

Rosas, A.L., J.D. Nosanchuk, and A. Casadevall. 2001. Passive immunization with melanin-binding monoclonal antibodies prolongs survival in mice with lethal *Cryptococcus neoformans* infection. *Infect Immun* 69:3410–3412.

Rosas, A.L., J.D. Nosanchuk, M. Feldmesser, G.M. Cox, H.C. McDade, and A. Casadevall. 2000. Synthesis of polymerized melanin by *Cryptococcus neoformans* in infected rodents. *Infect Immun* 68:2845–2853.

Różanowska, M., T. Sarna, E.J. Land, and T.G. Truscott. 1999. Free radical scavenging properties of melanin: interaction of eu- and pheo-melanin models with reducing and oxidising radicals. *Free Radic Biol Med* 26:518–525.

Saini, L.S., B. Galsworthy, M.A. John, and M.A. Valvano. 1999. Intracellular survival of *Burkholderia cepacia* complex isolates in the presence of macrophage cell activation. *Microbiology* 145:3465–3475.

Sakina, N.L., A.E. Dontsov, and M.A. Ostrovsky. 1985. Comparison of the antioxidative systems of retinal pigment epithelium of pigmented and albino animals. *Biokhimiya* (Moscow) 50:78–83.

Salas, S.D., J.E. Bennett, K.J. Kwon-Chung, J.R. Perfect, and P.R. Williamson. 1996. Effect of the laccase gene CNLAC1, on virulence of *Cryptococcus neoformans*. *J Exp Med* 184:377–386.

Sarna, T. 1992. Properties and function of the ocular melanin—A photobiophysical view. *J Photochem Photobiol B Biol* 12:215–258.

Sarna, T., B. Pilas, E.J. Land, and T.G. Truscott. 1986. Interaction of radicals from water radiolysis with melanin. *Biochim Biophys Acta* 883:162–167.

Sasaki, M., T. Horikoshi, H. Uchiwa, and Y. Miyachi. 2000. Up-regulation of tyrosinase gene by nitric oxide in human melanocytes. *Pigment Cell Res* 13:248–252.

Sava, V., B. Galkin, M.-Y. Hong, P.-C. Yang, and G. Huang. 2001. A novel melanin-like pigment derived from black tea leaves with immuno-stimulating activity. *Food Res Int* 34:337–343.

Schnitzler, N., H. Peltroche-Liacsahuanga, N. Bestier, J. Zundorf, R. Lutticken, and G. Haase. 1999. Effect of melanin and carotenoids of *Exophiala* (*Wangiella*) *dermatitidis* on phagocytosis, oxidative burst, and killing by human neutrophils. *Infect Immun* 67:94–101.

Schweitzer, A., R. Howell, Z. Jiang et al. 2009. Physico-chemical evaluation of rationally designed melanins as novel nature-inspired radioprotectors. *PLoS One* 4:e7229.

Schweitzer, A., E. Revskaya, P. Chu et al. 2010. Melanin-covered nanoparticles for protection of bone marrow during radiation therapy of cancer. *Int J Radiat Oncol Biol Phys* 78:1494–1502.

Sealy, R.C., J.S. Hyde, C.C. Felix et al. 1982. Novel free radicals in synthetic and natural pheomelanins: Distinction between dopa melanins and cysteinyldopa melanins by ESR spectroscopy. *Proc Natl Acad Sci USA* 79:2885–2889.

Seraglia, R., P. Traldi, G. Elli, A. Bertazzo, C. Costa, and G. Allegri. 1993. Laser desorption ionization mass spectrometry in the study of natural and synthetic melanins. I-tyrosine melanins. *Biol Mass Spectrom* 22:687–697.

Sidibe, S., F. Saal, A. Rhodes-Feuillette et al. 1996. Effects of serotonin and melanin on in vitro HIV-1 infection. *J Biol Regul Homeost Agents* 10:19–24.

Sigimura, H., X. Qin, and M. Boulineau. 2010. Polarizing plate with melanin. U.S. Pat. No. 7703916 B2.

Simon, J.D., D. Peles, K. Wakamatsu, and S. Ito. 2009. Current challenges in understanding melanogenesis: Bridging chemistry, biological control, morphology, and function. *Pigment Cell Melanoma Res* 22:563–579.

Slominski, A., C. Gomez-Sanchez, M. Foecking, and J. Wortsman. 1999. Metabolism of progesterone to DOC, corticosterone and 18OHDOC in cultured human melanoma cell. *FEBS Lett* 445:364–366.

Slominski, A. and G. Goodman-Snitkoff. 1992. Dopa inhibits induced proliferative activity of murine and human lymphocytes. *Anticancer Res.* 12:753–756.

Slominski, A., R. Paus, and M. Mihm. 1998. Inhibition of melanogenesis as an adjuvant strategy in the treatment of melanotic melanomas: Selective review and hypothesis. *Anticancer Res* 18:3709–3716.

Slominski, A., D. Tobin, S. Shibahara, and J. Wortsman. 2004. Melanin pigmentation in mammalian skin and its hormonal regulation. *Physiol Rev* 84:1155–1228.

Slominski, A., J. Wortsman, T. Luger, R. Paus, and S. Solomon. 2000. Corticotropin releasing hormone and proopiomelanocortin involvement in the cutaneous response to stress. *Physiol Rev* 80:979–1020.

Slominski, A., B. Zbytek, and R. Slominski. 2009. Inhibitors of melanogenesis increase toxicity of cyclophosphamide and lymphocytes against melanoma cells. *Int J Cancer* 124:1470–1477.

Slominski, A., B. Zbytek, A. Szczesniewski et al. 2005. CRH stimulation of corticosteroids production in melanocytes is mediated by ACTH. *Am J Physiol Endocrinol Metab* 288:E701–E706.

Smit, N., I. Le Poole, R. Van Der Wijngaard, A. Tigges, W. Westerhof, and P. Das. 1993. Expression of different immunological markers by cultured human melanocytes. *Arch Dermatol Res* 285:356–365.

Smit, N.P., A.A. Vink, R.M. Kolb et al. 2001. Melanin offers protection against induction of cyclobutane pyrimidine dimers and 6–4 photoproducts by UVB in cultured human melanocytes. *Photochem Photobiol* 74:424–430.

Stępień, K.B., J.P. Dworzański, B. Bilińska, M. Porębska-Budny, A.M. Hollek, and T. Wilczok. 1989. Catecholamine melanins. Structural changes induced by copper ions. *Biochim Biophys Acta* 997:49–54.

Stępień, K., A. Dzierżęga-Lęcznar, S. Kurkiewicz, and I. Tam. 2009. Melanin from epidermal human melanocytes: Study by pyrolytic GC/MS. *J Am Soc Mass Spectrom* 20:464–468.

Stępień, K., M. Porębska-Budny, A.M. Hollek, and T. Wilczok. 1992. The inhibiting effect of catecholamine-melanins on UV-induced lecithin peroxidation. *J Photochem Photobiol B Biol* 15:223–231.

Stępień, K. and I. Tam. 2009. Lipopolysaccharide-induced nitric oxide production in lightly and darkly pigmented human melanocytes. *Acta Biochim Pol* 56(Suppl. 3):139.

Stępień, K. and T. Wilczok. 1994. Antioxidant activity of model neuromelanins in the process of lipid peroxidation. *Curr Top Biophys* 18:135–138.

Stępień, K., A. Wilczok, A. Zajdel, A. Dzierżęga-Lęcznar, and T. Wilczok. 2000a. Peroxynitrite mediated linoleic acid oxidation and tyrosine nitration in the presence of synthetic neuromelanins. *Acta Biochim Pol* 47:931–940.

Stępień, K., A. Zajdel, G. Świerczek, A. Wilczok, and T. Wilczok.1998. Reduction of 13-hydroperoxy-9,11-octadecadienoic acid by dopamine-melanin. *Biochem Biophys Res Commun* 244:781–784.

Stępień, K., A. Zajdel, A. Wilczok et al. 2000b. Dopamine-melanin protects against tyrosine nitration, tryptophan oxidation and Ca^{2+}-ATPase inactivation induced by peroxynitrite. *Biochim Biophys Acta* 1523:189–195.

Sulzer, D., J. Boguslavsky, K.E. Larsen et al. 2000. Neuromelanin biosynthesis is driven by excess cytosolic catecholamines not accumulated by synaptic vesicles. *Proc Natl Acad Sci USA* 67:11869–11874.

Swan, G.A. 1974. Structure, chemistry, and biosynthesis of the melanins. *Fortschr Chem Org Naturst* 31:521–582.

Swope, V., D. Sauder, R. McKenzie et al. 1994. Synthesis of interleukin-1α and β by normal human melanocytes. *J Invest Dermatol* 102:749–753.

Tadokoro, T., N. Kobayashi, B.Z. Zmudzka et al. 2003. UV-induced DNA damage and melanin content in human skin differing in racial/ethnic origin. *FASEB J* 17:1177–1179.

Tam, I. and K. Stępień. 2010. Does pigmentation phenotype of melanocytes affect their ability to produce IL-8? *Acta Biochim Pol* 57(Suppl. 4):176.

Tam, I. and K. Stępień. 2011. Secretion of proinflammatory cytokines by normal human melanocytes in response to lipopolysaccharide. *Acta Biochim Pol* 58:507–511.

Tamta, H., N.D. Pugh, P. Balachandran, R. Moares, J. Sumiyanto, and D.S. Pasco. 2008. Variability in in vitro macrophage activation by commercially diverse bulk *Echinacea* plant material is due predominantly to bacterial lipoproteins and lipopolysaccharides. *J Agric Food Chem* 56:10552–10556.

Tolleson, W.H. 2005. Human melanocyte biology, toxicology, and pathology. *J Environ Sci Health* 23:105–161.

Tsai, H.F., Y.C. Chang, R.G. Washburn, M.H. Wheeler, and K.J. Kwon-Chung. 1998. The developmentally regulated *alb1* gene of *Aspergillus fumigatus*: Its role in modulation of conidial morphology and virulence. *J Bacteriol* 180:3031–3038.

Tsukamoto, K., I.J. Jackson, K. Urabe, P.M. Montague, and V.J. Hearing. 1992. A second tyrosinase-related protein, TRP-2, is a melanogenic enzyme termed DOPAchrome tautomerase. *EMBO J* 11:519–526.

Van Duin, D., A. Casadevall, and J.D. Nosanchuk. 2002. Melanization of *Cryptococcus neoformans* and *Histoplasma capsulatum* reduces their susceptibilities to amphotericin B and caspofungin. *Antimicrob Agents Chemother* 46:3394–3400.

Wakamatsu, K., D.N. Hu, S.A. McCormick, and S. Ito. 2008. Characterization of melanin in human iridal and choroidal melanocytes from eyes with various colored irides. *Pigment Cell Res* 21:97–105.

Wang, Y., P. Aisen, and A. Casadevall. 1995. *Cryptococcus neoformans* melanin and virulence: Mechanism of action. *Infect Immun* 63:3131–3136.

Wang, Y. and A. Casadevall. 1994. Susceptibility of melanized and nonmelanized *Cryptococcus neoformans* to nitrogen- and oxygen-derived oxidants. *Infect Immun* 62:3004–3007.

Wang, Y. and A. Casadevall. 1996. Melanin, melanin 'ghosts', and melanin composition in *Cryptococcus neoformans. Infect Immun* 64:2420–2424.

Wang, Z., J. Dillon, and E.R. Gaillard. 2006. Antioxidant properties of melanin in retinal pigment epithelial cells. *Photochem Photobiol* 82:474–479.

Weis, E., C.P. Shah, M. Lajous, J.A. Shields, and C.I. Shields. 2006. The association between host susceptibility factors and uveal melanoma: A meta-analysis. *Arch Ophthalmol* 124:54–60.

Wheeler, M.H. and A.A. Bell. 1988. Melanins and their importance in pathogenic fungi. *Curr Top Med Mycol* 2:338–387.

Wielgus, R.A. and T. Sarna. 2005. Melanin in human irides of different color and age of donor. *Pigment Cell Res* 18:454–464.

Wilczok, T., K. Stępień, A. Dzierzega–Lecznar, A. Zajdel, and A. Wilczok. 1999. Model neuromelanins as antioxidative agents during lipid peroxidation. *Neurotoxicity Res* 1:141–147.

Williamson, P.R. 1994. Biochemical and molecular characterization of the diphenol oxidase of *Cryptococcus neoformans*: Identification as a laccase. *J Bacteriol* 176:656–664.

Wilms, H., P. Rosenstiel, J. Sievers, G. Deuschl, L. Zecca, and R. Lucius. 2003. Activation of microglia by human neuromelanin is NF-κB dependent and involves p38 mitogen-activated protein kinase: Implications for Parkinson's disease. *FASEB J* 17:500–502.

Wlaschek, M., I. Tantcheva-Poor, L. Naderi et al. 2001. Solar UV irradiation and dermal photoaging. *J Photochem Photobiol B Biol* 63:41–51.

Yamaguchi, Y., S.G. Coelho, B.Z. Zmudzka et al. 2008. Cyclobutane pyrimidine dimer formation and p53 production in human skin after repeated UV irradiation. *Exp Dermatol* 17:916–924.

Yamaguchi, Y., K. Takahashi, B.Z. Zmudzka et al. 2006. Human skin responses to UV radiation: Pigment in the upper epidermis protects against DNA damage in the lower epidermis and facilitates apoptosis. *FASEB J* 20:E630-E639.

Yu, N., S. Zhang, F. Zuo, K. Kang, M. Guan, and L. Xiang. 2009. Cultured human melanocytes express functional toll-like receptors 2-4, 7, and 9. *J Dermatol Sci* 56:113–120.

Zaręba, M., A. Bober, W. Korytowski, L. Zecca, and T. Sarna. 1995. The effect of a synthetic neuromelanin on yield of free hydroxyl radicals generated in model systems. *Biochim Biophys Acta* 1271:343–348.

Zecca, L., P. Costi, C. Mecacci, S. Ito, M. Terreni, and S. Sonnino. 2000. Interaction of human substantia nigra neuromelanin with peptides and lipids. *J Neurochem* 74:1758–1765.

Zecca, L., C. Mecacci, R. Seraglia, and E. Parati. 1992. The chemical characterization of melanin contained in substantia nigra of human brain. *Biochim Biophys Acta* 1138:6–10.

Zecca, L., D. Tampellini, A. Gatti et al. 2002. The neuromelanin of human substantia nigra and its interaction with metals. *J Neural Transm* 109:663–672.

Zecca, L., D. Tampellini, M. Gerlach, P. Riederer, R.G. Fariello, and D. Sulzer. 2001. Substantia nigra neuromelanin: Structure, synthesis, and molecular behavior. *Mol Pathol* 54:414–418.

Zecca, L., H. Wilms, S. Geick et al. 2008. Human neuromelanin induces neuroinflammation and neurodegeneration in the rat substantia nigra: Implications for Parkinson's disease. *Acta Neuropathol* 116:47–55.

Zecca, L., F.A. Zucca, A. Albertini, E. Rizzio, and R.G. Fariello. 2006. A proposed dual role of neuromelanin in the pathogenesis of Parkinson's disease. *Neurology* 67(Suppl 2):S8–S11.

Zhang, W., K. Phillips, A. Wielgus et al. 2011. Neuromelanin activates microglia and induces degeneration of dopaminergic neurons: Implications for progression of Parkinson's disease. *Neurotox Res* 19:63–72.

Zucca, F.A., C. Bellei, S. Giannelli et al. 2006. Neuromelanin and iron in human locus coeruleus and substantia nigra during ageing: Consequences for neuronal vulnerability. *J Neural Transm* 113:757–767.

Zucca, F.A., G. Giaveri, M. Gallorini et al. 2004. The neuromelanin of human substantia nigra: Physiological and pathogenic aspects. *Pigment Cell Res* 17:610–617.

Zughaier, S.M., H.C. Ryley, and S.K. Jackson. 1999. A melanin pigment purified from an epidemic strain of *Burkholderia cepacia* attenuates monocyte respiratory burst activity by scavenging superoxide anion. *Infect Immun* 67:908–913.

10 Recent Acquisitions on Naturally Occurring Oxyprenylated Secondary Plant Metabolites

Francesco Epifano and Salvatore Genovese

CONTENTS

10.1 INTRODUCTION

Oxyprenylated secondary metabolites have been regarded, for several years, merely as biosynthetic intermediates of *C*-prenylated compounds and only in the last decade were characterized as phytochemicals able to exert interesting and effective biological activities. Considering the length of the carbon chain attached to the oxygen atom, three types of prenyloxy skeletons can be identified: those having 5 (isopentenyl), 10 (geranyl), or 15 (farnesyl) carbon atoms. Isopentenyloxy and geranyloxy chains are quite abundant in nature, while farnesyloxy ones are by far less common. The skeleton may consist only of carbon and hydrogen or may contain oxygen atoms, usually in the form of alcohols, ethers, carboxylic acids, or ketone functionalities, and less frequently nitrogen and halogen atoms. To date, about 350 oxyprenylated derivatives were isolated from natural sources, mainly plants, fungi, and bacteria, including marine organisms, and shown to exert a variety of valuable and promising biological activities. In the last 5 years, several new phytochemical and pharmacological data about the title secondary metabolites were reported in the literature. The aim of this review is to make a survey of the most recently published data and properties of these important and interesting class of natural products, some of which have been obtained from plants that have long been used for proven or supposed medical properties, according to some ancient ethnomedical traditions.

10.2 *O*-GERANYL DERIVATIVES

10.2.1 Anthraquinones

The genus *Vismia* is nowadays well recognized among the most important sources of geranyloxy anthranoids. In 2008, Mbaveng and coworkers isolated 3-geranyloxy-6-methyl-1,8-dihydroxyanthraquinone (**1**) from the leaves, stem bark, and roots of *Vismia guineensis* (Linn.) Choisy (Guttiferae) (Figure 10.1) (Mbaveng et al. 2008). The investigators tested antimicrobial activity of **1** and found it to have slight activity on *Mycobacterium smegmatis* (MIC = 39.06 μg/mL) and *M. tuberculosis* (MIC = 78.12 μg/mL). This compound (**1**) has also recently been isolated from *Cratoxylum glaucum* Korth. and *C. arborescens* (Vahl) Blume (Guttiferae) (Sim et al. 2011).

Two anthranoid dimers, named febriquinone (**2**) and adamabianthrone (**3**) (Figure 10.2), were isolated by Tsaffack and coworkers from the roots of *Psorospermum febrifugum* Spach and from the barks of *P. adamauense* Engl. (Guttiferae) together with known anthraquinones like bianthrone A1, vismione D, 3-geranyloxyemodinanthrone, and 3-geranyloxy-1,8-dihydroxy-6-methylanthraquinone (Tsaffack et al. 2009). The new isolates **2** and **3** showed fairly good antimicrobial activities on both Gram (+) and Gram (–) bacteria. In particular, febriquinone (**2**) exhibited an appreciable effect on *Bacillus cereus* (MIC = 9.76 μg/mL) and *Staphylococcus faecalis* (MIC = 9.76 μg/mL).

10.2.2 Cinnamic Acids

The most part of the reported data about oxyprenylated cinnamic acids concerns 4′-geranyloxyferulic acid (GOFA; **4**) (Figure 10.3). This compound (**4**) was isolated for the first time in 1966 from *Acronychia baueri* Schott (Fam. Rutaceae) (Prager and Thregold 1966).

In 2008, Tanaka and coworkers reported that a novel aminoacidic prodrug of GOFA, namely, 3-(4′-geranyloxy-3′-methoxyphenyl)-L-alanyl-L-proline (GAP; **5**) (Figure 10.4) was able to exert promising protective effect against colon carginogenesis in mice (Miyamoto et al. 2008).

This prodrug was conceived in such a way to be enzymatically cleaved once having reached the large bowel by the intestinal angiotensin converting enzyme (ACE) located in high concentrations in the brush border of colonocytes (Curini et al. 2005). The inhibitory effects of GAP (**5**) on colon carcinogenesis were investigated using male CD1 (ICR), firstly treated with a single intraperitoneal injection of azoxymethane (AOM) (10 mg/kg body weight) to induce colon cancer, and then administered with a 1% (w/v) solution of dextrane sodium sulfate (DSS) in drinking water for 7 days to promote the growth of neoplastic lesions. After 2 weeks of feeding basal diet, animals were given a diet containing GAP at two different concentrations, 0.01% and 0.05%, respectively, for 17 weeks. At the end of the study, animals were sacrificed and biological

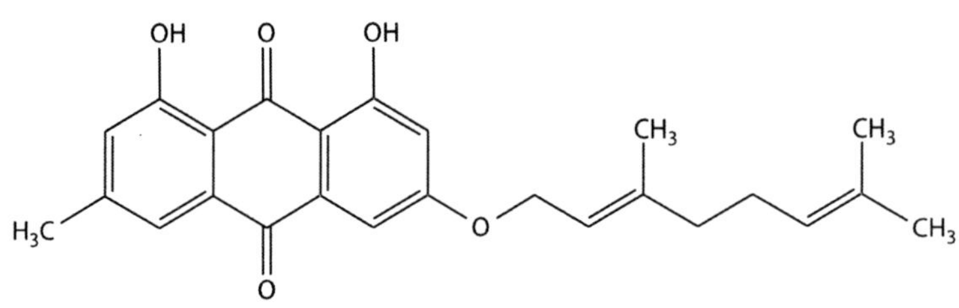

(1)

FIGURE 10.1 Structure of 3-geranyloxy-6-methyl-1,8-dihydroxyanthraquinone (**1**).

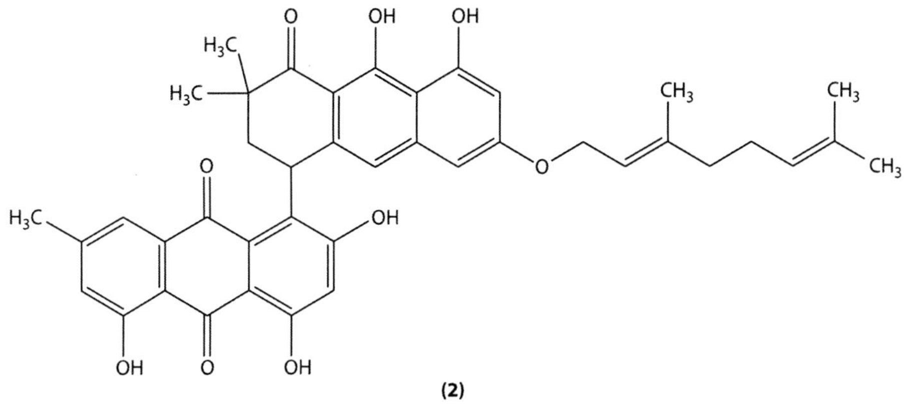

(2)

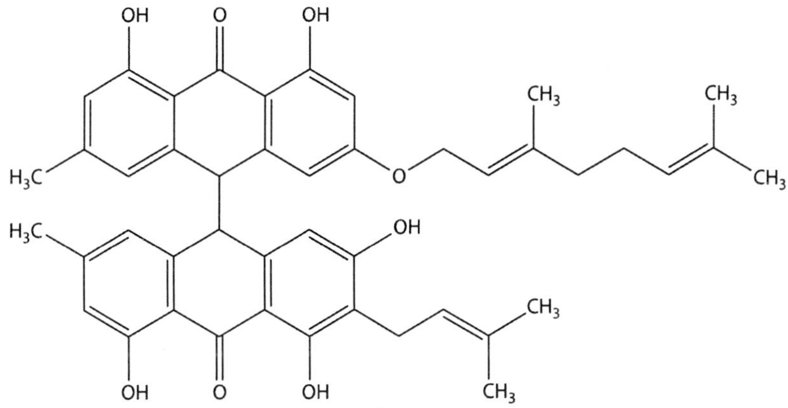

(3)

FIGURE 10.2 Structure of febriquinone (**2**) and adamabianthrone (**3**).

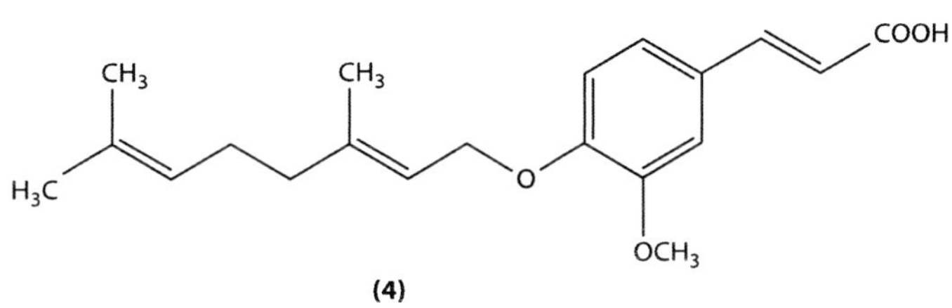

(4)

FIGURE 10.3 Structure of 4′-geranyloxyferulic acid (**4**).

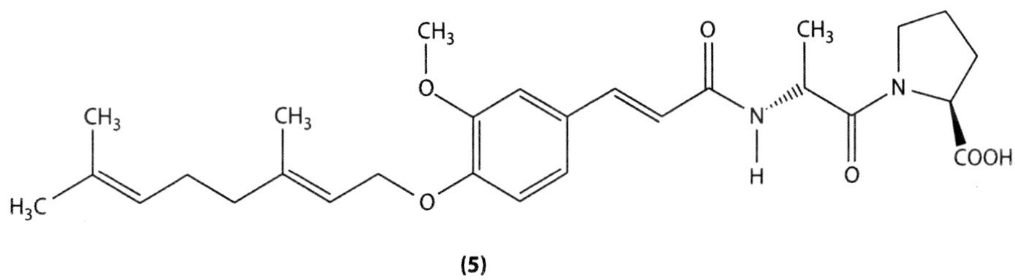

(5)

FIGURE 10.4 Structure of 3-(4′-geranyloxy-3′-methoxyphenyl)-L-alanyl-L-proline (**5**).

parameters relevant to cancer were measured. The development and the growth of colonic adenocarcinoma were significantly inhibited by GAP dietary feeding administration at the dose levels of 0.01% [60% incidence ($P < 0.05$)] and 0.05% [53% incidence ($P < 0.05$)]. These values were in both cases higher than those obtained for the AOM/DSS treated group of animals [95% incidence ($P < 0.05$)]. GAP feeding also provided lower indices of mitosis with respect to AOM/DSS only treated group. Finally, the two groups treated with GAP showed a great decrease of 8-hydroxy-2′-deoxyguanosine (8-OHdG)-positive cells, a decrease of urinary level of this metabolite, and an increase of immunoreactivity of an inducible form of heme oxygenase 1 (HO-1) in the colonic mucosa. It has to be kept in mind that both of these are indices of the oxidative stress and damage induced during inflammatory-based cancer growth and development. A similar approach was used for the synthesis and pharmacological assays of another prodrug of GOFA, namely, the one obtained by its inclusion into β-cyclodextrin (β-CD) (Tanaka et al. 2010). The complex GOFA/β-CD was rapidly synthesized by dissolving the parent acid in a suspension of β-CD in acetone at room temperature followed by evaporation of the solvent under vacuum to dryness. A pharmacological animal model similar to that described earlier was used to investigate the dietary feeding chemopreventive properties of this other prodrug. To this aim, animals were administered GOFA/β-CD at two dose levels (100 and 500 ppm, respectively). At the end of the study, the development of colonic adenocarcinoma was significantly inhibited by feeding with GOFA/β-CD at dose levels of 100 ppm (63% reduction, $P < 0.05$) and 500 ppm (83% reduction, $P < 0.001$), when compared to the AOM/DSS group. The dietary administration with GOFA/β-CD inhibited colonic inflammation and also modulated proliferation, apoptosis, and expression of several pro-inflammatory cytokines, such as nuclear factor-kappa B (NF-kB), tumor necrosis factor-α (TNF-α), Stat3, NF-E2-related factor 2 (Nrf2), interleukin (IL)-6, and IL-1β, all induced during adenocarcinomas development and growth. In particular, NF-kB decreased by 38.6% ($P < 0.001$) and 49.4% ($P < 0.001$), Nrf2 by 32.2% ($P < 0.01$) and 51.4% ($P < 0.01$), TNF-α by 21.7% ($P < 0.05$) and 43.8% ($P < 0.05$), Stat3 by 48.6% ($P < 0.001$) and 57.0% ($P < 0.001$), IL-1β by 42.4% ($P < 0.001$) and 49.2% ($P < 0.001$), and finally IL-6 by 31.4% ($P < 0.001$) and 41.6% ($P < 0.001$) when tested at the dose of 100 and 500 ppm, respectively.

GOFA (**4**) was shown to be an efficient anti-inflammatory agent (Epifano et al. 2007). In order to get further insights into these biological properties and to have pharmacologically active products designed in such a way to have a synergistic anti-inflammatory effects by means of chemical or enzymatic hydrolysis (e.g., lipases), Epifano and coworkers synthesized a series of esters in which the acid portion was represented by GOFA while the alcoholic one originated from natural, semisynthetic, or synthetic already known as *in vitro* and *in vivo* anti-inflammatory agents, both portions being able to potentially ensure such kind of effect. The pharmacological activity of the synthesized esters was evaluated using the Croton oil ear test in mice as a model of acute inflammation (Epifano et al. 2007). Each derivative was administered at the dose of 0.30 μmol/cm^2, and indomethacin at the same concentration was used as the control. GOFA induced a 41% edema inhibition, being slightly less active than indomethacin, which reduced the response by 62%. Among the eleven esters synthesized, three, the alcoholic portion of which was represented by paracetamol, guaiacol, and hydroquinone (Figure 10.5), showed effects ranging from 49% to 57% edema inhibition significantly higher than the parent acid and comparable to that of indomethacin, even though their phenol precursors were inactive (4%–13% inhibition).

The same group of compounds, as described previously, was tested for their antibacterial activity as well as inhibitory activity toward biofilm formation against two main oral pathogens, *Porphyromonas gingivalis* and *Streptococcus mutans* (Bodet et al. 2008). GOFA itself at the dose of 31.3 μg/mL (78.1 μM) led to an inhibition of biofilm formation by *P. gingivalis* of about 80%. Some of the esters however were seen to be more active than the parent acid. Guaiacol ester (**6b**) (Figure 10.5) was able to inhibit the biofilm formation of this latter microorganism at the lowest concentration tested (8.6 μM, $P < 0.05$). Also esters having 2-hydroxynaphtoquinone (**7**) and methyl vanillate (**8**) (Figure 10.6) as the alcoholic portions, showed an appreciable activity, although lower than GOFA.

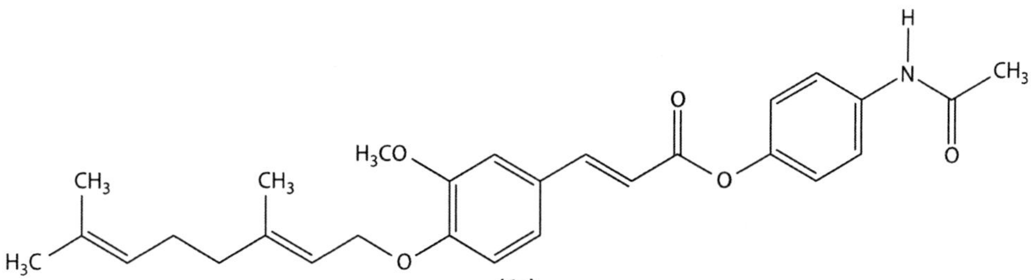

(6a)

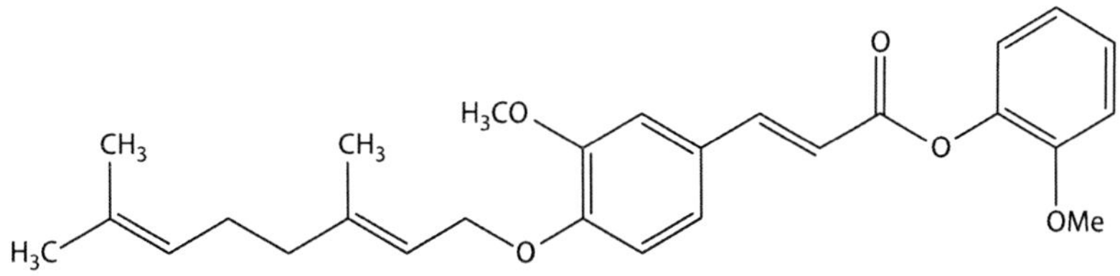

(6b)

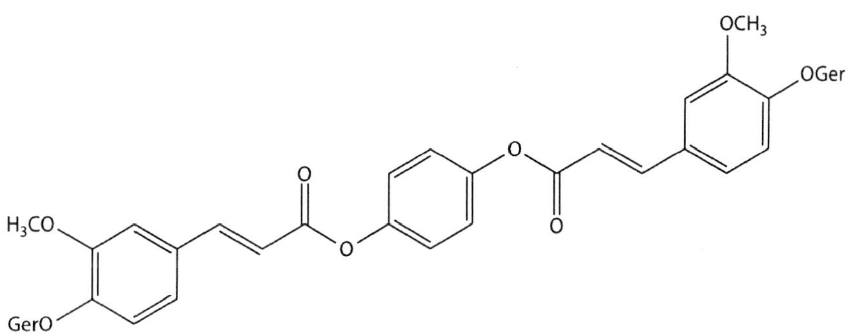

(6c)

FIGURE 10.5 Structure of GOFA esters with paracetamol (**6a**), guaiacol (**6b**), and hydroquinone (**6c**).

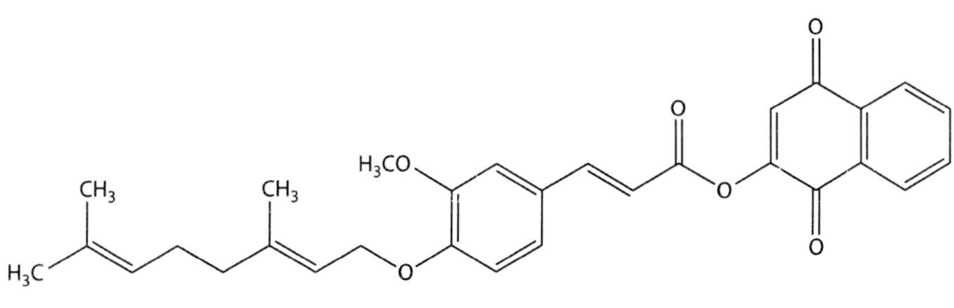

(7)

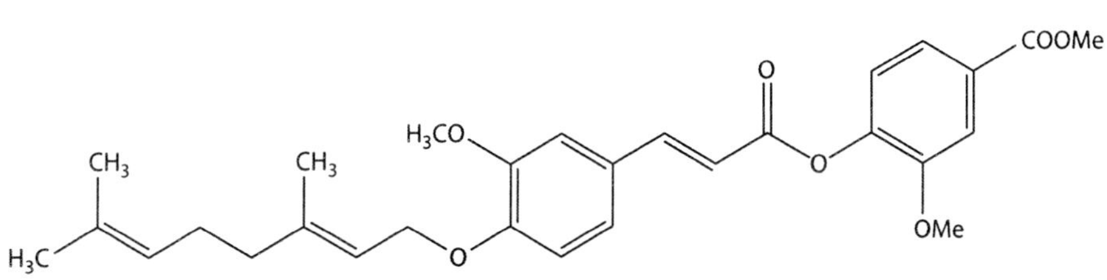

(8)

FIGURE 10.6 Structure of GOFA esters with 2-hydroxynaphtoquinone (**7**) and methyl vanillate (**8**).

GOFA and its three aforementioned esters (**6a**, **6b**, and **6c**) were able to reduce the growth of *P. gingivalis*, and this effect may be correlated with their capacity to inhibit biofilm formation. However, none of these compounds caused a significant decrease in viability of *P. gingivalis*, allowing the authors to hypothesize that the antibacterial effect could be bacteriostatic rather than bactericidal (Bodet et al. 2008). The earlier cited products were then tested as inhibitory agents of biofilm formation by *S. mutans*. GOFA at the lowest concentration tested (3.9 μg/mL; 9.8 μM) still had a significant inhibition of biofilm formation. The three ester derivatives, although able to reduce biofilm formation by *S. mutans* to some extent, were less effective than the parent molecule (Bodet et al. 2008).

The *in vivo* neuroprotective effects of GOFA were studied by Genovese and coworkers in 2009 using the mouse maximal electroshock-induced seizure model (MES-test) (Genovese et al. 2009). The MES-test was performed at different pre-treatment times (5, 15, 30, 60, and 120 min) using a group of eight animals after systemic intraperitoneal administration of GOFA at a dose of 300 mg/kg. Results obtained indicated that this prenyloxycinnamic acid was effective in protecting the animals against MES-induced seizures. In particular, the maximum anticonvulsant effect was obtained in the range 15–30 min after intraperitoneal administration, revealing absence of seizures in 7 of 8 and 6 of 8 animals, respectively. This effect tended to decrease with increasing time, being 2 of 8 animals and completely lacking (0 of 8 animals) after 60 and 120 min, respectively. The recorded ED_{50} at 15 min was 224.1 ± 10.7 mg/kg ($P < 0.05$).

For what concerns the mechanism of action of GOFA, until 2007 the only data at disposition was its capacity to inhibit cyclooxygenase-2 (COX-2) and the inducible form of nitric oxide synthase (*i*NOS) (Curini et al. 2006). Trying to get further insights, in 2007 Epifano and coworkers tested GOFA as *in vitro* inhibitors of prenyltransferases, known to play a pivotal role in the pathogenesis of several types of cancers (Epifano et al. 2007). At a concentration of 100 μM, GOFA selectively inhibited geranylgeranyl transferase I (GGTase I) (72.4%) while it was ineffective on farnesyl transferase (FTase) (12.7%). Genovese and coworkers investigated the efficacy of GOFA as activator of PPARs (Genovese et al. 2010). In the first series of experiments, GOFA was found to activate PPARα, PPARβ/δ, and PPARγ, the efficacy and selectivity being greater for PPARβ/δ. In order to reveal the effect of GOFA on PPARs β/δ, in a subsequent series of experiments, the investigators used wild-type mouse keratinocytes in which the activation of PPARs β/δ by selective ligands was known to evoke a significant increase in the expression of mRNA encoding *Angptl4* gene and PPARs β/δ-null cells. Culturing wild-type mouse keratinocytes with GOFA at a concentration of 10 μM led to a modest increase in the expression of *Angptl4* mRNA, an effect that was not found in PPARs β/δ-null cells. At a concentration of 100 μM, GOFA caused a marked increase in the expression of *Angptl4* mRNA comparable to that observed with GW0742, used as reference drug, and this change was still not observed in PPARs β/δ-null keratinocytes. Proliferation effects under the influence of GOFA were investigated in the human epithelial carcinoma cell line A431. Inhibition of cell growth was observed after 72 h following treatment with 100 μM GOFA, a result similar to that observed with GW0742. To determine if the observed inhibition of cell proliferation by GOFA was mediated by PPARs β/δ, cell proliferation was examined in wild-type and PPARs β/δ-null mouse primary keratinocytes. Ligand activation of PPARs β/δ with 1 μM GW0742 caused inhibition of cell proliferation in wild-type keratinocytes after 48 and 72 h of culture, and this effect was not found in similarly treated PPARs β/δ-null mouse primary keratinocytes. No changes in cell proliferation were recorded in keratinocytes following treatment with 25 μM GOFA. However, inhibition of cell proliferation was found in wild-type keratinocytes after 24 h of treatment with 100 μM GOFA and this effect was not observed in similarly treated PPARs β/δ-null keratinocytes. Moreover, inhibition of cell proliferation was found in primary keratinocytes lacking expression of PPARs β/δ. Thus, the authors made the hypothesis that the inhibitory effects of GOFA may be influenced by other PPARs β/δ-independent mechanisms.

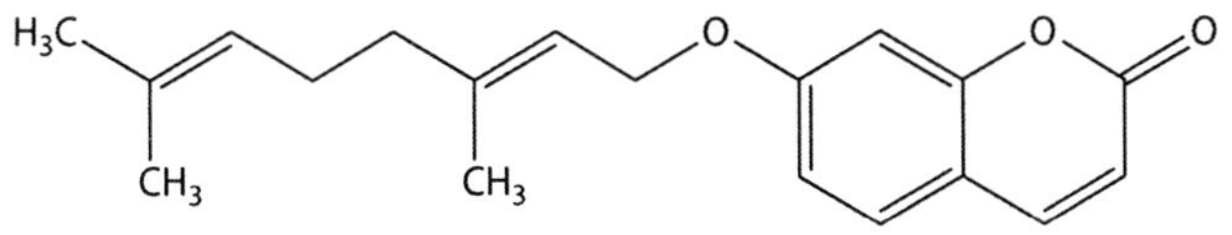

(9)

FIGURE 10.7 Structure of auraptene (**9**).

10.2.3 Coumarins

The pharmacological profile of one of the most abundant geranyloxycoumarin in nature, namely, auraptene (**9**) (Figure 10.7), has recently been reviewed (Genovese and Epifano 2011).

In addition to the reported data, it was found that auraptene (**9**), extracted from seeds of *Zosima absinthifolia* Link (Apiaceae), had a weak antifungal effect against the phytopathogenic fungus *Sclerotinia sclerotiorum* (30%), but a strong herbicide effect, being able, at a dose of 0.1 mg/mL, to entirely stunt seed germination and root and shoot growth in lettuce (Razavi et al. 2010). Recently, De Medina and coworkers shed light on the biological mechanism of action of auraptene (**9**), finding that this geranyloxycoumarin was able to inhibit acylCoA cholesterol acyltransferase and to modulate both estrogen receptors (ERs) α and β with binding affinity values of 7.8 and 7.9 μM, respectively (De Medina et al. 2010). Moreover, auraptene was able to modulate the transcription of both ERs via an ER-dependent reporter gene. The recorded effects correlated well with the control of growth and metastatic capacities of tumor cells. Like auraptene, geranyloxycoumarins and furanocoumarins are typically contained in edible *Citrus* fruits. The recent works by Dugo and coworkers allowed us to depict in more detail the geranyloxy- and furanocoumarins profile of *Citrus* fruits (Dugo et al. 2009; Costa et al. 2010).

10.2.4 Ketones

Oxyprenylated acetophenones were isolated for the first time from two *Melicope* spp., namely, *M. obscura* (Cordem) T.G. Hartley and *M. obtusifolia* ssp. *obtusifolia* var. *arborea* (Coode) T.G. Hartley (Andersen et al. 2007). These are 2,6-dihydroxy-4-geranyloxyacetophenone (**10a**), 4-geranyloxy-2,6,β-trihydroxyacetophenone (**10b**), 2,6-dihydroxy-4-geranyloxy-3-isopentenylacetophenone (**10c**), and 4-geranyloxy-3-isopentenyl-2,6,β-trihydroxyacetophenone (**10d**) (Figure 10.8).

Chemotaxonomic studies allowed us to reveal that these secondary metabolites could be regarded as markers of the Rutaceae subfamily Rutoideae, tribe Xanthoxyleae. The chemical synthesis of the

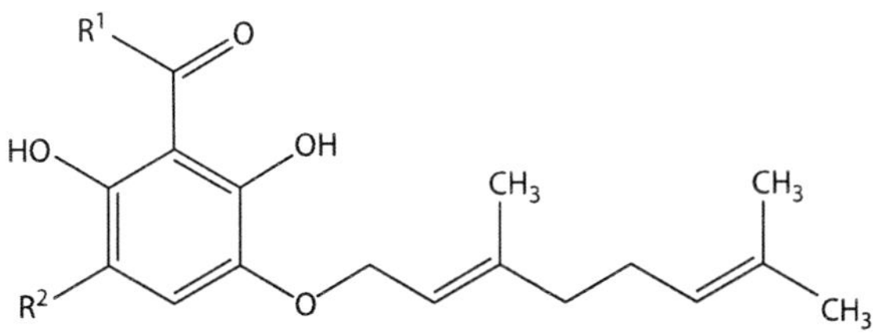

10a: $R^1=CH_3$, $R^2=H$
10b: $R^1=CH_2OH$, $R^2=H$
10c: $R^1=CH_3$, $R^2=$Isopentenyl
10d: $R^1=CH_2OH$, $R^2=$Isopentenyl

FIGURE 10.8 Structure of 2,6-dihydroxy-4-geranyloxyacetophenone (**10a**), 4-geranyloxy-2,6,β-trihydroxyacetophenone (**10b**), 2,6-dihydroxy-4-geranyloxy-3-isopentenylacetophenone (**10c**), and 4-geranyloxy-3-isopentenyl-2,6,β-trihydroxyacetophenone (**10d**).

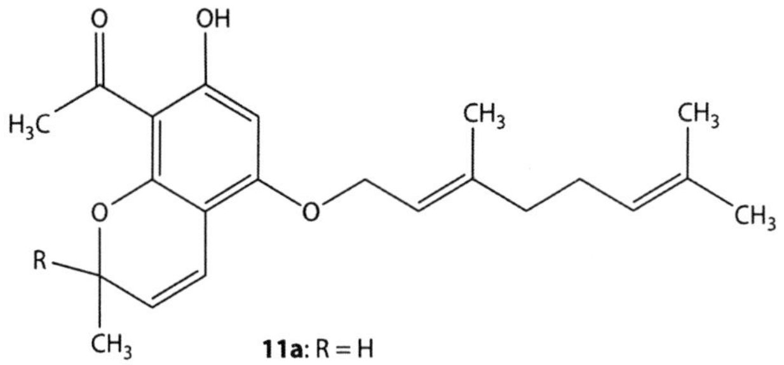

11a: R = H
11b: Isopentenyl

FIGURE 10.9 Further oxyprenylated acetophenones (**11a,b**) from *Melicope* spp.

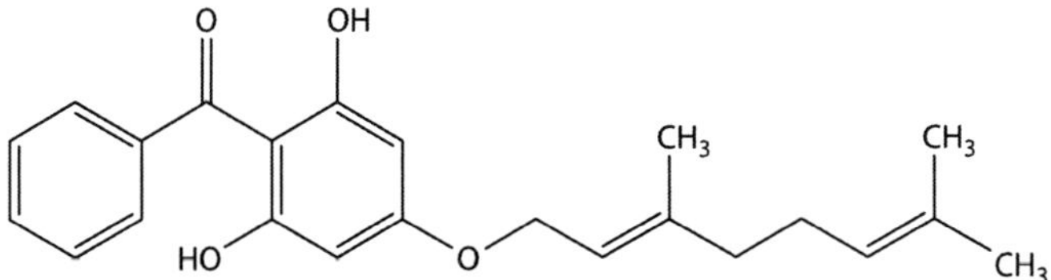

(12a)

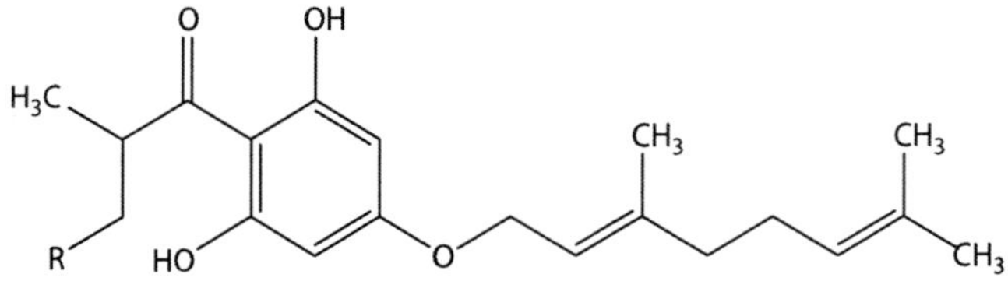

12b: R = H
12c: R = CH_3

FIGURE 10.10 Structures of 4-gerayloxy-2,6-dihydroxybenzophenone (**12a**), 4-geranyloxy-1-(2-methylpropanoyl)phloroglucinol (**12b**), and 4-geranyloxy-1-(2-methylbutanoyl)phloroglucinol (**12c**).

earlier listed acetophenones as well as of other two naturally occurring ones (**11a** and **11b**) (Figure 10.9) was recently reported by Xia et al. (2010).

Oxyprenylated phloroglucinol derivatives were isolated from the apolar extract of *Hyperichum densiflorum* Pusch. (Clusiaceae) (Henry et al. 2009). These were identified as 4-gerayloxy-2,6-dihydroxybenzophenone (**12a**), 4-geranyloxy-1-(2-methylpropanoyl)phloroglucinol (**12b**), and 4-geranyloxy-1-(2-methylbutanoyl)phloroglucinol (**12c**) (Figure 10.10).

These isolates (**12a–c**) were evaluated for a series of biological activities, including antitumor, anti-inflammatory, antioxidant, and antibacterial ones. All the compounds exhibited an appreciable *in vitro* growth inhibitory effect against a panel of human cancer cell lines, while showing a moderate anti-inflammatory activity, measured as COX-1 and COX-2 inhibition. 4-Geranyloxy-2,6-dihydroxybenzophenone (**12a**) revealed a good antioxidant effect (70% of reduction in lipid peroxidation test), while 4-geranyloxy-1-(2-methylpropanoyl)phloroglucinol (**12b**) and 4-geranyloxy-1-(2-methylbutanoyl)phloroglucinol (**12c**) recorded a worse effect in the same test. Finally, all these ketones showed a very good antibacterial activity against methicillin-resistant *Staphylococcus aureus* with IC_{50} values of 0.87, 1.14, and 1.80 μg/mL, respectively.

10.2.5 Quinones

A novel oxygeranylated 1,4-naphtoquinone, 7-geranyloxy-5-hydroxy-2-methoxy-6-methyl-1,4-napthoquinone, named flaviogeranin (**13**) (Figure 10.11) was isolated by Hayakawa and coworkers in 2010 from *Streptomyces* spp. strain RAC 226 (Hayakawa et al. 2010); the compound was tested as neuroprotective agent and exhibited very good activity. In fact, this quinone derivative was able

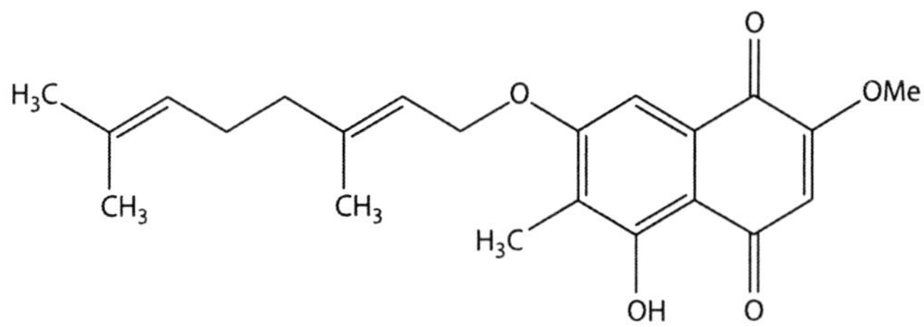

(13)

FIGURE 10.11 Structure of flaviogeranin (**13**).

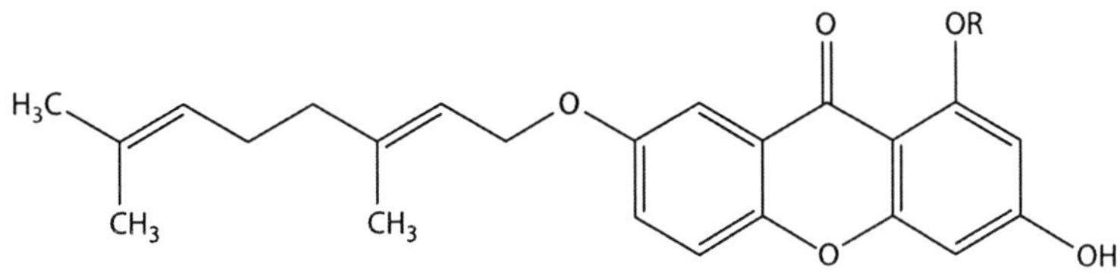

14a: R = H
14b: Ac

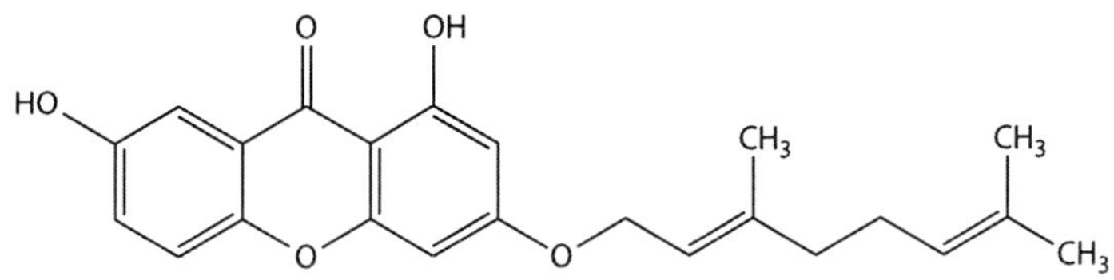

14c

FIGURE 10.12 Structure of geranyloxyxanthones (**14a–c**) from *Cratoxylum cochinchinense*.

to prevent neuronal cell death in C6 cells exposed to 100 mM glutamate (EC_{50} = 8.6 nM) and suppressed death in N18-RE-105 rat primary retina-mouse neuroblastoma hybrid cells exposed to glutamate 10 mM (EC_{50} = 360 nM).

10.2.6 Xanthones

Three novel oxygeranylated xanthones (**14a–c**) were isolated in 2009 by Bonnak and coworkers from the resin and green fruits of the Thai plant *Cratoxylum cochinchinense* Blume (Guttiferae) (Figure 10.12) (Bonnak et al. 2009). When tested as antimicrobial agents, all the isolates exhibited a significant activity against *Pseudomonas aeruginosa* (MIC = 4.7 μg/mL). Scanning electron microscopy studies revealed that they exerted this effect probably by interfering with the *de novo* formation of bacterial wall.

10.3 *O*-Isopentenyl Derivatives

10.3.1 Alkaloids

Several isopentenyloxy alkaloids are reported to be isolated from different plants. The South African shrub *Tecla gerrardi* I. Verd., commonly known as flaky cherry-orange (Rutaceae: Toddalioidae), provided an acridone alkaloid named tegerrardin B [(3-hydroxy-*N*-methyl-1-(γ,γ-dimethylallyloxy)acridone] (**15a**) that was characterized for the first time in nature, together with the already described furoquinoline alkaloids evoxine (**15b**) and 7-isopentenyloxy-γ-fagarine (**15c**) (Figure 10.13) (Waffo et al. 2007).

Among the isolates, evoxine (**15b**) showed a mild antimalarial activity against *Plasmodium falciparum* strain CQS D10, recording an IC_{50} value of 24.5 μM. The first evidence of the

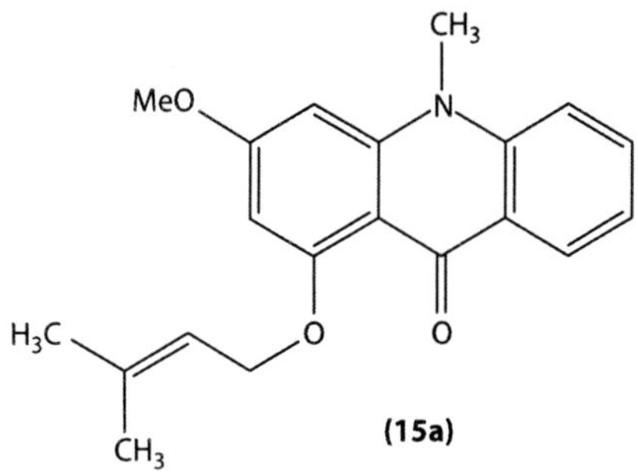

(15a)

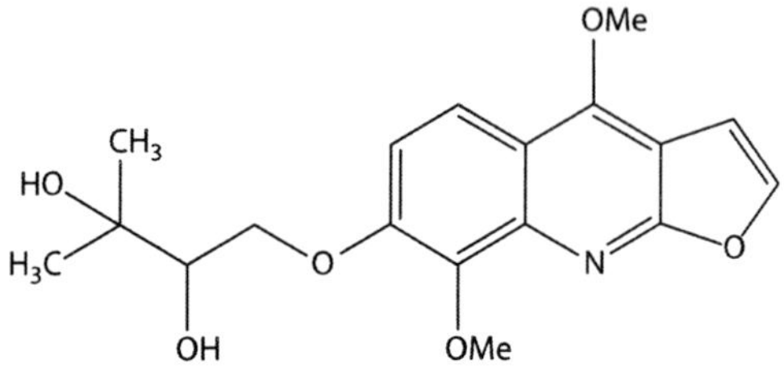

(15b)

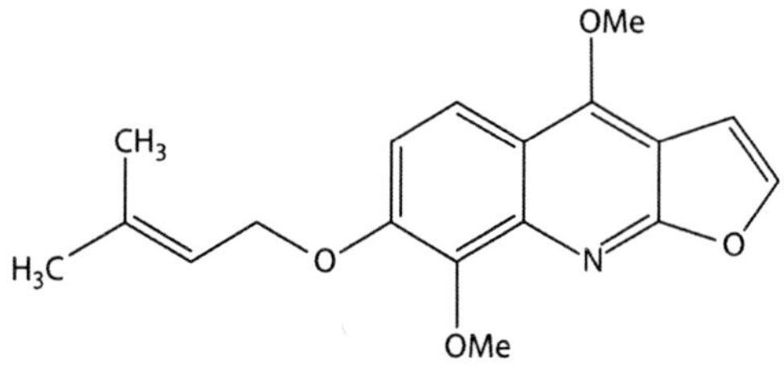

(15c)

FIGURE 10.13 Structure of tegerrardin B (**15a**), evoxine (**15b**), and 7-isopentenyloxy-γ-fagarine (**15c**).

isolation of chlorinated oxyprenylated furoquinoline alkaloids was given almost in the mean time but independently by two different research groups. Cao and coworkers extracted and purified 7-(2′-hydroxy-3′-chloroisopentenyloxy)-4,8-dimethoxyfuroquinoline (**16a**) and 6-(2′-hydroxy-3′-chloroisopentenyloxy)-4,7-dimethoxyfuroquinoline (**16b**) from the aerial parts of *Monnieria trifolia* (L.) Kuntze (Rutaceae) together with the known alkaloids 7-isopentenyloxy-γ-fagarine and tecleamatalesine B (**16c**) (Figure 10.14) (Cao et al. 2008).

Tested as anticancer agents *in vitro* on A2780 human ovarian cancer cell line, all these secondary metabolites showed a weak activity. The absolute configuration of 6-(2′-hydroxy-3′-chloroisopentenyloxy)-4,7-dimethoxyfuroquinoline (**16b**), isolated from the leaves of the ornamental shrub *Choisya ternata* H.B. & K., commonly known as Mexican orange, was then determined by Boyd and coworkers (Boyd et al. 2007). Finally in 2008, Varamini and coworkers extracted

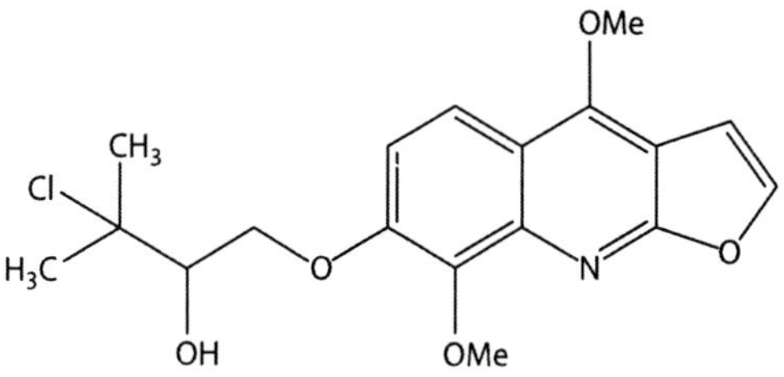

(16a)

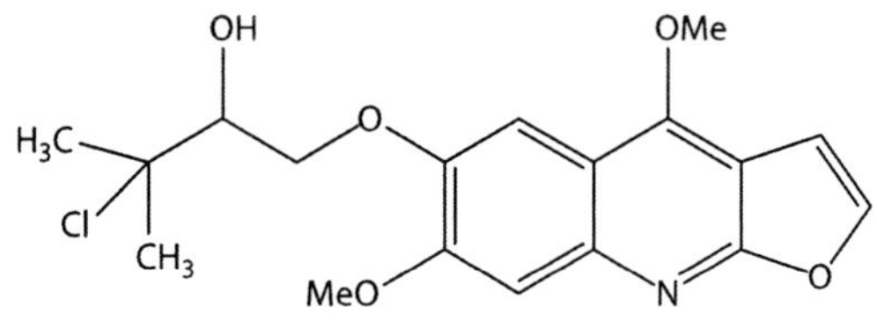

(16b)

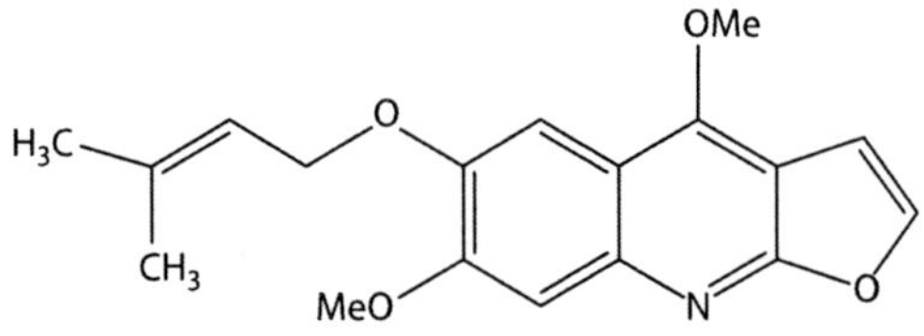

(16c)

FIGURE 10.14 Structure of 7-(2′-hydroxy-3′-chloroisopentenyloxy)-4,8-dimethoxyfuroquinoline (**16a**), 6-(2′-hydroxy-3′-chloroisopentenyloxy)-4,7-dimethoxyfuroquinoline (**16b**), and tecleamatalesine B (**16c**).

7-isopentenyloxy-γ-fagarine from *Haplophyllum canaliculatum* Boiss (Rutaceae) by means of a bioassay guided fractionation; the authors showed also that this alkaloid exerted *in vitro* an appreciable antitumor effect against Raji (IC_{50} = 1.50 μg/mL) and MCF-7 cells (IC_{50} = 15.50 μg/mL) (Varamini et al. 2008). Flow cytometry analysis performed with 7-isopentenyloxy-γ-fagarine also revealed that it was able to arrest the cell cycle at the sub-G1 phase in Raji and Jurkat cells in a dose-dependent manner.

10.3.2 Chalcones

Reddy and coworkers in 2008 extracted and structurally characterized a novel chalcone, namely, 2,3-dimethoxy-4′-isopentenyloxy-2′-hydroxychalcone (**17**) from the root bark of *Dalbergia sissoo* Roxb. (Figure 10.15) (Reddy et al. 2008).

Another chalcone named xinjiachalcone A (**18**) became part of the well-known secondary metabolites profile of licorice, having been isolated by Iwasaki and coworkers in 2009 (Figure 10.16) (Iwasaki et al. 2009).

The rare chalcone 4-hydroxycordoin (**19**; Figure 10.17) isolated from *Lonchocarpus neuroscapha* Benth. (Fabaceae) was recently investigated from a pharmacological point of view by Grenier and coworkers (Feldman et al. 2011; Messier et al. 2011). The authors showed that this compound exerted beneficial antibacterial effect against known periodontopathogens like *Prevotella intermedia* (MIC = 2.50 μg/mL), *Porphyromonas gingivalis* (MIC = 5.0 μg/mL), and *Fusobacterium nucleatum* (MIC = 40.0 μg/mL). Moreover, 4-hydroxycordoin at concentrations

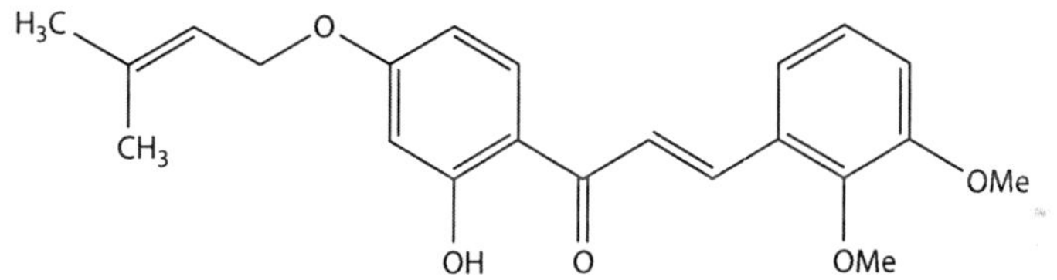

(17)

FIGURE 10.15 Structure of 2,3-dimethoxy-4′-isopentenyloxy-2′-hydroxychalcone (**17**).

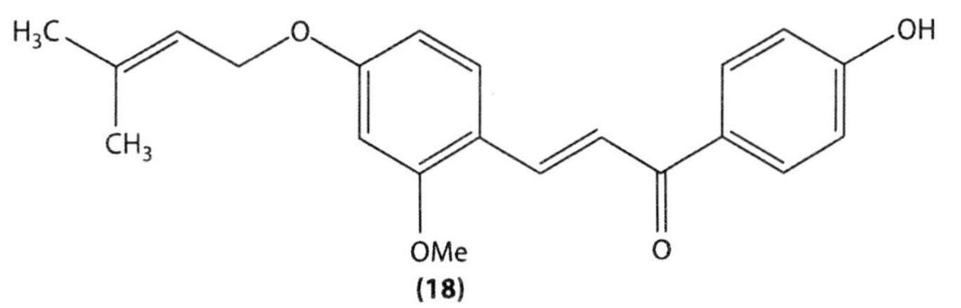

(18)

FIGURE 10.16 Structure of xinjiachalcone A (**18**).

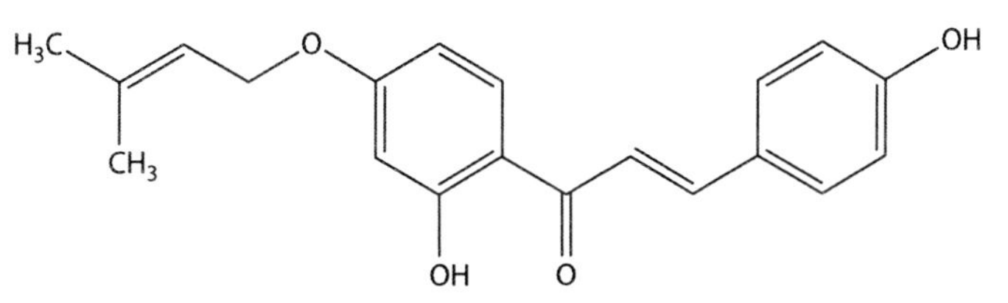

(19)

FIGURE 10.17 Structure of 4-hydroxycordoin (**19**).

of 1.0 and 5.0 μg/mL prevented the adhesion of *P. gingivalis* to the oral mucosa by 31% and 63%, respectively. In the same context, this chalcone was seen to markedly decrease the production of cytokines like IL-1β, TNF-α, and IL-6; and chemokines like IL-8 and CCL5 as well as of PGE_2 by LPS-stimulated macrophages. Finally, the same research group found that 4-hydroxycordoin exerted significant inhibitory effects on two main virulence factors of the fungus *Candida albicans*: biofilm formation (>85% at a concentration of 20 μg/mL) and yeast-hyphal transition (50–200 μg/mL).

10.3.3 Cinnamic Acids

Boropinic acid (**20**) is an isopentenyloxy cinnamic acid isolated in 2000 by Ito and coworkers from the leaves of the Australian plant *Boronia pinnata* Sm. (Rutaceae) (Figure 10.18) (Epifano et al. 2007).

Deepening the knowledge about inhibitory properties of this ferulic acid derivative against *Helicobacter pylori*, Touati and coworkers found that boropinic acid was active also *in vivo*, reducing the gastric mucosa colonization by *H. pylori* strain SS1 using the murine model of infection of C57BL/6 mice (Touati et al. 2009). The same research group found that boropinic acid exerted a topical anti-inflammatory effect, using the Croton oil–induced ear edema in mice as a model of acute inflammation, equal to the known LOX inhibitor nordihydroguaiaretic acid and about half of that of the NSAID indomethacin (Epifano et al. 2011). The structurally related isopentenyloxy-*p*-coumaric acid (**21**) (Figure 10.19), extracted from *Esenbeckia hieronimi* (Rutaceae), showed *in vivo* neuroprotective time and dose-dependent effects using the mouse maximal electroshock-induced seizures as a model of epilepsy (Genovese et al. 2009).

10.3.4 Coumarins

Oxyprenylated coumarins are among the most abundant prenyloxyphenylpropanoids occurring in nature. During the bioassay guided fractionation of the ethyl acetate extract of the leaves of *Melicope vitiflora* (F. Muell) T.G. Hartley (Rutaceae), O'Donnell and coworkers isolated 7-isopentenyloxycoumarin (**22a**) and two structurally related products deriving from the oxidation of one of the methyl group of the *O*-side chain, namely, 7-(3′-carboxybutoxy)coumarin (**22b**) and 7-(3′-carboxybutenoxy) coumarin (**22c**) (Figure 10.20) (O'Donnell et al. 2009).

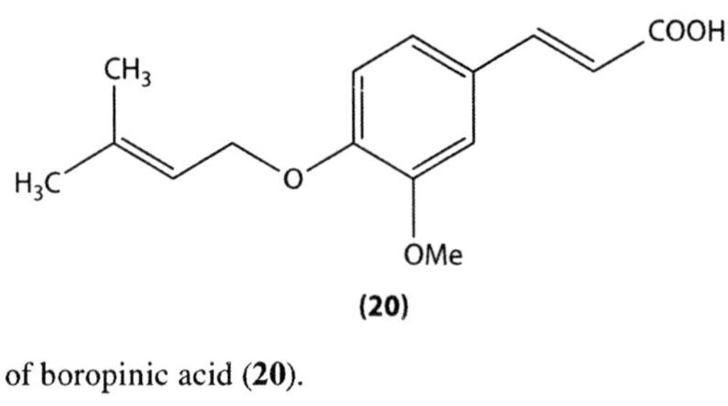

(20)

FIGURE 10.18 Structure of boropinic acid (**20**).

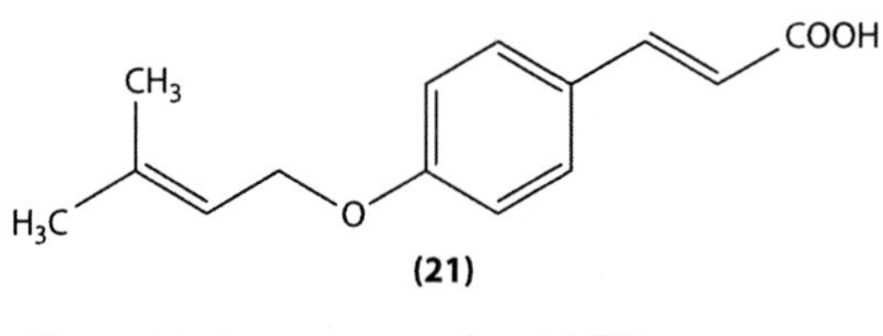

(21)

FIGURE 10.19 Structure of isopentenyloxy-*p*-coumaric acid (**21**).

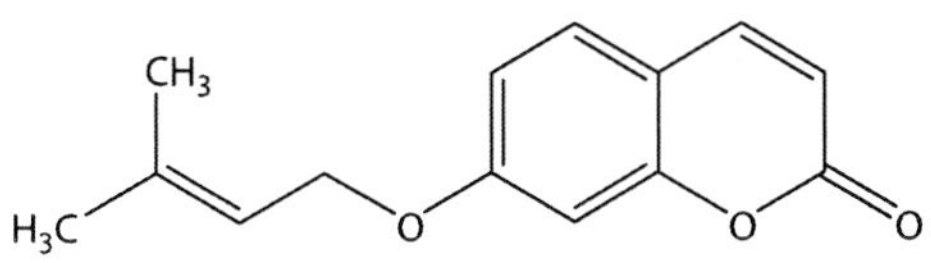

(22a)

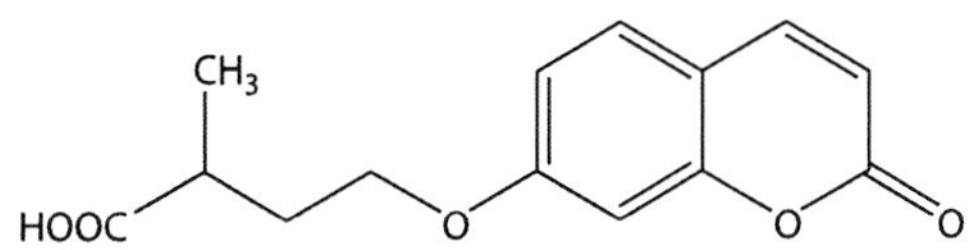

(22b)

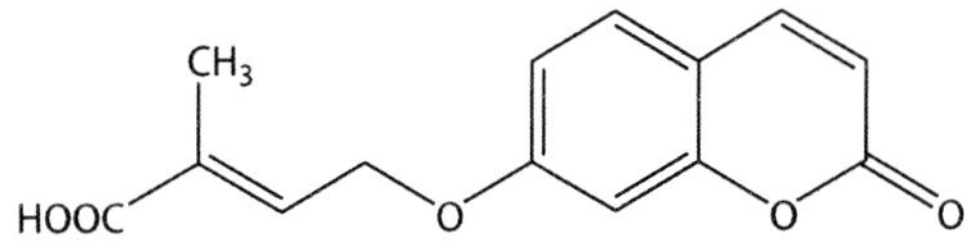

(22c)

FIGURE 10.20 Structure of 7-isopentenyloxycoumarin (**22a**), 7-(3′-carboxybutoxy)coumarin (**22b**), and 7-(3′-carboxybutenoxy)coumarin (**22c**).

In 2010, Razavi and coworkers isolated and structurally characterized from the *n*-hexane extract of the seeds of the Caucasian plant *Zosima absinthifolia* (Vent.) Link. (Apiaceae) imperatorin (**23**; Figure 10.21) and 7-isopentenyloxycoumarin (**22a**) (Razavi et al. 2010).

When tested as *in vitro* antifungal agents against the phytopathogenic fungus *Sclerotinia sclerotiorum*, both compounds exhibited a good level of activity, the effect of imperatorin being more pronounced. In fact, this latter product at the dose of 1 mg/mL was able to completely inhibit the growth of mycelia, while the activity of 7-isopentenyloxycoumarin (**22a**) was by far less (about 25% compared to imperatorin **23**). On the other hand, when tested as herbicides against lettuce seeds growth and development, 7-isopentenyloxycoumarin performed better than imperatorin being able to completely prevent the germination, root, and shoot growth. Studying the biotransformation of imperatorin to isoimperatorin (**24**; Figure 10.22) by the phytopathogenic fungus *Glomerella cingulata*, Marumoto and Miyazawa found that for the latter coumarin a cleavage of the lactone ring and a reduction of the conjugated double bond, yielding 6,7-furano-5-isopentenyloxydihydrocoumaric acid (**25**; Figure 10.23) occurred (Marumoto and Miyazawa 2010).

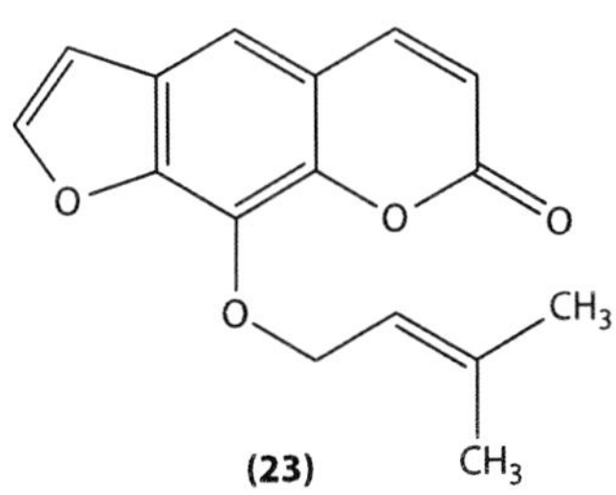

(23)

FIGURE 10.21 Structure of imperatorin (**23**).

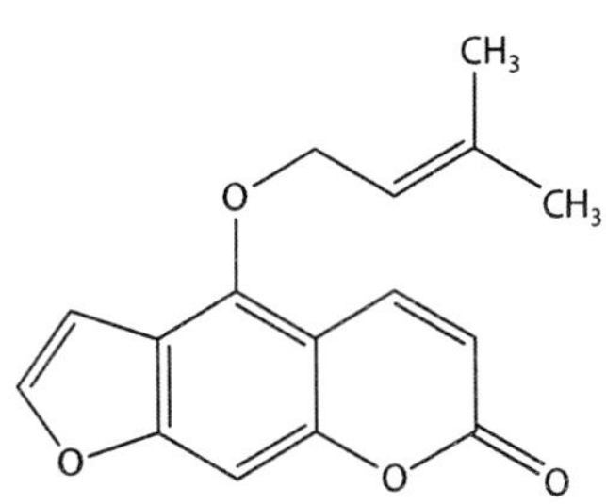

(24)

FIGURE 10.22 Structure of isoimperatorin (**24**).

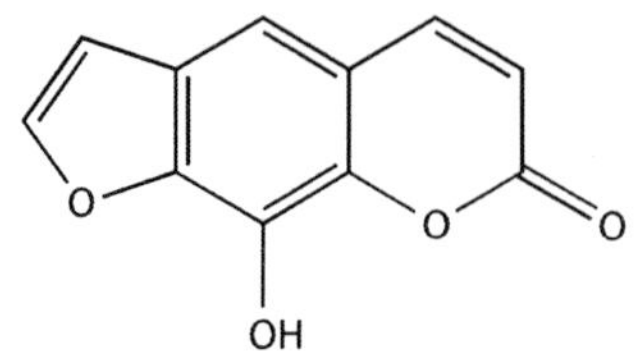

(25)

FIGURE 10.23 Structure of 6,7-furano-5-isopentenyloxydihydrocoumarin acid (**25**).

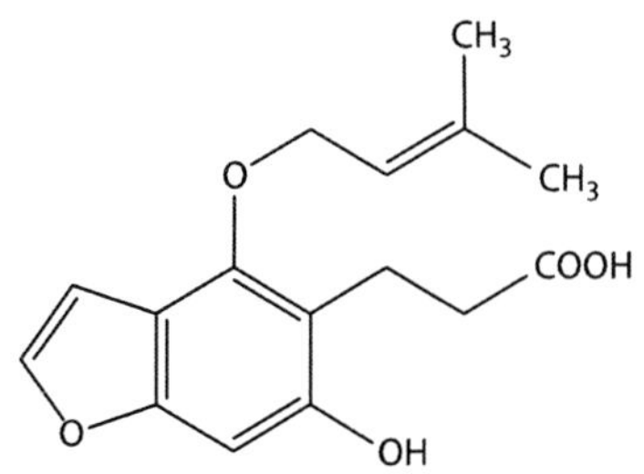

(26)

FIGURE 10.24 Structure of xanthotoxol (**26**).

The isomeric imperatorin was on the other hand dealkylated in quantitative yield to xanthotoxol (**26**; Figure 10.24).

Both products were investigated for their ability to inhibit *in vitro* the β-secretase (BACE1) resulting however in a very low activity for both coumarins. The first example of a double isopentenyloxy coumarin was recently reported by Fukuda et al. (2011). The authors isolated marianin A (**27**; Figure 10.25) from the culture extract of the fungus *Mariannaea camptospora* strain TAMA 118.

Marianin A (**27**) showed only a marginal antimicrobial activity against *Micrococcus luteus* (MIC = 15 μg/mL) and no activity on *Escherichia coli* and *Candida albicans*. The same pattern was recorded when marianin A was tested *in vitro* as growth inhibitory agent of cancer cell lines being the IC_{50} values of 34.0 and 39.0 μM on HeLa and MCF7 cells, respectively. Finally 7-isopentenyloxycoumarin was seen to exert both *in vitro* and *in vivo* remarkable neuroprotective effects. In the first case, this coumarin protected (50%) neuronal cells from cell death induced by glutamate (Epifano et al. 2008) while *in vivo* 7-isopentenyloxycoumarin showed a significant protection in animals of epileptic seizures induced by electroshock at different times (15–120 min) (Genovese et al. 2009).

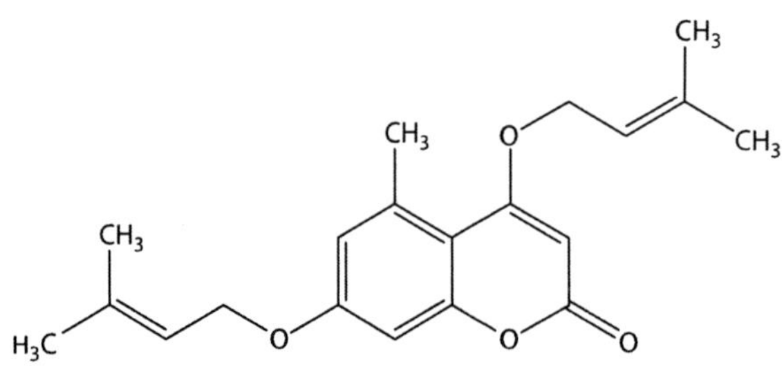

(27)

FIGURE 10.25 Structure of marianin A (**27**).

10.3.5 Flavonoids

Few novel isopentenyloxyflavonoids were discovered in the last 5 years. The methanol extract of the leaves of *Melicope triphylla* Merr. (Rutaceae), commonly known as Awadan, a shrub from south eastern Asia, afforded three flavones, namely, 3,5-dihydroxy-7-isopentenyloxy-8-methoxy-3′,4′-mehtylenedioxyflavone (**28a**), 5-hydroxy-3-isopentenyloxy-7-methoxy-3′,4′-mehtylenedioxyflavone (**28b**), and 5-hydroxy-7-isopentenyloxy-3,8-dimethoxy-3′,4′-mehtylenedioxyflavone (**28c**) (Figure 10.26) (Higa et al. 2010).

M. brandisiana Kurz (Leguminosae) afforded two novel isoflavones, namely, 4′-isopentenyloxy-5,7,2′,5′-tetramethoxyisoflavone (**29a**) and 7,4-diisopentenyloxygenistein (**29b**) (Figure 10.27) (Pancharoen et al. 2008).

Compounds **29a** and **29b** were isolated by Pancharoen and coworkers from the hexane extract of the flowers of the aforementioned plant. Another isoflavone 7-isopentenyloxy-5-hydroxy-4′-methoxyisoflavone (**30**; Figure 10.28) was isolated from the roots of *Dalbergia sissoo* Roxb. (Reddy et al. 2008).

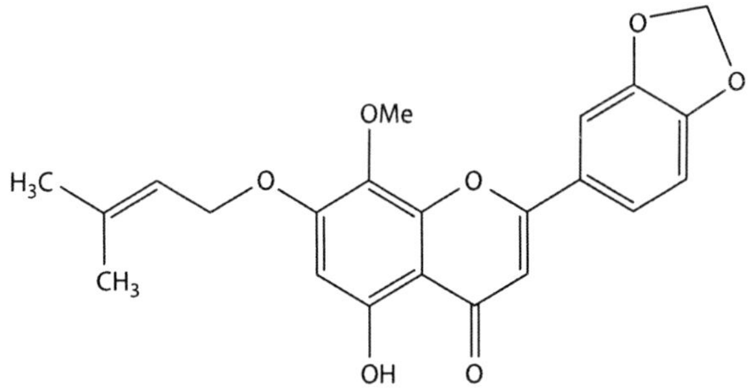

(28a)

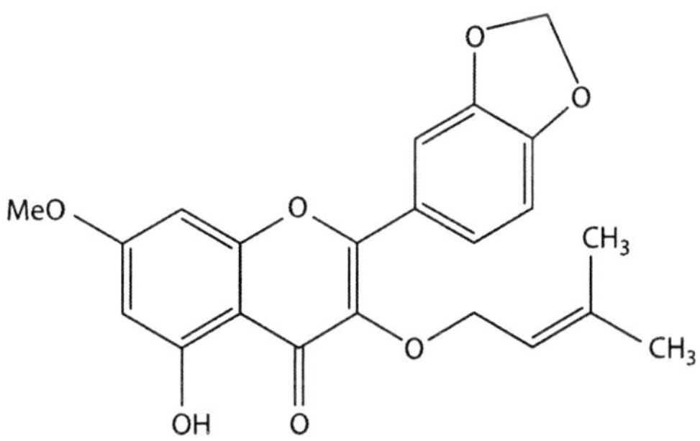

(28b)

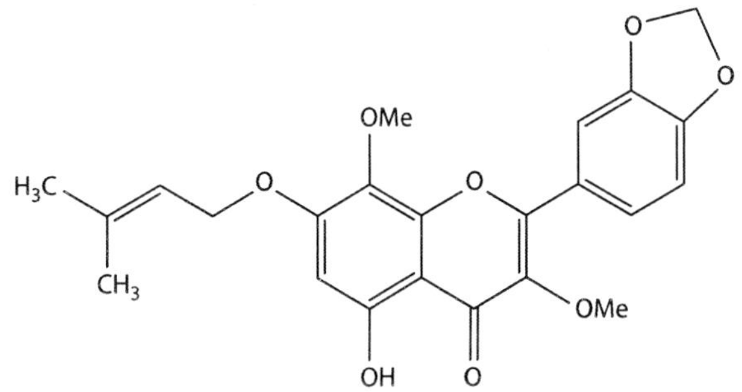

(28c)

FIGURE 10.26 Structure of 3,5-dihydroxy-7-isopentenyloxy-8-methoxy-3′,4′-mehtylenedioxyflavone (**28a**), 5-hydroxy-3-isopentenyloxy-7-methoxy-3′,4′-mehtylenedioxyflavone (**28b**), and 5-hydroxy-7-isopentenyloxy-3,8-dimethoxy-3′,4′-mehtylenedioxyflavone (**28c**).

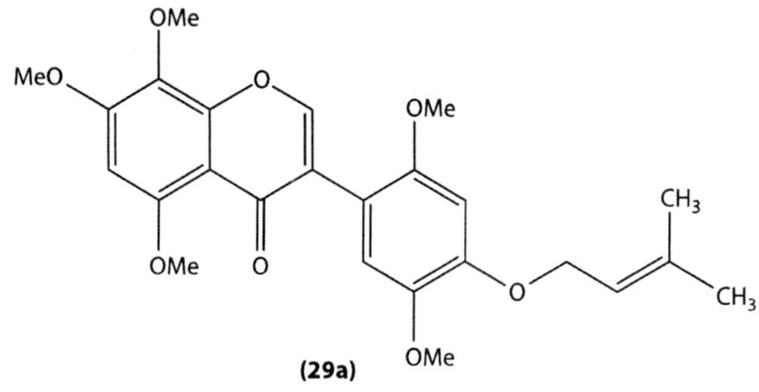

(29a)

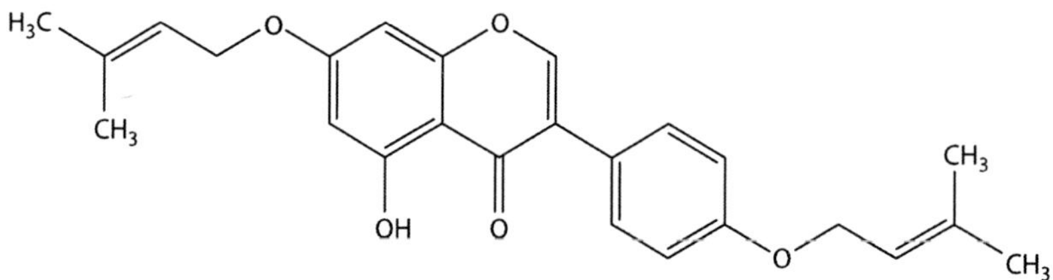

(29b)

FIGURE 10.27 Structure of 4′-isopentenyloxy-5,7,2′,5′-tetramethoxyisoflavone (**29a**) and 7,4-diisopentenyloxygenistein (**29b**).

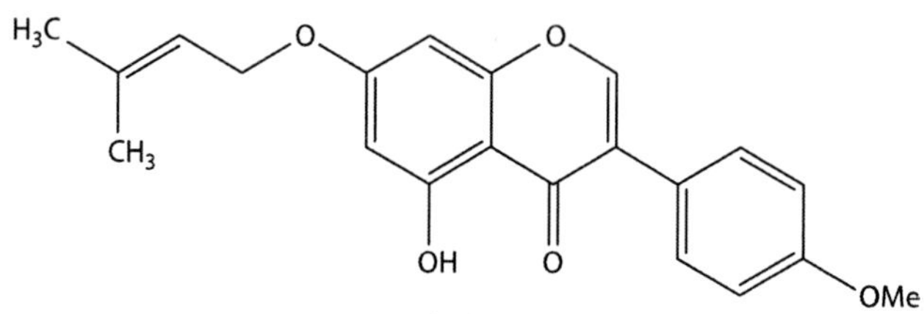

(**30**)

FIGURE 10.28 Structure of 7-isopentenyloxy-5-hydroxy-4′-methoxyisoflavone (**30**).

10.3.6 Phtalides

Novel skeletons structurally related to isopentenyloxyphenylpropanoids were discovered during the last 5 years. This is the case of phtalides; for example, 6-isopentenyloxy-4-methoxy-5-methylphtalide (**31**; Figure 10.29) was extracted and characterized by Demuner and coworkers in 2006 from the phytopathogenic fungus *Nimbya alternantherae* (Demuner et al. 2006). The authors revealed that the compound (**31**) acted as an herbicide being an inhibitor of the photosynthetic process in spinach thylakoids. In particular, it was able to uncouple ATP production.

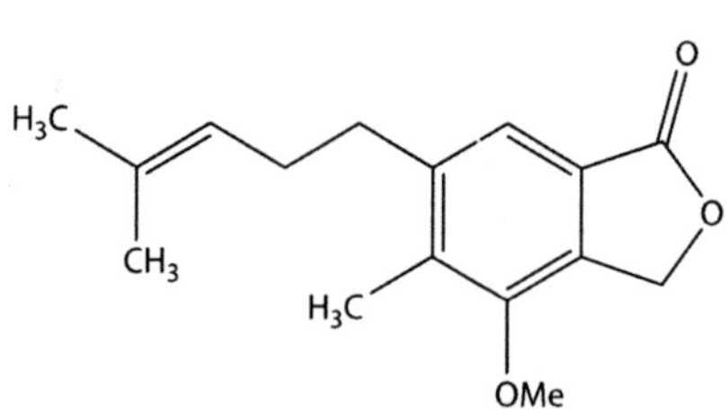

(31)

FIGURE 10.29 Structure of 6-isopentenyloxy-4-methoxy-5-methylphtalide (**31**).

10.4 PERSPECTIVES AND CONCLUSIONS

Only in the last 15 years naturally occurring secondary metabolites containing an *O*-prenyl side chain have been recognized as interesting and valuable biologically active phytochemicals. For these reasons, research on these secondary metabolites is a field of current and growing interest. Many of the previously (Epifano et al. 2007) and herein described oxyprenylated derivatives have been found in plants belonging to the family of Rutaceae and to a lesser extent in the families of Apiaceae, Guttiferae, Leguminosae, and few others. A peculiar feature of prenyloxy natural products is the low concentration at which in many cases they can be extracted and isolated from plant sources. This may be the main reason why these classes of natural compounds have not been fully considered about their pharmacological properties. Recently, the development of new high yielding procedures made possible the synthesis of some of these compounds in quantities more than sufficient to carry out more detailed studies on their pharmacological properties. Results of these investigations suggest that these secondary metabolites may represent in the next future a new frontier and a challenge for the development of novel anticancer, anti-inflammatory, neuroprotective, and antimicrobial compounds. On the basis of already reported (Epifano et al., 2007) and herein cited data about the synthesis and pharmacology of oxyprenylated derivatives, it is hopeful that in the next future more studies could be performed aimed at the search for prenyloxy phytochemicals from novel natural sources, to develop new environment friendly, cheap, and high yielding synthetic routes to obtain these compounds in large amounts and finally to get further insights and to depict in more detail their biological profile and mechanism of action.

ABBREVIATIONS

ACE	angiotensin converting enzyme
AOM	azoxymethane
ATP	adenosine triphosphate
BACE	β-secretase
CCL	chemokine (C–C motif) ligand
β-CD	β-cyclodextrin
CoA	coenzyme A
COX	cyclooxygenase
DSS	dextrane sodium sulfate
ED	effective dose
ER	estrogen receptor
FTase	farnesyl transferase
GAP	3-(4′-geranyloxy-3′-methoxyphenyl)-L-alanyl-L-proline
GGTase	geranylgeranyl transferase
GOFA	4′-geranyloxyferulic acid
HO-1	heme oxygenase 1
IC	inhibitory concentration
IL	interleukin
LOX	lipoxygenase
LPS	lipopolysaccharide
MES	maximal electroshock-induced seizure
MIC	minimum inhibitory concentration
NF-kB	nuclear factor-kappa B
NOS	nitric oxide synthase
Nrf2	NF-E2-related factor 2
NSAID	non steroidal anti-inflammatory drug
8-OHdG	8-hydroxy-2′-deoxyguanosine

PG prostaglandin
PPAR peroxisome proliferator-activated receptor
RNA ribonucleic acid
TNF-α tumor necrosis factor-α

REFERENCES

Andersen A., Smitt U.W., Simonsen H.T. et al. 2007. Prenylated acetophenones from *Melicope obscura* and *Melicope. obtusifolia* ssp. *obtusifolia* var. *arborea* and their distribution in Rutaceae. *Biochem. Syst. Ecol.* 35: 447–453.

Bodet C., Epifano F., Genovese S. et al. 2008. Effects of 3-(4′-geranyloxy-3′-methoxyphenyl)-2-*trans* propenoic acid and its ester derivatives on biofilm formation by two oral pathogens, *Porphyromonas gingivalis* and *Streptococcus mutans*. *Eur. J. Med. Chem.* 43: 1612–1620.

Bonnak N., Karalai C., Chantrapromma S. et al. 2009. Anti-*Pseudomonas aeruginosa* xanthones from the resin and green fruits of *Cratoxylum cochinchinense*. *Tetrahedron* 65: 3003–3013.

Boyd R.D., Sharma N.D., Loke P.L. et al. 2007. Synthesis, structure and stereochemistry of quinoline alkaloids from *Choisya ternata*. *Org. Biol. Chem.* 5: 2983–2991.

Cao S., Al-Rehaily A.J., Brodie P. et al. 2008. Furoquinoline alkaloids from *Ertela* (*Monnieria*) *trifolia* (L.) Kuntze from the Suriname rainforest. *Phytochemistry* 69: 553–557.

Costa R., Dugo P., Navarra M. et al. 2010. Study on the chimica composition variability of some processed bergamot (*Citrus bergamia*) essential oils. *Flav. Fragr. J.* 25: 4–12.

Curini M., Epifano F. and S. Genovese 2005. Synthesis of a novel prodrug of 3-(4′-geranyloxy-3′-methoxyphenyl)-2-trans-propenoic acid for colon delivery. *Bioorg. Med. Chem. Lett.* 15: 5049–5052.

Curini M., Epifano F., Genovese S. et al. 2006. 3-(4′-Geranyloxy-3′-methoxyphenyl)-2-*trans* propenoic acid: a novel promising cancer chemopreventive agent. *Anticancer Agents Med Chem.* 6: 571–577.

De Medina P., Genovese S., Paillasse M.R. et al. 2010. Auraptene is an inhibitor of cholesterol esterification and a modulator of estrogen receptors. *Mol. Pharmacol.* 78: 827–836.

Demuner A.J., Barboza L.C.A., Veiga T.A.M. et al. 2006. Phytotoxic constituents from *Nimbya alternantherae*. *Biochem. Syst. Ecol.* 34: 790–795.

Dugo P., Piperno A., Romeo R. et al. 2009. Determination of oxygen heterocyclic components in citrus products by HPLC with UV detection. *J. Agric. Food Chem.* 57: 8543–8551.

Epifano F., Curini M., Genovese S. et al. 2007a. Prenyloxyphenylpropanoids as novel lead compounds for the selective inhibition of geranylgeranyl transferase I. *Bioorg. Med. Chem. Lett.* 17: 2639–2642.

Epifano F., Genovese S., Menghini L. et al. 2007b. Chemistry and pharmacology of oxyprenylated secondary plant metabolites. *Phytochemistry* 68: 939–953.

Epifano F., Genovese S., Sosa S. et al. 2007c. Synthesis and anti-inflammatory activity of 3-(4′-geranyloxy-3′-methoxyphenyl)-2-*trans* propenoic acid and its ester derivatives. *Bioorg. Med. Chem. Lett.* 17: 5709–5714.

Epifano F., Molinaro G., Genovese S. et al. 2008. Neuroprotective effect of prenyloxycoumarins from edible vegetables. *Neurosci. Lett.* 443: 57–60.

Epifano F., Sosa S., Tubaro A. et al. 2011. Topical antiinflammatory activity of boropinic acid and its natural and semisynthetic derivatives. *Bioorg. Med. Chem. Lett.* 21: 769–772.

Feldman M., Tanabe S., Epifano F. et al. 2011. Antibacterial and antiinflammatory activities of 4-hydroxycordoin potential therapeutic benefits. *J. Nat. Prod.* 74: 26–31.

Fukuda T., Sudoh Y., Tsuchiya Y. et al. 2011. Marianins A and B, prenylated phenylpropanoids from *Mariannea camptospora*. *J. Nat. Prod.* 74: 1327–1330.

Genovese S., Epifano F., Curini M. et al. 2009. Prenyloxyphenylpropanoids as a novel class of anticonvulsive agents. *Bioorg. Med. Chem. Lett.* 19: 5419–5422.

Genovese S., Foreman J.E., Borland M.G. et al. 2010. A natural propenoic acid derivative activates peroxisome proliferator–activated receptor-β/δ (PPAR β/δ). *Life Sci.* 86: 493–498.

Genovese S. and Epifano F. 2011. Auraptene: A natural biologically active compound with multiple targets. *Curr. Drug Targets* 12: 381–386.

Hayakawa Y., Yamazak Y., Kukita M. et al. 2010. Flaviogeranin, a new neuroprotective compound from *Streptomyces* spp. *J. Antibiot.* 63: 379–380.

Henry G.E., Campbell M.S., Zelinsky A.A. et al. 2009. Bioactive acylphloroglucinol from *Hyperichum densiflorum*. *Phytotherapy Res.* 23: 1759–1762.

Higa M., Nakadomari E., Imamura M. et al. 2010. Isolation of four new flavonoids from *Melicope triphylla*. *Chem. Pharm. Bull.* 58: 1339–1342.

Iwasaki N., Baba M., Aishan H. et al. 2009. Studies of traditional folk medicines in Xinjang Uighur autonomous II. Research for chemical constituents of Xinjang licorice. *Heterocycles* 78: 1581–1587.

Marumoto S. and M. Miyazawa 2010. Biotransformation of isoimperatorin and imperatorin by *Glomerella cingulata* and β-secretase inhibitory activity. *Bioorg. Med. Chem.* 18: 455–459.

Mbaveng A.T., Kuete V., Nguemeving J.R. et al. 2008. Antimicrobial activity of the extracts and compounds obtained from *Vismia guineensis* (Guttiferae). *Asian J. Trad.* 3: 211–223.

Messier C., Epifano F., Genovese S. et al. 2011. Inhibition of *Candida albicans* biofilm formation and yeast-hyphal transitino by 4-hydroxycordoin. *Phytomedicine* 18: 380–383.

Miyamoto S., Epifano F., Curini M. et al. 2008. A novel prodrug of 4′-Geranyloxy-Ferulic Acid suppressess colitis-related colon carcinogenesis in mice. *Nutr. Cancer* 60: 675–684.

O'Donnell F., Ramachandran V.N., Smyth T.J.P. et al. 2009. An investigation of bioactive phytochemicals in the leaves of *Melicope vitiflora* by electrospray ionisation trap mass spectrometry. *Anal. Chim. Acta* 634: 115–120.

Pancharoen O., Athipornchai A., Panthong A. et al. 2008. Isoflavones and rotenoids from the leaves of *Millettia brandisiana. Chem. Pharm. Bull.* 56: 835–838.

Prager R.H. and H.M. Thregold 1966. Some neutral constituents of *Acronychia baueri. Aust. J. Chem.* 5: 451–453.

Razavi S.M., Imanzadeh G. and M. Davari 2010. Coumarins from *Zosima absinthifolia* seeds, with allelopatic effects. *EurAsia J. Biosci.* 4: 17–22.

Reddy R.V.N., Reddy N.P., Khalivulla S.I. et al. 2008. O-prenylated flavonoids from *Dalbergia sissoo. Phytochem. Lett.* 1: 23–26.

Sim W.C., Ee G.C.L., Lim C.J. et al. 2011. *Cratoxylum glaucum* and *Cratoxylum arborescens* (Guttiferae) two potential sources of antioxidant agents. *Asian J. Chem.* 23: 569–572.

Tanaka T., de Azevedo M.B.M., Duran N. et al. 2010. Colorectal cancer chemoprevention by 2-β-cyclodextrin inclusion compounds of auraptene and 4′-geranyloxyferulic acid. *Int. J. Cancer* 126: 830–840.

Touati E., Michel V., Correia M. et al. 2009. Boropinic acid a novel inhibito of *Helicobacter pylori* stomach colonization. *J. Antim. Chemoth.* 64: 210–211.

Tsaffack M., Nguemeving J.R., Kuete V. et al. 2009. Two new antimicrobial dimeric compounds: febriquinone, a vismione-anthraquinone coupled pigment and adamabianthrone, from two *Psorospermum* species. *Chem. Pharm. Bull.* 57: 1113–1118.

Varamini P., Javidnia K., Soltani M. et al. 2008. Cytotoxic activity and cell cycle analysis for quinoline alkaloids from *Haplophyllum canaliculatum* Boiss. *Planta Med.* 75: 1509–1516.

Waffo A.F.K., Coombes P.H., Crouch N.R. et al. 2007. Acridone and furoquinoline alkaloids from *Tecla gerrardi* (Rutaceae Toddalioideae) of Southern Africa. *Phytochemistry* 68: 663–667.

Xia C., Narasimhulu M., Li X. et al. 2010. New synthetic routes to biologically interesting geranylated acetophenones from *Melicope semecarpifolia* and their unnatural prenylated and farnesylated derivatives. *Bull. Kor. Chem. Soc.* 31: 664–669.

11 Role of Curcumin in Ameliorating Neuroinflammation and Neurodegeneration Associated with Alzheimer's Disease

Sumit Sarkar, Balmiki Ray, Jay Sharma,
Larry Schmued, and Debomoy K. Lahiri

CONTENTS

11.1 INTRODUCTION

Alzheimer's disease (AD) is the most common form of dementia in elderly people and the fifth leading cause of death for people who are 65 years or older (Alz. Assoc. Facts and Figures 2011). Neuropathologically, depositions of amyloid beta (Aβ) plaques in the brain interstitial and phosphorylation of microtubule-associated protein tau (MAPT) within axons are the hallmarks of AD (Hardy and Selkoe 2002). Aβ peptide, which is the proteolytic cleaved product of the transmembrane amyloid precursor protein (*APP*), is released by enzymatic cleavage by several secretase enzymes. *APP* is first cleaved by β-secretase (or *BACE-1*) to produce sAPPβ and a 99 amino acid fragment, which is further cleaved by γ-secretase to produce Aβ peptides (39–44 amino acids residue) (Sambamurti et al. 2002). Alternatively, *APP* can also be cleaved by another enzyme, α-secretase, to produce sAPPα and a 83 amino acid residue fragment (C83), which is further cleaved by γ-secretase to produce P3 fragment and precludes Aβ production (Lahiri et al. 2003; Marlow et al. 2003). Decreasing the levels of

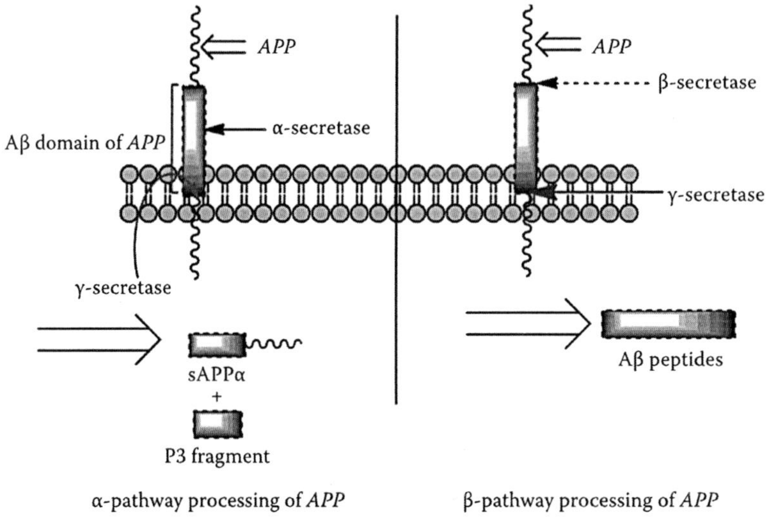

FIGURE 11.1 Schematic showing *APP* processing pathways by different secretase enzymes. In the left of the figure, *APP* is sequentially cleaved by α-secretase and γ-secretase enzymes to produce sAPPα and P3 fragment. Please note, α-secretase cleaves *APP* within its Aβ domain and hence, precludes production of Aβ peptides. In normal cells, *APP* is cleaved mainly by α-secretase pathway. In contrast to α-secretase pathway, *APP* can be cleaved by β-secretase enzyme and γ-secretase complex to produce sAPPβ, carboxyl-truncated fragments (not shown in the figure), and Aβ peptides (right side of the figure). *APP* cleavage by β-secretase is considered as a "minor" pathway except for some genetic conditions. *APP* pathway can be a target of several drugs and agents, including curcumin, the focus of this review.

APP and the activities of *BACE-1* and γ-secretase has already been identified as potential therapeutic strategies for the treatment of AD (Lahiri et al. 2007a; Imbimbo and Giardina 2011; Vassar and Kandalepas 2011). The schematic diagram in Figure 11.1 shows the major *APP* processing pathways.

It has been postulated that deposited Aβ peptides initiate inflammatory responses in the brain. Deposited Aβ can activate microglia, and the interaction between the latter and Aβ can produce reactive oxygen species and several cytokines, leading to neuronal damage (Ray and Lahiri 2009). Once neuroinflammation is set, several events can take place, including the activation of the proinflammatory transcription factor, nuclear factor kappa-light-chain-enhancer of activated B cell (NFκB). In the resting condition, NFκB stays within the cytoplasm of the cell binding with an inhibitor protein IκB. Once activated, NFκB is detached from the inhibitor molecule, enters the nucleus, and binds to and activates several proinflammatory genes. Further, NFκB can activate several genes directly related to the pathology of AD such as *APP*, *presenilin-1*, and *BACE-1*, leading to more Aβ production (for review see Ray and Lahiri 2009). Hence, apart from *APP*, *BACE-1*, and tau, regulation of NFκB activation is also being considered a rational strategy for therapeutic intervention in AD.

In addition to therapeutic interventions, preventive strategies for AD have also become a topic of research interest in recent years (Lahiri 2006; Camins et al. 2010; Frisardi et al. 2010). Epidemiological reports suggest that the elderly Indian populace have ~4.4-fold less incidence of AD when compared to a reference American populace (Chandra et al. 2001), and risk factors for AD (such as presence of APOE ε4 allele) also differ geographically (Murrell et al. 2006). Further, prevalence of AD in some Mediterranean countries was reported to be smaller than other European countries (Benedetti et al. 2002). Taken together, these facts suggest a strong environmental component for the development of dementias like AD (Lahiri et al. 2007b). Although nutritional components can be correlated with the etiology of AD, an extensive study with specific nutritional components, and how those can prevent AD, is mostly lacking.

Our laboratory is working to identify potential nutritional components to prevent and/or delay the onset/progression of AD (Ray et al. 2011b,c). In this chapter, we will discuss the potential preventive and/or curative roles of one of the plant-derived polyphenols, curcumin (diferuloylmethane), in AD. Curcumin exerts pleiotropic effects and has been reported to be effective in various disorders including cancer and neurodegenerative diseases (Aggarwal et al. 2006; Begum et al. 2008). By suppressing the activation of NFκB, defibrilling Aβ plaques, preserving neurons, upregulating neurotrophic factors, and facilitating neurogenesis, curcumin can emerge as a potential therapeutic/preventive agent in the treatment of several neurodegenerative disorders including AD.

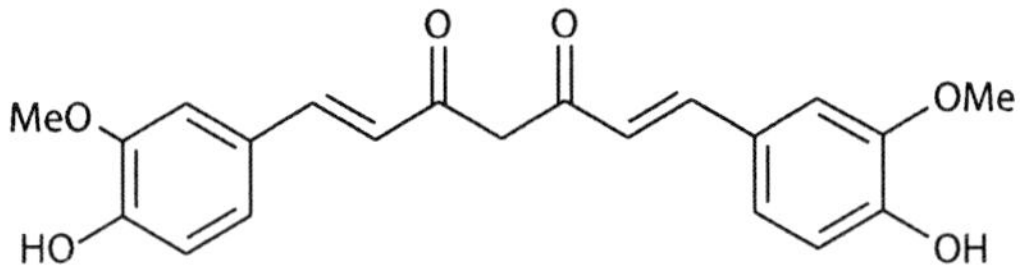

(Chemical structure of curcumin)

11.2 ROLE OF CURCUMIN IN NEURODEGENERATIVE DISEASES

Curcumin has been shown to exhibit beneficial activity against various neurological diseases, including AD (Lim et al. 2001), multiple sclerosis (Natarajan and Bright 2002), Parkinson's disease (PD) (Zbarsky et al. 2005), epilepsy (Sumanont et al. 2006), cerebral injury (Ghoneim et al. 2002), schizophrenia (Bishnoi et al. 2008), spongiform encephalopathy (Creutzfeldt–Jakob disease) (Hafner-Bratkovic et al. 2008), neuropathic pain (Sharma et al. 2006), and depression (Xu et al. 2005).

Before going into detail about the role of curcumin in modifying disease pathology as seen in AD, it is imperative to describe, in brief, the role of inflammation in AD pathology.

11.2.1 Inflammation and AD

The brain needs a constant supply of oxygen, as it consumes 20% of the body's oxygen despite having only 2% of the total body weight. With normal aging, the brain spontaneously accumulates several metal ions such as iron (Fe), zinc (Zn), and copper (Cu). However, the brain contains a rich amount of antioxidants that control and prevent the harmful reactive oxygen species (ROS) generated via Fenton chemistry that involves redox-active metal-ion reduction and activation of molecular oxygen (Smith et al. 1998).

Neuroinflammation plays a major role in the pathogenesis of many neurodegenerative diseases including AD. Although Aβ has been considered a key player in inducing AD pathogenesis (Walsh et al. 2002; Walsh and Selkoe 2004), it is not clear whether Aβ plaques and neurofibrillary tangles (NFT) are causative for AD. These doubts are fueled by a recent finding that the Aβ plaque burden poorly correlates with the progression and severity of dementia in AD. Moreover, transgenic animals that develop widespread Aβ plaque deposition in response to overexpression of *APP* mutations show only slight cognitive deficits (Braak and Braak 1998; Davis and Laroche 2003). It has also been shown that formation of NFT may more closely correlate with the decline in cognitive skills, but seem to occur as a late event subsequent to Aβ accumulation. However, some studies suggest that protofibrils and oligomers of Aβ 1–40 and Aβ 1–42, rather than the aggregated Aβ plaques, contribute to early dendritic and synaptic injury and thus contribute to neuronal dysfunction (Walsh et al. 2002).

While minor signs of neuroinflammation can be found in the normal aging brain, the AD brain faces a much stronger activation of inflammatory systems, indicating an increasing amount of immunostimulation present. A significant body of evidence suggests that Aβ peptides play a pivotal role as inducers of neuroinflammation.

Aβ itself has been shown to induce a local inflammatory type response, and fibrillar Aβ can bind to the complement factor C1 and hence potentially activate the classical complement pathway in an antibody-independent fashion (Rogers et al. 1992). Such activated complement factors could play an important role in the local recruitment and activation of microglial cells expressing the complement receptors CR3 and CR4 (Rozemuller et al. 1989). *In vitro* studies are also consistent with immunohistochemical data in AD brains, showing a weak immunostaining for early complement components in diffuse plaques that are composed of non- or low-grade fibrillar Aβ (Eikelenboom and Veerhuis 1996). Several studies indicate that extracellular deposition of Aβ in AD brains is one of the main triggers of inflammation. For example, Aβ activates microglia by binding to the receptor for advanced glycation end products (RAGE) (Yan et al. 1995) and to other scavenger receptors (Paresce et al. 1996). Moreover, the LPS receptor, CD14 also interacts with fibrillar Aβ (Fassbender et al. 2004), and microglia can kill Aβ 1–42 damaged neurons by a CD14-dependent process (Bate et al. 2004). Thus, the role of CD14 in Aβ-induced microglia activation strongly suggests that innate immunity is linked with AD pathology.

11.2.2 Role of Activated Glia

Microglia represent the brain's immune system and are known as a first-line defense when challenged by bacterial, viral, or fungal infection. Although these functions are important and beneficial, it is now clear that microglial activation may also be evoked by endogenous proteins and can significantly contribute to neuronal damage. Activated microglia upregulate the expression of a variety of surface proteins, including the major histocompatibility complex and complement receptors (Liu and Hong 2003). Once immunostimulated in response to neurodegenerative events, these microglial cells release an array of proinflammatory mediators including cytokines, ROS, complement factors, neurotoxic secretory products, free radical species, and nitric oxide (NO), all of which can contribute to neuronal dysfunction and cell death (Griffin et al. 1998). Several amyloid peptide species and *APP* can act as potent glial activators (Dickson et al. 1993; Barger and Harmon 1997; Schubert et al. 2000), and disruption of the *APP* gene and its proteolytic products delays and decreases microglial activation (DeGiorgio et al. 2002). It is interesting to note that some microglia activation may be beneficial as activated microglia is able to reduce Aβ accumulation by increasing its phagocytosis clearance and degradation (Frautschy et al. 1998; Qiu et al. 1998; Yan et al. 2003). Additionally, the secreted Aβ species (both the short 1–40 and long Aβ 1–42 peptides) are constitutively degraded by an insulin-degrading enzyme (IDE), a zinc metalloproteinase released by microglia and neural cells.

In addition to microglia, astrocytes also participate in Aβ clearance and degradation, provide trophic support to neurons, and form a protective barrier between Aβ deposits and neurons (Wyss-Coray et al. 2003; Koistinaho et al. 2004). The presence of astrocytes around the Aβ plaques in the AD brain suggests that these lesions generate chemotactic molecules that mediate astrocyte recruitment and hypertrophy. It has been shown that astrocytes throughout the entorhinal cortex of AD patients gradually accumulate Aβ 1–42 positive material, and the amount of this material correlates positively within the extent of local AD pathology. Aβ 1–42 within these astrocytes could be of neural origin and possibly accumulated by phagocytosis of locally degenerated dendrites and synapses (Nagele et al. 2003). Recent evidence also suggests that astroglial cells are able to phagocytize Aβ peptides, a process that may depend on their apolipoprotein E (ApoE) status. This work suggests that ApoE polymorphisms may influence the risk of developing AD by affecting astroglial Aβ phagocytosis (Niino et al. 2001).

11.2.3 Role of Nuclear Factor Kappa Beta

TNF-alpha has been found to be a major mediator of inflammation in most of the aforementioned diseases and its effect is regulated by the activation of transcription factor, NFκB. Under normal stable conditions, this transcription factor stays inactivated by IkB. However, upon activation, it enters the nucleus and increases the transcription of various inflammatory mediators. Several molecules have the ability to activate NFκB, which include TNFα, Aβ, and secreted *APP* (Barger and Harmon 1997; Guo et al. 1998). Gene mapping studies show that NFκB sites are present in the promoter regulatory region of the *APP*, PS-1, and *BACE-1* gene. Upon activation of NFκB, increased transcription of *APP* and *BACE-1* ensues, which subsequently leads to increased Aβ production. An increased level of NFκB in the brain has been observed in the presence of the APOE ε4 allele, compared with the activation in the presence of the APOE ε3 allele (Ophir et al. 2005). Thus, APOE ε4 might be playing a significant role in activating NFκB, leading to further damage. However, a recent APOE gene promoter study has shown that Aβ can also stimulate APOE through the NFκB-dependent pathway (Du et al. 2005).

11.2.4 Role of Peroxisome Proliferator-Activated Receptor G

Peroxisome proliferator-activated receptor G (PPARγ) is a ligand-dependent nuclear hormone receptor transcription factor and is a regulator of adipocyte differentiation. This factor has been implicated in the pathology of numerous diseases including obesity, diabetes, atherosclerosis, and cancer. Upon activation, PPARγ binds to peroxisome proliferator response element (PPRE) within the promoter regions of targeted genes of inflammatory mediators expressed by T cells, such as TNF alpha, IL-10, IFN-G, and IL-4, and also regulates their expression (Szczucinski and Losy 2007). Recent evidence suggests that PPARγ activation not only suppresses Aβ-mediated induction of microglial cells from producing proinflammatory cytokines, but also inhibits NFκB-mediated pathways by reducing its nuclear translocation (Heneka et al. 2005).

11.3 CURCUMIN IN *APP* PROCESSING

AD is believed to be primarily driven by the excessive production of Aβ, the principal component of senile plaques. Aβ is a 4 kDa peptide generated by a sequential proteolytic cleavage of the type I transmembrane protein, the *APP* (Sambamurti et al. 2002). A significant body of evidence suggests that curcumin, under both *in vivo* and *in vitro* conditions, can bind to amyloid plaques and inhibit Aβ aggregation (Hong et al. 2009), as well as the fibril and oligomer formation (Yang et al. 2005). Recently, Zhang et al. (2010) have described a possible cellular mechanism explaining how curcumin could decrease Aβ deposition and plaque formation. These investigators have shown that curcumin has the ability to decrease Aβ levels by reducing *APP* maturation, or possibly by preventing endocytosis from the plasma membrane. Immature *APP* is *N*-glycosylated in the ER, and a fraction of these molecules exit the ER and undergo *O*-glycosylation in the Golgi complex to become mature *APP* (Zhang et al. 2010). Mature *APP* is then transported onto the plasma membrane after which it can undergo endocytosis via clathrin-coated pits (Small and Gandy 2006). Their data also showed that curcumin can significantly change APPmature/APPimmature ratio and decrease the level of intermediate *APP* induced by Brefeldin A (BFA), an agent that disrupts the Golgi complex. Overall, these findings suggest that curcumin can affect *APP* metabolism at the level of the ER and the cumulative effect would be a significant decrease in both Aβ (1–40) and Aβ (1–42) levels.

11.4 CURCUMIN AND TAUOPATHIES

As stated earlier, the key hallmarks of AD pathology include brain depositions of Aβ-loaded plaques, intracellular NFT formation, and oxidative stress induced by impaired metabolic pathways and certain multivalent metals (Selkoe 1994). Several studies conducted in early 1980s suggest that microtubule-based axonal transport and synaptic function are impaired in AD (Grundke-Iqbal et al. 1986b). Microtubules are stabilized by the binding of the microtubule connection protein, tau (Grundke-Iqbal et al. 1986a). Basically, tau is a highly soluble microtubule-associated protein that plays an important role in the stabilization of axons. Tau, being rich in phosphorylation sites, makes it vulnerable to hyperphosphorylation, and a balance between its phosphorylation and dephosphorylation state denotes the presence of normal physiological condition. Abnormal hyperphosphorylation of tau leads to its impaired biological activity, resistance to degradation, induction of conformational changes, and promotion of paired helical filament (PHF), which is the principal component of NFT (Metcalfe and Figueiredo-Pereira 2010). Recent evidence suggests that although in a healthy brain, only two to three amino acid residues are phosphorylated; the accrual of phosphorylation with nearly nine phosphates per molecule leads to AD and other tauopathies (Medeiros et al. 2011). It has been reported recently that Aβ promotes neurite degeneration and microtubule disintegration by coordination with tau, and signs of degeneration disappeared in the absence of tau, which again underscores the role of tau in neurodegeneration (Metcalfe and Figueiredo-Pereira 2010). The significant increase in tau hyperphosphorylations, in which Aβ might play a key role, has also been reported in postmortem brain tissues obtained from AD patients (Wai et al. 2009). These observations suggest that preventing tau phosphorylation could protect cells against Aβ-induced neurotoxicity. In this context, it is noteworthy that curcumin plays a major role in reducing plaque deposition, proinflammatory cytokines, and pJNK expression in Tg2576 mice that lack tau pathology (Begum et al. 2008). Recent evidence also suggests that curcumin decreases tau hyperphosphorylation in PC12 cells (Park et al. 2008) and provides neuroprotective effect of curcumin against Aβ-induced toxicity. A recent study has shown that curcumin significantly reduced the phosphorylation of tau (Ma et al. 2009), and a combination of fish oil and curcumin treatment has shown a significant inhibition of tau phosphorylation, raising the possibility of therapeutic applications against AD. It is reasoned that docosahexaenoic acid (DHA, present in fish oil) and curcumin target different steps in the Aβ cascade. Further, curcumin has a direct anti-Aβ binding activity that can directly antagonize Aβ aggregation (Begum et al. 2008). Moreover, DHA competitively reduces availability of arachidonic acid substrate in phospholipids, whereas curcumin reduces activity of phospholipases, cyclooxigenases, and lipooxygenases enzymes, which produce proinflammatory lipid mediators. The combination of fish oil and curcumin also significantly prevented cognitive decline in 3xTg-AD mice (Ma et al. 2009). Four human clinical trials are ongoing to explore the potential use of curcumin against AD pathology (Baum et al. 2008). Further studies are warranted to unravel the factors that might be associated with tau hyperphosphorylation.

11.5 AD IS A SYNAPTIC FAILURE: CURCUMIN'S EFFECT ON PRE- AND POSTSYNAPTIC PROTEINS

Synaptic failure is a salient feature in individuals with AD and has shown to be correlated with the cognitive decline in AD patients (Scheff and Price 2001; Scheff et al. 2006). Indeed, synaptic loss is an early event in AD and synaptic contacts in both neocortex and hippocampus are lost in this neurodegenerative disorder. Recent studies suggest that levels of the presynaptic protein synaptophysin are significantly decreased and levels of postsynaptic density protein (PSD)-95 are also dramatically altered in AD brains compared to the control (Frautschy et al. 2001). A recent study has shown that in older female Sprague-Dawley rats, a significant loss of presynaptic protein synaptophysin was observed in the control animals, which were fed a normal diet, compared to the rats that received curcumin in their diet (Frautschy et al. 2001). In this study, Frautschy and coworkers also found increased levels of PSD-95 in the rats that had curcumin in their diet, compared to rats that received normal chow. Increases in both pre- and postsynaptic protein levels ultimately improved the synaptic transmission, which was evident when the Morris water maze test was conducted. In fact, curcumin-treated animals did significantly better in this maze test than the control rats that received the normal chow diet.

Further, curcumin not only reduces Aβ levels *in vivo* and prevents the presynaptic loss in AD patients, but also prevents the glutamate-induced excitotoxicity in rat cortical neurons (Wang et al. 2008). Mechanistically, curcumin increased the level of brain-derived neurotrophic factor (BDNF) via the TrkB receptor signaling pathway to prevent the toxicity generated due to glutamate induction.

11.6 CURCUMIN AND ITS EFFECT ON THE LEVELS OF BDNF AND NGF

In addition to curcumin's possible preventive role in AD and PD, curcumin has been reported to be effective in alleviating stress-induced disorders in rodents, possibly by exerting neuroprotection and neuroendocrine functions in the central nervous system (CNS) and BDNF is one of the most widely distributed neurotrophins in the CNS and plays an important role in neuronal survival and neurogenesis (Lee et al. 2002; Balaratnasingam and Janca 2012). Depletion of BDNF is strongly associated with several psychiatric and neurodegenerative disorders, including AD (Chu et al. 2011; Diniz and Teixeira 2011; Zhang et al. 2012). Further, acute stress in rats was observed to deplete CNS BDNF levels (Ray et al. 2011d), and curcumin treatment can alleviate stress by upregulating BDNF levels (Xu et al. 2007). A recent study (Wei et al. 2010) has demonstrated that when pigs were treated with curcumin, it can alleviate subacute stress response through modulation of hippocampal BDNF expression. Moreover, supplementing the diet with curcumin dramatically reduced oxidative damage and normalized the levels of BDNF, synapsin I, and CREB, which were altered following traumatic brain injury (Wu et al. 2006). Another study (Wang et al. 2008) has suggested that curcumin can be protective against glutamate excitotoxicity, seemingly mediated via the BDNF/TrkB signaling pathway.

Like BDNF, another neurotrophin, nerve growth factor (NGF) plays an important role in neural development, neuropreservation, and synaptic plasticity and also has been reported to be decreased in AD patients (Calissano et al. 2010). Recently, curcumin has been shown to increase the levels of NGF in different regions of the brain and can be effective in ameliorating AD pathologies (Hassanzadeh and Hassanzadeh 2012).

Postulated molecular targets of curcumin, in relation to AD, are depicted in Figure 11.2

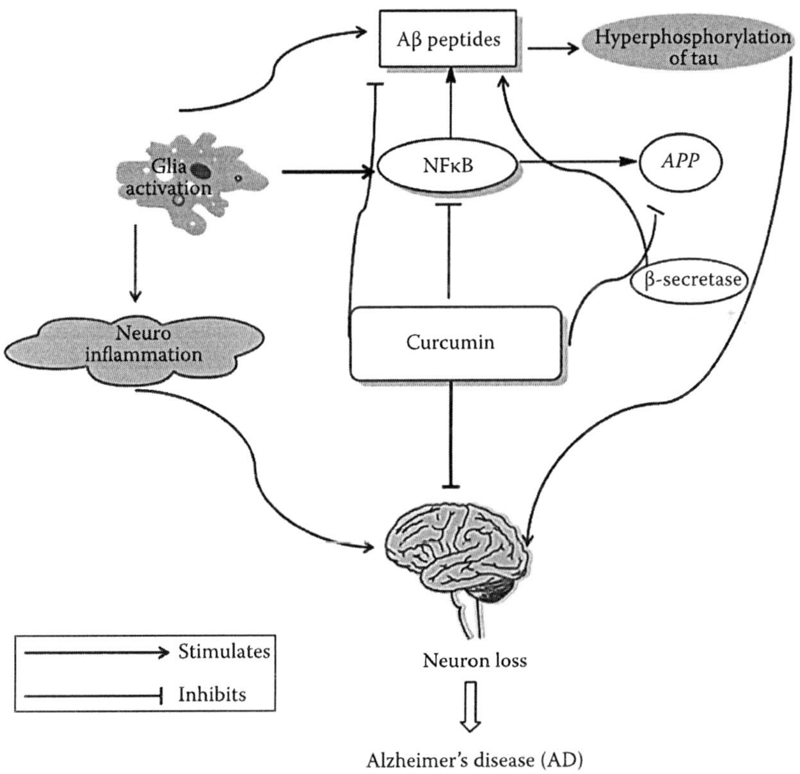

FIGURE 11.2 Molecular targets of curcumin are shown in this figure. The central role of curcumin's action is the inhibition of the proinflammatory transcription factor, NFκB, which is involved in the production of several cytochemokines, leading to neuronal loss. NFκB can increase the production of *APP* and β-secretase enzyme, which eventually causes more Aβ production. Curcumin has Aβ defibrillating properties (mainly by chelating action on metal ions) and can also prevent neuronal loss by preserving neurons. Neuropreservation by curcumin can be due to upregulating the levels of BDNF and NGF, among other factors. Curcumin has also been shown to stimulate neurogenesis in specific regions of the brain. Separate studies have also depicted that curcumin can prevent hyperphosphorylation of axonal protein tau (please see text for details). The hyperphosphorylation of tau protein is one of the cardinal features in AD. The schematic brain within this figure represents not normal but neurodegenerative AD brain.

11.7 CURCUMIN: BIOAVAILABILITY

Although clinical trials of curcumin have been ongoing to prevent various diseases including cancer and AD, poor oral absorption of curcumin in both humans and animals has raised several concerns about its clinical efficacy. Curcumin is a biphenolic compound that has hydroxyl groups at the *para*-position on the two aromatic rings that are connected by 1,3-diketone bridge that can undergo Michael addition, critical for some of the effects of curcumin but can contribute to its chemical instability in aqueous solution (Anand et al. 2007). Major limiting factors of curcumin include its low solubility in water and the fact that soluble curcumin molecules are highly unstable at physiological pH (Tonnesen 2002; Tonnesen et al. 2002). To date, several formulations of curcumin have been suggested to have more bioavailability than that of free curcumin. These include nanoparticle formulation (Bisht et al. 2008; Ray et al. 2011a), micelles (Mohanty et al. 2010), and curcumin combining with adjuvant (Anand et al. 2007; Sehgal et al. 2012). Large-scale clinical studies are needed to demonstrate the efficacy of these formulations.

11.8 CONCLUSION AND FUTURE DIRECTIONS

AD is a multifactorial disorder with many pathological sequelae. As mentioned, deposition of Aβ peptides in the brain is a hallmark of AD. However, whether Aβ is the sole triggering agent for the pathologies of AD is not clearly known. It was observed that deposition of Aβ takes place in the brain years, (even decades), before the appearance of clinical manifestations of AD (Morris et al. 1996), which may suggest that a second event or "hit" is necessary for the development of the disease. In this context, our laboratory has proposed the latent early-life associated regulation or LEARn in explaining the etiologies of several neurodegenerative chronic disorders, including AD (Lahiri and Maloney 2010). LEARn postulates that an initial insult or "hit," followed by a latent period in the pathogenesis of chronic disorders where the person stays symptomless, during the latent period, and the disease manifestation only occurs if a second insult or "hit" takes place in the later period of life. In AD, the second "hit" could be the physiological changes related to normal aging (Lahiri et al. 2009). It is possible that the activated inflammatory cascade in the brain can trigger other pathological events related to AD, and environmental factors, including diets, can have preventive roles (Lahiri et al. 2007). Once neuroinflammation sets in, it stimulates more Aβ production and the latter increases the production of cytokines, which intensifies the inflammation, thus creating a vicious cycle. Segregated approaches of decreasing Aβ load or ameliorating neuroinflammation alone did not produce desirable effects in clinical studies (Szekely and Zandi 2010; Sambamurti et al. 2011), which warrants effective therapy targeting multiple pathological cascades. Preclinical research work from all over the world has already established the effectiveness of curcumin in lowering Aβ loads, alleviating neuroinflammation, preserving neurons and synapses, and preventing hyperphosphorylation of tau proteins. However, because of curcumin's poor bioavailability and rapid biotransformation in blood, translation of curcumin's effectiveness in clinical settings has not been fully observed. Several formulations of curcumin, including encapsulation in nanoparticles, can be considered as potential ways to preserve curcumin's efficacy *in vivo*. Taken together, curcumin can have significant therapeutic potential in the treatment of several neurodegenerative disorders including AD and newer formulations of curcumin can open a new horizon in their treatments.

ABBREVIATIONS

Aβ	amyloid beta protein
AD	Alzheimer's disease
APP	amyloid precursor protein
APOE	apolipoprotein E

MAPT microtubule-associated protein tau
NFκB nuclear factor kappa-light-chain-enhancer of activated B cell
NFT neurofibrillary tangles
NO nitric oxide
PPAR γ peroxisome proliferator-activated receptor G
ROS reactive oxygen species

ACKNOWLEDGMENTS

This work was supported by grants from the Alzheimer's Association (IIRG) and NIH (AG18379 and AG18884) to DKL.

REFERENCES

Aggarwal, S., H. Ichikawa, Y. Takada, S. K. Sandur, S. Shishodia, and B. B. Aggarwal. 2006. Curcumin (diferuloylmethane) down-regulates expression of cell proliferation and antiapoptotic and metastatic gene products through suppression of IkappaBalpha kinase and Akt activation. *Mol Pharmacol* 69 (1):195–206.

Anand, P., A. B. Kunnumakkara, R. A. Newman, and B. B. Aggarwal. 2007. Bioavailability of curcumin: Problems and promises. *Mol Pharm* 4 (6):807–818.

Balaratnasingam, S. and A. Janca. 2012. Brain derived neurotrophic factor: A novel neurotrophin involved in psychiatric and neurological disorders. *Pharmacol Ther* 134:116–124.

Barger, S. W. and A. D. Harmon. 1997. Microglial activation by Alzheimer amyloid precursor protein and modulation by apolipoprotein E. *Nature* 388 (6645):878–881.

Bate, C., R. Veerhuis, P. Eikelenboom, and A. Williams. 2004. Microglia kill amyloid-beta1–42 damaged neurons by a CD14-dependent process. *Neuroreport* 15 (9):1427–1430.

Baum, L., C. W. Lam, S. K. Cheung, T. Kwok, V. Lui, J. Tsoh, L. Lam et al. 2008. Six-month randomized, placebo-controlled, double-blind, pilot clinical trial of curcumin in patients with Alzheimer disease. *J Clin Psychopharmacol* 28 (1):110–113.

Begum, A. N., M. R. Jones, G. P. Lim, T. Morihara, P. Kim, D. D. Heath, C. L. Rock et al. 2008. Curcumin structure-function, bioavailability, and efficacy in models of neuroinflammation and Alzheimer's disease. *J Pharmacol Exp Ther* 326 (1):196–208.

Benedetti, M. D., A. Salviati, S. Filipponi, M. Manfredi, L. De Togni, M. Gomez Lira, G. Stenta et al. 2002. Prevalence of dementia and apolipoprotein e genotype distribution in the elderly of buttapietra, verona province, Italy. *Neuroepidemiology* 21 (2):74–80.

Bishnoi, M., K. Chopra, and S. K. Kulkarni. 2008. Protective effect of Curcumin, the active principle of turmeric (*Curcuma longa*) in haloperidol-induced orofacial dyskinesia and associated behavioural, biochemical and neurochemical changes in rat brain. *Pharmacol Biochem Behav* 88 (4):511–522.

Bisht, S., G. Feldmann, J. B. Koorstra, M. Mullendore, H. Alvarez, C. Karikari, M. A. Rudek, C. K. Lee, and A. Maitra. 2008. *In vivo* characterization of a polymeric nanoparticle platform with potential oral drug delivery capabilities. *Mol Cancer Ther* 7 (12):3878–3888.

Braak, H., and E. Braak. 1998. Evolution of neuronal changes in the course of Alzheimer's disease. *J Neural Transm Suppl* 53:127–140.

Calissano, P., C. Matrone, and G. Amadoro. 2010. Nerve growth factor as a paradigm of neurotrophins related to Alzheimer's disease. *Dev Neurobiol* 70 (5):372–383.

Camins, A., F. X. Sureda, F. Junyent, E. Verdaguer, J. Folch, C. Beas-Zarate, and M. Pallas. 2010. An overview of investigational antiapoptotic drugs with potential application for the treatment of neurodegenerative disorders. *Expert Opin Investig Drugs* 19 (5):587–604.

Chandra, V., R. Pandav, H. H. Dodge, J. M. Johnston, S. H. Belle, S. T. DeKosky, and M. Ganguli. 2001. Incidence of Alzheimer's disease in a rural community in India: The Indo-US study. *Neurology* 57 (6):985–989.

Chu, C. L., C. K. Liang, M. Y. Chou, Y. T. Lin, C. C. Pan, T. Lu, L. K. Chen, and P. C. Chow. 2012. Decreased plasma brain-derived neurotrophic factor levels in institutionalized elderly with depressive disorder. *J Am Med Dir Assoc.*

Davis, S. and S. Laroche. 2003. What can rodent models tell us about cognitive decline in Alzheimer's disease? *Mol Neurobiol* 27 (3):249–276.

DeGiorgio, L. A., Y. Shimizu, H. S. Chun, B. P. Cho, S. Sugama, T. H. Joh, and B. T. Volpe. 2002. APP knockout attenuates microglial activation and enhances neuron survival in substantia nigra compacta after axotomy. *Glia* 38 (2):174–178.

Dickson, D. W., S. C. Lee, L. A. Mattiace, S. H. Yen, and C. Brosnan. 1993. Microglia and cytokines in neurological disease, with special reference to AIDS and Alzheimer's disease. *Glia* 7 (1):75–83.

Diniz, B. S. and A. L. Teixeira. 2011. Brain-derived neurotrophic factor and Alzheimer's disease: Physiopathology and beyond. *Neuromolecular Med* 13 (4):217–222.

Du, Y., X. Chen, X. Wei, K. R. Bales, D. T. Berg, S. M. Paul, M. R. Farlow, B. Maloney, Y. W. Ge, and D. K. Lahiri. 2005. NF-(kappa)B mediates amyloid beta peptide-stimulated activity of the human apolipoprotein E gene promoter in human astroglial cells. *Brain Res Mol Brain Res* 136 (1–2):177–188.

Eikelenboom, P. and R. Veerhuis. 1996. The role of complement and activated microglia in the pathogenesis of Alzheimer's disease. *Neurobiol Aging* 17 (5):673–680.

Fassbender, K., S. Walter, S. Kuhl, R. Landmann, K. Ishii, T. Bertsch, A. K. Stalder et al. 2004. The LPS receptor (CD14) links innate immunity with Alzheimer's disease. *FASEB J* 18 (1):203–205.

Frautschy, S. A., W. Hu, P. Kim, S. A. Miller, T. Chu, M. E. Harris-White, and G. M. Cole. 2001. Phenolic anti-inflammatory antioxidant reversal of Abeta-induced cognitive deficits and neuropathology. *Neurobiol Aging* 22 (6):993–1005.

Frautschy, S. A., F. Yang, M. Irrizarry, B. Hyman, T. C. Saido, K. Hsiao, and G. M. Cole. 1998. Microglial response to amyloid plaques in APPsw transgenic mice. *Am J Pathol* 152 (1):307–317.

Frisardi, V., F. Panza, D. Seripa, B. P. Imbimbo, G. Vendemiale, A. Pilotto, and V. Solfrizzi. 2010. Nutraceutical properties of Mediterranean diet and cognitive decline: Possible underlying mechanisms. *J Alzheimers Dis* 22 (3):715–740.

Ghoneim, A. I., A. B. Abdel-Naim, A. E. Khalifa, and E. S. El-Denshary. 2002. Protective effects of curcumin against ischaemia/reperfusion insult in rat forebrain. *Pharmacol Res* 46 (3):273–279.

Griffin, W. S., J. G. Sheng, M. C. Royston, S. M. Gentleman, J. E. McKenzie, D. I. Graham, G. W. Roberts, and R. E. Mrak. 1998. Glial-neuronal interactions in Alzheimer's disease: The potential role of a "cytokine cycle" in disease progression. *Brain Pathol* 8 (1):65–72.

Grundke-Iqbal, I., K. Iqbal, M. Quinlan, Y. C. Tung, M. S. Zaidi, and H. M. Wisniewski. 1986a. Microtubule-associated protein tau. A component of Alzheimer paired helical filaments. *J Biol Chem* 261 (13):6084–6089.

Grundke-Iqbal, I., K. Iqbal, Y. C. Tung, M. Quinlan, H. M. Wisniewski, and L. I. Binder. 1986b. Abnormal phosphorylation of the microtubule-associated protein tau (tau) in Alzheimer cytoskeletal pathology. *Proc Natl Acad Sci USA* 83 (13):4913–4917.

Guo, Q., N. Robinson, and M. P. Mattson. 1998. Secreted beta-amyloid precursor protein counteracts the pro-apoptotic action of mutant presenilin-1 by activation of NF-kappaB and stabilization of calcium homeostasis. *J Biol Chem* 273 (20):12341–12351.

Hafner-Bratkovic, I., J. Gaspersic, L. M. Smid, M. Bresjanac, and R. Jerala. 2008. Curcumin binds to the alpha-helical intermediate and to the amyloid form of prion protein—A new mechanism for the inhibition of PrP(Sc) accumulation. *J Neurochem* 104 (6):1553–1564.

Hardy, J. and D. J. Selkoe. 2002. The amyloid hypothesis of Alzheimer's disease: Progress and problems on the road to therapeutics. *Science* 297 (5580):353–356.

Hassanzadeh, P. and A. Hassanzadeh. 2012. The CB(1) Receptor-mediated endocannabinoid signaling and NGF: The novel targets of curcumin. *Neurochem Res* 37 (5):1112–1120.

Heneka, M. T., M. Sastre, L. Dumitrescu-Ozimek, A. Hanke, I. Dewachter, C. Kuiperi, K. O'Banion, T. Klockgether, F. Van Leuven, and G. E. Landreth. 2005. Acute treatment with the PPARgamma agonist pioglitazone and ibuprofen reduces glial inflammation and Abeta1–42 levels in APPV717I transgenic mice. *Brain* 128 (Pt 6):1442–1453.

Hong, H. S., S. Rana, L. Barrigan, A. Shi, Y. Zhang, F. Zhou, L. W. Jin, and D. H. Hua. 2009. Inhibition of Alzheimer's amyloid toxicity with a tricyclic pyrone molecule in vitro and in vivo. *J Neurochem* 108 (4):1097–1108.

Imbimbo, B. P. and G. A. Giardina. 2011. gamma-Secretase inhibitors and modulators for the treatment of Alzheimer's disease: Disappointments and hopes. *Curr Top Med Chem* 11 (12):1555–1570.

Koistinaho, M., S. Lin, X. Wu, M. Esterman, D. Koger, J. Hanson, R. Higgs, F. Liu, S. Malkani, K. R. Bales, and S. M. Paul. 2004. Apolipoprotein E promotes astrocyte colocalization and degradation of deposited amyloid-beta peptides. *Nat Med* 10 (7):719–726.

Lahiri, D. K. 2006. Where the actions of environment (nutrition), gene and protein meet: Beneficial role of fruit and vegetable juices in potentially delaying the onset of Alzheimer's disease. *J Alzheimers Dis* 10 (4):359–361; discussion 363–364.

Lahiri, D. K., D. Chen, B. Maloney, H. W. Holloway, Q. S. Yu, T. Utsuki, T. Giordano, K. Sambamurti, and N. H. Greig. 2007a. The experimental Alzheimer's disease drug posiphen [(+)-phenserine] lowers amyloid-beta peptide levels in cell culture and mice. *J Pharmacol Exp Ther* 320 (1):386–396.

Lahiri, D. K., M. R. Farlow, K. Sambamurti, N. H. Greig, E. Giacobini, and L. S. Schneider. 2003. A critical analysis of new molecular targets and strategies for drug developments in Alzheimer's disease. *Curr Drug Targets* 4 (2):97–112.

Lahiri, D. K. and B. Maloney. 2010. The "LEARn" (Latent Early-life Associated Regulation) model integrates environmental risk factors and the developmental basis of Alzheimer's disease, and proposes remedial steps. *Exp Gerontol* 45 (4):291–296.

Lahiri, D. K., B. Maloney, M. R. Basha, Y. W. Ge, and N. H. Zawia. 2007b. How and when environmental agents and dietary factors affect the course of Alzheimer's disease: The "LEARn" model (latent early-life associated regulation) may explain the triggering of AD. *Curr Alzheimer Res* 4 (2):219–228.

Lahiri, D. K., B. Maloney, and N. H. Zawia. 2009. The LEARn model: An epigenetic explanation for idiopathic neurobiological diseases. *Mol Psychiatry* 14 (11):992–1003.

Lee, J., W. Duan, and M. P. Mattson. 2002. Evidence that brain-derived neurotrophic factor is required for basal neurogenesis and mediates, in part, the enhancement of neurogenesis by dietary restriction in the hippocampus of adult mice. *J Neurochem* 82 (6):1367–1375.

Lim, G. P., T. Chu, F. Yang, W. Beech, S. A. Frautschy, and G. M. Cole. 2001. The curry spice curcumin reduces oxidative damage and amyloid pathology in an Alzheimer transgenic mouse. *J Neurosci* 21 (21):8370–8377.

Liu, B. and J. S. Hong. 2003. Role of microglia in inflammation-mediated neurodegenerative diseases: Mechanisms and strategies for therapeutic intervention. *J Pharmacol Exp Ther* 304 (1):1–7.

Ma, Q. L., F. Yang, E. R. Rosario, O. J. Ubeda, W. Beech, D. J. Gant, P. P. Chen et al. 2009. Beta-amyloid oligomers induce phosphorylation of tau and inactivation of insulin receptor substrate via c-Jun N-terminal kinase signaling: Suppression by omega-3 fatty acids and curcumin. *J Neurosci* 29 (28):9078–9089.

Marlow, L., R. M. Canet, S. J. Haugabook, J. A. Hardy, D. K. Lahiri, and K. Sambamurti. 2003. APH1, PEN2, and Nicastrin increase Abeta levels and gamma-secretase activity. *Biochem Biophys Res Commun* 305 (3):502–509.

Medeiros, R., D. Baglietto-Vargas, and F. M. LaFerla. 2011. The role of tau in Alzheimer's disease and related disorders. *CNS Neurosci Ther* 17 (5):514–524.

Metcalfe, M. J. and M. E. Figueiredo-Pereira. 2010. Relationship between tau pathology and neuroinflammation in Alzheimer's disease. *Mt Sinai J Med* 77 (1):50–58.

Mohanty, C., S. Acharya, A. K. Mohanty, F. Dilnawaz, and S. K. Sahoo. 2010. Curcumin-encapsulated MePEG/PCL diblock copolymeric micelles: A novel controlled delivery vehicle for cancer therapy. *Nanomedicine (Lond)* 5 (3):433–449.

Morris, J. C., M. Storandt, D. W. McKeel, Jr., E. H. Rubin, J. L. Price, E. A. Grant, and L. Berg. 1996. Cerebral amyloid deposition and diffuse plaques in "normal" aging: Evidence for presymptomatic and very mild Alzheimer's disease. *Neurology* 46 (3):707–719.

Murrell, J. R., B. Price, K. A. Lane, O. Baiyewu, O. Gureje, A. Ogunniyi, F. W. Unverzagt et al. 2006. Association of apolipoprotein E genotype and Alzheimer disease in African Americans. *Arch Neurol* 63 (3):431–434.

Nagele, R. G., M. R. D'Andrea, H. Lee, V. Venkataraman, and H. Y. Wang. 2003. Astrocytes accumulate A beta 42 and give rise to astrocytic amyloid plaques in Alzheimer disease brains. *Brain Res* 971 (2):197–209.

Natarajan, C. and J. J. Bright. 2002. Curcumin inhibits experimental allergic encephalomyelitis by blocking IL-12 signaling through Janus kinase-STAT pathway in T lymphocytes. *J Immunol* 168 (12):6506–6513.

Niino, M., K. Iwabuchi, S. Kikuchi, M. Ato, T. Morohashi, A. Ogata, K. Tashiro, and K. Onoe. 2001. Amelioration of experimental autoimmune encephalomyelitis in C57BL/6 mice by an agonist of peroxisome proliferator-activated receptor-gamma. *J Neuroimmunol* 116 (1):40–48.

Ophir, G., N. Amariglio, J. Jacob-Hirsch, R. Elkon, G. Rechavi, and D. M. Michaelson. 2005. Apolipoprotein E4 enhances brain inflammation by modulation of the NF-kappaB signaling cascade. *Neurobiol Dis* 20 (3):709–718.

Paresce, D. M., R. N. Ghosh, and F. R. Maxfield. 1996. Microglial cells internalize aggregates of the Alzheimer's disease amyloid beta-protein via a scavenger receptor. *Neuron* 17 (3):553–565.

Park, S. Y., H. S. Kim, E. K. Cho, B. Y. Kwon, S. Phark, K. W. Hwang, and D. Sul. 2008. Curcumin protected PC12 cells against beta-amyloid-induced toxicity through the inhibition of oxidative damage and tau hyperphosphorylation. *Food Chem Toxicol* 46 (8):2881–2887.

Qiu, W. Q., D. M. Walsh, Z. Ye, K. Vekrellis, J. Zhang, M. B. Podlisny, M. R. Rosner, A. Safavi, L. B. Hersh, and D. J. Selkoe. 1998. Insulin-degrading enzyme regulates extracellular levels of amyloid beta-protein by degradation. *J Biol Chem* 273 (49):32730–32738.

Ray, B., S. Bisht, A. Maitra, and D. K. Lahiri. 2011a. Neuroprotective and neurorescue effects of a novel polymeric nanoparticle formulation of curcumin (NanoCurc) in the neuronal cell culture and animal model: Implications for Alzheimer's disease. *J Alzheimers Dis* 23 (1):61–77.

Ray, B., N. B. Chauhan, and D. K. Lahiri. 2011b. The "aged garlic extract:" (AGE) and one of its active ingredients S-allyl-L-cysteine (SAC) as potential preventive and therapeutic agents for Alzheimer's disease (AD). *Curr Med Chem* 18 (22):3306–3313.

Ray, B., N. B. Chauhan, and D. K. Lahiri. 2011c. Oxidative insults to neurons and synapse are prevented by aged garlic extract and S-allyl-L-cysteine treatment in the neuronal culture and APP-Tg mouse model. *J Neurochem* 117 (3):388–402.

Ray, B., D. L. Gaskins, T. J. Sajdyk, J. P. Spence, S. D. Fitz, A. Shekhar, and D. K. Lahiri. 2011d. Restraint stress and repeated corticotrophin-releasing factor receptor activation in the amygdala both increase amyloid-beta precursor protein and amyloid-beta peptide but have divergent effects on brain-derived neurotrophic factor and pre-synaptic proteins in the prefrontal cortex of rats. *Neuroscience* 184:139–150.

Ray, B. and D. K. Lahiri. 2009. Neuroinflammation in Alzheimer's disease: Different molecular targets and potential therapeutic agents including curcumin. *Curr Opin Pharmacol* 9 (4):434–444.

Rogers, J., N. R. Cooper, S. Webster, J. Schultz, P. L. McGeer, S. D. Styren, W. H. Civin et al. 1992. Complement activation by beta-amyloid in Alzheimer disease. *Proc Natl Acad Sci USA* 89 (21):10016–10020.

Rozemuller, J. M., P. Eikelenboom, S. T. Pals, and F. C. Stam. 1989. Microglial cells around amyloid plaques in Alzheimer's disease express leucocyte adhesion molecules of the LFA-1 family. *Neurosci Lett* 101 (3):288–292.

Sambamurti, K., N. H. Greig, and D. K. Lahiri. 2002. Advances in the cellular and molecular biology of the beta-amyloid protein in Alzheimer's disease. *Neuromolecular Med* 1 (1):1–31.

Sambamurti, K., N. H. Greig, T. Utsuki, E. L. Barnwell, E. Sharma, C. Mazell, N. R. Bhat, M. S. Kindy, D. K. Lahiri, and M. A. Pappolla. 2011. Targets for AD treatment: Conflicting messages from gamma-secretase inhibitors. *J Neurochem* 117 (3):359–374.

Scheff, S. W. and D. A. Price. 2001. Alzheimer's disease-related synapse loss in the cingulate cortex. *J Alzheimers Dis* 3 (5):495–505.

Scheff, S. W., D. A. Price, F. A. Schmitt, and E. J. Mufson. 2006. Hippocampal synaptic loss in early Alzheimer's disease and mild cognitive impairment. *Neurobiol Aging* 27 (10):1372–1384.

Schubert, P., T. Morino, H. Miyazaki, T. Ogata, Y. Nakamura, C. Marchini, and S. Ferroni. 2000. Cascading glia reactions: A common pathomechanism and its differentiated control by cyclic nucleotide signaling. *Ann N Y Acad Sci* 903:24–33.

Sehgal, A., M. Kumar, M. Jain, and D. K. Dhawan. 2012. Piperine as an adjuvant increases the efficacy of curcumin in mitigating benzo(a)pyrene toxicity. *Hum Exp Toxicol.* 31 (5):473–482.

Selkoe, D. J. 1994. Alzheimer's disease: A central role for amyloid. *J Neuropathol Exp Neurol* 53 (5):438–447.

Sharma, S., S. K. Kulkarni, J. N. Agrewala, and K. Chopra. 2006. Curcumin attenuates thermal hyperalgesia in a diabetic mouse model of neuropathic pain. *Eur J Pharmacol* 536 (3):256–261.

Small, S. A. and S. Gandy. 2006. Sorting through the cell biology of Alzheimer's disease: Intracellular pathways to pathogenesis. *Neuron* 52 (1):15–31.

Smith, M. A., K. Hirai, K. Hsiao, M. A. Pappolla, P. L. Harris, S. L. Siedlak, M. Tabaton, and G. Perry. 1998. Amyloid-beta deposition in Alzheimer transgenic mice is associated with oxidative stress. *J Neurochem* 70 (5):2212–2215.

Sumanont, Y., Y. Murakami, M. Tohda, O. Vajragupta, H. Watanabe, and K. Matsumoto. 2006. Prevention of kainic acid-induced changes in nitric oxide level and neuronal cell damage in the rat hippocampus by manganese complexes of curcumin and diacetylcurcumin. *Life Sci* 78 (16):1884–1891.

Szczucinski, A. and J. Losy. 2007. Chemokines and chemokine receptors in multiple sclerosis. Potential targets for new therapies. *Acta Neurol Scand* 115 (3):137–146.

Szekely, C. A. and P. P. Zandi. 2010. Non-steroidal anti-inflammatory drugs and Alzheimer's disease: The epidemiological evidence. *CNS Neurol Disord Drug Targets* 9 (2):132–139.

Tonnesen, H. H. 2002. Solubility, chemical and photochemical stability of curcumin in surfactant solutions. Studies of curcumin and curcuminoids, XXVIII. *Pharmazie* 57 (12):820–824.

Tonnesen, H. H., M. Masson, and T. Loftsson. 2002. Studies of curcumin and curcuminoids. XXVII. Cyclodextrin complexation: Solubility, chemical and photochemical stability. *Int J Pharm* 244 (1–2):127–135.

Vassar, R. and P. C. Kandalepas. 2011. The beta-secretase enzyme BACE1 as a therapeutic target for Alzheimer's disease. *Alzheimers Res Ther* 3 (3):20.

Wai, M. S., Y. Liang, C. Shi, E. Y. Cho, H. F. Kung, and D. T. Yew. 2009. Co-localization of hyperphosphorylated tau and caspases in the brainstem of Alzheimer's disease patients. *Biogerontology* 10 (4):457–469.

Walsh, D. M., I. Klyubin, J. V. Fadeeva, M. J. Rowan, and D. J. Selkoe. 2002. Amyloid-beta oligomers: Their production, toxicity and therapeutic inhibition. *Biochem Soc Trans* 30 (4):552–557.

Walsh, D. M. and D. J. Selkoe. 2004. Deciphering the molecular basis of memory failure in Alzheimer's disease. *Neuron* 44 (1):181–193.

Wang, R., Y. B. Li, Y. H. Li, Y. Xu, H. L. Wu, and X. J. Li. 2008. Curcumin protects against glutamate excitotoxicity in rat cerebral cortical neurons by increasing brain-derived neurotrophic factor level and activating TrkB. *Brain Res* 1210:84–91.

Wei, S., H. Xu, D. Xia, and R. Zhao. 2010. Curcumin attenuates the effects of transport stress on serum cortisol concentration, hippocampal NO production, and BDNF expression in the pig. *Domest Anim Endocrinol* 39 (4):231–239.

Wu, A., Z. Ying, and F. Gomez-Pinilla. 2006. Dietary curcumin counteracts the outcome of traumatic brain injury on oxidative stress, synaptic plasticity, and cognition. *Exp Neurol* 197 (2):309–317.

Wyss-Coray, T., J. D. Loike, T. C. Brionne, E. Lu, R. Anankov, F. Yan, S. C. Silverstein, and J. Husemann. 2003. Adult mouse astrocytes degrade amyloid-beta in vitro and in situ. *Nat Med* 9 (4):453–457.

Xu, Y., B. Ku, L. Cui, X. Li, P. A. Barish, T. C. Foster, and W. O. Ogle. 2007. Curcumin reverses impaired hippocampal neurogenesis and increases serotonin receptor 1A mRNA and brain-derived neurotrophic factor expression in chronically stressed rats. *Brain Res* 1162:9–18.

Xu, Y., B. S. Ku, H. Y. Yao, Y. H. Lin, X. Ma, Y. H. Zhang, and X. J. Li. 2005. Antidepressant effects of curcumin in the forced swim test and olfactory bulbectomy models of depression in rats. *Pharmacol Biochem Behav* 82 (1):200–206.

Yan, S. D., S. F. Yan, X. Chen, J. Fu, M. Chen, P. Kuppusamy, M. A. Smith et al. 1995. Non-enzymatically glycated tau in Alzheimer's disease induces neuronal oxidant stress resulting in cytokine gene expression and release of amyloid beta-peptide. *Nat Med* 1 (7):693–699.

Yan, Q., J. Zhang, H. Liu, S. Babu-Khan, R. Vassar, A. L. Biere, M. Citron, and G. Landreth. 2003. Anti-inflammatory drug therapy alters beta-amyloid processing and deposition in an animal model of Alzheimer's disease. *J Neurosci* 23 (20):7504–7509.

Yang, F., G. P. Lim, A. N. Begum, O. J. Ubeda, M. R. Simmons, S. S. Ambegaokar, P. P. Chen et al. 2005. Curcumin inhibits formation of amyloid beta oligomers and fibrils, binds plaques, and reduces amyloid *in vivo*. *J Biol Chem* 280 (7):5892–5901.

Zbarsky, V., K. P. Datla, S. Parkar, D. K. Rai, O. I. Aruoma, and D. T. Dexter. 2005. Neuroprotective properties of the natural phenolic antioxidants curcumin and naringenin but not quercetin and fisetin in a 6-OHDA model of Parkinson's disease. *Free Radic Res* 39 (10):1119–1125.

Zhang, C., A. Browne, D. Child, and R. E. Tanzi. 2010. Curcumin decreases amyloid-beta peptide levels by attenuating the maturation of amyloid-beta precursor protein. *J Biol Chem* 285 (37):28472–28480.

Zhang, X. Y., J. Liang, D. C. Chen, M. H. Xiu, F. De Yang, T. A. Kosten, and T. R. Kosten. 2012. Low BDNF is associated with cognitive impairment in chronic patients with schizophrenia. *Psychopharmacology (Berl)* 222 (2):277–284.

12 Plant Metabolites
Inhibitors of NO Production

Marina P. Polovinka and Nariman F. Salakhutdinov

CONTENTS

12.1 INTRODUCTION

The discovery of the physiological and pathophysiological roles of nitric oxide (NO) in the 1980s became one of the most remarkable events in biology (Furchgott and Zavadski 1980; Palmer et al. 1987). From the chemical point of view, NO is an uncharged paramagnetic molecule. Its chemical and physiological properties are a result of its tendency to stabilize an unpaired electron (Stamler 1994; Kerwin et al. 1995). In contrast to oxygen radicals, the half lifetime of NO reaches several seconds depending on the type of tissues and physiological conditions (Kikuchi et al. 1993) as a result of which NO molecules can easily penetrate through biological membranes and interact with intracellular and extracellular structures that are located relatively far from the place where these molecules were produced, and readily react with other substances as well (Moncada et al. 1991). Nitric oxide plays a dual role in an organism: On the one hand, it diffuses into parasite cells and inhibits the key enzymes necessary to those cells, thereby exhibiting a protective effect against the parasite cells by destroying them; on the other hand, NO produced in an excess amount acts as a strong cytostatic, which causes appreciable harm to the organism itself under conditions of oxidative stress and production of active oxygen forms, mainly peroxynitrites (Schmidt and Walter 1994), and eventually takes part in the development of inflammatory processes.

Almost all endogenous NO is synthesized from L-arginine during the catabolism of L-arginine into L-citrulline by a family of cytochrome P450-like hemoproteins, i.e., nitric oxide synthases (NOSs) (Wang et al. 2005a). The isoforms of NOSs are the products of various genes: nNOS and eNOS are constitutive isoforms, while iNOS is an inducible isoform. In resting cells, the inducible isoform iNOS is not detected; for its expression, the activation of cells by lipopolysaccharides (LPS) or cytokines, e.g., interleukins 1,2,6 & 8, interferon-gamma (IFN-γ), tumor necrosis factor-α (TNF-α), etc., is required. In this context, it should be mentioned that the constitutive forms cause the production of NO in lower amounts (picomol), while the amount of NO synthesized under the action of iNOS may vary and reach higher values (nanomol).

The mechanism of NO formation from L-arginine is the same for all the three isoforms; the process proceeds via a two-stage oxidation reaction. In order to transform L-arginine into NO and L-citrulline, all the three isoforms of NOSs require the following coenzymes: the reduced form of nicotinamide adenine dinucleotide phosphate (NADPH), tetrahydrobiopterin (H_4B), flavin mononucleotide (FMN), and flavin adenine dinucleotide (FAD) (Griffith and Stuehr 1995; Woodward et al. 2009). NOS expression increases with activation of nuclear factor (NF-κB). The nuclear factor, κB, is a protein referred to the κ (kappa) group and brought to the active state by the action of lipopolysaccharides. NF-κB initiates transcription as a result of which the corresponding mRNA molecules are produced; these molecules enter to cytoplasm and take part in the process of synthesizing various proteins (including iNOS) on ribosomes (Bremner and Heinrich 2005). Three isoforms of NOSs are involved in various pathological processes, including Alzheimer's disease and stroke (nNOS), septic shock, arthritis and inflammatory processes (iNOS), formation of edemas, and endothelial damage (eNOS). Hence, selective inhibitors for various isoforms of NOSs are warranted (Babu and Griffith 1998).

In accordance with the mechanism of action, the NOS inhibitors can be divided into the following groups:

1. Compounds that prevent the transfer of L-arginine to active sites of enzymes
2. Compounds that inactivate the cofactors needed for NOS-catalyzed oxidation of L-arginine
3. Compounds that inhibit the electron transport, into which NADPH and flavins are involved, and the agents capable of interfering in the functions of heme
4. Compounds that inhibit the production of NO
5. Compounds that inhibit the activity of iNOS
6. Compounds that inhibit iNOS expression
7. Compounds that inhibit the activation of NF-κB

The inhibitors of NO-synthases can be divided according to their origin: natural, semisynthetic, and synthetic categories. Nowadays, the search for selective NOS inhibitors is conducted in all three areas. This review is devoted to the study of plant secondary metabolites, capable of inhibiting the production of NO and the activity of iNOS; the works presented in this review were mainly published within the period from 2000 to date.

12.2 PLANT METABOLITES AND INHIBITORS OF NITRIC OXIDE PRODUCTION

After the discovery of the important role of NO in inflammatory processes, a new tendency appeared in phytochemical studies, i.e., the works devoted to studying extracts and plant secondary metabolites for their NO- and iNOS-inhibitory activity. First of all, it is the studies of extracts that are obtained from fruits (Van Beharka et al. 2000; Tezuka et al. 2001; Meeteren et al. 2004; Jung et al. 2007a; Lin et al. 2008; Huang and Ho 2010), berries (Pergola et al. 2006; Lau et al. 2009), and vegetables (Wang et al. 2005b; Hwang et al. 2011) used for food or extracts of medicinal plants (Hong et al. 2002; Sutherland et al. 2006; Wang et al. 2008b; Lii et al. 2009; Lee et al. 2005, 2011; Mueller et al. 2010); the investigations are performed without isolating individual compounds or, in some cases, with a partial determination of the component composition by applying high-performance liquid chromatography, but always with establishing in details the mechanism of NO and iNOS inhibition (Hong et al. 2002; Kiemer et al. 2003; Lee and Jeon 2003; Kaszkin et al. 2004; Kim et al. 2004; Matheus et al. 2006; Sutherland et al. 2006; Jung et al. 2007b; Lee et al. 2007; Yen et al. 2008;

Chao et al. 2009; Ichikawa et al. 2009; Jung et al. 2009; Matsuda et al. 2009; Sheeba and Asha 2009; Sripanidkulchai et al. 2009; Jin et al. 2010; Ozer et al. 2010; Kang et al. 2011; Yu et al. 2011). The works devoted to these kinds of studies have been actively published in the last decade; the main purpose of these investigations is partly to make recommendations on the use of plant products for the prevention of various inflammatory and cancer diseases and, on the other hand, to reveal the extracts of medicinal plants that exhibit the highest NO- and iNOS-inhibitory activity. This, in turn, has motivated to undertake exhaustive research works directed toward the isolation of active plant metabolites and the determination of their structures including the establishment of the correlation between a structure and its activity concerned; the structure–activity relationship (SAR) draws significant importance in clinical chemistry when developing new drugs.

The most widespread experimental model for the primary investigation of the isolated metabolites for their NO-inhibitory activity is the mouse macrophages activated with lipopolysaccharide (LPS) (Nathan and Xie 1994; MacMicking et al. 1997; Alderton et al. 2001), which has an ability to induce iNOS expression and the formation of NO in cells. It should be noted that plant metabolites with NO-inhibiting activity are referred to in various classes of natural compounds such as terpenoids, phenolic compounds, alkaloids, and also their glycosides. The most representative group is the phenolic–phenolic compounds of various structural types: (1) simple phenolic compounds (C_6 compounds); (2) C_6–C_1 compounds (oxybenzoic acids and their derivatives); (3) C_6–C_3 phenolic compounds (phenylpropanoids, lignans, neolignans, coumarins, and their derivatives); (4) stilbenes (C_6–C_2–C_6 compounds); (5) C_6–C_3–C_6 compounds, cyclic and noncyclic—chalcones, flavonoids, and aurones; (6) C_6–C_4–C_6 compounds; (7) diarylheptanoids (C_6–C_7–C_6 compounds), etc.

12.2.1 Flavonoids

Plant flavonoids have already created a stir among the scientific community at large due their multidirectional biological activities, and at the same time they have been reported to possess promising capability of inhibiting NO production. More than 8000 individual flavonoid compounds of natural origin are known to date (Pietta 2000). These phenolic secondary metabolites may be divided into several structural subtypes; the primary amongst these being: flavones, flavonols, flavanones, flavanols, and anthocyanidins. Such flavonoid compounds are the constituents of fruits, berries, and vegetables; the average intake of flavonoids by humans is on the order of a few hundred milligrams per day in terms of the aglycon mass (Hollman and Katan 1999). Flavonoids are frequently called "molecules against oxidative stress," i.e., proinflammatory radical "scavengers." Products enriched with flavonoids are recommended for chronic diseases, including diseases caused by the excessive production of nitric oxide (NO).

Flavones, e.g., **1** Apigenin (Matsuda et al. 2003; Comalada et al. 2006; Tong et al. 2007; Kang et al. 2009), **2** Diosmetin (Comalada et al. 2006), and **3** Luteolin (Comalada et al. 2006; Lopez-Posadas et al. 2008; Wang et al. 2008a), which "work" in various cell models, *in vitro* and *in vivo*, are considered to be the most active NO-inhibiting metabolites from the flavonoid group (Matsuda et al. 2003). In order to understand how the structure influences the NO-inhibition activity, in Matsuda et al. (2003) the authors tested 73 flavonoids in the free and glycosylated forms, natural metabolites, and their methylated analogues: flavonones **1–12** and flavanones **13–17** (Table 12.1); flavonols **18–37** and flavanols **38–40** (Table 12.2); and isoflavones **41–48** (Table 12.3). According to the results obtained, flavonones **1–12** were found to be the most active among the compounds listed (Table 12.1). The IC_{50} values of compounds **1**, **2**, and **3** inhibiting the production of NO are 7.7, 8.9, and 20 μM, respectively; the IC_{50} values of Di- **4**, Tri- **5**, and tetra-*O*-methyl luteolins **6** are 11, 11, and 2.4 μM (Table 12.1). Thus, among other flavones considered, compounds **1**, **2**, and **6** exhibit the highest activity (IC_{50} < 10 μM). Flavanones **13–17** (Table 12.1) exhibit a lower activity, which indicates that the presence of the double bond between the C_2 and C_3 atoms in the ring-C is a very important factor for the occurrence of NO-inhibiting activity.

Among flavonols **18–37** (Table 12.2), only the completely methylated form, hexa-*O*-methyl mirecetin **36**, exhibited an activity of IC_{50} < 10 μM (IC_{50} < 7.4 μM, Table 12.2). All the other

TABLE 12.1
Data on the Inhibition of NO Production for Flavones 1–12 and Flavanones 13–17

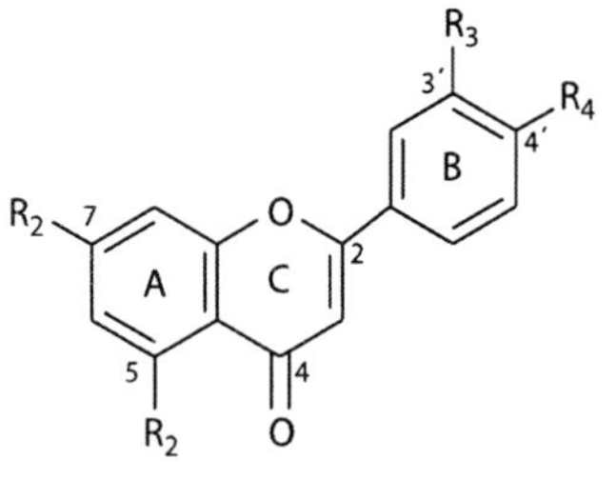

Flavones: **1–12**

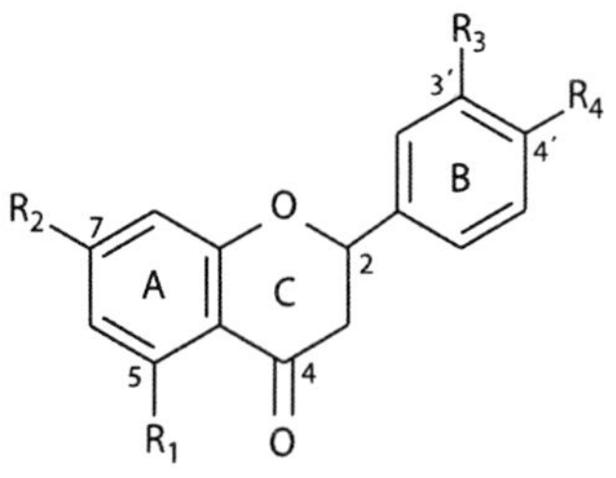

Flavanones: **13–17**

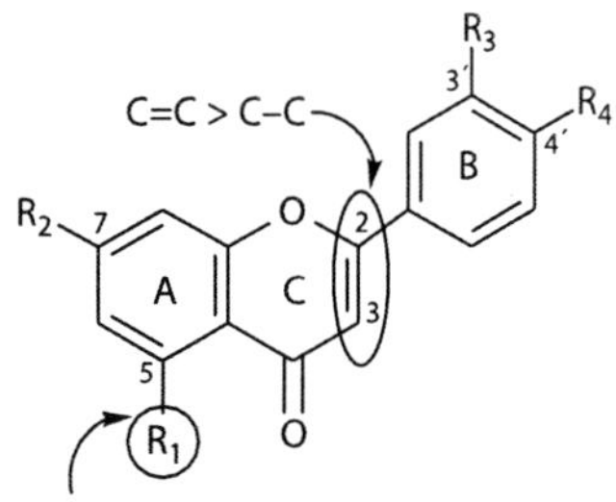

OCH$_3$ > OH

Compounds (Str. No.)	R_1	R_2	R_3	R_4	IC_{50} (μM)
Apigenin **1**	OH	OH	H	OH	7.7
Diosmetin **2**	OH	OH	OH	OCH_3	8.9
Luteolin **3**	OH	OH	OH	OH	20
4′,7-Dimethylluteolin **4**	OH	OCH_3	OH	OCH_3	11
3′,4′,7-Trimethylluteolin **5**	OH	OCH_3	OCH_3	OCH_3	11
3′,4′,5,7-Tetramethylmethylluteolin **6**	OCH_3	OCH_3	OCH_3	OCH_3	2.4
Apigenin, 7-*O*-Glc **7**	OH	O-Glc	H	OH	>100
Apigenin, 7-*O*-Rut **8**	OH	O-Rut	H	OH	>100
4′,7-Dihydroxyflavone **9**	H	OH	H	OH	14
3′,4′-Dihydroxyflavone **10**	H	H	OH	OH	23
3′,4′,7-Trihydroxyflavone **11**	H	OH	OH	OH	26
Luteolin, 7-*O*-Glc **12**	OH	O-Glc	OH	OH	>100
Liquiritigenin **13**	H	OH	H	OH	85
14	H	OCH_3	H	OH	38
Liquiritin **15**	H	OH	H	-O-Glc	>100
16	H	OCH_3	H	-O-Glc	>100
Eriodictyol **17**	OH	OH	OH	OH	>100

Source: Matsuda, H. et al., *Bioorgan. Med. Chem.*, 11, 1995, 2003.

flavonols and flavanols **38–40** (Table 12.2) considered in Matsuda et al. (2003) have high IC_{50} values characterizing the inhibition of NO within this experiment.

Let us compare the activities of flavones (Table 12.1) and flavonols (Table 12.2) corresponding to them. Apigenin **1** has a higher activity than kaempferol **20**; diosmetin **2** is more active than tamarixetin **26**; and luteolin **3** exhibits a higher activity than quercetin **22**. Thus, it can be concluded that the introduction of the OH group at C-3 in ring-C upon moving from flavones to flavonols leads to a decrease in the NO-inhibition activity. (This regularity remains valid for the other flavone/flavonol pairs, e.g., **4/28** and **5/30** compounds; see Tables 12.1 and 12.2, respectively).
Comparing the structural features of the flavonoids studied with the NO-inhibition activity as studied by Matsuda et al. (2003), the following conclusions can be made:

1. The activity of flavones is higher than the activity of flavonols corresponding to them (Tables 12.1 and 12.2). See the **1/20**, **2/26**, **3/22**, **4/28**, and **5/30** pairs.
2. The flavonoids containing the glycoside residue in the structure exhibit a lower NO-inhibition activity in all cases (Tables 12.1 through 12.3); the IC_{50} value is above 100 μM. Thus, the transition from aglycon to glycoside leads to a significant decrease in activity. See **1/7** and **1/8** pairs; in this case, it occurs independently on the structure of

TABLE 12.2
Data on the Inhibition of NO Production for Flavonols 18–37 and Flavanols 38–40

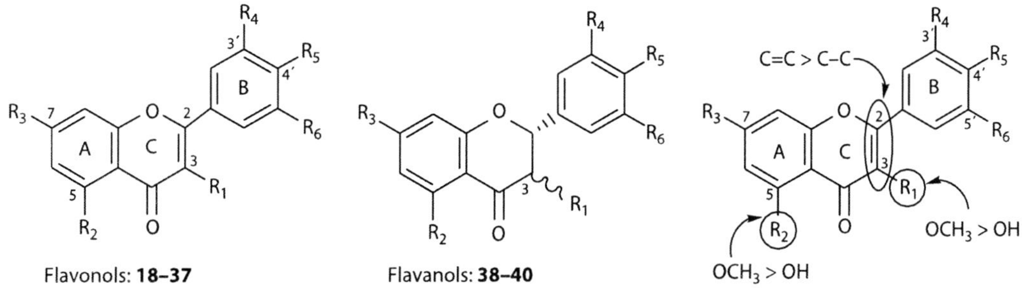

Flavonols: **18–37** Flavanols: **38–40**

Compounds (Str. No.)	R_1	R_2	R_3	R_4	R_5	R_6	IC_{50} (μM)
3-Hydroxyflavone **18**	OH	H	H	H	H	H	>10
Izalpinin **19**	OH	OH	OCH_3	H	H	H	>30
Kaempferol **20**	OH	OH	OH	H	OH	H	29
Kaempferol, -3-O-GlcA **21**	-O-GlcA	OH	OH	H	OH	H	>100
Quercetin **22**	OH	OH	OH	OH	OH	H	36
Rutin **23**	-O-Rut	OH	OH	OH	OH	H	>100
Quercetin, 3,7-di-*O*-Glc **24**	-O-Glc	OH	O-Glc	OH	OH	H	>100
Rhamnetin **25**	OH	OH	OCH_3	OH	OH	H	42
Tamarixetin **26**	OH	OH	OH	OH	OCH_3	H	25
27	OCH_3	OH	OCH_3	OH	OH	H	15
Ombuine **28**	OH	OH	OCH_3	OH	OCH_3	H	>30
Ayanin **29**	OCH_3	OH	OCH_3	OH	OCH_3	H	19
30	OH	OH	OCH_3	OCH_3	OCH_3	H	>10
31	OCH_3	OH	OCH_3	OCH_3	OCH_3	H	79
32	OCH_3	OCH_3	OCH_3	OCH_3	OCH_3	H	26
Myricetin **33**	OH	OH	OH	OH	OH	OH	99
34	OH	OH	OCH_3	OH	OCH_3	OH	24
35	OH	OH	OCH_3	OCH_3	OCH_3	OH	>10
36	OCH_3	OCH_3	OCH_3	OCH_3	OCH_3	OCH_3	7.4
Myricitrin **37**	-O-Rha	OH	OH	OH	OH	OH	>100
(+)-Catechin **38**	β-OH	OH	OH	OH	OH	H	>100
(−)-Epicatechin **39**	α-OH	OH	OH	OH	OH	H	>100
(−)-Epigallocatechin **40**	α-OH	OH	OH	OH	OH	OH	65

Source: Matsuda, H. et al., *Bioorgan. Med. Chem.*, 11, 1995, 2003.

glycoside. See **3/12** pair (Table 12.1); **20/21**, **22/23**, **22/24**, **33/37** pairs (Table 12.2); and **41/42**, **43/44**, **45/46** pairs (Table 12.3).

3. The activity of flavones (Table 12.1) is higher than the activity of flavanones corresponding to them (Table 12.1). See **3/17** and **9/13** pairs.
4. Flavones and flavonols containing only one hydroxyl group near the C-4′ atom in ring-B exhibit a higher activity than flavones and flavonols containing a larger number of OH groups in ring-B, including those that have two OH groups in the 3′ and 4′ positions. See **1/3** and **9/11** pairs (Table 12.1); and **20/22**, **20/25**, and **20/33** pairs (Table 12.2).
5. Flavonols containing the hydroxyl groups near the C-3′ and C-4′ atoms (Catechol type) exhibit a higher activity than those that contain three hydroxyl groups in the 3′, 4′, and 5′ positions (Pyrogallol type). See **22/33** compounds (Table 12.2).

TABLE 12.3
Data on the Inhibition of NO Production for Isoflavones 41–48

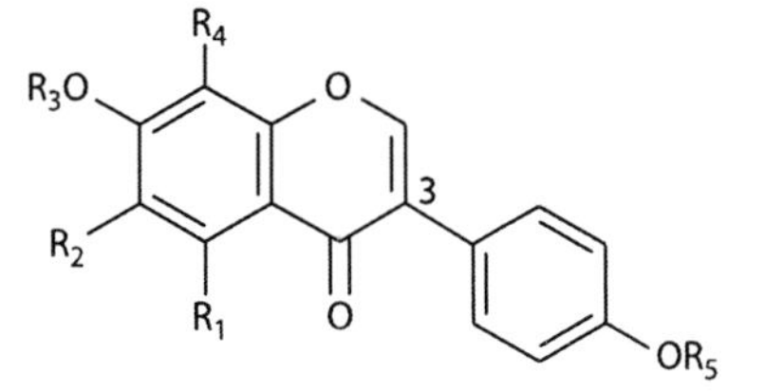

Isoflavones: **41–48**

Compounds (Str. No.)	R_1	R_2	R_3	R_4	R_5	IC_{50} (μM)
Daidzein **41**	H	H	H	H	H	33
Daidzin **42**	H	H	Glc	H	H	>100
Genistein **43**	OH	H	H	H	H	26
Genistin **44**	OH	H	Glc	H	H	>100
Tectorigenin **45**	OH	OCH_3	H	H	H	31
Tectoridin **46**	OH	OCH_3	Glc	H	H	>100
Biochanin A **47**	H	H	H	H	CH_3	30
Glycitein **48**	H	OCH_3	H	H	H	~100

Source: Matsuda, H. et al., *Bioorgan. Med. Chem.*, 11, 1995, 2003.

6. The presence of the OH group near the C-5 atom normally increases the activity. See the **1/9**, **3/11**, and **43/41** compounds.
7. If there is an OCH_3 group in the 3′, 5′, or 4′ position, the compounds exhibit higher NO-inhibitory activities. See compounds **2** (IC_{50} 8.9 μM), **6** (IC_{50} 2.4 μM), **36** (IC_{50} 7.4 μM), etc.
8. The activity of isoflavones is lower than the activity of corresponding flavones. In this review, we outline the data for the **1/43** pair (IC_{50} 7.7/26 μM, respectively).
9. The compounds containing the OCH_3 group near the C^3 exhibit a low cytotoxicity.

It was also established by Matsuda et al. (2003) that all flavonoids as studied inhibit iNOS expression without reducing the iNOS activity. Another important group of flavonoids is anthocyanidins (Figure 12.1). Anthocyanidins are aglycons of anthocyanins, natural pigments extracted from plants. Anthocyanidins are usually obtained from acidic plant extracts at low pH values; in this case, anthocyanidins are in the form of salts, in which an electron of the oxygen atom is involved into the heteroaromatic π-system of the benzpyrylium (chromenylium) cycle; the latter is a chromophore defining the color of these compounds.

The influence of anthocyanidins **49–53** and their glycosides isolated from the extracts of berries on the production of NO in *LPS/INF*-γ-activated macrophages was studied by Wang and Mazza (2002).

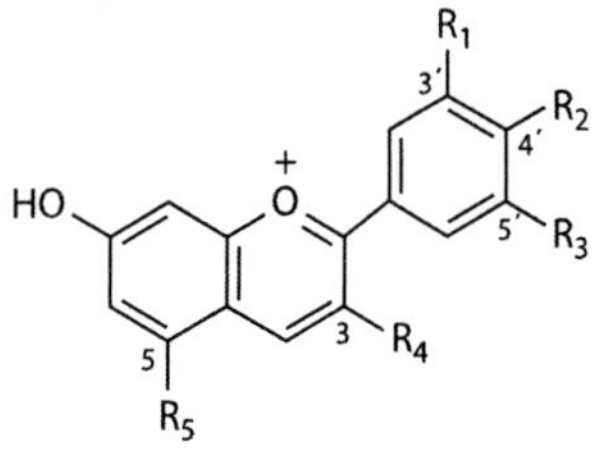

Anthocyanidins:
49: Pelargonidin: $R_1=H$, $R_2=OH$, $R_3=H$, $R_4=R_5=OH$
50: Delphinidin: $R_1=R_2=R_3=R_4=R_5=OH$
51: Peonidin: $R_1=OCH_3$, $R_2=OH$, $R_3=H$, $R_4=R_5=OH$
52: Malvidin: $R_1=OCH_3$, $R_2=OH$, $R_3=OCH_3$, $R_4=R_5=OH$
53: Cyanidin: $R_1=H$, $R_2=R_3=R_4=R_5=OH$

FIGURE 12.1 Anthocyanidin derivatives isolated from the extracts of berries.

The activity of this group of flavonoids was found to be lower than the activity of flavonols examined in the same work; in the experiments, anthocyanidins **49** and **50** exhibited the highest activity. Thus, at a concentration of 125 μM, compounds **49** and **50** inhibited the production of NO by 35%, compounds **51** and **52** inhibited by 30%, and compound **53** inhibited by 19%; for reference, kaempferol **20** and quercetin **22** (Table 12.2) at the same concentration inhibit the production of NO by 73% and 57%, respectively.

Every year, researchers usually report on more and more new compounds referred to the group of flavonoids; more than 450 new flavonoids were reported during the period from 2001 to 2003 (Veitch and Grayer 2008). Consequently, new works devoted to the study of biological activity of new plant metabolites have been appearing.

The flavonols, fisetin **54** (IC_{50} < 5 μM) and morin **55** (IC_{50} > 10 μM) (Figure 12.2), were evaluated to possess promising inhibitory activity against iNOS mRNA and the activation of the nuclear factor, NF-κB (Wang et al. 2006). From the ethanolic extract of *Agrimonia pilosa* Ledeb, a potent source of polyphenols, compounds **56–59** including three flavanols **56**, **58**, and **59** (Figure 12.2) in the free and glycolised forms were isolated (Taira et al. 2009); all the isolated compounds inhibited the production of NO but did not exhibit any cytotoxicity at the concentrations used. Compound **56** showed the highest activity. The investigators suggested that phenolic compounds are good radical "scavengers" and consequently exhibit the antioxidant properties (Taira et al. 2009). In this context,

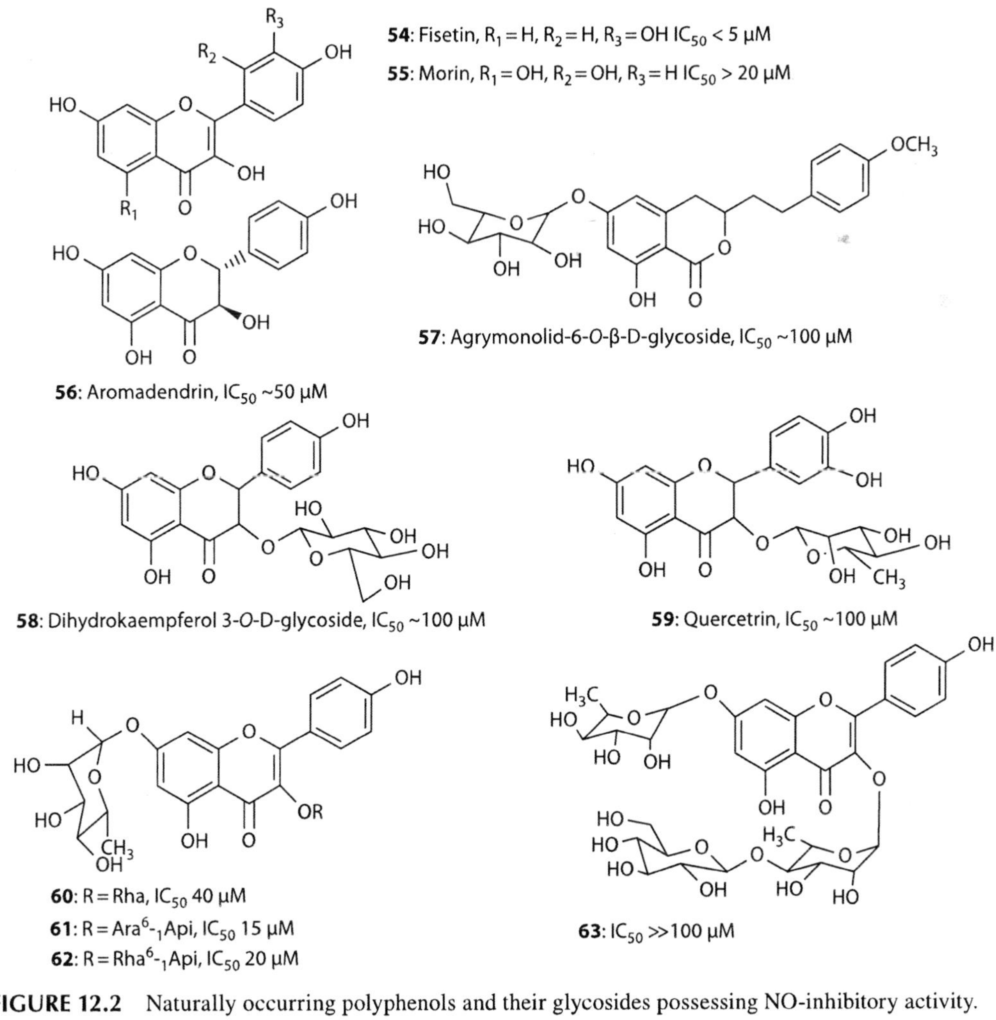

FIGURE 12.2 Naturally occurring polyphenols and their glycosides possessing NO-inhibitory activity.

in order to clarify the mechanism of action for compounds **56–59**, the same team carried out the experiments in LPS-induced macrophage cells and in the presence of the NO donor, 4-ethyl-2-hydroxyamino-5-nitro-3-hexenamide (NOR3). In all the cases, a decrease in the concentration of nitric oxide was observed; the experimental results prompted the investigators to suggest compounds **56–59** inhibit the production of NO in macrophages acting as traps for nitric oxide produced.

Four flavonol and kaempferol **20** glycosides, i.e., compounds **60–63** (Figure 12.2), were isolated from the methanolic extract of *Cinnamomum osmophloeum* Kaneh leaves, an endemic tree of Taiwan, by Fang et al. (2005). These compounds exhibit a dose-dependent inhibition of NO production in *LPS/γ-IFN*-activated macrophage cells. Among the compounds examined, compound **61**, kaempferol-3-*O*-β-D-apiofuranosyl-(1 → 2)-α-L-arabinofuranosyl-7-*O*-α-L-ramnopyranoside, was found to be the strongest inhibitor. For reference, at a concentration of 20 μM, compound **61** inhibited the production of NO by 69%, while compound **63** in the same concentration inhibited the production of NO by only 9%. The IC_{50} value was 40, 15, and 20 μM for compounds **60**, **61**, and **62**, respectively. In accordance with the data, the isolated glycosides can be ranked in the order of decreasing NO-inhibition activity as follows: **61** > **62** > **60** >> **63**. Although the investigators did not comment on the correlation between the structure and properties (Fang et al. 2005), following a comparison of the structures of compounds **60–63** that differ from each other by the structure of the glycoside residue, R, near the C-3 atom, it follows that the most active compound **61** contains two furanose cycles in the residue, compound **62** has one furanose cycle (apiofuranosyl, the same as in compound **61**), and the least active in this series compounds **60** and **63** (Figure 12.2) contain only pyranose residues in the radical. The most bulky glycoside residues are found in compound **63**, which corresponds to its low inhibition activity. It should be noted that the NO-inhibition activity of all glycosides **60–63** (Figure 12.2) is rather high; it is not in agreement with the data for other glycosides, including kaempferol **20** as presented by Matsuda et al. (2003).

By analyzing the data on the NO-inhibition activity of the most common flavonoids as discussed in literature (Wang and Mazza 2002; Matsuda et al. 2003; Wang et al. 2006; Puangpraphant et al. 2009), it may be said that a relative order of the activity among the compounds concerned persists. For instance, the activity of compound **20**, kaempferol, is higher than the activity of compound **22** in all experiments. However, the data on the NO-inhibition activity in LPS-activated macrophages differ quantitatively only for quercetin **22**: IC_{50} ~ 125 μM (Wang and Mazza 2002), IC_{50} < 10 μM (Wang et al. 2006), and IC_{50} = 11.6 μM (Puangpraphant et al. 2009). Thus, the most reliable results and conclusions concerning the structure–activity correlation can be obtained from the series of compounds with similar structures under the same conditions. Although works of this kind are rare, we would like to pay special attention to these particular works.

Acacia confusa Merr. is traditionally used in the folk medicine of Taiwan. In 2008, Wu and his coworkers reported the isolation of two flavonols, melanoxetin **64** and transilitin **65** (Figure 12.3), from the ethyl acetate extract of the plant-wood (Wu et al. 2008). The structure of transilitin **65** differs from the structure of melanoxetin **64** by the presence of the methylated OH group near the C-3 position; the inhibitory activity in this case decreases almost by a factor of two. For melanoxetin **64**, the activity is characterized by an IC_{50} value of 6.9 μM, which is comparable with quercetin **22** (IC_{50} 6.4 μM, compound **22** is used as a standard in the experiment), while for transilitin **65**, the IC_{50} value was determined as greater than 100 μM. Melanoxetin **64** exhibited high NO-inhibition activity and inhibited iNOS expression as well with IC_{50} of 50 μM. The investigators explained a decrease in the activity observed for compound **65** by an increase in the lipophilicity of its molecules with respect to compound **64**. These data do not correlate with the results from Matsuda et al. (2003) (flavonols **18–37**, Table 12.2), in which a substitution of the OH group near the C-3 atom of the ring-C normally led to a significant decrease in the NO-inhibition activity (IC_{50} > 100 μM) only in the case of introducing the glycoside residue. However, compounds **64** and **65** have the other type of substitution in ring A, i.e., they contain OH groups at the C-7 and C-8 positions; it might be an important factor in the inhibition activity.

The extracts from the stems of *Erycibe expansa*, a traditional plant in Thai medicine, yielded a number of isoflavones (Figure 12.3; Morikawa et al. 2006; Matsuda et al. 2007). Clycosin **66** and

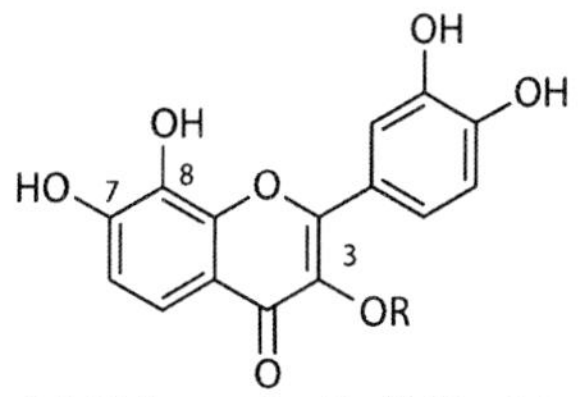

64: Melanoxetin, R = H, IC_{50} 6.9 μM
65: Transilitin, R = CH_3, IC_{50} >>100 μM

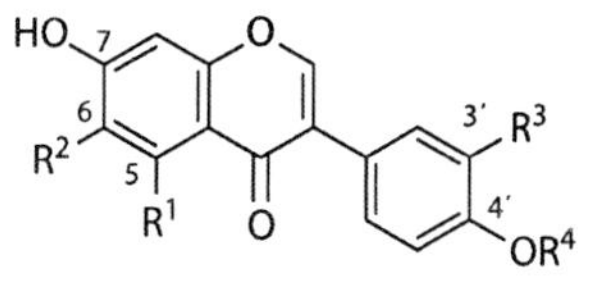

	R^1	R^2	R^3	R^4	IC_{50} μM
66: Clycosin:	H	H	OH	CH_3	13
67: Erythrinine B:	OH	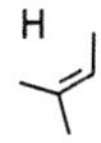	H	H	18
68:	OH	H	OCH_3	H	37
69: Orobol	OH	H	OH	H	44
70: Formononetin	H	H	H	CH_3	>100

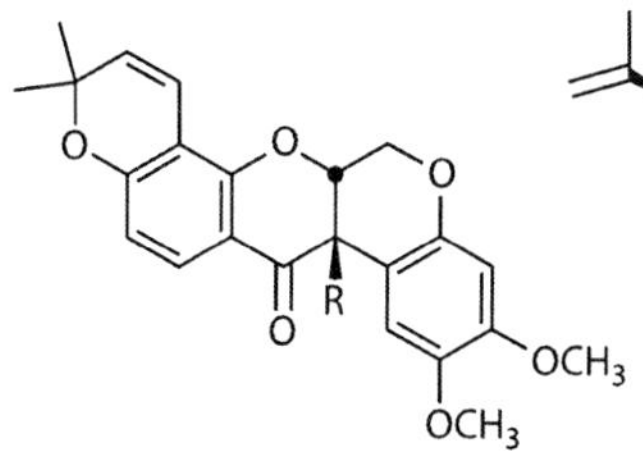

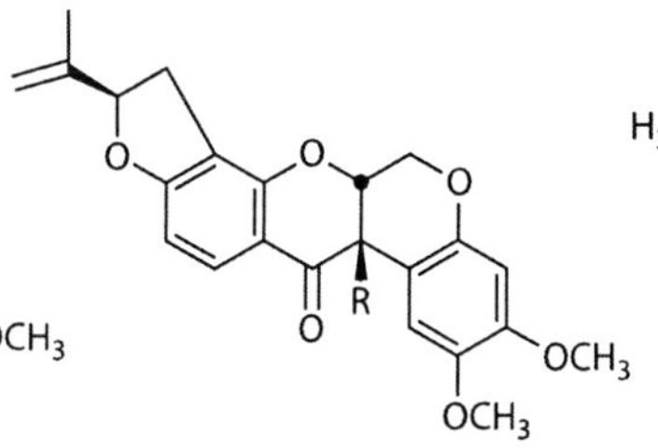

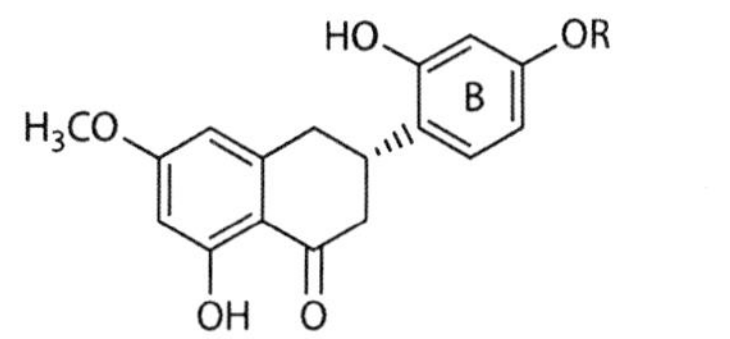

75: R = H, IC_{50} 35 μM
76: R = CH_3, IC_{50} >>50 μM

71: Deguelin; R = H, IC_{50} 26 μM
72: Tephrosin; R = OH, IC_{50} ~100 μM
73: Rotenone; R = H, IC_{50} 27 μM
74: 12a-Hydroxyrotenone; R = OH, IC_{50} ~100 μM

FIGURE 12.3 NO-Inhibitory naturally occurring flavonoids.

erythrinine B **67** exhibited the highest inhibitory activity with the IC_{50} values of 13 and 18 μM, respectively. In addition, two rotenoids with the isoflavane skeleton, deguelin **71** and rotenone **72**, were also found to be active having respective IC_{50} values of 26 and 27 μM. Rotenoids are naturally occurring compounds, whose structure is close to isoflavones and involves the *cis*-fused fragment of tetrahydrochromeno[3,4-b]chromene. Analysis of the structure–activity correlation shows that the introduction of the hydroxyl groups into the node positions of rotenones in the **71/72** and **73/74** pairs causes a significant decrease in the NO-inhibiting activity and an increase in the cytotoxicity.

The plants belonging to *Artocarpus* genus (mulberry family) grow in tropical and subtropical regions and are used against fever and malaria in the traditional folk medicine of Indonesia. Flavonoids isolated from the extracts of these plants were examined for various types of biological activity (Wei et al. 2005). Compound **75** (Figure 12.3) was found to be a good inhibitor of NO production in LPS-activated macrophages of the RAW 264.7 mouse; compound **76** (with methylated hydroxyl group in ring-B) exhibits a substantially lower inhibition activity (see **64/65** pairs, Figure 12.3). Compound **75** reduces the production of NO by inhibiting iNOS expression. The lowest activity of formononetin **70** (Figure 12.3) most likely can be explained by its structure: This compound is the most lipophilic isoflavonoid in the series (Wei et al. 2005).

12.2.2 Chalcones

Chalcones are compounds that can be considered as flavonoids containing an open pyran ring. The majority of the compounds from this group are found in plants in the form of glycosides. The following four chalcones were isolated from the extracts of *Alpinia pricei* Hayata roots: cardamonin **77**, flavokavain B **78**, and chalcones **79** and **80** (Lin et al. 2009; Figure 12.4); their NO-inhibition

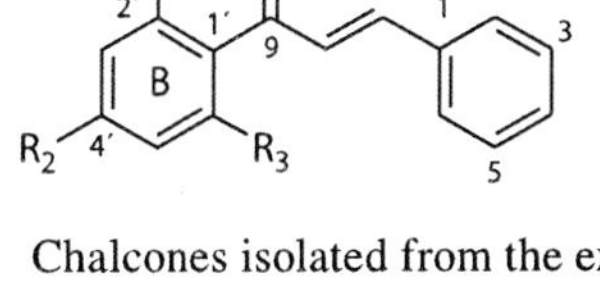

77: R_1 = OH, R_2 = OH, R_3 = OCH_3, IC_{50} 60.6 μM
78: R_1 = OH, R_2 = OCH_3, R_3 = OCH_3, IC_{50} 9.8 μM
79: R_1 = OCH_3, R_2 = OCH_3, R_3 = OCH_3, IC_{50} 79.0 μM
80: R_1 = OH, R_2 = OCH_3, R_3 = OH, IC_{50} 12.0 μM

FIGURE 12.4 Chalcones isolated from the extracts of *Alpinia pricei* Hayata roots.

activity was studied. Compound **78** had the highest activity; according to the mechanism established by the investigators, this compound exhibits a dose-dependent inhibition of iNOS expression and of NF-κB activation. The data obtained indicate that chalcones **78** and **80** exhibit a higher NO-inhibition activity; these compounds contain the phenolic hydroxyl group at C-2′ and the methoxy group at $C^{4'}$ position (Figure 12.4). Most likely, this kind of arrangement of the substituents in ring-B of chalcones (compounds **78** and **80**) may influence their inhibition activity.

12.2.3 Phenylpropanoids

The plants traditionally applied in folk medicine, which exhibits anti-inflammatory activity, are of interest. Many secondary metabolites having a NO-inhibition activity are isolated from the plants of the Ginger family; these plants are widely used in oriental medicine and oriental cooking.

Alpinia galanga SWARTZ is the genus of herbaceous plants from the Ginger family. The rhizome of *A. galanga* is widely used in folk medicine for preparing tinctures, e.g., as a remedy against stomach diseases in traditional Chinese medicine, as a remedy against tympanism, and also as antifungal and antipruritic agents in traditional Thai medicine, as additives in the cuisine of the South and Southeast Asian countries. The NO-inhibition activity of the metabolites from the extracts of this plant was also evaluated (Ando et al. 2005; Matsuda et al. 2005a,b; Morikawa et al. 2005).

The major components of the extracts from *A. galanga* are the phenolic compounds from the group of phenylpropanoids (Figure 12.5).

1′*S*-1′-Acetoxychavicol acetate **81**, a major component of the extracts from the rhizome of *A. galanga* (Ando et al. 2005; Matsuda et al. 2005a,b; Morikawa et al. 2005), was reported to inhibit the production of NO in LPS-activated peritoneal mouse macrophages by inhibiting β-interferon mRNA (Ando et al. 2005) as well as by inhibiting the activation of the nuclear factor, NF-κB (Morikawa et al. 2005). The correlation between the structure and the activity was studied not only for acetate **81**, but also for various natural and synthetic phenylpropanoids (C_6–C_3 phenolic compounds) (Matsuda et al. 2005). Let us consider the series of compounds **81–96** first; in these compounds, there is a double bond between the $C_{2'}$–$C_{3'}$ atoms in the propenyl substituent (Figure 12.5).

From the analysis of the data shown in Figure 12.5, the following conclusions can be made:

1. NO-inhibitory activity appears when the substituents occupy the *para*- or *ortho*-positions in the benzene ring. See compounds **81–83**, **85**, **90**, and **96**.
2. The highest inhibition activity is observed for the compounds that contain acetoxy groups both in the ring and near the C-1′ position as a propenyl fragment; in this case the *S*-configuration of the 1′-acetoxy group is preferable. See compounds **81–83** and **85**.
3. The substitution of the acetoxy group at the C-4 and/or C-1′ position for OH or H (Figure 12.3), i.e., both in the ring and the propenyl radical, leads to a significant decrease in the inhibition activity. See compounds **86–92**, **94**, and **95**. Compounds **93** and **96** are exceptions.

Among the phenylpropanoids, in which the double bond is located in the propenyl fragment between the $C_{1'}$–$C_{2'}$ atoms, only compounds **97–99** with the *para*-arrangement of the substituents in the ring exhibited NO-inhibiting activity (Figure 12.6; Matsuda et al. 2005).

On going from the diacetate **97** to the mono-acetate **100**, inhibitory activity receives a significant decrease (Figure 12.6); however, phenols **98** and **99** containing the OH group at the C-4 and alcoholic/aldehyde group at the C-3′ position exhibited the activity that is characterized by IC_{50} values of 72 and 20 μM, respectively. During the study of the NO-inhibition activity, phenethyl ester of caffeic acid, compound **101** with IC_{50} 15 μM was taken as a standard (Ando et al. 2005; Matsuda et al. 2005; Morikawa et al. 2005). Compound **101** is referred to phenylpropanoids as well; its small amounts can be found in propolis. This compound inhibits iNOS expression and the activation

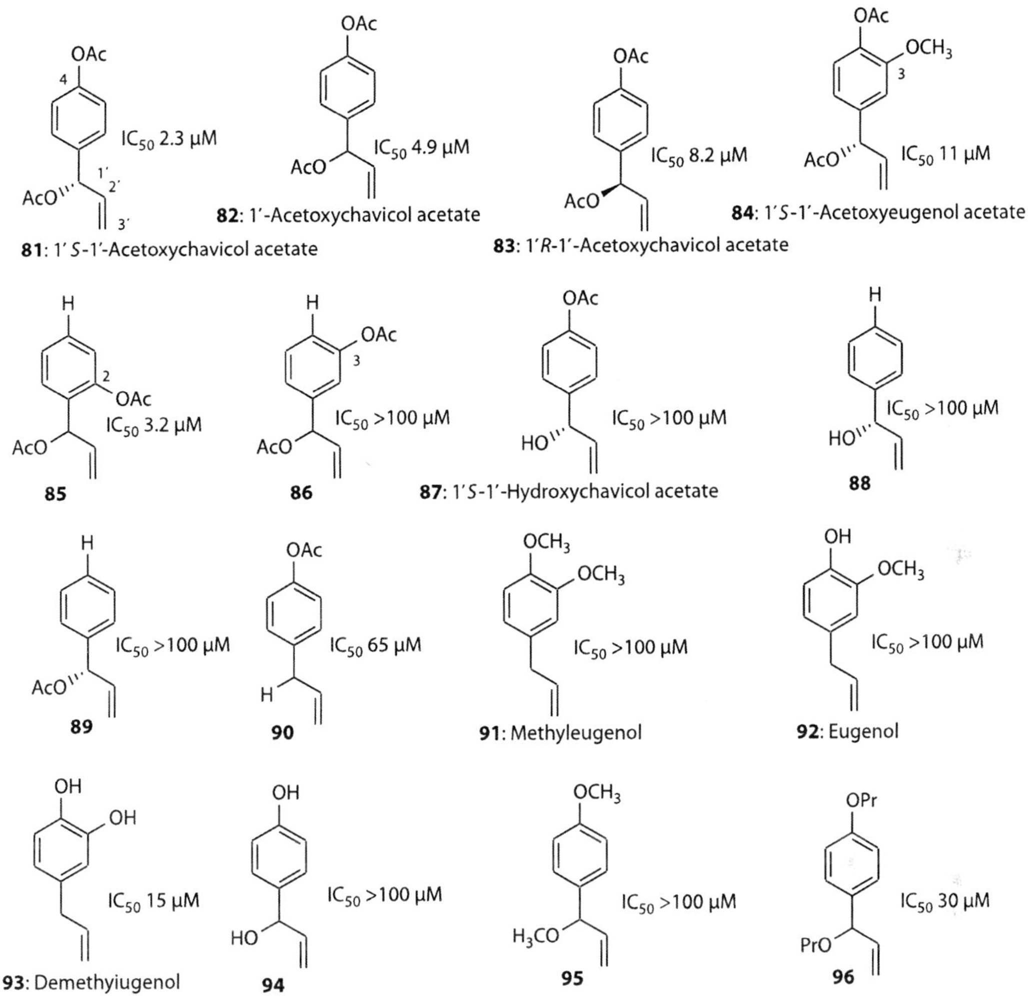

FIGURE 12.5 Major phenylpropanoids of *Alpinia galanga* possessing NO-inhibitory activity.

of NF-κB (Song et al. 2002) and is widely used as a standard in the experiments on studying the NO-inhibition activity in natural compounds. Caffeic acid **102** itself exhibits no inhibition activity (Matsuda et al. 2005; Figure 12.6).

12.2.4 Neolignans

Neolignans are the compounds that biogenetically related to phenylpropanoids. Along with the phenylpropanoids, new neolignans **103–106** (Figure 12.6) were also isolated from the 80%-water-acetone extract of *A. galanga* rhizome (Matsuda et al. 2005a,b). From the data as shown in Figure 12.6, it appears that the NO-inhibitory activity of neolignans **102–106** is lower than the activity of compound **81**, a major extract component; galanganol C **106** exhibits the highest NO-inhibition activity in the experiment (IC$_{50}$ 33 μM). Five glycosides of dibenzofuran neolignans **107–111** (Figure 12.7), isolated from *Coptis japonica*, were evaluated to possess inhibitory activity against the production of NO in activated macrophages with IC$_{50}$ values of 14–25 μM (Cho et al. 2000). All the compounds have close structures and all exhibit a high inhibition activity; however, compound **110** exhibited the best properties and inhibited the production of NO in macrophages with IC$_{50}$ 14 μM. All compounds **107–111** are glycosides, and it is not clear which peculiarities of the structure in compound **110** causes its higher inhibition activity.

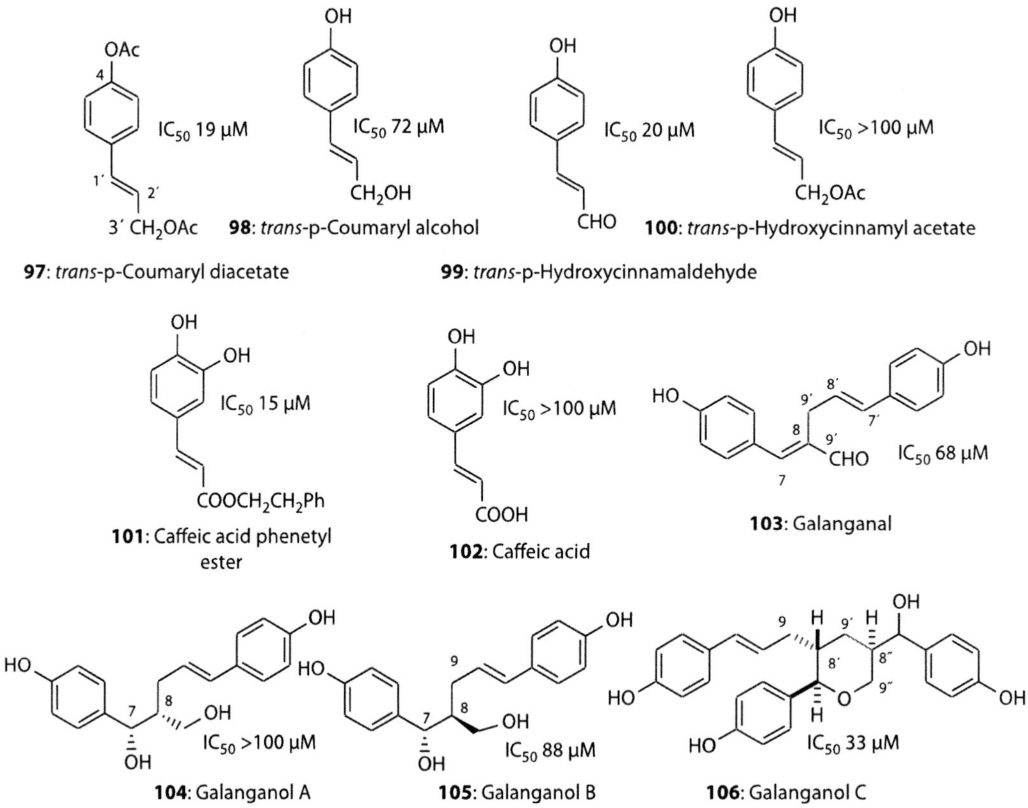

FIGURE 12.6 Structure-activity relationships within NO-inhibitory phenylpropanoid molecules.

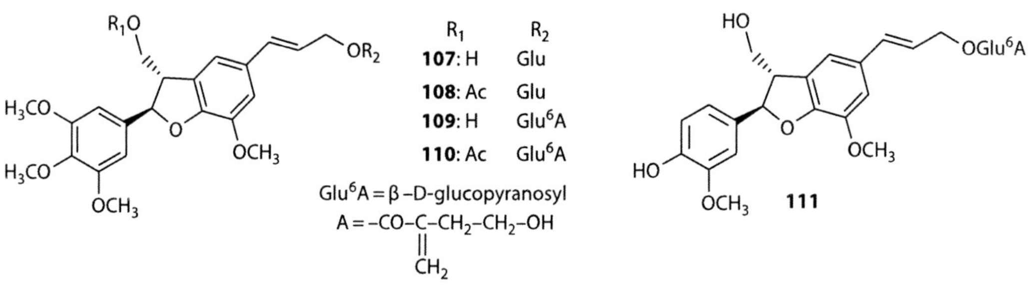

FIGURE 12.7 Dibenzofuran neolignan glycosides isolated from *Coptis japonica*.

12.2.5 Phenylbutanoids

Zingiber cassumunar, a plant from the Ginger family, is widely used in Southeast Asian countries. In Thailand, it is called "phlai" and finds application as a spice, and also in the treatment of asthma, bronchitis, and gastrointestinal disturbance. A methanolic extract from the rhizome of this plant have a NO-inhibitory activity in LPS-activated peritoneal macrophages of mice (Nakamura et al. 2009). Twenty-two compounds including the new compounds **112–121** (Figure 12.8) were isolated from this extract, most of which belong to the class of phenylbutanoids. Phlain I **112** and phlain II **113** (Figure 12.8) differ by the configuration of the substituents at C-1″; **112** exhibits NO-inhibitory activity (IC_{50} 24 μM), while such activity cannot be determined for **113** at the concentrations studied. Thus, the configuration of the substituents at the C-1″ is important and defines for such activity as well. The NO-inhibiting effect of compound **114** (Figure 12.8) is the same as in compounds

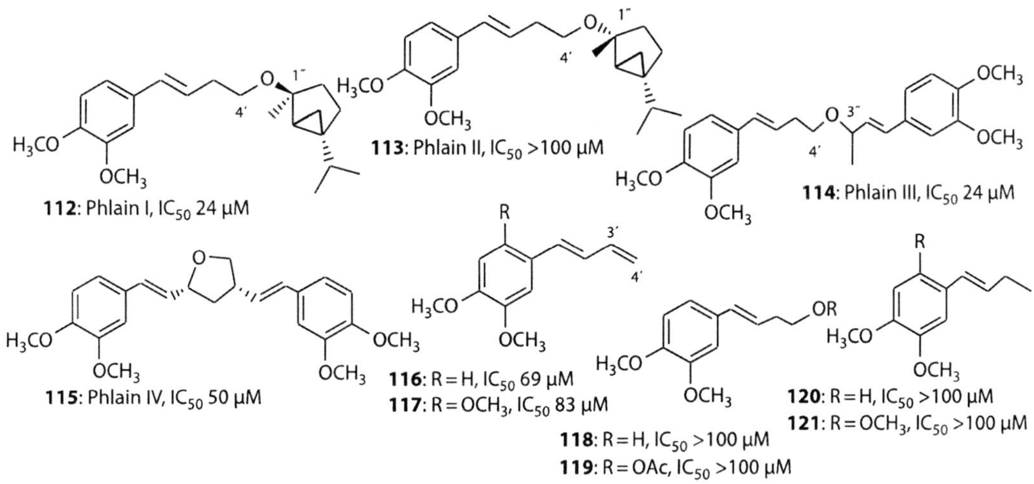

FIGURE 12.8 Phenylbutanoids isolated from *Zingiber cassumunar* and their comparative NO-inhibitory activity.

116–117; in this case, (*E*)-1-(3,4-dimethoxyphenyl)buta-1,3-dien **116** (IC_{50} 69 μM) and (*E*)-1-(2,4,5-trimethoxyphenyl)buta-1,3-dien **117** (IC_{50} 83 μM) inhibit the production of NO without exhibiting any cytotoxicity (Figure 12.8).

In the context of the structural peculiarities of phenylbutanoids affecting the activity, the presence and the absence of the terminal double bond in the butenyl radical (or the presence of the 1,3-dien fragment in the structure) should be noted. Here, compounds **116–121** can be compared (Figure 12.8). Compounds **116** and **117** that contain the terminal double bond in the butane fragment exhibit NO-inhibiting activity, whereas compounds **118–121** show no such activity in the range of concentrations studied.

12.2.6 Diarylheptanoids

A great deal of attention is given to diarylheptanoids (C_6–C_7–C_6 phenolic compounds) as inhibitors of NO production in activated macrophages. More and more new data on the biological activity of these compounds as anti-inflammatory agents have been published. By using various biochemical tests, the molecular mechanisms of inhibition of NO production were determined (Matsuda et al. 2001; Tao et al. 2002; Morikawa et al. 2003; Matsuda et al. 2006; Han et al. 2008; Li et al. 2010; Lai et al. 2011).

Compounds **122–125** (Figure 12.9) were isolated from the 80%-water-acetone extract of *Alpinia officinarum* (Ginger family) (Matsuda et al. 2006). Among these compounds, **125** and **124** exhibited a capability of inhibiting the production of NO (the IC_{50} values are 33 and 62 μM, respectively).

In order to make definite conclusions about the influence of the structure on the NO-inhibiting activity of diarylheptanoids, the investigation of compounds **122–134**, which were isolated from various plants applied in traditional Chinese medicine (Matsuda et al. 2001; Tao et al. 2002; Morikawa et al. 2003), was performed under the same conditions (Matsuda et al. 2006). Compounds **122–134** shown in Figure 12.9 were compared by the presence or absence of the enone fragment and conjugated double bonds in the seven-member bridge, and also of the substituents in the aromatic rings. First of all, it should be noted that compound **125**, which contains the enone fragment, showed the highest inhibitory activity among other diarylheptanoid components of the *A. officinarum* extract (Matsuda et al. 2006); compound **123** with the same substitution in the aromatic rings as in **125**, but a nonconjugated double bond in the carbonyl group at C-3, exhibited a lower activity (**125** [IC_{50} 33 μM] > **123** [IC_{50} > 100 μM]).

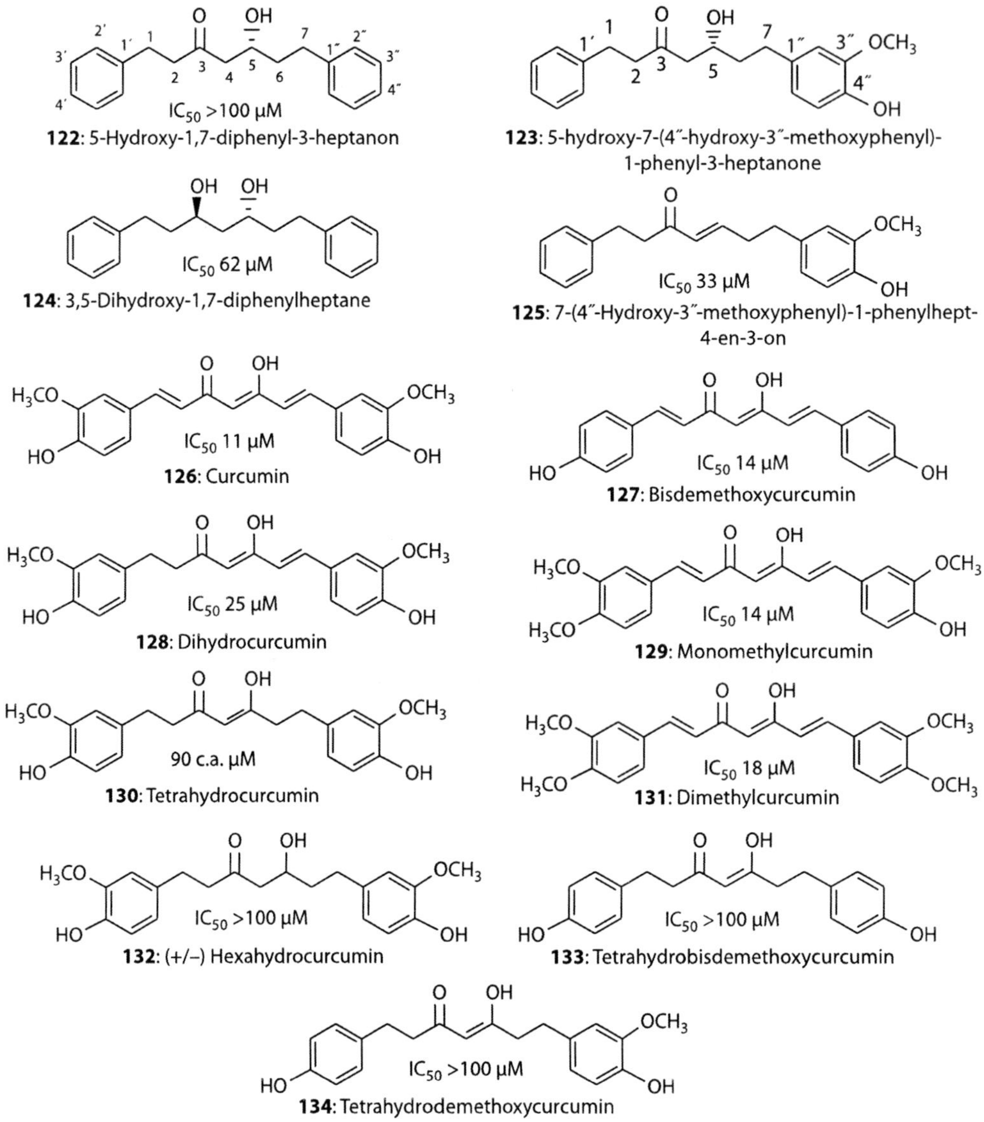

FIGURE 12.9 Diarylheptanoids isolated from *Alpinia officinarum* and their comparative NO-inhibitory activity.

Compound **122** containing unsubstituted aryl fragments and having the same structure of the aliphatic chain as in **123** also exhibited a low inhibition activity (IC_{50} > 100 μM). It is clear that the substituents in the aromatic rings do not influence the inhibitory activity, while the structure of the aliphatic fragment connecting two aromatic rings plays an important role in the inhibition. Thus, compounds **126**, **127**, **129**, and **131**, which contain the keto-group conjugated with three double bonds, exhibit the close and the high IC_{50} values, i.e., 11, 14, 14, and 18 μM, respectively. These diarylheptanoids have the same structure of the C_1–C_7 fragment, but differ by the substitution in the aromatic rings. It is appropriate to compare compounds **126**, **128**, **130**, and **132** (Figure 12.9) with each other. When the substitution of the aryl fragments is the same, and the number of double bonds between the C_1–C_7 atoms decreases, the NO-inhibiting activity decreases as well: IC_{50} 11, 25, 90, and >100 μM, respectively.

A low inhibition activity is observed for compounds **130**, **133**, and **134** (Figure 12.9), IC_{50} 90, >100, and >100 μM, which have the same structure of the C_1–C_7 fragment and differ by the substitution in the aromatic rings. Thus, the presence of the keto-enol fragment in the aliphatic chain does not cause an occurrence of a high NO-inhibition activity. The enone fragment without its conjugation with double bonds in compound **125** results in its rather high activity; however, the presence of the keto-enol fragments without the system of the conjugated double bonds in the C_1–C_7 bridge in compounds **130**, **133**, and **134** does not contribute to the NO-inhibition activity. It is suggested that the considered diarylheptanoids blocks the activation of NF-κB (Matsuda et al. 2006); no more precise information about the mechanism of inhibition was published.

5-*O*-Methylhirsutanonol **135**, isolated from *Alnus japonica*, Betulaceae (Figure 12.10), inhibited the production of NO depending in a dose-dependent manner (IC_{50} 14.5 μM), expression of iNOS proteins and iNOS mRNA, and also the activation of the nuclear factor, NF-κB (Han et al. 2008). This is true, the substance 135 acts as an inhibitor in various biochemical models. The inhibitory activity of compound **135** is higher than the activity of oregonin **136** (Figure 12.10) isolated from the same plant. Thus, the bulky hydrophilic substituent is supposed to reduce the inhibitory activity.

Twenty-two new diarylheptanoids were isolated from the extracts of *Curcuma kwangsiensis* (Ginger family) rhizome (Li et al. 2010); compounds **137–148** were examined for the inhibition of NO production in LPS-activated macrophages. For several compounds, the separation of

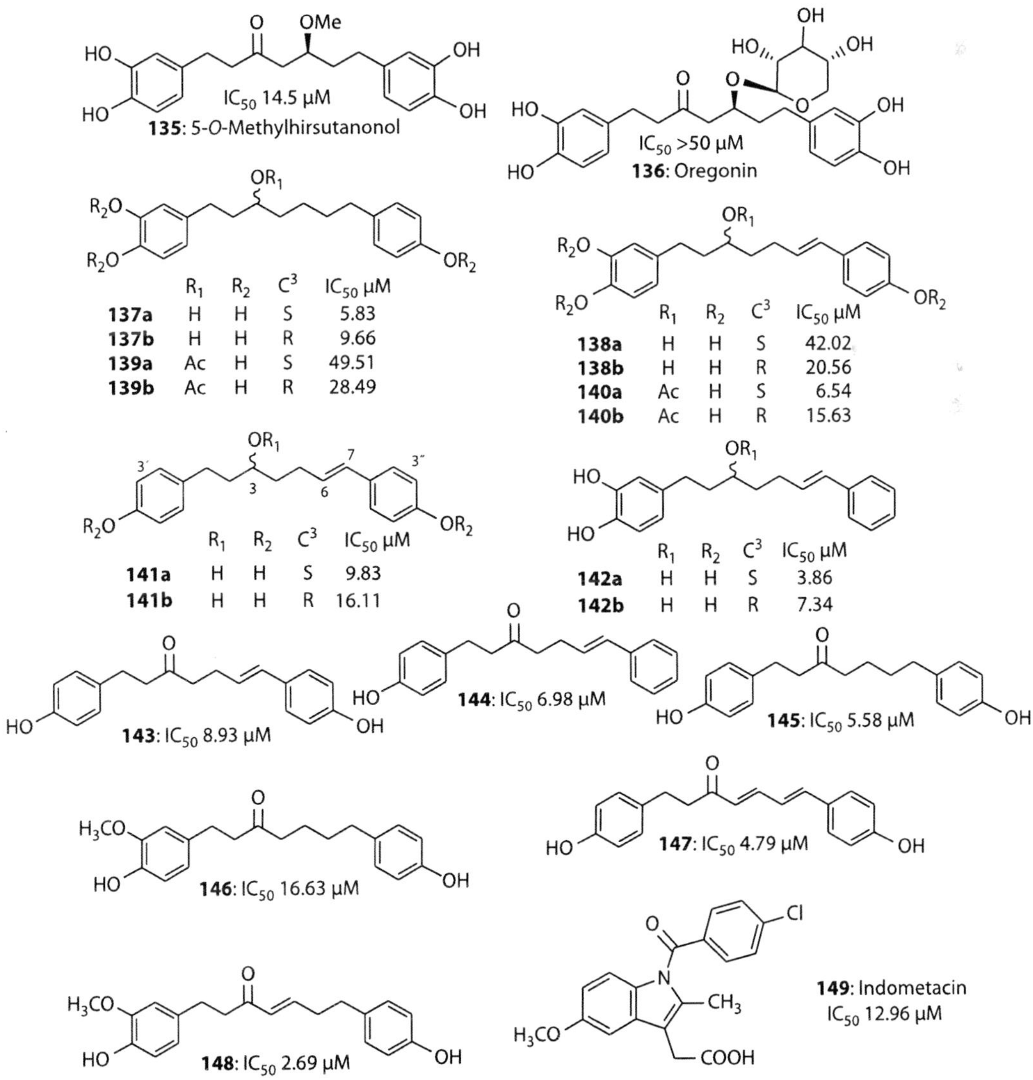

FIGURE 12.10 Structure-activity relationships within NO-inhibitory naturally occurring diarylheptanoids.

enantiomers was performed using chiral columns, and the conclusions about the structure–activity correlation were made with the asymmetric configuration of the carbon atom taken into account. The major part of the isolated compounds exhibit the high NO-inhibition activity, $IC_{50} < 15$ μM (Figure 12.10). Indometacin **149**, an anti-inflammatory drug (IC_{50} 12.96 μM), was used as a standard in the experiment; this substance is referred to the derivatives of indole acetic acid and has a pronounced analgetic activity found to be effective against rheumatoid arthritis, periarthritis, ankylosing spondylitis, osteoarthrosis, and podagra. All diseases outlined in the preceding text are believed to be associated with the excessive amounts of NO in organism (Li et al. 2010).

The inhibiting activity has almost no differences for the *S*/*R* enantiomers; it is high for both *S*- and *R*-isomers. However, it should be noted that in the majority of enantiomer pairs (except **138a/138b** and **139a/139b** pairs, Figure 12.10), the inhibition activity decreases upon moving from *S*- to *R*-enantiomer. The introduction of the substituents, OH or OCH_3 into the aryl fragments and the presence of the double bonds in the C_1–C_7 fragment have also no influence on the inhibitory activity. Although in each pair the activity differs almost by a factor of two, the authors assume it to be negligible, since all compounds **137–148** exhibit a rather high inhibition effect.

New data on the NO-inhibition activity of diarylheptanoids are continuingly published. Thus, 27 secondary metabolites were isolated from the extracts of *Alnus formosana* leaves with the purpose of studying their anti-inflammatory activity (Lai et al. 2011); 13 from those diarylheptanoids were newly found. Among the other components of the extract, new compound **150** and alnuside A **151** (Figure 12.11) exhibited the highest NO-inhibition activity in LPS-activated macrophages (IC_{50} 7.99 and 8.08 μM, respectively) without any cytotoxic effect. Let us compare compounds **150/151** with the pair of diarylheptanoids, **135/136** (Figure 12.10). While, in the **135/136** pair, the activity decreases almost by a factor of three upon replacing the methyl group for glycoside, in the **150/151** pair, the activity is almost the same upon replacing the *n*-butyl radical for the glycoside residue. It is difficult to make definite conclusions about how the substituent affects the activity.

The bark of *Acer nikoense* Maxim. (Aceraceae, grows in Japan) is used as folk medicine for the treatment of liver diseases and eye diseases. Cyclic diarylheptanoids **152–155** and acyclic diarylheptanoid **156** (Figure 12.11) were isolated from the extract of the *A, nikoense* Maxim. Bark (Morikawa et al. 2003); those compounds exhibited NO-inhibiting activity without any cytotoxic effect. Comparing the **152/153** and **154/155** pairs by the NO-inhibition activity, the authors indicated that biphenyl derivatives of diarylheptanoids, **154** and **155**, exhibit a higher activity than phenyl esters, **152** and **153**. Thus, the presence of the biphenyl fragment most likely affects the NO-inhibition activity.

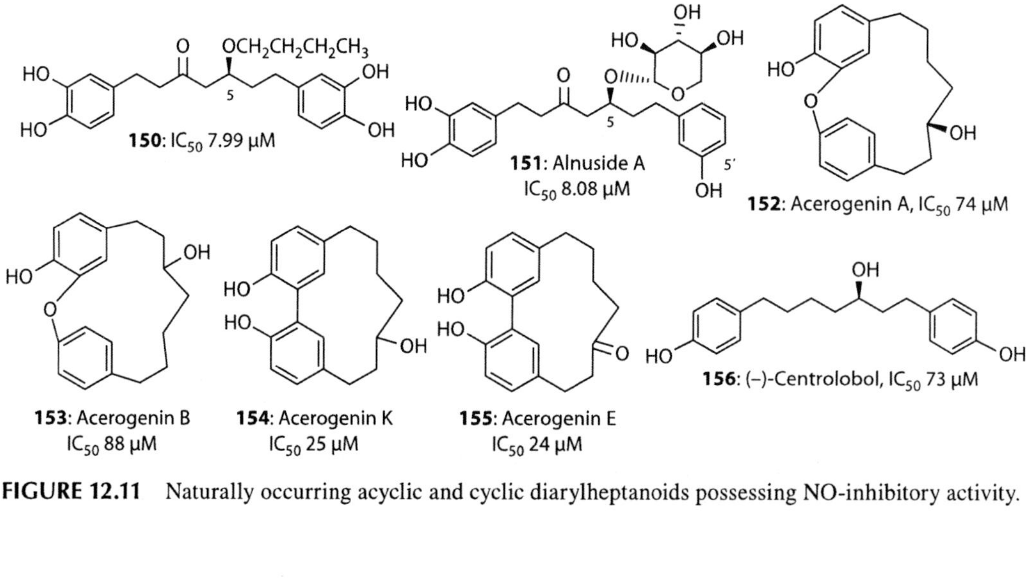

FIGURE 12.11 Naturally occurring acyclic and cyclic diarylheptanoids possessing NO-inhibitory activity.

12.2.7 Coumarins

More than 50 compounds were isolated from the extracts of *Angelica furcijuga*, a known medicinal plant widely used in Japanese folk medicine as a hepatoprotector, and an anti-inflammatory, anti-allergic, and hypotensive agent (Matsuda et al. 2005a,b; Yoshikawa et al. 2006). A large number of metabolites bear the coumarin skeleton. The coumarins of the khellactone-type, **157–167** (Figure 12.12), exhibited a significant activity with respect to the inhibition of NO production in LPS-activated macrophages (Matsuda et al. 2005; Yoshikawa et al. 2006); compounds **160** and **162–166** have $IC_{50} < 10$ μM and do not cause any toxic effect (Figure 12.12); and pterixin **161** and saxdorphin **167** showed IC_{50} values of 20 and 11 μM, respectively. The known iNOS inhibitor, L-NMMA, was used as a comparison sample; this compound exhibited a lower activity (IC_{50} 28 μM) under the experimental conditions in comparison to coumarins **160–167**.

Compounds **157** and **158** exhibited a much lower activity than coumarins **160–167**; most likely the acylation of both OH groups at the C-3′ and C-4′ atoms is required for an occurrence of NO-inhibitory activity. In addition, IC_{50} of isoepoxypteryxin **159** is almost four times less than the IC_{50} value of compound **160**. Thus, epoxidation of the double bond in the acyl radical causes a decrease in the activity.

The furocoumarins **168–170** (Figure 12.13) exhibited the NO-inhibitory activity in a significantly different manner than the others (Murakami et al. 1999). The furocoumarins **168–170** bear the skeleton of psoralen **171** (Figure 12.13); their activity varies depending on the structure of "R" group. Compounds **168** (Bergamottin) and **170** contain geranyl moiety in the structure; however, in **170**, there is a diol fragment over the $C_{6'}-C_{7'}$ atoms. The inhibition activity of compound **170** is almost 10 times less than the activity of furocoumarin **168**. In the structure of compound **169**, the diol fragment is located in the isoprenyl moiety, and the activity of compound **169** is lower than

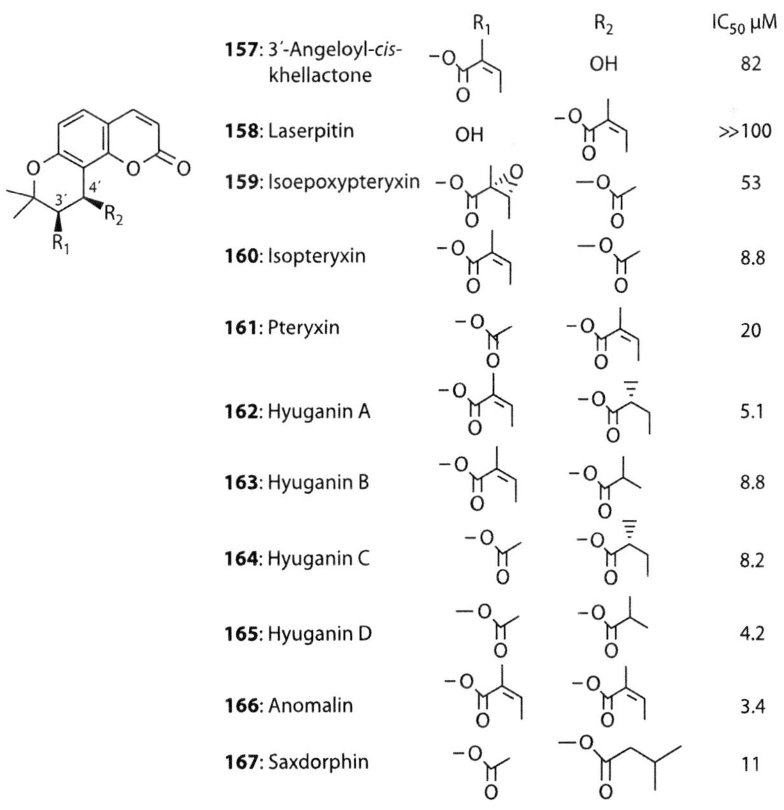

FIGURE 12.12 Khellactone-type coumarins exhibiting significant inhibition against NO production in LPS-activated macrophages.

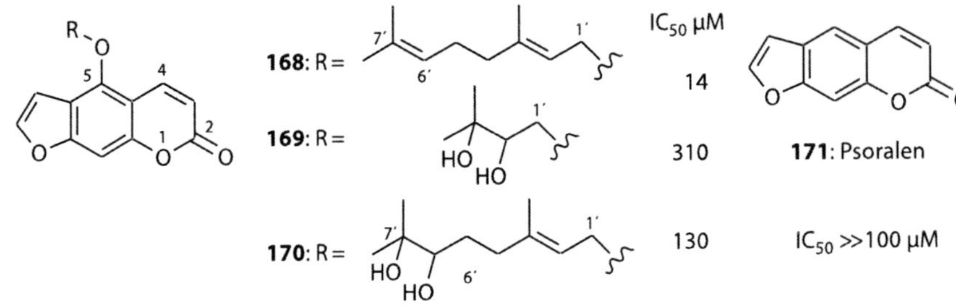

FIGURE 12.13 NO-inhibitory furocoumarins.

the activity of compound **168** more than in 20 times. Thus, the presence of the diol fragment in the moiety is supposed to influence the bioactivity of the substituted coumarins and leads to a decrease in the inhibition effect.

The known natural coumarins, **171–191** (Figures 12.13 and 12.14), isolated from the plants of Rutaceae exhibited no cytotoxicity during studying their NO-inhibition activity (Murakami et al. 1999). Compounds **171–191** were arbitrarily divided into three groups in accordance with their structural peculiarities. In the group of coumarins **172–177**, all compounds contain an isoprenyl fragment and alkylated OH groups in the structure (except compound **174**). In the group of coumarins **178–180**, the isoprenyl fragment forms a dimethyl chromene cycle. Compounds **182**, **183**, and **184** contain OH groups in the isoprenyl fragments. In the structures of compounds **185**, **190**, and **191**, the phenol groups are not alkylated. The investigation of the NO-inhibition was activity carried out at two concentrations of 10 and 50 μM (Murakami et al. 1999); the results indicated that, at a concentration of 10 μM, only dentatin **175** (Figure 12.14) inhibits the production of NO in LPS-activated macrophages by more than 50% (~80%), whereas the other coumarins exhibit a much lower activity. Thus, the compounds **168** > **172** > **173** > **174** (in the order of decreasing activity) inhibited the production of NO by 25%–30% at a concentration of 10 μM. Compounds **176**, **177**, **181**, and **178** had the activities close in value and inhibited the production of NO by ~15% at a concentration of 10 μM. More than half of all the compounds studied did not exhibit any inhibition capabilities at this concentration. In the experiment involving compound **170**, the production of NO increased. At a concentration of 50 μM, the inhibition of NO production by a little higher than 50% was observed for compounds **172** > **173** > **174** > **176** > **177** (in the order of decreasing activity); the concentration of NO was found to be abruptly increased to 20% in the experiments with compounds **190** and **191**. The analysis of the structures of the compounds exhibiting a high activity revealed the presence of the isoprenyl moiety and alkylated OH groups in the skeleton (except coumarin **174**). Thus, a higher NO-inhibition activity is observed for the coumarins containing bulky alkyl groups and alkylated OH functions in the structure; this kind of coumarin is also called prenylated coumarins.

12.2.8 Stilbenes

The other group of phytogenous phenolic compounds of plant origin capable of inhibiting the production of NO in activated macrophages is stilbenes (C_6–C_2–C_6 compounds). It was shown that pterostilbene **192** (Figure 12.15), *trans*-3,5-dimethoxy-4′-hydroxystilbene, a dimethyl analog of resveratrol **193**, isolated from *Vaccinium ashei* and *Vaccinium stamineum*, exhibits the anti-inflammatory activity similar to compound **193** and causes apoptosis of various cancer cells (Pan et al. 2008). By polymerase chain reaction (PCR) analysis, pterostilbene **192** was found to inhibit the production of NO (IC_{50} 25.3 μM) and to block the synthesis of iNOS mRNA in LPS-activated macrophages. Dihydrostilbenes (bibenzyls) **194–197** (Figure 12.15) isolated from *Dendrobium nobile* also exhibit iNOS-inhibitory activity (Zhang et al. 2007).

Compounds **194–197** (Figure 12.15) were evaluated to inhibit the production of NO without exhibiting a cytostatic effect. It should be noted that nobilin D **194** shows a higher NO-inhibitory

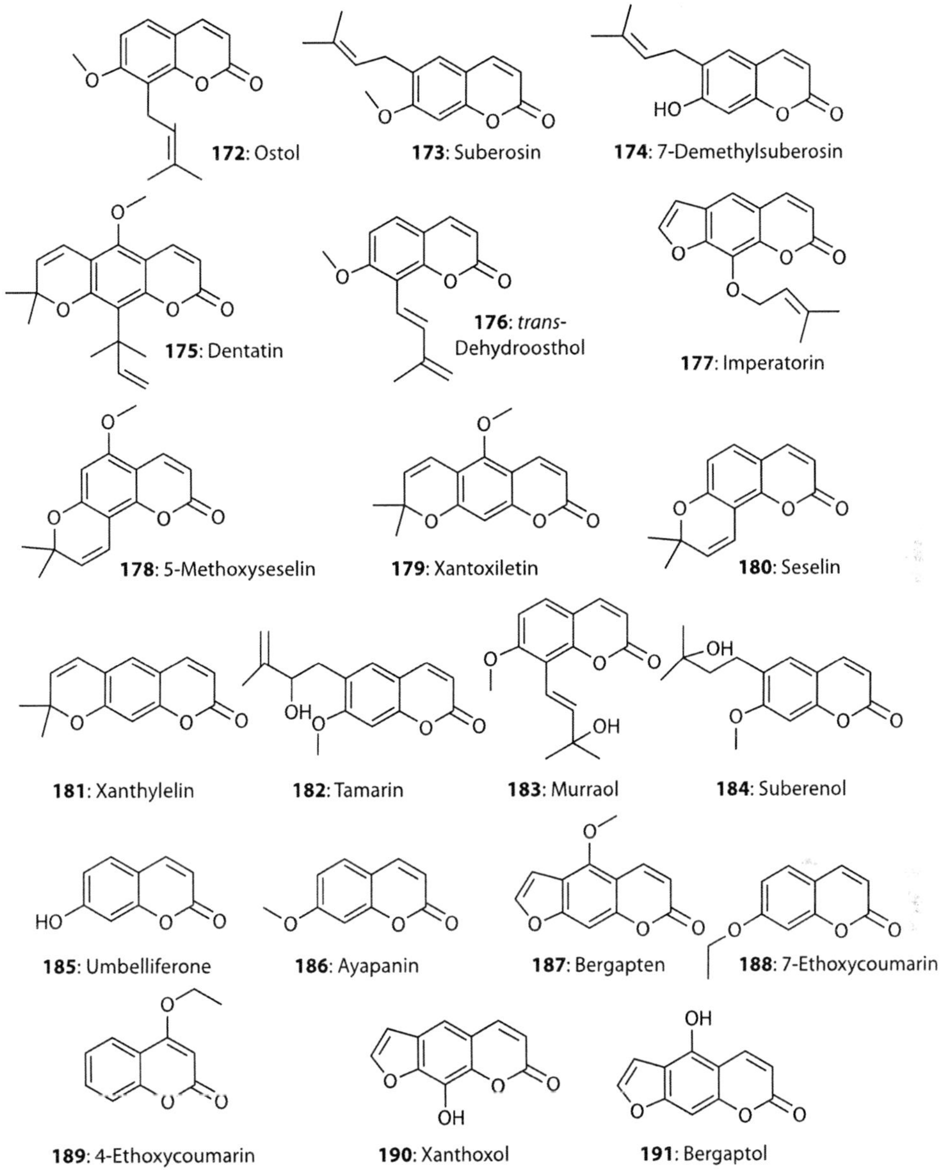

FIGURE 12.14 Naturally occurring NO-inhibitory coumarins.

activity than the comparison sample, resveratrol **193**, and higher by a factor of two more than dihydrostilbene **197**. The latter is most likely a result of the presence of the hydroxyl group in the α-position (R_4) in nobilin D **194**; this peculiarity distinguishes it from bibenzyls **195–197**. Thus, the presence of the OH group in the bridge connecting the aryl fragments is an important factor for an occurrence of inhibitory activity. It is believed that the biological activity of peanuts is defined by the presence of stilbenes in their composition, i.e., mostly resveratrol **193** and its derivatives (Stivala et al. 2001; Djoko et al. 2007). As can be seen from Figure 12.15, the stilbenes from peanuts **193**, **198**, and **199** differ from each other by the number of hydroxyl groups and the presence or absence of the isoprenyl fragment. It is reported that arahidin-1 **198** and piceatannol **199** have a higher NO-inhibition effect in LPS-activated macrophages than compound **193** (Djoko et al. 2007). Thus, it can be concluded that the combination of OH groups in the C-3′ and C-4′ positions affects an occurrence of inhibition activity.

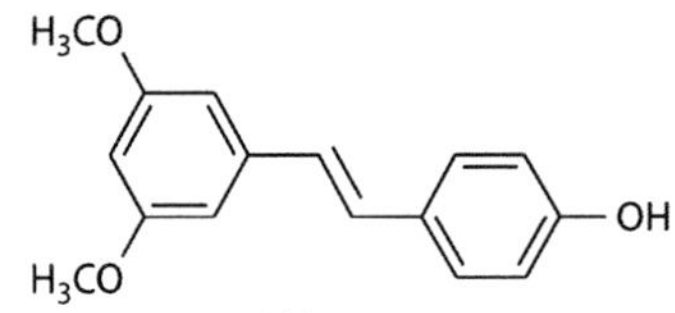

192

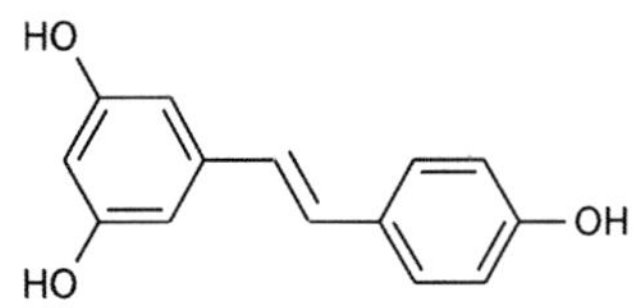

193: Resveratrol

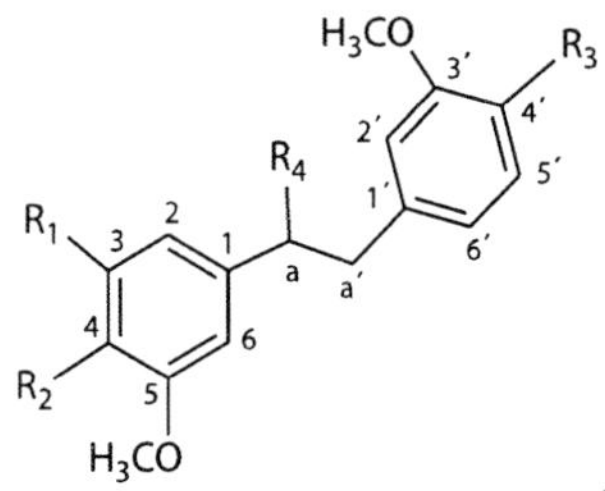

	IC_{50}, μM
193: Resveratrol	23.5
194: $R_1 = OCH_3$, $R_2 = R_3 = R_4 = OH$ nobilin D	15.3
195: $R_1 = R_2 = R_3 = OCH_3$, $R_4 = H$	48.2
196: $R_1 = OCH_3$, $R_2 = R_3 = OH$ $R_4 = H$	36.8
197: $R_1 = R_3 = H$ $R_2 = R_4 = OH$	32.9

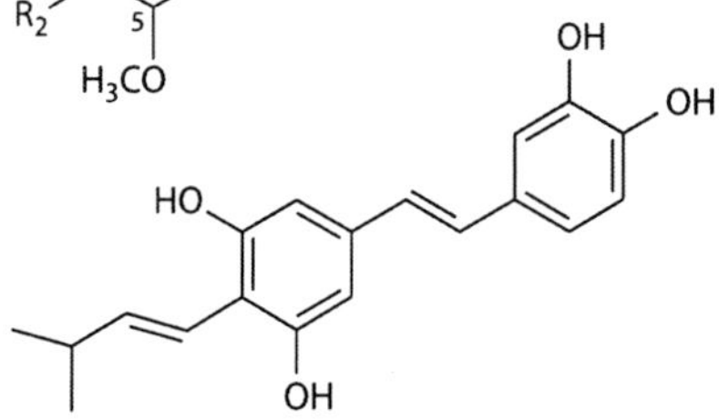

198: Arachidin-1, IC_{50} 15 μM

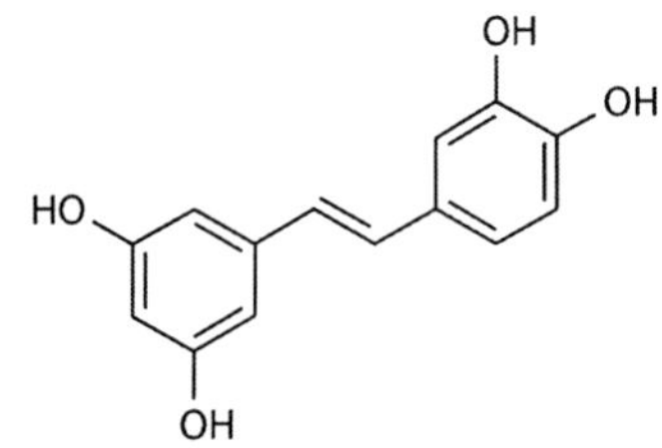

199: Piceatannol, IC_{50} 7.5 μM

FIGURE 12.15 Naturally occurring NO-inhibitory stilbene derivatives.

12.2.9 *Bis*-Bibenzyls

Bis-bibenzyls are macrocyclic dimer bibenzyls (**200–218**; Figure 12.16) biogenetically related to dihydrostilbenes. The natural source of *bis*-bibenzyls is liverworts, and their (**200–218**) NO-inhibitory activity has been evaluated by Harinantenaina et al. (2005). Marchantin A **200** has been found to be the strongest inhibitor in the series (IC_{50} 1.44 μM); the introduction of the hydroxyl group into the C-12 position (as can be seen by the example of marchantin B **201**) somewhat decreases the inhibition activity (marchantin B **201**, IC_{50} 4.1 μM). An additional hydroxyl group in the C-7′ position causes a decrease in the activity of compound **203** (IC_{50} 10.18 μM); methylation of this group also leads to a decrease in the activity by a factor of 40 (IC_{50} 62.16 μM) in comparison to marchantin A **200**. Perrotetin F **214** exhibits a high inhibition activity (IC_{50} 7.4 μM); the substituents in rings A and B in this compound are analogous to the substituents in marchantin A **200**.

The IC_{50} values for **206** and **207** containing methylated hydroxyl groups have been measured as 42.5 and 42.45 μM, respectively. This fact confirms a significant role of the nonsubstituted hydroxyl groups in the inhibition of NO production. However, riccardins A and F, **215** and **217**, which have one methoxy group in the C-11 and C-1′ positions, exhibit high inhibition activities (2.5 and 5 μM, respectively). Most likely, the reason behind this situation is a number of structural peculiarities of riccardins, i.e., the free hydroxyl group in the C-13′ position, C_{14}–$C_{12'}$-biphenyl bond, and the methoxy group in the C-11 or C-1′ position. The data shown in Figure 12.16 confirm this suggestion with the example of riccardin C **216**, which has a low inhibition activity characterized by an IC_{50} value of above 100 μM. It should be noted that the inhibition activity of isoplagiochin D **211** is three times lower than the activity of riccardin F **217**, despite the fact that rings B and D are substituted in the same way. A possible explanation of this behavior is the presence of the C_6–$C_{2'}$ biphenyl bond and the hydroxyl groups in rings A and C; altogether they provide a rigid conformation of compound **211**.

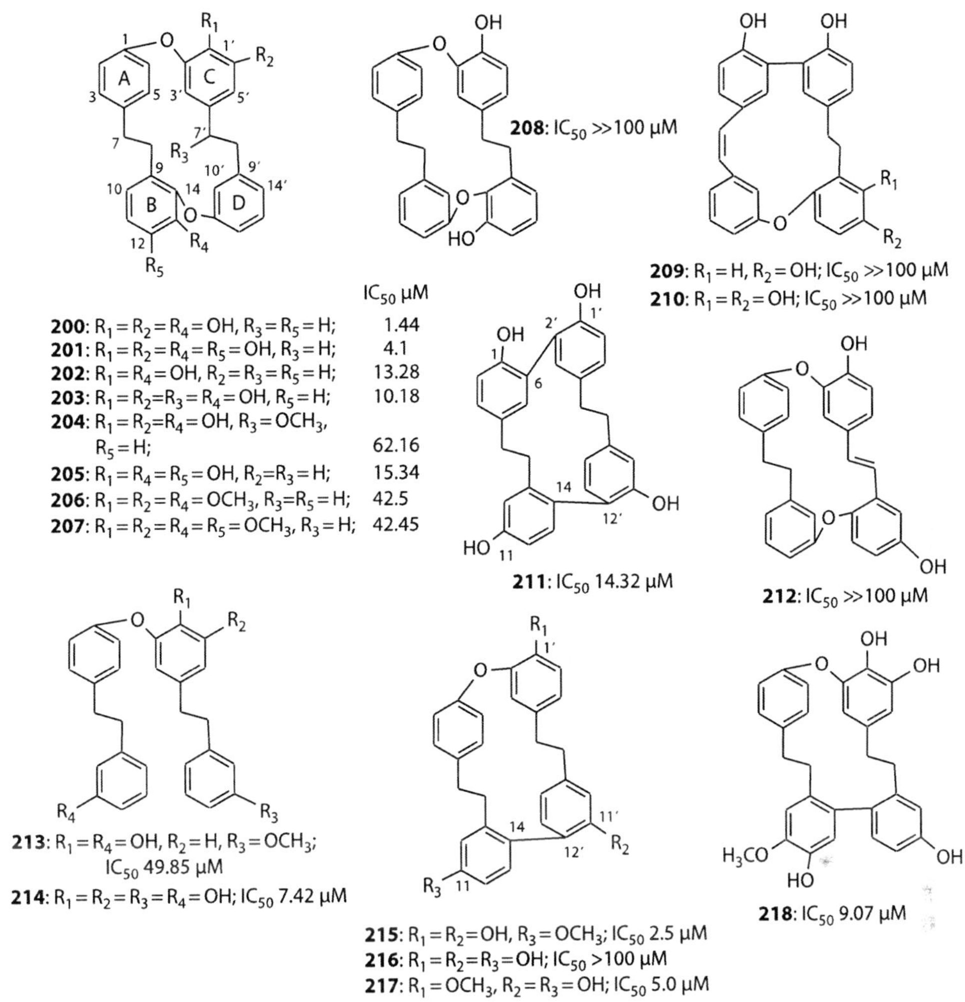

FIGURE 12.16 Naturally occurring NO-inhibitory bis-bibenzyls.

12.2.10 Terpenoids

The compounds isolated from *Laurus nobilis* are widely used in cookery and folk medicine. The extract metabolites **219**, **220**, and **221** (Figure 12.17) were found to exhibit the NO-inhibitory activity, but the molecular mechanism of this process still remains unclear (De Marino et al. 2004). Megastigmane glycosides **219**, **220**, and **221** isolated from the extracts of *Laurus* do not belong to terpenoid glycosides, since the aglycon skeleton contains 13 carbon atoms; however, the isolated glycosides exhibit a high biological activity. Compounds **220** and **221** have a special skeleton, which

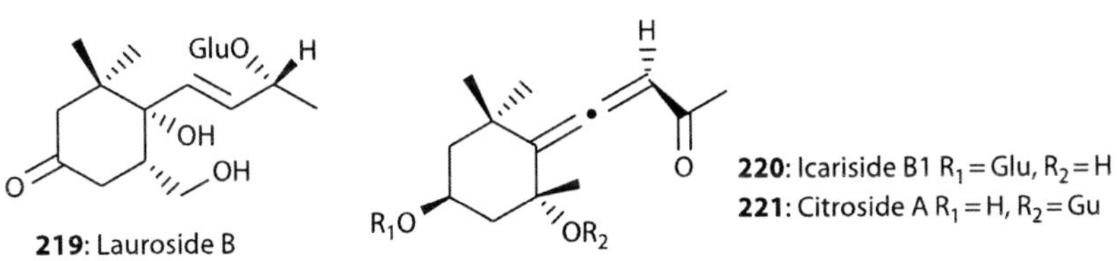

FIGURE 12.17 NO-inhibitory terepenoids isolated from *Laurus nobilis*.

involves the allene fragment; that is why they are considered in this review. Compounds **219–221** (Figure 12.17) inhibit the production of NO by 50% at a concentration of 10 μM.

It is noted in many phytochemical works (Rungeler et al. 1999; Castro et al. 2000; Siedle et al. 2004) that sesquiterpene lactones are found to be the major components of the extracts from the plants of the Asteraceae family. Extracts, tinctures, and decoctions from these plants are used in folk medicine as anti-inflammatory agents. Sesquiterpene lactones in low concentrations, from micromoles to nanomoles, inhibit the production of NO; a large number of these compounds do not exhibit any cytotoxic effect during inhibition. The mechanism of the action of sesquiterpene lactones involves the inhibition of the nuclear factor, NF-κB, with the IC_{50} value in the range of 5–10 μM. There are several works in which the attempts to determine the structure–activity correlation for the studied compounds were made, despite the complexity and diversity of lactone skeletons. Thus, analyzing the structure–activity correlation for 28 isolated lactones, Rungeler et al. (1999) commented that the high activity of compounds correlates with the presence of the following two fragments in the structure: α-methylene-γ-lactone fragment and α-, β- or γ-, δ-unsaturated carbonyl groups. An increase in the lipophilicity has no effect on the NF-κB-inhibition activity. A comprehensive study of the bioactivity of sesquiterpene lactones was conducted by Siedle et al. (2004), where the inhibitory activity with respect to the activation of the nuclear factor NF-κB was studied in 103 various sesquiterpene lactones belonging to six structural groups (44 germacranolides, 16 heliangolides, 22 guaianolides, 9 pseudoguaianolides, 2 hypocretenolide, and 10 eudesmanolides). Almost all compounds examined exhibited high levels of activity. The activation of the nuclear factor, NF-κB, initiates iNOS expression and other proinflammatory mediators.

Sesquiterpene lactones (**222–230**, Figure 12.18) isolated from *Artemisia sylvatica* exhibited NO-inhibition activity in LPS-activated macrophages characterized by the IC_{50} values in the range from 0.49 to 7.17 μM (Jin et al. 2004); the inhibition proceeds through the activation of the nuclear factor NF-κB without an occurrence of the cytostatic effect. The authors emphasize an importance of using extract components for treating inflammations of various etiologies. Lactones **222–230** contain the α-methylene-γ-lactone fragment in the structure; compounds **222**, **226**, **228**, **229**, and

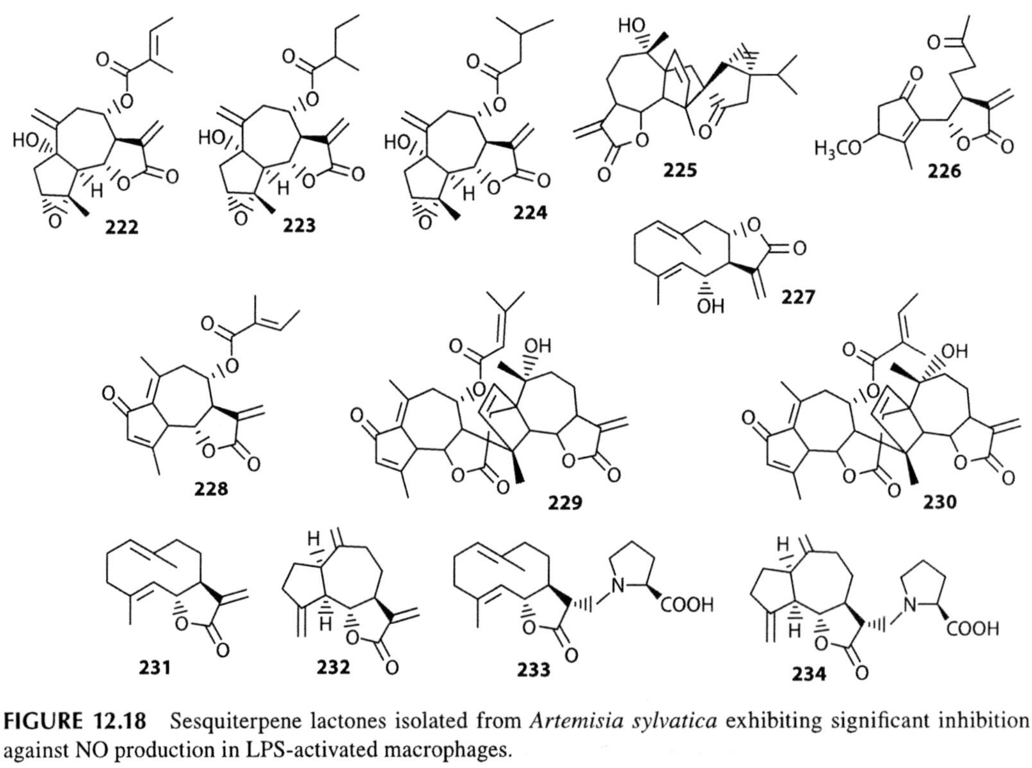

FIGURE 12.18 Sesquiterpene lactones isolated from *Artemisia sylvatica* exhibiting significant inhibition against NO production in LPS-activated macrophages.

230 also contain the enone fragment. Thus, the presence of these structural components plays an important role in the occurrence of inhibition activity (Rungeler et al. 1999).

Lactones **231–232** and the conjugates of these lactones with amino acids, **233–234** (Figure 12.18), the major components of the methanolic extract from *Saussurea lappa* Clarke roots, have the similar properties (Moore et al. 1994). Compounds **231–234** cause a reduction in the production of NO in LPS-activated peritoneal macrophages of mice by inhibiting NF-κB expression. A fact worthy of note is that during the formation of conjugates **233–234** (Figure 12.18), the methylene group in the active fragment of the lactone disappears; however, no decrease in the NO-inhibition activity was observed, and this is quite unclear from the standpoint of the postulated mechanism of inhibition (Rungeler et al. 1999; Castro et al. 2000; Jin et al. 2004; Siedle et al. 2004).

Balsamodendron mukul Hook. is a medicinal plant that grows in India, Sri Lanka, and in the north of Africa (Matsuda et al. 2004). The water-methanol extract from the resin of this plant also inhibited the production of NO in LPS-activated peritoneal macrophages of mice. The compounds **239–249** (Figure 12.19) of various structures (diterpenoids, triterpenoids, and steroids), isolated from this extract, exhibited the inhibition activity without any cytostatic effect. In particular, the triterpene alcohol, mirrhanol A **236**, and the diterpene alcohol, mukulol **249**, which are the major components of the extract, acted as selective dose-dependent inhibitors of iNOS expression. It is likely that the presence of these particular components defines the therapeutic effect of this plant. In the **235–237** series of compounds, compound **237**, whose structure contains the hydrocarbon substituent at C-9 position, has the lowest activity; hence, the presence of the polar group is important for an occurrence of activity. The presence of the OH group at the C-3 position in compounds **235–237** in place of the carbonyl group at the same carbon atom (compounds **238–241**) has no influence

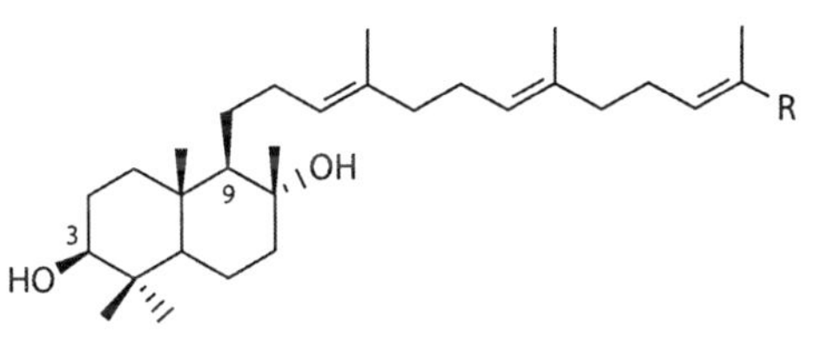

235: R = COOH; Myrrhanol B, IC_{50} 61 μM
236: R = CH_2OH; Mirrhanol A, IC_{50} 25 μM
237: R = CH_3; IC_{50} >>100 μM

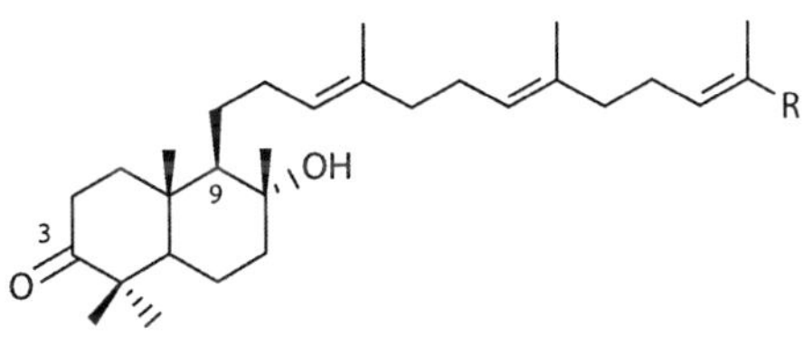

238: R = COOH; Myrrhanon B, IC_{50} >>100 μM
239: R = CH_2OH; Myrrhanon A, IC_{50} 35 μM
240: R = CH_3; IC_{50} >>100 μM
241: R = CH_2OAc; IC_{50} >>100 μM

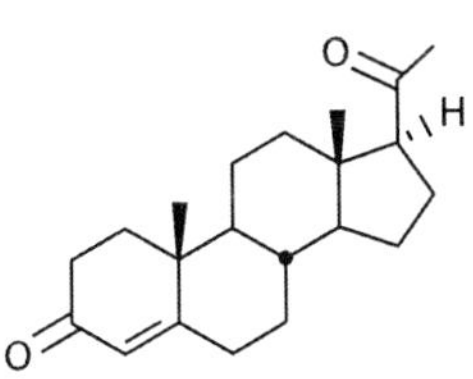

242: Progesterone, IC_{50} 11 μM

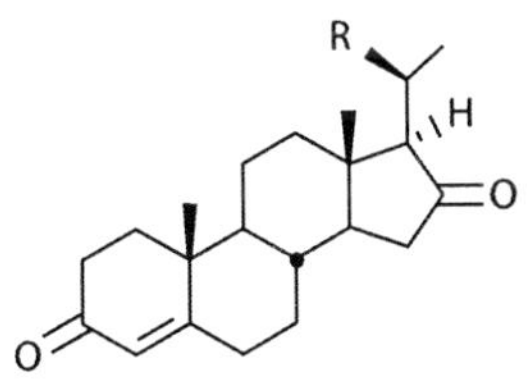

243: R = H, IC_{50} 40 μM
244: R = OAc, IC_{50} 56 μM

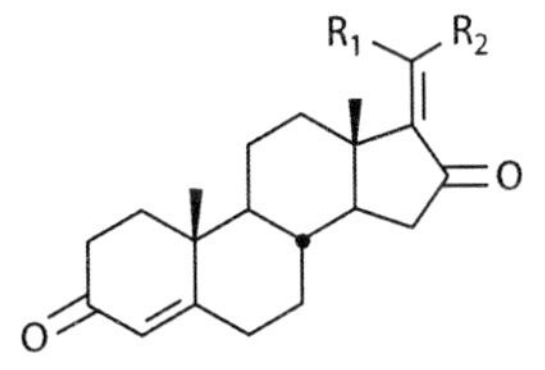

245: R_1 = CH_3, R_2 = H, IC_{50} 10 μM
246: R_1 = H, R_2 = CH_3, IC_{50} 8.2 μM

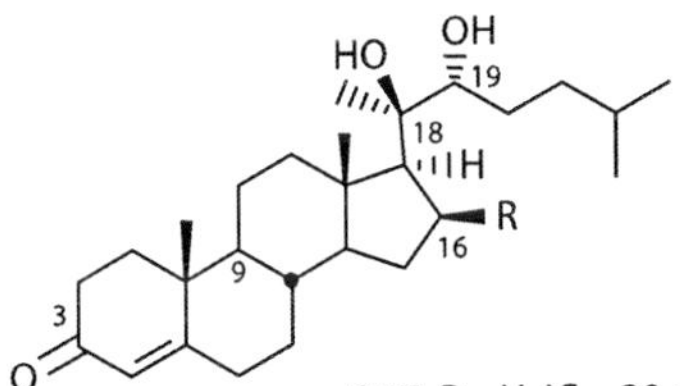

247: R = H, IC_{50} 20 μM
248: R = OH, IC_{50} >>100 μM

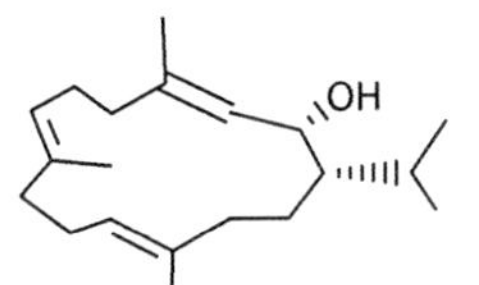

249: Mukulol, IC_{50} 24 μM

FIGURE 12.19 Terpenoid constituents of *Balsamodendron mukul* Hook.

on the inhibition activity. Most likely, the activity of compounds **235–241** is defined by the structure of the alkyl group at the C-9 position. It is interesting to note that compounds **243–246** having the skeleton of progesterone **242** exhibited a high degree of NO-inhibitory activity (Figure 12.19); the activity of compounds **245** and **246** is higher than the activity of compounds **243** and **244**. The main structural distinction of these pairs is the presence or absence of the exocyclic double bond, respectively; consequently, the configuration of the double bond is not so important for the occurrence of inhibition activity. Compounds **247** and **248** (Figure 12.19) are dammarane-type triterpene alcohols differing by the presence of the OH group at C-17 in the triterpene skeleton of compound **248**; the NO-inhibitory activity of triol **248** is over five times lower than the activity of diol **247**.

Isodon xerophilus is a perennial shrub from Yunnan, a Chinese province; the leaves of this plant are used as part of traditional Chinese medicine for the treatment of sore throats, inflammation, and flu.

Diterpenoids, i.e., xerophilusin A **250**, xerophilusin B **251**, longikaurin B **252**, and xerophilusin F **253** (Figure 12.20), so-called *ent*-kauranes, inhibited the production of NO in LPS-activated macrophages RAW 267.7 with the IC_{50} values of 0.60, 0.23, 0.44, and 0.67 μM, respectively, and also inhibited iNOS expression in these cells (Aquila et al. 2009). Compounds **250–253** were found to inhibit the activation of NF-κB as well. The inhibition activity of *ent*-kauranes with respect to the nuclear factor, NF-κB, is most likely defined by the presence of certain reaction sites in the structure. Thus, different authors pay attention to the *exo*-methylene group conjugated with the carbonyl group in the cyclopentanone fragment. This reactive group interacts with biological nucleophiles, such as the thiol group of cysteine, in the DNA-binding domain of the NF-κB subunits through the Michael-type reaction.

A high activity of *ent*-kaurane diterpenoids **254–257** (Figure 12.21) was noted by Giang et al. (2003); the sesquiterpene lactone, parthenolide **258**, used as a standard, showed the NO-inhibiting activity lower almost by an order of magnitude than diterpenoids **254–257** studied; the mechanism of inhibition for these compounds involves the suppression of NF-κB activation.

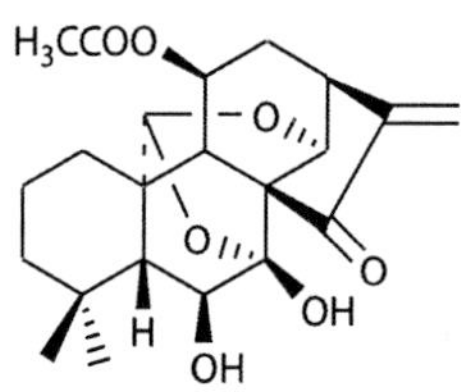

250: Xerophilusin A, IC_{50} 0.60 μM

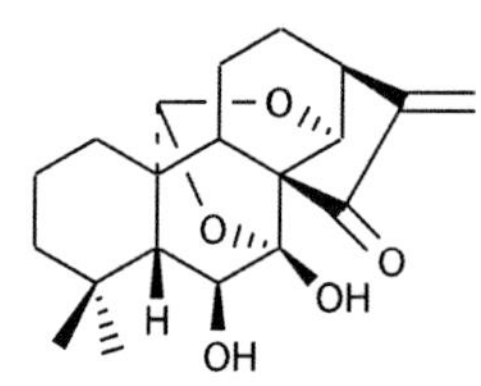

251: Xerophilusin B, IC_{50} 0.23 μM

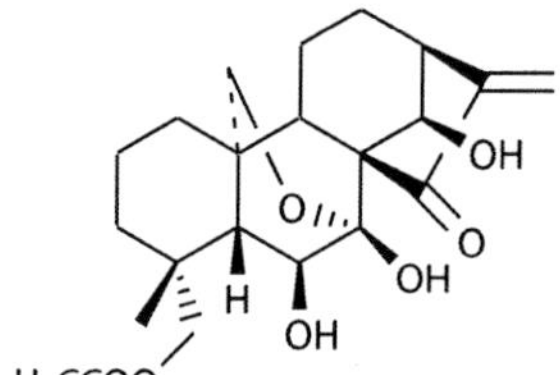

252: Longikaurin B, IC_{50} 0.44 μM

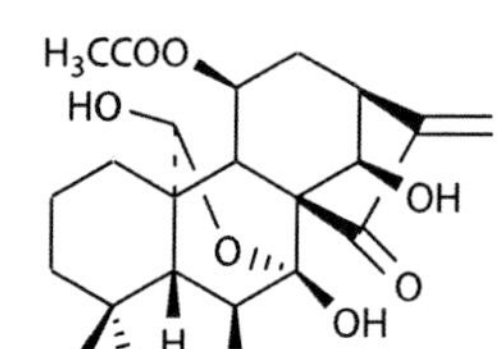

253: Xerophilusin F, IC_{50} 0.67 μM

FIGURE 12.20 *Ent*-kaurane diterpenoids isolated from *Isodon xerophilus* exhibiting significant inhibition against NO production in LPS-activated macrophages.

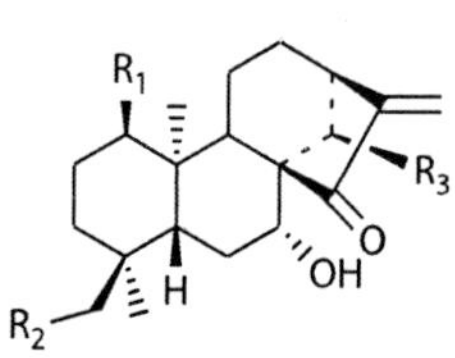

	R_1	R_2	R_3	IC_{50} μM
254:	H	H	OH	0.26
255:	H	OAc	H	0.21
256:	OAc	H	OH	0.47
257:	H	OAc	OH	0.15

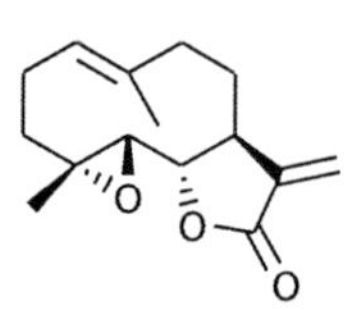

Parthenolide: **258**, IC_{50} 2.01 μM

FIGURE 12.21 Potent NO-inhibitory *ent*-kaurane diterpenoids.

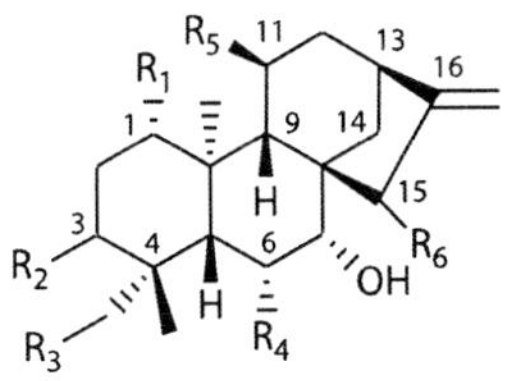

	R_1	R_2	R_3	R_4	R_5	R_6	IC_{50} μM
259:	H	β-OAc	OH	H	OH	=O	0.67
260:	OH	β-OAc	H	H	OH	=O	0.56
261:	H	α-OH	OAc	H	OH	β-OH	>10
262:	OH	β-OH	H	OH	OAc	=O	2.89
263:	H	=O	OAc	H	OH	=O	1.36
264:	H	=O	H	OH	OH	=O	1.24
265:	OH	β-OAc	H	H	OAc	=O	0.48
266:	OH	β-OAc	H	=O	OAc	=O	0.69
267:	H	β-OAc	H	OH	OH	=O	0.63
268:	OH	β-OAc	H	OH	OH	=O	2.52
269:	OH	β-OAc	H	OH	OAc	=O	0.94
270:	OH	β-OAc	H	OH	OAc	β-OH	>10

FIGURE 12.22 NO-inhibitory *ent*-kaurane diterpenoids isolated from *Isodon excisus*.

Isodon excisus (Labiatae) is a perennial plant commonly occurring in Korea, China, and Japan and is used in folk medicine for the treatment of gastrointestinal infections (Hong et al. 2007). *Ent*-kaurane diterpenoids **259–270** exhibiting a high NO-inhibition activity were isolated from the top of this plant (Figure 12.22). The mechanism of inhibition is the same as in the related *ent*-kauranes (Giang et al. 2003; Hong et al. 2007; Leung et al. 2005; Sun et al. 2006): The inhibition occurs through the suppression of NF-κB activation. A lower (relative to the other compounds of the group studied) NO-inhibition activity was observed for compounds **261** and **270**; this is most likely owing to the presence of four OH groups in the structure.

Three interesting compounds exhibiting a high NO-inhibition activity were isolated from the extracts of *Ferula fukanensis* roots: sesquiterpene coumarins **271–279** (Figure 12.23) (Motai et al. 2004), sesquiterpene phenylpropanoids **280–284** (Figure 12.24) (Motai and Kitanaka 2005a,b), and sesquiterpene chromones **285–289** (Figure 12.25) (Motai and Kitanaka 2005). The comparison of the structures of these compounds allows certain conclusions about the structure–activity correlation to be made. As seen from Figure 12.23, there is no keto-group in the sesquiterpene fragment in compounds **271** and **272**; both of these compounds do not show any inhibition activity (Figure 12.23). The inhibition activity of the other compounds is rather high (IC_{50} is in the range of 8.9–31.2 μM) and is comparable for the pairs of *cis*- and *trans*-isomers (by the C_2–C_3 bond in the furan cycle), i.e., **273/274**, **275/276**, and **277/278**; hence, it can be concluded that the configuration of the C-2 and C-3 atoms in the furan cycle have no effect on the inhibition activity. However, the inhibitory activity somewhat increases in the following order: **273/274** < **275/276** < **277/278**. From the comparison of the results, it can be concluded that the presence of the carbonyl group is an important factor for the occurrence of inhibition activity. If the carbonyl group is conjugated with the double bonds as observed in compounds **275/276**, the inhibition activity is somewhat higher than for compounds **273/274** containing the isolated carbonyl group. In these structural series of furo-coumarins, compounds **277/278**, in which the $C_{3'}$–$C_{4'}$ double bond has a Z-configuration, exhibited the highest activity. In compound **279**, furan cycle discloses, but owing to the fact that the enone fragment remains intact in the sesquiterpene molecule, the NO-inhibitory activity remains as well at rather high level (IC_{50} 19.5 μM, which is comparable with the IC_{50} values for compounds **273–278**).

Fukanedones A **280** and B **281** (Figure 12.24) do not inhibit the production of NO, since they do not have the carbonyl group in the sesquiterpene fragment (Motai and Kitanaka 2005). The other isolated compounds, **282–284**, exhibit NO-inhibitory activity in accordance with the values as shown in Figure 12.24; the cytotoxic effect of the compounds studied is negligible. Compounds **282** and **283**, which contain a keto-dienone fragment in the structure, have the highest inhibition

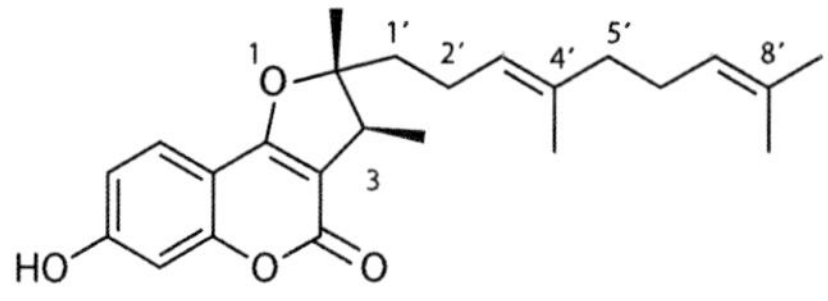

271

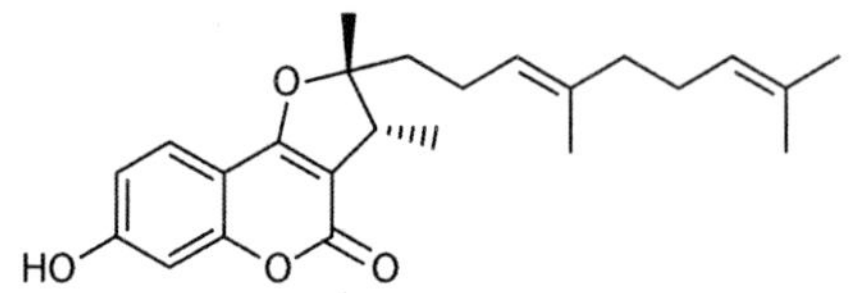

272

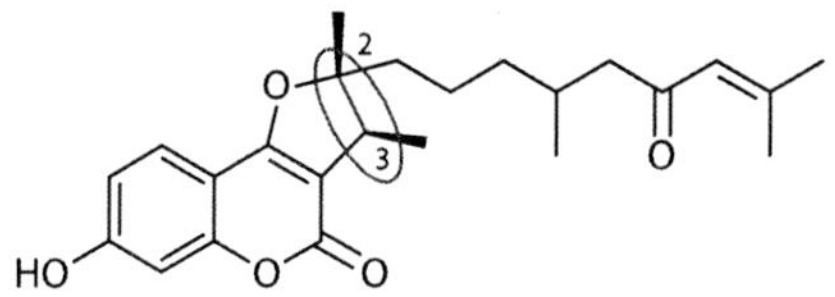

273: IC_{50} 24.6 μM

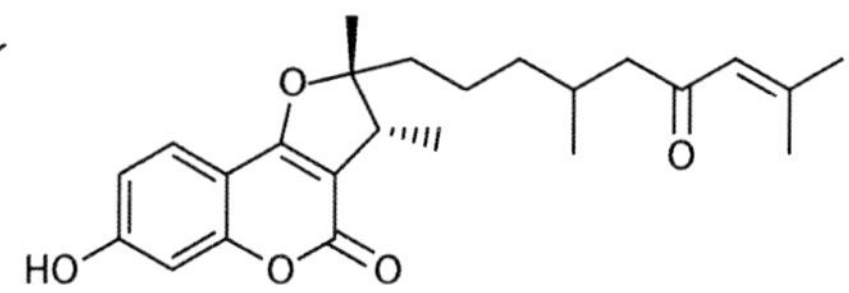

274: IC_{50} 31.2 μM

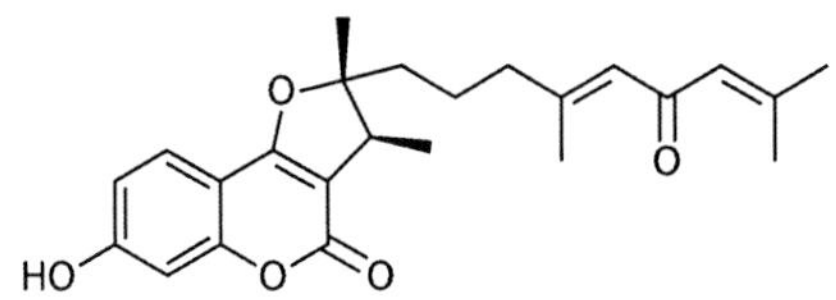

275: IC_{50} 13.0 μM

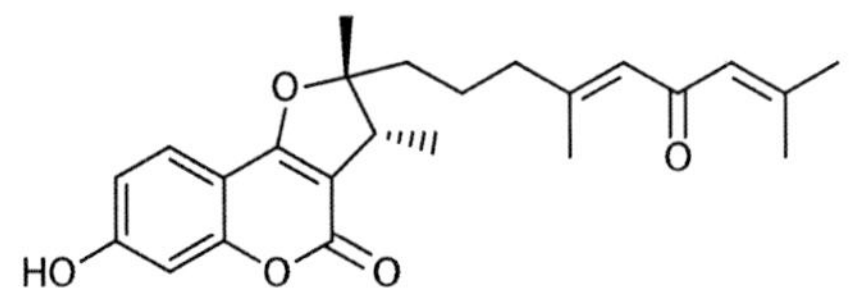

276: IC_{50} 16.2 μM

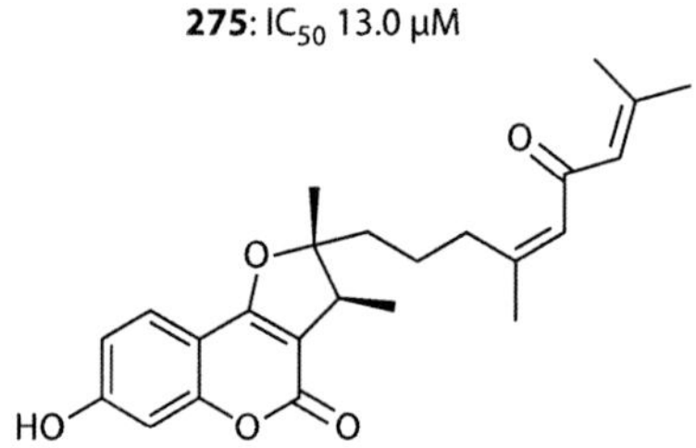

277: IC_{50} 11.1 μM

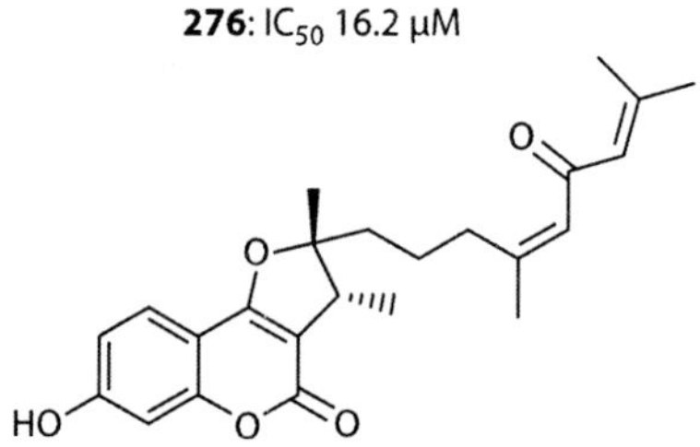

278: IC_{50} 8.9 μM

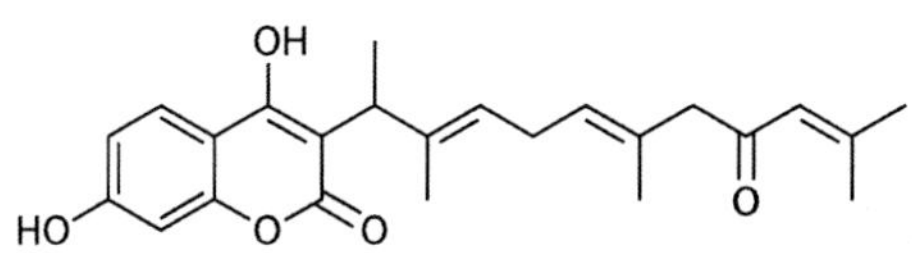

279: IC_{50} 19.5 μM

FIGURE 12.23 Sesquiterpene coumarins isolated from the extracts of *Ferula fukanensis* exhibiting a high NO-inhibition activity.

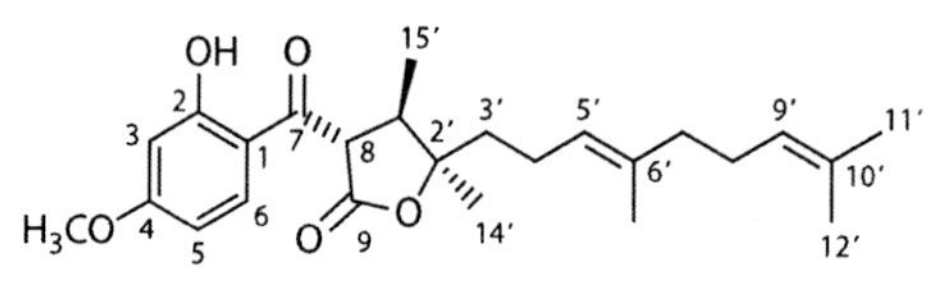

280: Fukanedone A

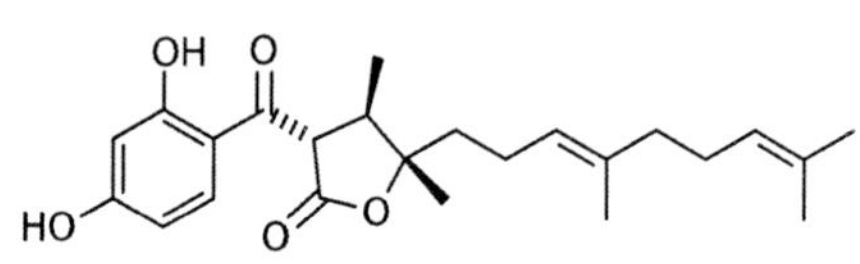

281: Fukanedone B

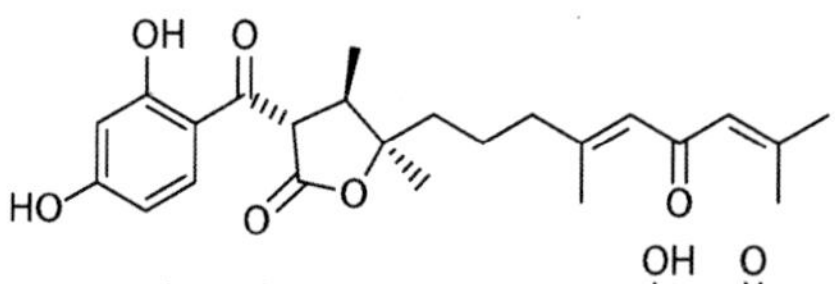

282: Fukanedone C, IC_{50} 6.7 μM

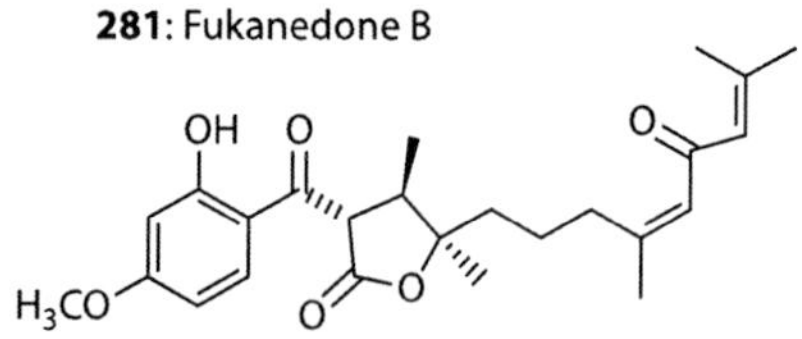

283: Fukanedone D, IC_{50} 27.8 μM

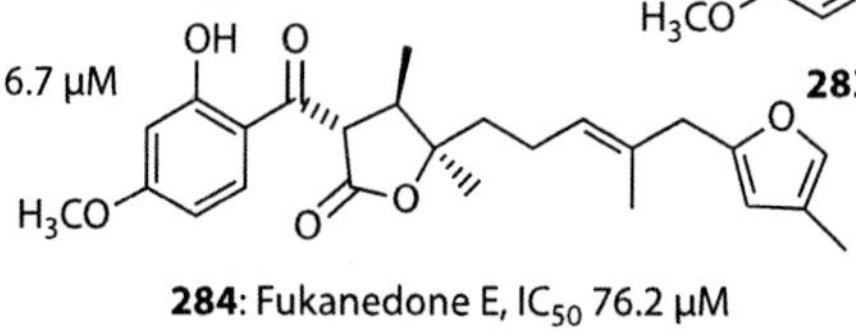

284: Fukanedone E, IC_{50} 76.2 μM

FIGURE 12.24 Sesquiterpene phenylpropanoids isolated from the extracts of *Ferula fukanensis* exhibiting a high NO-inhibition activity.

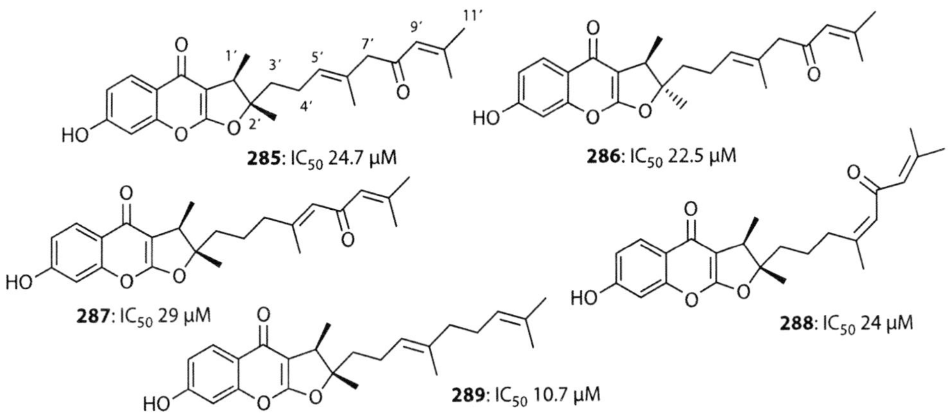

FIGURE 12.25 Sesquiterpene chromones isolated from the extracts of *Ferula fukanensis* exhibiting a high NO-inhibition activity.

activity (Figure 12.24); but in contrast to the previous work (Motai et al. 2004), compound **282** with the *E*-configuration exhibits the highest activity. It should be noted that the presence of an α,β-unsaturated keto-group in the sesquiterpene fragment is an important factor in the occurrence of NO-inhibition activity. When the furan cycle occurs between the $C_{8'}$–$C_{11'}$ atoms, the inhibition activity is decreased (compound **284**, Figure 12.24). The mechanism of action of compounds **282–284** involves inhibition of iNOS RNA expression in LPS-activated macrophages.

In the case of sesquichromones **258–289** (Figure 12.25), the NO-inhibition activity is high regardless of the structure of sesquiterpene radical (IC_{50} is in the range of 10.7–29 μM) (Motai and Kitanaka 2005). Moreover, the highest activity is found for compound **289**, the sesquiterpene fragment of which does not bear any carbonyl group; these data do not correlate with the data from Motai et al. (2004) and Motai and Kitanaka (2005a,b). From the analysis of these data, it can be concluded that, in the case of sesquichromones **285–289**, the chromone skeleton itself has the most influence on the NO-inhibition activity.

12.2.11 Alkaloids

During studying the 80%-water-methanol extract from *Crinum yemense* (Amaryllidaceae), it was shown that the extract inhibits the production of NO in the LPS-activated macrophage medium. The following alkaloids isolated from this extract also inhibit the production of NO (Abdel-Halim et al. 2004): yemenine A **290**, (+)-crinamine **291**, (+)-6-hydroxycrinamine **292**, and (–)-licorine **293** (Figure 12.26) (the IC_{50} values are 4.9, 1.8, 5.4, and 2.5 μM, respectively). By the mechanism of action, these compounds were referred to the agents inhibiting iNOS expression. Eucophylline **294** (Figure 12.26), a new tetracyclic vinyl-chinoline-type alkaloid isolated from the extract of *Leuconotis eugenifolius* along with leucophyllidine **295** (Deguchi et al. 2010). During the examination of these compounds for the NO-inhibition activity, it was found that compound **295** exhibits a high dose-dependent inhibition activity, while compound **294** does not (Figure 12.26).

Recently, Chen et al. (2010) reported more than 20 β-carboline-type alkaloids from the extracts of *Stellaria dichotoma* var. *Lanceolata* roots, among which 13 were new; five compounds (**296–300**) are the major components of the extract and exhibit high inhibition activities (IC_{50} are shown in Figure 12.27) comparable with the standard, aminoguanidine (IC_{50} 4.6 μM). *Isatis indigotica* Fort is referred to Cruciferaceae family and is a natural source of indigo. The extract of *Isatis* roots is widely used in the traditional Chinese medicine against acute and chronic diseases, such as flu, viral pneumonia, and hepatitis. Isaindigotone **301** (Figure 12.27) isolated from the chloroform-butanol fraction of the extract from roots (Wu et al. 1997) inhibits the production of NO in LPS-activated macrophages RAW 264.7 with $IC_{50} > 10$ μM (Molina et al. 2001). The synthetic analogs

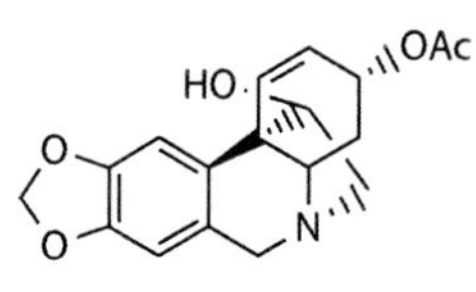

290: Yemenine A, IC_{50} 4.9 μM

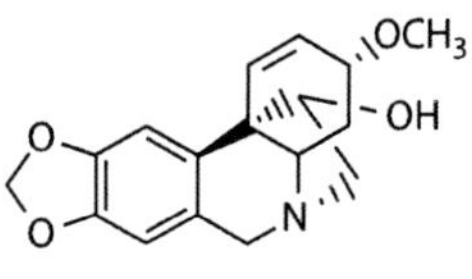

291: (+)-Crinamine, IC_{50} 1.8 μM

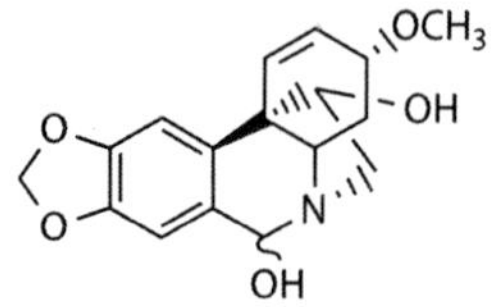

292: (+)-6-Hydroxycrinamine, IC_{50} 5.4 μM

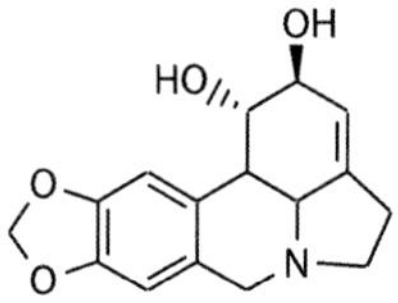

293: (−)-Licorine, IC_{50} 2.5 μM

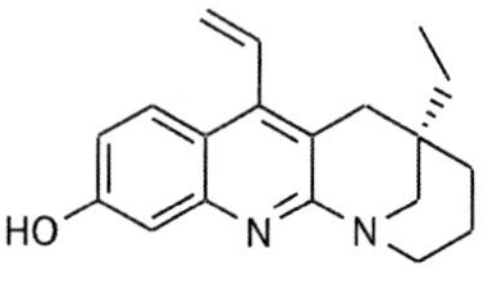

294

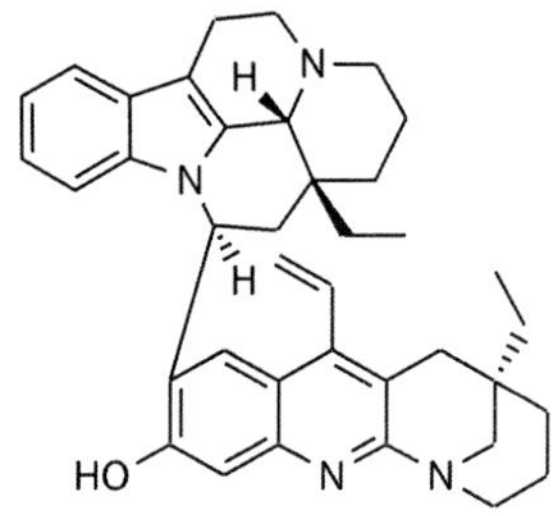

295: IC_{50} 7.1 μM

FIGURE 12.26 Naturally occurring alkaloids with NO-inhibitory activity.

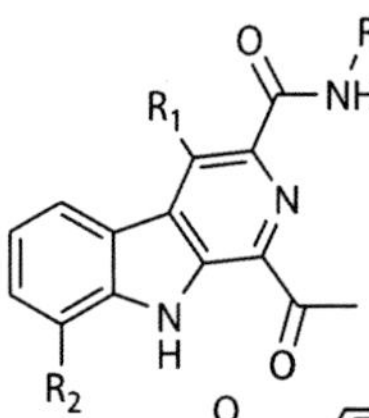

296: R_1 = H, R_2 = H, R_3 = (E)–CH = $CHCO_2CH_3$, IC_{50} 17.3 μM
297: R_1 = H, R_2 = H, R_3 = H, IC_{50} 19.3 μM
298: R_1 = H, R_2 = H, R_3 = (Z)–CH = $CHCO_2CH_3$, IC_{50} 18.6 μM

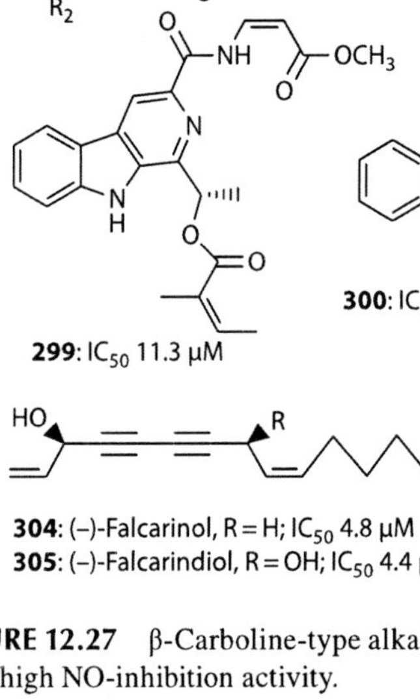

299: IC_{50} 11.3 μM

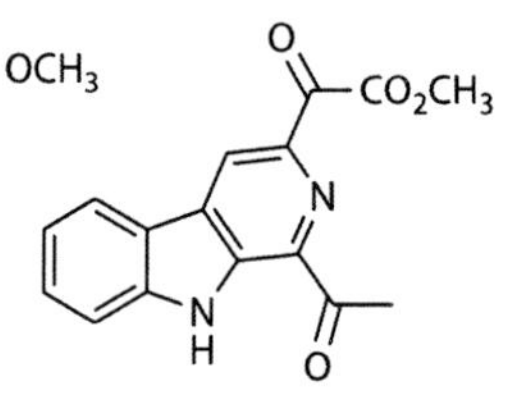

300: IC_{50} 17.9 μM

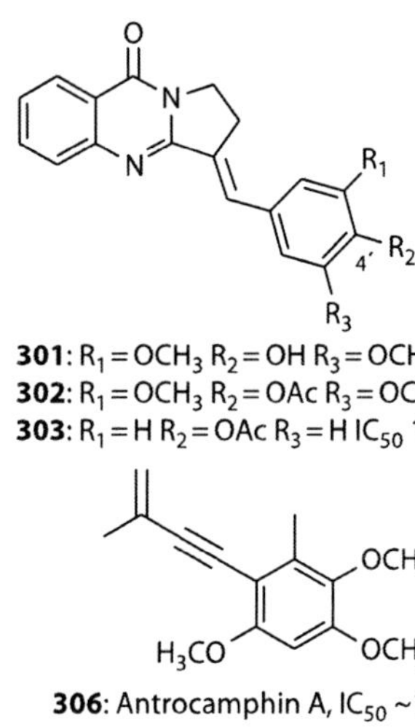

301: R_1 = OCH_3 R_2 = OH R_3 = OCH_3 IC_{50} >10 μM
302: R_1 = OCH_3 R_2 = OAc R_3 = OCH_3 IC_{50} 4.2 μM
303: R_1 = H R_2 = OAc R_3 = H IC_{50} 1.8 μM

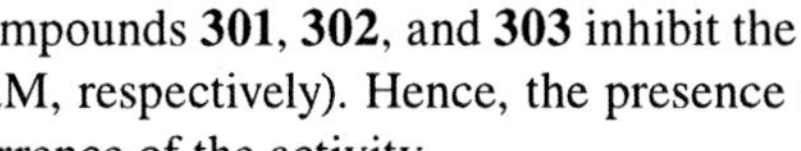

304: (−)-Falcarinol, R = H; IC_{50} 4.8 μM
305: (−)-Falcarindiol, R = OH; IC_{50} 4.4 μM

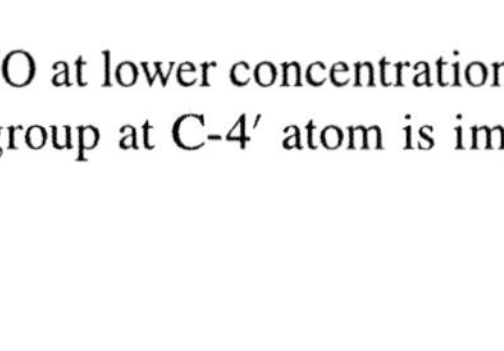

306: Antrocamphin A, IC_{50} ~15 μM

FIGURE 12.27 β-Carboline-type alkaloids isolated from *Stellaria dichotoma* var. *Lanceolata* roots exhibiting a high NO-inhibition activity.

of compounds **301**, **302**, and **303** inhibit the production of NO at lower concentrations (IC_{50} 4.2 and 1.8 μM, respectively). Hence, the presence of the acetoxy group at C-4′ atom is important for the occurrence of the activity.

12.2.12 Acetylenes

Diacetylenes **304** and **305** (Figure 12.27) were isolated from the extracts of *A. furcijuga*, a well-known medicinal plant widely used in the folk medicine of Japan as a hepatoprotector and anti-inflammatory, anti-allergic, and hypotensive agents (Yoshikawa et al. 2006). Compounds **304** and **305** exhibited a high activity as inhibitors of iNOS. The major component of the extract from *Taiwanofungus camphorates* used in the Taiwanese medicine for treating liver cancer (Hsieh et al. 2010) can be given as an example of the acetylene-type compound having a high inhibition activity. Antrocamphin A **306** (Figure 12.27) inhibits the production of NO at a concentration of $IC_{50} \sim 15\ \mu M$, which is higher than the activity of the standard, quercetin **22** under given conditions; it suppresses iNOS expression by inhibiting the activation of NF-κB. In the series of compounds **304–306**, there is a conjugated system of double and triple bonds (**304** and **305**), including the aromatic ring **306**, which might have an important influence on the occurrence of NO-inhibiting activity.

12.3 CONCLUSIONS

Plant secondary metabolites possess promising nitric-oxide-inhibitory potential as reflected from the detailed discussion in this chapter. The following conclusions about the correlation between the structure and the NO-inhibition activity can be inferred:

1. Upon moving from the phenolic compounds to the corresponding glycosides, a decrease in the inhibition activity is normally observed.
2. Flavonoid-type metabolites, in which the hydroxyl functions (phenolic or alcoholic) are "shielded" with alkyl groups, have a higher NO-inhibitory activity; as explained by the investigators of various works, an increase in the "lipophilicity" of molecules causes an increase in the inhibition activity. This is valid in the case of coumarins as well, and it is shown that molecules containing isoprenyl moiety exhibit high inhibition activity.
3. Arylpropanoids and diarylheptanoids form a large group of compounds having a high inhibitory activity against NO; for diarylheptanoids, the inhibitory activity increases in the case if there is an enone fragment in the seven-member bridge, especially if this fragment is conjugated with double bonds. Thus, the higher the length of conjugation in the seven-member bridge, the higher the inhibition activity.
4. Unfortunately, the structure–activity correlation cannot always be explained even in the groups of structure-related metabolites.

Analysis of the current status in the problem related to the search for new selective iNOS inhibitors among natural compounds shows the following:

1. A large number of natural compounds, isolated from the extracts of plants that are used in folk medicine against various inflammatory diseases, were examined for the inhibition of NO production LPS- or γ-IFN-activated macrophages.
2. The inhibitory mechanism of NO production in cells is associated with both the inhibition of iNOS expression and the inhibition of NF-κB activation, and in the case of the metabolites, the inhibition of NF-κB activation is more common.
3. Frequently, secondary plant metabolites have a very high inhibitory activity; hence, there are two possible research areas: (a) searching for natural inhibitors that involve the determination of the major components of the extracts, the development of the methods for their isolation, and the chemical modification of available natural compounds with the peculiarities of structure–activity correlation taken into account, and (b) searching for natural inhibitors having a high and unique activity; if the isolation of these compounds is not possible in sufficient amounts due their low contents, the synthesis from the available substrates is performed.
4. Analysis of the structure-activation correlation performed for secondary plant metabolites is very important for determining the main direction of chemical modification of natural skeletons, since the purpose of these transformations is to obtain compounds having a higher inhibition activity and low toxicity.

The diversity of the nature is inexhaustible, hence the investigation of plant extracts, the isolation of bioactive compounds from them, and the analysis of the structure-activation correlation expand the outlook of chemists-organics and turn their steps to obtaining biologically active compounds with a given activity and to the synthesis of compounds that can be used for creating new drugs with a high activity and low toxicity.

REFERENCES

Abdel-Halim, O. B., Morikawa, T., Ando, S. et al. 2004. New crinine-type alkaloids with inhibitory effect on induction of inducible nitric oxide synthase from *Crinum yemense*. *J Nat Prod* 67:1119–1124.

Alderton, W. K., Cooper, C. E., and Knowles, R. G. 2001. Nitric oxide synthases: Structure, function and inhibition. *Biochem J* 357:593–615.

Ando, S., Matsuda, H., Morikawa, T., and Yoshikawa, M. 2005. 1′*S*-1′-Acetoxychavicol acetate as a new type inhibitor of interferon-β production in lipopolysaccharide-activated mouse peritoneal macrophages. *Bioorgan Med Chem* 13:3289–3294.

Aquila, S., Weng, Z. Y., Zeng, Y. Q. et al. 2009. Inhibition of NF-κB activation and iNOS induction by ent-kaurane diterpenoids in LPS-stimulated RAW264.7 murine macrophages. *J Nat Prod* 72:1269–1272.

Babu, B. R. and Griffith, O. W. 1998. Design of isoform-selective inhibitors of nitric oxide synthase. *Curr Opin Chem Biol* 2:491–500.

Beharka, A. A., Han, S. N., Adolfsson, O. et al. 2000. Long-term dietary antioxidant supplementation reduces production of selected inflammatory mediators by murine macrophages. *Nutr Res* 20: 281–296.

Bremner, P. and Heinrich M. 2005. Natural products and their role as inhibitors of the pro-inflammatory transcription factor NF-κB. *Phytochem Rev* 4:27–37.

Castro, V., Rungeler, P., Murillo, R. et al. 2000. Study of sesquiterpene lactones from *Milleria quinqueflora* on their anti-inflammatory activity using the transcription factor NF-_B as molecular target. *Phytochemistry* 53:257–263.

Chao, J., Lu, T. C., Liao, J. W. et al. 2009. Analgesic and anti-inflammatory activities of ethanol root extract of *Mahonia oiwakensis* in mice. *J Ethnopharmacol* 125:297–303.

Chen, Y. F., Kuo, P. C., Chan, H. H. et al. 2010. β-Carboline alkaloids from *Stellaria dichotoma* var. *lanceolata* and their anti-inflammatory activity. *J Nat Prod* 73:1993–1998.

Cho, J. Y., Baik, K. U., Yoo, E. S. et al. 2000. In vitro antiinflammatory effects of neolignan woorenosides from the rhizomes of *Coptis japonica*. *J Nat Prod* 63:1205–1209.

Comalada, M., Ballester, I., Bailon, E. et al. 2006. Inhibition of pro-inflammatory markers in primary bone marrow-derived mouse macrophages by naturally occurring flavonoids: Analysis of the structure-activity relationship. *Biochem Pharmacol* 72:1010–1021.

De Marino, S., Borbone, N., Zollo, F. et al. 2004. Megastigmane and phenolic components from Laurus nobilis L. leaves and their inhibitory effects on nitric oxide production. *J Agric Food Chem* 52:7525–7531.

Deguchi, J., Shoji, T., Nugroho, A. E. et al. 2010. Eucophylline, a tetracyclic vinylquinoline alkaloid from *Leuconotis eugenifolius*. *J Nat Prod* 73:1727–1729.

Djoko, B., Chiou, R. Y. Y., Shee, J. J., and Liu, Y. W. 2007. Characterization of immunological activities of peanut stilbenoids, arachidin-1, piceatannol, and resveratrol on lipopolysaccharide-induced inflammation of RAW 264.7 macrophages. *J Agric Food Chem* 55:2376–2383.

Fang, S. H., Rao, Y. K., and Tzeng Y. M. 2005. Inhibitory effects of flavonol glycosides from *Cinnamomum osmophloeum* on inflammatory mediators in LPS/IFN-gamma-activated murine macrophages. *Bioorgan Med Chem* 13:2381–2388.

Furchgott, R. F. and Zavadski, J. W. 1980. The obligatory role of endothelial cells in the relaxation of vascular smooth muscle by acetylholine. *Nature* 286:373–376.

Giang, P. M., Jin, H. Z., Son, P. T. et al. 2003. *ent*-kaurane diterpenoids from *Croton tonkinensis* inhibit LPS-induced NF-KB activation and NO production. *J Nat Prod* 66:1217–1220.

Griffith, O. and Stuehr, D. 1995. NO synthases: Properties and catalytic mechanism. *Annu Rev Physiol* 57:707–736.

Han, J. M., Lee, W. S., Kim, J. R. et al. 2008. Effect of 5-*O*-methylhirsutanonol on nuclear factor- κB-dependent production of NO and expression of iNOS in lipopolysaccharide-induced RAW264.7 cells. *J Agric Food Chem* 56:92–98.

Harinantenaina, L., Quang, D. N., Takeshi, N. et al. 2005. Bis(bibenzyls) from liverworts inhibit LPS induced iNOS in 264.7 RAW cells. *J Nat Prod* 68(12):1779–1781.

Hollman, P. C. H. and Katan, M. B. 1999. Dietary flavonoids: Intake, health effects and bioavailability. *Food Chem Toxicol*. 37:937–942.

Hong, C. H., Hur, S. K., Oh, O. J. et al. 2002. Evaluation of natural products on inhibition of inducible cyclooxygenase (COX-2) and nitric oxide synthase (iNOS) in cultured mouse macrophage cells. *J Ethnopharmacol* 83:153–159.

Hong, S. S., Lee, S. A., Han, X. H. et al. 2007. Kaurane diterpenoids from *Isodon excisus* inhibit LPS-induced NF-KB activation and NO production in macrophage RAW264.7 cells. *J Nat Prod* 70:632–636.

Hsieh, Y. H., Chu, F. H., Wang, Y. S. et al. 2010. Antrocamphin A, an anti-inflammatory principal from the fruiting body of *Taiwanofungus camphoratus*, and its mechanisms. *J Agric Food Chem* 58:3153–3158.

Huang, Y. S. and Ho, S. C. 2010. Polymethoxy flavones are responsible for the anti-inflammatory activity of citrus fruit peel. *Food Chem* 119:868–873.

Hwang, Y. P., Choi, J. H., Yun, H. J. et al. 2011. Anthocyanins from purple sweet potato attenuate dimethylnitrosamine-induced liver injury in rats by inducing Nrf2-mediated antioxidant enzymes and reducing COX-2 and iNOS expression. *Food Chem Toxicol* 49:93–99.

Ichikawa, T., Li, J., Nagarkatti, P. et al. 2009. American ginseng preferentially suppresses STAT/iNOS signaling in activated macrophages. *J Ethnopharmacol* 125:145–150.

Jin, H. Z., Lee, J. H., Lee, D. et al. 2004. Inhibitors of the LPS-induced NF-kB activation from Artemisia sylvatica. *Phytochemistry* 65:2247–2253.

Jin, M., Suh, S. J., Yang, J. H. et al. 2010. Anti-inflammatory activity of bark of *Dioscorea batatas* DECNE through the inhibition of iNOS and COX-2 expressions in RAW264.7 cells via NF-κB and ERK1/2 inactivation. *Food Chem Toxicol* 48:3073–3079.

Jung, K. H., Ha, E., Kim, M. J. et al. 2007a. Suppressive effects of nitric oxide (NO) production and inducible nitric oxide synthase (iNOS) expression by *Citrus reticulata* extract in RAW 264.7 macrophage cells. *Food Chem Toxicol* 45:1545–1550.

Jung, C. H., Jung, H., Shin, Y. C. et al. 2007b. *Eleutherococcus senticosus* extract attenuates LPS-induced iNOS expression through the inhibition of Akt and JNK pathways in murine macrophage. *J Ethnopharmacol* 113:183–187.

Jung, H. W., Yoon, C. H., Park, K. M. et al. 2009. Hexane fraction of *Zingiberis* Rhizoma Crudus extract inhibits the production of nitric oxide and proinflammatory cytokines in LPS-stimulated BV2 microglial cells via the NF-kappaB pathway. *Food Chem Toxicol* 47:1190–1197.

Kang, H. K., Ecklund, D., Liu, M., and Datta, S. K. 2009. Apigenin, a non-mutagenic dietary flavonoid, suppresses lupus by inhibiting autoantigen presentation for expansion of autoreactive Th1 and Th17 cells. *Arthritis Res Ther* 11:R59.

Kang, H., Kwon, S. R., and Choi, H. Y. 2011. Inhibitory effect of *Physalis alkekengi* L. var. *franchetii* extract and its chloroform fraction on LPS or LPS/IFN-γ-stimulated inflammatory response in peritoneal macrophages. *J Ethnopharmacol* 135:95–101.

Kaszkin, M., Beck, K. F., Koch, E. et al. 2004. Downregulation of iNOS expression in rat mesangial cells by special extracts of *Harpagophytum procumbens* derives from harpagoside-dependent and independent effects. *Phytomedicine* 11:585–595.

Kerwin, J. F., Lancaster, J. R., and Feldman, P. L. 1995. Nitric oxide: A new paradigm for second messengers. *J Med Chem* 38:4343–4362.

Kiemer, A. K., Hartung, T., Huber, C., and Vollmar, A. M. 2003. *Phyllanthus amarus* has anti-inflammatory potential by inhibition of iNOS, COX-2, and cytokines via the NF-κB pathway. *J Hepatol* 38:289–297.

Kikuchi, K., Nagano, T., Hayakawa, H. et al. 1993. Real time measurement of nitric oxide produced ex vivo by luminol-H202 chemiluminescence method. *J Biol Chem* 268:23106–23110.

Kim, M. J., Kim, H. N., Kang, K. S. et al. 2004. Methanol extract of *Dioscoreae* Rhizoma inhibits pro-inflammatory cytokines and mediators in the synoviocytes of rheumatoid arthritis. *Int Immunopharmacol* 4:1489–1497.

Lai, Y. C., Chen, C. K., Lin, W. W., and Lee, S. S. 2011. A comprehensive investigation of anti-inflammatory diarylheptanoids from the leaves of *Alnus formosana*. *Phytochemistry* 73: 84–94.

Lau, F. C., Joseph, J. A., McDonald, J. E., and Kalt, W. 2009. Attenuation of iNOS and COX2 by blueberry polyphenols is mediated through the suppression of NF-κB activation. *J Funct Food* 1: 274–283.

Lee, K. Y. and Jeon, Y. J. 2003. Polysaccharide isolated from *Poria cocos* sclerotium induces NF-κB/Rel activation and iNOS expression in murine macrophages. *Int Immunopharmacol* 3:1353–1362.

Lee, M. H., Lee, J. M., Jun, S. H. et al. 2007. The anti-inflammatory effects of *Pyrolae herba* extract through the inhibition of the expression of inducible nitric oxide synthase (iNOS) and NO production. *J Ethnopharmacol* 112:49–54.

Lee, M. Y., Lee, J. A., Seo, C. S. et al. 2011. Anti-inflammatory activity of *Angelica dahurica* ethanolic extract on RAW264.7 cells via upregulation of heme oxygenase-1. *Food Chem Toxicol* 49:1047–1055.

Lee, J. S., Oh, T. Y., Kim, Y. K. et al. 2005. Protective effects of green tea polyphenol extracts against ethanol-induced gastric mucosal damages in rats: Stress-responsive transcription factors and MAP kinases as potential targets. *Mutat Res Fund Mol* 579: 214–224.

Leung, C. H., Grill, S. P., Lam, W. et al. 2005. Novel mechanism of inhibition of NF-kappaB DNA-binding activity by diterpenoids isolated from Isodon Rubescens. *Mol Pharmacol* 68:286–297.

Li, J., Zhao, F., Li, M. Z. et al. 2010. Diarylheptanoids from the Rhizomes of *Curcuma kwangsiensis*. *J Nat Prod* 73:1667–1671.

Lii, C. F., Chen, H. W., Yun, W. T., and Liu, K. L. 2009. Suppressive effects of wild bitter gourd (*Momordica charantia* Linn. var. *abbreviata* ser.) fruit extracts on inflammatory responses in RAW 264.7 macrophages. *J Ethnopharmacol* 122:227–233.

Lin, C. C., Hung, P. F., and Ho, S. C. 2008. Heat treatment enhances the NO-suppressing and peroxynitrite-intercepting activities of kumquat (*Fortunella margarita* Swingle) peel. *Food Chem* 109:95–103.

Lin, C. T., Kumar, K. J. S., Tseng, Y. H. et al. 2009. Anti-inflammatory Activity of Flavokawain B from *Alpinia pricei* Hayata. *J Agric Food Chem* 57:6060–6065.

Lopez-Posadas, R., Ballester, I., Abadia-Molina, A. C. et al. 2008. Effect of flavonoids on rat splenocytes, a structure-activity relationship study. *Biochem Pharmacol* 76:495–506.

MacMicking, J., Xie, Q. W., and Nathan, C. 1997. Nitric oxide and macrophage function. *Annu Rev Immunol* 15:323–350.

Matheus, M. E., Fernandes, S. B. O., Silveira, C. S. et al. 2006. Inhibitory effects of *Euterpe oleracea* Mart. on nitric oxide production and iNOS expression. *J Ethnopharmacol* 107:291–296.

Matsuda, H., Ando, S., Kato, T., Morikawa, T., and Yoshikawa, M. 2006. Inhibitors from the rhizomes of Alpinia officinarum on production of nitric oxide in lipopolysaccharide-activated macrophages and the structural requirements of diarylheptanoids for the activity. *Bioorgan Med Chem* 14:138–142.

Matsuda, H., Ando, S., Morikawa, T. et al. 2005a. Structure–activity relationships of 1′*S*-1′-acetoxychavicol acetate for inhibitory effect on NO production in lipopolysaccharide-activated mouse peritoneal macrophages. *Bioorg Med Chem Lett* 15:1949–1953.

Matsuda, H., Kiyohara, S., Sugimoto, S. et al. 2009. Bioactive constituents from Chinese natural medicines. XXXIII. Inhibitors from the seeds of *Psoralea corylifolia* on production of nitric oxide in lipopolysaccharide-activated macrophages. *Biol Pharm Bull* 32:147–149.

Matsuda, H., Morikawa, T., Ando, S. et al. 2003. Structural requirements of flavonoids for nitric oxide production inhibitory activity and mechanism of action. *Bioorgan Med Chem* 11:1995–2000.

Matsuda, H., Morikawa, T., Ando, S. et al. 2004. Absolute stereostructures of polypodane- and octanordammarane-type triterpenes with nitric oxide production inhibitory activity from guggul-gum resins. *Bioorgan Med Chem* 12:3037–3046.

Matsuda, H., Morikawa, T., Ohgushi, T. et al. 2005b. Inhibitors of nitric oxide production from the flowers of *Angelica furcijuga*: Structures of hyuganosides IV and V. *Chem Pharm Bull* 53:387–392.

Matsuda, H., Morikawa, T., Toguchida, I. et al. 2001. Medicinal foodstuffs. XXVIII. Inhibitors of nitric oxide production and new sesquiterpenes, zedoarofuran, 4-epicurcumenol, neocurcumenol, gajutsulactones A and B, and zedoarolides A and B, from *Zedoariae Rhizoma*. *Chem Pharm Bull* 49:1558–1566.

Matsuda, H., Yoshida, K., Miyagawa, K. et al. 2007. Rotenoids and flavonoids with anti-invasion of HT1080, anti-proliferation of U937, and differentiation-inducing activity in HL-60 from *Erycibe Expansa*. *Bioorgan Med Chem* 15:1539–1546.

Molina, P., Tarraga, A., Gonzalez-Tejero, A. et al. 2001. Inhibition of leukocyte functions by the alkaloid isaindigotone from *Isatis indigotica* and some new synthetic derivatives. *J Nat Prod* 64:1297–1300.

Moncada, S., Palmer, R. M. J., and Higgs, E. A. 1991. Nitric oxide: Physiology, pathophysiology and pharmacology. *Pharmacol Rev* 43:109–142.

Moore, W. M., Webber, R. K., Jerome, G. M. et al. 1994. L-N6-(1-Iminoethyl)-lysine: A selective inhibitor of inducible nitric oxide synthase. *J Med Chem* 37:3886–3888.

Morikawa, T., Ando, S., Matsuda, H. et al. 2005. Inhibitors of nitric oxide production from the rhizomes of *Alpinia galanga*: Structures of new 8–9′ linked neolignans and sesquineolignan. *Chem Pharm Bull* 53:625–630.

Morikawa, T., Tao, J., Toguchida, I. et al. 2003. Structures of new cyclic diarylheptanoids and inhibitors of nitric oxide production from Japanese folk medicine *Acer nikoense*. *J Nat Prod* 66:89–91.

Morikawa, T., Xu, F., Matsuda, H., and Yoshikawa, M. 2006. Structures of new flavonoids, erycibenins D, E, and F, and NO production inhibitors from *Erycibe expansa* originating in Thailand. *Chem Pharm. Bull.* 54:1530–1534.

Motai, T., Daikonya, A., and Kitanaka, S. 2004. Sesquiterpene coumarins from *Ferula fukanensis* and nitric oxide production inhibitory effects. *J Nat Prod* 67:432–436.

Motai, T. and Kitanaka, S. 2005a. Sesquiterpene phenylpropanoids from *Ferula fukanensis* and their nitric oxide production inhibitory effects. *J Nat Prod* 68:365–368.

Motai, T. and Kitanaka, S. 2005b. Sesquiterpene chromones from *Ferula fukanensis* and their nitric oxide production inhibitory effects. *J Nat Prod* 68:1732–1735.

Mueller, M., Hobiger, S., and Jungbauer, A. 2010. Anti-inflammatory activity of extracts from fruits, herbs and spices. *Food Chem* 122:987–996.

Murakami, A., Gao, G., Kim, O. K. et al. 1999. Identification of coumarins from the fruit of *Citrus hystrix* DC as inhibitors of nitric oxide generation in mouse macrophage RAW 264.7 Cells. *J Agric Food Chem* 47:333–339.

Nakamura, S., Iwami, J., Matsuda, H. et al. 2009. Structures of new phenylbutanoids and nitric oxide production inhibitors from the rhizomes of *Zingiber cassumunar. Chem Pharm Bull* 57:1267–1272.

Nathan, C. and Xie, Q. W. 1994. Nitric oxide synthases: Roles, tolls, and controls. *Cell* 78:915–918.

Ozer, L., El-On, J., Golan-Goldhirsh, A., and Gopas, J. 2010. Leishmania major: Anti-leishmanial activity of *Nuphar lutea* extract mediated by the activation of transcription factor NF-κB. *Exp Parasitol* 126:510–516.

Palmer, R. M. J., Ferigge, A. G., and Moncada, S. 1987. Nitric oxide release accounts for the biological activity of endothelium derived relaxing factor. *Nature* 327:524–526.

Pan, M. H., Chang, Y. H., Tsai, M. L. et al. 2008. Pterostilbene suppressed lipopolysaccharide-induced upexpression of iNOS and COX-2 in murine macrophages. *J Agric Food Chem* 56:7502–7509.

Pergola, C., Rossi, A., Dugo, P. et al. 2006. Inhibition of nitric oxide biosynthesis by anthocyanin fraction of blackberry extract. *Nitric Oxide* 15(1):30–39.

Pietta, P. G. 2000. Flavonoids as antioxidants. *J Nat Prod* 63:1035–1042.

Puangpraphant, S., Berhow, M. A., and Gonzalez de Mejia, E. 2009. Saponins in yerba mate tea (*Ilex paraguariensis* A. St.-Hil) and quercetin synergistically inhibit iNOS and COX-2 in lipopolysaccharide-induced macrophages through NF-κB pathways. *J Agric Food Chem* 57:8873–8883.

Rungeler, P., Castro, V., Mora, G. et al. 1999. Inhibition of transcription factor NF-κB by sesquiterpene lactones: A proposed molecular mechanism of action. *Bioorgan Med Chem* 7:2343–2352.

Schmidt, H. H. and Walter, U. 1994. NO at work. *Cell* 78:919–925.

Sheeba, M. S. and Asha, V. V. 2009. *Cardiospermum halicacabum* ethanol extract inhibits LPS induced COX-2, TNF-α and iNOS expression, which is mediated by NF-κB regulation, in RAW264.7 cells. *J Ethnopharmacol* 124:39–44.

Siedle, B., Garcia-Pineres, A. J., Murillo, R. et al. 2004. Quantitative structure-activity relationship of sesquiterpene lactones as inhibitors of the transcription factor NF-κB. *J Med Chem* 47:6042–6054.

Song, Y. S., Park, E. H., Hur, G. M. et al. 2002. Caffeic acid phenethyl ester inhibits nitric oxide synthase gene expression and enzyme activity. *Cancer Lett* 175:53–61.

Sripanidkulchai, B., Junlatat, J., Wara-aswapati, N., and Hormdee, D. 2009. Anti-inflammatory effect of *Streblus asper* leaf extract in rats and its modulation on inflammation-associated genes expression in RAW 264.7 macrophage cells. *J Ethnopharmacol* 124:566–570.

Stamler, J. S. 1994. Redox signaling: Nitrisylation and related target interactions of nitric oxide. *Cell* 78:931–936.

Stivala, L. A., Savio, M., Carafoli, F. et al. 2001. Specific structural determinants are responsible for the antioxidant activity and the cell cycle effects of resveratrol. *J Biol Chem* 276:22586–22594.

Sun, H. D., Huang, S. X., and Han, Q. B. 2006. Diterpenoids from *Isodon* species and their biological activities. *Nat Prod Rep* 23:673–698.

Sutherland, B. A., Rahman, R. M. A., and Appleton, I. 2006. Mechanisms of action of green tea catechins, with a focus on ischemia-induced neurodegeneration. *J Nutr Biochem* 17:291–306.

Taira, J., Nanbu, H., and Ueda, K. 2009. Nitric oxide-scavenging compounds in *Argimonia pilosa* Lebed on LPS-induced RAW 264.7 macrophages. *Food Chem* 115:1221–1227.

Tao, J., Morikawa, T., Toguchida, I. et al. 2002. Inhibitors of nitric oxide production from the bark of *Myrica rubra* structures of new biphenyl type diarylheptanoid glycosides and taraxerane type triterpene. *Bioorgan Med Chem* 10:4005–4012.

Tezuka, Y., Irikawa, S., Kaneko, T. et al. 2001. Screening of Chinese herbal drug extracts for inhibitory activity on nitric oxide production and identification of an active compound of *Zanthoxylum bungeanum. J Ethnopharmacol* 77:209–217.

Tong, X., Van Dross, R.T., Abu-Yousif, A. et al. 2007. Apigenin prevents UVB-induced cyclooxygenase 2 expression: Coupled mRNA stabilization and translational inhibition. *Mol Cell Biol* 27:283–296.

Van Meeteren, M. E., Hendriks, J. J., Dijkstra, C. D. et al. 2004. Dietary compounds prevent oxidative damage and nitric oxide production by cells involved in demyelinating disease. *Biochem Pharmacol* 67:967–975.

Veitch, N. C. and Grayer, R. J. 2008. Flavonoids and their glycosides, including anthocyanins. *Nat Prod Rep* 55:555–611.

Wang, P. G., Cai, T. B., and Taniguchi, N., eds. *Nitric Oxide Donors: For Pharmaceutical and Biological Applications*. Wiley-VCH, Weinheim, Germany, 2005a.

Wang, B. S., Chen, J. H., Liang, Y. C., and Duh, P. D. 2005b. Effects of Welsh onion on oxidation of low-density lipoprotein and nitric oxide production in macrophage cell line RAW 264.7. *Food Chem* 91(1):147–155.

Wang, G. J., Chen, Y. M., Wang, T. M. et al. 2008a. Flavonoids with iNOS inhibitory activity from *Pogonatherum crinitum. J Ethnopharmacol* 8:71–78.

Wang, J. and Mazza, G. 2002. Inhibitory effects of anthocyanins and other phenolic compounds on nitric oxide production in LPS/IFN-γ-Activated RAW 264.7 macrophages. *J Agric Food Chem* 50:850–857.

Wang, L., Tu, Y. C., Lian, T. W. et al. 2006. Distinctive antioxidant and antiinflammatory effects of flavonols. *J Agric Food Chem* 54:9798–9804.

Wang, B. S., Yu, H. M., Chang, L. W. et al. 2008b. Protective effects of pu-erh tea on LDL oxidation and nitric oxide generation in macrophage cells. *Food Sci Technol* 1:1122–1132.

Wei, B. L., Weng, J. R., Chiu, P. H. et al. 2005. Antiinflammatory flavonoids from *Artocarpus heterophyllus* and *Artocarpus communis. J Agric Food Chem* 53:3867–3871.

Woodward, J. J., Chang, M. M., Martin, N. I., and Marletta, M. A. 2009. The second step of the nitric oxide synthase reaction: Evidence for ferric-peroxo as the active oxidant. *J Am Chem Soc* 131:297–305.

Wu, X., Qin, G., Cheung, K. K., and Cheng, K. F. 1997. New alkaloids from Isatis indigotica. *Tetrahedron* 53:13323–13328.

Wu, J. H., Tung, Y. T., Chein, S. C. et al. 2008. Effect of phytocompounds from the heartwood of *Acacia confusa* on inflammatory mediator production. *J Agric Food Chem* 56:1567–1573.

Yen, G. C., Duh, P. D., Huang, D. W. et al. 2008. Protective effect of pine (*Pinus morrisonicola* Hay.) needle on LDL oxidation and its anti-inflammatory action by modulation of iNOS and COX-2 expression in LPS-stimulated RAW 264.7 macrophages. *Food Chem Toxicol* 46:175–185.

Yoshikawa, M., Nishida, N., Ninomiya, K. et al. 2006. Inhibitory effects of coumarin and acetylene constituents from the roots of *Angelica furcijuga* on D-galactosamine/lipopolysaccharide-induced liver injury in mice and on nitric oxide production in lipopolysacchride-activated mouse peritoneal macrophages. *Bioorgan Med Chem* 14:456–463.

Yu, T., Lee, Y. J., Yang, H. M. et al. 2011. Inhibitory effect of *Sanguisorba officinalis* ethanol extract on NO and PGE2 production is mediated by suppression of NF-κB and AP-1 activation signaling cascade. *J Ethnopharmacol* 134:11–17.

Zhang, X., Xu, J. K., Wang, J. et al. 2007. Bioactive bibenzyl derivatives and fluorenones from *Dendrobium nobile. J Nat Prod* 70: 24–28.

13 X-Ray Structural Behavior of Some Significant Bioactive Steroids and Their Chemistry in the Crystal Packing and Related Matters

Vivek K. Gupta

CONTENTS

13.1 INTRODUCTION

Steroids constitute a very large group of natural and synthetic compounds with a broad range of biological activities. The basic steroid nucleus consists of three fused cyclohexane rings and one cyclopentane ring as shown in Figure 13.1, which illustrates the standard steroid numbering and ring nomenclature.

The diversity of steroids results primarily from variation in the side chains R_1, R_2, and R_3, and secondarily from differences in nuclear substitution and in the degree of unsaturation. R_1 and R_2 are generally methyl groups, which may occasionally be oxygenated. R_2 is absent in the estrogenic hormones and other steroids having ring A and/or B aromatic. The side chain R_3 may comprise 2, 4, 5, 8, 9, or 10 carbon atoms: if it is absent, the position is usually oxygenated. There are four rings in a steroid skeleton and hence there are three fusion points. A/B, B/C, and C/D rings share two carbons each (fusion). Every fusion center can either be *trans*- or *cis*-fused. In discussing intermolecular interaction, it has become customary to refer to the head and the tail of the steroid. The head–tail designation usually refers to C_3 and C_{17} respectively or to substituents on these positions. The most important classes of steroids include

1. Sterols
2. Bile acids
3. Corticosteroids
4. Sex hormones
5. Saponins
6. Withanolides

13.1.1 Sterols

Sterols are crystalline alcohols occurring in animal, plant oils, and fats freely or esters of higher fatty acids. On the basis of the sources, they are grouped into the following classes:

13.1.1.1 Zoosterols

Zoosterols are the compounds obtained from animal fats. Cholesterol belongs to this class. It is the most important zoosterol and occurs widely either freely or as esters in nearly all animal cells

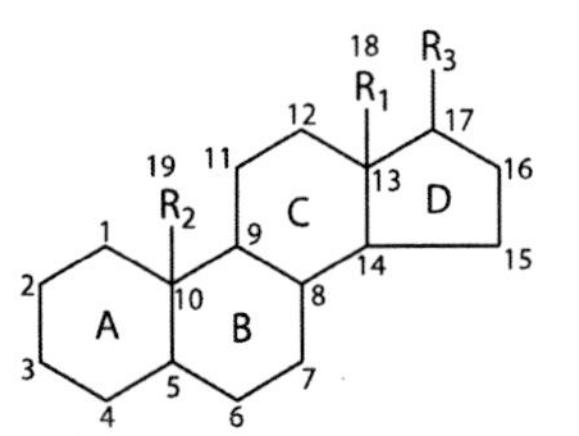

FIGURE 13.1 Steroid atomic numbering.

especially in brain, spinal cord, and human gallstones. The chief commercial sources of cholesterol are fish liver oil, brain, and spinal cord of cattle.

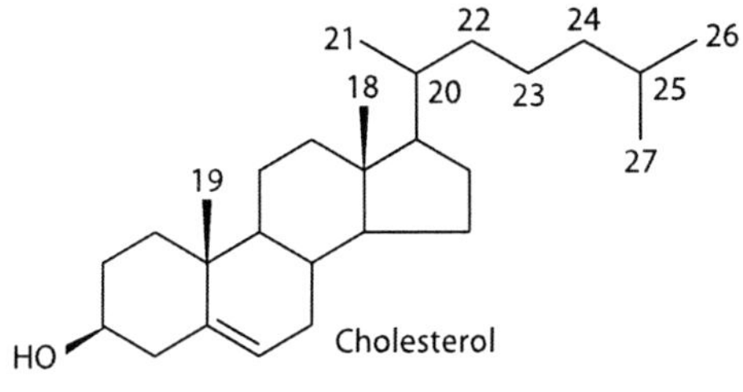

Cholesterol

13.1.1.2 Phytosterols

Phytosterols are sterols obtained from vegetable oils and fats. Stigmasterol is the most important member of this group. It occurs exclusively in plants and its chief source is soybean oil. Its structure differs from that of cholesterol in the presence of a second double bond at C_{22}–C_{23} and an ethyl substituent at C_{24}.

13.1.1.3 Mycosterols

Mycosterols are the sterols occurring in fungi and yeast, for example, ergosterol. The main structural feature of ergosterol is that it has the same structure as cholesterol except that it has, in addition, two more double bonds, one at C_7_C_8 position and the other at C_{22}–C_{23}, and a methyl substituent at C_{24}.

13.1.2 Bile Acids

Bile acids (C_{24} compounds) are hydroxy derivatives of cholanic acid. They have a 3α-hydroxyl group (with one exception), other hydroxyl groups may be present at C_6, C_7, C_{12}, and occasionally at other carbon atoms. A significant fraction of the body's cholesterol is used to form bile acids. Oxidation in the liver removes a portion of the C_8H_{17} side chain, and additional hydroxyl groups are introduced at various positions on the steroid nucleus. Cholic acid is the most abundant of the bile acids. Bile acids act as emulsifying agents to aid the digestion of fats.

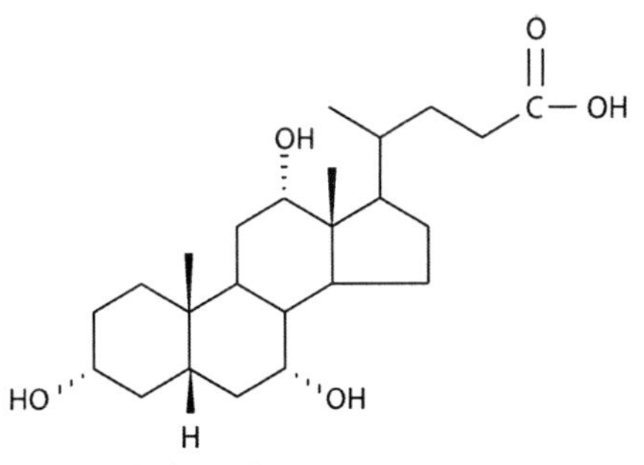

Cholic acid

13.1.3 Corticosteroids

The outer layer, or cortex, of the adrenal gland is the source of a large group of substances known as corticosteroids. Like the bile acids, they are derived from cholesterol by oxidation, with cleavage of a portion of the alkyl substituent on the D ring. Cortisol is the most abundant of the corticosteroids, but cortisone is probably the best known. Cortisone is commonly prescribed as an antiinflammatory drug, especially in the treatment of rheumatoid arthritis.

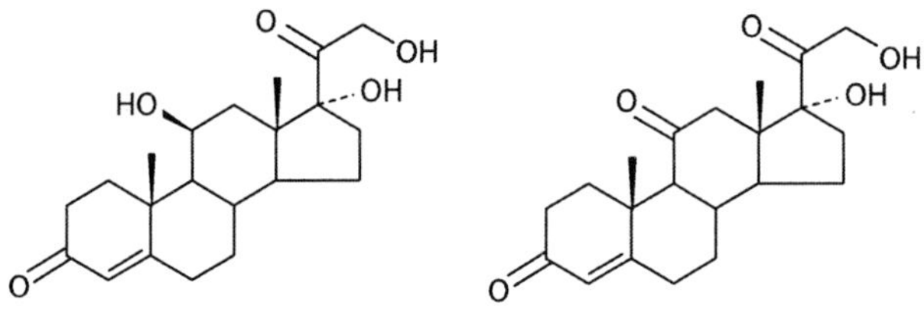

Cortisol Cortisone

Corticosteroids exhibit a wide range of physiological effects. One important function is to assist in maintaining the proper electrolyte balance in body fluids. Though natural and synthetic corticosteroids are both potent antiinflammatory compounds, the synthetics exert a stronger effect. Oral forms of corticosteroids are used to treat numerous autoimmune and inflammatory conditions, including asthma, bursitis, skin disorders, tendinitis, ulcerativecolitis, and others. They are also used to treat severe allergic reactions and to prevent rejection after organ transplant. Dexamethasone, a synthetic corticosteroid, is similar to a natural hormone produced by adrenal glands. It often is used to replace this chemical when body does not make enough of it. It relieves inflammation (swelling, heat, redness, and pain) and is used to treat certain forms of arthritis, skin, blood, kidney, eye, thyroid, intestinal disorders, allergies, and asthma.

13.1.4 Sex Hormones

Hormones are the chemical messengers of the body. They are synthesized and secreted into the bloodstream by the endocrine glands and regulate biological processes. The sex glands—testes in males, ovaries in females—secrete a number of hormones that are involved in sexual development and reproduction. The sex hormones can be classified into three major groups:

1. Female sex hormones (estrogens)
2. Male sex hormones (androgens)
3. Pregnancy hormones (progestins)

13.1.4.1 Female Sex Hormones (Estrogens)

In 1929, the first sex hormone (estrone) was isolated from the urine of pregnant women. Later a much more potent estrogen, estradiol, was isolated. Estradiol is the true female sex hormone, and estrone is a metabolized form of estradiol that is excreted. Estradiol is secreted by the ovaries and promotes the development of the secondary female characteristics that appear at the onset of puberty.

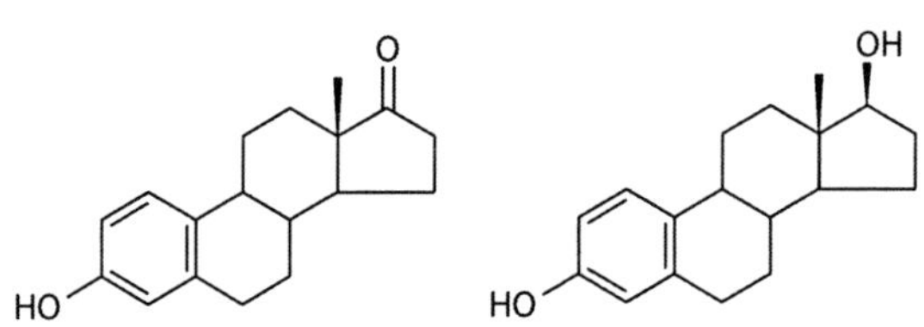

Estrone Estradiol

13.1.4.2 Male Sex Hormones (Androgens)

In 1931, the first androgen (androsterone) was isolated by extracting male urine. Soon afterward in 1935, another male sex hormone, testosterone, was isolated from bull testes. Testosterone is the true male sex hormone and that androsterone is a metabolized form of testosterone that is extracted in the urine.

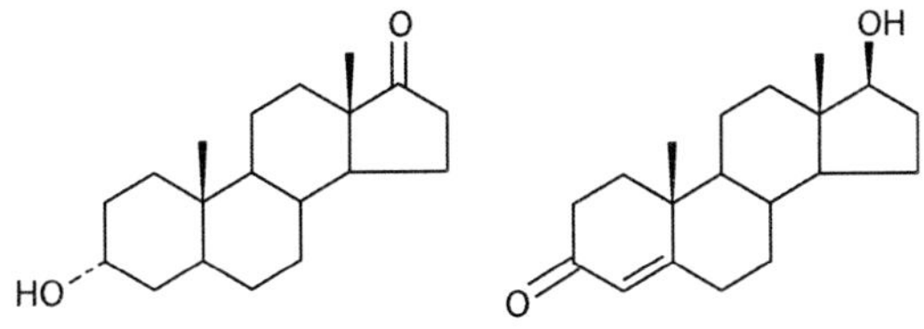

Androsterone Testosterone

Testosterone, secreted by the testes, is the hormone that promotes the development of secondary male characteristics: the growth of facial and body hair, the deepening of the voice, muscular development, and the maturation of the male sex organs.

13.1.4.3 Pregnancy Hormones (Progestins)

Progesterone is the most important progestin (pregnancy hormone). After ovulation occurs, the remnant of the ruptured ovarian follicle (called the corpus luteum) begins to secrete progesterone. This hormone prepares the lining of the uterus for implantation of the fertilized ovum, and continued progesterone secretion is necessary for the completion of pregnancy. Progesterone is secreted by the placenta after secretion by the corpus luteum declines. Progesterone also suppresses ovulation, and it is the chemical agent that apparently accounts for the fact that pregnant women do not conceive again while pregnant. It was this observation that led to the search for synthetic progestins, such as norethindrone, that could be used to "turn off" ovulation. By inducing temporary infertility, synthetic progestins form the basis of most oral contraceptive agents.

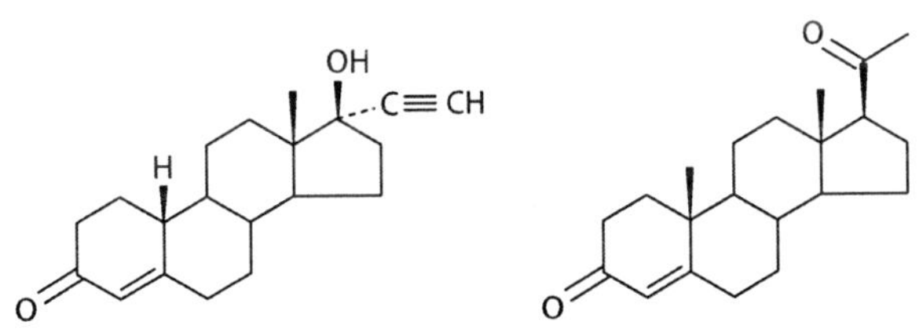

Norethindrone Progesterone

To give an idea of how small molecular differences can have large differences when affecting humans and animals, two hormones are shown below: One is testosterone, the "male" hormone, and the other is estradiol, the potent "female" hormone. Even though the molecular differences are very small and the hormones look very similar, testosterone tells the body it is male, and estradiol tells the body it is female. So, a small difference in molecular structure causes the difference between male and female.

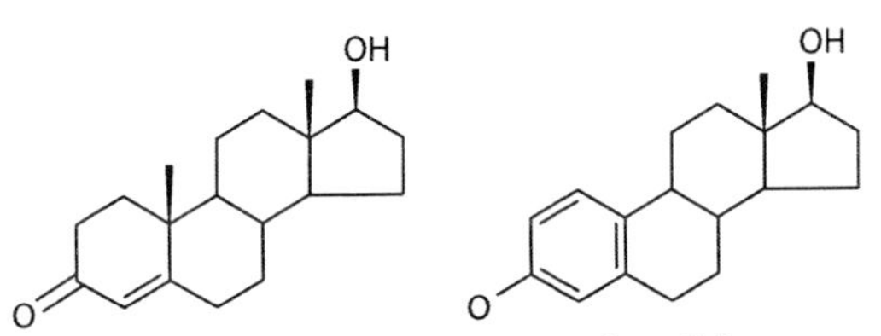

Testosterone Estradiol

With the exception of retinoic acid, the steroid hormones are all derived from cholesterol. Moreover, with the exception of vitamin D, they all contain the same cyclopentanophenanthrene ring and atomic numbering system as cholesterol. The conversion of C_{27} cholesterol to the 18-, 19-, and 21-carbon steroid hormones (designated by the nomenclature C with a subscript number indicating the number of carbon atoms, e.g., C_{19} for androstanes) involves the rate-limiting, irreversible cleavage of a 6-carbon residue from cholesterol, producing pregnenolone (C_{21}) plus isocaproaldehyde. Common names of the steroid hormones are widely recognized, but systematic nomenclature is gaining acceptance and familiarity with both nomenclatures is increasingly important. Steroids with 21 carbon atoms are known systematically as pregnanes, whereas those containing 19 and 18 carbon atoms are known as androstanes and estranes, respectively.

13.1.5 Saponins

Saponins are high-molecular-weight glycosides, consisting of a sugar moiety linked to a *triterpene* or *steroid* aglycone. The name "saponin" comes from the Latin word *sapo* (Soap). Saponins are widely distributed in the plant kingdom. Even by 1927, Kofler had listed 472 saponin-containing plants (Kofler, 1927) and it is now known that over 90 families contain saponins. Saponins occur in plants that are used as human food: soybeans, chick peas, peanuts, mung beans, broad beans, kidney beans, lentils, garden peas, spinach, oats, aubergines, asparagus, fenugreek, garlic, sugar beet, potatoes, green peppers, tomatoes, onions, tea, cassava, yams (Birk and Peri, 1980; Oakenfull, 1981; Price et al., 1987). The aglycone or non-saccharide portion of the saponin molecule is called the *genin* or *sapogenin*. Depending on the type of genin present, the saponins can be divided into three major classes:

1. Steroid sapogenins
2. Steroid alkaloid sapogenins
3. Triterpene sapogenins

13.1.5.1 Steroid Sapogenins

Over 100 steroid sapogenins are known and most are derived from the spirostan or furostan skeleton. In all cases, the C_{18} and C_{19} angular methyl groups are β-orientated and the C_{21} methyl group has the α-configuration. There is sometimes a 5,6-double bond. The sapogenins are mostly hydroxylated at C_3.

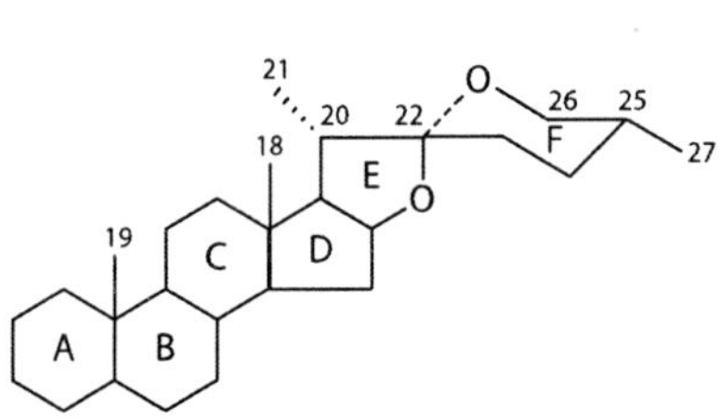

Spirostan

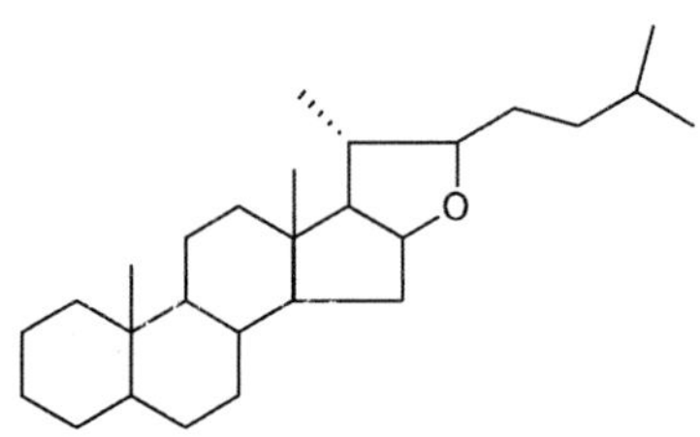

Furostan

Spirostans are characterized by the existence of a ketospiroketal moiety (rings E/F) and may be subdivided into a 25*S* or a 25*R* series. The 25*S* series (e.g., yamogenin) and 25*R* series (e.g., diosgenin) were formerly referred to as neosapogenins or isosapogenins, respectively. The C_{25} methyl group is axially oriented in neosapogenins and equatorially oriented in isosapogenins.

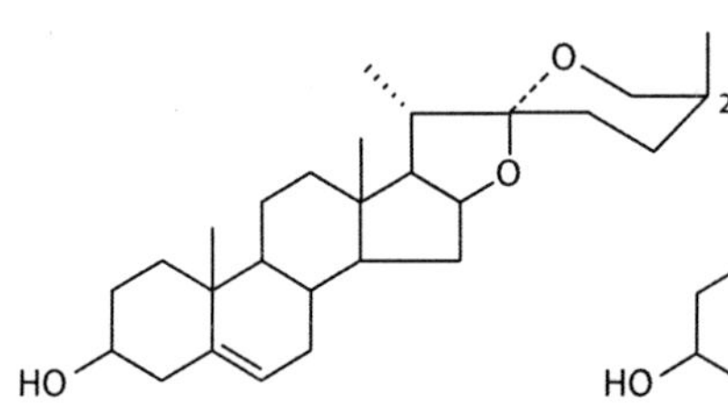

Yamogenin (25*S*)

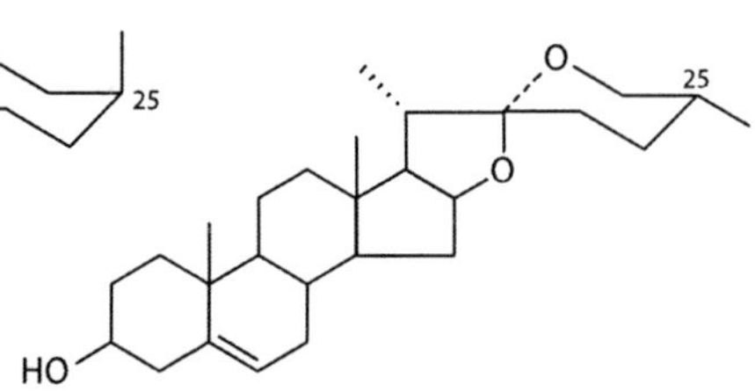

Diosgenin (25*R*)

13.1.5.2 Steroid Alkaloid Sapogenins

There are two classes of steroid alkaloid sapogenin: the spirosolans and the solanidans.

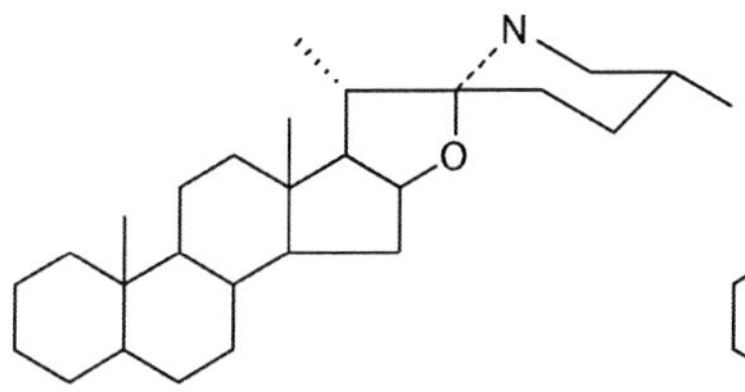

Spirosolan

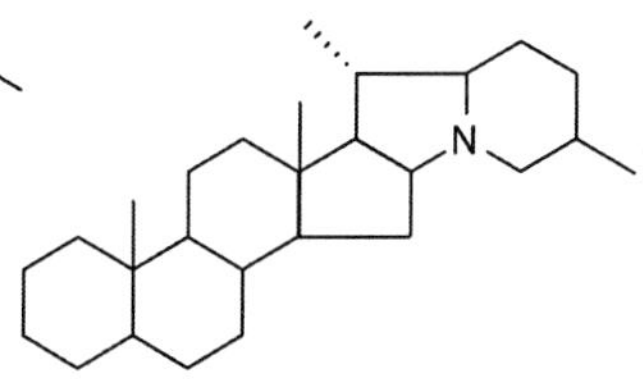

Solanidan

13.1.5.3 Triterpene Sapogenins

The triterpene sapogenins can be divided into three main classes, depending on whether they have a β-amyrin (oleanane-type), α-amyrin (ursane-type), or lupeol skeleton.

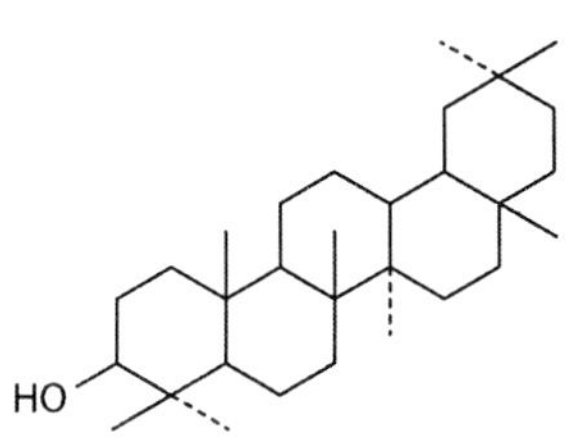

β-Amyrin type (oleanane-type)

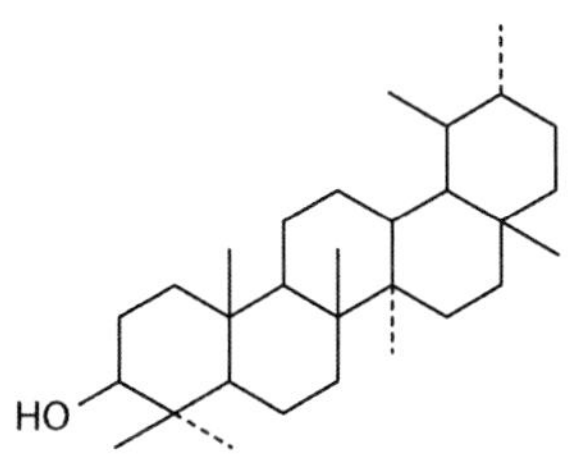

α-Amyrin type

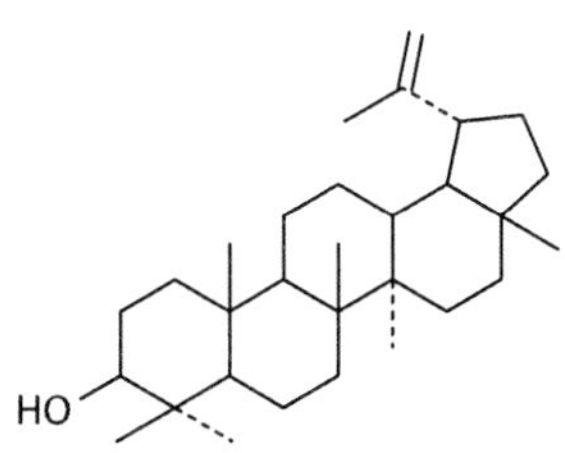

(ursane-type) Lupeol type

13.1.6 Withanolides

The C_{28}-steroidal lactones characterized by a nine carbon side chain with a six membered lactone ring were designated as "withanolides." Genuine interest arose in withanolides when it was found that they are able to exhibit antitumor activity in a number of animal studies (Chakraborti et al., 1974; Umadevi et al., 1992). In addition, cytotoxicity, immunosuppressive, antimicrobial, hepatoprotective, insect antifeedant and antiinflammatory properties were observed (Budhiraja et al., 1984; Gil et al., 1997; Furmanowa et al., 2001). Several review articles on withanolides have appeared since the isolation of withaferin A from the leaves of *Withania somnifera* (Kundu et al., 1976; Glotter et al., 1978; Budhiraja and Sudhir, 1987; Ray, 1989; Glotter, 1991; Singh and Kumar, 1998). Today there is much interest in natural products with anticancer activity. Withanolides are considered as potential candidates as far as treatment of cancer is concerned.

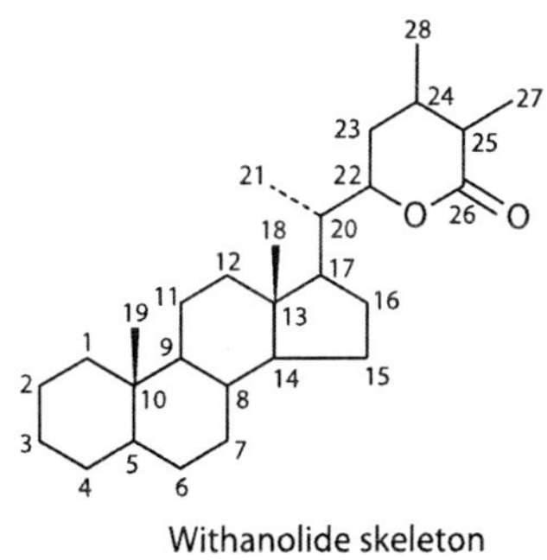

Withanolide skeleton

13.2 X-RAY CRYSTAL STRUCTURE OF STEROIDS CONCERNED

Crystal and molecular structure determinations have been reported for single crystals of 11 steroids (1 Sterol, 5 Withanolides, 3 Pregnanes, 1 Steroid sapogenin, and 1 Androstane). None of these steroids contain more than one crystallographically independent molecule in the asymmetric unit.

13.3 EXPERIMENTAL

For obtaining x-ray diffraction quality single crystals, solvent loss technique was employed. Three dimensional x-ray diffraction intensity data from single crystal samples were collected on Enraf-Nonius CAD-4 diffractometer and Brucker SMART APEX CCD area-detector diffractometer. The crystal structures were solved by direct methods and refined by standard least-squares methods. The computer programs used for structure solution are SHELXS97 (Sheldrick, 1997a); the refinements were carried out by using SHELXL97 program (Sheldrick, 1997b). ORTEP-3 for Windows (Farrugia, 1997) software was used for making the thermal ellipsoids. Geometrical calculations were performed using PLATON (Spek, 1999) and PARST (Nardelli, 1995) software.

13.4 RESULTS AND DISCUSSION

13.4.1 Crystal Structure of Z-Guggulsterone: A Sterol

Guggul is the yellowish resin (or gum) that is produced by the *Commiphora mukul*, a small, thorny plant that grows in dry areas of India, Pakistan, and Afghanistan. Guggul is also referred as gugglesterone, guggul gum, guggal, guggul, gugulu, and gum gugal. Guggul is used in the treatment of arthritis, skin diseases, pains in the nervous system, obesity, digestive problems, infections in the mouth, and menstrual problems. The resin has been used for centuries as part of India's traditional medicine called Ayurveda. This resinous sap is processed and purified, and then standardized for a given amount of its active constituents—*E*- and *Z*-guggulsterones. These two compounds are plant sterols with a high degree of human bioactivity and have been shown in studies to affect many biological processes including thyroid metabolism, cholesterol management, and dermal (skin) function. In each of these areas, Guggulsterones were shown in studies to be highly effective modulators with near drug-like potency. Guggulsterones stimulate the thyroid gland, which in turn produces more thyroid hormones such as thyroxin. Guggul is also an antioxidant, which helps stop the oxidization of cholesterol and the subsequent hardening of the arteries. Since guggul supports hardening of the arteries, which may impede blood flow to and from the penis, guggul may possibly be the treatment for impotence many men are looking for (Nadkarni, 1954; Patil et al., 1972; Nityanand et al., 1973; Kuppuranjan, 1978; Bordia, 1979; Mester et al., 1979; Singh et al., 1982; Tripathy et al., 1985, 1988; Satyavati, 1991; Urizar et al., 2002; Wu et al., 2002).

E- and *Z*-Guggulsterones have been isolated from the gum resin of *Commiphora mukul* by Patil et al. (1972) along with a number of other compounds. This was the first report of their occurrence in nature. The synthesis and stereochemistry of these compounds was reported much before their isolation by Benn et al. (1964). The compounds isolated from *Commiphora mukul* were found identical to the synthesized compounds in all respects. For the present study guggulsterone Z [4,17(20)-(*trans*)-pregnadiene-3,16-dione, $C_{21}H_{28}O_2$] was synthesized by the method described earlier (Benn et al., 1964). 16,17-Epoxypregnenolone was refluxed with hydrazine hydrate to obtain a mixture of isomeric diols, 5,17(20)-(*cis*)-pregnadiene-3β,16α-diol and 5,17(20)-(*trans*)-pregnadiene-3β,16α-diol. The mixture of diols was subjected to oppenaure oxidation using toluene, cyclohexanone, and aluminum isopropoxide, which yielded a mixture of dienones. This mixture of dienones was chromatographed over a column of natural alumina. Elution with hexane: ethyl acetate (9:1) gave the *trans* isomer, 4,17(20)-(*trans*)-pregnadiene-3,16-dione (Z-guggulsterone as designated by Patil et al., 1972) followed by the *cis* isomer, 4,17(20)-(*cis*)-pregnadiene-3,16-dione (*E*-guggulsterone as designated by Patil et al., 1972) (Gupta et al., 2006). The chemical structures of *Z*- and *E*-guggulsterones are shown in Figure 13.2 and an *ORTEP* view of the molecule indicating atom numbering scheme is shown in Figure 13.3 (Farrugia, 1997).

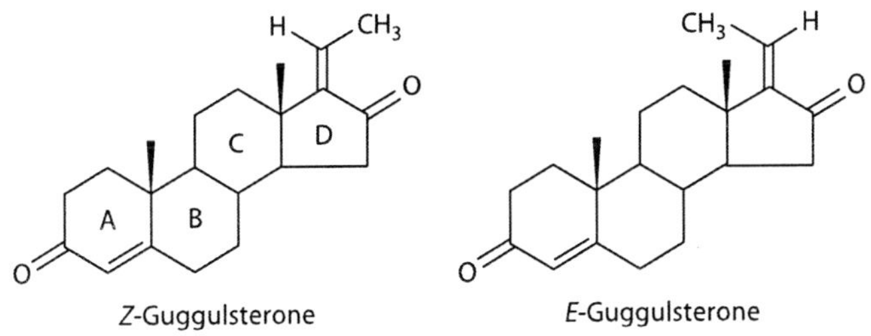

FIGURE 13.2 Chemical structures of *Z*- and *E*-guggulsterones.

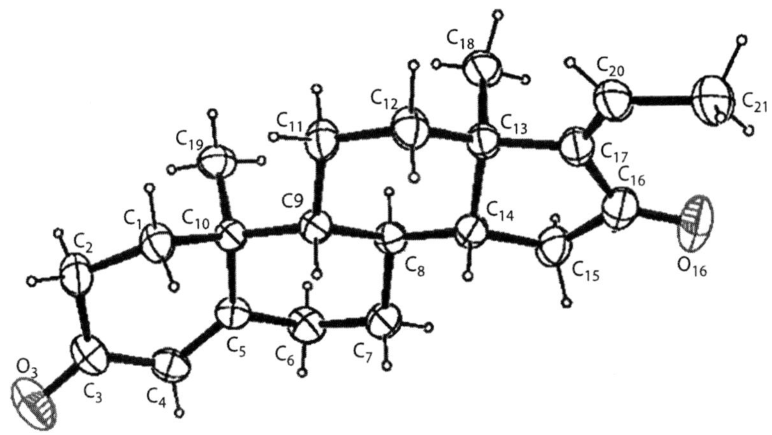

FIGURE 13.3 ORTEP view of the molecule with displacement ellipsoids drawn at the 50% probability level. H atoms are shown as small spheres of arbitrary radii.

Mean bond lengths [C(sp^3)-C(sp^3) = 1.534(3); C(sp^3)-C(sp^2) = 1.508(3); C(sp^2)-C(sp^2) = 1.468(3); C(sp^2) = C(sp^2) = 1.338(3) Å] are comparable to the theoretical values as reported by Allen et al. (1987), although the bond C_9–C_{10} = 1.567(2) Å shows significant deviation from the mean value. An examination of other published steroid structures seems to suggest that this lengthening is related to the presence of a double bond in ring A. The length of the C_9–C_{10} bond in some of the steroid structures having double bond in ring A ranges from 1.553 to 1.570 Å with a mean value of 1.563 Å (Roberts et al., 1973; Eggleston et al., 1990; Gupta et al., 1994; Sarkhel et al., 2001; Thamotharan et al., 2004). In steroids with a fully saturated or fully unsaturated ring A, the C_9–C_{10} bond does not show such a systematic lengthening (Weeks et al., 1971; Duax et al., 1989; Ribar et al., 1993; Starova et al., 2003; Matsumoto et al., 2004).

The presence of double bond at C_4–C_5 and =O at C_3 imposes a distorted 1α-*sofa* conformation on ring A, with asymmetry parameters $\Delta C_s(C_1–C_4)$ = 9.40, $\Delta C_2(C_1–C_2)$ = 15.35 (Duax and Norton, 1975). The overall shape of ring B is still approximately the *chair* conformation typical for totally saturated six membered rings. Distortion from that ideal form can be expressed by the loss of mirror symmetry through atoms C_6 and C_9 [$\Delta C_s(C_6–C_9)$ = 8.14] with the retention of perpendicular rotational symmetry [$\Delta C_2(C_5–C_{10})$ = 1.83]. The best mirror plane passes through C_7 and C_{10}, with $\Delta C_s(C_7)$ = 3.34. Ring C has a distorted *chair* form [$\Delta C_s(C_{11})$ = 2.75, $\Delta C_2(C_8–C_{14})$ = 8.10]. The conformation of D ring is intermediate between 13β,14α-*half chair* [$\Delta C_2(C_{13}–C_{14})$ = 8.71] and 14α-*envelope* [$\Delta C_s(C_{14})$ = 8.71)] with pseudorotation parameters Δ = −15.57° and φ_m = −40.98° (Altona et al., 1968). The $C_3 \ldots C_{16}$ distance, which is a measure of the length of the steroid nucleus, is 8.870 Å. The distance between terminal atoms O_3 and C_{21} is 12.028 Å. The $C_{19}–C_{10} \ldots C_{13}–C_{18}$ pseudo-torsion angle, which gives a measure of the molecular twist, is 7.2°. The B/C and C/D ring junctions approach *trans* characteristics about the C_8–C_9 and C_{13}–C_{14} bonds, respectively, whereas

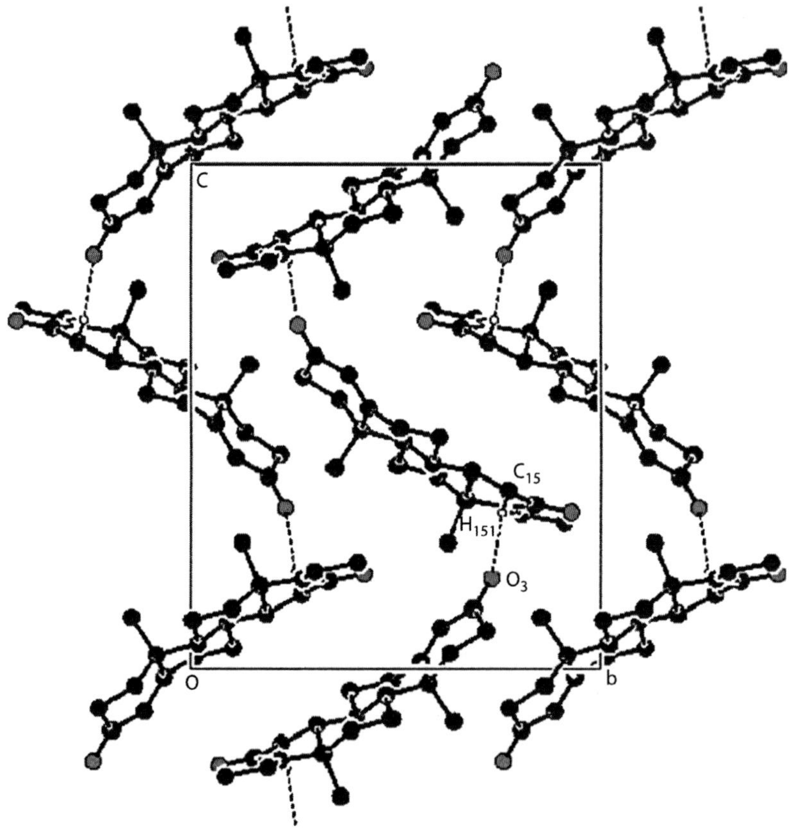

FIGURE 13.4 A view of the unit cell packing structure illustrating the C–H...O intermolecular interactions. H atoms have been omitted for clarity, except those involved in hydrogen bonding.

the A/B ring junction is quasi-*trans* (Bucourt, 1974). This quasi characteristic of the A/B *trans* ring junction is due to the existence of the trigonal atom C_5.

Figure 13.4 illustrates the packing of the molecules and their hydrogen-bonding arrangement. Atom O_3 was found to have close C–H...O contact [distance 2.55(3) Å to H_{151}] to neighboring molecules related by screw axes in c-direction forming zigzag chains. This distance lies within the 2.7 Å range we usually employ for nonbonded H...O packing interactions (Steiner, 1997). Using compiled data for a large number of C–H...O contacts, Steiner and Desiraju (1998) found significant statistical directionality even as far out as 3.0 Å, and concluded that these are legitimately viewed as "weak hydrogen bonds," with a greater contribution to packing forces than simple van der Waals attractions. Chains of hydrogen-bonded molecules are packed with van der Waals contacts. The crystallographic data are summarized in Table 13.1.

13.4.2 Crystal Structure of Withaferin A (5β,6β-Epoxy-4β,27-Dihydroxy-1-Oxo-22*R*-Witha-2,24-Dienolide): A Withanolide

In view of the wide applications in indigenous or traditional systems of medicine as well as in folk medicines, the plant *W. somnifera* has attracted attention of phytochemists all over the world since a long time in the study of its constituents. Withaferin A is the most active withanolide contained in the leaves of *W. somnifera* and has been isolated from the 95% alcohol extract of the leaves of *W. somnifera*. The antibacterial activity of Withaferin A was established long before the structure of this compound was fully clarified (Kurup, 1956). Withaferin A has also shown significant anticancer activity (Sing et al., 1998).

TABLE 13.1
Crystal and Experimental Data

CCDC no	261919
Crystal description	Colorless rectangular
Crystal size	0.18 × 0.16 × 0.13 mm
Empirical formula	$C_{21}H_{28}O_2$
Formula weight	312.43
Radiation, wavelength	Mo $K\alpha$, 0.71073 Å
Unit cell dimensions	a = 7.908(2) Å, b = 13.611(3) Å, c = 16.309(4) Å
Crystal system	Orthorhombic
Space group	$P2_12_12_1$
Unit cell volume	1755.4(7) $Å^3$
Density (calculated)	1.182 Mgm^{-3}
No. of molecules per unit cell, Z	4
Temperature	100 K
Absorption coefficient(μ)	0.074 mm^{-1}
F(000)	680
Refinement of unit cell	999 reflections (2.04° < θ < 28.23°)
Scan mode	φ and ω scans
θ range for entire data collection	2.50° < θ < 28.30°
Reflections collected/unique	11819/4307
Reflections observed [I > 2σ(I)]	3667
Range of indices	h = −10 to 7, k = −18 to 15, l = −21 to 19
R_{int}	0.0242
R_{sigma}	0.0303
No. of parameters refined	320
Final R-factor	0.0578
$wR(F^2)$	0.1430
Weight	$1/[\sigma^2(F_o^2) + (0.1106P)^2 + 0.00P]$
	where $P = \left[F_o^2 + 2F_c^2\right]/3$
Goodness-of-fit	0.974
$(\Delta/\sigma)_{max}$	−0.047 (for z H182)
Final residual electron density	$-0.191 < \Delta\rho < 0.293$ $eÅ^{-3}$

Air dried and powdered leaves (500 g) of *W. somnifera* were extracted with 95% ethanol at room temperature while stirring for 2 h. The extract was filtered through muslin cloth followed by centrifugation. The marc was again extracted twice as mentioned earlier. All the three extracts were pooled and concentrated to 1/8th of its volume. Syrupy solution was diluted with water and the resulting suspension was extracted sequentially with $CHCl_3$, EtOAc, and *n*-BuOH. The $CHCl_3$ extract was subjected to chromatography over silica gel (60–120 mesh) and the elution was carried in increasing polarity with $CHCl_3$, 2% MeOH in $CHCl_3$, 5% MeOH in $CHCl_3$ and MeOH. Fractions got eluted in 2% MeOH in $CHCl_3$ were pooled, concentrated, the residue after crystallization from EtOAc yielded withaferin A (200 mg) (Figure 13.5), mp 252°C–253°C (Bandhoria et al., 2006d).

An *ORTEP* view of the molecule indicating atom numbering scheme is shown in Figure 13.6. The mean bond lengths are $C(sp^3)–C(sp^3)$ = 1.532(3); $C(sp^3)–C(sp^2)$ = 1.501(3); $C(sp^2)–C(sp^2)$ = 1.476(4); $C(sp^2) = C(sp^2)$ = 1.329(4); $C(sp^3)–O$ = 1.445(3); C = O = 1.214(3) Å. The shortest and the longest $C(sp^3)–C(sp^3)$ bond distances are observed in ring B to which the epoxide is fused [$C_5–C_6$ = 1.467(3) and $C_9–C_{10}$ = 1.579(3) Å]. For the similar withanolides having epoxide at the same position,

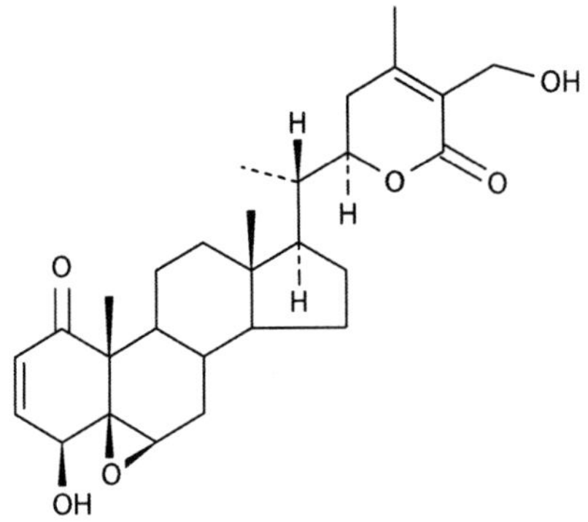

FIGURE 13.5 Chemical structure of withaferin A.

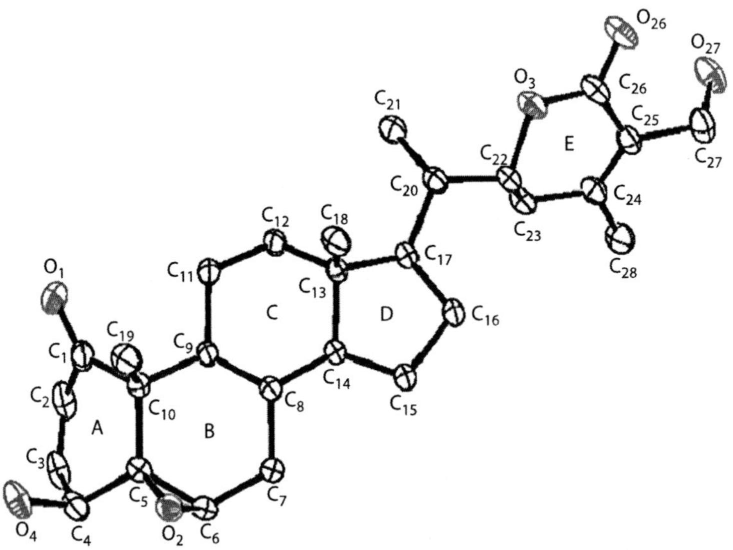

FIGURE 13.6 *ORTEP* plot of the molecule with 40% probability thermal ellipsoids. All H atoms have been omitted for clarity.

variations in ring B C(sp^3)–C(sp^3) bond lengths have been reported (Parvez et al., 1988, 1990). The two C–O epoxy bond lengths [C_5–O_2 1.438(3), C_6–O_2 1.437(3) Å] are the same.

Ring A has a 1α,4α-*twist boat* conformation with C_1 and C_4 −0.346(3) and −0.576(2) Å, respectively, from the C_2, C_3, C_5, C_{10} plane. Ring B is *cis* fused to ring A and has a *half chair* conformation with 8β,9α orientation [$\Delta C_2(C_8–C_9) = 7.9$]. The overall shape of ring C is close to the *chair* conformation. Distortion from the ideal form could be expressed as due to the loss of rotational symmetry [$\Delta C_2(C_{11}–C_{12}) = 12.14$] with the retention of perpendicular mirror symmetry through atoms C_9 and C_{13} [$\Delta C_s(C_9) = 2.06$]. The conformation of D ring is 13α-*envelope* [$\Delta C_s(C_{13}) = 4.02$] with pseudo-rotation parameters $\Delta = 27.96°$ and $\varphi_m = 48.02°$. The δ-lactone ring E adopts 22α-*distorted sofa* conformation [$\Delta C_s(C_{22}) = 9.07$]. Ring E makes a dihedral angle of 59.0(1)° with the plane of the steroid nucleus. The distance between the terminal atoms O_4 and O_{27} is 16.468 Å and the C_3…C_{16} distance, which is a measure of the length of the steroid nucleus, is 8.471 Å. The pseudo-torsion angle C_{19}–C_{10}…C_{13}–C_{18} is 1.9°, indicating that the steroid nucleus is untwisted. This is attributed to the short intramolecular C–H…O contacts [C_{11}–H_{112}…O_1 3.056(3) and C_{19}–H_{191}…O_1 2.808(4) Å] present in the molecule. The B/C and C/D ring junctions approach *trans* characteristics about the C_8–C_9 and C_{13}–C_{14} bonds, respectively.

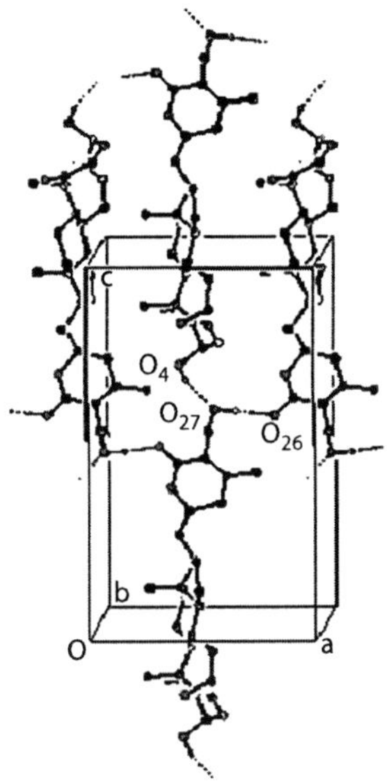

FIGURE 13.7 Part of the crystal structure, showing the formation of molecular chains along the c-axis.

The crystal structure of withaferin A is dictated by two intermolecular hydrogen bonds. Atom O_4 of the hydroxy group in the molecule at (x, y, z) acts as a hydrogen-bond donor to hydroxy atom 027 in the molecule at (x, y, z + 1), producing a chain. Chains of molecules are packed together to form well-defined layers. Molecules within the layers are arranged in an antiparallel manner and are stabilized by the second hydrogen bond ($O_{27}-H_{27}O...O_{26}$) (Figure 13.7). The crystallographic data are summarized in Table 13.2.

13.4.3 Crystal Structure of Withanone (6α,7α-Epoxy-5α,17α, Dihydroxy-1-Oxo-22*R*-Witha-2,24-Dienolide): A Withanolide

Withanone was isolated from *W. somnifera* leaves. Air dried and powdered leaves (500 g) of *W. somnifera* were extracted with 95% ethanol at room temperature while stirring for 2 h. The extract was filtered through muslin cloth followed by centrifugation. The marc was again extracted twice as mentioned earlier. All the three extracts were pooled and concentrated to 1/8th of its volume. Syrupy solution was diluted with water and the resulting suspension was extracted sequentially with $CHCl_3$, EtOAc, and *n*-BuOH. The $CHCl_3$ extract was subjected to chromatography over silica gel (60–120 mesh) and the elution was carried in increasing polarity with $CHCl_3$, 2% MeOH in $CHCl_3$, 5% MeOH in $CHCl_3$ and MeOH. Fractions got eluted in $CHCl_3$ were pooled, concentrated and the residue yielded withanone (150 mg) (Figure 13.8), mp 275°C–276°C (Bandhoria et al., 2006e).

An *ORTEP* view of the molecule indicating atom numbering scheme is shown in Figure 13.9. Mean bond lengths are: $C(sp^3)-C(sp^3) = 1.532(6)$; $C(sp^3)-C(sp^2) = 1.506(7)$; $C(sp^2)-C(sp^2) = 1.459(7)$; $C(sp^3)-O = 1.432(5)$; $C(sp^2) = O = 1.225(6)$ Å. The shortest $C(sp^3)-C(sp^3)$ bond distance is observed in ring B to which the epoxide is fused [$C_6-C_7 = 1.461(6)$ Å]. In ring A, the double bond imposes a 10β,5α–*half-chair* conformation on ring A, with asymmetry parameter $\Delta C_2(C_5-C_{10}) = 6.13$. The conformation of ring B is intermediate between 9α,10β-*half-chair* and 10β-*sofa*, with asymmetry parameters: $\Delta C_2(C_9-C_{10}) = 11.96$, $\Delta C_s(C_{10}) = 13.58$. The overall shape of ring C is close to the *chair* conformation. Distortion from the ideal form could be expressed as due to the loss of mirror symmetry through atoms C_{11} and C_{14} [$\Delta C_s(C_{14}) = 11.0$], with the retention of perpendicular rotational symmetry [$\Delta C_2(C_{12}-C_{13}) = 1.43$]. The conformation of ring D is intermediate between 13β,14α-*half chair* [$\Delta C_2(C_{13}-C_{14}) = 8.98$] and 14α-*envelope* [$\Delta C_s(C_{14}) = 11.73$] with pseudorotation parameters $\Delta = 13.24°$ and $\varphi_m = 47.93°$. The δ-lactone ring E adopts 22β-*sofa* conformation

TABLE 13.2
Crystal and Experimental Data

CCDC no	285229
Crystal description	Colorless rectangular
Crystal size	0.3 × 0.2 × 0.2 mm
Empirical formula	$C_{28}H_{38}O_6$
Formula weight	470.58
Radiation, wavelength	Cu $K\alpha$, 1.5418 Å
Unit cell dimensions	a = 10.697(1) Å, b = 12.344(2) Å, c = 18.714(2) Å
Crystal system	Orthorhombic
Space group	$P2_12_12_1$
Unit cell volume	2471.1(5) $Å^3$
Density (calculated)	1.265 Mgm^{-3}
No. of molecules per unit cell, Z	4
Temperature	293(2) K
Absorption coefficient	0.707 mm^{-1}
Absorption correction	ψ-scan (T_{min} = 0.9373 and T_{max} = 0.9807)
Extinction coefficient	0.0029(5)
F(000)	1016
Refinement of unit cell	25 reflections (12° < θ < 24°)
Scan mode	ω/2θ
θ range for entire data collection	4.29° < θ < 67.90°
Reflections collected/unique	2639/2552
Reflections observed [I > 2σ(I)]	2471
Range of indices	h = 0–12, k = 0–14, l = −22 to 22
R_{int}	0.0096
R_{sigma}	0.0107
No. of parameters refined	460
Final R	0.0382
$wR(F^2)$	0.1156
Weight	$1/[\sigma^2(F_o^2) + (0.1068\,P)^2 + 0.1682\,P]$ where $P = \left[F_o^2 + 2F_c^2\right]/3$
Goodness-of-fit	1.003
$(\Delta/\sigma)_{max}$	−0.028 (for X H112)
Final residual electron density	$-0.150 < \Delta\rho < 0.265$ $eÅ^{-3}$

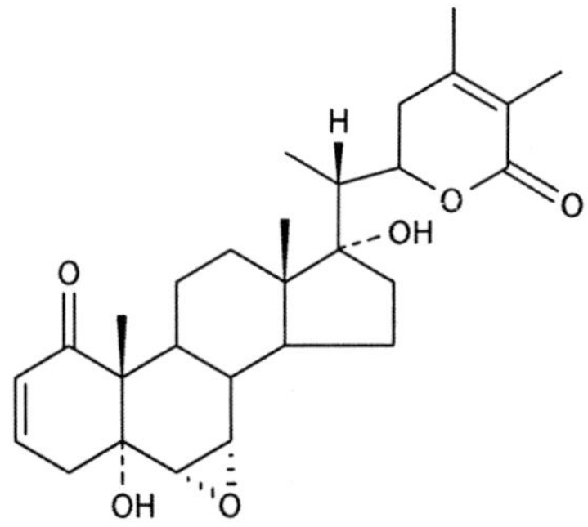

FIGURE 13.8 Chemical structure of Withanone.

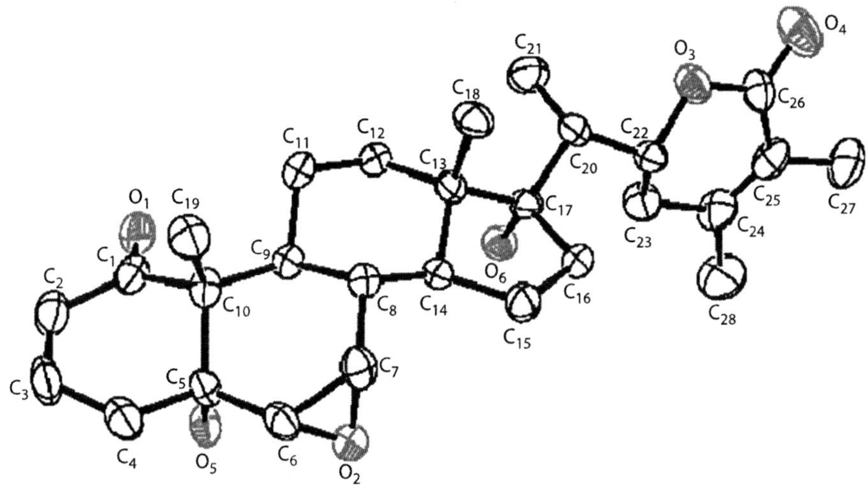

FIGURE 13.9 *ORTEP* view of the molecule, showing the atom-labeling scheme. Displacement ellipsoids are drawn at the 50% probability level and H atoms have been omitted for clarity.

with atom C_{22} disposed 0.605(5) above the plane defined by other five ring atoms $[\Delta C_s(C_{22}) = 5.16]$. Ring E makes a dihedral angle of 73.9(1)° with the plane of the steroid nucleus. The ring junctions A/B, B/C, and C/D are *trans* fused about the C_5–C_{10}, C_8–C_9, and C_{13}–C_{14} bonds, respectively. The distance between the terminal atoms C_3 and C_{27} is 15.238 Å. The length of the steroid nucleus is 8.948 Å (C_3...C_{16}) and is more in comparison to withaferin A. This is attributed to the A/B ring junction, which is *trans* fused in withanone and *cis* fused in withaferin A. The pseudo-torsion angle C_{19}–C_{10}...C_{13}–C_{18} is 1.8°, indicating that the steroid nucleus is untwisted.

Packing view of the molecules in the unit cell viewed down the a-axis is shown in Figure 13.10. The characteristic pattern observed in packing diagram is the appearance of twisted chains of

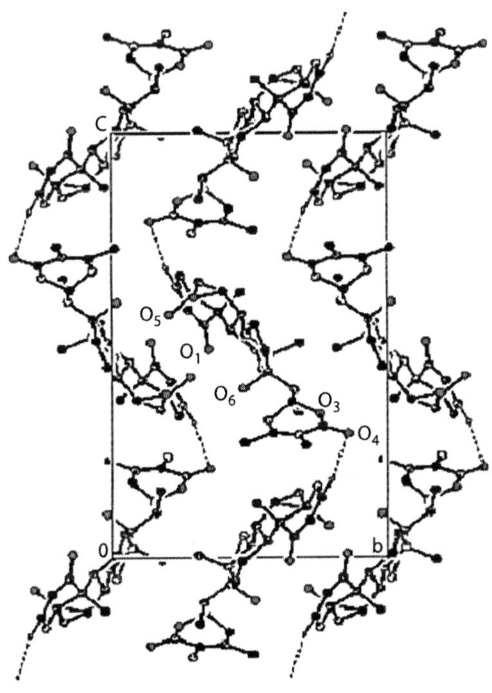

FIGURE 13.10 Appearance of chains of molecules that are hydrogen bonded.

TABLE 13.3
Crystal and Experimental Data

CCDC number	272526
Crystal description	Colorless irregular
Crystal size	0.3 × 0.2 × 0.3 mm
Empirical formula	$C_{28}H_{38}O_6$
Formula weight	470.58
Radiation, wavelength	Mo $K\alpha$, 0.71073 Å
Unit cell dimensions	a = 9.191(10) Å, b = 12.858(6) Å, c = 21.400(16) Å
Crystal system	Orthorhombic
Space group	$P2_12_12_1$
Unit cell volume	2529(3) $Å^3$
Density (calculated)	1.236 Mgm^{-3}
No. of molecules per unit cell, Z	4
Temperature	293(2) K
Absorption coefficient	0.086 mm^{-1}
Absorption correction	ψ-scan (T_{min} = 0.9564 and T_{max} = 0.9970)
F(000)	1016
Refinement of unit cell	25 (6° < θ < 12°)
Scan mode	ω/2θ
θ range for entire data collection	2.41° < θ < 24.97°
Reflections collected/unique	3158/3076
Reflections observed (I > 2σ(I))	1742
Range of indices	h = 0–10, k = 0–15, l = −22 to 25
R_{int}	0.0653
R_{sigma}	0.0665
No. of parameters refined	312
Final R-factor	0.0603
$wR(F^2)$	0.1427
Weight	$1/[\sigma^2(F_o^2) + (0.1064P)^2 + 0.00\ P]$
	where $P = \left[F_o^2 + 2F_c^2\right]/3$
Goodness-of-fit	0.897
$(\Delta/\sigma)_{max}$	0.586 (for U11 H5′)
Final residual electron density	$-0.203 < \Delta\rho < 0.288$ $eÅ^{-3}$

molecules packed together to form layers. C_3–H_3...O_4 weak hydrogen bond binds adjacent links in these chains. The adjacent chain links are rotationally related. Chains of hydrogen-bonded molecules are parallel to c-axis. The packing of the chains in the crystal is further stabilized into a three-dimensional network by strong O–H...O and C–H...O hydrogen bonds. In the packing diagram it can also be seen that the δ-lactone ring E lies in layers perpendicular to the c-axis. The crystallographic data are summarized in Table 13.3.

13.4.4 Crystal Structure of 6α,7α:24α,25α-Diepoxy-5α,12α, Dihydroxy-1-Oxo-20*S*, 22*R*-Witha-2-Enolide Methanol Solvate: A Withanolide

Phytochemical investigations of *Datura quercifolia* plant, growing in Jammu and Kashmir State of India at high altitudes, led to the isolation and characterization of several datura lactones which are of withanolide skeleton (Dhar et al., 1976; Kalla et al., 1979; Qurishi et al., 1979). Datura lactones differ from other withanolides in having a rare epoxide functionality in the lactone ring and are used for the treatment of various conditions, such as infections, organ transplantation, cancer, rheumatoid arthritis, etc. (Sany, 1987; Bartlett et al., 1991; Gonsette, 1996).

Powdered leaves of *Datura quercifolia* were extracted with toluene while stirring. The extract was concentrated to dryness under reduced pressure and was subjected to column chromatography over silica gel on a glass column of 1½ in. dia. The column was eluted with solvents by gradually increasing the percentage of MeOH in $CHCl_3$. In all 105 fractions of 100 mL each were collected and pooled on the bases of TLC patterns using $CHCl_3$: MeOH (9:1) as developing solvent. Spots were visualized by spraying with freshly prepared cerric-ammonium sulfate. Fractions 23–29 showing same TLC pattern were pooled, dried, and subjected to further chromatographic resolution using 100–200 mesh SiO_2 gel column (1:20 ratio) and eluted with $CHCl_3$: MeOH mixtures of increasing polarity. In all 60 fractions of 200 mL each were collected. Fractions 37–44 were pooled on the bases of TLC ($CHCl_3$: MeOH = 9:1) and again subjected to column chromatography. Thirty fractions of 100 mL each were collected. Fractions 23–28 were concentrated under reduced pressure. Residue on crystallization from MeOH yielded the title compound (Figure 13.11), m.p. 261°C–262°C (Bandhoria et al., 2006f).

An ORTEP view of the molecule indicating atom numbering scheme is shown in Figure 13.12. The mean bond lengths are: C(sp^3)–C(sp^3) = 1.523(7); C(sp^3)–C(sp^2) = 1.501(7); C(sp^3)–O = 1.446(6);

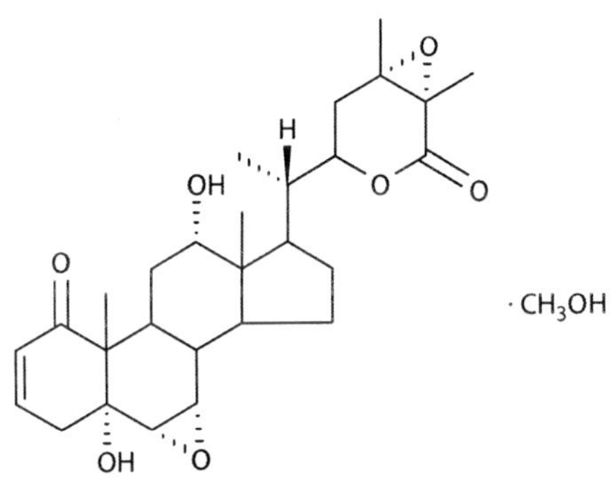

FIGURE 13.11 Chemical structure of 6α,7α:24α,25α-diepoxy-5α,12α,dihydroxy-1-oxo-20*S*, 22*R*-witha-2-enolide.

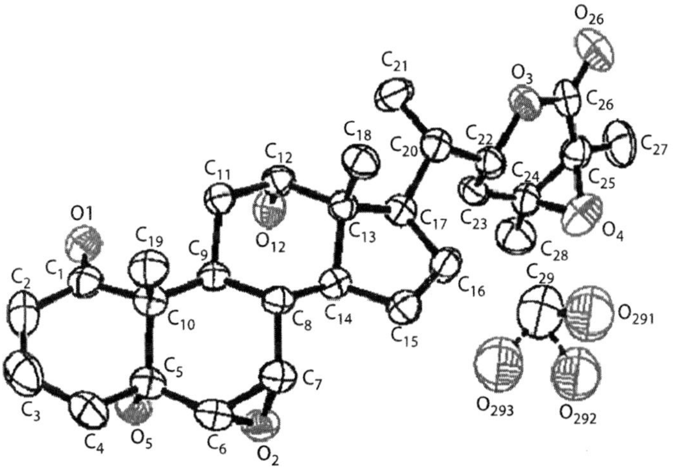

FIGURE 13.12 *ORTEP* view of the molecule with displacement ellipsoids drawn at 40% probability level. H atoms have been omitted for clarity.

$C(sp^2) = O = 1.207(6)$ Å. The shortest $C(sp^3)–C(sp^3)$ bond lengths are observed in rings B and E to which epoxide is fused at C_6, C_7 and C_{24}, C_{25} atoms [$C_6–C_7 = 1.456(7)$ and $C_{24}–C_{25} = 1.478(7)$ Å]. Ring A is highly distorted from the normal *chair* conformation, assuming instead an 5α,10β–*half-chair* conformation due to the localization of double bond at $C_2 = C_3$ position [$C_2 = C_3 = 1.335(8)$ Å] with asymmetry parameter [$\Delta C_2(C_2–C_3) = 4.88$]. Ring B assumes 9α,10β–*half chair* conformation instead of *chair* conformation due to epoxide fused at C_6 and C_7. The best rotational axis for this ring passes through the $C_6–C_7$ and $C_9–C_{10}$ bonds with the asymmetry parameter $\Delta C_2(C_9–C_{10}) = 3.08$. Ring C has a *chair* conformation. Rotational symmetry is dominant; a pseudo-C_2 axis intercepts the $C_8–C_9$ and $C_{12}–C_{13}$ bonds with the asymmetry parameter [$\Delta C_2(C_8–C_9) = 2.20$]. The best mirror plane for this ring passes through C_8 and C_{12} atoms with asymmetry parameter [$\Delta C_s(C_8) = 3.47$]. The five-membered ring D adopts a conformation approximately halfway between that of a 13β,14α-*half-chair* [$\Delta C_2(C_{13}–C_{14}) = 7.50$] and a 13β-*envelope* [$\Delta C_s(C_{13} = 12.74)$] with the phase angle of pseudorotation $\Delta = 10.87°$ and maximum angle of torsion $\varphi_m = 48.01°$. The conformation of ring E is C_{25}, O_3 diplanar [$\Delta C_2(C_{22}–C_{23}) = 7.25$]. The average of the torsion angles in this ring is 25.8(6)°. Ring E makes a dihedral angle of 73.1(1)° with the plane of the steroid nucleus. The $C_{19}–C_{10}...C_{13}–C_{18}$ pseudo-torsion angle, which gives a measure of the molecular twist, is 3.3°. The $C_3...C_{16}$ distance, which is a measure of the length of the steroid nucleus, is 8.987 Å. The distance between the terminal atoms is 15.256 Å ($C_3...C_{27}$). The geometry of rings is *trans* at the A/B, B/C, and C/D ring junctions. The packing of the molecules in the unit cell is shown in Figure 13.13. From the figure it is evident that the molecules related by two-fold screw are packed in interpenetrating layers. The crystal structure is stabilized by the presence of O–H...O and C–H...O intra- and intermolecular hydrogen bonds. The crystallographic data are summarized in Table 13.4.

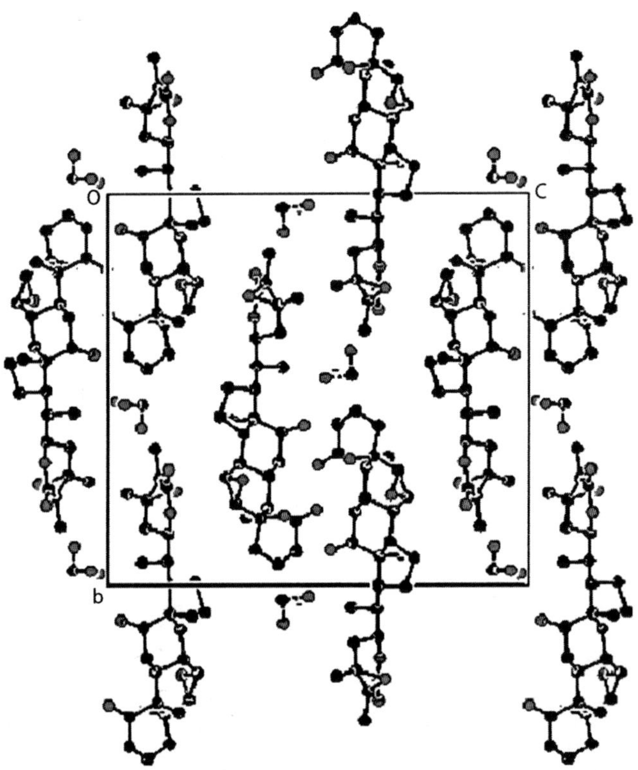

FIGURE 13.13 Packing of the molecules in the unit cell down a-axis.

TABLE 13.4
Crystal and Experimental Data

CCDC no	293742
Crystal description	Light green rectangular
Crystal size	0.3 × 0.2 × 0.2 mm
Empirical formula	$C_{28}H_{38}O_7 \cdot CH_3OH$
Formula weight	259.31
Radiation, wavelength	Mo $K\alpha$, 0.71073 Å
Unit cell dimensions	a = 6.916(4) Å, b = 19.199(2) Å, c = 20.138(5) Å
Crystal system	Orthorhombic
Space group	$P2_12_12_1$
Unit cell volume	2674(2) Å^3
Density (calculated)	1.288 Mgm^{-3}
No. of molecules per unit cell, Z	4
Temperature	293(2) K
Absorption coefficient	0.093 mm^{-1}
Absorption correction	ψ-scan (T_{min} = 0.9666 and T_{max} = 0.9872)
Extinction coefficient	0.0025(9)
F(000)	1120
Refinement of unit cell	25 reflections (9° < θ < 14°)
Scan mode	ω/2θ
θ range for entire data collection	2.12° < θ < 24.97°
Reflections collected/unique	2691/2690
Reflections observed (I > 2σ(I))	1628
Range of indices	h = 0–8, k = 0–22, l = 0–23
R_{int}	0.0060
R_{merge}	0.0524
No. of parameters refined	350
Final R-factor	0.0462
$wR(F^2)$	0.1203
Weight	$1/[\sigma^2(F_o^2) + (0.0838P)^2 + 0.5979\,P]$ where $P = \left[F_o^2 + 2F_c^2\right]/3$
Goodness-of-fit	1.005
$(\Delta/\sigma)_{max}$	−0.001 for y C_{12}
Final residual electron density	$-0.194 < \Delta\rho < 0.217$ eÅ^{-3}

13.4.5 Crystal Structure of (20*R*,22*R*)-6α,7α-Epoxy-5α,27-Dihydroxy-1-Oxowitha-2,24-Dienolide: A Withanolide

Powdered leaves of *W. somnifera* were extracted with 95% ethanol while refluxing. The alcoholic extract was concentrated to its 1/8th volume under reduced pressure and diluted with water. Resulting suspension was extracted sequentially with $CHCl_3$, EtOAc, and *n*-BuOH. The $CHCl_3$ extract was subjected to chromatography over silica gel (60–120 mesh) and the elution was carried in increasing polarity with $CHCl_3$, 2% MeOH in $CHCl_3$, 5% MeOH in $CHCl_3$ and MeoH. Fractions got eluted in 5% MeOH in $CHCl_3$ were pooled, concentrated, the residue after crystallization from MeOH yielded 27-hydroxywithanolide B (Figure 13.14) (105 mg, mp 292°C–294°C) (Gupta et al., 2008).

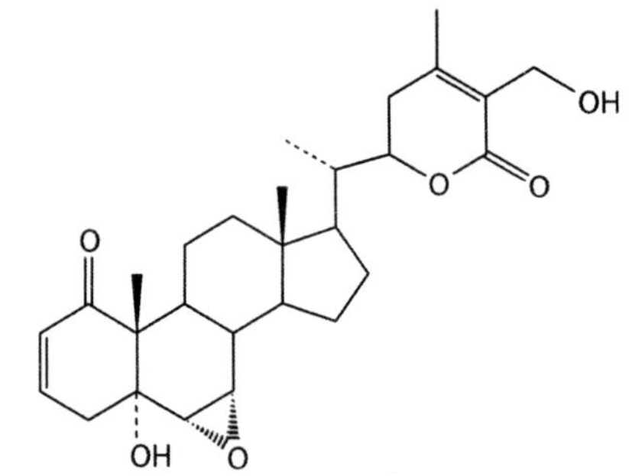

FIGURE 13.14 Chemical structure of (20*R*,22*R*)-6α,7α–Epoxy-5α,27-dihydroxy-1-oxowitha-2,24-dienolide.

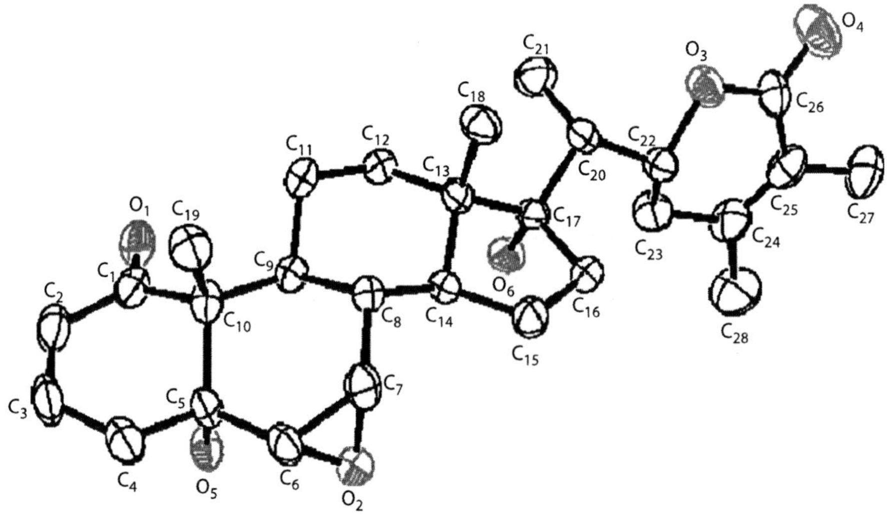

FIGURE 13.15 ORTEP view of the molecule with displacement ellipsoids drawn at 50% probability level.

An *ORTEP* view of the title compound with atomic labeling is shown in Figure 13.15. Mean bond lengths are: C(sp^3)–C(sp^3) = 1.532(6); C(sp^3)–C(sp^2) = 1.506(7); C(sp^2)–C(sp^2) = 1.459(7); C(sp^3)–O = 1.432(5); C(sp^2) = O = 1.225(6) Å. In ring A, the $C_2 = C_3$ distance of 1.336(7) Å confirms the localization of a double bond at this position. This double bond imposes a 10β,5α–half-chair conformation on ring A, with asymmetry parameters: $\Delta C_2(C_5–C_{10}) = 6.13$. Ring B is intermediate between a 9α-10β-half-chair and a 10β-sofa, with asymmetry parameters: $\Delta C_2(C_9–C_{10}) = 11.96$, $\Delta C_s(C_{10}) = 13.58$. The overall shape of ring C is close to the chair conformation. Distortion from the ideal form could be expressed as due to the loss of mirror symmetry through atoms C_{11} and C_{14} [$\Delta C_s(C_{14}) = 11.0$], with the retention of perpendicular rotational symmetry [$\Delta C_2(C_{12}–C_{13}) = 1.43$]. Distortion in ring C is because of strain at the junction with the five-membered D ring. The conformation of the D ring is intermediate between 13β,14α-half chair [$\Delta C_2(C_{13}–C_{14}) = 8.98$] and 14α-envelope [$\Delta C_s(C_{14} = 11.73)$] with pseudorotation parameters $\Delta = 13.24$ and $\varphi_m = 47.93$. The δ-lactone ring E adopts a 22β-*sofa* conformation. The asymmetry parameter, $\Delta C_s(C_{22}) = 5.16$. Ring E makes a dihedral angle of 73.9(1)° with the plane of the steroid nucleus.

The distance between the terminal atoms C_3 and C_{27} is 15.238 Å and the $C_3 \ldots C_{16}$ distance, which is a measure of the length of the steroid nucleus, is 8.948 Å. The pseudo-torsion angle $C_{19}–C_{10} \ldots C_{13}–C_{18}$ is 1.8°, indicating that the steroid nucleus is untwisted. Packing view of the molecules in the unit cell viewed down the a-axis is shown in Figure 13.16. The characteristic pattern observed in packing diagram is the appearance of twisted chains of molecules packed together to form-well

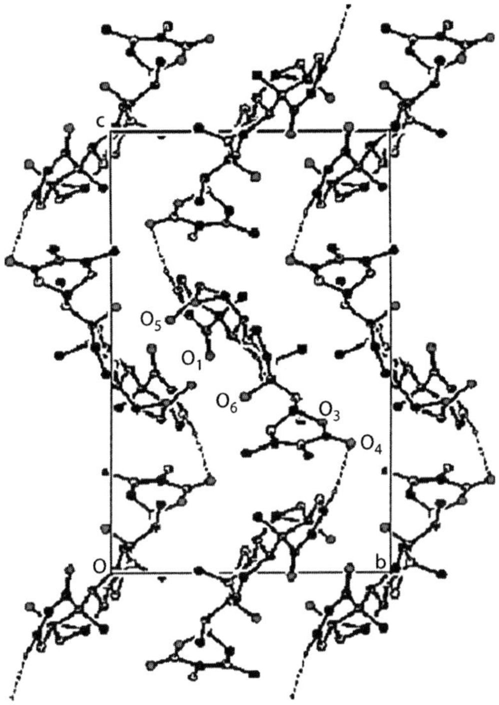

FIGURE 13.16 Appearance of chains of molecules that are hydrogen bonded.

defined layers. C_3–H_3...O_4 weak hydrogen bond binds adjacent links in these chains. The adjacent chain links are rotationally related. Chains of hydrogen-bonded molecules are parallel to the c-axis. The packing of the chains in the crystal is further stabilized into a three-dimensional network by strong O–H...O and C–H...O hydrogen bonds. In the packing diagram it can also be seen that the δ-lactone ring E lies in layers perpendicular to the c-axis. The crystallographic data are summarized in Table 13.5.

13.4.6 Crystal Structure of 6α,7α-Epoxy-5α,17α,27-Trihydroxy-1-Oxo-22*R*-Witha-2,24-Dienolide Monohydrate: A Withanolide

Air dried and powdered leaves (500 g) of *W. somnifera* were extracted with 95% ethanol at room temperature while stirring for 2 h. The extract was filtered through muslin cloth followed by centrifugation. The marc was again extracted twice as mentioned earlier. All the three extracts were pooled and concentrated to 1/8th of its volume. Syrupy solution was diluted with water and the resulting suspension was extracted sequentially with chloroform (four times), by ethyl acetate (four times) and *n*-butanol (10 times) in a separating funnel. The concentrated chloroform extract was subjected to chromatography over silica gel (60–120 mesh) and the elution was carried in increasing polarity with $CHCl_3$, 2% MeOH in $CHCl_3$, 5% MeOH in $CHCl_3$ and MeOH. Fractions got eluted in $CHCl_3$ were pooled, concentrated and the residue on crystallization from methanol yielded 6α,7α-epoxy-5α,17α,27-trihydroxy-1-oxo-22*R*-witha-2,24-dienolide (40 mg), m.p. 242°C–243°C (Figure 13.17) (Gupta et al., 2011).

An *ORTEP* view of the molecule indicating atom numbering scheme is shown in Figure 13.18. The mean bond lengths are: [C(sp^3)–C(sp^3) = 1.525(2); C(sp^3)–C(sp^2) = 1.484(3); C(sp^2)–C(sp^2) = 1.327(3); C(sp^3)–O = 1.443(2); C(sp^2) = O = 1.217(2) Å]. The shortest and the longest C(sp^3)–C(sp^3) bond distances are observed in ring B to which the epoxide is fused [C_6–C_7 = 1.464(2) and C_9–C_{10} = 1.545(2) Å].

In ring A, the C_2=C_3 double bond imposes a 10α,5β–*half-chair* conformation on ring A, with asymmetry parameter $\Delta C_2(C_5$–$C_{10})$ = 2.77. The conformation of ring B is 9β,10α-*half-chair*, with

TABLE 13.5
Crystal Data and Experimental Details

CCDC number	272526
Crystal description	Colorless (irregular)
Chemical formula	$C_{28}H_{38}O_6$
Molecular weight	470.58
Cell parameters	a = 9.191(10) Å, b = 12.858(6) Å, c = 21.400(16) Å
Unit cell volume	2529(3) Å^3
Crystal system	Orthorhombic
Space group	$P2_12_12_1$
Density (calculated)	1.236 Mgm^{-3}
No. of molecules per unit cell, Z	4
Radiation, wavelength	Mo*K*α, 0.71073 Å
Temperature	293(2) K
Absorption coefficient(μ), correction	0.086 mm^{-1}; Psi-scan
Max. and Min. transmission	0.9970 and 0.9564
F(000)	1016
θ range for entire data collection	2.41° < θ < 24.97°
No. of measured reflections	3182
No. of unique reflections	3076
No. of observed reflections	1742 [$F_o > 4\sigma(F_o)$]
No. of parameters refined	312
Final R-factor	0.0603
wR(F^2)	0.1427
Weight	$1/[\sigma^2(F_o^2) + (0.1064\,P)^2 + 0.00\,P]$ where $P = \left[F_o^2 + 2F_c^2\right]/3$
Goof (S) on F^2	0.897
Final residual electron density	$-0.203 < \Delta\rho < 0.288$ eÅ^{-3}
$(\Delta/\sigma)_{max}$ in the final cycle	0.586 (for U11 H5′)

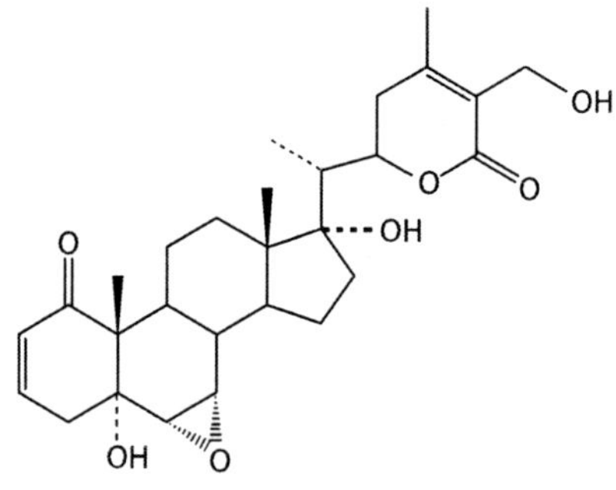

FIGURE 13.17 Chemical structure of 6α,7α-epoxy-5α,17α,27-trihydroxy-1-oxo-22*R*-witha-2,24-dienolide monohydrate.

asymmetry parameters: $\Delta C_2(C_9–C_{10}) = 6.33$. Ring C is in *chair* conformation [$\Delta C_s(C_8–C_{12}) = 2.82$, $\Delta C_2(C_{11}–C_{12}) = 3.77$]. The conformation of ring D is intermediate between 13β,14α-*half chair* [$\Delta C_2(C_{13}–C_{14}) = 10.31$] and 13β-*envelope* [$\Delta C_s(C_{13}) = 10.63$] with pseudorotation parameters $\Delta = -15.28°$ and $\varphi_m = 47.93°$. The δ-lactone ring E adopts distorted *sofa* conformation [$\Delta C_s(C_{22}) = 4.68$]. Ring E makes a dihedral angle of 83.7(1)° with the plane of the steroid nucleus. The distance between the

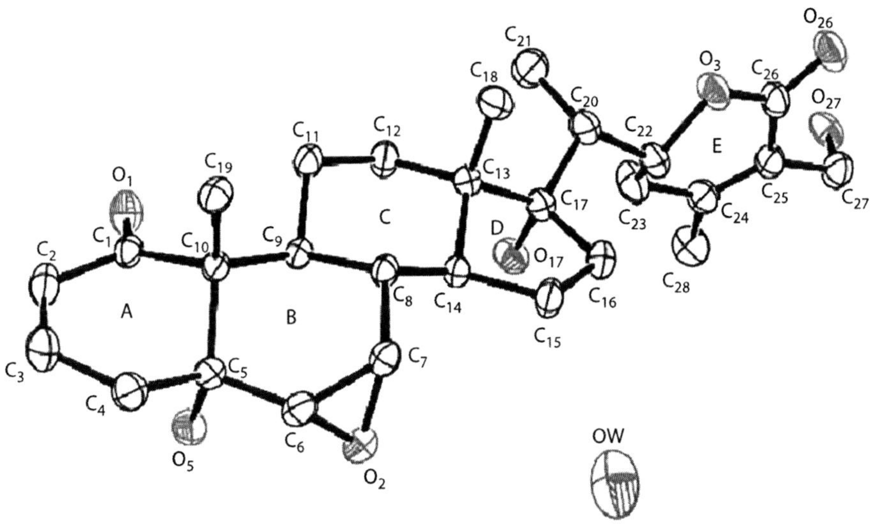

FIGURE 13.18 *ORTEP* view of the molecule, showing the atom-labeling scheme. Displacement ellipsoids are drawn at the 50% probability level. H atoms have been omitted for clarity.

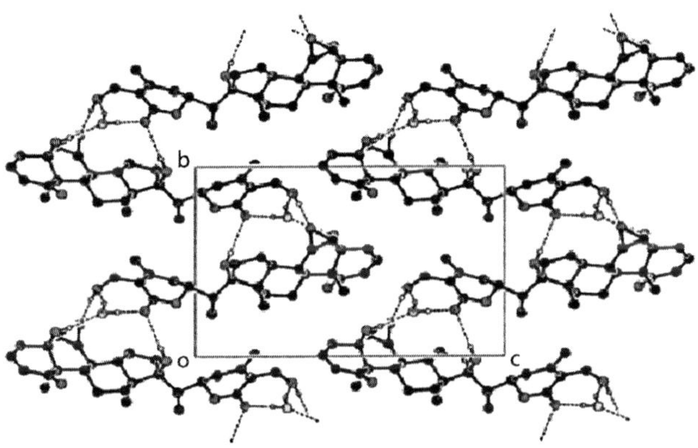

FIGURE 13.19 Appearance of twisted chains of molecules that are hydrogen bonded.

terminal atoms C_3 and C_{27} is 15.362 Å and the $C_3...C_{16}$ distance, which is a measure of the length of the steroid nucleus, is 8.994 Å. The pseudo-torsion angle $C_{19}–C_{10}...C_{13}–C_{18}$ is −1.5°, indicating that the steroid nucleus is untwisted. This is attributed to the short intramolecular O–H...O and C–H...O contacts [$O_5–H_5...O_2$, $C_9–H_9...O_5$, $C_{14}–H_{14}...O_{17}$; $C_{11}–H_{112}...O_1$; $C_{12}–H_{121}...O_{17}$] present in the molecule. The ring junctions A/B, B/C, and C/D are *trans* fused about the $C_5–C_{10}$, $C_8–C_9$, and $C_{13}–C_{14}$ bonds, respectively. Packing view of the molecules in the unit cell viewed down the a-axis is shown in Figure 13.19. The presence of hydroxyl groups in the molecule and of water solvent results in intermolecular hydrogen bonding. The characteristic pattern observed in packing diagram is the appearance of twisted chains of molecules. The adjacent chain links are rotationally related. The packing of the chains in the crystal is further stabilized into a three-dimensional network by weak C–H...O hydrogen bonds. The crystallographic data are summarized in Table 13.6.

13.4.7 Crystal Structure of 16-Dehydropregnenolone Acetate (16-DPA): A Pregnane

16-Dehydropregnenolone Acetate (3β-acetoxy-pregna-5,16-dien-20-one, $C_{23}H_{32}O_3$) is a key intermediate for the production of steroidal hormone drugs like betamethesone, dexamethasone, beclomethasone, eluticasone, prednicarlate as well as other sex hormones. Diosgenin, a primary material

TABLE 13.6
Crystal and Experimental Data

CCDC no.	734187
Crystal description	Irregular (colorless)
Crystal size	0.3 × 0.2 × 0.2 mm
Empirical formula	$C_{28}H_{38}O_7 \cdot H_2O$
Formula weight	504.60
Radiation, wavelength	Mo $K\alpha$, 0.71073 Å
Unit cell dimensions	a = 6.4540(2) Å, b = 11.3656(4) Å, c = 17.4982(5) Å β = 90.730(2)°
Crystal system	Monoclinic
Space group	$P2_1$
Unit cell volume	1283.45(7) Å^3
Density (calculated)	1.306 Mgm^{-3}
No. of molecules per unit cell, Z	2
Temperature	293(2) K
Absorption coefficient	0.095 mm^{-1}
Absorption correction	multi-scan (T_{min} = 0.9122 and T_{max} = 0.9741)
F(000)	544
Refinement of unit cell	5833 (2.5° < θ < 28.8°)
Scan mode	ω and φ scan
θ range for entire data collection	1.16° < θ < 29.77°
Reflections collected/unique	31300/7247
Reflections observed (I > 2σ(I))	6122
Range of indices	h = −8 to 9, k = −15 to 15, l = −24 to 24
R_{int}	0.0319
R_{sigma}	0.0310
No. of parameters refined	356
Final R-factor	0.0419
wR(F^2)	0.1049
Weight	$1/[\sigma^2(F_o^2) + (0.0693\ P)^2 + 0.0416\ P]$ where $P = \left[F_o^2 + 2F_c^2\right]/3$
Goodness-of-fit	1.041
$(\Delta/\sigma)_{max}$	0.001 (for z C_3)
Final residual electron density	−0.207 < $\Delta\rho$ < 0.275 eÅ^{-3}

for synthesizing steroid hormone drugs, is a raw material of 16-dehydropregnenolone acetate. It synthesizes the hydrocortisone, prednisone, norethindronum, fluocinolone, and dexamethasone etc. and kinds of steroid hormone drugs. A mixture of diosgenin (4.9 g) and acetic anhydride (10 mL) was heated at 195°C at elevated pressure for an hour to give pseudodiosgenin diacetate. The resulting solution was brought to room temperature and diluted with water (1.5 mL) and acetic acid (10 mL). The mixture was cooled to 10°C and to this was added a solution of chromic anhydride (2 g) in acetic acid (10 mL) with stirring maintaining the temperature between 10°C and 15°C and the stirring was continued for 3 h at this temperature. This resulted in the oxidation of pseudodiosgenin diacetate to yield the ester. Sodium bisulfite (2 g) in water (5 mL) was added to the mixture which was then refluxed for an hour, cooled and poured in water (200 mL). The precipitated solid was filtered, washed with water and dried at room temperature. Repeated crystallization from methanol gave 16-Dehydropregnenolone Acetate as crystals (2.3 g), m.p. 171°C–172°C (Bandhoria et al., 2006b) (Figure 13.20).

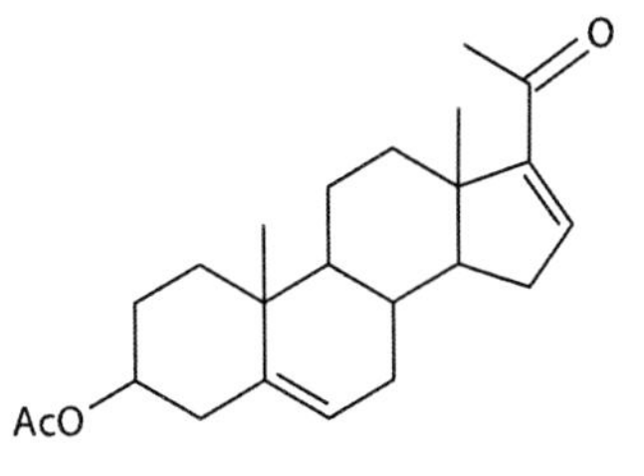

FIGURE 13.20 Chemical structure of 16-dehydropregnenolone acetate.

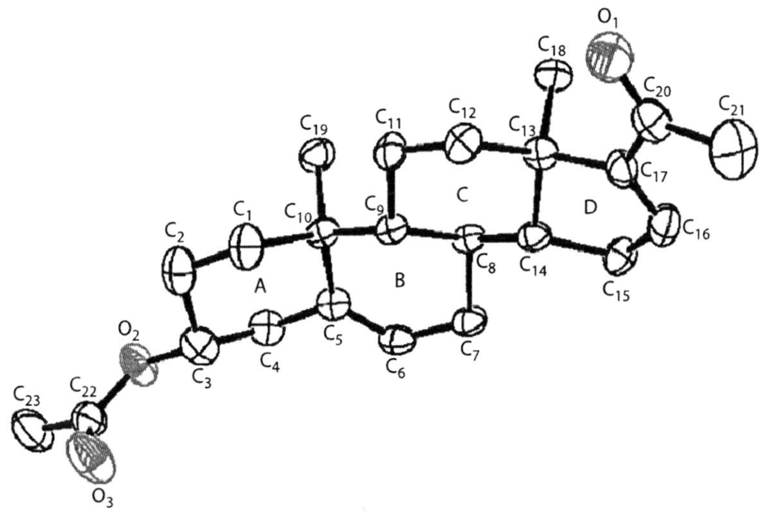

FIGURE 13.21 *ORTEP* view of the molecule with displacement ellipsoids drawn at 50% probability level.

An *ORTEP* view of the molecule indicating atomic numbering scheme is shown in Figure 13.21. Mean bond lengths are [$C(sp^3)–C(sp^3) = 1.532(9)$; $C(sp^3)–C(sp^2) = 1.499(9)$; $C(sp^2) = C(sp^2) = 1.325(9)$ Å]. The overall shape of ring A is very close to the *chair* conformation typical for totally saturated A rings. However, some distortion from the ideal form could be expressed as due to the loss of rotational symmetry, with rotational axis passing through $C_1–C_2$ and $C_4–C_5$ bonds [$\Delta C_2(C_1–C_2) = 12.75$], with the retention of perpendicular mirror symmetry [$\Delta C_s(C_3) = 2.32$]. In ring B, the $C_5{=}C_6$ distance of 1.318(9) Å confirms the localization of a double bond at this position. This double bond imposes 9β-8α *half-chair* conformation on ring B with the rotational axis bisecting $C_5–C_6$ and $C_8–C_9$ bonds and with the asymmetry parameter [$\Delta C_2(C_5–C_6) = 5.20$]. Ring C has a distorted *chair* form [$\Delta C_s(C_8) = 8.60$, $\Delta C_2(C_8–C_9) = 10.45$]. The conformation of ring D is 14α-*envelope* ($\Delta C_s(C_{14} = 4.16)$) with pseudorotation parameters $\Delta = -25.22°$ and $\varphi_m = -37.53°$. The conformation of O1 relative to C_{13} is assigned one of the qualitative descriptors, *syn*periplanar, syn-clinal, anticlinal, or antiperiplanar in accordance with the definition of Klyne and Prelog (1960). In the present study, the carbonyl oxygen atom O1 is oriented over the D-ring syn-periplanar to the $C_{13}–C_{17}$ bond [$C_{13}–C_{17}–C_{20}–O_1 = -5.0(9)°$]. The conformation of the side chain at C_{17} is further illustrated in Figure 13.22.

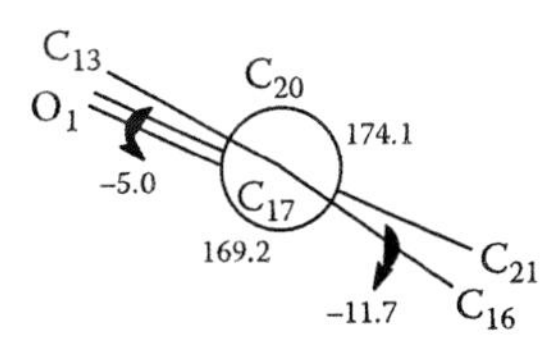

FIGURE 13.22 Newman projection with torsion angles about the $C_{17}–C_{20}$ bond (angles given in degrees).

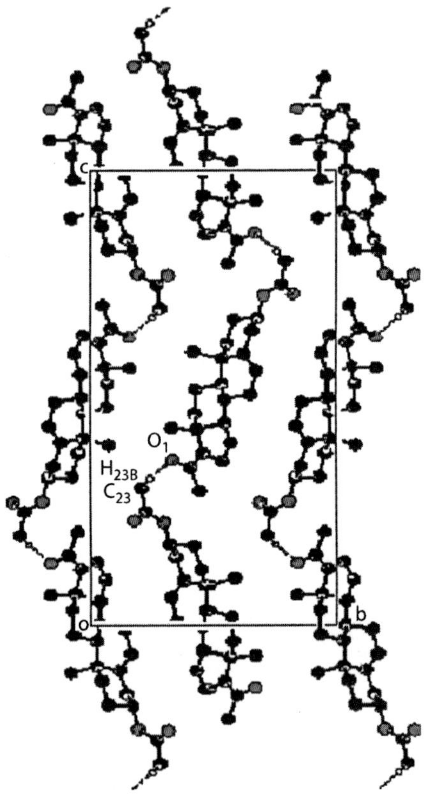

FIGURE 13.23 Appearance of chains of molecules that are head-to-tail hydrogen bonded.

The $C_3...C_{16}$ distance, which is a measure of the length of the steroid nucleus, is 8.739 Å. The distance between terminal atoms C_{21} and C_{23} is 14.543 Å. The C_{19}–C_{10}...C_{13}–C_{18} pseudo-torsion angle, which gives a measure of the molecular twist, is 9.5°. The A/B ring junction is *quasi-trans*, while ring systems B/C and C/D are *trans* fused about the C_8–C_9 and C_{13}–C_{14} bonds, respectively.

Packing view of the molecules in the unit cell viewed down the a-axis is shown in Figure 13.23. Atom C_{23} acts as a donor for a weak intermolecular C–H...O hydrogen bond with carbonyl atom O_1 of an adjacent molecule. This interaction links the molecules head-to-tail into twisted chains that run along the c-axis. The adjacent chain links are rotationally related. Chains of hydrogen-bonded molecules are packed with van der Waals contacts to form layers. In the packing diagram it can also be seen that the acetate groups lie in layers perpendicular to the c-axis. The crystallographic data are summarized in Table 13.7.

13.4.8 Crystal Structure of 3β-Acetoxy-17α-Hydroxy-16α-Methylallopregnan-20-One Hemihydrates: A Pregnane

3β-acetoxy-17α-hydroxy-16α-methylallopregnan-20-one, $C_{24}H_{38}O_4 \cdot 0.5H_2O$, was prepared from 16-dehydropregnenolone acetate by the method described earlier (Marker and Crooks 1942; Oliveto et al., 1958). 16-dehydropregnenolone acetate was treated with methylmagnesium iodide in the presence of cuprous chloride in dry tetrahydrofuran to obtain 3β-hydroxy-16α-methylpregn-5-en-20-one, which was hydrogenated over palladium on charcoal to yield 3β-hydroxy-16α-methylallopregnan-20-one. Enolacetylation at C_{20}, treatment with perbenzoic acid and subsequent alkaline hydrolysis gave 3β-17α-dihydroxy-16α-methylallopregnan-20-one, which was acetylated using acetic anhydride/pyridine to obtain 3β-acetoxy-17α-hydroxy-16α-methylallopregnan-20-one (Figure 13.24) (Bandhoria et al., 2006c).

TABLE 13.7
Crystal and Experimental Data

CCDC no.	259203
Crystal description	Colorless plate
Crystal size	0.3 × 0.2 × 0.2 mm
Empirical formula	$C_{23}H_{32}O_3$
Formula weight	356.49
Radiation, wavelength	Mo $K\alpha$, 0.71073 Å
Unit cell dimensions	a = 6.031(4) Å, b = 12.481(2) Å, c = 27.162(5) Å
Crystal system	Orthorhombic
Space group	$P2_12_12_1$
Unit cell volume	2044.6(14) $Å^3$
Density (calculated)	1.158 Mgm^{-3}
No. of molecules per unit cell, Z	4
Temperature	293(2) K
Absorption coefficient (μ)	0.075 mm^{-1}
Absorption correction	ψ-scan (T_{min} = 0.9664 and T_{max} = 0.9982)
Extinction coefficient	0.002(4)
F(000)	776
Refinement of unit cell	25 reflections (6° < θ < 12°)
Scan mode	ω/2θ
θ range for entire data collection	2.22° < θ < 24.96°
Reflections collected/unique	2097/2097
Reflections observed (I > 2σ(I))	1291
Range of indices	h = 0–7, k = 0–14, l = 0–32
R_{int}	0.0000
R_{sigma}	0.0408
No. of parameters refined	340
Final R-factor	0.0597
$wR(F^2)$	0.1748
Weight	$1/[\sigma^2(F_o^2) + (0.1774\ P)^2 + 0.00\ P]$ where $P = \left[F_o^2 + 2F_c^2\right]/3$
Goodness-of-fit	1.056
$(\Delta/\sigma)_{max}$	0.153 (for y H_{11})
Final residual electron density	$-0.243 < \Delta\rho < 0.244$ $eÅ^{-3}$

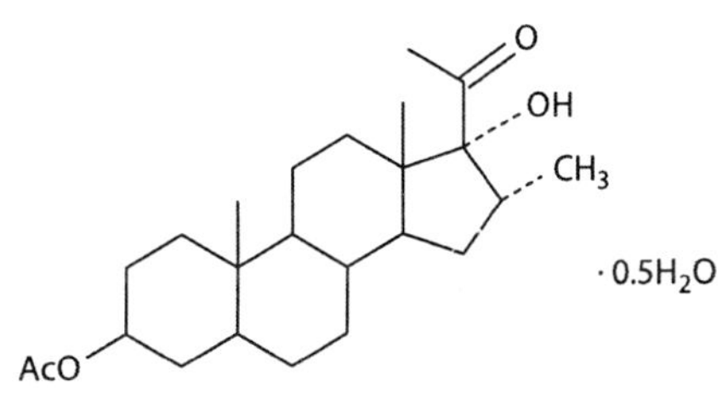

FIGURE 13.24 Chemical structure of 3β-acetoxy-17α-hydroxy-16α-methylallopregnan-20-one.

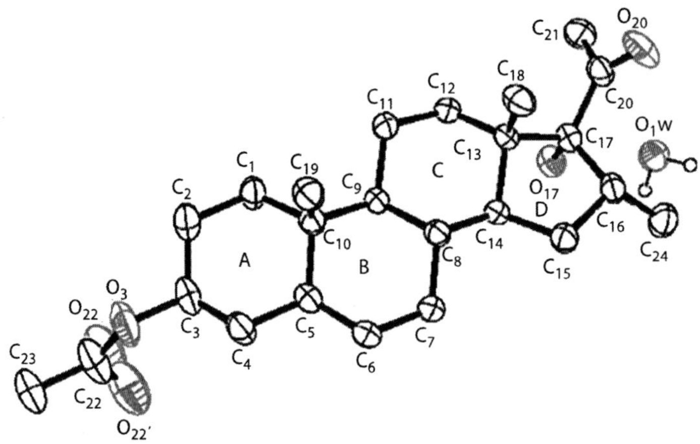

FIGURE 13.25 *ORTEP* view of the molecule with displacement ellipsoids drawn at 50% probability level.

An *ORTEP* view of the molecule indicating atomic numbering scheme is shown in Figure 13.25. The average C(sp^3)–C(sp^3) bond lengths in rings A, B, and C are 1.530(5), 1.537(5), and 1.535(5) Å, respectively. The average value of all the C(sp^3)–C(sp^3) bond lengths in the molecule is 1.538(5) Å. The shortening in the C_2–C_3 bond length 1.498(7) Å may be associated with the attachment of the acetoxy group to C_3. The interior angle C_{17}–C_{13}–C_{14} 98.4(3)° of ring D is significantly smaller than the other endocyclic bond angles. This effect has been observed in other pregnane steroids, for example, 11-ketoprogesterone (Gupta et al., 1994), 3β, 20-diacetoxy-16α-methylallopregn-17(20)-ene (Singh et al., 1994), which have angles having corresponding values of 99.7(2)°, 101.1(3)°. Further examples may be found in Duax and Norton (1975) and Griffin et al. (1984). In the 3β-acetoxy group, the C_{22}–O_{22} bond length approaches the value of a delocalized double bond. Furthermore, the structure refinement shows that this group is disordered over two positions in such a way that the 43% occupied site appears to have a normal C_{22}–O_{22} double bond and the 57% occupied site appears to have a longer C_{22}–$O_{22'}$ bond relative to the standard value. Both disordered components have a small distortion from the expected coplanarity of an acetoxy group (sums of valence angles around C_{22} are 355.0(9)° and 354.1(7)°). Similar distortion from the expected coplanarity is also observed in steroids having disordered acetoxy group at C_3 (Andrade et al., 1999). A comparison of C_3–O_3–C_{22}–O_{22} torsion angle (°) in some steroid structures having a saturated A ring and a 3-acetoxy substituent shows that the 3-acetoxy group is planar in steroids having no disorder in the group (Lindeman et al., 1992; Paixao et al., 2004). The C_3–O_3 bond is oriented equatorially and is (+)anticlinal to the C_2–C_3 bond. The dihedral angle between the plane of the acetoxy group and the mean molecular plane is 84.4(3)°, showing that this group is twisted around the C_3–O_3 bond.

Ring A has a highly symmetrical *chair* conformation. All asymmetry parameters are below 5.16. Mirror symmetry is dominant with the best C_s plane passing through C_3 and C_{10}. The asymmetry parameters in ring A are: $\Delta C_s(C_3) = 0.47$, $\Delta C_2(C_3–C_4) = 2.46$, $\Delta C_2(C_2–C_3) = 3.15$. Ring B has an ideal *chair* conformation typical for totally saturated B rings with the best rotational axis bisecting the C_5–C_{10} and C_7–C_8 bonds and with the asymmetry parameter $\Delta C_2(C_5–C_{10}) = 2.93$. The best mirror plane for this ring passes through C_7 and C_{10} [$\Delta C_s(C_7) = 0.31$]. The overall shape of ring C is close to the *chair* conformation. However, some distortion from the ideal form could be expressed as due to the loss of rotational symmetry with rotational axis passing through C_8–C_{14} and C_{11}–C_{12} bonds [$\Delta C_2(C_8–C_{14}) = 8.47$], with the retention of perpendicular mirror symmetry [$\Delta C_s(C_9) = 1.71$]. The conformation of ring D is perfect 13β-*envelope* [$\Delta C_s(C_{13}) = 1.01$] with pseudorotation parameters $\Delta = 38.16°$ and maximum angle of torsion $\varphi_m = 49.20°$. From the numerous pregnanes studied crystallographically thus far, it has been observed that the rotation of the side chain at C_{17} is hindered despite apparent freedom of rotation and the C_{16}–C_{17}–C_{20}–O_{20} torsion angle is between 0° and −46°

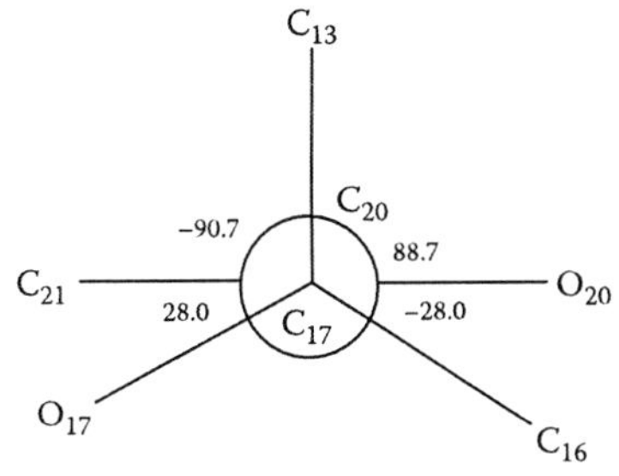

FIGURE 13.26 Newman projection with torsion angles about the C_{17}–C_{20} bond (angles in degrees).

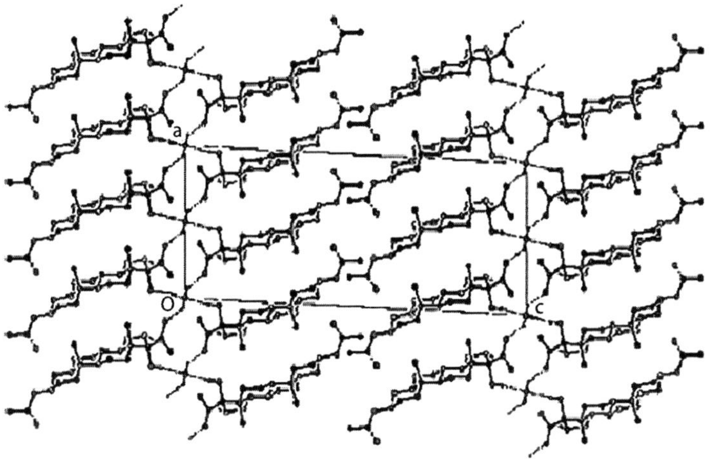

FIGURE 13.27 Appearance of supramolecular chains of molecules that are hydrogen bonded.

(Duax et al., 1981). In the title compound the torsion angle is –28.0(5)°. The carbonyl oxygen atom O_{20} is oriented over the D-ring (+) *syn*-clinal to the C_{13}–C_{17} bond [C_{13}–C_{17}–C_{20}–O_{20} = 88.7(5)°]. The conformation of the side chain at C_{17} is further illustrated in Figure 13.26.

The distance between the terminal atoms C_{23} and O_{20} is 14.306 Å and the pseudo-torsion angle C_{19}–C_{10}...C_{13}–C_{18} is 2.3°, showing that the molecule has a negligible twist. The C_3...C_{16} distance, which is a measure of the length of the steroid nucleus, is 9.046 Å. All rings of the steroid skeleton are *trans* connected. The presence of hydroxy group in the molecule and of water solvent molecules results in intermolecular hydrogen bonding, which links the molecules into infinite supramolecular chains running along the [100] direction in a head-to-head fashion (Figure 13.27). There are no hydrogen bonds between separate chains. In the packing diagram it can also be seen that the acetoxy groups lie in layers perpendicular to the c-axis. The crystallographic data are summarized in Table 13.8.

13.4.9 Crystal Structure of 3β-Hydroxy-16α-Methylpregn-5-en-20-One: A Pregnane

3β-Hydroxy-16α-methylpregn-5-en-20-one was prepared from 16-dehydro-pregnenolone acetate by the method described earlier (Marker and Crooks, 1942; Oliveto et al., 1958). 16-dehydro-pregnenolone acetate was treated with methylmagnesium iodide in presence of cuprous chloride in dry tetrahydro-furan. The product obtained was purified by repeated crystallization to yield 3β-hydroxy-16α-methylpregn-5-en-20-one (Figure 13.28) (Gupta et al., 2011).

An *ORTEP* view of the molecule indicating atomic numbering scheme is shown in Figure 13.29. Mean bond lengths are: [C(sp^3)–C(sp^3) = 1.533(4) Å; C(sp^3)–C(sp^2) = 1.510(4) Å]. Ring A is in *chair* conformation with the best rotational axis bisecting the C_2–C_3 and C_5–C_{10} bonds [asymmetry parameter $\Delta C_2(C_2–C_3)$ = 2.53]. The best mirror plane for this ring passes through C_2 and C_5

TABLE 13.8
Crystal and Experimental Data

CCDC number	270026
Crystal description	Transparent plate
Crystal size	0.3 × 0.2 × 0.2 mm
Empirical formula	$C_{24}H_{38}O_4 \cdot 0.5H_2O$
Formula weight	399.55
Radiation, Wavelength	Mo $K\alpha$, 0.71073 Å
Unit cell dimensions	a = 10.665(2) Å, b = 7.497(1) Å, c = 28.200(4) Å
	β = 92.74(2)°
Crystal system	Monoclinic
Space group	C_2
Unit cell volume	2252.1(7) $Å^3$
Density (calculated)	1.178 Mgm^{-3}
No. of molecules per unit cell, Z	4
Temperature	293(2) K
Absorption coefficient	0.079 mm^{-1}
Absorption correction	ψ-scan (T_{min} = 0.9969 and T_{max} = 0.9469)
Extinction coefficient	0.0030(8)
F(000)	876
Refinement of unit cell	25 reflections (4° < θ < 13°)
Scan mode	ω/2θ
θ range for entire data collection	2.17° < θ < 24.96°
Reflections collected/unique	2267/2137
Reflections observed (I > 2σ(I))	1579
Range of indices	h = −0 to 12, k = 0 to 8, l = −33 to 33
R_{int}	0.0140
R_{sigma}	0.0287
No. of parameters refined	267
Final R-factor	0.0419
wR(F^2)	0.1124
Weight	$1/[\sigma^2(F_o^2) + (0.0695\ P)^2 + 1.7475\ P]$
	where $P = \left[F_o^2 + 2F_c^2\right]/3$
Goodness-of-fit	0.962
$(\Delta/\sigma)_{max}$	−0.247 (for y O_3)
Final residual electron density	$-0.197 < \Delta\rho < 0.276$ $eÅ^{-3}$

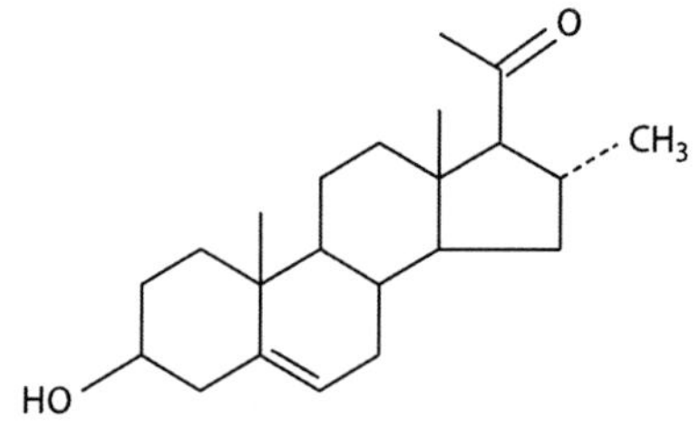

FIGURE 13.28 Chemical structure of 3β-hydroxy-16α-methylpregn-5-en-20-one.

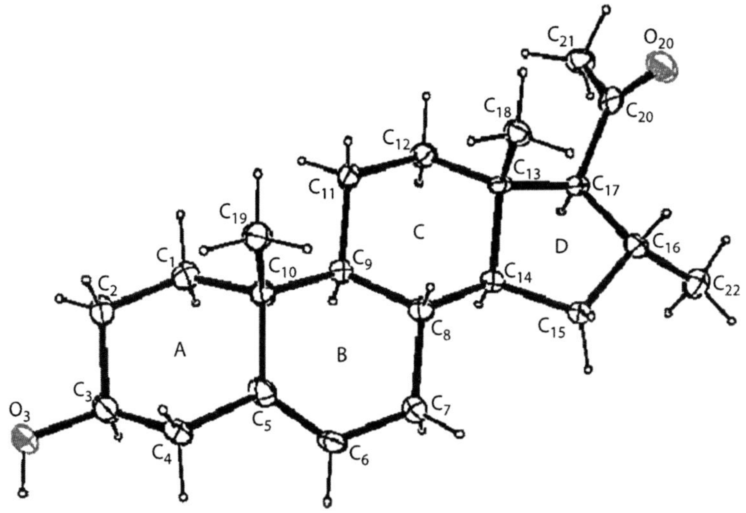

FIGURE 13.29 *ORTEP* view of the molecule with displacement ellipsoids drawn at 50% probability level. H atoms are shown as small spheres of arbitrary radii.

with asymmetry parameter $[\Delta C_s(C_2) = 3.51]$. In ring B, the $C_5 = C_6$ distance of 1.330(5) Å confirms the localization of a double bond at this position. This double bond imposes an 8β,9α–*half-chair* conformation on ring B with the rotational axis bisecting C_5–C_6 and C_8–C_9 bonds and with the asymmetry parameter $[\Delta C_2(C_5–C_6) = 2.21]$. The overall shape of ring C is very close to the *chair* conformation. Distortion from the ideal form could be expressed as due to the loss of rotational symmetry through C_8–C_{14} and C_{11}–C_{12} bonds $[\Delta C_2(C_8–C_{14}) = 12.37]$, with the retention of perpendicular mirror symmetry $[\Delta C_s(C_9) = 2.52]$. The D-ring has a nearly 13β-*envelope* conformation described in terms of Altona's notation by $\Delta = -28.20°$ and $\varphi_m = -48.15°$ (1968). Alternatively, the slight deviation from mirror symmetry across a plane bisecting the C_{15}–C_{16} bond and containing the C_{13} atom is indicated by the small magnitude of the asymmetry parameter $[\Delta C_s(C_{13}) = 3.91]$. The $C_3...C_{16}$ distance, which is a measure of the length of the steroid nucleus, is 9.035 Å. The distance between terminal atoms O_3 and O_{20} is 11.931 Å. It was found that the equilibrium geometry of the isolated molecule features a sizeable twist of the steroid nucleus, as measured by the pseudo-torsion angle C_{19}–$C_{10}...C_{13}$–C_{18} of 8.0°. The A/B ring junction is *quasi-trans*, while ring systems B/C and C/D are *trans* fused about the C_8–C_9 and C_{13}–C_{14} bonds, respectively.

In the present study, the carbonyl oxygen atom O_{20} is oriented over the D-ring (+) syn-clinal to the C_{13}–C_{17} bond $[C_{13}–C_{17}–C_{20}–O_{20} = 87.3(4)°]$. The conformation of the side chain at C_{17} is further illustrated in Figure 13.30.

In the crystal structure of the title compound, head-to-tail hydrogen-bonded extended chains are observed. Atom C_{21} of the methyl group in the molecule at (x, y, z) acts as a hydrogen-bond donor to hydroxy atom O_3 in the molecule at (x, y, z + 1), producing a chain. The adjacent chain

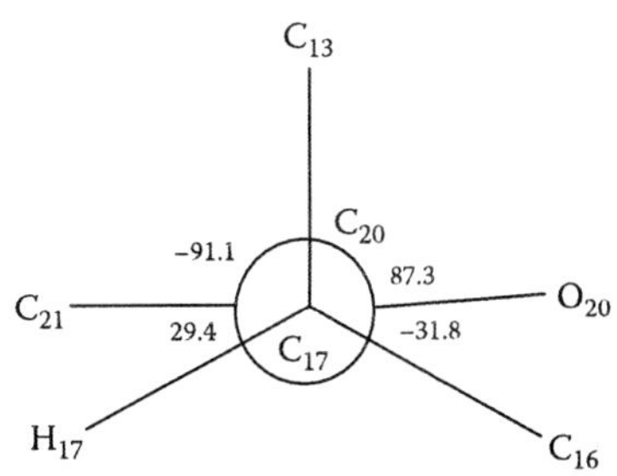

FIGURE 13.30 Newman projection with torsion angles about the C_{17}–C_{20} bond (angles given in degrees).

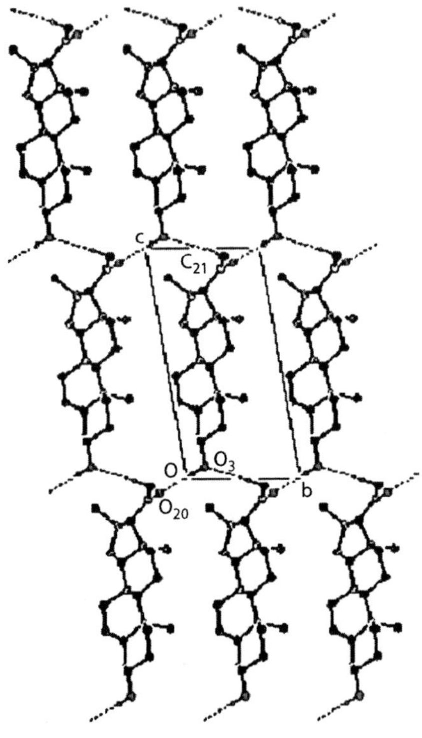

FIGURE 13.31 Appearance of chains of molecules that are hydrogen bonded. For clarity, only H atoms involved in hydrogen bonding have been included.

links are translationally related. The extended chains are packed together to form layers. Molecules within the layers are arranged parallel to each other and are stabilized by the second hydrogen bond (O_3–$H_3O\ldots O_{20}$) (Figure 13.31). Atom O_3 of the hydroxy group in the molecule at (x, y, z) acts as a hydrogen bond donor to carbonyl atom O20 in the molecule at (x, y − 1, z − 1).

13.4.10 Crystal Structure of (25*R*)-Spirost-5-en-3β-Acetate: A Steroid Sapogenin

The compound [(25*R*)-Spirost-5-en-3β-acetate, $C_{29}H_{44}O_4$] was synthesized by acetylation of diosgenin. Diosgenin is a steroidal sapogenin, which is extracted from the roots of Wild Yam (*Dioscorea deltoidea*), an important plant for the pharmaceutical industry used mainly as a source of steroidal harmones. Diosgenin is used for making intermediates of residual hormone medicines such as cortisone, floucinoline acetonide, progesterone, testosterone, etc., Diosgenin (100 mg) was subjected to microwave irradiation in 10 mL acetic anhydride for 20 min at 30% power. The contents of reaction mixture were cooled and kept overnight in the refrigerator. The crystals formed were filtered and recrystallized from acetic acid to yield (25*R*)-Spirost-5-en-3β-acetate, m.p. 193°C–195°C (Figure 13.32) (Bandhoria et al., 2006a).

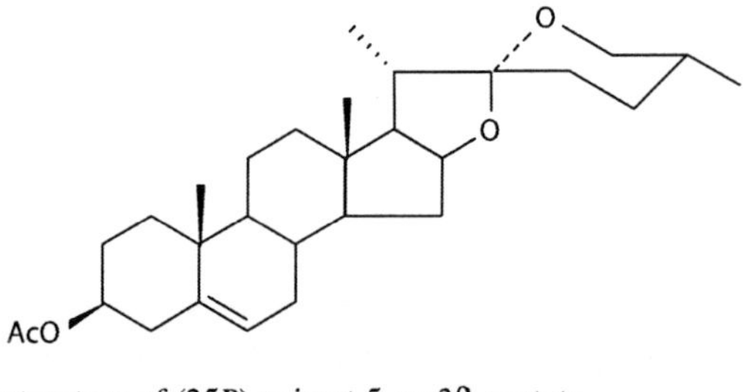

FIGURE 13.32 Chemical structure of (25*R*)-spirost-5-en-3β-acetate.

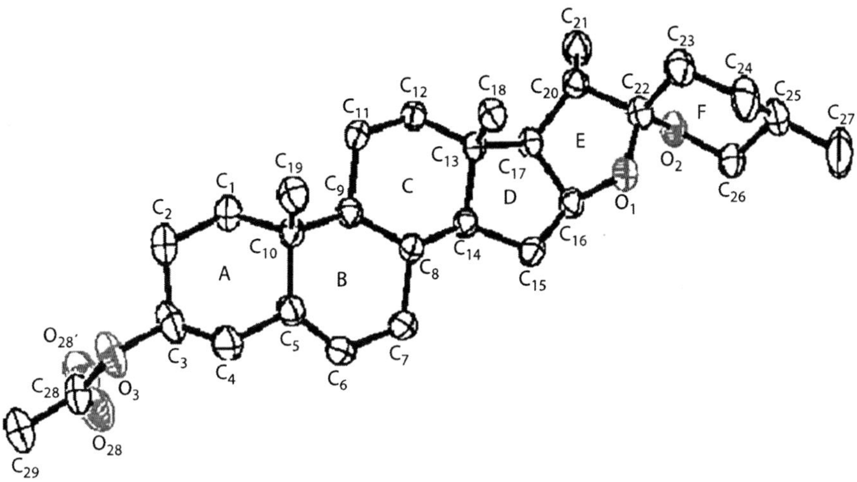

FIGURE 13.33 *ORTEP* view of the molecule with thermal ellipsoids drawn at 40% probability level.

An *ORTEP* view of the molecule indicating atomic numbering scheme is shown in Figure 13.33. The mean bond lengths are $C(sp^3)–C(sp^3) = 1.533(5)$; $C(sp^3)–C(sp^2) = 1.509(4)$; $C(sp^3)–O = 1.420(4)$ Å. The conformation of A ring is approximately the *chair*, typical for totally saturated A rings. Distortions from the ideal chair form can be expressed by the loss of rotational symmetry through bonds $C_1–C_2$ and $C_4–C_5$ [$\Delta C_2(C_1–C_2) = 13.70$] with a retention of perpendicular mirror symmetry [$\Delta C_s(C_3–C_{10}) = 1.88$]. Due to the $C_5=C_6$ double bond, the environment of atom C_5 is *planar*. Hence, ring B is highly distorted from the normal chair conformation, assuming instead an 8β,9α-*half-chair* conformation. Atoms C_8 and C_9 are displaced to opposite sides by 0.395(4) and 0.364(3) Å, respectively, from the mean $C_{10}/C_5/C_6/C_7$ plane. Ring C has a distorted *chair* form [$\Delta C_s(C_8) = 8.31$, $\Delta C_2(C_8–C_{14}) = 11.14$]. The conformation of D ring is 13β,14α-*half-chair* [$\Delta C_2(C_{13}–C_{14}) = 2.45$] with pseudorotation parameters $\Delta = 3.18°$ and $\varphi_m = 47.91°$. The conformation of ring E is *envelope* [$\Delta C_s(O_1) = 3.71$] with pseudorotation parameters $\Delta = -27.30°$ and $\varphi_m = -41.16°$. Ring F has a symmetrical *chair* conformation in which rotational symmetry is dominant, with asymmetry parameters [$\Delta C_2(C_{23}–C_{24}) = 0.45$, $\Delta C_s(O_2–C_{24}) = 3.11$].

In the 3β-acetoxy group, both disordered components have distortion from the expected coplanarity of an acetoxy group [sums of valence angles around C_{28} are 358.7(5)° and 352.8(7)°]. Similar distortion from the expected coplanarity is also observed in steroids having disordered acetoxy group at C_3. The distance between the terminal atoms C_{29} and C_{27} is 17.977 Å and the $C_3 \ldots C_{16}$ distance, which is a measure of the length of the steroid nucleus, is 8.921 Å. The pseudo-torsion angle $C_{19}–C_{10} \ldots C_{13}–C_{18}$ is 9.2°, indicating that the steroid nucleus is slightly twisted. The molecule is slightly convex toward the β side, with an angle of 10.7(2)° between the $C_{10}–C_{19}$ and $C_{13}–C_{18}$ vectors. The A/B ring junction is *quasi-trans*, while ring junctions B/C and C/D are *trans* fused about the $C_8–C_9$ and $C_{13}–C_{14}$ bonds, respectively. The D/E ring junction is *cis*. Owing to the absence of any strong donor group, cohesion of molecules is mainly achieved by van der Waals interactions and a weak hydrogen bond involving CH group [$C_{29}–H_{29}C \ldots O_3$ (3.594(6) Å, 166.9(3)°); $H_{29}C \ldots O_3$ 2.652(3) Å; Symmetry: $-x + 2$, $y - 1/2$, $-z$]. Molecules in the unit cell are packed together to form well-defined layers. Molecules within the layers are arranged in an antiparallel manner (Figure 13.34). The crystallographic data are summarized in Table 13.9.

13.4.11 Crystal Structure of 3β-Hydroxy-17-Oximinoandrost-5-ene Monohydrate

The compound, 3β-hydroxy-17-oximinoandrost-5-ene monohydrate, was synthesized by dissolving dehydroepiandrosterone (1 g) in a mixture of methanol (10 mL) and anhydrous pyridine (10 mL); 250 mg of hydroxylamine hydrochloride was added to it. The solution was refluxed on a water

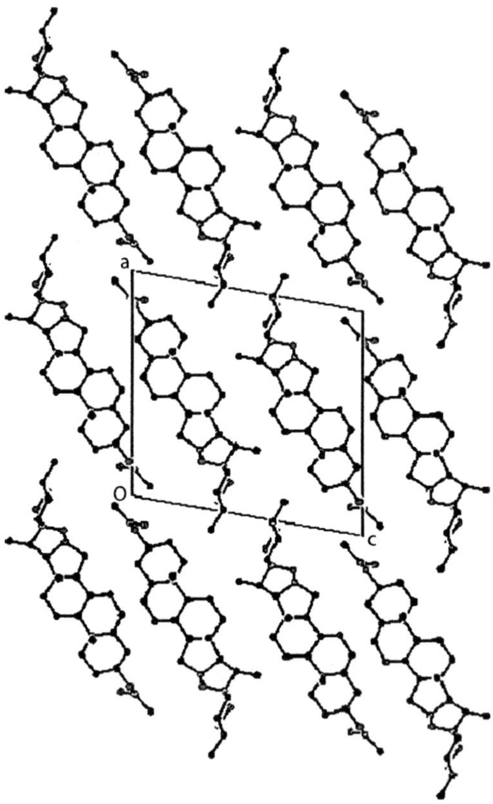

FIGURE 13.34 The crystal packing showing the formation of molecular layers.

bath for 30 min. The reaction mixture was cooled and the separated solid was filtered, washed free of pyridine, and dried under a vacuum. Recrystallization from aq. methanol gave 3β-hydroxy-17-oximinoandrost-5-ene (dehydroepiandrosterone oxime) (760 mg), m.p. 205°C–206°C (Figure 13.35) (Kanwal et al., 2007).

An *ORTEP* view of the molecule indicating atomic numbering scheme is shown in Figure 13.36. The mean bond lengths are: C(sp^3)–C(sp^3) = 1.535(4) Å; C(sp^3)–C(sp^2) = 1.506(4) Å. Ring A is in *chair* conformation with the best rotational axis bisecting the C_2–C_3 and C_5–C_{10} bonds [asymmetry parameter $\Delta C_2(C_2–C_3)$ = 2.15]. The best mirror plane for this ring passes through the C_3 and C_{10} atoms with asymmetry parameter [$\Delta C_s(C_3)$ = 2.87]. In ring B, due to the presence of a double bond at C_5–C_6, the environment of atom C_5 is *planar.* Hence, ring B is highly distorted from the normal *chair* conformation, assuming instead an 8α,9β–*half-chair* conformation with the rotational axis bisecting the C_5–C_6 and C_8–C_9 bonds and with the asymmetry parameter [$\Delta C_2(C_5–C_6)$ = 1.27]. Atoms C_8 and C_9 are displaced to opposite sides by −0.378(3) and 0.347(3) Å, respectively, from the mean plane ($C_{10}/C_5/C_6/C_7$). The overall shape of ring C is very close to the *chair* conformation. A distortion from the ideal form could be expressed as being due to a loss of rotational symmetry through the C_8–C_{14} and C_{11}–C_{12} bonds [$\Delta C_2(C_8–C_{14})$ = 6.08], with the retention of perpendicular mirror symmetry [$\Delta C_s(C_{13})$ = 2.59]. Atoms C_9 and C_{13} are situated at 0.636(3) and −0.702(3) Å, respectively, above and below the plane defined by the other four ring atoms ($C_8/C_{14}/C_{12}/C_{11}$). The conformation of D ring is 13α, 14β—distorted *half chair* [$\Delta C_2(C_{13}–C_{14})$ = 6.93] with the phase angle of pseudorotation Δ = −11.28° and maximum angle of torsion φ_m = 45.62°. Atoms C_{13} and C_{14} are disposed −0.258(3) and 0.466(3) Å, respectively, below and above the plane defined by the other three ring atoms.

TABLE 13.9
Crystal and Experimental Data

CCDC no	261918
Crystal description	Colorless rectangular plate
Crystal size	0.3 × 0.2 × 0.2 mm
Empirical formula	$C_{29}H_{44}O_4$
Formula weight	456.64
Radiation, Wavelength	Mo $K\alpha$, 0.71073 Å
Unit cell dimensions	a = 14.559(2) Å, b = 6.212(3) Å, c = 14.847(7) Å
	β = 100.22(2)°
Crystal system	Monoclinic
Space group	$P2_1$
Unit cell volume	1321.5(9) $Å^3$
Density (calculated)	1.148 Mgm^{-3}
No. of molecules per unit cell, Z	2
Temperature	293(2) K
Absorption coefficient(μ)	0.074 mm^{-1}
Absorption correction	ψ-scan (T_{min} = 0.9567 and T_{max} = 0.9983)
Extinction coefficient	0.007(3)
F(000)	500
Refinement of unit cell	25 reflections (10° < θ < 15°)
Scan mode	$\omega/2\theta$
θ range for entire data collection	2.16° < θ < 24.97°
Reflections collected/unique	2666/2561
Reflections observed (I > 2σ(I))	2176
Range of indices	h = 0–17, k = 0–7, l = −17 to 17
R_{int}	0.0341
R_{sigma}	0.0214
No. of parameters refined	303
Final R	0.0513
$wR(F^2)$	0.1419
Weight	$1/[\sigma^2(F_o^2) + (0.1085\ P)^2 + 0.2339\ P]$
	where $P = \left[F_o^2 + 2F_c^2\right]/3$
Goodness-of-fit	1.050
$(\Delta/\sigma)_{max}$ in the final cycle	−0.002 (for z O_{28})
Final residual electron density	$-0.221 < \Delta\rho < 0.406\ eÅ^{-3}$

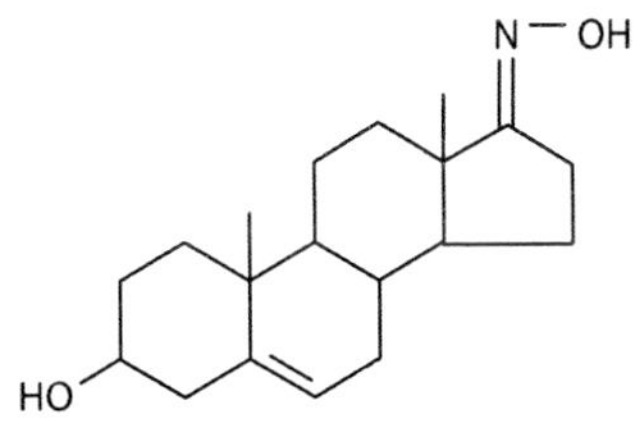

FIGURE 13.35 Chemical Structure of 3β-hydroxy-17-oximinoandrost-5-ene.

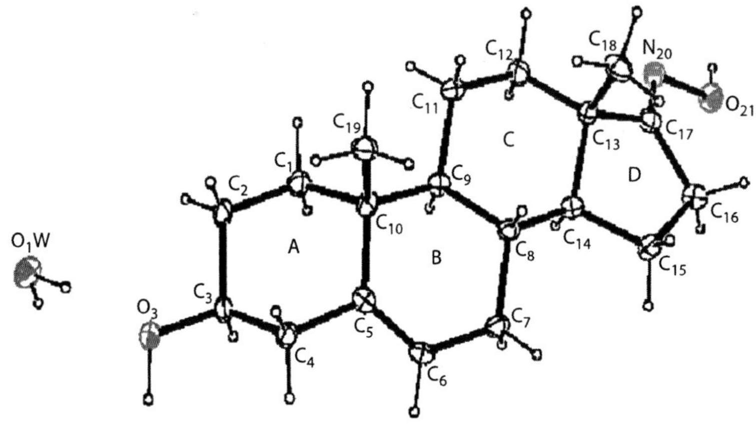

FIGURE 13.36 ORTEP view of the molecule with displacement ellipsoids drawn at the 50% probability level. H atoms are shown as small spheres of arbitrary radii.

The $C_3...C_{16}$ distance, which is a measure of the length of the steroid nucleus, is 8.917 Å. The distance between terminal atoms O_3 and O_{21} is 11.936 Å. The C_{19}–$C_{10}...C_{13}$–C_{18} pseudo-torsion angle is −12.5°, indicating that the steroid nucleus is twisted. The B/C and C/D ring junctions approach *trans* characteristics about the C_8–C_9 and C_{13}–C_{14} bonds, respectively, whereas the A/B ring junction is quasi-*trans*.

The molecular packing down the b-axis is illustrated in Figure 13.37. The presence of hydroxy groups in the molecule and of water solvent results in inter-molecular hydrogen bonding. O–H...O hydrogen bonds connect the parent steroid molecule to its neighbors via water solvent ($O_3...O_1W$ and $O_{21}...O_1W$) forming dimers. Further, $O_3...O_{21}$, $O_3...N_{20}$, O_1W-$H_{111}...O_3$ and O_1W–$H_{112}...O_3$ intermolecular hydrogen bonds link the dimers to form dimer chains. The dimer chains are packed together to form layers. Dimers within the layers are arranged parallel to each other. The crystallographic data are summarized in Table 13.10.

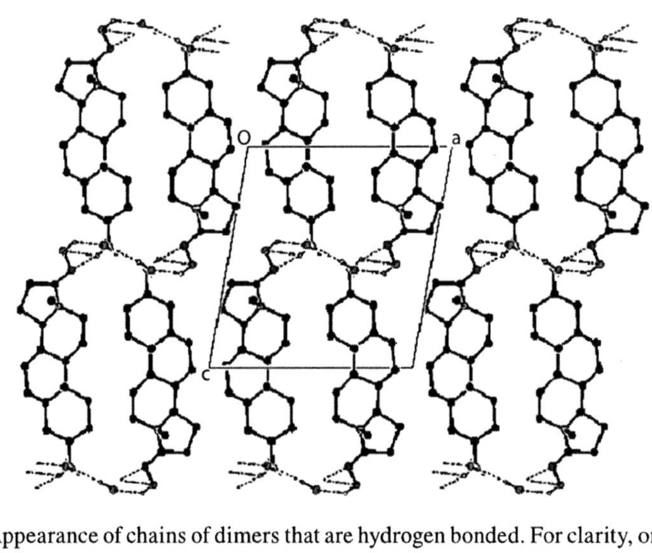

FIGURE 13.37 Appearance of chains of dimers that are hydrogen bonded. For clarity, only H atoms involved in hydrogen bonding are included.

TABLE 13.10
Crystal and Experimental Data

CCDC deposition no	649354
Crystal description	Irregular (transparent)
Crystal size	0.3 × 0.2 × 0.2 mm
Empirical formula	$C_{19}H_{29}NO_2 \cdot H_2O$
Formula weight	321.45
Wavelength	0.71073 Å
Crystal system	Monoclinic
Space group, Z	$P2_1$, 2
Temperature	100 K
Unit cell dimensions	
a	11.632(3) Å
b	6.255(2) Å
c	12.155(3) Å
β	100.488(4)°
Volume(V)	869.6(4) $Å^3$
D_x	1.228 Mgm^{-3}
Absorption coefficient	0.082 mm^{-1}
Reflections collected/unique	5758/3967
Reflections observed (I > 2σ(I))	3421
θ_{max}	28.35°
No. of parameters refined	220
Final R-factor	0.0549
$wR(F^2)$	0.1389
Goodness-of-fit	1.042
$(\Delta\rho)_{max}$	0.295 $eÅ^{-3}$
$(\Delta\rho)_{min}$	−0.302 $eÅ^{-3}$

13.5 CONCLUSIONS

1. None of the steroids, whose crystal structures have been reported here, contain more than one crystallographically independent molecule in the asymmetric unit.
2. In steroids with acetoxy group positioned at C_3, the C_2–C_3 bond length is significantly shorter than the expected value for a $C(sp^3)$–$C(sp^3)$ bond length.
3. The $C(sp^3)$–$C(sp^3)$ bond lengths in epoxide fused rings deviate significantly from the standard value.
4. Some of the bond angles show significant deviations from the ideal tetrahedral value. These deviations in steroids are a result of strain induced by the fusion of five- and six-membered rings, side chains, and bond unsaturations.
5. Steroid structures having a saturated A ring and a 3-acetoxy substituent show that the 3-acetoxy group is planar in steroids having no disorder in the group. Distortion from the coplanarity is observed in steroids having disordered acetoxy group at C_3.
6. The presence of a double bond in a six-membered ring imposes *half-chair* conformation.
7. The fusion of epoxide to six-membered ring changes the conformation from *chair* (typical for totally saturated rings) to *half-chair/sofa.*
8. The presence of solvent molecules helps in stabilizing the crystal structure through a network of intra- and intermolecular hydrogen bonds.

9. The characteristic pattern observed in steroid packing is the appearance of either extended chains or twisted chains. Head-to-tail hydrogen bonding is commonly observed to bind adjacent links in these chains.
10. Extended chains and twisted chains are packed together to form layers. The layers are much better defined in those structures having extended chains rather than twisted chains.
11. The 17-side chain orientation in pregnane derivatives is normally restricted with respect to the D ring despite apparent freedom of rotation. In the present study, in pregnane structures having no epoxide link on ring D or unsaturation at C_{16}–C_{17}, the C_{16}–C_{17}–C_{20}–O_{20} torsion angle is observed to be between −28.0° and −31.8°, that is, synperiplanar to the C_{16}–C_{17} bond. In 16-en-20-one pregnane structure, the carbonyl at C_{20} is antiperiplanar to the unsaturated C_{16}–C_{17} bond.
12. In pregnane derivatives having unsaturation at C_{16}–C_{17}, an unusual conformation of the substituent group at C_{17} is observed where the C_{13}–C_{17} bond almost eclipses the C_{20}–O_{20} bond. The unusual eclipsed conformation is responsible for the relatively large value of the pseudo-torsion angle C_{19}–C_{10}...C_{13}–C_{18}, which measures the twist of the molecule.

ACKNOWLEDGMENTS

The author is grateful to Dr. B. D. Gupta, Dr. K. A. Suri, Dr. N. K. Satti and Dr. D. K. Gupta, Indian Institute of Integrative Medicine, Canal Road, Jammu, India, for providing the samples for which the three-dimensional crystal structures have been reported.

REFERENCES

Allen, F.H., Kennard, O., Watson, D.G., Brammer, L., Orpen, A.G., and Taylor, R. 1987. Tables of bond lengths determined by x-ray and neutron diffraction. Part 1. Bond lengths in organic compounds. *J Chem Soc Perkin Trans* 2: S1–S19.

Altona, C., Geise, H.J., and Romers, C. 1968. Geometry and conformation of ring D in some steroids from x-ray structure determinations. *Tetrahedron* 24: 13–32.

Andrade, L.C.R., Paixao, J.A., de Almeida, M.J., Tavares da silva, E.J., Sae Melo, M.L., and Campos NeVes, A.S. 1999. 3,17-Dioxo-4-oxaandrostane-5α-carbaldehyde. *Acta Cryst* C55: 637–639.

Bandhoria, P., Gupta, V.K., and Gupta, D.K. 2006a. Crystal structure of (25R)-Spirost-5-en-3β-acetate. *Anal Sci* 22: 91–92.

Bandhoria, P., Gupta, V.K., Gupta, D.K., Jain, S.M., and Varghese, B. 2006b. Crystal structure of 3β-acetoxy-pregna-5,16-dien-20-one (16 DPA). *J Chem Crystallogr* 36: 161–166.

Bandhoria, P., Gupta, V.K., Gupta, B.D., and Varghese, B. 2006c. Crystal structure of 3β-acetoxy-17α-hydroxy-16α-methylallopregnan-20-one hemihydrates. *J Chem Crystallogr* 36:427–433.

Bandhoria, P., Gupta, V.K., Kumar, P., Satti, N.K., Dutt, P., and Suri, K.A. 2006d. Crystal structure of 5β,6β-epoxy-4β,27-dihydroxy-1-oxo-22R-witha-2,24-dienolide isolated from *Withania somnifera* Leaves. *Anal Sci* 22: 89–90.

Bandhoria, P., Gupta, V.K., Kumar, P., Satti, N.K., Dutt, P., and Suri, K.A. 2006e. 6α,7α-Epoxy-5α,17α,dihydroxy-1-oxo-22R-witha-2,24-dienolide in leaves of *Withania somnifera*: Isolation and its crystal structure. *J Chem Crystallogr* 36:153–159.

Bandhoria, P., Gupta, V.K., Sharma, V.K., Satti, N.K., Dutt, P., and Suri, K.A. 2006f. Crystal Structure of 6α,7α:24 α,25α-diepoxy-5α,12α,dihydroxy-1-oxo-20S,22R-witha-2-enolide isolated from *Datura quercifolia* leaves. *Anal Sci* 22: 169–170.

Bartlett, R.R., Dimitrijevic, M., Mattar, A.T., Zielinski, T., German, T., Rude, E., Thoenes, G.H., Kuchle, C.C., Schorlemmer, H.U., and Bremer, E. 1991. Leflunomide (HWA 486), a novel immunomodulating compound for the treatment of autoimmune disorders and reactions leading to transplant rejection. *Agents Actions* 32: 10–21.

Benn, W.R. and Dodson, R.M. 1964. The synthesis and stereochemistry of isomeric 16-hydroxy-17(20)-pregnenes. *J Org Chem* 29: 1142–1148.

Birk, Y. and Peri, I. 1980. Saponins. In *Toxic Constituents of Plant Foodstuffs*, 2nd edn., vol. 1, pp. 161–182, ed. I.E. Liener New York: Academic press.

Bordia, A. and Chuttani, S.K. 1979. Effect of gum guggulu on fibrinolysis and platelet adhesiveness in coronary heart disease. *Ind J Med Res* 70: 992–996.

Bucourt, R. 1974. The torsion angle concept in conformational analysis. In *Topics in Stereochemistry*, vol. 8, p. 159, eds. E.L. Eliel and N.L. Allinger. New York: Interscience.

Budhiraja, R.D. and Sudhir, S. 1987. Review of biological activity of withanolides. *J Sci Ind Res* 46: 488–491.

Budhiraja, R.D., Sudhir, S., and Garg, K.N. 1984. Pharmacological studies on leaves of *Withania somnifera. Planta Med* 50: 134–136.

Chakraborti, S.K., De, B.K., and Bandyopadhyay, T. 1974. Variation in the antitumor constituents of *Withania somnifera. Experientia* 30: 852–853.

Dhar, K.L. and Kalla, A.K. 1976. HA12-oxowithanolidefrom *Datura quercifoliaH Phytochemistry* 15: 339–340.

Duax, W.L., Griffin, J.F., Strong, P.D., and Wood, K.J. 1989. 11β-Hydroxy-9β-estrone *Acta Cryst C* 45: 930–932.

Duax, W.L. and Norton, D.A. 1975. *Atlas of Steroid Structures*, vol. 1. New York: Plenum.

Duax, W.L., Griffin, J.F., and Rohrer, D.C. 1981. Conformation of progesterone side chain: Conflict between X-ray data and force-field calculations. *J Am Chem Soc* 103: 6705–6712.

Eggleston, D.S. and Lan-Hargest, H.Y. 1990. 6α- and 6β-Trifluoromethyl-substituted androstenedione. *Acta Cryst C* 46: 1686–1691.

Farrugia, L., J. 1997. ORTEP-3 for windows—A version of ORTEP-III with a graphical user interface (GUI). *J Appl Cryst* 30: 565.

Furmanowa, M., Gajdzis-Kuls, D., Ruszkowska, J., Czarnocki, Z., Obidoska, G., Sadowska, A., Rani, R., and Upadhyay, S.N. 2001. In vitro propagation of *Withania somnifera* and isolation of withanolides with immunosuppressive activity. *Planta Med* 67(2): 146–149.

Gil, R.R., Misico, R.I., Sotes, I.R., Oberti, J.C., Veleiro, A.S., and Burton, G. 1997. 16-Hydroxylated withanolides from *Exodeconus maritimusH. J Nat Prod* 60: 568–572.

Glotter, E. 1991. Withanolides and related ergostane-type steroids. *Nat Prod Rep* 8: 415–440.

Glotter, E., Kirson, I., Lavie, D., and Abraham, A. 1978. The Withanolides—A group of natural steriods. *Bio-Organic Chemistry*, vol. 2, p. 57, ed. E.E. van Tamelen. New York: Academic Press.

Gonsette, R.E. (1996) Introductory remarks: Immunosuppressive and immunomodulating drugs, where and how do they act? *Mult Scler* 1: 306–312.

Griffin, J.F., Duax, W.L., and Weeks, C.M. 1984. *Atlas of Steroid Structures*, vol. 2, p. 21. New York: Plenum.

Gupta, V.K., Bandhoria, P., and Gupta, B.D. 2011a. 3β-Hydroxy-16α-methylpregn-5-en-20-one. *J Chem Crystallogr* 51: 265–270.

Gupta, V.K., Bandhoria, P., Gupta, B.D., and Gupta, K.K. 2006. Crystal structure of guggulsterone Z. *Crystallogr Rep* 51: 265–270.

Gupta, V.K., Lal, M.M., Satti, N.K., Dutt, P., Sharma, P., Amina, M., and Suri, K.A. 2011b. Isolation and crystal structure of 6α,7α-Epoxy-5α,17α,27-trihydroxy-1-oxo-22R-witha-2,24-dienolide monohydrate-A Withasteroid from *Withania somnifera* leaves. *J Chem Crystallogr* 41:1064–1070.

Gupta, V.K., Mahajan, S., Satti, N.K., Suri, K.A., and Qazi, G.N. 2008. (20*R*,22*R*)-6α,7α-Epoxy-5α,27-dihydroxy-1-oxowitha-2,24-dienolide in leaves of *Withania somnifera*: Isolation and its crystal structure. *J Chem Crystallogr* 38: 769–773.

Gupta, V.K., Kant, R., Goswami, K.N., Mazumdara, S.K., and Bhutani, K.K. 1994. 11-ketoprogesterone. *Acta Cryst C* 50: 798–801.

Kalla, A.K., Raina, M.L., Dhar, K.L., Qurishi, M.A., and Snatzke, G. 1979. Revised structures of datural actone and 12-oxowithanolide. *Phytochemistry* 18: 637–640.

Kanwal, P., Gupta, V.K., and Gupta, B.D. 2007. Crystal structure of 3β-hydroxy-17-oximinoandrost-5-ene monohydrate. *Anal Sci* 23: 239–240.

Klyne, W. and Prelog, V. 1960, Description of steric relationships across single bonds. *Experientia* 16: 521–523.

Kofler, L. 1927. *Die Saponine*. Vienna, Austria: Julius Springer Verlag.

Kundu, A.B., Mukherjee, A., and Dey, A.K. 1976. Recent development in the chemistry of withanolides. *J Sci Ind Res* 35: 616–626.

Kuppuranjan, K., Rajagopalan, S.S., Koteswara Rao, T., and Sitaraman, R. 1978. Effect of guggulu (*Commiphora mukul* Engl) on serum, lipids in obese, hypercholesterolemic and hyperlipemic cases. *J Assoc Physicians Ind* 26: 367–369.

Kurup, P.A. 1956. Antibiotic principle of the leaves of *Withania somnifera. Curr Sci* 25: 57–58.

Lindeman, S.V., Alexanyan, M.S., Struchkov, Y.T., Thaper, R.K., Reshetova, I.G., and Kamernitzky, A.V. 1992. Structure of (22S)-3-acetoxy-20-(3-isopropylisoxazolin-5-yl)-4,4,14-trimethylpregn-8(9)-ene. *Acta Cryst C* 48: 290–292.

Marker, R.E. and Crooks, H.M. 1942. Sterols. CXLIV. Some 16-Alkyl-pregnenolones and progesterones. *J Am Chem Soc* 64: 1280–1281.

Matsumoto, T., Watanabe, M., Matsumoto, T., Mataka, S., and Thiemann, T. 2004. 3-Benzyloxy-16-[(*N*-methyl-*N*-phenylamino)methylidene]estra-1,3,5(10)-trien-17-one. *Acta Cryst C* 60: 813–814.

Mester, M., Mester, L., and Nityanand, S. 1979. Inhibition of platelet aggregation by guggulu steroids. *Planta Medica* 37: 367–369.

Nadkarni, A.K. 1954. *Indian Materia Medica*, vol. 1. Bombay, India: Popular Book Depot.

Nardelli, M. 1995. PARST95-An update to PARST. A system of Fortran routines for calculating molecular structure parameters from the results of the crystal structure analysis. *J Appl Cryst* 28: 659.

Nityanand, S. and Kapoor, N.K. 1973. Cholesterol lowering activity of the various fractions of the guggal. *Ind J Exp Biol* 11: 395–396.

Oakenfull, D. 1981. Saponins in foods—A review. *Food Chem* 6: 19–40.

Oliveto, E.P., Rausser, R., Weber, L., Nussbaum, A.L., Gebert, W., Coniglio, C.T., Hershberg, E.B. et al.1958. 16-Alkylated corticoids. II. 9α-fluoro-16α-methylprednisolone-21-acetate. *J Am Chem Soc* 80: 4431–4431.

Paixao, J.I.F., Salvador, J.A.R., Paixao, J.A., Beja, A.M., Silva, M.R., and Gonsalves, A.M.d'A.R. 2004. 6β-Azido-7α-hydroxy-17-oxo-5α-androstan-3β-yl acetate. *Acta Cryst C* 60: 630–632.

Parvez, M., Fazardo, V., and Shamma, M. 1988. (+)-Jaborosalactone M, a hemiketal withanolide from *Jaborosa magellanica. Acta Cryst C* 44: 553–555.

Parvez, M., Fazardo, V., and Shamma, M. 1990. (+)-Jaboromagellone, a new withanolide from *Jaborosa magellanica. Acta Cryst C* 46: 1850–1853.

Patil, V.D., Nayak, U.R., and Dev, S. 1972. Chemistry of ayurvedic crude drugs—I: Guggulu(resin from *Commiphora mukul*)—1:Steroidal constituents. *Tetrahedron* 28: 2341–2352.

Price, K.R., Johnson, I.T., and Fenwick, G.R. 1987. The chemistry and biological significance of saponins in foods and feeding stuffs. *C R Food Sci Nutr* 26: 27–135.

Qurishi, M.A., Dhar, K.L., and Atal, C.K. 1979. A novel withanolide from *Datura quercifolia. Phytochemistry* 18: 283–284.

Ray, A.B. 1989. Recent progress in with asteroids. In: *Frontiers in Applied Chemistry*, ed. A.K. Biswas. New Delhi, India: Narosa.

Ribar, B., Stankovic, S., Meszaros, C., Miljkovic, D., Pejanovic, V., and Petrovic, J. 1993. Structure of 3-methoxy-6α,17β-dihydroxyestra-1,3,5(10)-trien-7-one oxime. *Acta Cryst C* 49: 270–273.

Roberts, P.J., Pettersen, R.C., Sheldrick, G.M., Isaacs, N.W., and Kennard, O. 1973. Crystal and molecular structure of 17β-hydroxyandrost-4-en-3-one (testosterone). *J Chem Soc Perkin Trans 2* 1978–1984.

Sany, J. 1987. Prospects in the immunological treatments of rheumatoid arthritis. *Scand J Rheumatol Suppl* 66: 129–136.

Sarkhel, S., Yadava, U., Prakas, P., Jain, G.K., Singh, S., and Maulik, P.R. 2001. Guggulsterone E, a lipid-lowering agent from *Commiphora mukul. Acta Cryst* E57: 285–286.

Satyavati, G.V. 1991. Guggulipid: A promising hypolipidaemic agent from gum guggul (*Commiphora mukul*). In: *Economic and Medicinal Plant Research*, vol. 5, Plants and Traditional Medicine, pp. 47–82, eds. H. Wagner, N.R. Farnsworth. New York: Academic Press.

Sheldrick, G.M. 1997a. SHELXS97, Program for the Solution of Crystal Structures, University of Gottingen, Gottingen, Germany.

Sheldrick, G.M. 1997b. SHELXL97, Program for the Refinement of Crystal Structures, University of Gottingen, Gottingen, Germany.

Singh, A., Gupta, V.K., Rajnikant, and Goswami, K.N. 1994. Structure of 3β, 20-Diacetoxy-16α-methylallopregn-17(20)-ene. *Cryst Res Technol* 29: 837–842.

Singh, S. and Kumar, S. 1998. *Withania somnifera*: The Indian Ginseng *Ashwagandha,* pp. 293. Central Institute of Medicinal and Aromatic Plants (CIMAP), Lucknow, India.

Singh, A.K., Prasad, G.C., and Tripathi, S.N. 1982. In vitro studies on thyrogenic effect of *Commiphora mukul* (guggulu). *Ancient Sci Life* 2: 23–28.

Spek, A.L. 1999. PLATON for Windows, September 1999 Version. Utrecht, the Netherlands: University of Utrecht.

Starova, G.L., Egorov, M.S., Vasiljeva, E.S., and Shavva, A.G. 2003. 17β-Ethoxy-3-methoxy-8-isoestra-1,3,5(10)-triene. *Acta Cryst C* 59: 451–453.

Steiner, T. 1997. Unrolling the hydrogen bond properties of C–H ··· O interactions. *Chem Commun* 727–734.

Steiner, T. and Desiraju, G.R. 1998. Distinction between the weak hydrogen bond and the van der Waals interaction. *Chem Commun* 891–892.

Thamotharan, S., Parthasarathi, V., Dubey, S., Jindal, D.P., and Linden, A.H. 2004. 16-(4-Isopropylbenzylidene) androst-4-ene-3,17-dione. *Acta Cryst C* 60: o110–o112.

Tripathy, Y.B., Malhotra, O.P., and Tripathy, S.N. 1985. Thyroid-stimulating actions of (Z)guggulsterone obtained from *Commiphora mukul. Planta Med* 50: 78–80.

Tripathy, Y.B., Tripathy, P., Malhotra, O.P., and Tripathy, S.N. 1988. Thyroid stimulating action of (Z) guggulsterone. Mechanism of action. *Planta Med* 271–276.

Umadevi, P., Akagi, K., Ostapenko, V., Tanaka, Y., and Sugahara, T. 1996. Withaferin A: A new radiosensitizer from the Indian medicinal plant *Withania somnifera. Int J Radiat Biol* 69: 193–197.

Urizar, N.L., Liverman, A.B., Dodds, D.N.T., Silva, F.V., Ordentlich, P. Yan, Y., Gonzzalez, F.J., Heyaman, R.A., Mangelsdorf, D.J., and Moore D.D. 2002. A natural product that lowers cholesterol as an antagonist ligand for FXR. *Science* 296: 1703–1706.

Weeks, C.M., Cooper, A., and Norton, D.A. 1971. The crystal and molecular structure of 3β-chloro-5-androsten-17β-ol. *Acta Cryst* B27: 531–538.

Wu, J., Xia, C., Meier, J., Li, S., Hu, X., and Lala, D.S. 2002. The hypolipidemic natural product guggulsterone acts as an antagonist of the bile acid receptor. *Mol Endocrinol* 16: 1590–1597.

14 Three-Dimensional Structure of Xanthones

Luís Gales and Ana M. Damas

CONTENTS

14.1 INTRODUCTION

Xanthone derivatives are involved in a multiplicity of pharmacological activities, which reflects their large structural and chemical variety. Among the documented activities are tuberculostatic, antimycotic, antimalarial (Riscoe et al. 2005), antiplatelet, antithrombotic, anti-inflammatory, antiallergic, antitumor, antimutagenic, and antioxidant (Pinto et al. 2005; Rodrigues et al. 2010). Xanthone pharmacological research has been particularly successful in cancer (Pinto et al. 2005). In fact, several xanthone derivatives have been described as antitumor agents. They are the xanthone precursor as well as oxygenated, sulfonated, glycosylated, and prenylated xanthones. These derivatives show "*in vitro*" growth-inhibitory activity on a remarkable range of tumor cell lines such as leukemia, multiple myeloma, breast adenocarcinoma, melanoma, hepatoma, glioma, neuroblastoma, pheochromocytoma, fibroblasts tumor cells, fibrosarcoma, epithelial tumor cells, Friend tumor cells, and carcinomas from many origins such as oral squamous cell, colon, ovarian, uterine, prostate, lung, liver, stomach, renal, pancreatic, CNS, colorectal, bladder, adrenocortical, and nasopharynx epidermoid (Pinto et al. 2005).

More than 1000 natural xanthones have been identified from 20 families of higher plants, fungi, and lichens (Vieira and Kijjoa 2005; El-Seedi et al. 2009). They can be classified according to the nature of the substituents present in the three-ring scaffold. There are simple oxygenated xanthones, glycosylated xanthones, prenylated xanthones and their derivatives, xanthone dimers, xanthonolignoids, and miscellaneous xanthones. However, some synthetic strategies have been implemented in order to obtain more complex derivatives or even to produce the bioactive natural xanthones. While some xanthones were synthesized with a simple hydroxyl, methoxyl, methyl, or carboxyl, others have more complex substituents such as epoxide, azole, methylidene-butyrolactone, aminoalcohol, sulfamoyl, methylthiocarboxylic acid, and dihydropyridine in their scaffold. There are three traditional methods for the synthesis of simple xanthones: the Grover, Shah, and Shah (GSS) reaction, the synthesis via benzophenone, and the synthesis via diphenyl ethers intermediates. Additional information about the synthesis pathways can be found in some review articles (Sousa and Pinto 2005; El-Seedi et al. 2010).

The variety of the reported xanthone bioactivities is obviously related to their molecular structure. It is thus essential to elucidate the structure of the new isolated xanthones as well as those obtained from synthetic pathways. Currently, there are two techniques available to perform this structure elucidation: x-ray crystallography and nuclear magnetic resonance (NMR) spectroscopy. The NMR results are out of the scope of this chapter; the main NMR features of the most common and important classes of xanthones can be found elsewhere (Silva and Pinto 2005). The next sections will review the available crystal structures of xanthones and xanthone derivatives. This information is essential to understand the activity of the compounds and hopefully, it will be valuable for the design of new and more potent xanthone derivatives.

14.2 XANTHONE CRYSTAL STRUCTURES

14.2.1 Crystallization and Crystal Packing

Suitable crystals for structure elucidation using x-ray diffraction are usually obtained by slow evaporation of their mother solvents. They vary in size, shape, and color. Crystals are essentially colorless or yellow and the shape description includes needlelike, block, plate, and others. They diverge in the crystallographic system, although the most common are triclinic or monoclinic. Currently, there are 135 structures deposited in the Cambridge Crystallographic Data Centre (CCDC) from 131 different xanthone derivatives and these data more than double the structures available and are described in a previous review in 2005 (Gales and Damas 2005).

Xanthones are planar and highly symmetric molecules that are more efficiently packed into columns with their planes parallel to one another and about 3.5 Å apart. The packing of the molecules is governed essentially by van der Waals forces. However, when hydroxyl or amino substitutions are present, stabilizing intermolecular hydrogen bonds are frequent. Occasionally, a solvent molecule becomes trapped in the crystal framework. In these cases, the most common molecules are water and methanol.

14.2.2 Crystal Structure of Xanthone

There are four different entries in the CCDC for the xanthone structure that were deposited in 1956 (Toussaint 1956), 1969 (Biswas et al. 1969), 1982 (Biswas et al. 1982), and 1990 (Onuma et al. 1990). The most recent structure from Onuma and colleagues is the most accurate (Onuma 1990). However, the crystallographic packing is an exception since there are four molecules in the asymmetric unit and they are not stacked. The compound has two benzenoid and one pyranoid rings as shown in Scheme 14.1.

The xanthone molecule adopts a much flattened boat conformation: the aromatic ring planes form angles of 3.7° and 2.0° with the plane of the pyranoid. The central pyranoid ring has a partial aromatic character: the C(4a)–O(10)–C(10a) angle is 119.4(6)° and the C(4a)–O(10) and C(10a)–O(10) bond lengths are 1.35(1) Å and 1.37(1) Å, which are values slightly shorter than those observed for

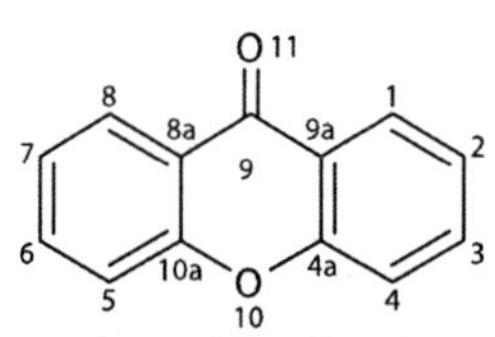

SCHEME 14.1 Xanthone.

diaryl ethers (Car-O-Car): 1.384(14) Å Allen et al. 1987. Also, the C(8a)–C(9) and the C(9)–C(9a) bonds are shorter than the corresponding bonds in acetone. Thus, it appears that the p_z electrons of atoms O(10) and C(9) are used for conjugation conferring an aromatic character to the central ring.

The O(11) atom deviates 0.13 Å from the central pyranoid ring plane. This is not a general feature of xanthonic derivatives; it probably results from intermolecular repulsion between O(11) and C(6) in the xanthone crystal packing.

Xanthone substituents may introduce small distortions in the skeleton structure, mostly due to steric limitations associated with the substituents packing. The most common deviation is a slight twisting of the three-ring system along its longitudinal axis to adopt a propeller conformation. A detailed analysis of the xanthone derivatives will be presented next. They were classified as hydroxylated and methoxylated xanthones, glycosidic xanthones, prenylated and related xanthones, xanthones containing a bis-dihydrofuran ring system, halogenated xanthones, xanthones containing a crown ether, and xanthones forming metal complexes.

14.2.3 Hydroxylated and Methoxylated Xanthones

This group of xanthonic compounds is associated with many pharmacological activities such as antidepressant, antitumor, antimicrobial, antifungal, anti-inflammatory, antiviral, cardiotonic, hypoglycemic, antihepatotoxic, and immunomodulatory (Pinto et al. 2005). It was found that they may inhibit many enzymatic systems: acetylcholinesterase, angiotensin-I-converting enzyme, catalase, cyclooxygenases (COXs), cyclic AMP-phosphodiesterase, cyclic GMP-phosphodiesterase, glutamic oxaloacetic transaminase, glutathione-S-transferase, glutathione peroxidase, glutamic acid decarboxylase, and hypoxanthine–xanthine oxidase (Pinto et al. 2005). The relevant feature associated with the biological activity is the position of the substituents. Thus, a number of systematic studies covering a large number of substituents' positions and a variety of pharmacological applications were already addressed. In particular, a series of xanthone derivatives were investigated in relation to their antiplatelet effect and 1,3,5,6-tetrahydroxyxanthone showed potent and significant inhibitory effect on collagen-induced platelet aggregation (Lin et al. 1992); the inhibitory activity toward monoamine-oxidase A (MAO-A) was also investigated and 1,5-dihydroxy-3-methoxyxanthone with an IC_{50} of 40 nM for MAO-A emerged as the most active inhibitor; complementary computational studies revealed the importance of an OH substituent in position 1 or 5 in the inhibitory activity toward MAO-A (Guerre et al. 2001); the modulatory activity of 3,4-dihydroxyxanthone on several isoforms of protein kinase C (PKC) was evaluated using an "*in vivo*" yeast phenotypic assay. PKC inhibition caused by this derivative was confirmed using an "*in vitro*" kinase assay and differences on its potency toward the distinct PKC isoforms were observed (Guerre et al. 2001).

There are 17 crystal structures available for simple oxygenated xanthones and many others sharing other substituents. The essential feature of this class of compounds is that the substituents adopt a nearly coplanar conformation in relation to the xanthone skeleton.

In hydroxylated xanthones, the rotation of the OH group around the C–OH bond is usually less than 10° in relation to the plane of the aromatic ring. However, when there is a substituent in a position adjacent to the position of the hydroxyl group and no stabilizing interactions are formed between them, the hydrogen from OH remains in the aromatic plane, but oriented far away from the other substituent. When the –OH is bound to C(1) or C(8), a strong hydrogen bond to O(11) is always established.

The methoxy substituents also adopt a coplanar conformation relative to the three-ring system, allowing maximum overlap of the unshared oxygen electrons with the aromatic π electron cloud. This conformation creates a close approach between the methoxy carbon and one of the adjacent aromatic carbons (Cadj). To relieve the resulting strain, there is an opening of the Cadj-C-OMe bond angle from its nominal value of 120° to values between 122° and 130°. When the substitution occurs at C(1) or C(8), there is a repulsion between the methoxy oxygen and O(11). To increase the distance between the two oxygen atoms, there is an opening of the angles C(9a)-C(1)-OMe or C(8a)-C(8)-OMe.

On the other hand, it is frequently observed, when the methoxy group is hindered by adjacent substituents, that the carbon atom of OMe is well out of the aromatic plane due to a rotation along the C–OMe bond (Table 14.1).

14.2.4 Glycosidic Xanthones

Some glycosidic xanthones show an inhibitory effect or contribute to a decrease of the activity of several enzyme systems such as aldose reductase, catalase, cyclooxygenases (COXs), α-glycosidase, isomaltase, creatine kinase, IκB kinase, monoamine oxidases (MAOs) A and B, nitric oxide synthase, sucrase, superoxide dismutase, and reductase (Pinto et al. 2005). They also modulate other cellular systems such as the proliferation of lymphocytes, the phagocytic activity, and the ROS and NO production in macrophages (Pinto et al. 2005). Finally, they have been associated with an antitumor activity through apoptosis induction via active caspase 3 pathways (Peng et al. 2004) and by transforming growth factor-β (TGF-β) gene expression regulation (Leiro et al. 2003).

Glycosidic xanthones can be considered structurally related to anthracyclines (Arcamone 1981), which remain one of the most effective agents for the treatment of solid tumors. One of the glycosides revealed structural changes that are pertinent to possible DNA interaction (William Lown and Sondhi 1985). The hydroxyl at C(4) is hydrogen bonded to the flanking carbonyl in doxorubicin and related anthracycline antitumor drugs. In the xanthone glycoside, it interacts with the oxygen of the sugar. The formation of this new hydrogen bond implies the alteration in the dihedral angle between the aglycone and the sugar ring and a slight bend of the chromophore (William Lown and Sondhi 1985).

Another glycoside xanthone, which has moderate activity against several tumor cell lines, is the antibiotic FD-594 (Eguchi et al. 1999). The absolute structure determined by x-ray diffraction revealed that the compound has a glycosylated pyrano[4′,3′:6,7]naphtha[1,2-b]xanthenes skeleton and that the aglycon part has (3*R*,6*S*,7*S*) configuration. In the crystal structure, there are no stabilizing interactions between the xanthone skeleton and the aglycon part of the compound. This may contribute to the solvent-dependent atropisomerism of this natural product, as assessed by circular dichroism (CD) and NMR spectroscopy studies performed in two different solvent systems (Eguchi et al. 1999) (Table 14.2).

14.2.5 Prenylated and Related Xanthones

Prenylated xanthones are also implicated in numerous biological activities. The interference of this class of compounds in enzyme activity is well documented. Inhibition of acetylcholinesterase, calcium-ATPase, cyclooxygenases (COXs), cyclic AMP-binding phosphatase, cyclic AMP-phosphodiesterase, calcium-dependent protein kinase, cyclic AMP-dependent protein kinase, IκB kinase, myosin light chain kinase (MLCK), PKC, monoamine oxidases (MAOs) A and B, aspartic protease, HIV-1 protease, sphingomyelinases, and topoisomerases I and II was observed (Pinto et al. 2005). On the other hand, activation of Jun N-terminal kinase/stress-activated protein kinase (JNK/SAPK) and caspase-9 was also observed (Pinto et al. 2005).

Other cellular systems are also prone to prenylated xanthone activity such as 5-HT 2A receptor, the human complement system, histamine H1 receptor, lymphocytes, platelet-activating factor receptors, and prostaglandin D2(DP), E1, and E2 receptors (Pinto et al. 2005).

Overall, they have been associated to antibacterial, antimalarial, antifungal, antiplatelet/anticoagulants, anti-inflammatory, and antitumor activities. In the context of cancer, it has been shown that they are involved in several mechanisms such as apoptosis induction via active caspase 3 pathways, DNA synthesis suppression, kinases modulation, prostaglandin (PG) E2 receptors blocking, sphingomyelinases inhibition, and topoisomerases I and II inhibition (Pinto et al. 2005).

The crystal structures of the prenylated xanthones reveal that the isoprenyl group may adopt multiple conformations, but it usually comes out of xanthone skeleton plane, even when it is not

TABLE 14.1
Simple Hydroxylated and Methoxylated Xanthones

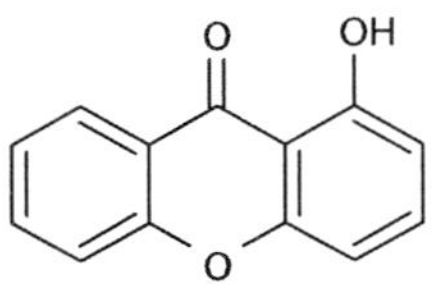 1-Hydroxyxanthone	Formula: $C_{13}H_8O_3$ Temperature: 298 K Density: 1.465 g/cm^3 Space group: $P2_1/c$ R-factor (%): 4.92 Ref: (Corrêa et al. 2010)	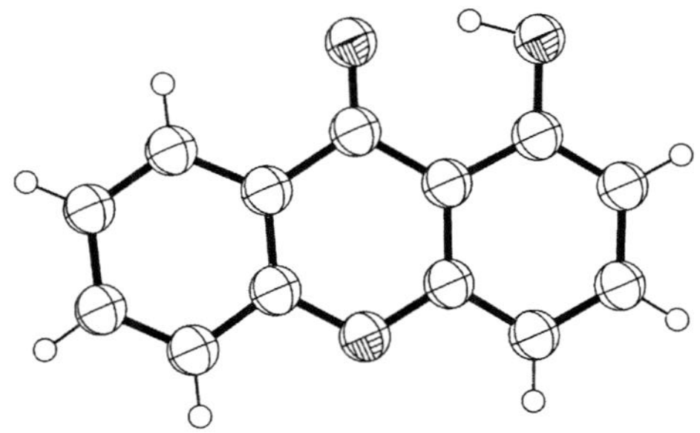
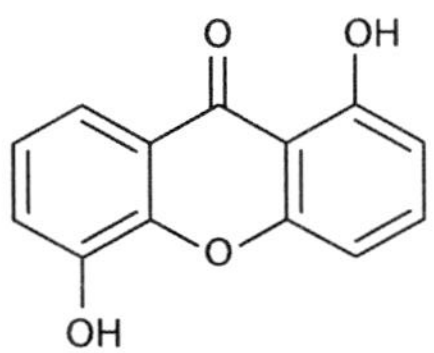 1,5-Dihydroxyxanthone	Formula: $C_{13}H_8O_4$ Temperature: 293 K Density: 1.598 g/cm^3 Space group: Pc R-factor (%): Ref: (Kabaleeswaran et al. 2003)	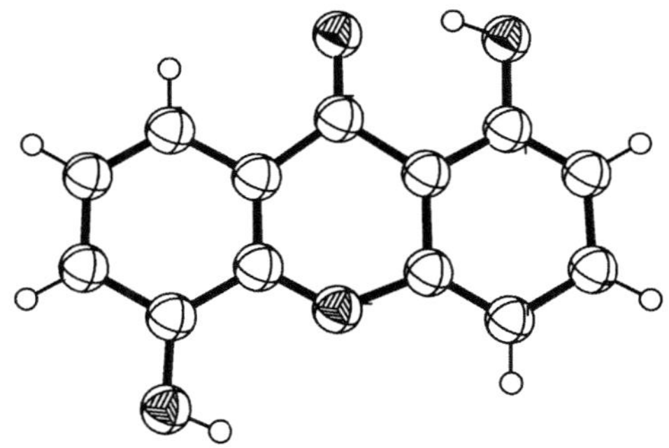
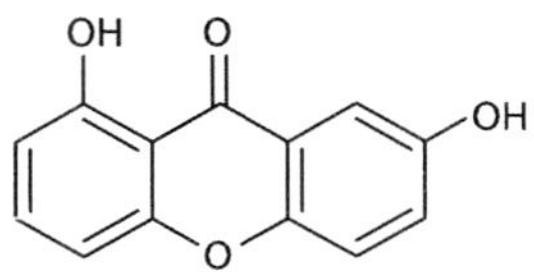 1,7-Dihydroxyxanthone	Formula: $C_{13}H_8O_4$ Temperature: 293 K Density: 1.543 g/cm^3 Space group: $P2_12_12_1$ R-factor (%): 3.15 Ref: (Kato et al. 2005)	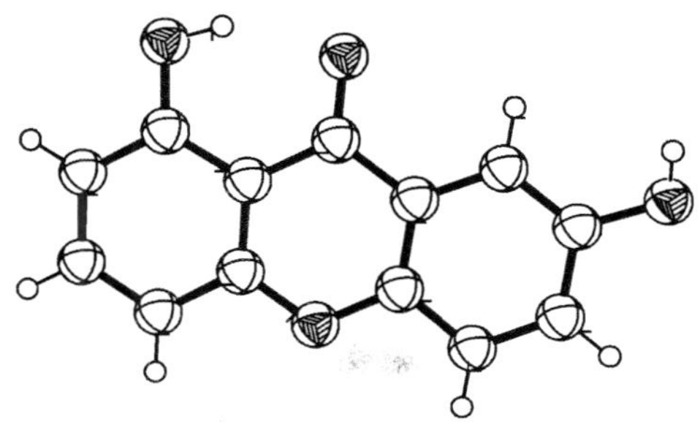

(continued)

TABLE 14.1 (continued)
Simple Hydroxylated and Methoxylated Xanthones

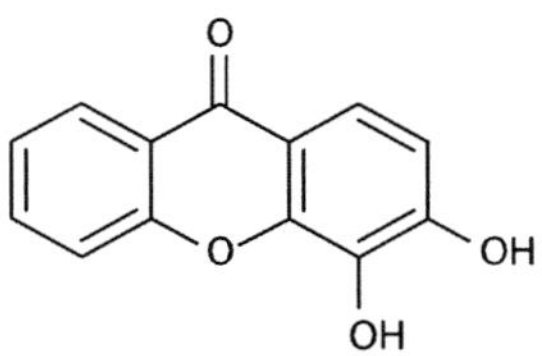 3,4-Dihydroxyxanthone	Formula: $C_{13}H_8O_4 \cdot 3H_2O$ Temperature: 291 K Density: 1.488 g/cm^3 Space group: P-1 R-factor (%): 4.99 Ref: (Gales et al. 2005)	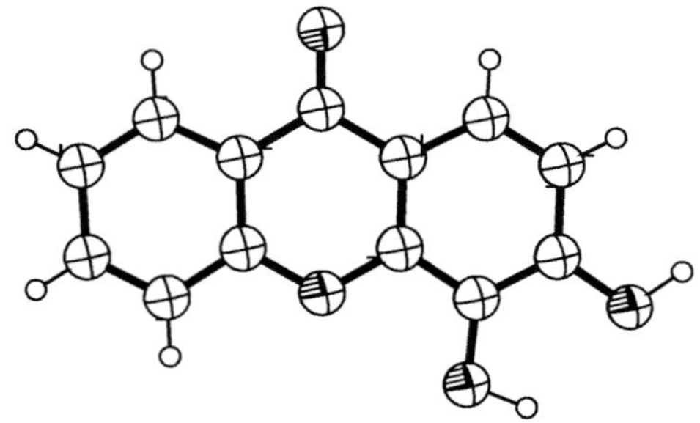
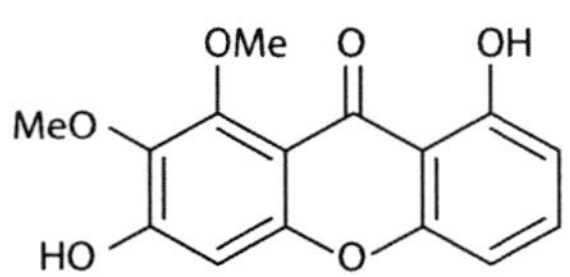 1,6-Dihydroxy-7,8-dimethoxyxanthone	Formula: $C_{15}H_{12}O_6$ Temperature: 295 K Density: 1.499 g/cm^3 Space group: A2/a R-factor (%): 7.60 Ref: (Vijayalakshmi et al. 1987)	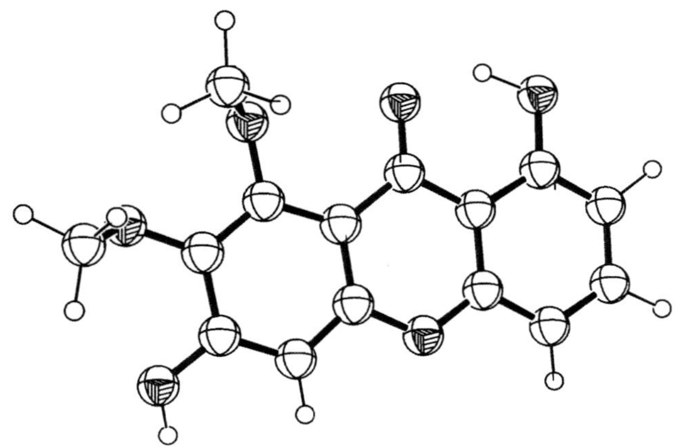

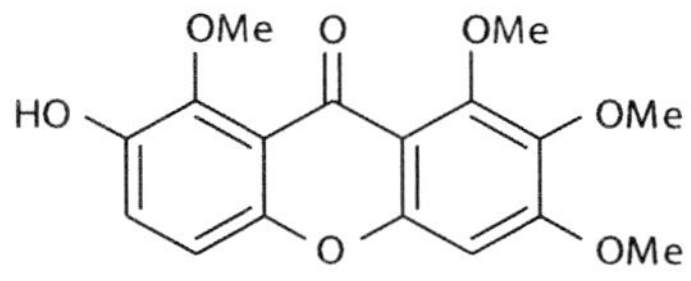

7-Hydroxy-1,2,3,8-tetramethoxyxanthone

Formula: $C_{17}H_{16}O_7$
Temperature: 295 K
Density: 1.414 g/cm^3
Space group: $P2_1/c$
R-factor (%): 5.27
Ref: (Kijjoa et al. 1998)

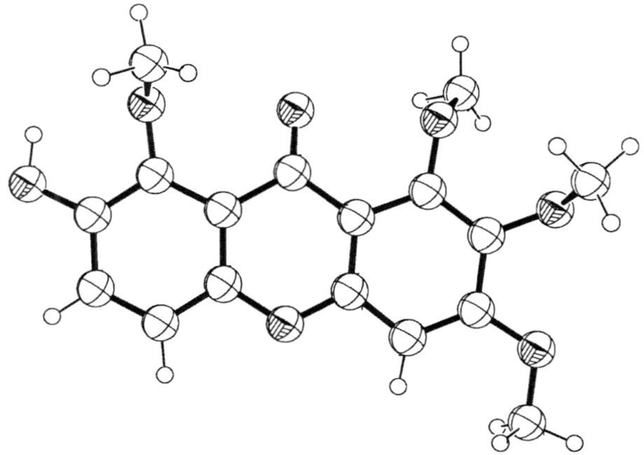

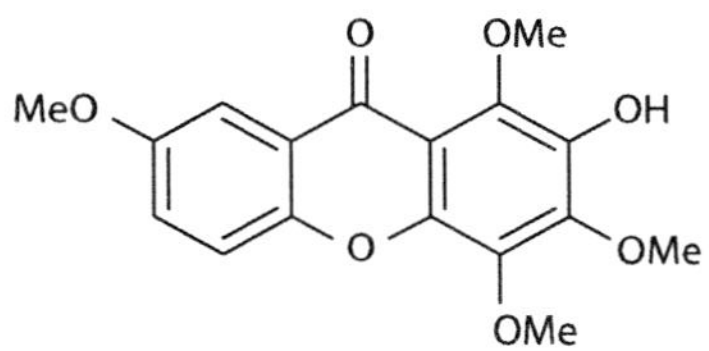

2-Hydroxy-1,3,4,7-tetramethoxyxanthone

Formula: $C_{17}H_{16}O_7$
Temperature: 295 K
Density: 1.429 g/cm^3
Space group: $P2_1/c$
R-factor (%): 5.40
Ref: (Stout et al. 1969)

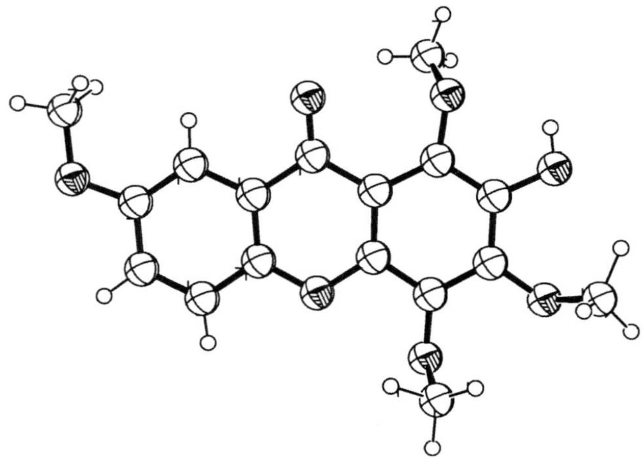

(continued)

TABLE 14.1 (continued)
Simple Hydroxylated and Methoxylated Xanthones

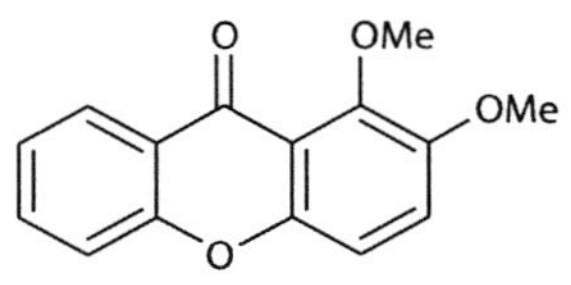 1,2-Dimethoxyxanthone	Formula: $C_{15}H_{12}O_4$ Temperature: 293 K Density: 1.407 g/cm^3 Space group: P2$_1$/n R-factor (%): 5.20 Ref: (Gales et al. 2001)	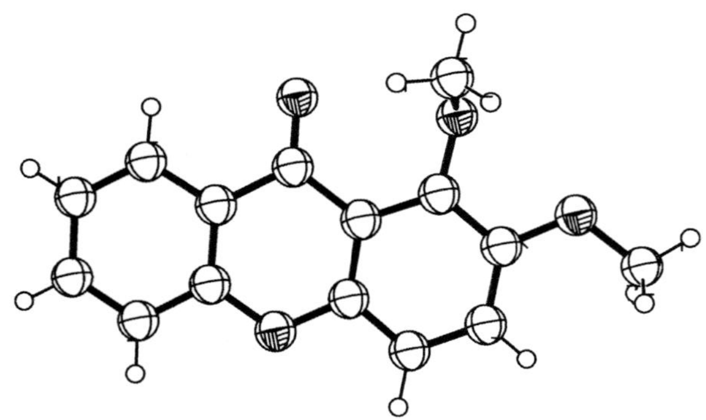
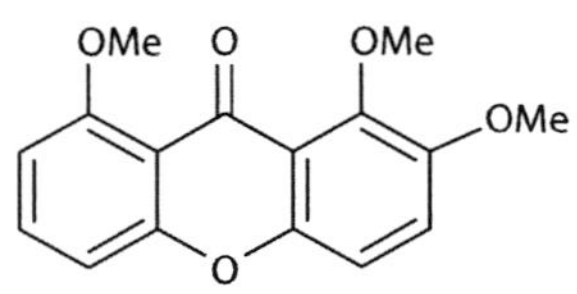 1,2,8-Trimethoxyxanthone	Formula: $C_{16}H_{14}O_5$ Temperature: 293 K Density: 1.411 g/cm^3 Space group: P2$_1$2$_1$2$_1$ R-factor (%): 4.20 Ref: (Gales et al. 2001)	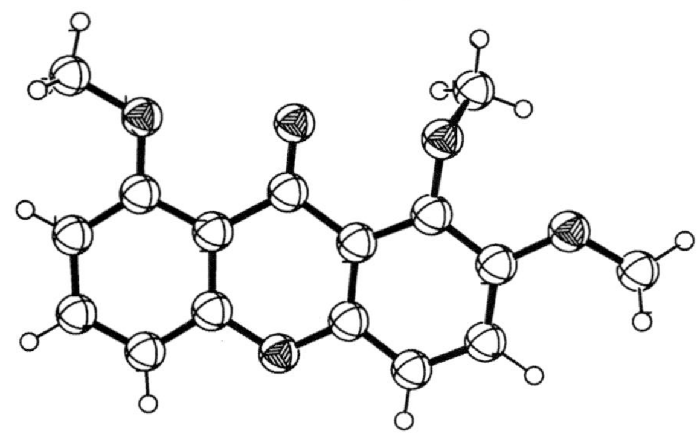

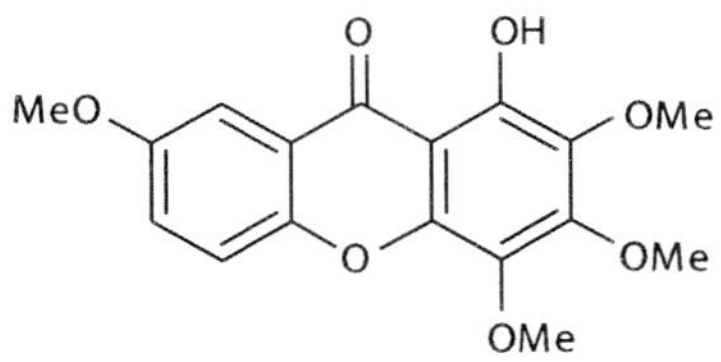

1-Hydroxy-2,3,4,7-tetramethoxyxanthone

Formula: $C_{17}H_{16}O_7$
Temperature: 293 K
Density: 1.460 g/cm³
Space group: $P2_1/n$
R-factor (%): 5.20
Ref: (Shi et al. 2004a)

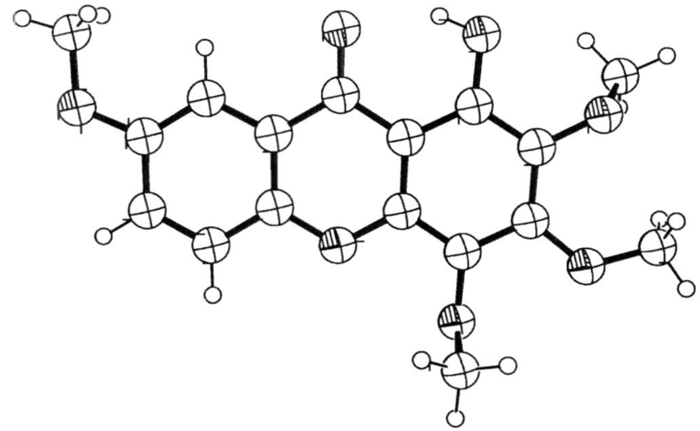

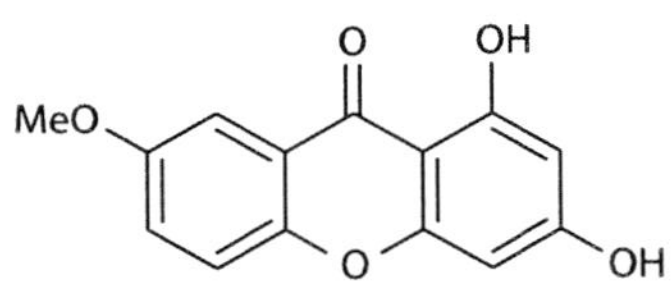

1,3-Dihydroxy-7-methoxyxanthone

Formula: $C_{14}H_{10}O_5$
Temperature: 120 K
Density: 1.624 g/cm³
Space group: P-1
R-factor (%): 3.38
Ref: (Evans et al. 2004)

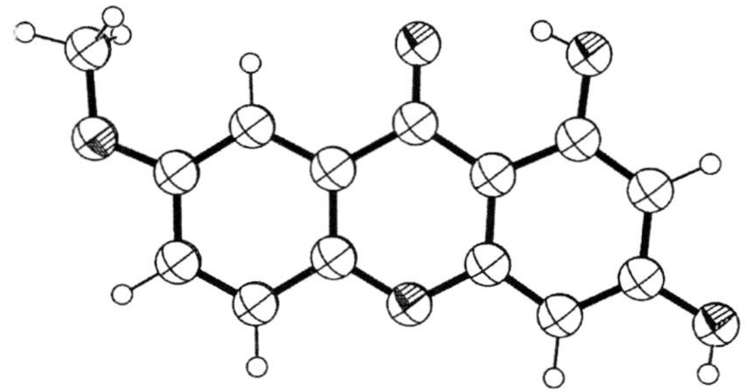

(continued)

TABLE 14.1 (continued)
Simple Hydroxylated and Methoxylated Xanthones

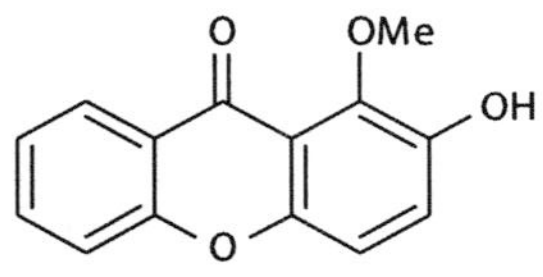 2-Hydroxy-1-methoxyxanthone	Formula: $C_{14}H_{10}O_4 \cdot H_2O$ Temperature: 295 K Density: 1.459 g/cm^3 Space group: P2$_1$/c R-factor (%): 5.34 Ref: (Chen et al. 2009)	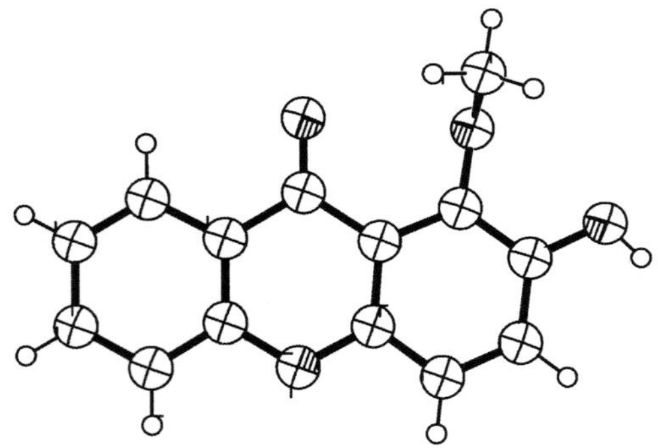
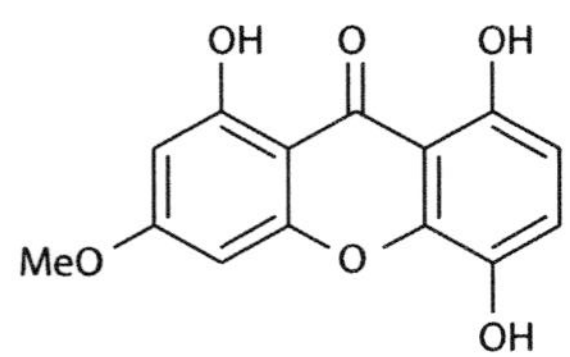 1,5,8-Trihydroxy-5-methoxyxanthone	Formula: $C_{14}H_{10}O_6$ Temperature: 293 K Density: 1.619 g/cm^3 Space group: P-1 R-factor (%): 4.05 Ref: (Shi et al. 2004b)	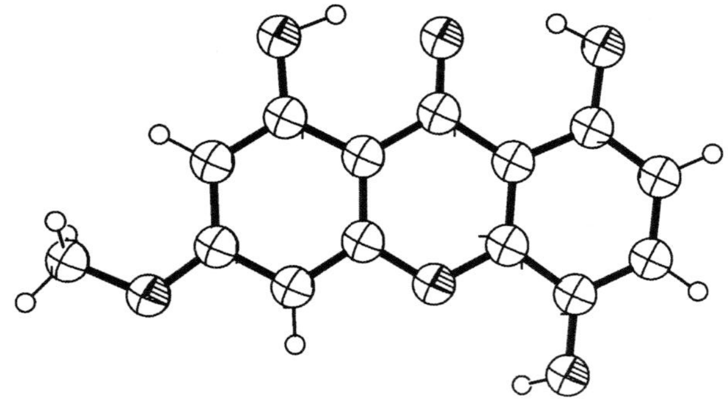

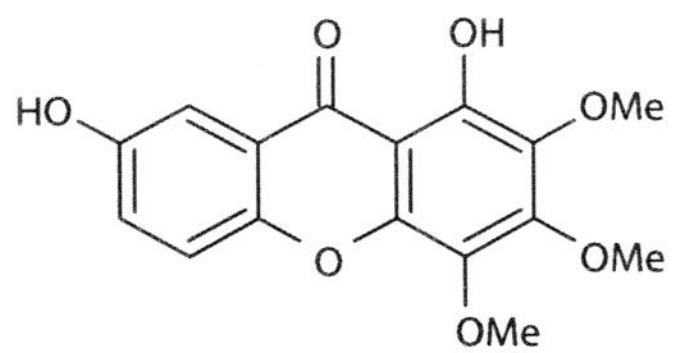 1,7-Dihydroxy-2,3,4-trimethoxyxanthone	Formula: $C_{16}H_{14}O_7 \cdot H_2O$ Temperature: 298 K Density: 1.501 g/cm^3 Space group: $P2_1/c$ R-factor (%): 4.60 Ref: (Yu et al. 2008)	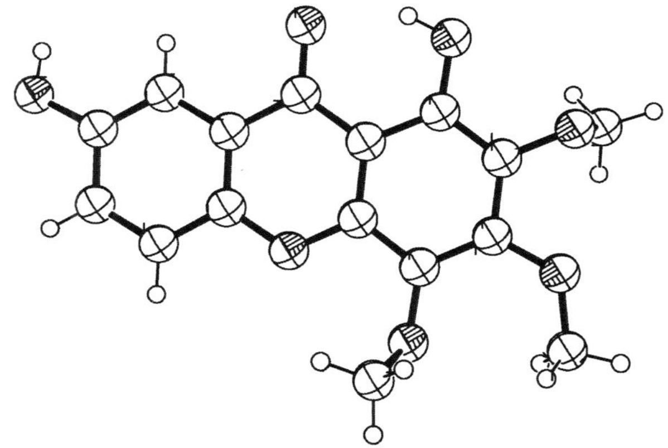
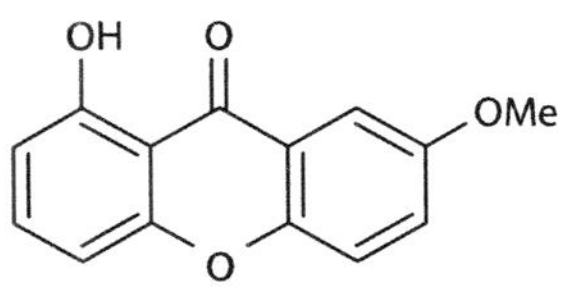 1-Hydroxy-7-methoxyxanthone	Formula: $C_{14}H_{10}O_4$ Temperature: 294 K Density: 1.451 g/cm^3 Space group: $P2_12_12_1$ R-factor (%): 3.49 Ref: (Corrêa et al. 2010)	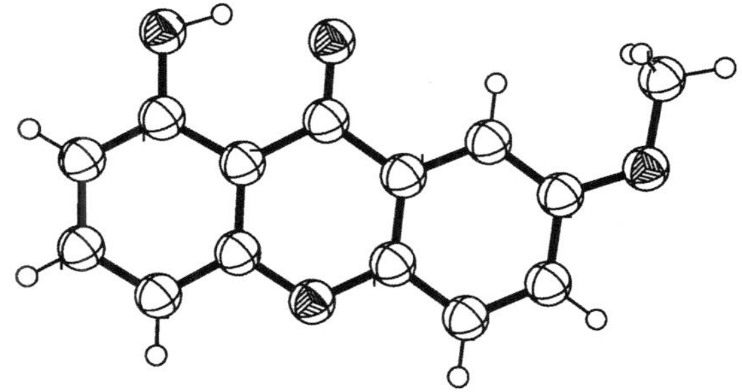

(continued)

TABLE 14.1 (continued)
Simple Hydroxylated and Methoxylated Xanthones

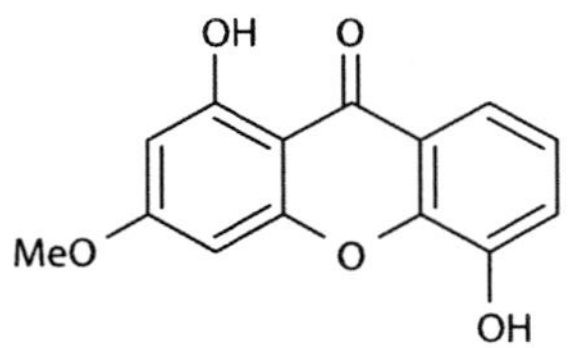

1,5-Dihydroxy-3-methoxyxanthone

Formula: $C_{14}H_{10}O_5$
Temperature: 298 K
Density: 1.564 g/cm^3
Space group: $P2_1/c$
R-factor (%): 4.09
Ref: (Corrêa et al. 2010)

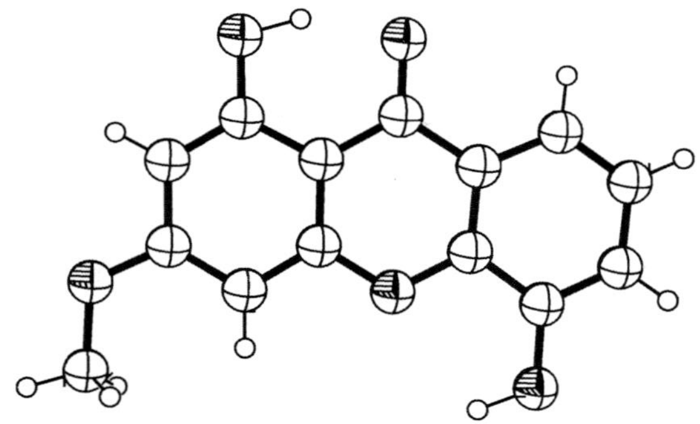

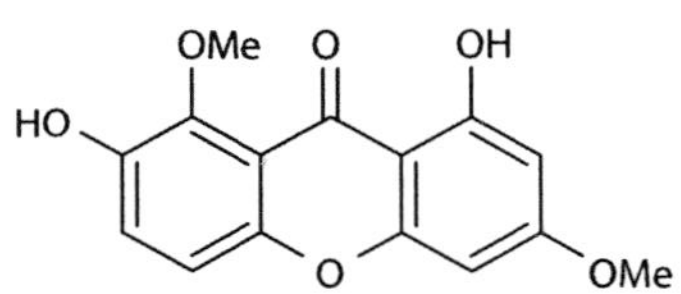

2,8-Dihydroxy-1,6-dimethoxyxanthone

Formula: $C_{15}H_{12}O_6$
Temperature: 298 K
Density: 1.540 g/cm^3
Space group: $P2_12_12_1$
R-factor (%): 4.28
Ref: (Corrêa et al. 2010)

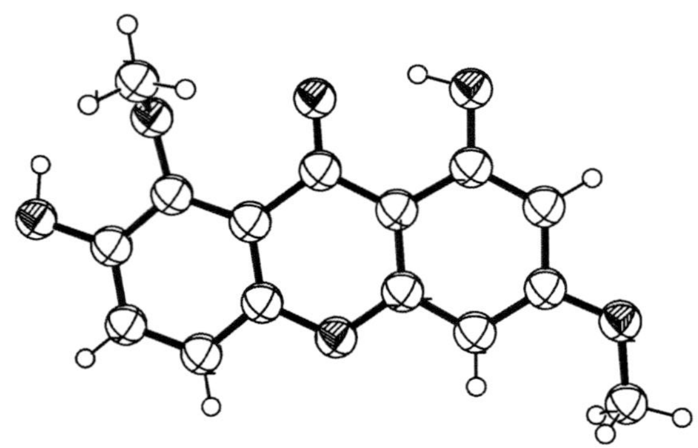

TABLE 14.2
Selected Glycosidic Xanthones

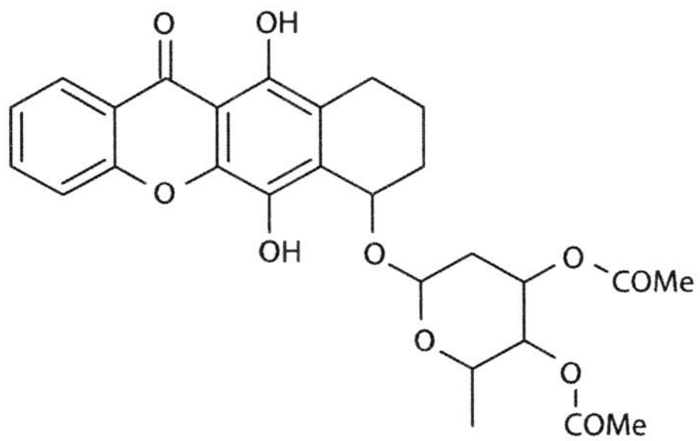 6,7,11-Trihydroxy-xantho(2,3-g) tetralinyl-3′,4′-di-O-acetyl-2′,6′-deoxy-→-L-arabino-hexopyranose	Formula: $C_{27}H_{28}O_{10}$ Temperature: 295 K Density: 1.343 g/cm³ Space group: $P2_1$ R-factor (%): 8.90 Ref: (William Lown and Sondhi 1985)	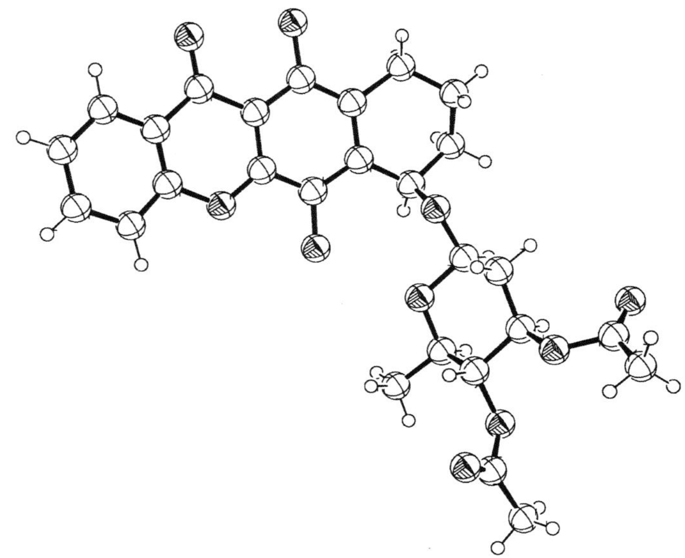
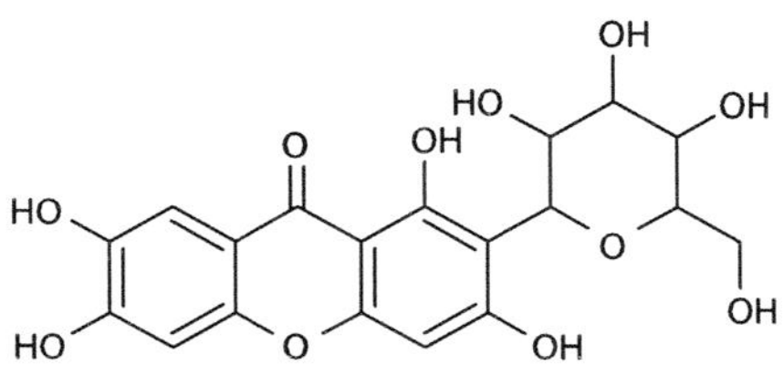 Mangiferin hydrate	Formula: $C_{19}H_{18}O_{11} \cdot 2.5\ H_2O$ Temperature: 150 K Density: 1.625 g/cm³ Space group: P1 R-factor (%): 4.71 Ref: (Da Cruz et al. 2008)	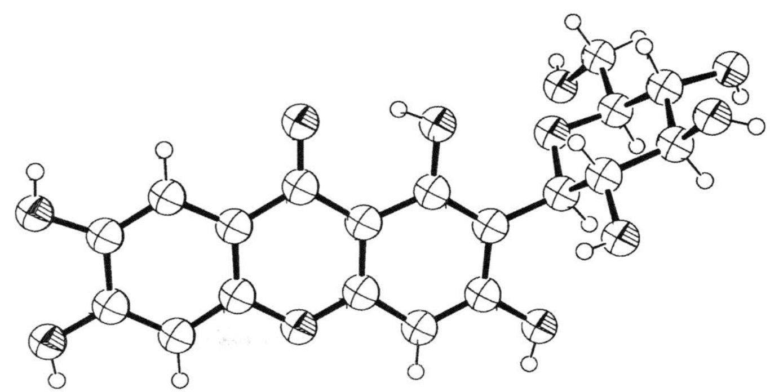

(continued)

TABLE 14.2 (continued)
Selected Glycosidic Xanthones

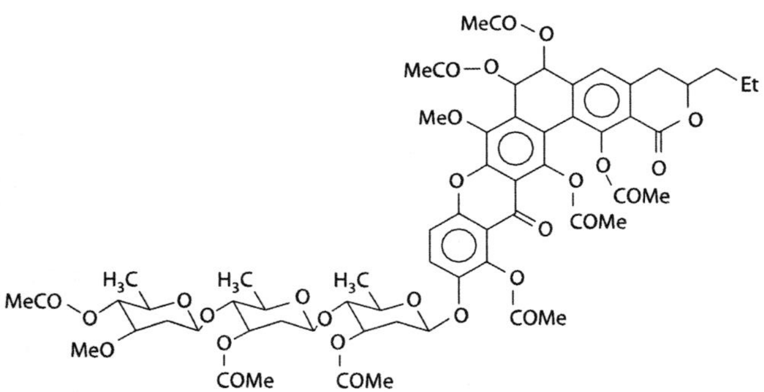

FD-594 octaacetate

Formula: $C_{63}H_{72}O_{28}$
Temperature: 193 K
Density: 1.225g/cm³
Space group: C2
R-factor (%): 4.78
Ref: (Eguchi et al. 1999)

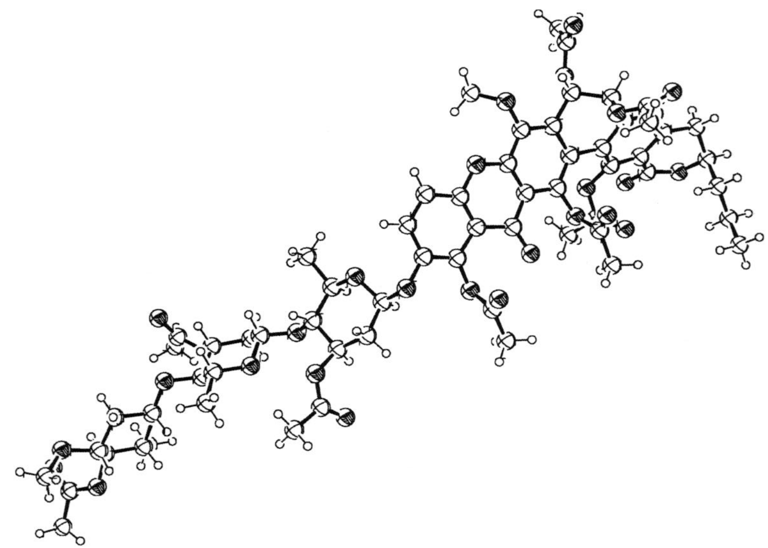

hindered by neighboring aromatic substituents. The corresponding dihedral angle is 111.8° in emericellin, 95.4° in epishamixanthone, and 111.5° in garcinone.

A series of xanthone derivatives have been evaluated for their binding affinity to transthyretin (Maia et al. 2005). Transthyretin is a plasma protein involved in the transport of thyroxine (T4) and also implicated in amyloid diseases. Till now, the only efficient therapy available is liver transplant when performed in an early phase of the onset of the disease symptoms. Alternatives are desirable and in that sense, several compounds have been proposed to impair amyloid formation through the stabilization of the native tetrameric fold of the protein (Almeida et al. 2005). Using competition-binding studies with the protein natural ligand T4, one prenylated xanthone (1,5-dihydroxy-6-(4-hydroxy-3-methylbutyl)-xanthone) with a nanomolar affinity toward transthyretin, which is remarkable and makes this class of compounds a promising template for the design of new candidates for a small molecule therapeutic approach against transthyretin-related amyloidosis.

The cyclization of an isoprenoid side chain results in compounds comprising four 6-membered rings. These compounds were found to have antibacterial, antimalarial, and antiretroviral activity (Pinto et al. 2005). The xanthone skeleton remains essentially planar and the extra ring adopts a half-chair conformation. The angle between the skeleton plane and the extra ring is always lower than 8°. Garcinone B (Ravikumar and Rajan 1987) and isojacareubin (Ishiguro et al. 1993) have an extra ring angularly fused; the γ-pyrone ring also assumes a half-chair conformation in these natural compounds (Table 14.3).

14.2.6 Xanthones Containing a Bis-Dihydrofuran Ring System

Sterigmatocystin (Kukuyama et al. 1978) contains a bis-dihydrofuran ring system and is structurally similar to the potent carcinogen aflatoxin B1. Despite being acutely toxic to the liver of most animals that were tested, this xanthonic compound is also a potent carcinogenic. Much interest has been dedicated to this compound and to a number of closely related ones that occur in nature. A total of six crystal structures of sterigmatocystin derivatives were already determined. The skeleton of the xanthone ring system is essentially planar, although it can be slightly twisted along its longitudinal axis or assume a bent form. The hydrofuran fused to the xanthone ring system assumes a half-chair conformation, while the terminal five-membered ring is either planar or takes an envelope conformation.

Sterigmatin (Fukuyama et al. 1975b) has the two-ring system fused to the C(2) and C(3) atoms of the xanthone skeleton. It also assumes a half-chair conformation like sterigmatocystin derivatives; the terminal five-membered ring, which is essentially planar, forms dihedral angles of approximately 65.0° with the plane of the xanthone skeleton (Table 14.4).

14.2.7 Halogenated Xanthones

Bromoalkoxyxanthones have been identified as interesting scaffolds for the design of potential anticancer drugs as they inhibit the growth of various human tumor cell lines (Sousa et al. 2009). The structure of 1-((6-bromohexyl)oxy)-9H-xanthen-9-one was elucidated and revealed that the 6-bromohexyl side chain is out of the plane of the xanthone skeleton. The structure–activity relationship of the bromoalkoxyxanthone was not explored; however, DNA unwinding experiments combined with molecular modeling studies concerning the binding of the bromoalkoxyxanthone to DNA or to estrogen receptors might disclose some mechanistic aspects in the future.

On the other hand, a series of chlorinated xanthones were identified in a wide variety of lichens and the structure of three of these compounds (demethylchodatin, thiomelin diacetate, and hexachloroxanthone) was determined by x-ray diffraction (Table 14.5). They revealed a small twist along the longitudinal axis of the xanthone skeleton, probably due to steric factors associated with the chlorine substitution. The x-ray analysis of the triacetate form of the antibiotic lysolipin, which has a chlorine atom bound to C(3), allowed the elucidation of this rather complicated polycyclic structure (Dobler and Keller-Schierlein 1977). The xanthone skeleton of lysolipin remains essentially planar, although the four-ring system attached to this structure is slightly twisted, taking a propeller-like conformation.

TABLE 14.3
Selected Prenylated and Related Xanthones

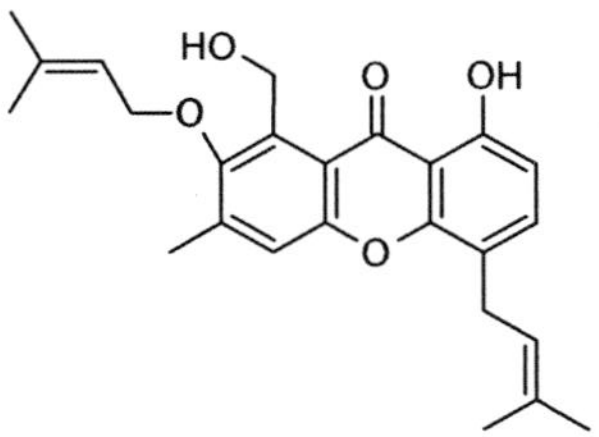 Emericellin	Formula: $C_{25}H_{28}O_5$ Temperature: 295 K Density: 1.268 g/cm^3 Space group: P-1 R-factor (%): 8.60 Ref: (Fukuyama et al. 1975c)	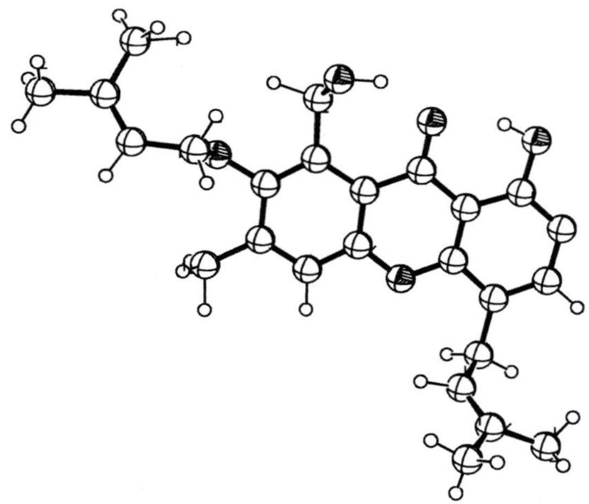
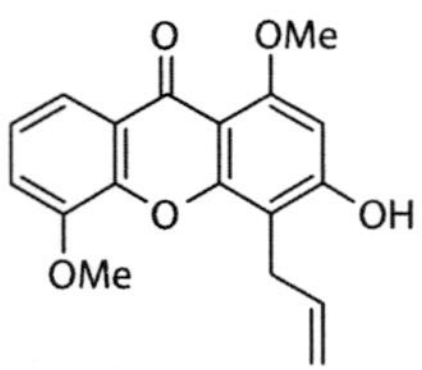 4-Allyl-1,5-dimethoxy-3-hydroxyxanthone	Formula: $C_{18}H_{16}O_5$ Temperature: 295 K Density: 1.340 g/cm^3 Space group: $P2_1/c$ R-factor (%): 8.80 Ref: (Ho et al. 1987)	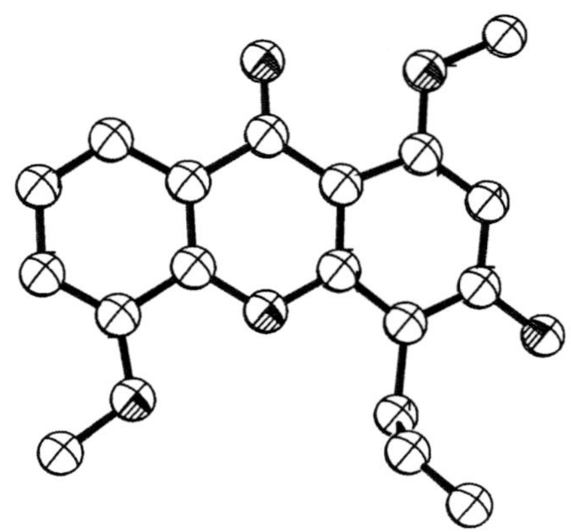

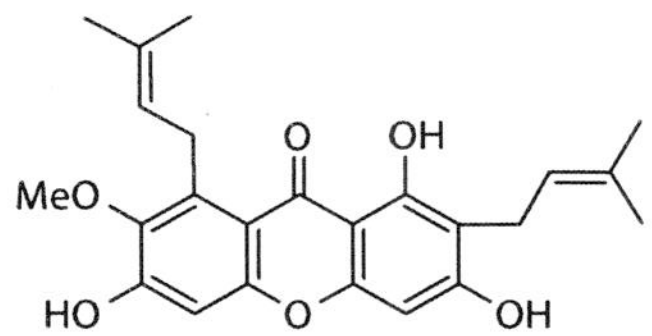

Mangostin dimethylsulfoxide solvate

Formula: $C_{24}H_{26}O_6 \cdot C_2H_6OS$
Temperature: 295 K
Density: 1.264 g/cm³
Space group: P-1
R-factor (%): 8.28
Ref: (Malathi et al. 2000)

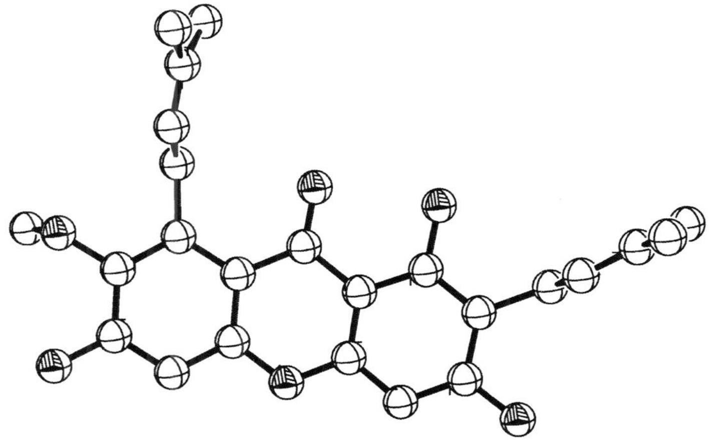

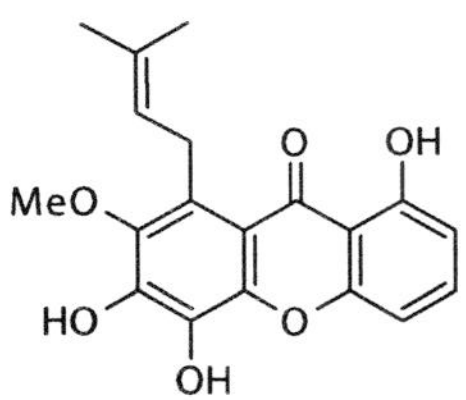

1,5,6-Trihydroxy-7-methoxy-8-(3-methyl-2-butenyl)-xanthone

Formula: $C_{19}H_{18}O_6$
Temperature: 295 K
Density: 1.338 g/cm³
Space group: $P2_1/c$
R-factor (%): 8.20
Ref: (Stout et al. 1963)

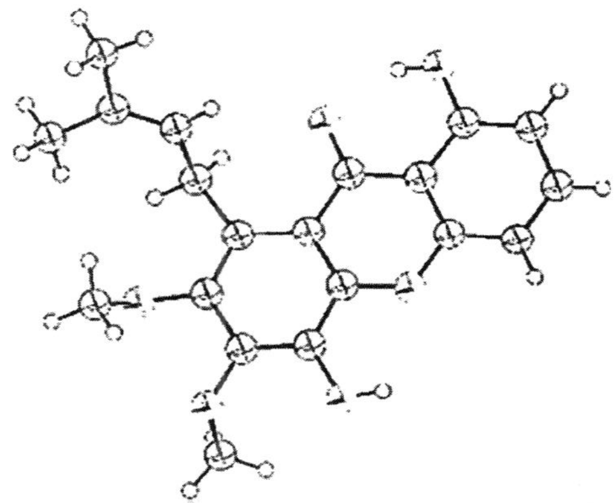

(continued)

TABLE 14.3 (continued)
Selected Prenylated and Related Xanthones

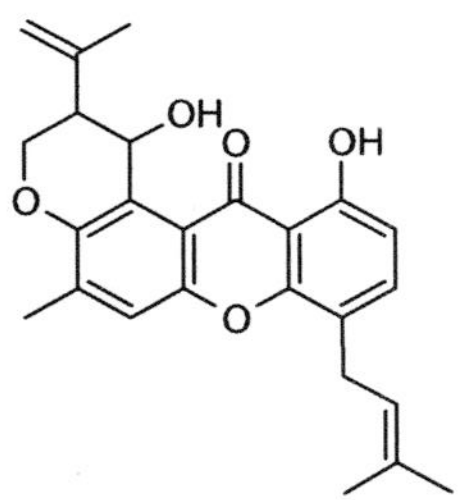

Epishamixanthone

Formula: $C_{25}H_{26}O_5$
Temperature: 295 K
Density: 1.283 g/cm^3
Space group: $P2_1$
R-factor (%): 5.80
Ref: (Fukuyama et al. 1978)

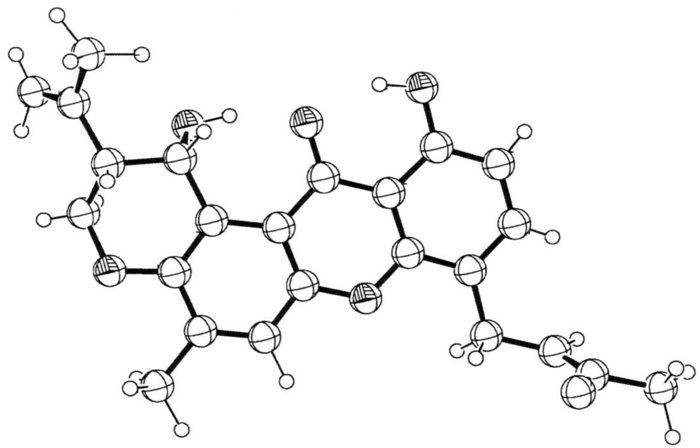

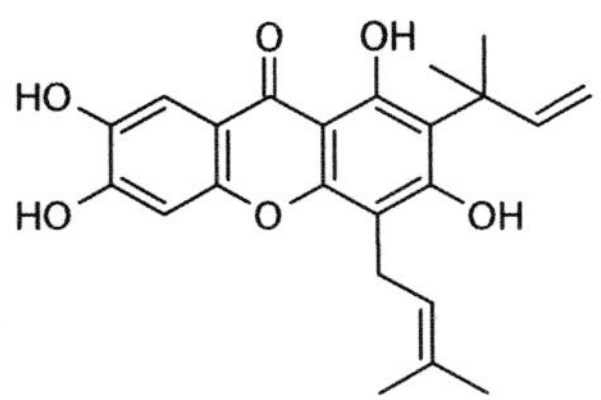

Macluraxanthone B

Formula: $C_{23}H_{24}O_6$
Temperature: 173 K
Density: 1.382 g/cm^3
Space group: $P2_1/c$
R-factor (%): 6.40
Ref: (Lee et al. 2005)

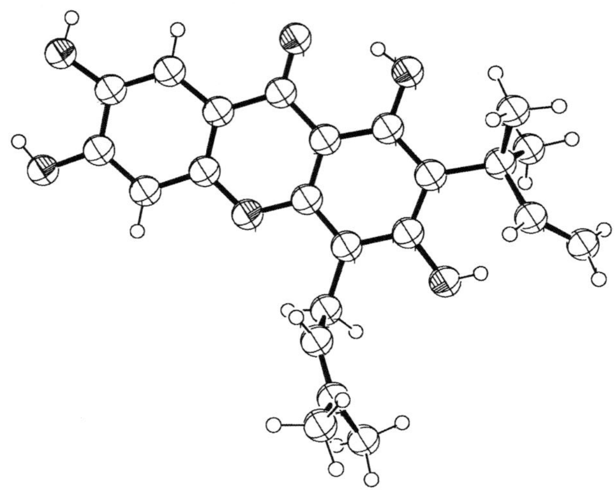

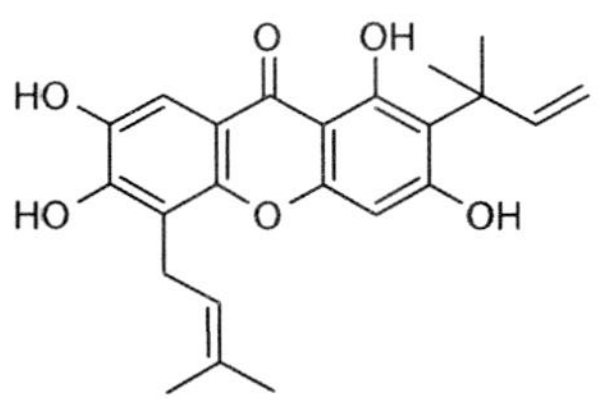

Macluraxanthone B

Formula: $C_{23}H_{24}O_6$
Temperature: 173 K
Density: 1.365 g/cm³
Space group: P-1
R-factor (%): 6.27
Ref: (Lee et al. 2005)

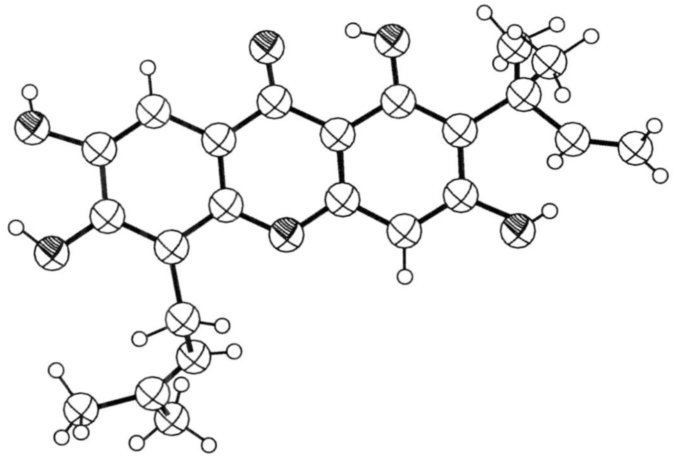

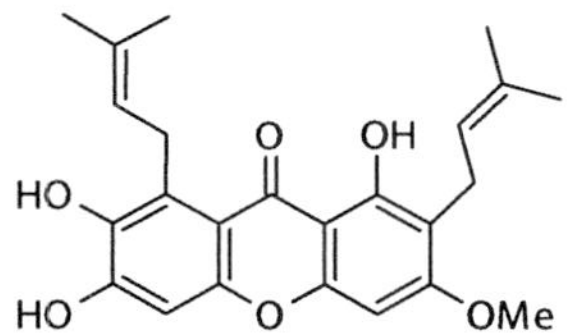

Prunifoxanthone A chloroform solvate

Formula: $C_{24}H_{26}O_6 \cdot HCCl_3$
Temperature: 100 K
Density: 1.428 g/cm³
Space group: $P2_1/c$
R-factor (%): 6.34
Ref: (Boonnak et al. 2005)

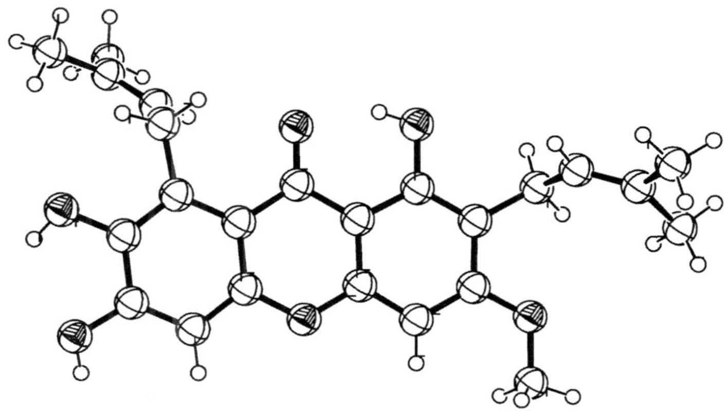

(continued)

TABLE 14.3 (continued)
Selected Prenylated and Related Xanthones

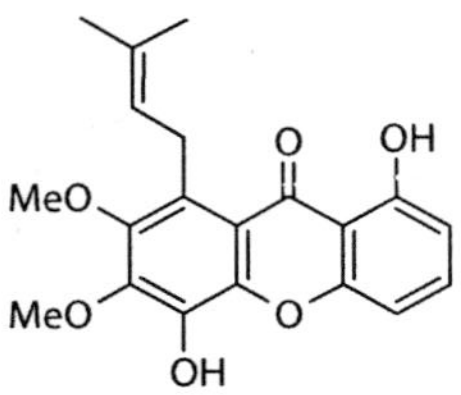 4,8-Dihydroxy-2,3-dimethoxy-1-(3-methyl-2-butenyl)-xanthone	Formula: $C_{20}H_{20}O_6$ Temperature: 100 K Density: 1.418 g/cm^3 Space group: $P2_1/c$ R-factor (%): 4.87 Ref: (Boonnak et al. 2007a)	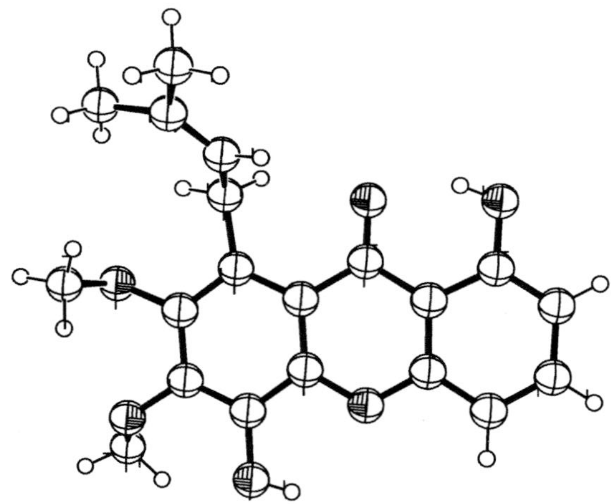
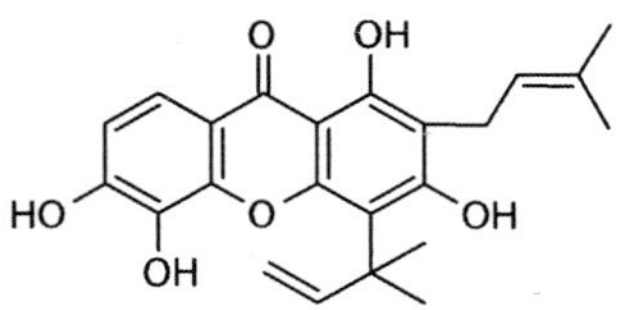 Gerontoxanthone monohydrate	Formula: $C_{23}H_{24}O_6 \cdot CH_4O$ Temperature: 297 K Density: 1.302 g/cm^3 Space group: $P2_1/c$ R-factor (%): 4.80 Ref: (Boonnak et al. 2007b)	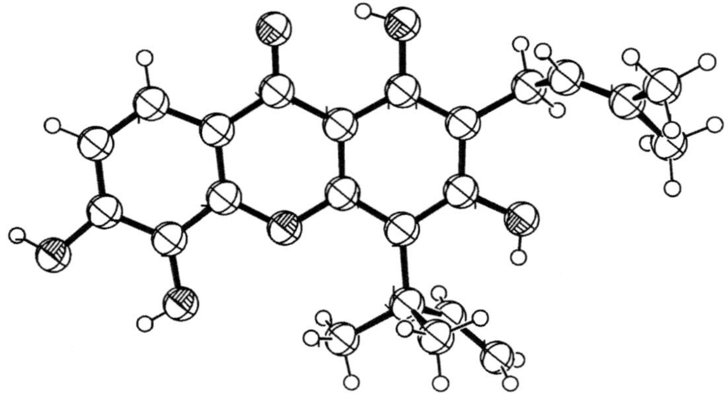

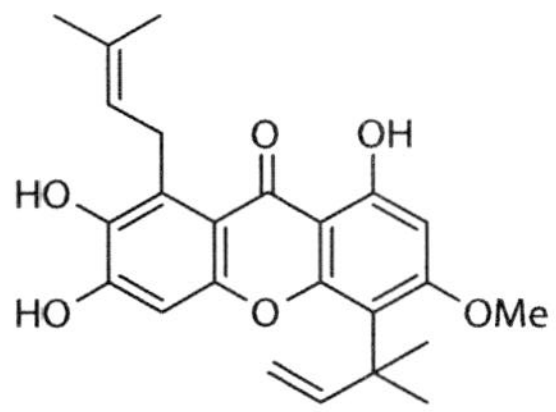

Cudratricusxanthone F hemihydrate

Formula: $C_{24}H_{26}O_6 \cdot H_2O$
Temperature: 173 K
Density: 1.307 g/cm³
Space group: $P2_1/c$
R-factor (%): 7.76
Ref: (Seo et al. 2007)

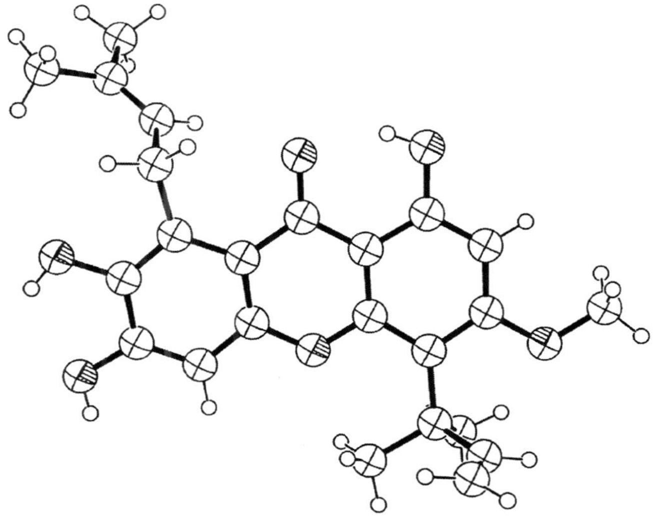

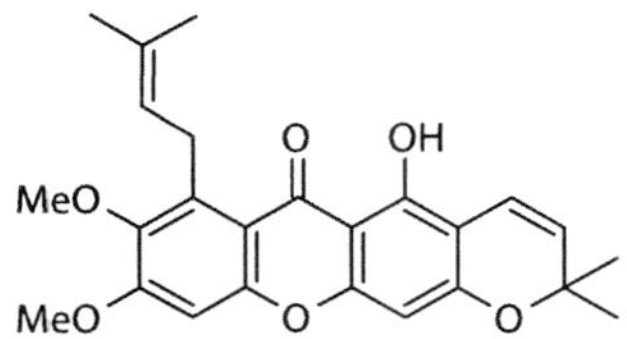

5-Hydroxy-8,9-dimethoxy-2,2-dimethyl-7-(3methyl-2-butenyl)2H,6H-pyrano(3,2) xanthone

Formula: $C_{25}H_{26}O_6$
Temperature: 295 K
Density: 1.369 g/cm³
Space group: $P2_1/c$
R-factor (%): 6.50
Ref: (Ravikumar et al. 1987)

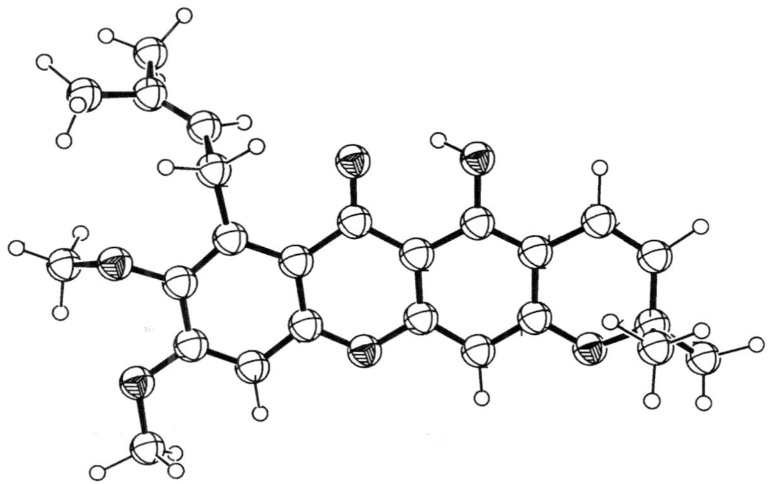

(continued)

TABLE 14.3 (continued)
Selected Prenylated and Related Xanthones

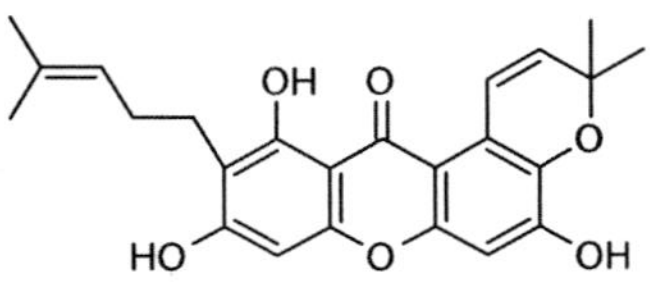

Garcinone B

Formula: $C_{23}H_{22}O_6 \cdot H_2O$
Temperature: 295 K
Density: 1.347 g/cm^3
Space group: P-1
R-factor (%): 6.20
Ref: (Ravikumar and Rajan 1987)

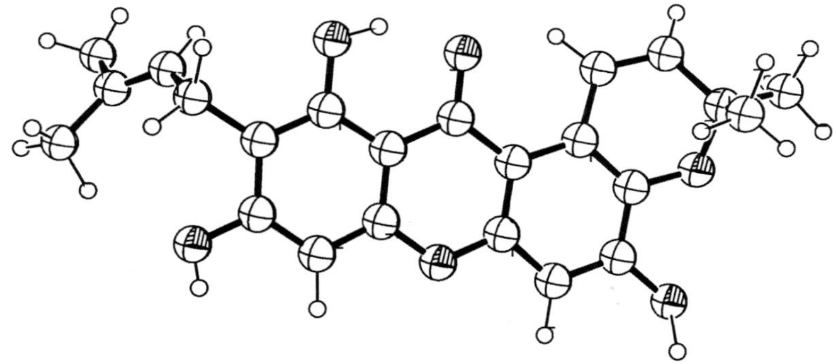

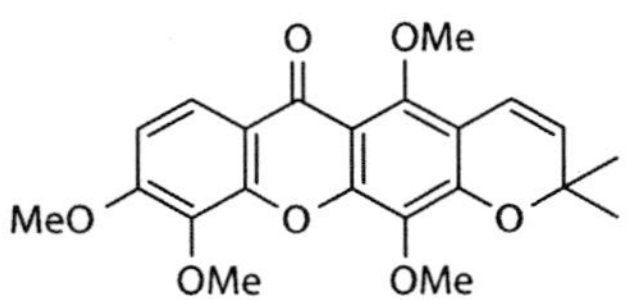

Dulxanthone E

Formula: $C_{22}H_{22}O_7$
Temperature: 295 K
Density: 1.360 g/cm^3
Space group: $P2_1/c$
R-factor (%): 4.30
Ref: (Kosela et al. 1999)

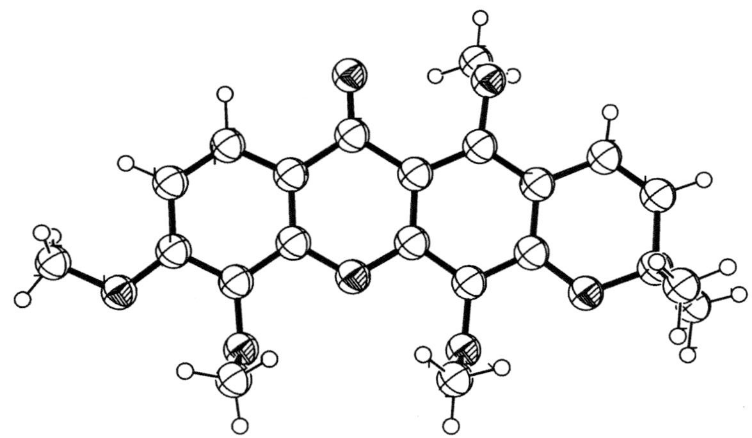

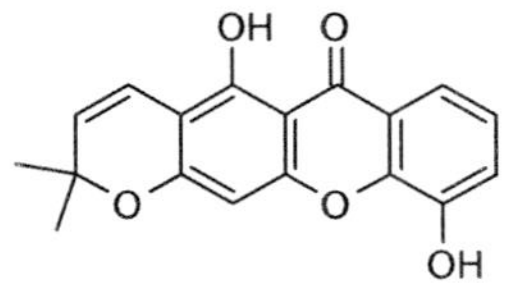 6-Deoxyjacareubin	Formula: $C_{18}H_{14}O_5$ Temperature: 293 K Density: 1.424 g/cm^3 Space group: P-1 R-factor (%): 5.14 Ref: (Doriguetto et al. 2001)	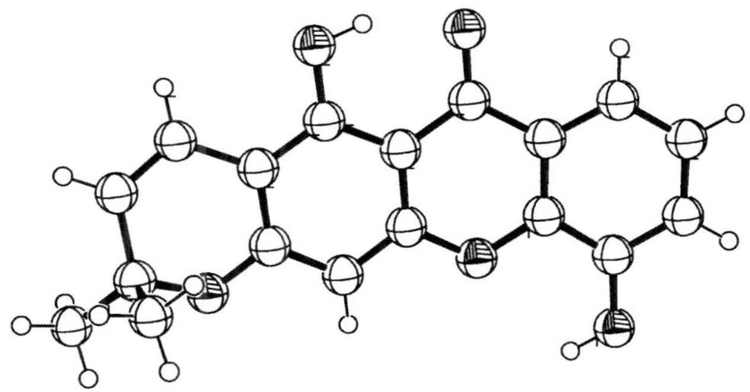
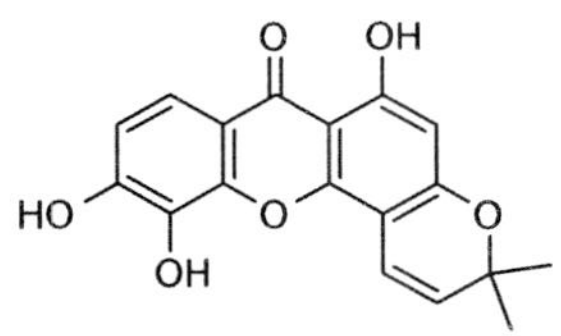 Isojacreubin	Formula: $C_{18}H_{14}O_6$ Temperature: 295 K Density: 1.482 g/cm^3 Space group: C2/c R-factor (%): 5.00 Ref: (Ishiguro et al. 1993)	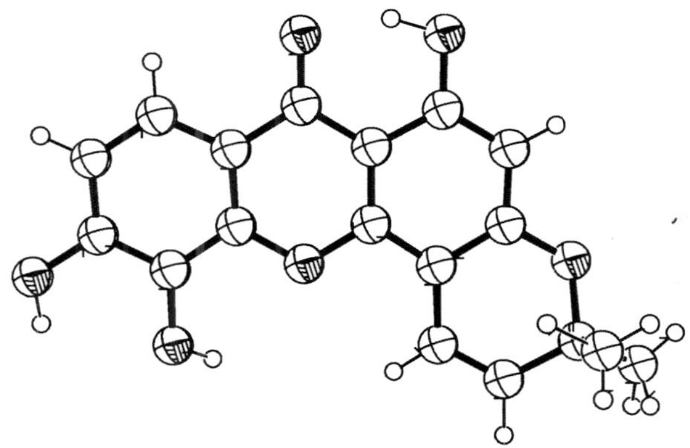

(continued)

TABLE 14.3 (continued)
Selected Prenylated and Related Xanthones

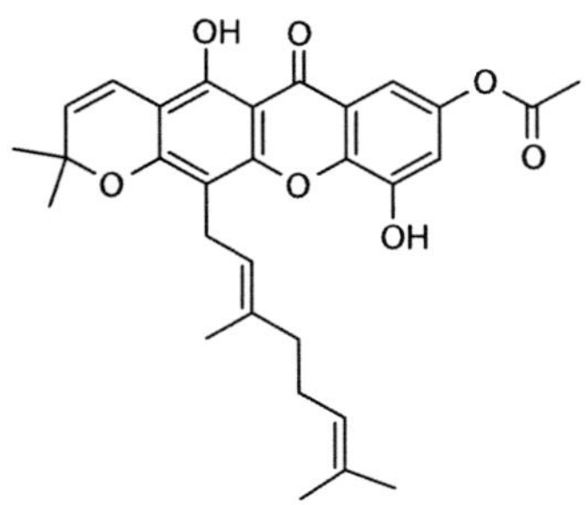

7-Acetoxycochinchinone I

Formula: $C_{30}H_{32}O_6$
Temperature: 100 K
Density: 1.282 g/cm^3
Space group: P-1
R-factor (%): 9.47
Ref: (Chantrapromma et al. 2009)

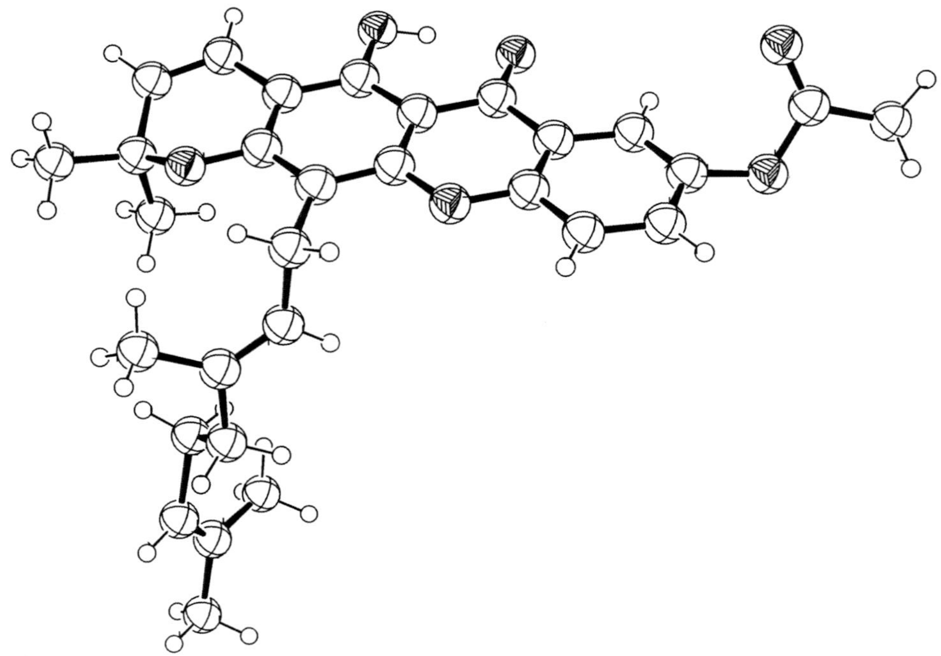

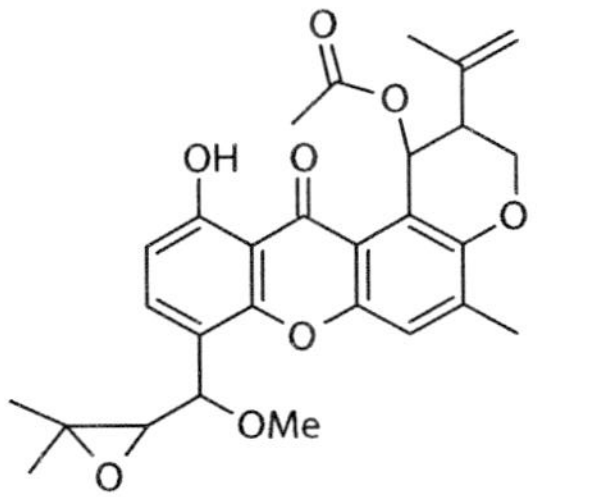 14-Methoxytajixanthone-25-acetate	Formula: $C_{28}H_{30}O_8$ Temperature: 293 K Density: 1.277g/cm^3 Space group: $P2_1$ R-factor (%): 5.59 Ref: (Liangsakul et al. 2009)	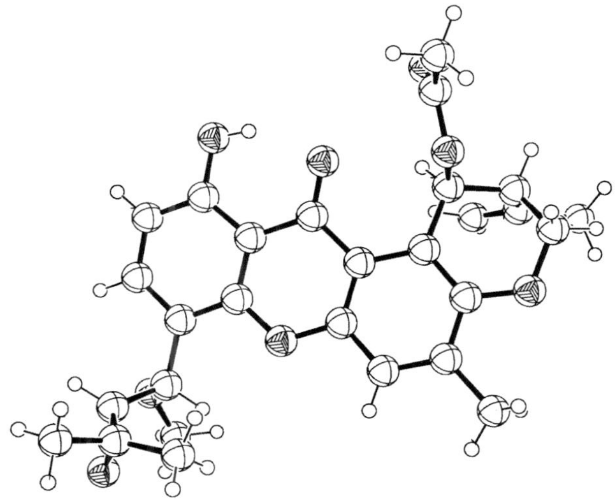
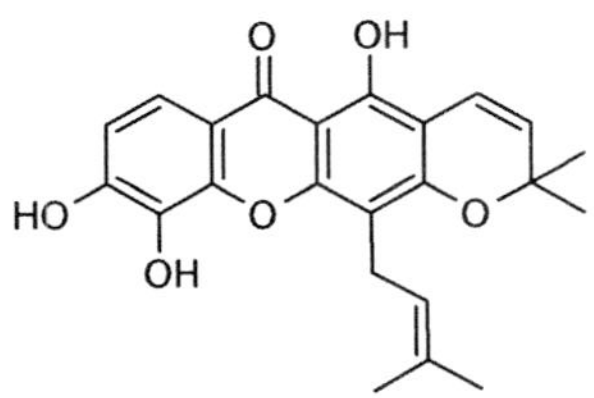 Xanthone V1	Formula: $C_{23}H_{22}O_6 \cdot CH_4O$ Temperature: 293 K Density: 1.286 g/cm^3 Space group: $P2_1/c$ R-factor (%): 8.68 Ref: (Chantrapromma et al. 2005)	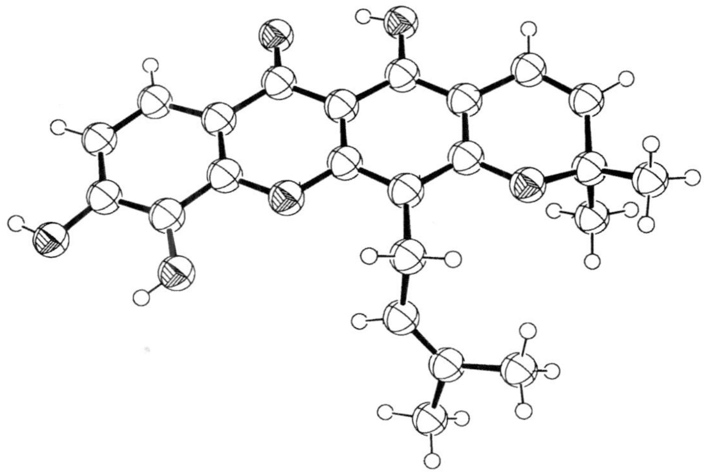

(continued)

TABLE 14.3 (continued)
Selected Prenylated and Related Xanthones

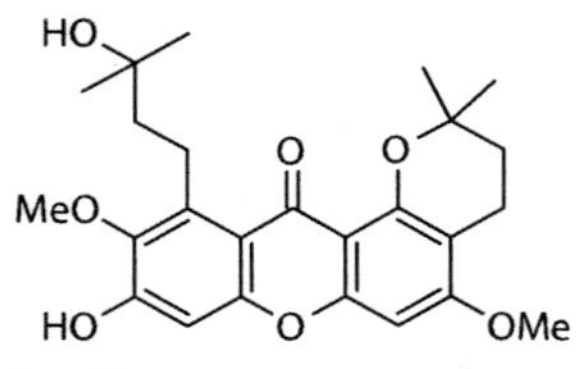

Pruniflorone A

Formula: $C_{25}H_{30}O_7$
Temperature: 100 K
Density: 1.305 g/cm^3
Space group: P2_1/n
R-factor (%): 7.59
Ref: (Boonnak et al. 2006)

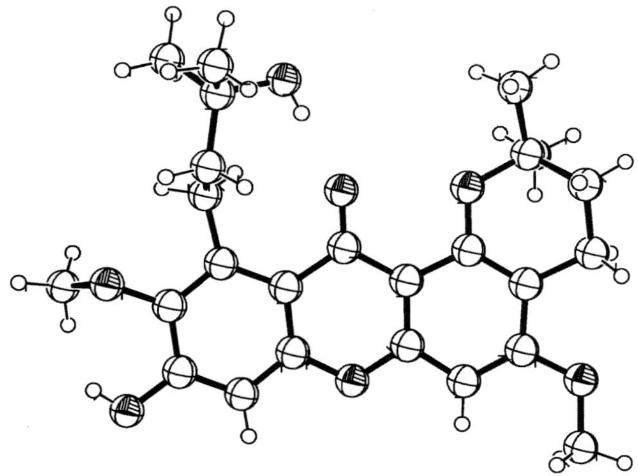

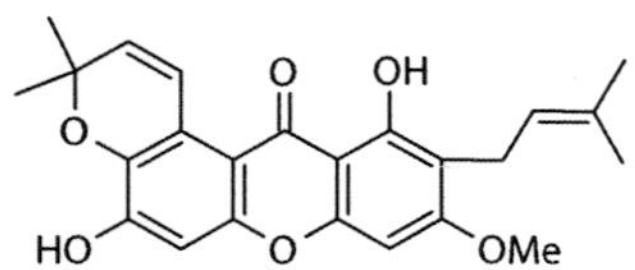

Dulxisxanthone F

Formula: $C_{24}H_{24}O_6$
Temperature: 100 K
Density: 1.351 g/cm^3
Space group: P-1
R-factor (%): 10.64
Ref: (Boonnak et al. 2006)

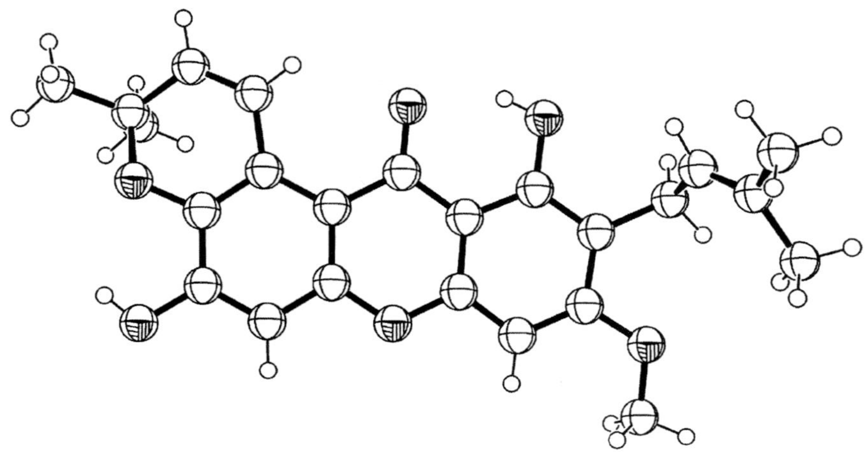

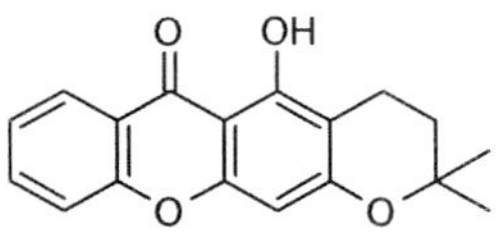 1-Hydroxy-6,6′-dimethyl-4′,5′-dihydropyrano(2′,3′:3,2) xanthone	Formula: $C_{18}H_{16}O_4$ Temperature: 293 K Density: 1.363 g/cm^3 Space group: C2/c R-factor (%): 4.00 Ref: (Castanheiro et al. 2007)	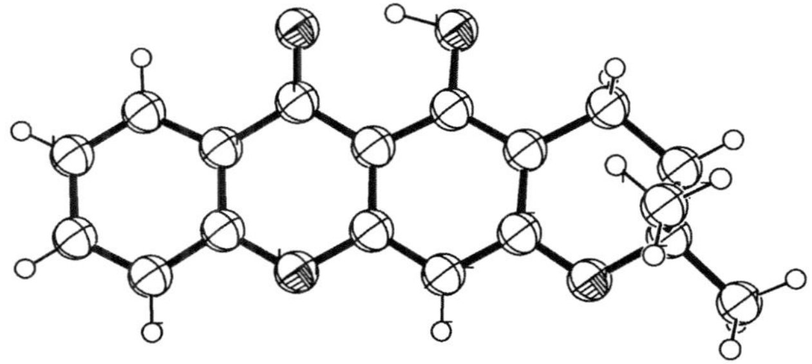
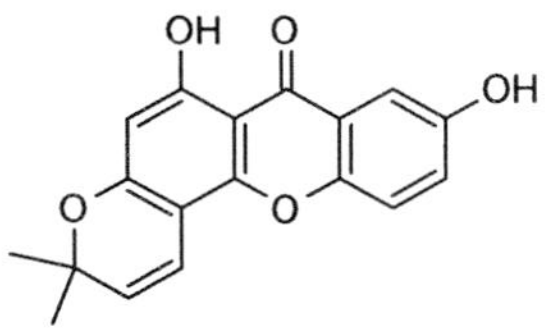 Nigrolineaxanthone F	Formula: $C_{18}H_{14}O_5$ Temperature: 293 K Density: 1.442 g/cm^3 Space group: C2/c R-factor (%): 4.24 Ref: (Mondal et al. 2006)	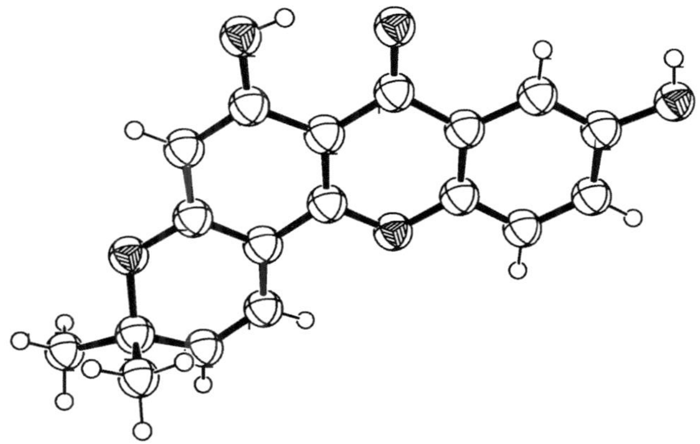
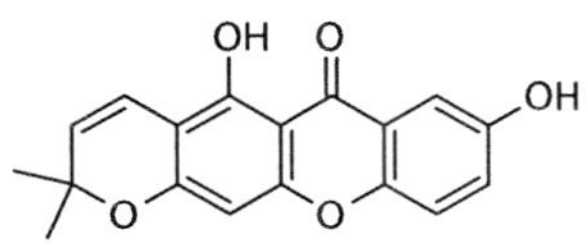 Osajaxanthone	Formula: $C_{18}H_{14}O_5$ Temperature: 293 K Density: 1.452 g/cm^3 Space group: $P2_1/c$ R-factor (%): 3.38 Ref: (Mondal et al. 2006)	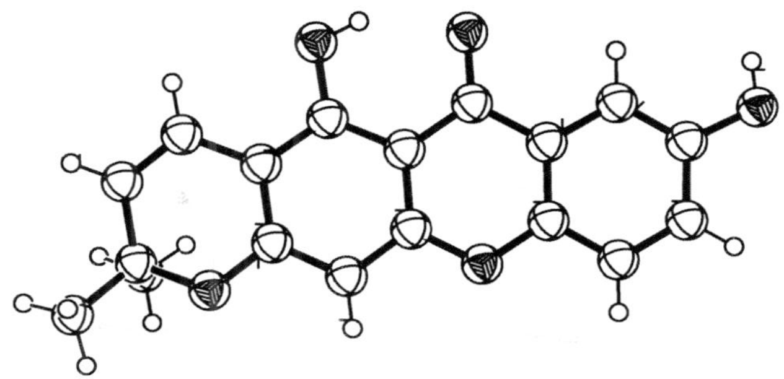

(continued)

TABLE 14.3 (continued)
Selected Prenylated and Related Xanthones

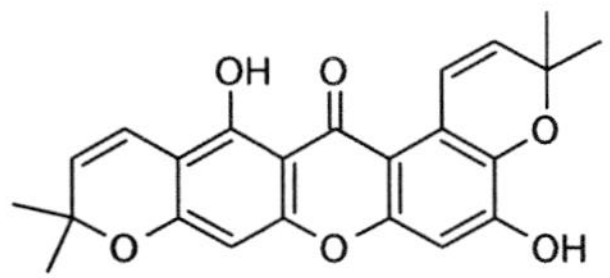 Brasilixanthone	Formula: $C_{23}H_{20}O_6$ Temperature: 100 K Density: 1.436 g/cm^3 Space group: Pc R-factor (%): 10.26 Ref: (Chantrapromma et al. 2010)	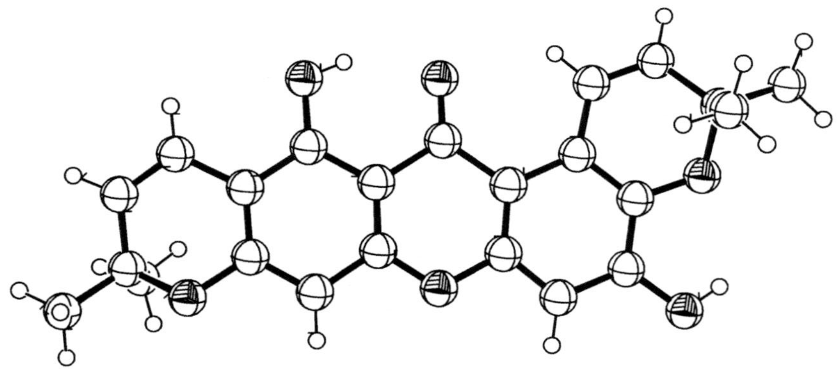
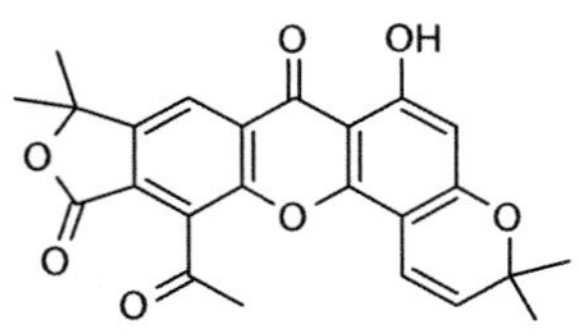 12-Acetyl-6-hydroxy-3,3,9,9-tetramethyl-3H,7H-furo[3,4-b]pyrano[3,2-h]xanthone	Formula: $C_{24}H_{20}O_7$ Temperature: 150 K Density: 1.443 g/cm^3 Space group: C2/c R-factor (%): 3.88 Ref: (Ee et al. 2010)	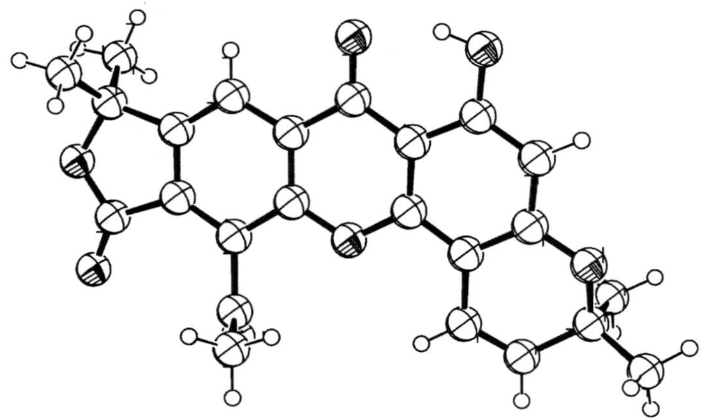

TABLE 14.4
Selected Xanthones Containing a Bis-Dihydrofuran Ring System

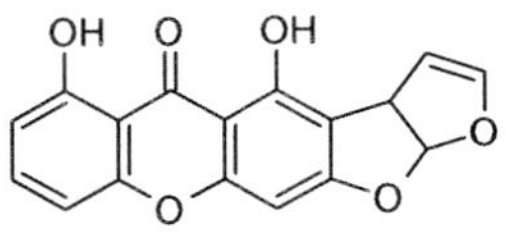 Sterigmatin	Formula: $C_{17}H_{10}O_6$ Temperature: 295 K Density: 1.555 g/cm^3 Space group: P1 R-factor (%): 3.90 Ref: (Fukuyama et al. 1975b)	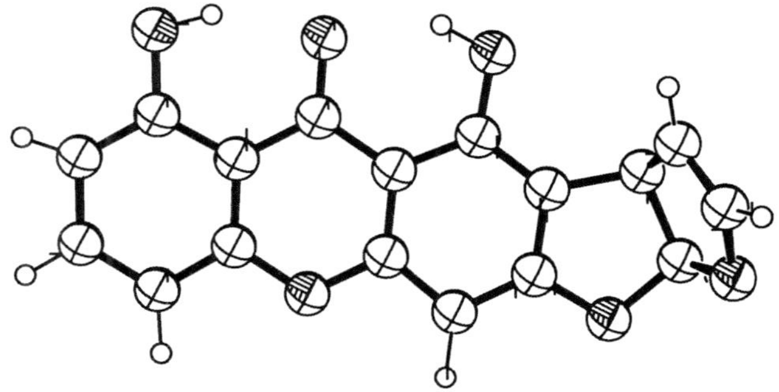
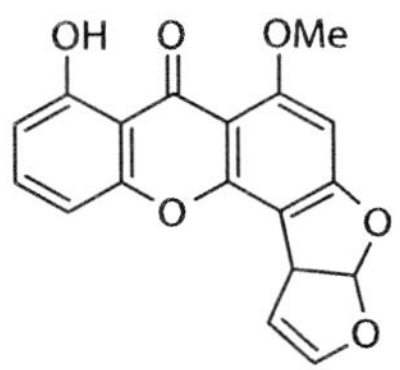 Sterigmatocystin	Formula: $C_{18}H_{12}O_6$ Temperature: 295 K Density: 1.504 g/cm^3 Space group: P2$_1$ R-factor (%): 3.90 Ref: (Fukuyama et al. 1978)	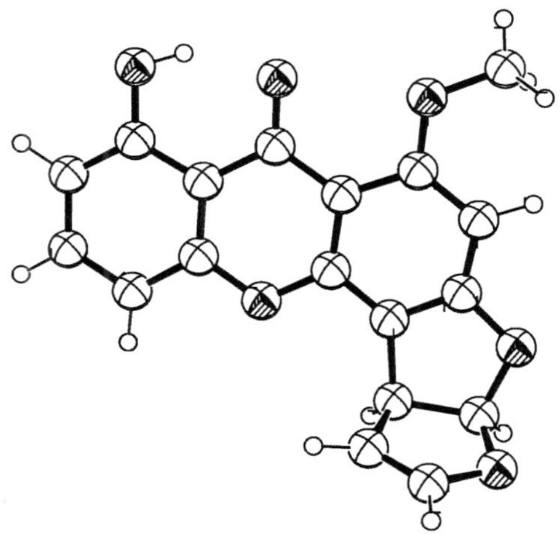

(continued)

TABLE 14.4 (continued)
Selected Xanthones Containing a Bis-Dihydrofuran Ring System

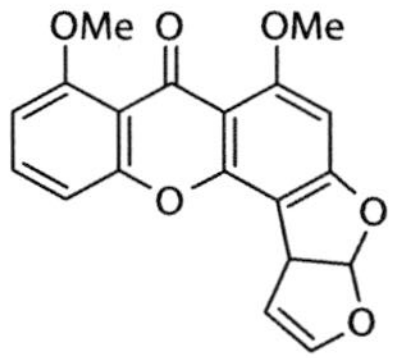

O-Methylsterigmatocystin

Formula: $C_{19}H_{14}O_6$
Temperature: 295 K
Density: 1.486 g/cm^3
Space group: $P2_1$
R-factor (%): 5.50
Ref: (Fukuyama 1978)

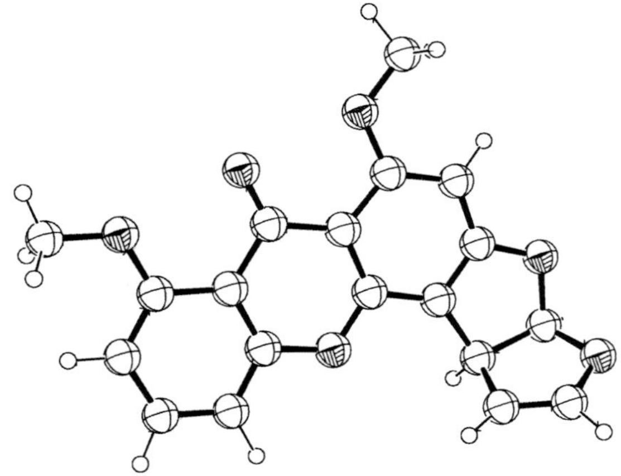

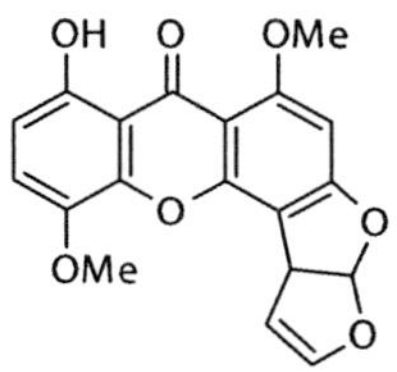

5-Methoxy-sterigmatocystin

Formula: $C_{19}H_{14}O_7$
Temperature: 295 K
Density: 1.520 g/cm^3
Space group: $P2_12_12_1$
R-factor (%): 6.00
Ref: (Smith and Duax 1975)

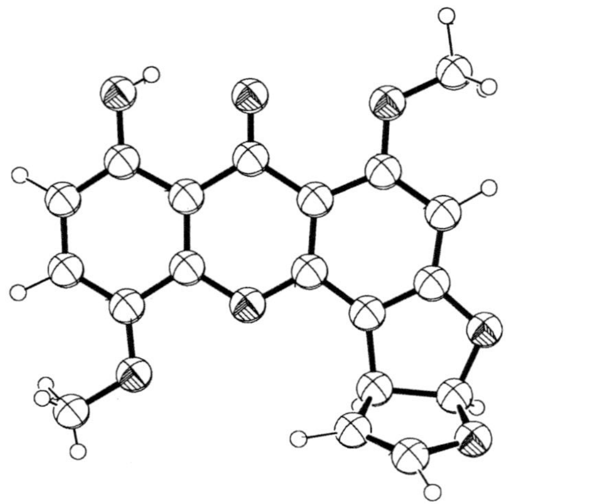

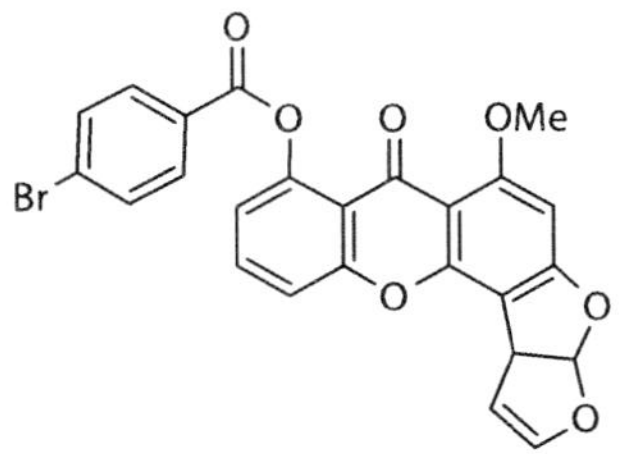

Sterigmatocystin p-bromobenzoate

Formula: $C_{25}H_{15}BrO_7$
Temperature: 295 K
Density: 1.635 g/cm^3
Space group: $P2_1$
R-factor (%): 5.50
Ref: (Fukuyama et al. 1975a)

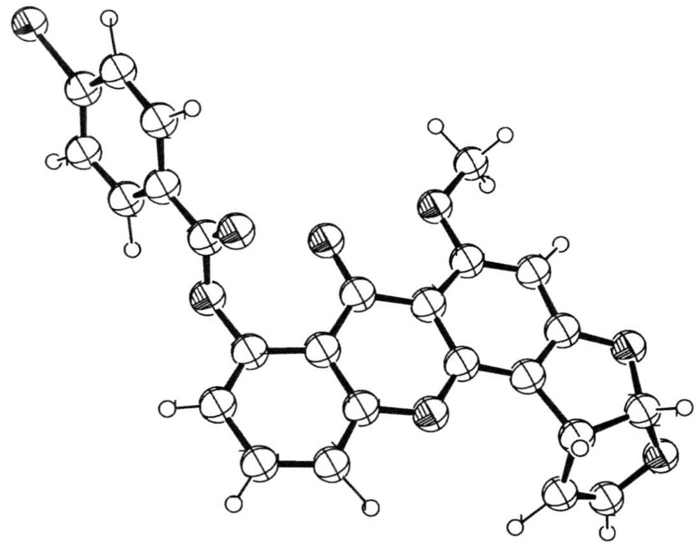

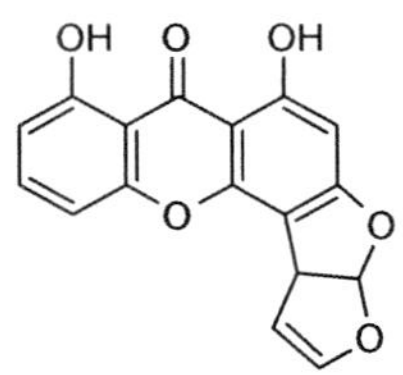

Demethyl-sterigmatocystin

Formula: $C_{17}H_{10}O_6$
Temperature: 295 K
Density: 1.533 g/cm^3
Space group: C2
R-factor (%): 6.80
Ref: (Yoshida et al. 1979)

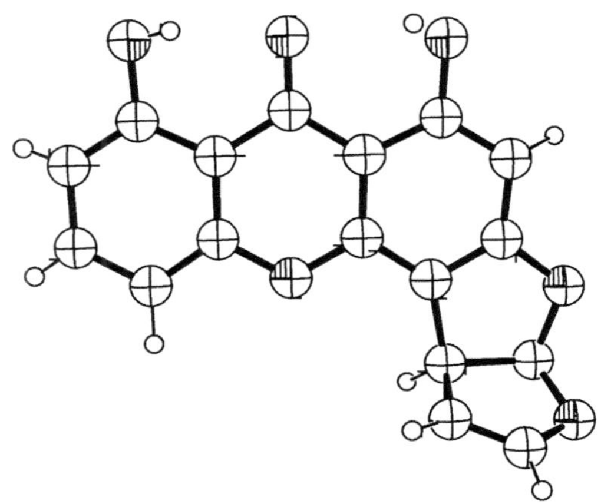

TABLE 14.5
Selected Halogenated Xanthones

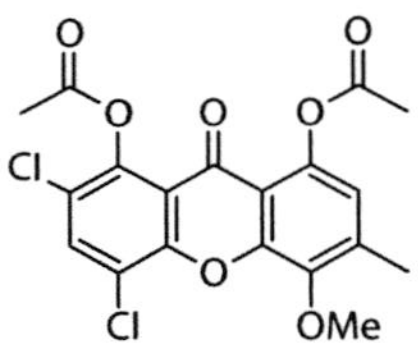

Thiomelin diacetate

Formula: $C_{19}H_{14}Cl_2O_7$
Temperature: 295 K
Density: 1.532 g/cm^3
Space group: $P2_1/c$
R-factor (%): 4.60
Ref: (Elix et al. 1987)

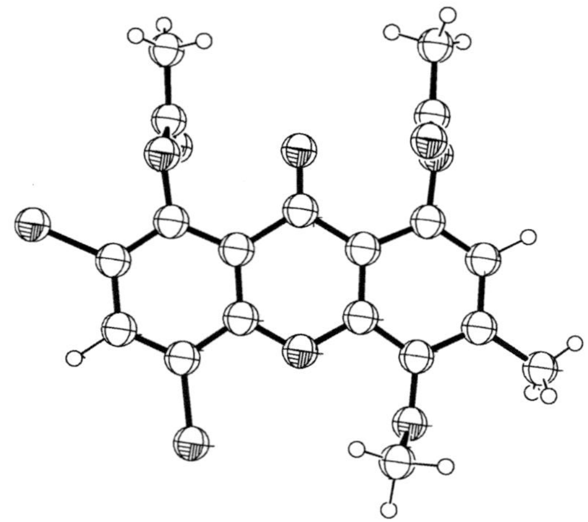

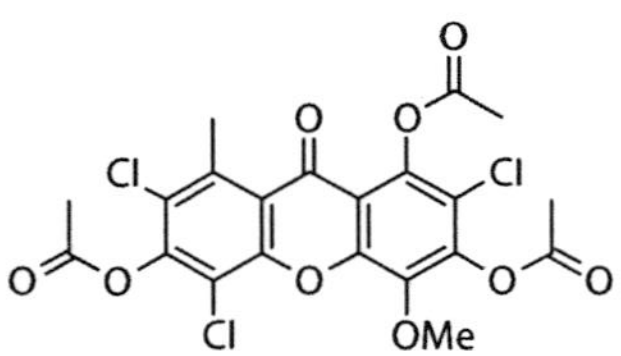

Demethylchodatin tri-O-acetate

Formula: $C_{21}H_{15}\,Cl_3O_9$
Temperature: 295 K
Density: 1.577 g/cm^3
Space group: $P2_1/a$
R-factor (%): 3.30
Ref: (Elix et al. 1994)

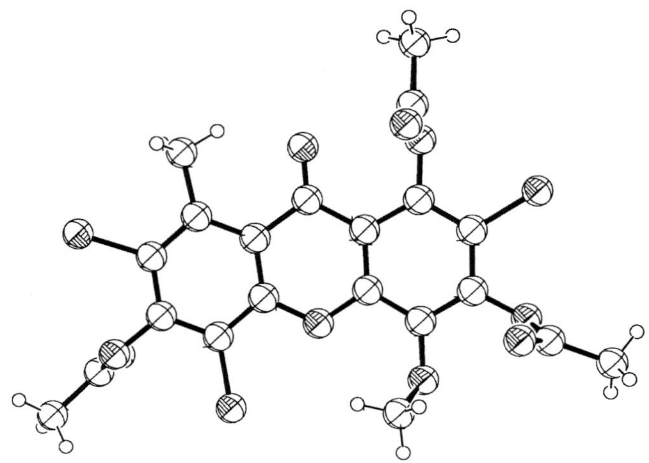

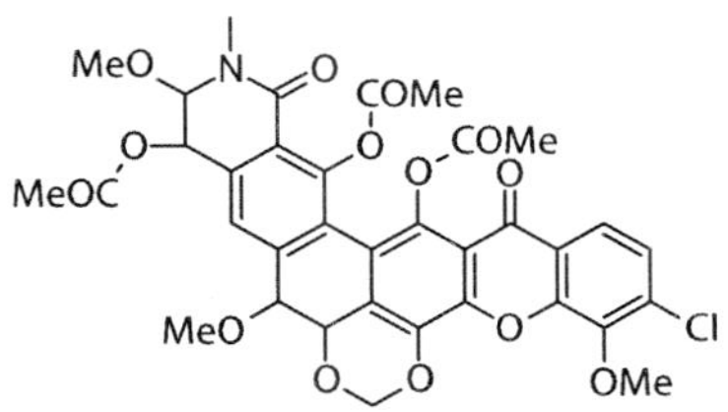

Lysolipin triacetate acetone solvate

Formula: $C_{35}H_{30}ClO_{14} \cdot C_3H_6O$
Temperature: 295 K
Density: 1.381 g/cm³
Space group: $P2_1$
R-factor (%): 6.70
Ref: (Dobler and Keller-Schierlein 1977)

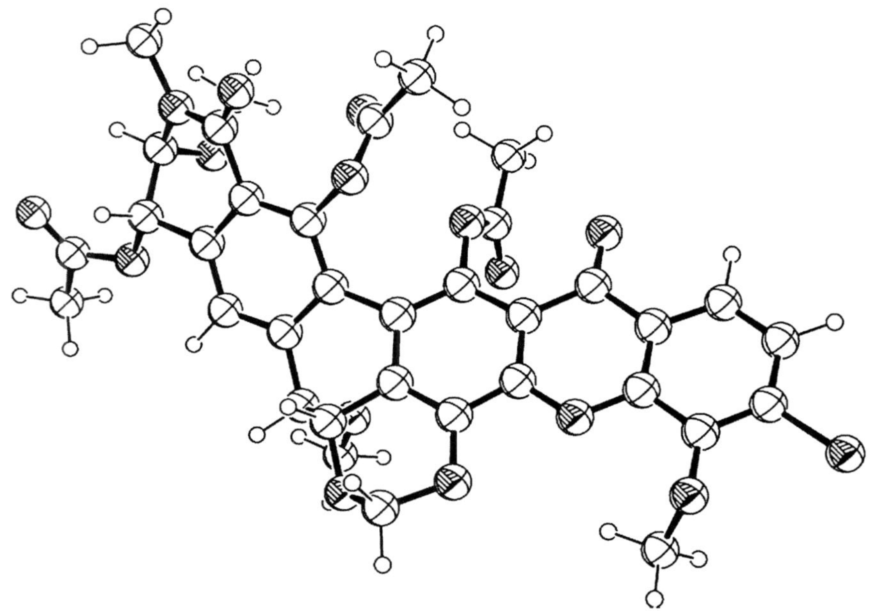

(*continued*)

TABLE 14.5 (continued)
Selected Halogenated Xanthones

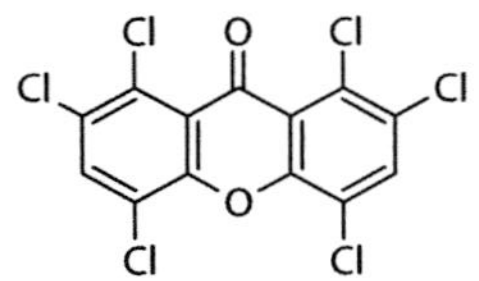 1,3,4,6,7,9-Hexachloroxanthone	Formula: $C_{13}H_2Cl_6O_2$ Temperature: 295 K Density: 1.931 g/cm^3 Space group: $P2_1/n$ R-factor (%): 6.10 Ref: (Soderholm et al. 1976)	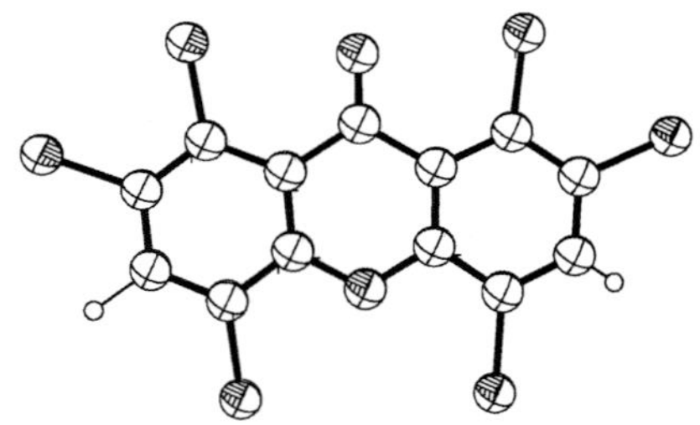
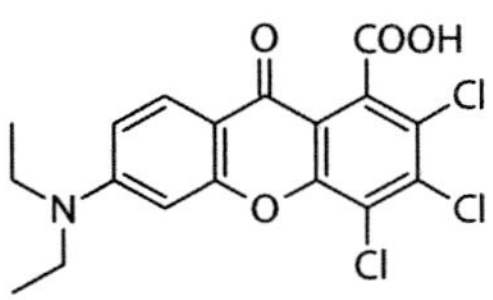 2,3,4-Trichloro-6-(diethylamino)-9-oxo-9H-xanthene-1-carboxylic acid	Formula: $C_{18}H_{14}Cl_3NO_4 \cdot CH_4O$ Temperature: 294 K Density: 1.496 g/cm^3 Space group: P-1 R-factor (%): 3.96 Ref: (Zhang et al. 2007)	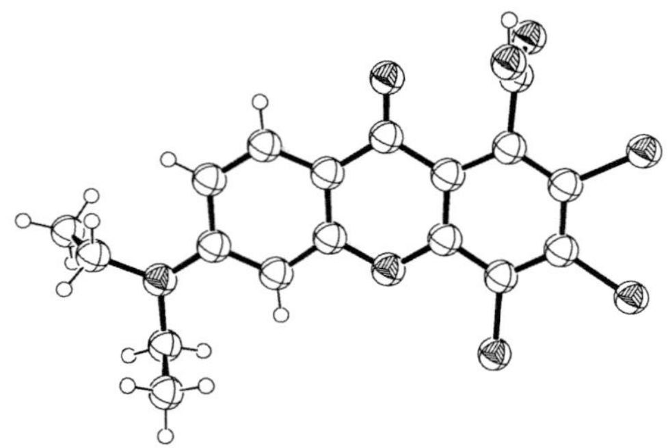

14.2.8 Xanthones Containing a Crown Ether

Xanthonic compounds have shown their utility outside the pharmacologic and biological fields. Xanthones containing crown ethers have been investigated to catalyze hydrolyses of simple carboxylic acid derivatives. The xanthone skeleton adopts a flattened boat conformation in the crystal structures. Angles between the aromatic planes range from 8° to 17°, which direct the carbonyl oxygen out of the mean plane of the macrocyclic array.

The crown ether sensors also attract considerable attention as receptors for the alkali metal and alkaline earth metal cations to form the host–guest complexes. A series of sodium(I), magnesium(II), and aluminum(III) xanthone–crown ether complexes were synthesized in order to investigate their anion sensing capabilities (Shen et al. 2008b). It was observed that the magnesium(II) xanthone–crown ether complex acts as a good fluorescent and colorimetric detector, with high sensitivity and selectivity to HSO_4^-. Later, it was found that the zinc(II) xanthone–crown ether could also selectively detect HSO_4^- over other anions by fluorescence and UV–Vis absorption spectroscopy (Shen et al. 2008a).

In 2009, a series of sandwich, monomeric, dimeric, and polymeric complexes supported with 1,8-xanthone-18-crown-5 (L) were synthesized. The crystal structures of six complexes show the strong coordination of xanthone-18-crown-5 carbonyl oxygen with alkaline earth metal cation, which results in high fluorescent increase in the alkaline earth metal complexes (Wu et al. 2009).

Another field of application is in the synthesis of macrocyclic aromatic ethers and in the chemistry of their interconversion with polymeric materials. Low-viscosity macrocyclic precursors may undergo ring-opening polymerization *in situ* leading to high-molar-mass polymers without generating by-products (Kogan and Biali 2007). X-ray diffraction analysis shows that replacement of the dibenzofuran unit by a xanthone brings about a significant relaxation of the macrocyclic ring strain, reflected in an increase in the angle subtended by the two biphenylene-to-carbonyl bonds from 115° to 138°. However, the twisting of the biphenylene unit is essentially unchanged (Table 14.6).

14.2.9 Xanthone: Metal Complexes

The formation of complexes, where the C=O from the pyranoid ring coordinates with a metal atom, has been studied due to the importance of these compounds in understanding the donor–acceptor interactions. Moreover, a xanthone was included in the analysis of the molecular packing of more than 200 tetraaryl porphyrin-based lattices clathrates, which have interesting properties concerning the capability to accommodate a wide range of guest molecules and therefore may be used in the design of new molecular solids (Byrn et al. 1993). The molecular structures of the complexes show that the xanthone skeleton remains essentially planar and that there is a lengthening of the C=O bond, depending on the nature of the metal atom. For example, the change in C=O bond length is larger on complexation with trichloro-aluminum(III) (Boucher et al. 1999) than with indium(III) iodide (Brown and Tuck 2000).

As described in the previous section, xanthone:metal complexes have been investigated for anion sensing. Of particular interest is a xanthone-based chemosensor that bears two 2,2′-dipicolylamine–zinc(II) sites and displays changes in the excitation spectrum at three wavelengths upon binding phosphates (Ojida et al. 2006). These changes in the signals arise from the coordination rearrangement of the xanthone fluorophore around the zinc centers, which modulates the photophysical property of the chemosensor, thus enabling ratiometric fluorescent analysis (Table 14.7).

TABLE 14.6
Selected Xanthones Containing a Crown Ether

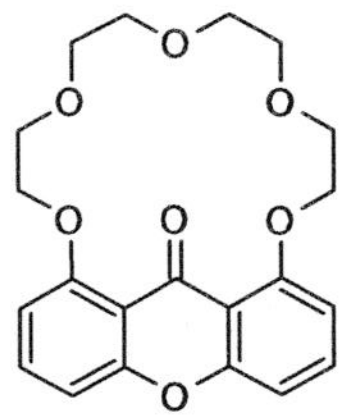 1,8-((3,6,9-Trixaundecane-1,11-diyl) dioxy)xanthone	Formula: $C_{21}H_{22}O_7$ Temperature: 298 K Density: 1.566 g/cm^3 Space group: Pnna R-factor (%): 8.21 Ref: (Shen et al. 2008b)	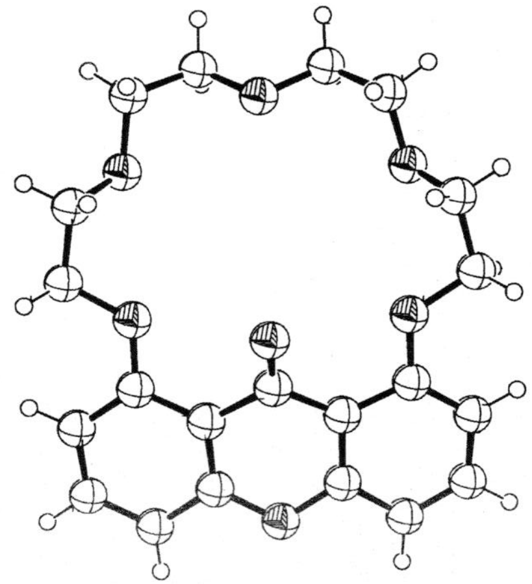
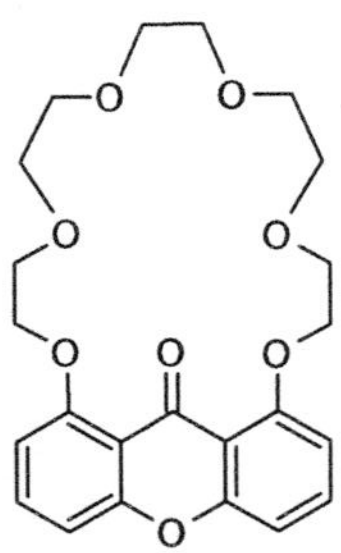 1,8-((3,6,9,12-Tetraxaundecane-1,14-diyl)dioxy)xanthone	Formula: $C_{23}H_{26}O_8$ Temperature: 295 K Density: 1.363 g/cm^3 Space group: $P2_1/n$ R-factor (%): 5.00 Ref: (Beddoes et al. 1996)	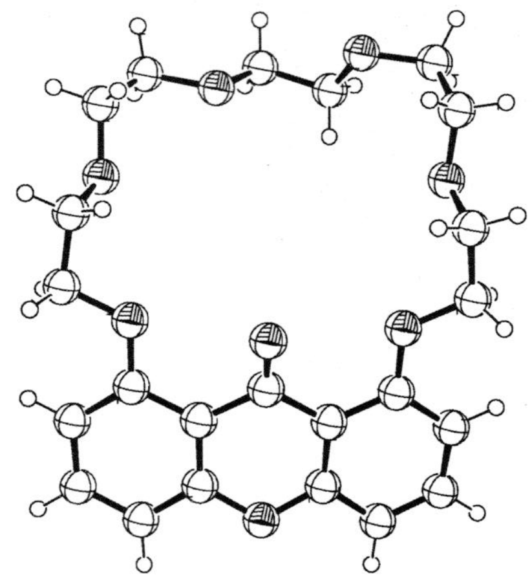

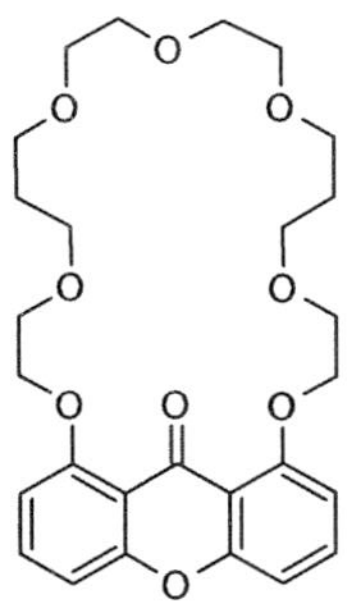

1,8-((3,6,9,12,15-Pentaoxaheptadecane-1,17-diyl)dioxy)xanthone

Formula: $C_{25}H_{30}O_9$
Temperature: 295 K
Density: 1.283 g/cm^3
Space group: Pccn
R-factor (%): 6.40
Ref: (Beddoes et al. 1996)

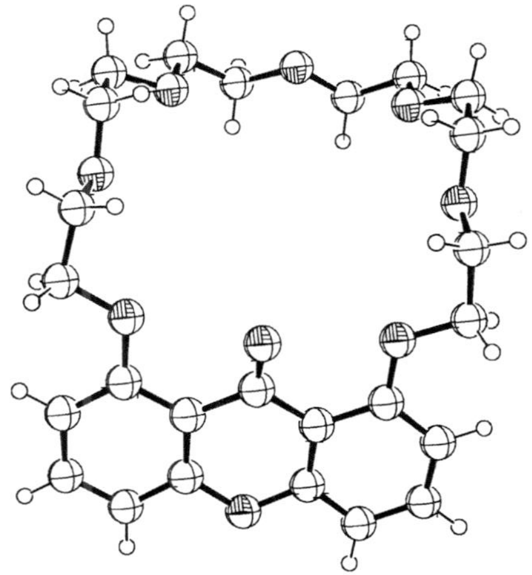

TABLE 14.7
Xanthone: Metal Complexes

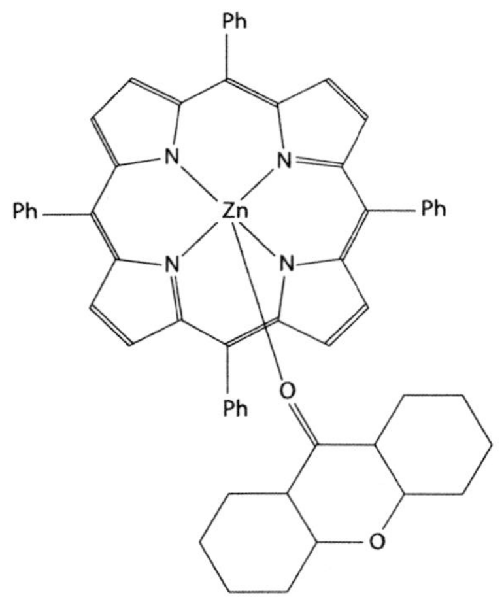

Tetraphenylporphyrin-(9-xanthone)-zinc(II)

Formula: $C_{57}H_{36}N_4O_2Zn$
Temperature: 295 K
Density: 1.360 g/cm^3
Space group: P-1
R-factor (%): 4.10
Ref: (Byrn et al. 1993)

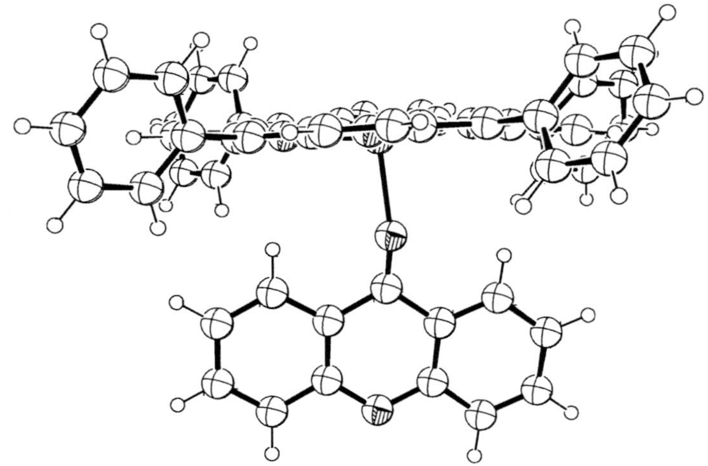

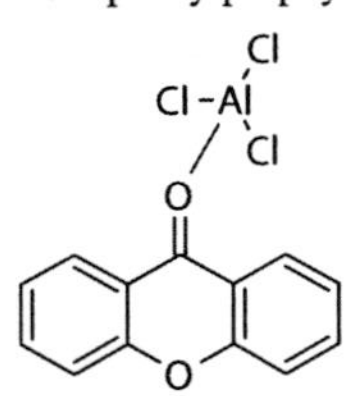

Trichloro-(xanthone-O)-aluminum(III)

Formula: $C_{13}H_8AlCl_3O_2$
Temperature: 295 K
Density: 1.568 g/cm^3
Space group: Pbca
R-factor (%): 4.30
Ref: (Boucher et al. 1999)

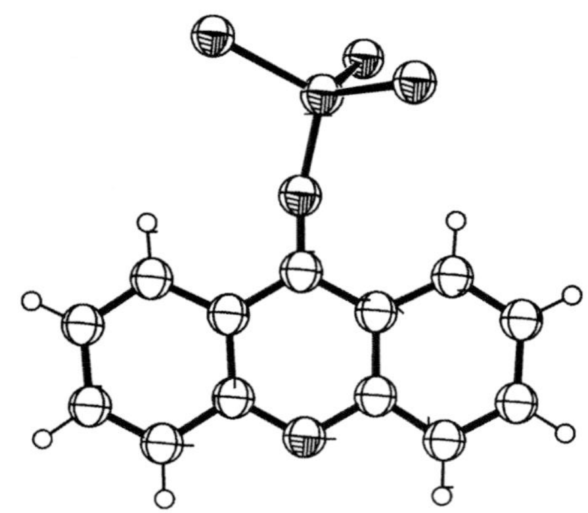

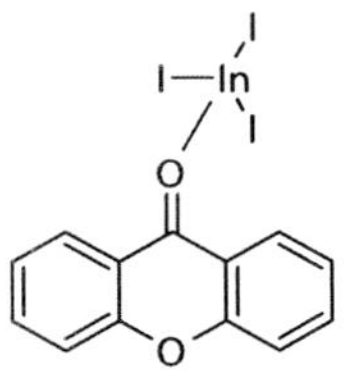

Triiodo-(xanthone-O)-indium(III)

Formula: $C_{13}H_8I_3InO_2$
Temperature: 295 K
Density: 2.665 g/cm^3
Space group: P-1
R-factor (%): 5.80
Ref: (Brown and Tuck 2000)

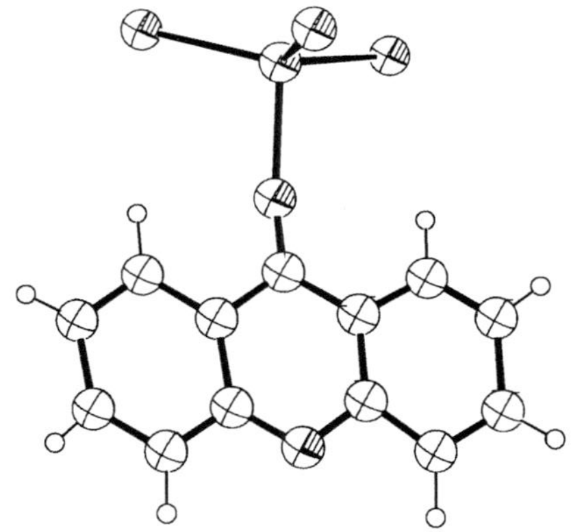

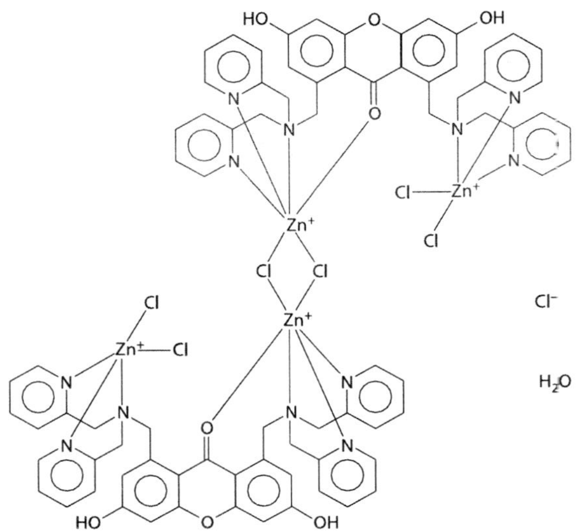

bis(μ2-Chloro)-bis(m2–1,8-bis((2,2′dipicolylamino) methyl)-3,6-dihydroxy-9H-xanthen-9-one)-tetrachloro-tetra-zinc(ii)

Formula:
$C_{78}H_{68}Cl_6N_{12}O_8Zn_4^{4+}$, $4(Cl^-)$, $7(H_2O)$
Temperature: 123 K
Density: 1.481 g/cm^3
Space group: $P2_1/n$
R-factor (%): 7.60
Ref: (Ojida et al. 2006)

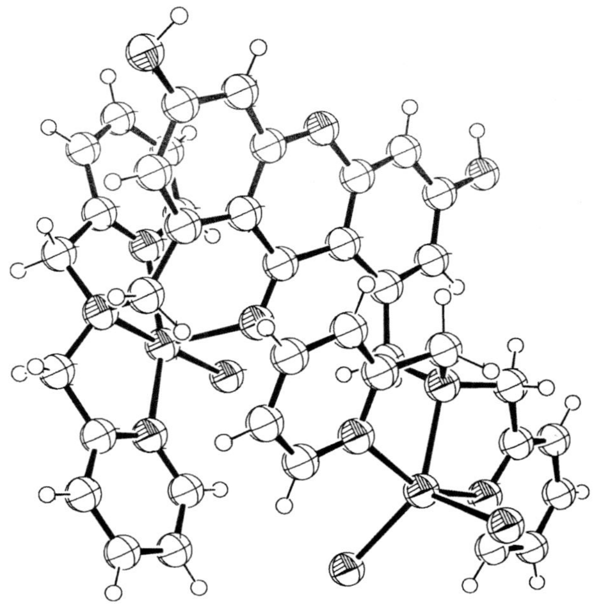

14.3 CONCLUSIONS

The crystal structures of all these xanthones reveal that the three-ring system is mostly planar. The central pyranoid ring has, in the majority of the cases, an aromatic character, as evaluated by the bond lengths and angles determined for the central ring. However, small deviations from planarity are frequent, mainly due to steric factors associated with substituents.

The xanthone three-ring system may accommodate a large number of substituents, which depending on their chemical nature and position on the aromatic rings, lead to a myriad of biological activities of its derivatives. Moreover, outside the medicinal chemistry field, xanthone: metal complexes involving the coordination of the C=O from the pyranoid ring have found interesting applications in chemical sensing of anions.

Further studies concerning the biological activities of xanthone derivatives need to be performed at the molecular level. It is important to know which proteins or macromolecular assemblies interact with xanthones in order to determine the three-dimensional structure of such complexes and thereby design derivatives with improved activity.

REFERENCES

Allen F.H., O. Kennard, D.G. Watson et al. 1987. Bond lengths in organic compounds. *J. Chem. Soc., Perkin Trans.* 212: S1–9.

Almeida M.R., L. Gales, A.M. Damas et al. 2005. Small transthyretin (TTR) ligands as possible therapeutic agents in TTR amyloidoses. *Curr. Drug Targets CNS Neurol. Disord.* 4: 587–596.

Arcamone F. 1981. *Doxorubicin Anticancer Antibiotics*. New York: Academic Press.

Beddoes R.S., B.G. Cox, O.S. Mills et al. 1996. Structures and complexing properties of crown ethers incorporating 1,8-dioxyxanthones. *J. Chem. Soc. Perkin Trans.* 2 10: 2091–2098.

Biswas S.C.S., and R.K. Sen. 1969. Crystal and molecular structure of xanthane. *Indian J. Pure Appl. Phys.* 7: 408.

Biswas S.C.S. and R.K. Sen. 1982. X-ray crystallographic studies of xanthones. *Indian J. Pure Appl. Phys.* 20: 414–415.

Boonnak N., H.K. Fun and S. Chantrapromma, et al. 2007b. Gerontoxanthone I methanol solvate. *Acta Crystallogr. E* 63: o3958–o3959.

Boonnak N., S. Chantrapromma, H.K. Fun et al. 2005. 1,6,7-Trihydroxy-3-methoxy-2,8-bis(3-methyl2-butenyl)-9H-xanthen-9-one chloroform solvate. *Acta Crystallogr. E* 61: o4376–o4378.

Boonnak N., S. Chantrapromma, H.K. Fun et al. 2007a. 4,8-Dihydr-oxy-2,3-dimeth-oxy-1-(3-methyl-but-2-en-yl)-9H-xanthen-9-one. *Acta Crystallogr. E* 63: o4903–o4904.

Boonnak N., C. Karalai, S. Chantrapromma et al. 2006. Bioactive prenylated xanthones and anthraquinones from *Cratoxylum formosum* ssp. pruniflorum. *Tetrahedron* 62: 8850–8859.

Boucher D.L., M.A. Brown, B.R. McGarvey et al. 1999. Spectroscopic and crystallographic studies of adducts of aluminium trichloride with cyclic ketones, para-quinones and ortho-quinones. *J. Chem. Soc. Dalton Trans.* 19: 3445–3450.

Brown M.A. and D.G. Tuck. 2000. Crystallographic and spectroscopic studies of adducts of indium (III) iodide with two cyclic ketones. *Can. J. Chem.* 78: 536–541.

Byrn M.P., C.J. Curtis, Y. Hsiou et al. 1993. Porphyrin sponges: Conservation of host structure in over 200 porphyrin-based lattice clathrates. *J. Am. Chem. Soc.* 115: 9480–9497.

Castanheiro R.A.P., M.M.M. Pinto, A.M.S. Silva et al. 2007. Dihydroxyxanthones prenylated derivatives: Synthesis, structure elucidation, and growth inhibitory activity on human tumor cell lines with improvement of selectivity for MCF-7. *Bioorg. Med. Chem.* 15: 6080–6088.

Chantrapromma S., N. Boonnak, and H.K. Fun. 2009. 7-Acetoxycochinchinone i. *Acta Crystallogr. E* 65: o2223–o2224.

Chantrapromma S., N. Boonnak, H.K. Fun et al. 2005. 5,9,10-Trihydroxy-2,2-dimethyl-12-(3-methyl-but-2-enyl)-2H,6H-pyrano[3,2-b] xanthen-6-one methanol solvate. *Acta Crystallogr. E* 61: o2136–o2138.

Chantrapromma S., N. Boonnak, H.K. Fun et al. 2010. Brasilixanthone. *Acta Crystallogr. E* 66: o2066–o2067.

Chen G., J. Zhao, C. Cen et al. 2009. 2-Hydr-oxy-1-methoxyxanthen-9-one monohydrate. *Acta Crystallogr. E* 65: o2798.

Corrêa R.S., M.H. dos Santos, T.J. Nagem et al. 2010. On the relationships between molecular conformations and intermolecular contacts toward crystal self-assembly of mono-, di-, tri-, and tetra-oxygenated xanthone derivatives. *Struct. Chem.* 21: 555–563.

Da Cruz Jr J.W., L.R. De Moraes, M.H. Dos Santos et al. 2008. Crystalline structure of mangiferin, a C-glycosyl-substituted 9H-xanthen-9-one isolated from the stem bark of *Mangifera indica*. *Helv. Chim. Acta* 91: 144–154.

Dobler M. and W. Keller-Schierlein. 1977. Metabolites of microorganisms. 162nd Communication. The crystal and molecular structure of lysolipin I. *Helv. Chim. Acta* 60: 178–185.

Doriguetto A.C., M.H. Santos, J.A. Ellena et al. 2001. 6-deoxyjacareubin. *Acta Crystallogr. C* 57: 1095–1097.

Ee G.C.L., S.H. Teo, H.C. Kwong et al. 2010. 12-Acetyl-6-hydroxy-3,3,9,9-tetramethylfuro[3,4-b]pyrano[3,2-h]-xanthene-7, 11(3H,9H)-dione. *Acta Crystallogr. E* 66: o3331–o3332.

Eguchi T., K. Kondo, K. Kakinuma et al. 1999. Unique solvent-dependent atropisomerism of a novel cytotoxic naphthoxanthene antibiotic FD-594. *J. Org. Chem.* 64: 5371–5376.

El-Seedi H.R., M.A. El-Barbary, D.M.H. El-Ghorab et al. 2010. Recent insights into the biosynthesis and biological activities of natural xanthones. *Curr. Med. Chem.* 17: 854–901.

El-Seedi H.R., D.M.H. El-Ghorab, M.A. El-Barbary et al. 2009. Naturally occurring xanthones; Latest investigations: Isolation, structure elucidation and chemosystematic significance. *Curr. Med. Chem.* 16: 2581–2626.

Elix J., K. Gaul, M. Sterns et al. 1987. The Structure of the novel lichen xanthone thiomelin and its cogenors. *Aust. J. Chem.* 40: 1169–1178.

Elix J., F. Robertson, J. Wardlaw et al. 1994. Isolation and structure determination of demethylchodatin 8212; a new lichen xanthone. *Aust. J. Chem.* 47: 2291–2295.

Evans I.R., J.A.K. Howard, K. Šavikin-Fodulovіç et al. 2004. Isogentisin (1,3-dihydroxy-7-methoxyxanthone). *Acta Crystallogr. E* 60: o1557–o1559.

Fukuyama K., K. Hamada, T. Tsukihara, and Y. Katsube. 1978. Determination of absolute configuration of epishamixanthone, a metabolite of *Aspergillus rugulosus*, by anomalous scattering of light atoms. *Bull. Chem. Soc. Jpn.* 51: 37–44.

Fukuyama K., T. Tsukihara, and Y. Katsube. 1975a. The crystal and molecular structure of the p bromobenzoate of sterigmatocystin. *Bull. Chem. Soc. Jpn.* 48: 1980–1983.

Fukuyama K., T. Tsukihara, Y. Katsube, T. Hamasaki, Y. Hatsuda, N. Tanaka, T. Ashida, and M. Kakudo. 1975b. The crystal and molecular structure of Sterigmatin, a metabolite of *Aspergillus versicolor*. *Bull. Chem. Soc. Jpn.* 48: 1639–1640.

Fukuyama K., T. Tsukihara, and S. Kishida. 1975c. The crystal and molecular structure of emericellin, a metabolite of *Aspergillus nidulans*. *Bull. Chem. Soc. Jpn.* 48: 2947–2948.

Gales L. and A.M. Damas. 2005. Xanthones—A structural perspective. *Curr. Med. Chem.* 12: 2499–2515.

Gales L., M.E.d. Sousa, M.M.M. Pinto et al. 2001. Naturally occurring 1,2,8-trimethoxyxanthone and biphenyl ether intermediates leading to 1,2-dimethoxyxanthone. *Acta Crystallogr. C* 57: 1319–1323.

Gales L., M.E. Sousa, M.M.M. Pinto et al. 2005. 3,4-Dihydroxy-9H-xanthen-9-one trihydrate. *Acta Crystallogr. E* 61: o2213–o2215.

Guerre C., U. Thull, P. Gaillard et al. 2001. Natural and synthetic xanthones as monoamine oxidase inhibitors: Biological assay and 3D-QSAR. *Helv. Chim. Acta* 84: 552–570.

Ho D.K., A.T. McKenzie, S.R. Byrn et al. 1987. O5-methyl-(±)-(2′R,3′,S)-psorospermin. *J. Org. Chem.* 52: 342–347.

Ishiguro K., S. Nagata, H. Fukumoto et al. 1993. An isopentenylated flavonol from *Hypericum japonicum*. *Phytochemistry* 32: 1583–1585.

Kabaleeswaran V., R. Malathi, and S.S. Rajan. 2003. Rotational disorder in 1,5-dihydroxy xanthone. *J. Chem. Crystallogr.* 33: 233–237.

Kato L., C.M. Alves De Oliveira, I. Vencato et al. 2005. Crystal structure of 1,7-dihydroxyxanthone from *Weddellina squamulosa* Tul. *J. Chem. Crystallogr.* 35: 23–26.

Kijjoa A., M. José, T.G. Gonzalez et al. 1998. Xanthones from *Cratoxylum maingayi*. *Phytochemistry* 49: 2159–2162.

Kogan K. and S.E. Biali. 2007. Intramolecular SNAr reactions in a large-ring Ketocalix[6]arene. *Org. Lett.* 9: 2393–2396.

Kosela S., L.H. Hu, S.C. Yip et al. 1999. Dulxanthone E: A pyranoxanthone from the leaves of *Garcinia dulcis*. *Phytochemistry* 52: 1375–1377.

Lee B.W., S.W. Gal, K.M. Park et al. 2005. Cytotoxic xanthones from *Cudrania tricuspidata*. *J. Nat. Prod.* 68: 456–458.

Leiro J.M., E. Álvarez, J.A. Arranz et al. 2003. In vitro effects of mangiferin on superoxide concentrations and expression of the inducible nitric oxide synthase, tumour necrosis factor-α and transforming growth factor-β genes. *Biochem. Pharmacol.* 65: 1361–1371.

Liangsakul J., S. Srisurichan, N. Muangsin et al. 2009. 8-[(3,3-Dimethyl-oxiran-2-yl)methoxy-meth-yl]-11-hydr-oxy-2-isopropenyl-5- methyl-12-oxo-1,2,3,12-tetra-hydro-pyrano[3,2-a]xanthen-1-yl acetate. *Acta Crystallogr. E* 65: o2558–2559.

Lin C.N., S.S. Liou, F.N. Ko et al. 1992. γ-Pyrone compounds. II: Synthesis and antiplatelet effects of tetraoxygenated xanthones. *J. Pharm. Sci.* 81: 1109–1112.

Maia F., M.d.R. Almeida, L. Gales et al. 2005. The binding of xanthone derivatives to transthyretin. *Biochem. Pharmacol.* 70: 1861–1869.

Malathi R., V. Kabaleeswaran, and S.S. Rajan. 2000. Structure of mangostin. *J. Chem. Crystallogr.* 30: 203–205.

Mondal M., V.G. Puranik, and N.P. Argade. 2006. Facile synthesis of 1,3,7-trihydroxyxanthone and its regioselective coupling reactions with prenal: Simple and efficient access to osajaxanthone and nigrolineaxanthone F. *J. Org. Chem.*71: 4992–4995.

Ojida A., H. Nonaka, Y. Miyahara et al. 2006. Bis(Dpa-ZnII) appended xanthone: Excitation ratiometric chemosensor for phosphate anions. *Angew. Chem. Int. Edn.* 45: 5518–5521.

Onuma S., K. Iijima, and I. Oonishi. 1990. Structure of xanthone. *Acta Crystallogr C* 46: 1725–1730.

Peng Z.G., J. Luo, L.H. Xia et al. 2004. CML cell line K562 cell apoptosis induced by mangiferin. *J. Exp. Hematol.* 12: 590–594.

Pinto M.M.M., M.E. Sousa, and M.S.J. Nascimento. 2005. Xanthone derivatives: New insights in biological activities. *Curr. Med. Chem.* 12: 2517–2538.

Ravikumar K. and S.S. Rajan. 1987. Structure of garcinone B. *Acta Crystallogr. C* 43: 1927–1929.

Ravikumar K., S.S. Rajan, and V.M. Padmanabhan. 1987. Structure of 5-hydroxy-8,9-dimethoxy-2,2-dimethyl-7-(3-methyl-2-butenyl)-2H,6H-pyrano[3,2-b]xanthen-6-one. *Acta Crystallogr. C* 43: 553–555.

Riscoe M., J.X. Kelly, and R. Winter. 2005. Xanthones as antimalarial agents: Discovery, mode of action, and optimization. *Curr. Med. Chem.* 12: 2539–2549.

Rodrigues T., F. Lopes, and R. Moreira. 2010. Inhibitors of the mitochondrial electron transport chain and de novo pyrimidine biosynthesis as antimalarials: The present status. *Curr. Med. Chem.* 17: 929–956.

Seo E.J., M.J. Curtis-Long, B.W. Lee et al. 2007. Xanthones from *Cudrania tricuspidata* displaying potent α-glucosidase inhibition. *Bioorg. Med. Chem. Lett.* 17: 6421–6424.

Shen R., X. Pan, H. Wang et al. 2008a. Anion recognition by a novel zinc(Π) xanthone-crown ether complex. *Inorg. Chem. Commun.* 11: 318–322.

Shen R., X. Pan, H. Wang et al. 2008b. Selective colorimetric and fluorescent detection of HSO4- with sodium(i), magnesium(ii) and aluminium(iii) xanthone-crown ether complexes. *Dalton Trans.* 3574–3581.

Shi G.F., R.H. Lu, Y.S. Yang et al. 2004a. 1-Hydroxy-2,3,4,7-tetramethoxyxanthone from *Swertia Chirayita. Acta Crystallogr. E* 60: o878–o880.

Shi G.F., R.H. Lu, Y.S. Yang et al. 2004b. Isolation and crystal structure of xanthones from *Swertia chirayita. Jiegou Huaxue* 23: 1164–1168.

Silva A.M.S. and D.C.G.A. Pinto. 2005. Structure elucidation of xanthone derivatives: Studies of nuclear magnetic resonance spectroscopy. *Curr. Med. Chem.* 12: 2481–2497.

Smith G.D. and W.L. Duax 1975. 5-Methoxy-sterigmatocystin. *Cryst. Struct. Commun.* 4: 697.

Soderholm M., U. Sonnerstam, R. Norrestam et al. 1976. Structural studies of polychlorinated hydrocarbons. II. Hexachloroxanthene and hexachloroxanthone. *Acta Crystallogr. B* 32: 3013–3018.

Sousa E., A. Paiva, N. Nazareth et al. 2009. Bromoalkoxyxanthones as promising antitumor agents: Synthesis, crystal structure and effect on human tumor cell lines. *Eur. J. Med. Chem.* 44: 3830–3835.

Sousa M.E. and M.M.M. Pinto. 2005. Synthesis of xanthones: An overview. *Curr. Med. Chem.* 12: 2447–2479.

Stout G.H., T.S. Lin, and I. Singh 1969. Xanthones of the Gentianaceae-III. The crystal and molecular structure of 2-hydroxy-1,3,4,7-tetramethoxyxanthone. *Tetrahedron* 25: 1975–1983.

Stout G.H., V.F. Stout, and M.J. Welsh 1963. Celebixanthone. A combined chemical and crystallographic structure proof. *Tetrahedron* 19: 667–676.

Toussaint J. 1956. Étude radiocristallographique de l'hétérocycle γ-pyrone I. Structure moléculaire et cristalline de la diméthyl-2-6 γ-thiopyrone. *Bull. Soc. Chim. Belg.* 65: 213–228.

Vieira L.M.M. and A. Kijjoa 2005. Naturally-occurring xanthones: Recent developments. *Curr. Med. Chem.* 12: 2413–2446.

Vijayalakshmi J., S.S. Rajan, and R. Srinivasan 1987. The structure of 1,6-dihydroxy-7,8-dimethoxyxanthone. *Acta Crystallogr. C* 43: 2108–2110.

William Lown J. and S.M. Sondhi 1985. Glycosidic coupling of regiospecifically synthesized xantho[2,3-g] tetralin aglycones to afford moderately antileukemic but redox inactive structures related to anthracyclines. *J. Org. Chem.* 50: 1413–1418.

Wu J., X. Pan, L. Yao et al. 2009. Synthesis and x-ray crystallography of diverse metal complexes derived from xanthone-crown ether. *Supramol. Chem.* 21: 707–716.

Yoshida O., N. Tanaka, T. Ashida et al. 1979. Structure of demethylsterigmatocystin. *Acta Crystallogr. B* 35: 1266–1268.

Yu P., X. Shen, C. Hu et al. 2008. 1,7-Dihydr-oxy-2,3,4-trimeth-oxy-9H-xanthen-9-one monohydrate from *Halenia elliptica*. *Acta Crystallogr. E* 64: o651–o652.

Zhang H., B. Sun, X. Zhao et al. 2007. 2,3,4-Trichloro-6-(diethylamino)-9-oxo-9H-xanthene-1-carboxylic acid. *Acta Crystallogr. E* 63: 0863–0864.

15 Gambogic Acid
*A Caged Prenylated Garcinia Xanthone Potent Anticancer Agent of Pharmaceutical Promise**

Goutam Brahmachari

CONTENTS

15.1 INTRODUCTION

Gambogic acid (GA; m.f. $C_{38}H_{44}O_8$; $[\alpha]^{20}_{D}$–714.1^0 [*c* 0.17, $CHCl_3$]; Figure 15.1), a "caged prenylated xanthone," is the principal acidic component of the pigment gamboge, the dried resin of various *Garcinia* species including *Garcinia morella* and *Garcinia hanburyi* Hook.f (family: Clusiaceae). *G. hanburyi* mainly grows in South China, Cambodia, Vietnam, and Thailand (Yang et al. 1994) and is used as a folk medicine to treat infections and tumors (Asano et al. 1996; Han et al. 2006, 2009; Reutrakul et al. 2007; Tao et al. 2009). The xanthone molecule was isolated from *G. hanburyi* for the first time in 1949 (Land and Katz 1949, Ren et al. 2011) and its planar structure was deduced through several chemical reactions and NMR spectroscopy in 1965 (Ollis et al. 1965) and also confirmed later on by X-ray diffraction analysis in the form of a pyridine salt (Weakley et al. 2001). The ^1H and Carbon-13 Nuclear Magnetic Response (^{13}CNMR) spectral data of the compound were

* This chapter is dedicated to Late Santosh K. Brahmachari—In memory to my beloved father.

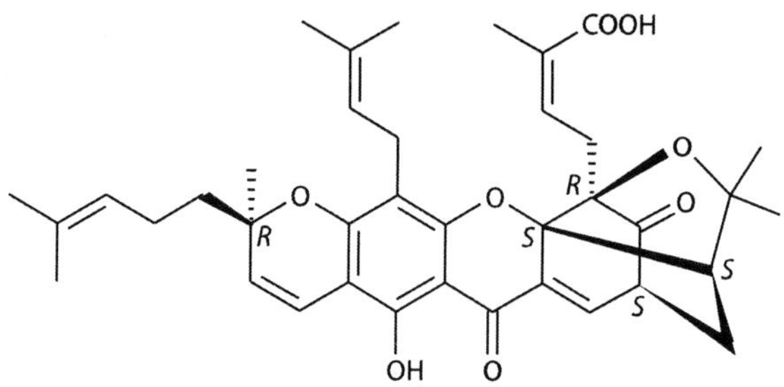

FIGURE 15.1 (–) Gambogic acid (GA; **1**).

assigned with the aid of HMBC and ROESY techniques as well (Lin et al. 1993). Partial absolute configuration was determined by a series of chemical degradations by Cardillo and Merlini (1967). However, its complete absolute configuration has recently been confirmed by Ren et al. (2011) by comparison of physical and spectroscopic data, especially experimental and calculated electronic circular dichroism, with those of (–)-morellic acid, an analog of (–)-gambogic acid.

Gambogic acid has already been established as a potent anticancer agent; the test compound has been found to exhibit promising activity against a variety of human cancer cell lines as well. Gambogic acid has already been subjected to a phase I clinical trial as an anticancer agent in the People's Republic of China, with a dose regimen developed for subsequent phase II testing (Zhou and Wang 2007). The present chapter deals with the detailed studies on anticancer activity of gambogic acid, its mode of action, semi-synthesis and structure–activity relationship (SAR), bioavailability, toxicology, pharmacokinetics, and metabolism.

15.2 ANTICANCER STUDIES WITH GAMBOGIC ACID: PHARMACEUTICAL POTENTIAL AND MODE OF ACTION

Gambogic acid (GA; **1**), a natural product isolated from the resin of *G. hurburyi* tree, bears a unique 4-oxatricyclo[4.3.1.0]decan-2-one scaffold in its molecule and has already been demonstrated as a potent antitumor agent by numerous workers; the outcome of such extensive works has evoked tremendous promise in anticancer drug research. In order to study the SAR of GA (**1**), several derivatives of this compound have also been synthesized by modifying the functional groups at different positions in its molecule and their relative efficacies have been compared; these studies are to be presented in the next section in detail. However, the mode of action of GA's anticancer activity is still of much discussion. Multiple mechanism of actions have been proposed by research groups worldwide, such as apoptosis induction (Zhang et al. 2004; Zhao et al. 2004), cell cycle regulation (Zhao et al. 2008), telomerase depression (Guo et al. 2006; Yi et al. 2008), angiogenesis inhibition (Lu et al. 2007), reactive oxygen species (ROS) generation (Nie et al. 2009), etc. In addition, Wang et al. (2009) reported a comprehensive proteome profiling that deduced stathmin-1 (STMN1) as the target of GA. Recently, it has also been reported that the xanthone molecule inhibited nuclear factor kappa-light-chain-enhancer of activated B cell (NF-κB) signaling pathway (Pandey et al. 2007; Palempalli et al. 2009) and induced apoptosis through its interaction with the transferring receptor (Kasibhatla et al. 2005).

Zhang et al. (2004) demonstrated gambogic acid (GA; **1**) as a potent apoptosis inducer with a different mechanism of action independent of cell cycle by using cell- and caspase-based high-throughput screening assays. Gambogic acid was found to have an EC_{50} of 0.78 μM in the caspase activation assay in human ductal breast epithelial tumor (T47D) cell lines. The apoptosis-inducing activity of GA was further characterized by a nuclear fragmentation assay and flow cytometry analysis in human breast tumor cells T47D. The drug candidate (**1**) was found to induce apoptosis independent of cell cycle, which is different from paclitaxel that arrests cells in the G2/M phase; it was found to induce apoptosis by an effective activation of caspases in T47D cells (Zhang et al. 2004). Zhao et al. (2004)

also observed that gambogic acid (GA) can selectively induce apoptosis and can regulate expressions of *Bax* and *Bcl-2* protein in human gastric carcinoma MGC-803 cells. The test compound potently inhibited (24, 48, 72 h) the growth of MGC-803 cells (by MTT) with the IC_{50} value of 0.96 μg/mL at 48 h. Interestingly, no influence was observed on body weight, number of white blood cells in blood or karyote in marrow of rats after GA was injected intravenously, thereby suggesting GA does not affect normal cells, but that it can induce apoptosis in tumor cells selectively. The investigators observed a huge quantity of apoptotic cells and increasing G2/M phase cells using flow cytometry, and a significant percentage of early apoptotic cells were also observed by Annexin-V/PI double staining assay. The increase of *Bax* gene and the decrease of *Bcl-2* gene expressions were detected by immunohistochemistry. Overexpression of *Bax* usually accelerates cell death (Mehta et al. 2002; Mertens et al. 2002), while overexpression of antiapoptotic proteins such as *Bcl-2* represses the death function of *Bax* (Tilli et al. 2002). Thus, the ratio of Bcl-2/Bax might be one a critical factor of a cell's threshold for undergoing apoptosis (Petersson et al. 2002). In the present study, the investigators deduced that *Bax* leads to the release of cytochrome c, which is restrained by binding to *Bcl-2*; hence the activation of cytochrome c is suppressed, and subsequently apoptosis occurs. Activation of *Bax* and suppression of *Bcl-2* (i.e., lowering the ratio of *Bcl-2/Bax*) might be one of its molecular mechanisms, which contributes to the high apoptosis rate of MGC-803 cancer cells treated with GA (Zhao et al. 2004).

Liu et al. (2005) studied the effects of GA on proliferation and apoptosis of a human gastric cancer line BGC-823 both *in vitro* and *in vivo*. GA was found to inhibit the growth of BGC-823 cells in a dose-dependent manner, and an inhibition up to 99.07% was observed at a concentration of 4.40 μM. The IC_{50} values were determined for GA on its exposure to BGC-823 cells at 24, 48, and 72 h as 1.02 ± 0.05, 1.41 ± 0.20, and 1.14 ± 0.19 μM, respectively. The investigators further demonstrated (by means of Annexin-V/PI double-staining flow cytometry assay) that the cell-death is associated with the apoptosis induced by GA, and the apoptotic population of BGC-823 cells was measured about 12.96% and 24.58%, respectively, when cells were incubated with 1.2 μM GA for 48 and 72 h. The molecular mechanism of such apoptosis was also traced to be induced by upregulating *Bax* gene and downregulating *Bcl-2* gene; hence, the anticancer effect of GA may thus be associated with its ability to regulate the expression of apoptotic-related genes (Liu et al. 2005).

Potent *in vitro* antitumor activity of GA was also reported by Chen (1980) and Dong et al. (1988). Wu et al. (2004) demonstrated promising *in vivo* and *in vitro* inhibition of the proliferation of human lung carcinoma SPC-A1 cells exerted by GA in dose-dependent and time-dependent manners; the investigators also indicated that the telomerase activity inhibition by GA was one of the possible causes of the low toxicity and high selectivity of the drug. Telomerase activity and human telomerase reverse transcriptase (hTERT) mRNA expression were both found to be decreased significantly, when cells were exposed to GA for 24, 48, and 72 h (for 24 h, $p < 0.05$, and for 48 h and 72 h, $p < 0.01$). The experimental results inferred that gambogic acid could inhibit the growth of SPC-A1 cells and its tumor xenografts, and when treated with GA for a period of time, telomerase activity and expression of hTERT mRNA in the tumor cells were both inhibited significantly. The downregulating telomerase activity of GA by modifying partly the expression of hTERT mRNA in SPC-A1 cells might thus be one possible mechanism for the inhibitory activity of the drug candidate in the cells (Wu et al. 2004). Guo et al. (2004) further demonstrated that GA displayed an inhibitory effect on the growth of transplantation tumor SMMC-7721 in nude mice and also potently inhibited the proliferation of human hepatoma SMMC-7721 cells *in vitro*, and such growth inhibitory effect might also be related to its inhibition of telomerase activity (Guo et al. 2004). The activation of human telomerase, a process regulated by the hTERT, is regarded as a crucial step during cellular immortalization and malignant transformation. Later on, the same group (Guo et al. 2006) further demonstrated that GA inhibited telomerase activity by downregulating the expression of the *hTERT* gene (a target of oncogene cellular myelocytomatosis [*c-MYC*] activity, which is a ubiquitous transcription factor involved in the control of cell proliferation and differentiation). It was shown that GA treatment of a human hepatoma cell line SMMC-7721 significantly reduces the expression of *c-MYC* in a time- and concentration-dependent manner accompanied with the downregulation of the *hTERT* transcription and the ultimate reduction in telomerase activity; treatment of SMMC-7721 cells with 1.2 μM of GA for a period of 24, 48, and 72 h was found to decrease the telomerase activity to 68.5%, 42.3%, and 12.5%, respectively (Guo et al. 2006). Yu et al. (2006) also demonstrated significant GA-induced inhibition of cell proliferation of two human gastric carcinoma cell lines, MGC-803 and SGC-7901, which eventually reduced the expression of *c-MYC* in a time- and concentration-dependent manner accompanied with the downregulation of the *hTERT* transcription and the ultimate reduction in telomerase activity. From their detailed study, Zhao et al. (2008) also demonstrated that GA repressed telomerase activity not only by repressing *hTERT* transcriptional activity via downregulation of *c-MYC* expression in BGC-823 human gastric carcinoma cells, but also by modifying phosphorylation of *hTERT* protein via the deactivation of *AKT* (also known as protein kinase B [PKB]).

Yang et al. (2007) carried out a detailed investigation on selective anticancer activity of GA by comparing different apoptotic induction of the drug on human normal embryo hepatic L02 cells and human hepatoma SMMC-7721 cells by detecting growth inhibition, observing morphological changes, and the expressions of the relative apoptotic proteins (*Bax, Bcl-2*, and caspase-3). The results indicated that GA could selectively induce apoptosis of SMMC-7721 cells, while had relatively less effect on L02 cells; this is attributed to the observation that SMMC-7721 cells had higher GA binding activity than the L02 cells; the retention time of GA in grafted tumor was found to be longer than in liver, renal, and other organs. Hence, the investigators suggested that the selective anticancer activity of GA could be due to its significant apoptotic-inducing effects as well as its higher distribution and longer retention time in tumor cells compared to the normal cells, thereby indicating GA as a promising and effective anticancer drug candidate with low toxicity to normal tissue (Yang et al. 2007).

Yu et al. (2007) offered an insight into the mechanistic pathway for GA-treated inhibition against the proliferation of human gastric carcinoma BGC-823 cells; their experimental observation led to the conclusion that the antitumor activity was associated with the decreased production of cyclin-dependent kinase 7 (CDK7) mRNA and protein, which in turn, resulted in the reduction of CDK7 kinase activity. The reduced CDK7 kinase activity is responsible for the inactivation of cell division cycle 2 (CDC2)/p34 kinase and the irreversible G2/M phase cell-cycle arrest of the cancer cells.

A potent antiangiogenic activity of GA *in vitro* and *in vivo* was reported by Lu et al. (2007) for the first time; the drug inhibits angiogenesis through suppressing vascular endothelial growth factor (VEGF)-induced tyrosine phosphorylation of KDR/Flk-1. The investigators showed that GA inhibited the VEGF-stimulated proliferation, migration, and tube formation of human umbilical vein endothelial cells (HUVECs) as well as microvessel sprouting from rat aortic rings *in vitro*. In addition, GA inhibited vessel growth in matrigel plugs and chicken chorioallantoic membrane (CAM) *in vivo* and transplanted tumor in mice. This inhibition of receptor phosphorylation was also found to be correlated with a significant decrease in VEGF-triggered phosphorylated forms of extracellular-signal-regulated kinase (ERK), AKT and p38, making GA as a structurally novel angiogenesis inhibitor and a novel antitumor agent (Lu et al. 2007).

Since *hTERT* is also regulated by NF-κB, it might be possible that *hTERT* is downregulated by the downregulation of NF-κB (Akiyama et al. 2003). In 2007, Pandey et al. carried out a thorough investigation on the effects of GA on NF-κB-mediated cellular responses and NF-κB-regulated gene products in human leukemia cancer cells. The investigators demonstrated that the drug candidate inhibits NF-κB signaling pathway and potentiates apoptosis through its interaction with the transferring receptor. It was observed that GA is capable to suppress NF-κB activation induced by various inflammatory agents and carcinogens and also induced by tumor-necrosis factor receptor-1 (TNFR1), tumor necrosis factor receptor-associated death domain (TRADD), tumor necrosis factor receptor-associated factor (TRAF2), NF-κB-inducing kinase (NIK), transforming growth factor-β-activated kinase (TAK1)/TAB1, and IκBα kinase β (IKKβ). The drug prevented NF-κB activation by other agents also by suppressing IKK activation.

The experimental results demonstrated that treatment of cells with the drug candidate was found to enhance apoptosis induced by tumor necrosis factor (TNF) and chemotherapeutic agents and also found to inhibit the expression of gene products involved in antiapoptosis (*IAP1* and *IAP2*, *Bcl-2*, *Bcl-xL*, and *TRAF1*), proliferation (cyclin D1 and *c-MYC*), invasion (*COX-2 and MMP-9*), and angiogenesis (VEGF), all of which are known to be regulated by NF-κB. The overall results provide an insight into the molecular basis for the antiproliferative and anti-inflammatory effects of GA (Pandey et al. 2007).

In a recent study carried out by Zhang et al. (2010), it has been demonstrated that GA inhibits heat shock protein 90 (Hsp90) and downregulates TNF-α/NF-κB in HeLa cells. The drug was evaluated to inhibit HeLa cells proliferation in a dose-dependent manner with an IC_{50} of 0.69 ± 0.22 μM and also to induce HeLa cells apoptosis in a concentration-dependent manner and time of exposure. It has been reported that several antiapoptotic genes are regulated by NF-κB, including *XIAP* (x-linked-inhibitor of apoptosis protein), surviving *Bcl-2*, and *Bcl-xL* (Chen et al. 2001); in the present study with HeLa cells treated with GA, NF-κB activation was found to be inhibited along with the marked decrease in *XIAP* expression levels and also in the ratio of Bcl-2/Bax triggering the cell apoptosis. Thus, the present study of Zhang et al. (2010) offers GA as a promising therapeutic agent for cancer, and also provides a useful probe to increase understanding of the biological functions of Hsp90. Potent Hsp90 inhibition efficacy of GA has also been reported by Davenport et al. (2011); they showed that the drug inhibits cell proliferation and accelerates degradation of Hsp90 client proteins in cultured cells.

Kasibhatla et al. (2005) showed that GA significantly inhibited the growth of several cancer cell lines such as T47D (breast; GI_{50} 630 nM), ZR751 (breast; GI_{50} 400 nM), HL60 (lymphocyte; GI_{50} 115 nM), Jurkat (lymphocyte; GI_{50} 168 nM), Calu1 (non-small cell lung; GI_{50} 550 nM) and MES (uterine; GI_{50} 300 nM). It is reported that transferrin receptor (TfR) becomes prominently overexpressed in different types of cancers (Szekeres et al. 2002; Ryschich et al. 2004). The present investigators unearthed a possible link between TfR and the rapid activation of GA-induced apoptosis of tumor cells; GA interferes with TfR internalization leading to the generation of rapid signaling for apoptosis. The binding site of GA on TfR is independent of the transferrin binding site, and GA binding to TfR potentially inhibits TfR internalization, thereby inducing a unique signal leading to rapid apoptosis of tumor cells. Such promising experimental results of Kasibhatla et al. (2005) offer an additional approach for targeting the TfR and its use in cancer therapy. The antiproliferative activity of GA was further confirmed by Qin et al. (2007) in a panel of human tumor cells and multidrug resistant (MDR) cells. However, it was found that GA exhibited almost equipotent cytotoxicity against T47D (high TfR cell surface expression) and HMEC (negative TfR cell surface expression); these results suggested that transferring receptor might not be the dominant target of GA in the cell lines as they tested for. The investigators demonstrated topoisomerase (Topo) IIα as a cellular target for GA to exert its antiproliferative activity. GA significantly inhibited the catalytic activity of Topo IIα by binding to the ATPase domain, thereby preventing DNA cleavage and inhibiting ATP hydrolysis. The overall experimental outcome would thus help in understanding GA as a potential antitumor candidate for clinical development as well as facilitate designing chemotherapy regimens in clinical application (Larsen et al. 2003; Qin et al. 2007).

GA is capable to restore docetaxel sensitivity in gastric cancer cell lines; Wang et al. (2008) demonstrated that GA could reverse docetaxel resistance in BGC-823/Doc cells *in vitro* in a dose-dependent manner. GA-treatment enhanced docetaxel-induced apoptosis in BGC-823/Doc cells by potentiating the ability of docetaxel to disturb microtubule function for G2/M arrest; the investigators concluded that GA exerts such activity through downregulation of survivin, a member of the inhibitor of apoptosis protein (IAP) family (Zaffaroni et al. 2002; Altieri 2003). BGC-823/Doc cell line was evaluated to be 56.3 times more resistant to docetaxel than the parental BGC-823 cell line. The present investigators showed that GA, at concentrations of 0.05, 0.1, and 0.2 μM, is able to decrease the IC_{50} of docetaxel *in vitro* in a dose-dependent manner; however, these findings demand for further exploration *in vivo* and in clinic.

Gu et al. (2008a,b) showed that GA induces tumor cell apoptosis by T lymphocyte activation in H_{22} transplanted mice using cDNA microarray technique; on the basis of their detailed experimental results the investigators suggested that GA could also exert its antitumor activity through a mechanistic path involving activation in the chain of tumor peptides-MHCII-CD4$^+$ T lymphocytes, which in turn triggers CD8$^+$ T lymphocytes to induce tumor cell apoptosis *in vivo*.

Yi et al. (2008) demonstrated that GA acts as an inhibitor of vascular endothelial growth factor receptor 2 (VEGFR2) and its downstream protein kinases, cellular-sarcoma (c-Src), focal adhesion kinase (FAK) and AKT, thereby creating an obstacle toward the signaling pathway that eventually results in the inhibition angiogenesis and tumorigenesis. However, in a xenograft prostate tumor model GA was found to be more effective (at nM concentration) in activating apoptosis and inhibiting migration and proliferation in HUVECs than that in PC3 cancer cells, which was also in agreement with previous reports that HUVECs express VEGFR2 but human prostate cancer cells (PC3) do not express it (Kitagawa et al. 2005). Hence, GA might be a potential drug candidate in cancer therapy through angioprevention with low chemotoxicity. Tang et al. (2011) reported that GA can significantly suppress the cell proliferation and also induce the apoptosis of human prostate cancer PC3 cells.

Qiang et al. (2008) also evaluated from their *in vitro* and *in vivo* studies that GA may be of potential use in treatment of glioblastoma by apoptotic induction and antiangiogenic effects; the drug exhibited potent anticancer activity for glioblastoma in rat brain microvascular endothelial cells (rBMEC) effecting apoptosis of rat C_6 glioma cells *in vitro* in a concentration-dependent manner by triggering the intrinsic mitochondrial pathway of apoptosis *in vitro* and also *in vivo* reduction of tumor volumes (on administration of GA i.v. injection once a day for 2 weeks) by antiangiogenesis and apoptotic induction of the cancer cells.

GA was reported to induce apoptosis and growth inhibition in a panel of cells expressing or not expressing p53 (Gu et al. 2008a,b; Rong et al. 2009), and in both the cases GA-mediated antitumor effects are due to the suppression of MDM2. Rong et al. (2009) observed that GA could inhibit cell growth, induce apoptosis and cell cycle arrest in G2/M phase in cancer cells with or without functional p53. It is reported that high MDM2 levels are associated with poor prognosis and resistance to chemotherapy (Lundgren et al. 1997). Again, p53 is genetically mutated in over 50% of human cancers; these novel findings offer implication for clinical application of GA with the comparative advantages over currently used anticancer drugs in the market. The same group of investigators further proved that GA triggers DNA damage signaling by activating and stabilizing p53 through an ATR/Chk1 pathway, which eventually results in the post-translational modification of p53 including phosphorylation at specific sites that are necessary for activation of p53 and its downstream target $p21^{Waf1/CIP1}$ (Rong et al. 2010).

Stathmin 1 (STMN1), an abundant cytoplasmic tubulin-binding phosphoprotein, is found to play a crucial role in tumor cell proliferation and migration (Rubin and Atweh 2004; Benlhabib and Herrera 2006; Nogales and Wang 2006). It is highly overexpressed in primary acute leukemia cells as well as high proliferative breast, prostate, or hepatocellular carcinoma (HCC) (Cassimeris 2002; Rubin and Atweh 2004). Mutation or dysfunction of STMN may cause an inability to bind to tubulin, thus leading to constant microtubule assembly and uncontrolled cell cycles, which result in continuous abnormal cell growth causing tumor formation (Jourdain et al. 1997; Baldassarre et al. 2005; Clement et al. 2005). Wang et al. (2009) demonstrated that GA significantly inhibited the proliferation of HCC cells in a dose-dependent manner; the IC_{50} for the drug at 48 h post-treatment was determined as 1.0 μM. GA could downregulate the expression of STMN1 in HCC cells, and the investigators suggested that downregulation of STMN1 might be, at least by part, the underlying therapeutic mechanism of GA (Wang et al. 2009).

Earlier works on mechanistic aspects of GA-induced antitumor activity by several workers as discussed in earlier sections led to conceive an idea that mitochondrial pathway might play an important role by which GA exerts its anticancer activity. In their detailed studies on the effect of GA on induction of reactive oxygen species (ROS) accumulation and triggering the mitochondrial signaling pathway in human hepatoma SMMC-7721 cells, Nie et al. (2009) demonstrated that GA-induced ROS accumulation and collapse of mitochondrial membrane potential in SMMC-7721 cells in a concentration-dependent manner and subsequently induced the release of cytochrome c and apoptosis-inducing factor (AIF) from mitochondria to cytosol, which inhibited ATP generation and induced apoptosis in the cells. Moreover, GA elevated the phosphorylation of c-Jun-N-terminal protein kinase (JNK) and p38, which was the downstream effect of ROS accumulation; hence accumulation of ROS might play a crucial role in GA-induced mitochondrial signaling pathway. These findings warrant future investigation to address the detailed molecular mechanism of GA interacting with mitochondria and stimulating ROS generation.

Regulation of cell–cell (intercellular) and cell–matrix adhesion is regarded as a useful strategy for cancer progression because cell adhesion plays an important role in the steps of cancer metastasis (Pignatelli and Stamp 1995; Glinsky 1998; Ellenrieder et al. 1999; Posey et al. 2001; Bogenrieder and Herlyn 2003; Dolle et al. 2006; Zhang et al. 2009). Very recently, Li et al. (2011) showed that gambogic acid strongly inhibited the adhesion of human cancer cells to fibronectin; at a dose of 1 μM, GA was found to suppress MDA-MB-231, HCT-116, and HepG2 cells adhesion to fibronectin *in vitro* by 60% ± 9% (n = 3, $P < 0.01$), 52% ± 11% (n = 3, $P < 0.01$), and 39% ± 10% (n = 3, $P < 0.05$), respectively, and most importantly, the drug (up to 1 μM) did not influence the cell viability. The investigators demonstrated that such inhibitory effect is due to suppressing integrin β1 and membrane lipid rafts, thus inhibiting integrin β1 clustering and the lipid raft-associated integrin signaling pathway—the results obviously provided a new understanding of the antimetastatic activity of GA, which would in turn help in understanding molecular mechanism of GA-induced anticancer effects.

Gambogic acid (GA) could also be used safely in combination with proteasome inhibitors wherein the drug plays a synergistic role in proteasome inhibitor-induced apoptosis and growth inhibition in cancer cells, thus representing a compelling anticancer strategy. Huang et al. (2011) demonstrated that GA activates the caspase pathway, thus sensitizing cancer cells to growth inhibition and apoptosis induction by proteasome inhibitors (viz. MG132 and MG262) *in vitro* and *in vivo*; these findings indicate GA as a potent chemosensitizer in the treatment of human cancers.

Prasad et al. (2011) evaluated that GA could inhibit growth and induce apoptosis in various human tumor cells through suppression of both inducible and constitutive of STAT3, a transcription factor, associated with proliferation, survival, and metastasis of cancer cells. The investigators observed that GA inhibited STAT3 phosphorylation at both tyrosine residue 705 and serine residue 727 via suppressing the activation of protein tyrosine kinases Janus-activated kinase 1 (JAK1) and JAK2. It was further demonstrated that GA inhibits signal transducer and activator of transcription 3 (STAT3) phosphorylation through activation of protein tyrosine phosphatase Src homology phosphatase-1 (SHP-1); in addition, the present investigators also showed that the application of the drug downregulated the expression of STAT3-regulated antiapoptotic (Bcl-2, Bcl-xL, and Mcl-1), proliferative (cyclin D1), and angiogenic (VEGF) proteins, which are usually correlated with suppression of proliferation and induction of apoptosis.

P53 mutation is one of the most frequent alterations in human malignancy, and hence, this may serve as a potential target for drug development in overcoming drug resistance in cancer therapy (Hainaut and Hollstein 2000). Wang et al. (2011) demonstrated that GA downregulates mutant p53 at post-transcription level through destabilization and degradation of the protein. From their detailed studies, it appeared that GA accelerates the degradation of mutant p53 through chaperones-assisted ubiquitin/proteasome degradation pathway in cancer cells. Mutant p53 was found to be ubiquitinated and it was the chaperones related ubiquitin ligase carboxy terminus of Hsp70-interacting protein (CHIP) rather than MDM2 involved in the degradation of mutant p53; besides, GA was also found to inhibit Hsp90/mutant p53 complex formation and to accelerate interaction of mutant p53 with heat shock protein 70 (Hsp70) (Wang et al. 2011).

Such extensive studies with a particular compound on its anticancer profile, thus, eventually mark gambogic acid as a promising "lead candidate" in the on-going anticancer drug discovery program.

15.3 SEMI-SYNTHETIC STUDIES WITH GAMBOGIC ACID: COMPARATIVE ANTICANCER POTENTIAL AND STRUCTURE–ACTIVITY RELATIONSHIP

In order to study the SAR of gambogic acid (**1**), several derivatives (**2–16**) of this compound including a number of esters and amides were synthesized by Zhang et al. (2004) and Chantarasriwong et al. (2009) as shown in Figure 15.2. All the synthetic derivatives of GA were tested by the high-throughput screening (HTS) caspase activation assay using three tumor cell lines, T47D, ZR751, and dihydrolipoamide dehydrogenase (DLD-1), and the results were compared with those of GA. Table 15.1 summarizes the EC_{50} data of these compounds.

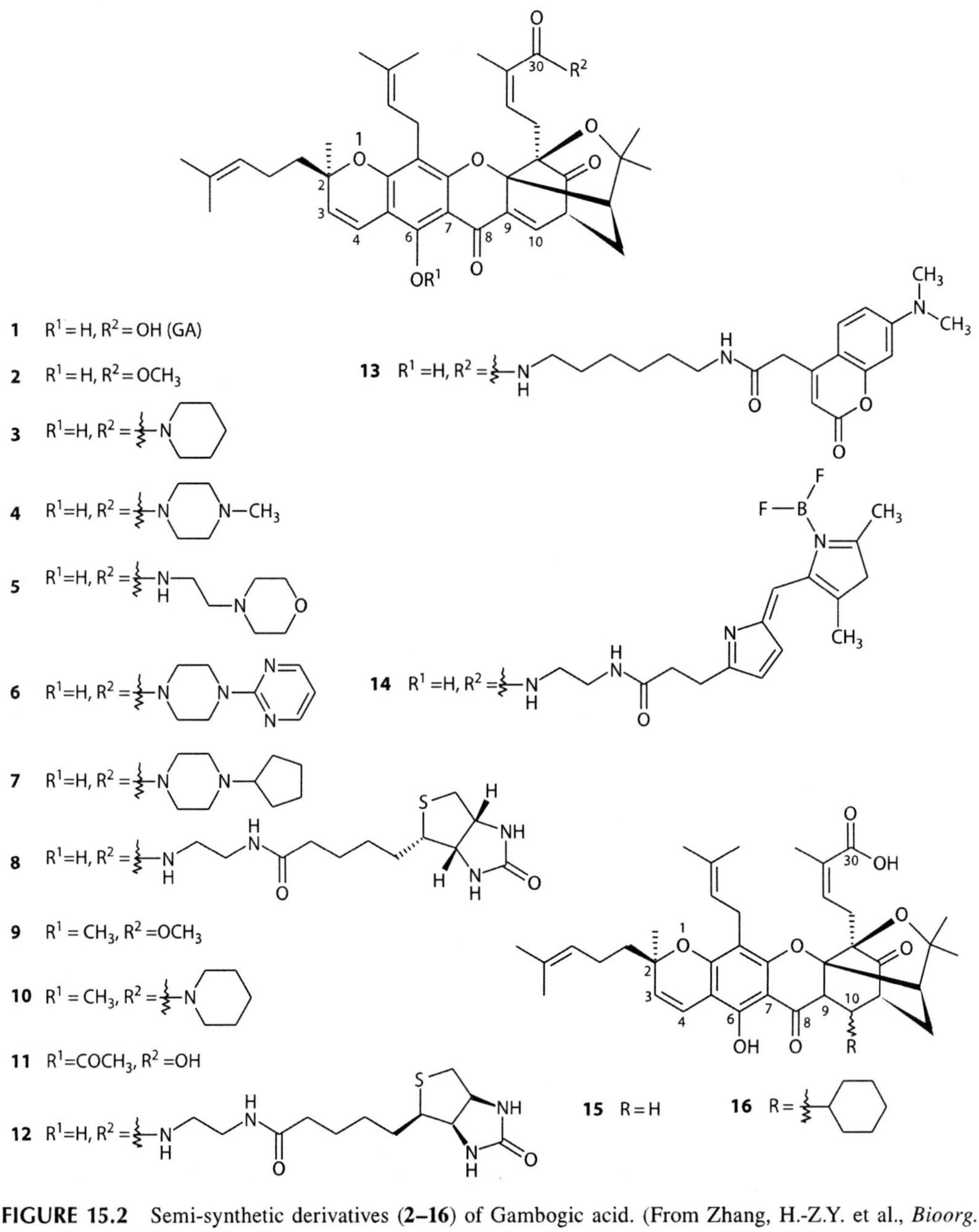

FIGURE 15.2 Semi-synthetic derivatives (**2–16**) of Gambogic acid. (From Zhang, H.-Z.Y. et al., *Bioorg. Med. Chem.*, 12, 309, 2004; Chantarasriwong, O. et al., *Org. Biomol. Chem.*, 7, 4886, 2009.)

TABLE 15.1
Comparative Antitumor Activity of Gambogic Acid (1) and Its Semi-Synthetic Derivatives (2–16)

Compounds	EC_{50} (μM)		
	T47D	**ZR751**	**DLD-1**
1 (GA)	0.78 ± 0.02	1.64 ± 0.04	0.89 ± 0.01
2	0.44 ± 0.01	1.29 ± 0.07	0.64 ± 0.02
3	0.21 ± 0.01	0.78 ± 0.01	Not determined
4	1.39 ± 0.15	3.85 ± 0.56	2.70 ± 0.01
5	0.68 ± 0.05	1.76 ± 0.15	1.39 ± 0.03
6	0.24 ± 0.01	0.78 ± 0.05	0.51 ± 0.01
7	0.38 ± 0.01	1.24 ± 0.04	0.67 ± 0.02
8	0.56 ± 0.03	1.14 ± 0.04	1.86 ± 0.05
9	0.80 ± 0.42	2.31 ± 1.00	1.37 ± 0.03
10	0.42 ± 0.09	1.01 ± 0.14	0.43 ± 0.02
11	0.51 ± 0.05	1.39 ± 0.03	1.00 ± 0.01
15	>10	>10	>10
16	>10	>10	>10

Source: Zhang, H.-Z.Y. et al., *Bioorg. Med. Chem.*, 12, 309, 2004; Chantarasriwong, O. et al., *Org. Biomol. Chem.*, 7, 4886, 2009.

Gambogic acid (**1**) was found to activate caspases and induce apoptosis with an EC_{50} value of 0.78, 1.64, and 0.89 μM in T47D, ZR751, and DLD-1 cell lines, respectively. The 30-carboxy group was found to tolerate a variety of modifications (viz. **2–8**) without much effect on apoptosis-inducing activity. Derivatives with modification of the 6-hydroxy group by either methylation or acylation produced compounds **9**, **10**, and **11**; these compounds were found to have similar activity as the corresponding 6-hydroxy compound. However, the experimental results demonstrated that the 9,10 carbon–carbon double bond of the α,β-unsaturated ketone moiety is important for the biological activity of GA. A similar observation was also reported by Chantarasriwong et al. (2009); from their semi-synthetic studies it was again clarified that the carboxylic acid of GA can be functionalized without loss of its activity. The amide derivatives (**12–14**; Figure 15.2) of GA showed similar growth inhibitory activity against a multidrug-resistant promyelocytic leukemia cell line, HL-60, as that of GA with respective IC_{50} values of 1.1, 0.3, 0.6, and 0.5 μM. The overall results from the SAR studies of gambogic acid can be summarized as follows:

1. The 6-hydroxy group within the molecule is not important for the apoptosis-inducing activity of gambogic acid; however, it is involved in a strong intra-molecular hydrogen bonding with the C-8 carbonyl group.
2. The 30-carboxyl group was found to tolerate many modifications (esters, amides) and even large modifications are also well tolerated, thereby indicating that the area around the carboxy moiety is not probably involved in important interactions of gambogic acid with biological targets. It is evident from the crystal structure (Weakley et al. 2001) of gambogic acid that the xanthone ring structure is planar with two different faces, the top and bottom. The two prenyl chains and the polycyclic ring on top consist of the "hydrophobic face," while the carboxylic acid and the carbonyl in the polycyclic ring on the bottom consist of the "hydrophilic face." From SAR studies, it can be inferred that the "hydrophilic face" is less important for biological activity of GA than the "hydrophobic face."

3. The double bond between C-9 and C-10 of the α,β-unsaturated ketone moiety is important for the biological activity of gambogic acid. Semi-synthetic derivatives (**15** and **16**) of GA without the 9,10 carbon–carbon double bond were found to be inactive in the caspase activation assay in all three cancer cell lines tested at concentrations up to 10 μM. These compounds are more than 10-fold less active than GA in T47D cells, indicating that the 9,10-double bond in the α,β-unsaturated ketone is critical for activity. The investigators suggested that gambogic acid–induced apoptosis proceeds in two steps: first, gambogic acid binds to its molecular target(s), and thereafter a nucleophile presented in the target attacks the carbon–carbon double bond in the α,β-unsaturated ketone. Such Michael addition results in the covalent attachment of GA to its target, which then activates the apoptosis signal leading to the activation of caspase cascade and cell death (Zhang et al. 2004).

From these results, putative pharmacophoric data for cytotoxic activity and apoptosis induction are summarized in Figure 15.3.

That the double bond between C-9 and C-10 of the α,β-unsaturated ketone moiety is important for the biological activity of gambogic acid also received support from the works of Han and his group (2005). It was observed that GA is quite stable when dissolved in acetone, acetonitrile, and chloroform, even when acids were added. However, the investigators demonstrated the formation of a new derivative (~7% yield) on storing GA in methanolic solution for a week at room temperature; addition of alkalis enhanced the increase in the rate of this chemical transformation. The new derivative was characterized as gambogoic acid (GOA; **17**; Figure 15.4) proposed to be the product of nucleophilic addition of methanol to the olefinic bond at C-10 of GA. Furthermore, when these two compounds were tested for their cytotoxicity using five tumor cell lines including Jurkat, K562, HL60, MCF7, and MDA-MB-468 cell lines, GOA (**17**) showed significantly weaker inhibitory effects than GA. The respective IC_{50} (μg/mL) values were determined as 0.32 ± 0.03, 0.28 ± 0.04, 0.17 ± 0.03, 0.24 ± 0.02, and 0.33 ± 0.05 for GA, and 0.71 ± 0.03, 0.65 ± 0.07, 0.83 ± 0.1, 0.47 ± 0.02, and 0.54 ± 0.02 for GOA. It was therefore deduced that the α,β-unsaturated carbonyl

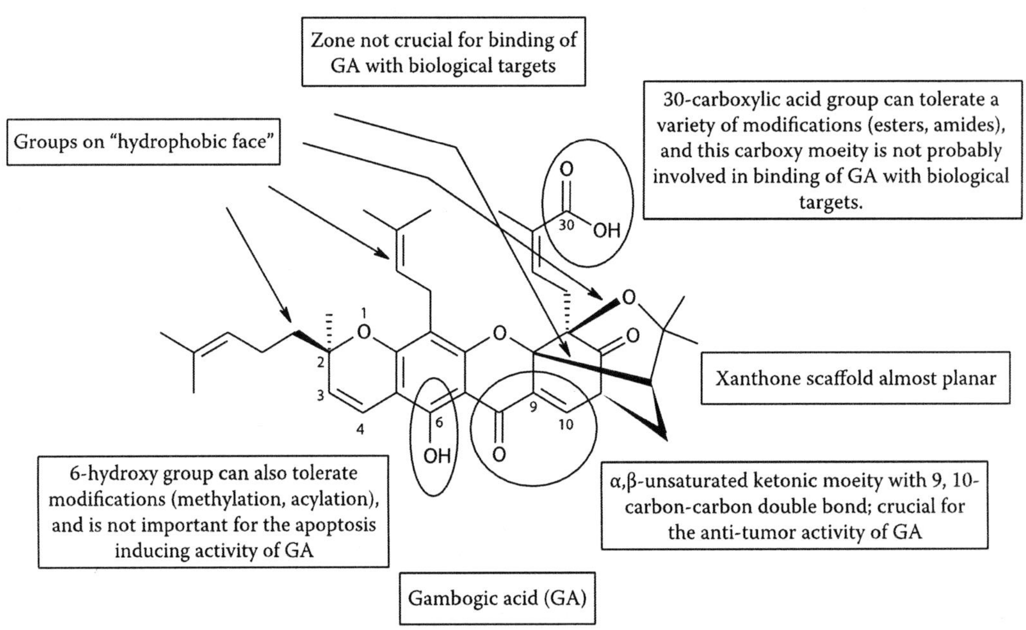

FIGURE 15.3 Putative pharmacophoric data for cytotoxic and apoptosis induction activities for derivatives of gambogic acid (**1**).

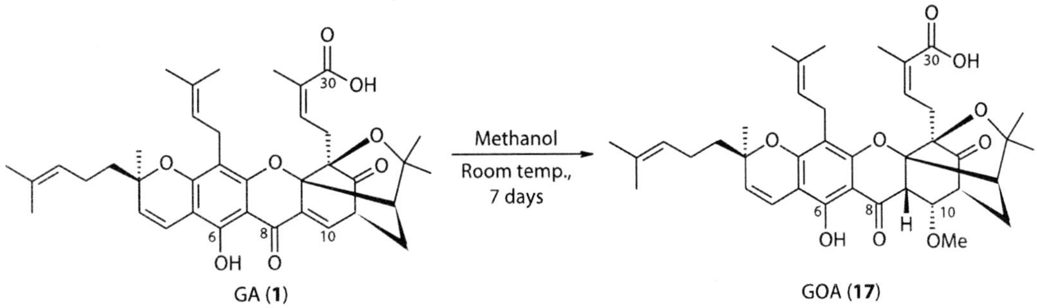

FIGURE 15.4 Conversion of gambogic acid (**1**) to gambogic acid (**17**). (From Han, Q.-B. et al., *Biol. Pharm. Bull.*, 28, 2335, 2005.)

moiety at C-10 contributed to the cytotoxicity of GA (Han et al. 2005). Similar conclusion for the structural requirement of GA to exhibit antitumor activity was also drawn by Kuemmerle et al. (2008) from their comparative antitumor studies of GA with some synthetic caged xanthone-type entities.

GA3 (**19**), the semi-synthetic derivative of GA (Figure 15.5), exhibited better water solubility with effective antitumor potential against diversified human cancer cell lines with a mean IC_{50} value of 2.15 μM; the derivative was also found to be effective against multidrug resistant (MDR) cells, with an average resistance factor much lower than that of the reference drug, doxorubicin (Xie et al. 2009).

Wang et al. (2010) and He et al. (2012) prepared a number of semi-synthetic derivatives (**20–34**) (Figure 15.6) of gambogic acid and compared relative anticancer activities of most of them. Table 15.2 summarizes all these experimental data that also offer an essence of comparative SARs among the semi-synthetic derivatives and the parent compound, gambogic acid.

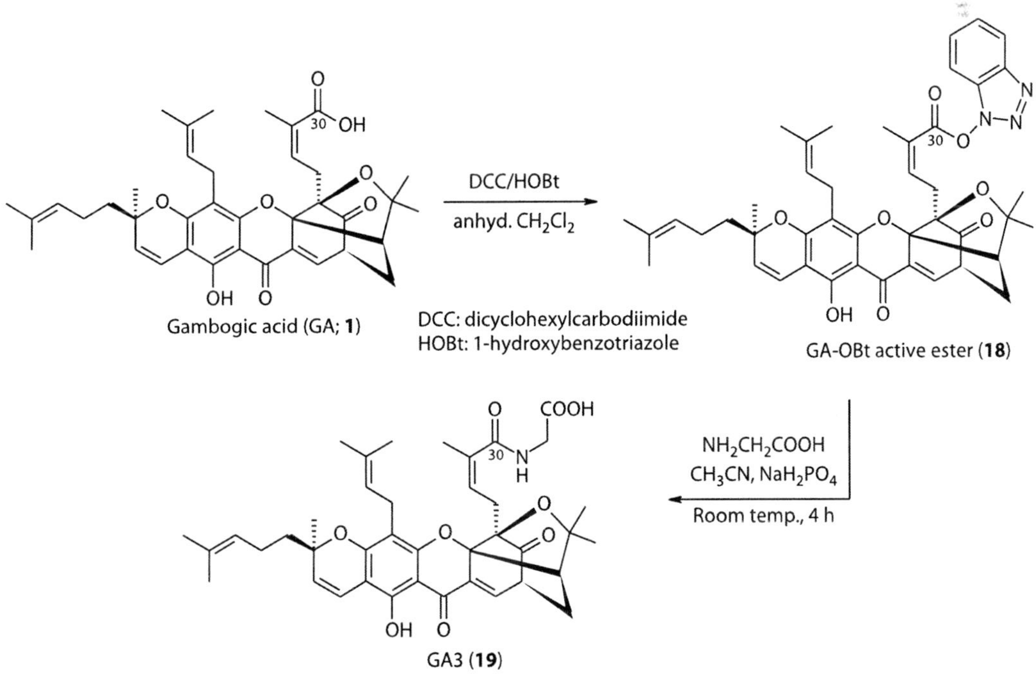

FIGURE 15.5 Conversion of gambogic acid (**1**) to GA3 (**19**). (From Xie, H. et al., *Acta Pharmacol. Sin.*, 30, 346, 2009.)

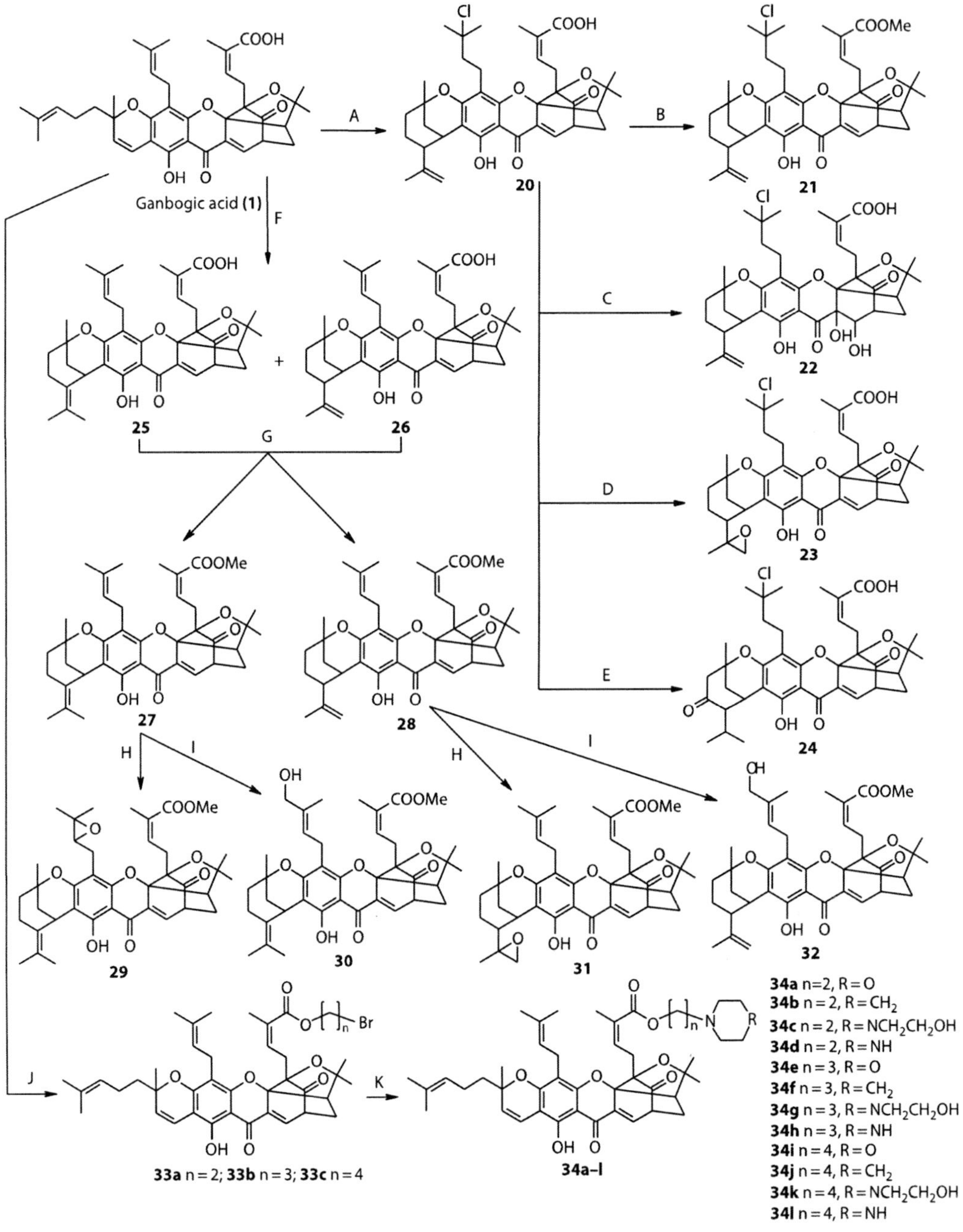

Reagents and conditions: A: HOAc, HCl(q), 100°C, 1 h (90%); B: K_2CO_3, MeI, $(Me)_2CO$, RT, 0.5 h (25%); C: $KMnO_4$, EtOH, RT, 6 h (30%); D: *m*-CPBA, CH_2Cl_2, RT, 7.5 h (40%); E: SeO_2, *t*-BuOOH, CH_2Cl_2, RT, 11 h (25%); F: HAc-CH_3SO_3H, 70°C, 30 min (6+7, 64%): G: K_2CO_3, MeI, $(Me)_2CO$, RT, 1 h (8, 40% and 9, 48%); H: *m*-CPBA, CH_2Cl_2, RT, 45 min (10, 73% and 11, 80%); I: SeO_2, *t*-BuOOH , CH_2Cl_2, RT, 5 h (12, 39% and 13, 39%); J: Br-$(CH_2)_n$-Br (n = 2–4), DMF, K_2CO_3, RT, 2 h; K: Piperazidine, *N*-hydroxyethyl piperazidine, Piperidine or morpholine, K_2CO_3, CH_3CN, RT, 10–24 h.

FIGURE 15.6 Semi-synthetic derivatives (**20–34**) of gambogic acid (**1**). (From Wang, J. et al., *Eur. J. Med. Chem.*, 45, 4343, 2010; He, L. et al., *Bioorg. Med. Chem. Lett.*, 22, 289, 2012.)

TABLE 15.2
Comparative Anticancer Activities of Gambogic Acid and Its Semi-Synthetic Derivatives against the Cancer Cell Lines Tested

Compounds	IC$_{50}$ Values (μM)							
	A549	BGC-823	HT-29	Bel-7402	SMMC-7721	Bel-7404	QGY-7701	HepG2
1(GA)	3.65 ± 0.97	2.83 ± 0.71	3.52 ± 0.82	2.08 ± 0.45	1.13	4.70	0.24	1.17
20	1.24 ± 0.31	4.0 ± 1.12	1.64 ± 0.41	18.1 ± 4.6	—	—	—	—
21	4.63 ± 1.21	1.98 ± 0.42	4.55 ± 1.15	2.63 ± 0.65	—	—	—	—
22	>100	>100	>100	>100	—	—	—	—
23	**3.18 ± 0.70**	**0.28 ± 0.05**	**1.33 ± 0.21**	**0.44 ± 0.09**	—	—	—	—
24	27.60 ± 8.02	2.93 ± 0.75	5.60 ± 1.35	4.12 ± 1.06	—	—	—	—
26	1.12 ± 0.29	0.46 ± 0.12	1.34 ± 0.30	6.48 ± 1.73	—	—	—	—
27	5.55 ± 0.97	6.62 ± 1.58	0.54 ± 0.15	7.86 ± 1.60	—	—	—	—
28	1.79 ± 0.43	2.61 ± 0.68	0.81 ± 0.23	4.56 ± 1.14	—	—	—	—
29	2.34 ± 0.45	2.53 ± 0.48	0.12 ± 0.02	8.10 ± 1.85	—	—	—	—
30	1.56 ± 0.34	1.42 ± 0.34	0.14 ± 0.03	11.0 ± 2.67	—	—	—	—
31	2.42 ± 0.48	0.92 ± 0.20	0.49 ± 0.09	4.38 ± 0.86	—	—	—	—
32	0.98 ± 0.22	0.62 ± 0.17	0.24 ± 0.05	13.0 ± 3.23	—	—	—	—
34a	—	—	—	**0.59**	**1.59**	**1.99**	**0.41**	**0.94**
34b	—	—	—	1.72	6.52	7.23	0.70	3.18
34c	—	—	—	6.36	8.57	10.24	1.86	9.90
34d	—	—	—	19.71	28.40	31.13	5.43	20.07
34e	—	—	—	**0.045**	**0.73**	**1.25**	**0.12**	**0.067**
34f	—	—	—	**0.086**	**1.06**	**1.42**	**0.43**	**0.15**
34g	—	—	—	3.12	6.61	8.08	0.96	3.25
34h	—	—	—	18.06	26.21	27.62	4.38	19.06
34i	—	—	—	2.13	1.54	2.11	0.53	1.51
34j	—	—	—	3.14	4.53	6.46	4.36	5.59
34k	—	—	—	1.102	3.52	9.65	2.56	1.16
34l	—	—	—	16.67	23.46	28.62	4.12	24.12

Source: Wang, J. et al., *Eur. J. Med. Chem.*, 45, 4343, 2010; He, L. et al., *Bioorg. Med. Chem. Lett.*, 22, 289, 2012.

Note: Compounds **23**, **34a**, **e**, **f** showed better anticancer profiles compared to gambogic acid (GA) with a hope to become potential leads in drug discovery program.

15.4 GAMBOGIC ACID: STUDIES WITH DRUG-DELIVERY SYSTEMS

Although gambogic acid has been established to possess potent anticancer efficacy, the major problems for clinical applications of this active chemical agent are developed due to its poor aqueous solubility (less than 0.5 μg/mL), rapid plasma clearance, and wide distribution *in vivo*, which would cause low bioavailability of the drug (Liu et al. 2006; Hao et al. 2007a,b). To improve the water solubility, certain solubilizers such as L-arginine or Cremophor EL were employed earlier (Dai 2003; You et al. 2003); however, these agents may cause a series of side effects such as hypersensitivity reactions, nephrotoxicity, neurotoxicity, and cardiotoxicity. Besides, rapid plasma elimination of the active drug molecule cannot be avoided by these formulations (Han et al. 2001; Ding et al. 2012). Afterward an alternative way of polymeric carrier to GA was constructed by amphipathic chitosan derivatives (*N*-octyl-*O*-sulfate chitosan, NOSC) wherein the drug molecule became encapsulated in micelle—compared with GA-Larginine formulation, NOSC-encapsulated drug showed higher drug-loading rate (29.8% ± 0.17%) and improved entrapment efficiency (63.8% ± 0.52%) (Zhu et al. 2008; Qu et al. 2009). Very recently, more efficacious polyethylene glycol (PEG)-prodrug and magnetic nanoparticle-embedded prodrug of gambogic acid have been reported; these formulations are described in brief.

15.4.1 Polyethylene Glycol Prodrug of Gambogic Acid: Efficacy and Bioavailability

To attain better bioavailability of gambogic acid, Ding et al. (2012) recently prepared thirty poly(ethylene glycol)-amino acid (or dipeptide)-gambogic acid (PEG-spacer-GA) conjugates using different amino acid and dipeptide as spacers, and also studied their pharmacokinetics, biodistribution, bioavailability, and cell cytotoxicity. The PEG-GA conjugates were found to show satisfactory water solubility ($1.2 \times 10^3 - 4.5 \times 10^5$ times of GA solubility) compared with gambogic acid. Based on their experimental results *in vitro*, the molecular weight of polymeric carrier and the choice of spacers were found to play determining role in the solubility, drug content, and drug release behavior of polymeric prodrugs. The investigators demonstrated from their *in vivo* studies that employment of the polymeric conjugation strategy remarkably improves circulatory retention time and bioavailability, as well as reduces peripheral toxicity in comprising with GA and its Cremophor EL formulation (Ding et al. 2012). Besides, the liver target character of PEG-GA conjugates made them potential prodrugs for liver cancer treatment as well. Hence, the polymeric conjugation method, by rational design and component selection, would solve many problems of insoluble drug molecules, and at the same time achieve excellent properties of drug delivery systems.

15.4.2 Gambogic Acid-Loaded Magnetic Fe_3O_4 Nanoparticles (GA-MNP-Fe_3O_4): Efficacy and Bioavailability

E26 transformation-specific sequence-1 (ETS1), a proto-oncoprotein transcription factor, plays important roles in both carcinogenesis and the progression of a wide range of malignancies. Wang et al. (2012) evaluated the efficacy of a drug delivery system for gambogic acid through GA-loaded magnetic Fe_3O_4 nanoparticles (GA-MNP-Fe_3O_4) on the suppression of ETS1-mediated cell proliferation and migration in Panc-1 pancreatic cancer cells; it was observed that treatment with GA-MNP-Fe_3O_4 suppressed cancer cell proliferation as well as prevented cells from migrating effectively; Panc-1 pancreatic cancer cells on treatment with the drug system showed significantly decreased expression of ETS1 and also its downstream target genes for cyclin D1, urokinase-type plasminogen activator (u-PA), and VEGF.

Anticancer efficacy of GA-MNP-Fe_3O_4 was found to be enhanced in comparison to GA alone; the *in vitro* IC_{50} value of free GA for the cancer cells was 1.29 μM, whereas the value for GA-MNP-Fe_3O_4 was determined as 0.52 μM. Hence, the lower IC_{50} of GA-MNP-Fe_3O_4 for the drug delivery system could improve the therapeutic efficacy without high usage of GA to inhibit cancer cell proliferation. The increased antiproliferative effect may be attributed to improved GA cellular uptake by the GA-MNP-Fe_3O_4 drug delivery system, which increases the water solubility of GA through the endocytosis pathway, thereby inducing the release of GA from the MNP-Fe_3O_4 in cancer cells to promote efficient cell killing (Yoo et al. 2000; Bareford and Swaan 2007; Wang et al. 2012). Application of GA-MNP-Fe_3O_4 nanoparticles targeting ETS1 might be a promising strategy for better pancreatic cancer care. The same group of investigators (Wang et al. 2011) also demonstrated earlier that this delivery system of GA enhances its chemotherapeutic efficiency against Capan-1 pancreatic cancer cells by inducing apoptosis, and the synergistic effect may be due to regulation of various antiapoptotic and proapoptotic gene products, including Bax, Bcl-2, caspase-9, and caspase-3. Synergistic effect of magnetic nanoparticles of Fe_3O_4 with GA was first observed by Chen et al. (2009); it was found that MNPs-Fe_3O_4 can promote apoptosis induction of GA *in vitro* in K562 leukemia cells. From their detailed experimental results, the investigators suggested that the synergistic effect of metal nanoparticle–GA composite on apoptosis induction is due to the regulation of various proliferative and antiapoptotic gene products including *caspase-3, Bax, Bcl-2,* NF-κB, and survivin. Thus, a combination of MNPs-Fe_3O_4 and GA may find an efficient and less toxic method in cancer therapy.

15.5 GAMBOGIC ACID: PHARMACOKINETICS, TOXICOLOGY, AND METABOLISM

15.5.1 Pharmacokinetics

Hao et al. (2007b) carried out pharmacokinetic evaluation of gambogic acid in dog plasma by using a rapid high-performance liquid chromatography (HPLC) method; mean plasma concentration (AUC)–time profile of GA after intravenous (i.v.) administration of 0.5, 1, and 2 mg/kg GA to 12 dogs was determined. The elimination half-life ($t_{1/2}$) values were estimated to be 57.9, 59.2, and 60.9 min and the mean AUC_t values were 49.5, 116.6, and 250.9 μg min/mL, respectively. The AUC was found to increase with increasing doses for i.v. administration; the pharmacokinetics of GA in dog was shown to be linear for the dose range studied. In another study, Hao et al. (2007a) evaluated pharmacokinetics, tissue distribution and excretion of GA on its intravenous (i.v.) bolus administration in rats at the doses of 1, 2, and 4 mg/kg. The elimination half-life ($t_{1/2}$) values for GA were determined as 14.9, 15.7, and 16.1 min, while the mean area under concentration–time curve (AUC_t) values were 54.2, 96.1, and 182.4 μg min/mL, respectively. GA was found to reach its maximal concentration in all tissues at 5 min post-dose; however, no GA was detected in urine after i.v. administration. GA had a limited tissue distribution, with the highest concentrations being found in the liver; from their detailed study, the present investigators suggested that GA is rapidly eliminated from the blood and transferred to the tissues. It was also observed that most of the drug appeared to be excreted into the bile within 16 h of i.v. administration (Hao et al. 2007a). In their earlier study, Hao et al. (2005) also showed that GA is rapidly eliminated from the rat plasma, extensively distributed and metabolized in rats after i.v. administration, and the drug is mainly excreted in the bile in both parent and metabolite forms.

15.5.2 Toxicology

Zhao et al. (2010) studied general pharmacological toxicity of gambogic acid (GA) on the dog cardiovascular and respiratory system and the mouse central nervous system (CNS); experimental observations revealed that administration of the drug did not cause any toxic symptoms on blood pressure (mean arterial pressure), heart rate, and respiratory frequency. However, slight adverse effects on the mouse CNS were noticed on administration of a high dose of GA; evidence of maternal and developmental toxicity was also observed in a dose-dependent manner. The maternal body-weight gain, as well as the birth weights and live birth index, were decreased significantly in the treatment groups. The investigators reported inhibitory effects of GA on fetal skeletal development, although no obvious effects of the drug on external alterations and visceral alterations were observed (Zhao et al. 2010). From the detailed experimental results of Qi et al. (2008) carried out with rats, it was found that the toxicity targets of GA in the experimental animals are the kidney and liver; it was demonstrated that rats treated with high dose (120 mg/kg) of GA for a long time can lead to the damage on the kidney and liver. A safe and non-toxic dosage of GA administered via intragastric administration at a frequency of one treatment every two days for a total of 13 weeks is 60 mg/kg. It was also estimated that this dosage is approximately 18.0 (body weight) or 9.6 (body surface area) times higher than that of the dose (200 mg/60 kg, every other day) used for human trials (Qi et al. 2008).

15.5.3 Metabolism

Liu et al. (2006) studied the metabolism of gambogic acid and the effects of selective cytochrome P-450 (CYP450) inhibitors on the metabolism of GA in rat liver microsomes *in vitro*. GA was determined to be rapidly metabolized in rat liver microsomes. Two phase I metabolites of GA such as 27-hydroxygambogic acid (M_1; **35**) and 27,28-epoxygambogic acid (M_2; **36**), tentatively presumed to be the hydration metabolite and epoxide metabolite of GA, respectively, were identified in rat liver microsomes (Figure 15.7). The investigators suggested that M_1 (**35**) is crucial for the elimination of GA, and cytochrome P-450 1A2 is the major rat CYP involved in the metabolism of GA (Liu et al. 2006).

However, Feng et al. (2007) suggested from their detailed investigation on the metabolites of GA in rat bile *in vivo* that these two Phase I metabolites actually are 10-hydroxygambogic acid (MT1; **37**) and

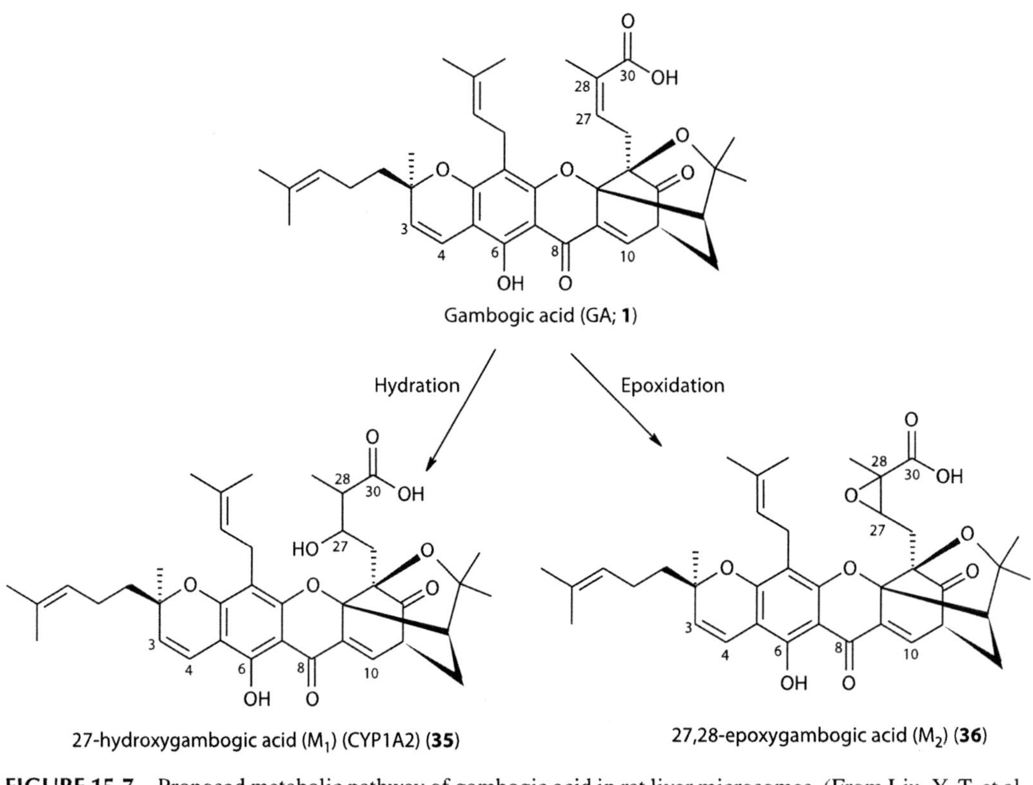

FIGURE 15.7 Proposed metabolic pathway of gambogic acid in rat liver microsomes. (From Liu, Y.-T. et al., *Acta Pharmacol. Sin.*, 27, 1253, 2006.)

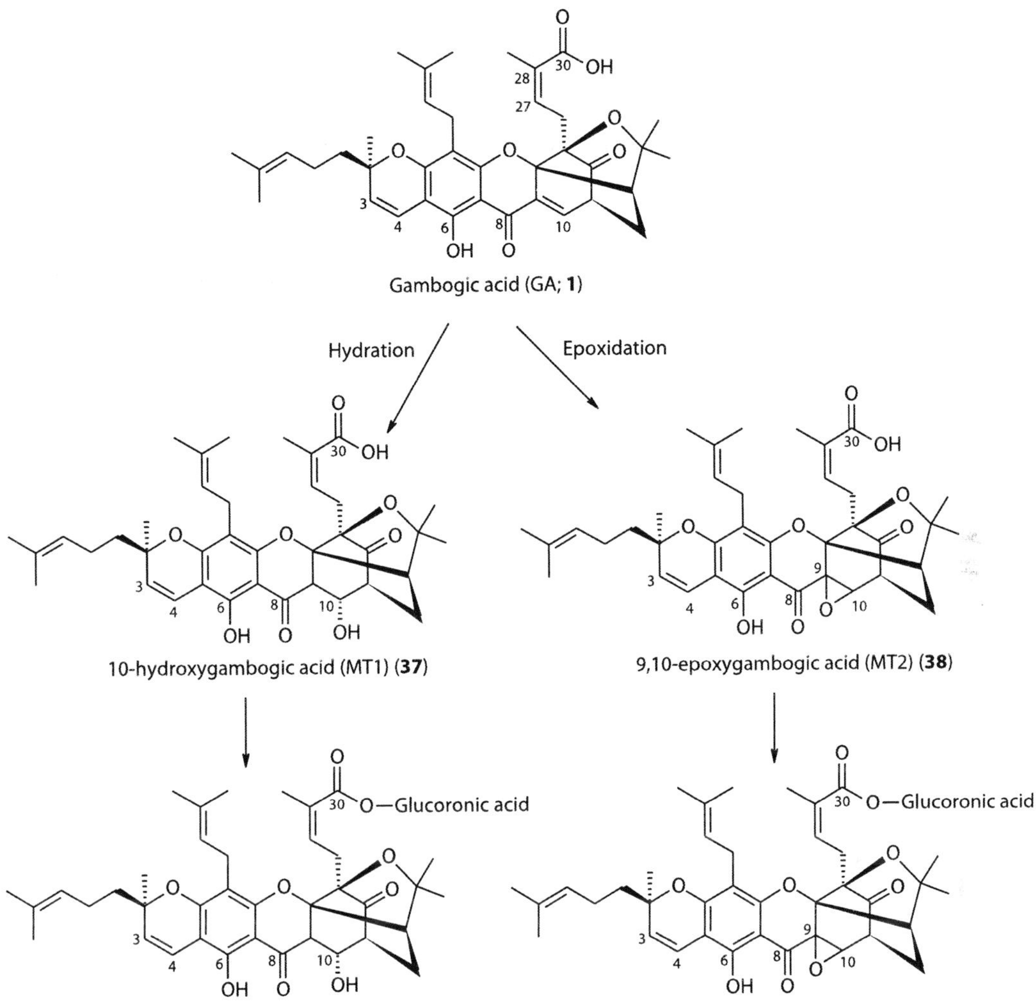

FIGURE 15.8 Proposed metabolic pathway of gambogic acid in rat bile *in vivo*. (From Feng, F. et al., *J. Chromatogr. B*, 860, 218, 2007.)

9,10-epoxygambogic acid (MT2; **38**), and two other phase II metabolites are their corresponding glucuronide conjugates, 10-hydroxylgambogic acid-30-*O*-glucuronide (MT3; **39**) and 9,10-epoxygambogic acid-30-*O*-glucuronide (MT4; **40**) (Figure 15.8).

Along with 10-hydroxygambogic acid (**37**) and 9, 10-epoxy-gambogic acid (**38**), one more metabolite, 3,4-dihydrogambogic acid was also identified in rat bile by Zhang et al. (2009). Yang et al. (2010) identified 10-hydroxygambogic acid (**37**) as the major circulating metabolite of GA in human. Very recently, Yang et al. (2011) concluded that although GA is metabolized by a variety of routes including monooxidation, hydration, epoxidation, glutathionylation, glucuronidation, and glucosidation in the liver of rats, a large amount of GA is excreted into the intestinal tract through bile as the unchanged form, and hence bile excretion process is primarily responsible for removing GA from the systemic circulation. The investigators also proposed for a rare intestinal metabolic pathway for the final disposition of GA in rats, which mainly involves Michael addition reaction of sulfite ion in the intestinal contents to the metabolic soft spot, 9,10-carbon–carbon bond of α,β-unsaturated ketone, of gambogic acid from biliary excretion (Yang et al. 2011).

15.6 CONCLUDING REMARKS

This chapter overviews promising anticancer potential of gambogic acid (GA), a "caged prenylated xanthone"—the major chemical constituent of gamboge, the dried resin of *Garcinia* species (Clusiaceae). Besides detailed anticancer properties, the mode of action, semi-synthesis and SAR, pharmacokinetics, toxicology, metabolism, and bioavailability of this drug molecule have also been discussed herein. The potential role of gambogic acid in exhibiting anticancer efficacies has created a stir among the scientific community at large to undertake extensive research for exploring the possibility of its prospective use as a "lead molecule" in the on-going drug discovery process against cancerous diseases. Semi-synthetic derivatives of the compound also showed better activity profiles and bioavailability in certain situations. It is expected that interest in this molecule grows even more, and further research on this molecule is strongly recommended to assess its potential to become a useful anticancer drug in near future.

ABBREVIATIONS

A549	adenocarcinomic human alveolar basal epithelial cells
AIF	apoptosis-inducing factor
ATP	adenosine triphosphate
Bcl-2	B cell lymphoma-2
Bcl-xL	B cell lymphoma-xL
BGC-823	gastric carcinoma cell line
CAM	chicken chorioallantoic membrane
CDC2	cell division cycle 2
CDK7	cyclin-dependent kinase 7
CHIP	carboxy terminus of Hsp70-interacting protein
c-MYC	cellular myelocytomatosis
CNS	central nervous system
COX-2	cyclooxygenase-2
c-Src	cellular-sarcoma
CYP450	cytochrome P-450
DLD-1	dihydrolipoamide dehydrogenase
DNA	deoxyribonucleic acid
EC_{50}	half maximal effective concentration
ERK	extracellular-signal-regulated kinase
ETS1	E26 transformation-specific sequence-1
FAK	focal adhesion kinase
FLK-1	fetal liver kinase 1
GA	gambogic acid
GA-MNP-Fe_3O_4	GA-loaded magnetic Fe_3O_4 nanoparticles
HCC	hepatocellular carcinoma
HL60	human promyelocytic leukemia cells
HMBC	heteronuclear multiple bond correlation
HMEC	human mammary epithelial cells
HPLC	high-performance liquid chromatography
Hsp70	heat shock protein 70
Hsp90	heat shock protein 90
hTERT	human telomerase reverse transcriptase
HTS	high-throughput screening
HUVECs	human umbilical vein endothelial cells
IAP	inhibitor of apoptosis proteins

IKKβ	IκBα kinase β
i.v.	intravenous
JAK	Janus-activated kinase
JNK	Jun-*N*-terminal protein kinase
KDR	kinase insert domain receptor
MCF7	Michigan cancer foundation-7
Mcl-1	myeloid cell leukemia sequence 1
MDR	multidrug resistant
MES	mouse embryonic stem
MMP-9	matrix metalloproteinase-9
NF-κB	nuclear factor kappa-light-chain-enhancer of activated B cell
NIK	NF-κB-inducing kinase
NOSC	*N*-octyl-*O*-sulfate chitosan
PC3	prostate cancer-3
PEG	polyethylene glycol
rBMEC	rat brain microvascular endothelial cells
ROESY	rotating-frame overhauser effect spectroscopy
ROS	reactive oxygen species
SAR	structure-activity relationship
SHP-1	Src homology phosphatase-1
SPC-A1	secretory pathway $Ca2^{+}$-ATPase isoform 1
STAT3	signal transducer and activator of transcription 3
STMN1	stathmin-1
T47D	human ductal breast epithelial tumor cell line
TAB1	TAK1, Binding protein
TAK1	transforming growth factor-B-activated kinase
TfR	transferrin receptor
TNF	tumor-necrosis factor
TNFR1	tumor-necrosis factor receptor-1
Topo	topoisomerase
TRADD	tumor necrosis factor receptor-associated death domain
TRAF2	tumor necrosis associated factor-2
u-PA	urokinase-type plasminogen activator
VEGF	vascular endothelial growth factor
VEGFR2	vascular endothelial growth factor receptor-2
XIAP	X-linked-inhibitor of apoptosis protein

ACKNOWLEDGMENT

Fruitful and valuable works with gambogic acid carried out by numerous researchers worldwide, upon which the present chapter is based, are being deeply acknowledged herein.

REFERENCES

Akiyama, M., Hideshima, T., Hayashi, T. et al. 2003. Nuclear factor-kappaB p65 mediates tumor necrosis factor alpha-induced nuclear translocation of telomerase reverse transcriptase protein. *Cancer Res.* 63: 18–21.

Altieri, D.C. 2003. Survivin, versatile modulation of cell division and apoptosis in cancer. *Oncogene* 22: 8581–8589.

Asano, J., Chiba, K., Tada, M., and Yoshii, T. 1996. Cytotoxic xanthones from *Garcinia hanburyi. Phytochemistry* 41: 815–820.

Baldassarre, G., Belletti, B., Nicoloso, M. S. et al. 2005. p27(Kip1)-stathmin interaction influences sarcomacell migration and invasion. *Cancer Cell* 7: 51–63.

Bareford, L. A. and Swaan, P. W. 2007. Endocytic mechanisms for targeted drug delivery. *Adv. Drug Deliv. Rev.* 59: 748–758.

Benlhabib, H. and Herrera, J. E. 2006. Expression of the Op18 gene is maintained by the CCAAT-binding transcription factor NF-Y. *Gene* 377: 177–185.

Bogenrieder, T. and Herlyn, M. 2003. Axis of evil: Molecular mechanisms of cancer metastasis. *Oncogene* 22: 6524–6536.

Cardillo, G. and Merlini, L. 1967. Absolute configuration of carbon 2 in the chromene ring of gambogic acid. *Tetrahedron Lett.* 27: 2529–2530.

Cassimeris, L. 2002. The oncoprotein 18/stathmin family of microtubule destabilizers. *Curr. Opin. Cell Biol.* 14: 18–24.

Chantarasriwong, O., Cho, W. C., Batova, A. et al. 2009. Evaluation of the pharmacophoric motif of the caged *Garcinia* xanthones. *Org. Biomol. Chem.* 7: 4886–4894.

Chen, B. R. 1980. Investigation of the active anticancer constituents of gamboge. *J. Jiangxi Med. College* 2: 1–7.

Chen, F., Castranova, V., and Shi, X. 2001. New insights into the role of nucleus factor kappa B in cell growth regulation. *Am. J. Pathol.* 159: 387–397.

Chen, B. A., Liang, Y. Q., Wu, W. W. et al. 2009. Synergistic effect of magnetic nanoparticles of Fe_3O_4 with gambogic acid on apoptosis of K562 leukemia cells. *Int. J. Nanomedicine* 4: 251–259.

Clement, M. J., Jourdain, I., Lachkar, S. et al. 2005. N-terminal stathmin-like peptides bind tubulin and impede microtubule assembly. *Biochemistry* 44: 14616–14625.

Dai, J. G. 2003. The preparation of a kind of gamboic acid injection. CN Patent No. 03131511.9.

Davenport, J., Manjarrez, J. R., Peterson, L., Krumm, B., Blagg, B. S. J., and Matts, R. L. 2011. Gambogic Acid, a natural product inhibitor of Hsp90. *J. Nat. Prod.* 74: 1085–1092.

Ding, Y., Zhang, P., Tang, X.-Y. et al. 2012. PEG prodrug of gambogic acid: Amino acid and dipeptide spacer effects. *Polymer* 53: 1694–1702.

Dolle, L., Depypere, H. T., and Bracke, M. E. 2006. Anti-invasive/anti-metastasis strategies: new roads, new tools and new hopes. *Curr. Cancer Drug Targets* 6: 729–751.

Dong, C., Jin, T. Y., Lu, F. D. et al. 1988. The anticancer activity of gambogic acid *in vitro*. *Bull. Chin. Pharm.* 23: 89–90.

Ellenrieder, V., Adler, G., and Gress, T. M. 1999. Invasion and metastasis in pancreatic cancer. *Ann. Oncol.* 10: 46–50.

Feng, F., Liu, W., Wang, Y., Guo, Q., and You, Q. 2007. Structure elucidation of metabolites of gambogic acid in vivo in rat bile by high-performance liquid chromatography-mass spectrometry and high-performance liquid chromatography-nuclear magnetic resonance. *J. Chromatogr. B* 860: 218–226.

Glinsky, G. V. 1998. Anti-adhesion cancer therapy. *Cancer Metast. Rev.* 17: 177–185.

Gu, H. Y., Wang, X. T., Rao, S. et al. 2008a. Gambogic acid mediates apoptosis as a p53 inducer through down-regulation of mdm2 in wild-type p53-expressing cancer cells, *Mol. Cancer Ther.* 7: 3298–3305.

Gu, H., You, Q., Liu, W. et al. 2008b. Gambogic acid induced tumor cell apoptosis by T lymphocyte activation in H22 transplanted mice. *Int. Immunopharmacol.* 8: 1493–1502.

Guo, Q. L., Lin, S. S., You, Q. D. et al. 2006. Inhibition of human telomerase reverse transcriptase gene expression by gambogic acid in human hepatoma SMMC-7721 cells. *Life Sci.* 78: 1238–1245.

Guo, Q.-L., You, Q.-D., Wu, Z.-Q., Yuan, S.-T., and Zhao, L. 2004. General gambogic acids inhibited growth of human hepatoma SMMC-7721 cells in vitro and in nude mice. *Acta Pharmacol. Sin.* 25: 769–774.

Hainaut, P. and Hollstein, M. 2000. p53 and human cancer: The first ten thousand mutations. *Adv. Cancer Res.* 77: 81–137.

Han, Q.-B., Cheung, S., Tai, J., Qiao, C.-F., Song, J.-Z., and Xu, H.-X. 2005. Stability and cytotoxicity of gambogic acid and its derivative, gambogic acid. *Biol. Pharm. Bull.* 28: 2335–2337.

Han, Q.-B., Wang, Y.-L., Yang, L. et al. 2006. Cytotoxic polyprenylated xanthones from the resin of *Garcinia hanburyi*. *Chem. Pharm. Bull.* 54: 265–267.

Han, J. Y., Xiao, J., Wang, H., Chang, H. Y., and Ma, P. S. 2001. Measurement and correlation of taxol solubility in methanol, ethanol and methanol-water systems. *J. Chem. Ind. Eng.* 52: 64–67.

Han, Q.-B. and Xu, H.-X. 2009. Caged *Garcinia* xanthones: Development since 1937. *Curr. Med. Chem.* 16: 3775–3796.

Hao, K., Liu, X.-Q., and Wang, G.-J. 2005. Pharmacokinetics of gambogic acid in rats. *J. China Pharma. Univ.* 36: 338–341.

Hao, K., Liu, X.-Q., Wang, G. J., and Zhao, X. P. 2007a. Pharmacokinetics, tissue distribution and excretion of gambogic acid in rats. *Eur. J. Drug Metab. Pharmacokinet.* 32: 63–68.

Hao, K., Zhao, X.-P., Liu, X.-Q., and Wang, G.-J. 2007b. Determination of gambogic acid in dog plasma by high-performance liquid chromatography for a pharmacokinetic study. *Biomed. Chromatogr.* 21: 279–283.

He, L., Ling, Y., Fu, L., Yin, D., Wang, X., and Zhang, Y. 2012, Synthesis and biological evaluation of novel derivatives of gambogic acid as anti-hepatocellular carcinoma agents. *Bioorg. Med. Chem. Lett.* 22: 289–292.

Huang, H., Chen, D., Li, S. et al. 2011. Gambogic acid enhances proteasome inhibitor-induced anticancer activity. *Cancer Lett.* 301: 221–228.

Jourdain, L., Curmi, P., Sobel, A., Pantaloni, D., and Carlier, M. F. 1997. Stathmin: A tubulin-sequestering protein which forms a ternary T2S complex with two tubulin molecules. *Biochemistry* 36: 10817–10821.

Kasibhatla, S., Jessen, K. A., Maliartchouk, S. et al. 2005. A role for transferring receptor in triggering apoptosis when targeted with gambogic acid. *Proc. Natl. Acad. Sci. USA* 102: 12095–12100.

Kitagawa, Y., Dai, J., Zhang, J. et al. 2005. Vascular endothelial growth factor contributes to prostate cancermediated osteoblastic activity. *Cancer Res.* 65: 10921–10929.

Kuemmerle, J., Jiang, S., Tseng, B., Kasibhatla, S., Drewe, J., and Cai, S. X. 2008. Synthesis of caged 2,3,3a,7a-tetrahydro-3,6-methanobenzofuran-7(6*H*)-ones: Evaluating the minimum structure for apoptosis induction by gambogic acid. *Bioorg. Med. Chem.* 16: 4233–4241.

Land, M. and Katz, A. 1949. Beitrag zur chemie des gummigutt *Pharm. Acta Helv.* 24: 387–401, PMID: 15398575.

Larsen, A. K., Escargueil, A. E., and Skladanowskim, A. 2003. Catalytic topoisomerase II inhibitors in cancer therapy. *Pharmacol. Ther.* 99: 167–181.

Li, C., Lu, N., Qi, Q. et al. 2011. Gambogic acid inhibits tumor cell adhesion by suppressing integrin b1 and membrane lipid rafts-associated integrin signaling pathway. *Biochem. Pharmacol.* 82: 1873–1883.

Lin, L. J., Lin, L. Z., Pezzuto, J. M., Cordell, G. A., and Ruangrungsi, N. 1993. Isogambogic acid and isomorellinol from *Garcinia hanburyi*. *Magn. Reson. Chem.* 31: 340–347.

Liu, W., Guo, Q.-L., You, Q.-D., Zhao, L., Gu, H.-Y., and Yuan. S.-T. 2005. Anticancer effect and apoptosis induction of gambogic acid in human gastric cancer line BGC-823. *World J. Gastroenterol.* 11: 3655–3659.

Liu, Y.-T., Hao, K., Liu, X.-Q., and Wang, G.-J. 2006. Metabolism and metabolic inhibition of gambogic acid in rat liver microsomes. *Acta Pharmacol. Sin.* 27: 1253–1258.

Lu, N., Yang, Y., You, Q. D. et al. 2007. Gambogic acid inhibits angiogenesis through suppressing vascular endothelial growth factor-induced tyrosine phosphorylation of KDR/Flk-1. *Cancer Lett.* 258: 80–89.

Lundgren, K., de Oca Luna, R. M., McNeill, Y. B. et al. 1997. Targeted expression of MDM2 uncouples S phase from mitosis and inhibits mammary gland development independent of p53. *Genes Dev.* 11: 714–725.

Mehta, U., Kang, B. P., Bansal, G., and Bansal, M. P. 2002. Studies of apoptosis and bcl-2 in experimental atherosclerosis in rabbit and influence of selenium supplementation. *Gen. Physiol. Biophys.* 21: 15–29.

Mertens, H. J., Heineman, M. J., and Evers, J. L. 2002. The expression of apoptosisrelated proteins bcl-2 and ki67 in endometrium of ovulatory menstrual cycles. *Gynecol. Obstet. Invest.* 53: 224–230.

Nie, F., Zhang, X., Qi, Q. et al. 2009. Reactive oxygen species accumulation contributes to gambogic acid-induced apoptosis in human hepatoma SMMC-7721 cells. *Toxicology* 260: 60–67.

Nogales, E. and Wang, H. W. 2006. Structural intermediates in microtubule assembly and disassembly: How and why? *Curr. Opin. Cell Biol.* 18: 179–184.

Ollis, W. D., Ramsay, M. V. J., Sutherland, I. O., and Mongkolsuk, S. 1965. The constitution of gambogic acid. *Tetrahedron* 21: 1453–1470.

Palempalli, U. D., Gandhi, U., Kalantari, P. et al. 2009. Gambogic acid covalently modifies IκB kinase-β subunit to mediate suppression of lipopolysaccharide-induced activation of NF-κB in macrophages. *Biochem. J.* 419: 401–409.

Pandey, M. K., Sung, B., Ahn, K. S., Kunnumakkara, A. B., Chaturvedi, M. M., and Aggarwal, B. B. 2007. Gambogic acid, a novel ligand for transferrin receptor, potentiates TNF-induced apoptosis through modulation of the nuclear factor-κB signaling pathway. *Blood* 110: 3517–3525.

Petersson, F., Dalgleish, A. G., Bissonnette, R. P., and Colston K. W. 2002. Retinoids cause apoptosis in pancreatic cancer cells via activation of RAR-γ and altered expression of Bcl-2/Bax. *Br. J. Cancer*, 87: 555–561.

Pignatelli, M. and Stamp, G. 1995. Integrins in tumour development and spread. *Cancer Surv.* 24: 113–127.

Posey, J. A., Khazaeli, M. B., DelGrosso, A. et al. 2001. A pilot trial of Vitaxin, a humanized anti-vitronectin receptor (anti alpha v beta 3) antibody in patients with metastatic cancer. *Cancer Biother. Radiopharm.* 16: 125–132.

Prasad, S., Pandey, M. K., Yadav, V. R., and Aggarwal, B. B. 2011. Gambogic acid inhibits STAT3 phosphorylation through activation of protein tyrosine phosphatase SHP-1: Potential Role in proliferation and apoptosis. *Cancer Prev. Res.* 4: 1084–1094.

Qi, Q., You, Q., Gu, H. et al. 2008. Studies on the toxicity of gambogic acid in rats. *J. Ethnopharmacol.* 117: 433–438.

Qiang, L., Yang, Y., You, Q.-D. et al. 2008. Inhibition of glioblastoma growth and angiogenesis by gambogic acid: An in vitro and in vivo study. *Biochem. Pharmacol.* 75: 1083–1092.

Qin, Y., Meng, L., Hu, C. et al. 2007. Gambogic acid inhibits the catalytic activity of human topoisomerase IIA by binding to its ATPase domain. *Mol. Cancer Ther.* 6: 2429–2440.

Qu, G., Zhu, X., Zhang, C., and Ping, Q. N. 2009. Modified chitosan derivative micelle system for natural antitumor product gambogic acid delivery. *Drug Deliv.* 16: 363–370.

Ren, Y., Yuan, C., and Chai, H.-B. et al. 2011. Absolute configuration of (−)-gambogic acid, an antitumor agent. *J. Nat. Prod.* 74: 460–463.

Reutrakul, V., Anantachoke, N., Pohmakotr, M. et al. 2007. Cytotoxic and anti-HIV-1 caged xanthones from the resin and fruits of *Garcinia hanburyi. Planta Med.* 73: 33–40.

Rong, J.-J., Hu, R., Song, X.-M. et al. 2010. Gambogic acid triggers DNA damage signaling that induces $p53/p21^{Waf1/CIP1}$ activation through the ATR-Chk1 pathway. *Cancer Lett.* 296: 55–64.

Rong, J.-J., Hu, R., Qi, Q. et al. 2009. Gambogic acid down-regulates MDM2 oncogene and induces $p21^{Waf1/CIP1}$ expression independent of p53. *Cancer Lett.* 284: 102–112.

Rubin, C. I. and Atweh, G. F. 2004. The role of stathmin in the regulation of the cell cycle. *J. Cell Biochem.* 93: 242–250.

Ryschich, E., Huszty, G., Knaebel, H., Hartel, M., Buchler, M., and Schmidt, J. 2004. Transferrin receptor is a marker of malignant phenotype in human pancreatic cancer and in neuroendocrine carcinoma of the pancreas. *Eur. J. Cancer* 40: 1418–1422.

Szekeres, T., Sedlak, J., and Novotny, L. 2002. Benzamide riboside, a recent inhibitor of inosine 5′-monophosphate dehydrogenase induces transferrin receptors in cancer cells. *Curr. Med. Chem.* 9: 759–764.

Tang, D., Lu, L., Zeng, F.-Q., He, J., Jiang, G.-S., and Wang, Z.-D. 2011. Gambogic acid inhibits cell proliferation and induces apoptosis of human prostate cancer PC-3 cells *in vitro. Tumor* 31: 688–692.

Tao, S.-J., Guan, S.-H., Wang, W. et al. 2009. Cytotoxic polyprenylated xanthones from the resin of *Garcinia hanburyi. J. Nat. Prod.* 72: 117–124.

Tilli, C. M., Stavast-Koey, A. J., Ramaekers, F. C., and Neumann, H. A. 2002. Bax expression and growth behavior of basal cell carcinomas. *J. Cutan. Pathol.* 29: 79–87.

Wang, X., Chen, Y., Han, Q.-B. et al. 2009. Proteomic identification of molecular targets of gambogic acid: Role of stathmin in hepatocellular carcinoma. *Proteomics* 9: 242–253.

Wang, J., Ma, J., You, Q., Zhao, Li., Wang, F., and Li, C. 2010. Studies on chemical modification and biology of a natural product, gambogic acid (II): Synthesis and bioevaluation of gambogellic acid and its derivatives from gambogic acid as antitumor agents. *Eur. J. Med. Chem.* 45: 4343–4353.

Wang, T., Wei, J., Qian, X., Ding, Y., Yu, L., and Liu, B. 2008. Gambogic acid, a potent inhibitor of survivin, reverses docetaxel resistance in gastric cancer cells. *Cancer Lett.* 262: 214–222.

Wang, C., Zhang, H., Chen, Y., Shi, F., and Chen, B. 2012. Gambogic acid-loaded magnetic Fe_3O_4 nanoparticles inhibit Panc-1 pancreatic cancer cell proliferation and migration by inactivating transcription factor ETS1. *Int. J. Nanomedicine* 7: 781–787.

Wang, C., Zhang, H., Chen, B., Yin, H., and Wang, W. 2011. Study of the enhanced anticancer efficacy of gambogic acid on Capan-1 pancreatic cancer cells when mediated via magnetic Fe_3O_4 nanoparticles. *Int. J. Nanomedicine* 6: 1929–1935.

Wang, J., Zhao, Q., Qi, Q. et al. 2011. Gambogic acid-induced degradation of mutant p53 is mediated by proteasome and related to CHIP. *J. Cell. Biochem.* 112: 509–519.

Weakley, T. J. R., Cai, S. X., Zhang, H.-Z., and Keana, J. F. W. 2001. Crystal structure of the pyridine salt of gambogic acid. *J. Chem. Cryst.* 31: 501–505.

Wu, Z.-Q., Guo, Q.-L., You, Q.-D., Zhao, L., and Gu, H.-Y. 2004. Gambogic acid inhibits proliferation of human lung carcinoma SPC-A1 cells in vivo and in vitro and represses telomerase activity and telomerase reverse transcriptase mRNA expression in the cells. *Bio. Pharma. Bull.* 27: 1769–1774.

Xie, H., Qin, Y.-X., Zhou, Y.-L. et al. 2009. GA3, a new gambogic acid derivative, exhibits potent antitumor activities in vitro via apoptosis-involved mechanisms. *Acta Pharmacol. Sin.* 30: 346–354.

Yang, J., Ding, L., Hu, L. et al. 2011. Metabolism of gambogic acid in rats: A rare intestinal metabolic pathway responsible for its final disposition. *Drug Meta. Dispos.* 39: 617–626.

Yang, J., Ding, L., Jin, S., Liu, X., Liu, W., and Wang, Z. 2010. Identification and quantitative determination of a major circulating metabolite of gambogic acid in human. *J. Chromatogr. B* 878: 659–666.

Yang, Q. Z., Jia, S. J., and Dh, L. 1994. The neoteric study of Chinese traditional drug, gamboge. *Chin. J. Clin. Oncol.* 21: 464–465.

Yang, Y., Yang, L., You, Q.-D. et al. 2007. Differential apoptotic induction of gambogic acid, a novel anticancer natural product, on hepatoma cells and normal hepatocytes. *Cancer Lett.* 256: 259–266.

Yi, T., Yi, Z., Cho, S.-G. et al. 2008. Gambogic acid inhibits angiogenesis and prostate tumor growth by suppressing vascular endothelial growth factor receptor 2 signaling. *Cancer Res.* 68: 1843–1850.

Yoo, H. S., Lee, K. H., Oh, J. E., and Park, T. G. 2000. In vitro and in vivo anticancer activities of nanoparticles based on doxorubicin-PLGA conjugates. *J. Control. Release* 68: 419–431.

You, Q. D., Guo. Q. L., Ke, X., Xiao, W., Dai, L. L., and Lin, Y. 2003. The preparation of gamboic acid and its compound. CN Patent No. 03132386.3.

Yu, J., Guo, Q. L., You, Q. D. et al. 2006. Repression of telomerase reverse transcriptase mRNA and hTERT promoter by gambogic acid in human gastric carcinoma cells. *Cancer Chemother. Pharmacol.* 58: 434–443.

Yu, J., Guo, Q. L., You, Q. D. et al. 2007. Gambogic acid-induced G2/M phase cell-cycle arrest via disturbing CDK7-mediated phosphorylation of CDC2/p34 in human gastric carcinoma BGC-823 cells. *Carcinogenesis* 28: 632–638.

Zaffaroni, N., Pennati, M., Colella, G. et al. 2002. Expression of the anti-apoptotic gene survivin correlates with taxol resistance in human ovarian cancer, *Cell. Mol. Life Sci.* 59: 1406–1412.

Zhang, H.-Z., Kasibhatla, S., Wang, Y. et al. 2004. Discovery, characterization and SAR of gambogic acid as a potent apoptosis inducer by a HTS assay. *Bioorg. Med. Chem.* 12: 309–317.

Zhang, C., Liu, Y., Gao, Y. et al. 2009. Modified heparins inhibit integrin alpha(IIb)beta(3) mediated adhesion of melanoma cells to platelets in vitro and in vivo. *Int. J. Cancer* 125: 2058–2065.

Zhang, L., Yi, Y., Chen, J. et al. 2010. Gambogic acid inhibits Hsp90 and deregulates TNF-α/NF-κB in HeLa cells. *Biochem. Biophys. Res. Commun.* 403: 282–287.

Zhang, L., You, Q., Liang, Y., Liu, W., Guo, Q., and Wang, J. 2009. Identification of gambogic acid metabolites in rat bile by liquid chromatography-tandem mass spectrometry-ion trap-time-off-light. *Chin. J. Nat. Med.* 7: 376–380.

Zhao, L., Guo, Q. L., You, Q. D., Wu, Z. Q., and Gu, H. Y. 2004. Gambogic acid induces apoptosis and regulates expressions of Bax and Bcl-2 protein in human gastric carcinoma MGC-803 cells. *Biol. Pharm. Bull.* 27: 998–1003.

Zhao, Q., Yang, Y., Yu, J. et al. 2008. Posttranscriptional regulation of the telomerase hTERT by gambogic acid in human gastric carcinoma 823 cells. *Cancer Lett.* 262: 223–231.

Zhao, L., Zhen, C., Wu, Z., Hu, R., Zhou, C., and Guo, Q. 2010. General pharmacological properties, developmental toxicity, and analgesic activity of gambogic acid, a novel natural anticancer agent. *Drug Chem. Toxicol.* 33: 88–96.

Zhou, Z. T. and Wang, J. W. 2007. Phase I human tolerability trial of gambogic acid. *Chin. J. New Drugs* 16: 79–83.

Zhu, X., Zhang, C., Wu, X., Tang, X., and Ping, Q. 2008. Preparation, physical properties, and stability of gambogic acid-loaded micelles based on chitosan derivatives. *Drug Dev. Ind. Pharm.* 34: 2–9.

16 Neuroplasticity as a New Approach to the Pathophysiology of Depression and the Role of Modern Antidepression Drugs

Gianluca Serafini, Maurizio Pompili, and Paolo Girardi

CONTENTS

16.1 INTRODUCTION

Major depressive disorder (MDD) is a common mental illness affecting approximately 2.5% of the general population. MDD is one of the leading causes of disability and it has been suggested to become the second highest burden of disease (measured in disability-adjusted life years) by 2020 (WHO 2007). MDD has negative social consequences in terms of unemployment and psychosocial impairment (Anderson et al. 2011). The pathophysiology of depression has been suggested to involve both external social stressors and internal genetic vulnerability.

Among all biological theories postulated about MDD, an impairment of neuroplasticity and cellular resilience has been also suggested (Duman et al. 1999). Resiliency depends on the general ability to adapt and react to stressful life events whereas neuroplasticity generally refers to a collection of molecular, cellular, and systemic events which are critical for neuronal adaptation. The existence of neuroplasticity-related changes derived from several lines of evidence showing that neural circuits and connections undergo lifelong modifications and reorganizations in response to external or internal environmental stimuli. Generally, enhanced axonal outgrowth and collateral sprouting on the presynaptic site may lead to the formation of novel synapses whereas the existing synapses can be deleted through terminal retrograde degeneration (for more details, see Fuchs 2009). Adult neurogenesis involves precursors of cell proliferation, migration, and differentiation, mainly occurring in the dentate gyrus of the hippocampus (Eriksson et al. 1998). For example, several neurotoxic agents such as chronic stress, excessive concentrations of glutamate, biogenic amines, and glucocorticoids may affect the morphology of hippocampal CA3 pyramidal neurons and also pyramidal cells in the prefrontal cortex. It has been demonstrated that neural cells may react to chronic stress debranching apical dendrites or with spine loss; these specific changes are closely associated to daily periods of resting and activity (Peter-Cruz 2007, 2009).

Interestingly, it has been demonstrated that some ADs may increase neurotrophin signaling, promoting neuronal and synaptic remodeling as well as the formation of new neurons in the hippocampus and prefrontal cortex (Duman 2004b; McEwen 2004; Castren et al. 2007; Sairanen et al. 2007; Bessa et al. 2009). In recent years, it has been suggested that monoamine deficits may conceptualize only part of the depression pathogenesis; therefore, additional theories have reported that modern ADs may act enhancing neuroplasticity mechanisms and renewing the impairment in neural circuits contributing to their normalization (Duman et al. 1999; Manji et al. 2000; Manji and Duman 2001; Pittenger and Duman 2008). Although the exact modifications that ADs may induce at the synaptic level are still unclear, antidepressant action may promote neuronal connectivity and strengthen specific synapses or normalize glutamatergic tone, the alterations of which are supposed to be underlying depression (Reznikov et al. 2009). Pharmacological manipulation of glutamatergic system in animal models has been shown to reduce stress-induced morphological changes in the hippocampus (McEwen and Chattarji 2004; Reagan et al. 2004; Kasper and McEwen 2008), and some ADs have been reported to regulate glutamatergic transmission through the inhibition of stress-induced morphological changes in both the hippocampus and amygdala (McEwen and Chattarji 2004).

Considering this novel and fascinating background, this chapter aims to critically review the current literature about neuroplasticity, depression, stress-related neuropathological changes, and antidepressant treatment, based on the assumption that ADs are effective in promoting neuroplasticity mechanisms and neurogenesis.

16.2 NEUROPLASTICITY AND RESILIENCY

Generally, resiliency is the ability of an organism to adapt and react to stressful life events and environmental situations. This ability is mediated by the involvement of several brain areas such as the hippocampus, amygdala, and prefrontal cortex playing a key role in either cognitive or affective domains and requiring the involvement of specific neurotransmitter molecules.

Indeed, neuroplasticity is a general term indicating the neural framework in which all the different internal events at either the molecular and systemic levels determine neuronal modifications (Reznikov et al. 2009). Overall, neuroplasticity is actually conceptualized as the ability of each neural component and structure to react in response to external or internal stimuli. At the neuronal level, neuroplasticity refers to the complex cascade of neurophysiological, neurochemical, and neurohistological changes determining synaptic strengthening, dendritic growth, branching, and new synapse formation. Mainly, neuroplasticity is closely associated with neurogenesis, gliogenesis, enhancement of vascular endothelial support, and finally with the inhibition of apoptotic mechanisms.

16.3 DEPRESSION AND NEUROPLASTICITY: UNDERSTANDING THE NEUROBIOLOGY OF DEPRESSION THROUGH THE EFFECT OF STRESS

In recent decades, the view that the brain is a static structure in which electrical and chemical information are processed within a fixed system of neuronal circuits has been widely debated. Neural circuits, brain nuclei, neurons, and synaptic connections undergo several lifelong modifications and relevant adaptations due to environmental stimuli. Continuous modifications often induced by steroid hormones such as increased axonal growth and collateral sprouting determine the development of new synapses and retrograde elimination of the pre-existing ones as well as changes in the dimensions of the dendritic tree and spine density influencing the number of postsynaptic sites (Woolley 1999; Carvalho et al. 2010). Popov and Bochorova (1992) have found that specific and multifaceted structural changes at the synaptic level may be induced by mossy fibers and hippocampal pyramidal neurons. Also, hippocampal CA3 pyramidal neurons undergo dendritic shrinkage after chronic stress induced by corticosterone (Woolley et al. 1990; Magariños et al. 1996; Sousa et al. 2000).

Interestingly, the fact that changes in the regulation of endocrine systems, learning and memory, brain histology, and behavior show strong similarities among stressed animals and depressed human subjects, confirmed the usefulness of animal models to describe the effect of stress and antidepressant mechanisms in humans (Fuchs 2004). Neurohistological changes in stress and antidepressant response are also site specific (Andrade and Rao 2010). The reduction in hippocampal volume is presumably the most common finding in depressed subjects and the duration of depressive episodes is known to be closely related to the volumetric hippocampal modifications (Sheline et al. 2003; Lorenzetti et al. 2009). Somatodendritic, axonal, and synaptic components may inhibit adult neurogenesis, and changes in glial cell number have been identified to be one of the most relevant finding underlying hippocampal volume loss. However, the precise molecular mechanism explaining volumetric hippocampal modifications is now unclear. Recent evidence (Czéh and Lucassen 2007) suggested that both hypercholesterolemia (no postmortem neuronal loss was found in the brain tissue) and apoptosis (evident only in a small hippocampal area) have not been identified as possible neurotoxic agents. Although it has been suggested that reduced adult neurogenesis is involved in the pathophysiology of major affective disorders, Reif et al. (2006), in a small sample of subjects, did not find any significant change in neurogenesis of depressed patients such as evidence of neural stem cell proliferation. Moreover, the small hippocampal volume may also be found in several neuropsychiatric disorders such as anxiety disorders, schizophrenia, and many neurological diseases. It is presumable that hippocampal neuropathological changes and volume reduction associated with depression should be considered in a broader context (Fuchs 2009).

The hippocampus is not the only cerebral area subject to structural modifications. Several structural changes have been noticed in the rat prefrontal cortex, a brain area in which a retraction of dendrites and a spine loss induced by chronic stress and associated with daily periods resting and activities were observed (Perez-Cruz et al. 2007, 2009). Postmortem histopathological studies (Rajkowska et al. 1999; Cotter et al. 2002) have reported reduced neuronal density, smaller neuronal somata, and a relevant reduction in the prefrontal cortical thickness. The hippocampal volume reduction may be observed in postmortem animal models of both stress and depression (D'Sa and Duman 2002; Pittenger and Duman 2008). In animal models, stress-induced hippocampal neuropathological changes may be summarized as follows: loss of dendritic spines; decrease in the number and length of dendrites; loss of synapses; loss of glia and impaired neurogenesis (D'Sa and Duman 2002; Sheline 2004; Pittenger and Duman 2008; Fuchs 2009; Gorwood 2009; Jay et al. 2009). The retraction of dendrites and synapses determine a reduction of connectivity, multiple impairments of neurons and loss of glia, consequent reduction of neurotransmission, decreased neurogenesis, and presumably apoptosis (Andrade and Rao 2010). Considering that the hippocampus is a key structure involved in learning and memory, as a result of hippocampal impairment, stressed animals and depressed humans may be affected by impaired learning and memory.

Stress-induced neurohistological changes do not simply interfere with hippocampal functioning but they also affect functioning of other downstream areas. A stress-induced inhibition of cell proliferation and gliogenesis may also be observed in the whole prefrontal cortex whereas a stress-induced dendritic reorganization in pyramidal neurons has been reported in the medial prefrontal cortex. In animal models, stress-induced neurohistological changes in the prefrontal cortex determine loss of dendritic spines, atrophy of the dendritic tree, loss of synapses, decreased number and size of glia (D'Sa and Duman 2002; Pittenger and Duman 2008; Fuchs 2009; Jay 2009). Postmortem studies in depressed subjects showed a decrease in neuronal and glial cells (both in number and size), and overall cortical thickness (D'Sa and Duman 2002). A glial cells loss was found not only in the amygdala but also in limbic and extralimbic structures, prefrontal, orbitofrontal, and cingulated cortices of depressed individuals. The prefrontal cortex plays a key role in cognitive functions such as attention, concentration, learning, and memory.

Moreover, structural modifications have been described in the amygdala where an enhanced dendritic arborization (but not an increase in all classes of amygdaloid neurons) has been shown by Vyas et al. (2002). After chronic stress, they observed an enhanced dendritic arborization in the basolateral nucleus of the amygdala and specifically in the pyramidal and excitatory projections of the stellate neurons. Also in animal models, stress-induced neurohistological changes in the amygdala include increased dendritic arborization and synaptogenesis (Pittenger and Duman 2008; Jay 2009). No alteration was reported in neuronal amygdalar number although a reduced number of glial cells has been demonstrated in depressed patients (Bowley et al. 2002). Glial cells are mostly present in the adult brain, providing a relevant structural framework. In recent decades, it has been found that abnormalities in glial cells functions are strongly involved in determining multiple impairments of structural plasticity (Coyle and Schwarcz 2000; Cotter et al. 2001). Several studies (Sheline et al. 1998; Bremner et al. 2000) using magnetic resonance imaging (MRI) showed an altered amygdalar core. Overall, an increased amygdalar volume determining not only structural but also functional impairments has been described in either stressed animals and/or depressed subjects. Stress-induced neurohistological modifications in the amygdala were not reversed after some weeks but required longer periods (Pittenger and Duman 2008). The amygdala plays a key role in social and emotional learning and, particularly, in emotions such as anxiety and fear. Therefore, morphological modifications occurring after exposure to chronic stress are not limited to hippocampus but are also extended to the amygdala and prefrontal cortex. In addition, chronic stress changes on dendrites and spines influence the expression of several synaptic molecules resulting crucial for the information transfer between neurons. Cooper et al. (2009) showed that the expression of M6a, particularly the splice variant M6a-Ib, a glycoprotein that appears to be located in the axonal plasma membrane of glutamatergic neurons may be differently regulated by stress. Chronic stress may differently induce the expression of M6a-Ib in a region-dependent manner, specifically down-regulating M6a-Ib in the dentate gyrus granule neurons and CA3 pyramidal neurons, and up-regulating M6a-Ib in the medial prefrontal cortex. This different regulation of targeted glycoproteins induced by chronic stress presumably leads to reduced axonal output in hippocampal neurons also altering the integrity of axons and the information transfer between neurons in different brain regions. Finally, chronic stress may also affect neuron–glia communication, inducing a remodeling of hippocampal dendrites and increased expression of the GLT-1 glial glutamate transporter in the dentate gyrus and CA3 hippocampal neurons (Reagan et al. 2004).

Overall, pathological stress and depression result in abnormalities in neuroplasticity response characterized by an abnormal increased activity in the amygdala and an impaired hippocampal and prefrontal cortex functioning.

16.4 EVOLUTION OF ADs

Each class of modern ADs was empirically discovered. Enzyme inhibitors (mainly monoamine oxidase inhibitors (MAOI)), uptake blockers tricyclics (TCAs), noradrenergic and dopaminergic inhibitors (NDRIs), noradrenergic selective reuptake inhibitors (NARIs), serotonergic and noradrenergic reuptake inhibitors (SNRIs), and selective serotonergic reuptake inhibitors (SSRIs) were randomly discovered by investigating patients with tuberculosis and schizophrenia, respectively. The third class of ADs (receptor-acting drugs) derived from animal studies showed similar behavioral effects although possessing differential mechanisms of action. This actual empirical classification of ADs undergoes major refinements due to the continuous advances in the field of psychopharmacology. Since the discovery of the first AD in the 1950s, deficits in synaptic levels of serotonin, noradrenaline, and dopamine have been considered as a prominent biological feature underlying depression (see the well-known "monoaminergic hypothesis" of depression; Nestler et al. 2002). After the introduction of TCAs, in order to reduce the emergence of several side effects associated with these compounds, newer and more selective ADs appeared. However, although neurotransmitter dysfunctions are undoubtedly involved in the pathophysiology of depression, it is widely recognized that these deficits alone are not sufficient to explain the complex modifications related to mood disturbances and the different mechanisms of action of AD agents. One of the most relevant criticism about this model is the gap between the rapid occurrence of AD-induced synaptic effects (due to the blockade of reuptake of monoamines and their persistence in the synaptic cleft, resulting in increased postsynaptic receptor stimulation and neurotransmission) and the delay in clinical improvement of depressive symptoms. In fact, an antidepressant response takes weeks or longer to develop and induce response and remission (Calabrese et al. 2011).

Additional hypotheses have been suggested (Duman 2004; Andrade and Rao 2010) postulating the occurrence of a compensatory down-regulation of the facilitatory postsynaptic receptors (resulting in decreased activity of postsynaptic neurons) and a down-regulation of the inhibitory presynaptic receptors (resulting in increased activity of the presynaptic neurons requiring days or weeks). For example, TCAs induce a down-regulation of the presynaptic alpha-2-adrenoceptors and β postsynaptic adrenoceptors as well as SSRIs determine a down-regulation of the presynaptic 5-HT_{1a} receptors and postsynaptic 5-HT_2 receptors. However, these additional hypotheses result still incomplete (Andrade and Rao 2010). Although monoamines are recognized as having a distinct role in the pathophysiology of depression and antidepressant action, recent theories postulating the existence of an impairment in neuroplasticity and cellular resiliency gained growing consensus among researchers. According to neuroplasticity model underlying pathological stress and depression, dysfunctional neurohistological changes in the hippocampus, prefrontal cortex, amygdala, and other related brain structures may occur. Notably, by the initial investigation of receptor changes clinicians moved to consider downstream mechanisms and the regulation of intracellular signal transduction molecules (Coyle and Duman 2003; Manji et al. 2003). Consequently, according to the new neuroplastic/neurotrophic hypothesis it has been suggested that even the delayed antidepressant action may be due to difficulties in the adaption of post-receptors signaling cascade, and to the delayed remodeling of synaptic and cellular abnormalities (Duman and Monteggia 2006; Calabrese et al. 2009). ADs may protect against and reverse those neurohistological changes (Pittenger and Duman 2008). Accordingly, the insufficient response to antidepressant treatment (Frodl et al. 2006) has been hypothesized to the limited ability of drugs to modulate systems involved in neuroplasticity.

Based on the fact that stress-related neuropathological changes represent a key element in the emergence of depression presumably related to a genetic predisposition, our aim is to test whether and to what extent ADs may influence (reverse) stress-related changes.

16.5 ANTIDEPRESSANTS AND NEUROPLASTICITY

Several lines of evidence demonstrate that some ADs may reverse neuroplasticity and neurogenesis modifications induced by chronic stress. In animal models, ADs may reverse and remodel many of the stress-induced neurohistological changes. It is possible to speculate that by reversing the neurohistological effects of stress in animal models, ADs may reverse depression in human subjects. There is evidence that treatment with modern ADs significantly improves both hippocampal shrinkage (Fuchs et al. 2004; Dranovsky and Hen 2006; Czéh and Lucassen 2007) and function (e.g., cognitive functions) (Yan et al. 2011). Therefore, most of the abnormalities observed in neuroplasticity are presumably reversible after short-term administration of ADs being not determined by neurodegenerative processes. Interestingly, Rocher et al. (2004) suggested that both tianeptine (see Figure 16.1) and fluoxetine (see Figure 16.2) may reverse the inhibition of long-term potentiation (an interesting

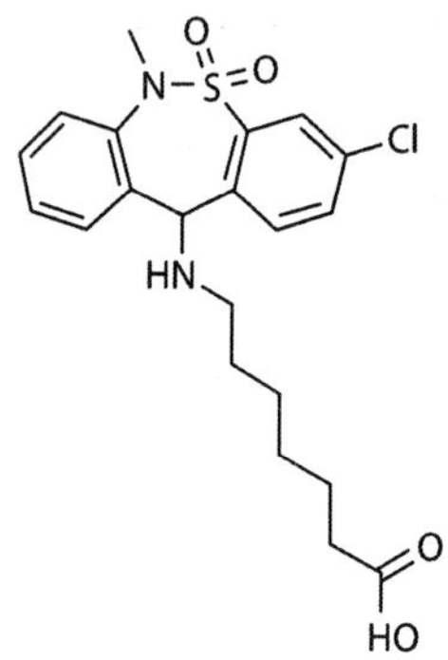

FIGURE 16.1 Chemical structure of tianeptine.

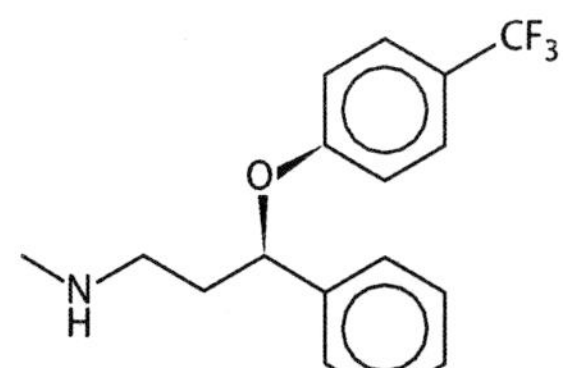

FIGURE 16.2 Chemical structure of fluoxetine.

prototype of synaptic plasticity) not only in the hippocampus but also in the prefrontal cortex. Czéh et al. (2006) pointed out that fluoxetine may inhibit stress-related reduced number of hippocampal astrocytes, reversing morphological modifications observed in the somal volume. Although treatment with some ADs has been demonstrated to rectify some neuroplasticity dysfunctions, the exact mechanism underlying this phenomenon is still unclear (Vermetten et al. 2003; Berton and Nestler 2006). Evidence suggests a central role of both N-Methyl-D-aspartate (NMDA) and α-amino-3-hydroxy-5-methyl-4-isoxazolepropionic acid (AMPA) glutamate receptors activation in inducing morphological changes, regulating neuroplasticity such as dendritic length and branching, spine density, and volume in several brain regions, specifically in the hippocampal dentate gyrus (Fifkova et al. 1977; Halpain et al. 1998).

Glutamate, in certain concentrations and presumably under influence of elevated glucocorticoids levels, mediates structural remodeling of neurons, leading to reversible modifications such as reduced neurogenesis, neuronal shrinkage, and decreased growth (Virgin et al. 1991). Several authors (Watanabe et al. 1992; Magariños and McEwen 1995) showed that the inhibition of glutamate release by NMDA receptors prevents this remodeling. Of particular interest is the evidence suggesting that suicidal ideation in depressed subjects is associated with genes encoding ionotropic glutamate receptors (Laje et al. 2007), and that some ADs and electroconvulsive shock therapy may reverse glutamate impairment in the anterior cingulated cortex of depressed individuals (Pfleiderer et al. 2003). These findings suggest a crucial role for glutamate abnormalities in depression and enhance the preliminary evidence that administration of ADs and electroconvulsive shock therapy may reverse neuroplasticity alterations specifically acting on glutamate dysregulation. Drugs having mood stabilization properties and additionally modulating glutamate release may mediate morphological plasticity abnormalities (McEwen and Chattarji 2004) and stress-induced morphological hippocampal together with amygdalar changes, which are reduced by AD manipulation (McEwen and Chattarji 2004; Reagan et al. 2004). Specifically, Malberg et al. (2000) found that antidepressant treatment prevented the retraction of apical dendrites of hippocampal CA3 pyramidal neurons and the increased granule cell proliferation. While tianeptine may prevent glutamate efflux in the basolateral nucleus of the amygdala, this effect seems not to be induced with the administration of fluoxetine. Therefore, Reznikov et al. (2007) postulated that the impact of ADs in mediating the stress-induced neuropathological changes was quite specific. Interestingly, Emery et al. (2005) found that neural stem cells involved in the proliferation and differentiation of adult new neurons extended axons to the CA3 region only after 2 weeks of ADs administration, explaining at least partially the delayed action of ADs.

Additionally, recent studies have shown that glucocorticoids are involved in the neurogenic action of ADs (Huang and Herbert 2006; David et al. 2009). The potential role of glucocorticoids in antidepressant-induced neurogenesis is also consistent with the evidence that ADs regulate the function of the glucocorticoid receptor (GR) (Pariante et al. 1997, 2003a,b; Funato et al. 2006; Pariante and Lightman 2008; Anacher et al. 2010). In a recent study (Anacker et al. 2011), it has been identified for the first time that antidepressant-induced changes in neurogenesis are dependent on the GR. Particularly, the SSRI sertraline enhances neuronal differentiation and promotes neuronal maturation of human hippocampal progenitor cells through a GR-dependent mechanism associated with GR phosphorylation via protein chinase-A signaling. The authors concluded that this effect (only observed with sertraline) is present during the proliferation phase, but suggested a complex regulation of neurogenesis by ADs, with different GR-dependent mechanisms leading to enhanced cell proliferation without changes in neuronal differentiation, or enhanced neuronal differentiation in the presence of decreased cell proliferation.

However, what are the molecular mechanisms underlying ADs regulation of neuroplasticity and neurogenesis? Svenningsson et al. (2007) suggested that ADs may induce a phosphorilation of the AMPA receptors in two main sites of the subunit GluR1: Ser831, which is phosphorilated by protein kinase C or CaMK-II determining elevations in hippocampal currents (Barria et al. 1997) and Ser845 which appears crucial for protein chinase-A amplification of peak current by the GluR1 receptors (Roche et al. 1996). The TCA imipramine and the SSRI fluoxetine actually increases phosphorylation at Ser845 on the subunit GluR1 (Svenningsson et al. 2002; Du et al. 2007) whereas tianeptine may reverse stress-induced changes in glutamate receptors expression (Reagan et al. 2004), stress-induced impairment in neurogenesis (Czéh et al. 2001), and reduced stress-induced apoptosis in the hippocampus and temporal cortex (Lucassen et al. 2004). In addition, other mechanisms of action for ADs have been hypothesized. ADs of different classes act enhancing phosphorilation at the *c*-AMP regulatory element-binding protein (CREB) (Manji et al. 2001) and CaMK-II (Popoli et al. 1995) as well as electroconvulsive shock therapy has been demonstrated to increase hippocampal CREB phosphorilation (Jeon et al. 1997). Specific neurotrophic factors such as brain-derived neurotrophic factor (BDNF) binding to tyrosine kinase (TrK) receptors may activate intracellular cascades involving *c*AMP-dependent protein kinase A (PKA), mitogen-activated protein kinase (MAPK), calcium/calmodulin-dependent protein kinase II (CaMK-II) and also transcription factors such as CREB. This last transcription factor is involved in the synthesis of different enzymes and proteins involved in the structural mechanisms underlying neuroplasticity. CREB and BDNF are among the most important effectors in neuroplasticity (D'Sa and Duman 2002; Pittenger and Duman 2008). Additionally, ADs may increase the activity of *c*-Fos, considered as a marker of biochemical activity (Beck 1995); some ADs such as tianeptine may reduce *c*-Fos levels, reversing its previous stress-induced increase (Duncan et al. 1996). Tianeptine has been demonstrated to prevent impairments in stress-induced amygdalar and prefrontal changes (Watanabe et al. 1992; Vouimba et al. 2006) as well as synaptic plasticity dysfunctions also preventing the reduction of length and branching of apical dendrites of the hippocampal CA3 neurons exposed to stress (Watanabe et al. 1992; Magariños et al. 1999). Rocher et al. (2004) have suggested that fluoxetine possesses a similar slower activity, blocking the effect of stress in the prefrontal cortex.

The final result of all these intracellular signaling cascades is a stimulation of neurogenesis in the dentate gyrus, including an increase in glial cells in the complexity of dendritic branching and new synaptic connections.

16.6 NOVEL ANTIDEPRESSION DRUGS AND NEUROPLASTICITY

Over the last few years, current literature has also focused on the neurotrophic actions of most recent ADs such as agomelatine (see Figure 16.3) mediating different therapeutic mechanisms compared to other SSRIs and TCAs (Banasr et al. 2006; Soumier et al. 2009) and presumably promoting hippocampal neurogenesis under basal conditions (Hanoun et al. 2004; Païzanis et al. 2010). Agomelatine has been found to selectively increase cell proliferation and neurogenesis in the ventral hippocampus, and to enhance the survival of newly generated cells throughout the entire hippocampus in rats under basal and stressful conditions (Daszuta et al. 2005; Banasr et al. 2006). Following the initial observation about melatonin that has been shown to increase hippocampal cell proliferation in maternally separated pups (Kim et al. 2004), Banasr et al. (2006) demonstrated that agomelatine expressed its neurotrophic properties in the ventral portion of the dentate gyrus after chronic (21 days), but not acute (4 h) or subchronic (8 days) administration, presumably through a joint effect of melatonin agonism and $5HT_{2C}$ antagonism playing a crucial role to facilitate all stages of neurogenesis and promote cell survival (Soumier et al. 2009). Other evidence (AlAhmed and Herbert 2010; Dagyte et al. 2010, 2011b) suggested that agomelatine may stimulate adult neurogenesis in the hippocampal dentate gyrus reducing the increase in glutamate release induced by acute stress in both the prefrontal and frontal cortex (Tardito et al. 2010). Several preclinical studies have reported that agomelatine can modulate the expression of various depression-related molecules such as BDNF, basic fibroblast-growth factor-2, and activity regulated cytoskeleton-associated protein (Calabrese et al. 2010) also activating several cellular signals implicated in the action of ADs such as extracellular signal-regulated kinase 1/2 (ERK1/2), protein kinase B (Akt), and glycogen synthase kinase 3β (GSK3β) (Banasr et al. 2006; Conboy et al. 2009; Soumier et al. 2009; Molteni et al. 2010).

Recently, Morley-Fletcher et al. (2011) found that a 3- and 6-week treatment with agomelatine (40–50 mg/kg daily) may rehabilitate in adult prenatal restraint stress rats the hippocampal levels of phosphorylated *c*AMP-responsive p-CREB and metabotropic glutamate receptors (mGluRs) 2/3 and mGluR5 as well as the reduced neurogenesis in the ventral hippocampus, a brain structure mainly involved in encoding memories associated to stress and emotions. In addition, Dagyte et al. (2011a) found that agomelatine normalized the stress-affected neuronal activity and promoted neurogenesis in the hippocampus of rats exposed to chronic footshock stress. They showed that chronic stress reduced *c*-Fos (a cellular proto-oncogene) expression in the hippocampal dentate gyrus, but chronic agomelatine treatment normalized neuronal activity in chronically stressed rats enhancing hippocampal cell proliferation and cell survival.

More recently, Dagyte et al. (2011b) found that chronic stress increased total synapsin I (SynI) (a regulator of synaptic transmission and plasticity) expression in all layers of the medial prefrontal cortex whereas agomelatine treatment administered for 3 weeks eliminated some of these effects. Chronic agomelatine administration reduced the fraction of phosphorylated SynI in all layers of the medial prefrontal cortex as well as selectively in the outer and middle molecular layers of the hippocampal dentate gyrus. In order to better understand agomelatine effects in the stress-compromised brain, Dagyte et al. (2011a) also investigated hippocampal neurogenesis in adult male rats subjected to different mild stressors for 5 weeks and treated with agomelatine during the last 3 weeks of the stress period. They reported that chronic mild stress significantly decreased the newborn cell survival and double cortin expression in the dentate gyrus but these changes can be reversed with

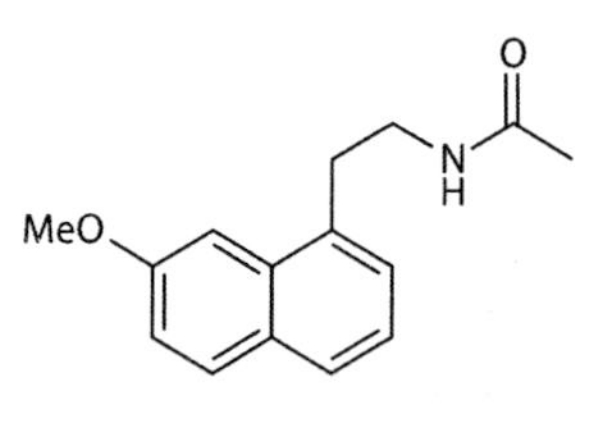

FIGURE 16.3 Chemical structure of agomelatine.

agomelatine that completely normalized stress-affected cell survival and partly reduced double cortin expression. AlAhmed and Herbert (2010) have found that agomelatine through an intact diurnal corticosterone rhythm may promote, presumably through its antagonism of the $5HT_{2C}$ receptor, progenitor cell mitosis in the dentate gyrus.

Agomelatine was also reported to reverse abnormal hippocampal neurogenesis and BDNF expression in transgenic mice with impaired glucocorticoid receptors (Païzanis et al. 2010). Chronic agomelatine, but not fluoxetine, increased survival of newly formed cells in the ventral part of the hippocampus without changing their phenotypic differentiation into neurons and by promoting cell proliferation and BDNF messenger RNA expression. Moreover, agomelatine has been reported to have an antidepressant-like activity in specific models of learned helplessness (Barden et al. 2005; Bertaina-Anglade et al. 2006), chronic mild stress (Papp et al. 2003), and forced swim test (Bourin et al. 2004). The modulation of BDNF, whose gene, protein expression, and function were found to be defective in mood disorders (Groves 2007; Martinowich et al. 2007; Kozisek et al. 2008; Calabrese et al. 2009), is a key element in long-term adaptive changes and reversion/normalization of neural defects induced by ADs (Calabrese et al. 2011b). Molteni et al. (2010) showed that the expression of BDNF messenger RNA (mRNA) levels in the prefrontal cortex may be up-regulated preventing the circadian down-regulation of the neurotrophin after an acute injection of agomelatine presumably through the functional interaction between melatonergic MT1/MT2 and 5-HT2C receptors. Additionally, Calabrese et al. (2011a), investigating the effects on the mRNA and protein expression of the BDNF of chronic agomelatine treatment compared to those of venlafaxine, found that only agomelatine produced major transcriptional changes in the hippocampus and increased levels of BDNF in the hippocampus and prefrontal cortex. Considering the different effect on mRNA levels and the similar cumulative effects on BDNF levels in the hippocampus and prefrontal cortex, the authors suggested that different modulatory mechanisms may be induced in these two regions by agomelatine.

Finally, agomelatine has been also found to induce neuroprotection and neuroplasticity mechanisms (Gressens et al. 2008) in newborn rats. Both melatonin (administered 2 h by the acute lesion) and agomelatine (up to 8 h after) may partially recover white matter periventricular cysts previously induced by intraperitoneal ibotenate injection, a glutamatergic-like agent. Specifically, although agomelatine and melatonin did not prevent the initial manifestation of white matter lesions, they did promote secondary lesion repair as some axonal markers supported the hypothesis that melatonin induced axonal regrowth or sprouting. Neurocognitive and antidepressant actions of agomelatine compared to imipramine, melatonin, and fluoxetine have been also showed in "depressed" rodents using the forced swimming test (Bourin et al. 2004) as well as an improvement of mice circadian system through the phase response curve record and motor activity initiative (Van Reeth et al. 1997, 1998).

16.7 CONCLUSION: CLINICAL IMPLICATIONS AND FUTURE DIRECTIONS

Stress and depression are associated with loss of dendritic spines, dendritic atrophy and loss of synapses, decrease of glial cells both in number and size in the hippocampus and prefrontal cortex. Consequently, the hippocampus, prefrontal cortex, and related downstream structures were impaired not only structurally but also functionally. Stress is also associated with increased dendritic arborization and new synapse formation in the amygdala showing an increased volume and an abnormal functioning. The site-specific neurohistological changes in these brain areas explain most of depressive clinical correlates such as anhedonia, loss of motivation, anxiety, fear, and other cognitive dysfunctions.

In recent decades, the scenario of mood disorders and the proposed mechanisms of action of most ADs have undergone significant refinements. The first conceptualization was that regarding the existence of a brain chemical imbalance that, however, was enriched by the most recent neuroplasticity hypothesis. Neuroplasticity theory postulates the existence of structural and functional modifications in neural circuits with the final result to induce adaptations to the different environmental stimuli (see Figure 16.1 for more details). Neuroplasticity, no doubt, provides a new approach for the study of the complex pathophysiology underlying depressive illness but also for the development of new pharmacological active agents. Currently our knowledge does not allow us to conclude whether neuroplasticity and neurogenesis modifications represent the cause or result of the neuropathological processes related to depression. Another criticism is that neuroplasticity theory is not able to explain why ketamine (Zarate et al. 2006; Anh et al. 2010), scopolamine (Furey and Drevets 2006; Drevets and Furey 2010), and electroconvulsive shock therapy (Andrade et al. 1990) exert relevant antidepressant properties.

Future longitudinal studies including larger samples of subjects should elucidate this crucial point allowing a more detailed understanding of the pathophysiology of mood disorders. However, we know that ADs are associated with the induction of neuroplasticity, neurogenesis, gliogenesis, dendritic arborization, and new synapse formation in both the hippocampus and prefrontal cortex. Glutamate and both NMDA and AMPA activation are thought to play a crucial role in morphological changes regulating neuroplasticity. Depressive illness and stress-related modifications may

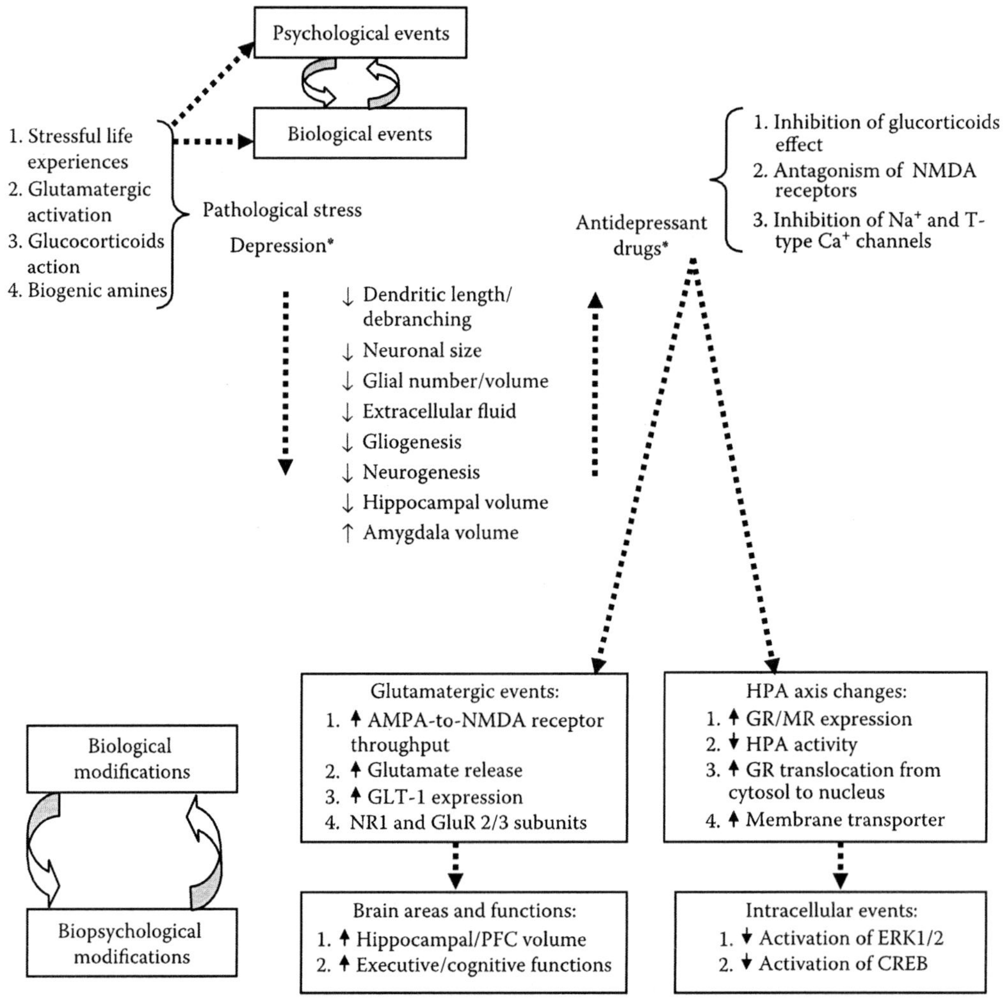

FIGURE 16.4 Proposed mechanisms leading to structural and functional modifications underlying pathological stress and depression together with the effect of antidepressant treatments on the most relevant stress-induced changes. **Source*: Modified by Fuchs (2009) and Calabrese et al. (2011a,b).

affect glutamate receptors and glutamatergic neurotransmitter system whose alterations, however, may be reversed by the administration of modern ADs.

As suggested by Kasper and McEwen (2008), ADs may reverse structural and functional modifications underlying depression promoting neuroplasticity mechanisms and presumably preventing the illness progression. Overall, such ADs may reverse stress-induced loss by reducing the retraction of hippocampal neurons (neuroplasticity) or increasing cell survival and functions (neurogenesis) (Figure 16.4). However, drugs may reverse the structural and functional consequences of stress in a site-specific manner (different changes induced in the hippocampus, prefrontal cortex, amygdala). This presumably explains why, although most of depressive clinical manifestations may be reversed with the ADs administration, the vulnerability to stress therefore remains (Diamond et al. 2004) providing a rationale for the required maintenance AD therapy also after the initial successful treatment of depression.

ABBREVIATIONS

ADs	antidepression drugs
AMPA	α-amino-3-hydroxy-5-methyl-4-isoxazolepropionic acid
BDNF	brain-derived neurotrophic factor
CaMKII	calcium/Calmodulin-dependent protein kinase II
CREB	*c*-AMP regulatory element-binding protein
ERK1/2	extracellular signal-regulated kinase
GLT	glutamate transporter
GR	glucocorticoid receptor
GSK3β	glycogen synthase kinase 3β
HPA	hypotalamic pituitary adrenal
MAOI	monoaminooxidase inhibitors
MAPK	mitogen-activated protein kinase
MDD	major depressive disorder
mGluRs	metabotropic glutamate receptors
MR	mineralocorticoid receptor
MRI	magnetic resonance imaging
mRNA	messenger RNA
NARIs	noradrenergic selective reuptake inhibitors
NDRIs	noradrenergic and dopaminergic inhibitors
NMDA	N-methyl-D-aspartate
NR1	NMDA NR1 subunit
PFC	prefrontal
PKA	protein kinase A
SNRIs	noradrenergic reuptake inhibitors
SSRIs	selective serotonergic reuptake inhibitors
SynI	synapsin I
TCAs	tricyclics
TrK	tyrosine kinase

REMARKS ON THE METHODOLOGY

In order to provide a critical review about neuroplasticity, depression, stress-related neuropathological changes, and antidepressant treatment, we performed a careful Pubmed/Medline, Scopus, PsycLit, and PsycInfo search to identify all papers and book chapters during the period between 1980 and 2010. The search used the following terms: "Depression" OR "Affective Disorders" OR "Mood Disorders" AND "Neuroplasticity" OR "Neurogenesis" OR "Synaptic plasticity" AND

"Antidepressants" OR "Antidepressant drugs" OR "Antidepressant medications" OR "Antidepressant agents" AND "Treatment" OR "Intervention" OR "Future implications." Reference lists of the articles included in the review were manually checked for relevant studies. Included papers were restricted to those in English. Where a title or abstract seemed to describe a study eligible for inclusion, the full article was obtained and examined to assess its relevance based on the inclusion criteria. Two independent researchers conducted a two-step literature search. Any discrepancies between the two reviewers who, blind to each other, examined the studies for the possible inclusion were resolved by consultations with a senior author.

REFERENCES

Aan het Rot, M., Collins, K.A., Murrough, J.W. et al. 2010. Safety and efficacy of repeated-dose intravenous ketamine for treatment-resistant depression. *Biol. Psychiatry* 67:139–145.

AlAhmed, S. and Herbert, J. 2010. Effect of agomelatine and its interaction with the daily corticosterone rhythm on progenitor cell proliferation in the dentate gyrus of the adult rat. *Neuropharmacology* 59:375–379.

Anacker, C., Zunszain, P.A., Carvalho, L.A., and Pariante, C.M. 2010. The glucocorticoid receptor: Pivot of depression and of antidepressant treatment? *Psychoneuroendocrinology* 36:415–425.

Anacker, C., Zunszain, P.A., Cattaneo, A. et al. 2011. Anti depressants increase human hippocampal neurogenesis by activating the glucocorticoid receptor. *Mol. Psychiatry* 16(7):738–750.

Andersen, I., Thielen, K., Bech, P., Nygaard, E., and Diderichsen, F. 2011. Increasing prevalence of depression from 2000 to 2006. *Scand. J. Public Health* 39(8):857–863.

Andrade, C., Gangadhar, B.N., and Channabasavanna, S.M. 1990. Further characterization of mania as a side effect of ECT. *Convul. Ther.* 6:318–319.

Andrade, C. and Rao, N.S. 2010. How antidepressant drugs act: A primer on neuroplasticity as the eventual mediator of antidepressant efficacy. *Indian J. Psychiatry* 52(4):378–386.

Banasr, M., Soumier, A., Hery, M., Mocaer, E., and Daszuta, A. 2006. Agomelatine, a new antidepressant, induces regional changes in hippocampal neurogenesis. *Biol. Psychiatry* 59:1087–1096.

Barden, N., Shink, E., Labbé, M., Vacher, R., Rochford, J., and Mocaër, E. 2005. Antidepressant action of agomelatine (S 20098) in a transgenic mouse model. *Prog. Neuropsychopharmacol. Biol. Psychiatry* 29:908–916.

Barria, A., Muller, D., Derkach, V., Griffith, L.C., and Soderling, T.R. 1997. Regulatory phosphorylation of AMPA-type glutamate receptors by CaM-KII during long-term potentiation. *Science* 276(5321):2042–2045.

Beck, C.H. 1995. Acute treatment with anti depressant drugs selectively increases the expression of c-fos in the rat brain. *J. Psychiatry Neurosci.* 20(1):25–32.

Bertaina-Anglade, V., la Rochelle, C.D., Boyer, P.A., and Mocaër, E. 2006. Antidepressant-like effects of agomelatine (S 20098) in the learned helplessness model. *Behav. Pharmacol.* 17:703–713.

Berton, O. and Nestler, E.J. 2006. New approaches to antidepressant drug discovery: Beyond monoamines. *Nat. Rev. Neurosci.* 7(2):137–151.

Bessa, J.M., Ferreira, D., Melo, I. et al. 2009. The mood-improving actions of antidepressants do not depend on neurogenesis but are associated with neuronal remodeling. *Mol. Psychiatry* 14:764–773.

Bourin, M., Mocaër, E., and Porsolt, R. 2004. Antidepressant-like activity of S 20098 (agomelatine) in the forced swimming test in rodents: Involvement of melatonin and serotonin receptors. *J. Psychiatry Neurosci.* 29:126–133.

Bowley, M.P., Drevets, W.C., Ongür, D., and Price, J.L. 2002. Low glial numbers in the amygdala in major depressive disorder. *Biol. Psychiatry* 52(5):404–412.

Bremner, J.D., Narayan, M., Anderson, E.R., Staib, L.H., Miller, H.L., and Charney, D.S. 2000. Hippocampal volume reduction in major depression. *Am. J. Psychiatry* 157(1):115–118.

Calabrese, F., Molteni, R., Cattaneo, A. et al. 2010. Long-Term duloxetine treatment normalizes altered brain-derived neurotrophic factor expression in serotonin transporter knockout rats through the modulation of specific neurotrophin isoforms. *Mol. Pharmacol.* 77(5):846–853.

Calabrese, F., Molteni, R., Gabriel, C., Mocaer, E., Racagni, G., and Riva, M.A. 2011a. Modulation of neuroplastic molecules in selected brain regions after chronic administration of the novel antidepressant agomelatine. *Psychopharmacology* 215(2):267–275.

Calabrese, F., Molteni, R., Racagni, G., and Riva, M.A. 2009. Neuronal plasticity: A link between stress and mood disorders. *Psychoneuroendocrinology* 34(Suppl. 1):S208–S216.

Calabrese, F., Molteni, R., and Riva, M.A. 2011b. Antistress properties of antidepressant drugs and their clinical implications. *Pharmacol. Ther.* 132:39–56.

Carvalho, L.A., Garner, B.A., Dew, T., Fazakerley, H., and Pariante, C.M. 2010. Antidepressants, but not antipsychotics, modulate GR function in human whole blood: An insight into molecular mechanisms. *Eur. Neuropsychopharmacol.* 20(6):379–387.

Castren, E., Voikar, V., and Rantamaki, T. 2007. Role of neurotrophic factors in depression. *Curr. Opin. Pharmacol.* 7:18–21.

Conboy, L., Tanrikut, C., Zoladz, P.R. et al. 2009. The antidepressant agomelatine blocks the adverse effects of stress on memory and enables spatial learning to rapidly increase neural cell adhesion molecule (NCAM) expression in the hippocampus of rats. *Int. J. Neuropsychopharmacol.* 12:329–341.

Cooper, B., Fuchs, E., and Flügge, G. 2009. Expression of the axonal membrane glycoprotein M6a is regulated by chronic stress. *PLoS One* 4(1):e3659.

Cotter, D., Mackay, D., Chana, G., Beasley, C., Landau, S., and Everall, I.P. 2002. Reduced neuronal size and glial cell density in area 9 of the dorsolateral prefrontal cortex in subjects with major depressive disorder. *Cereb. Cortex* 12(4):386–394.

Cotter, D.R., Pariante, C.M., and Everall, I.P. 2001. Glial cell abnormalities in major psychiatric disorders: The evidence and implications. *Brain Res. Bull.* 55(5):585–595.

Coyle, J.T. and Duman, R.S. 2003. Finding the intracellular signaling pathways affected by mood disorder treatments. *Neuron* 38:157–160.

Coyle, J.T. and Schwarcz, R. 2000. Mindglue: Implications of glial cell biology for psychiatry. *Arch. Gen. Psychiatry* 57(1):90–93.

Czéh, B. and Lucassen, P.J. 2007. What causes the hippocampal volume decrease in depression? Are neurogenesis, glial changes and apoptosis implicated? *Eur. Arch. Psychiatry Clin. Neurosci.* 257(5):250–260.

Czéh, B., Michaelis, T., Watanabe, T. et al. 2001. Stress-induced changes in cerebral metabolites, hippocampal volume, and cell proliferation are prevented by antidepressant treatment with tianeptine. *Proc. Natl Acad. Sci. USA* 98(22):12796–12801.

Czéh, B., Simon, M., Schmelting, B., Hiemke, C., and Fuchs, E. 2006. Astroglial plasticity in the hippocampus is affected by chronic psychosocial stress and concomitant fluoxetine treatment. *Neuropsychopharmacology* 31(8):1616–1626.

Dagyte, G., Crescente, I., Postema, F. et al. 2011a. Agomelatine reverses the decrease in hippocampal cell survival induced by chronic mild stress. *Behav. Brain Res.* 218(1):121–128.

Dagyte, G., Luiten, P.G., DeJager, T. et al. 2011b. Chronic stress and antidepressant agomelatine induce region-specific changes in synapsin I expression in the rat brain. *J. Neurosci. Res.* 89(10):1646–1657.

Dagyte, G., Trentani, A., Postema. F. et al. 2010. The novel antidepressant agomelatine normalizes hippocampal neuronal activity and promotes neurogenesis in chronically stressed rats. *CNS Neurosci. Ther.* 16:195–207.

Daszuta, A., Ban, M. Sr., Soumier, A., Hery, M., and Mocaer, E. 2005. Depression and neuroplasticity: Implication of serotoninergic systems. *Therapie* 60(5):461–468.

David, D.J., Samuels, B.A., Rainer, Q. et al. 2009. Neurogenesis-dependent and -independent effects of fluoxetine in an animal model of anxiety/depression. *Neuron* 62:479–493.

Diamond, D.M., Campbell, A., Park, C.R., and Vouimba, R.M. 2004. Preclinical research on stress, memory, and the brain in the development of pharmacotherapy for depression. *Eur. Neuropsychopharmacol.* 14:S491–S495.

Dranovsky, A. and Hen, R. 2006. Hippocampal neurogenesis: Regulation by stress and antidepressants. *Biol. Psychiatry* 59(12):1136–1143.

Drevets, W.C. and Furey, M.L. 2010. Replication of scopolamine's antidepressant efficacy in major depressive disorder: A randomized, placebo-controlled clinical trial. *Biol. Psychiatry* 67:432–438.

D'Sa, C. and Duman, R.S. 2002. Antidepressants and neuroplasticity. *Bipolar Disord.* 4:183–194.

Du, J., Suzuki, K., Wei, Y. et al. 2007. The anticonvulsants lamotrigine, riluzole, and valproate differentially regulate AMPA receptor membrane localization: Relationship to clinical effects in mood disorders. *Neuropsychopharmacology* 32(4):793–802.

Duman, R.S. 2004a. Introduction: Theories of depression—From monoamines to neuroplasticity. In *Neuroplasticity: A New Approach to the Pathophysiology of Depression*, eds. J.P. Olie, J.A. Costa e Silve, and J.P. Macher, pp. 1–11. London, U.K.: Science Press Ltd.

Duman, R.S. 2004b. Role of neurotrophic factors in the etiology and treatment of mood disorders. *Neuromol. Med.* 5:11–25.

Duman, R.S., Malberg, J., and Thome, J. 1999. Neural plasticity to stress and antidepressant treatment. *Biol. Psychiatry* 46(9): 1181–1191.

Duman, R.S. and Monteggia, L.M. 2006. A neurotrophic model for stress-related mood disorders. *Biol. Psychiatry* 59:1116–1127.

Duncan, G.E., Knapp, D.J., Johnson, K.B., and Breese, G.R. 1996. Functional classification of anti depressants based on antagonism of swimstress-induced fos-like immunoreactivity. *J. Pharmacol. Exp. Ther.* 277(2):1076–1089.

Emery, D.L., Fulp, C.T., Saatman, K.E., Schütz, C., Neugebauer, E., and McIntosh, T.K. 2005. Newly born granule cells in the dentate gyrus rapidly extend axons into the hippocampal CA3 region following experimental brain injury. *J. Neurotrauma* 22(9):978–988.

Eriksson, P.S., Perfilieva, E., Bjork-Eriksson, T. et al. 1998. Neurogenesis in the adult human hippocampus. *Nat. Med.* 4:1313–1317.

Fifková, E. and VanHarreveld, A. 1977. Long-lasting morphological changes in dendritic spines of dentate granular cells following stimulation of the entorhinal area. *J. Neurocytol.* 6(2):211–230.

Frodl, T., Schaub, A., Banac, S. et al. 2006. Reduced hippocampal volume correlates with executive dysfunctioning in major depression. *J. Psychiatry Neurosci.* 31:316–323.

Fuchs, E. 2004. Animal models of depression. In *Neuroplasticity: A New Approach to the Pathophysiology of Depression*, eds. J.P. Olie, J.A. Costa e Silve, and J.P. Macher, pp. 39–50. London, U.K.: Science Press Ltd.

Fuchs, E. 2009. Neuroplasticity: A new approach to the pathophysiology of depression. In *Neuroplasticity: New Biochemical Mechanisms*, eds. J.A. Costa e Silva, J.P. Macher, and J.P. Olié, pp. 1–12. London, U.K.: Current Medicine Group.

Fuchs, E., Czéh, B., Kole, M.H., Michaelis, T., and Lucassen, P.J. 2004. Alterations of neuroplasticity in depression: The hippocampus and beyond. *Eur. Neuropsychopharmacol.* 14(Suppl 5):S481–S490.

Funato, H., Kobayashi, A., and Watanabe, Y. 2006. Differential effects of antidepressants on dexamethasone-induced nuclear translocation and expression of glucocorticoid receptor. *Brain Res.* 1117:125–134.

Furey, M.L. and Drevets, W.C. 2006. Antidepressant efficacy of the antimuscarinic drug scopolamine: A randomized, placebo-controlled clinical trial. *Arch. Gen. Psychiatry* 63:1121–1129.

Gorwood, P. 2009. Clinical consequences of the role of glutamate and neuroplasticity in depressive disorder. In *Neuroplasticity: New Biochemical Mechanisms*, eds. J.A. Costa e Silve, J.P. Macher, and J.P. Olie, pp. 57–68. London, U.K.: Current Medical Group.

Gressens, P., Schwendimann, L., Husson, I. et al. 2008. Agomelatine, a melatonin receptor agonist with 5-HT(2C) receptor antagonist properties, protects the developing murine white matter against excitotoxicity. *Eur. J. Pharmacol.* 588(1):58–63.

Groves, J.O. 2007. Is it time to reassess the BDNF hypothesis of depression? *Mol. Psychiatry* 12:1079–1088.

Halpain, S., Hipolito, A., and Saffer, L. 1998. Regulation of F-actin stability in dendritic spines by glutamate receptors and calcineurin. *J. Neurosci.* 18(23):9835–9844.

Hanoun, N., Mocaer, E., Boyer, P.A., Hamon, M., and Lanfumey, L. 2004. Differential effects of the novel antidepressant agomelatine (S 20098) versus fluoxetine on 5-HT1A receptors in the rat brain. *Neuropharmacology* 47:515–526.

Huang, G.J. and Herbert, J. 2006. Stimulation of neurogenesis in the hippocampus of the adult rat by fluoxetine requires rhythmic change in corticosterone. *Biol. Psychiatry* 59:619–624.

Jay, T. 2009. Cellular plasticity and the pathophysiology of depression. In *Neuroplasticity: New Biochemical Mechanisms*, eds. J.A. Costa e Silve, J.P. Macher, and J.P. Olie, pp. 41–56. London, U.K.: Current Medical Group.

Jeon, S.H., Seong, Y.S., Juhnn, Y.S. et al. 1997. Electro convulsive shock increases the phosphorylation of cyclic AMP response element binding protein at Ser-133 inrathippocampus but not in cerebellum. *Neuropharmacology* 36(3):411–414.

Kasper, S. and McEwen, B.S. 2008. Neurobiological and clinical effects of the antidepressant tianeptine. *CNS Drugs* 22(1):15–26.

Kim, M.J., Kim, H.K., Kim, B.S., and Yim, S.V. 2004. Melatonin increases cell proliferation in the dentate gyrus of maternally separated rats. *J. Pineal. Res.* 37:193–197.

Kozisek, M.E., Middlemas, D., and Bylund, D.B. 2008. Brain-derived neurotrophic factor and its receptor tropomyosin-related kinase B in the mechanism of action of antidepressant therapies. *Pharmacol. Ther.* 117:30–51.

Laje, G., Paddock, S., Manji, H. et al. 2007. Genetic markers of suicidal ideation emerging during citalopram treatment of major depression. *Am. J. Psychiatry* 164(10):1530–1538.

Lorenzetti, V., Allen, N.B., Fornito, A., and Yücel, M. 2009. Structural brain abnormalities in major depressive disorder: A selective review of recent MRI studies. *J. Affect. Disord.* 117(1–2):1–17.

Lucassen, P.J., Fuchs, E., and Czéh, B. 2004. Antidepressant treatment with tianeptine reduces apoptosis in the hippocampal dentate gyrus and temporal cortex. *Biol. Psychiatry* 55(8):789–796.

Magariños, A.M., Deslandes, A., and McEwen, B.S. 1999. Effects of antidepressants and benzodiazepine treatments on the dendritic structure of CA3 pyramidal neurons after chronic stress. *Eur. J. Pharmacol.* 371(2–3):113–122.

Magariños, A.M. and McEwen, B.S. 1995. Stress-induced atrophy of apical dendrites of hippocampal CA3 neurons: Involvement of glucocorticoid secretion and excitatory amino acid receptors. *Neuroscience* 69(1):89–98.

Magariños, A.M., McEwan, B.S., Flugge, G., and Fuchs, E. 1996. Chronic psychosocial stress causes apical dendritic atrophy of hippocampal CA3 pyramidal neurons in subordinate tree shrews. *J. Neurosci.* 16:3534–3540.

Malberg, J.E., Eisch, A.J., Nestler, E.J., and Duman, R.S. 2000. Chronic antidepressant treatment increases neurogenesis in adult rat hippocampus. *J. Neurosci.* 20(24):9104–9110.

Manji, H.K., Drevets, W.C., and Charney, D.S. 2001. The cellular neurobiology of depression. *Nat. Med.* 7(5):541–547.

Manji, H.K. and Duman, R.S. 2001. Impairments of neuroplasticity and cellular resilience in severe mood disorders: Implications for the development of novel therapeutics. *Psychopharmacol. Bull.* 35(2):5–49.

Manji, H.K., Moore, G.J., Rajkowska, G., and Chen, G. 2000. Neuroplasticity and cellular resilience in mood disorders. *Mol. Psychiatry* 5(6):578–593.

Manji, H.K., Quiroz, J.A., Sporn, J. et al. 2003. Enhancing neuronal plasticity and cellular resilience to develop novel, improved therapeutics for difficult-to-treat depression. *Biol. Psychiatry* 53:707–742.

Martinowich, K., Manji, H., and Lu, B. 2007. New insights into BDNF function in depression and anxiety. *Nat. Neurosci.* 10:1089–1093.

McEwen, B.S. and Chattarji, S. 2004. Molecular mechanisms of neuroplasticity and pharmacological implications: The example of tianeptine. *Eur. Neuropsychopharmacol.* 14(Suppl 5):S497–S502.

Molteni, R., Calabrese, F., Pisoni, S. et al. 2010. Synergistic mechanisms in the modulation of the neurotrophin BDNF in the rat prefrontal cortex following acute agomelatine administration. *World J. Biol. Psychiatry* 11:148–153.

Morley-Fletcher, S., Mairesse, J., Soumier, A. et al. 2011. Chronic agomelatine treatment corrects behavioral, cellular, and biochemical abnormalities induced by prenatal stress in rats. *Psychopharmacology* 217(3):301–313.

Nestler, E.J., Barrot, M., DiLeone, R.J., Eisch, A.J., Gold, S.J., and Monteggia, L.M. 2002. Neurobiology of depression. *Neuron* 34:13–25.

Païzanis, E., Renoir, T., Lelievre, V. et al. 2010. Behavioural and neuroplastic effects of the new-generation antidepressant. *Int. J. Neuropsychopharmacol.* 13:759–774.

Papp, M., Gruca, P., Boyer, P.A., and Mocaer, E. 2003. Effect of agomelatine in the chronic mild stress model of depression in the rat. *Neuropsychopharmacology* 28:694–703.

Pariante, C.M., Hye, A., Williamson, R., Makoff, A., Lovestone, S., and Kerwin, R.W. 2003a. The antidepressant clomipramine regulates cortisol intracellular concentrations and glucocorticoid receptor expression in fibroblasts and rat primary neurones. *Neuropsychopharmacology* 28:1553–1561.

Pariante, C.M., Kim, R.B., Makoff, A., and Kerwin, R.W. 2003b. Antidepressant fluoxetine enhances glucocorticoid receptor function in vitro by modulating membrane steroid transporters. *Br. J. Pharmacol.* 139:1111–1118.

Pariante, C.M. and Lightman, S.L. 2008. The HPA axis in major depression: Classical theories and new developments. *Trends Neurosci.* 31: 464–468.

Pariante, C.M., Pearce, B.D., Pisell, T.L., Owens, M.J., and Miller, A.H. 1997. Steroid-independent translocation of the glucocorticoid receptor by the antidepressant desipramine. *Mol. Pharmacol.* 52:571–581.

Perez-Cruz, C., Müller-Keuker, J.I., Heilbronner, U., Fuchs, E., and Flügge, G. 2007. Morphology of pyramidal neurons in the rat prefrontal cortex: Lateralized dendritic remodeling by chronic stress. *Neural Plast.* 46276.

Perez-Cruz, C., Simon, M., Czéh, B., Flügge, G., and Fuchs, E. 2009. Hemispheric differences in basilar dendrites and spines of pyramidal neurons in the rat prelimbic cortex: Activity- and stress-induced changes. *Eur. J. Neurosci.* 29(4):738–747.

Pfleiderer, B., Michael, N., Erfurth, A. et al. 2003. Effective electroconvulsive therapy reverses glutamate/glutamine deficit in the left anterior cingulum of unipolar depressed patients. *Psychiatry Res.* 122(3):185–192.

Pittenger, C. and Duman, R.S. 2008. Stress, depression, and neuroplasticity: A convergence of mechanisms. *Neuropsychopharmacology* 33:88–109.

Popoli, M., Vocaturo, C., Perez, J., Smeraldi, E., and Racagni, G. 1995. Presynaptic Ca^{2+}/calmodulin-dependent protein kinase II: Autophosphorylation and activity increase in the hippocampus after long-term blockade of serotonin reuptake. *Mol. Pharmacol.* 48(4):623–629.

Popov, V.I. and Bochorova, L.S. 1992. Hibernation-induced structural changes in synaptic contacts between mossy fibres and hippocampal pyramidal neurons. *Neuroscience* 48:53–62.

Rajkowska, G., Miguel-Hidalgo, J.J., Wei, J. et al. 1999. Morphometric evidence for neuronal and glial prefrontal cell pathology in major depression. *Biol. Psychiatry* 45(9):1085–1098.

Reagan, L.P., Rosell, D.R., Wood, G.E. et al. 2004. Chronic restraint stress up-regulates GLT-1 mRNA and protein expression in the rat hippocampus: Reversal by tianeptine. *Proc. Natl Acad. Sci. USA* 101(7):2179–2184.

Reif, A., Fritzen, S., Finger, M. et al. 2006. Neural stem cell proliferation is decreased in schizophrenia, but not in depression. *Mol. Psychiatry* 11(5):514–522.

Reznikov, L.R., Fadel, J.R., and Reagan, L.P. 2009. Glutamate-mediated neuroplasticity deficits in mood disorders. In *Neuroplasticity: New Biochemical Mechanisms*, eds. J.A. Costa e Silva, J.P. Macher, and J.P. Olié, pp. 13–26. London, U.K.: Current Medicine Group.

Reznikov, L.R., Grillo, C.A., Piroli, G.G., Pasumarthi, R.K., Reagan, L.P., and Fadel, J. 2007. Acute stress-mediated increases in extracellular glutamate levels in the rat amygdala: Differential effects of antidepressant treatment. *Eur. J. Neurosci.* 25(10):3109–3114.

Roche, K.W., O'Brien, R.J., Mammen, A.L., Bernhardt, J., and Huganir, RL. 1996. Characterization of multiple phosphorylation sites on the AMPA receptor GluR1subunit. *Neuron* 16(6):1179–1188.

Rocher, C., Spedding, M., Munoz, C., and Jay, T.M. 2004. Acute stress-induced changes in hippocampal/prefrontal circuits in rats: Effects of antidepressants. *Cereb. Cortex* 14(2):224–229.

Sairanen, M., O'Leary, O.F., Knuuttila, J.E., and Castren, E. 2007. Chronic antidepressant treatment selectively increases expression of plasticity-related proteins in the hippocampus and medial prefrontal cortex of the rat. *Neuroscience* 144:368–374.

Sheline, Y.I. 2004. Consequences of depression in the hippocampus and other brain regions. In *Neuroplasticity: A New Approach to the Pathophysiology of Depression*, eds. J.P. Olie, J.A. Costa e Silve, and J.P. Macher, pp. 25–37. London, U.K.: Science Press Ltd.

Sheline, Y.I., Gado, M.H., and Kraemer, H.C. 2003. Untreated depression and hippocampal volume loss. *Am. J. Psychiatry* 160(8):1516–1518.

Sheline, Y.I., Gado, M.H., and Price, J.L. 1998. Amygdala core nuclei volumes are decreased in recurrent major depression. *Neuroreport* 9(9):2023–2028.

Soumier, A., Banasr, M., Lortet, S. et al. 2009. Mechanisms contributing to the phase-dependent regulation of neurogenesis by the novel antidepressant, agomelatine, in the adult rat hippocampus. *Neuropsychopharmacology* 34:2390–2403.

Sousa, N., Lukoyanov, N.V., Madeira, M.D., Almeida, O.F., and Paula-Barbosa, M.M. 2000. Reorganization of the morphology of hippocampal neurites and synapses after stress-induced damage correlates with behavioral improvement. *Neuroscience* 97(2):253–266.

Svenningsson, P., Bateup, H., Qi, H. et al. 2007. Involvement of AMPA receptor phosphorylation in antidepressant actions with special reference to tianeptine. *Eur. J. Neurosci.* 26(12):3509–3517.

Svenningsson, P., Tzavara, E.T., Witkin, J.M., Fienberg, A.A., Nomikos, G.G., and Greengard, P. 2002. Involvement of striatal and extrastriatal DARPP-32 in biochemical and behavioral effects of fluoxetine (Prozac). *Proc. Natl Acad. Sci. USA* 99(5):3182–3187.

Tardito, D., Milanese, M., Bonifacino, T. et al. 2010. Blockade of stress-induced increase of glutamate release in the rat prefrontal/frontal cortex by agomelatine involves synergy between melatonergic and 5-HT_{2C} receptor-dependent pathways. *BMC Neurosci.* 11:68.

Van Reeth, O., Olivares, E., Turek, F.W., Granjon, L., and Mocaer, E. 1998. Resynchronisation of a diurnal rodent circadian clock accelerated by a melatonin agonist. *Neuroreport* 9(8): 1901–1905.

Van Reeth, O., Olivares, E., Zhang, Y. et al. 1997. Comparative effects of a melatonin agonist on the circadian system in mice and Syrian hamsters. *Brain Res.* 762(1–2): 185–194.

Vermetten, E., Vythilingam, M., Southwick, S.M., Charney, D.S., and Bremner, J.D. 2003. Long-term treatment with paroxetine increases verbal declarative memory and hippocampal volume in posttraumatic stress disorder. *Biol. Psychiatry* 54(7):693–702.

Virgin, C.E. Jr., Ha, T.P., Packan, D.R. et al. 1991. Glucocorticoids inhibit glucose transport and glutamate uptake in hippocampal astrocytes: Implications for glucocorticoid neurotoxicity. *J. Neurochem.* 57(4):1422–1428.

Vouimba, R.M., Muñoz, C., and Diamond, D.M. 2006. Differential effects of predator stress and the antidepressant and leptin on physiological plasticity in the hippocampus and basolateral amygdala. *Stress* 9(1):29–40.

Vyas, A., Mitra, R., Shankaranarayana Rao, B.S., and Chattarji, S. 2002. Chronic stress induces contrasting patterns of dendritic remodeling in hippocampal and amygdaloid neurons. *J. Neurosci.* 22(15):6810–6818.

Watanabe, Y., Gould, E., Daniels, D.C., Cameron, H., and McEwen, B.S. 1992. Tianeptine attenuates stress-induced morphological changes in the hippocampus. *Eur. J. Pharmacol.* 222(1):157–162.

Wooley, C.S. 1999. Effects of estrogens in the CNS. *Curr. Opin. Neurobiol.* 9:349–354.

Woolley, C.S., Gould, E., and McEwan, B.S. 1990. Exposure to excess glucocorticoids alters dendritic morphology of adult hippocampal pyramidal neurons. *Brain Res.* 531:225–231.

World Health Organization. 2007. *Information on Mental Disorders Management: Depression*. Geneva, Switzerland: WHO.

Yan, H.C., Cao, X., Gao, T.M., and Zhu, X.H. 2011. Promoting adult hippocampal neurogenesis: A novel strategy for antidepressant drug screening. *Curr. Med. Chem.* 18(28):4359–4367.

Zarate, C.A. Jr., Singh, J.B., Carlson, P.J. et al. 2006. A randomized trial of an N-methyl-D-aspartate antagonist in treatment-resistant major depression. *Arch. Gen. Psychiatry* 63:856–864.

17 Statins
Fermentation Products for Cholesterol Control in Humans

Vincent Gullo and Arnold L. Demain

CONTENTS

17.1 INTRODUCTION

Only 25% of the cholesterol in the human body comes from the diet. The remaining 75% is synthesized by the body, mainly in the liver. Many people cannot control their cholesterol at a healthy level by diet alone but must depend on hypocholesterolemic drugs. High blood cholesterol leads to atherosclerosis, which is a causal factor in many types of coronary heart disease, a leading cause of human death. Statins reduce cardiovascular events including myocardial infarction, stroke, and death (Veillard and Mach 2002). The statins inhibit *de novo* synthesis of cholesterol in the liver, the major source of blood cholesterol. Statins are successful because they reduce total plasma cholesterol by 20%–40%, whereas the previously used fibrates only reduced it by 10%–15% (Knowles and Gromo 2003). A large segment of the U.S. pharmaceutical business is for cholesterol-lowering drugs, with worldwide sales reaching $32.4 billion in 2006 (Branca and Sannes 2006). Lipitor is the leading drug of the pharmaceutical industry, with worldwide sales in 2008 of $13.7 billion (Van Arnum 2009). Zocor generated worldwide annual sales of $7.2 billion in 2002 (Downton and Clark 2003) and Crestor sales were $5.7 billion in 2010 (Hirschler 2011).

It is quite remarkable that a microbial natural product, an antifungal antibiotic (Brown et al. 1976), has extended the lives of millions of people by lowering their cholesterol levels. The statins are microbially produced enzyme inhibitors, inhibiting 3-hydroxy-3-methylglutaryl-coenzyme A reductase, the regulatory and rate-limiting enzyme of cholesterol biosynthesis in liver.

17.2 DISCOVERY AND DEVELOPMENT OF THE STATINS

The discovery and development of the statins present a fascinating story (Endo 2010). Akira Endo, a scientist of the pharmaceutical company Sankyo, took a sabbatical leave in the mid-1960s and joined the laboratory of the famous biochemist, Bernard Horecker, at the Albert Einstein Medical College in New York City. His goal was to study phospholipid metabolism. He became fascinated with cholesterol biosynthesis, and upon his return to the company in 1968, screened 6000 fungal extracts for inhibition of cholesterol biosynthesis in rat liver membranes. He found two closely related active compounds, i.e., ML-236A and 236B (Figure 17.1), produced by *Penicillium citrinum* (Endo et al. 1976a,b). The structure of ML-236A differs from ML-236B in lacking the sec-butyl ester side chain. ML-236B, the more potent inhibitor of 3-hydroxy-3-methylglutaryl coenzyme A reductase, $IC_{50} = 2.6 \times 10^{-8}$ M, was named compactin (mevastatin) (Endo et al. 1976a,b). In 1976, Sankyo prepared a patent application on compactin but did not commercialize the drug. Also, in 1976, the British company, Glaxo, reported on an antifungal agent from *Penicillium compactum;* it also was compactin (Brown et al. 1976).

In 1976, Dr. H. Boyd Woodruff, a former Director of the Microbiology Department at Merck in New Jersey, was functioning as the company's representative in Japan. He heard of the work of Endo and requested a sample of compactin. Merck and Sankyo signed a confidentiality agreement and a compactin sample was given to Merck. Merck evaluated compactin in cultured mammalian cells, rats, and dogs, and by 1978 had obtained promising results. Endo reported on the discovery of monocolin K (lovastatin, mevinolin) (Figure 17.2) from *Monascus ruber* (Endo 1979) and its inhibitory activity against 3-hydroxy-3-methylglutaryl coenzyme A reductase (Endo 1980a,b). It was patented in Japan but without structure elucidation. At the same time, Merck scientists, who had been screening for new inhibitors, discovered lovastatin produced by *Aspergillus terreus* (Alberts et al. 1980). Its structure was similar to compactin but contained an additional methyl group on the hexahydronaphthalene portion of the molecule. Merck filed for a patent containing their findings, including the structure of lovastatin. The company received a U.S. patent in 1980. Sankyo, in their studies with compactin in dogs, apparently noted lymphoma at high doses and, in 1980, stopped further clinical development (Endo 2010). Merck also stopped the development of lovastatin at that

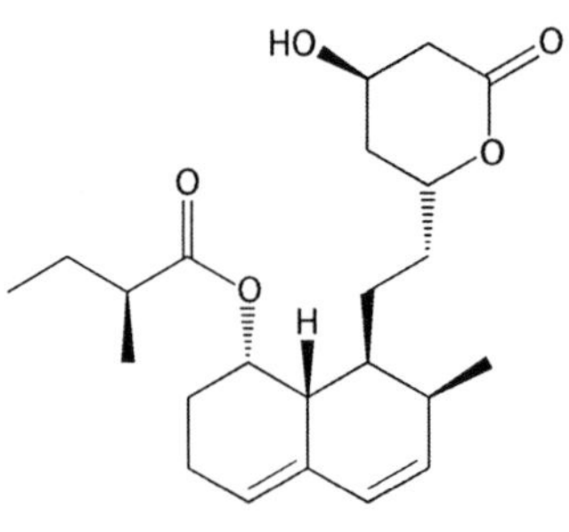

FIGURE 17.1 Structure of ML-236B (compactin).

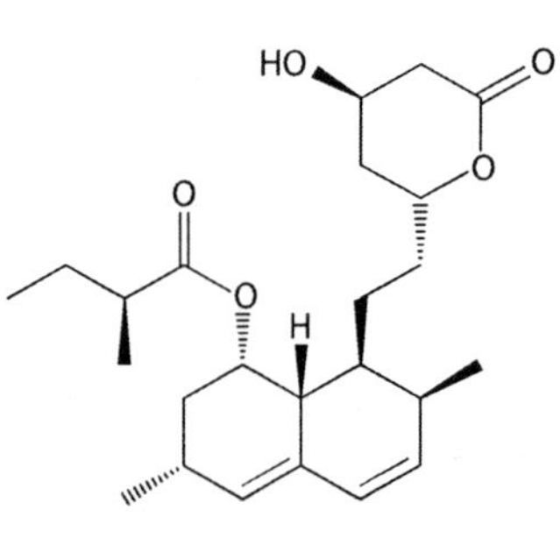

FIGURE 17.2 Structure of lovastatin (Mevacor®).

time and this inactivity continued for 2–3 years. However, since Merck observed no significant toxicity with lovastatin, they decided to resume their efforts. Further clinical tests on lovastatin progressed rapidly. The drug was approved by the FDA in 1987 after clinical tests in humans showed a lowering of total blood cholesterol of 18%–34%, a 19%–39% decrease in low-density lipoprotein cholesterol ("bad cholesterol") and a slight increase in high-density lipoprotein cholesterol ("good cholesterol"). Mevacor® (mevinolin, monocolin K, lovastatin) became the first statin on the market. The history of the medical aspects of the statins has been reviewed by Tobert (2003).

17.3 MICROBIAL PRODUCTION OF STATINS

Natural statins are produced by many fungi: *Aspergillus terreus* and species of *Monascus*, *Penicillium*, *Doratomyces*, *Eupenicillium*, *Gymnoascus*, *Hypomyces*, *Paecilomyces*, *Phoma*, *Trichoderma*, and *Pleurotis* (Manzoni and Rollini 2002).

17.3.1 Compactin

Compactin (Figure 17.1) is produced by species of *Penicillium* including *P. citrinum* (Endo et al. 1976a,b; Endo 1985), *P. brevicompactum* (Brown et al. 1976), and *P. cyclopium* (Bazaraa et al. 1998). Other producing strains include *Paecillomyces* sp., *Eupenicillium* sp, *Hypomyces chrysospermus*, *Trichoderma longibrachiatum*, and *Trichoderma pseudokoningii* (Endo et al. 1986). Statistical design studies of the medium components resulted in the production of 400 mg/L by *Penicillium* sp. IDR-629 (Konya et al. 1998). Control of morphology was found to be important in the seed stage, i.e., subsequent production was favored by increasing the level of free mycelia and decreasing the level of mycelial pellets in the seed stage (Hosobuchi et al. 1993a), and increasing the presence of small pellets in the fermentation stage (Hosobuchi et al. 1993c). Strain improvement of *P. citrinum* involving nitrosoguanidine (NTG) mutagenesis and isolation of nystatin-resistant mutants led to an improved strain, which performed well in fed-batch fermentation with continuous feeding of glycerol or glycerol plus maltose (Hosobuchi et al. 1993b). Further studies using statistical methods with *P. citrinum* resulted in a chemically defined medium capable of supporting the production of 456 mg/L of compactin (Chakravarti and Sahai 2002). A genetic transformation system was developed for *P. citrinum* (Nara et al. 1993). Cloning of genes encoding polyketide synthases in *P. citrinum* was used to identify a compactin biosynthetic gene cluster containing nine genes (Abe et al. 2002b). Increased dosage of some of the genes of the cluster was found to increase compactin production (Abe et al. 2002a). With all of the improvements mentioned earlier, the titer of compactin in *P. citrinum* was thought to be about 5 g/L (Manzoni and Rollini 2002).

17.3.2 Lovastatin (Mevacor®)

Lovastatin (Figure 17.2) is produced by *Monascus* species including *M. ruber* (Endo 1979), *M. purpureus*, *M. vitreus*, *M. pilosus* (Miyake et al. 2006a,b), and *M. pubigerus* (Negishi et al. 1986); *Aspergillus* species including *A. terreus* (Alberts et al. 1980) and *A. flavipes* (Valera et al. 2005); and *Pleurotis* species *P. ostreatus*, *P. sapidus*, and *P. saca* (Gunde-Cimerman et al. 1993). It is five times more active than compactin against 3-hydroxy-3-methylglutaryl-coenzyme A *in vitro* and when tested in rats *in vivo* (Endo 1980a,b).

Lovastatin is biosynthesized from acetate and methionine by a polyketide pathway (Moore et al. 1985). The route involves formation, from two polyketide pathways, of two polyketide chains (4-carbons and 18-carbons) of acetate units coupled head to tail, each chain bearing a methyl group derived from methionine. Separate polyketide synthases are involved (Hendrickson et al. 1999). The larger chain is a highly reduced backbone known as a triol. Oxygen atoms on the main chain are derived via aerobic oxidation of a deoxygenated precursor. The chains are joined through an ester linkage.

Of importance in the *A. terreus* lovastatin fermentation are pH control, slow use of the carbon source, and oxygen uptake rate (Buckland et al. 1989). Early addition of linoleic acid to *A. terreus* had a major positive effect on lovastatin production (Sorrentino et al. 2010). It is thought that this is due to the formation of oxylipins, which are active in quorum sensing. The addition increased the transcriptional levels of the biosynthetic genes *lovB* and *lovF*.

The production of lovastatin in *M. pilosus* is controlled by glucose repression (Miyake et al. 2006a). Use of a favorable mixture of 1% maltose plus 7% glycerol yielded 444 mg/L of lovastatin. A mutant insensitive to glucose repression produced 725 mg/L. Involved in such control is the cAMP signaling pathway (Miyaki et al. 2006b). Addition of cAMP decreased production from 379 to 5 mg/L. A mutant (strain MK-1) insensitive to cAMP addition was isolated and found to produce 620 mg/L. Although lovastatin production by *A. terreus* is negatively affected by glucose, lactose is much more favorable (Lai et al. 2007). Titer reached in the lactose-containing medium by ATCC strain 20,542 was 953 mg/L.

Metabolic engineering studies showed that the transcription factor gene *lovE* regulates the coordinate expression of many *lov* genes and high expression of it results in increased titers of lovastatin (Askenazi et al. 2003). Fermentation studies have shown superiority of solid-state fermentation (SSF) of *A. terreus* over liquid submerged fermentation, and the level of *lovE* was found to be 4.6-fold higher in SSF (Barrios-Gonzalez et al. 2008). The lovastatin gene cluster in *M. pilosus* has been cloned and characterized (Chen et al. 2008). Titers of lovastatin in *A. terreus* reached 7–8 g/L upon optimizing glucose and soybean meal concentrations in a 15 L fermentor (Wu et al. 2007). The addition of trisodium citrate led to the production of over 9 g/L (Zhang et al. 2006).

Red yeast has been used as a traditional Chinese food and medicine since 800 A.D. (Ma et al. 2000). It is also known as red koji or Hongqu. The fermentation organism is *Monascus purpureus* whose pigments give the food its characteristic color. The organism is used for the production of rice, red wine, red soy bean cheese, meat, fish and is authorized for food use in China and Japan. The orange pigments, monascorubin and rubropunctatin, have both antibacterial and antifungal activities (Martinkova et al. 1995). Li Shizhen, the noted pharmacologist of the Ming Dynasty (1368–1644), reported the favorable effects of red rice on blood circulation. More recent work has shown that red rice lowers blood-lipid levels due to its content of statins, especially lovastatin, and clinical trials demonstrated the lowering of cholesterol in humans (Heber et al. 1999).

17.3.3 Pravastatin (Pravacol®)

Although compactin was never used medically, Sankyo devised a bioconversion of it by hydroxylation, yielding pravastatin (Figure 17.3), which became commercialized in 1989 and was then licensed to Bristol-Myers Squibb. This bioconversion, induced by compactin, is carried out industrially using actinomycetes.

Pravastatin is the 3β-hydroxy derivative of compactin (Figure 17.3) and is more active than compactin in inhibition of HMG-CoA reductase (Hosobuchi et al. 1993b,d). Industrial production of pravastatin was originally carried out by the bioconversion of compactin using *Streptomyces*

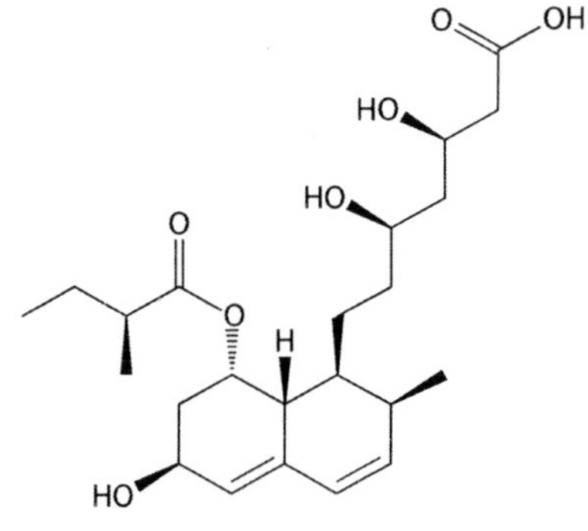

FIGURE 17.3 Structure of pravastatin (Pravacol®).

carbophilus (Serizawa and Matsuoka 1991) whose activity was due to a 2-component cytochrome P450 monooxygenase system induced by compactin. Screening of an actinomycete collection revealed three additional strains capable of carrying out the conversion, i.e., *Actinomadura* sp., *Streptomyces tanashiensis*, and *Streptomyces anulatus* (Yashphe et al. 1997). Further studies on the *Actinomadura* sp. strain 2966 indicated that it was a system quite different from that of *S. carbophilus* and was a unique hydroxylase, e.g., it did not require compactin induction (Peng et al. 1997). Examination of the *Actinomadura* strain's constitutive activity in cell-free extracts showed additional differences, i.e., it required NADPH as coenzyme and Mg^{2+} as cofactor, was stimulated by ATP and ascorbic acid, and was not inactivated by CO (Peng and Demain 1998).

Although pravastatin is commercially produced by bioconversion of compactin, certain strains of *Aspergillus* and *Monascus* can also produce pravastatin directly (Manzoni et al. 1998, 1999).

17.4 SEMI-SYNTHETIC STATIN: SIMVASTATIN (ZOCOR®)

Merck produces simvastatin (Figure 17.4) by a synthetic multistep process starting with lovastatin in which the 2-methylbutanoate side-chain of lovastatin is chemically modified to 2,2-dimethylbutanoate. Simvastatin, a more potent analog of lovastatin (Illingworth and Tobert 1994), was launched in 1988 and its bioconversion process was described by Gbewonyo et al. (1991). Simvastatin can also be made with an *Escherichia coli* strain over-expressing acyltransferase LovD in the presence of a cell-membrane permeable thioester, i.e., α-dimethylbutyryl-*S*-methyl mercaptopropionate (Xie et al. 2007; Xie and Tang 2007). LovF is an acyltransferase usually converting monocolin J to lovastatin by transfer of a 2-methylbutyryl group in the normal lovastatin biosynthetic pathway. The enzyme was first incorporated genetically into *E. coli* and its activity was increased by directed evolution (Gao et al. 2009). There were seven amino acid changes in the enzyme, resulting in a mutant with an 11-fold increase in activity. The modified cells were then used for the bioconversion. The whole-cell procedure converts monocolin J acid to simvastatin acid in high yields, i.e., over 99%.

17.5 SYNTHETIC STATINS

17.5.1 Atorvastatin (Lipitor®)

In 1981, Bruce D. Roth was a postdoctoral fellow in the Chemistry Department of the University of Rochester. He worked on the synthesis of the statins that Endo had isolated in the 1970s. Two years later, Roth led an 18-person group at Parke-Davis working on the synthesis of synthetic statins. Several groups (Stokker et al. 1986; Roth et al. 1990) demonstrated that the hexahydronaphthalene portion of the natural statins could be replaced with different ring systems and that the mevalonolactone was the key moiety for biological activity. Starting with a pyrrole template, the Parke-Davis team in 1985 synthesized atorvastatin (Figure 17.5), which demonstrated a superior profile as compared with the naturally derived inhibitors of 3-hydroxy-3-methylglutaryl-coenzyme A reductase (Roth 2002). They compared atorvastatin in a clinical trial versus fluvastatin, lovastatin, pravastatin,

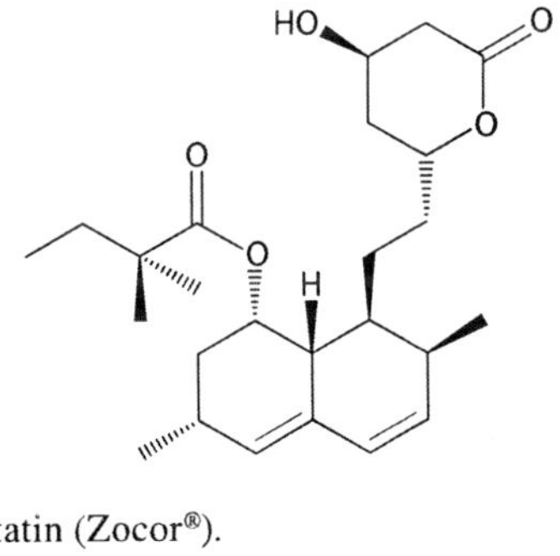

FIGURE 17.4 Structure of simvastatin (Zocor®).

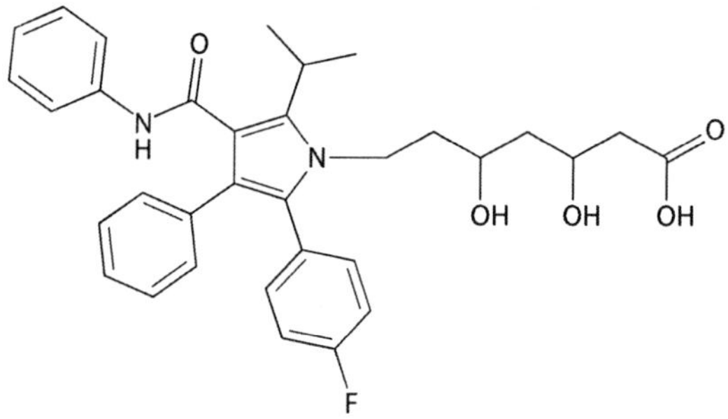

FIGURE 17.5 Structure of atorvastatin (Lipitor®).

and simvastatin; atorvastatin showed the best results (Jones et al. 1998). The FDA approved Lipitor® in January 1997. Parke-Davis decided to co-market Lipitor with Pfizer in 1996. By mid-1998, atorvastatin had 18% of the statin market as compared to 37% by simvastatin. Pfizer, in 2000, purchased Warner-Lambert, the parent of Parke-Davis, and became the sole owner of atorvastatin, which became the leading selling drug in the world.

17.5.2 Other Synthetics

Other important synthetic statins include fluvastatin (Lescol®) and rosuvastatin (Crestor®). For both compounds, the structural modification is in the hexahydronaphthalene portion of the natural statins with the critical mevalonolactone moiety in place.

17.6 OTHER SIGNIFICANT ACTIVITIES OF STATINS

Statins lower elevated *C*-reactive protein (CRP) levels, an anti-inflammatory effect, independent of their effect on cholesterol (Youssef et al. 2002; Chan et al. 2004). This is important since half of all myocardial infarctions occur in patients with normal LDL levels. High CRP is associated with the inflammatory response in atherosclerosis and is a predictor of future cardiovascular mortality. A large clinical trial (McCarey et al. 2004) suggested that atorvastatin is potentially a candidate for rheumatoid arthritis (RA). A recent study (Tang et al. 2011) describes improvement in the number of regulatory T cells as a mechanism for improved RA disease status. The activity of statins as anti-inflammatory agents may also prove useful against sepsis, which kills 210,000 people in the United States each year (Makris et al. 2010). Not only are statins active against atherosclerosis, the most common cause of death in Western countries, but they also improve endothelial function, anti-atherothrombosis, anti-proliferation, and anti-migration activities (Veillard and Mach 2002).

Statins were reported to reduce the occurrence of Alzheimer's disease (AD). A close relationship exists between high cholesterol levels in humans and AD (Solomon and Kivipelto 2009). Individuals who used statins to lower cholesterol had a 70% lower level of AD (Jick et al. 2000; Rockwood et al. 2002; Solomon et al. 2010). A new statin derivative (NST 0037) has both hypocholesterolemic and neuroprotective activity, the latter being detected using an oxidative stress-induced neuron cell death model with neuroblastoma cells (Campoy et al. 2010). Additional studies linking high cholesterol to AD have been described (Refolo et al. 2000; Puglielli et al. 2003; Tan et al. 2003; Wolozin et al. 2006; Harris and Milton 2010; McGuiness and Passmore 2010). The neuroprotective effect of statins has been demonstrated in an *in vitro* model of AD using primary cultures of cortical neurons (Fonseca et al. 2009). The effect did not appear to be due to cholesterol lowering but rather to the reduction in formation of isoprenyl intermediates of the cholesterol biosynthetic process.

Statins may also prevent stroke and reduce the development of peripheral vascular disease (Menge et al. 2005). They are also showing beneficial effects in multiple sclerosis and cancer (Wekerle 2002; Youssef et al. 2002; Fogarty 2003; Stuve et al. 2003; Ho and Pan 2009). Experiments with oral statins showed efficacy in a mouse model of multiple sclerosis. The effect appears to be independent of cholesterol lowering. Other activities being studied are stimulation of bone formation and anti-oxidation (Wrigley 2004). The applications of statins against multiple sclerosis, AD, and ischemic stroke have not yet been approved since more clinical studies are required (Menge et al. 2005).

Fluvastatin has unexpected antiviral activity against RNA viruses and acts against chronic hepatitis C virus in humans (Bader et al. 2008). Statins may also be useful against hepatitis B virus (HBV), which infects 400 million people and is the most common infectious disease agent in the world. The virus causes hepatocellular cancer that is thought to be the number one cause of cancer death globally (Bader and Korba 2010). One method of treating HBV has involved the use of nucleoside analogs such as lamivudine, adefovir, tenofovir, entecavir, and telbuvidine, all of which are FDA approved. Unfortunately, these nucleic acid antagonists only work on 11%–17% of patients in any 1 year. Thus, they are given over the course of many years. Simvastatin was shown to act *in vitro* against HBV and is synergistic with the nucleoside analogs.

17.7 CONCLUSIONS

The statin family of drugs has contributed significantly in the treatment of cardiovascular disease and extending lives of millions of people. Since the initial discovery of the microbial product, compactin by Endo in 1976, a great deal of effort has been undertaken to discover natural analogs from other microbial sources. To optimize the production of this family of compounds in fermentation, many approaches have been successfully applied. For example, media optimization, mutagenesis, and genetic modifications have been utilized to increase production in various strains that biosynthesize the statins. In addition, improved analogs, such as pravastatin, have been prepared by bioconversion. In recent years, the number one selling drug, Lipitor, was chemically synthesized based on the key structural feature, the mevalonolactone. From the first marketed statin, Mevacor®, to the present, the statins have made a remarkable contribution to human health. Based on recent studies demonstrating other significant activities for the statins, such as anti-inflammatory activity, neuroprotection in AD, or antiviral activity against HBV, more of the story may yet unfold.

ABBREVIATIONS

AD	Alzheimer's disease
ATP	adenosine triphosphate
cAMP	cyclic adenosine monophosphate
CO	carbon monoxide
CRP	C-reactive protein
HBV	hepatitis B virus
HMG-CoA reductase	3-hydroxy-3-methylglutaryl coenzyme A reductase
IC_{50}	half maximal inhibitory concentration
ML-236B	compactin, mevastatin
NADPH	nicotinamide adenine dinucleotide phosphate
RA	rheumatoid arthritis

REFERENCES

Abe, Y., T. Suzuki, T. Mizuno et al. 2002a. Effect of increased dosage of the ML-236B (compactin) biosynthetic gene cluster on ML236B production in *Penicillium citrinum*. *Mol Genet Genomics* 268:130–137.

Abe, Y., T. Suzuki, C. Ono, K. Iwamoto, M. Hosobuchi, and H. Yoshikawa. 2002b. Molecular cloning and characterization of an ML-236B (compactin) biosynthetic gene cluster in *Penicillium citrinum*. *Mol Genet Genomics* 267:636–646.

Alberts, A.W., J. Chen, G. Kuron et al. 1980. Mevinolin: A highly potent competitive inhibitor of hydroxymethylglutaryl-coenzyme A reductase and a cholesterol-lowering agent. *Proc Natl Acad Sci USA* 77:3957–3961.

Askenazi, M., E.M. Driggers, D.A. Holtzman et al. 2003. Integrating transcriptional and metabolite profiles to direct the engineering of lovastatin-producing fungal strains. *Nat Biotechnol* 21:150–156.

Bader, T., J. Fazili, M. Madhoun et al. 2008. Fluvastatin inhibits hepatitis C replication in humans. *Am J Gastroenterol* 103:1383–1389.

Bader, T. and B. Korba. 2010. Simvastatin potentiates the anti-hepatitis B virus activity of FDA-approved nucleoside analogue inhibitors in vitro. *Antiviral Res* 86:241–245.

Barrios-Gonzalez, J., J.G. Banos, A.A. Covarrubias, and A. Garay-Arroyo. 2008. Lovastatin biosynthetic genes of *Aspergillus terreus* are expressed differentially in solid-state and in liquid submerged fermentation. *Appl Microbiol Biotechnol* 79:179–186.

Bazaraa, W.A., M.K. Hamdy, and R. Toledo. 1998. Bioreactor for continuous synthesis of compactin by *Penicillium cyclopium*. *J Ind Microbiol Biotechnol* 21:192–202.

Branca, M.A. and L. Sannes. 2006. Battle of the super statins. *Pharma DD* 1(2):12.

Brown, A.G., T.C. Smale, T.J. King, R. Hasenkamp, and R.H. Thompson. 1976. Crystal and molecular structure of compactin, a new antifungal metabolite from *Penicillium brevicompactum*. *J Chem Soc Perkins Trans* I:1165–1170.

Buckland, B., K. Gbewonyo, T. Hallada, L. Kaplan, and P. Masurekar. 1989. Production of lovastatin, an inhibitor of cholesterol accumulation in humans. In: *Novel Microbial Products for Medicine and Agriculture*, eds. A.L. Demain, G.A. Somkuti, J.C. Hunter-Cevera, and H.W. Rossmoore, pp. 161–169. Amsterdam, the Netherlands: Elsevier.

Campoy, S., S. Sierra, B. Suarez et al. 2010. Semisynthesis of novel monacolin J derivatives: Hypocholesterolemic and neuroprotective activities. *J Antibiot* 63:499–505.

Chakravarti, R. and V. Sahai. 2002. Optimization of compactin production in chemically defined production medium by *Penicillium citrinum* using statistical methods. *Proc Biochem* 38:481–486.

Chan, K.Y., E.S. Boucher, P.J. Gandhi, and M.A. Silva. 2004. HMG-CoA reductase inhibitors for lowering elevated levels of C-reactive protein. *Am J Health Syst Pharm* 61:1676–1681.

Chen, Y.-P., C.-P. Tseng, L.-L. Liaw et al. 2008. Cloning and characterization of monocolin K biosynthetic gene cluster from *Monascus pilosus*. *J Agric Food Chem* 56:5639–5646.

Downton, C. and I. Clark. 2003. Statins-the heart of the matter. *Nat Rev Drug Disc* 2:343–344.

Endo, A. 1979. Monocolin K, a new hypocholesterolemic agent produced by a *Monascus* species. *J Antibiot* 32:852–854.

Endo, A. 1980a. Biological and biochemical aspects of ML-236B (compactin) and monocolin K, specific competitive inhibitors of 3-hydroxy-3-methylglutaryl coenzyme A reductase. In: *Atherosclerosis V, Proceedings of 5th International Symposium on Atherosclerosis*, eds. A.M. Gotto Jr, L.C. Smith, and B. Allen, pp. 152–155. Heidelberg, Germany: Springer-Verlag.

Endo, A. 1980b. Monocolin K, a new hypocholesterolemic agent that specifically inhibits 3-hydroxy-3-methylglutaryl coenzyme A reductase. *J Antibiot* 33:334–336.

Endo, A. 1985. Compactin (ML-236B) and related compounds as potential cholesterol-lowering agents that inhibit HMG-CoA reductase. *J Med Chem* 28:401–405.

Endo, A. 2010. A historical perspective on the discovery of statins. *Proc Jpn Acad Ser B* 86:484–492.

Endo, A., K. Hasumi, A. Yamada, R. Shimoda, and H. Takeshima. 1986. The synthesis of compactin (ML-236B) and monocolin K in fungi. *J Antibiot* 39:1609–1610.

Endo, A., M. Kuroda, and K. Tanzawa. 1976a. Competitive inhibition of 3-hydroxyglutaryl coenzyme A reductase by ML-236A and ML-236B fungal metabolites, having hypocholesterolemic activity. *FEBS Lett* 72:323–326.

Endo, A., M. Kuroda, and Y. Tsujita. 1976b. ML-236A, ML-236B, and ML-236C, new inhibitors of cholesterogenesis produced by *Penicillium citrinum*. *J Antibiot* 29:1346–1348.

Fogarty, M. 2003. Evidence suggests beneficial secondary effects from these cholesterol-lowering drugs. *Scientist* 17(22):22–33.

Fonseca, A.C.R.G., T. Proenca, R. Resende, C.R. Oliviera, and C.M.F. Pereira. 2009. Neuroprotective effects of statins in an in vitro model of Alzheimer's disease. *J Alzheimer's Dis* 17:503–517.

Gao, X., X. Xie, I. Pashkov et al. 2009. Directed evolution and structural characterization of a simvastatin synthase. *Chem Biol* 16:1064–1074.

Gbewonyo, K., B.C. Buckland, and M.D. Lilly. 1991. Development of a large-scale continuous substrate feed process for the biotransformation of simvastatin by *Nocardia* sp. *Biotechnol Bioeng* 37:1101–1107.

Gunde-Cimerman, N., A. Plemenitas, and A. Cimerman. 1993. *Pleurotus* fungi produce mevinolin, an inhibitor of HMG CoA reductase. *FEMS Microbiol Lett* 113:333–338.

Harris, J.R. and N.G. Milton. 2010. Cholesterol in Alzheimer's disease and other amyloidogenic disorders. *Subcell Biochem* 51:47–75.

Heber, D., I. Yip, J.M. Ashley, D.A. Elashoff, R.M. Elashoff, and V.L.M. Go. 1999. Cholesterol-lowering effects of a proprietary Chinese red-yeast-rice dietary supplement. *Am J Clin Nutr* 69:231–236.

Hendrickson, L., C.R. Davis, C. Roach et al. 1999. Lovastatin biosynthesis in *Aspergillus terreus*: Characterization of blocked mutants, enzyme activities and a multifunctional polyketide synthase gene. *Chem Biol* 6:429–439.

Hirschler, B. 2011. AstraZeneca back in court to defend Crestor patent. *Reuters*, Oct. 5.

Ho, B.-Y. and T.-M. Pan. 2009. The *Monascus* metabolite monocolin K reduces tumor progression and metastasis of Lewis lung carcinoma cells. *J Agric Food Chem* 57:8258–8265.

Hosobuchi, M., F. Fukui, H. Matsukawa, T. Suzuki, and H. Yoshikawa. 1993a. Morphology control of preculture during production of ML-236B, a precursor of pravastatin sodium, by *Penicillium citrinum. J Ferm Bioeng* 76:476–481.

Hosobuchi, M., K. Kurosawa, and H. Yoshikawa. 1993b. Application of computer to monitoring and control of fermentation process: Microbial conversion of ML-236B Na to pravastatin. *Biotechnol Bioeng* 42:815–820.

Hosobuchi, M., K. Ogawa, and H. Yoshikawa. 1993c. Morphology study in production of ML-236B, a precursor of pravastatin sodium, by *Penicillium citrinum. J Ferm Bioeng* 76:470–475.

Hosobuchi, M., T. Shiori, J. Ohyama, M. Arai, S. Iwado, and H. Yoshikawa. 1993d. Production of ML-236B, an inhibitor of 3-hydroxy-3-methylglutaryl CoA reductase by *Penicillium citrinum*: Improvement of strain and culture conditions. *Biosci Biotech Biochem* 57:1414–1419.

Illingworth, D.R. and J.A. Tobert. 1994. A review of clinical trials comparing HMG-CoA reductase inhibitors. *Clin Ther* 16:366–385.

Jick, H., G.L. Zomberg, S.S. Jick, S. Seshadri, and D.A. Drachman. 2000. Statins and the risk of dementia. *Lancet* 356:1627–1631.

Jones, P., S. Kafonek, I. Laurora, and D. Hunninghake. 1998. Comparative dose efficiency study of atorvastatin versus simvastatin, pravastatin, lovastatin, and fluvastatin in patients with hypercholesterolemia (The CURVES Study). *Am J Cardiol* 81:582–587.

Knowles, J. and G. Gromo. 2003. Target selection in drug discovery. *Nat Rev* 2:63–69.

Konya, A., A. Jekkel, J. Suto, and J. Salat. 1998. Optimization of compactin fermentation. *J Ind Microbiol Biotechnol* 20:150–152.

Lai, L.S., C.S. Hung, and C.-C. Lo. 2007. Effects of lactose and glucose on production of itaconic acid and lovastatin by *Aspergillus terreus* ATCC 20542. *J Biosci Bioeng* 104:9–13.

Ma, J., Y. Ji, Q. Ye et al. 2000. Constituents of red yeast rice, a traditional Chinese food and medicine. *J Agric Food Chem* 48:5220–5225.

Makris, G.C., G. Geroulakos, M.C. Makris, D. Mikhailidis, and M.E. Falagas. 2010. The pleiotropic effects of statins and omega-3 fatty acids against sepsis: A new perspective. *Expert Opin Invest Drugs* 19:809–814.

Manzoni, M., S. Bergomi, M. Rollini, and N. Cavazzoni. 1999. Production of statins by filamentous fungi. *Biotechnol Lett* 21:253–257.

Manzoni M. and M. Rollini. 2002. Biosynthesis and biotechnological production of statins by filamentous fungi and application of these cholesterol-lowering drugs. *Appl Microbiol Biotechnol* 58:555–564.

Manzoni, M., M. Rollini, M.S. Bergomi, and V. Cavazzoni. 1998. Production and purification of statins from *Aspergillus terreus* strains. *Biotechnol Technol* 12:529–532.

Martinkova, L., P. Juzlova, and D. Vesely. 1995. Biological activity of polyketide pigments produced by the fungus *Monascus. J Appl Bacteriol* 79:609–616.

McCarey, D.W., I.B. McInnes, R. Madhok, R. Hampson, O. Scherbakov, I. Ford, H.A. Capell, and N. Sattar. 2004. Trial of atorvastatin in rheumatoid arthritis (TARA): Double-blind randomized placebo-controlled trial. *Lancet* 363:2015–2021.

McGuiness, B. and P. Passmore. 2010. Can statins prevent or help treat Alzheimer's disease? *J Alzheimers Dis* 20:925–933.

Menge, T., H.-P. Hartung, and O. Stueve. 2005. Statins—A cure-all for the brain? *Nat Rev/Neurosci* 6:325–331.

Miyake, T., K. Uchitomi, M.-Y. Zhang et al. 2006a. Effects of the principal nutrients on lovastatin production by *Monascus pilosus. Biosci Biotechnol Biochem* 70:1154–1159.

Miyake, T., M.-Y. Zhang, I. Kono, N. Nozaki, and H. Sammoto. 2006b. Repression of secondary metabolite production by exogenous cAMP in *Monascus. Biosci Biotechnol Biochem* 70:1521–1523.

Moore, R.N., G.J.K. Chan, A.M. Hogg, T.T. Nakashima, and J.C. Vederas. 1985. Biosynthesis of the hypocholesterolemic agent mevinolin by *Aspergillus terreus*. Determination of the origin of carbon, hydrogen, and oxygen atoms by 13C NMR and mass spectrometry. *J Am Chem Soc* 107:3694–3701.

Nara, F., I. Watanabe, and N. Serizawa. 1993. Development of a transformation system for filamentous, ML-236B (compactin)-producing fungus *Penicillium citrinum. Curr Genet* 23:28–32.

Negishi, S., Z. Cai-Huang, K. Husuni, S. Murakawa, and A. Endo. 1986. Productivity of monacolin K (mevinolin) in the genus *Monascus. Hakkokogaku Kaishi* 64:509–512.

Peng, Y. and A.L. Demain. 1998. Properties of the hydroxylase in *Actinomadura sp* cells converting compactin to pravastatin. *J Ind Microbiol Biotechnol* 20:373–375.

Peng, Y., J. Yashphe, and A.L. Demain. 1997. Biotransformation of compactin to pravastatin by *Actinomadura* sp. 2966. *J Antibiot* 50:1032–1035.

Puglielli, L., R.E. Tanzi, and D.M. Kovacs. 2003. Alzheimer's disease: The cholesterol connection. *Nat Neurosci* 6:345–351.

Refolo, L.M., B. Malester, J. LaFrancois et al. 2000. Hypercholesterolemia accelerates the Alzheimer's amyloid pathology in a transgenic mouse model. *Neurobiol Dis* 7:321–331.

Rockwood, K., S. Kirkland, D.B. Hogan et al. 2002. Use of lipid-lowering agents, indication bias, and the risk of dementia in community-dwelling elderly people. *Arch Neurol* 59:223–227.

Roth, B.D. 2002. The discovery and development of atorvastatin, a potent novel hypolipidemic agent. *Prog Med Chem* 40:1–22.

Roth, B.D., D.F. Ortwine, M.L. Hoefle et al. 1990. Inhibitors of cholesterol biosynthesis. 1.Trans-6-(2-Pyrrol-1-ylethyl)-4-hydroxypyran-2-ones, a novel series of HMG-CoA reductase inhibitors. 1. Effects of structural modification at the 2- and 5-positions of the pyrrole nucleus. *J Med Chem* 33:21–31.

Serizawa, N. and T. Matsuoka. 1991. A two-component-type cytochrome P-450 monoxygenase system in prokaryotes that catalyzes hydroxylation of ML-236B to pravastatin, a tissue-selective inhibitor of 3-hydroxy-3-methylglutaryl coenzyme A reductase. *Biochim Biophys Acta* 1084:35–40.

Solomon, A. and M. Kivipelto. 2009. Cholesterol-modifying strategies for Alzheimer's disease. *Expert Rev Neurother* 9:695–709.

Solomon, A., R. Sippola, H. Soininen et al. 2010. Lipid-lowering treatment is related to decreased risk of dementia: A population-based study (FINRISK). *Neurodegener Dis* 7:180–182.

Sorrentino F., I. Roy, and T. Keshavarz. 2010. Impact of linoleic acid supplementation on lovastatin production in *Aspergillus terreus* cultures. *Appl Microbiol Biotechnol* 88:65–73.

Stokker, G.E., A.W. Alberts, P.S. Anderson et al. 1986. 3-Hydroxy-3-methylglutaryl-coenzyme A reductase inhibitors. 3. 7-(3,5-disubstituted-[1-1′-biphenyl]-2-yl)-3,5-dihydroxy-6-heptenoic acids and their lactone derivatives. *J Med Chem* 29:170–181.

Stuve, O., S. Youssef, S. Dunn, A.J. Stavin, L. Steinmann, and S.S. Zamvil. 2003. The potential therapeutic role of statins in central nervous system autoimmune disorders. *Cell Mol Life Sci* 60:2483–2491.

Tan, S.Z., S. Seshadri, A. Beiser et al. 2003. Plasma total cholesterol level as a risk factor for Alzheimer disease: The Framingham study. *Arch Intern Med* 163:1053–1057.

Tang, T.-T., Y. Song, Y.-J. Ding et al. 2011. Atorvastatin upregulates regulatory T cells and reduces clinical disease activity in patients with rheumatoid arthritis. *J Lipid Res* 52:1023–1032.

Tobert, J.A. (2003) Lovastatin and beyond: The history of the HMG-COA reductase inhibitors. *Nat Rev Drug Disc* 2:517–526.

Valera, H.R., J. Gomes, S. Lakshmi, R. Guraraja, S. Suryanarayan, and D. Kumar. 2005. Lovastatin production by solid state fermentation using *Aspergillus flavipes*. *Enzyme Microb Technol* 37:521–526.

Van Arnum, P. 2009. Shifting fortunes in API market growth. *Pharm Technol Suppl* 36:19–21.

Veillard, N.R. and F. Mach. 2002. Statins: The new aspirin? *Cell Mol Life Sci* 59:1771–1786.

Wekerle, H. 2002. Tackling multiple sclerosis. *Nature* 420:39–40.

Wolozin, B., J. Manger, R. Bryant, J. Cordy, R.C. Green, and A. McKee. 2006. Re-assessing the relationship between cholesterol, statins and Alzheimer's disease. *Acta Neurol Suppl* 185:63–70.

Wrigley, S.K. 2004. Pharmacologically active agents of microbial origin. In: *Microbial Diversity and Bioprospecting*, ed. A.T. Bull, pp. 356–374. Washington, DC: ASM Press.

Wu, B, C.H. Chen, and L. Yang. 2007. Optimization of medium components for lovastatin production and scale up in 15L bioreactor. *Chi J Antibiot* 32:409–413.

Xie, X. and Y. Tang. 2007. Efficient synthesis of simvastatin by use of whole-cell biocatalysis. *Appl Environ Microbiol* 73:2054–2060.

Xie, X., W.W. Wong, and Y. Tang. 2007. Improving simvastatin bioconversion in *Escherichia coli* by deletion of *bioH*. *Metab Eng* 9:379–386.

Yashphe, J., J. Davis, Y. Peng, S.H. Bok, and A.L. Demain. 1997. New microorganisms which convert compactin to pravastatin. *Actinomycetologica* 11:20–25.

Youssef, S., O. Stueve, J.C. Patarroyo et al. 2002. The HMG-CoA reductase inhibitor, atorvastatin, promotes a Th2 bias and reverses paralysis in central nervous system autoimmune disease. *Nature* 420:78–84.

Zhang, Y.-J., H.-F. Hu, and B.-Q. Zhu. 2006. Effects of precursors on lovastatin biosynthesis. *Chi J Antibiot* 9:529–531.

18 Molecular Aspects of Fungal Bioactive Polyketides

Ira Bhatnagar and Se-Kwon Kim

CONTENTS

18.1 INTRODUCTION: FUNGI AS PRODUCERS OF BIOLOGICALLY ACTIVE SECONDARY METABOLITES

Secondary metabolites often have obscure or unknown functions in organisms but have considerable importance for mankind due to their broad range of useful antibiotic, pharmaceutical as well as toxic activities (Yu and Keller 2005). The products of fungal metabolic pathways include important pharmaceuticals such as Penicillin, cyclosporine and statins, as well as potent poisons including aflatoxins (e.g., aflatoxin B1) and trichothecenes (Keller et al. 2005). Harold Raistrick was the first to initiate systematic study of fungal secondary metabolites in 1922 and was successful in characterizing more than 200 fungal metabolites (Keller et al. 2005). However, the major breakthrough in fungal metabolite research was achieved in 1929, when the first broad-spectrum antibiotic Penicillin was discovered from fungus *Penicillium notatum* (alias *Penicillium chrysogenum*) by Alexander Fleming (Fleming 1929, Bennett 2001). The discovery of Penicillin and its clinical use initiated extensive screening programs for microbial bioactive metabolites (Keller et al. 2005). More than half of the molecules isolated and characterized between 1993 and 2001 show antibacterial, antifungal, or antitumour activity (Pelaez 2005), thereby illustrating that natural products are the most important source of anticancer and anti-infective agents (Da Rocha et al. 2001).

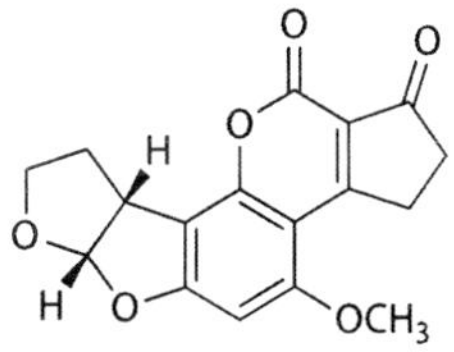

Aflatoxin B1

Up to date, there have been dozens of fungal secondary metabolites that are used as antibiotic, antitumor, immunosuppressive, hypocholesterolemic, antimigraine, and antiparasitic agents. In addition, new active fungal secondary metabolites have been discovered with pharmacological activities both in terrestrial (Bentley 1997, Demain 2006) and marine environments (Saleem et al. 2007, Blunt et al. 2008), and some of them have already entered clinical testing. The most prolific sources of fungal secondary metabolites belong to the genus *Aspergillus* and *Penicillium*.

18.1.1 Fungal Secondary Metabolites as Approved Pharmaceuticals

Fleming's discovery led to the first successful chemotherapeutic produced by microbes that has initiated the golden age of antibiotics. This discovery opened the way for the development of many other antibiotics, and up to date Penicillin has still remained the most active and the least toxic compound among many others (Demain 2006). Penicillins together with cephalosporins belong to the group of β-lactam compounds. There are several types of Penicillins, e.g., F, G, K, N, and V. Penicillin V and G are active against most aerobic Gram-positive organisms. Penicillin G is one of the most widely used antibiotic agents today and is used against Streptococcal, Staphylococcal, and Meningococcal infections. In the industrial scale Penicillin G is produced by fermentation of *P. chrysogenum* (Laich et al. 2002). Zearalenone is another example of a mycotoxin with pharmacologically useful properties. It is a polyketide that is synthesized entirely from acetate–malonate units (Dewick 2001). Zearalenone is produced by several *Fusarium* species (Lysoe et al. 2006). It resembles 17β-estradiol, the principal hormone produced by the human ovary, to allow binding to the estrogen receptors in mammalian target cells. The reduced form of Zearalenone, i.e., α-Zearalenol, has revealed to increase estrogenic activity (Shier et al. 2001). A synthetic commercial formulation called Zeranol has been successfully marketed for use as an anabolic agent for both sheep and cattle (Hodge et al. 1966). Zearalenone has also been applied for the treatment of postmenopausal symptoms in women (Utian 1973), and both Zearalenol and Zearalenone have been patented as oral contraceptives (Hidy et al. 1977). In summary, the Zearalenone family of metabolites is an example of both potentially harmful metabolites and promising pharmaceutical candidate.

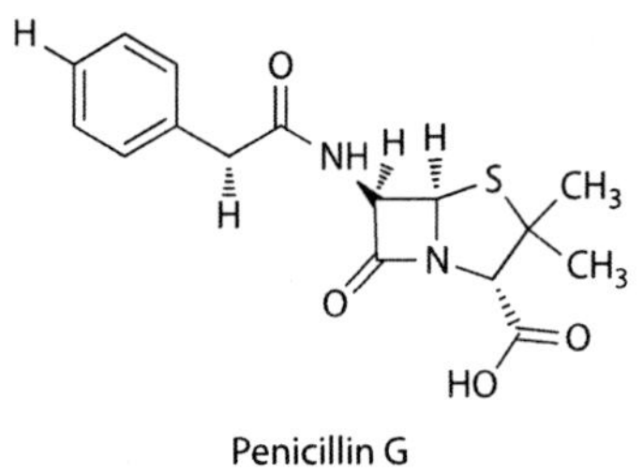

Penicillin G

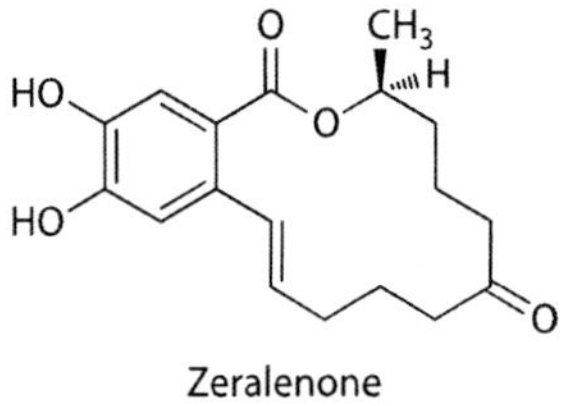

Zeralenone

Another fungal secondary metabolite, cyclosporin A, was originally discovered as a narrow-spectrum antifungal metabolite produced by the fungus *Tolypocladium infatum* (Borel et al. 1976). This fungal secondary metabolite is associated with reduction of cytokine formation and inhibition of activation and/or maturation of various cell types. This includes as well those cells involved in cell-mediated immunity, thus, cyclosporin A is used as an immunosuppressant in human transplantation surgery and the treatment of autoimmune diseases (Faulds et al. 1993).

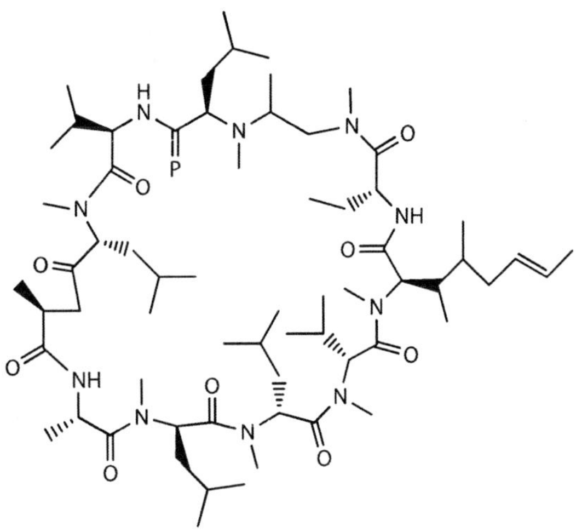

Cyclosporin A

A very old broad-spectrum compound, mycophenolic acid, first discovered in 1896 and never commercialized as an antibiotic, has recently been developed as a new immunosuppressant (Bookstein et al. 1990). Before being developed for an approved immunosuppressant, this organic acid was used to treat psoriasis (Demain 1999). 5-Methylorsellinic acid, but not orsellinic acid, is a precursor of mycophenolic acid in *Penicillium brevicompactum* (Dewick 2001). The members of the statin family of secondary metabolites are potent inhibitors of 3-hydroxy-3-methylglutaryl-coenzyme A (HMG-CoA) reductase, the key enzyme in cholesterol biosynthesis in humans (Royer et al. 2003). Besides their main cholesterol-lowering effect, members of the statin family have also strong antifungal activities, especially against yeasts (Demain 1999). Brown et al. discovered in 1976 the first member of this group—compactin (i.e., ML-236B)—as an antibiotic product of *P. brevicompactum*.

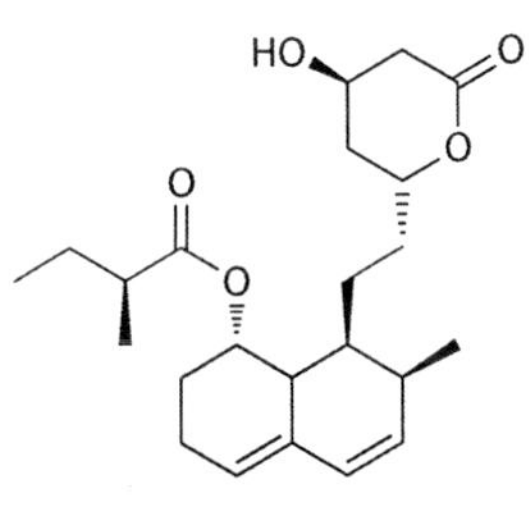

Compactin

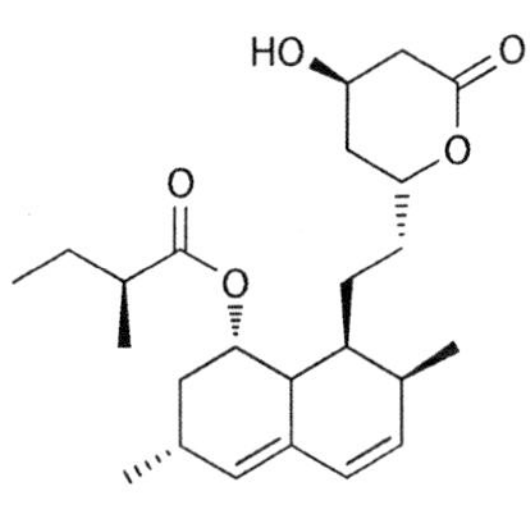

Lovastatin

Independently, in the same year, Endo et al. discovered compactin in broths of *Penicillium citrinum* as an inhibitor of HMG-CoA. Few years later Endo and Alberts independently discovered the more active methylated form of compactin known as lovastatin (Monacolin K or Mevinolin) in broths of *Monascus ruber* and *Aspergillus terreus,* respectively (Moore et al. 1985). It is important to emphasize that natural statins and their derivatives are an example of multibillion dollar drugs arising from fungal secondary metabolites (McAlpine 1998). One such derivative is pravastatin, which is produced by bioconversion of compactin (Peng and Demain 1998). Yet another compound of pharmaceutical importance is a diketopiperazine known as plinabulin (NPI-2358), isolated from a marine alga associated *Aspergillus* sp. CNC-139. This compound also inhibits tubulin assembly and acts as a vasculature disrupting agent that destabilizes the tumor vascular endothelial architecture and leads to cell damage. This compound is presently under phase II clinical trials (Bhatnagar and Kim 2010a).

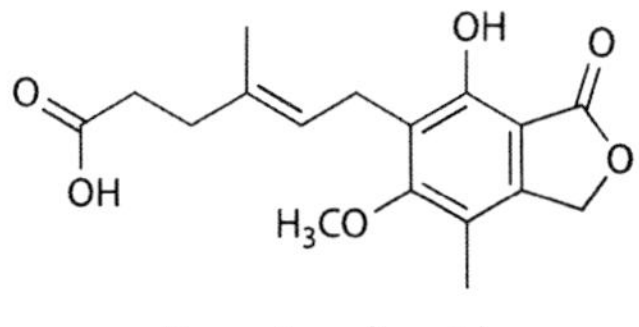

Mycophenolic acid

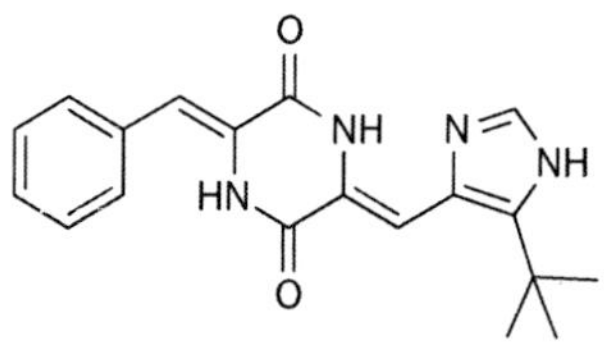

Plinabulin

18.2 MARINE-DERIVED FUNGAL POLYKETIDE METABOLITES

In search for novel and bioactive molecules for drug development, marine-derived natural resources have become an important research area. Although almost three quarters of the Earth's surface is occupied by seas and oceans, the isolation from soil was rather a common method to get fungal isolates. However, since fungal strains from terrestrial sources often yielded already known secondary metabolites, it was obvious that new sources are needed and therefore marine fungi have become an important source for isolation of pharmacologically active metabolites (Haefner 2003). The group of marine-derived fungi include obligate marine fungi, which grow and sporulate exclusively in the marine or estuarine habitat, as well as facultative marine fungi, which grow equally well in marine, freshwater, and terrestrial environments (Kohlmeyer et al. 2004). Interestingly, all known marine fungal products have been isolated from cultured organisms, though, up to now it is estimated that fewer than 1% of all microorganisms, including fungi, have been successfully cultured (Torsvik et al. 1990). Terrestrial fungi produce many therapeutically significant molecules, since marine organisms live in an environment significantly different from those of terrestrial organisms, it is reasonable to expect that their secondary metabolites will differ considerably. Although these natural resources have only recently been explored for natural products, there are currently over 15 fungal marine-derived secondary metabolites in clinical trials (Saleem et al. 2007, Bhatnagar and Kim 2010b). Moreover, it has been shown that among the nearly 300 new natural products isolated from marine-derived fungi (Bugni and Ireland 2004), many of them are polyketides (Saleem et al. 2007, Mayer et al. 2007). Therefore, metabolites of this class might be the largest part of the secondary metabolites derived from marine as well as terrestrial microorganisms.

Evidentially, polyketides isolated from marine-derived fungi represent an interesting group of bioactive substances that display myriads of effects on cell systems such as antioxidant, antibiotic, cytotoxic, neurotrophic, and antiproliferative activities. Furthermore, polyketides can act as inhibitors of microtubule assembly, be responsible for inhibition of DNA polymerase, or have antagonistic effect on receptors in the central nervous system (CNS). Their structural complexity can be quite impressive, which complicates the establishment of commercially viable synthesis (Saleem et al. 2007).

18.3 POLYKETIDE BIOSYNTHESIS

More than hundred years ago, Collie coined the term "polyketide" for natural products derived from simple two-carbon acetate building blocks (Fujii et al. 2004). This proposal was later proved experimentally by Birch who used isotopically labeled acetate in the study of 6-methylsalicylic acid (6-MSAS) biosynthesis in fungi and showed that it was formed from four acetate units. Then, Lynen and his coworkers succeeded in detecting MSAS activity in a cell-free extract of *P. patulum*, the first demonstration of polyketide synthase (PKS) function *in vitro*. These chemical and biochemical experiments with fungi established the concepts of "polyketide biosynthesis" and "PKS" (Fujii et al. 2004). Nowadays it is obvious that polyketides represent the largest family of structurally diverse secondary metabolites synthesized in both prokaryotic and eukaryotic organisms (Simpson 1995). The biological activities associated with polyketides encompass, e.g., antibacterial, antiviral, antitumor, antihypertensive activities, as well as immunosuppressant and mycotoxin compounds.

18.3.1 Molecular Background to Understand Polyketide Biosynthesis

Independent of their structural diversity, all polyketides have a common biosynthetic origin. They are derived from highly functionalized carbon chains whose assembly mechanism has close resemblance to the fatty acid biosynthetic pathway (O'Hagan 1991). The assembly process is controlled by multifunctional enzyme complexes called PKS (Simpson 1995). The core of the PKS function is the synthesis of long chains of carbon atoms through repetitive Claisen condensation reactions of small organic acids (such as acetic and malonic acid) via a ketosynthase (KS) enzyme activity. The building units, acetate, propionate, malonate, or methylmalonate, are activated units in the form of CoA esters, such as acetyl-CoA and malonyl-CoA, before involvement in the assembly of the polyketide chain. The most common starter-unit acetyl-CoA with two carbon atoms is condensed with a malonyl-CoA, with three carbons, to give a chain of four carbon atoms with loss of one carbon dioxide. Only two carbons are included into the chain in each round of condensation with malonyl-CoA. If the extender unit is methylmalonyl-CoA, the "extra" carbon forms a methyl side branch to keep the original extension speed in the main chain (Hopwood 2004). Each condensation is followed by a cycle of optional modifying reactions that involve the enzymes ketoreductase (KR), dehydratase (DH), and enoylreductase (ER) in the subsequent reduction steps. At this stage, a major difference between fatty acid and polyketide biosynthesis becomes apparent. Fatty acid synthases (FAS) catalyze the full reduction of each β-keto moiety prior to further chain extension in every cycle. The polyketide biosynthesis, however, shows a higher degree of complexity due to full or partial omission of reduction steps following condensation and thus affecting function: β-keto (no reduction), β-hydroxy (keto reduction), enoyl (keto reduction and dehydration), to alkyl (keto reduction, dehydration, and enoyl reduction). This control of β-keto reduction is the key feature of the reducing PKS (R-PKS) that differentiates these enzymes from FAS and that leads to a great structural diversity among polyketide compounds (Hutchinson and Fujii 1995, Hopwood 1997). The ability to use different chain starter-units (such as acetate, benzoate, cinnamate, and/or amino acids) and alternate extender units (malonate, methylmalonate, and ethylmalonate) by bacterial synthases gives rise to further structural diversity among the polyketides. The assembled polyketide chain can also undergo further modifications such as cyclization, reduction or oxidation, alkylation, and rearrangements after release from PKS.

18.3.2 Types of PKSs

In analogy to the classification of FASs, PKSs have traditionally been subdivided into two main categories (Hopwood and Sherman 1990). The first category encompasses multifunctional modular systems that are responsible for the biosynthesis of macrolactones, polyenes, and polyethers and are designated as Type I PKS (Staunton and Weissman 2001). The fully dissociable complex of small, discrete monofunctional proteins that catalyze the biosynthesis of bacterial aromatic polyketides

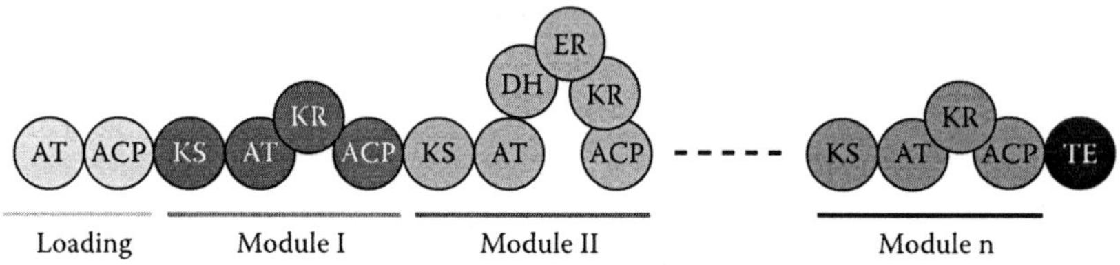

FIGURE 18.1 Modular type I PKS found in bacteria. KS, keto acyl synthase; AT, acyl transferase; KR, keto reductase; ACP, acyl carrier protein; DH, dehydratase; ER, enoyl reductase; TE, thioesterase.

is termed Type II PKS (Shen 2000). In the past decade, as cloning and sequencing of PKS genes were advancing especially since the discovery of fungal and plant PKSs, these categories of PKSs were redefined and enriched by Type III PKS and the expansion of Type I PKS into two subclasses, modular Type I PKSs (Figure 18.1) and iterative Type I PKSs (Shen 2003). The modular Type I PKSs representing bacterial systems are large multifunctional polypeptides arranged in a modular fashion with each module being responsible for one round of chain extension and subsequent β-keto processing. Particularly, each active site in modular Type I PKS is used only once during polyketide biosynthesis (Staunton and Weissman 2001).

The iterative Type I PKSs (Figure 18.2) are responsible for the biosynthesis of fungal metabolites such as 6-methylsalicyclic acid (Beck et al. 1990) and lovastatin (Hendrickson et al. 1999, Kennedy et al. 1999). The iterative Type I PKS has only one multidomain protein, in which all the enzyme activities are covalently bound together. The single multifunctional protein is used iteratively to catalyze multiple rounds of chain elongation and appropriate β-keto processing (Staunton and Weissman 2001).

In the iterative Type II PKSs (Figure 18.3), the active site for each biosynthetic step is encoded in a single gene. There is only one set of a heterodimeric KS (KSα-KSβ) and an acyl carrier protein (ACP) that have to operate a specific number of times in building a polyketide chain in correct length and subsequent cyclization, reduction, and aromatization are performed by cyclase (CYC), KR, and aromatase (ARO), respectively. In some Type II PKSs, the malonyl-CoA ACP acyl transferase (MAT), which catalyzes acyl transfer between malonyl-CoA and the ACP, is missing and is possibly shared between the PKS and the housekeeping FAS (Revill et al. 1995). The Type II PKSs usually catalyze the biosynthesis of a broad range of polyfunctional aromatic natural products and are so far restricted to bacteria (Shen 2000).

In contrast to the Type I and II PKSs that are composed of KSs and accessory enzymes, the Type III PKSs are dimers of KS-like enzymes (more precisely homodimers) that accomplish a complex set of reactions, such as priming of a starter-unit, decarboxylative condensation of extender units, ring closure, and aromatization of the polyketide chain, in a multifunctional active site pocket (Funa et al. 2007). Chalcone synthases, the most well-known representatives of this family, are ubiquitous in higher plants and provide the starting material for a diverse set of biologically important

FIGURE 18.2 Iterative Type I PKS found in fungi. KS, keto acyl synthase; MT, methyl transferase; ER, enoyl reductase; DH, dehydratase; KR, keto reductase; ACP, acyl carrier protein.

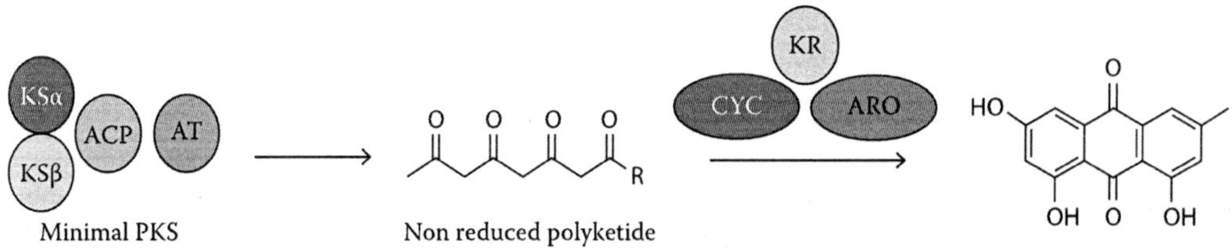

FIGURE 18.3 Iterative Type II PKS found in bacteria. KSα-KSβ, heterodimeric ketosynthase; ACP, acyl carrier protein; AT, acyl transferase; KR, keto reductase; CYC, cyclase; ARO, aromatase.

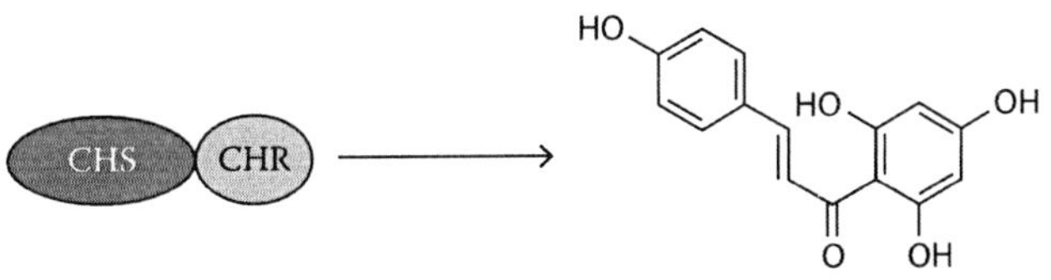

FIGURE 18.4 Chalcone synthase Type III PKS found in plants, bacteria and some fungi. CHS, chalcone synthase (a homodimer of identical KS monomeric domains); CHR, chalcone reductase.

phenylpropanoid metabolites (Schroder 1999). Type III PKSs (Figure 18.4) were traditionally associated with plants but recently discovered in a number of bacteria (Austin and Noel 2003) as well as in fungi (Funa et al. 2007, Seshime et al. 2005).

To summarize, modular Type I PKSs consist of multidomain proteins forming a modular unit for each condensation cycle. In iterative Type I PKSs, one copy of each active domain, KS, MAT, ACP and optional activities for reduction, KR, DH, and ER are assembled in one protein and iteratively used during the biosynthesis. In contrast, active sites of Type II PKSs are encoded in different genes and act in an iterative fashion. Type III CHS-like PKS have a simple architecture like CHS (a homodimer of identical KS monomeric domains) with an optional CHR, TE, CYC, ARO (He 2005).

18.3.2.1 Fungal PKSs

Up to date, only a few fungal PKS genes have been isolated as compared to the large number of isolated bacterial PKS genes. In general, the fungal PKSs are iterative Type I enzymes. The noniterative fungal Type I PKSs perform only one condensation cycle and result in the production of a diketide (e.g., LDKS—lovastatin diketide synthase). Each one that has been characterized so far is encoded by a gene that resides in a gene cluster, along with a PKS gene encoding an iterative PKS (Hendrickson et al. 1999, Abe et al. 2002). Recent genome projects for *Neurospora crassa* (Galagan et al. 2003) and *A. oryzae* (Machida et al. 2005) predict the presence of Type III PKS genes in these filamentous fungi (Funa et al. 2007, Seshime et al. 2005). On the other hand, fungal modular Type I and fungal Type II systems have not yet been observed. The minimal domain structure of fungal PKSs consists of KS, AT, and ACP domains. The KS domain is the most highly conserved domain in Type I PKSs and FASs (Kroken et al. 2003). The optional β-keto processing reactions may be catalyzed by KR, DH, and ER domains, in a stepwise fashion. Further accessory domains are represented by CYC and methyl transferase (MT) activities (Schumann and Hertweck 2006). The *C*-methylation takes place during polyketide chain formation due to activity of this intrinsic MT domain. It is important to note that in the fungal polyketide biosynthesis no methylmalonyl elongation units are employed, in contrary to the numerous examples of the bacterial Type I PKSs. As mentioned previously, the fungal iterative PKS can use each active site in an iterative way during chain assembly and determine the degree of reduction and *C*-methylation within each elongation round. It is fascinating, but a still-not-resolved mystery, how a single set of active domains determines chain length, degree of reduction, and timing of *C*-methylation at a particular step in the pathway (Schumann and Hertweck 2006).

18.3.2.2 Classification of Fungal PKSs

According to their architecture and the presence or absence of additional β-keto-processing domains, fungal PKSs are grouped into the nonreducing (NR PKS), partially reducing (PR PKS), and highly reducing PKSs (HR PKS) (Schumann and Hertweck 2006). Recent phylogenetic studies on the basis of KS amino acid sequences have provided valuable insights into the evolutionary relationship between different types of fungal PKS (Nicholson et al. 2001, Bingle et al. 1999). Kroken et al. showed that amino acid sequences of fungal KS domains cluster according to the degree of reduction of their products into reducing (β-keto reductive domains: KR, ER, DH) and NR PKSs (no β-keto reduction), each type being further divided into four subclades (Kroken et al. 2003).

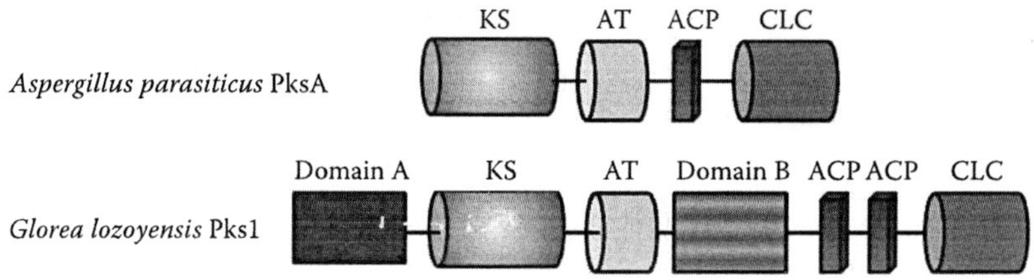

FIGURE 18.5 The general architecture of NR PKS genes in fungi: Domain A, which may be the starter-unit; KS, β ketoacyl synthase; AT, acyl transferase; Domain B (PT), product template; ACP, acyl carrier protein; CLC, claisen cyclase. There is a possibility that other domains can be included as well at the C-terminus after TE/CLC (e.g., C-MeT).

18.3.2.2.1 Nonreducing PKSs

NR PKS are shown to be responsible for the biosynthesis of nonreduced polyketides such as 1,3,6,8-tetrahydroxynaphthalene, norsolorinic acid (NA), and naphthopyrone (YWA1) that require no β-keto reductive steps during their biosynthesis (Figure 18.5). In all cases, known genes for these synthases encode Type I iterative PKS proteins (Cox 2007). The main characteristic of NR PKSs is that they do not contain β-keto-processing domains in their multidomain organization. At the N-terminus, a domain is present that appears to mediate the loading of a starter-unit and is thus named starter-unit-ACP transacylase (SAT) component. It is assumed that the starter-unit is derived from corresponding FAS, another PKS or an acyl-CoA. The SAT domain is followed by typical KS and AT domains responsible for chain extension and malonate loading. Beyond the AT there is a conserved domain designated as a product template (PT) with a not-yet-proven function. Nevertheless, a sequence analysis of this domain suggested that it may be involved in the control of chain-length. The PT domain is followed by one or more ACP domains. Some NR PKSs appear to terminate after the ACP, but many feature a diverse range of different domains including Claisen-cyclase–thioesterases (CLC–TE), MT, and reductases (R). Although not described in the literature, a sequence analysis of the *M. purpureus pksCT* sequence (Shimizu et al. 2005) showed that it has a C-terminal thioester reductase domain. Similar domains were found in the nonribosomal peptide synthase (NRPS) systems with reductase domains as chain release mechanisms resulting in an aldehyde or primary alcohol. Very recently, by joint efforts Cox and Simpsons demonstrated the role of the terminal reductase domain in product release via heterologous expression of MOS in *A. oryzea* (Bailey et al. 2007). In sum, it appears that these synthases are equipped with an N-terminal loading component, a central chain extension component consisting of KS, AT, and ACP domains with a possible control over a number of extensions, and a C-terminal processing component.

18.3.2.2.2 Partially Reducing PKSs

Less is known about the enzymology of the PR PKS (Cox 2007). The domain structure is much closer to mammalian FASs, with an N-terminal KS followed by AT and DH domains. A so-called "core" domain follows the DH, and this is followed by a KR domain. A typical PR PKS terminates with an ACP domain as, e.g., for MSAS (Figure 18.6). The domain structure differs considerably from the NR PKS in such a way that there is no SAT or PT domain, and the PKS terminates after the ACP with obviously no requirement for a CLC–TE domain responsible for offloading of the product. Although a number of PR PKS genes are known from genome sequencing projects, only

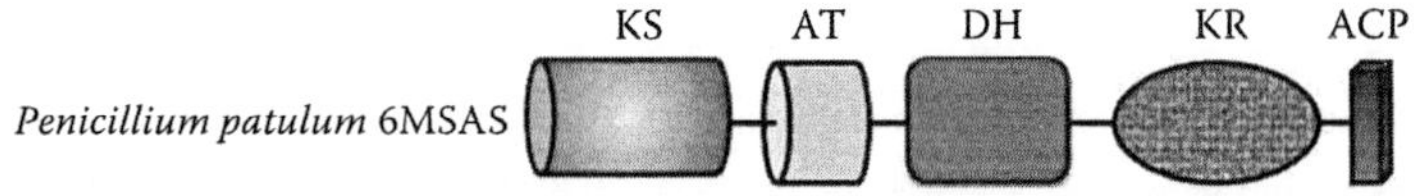

FIGURE 18.6 Domain organization of MSAS. Domain architecture of MSAS encoded by *P. patulum* 6MSAS. KS, β ketoacylsynthase; AT, acyl transferase; DH, dehydratase; KR, ketoreductase; ACP, acyl carrier protein.

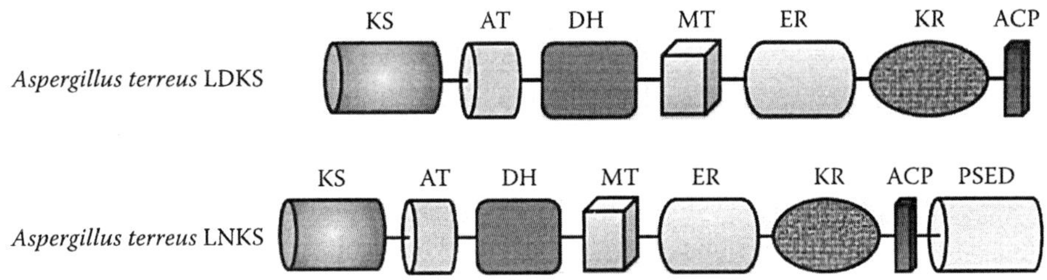

FIGURE 18.7 General domain organization of HR PKSs. General domain architecture of HR PKSs. KS, β ketoacylsynthase; AT, acyl transferase; DH, dehydratase; KR, ketoreductase; ER, enoyl reductase as optional; ACP, acyl carrier protein; MT, methyltransferase.

three genes have been matched to their chemical products—in all cases the tetraketide 6MSA (e.g., a single round of KR and DH). The first MSAS to be discovered was from *P. patulum* (Beck et al. 1990). The Ebizuka group have worked with the *atX* gene from *A. terreus* (Fujii et al. 1996) and most recently Tkacz et al. have described an MSAS gene (*pks2*) isolated from *Glarea lozoyensis* (Lu et al. 2005). Both *P. patulum* and *A. terreus* MSAS form homo-tetramers (Cox 2007). A short region of the core domain was identified by Fujii et al., the presence of this region proved to be essential for successful complementation among diverse deletion mutants of *atX* gene. It was hypothesized that this region of 122 amino acids probably forms a motif required for subunit–subunit interaction. Interestingly, this core sequence is present in other fungal PR PKS and in the bacterial PKS such as *CalO5* from calicheamicin biosynthesis (Moriguchi et al. 2006, 2008).

18.3.2.2.3 Highly Reducing PKSs

The HR PKSs is the third class of fungal PKSs that produce complex, highly reduced compounds such as lovastatin, T-toxin, fumonisin B1, and squalestatin. These PKSs have an N-terminal KS domain, followed by AT and DH domains (Figure 18.7). In many cases, the DH is followed by a MT domain. Some HR PKSs possess an ER domain, in others there is a roughly equivalent length of sequence without known function. An ER domain is succeeded by a KR domain, and finally the PKS often terminates with an ACP. The *lovB*, gene that encodes for LNKS involved in lovastatin biosynthesis, appears to encode one part of an NRPS condensation (C) domain immediately downstream of the ACP. It was proposed that this domain plays role in product release. In general, in HR PKS there seems to be no domains that are similar to the PT or SAT domains of the NR PKS, as well as no "core domain" of the PR PKS (Cox 2007).

18.4 FUTURE PERSPECTIVES IN FUNGAL POLYKETIDE RESEARCH

As a consequence of recent fungal genome sequencing projects, the data on as-yet unexplored putative PKS gene clusters have been considerably enlarged (Kroken et al. 2003, Shumann and Hertweck 2006, Galagan et al. 2005). The opportunities that have been opened for work in the PKS research area by means of "combinatorial biosynthesis" and "metabolic engineering" approaches are more than impressive. Combinatorial biosynthesis makes use of more than one biosynthetic pathway and is able to generate hybrid polyketide products with improved or even novel bioactivities. Additionally, metabolic pathway engineering is an approach to optimize the production of a single polyketide in a suitable heterologous host (Wilkinson and Micklefield 2007). Based on recent pioneering work it seems to be manageable to engineer fungal polyketide biosynthesis pathways as well (Schumann and Hertweck 2006). The experiments that involved domain swapping have been recently done by Du et al. in order to closely investigate programming in HR PKS systems (Zhu et al. 2006). Very recently, Hertweck and collaborators managed to express a silent PKS-NRPS gene cluster from the genome of *A. nidulans* via ectopic expression of a specific regulatory gene (Bergmann et al. 2007). Nevertheless, very few conclusions have been drawn from such experiments and many more questions have arisen so far. In addition, it is extremely difficult to provide sufficient amounts of active substances from fungi due to their limited levels of biosynthesis. Biosynthesis of bioactive marine-derived polyketides is dependent on many factors, one of which is certainly the rare occurrence of fungi themselves, especially those from a marine environment. The establishment of fermentation procedures for such microorganisms would be an option, even though good knowledge on the biosynthetic as well as the genetic background of a particular strain is an essential prerequisite. Therefore, the research on marine fungi and the use of biotransformation and biotechnological methods may help to obtain potent candidates among polyketides for clinical use.

Even though many molecular tool kits for fungi have been developed over the last few decades, it will still take a while until fungal PKS research can follow up the speed at which achievements are currently accumulating for bacterial PKS systems. Apparently, the monomodular-iterative nature of fungal PKS systems, together with the absence of sexual systems for many producer fungal species, is regarded as constantly challenging genetic work on these gene cluster systems. In the end, taking into account the moderate size of the fungal research community, tool development and improvement remain a constant challenge in order to clarify many fundamental aspects of fungal PKS systems.

ACKNOWLEDGMENTS

The authors are thankful to the Marine Bioprocess Research Center of the Marine Bio 21 center funded by the "Ministry of Land, Transport and Maritime," Republic of South Korea, for providing the lab space and grant. Ira Bhatnagar is thankful to CSIR, India, for her sabbaticals to Korea.

ABBREVIATIONS

ACP	acyl carrier protein
AT	acyl transferase
ARO	aromatase(s)
C	condensation
CHR	chalcone reductase
CHS	chalcone synthase
CLC-TE	claisen-cyclase-thioesterases
CYC	cyclase
DH	dehydratase
ER	enoyl reductase
FAS	fatty acyl synthase
HR PKS	highly reducing polyketide synthase
KR	ketoacyl reductase
KS	ketoacyl synthase
KSα-KSβ	heterodimeric ketosynthase
LDKS	lovastatin diketide synthase
LNKS	lovastatin nonaketide synthase
MAT	malonyl-CoA ACP acyl transferase
MT	methyl transferase
6-MSA	6-methylsalicylic acid
NA	norsolorinic acid
NR PKS	nonreduced polyketide synthase
NRPS	non-ribosomal peptide synthetase
PKS	polyketide synthase

PR PKS	partially reducing polyketide synthase
PT	product template
R	reductase
SAT	starter-unit-ACP transacylase
TE	thioesterase
YWA1	naphthopyrone

REFERENCES

Abe, Y., T. Suzuki, C. Ono, K. Iwamoto, M. Hosobuchi, and H. Yoshikawa. 2002. Molecular cloning and characterization of an ML-236B (compactin) biosynthetic gene cluster in *Penicillium citrinum*. *Mol Genet Genomics* 267:636–646.

Austin, M.B. and J.P. Noel. 2003. The chalcone synthase superfamily of type III polyketide synthases. *Nat Prod Rep* 20:79–110.

Bailey, A.M., R.J. Cox, K. Harley, C.M. Lazarus, T.J. Simpson, and E. Skellam. 2007. Characterisation of 3-methylorcinaldehyde synthase (MOS) in *Acremonium strictum*: First observation of a reductive release mechanism during polyketide biosynthesis. *Chem Commun* 39:4053–4055.

Beck, J., S. Ripka, A. Siegner, E. Schiltz, and E. Schweiser. 1990. The multifunctional 6-methylsalicylic acid synthase gene of *Penicillium patulum*. Its gene structure relative to that of other polyketide synthases. *Eur J Biochem* 192:487–498.

Bennett, J. 2001. Alexander Fleming and the discovery of Penicillin. *Adv Appl Microbiol* 49:163–184.

Bentley, R. 1997. Microbial secondary metabolites play important roles in medicine; prospects for discovery of new drugs. *Perspect Biol Med* 40:364–394.

Bergmann, S., J. Schumann, K. Scherlach, C. Lange, A.A. Brakhage, and Hertweck C. 2007. Genomics-driven discovery of PKS-NRPS hybrid metabolites from *Aspergillus nidulans*. *Nat Chem Biol* 3:213–217.

Bhatnagar, I. and S.-K. Kim. 2010. Immense essence of excellence: Marine microbial bioactive compounds. *Mar Drugs* 8:2673–2701.

Bhatnagar, I. and S.-K. Kim. 2010. Marine antitumor drugs: Status, shortfalls and strategies. *Mar Drugs* 8:2702–2720.

Bingle, L.E., T.J. Simpson, and C.M. Lazarus. 1999. Ketosynthase domain probes identify two subclasses of fungal polyketide synthase genes. *Fungal Genet Biol* 26:209–223.

Blunt, J.W., B.R. Copp, W.-P. Hu, M.H. Munro, P.T. Northcote, and M.R. Prinsep. 2008. Marine natural products. *Nat Prod Rep* 25:35–94.

Bookstein, R., C.C. Lai, H. To, and W.H. Lee. 1990. PCR-based detection of a polymorphic BamHI site in intron 1 of the human retinoblastoma (RB) gene. *Nucleic Acids Res* 18:1666.

Borel, J.F., C. Feurer, H.U. Gubler, and H. Stahelin. 1976. Biological effects of cyclosporin A: A new antilymphocytic agent. *Agents Actions* 6:468–475.

Bugni, T.S. and C.M. Ireland. 2004. Marine-derived fungi: A chemically and biologically diverse group of microorganisms. *Nat Prod Rep* 21:143–163.

Cox, R.J. 2007. Polyketides, proteins and genes in fungi: Programmed nano-machines begin to reveal their secrets. *Org Biomol Chem* 5:2010–2026.

Da Rocha, A.B., R.M. Lopes, and G. Schwartsmann. 2001. Natural products in anticancer therapy. *Curr Opin Pharmacol* 1:364–369.

Demain, A.L. 1999. Pharmaceutically active secondary metabolites of microorganisms. *Appl Microbiol Biotechnol* 52:455–463.

Demain, A.L. 2006. From natural products discovery to commercialization: A success story. *J Ind Microbiol Biotechnol* 33:486–495.

Dewick, P.N. 2001. The acetate pathway: Fatty acids and polyketides. In: *Medicinal Natural Products: A Biosynthetic Approach* (2nd edn.), ed. Dewick, P.N., pp. 35–117. John Wiley & Sons Ltd., West Sussex, U.K.

Faulds, D., K.L. Goa, and P. Benfield. 1993. Cyclosporin. A review of its pharmacodynamic and pharmacokinetic properties, and therapeutic use in immunoregulatory disorders. *Drugs* 45:953–1040.

Fleming, A. 1929. On the bacterial action of cultures of a *Penicillium*, with special reference to their use in the isolation of *B. influenzae*. *Br J Exp Pathol* 10:226–236.

Fujii, I., Y. Ono, H. Tada, K. Gomi, Y. Ebizuka, and U. Sankawa. 1996. Cloning of the polyketide synthase gene *atX* from *Aspergillus terreus* and its identification as the 6-methylsalicylic acid synthase gene by heterologous expression. *Mol Gen Genet* 253:1–10.

Fujii, I., A. Watanabe, and Y. Ebizuka. 2004. More functions for multifunctional polyketide synthases. In: *Advances in Fungal Biotechnology for Industry, Agriculture, and Medicine*, eds. Tkacz, J.S. and L. Lange, pp. 97–125. Kluwer Academic/Plenum Publishers, New York.

Funa, N., T. Awakawa, and S. Horinouchi. 2007. Pentaketide resorcylic acid synthesis by type III polyketide synthase from *Neurospora crassa. J Biol Chem* 282:14476–14481.

Galagan, J.E., S.E. Calvo, K.A. Borkovich, E.U. Selker, N.D. Read, D. Jaffe et al. 2003. The genome sequence of the filamentous fungus *Neurospora crassa. Nature* 422:859–868.

Galagan, J.E., S.E. Calvo, C. Cuomo, L.J. Ma, J.R. Wortman, S. Batzoglou et al. 2005. Sequencing of *Aspergillus nidulans* and comparative analysis with *A. fumigatus* and *A. oryzae. Nature* 438:1105–1115.

Haefner, B. 2003. Drugs from the deep: Marine natural products as drug candidates. *Drug Discov Today* 8:536–544.

He, J. 2005. Molecular analysis of the aureothin biosynthesis gene cluster from *Streptomyces thioluteus* HKI-227; new insights into polyketide assembly. Friedrich-Schiller-Universität Jena, Jena, LA.

Hendrickson, L., C.R. Davis, C. Roach, D.K. Nguyen, T. Aldrich, P.C. McAda et al. 1999. Lovastatin biosynthesis in *Aspergillus terreus*: Characterization of blocked mutants, enzyme activities and a multifunctional polyketide synthase gene. *Chem Biol* 6:429–439.

Hidy, P.H., R.S. Baldwin, R.L. Greasham, C.L. Keith, and J.R. McMullen. 1977. Zearalenone and some derivatives: Production and biological activities. *Adv Appl Microbiol* 22:59–82.

Hodge, E.G., P.H. Hidy, and H.J. Wehrmeister. 1966. Inventors; estrogenic compounds and animal growth promoters. U.S. Patent 3239345.

Hopwood, D.A. 1997. Genetic contributions to understanding polyketide synthases. *Chem Rev* 97:2465–2498.

Hopwood, D.A. 2004. Cracking the polyketide code. *PloS Biol* 2:166–169.

Hopwood, D.A. and D.H. Sherman. 1990. Molecular genetics of polyketides and its comparison to fatty acid biosynthesis. *Annu Rev Genet* 24:37–66.

Hutchinson, C.R. and I. Fujii. 1995. Polyketide synthase gene manipulation: A structure-function approach in engineering novel antibiotics. *Annu Rev Microbiol* 49:201–238.

Keller, N.P., G. Turner, and J.W. Bennett. 2005. Fungal secondary metabolism—From biochemistry to genomics. *Nat Rev Microbiol* 3:937–947.

Kennedy, J., K. Auclair, S.G. Kendrew, C. Park, J.C. Vederas, and C.R. Hutchinson. 1999. Modulation of polyketide synthase activity by accessory proteins during lovastatin biosynthesis. *Science* 284:1368–1372.

Kohlmeyer, J., B. Volkmann-Kohlmeyer, and S.Y. Newell. 2004. Marine and estaurine mycelia *Eumycota and Oomycota*. In: *Biodiversity of Fungi: Inventory and Monitoring Methods,* eds. Mueller, G.M., G.F. Bills, and M.S. Foster, pp. 533–546. Elsevier Academic Press, Amsterdam, the Netherlands.

Kroken, S., N.L. Glass, J.W. Taylor, O.C. Yoder, and B.G. Turgeon. 2003. Phylogenomic analysis of type I polyketide synthase genes in pathogenic and saprobic ascomycetes. *Proc Natl Acad Sci USA* 100:15670–15675.

Laich, F., F. Fierro, and J.F. Martin. 2002. Production of Penicillin by fungi growing on food products: Identification of a complete Penicillin gene cluster in *Penicillium griseofulvum* and a truncated cluster in *Penicillium verrucosum. Appl Environ Microbiol* 68:1211–1219.

Lu, P., A. Zhang, L.M. Dennis, A.M. Dahl-Roshak, Y.Q. Xia, B. Arison et al. 2005. A gene (pks2) encoding a putative 6-methylsalicylic acid synthase from *Glarea lozoyensis. Mol Genet Genomics* 273:207–216.

Lysoe, E., S.S. Klemsdal, K.R. Bone, R.J. Frandsen, T. Johansen, U. Thrane et al. 2006. The *PKS4* gene of *Fusarium graminearum* is essential for zearalenone production. *Appl Environ Microbiol* 72:3924–3932.

Machida, M., K. Asai, M. Sano, T. Tanaka, T. Kumagai, G. Terai et al. 2005. Genome sequencing and analysis of *Aspergillus oryzae. Nature* 438:1157–1161.

Mayer, K.M., J. Ford, and G.R. Macpherson. 2007. Exploring the diversity of marine-derived fungal polyketide synthases. *Can J Microbiol* 53:291–302.

McAlpine, J. 1998. Unnatural natural products by genetic manipulation. In: *Natural Products II: New Technologies to Increase Efficiency and Speed.* eds. Sapienza, D.M., W. Mori, and L.M. Savage, pp. 251–278. International Business Communications, Southborough, MA.

tMoore, R.N., G. Bigam, J.K. Chan, A.M. Hogg, T.T. Nakashima, and J.C. Vederas. 1985. Biosynthesis of the hypocholesterolemic agent mevinolin by *Aspergillus terreus*. Determination of the origin of carbon, hydrogen, and oxygen atoms by ^{13}C NMR and mass spectrometry. *J Am Chem Soc* 107:3694–3701.

Moriguchi, T., Y. Ebizuka, and I. Fujii. 2006. Analysis of subunit interactions in the iterative type I polyketide synthase ATX from *Aspergillus terreus. Chembiochem* 7:1869–1874.

Moriguchi, T., Y. Ebizuka, and I. Fujii. 2008. Domain-domain interactions in the iterative type I polyketide synthase ATX from *Aspergillus terreus. Chembiochem* 9:1207–1212.

Nicholson, T.P., B.A. Rudd, M. Dawson, C.M. Lazarus, T.J. Simpson, and R.J. Cox. 2001. Design and utility of oligonucleotide gene probes for fungal polyketide synthases. *Chem Biol* 8:157–178.

O'Hagan, D. 1991. *The Polyketide Metabolites*. Ellis Horwood Limited, Chichester, West Sussex, U.K.

Pelaez, F. 2005. Biological activities of fungal metabolites. In: *Handbook of Industrial Mycology*, ed. An, Z., pp. 49–92. Marcel Dekker, New York.

Peng, Y.L. and A.L. Demain. 1998. A new hydroxylase system in *Actinomadura* sp cells converting compactin to pravastatin. *J Ind Microbiol Biotechnol* 20:373–375.

Revill, W.P., M.J. Bibb, and D.A. Hopwood. 1995. Purification of a malonyltransferase from *Streptomyces coelicolor* A3(2) and analysis of its genetic determinant. *J Bacteriol* 177:3946–3952.

Royer, J.C., K.T. Madden, T.C. Norman, and K.F. LoBuglio. 2004. *Penicillium* genomics. In: *Applied Mycology and Biotechnology*, eds. Dilip, K.A. and G.K. George, Vol. 4, pp. 285–293. Elsevier, Amsterdam, the Netherlands.

Saleem, M., M.S. Ali, S. Hussain, A. Jabbar, M. Ashraf, and Y.S. Lee. 2007. Marine natural products of fungal origin. *Nat Prod Rep* 24:1142–1152.

Schroder, J. 1999. Probing plant polyketide biosynthesis. *Nat Struct Biol* 6:714–716.

Schumann, J. and C. Hertweck. 2006. Advances in cloning, functional analysis and heterologous expression of fungal polyketide synthase genes. *J Biotechnol* 124:690–703.

Seshime, Y., P.R. Juvvadi, I. Fujii, and K. Kitamoto. 2005. Discovery of a novel superfamily of type III polyketide synthases in *Aspergillus oryzae*. *Biochem Biophys Res Commun* 331(1):253–260.

Shen, B. 2000. Biosynthesis of aromatic polyketides. *Top Curr Chem* 209:1–51.

Shen, B. 2003. Polyketide biosynthesis beyond the type I, II and III polyketide synthase paradigms. *Curr Opin Chem Biol* 7:285–295.

Shier, W.T., A.C. Shier, W. Xie, and C.J. Mirocha. 2001. Structure-activity relationships for human estrogenic activity in zearalenone mycotoxins. *Toxicon* 39:1435–1438.

Shimizu, T., H. Kinoshita, S. Ishihara, K. Sakai, S. Nagai, and T. Nihira. 2005. Polyketide synthase gene responsible for citrinin biosynthesis in *Monascus purpureus*. *Appl Environ Microbiol* 71:3453–3457.

Cox, R.J. and T.J. Simpson. 2009. Fungal type I polyketide synthases. In: *Methods in Enzymology*, ed. David, A.H., Vol. 459, pp. 49–78. Academic Press, Amsterdam, the Netherlands.

Staunton, J. and K.J. Weissman. 2001. Polyketide biosynthesis: A millennium review. *Nat Prod Rep* 18:380–416.

Torsvik, V., K. Salte, R. Sorheim, and J. Goksoyr. 1990. Comparison of phenotypic diversity and DNA heterogeneity in a population of soil bacteria. *Appl Environ Microbiol* 56:776–781.

Utian, W.H. 1973. Comparative trial of P1496, a new non-steroidal oestrogen analogue. *Br Med J* 1:579–581.

Wilkinson, B. and J. Micklefield. 2007. Mining and engineering natural-product biosynthetic pathways. *Nat Chem Biol* 3:379–386.

Yu, J.H. and N. Keller. 2005. Regulation of secondary metabolism in filamentous fungi. *Annu Rev Phytopathol* 43:437–458.

Zhu, X., F. Yu, R.S. Bojja, K. Zaleta-Rivera, and L. Du. 2006. Functional replacement of the ketosynthase domain of FUM1 for the biosynthesis of fumonisins, a group of fungal reduced polyketides. *J Ind Microbiol Biotechnol* 33:859–868.

19 Marine Microalgal Metabolites
A Promising Source of Pharmaceuticals

S.W.A. Himaya and Se-Kwon Kim

CONTENTS

19.1 INTRODUCTION

Microalgae are microscopic organisms with a worldwide distribution in both freshwater and marine environments (Chacón-Lee and González-Mariño 2010). They are photosynthetic in nature and effectively utilize the solar energy to produce complex organic metabolites. Among all microalgal classes, marine microalgae have adapted to adverse conditions throughout their evolution due to the stress conditions in their habitat such as temperature variations, salinity, osmotic pressure, ultraviolet rays, and pH differences, and therefore marine microalgae have the potency to produce novel metabolites (Tandeau-de-Marsac and Houmard, 1993). Microalgae have been treated as a source of fish feed for years, until scientists understood the great potential in microalgae in many applications such as biofuel, bio-ethanol, nutrition supplement, and pharmaceutical agents. Due to recent identification of this great potential, much study has not been conducted till date exploring bioactive compounds from microalgae. However, the limited studies on this aspect have already established that the microalgal metabolites are highly bioactive. Being the primary producers of the sea, it is generally believed that many of the active compounds isolated from marine resources have actually been synthesized in microalgae and then passed into other members in food chain.

Considering the novelty in their structures and wide array of selective molecular targets, marine microalgal-derived metabolites such as carotenoids, fatty acids, active peptides, natural compounds, and polysaccharides raise a potent pharmaceutical interest (Guedes et al. 2011b; Lorenz and Cysewski 2000). Identification of pharmaceutical agents from marine algae is advantageous not only because of this wide spectrum of active compounds but also due to environmental considerations. As far as the environment factors are concerned, production of "green natural metabolites" is gaining more attention over scarce marine resources due to the culturability and reproducibility. In this regard, marine microalgae have an added advantage as the primary owners of most active marine ingredients, which can be readily cultured for the continuous production of biologically active metabolites in larger amounts (Kim et al. 2007). Accordingly, the current chapter presents an overview of the natural metabolites isolated from microalgae as potential candidates for the pharmaceutical industry, with a special focus on pharmacological potential of marine microalgae-derived carotenoids, fatty acids, peptides, and polysaccharides. Furthermore, the economic feasibility and future perspectives are also discussed in brief.

19.2 CAROTENOIDS FROM MARINE MICROALGAE

Carotenoids are a class of terpenoid pigments, derived from a 40-carbon polyene chain, which provides distinctive molecular structures of carotenoids (Guedes et al. 2011a). This polyene chain can be substituted with cyclic groups and oxygen-containing functional groups in the biosynthesis of different carotenoid compounds. The oxygenated derivatives of carotenoids are denoted as xanthophylls. Xanthophylls are again classified depending on the nature of oxygen present, such as lutein (oxygen is present as –OH), canthaxanthin (as oxy-groups), and astaxanthin (combination of both –OH and oxy-groups) (Del Campo et al. 2007; Guedes et al. 2011a). There are a number of pharmaceutically potent carotenoids, such as β-carotene, astaxanthin, cantaxanthin, lutein, and violaxanthin, identified from marine microalgal species (Plaza et al. 2009). These carotenoids are potent antioxidants and this activity is well described by relating to its complex ring structure, which has the ability to absorb the energy of oxidative radical species (Guerin et al. 2003). And also, it has been reported that β-carotene enhances the immunity via facilitating the monocyte function to increase the number of surface molecules expressed (Hughes et al. 2000). Based on this antioxidative and immune modulatory power of carotenoids, they can extend their activity in several disease conditions, which have a direct relation to oxidative stress and chronic inflammation such as cancer, cardiovascular disease, rheumatoid arthritis, and several neurodegenerative diseases (Abe et al. 2005).

β-Carotene (Figure 19.1A), the most common carotenoid, has wide range of applications in food, cosmetic, and pharmaceutical industries, which raises the necessity of continuous production of natural β-carotene in large amounts (Hejazi et al. 2004). Green unicellular marine microalgae *Dunaliella salina* has been identified as one of the most promising source of β-carotene with 14% of dry weight being β-carotene at appropriate culture conditions (Metting 1996). The particular growth conditions of *Dunaliella salina*, which yields highest amount of β-carotene at industrial

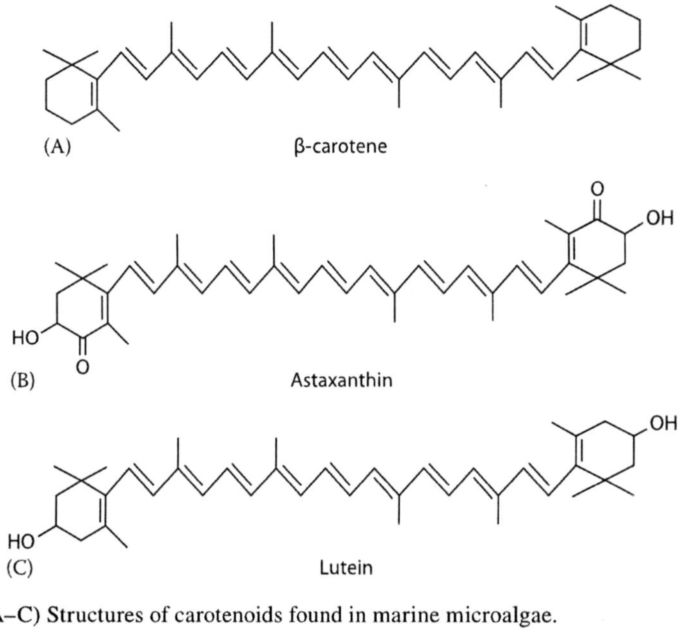

FIGURE 19.1 (A–C) Structures of carotenoids found in marine microalgae.

scale (100 g/kg dry weight), and specific extraction methods of β-carotene have been extensively studied (Denery et al. 2004; Mojaat et al. 2008a). The most appropriate technique for the extraction of β-carotene was found as the super critical fluid extraction (SFE) method. Furthermore, Jaime et al. (2007) has reported that this SFE method yields β-carotene with highest antioxidant capacity. With the development of optimum culture and extraction conditions, *Dunaliella salina* is commercially cultured as a source of β-carotene (Guedes et al. 2011a). SFE of metabolites from *Haematococcus pluvialis* has also produced β-carotene and astaxanthin (Figure 19.1B), where astaxanthin was the main product (about 75%) (Nobre et al. 2006). Up to date, *Haematococcus pluvialis* is considered the highest astaxanthin, producing microalgae with approximately 1.5%–3% by weight, and therefore, it has been exploited commercially (Machmudah et al. 2006). This astaxanthin from *Haematococcus pluvialis* has shown protective effects against colon cancer progression by upregulating the tumor suppressive gene p53 and inhibiting the cell cycle protein cyclin D1 in colon cancer cell models (Mojaat et al. 2008b). In addition, astaxanthin from *chlorella* has shown to stimulate immune responses and this aspect could be used against many immune system-related complications (Plaza et al. 2009). Interestingly, it has been reported that astaxanthin bears the highest antioxidative potential compared to other carotenoids and commercially available antioxidative agents such as vitamin E (Shimidzu et al. 1996). This promising antioxidative potential is attributed to the presence of both hydroxyl and oxy groups in the structure, which can readily quench oxygen radicals.

Lutein (Figure 19.1C) is a highly polar carotene compound corresponding to the presence of hydroxyl groups on the cyclic ring structure, and as a result, it bears a higher antioxidative potential, which could be effectively harvested to promote health and protect against chronic disease conditions. Specifically, lutein has the ability to protect the aging eye-related complications such as cataract and macular degeneration via retarding the pathological mechanisms underlying these diseases (Mares-Perlman et al. 2002). As a result of this potent activity, extraction of lutein from natural sources is explored widely. Among them, marigold is the most widely used source. However, mass plantation of marigold requires large areas and also it is climate dependent, and hence, the scientists have been motivated to find alternative sources (Wu et al. 2007). In this exploration, heterotrophically cultivated *Chlorella pyrenoidosa* was discovered as the most promising source of lutein with 2–4 mg/g dry weight, which is comparable to that of marigold (Shi et al. 1997). Wu et al. (2007) has published SPE as the most appropriate method for optimal extraction of lutein, which resulted in 87.0% extractive of lutein under the optimized conditions of 25 MPa and modified CO_2 with 50% ethanol. In addition, it was found later that *Muriellopsis* sp. has high lutein content up to 35 mg/L under specific culture conditions of 20–40 mM $NaNO_3$, 2–100 mM NaCl, 460 mmol photon/m^2/s, pH 6.5, and 28°C temperature (Del Campo et al. 2000). Corresponding to high growth rate and higher cell density, the culture of *Muriellopsis* sp. as the best source of lutein is practiced commercially (Eonseon et al. 2003). Other than these three most studied carotenoids of microalgae, some other less-known carotenoids such as cantaxanthin from *Chlorella vulgaricus* and *Haematococcus pluvialis* and violaxanthin from *Chlorella ellipsoidea* have been reported to possess pharmaceutical potential due to their antioxidative, immunomodulatory, and cancer prevention effects (Plaza et al. 2009).

19.3 POLYUNSATURATED FATTY ACIDS FROM MARINE MICROALGAE

Polyunsaturated fatty acids (PUFAs), especially ω-3 fatty acids, are essential for normal metabolism and cannot be synthesized by the human body as human body does not synthesize α-linolenic acid (ALA), which is the precursor for other ω-3 fatty acids. Therefore, external supplement of PUFA is required to maintain the physiological functions of the body. Microalgae have been identified as a rich source of long-chain PUFAs where it accounts for 10%–20% of cell weight in some species. Biosynthesis of PUFA in microalgae has been described by Tonon et al. (2004), which involves a series of desaturation and elongation steps starting with oleic acid. As a commercially demanding source of PUFA, design of optimum culture conditions to yield maximum amount of PUFA in a cost-effective manner is essential. Environmental conditions have a significant influence on the PUFA compositions of microalgae, and therefore, external factors such as the medium composition, nitrogen source, pH, incident light intensity, degree of aeration, and temperature should be optimally manipulated depending on the species (Tonon et al. 2002; Volkman et al. 1999). Use of ammonium or nitrate as the nitrogen source has produced highest level of PUFA in *Chlorella minutissima* (44.4% of the total fatty acids) and *Tetraselmis gracilis* (46.6% of the total fatty acids) (Lourenço et al. 2002). Interestingly, the presence of PUFA in marine microalgae is higher compared to freshwater microalgal species as they produce higher amounts of PUFA to survive in marine environments compared to the freshwater counterparts (Bell et al. 1986).

Besides their functions in regulating the normal metabolism, several PUFAs such as eicosapentaenoic acid (EPA), docosahexaenoic acid (DHA), and alpha-linolenic acid (ALA) (Figure 19.2) have proven to possess vast array of health benefits, such as prevention of coronary heart diseases, hypertriglyceridemia, blood platelet aggregation, atherosclerosis, general inflammation, hypertension, type II diabetes, ocular diseases, arthritis, cystic fibrosis, and several carcinomas (Guil-Guerrero et al. 2001; Shahidi and Wanasundara 1998; Wu and Bechtel 2008). In 2004, U.S. Food and Drug Administration has declared a "qualified health claim" status to EPA and DHA (FDA 2004). Therefore, both EPA and DHA have been extensively used as therapeutics to prevent aforementioned adverse health conditions (Mazza et al. 2007; Mullen et al. 2010). In consideration to the promising pharmacological properties, worldwide demand for these PUFAs is increasing rapidly. The principal dietary source of DHA and EPA is fish and fish by-products. However, due to unpleasant fishy odor and as a declining resource, there is a serious commercial and environmental issue related to the continued exploitation of fish in order to meet the demands of the expanding market (Medina et al. 1999; Tonon et al. 2002). And also the fish oil can be contaminated with trace pollutants such as mercury, polychlorinated biphenyls, and dioxins from their habitats (Doughman et al. 2007). More importantly, the realization of the fact that fish itself do not have the ability to synthesize DHA and EPA, and microalgae in their diet is the primary source of fatty acids for fish (Guedes et al. 2011c) arises the interest to explore marine microalgae as an alternative safe source of PUFA, especially EPA and DHA.

Studies on evaluating the EPA and DHA content of microalgae have reported an immense potential of several microalgal species as a primary source of ALA, EPA, and DHA. *Nannochloropsis* spp. (Fuentes et al. 2000; Wu et al. 2000), *Porphyridium cruentum* (Fuentes et al. 2000; Guerrero et al. 2001), *Phaeodactylum tricornutum* (Grima et al. 2003; Pérez et al. 2001), and *Chaetoceros calcitrans* (Delaporte et al. 2003) are identified as the best EPA-producing microalgae, while *Isochrysis galbana* (Qi et al. 2001), *Crypthecodinium* spp. (Atalah et al. 2007), and *Schyzotrichium* spp. (Mattos et al. 2004) have been reported as promising sources of DHA. Furthermore, *Pavlova lutheri* and

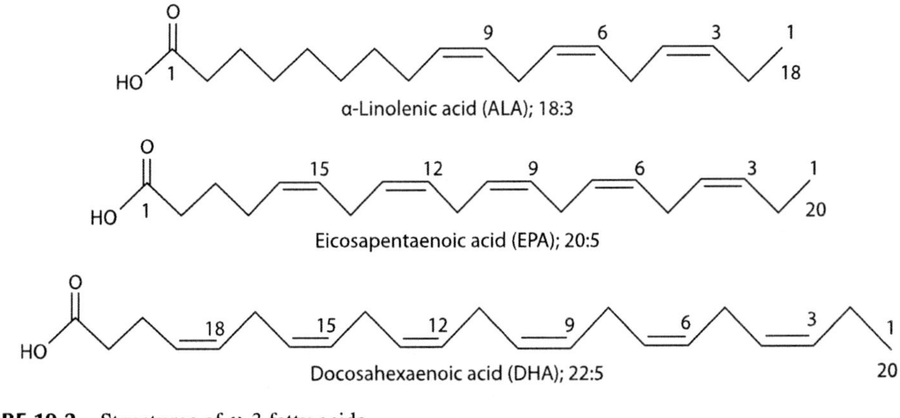

FIGURE 19.2 Structures of ω-3 fatty acids.

Thalassiosira pseudonana have been identified as a rich source of both EPA and DHA (Bigogno et al. 2002; Guedes et al. 2011c; Meireles et al. 2002, 2003). Commercialization of these species as sources of EPA and DHA should be followed by extensive studies on the economically feasible culture and extraction conditions.

19.4 PROTEINS AND BIOACTIVE PEPTIDES FROM MARINE MICROALGAE

Demand for protein as a dietary supplement has been increasing due to change in lifestyle. To supply for this demand, natural sources with higher protein content are being searched as ideal candidates. In this regard, microalgal proteins gained much attention as single-cell proteins. The protein content of some microalgal species including *Chlorella*, *Spirulina*, *Scenedesmus*, *Dunaliella*, *Micractinium*, *Oscillatoria*, *Chlamydomonas*, and *Euglena* has accounted for more than 50% of the dry weight, proving that these microalgae are promising protein sources (Becker 2007). Microalgal proteins are rich in essential amino acids, specifically lysine (Kuhad et al. 1997). Due to high lysine content, microalgae proteins can be effectively used as a supplement in cereal products. The biological value, a measure of absorbed proteins, is considerably high in some microalgae species such as *Spirulina* sp. and *Chlorella* sp., where it is 77.6% and 71.6%, respectively (Scrimshaw and Murray 1995). However, microalgal proteins and peptides could not be harvested directly due to the presence of cell wall. To overcome this constraint, hydrolysis of the cell wall with commercial or bacterial enzymes has been used widely. In a recent study, hydrolysis of *Chlorella* by *Cellulomonas* sp. YJ5 cellulases has resulted in increased level of soluble proteins, which bears higher reducing power (Yin et al. 2010). Besides hydrolysis, proteins in its natural form have been extracted by bead milling technique from microalgae *Tetraselmis* sp. This method has yielded 64% (w/w) proteins (Schwenzfeier et al. 2011).

Corresponding to the specific amino acid composition, microalgal proteins have gained attention as a promising source to isolate bioactive peptides with pharmaceutical applications (Table 19.1). Bioactive peptides are 2–20 amino acids long protein fragments; generally remain latent within the parent protein molecule (Himaya et al. 2012). Once released by hydrolysis, the active peptides exert physiological hormone-like effects on physiological complications of human body

TABLE 19.1
Bioactive Peptides Isolated from Marine Microalgae

The Peptide Sequence	Activity	Source	References
Val-Glu Cys-Tyr-Gly-Pro-Asn-Arg-Pro-Gln-Phe	ACE inhibitory, antioxidant, anticancer	*Chlorella vulgaris* protein waste	Sheih et al. (2009a,b, 2010)
Ile-Val-Val-Glu Ala-Phe-Leu Phe-Ala-Leu Ala-Glu-Leu Val-Val-Pro-Pro-Ala	ACE inhibitory	*Chlorella vulgaris* peptic hydrolysate	Suetsuna and Chen (2001)
Ile-Ala-Glu, Ile-Ala-Pro-Gly and Val-Ala-Phe ValAla-Phe	ACE inhibitory	*Spirulina platensis* peptic hydrolysate	Suetsuna and Chen (2001)
Pro-Gly-Trp-Asn-Gln-Trp-Phe-Leu	Liver protection against ethanol,	*Navicula incerta*	Kang et al. (2011, 2012)
Val-Glu-Val-LeuPro-Pro-Ala-Glu-Leu	Anti-fibrosis	Papain hydrolysate	
Met-Pro-Gly-Pro-Leu-Ser-Pro-Leu	Antioxidant Wound healing	*Pavlova lutheri* Microbial fermentation	Ryu et al. (2012)

beyond their nutritional value (Erdmann et al. 2008). Recently, the reports on hydrolysis of microalgal protein to obtain active peptides with pharmacological potential are increasing. The high protein amounts present in different *chlorella* species has attracted the attention of researchers to use it as a protein source to obtain active peptides. Therefore, hydrolysis of the *chlorella* protein, either using enzymes or microbes, was used as a practical approach. The hydrolysates and the peptides isolated from these hydrolysates have shown beneficial functions in promoting human health. *Chlorella vulgaris* is a well-established edible microalga in Japan. However, during processing, large amount of protein is discarded as waste. This protein waste consists of over 50% protein, and was utilized to produce antioxidative and angiotensin converting enzyme (ACE) inhibitory peptides following enzyme hydrolysis. Peptic hydrolysate of this waste protein of *C. vulgaris* has yielded antioxidant and ACE inhibitory peptide, Val-Glu Cys-Tyr-Gly-Pro-Asn-Arg-Pro-Gln-Phe. The peptide has quenched a variety of free radicals, including hydroxyl radicals (IC_{50} = 8.3 ± 0.15 μM), superoxide radicals (IC_{50} = 7.5 ± 0.12 μM), DPPH radicals (IC_{50} = 23.0 ± 1.8 μM) and ABTS radicals (IC_{50} = 9.8 ± 0.5 μM), and performed more efficiently than that observed for commercially available antioxidants BHT and Trolox (Sheih et al. 2009b). Furthermore, the peptide has effectively protected lung fibroblast cells (WI-38) from oxidative damage (Sheih et al. 2010). And also the peptide has shown a potent ACE inhibitory activity with 29.6 μM IC_{50} value (Sheih et al. 2009a). Inhibitory kinetics has revealed that the peptide binds to the ACE in a noncompetitive manner. Moreover, the peptic digest of *Chlorella vulgaris* cells has produced ACE inhibitory peptides Ile-Val-Val-Glu, Ala-Phe-Leu, Phe-Ala-Leu, Ala-Glu-Leu, and Val-Val-Pro-Pro-Ala with IC_{50} values of 315.3, 63.8, 26.3, 57.1, and 79.5 μM, respectively (Suetsuna and Chen, 2001). The same study reports about ACE inhibitory peptides from the peptic hydrolysate of microalgae *Spirulina platensis,* Ile-Ala-Glu, Ile-Ala-Pro-Gly, and Val-Ala-Phe with IC_{50} values of 34.7, 11.4, and 35.8 μM, respectively. Collectively, it can be observed that the use of pepsin as the hydrolyzing enzyme has shown the highest potency to break down the microalgal proteins into active peptide fractions over other commercially available enzymes. The antioxidative and ACE inhibitory activity of a peptide is strongly depending on the amino acid composition. ACE inhibitory activity of a peptide is influenced by the C-terminal tri-peptide sequence as these can interact with the subsites of ACE; S_1, $S_1'$and S_2' (Ondetti and Cushman 1982; Qian et al. 2007). Therefore, hydrophobic amino acid residues with aromatic or branched side chains are preferred at the C-terminal position (Byun and Kim 2002). The presence of branched amino acids at the *N*-terminal position is also facilitating the ACE inhibitory activity. Increased hydrophobicity of the peptide enhances the antioxidant activity of a peptide, as it allows the peptide to reach hydrophobic targets like cell membranes (Hsu 2010). It is evident that these isolated peptides from microalgae contain the amino acid sequences that enhance their activity. Abundance of these amino acids in microalgae would be facilitating the isolation of potent peptides from microalgal proteins. Further studies on biological activities of these isolated peptides would lead to more interesting findings.

The use of microalgae *Navicula incerta* and *Pavlova lutheri* to isolate active peptides is researched in our laboratory. The protein content of *Navicula incerta* was around 50% and the amino acid composition analysis showed that the most abundant (51%) amino acids present were lysine and arginine (Kang et al. 2011). The papain hydrolysate of this protein resulted in two potent peptides Pro-Gly-Trp-Asn-Gln-Trp-Phe-Leu and Val-Glu-Val-LeuPro-Pro-Ala-Glu-Leu. Both peptides have shown potent liver protective activities against ethanol-induced damage by downregulating cytochrome P-450 2E1 gene expression (Kim and Kang 2011). In addition, both the peptides and the hydrolysate have clearly reduced the amount of pro-collagen, which is an important indicator for hepatic fibrosis (Kang et al. 2012). Furthermore, *Pavlova lutheri* protein hydrolyzed by the yeast *Candida rugopelliculosa* has resulted in a peptide with antioxidant and wound healing properties (Ryu et al. 2012). This is the first report on extracting an active peptide from microalgae protein with microbial fermentation. This method can be suggested to commercialize over enzymatic hydrolysis due to cost effectiveness of the yeast compared to the commercial enzymes. Further research should be conducted to explore microalgal proteins as a source of bioactive peptide with pharmaceutical potential. In addition to the culture techniques, hydrolysis methods and peptide extraction techniques should be standardized in order to produce commercial therapeutics.

19.5 POLYSACCHARIDES

Marine polysaccharides such as carrageenan and agar are widely used commercial products. However, medicinal potential of polysaccharides isolated from microalgae is rarely studied. Recently, considerable attention has been rendered to explore the pharmaceutical potential of marine polysaccharides. This has eventually resulted that marine algal-derived polysaccharides possess various pharmaceutical properties such as antitumor, antivirus, anti-hyperlipidemia, and anticoagulant activities, and these chemical entities would serve as novel lead candidates in generating new drugs. Among microalgae, red microalgal species *Porphyridium spirulina* and *Rhodella reticulate* are known to produce sulfated polysaccharides that bear anti-inflammatory, antiviral, and antioxidant properties (Arad and Levy-Ontman 2010; Matsui et al. 2003). The isolated polysaccharides are composed of neutral monosaccharides (xylose, glucose, and galactose), glucoproteins, and sulfate groups. *Chlorella pyrenoidosa*, which has been named as a green healthy food by FAO, is being used in healthy diets (Robledo and Freile Pelegrin 1997). Shi et al. (2007) have isolated a sulfated polysaccharide from this species and commercially feasible extraction procedure has been optimized as 400 W of ultrasound for 800 s, and then followed by incubation in a water bath at 100°C for 4 h in the presence of 80% ethanol. As a consequence of shown potency of microalgal polysaccharides as pharmaceutical agents, future studies should be conducted in order to realize the full potential of this unexplored resource.

19.6 STEROLS

Microalgae originated sterols show a greater promise as a chemotaxonomic marker for differentiating between microalgal groups due to their uniqueness (Patterson et al. 1994). In particular, most dinoflagellates contain higher amounts of 4α-methyl sterols such as dinosterol, which are rarely found in other classes of microalgae (Mansour et al. 1999). Furthermore, dinoflagellate sterols are characterized with fully saturated ring system (stanols). In the same manner, most diatoms contain 4-desmethylsterols, such as cholesterol, 24-methylcholesta-5,22E-dien-3β-ol, and 24-methylenecholesterol (Volkman et al. 1993). Microalgal sterols can be divided into two groups, the normal sterols that function as membrane lipids, and the unusual, mostly toxic, sterols that are possibly used as defense materials (Giner and Wikfors 2011). This feature can be easily related to the potential of these toxic sterols as pharmaceutical materials such as anticancer and antitumor agents. However, to date, possible use of marine sterols as anticancer agents is rarely studied. As an initiative research, our research group has studied on anti-hepatocarcinoma sterol, stigmasterol, isolated from *Navicula incerta*, and it has shown promising antiproliferative effects against hepatocarcinoma cells via upregulating apoptosis (Kim 2012). Further studies are recommended to explore marine microalgal sterols as effective anticancer agents.

19.7 COMMERCIAL SCALE PRODUCTION, APPLICATIONS, AND FUTURE PROSPECTS

Throughout this chapter, the potential of microalgae as a producer of wide range of pharmaceutical products has been discussed. These potent metabolites can be readily used as novel candidates or can be used to replace existing sources with limitations, such as omega-3 fatty acids from fish oil. To harvest this potency, commercial cultivation and production of these agents should be promoted. Most commercialized microalgal products are carotenoids, based on their high potency as antioxidants over synthetic and other naturally existing antioxidants. Microalgal carotenoids are gaining a strong market demand, where the price of microalgal β-carotene is around 700 €/kg, and, comparatively, its synthetic counterpart would hardly reach to half that figure (Guedes et al. 2011a). Natural β-carotene is preferred as a health-promoting agent as it consists of both *trans* and *cis* isomers where the latter possesses anticancer features. This natural composition cannot be obtained via chemical synthesis (Demming-Adams and Adams 2002). Due to high demand in the market, several leading companies in the world have already started the production of carotenoids in a commercial scale using different techniques such as extraction of astaxanthin from *Haematococcus* by Cyanotech (Hawaii), Mera Pharmaceutical (Hawaii), and Fuji Health Science (Japan), β-carotene production from *Dunaliella* by Betatene (Australia), Western Biotechnology (Australia), and AquaCarotene (Australia), Cyanotech (Hawaii), and Inner Mongolia Biological Engineering (China). Commercial production of microalgal biomass, producing valuable metabolites, is continuously increasing as the research on microalgal metabolites is rising (Del Campo et al. 2007; Guedes et al. 2011a). Carotenoid mixtures extracted from *Dunaliella salina* are commercially available under different brands as supplements.

Even though microalgae present a promising potential as high-quality pure fatty acids, only DHA is commercially available (Spolaore et al. 2006). Species such as *Pavlova lutheri, Isochrysis galbana*, and *Porphyridium purpureum* have a great potential to be cultured as both EPA- and DHA-producing microalgae (Meireles et al. 2002). Despite the useful features discussed in this chapter, microalgae are in general expensive to produce; therefore, cost-efficient culture and extraction methods should be standardized for the mass cultivation and isolation of active ingredients.

19.8 CONCLUDING REMARKS

Microalgae are an untapped resource of potential pharmaceutical compounds. Nevertheless, the awareness and knowledge about this resource is dramatically increasing, which can be evidently concluded by the increasing publications on active ingredients of marine microalgae. To exploit this resource, multidisciplinary studies including optimization of culture conditions, standardization of extraction methods, biotechnological applications, and bioactivity screening should be conducted. Even though exploitation of marine microalgae as a pharmaceutical resource is challenging, the advances in technologies on culture techniques and molecular biology would ultimately result in phenomenal success.

REFERENCES

Abe, K., Hattor, H., and Hiran, M. 2005. Accumulation and antioxidant activity of secondary carotenoids in the aerial microalga *Coelastrella striolata* var. *multistriata*. *Food Chem* 100:656–661.

Arad, S.M. and Levy-Ontman, O. 2010. Red microalgal cell-wall polysaccharides: Biotechnological aspects. *Curr Opin Biotechnol* 21:358–364.

Atalah, E., Hernández-Cruz, C.M., Izquierdo, M.S. et al. 2007. Two microalgae *Crypthecodinium cohnii* and *Phaeodactylum tricornutum* as alternative source of essential fatty acids in starter feeds for seabream (*Sparus aurata*). *Aquaculture* 270:178–185.

Becker, E.W. 2007. Micro-algae as a source of protein. *Biotech Adv* 25:207–210.

Bell, M.V., Henderson, R.J., and Sargent, J.R. 1986. The role of polyunsaturated fatty acids in fish. *Comp Biochem Physiol B* 83:711–719.

Bigogno, C., Khozin-Goldberg, I., and Cohen, Z. 2002. Accumulation of arachidonic acidrich triacylglycerols in the microalga *Parietochloris incisa* (Trebuxiophyceae, Chlorophyta). *Phytochemistry* 60:135–143. *Bioresour Technol* doi:10.1016/j.biortech.2011.12.014.

Byun, H.G. and Kim, S.K. 2002. Structure and activity of angiotensin I converting enzyme inhibitory from Alaskan Pollack skin. *J Biochem Mol Biol* 35:239–243.

Chacón-Lee, T.L. and González-Mariño, G.E. 2010. Microalgae for "healthy" foods—Possibilities and challenges. *Compr Rev Food Sci F* 9:655–675.

Del Campo, A.J., García-González, M., and Guerrero, M.G. 2007. Outdoor cultivation of microalgae for carotenoid production: Current state and perspectives. *Appl Microbiol Biotechnol* 74: 1163–1174.

Del Campo, J.A., Moreno, J., Rodríguez, H. et al. 2000. Carotenoid content of chlorophycean microalgae. Factors determining lutein accumulation in *Muriellopsis* sp. (Chlorophyta). *J Biotechnol* 76:51–59.

Delaporte, M., Soudant, P., Moal, J. et al. 2003. Effect of a mono-specific algal diet on immune functions in two bivalve species *Crassostrea gigas* and *Ruditapes philippinarum*. *J Exp Biol* 206:3053–3064.

Demming-Adams, B. and Adams, W.W. 2002. Antioxidants in photosynthesis and human nutrition. *Science* 298:2149–2153.

Denery, J.R., Dragull, K., Tang, C.S. et al. 2004. Pressurized fluid extraction of carotenoids from Haematococcus pluvialis and Dunaliella salina and kavaloctones from Piper methysticum. *Anal Chim Acta* 501:175–181.

Doughman, S.D., Krupanidhi, S., and Sanjeevi, C.B. 2007. Omega-3 fatty acids for nutrition and medicine: Considering microalgae oil as a vegetarian source of EPA and DHA. *Curr Diabet Rev* 3:198–203.

Eonseon, J., Polle, J.E.W., Lee, H.K. et al. 2003. Xanthophylls in microalgae: from biosynthesis to biotechnological mass production and application. *Microb Biotechnol* 13:165–174.

Erdmann, K., Cheung, B.W.Y., and Schröder, H. 2008. The possible roles of food derived bioactive peptides in reducing the risk of cardiovascular disease. *J Nutr Biochem* 19:643–654.

FDAannouncesqualifiedhealthclaimsforomega-3fattyacids (Press release). United States Food and Drug Administration. September 8, 2004. Retrieved 2006-07-10.

Fuentes, M.M.R., Fernández, G.G.A., Pérez, J.A.S. et al. 2000. Biomass nutrient profiles of the microalga Porphyridium cruentum. *Food Chem* 70:345–353.

Giner, J.L. and Wikfors, G.H. 2011. "Dinoflagellate Sterols" in marine diatoms. *Phytochemistry* 72:1896–1901.

Grima, E.M., Belarbi, H., Acién-Fernández, F.G. et al. 2003. Recovery of microalgal biomass and metabolites: process options and economics. *Biotechnol Adv* 20:491–515.

Guedes, A.C., Amaro, H.M., Barbosa, C.R. et al. 2011c. Fatty acid composition of several wild microalgae and cyanobacteria, with a focus on eicosapentaenoic, docosahexaenoic and α-linolenic acids for eventual dietary uses. *Food Res Int* 44:2721–2729.

Guedes, A.C., Amaro, H.M., and Malcata, F.X. 2011a. Microalgae as sources of high added-value compounds—A brief review of recent work. *Biotechnol Prog* 27:597–613.

Guedes, A.C., Amaro, H.M., and Malcata, F.X. 2011b. Microalgae as sources of carotenoids. *Mar Drugs* 9:625–644.

Guerin, M., Huntley, M.E., and Olaizola, M. 2003. Haematococcus astaxanthin: Applications for human health and nutrition. *Trends Biotechnol* 21:210–215.

Guerrero, J.L.G., Belarbi, H., and Rebolloso-Fuentes, M.M. 2001. Eicosapentaenoic and arachidonic acids purification from the red microalga Porphyridium cruentum. *Bioseparation* 9:299–306.

Guil-Guerrero, J.L., Belarbi, H., and Rebolloso-Fuentes, M.M. 2001. Eicosapentaenoic and arachidonic acids purification from the red microalga *Porphyridium cruentum Bioseparation* 9:299–306.

Hejazi, M.A., Kleinegris, D., and Wijffels, R.H. 2004. Mechanism of extraction of ß-carotene from microalga Dunaliella salina in two-phase bioreactors. *Biotechnol Bioeng* 88:593–600.

Himaya, S.W.A., Ngo, D.H., Ryu, B. et al. 2012. An active peptide purified from gastrointestinal enzyme hydrolysate of Pacific cod skin gelatin attenuates angiotensin-1 converting enzyme (ACE) activity and cellular oxidative stress. *Food Chem* doi:10.1016/j.foodchem.2011.12.020.

Hsu, K.C. 2010. Purification of antioxidative peptides prepared from enzymatic hydrolysates of tuna dark muscle by-product. *Food Chem* 122:42–48.

Hughes, D.A., Wright, A.J.A., Finglas, P.M. et al. 2000. Effects of lycopene and lutein supplementation on the expression of functionally associated surface molecules on blood monocytes from healthy male non-smokers. *J Infect Dis* 182:S11–S15.

Jaime, L. Mendiola, J.A. Ibáñez, E. et al. 2007. ß-Carotene isomer composition of sub- and supercritical carbon dioxide extracts. Antioxidant activity measurement. *J Agric Food Chem* 55:10585–10590.

Kang, K.H., Qian, Z.J., Ryu, B. et al. 2011. Characterization of growth and protein contents from microalgae navicula incerta with the investigation of antioxidant activity of enzymatic hydrolysates. *Food Sci Biotechnol* 20:183–191.

Kang, K.H., Qian, Z.J., Ryu, B. et al. 2012. Protective effects of protein hydrolysate from marine microalgae Navicula incerta onethanol-inducedtoxicityinHepG2/CYP2E1cells. *Food Chem* 132:677–685.

Kim, Y.S. 2012. Effect of stigmasterol isolated from microalgae Navicula incerta on apoptosis in human hepatoma HepG2 cells. MSc dissertation. *Pukyong National University*, Busan, South Korea.

Kim, S.K. and Kang, K.H. 2011. Medicinal effects of peptides from marine microalgae. *Adv Food Nutr Res* 64:313–323.

Kim, M.K. Park, J.W., Park, C.S. et al. 2007. Enhanced production of *Scenedesmus* spp. (green microalgae) using a new medium containing fermented swine wastewater. *Biores Technol* 98:2220–2228.

Kuhad, R.C., Singh, A., Tripathi, K.K. et al. 1997. Microorganisms as an alternative source of protein. *Nutr Rev* 55:65–75.

Lorenz, T.R. and Cysewski, G.R. 2000. Commercial potential for *Haematococcus* microalgae as a natural source of astaxanthin. *Trends Biotechnol* 18:160–167.

Lourenço, S.O., Barbarino, E., Mancini-Filho, J. et al. 2002. Effects of different nitrogen sources on the growth and biochemical profile of 10 marine microalgae in batch culture: An evaluation for aquaculture. *J Phycol* 41:158–168.

Machmudah, S., Shotipruk, A., Goto, M. et al. 2006. Extraction of astaxanthin from Haematococcus pluvialis using supercritical CO_2 and ethanol as entrainer. *Ind Eng Chem Res* 45:3652–3657.

Mansour, M.P., Volkman, J.K., Jackson, A.E. et al. 1999. The fatty acid and sterol composition of five marine dinoflagellates. *J Phycol* 35:710–720.

Mares-Perlman, J.A., Millen, A.E., Ficek, T.L. et al. 2002. The body of evidence to support a protective role for lutein and zeaxanthin in delaying chronic disease. Overview. *Am Soc Nutr Sci* 518S–524S.

Matsui, M.S., Muizzuddin, N., Arad, S. et al. 2003. Sulfated polysaccharides from red microalgae have antiinflammatory properties in vitro and in vivo. *Appl Biochem Biotechnol* 104:13–22.

Mattos, R., Staples, C.R., Arteche, A. et al. 2004. The effects of feeding fish oil on uterine secretion of $PGF_2\alpha$, milk composition, and metabolic status of periparturient Holstein cows. *J Dairy Sci* 87:921–932.

Mazza, M., Pomponi, M., Janiri, L. et al. 2007. Omega-3 fatty acids and antioxidants in neurological and psychiatric diseases: An overview. *Prog Neuropsychopharmacol Biol Psychiatry* 31:12–26.

Medina, A.R., Cerdán, L.E., Giménez, A.G. et al. 1999. Lipase-catalyzed esterification of glycerol and polyunsaturated fatty acids from fish and microalgae oils. *J Biotechnol* 70:379–391.

Meireles, L.A., Guedes, A.C., and Malcata, F.X. 2002. Increase of the yields of eicosapentaenoic and docosahexaenoic acids by the microalga *Pavlova lutheri* following random mutagenesis. *Biotechnol Bioeng* 81:50–55.

Meireles, L.A., Guedes, A.C., and Malcata, F.X. 2003. Lipid class composition of the microalga *Pavlova lutheri*: Eicosapentaenoic and docosahexaenoic acids. *Agric Food Chem* 51:2237–2241.

Metting, F.B. 1996. Biodiversity and application of microalgae. *J Int Microbiol* 17:477–489.

Mojaat, M., Faucault, A., Pruvost, J., and Legrand, J. 2008a. Optimal selection of organic solvents for biocompatible extraction of ß-carotene from *Dunaliella salina. J Biotechnol* 133:433–441.

Mojaat, M., Pruvost, J., Foucault, A. et al. 2008b. Effect of organic carbon sources and Fe^{2+} ions on growth and ß-carotene accumulation by *Dunaliella salina. Biochem Eng J* 39:177–184.

Mullen, A., Loscher, C.E., and Roche, H.M. 2010. Anti-inflammatory effects of EPA and DHA are dependent upon time and dose-response elements associated with LPS stimulation in THP-1-derived macrophages. *J Nutr Biochem* 21:444–450.

Nobre, B., Marcelo, F., Passos, R. et al. 2006. Supercritical carbon dioxide extraction of astaxanthin and other carotenoids from the microalga Haematococcus pluvialis. *Eur Food Technol* 223:787–790.

Ondetti, M.A. and Cushman, D.W. 1982. Enzyme of the renin–angiotensin system and their inhibitors. *Annu Rev Biochem* 51:283–308.

Patterson, G.W., Tsitsa-Tzardis, E., Wikfors, G.H. et al. 1994. Sterols of eustigmatophytes. *Lipids* 29:661–664.

Pérez, A.N., Rebolloso-Fuentes, M.M., Ramos-Miras, J.J. et al. 2001. Biomass nutrient profiles of the microalga Phaeodactylum tricornutum. *J Food Biochem* 25:57–76.

Plaza, M., Herrero, M., Cifuentes, A. et al. 2009. Innovative natural functional ingredients from microalgae. *J Agric Food Chem* 57:7159–7170.

Qi, B., Beaudoin, F., Fraser, T. et al. 2001. Identification of a cDNA encoding a novel C18-Δ^9 polyunsaturated fatty acid specific elongating activity from the docosahexaenoic acid (DHA)-producing microalga, Isochrysis galbana. *FEBS Lett* 510:159–165.

Qian, Z.J., Jung, W.K., Lee, S.H., Byun, H.G., and Kim, S.K. 2007. Antihypertensive effect of an angiotensin I-converting enzyme inhibitory peptide from bullfrog (*Rana catesbeiana* Shaw) muscle protein in spontaneously hypertensive rats. *Process Biochem* 42:1443–1448.

Robledo, D. and Freile Pelegrin, Y. 1997. Chemical and mineral composition of six potentially edible seaweed species of Yucatan. *Bot Mar* 40:301–306.

Ryu, B., Kang, K.H., Ngo, D.H. et al. 2012. Statistical optimization of microalgae *Pavlova lutheri* cultivation conditions and its fermentation conditions by yeast, *Candida rugopelliculosa. Bioresour Technol* 107:307–313.

Schwenzfeier, A., Wierenga, P.A., and Gruppen, H. 2011. Isolation and characterization of soluble protein from the green microalgae *Tetraselmis* sp. *Bioresour Technol* 102:9121–9127.

Scrimshaw, N.S. and Murray, E.B. 1995. Nutritional value and safety of "single-cell protein." In: *Biotechnology*, 2nd edn. Eds. H.J. Rehm and G. Reed, pp. 221–237. Weinheim, Germany: VCH.

Shahidi, F. and Wanasundara, U.N. 1998. Omega-3 fatty acid concentrates: Nutritional aspects and production technologies. *Trends Food Sci Technol* 9:230–240.

Sheih, I.C., Fang, T.J., and Wu, T.K. 2009a. Isolation and characterisation of a novel angiotensin I-converting enzyme (ACE) inhibitory peptide from the algae protein waste. *Food Chem* 115:279–284.

Sheih, I.C., Fang, T.J., Wu, T.K. et al. 2010. Anticancer and antioxidant activities of the peptide fraction from algae protein waste. *J Agric Food Chem* 58:1202–1207.

Sheih, I.C., Wu, T.K., and Fang, T.J. 2009b. Antioxidant properties of a new antioxidative peptide from algae protein waste hydrolysate in different oxidation systems. *Bioresour Technol* 100:3419–3425.

Shi, X.M., Chen, F., Yuan, J.P. et al. 1997. Heterotrophic production of lutein by selected Chlorella strains. *J Appl Phycol* 9:445–450.

Shi, Y., Sheng, J., Yang, F. et al. 2007. Purification and identification of polysaccharide derived from Chlorella pyrenoidosa. *Food Chem* 103:101–105.

Shimidzu, N., Goto, M., Miki, W. 1996. Carotenoids as singlet oxygen quenchers in marine organisms. *Fish Sci* 62:134.

Spolaore, P., Joannis-Cassan, C., Duran, E. et al. 2006. Commercial applications of microalgae. *J. Biosci Bioeng* 101:87–96.

Suetsuna, K. and Chen, J.R. 2001. Identification of antihypertensive peptides from peptic digest of two microalgae, *Chlorella vulgaris* and *Spirulina platensis. Mar Biotechnol* 3:305–309.

Tandeau-de-Marsac, N. and Houmard, J. 1993. Adaptation of cyanobacteria to environmental stimuli: New steps towards molecular mechanisms. *FEMS Microb Rev* 104:119–190.

Tonon, T., Harvey, D., Larson, T.R. et al. 2002. Long chain polyunsaturated fatty acid production and partitioning to triacylglycerols in four microalgae. *Phytochem* 61:15–24.

Tonon, T., Harvey, D., Qing, R. et al. 2004. Identification of a fatty acid Δ^{11}-desaturase from the microalga Thalassiosira pseudonana. *FEBS Lett* 563:28–34.

Volkman, J.K., Barrett, S.M., and Blackburn, S.I. 1999. Fatty acids and hydroxy fatty acids in three species of freshwater eustigmatophytes. *J Phycol* 35:1005–1012.

Volkman, J.K., Brown, M.R., Dunstan, G.A. et al. 1993. Biochemical composition of marine microalgae from the class Eustigmatophyceae. *J Phycol* 29:69–78.

Wu, T.H. and Bechtel, P.J. 2008. Salmon by-product storage and oil extraction. *Food Chem* 111:868–871.

Wu, Z., Wu, S., and Shi, X. 2007. Supercritical fluid extraction and determination of lutein in heterotrophically cultivated Chlorella pyrenoidosa. *J Food Process Eng* 30:174–185.

Wu, Z.C., Zmora, O., Kopel, R. et al. 2000. An industrial-size flat plate glass reactor for mass production of *Nannochloropsis* sp. (Eustigmatophyceae). *Aquaculture* 195:35–49.

Yin, L.J., Jiang, S.T., Pon, S.H. et al. 2010. Hydrolysis of *Chlorella* by *Cellulomonas* sp. YJ5 cellulases and its biofunctional properties. *J Food Sci* 75:H317–H323.

20 Rosmarinic Acid
Biological, Pharmacological, and In Vitro Plant Cell Culture Approximation

Juan C. Luis, Maria Y. González-Padrón, Raquel M. Pérez, Ignacio F. Viera, and Francisco V. González

CONTENTS

20.1 INTRODUCTION

One characteristic of the higher plants is their ability to synthesize an enormous variety of organic molecules known as "secondary metabolites." These molecules are widely used with an important economic impact (Balandrin et al., 1985). Therefore, chemical and biotechnological strategies have been used for the production of these compounds. However, many of these organic molecules have complex structures and sometimes can be chiral, which are difficult, expensive, or even impossible to synthesize (Yeoman and Yeoman, 1996). But, why are they called secondary metabolites? In a living plant, there are a coordinated series of enzyme facilitated chemical reactions known as metabolism. These partial reactions are organized together to form metabolic pathways for the synthesis and utilization of molecular species such as sugars, amino acids, fatty acids, nucleotides, and the polymers derived from them (polysaccharides, proteins, lipids, RNA, DNA, etc.). This collection of processes is known as primary metabolism and the compounds involved, which are essential for the survival of the plant, are described as "primary metabolites." In addition, all plants also use other metabolic pathways, producing compounds, which initially might not be of any obvious use to the organism. These are "secondary metabolites," and the reactions involved in their biosynthesis and utilization constitute the secondary metabolism. However, the dividing line between primary and secondary metabolism is not so distinct, because many of the intermediates in primary metabolism are also intermediates of the secondary metabolism. In fact, the overlapping role of many

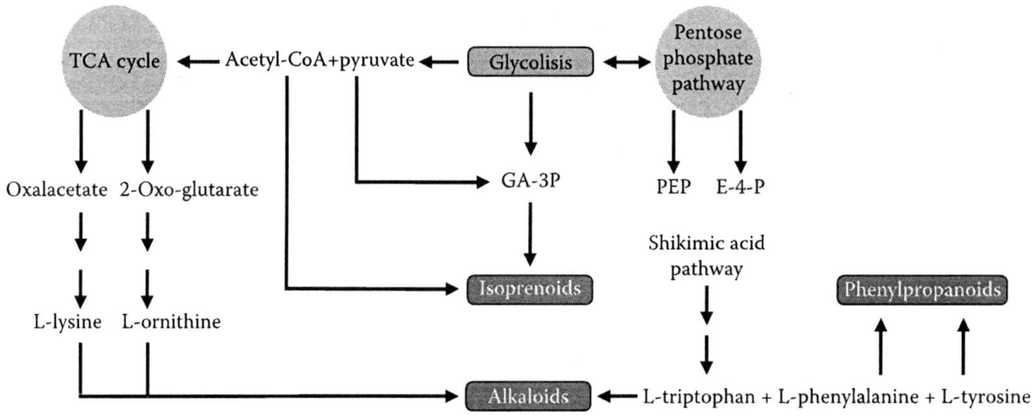

FIGURE 20.1 Metabolic roots of plant secondary metabolites. Plant secondary metabolites can be tracked back to common primary metabolites from which they are ultimately originated. In all cases, the fundamental carbon and nitrogen come from simple sugars and amino acids. GA-3P, glyceraldehyde-3-phosphate; E-4-P, erythrose-4-phosphate; PEP, phosphoenolpyruvate.

compounds ensures a close interconnection between primary and secondary metabolism because many of the small molecules generated in the primary metabolism are building blocks for all the secondary metabolic pathways (Figure 20.1).

20.2 FUNCTIONS OF SECONDARY METABOLITES

Traditionally "secondary metabolites" were considered waste products or those without function at all (Wink, 1988). However, this idea has evolved during the twentieth century based on their biological activities. According to the new ideas, these compounds have evolved during the evolution as defense against plant pathogen microorganisms, against herbivores, and against competing plants. Another function can be attracting animals for pollination or seed dispersal. Secondary metabolites playing the defense role are often not directed against a single organism, but generally to a variety of potential enemies, or in some cases they may combine different roles. Some examples are anthocyanins and volatile terpenes (essential oils), which can be attractants in terms of color and adore, but also insecticidal and antimicrobial. Even more interesting is the capacity of some plants to condense in one compound, playing two different roles. An example is the flavonoid catechin, which unravels part of the mystery of the allelopathic capacity of *Centaurea maculosa* (Bais et al. 2002b). The roots of this plant exude (±) catechin, but only the (–) catechin enantiomer was found phytotoxic, while (+) catechin had the antibacterial activity against root-infesting pathogens, which (–) catechin did not show. Nevertheless, in the last 40 years, new roles have been described for secondary metabolites such as nitrogen transport and storage (Rosenthal, 1982), UV-B protection (Harborne, 1982), antioxidants or scavengers of reactive oxygen species when plants are under stress conditions (Grace and Logan, 2000).

20.3 ROSMARINIC ACID: PRESENCE IN THE PLANT KINGDOM

Research in Lamiaceae secondary metabolites elucidation started around 1950, and the compound named rosmarinic acid (RA; Figure 20.2) was isolated from *Rosmarinus officinalis* (Lamiaceae) and identified for the first time in 1958 (Petersen et al., 2009). RA was later isolated from many species within the Lamiaceae and Boraginaceae families, so becoming one of the active components of several medicinal plants within these families (e.g., *Salvia officinalis*, *Thymus vulgaris*, *Melissa officinalis*, *Symphytum officinale*, among others). However, not all species belonging to the

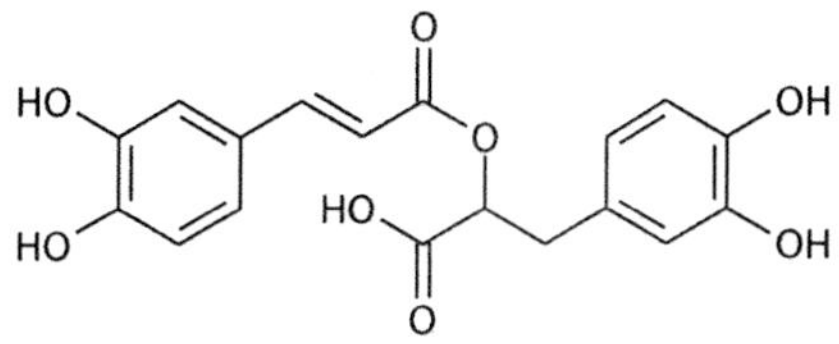

FIGURE 20.2 Structure of RA.

TABLE 20.1
RA Presence in the Plant Kingdom

Division	Class	Family
Briophyta	Anthocerotopsida	Anthocerotaceae
Pteridophyta	Filicopsida	Blechnaceae
Spermatophyta	Magnoliopsida	*Monocotyledonous*: Araceae, Potamogetonaceae, Zosteraceae, Cannaceae, Marantaceae, Meliantaceae *Dicotyledonous*: Onagraceae, Celastraceae, Rosaceae, Cucurbitaceae, Malvaceae, Sterculiaceae, Tiliaceae, Rubiaceae, Lamiaceae, Plantaginaceae, Acanthaceae Scrophulariaceae, Boraginaceae, Hydrophyllaceae, Apiaceae, Araliaceae, Asteraceae, Dipsacaceae

Source: Petersen, M. et al., *Phytochemistry*, 70, 1663, 2009.

Lamiaceae and Boraginaceae families contain RA; in fact, RA is mainly restricted to the subfamily Nepetoideae (Petersen et al., 2009). Outside the Lamiaceae and Boraginaceae families RA is a quite common secondary metabolite found in many species of land plants (Table 20.1), not being therefore recommended as a chemotaxonomical marker.

20.4 ROSMARINIC ACID: BIOSYNTHESIS

The biosynthesis of RA (see Figure 20.3 for molecules structures and enzymes involved) has been studied with intermediates and enzymes involved for almost 10 years, in different plant species from Lamiaceae and Boraginaceae families. The biosynthesis starts with the aromatic amino acids L-phenylalanine and L-tyrosine, which are transformed into the intermediary products 4-coumaroyl-CoA and 4-hydroxyphenyllactic acid, respectively. PAL, C_4H, and 4CL, enzymes of the general phenylpropanoid pathway, are responsible for L-phenylalanine transformation, while tyrosine is transaminated by TAT to produce 4-hydroxyphenylpyruvic acid, which is later reduced to 4-hydroxyphenyllactic acid by HPPR.

The two intermediary products are coupled by ester formation and with release of coenzyme A and the formation of 4-coumaroyl-4′-hydroxyphenyllactic acid. This reaction is catalyzed by the so-called rosmarinic acid synthase (RAS). Finally, the 3 and 3′ hydroxyl groups are introduced by cytochrome P450–dependent monooxygenase reactions, which have been characterized in protein and microsome preparations of suspension cells of *Coleus blumei* (Petersen et al., 1993, 1997).

20.5 ROSMARINIC ACID: ROLE IN PLANTS

The role of RA in plants could be determined, partly, by its distribution in the plant cells. In fact, hydroxycinnamic acids and esters are synthesized in plant cells via the phenylpropanoid pathway localized in the cytosol, with the final stages of biosynthesis and accumulations in the vacuole. They are present in leaves, petals, sepals, stems, and roots, which is consistent with the results

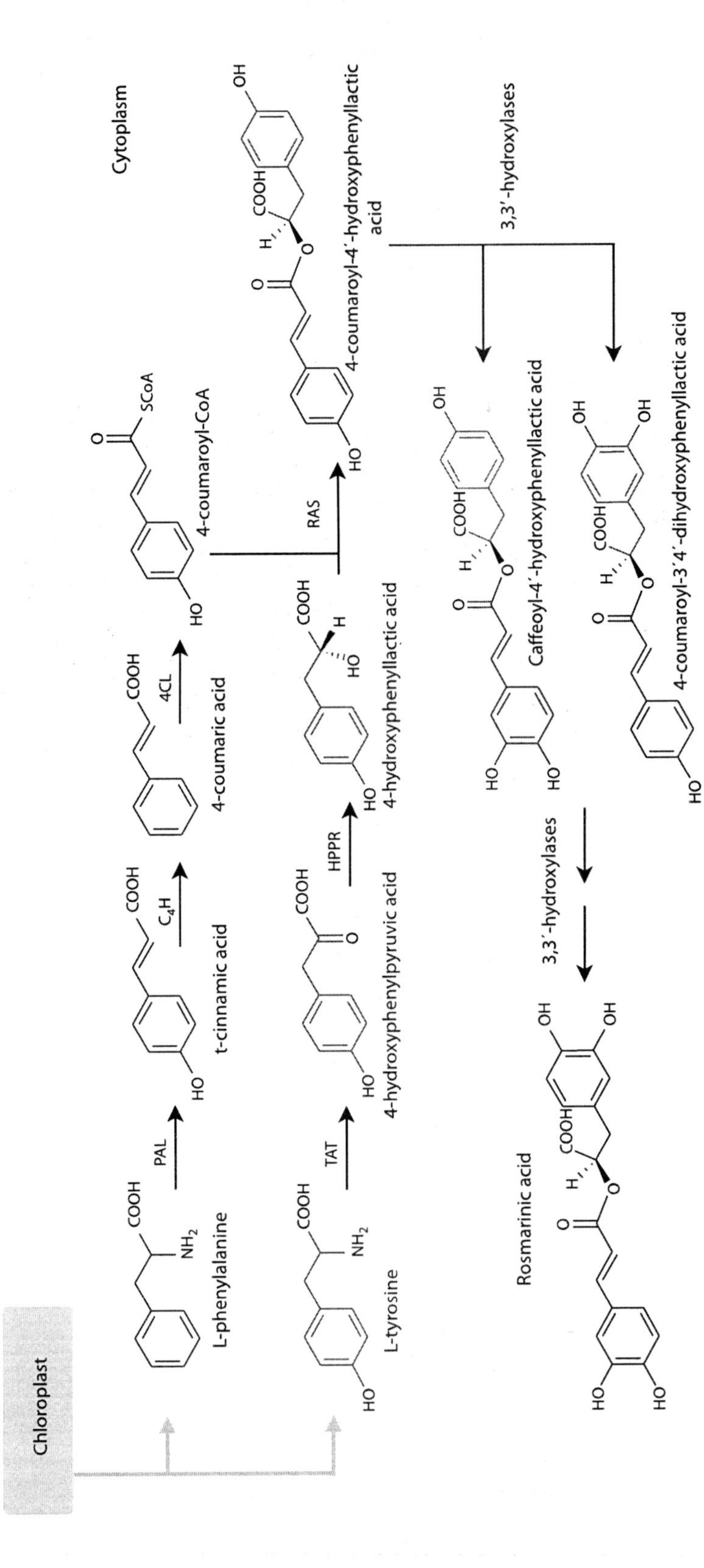

FIGURE 20.3 Proposed biosynthetic pathway for RA. The involved enzymes are PAL, phenylalanine ammonia lyase; C_4H, cinnamic acid 4-hydroxylase; 4CL, 4-coumaric acid CoA-ligase; TAT, tyrosine aminotransferase; HPPR, hydroxyphenylpyruvate reductase; RAS, "rosmarinic acid synthase."

obtained for RA (Luis and Johnson, 2005). In addition, certain phenylpropanoids, for example, have been shown to act as powerful one-electron scavengers of free radicals as well as two-electron donors to the hydrogen peroxide (Grace and Logan, 2000). Furthermore, it has been shown that those environmental stresses such as cold temperatures and UV radiation stimulate the production of phenylpropanoid compounds such as flavonoids, anthocyanins, and hydroxycinnamic acids (Grace and Logan, 2000). In fact, chilling temperatures and enhanced UV-B radiation levels induced high levels of RA in *R. officinalis* plants (Luis et al., 2007a,b). RA has also shown a potent antimicrobial activity against a good array of microorganisms. The mechanism of action of RA included changes in their morphology or cell wall modifications, together with nucleoid damage, increase of spatial division, and condensation of the genetic material (Bais et al., 2002a). The presence of RA in roots of the Lamiaceae species *R. officinalis* and *Ocimum basilicum* suggests the development of strategies such as exudation of RA to combat potential infections from soil-born microorganisms.

20.6 ROSMARINIC ACID: PHARMACOLOGY

RA as a phenolic compound has a well-documented antioxidant activity. The antioxidant activity of RA is stronger than that of vitamin E when hydrophilic systems were used. RA helps to prevent cell damage caused by free radicals, thereby reducing the risk of cancer and atherosclerosis. In addition, RA has anti-inflammatory and anti-allergic activity. A study by Sanbongi et al. (2004) has shown that the oral administration of RA is an effective intervention for allergic asthma. Another study by Youn et al. (2003) demonstrated that RA suppressed synovitis in mice and that it may be beneficial for the treatment of rheumatoid arthritis. Unlike antihistamines, RA prevents the activation of immune responder cells, which cause swelling and fluid formation. RA is also used for food preservation. In Japan the *Perilla* extracts, rich in RA, are used to garnish and improve the shelf life of fresh seafood. Finally, RA has shown in the last 10 years a good array of potential new biological activities (Table 20.2).

20.7 BIOTECHNOLOGICAL APPROACHES TO RA PRODUCTION

Nowadays, chemical synthesis of natural products is possible and sometimes commercially feasible, especially for those with relatively simple chemical structures. However, so often many natural products have a complex chemical structure, which usually includes multiple rings and chiral

TABLE 20.2
RA Biological Activities from Results Published in the Last 10 Years

Biological Activity	Reference
Inhibition of snake venom–induced hemorrhage	Aung et al. (2010)
Enhance cognitive performances	Park et al. (2010)
Inhibition of bone metastasis from breast carcinoma	Xu et al. (2010)
Antifibrotic	Li et al. (2010)
Anti-inflammatory and wound healing	Geller et al. (2010)
Anxiolytic	Awad et al. (2009)
Cardioprotective	Psotova et al. (2005)
Anti-apoptotic and antioxidant	Gao et al. (2005)
Anti-HIV activities (*in vitro*)	Bailly and Cotelle (2005)
Anti-allergic	Makino et al. (2003)

centers, making it difficult and sometimes impossible for the synthetic production. Unfortunately many natural products fall into this group, and RA is one of them. Direct extraction from harvested plant material is a commercial option, if mass cultivation is possible. However, in many cases the inefficiency and the cost of the extraction procedure provoke the search for an alternative supply. When natural supply is limited due to the combination of the mentioned circumstances and new ones such as slow growth, *in vitro* culture becomes an attractive alternative.

In vitro undifferentiated or differentiated cultures can be initiated from most of the plant species, including members of Boraginaceae and Lamiaceae species (Table 20.3). Suspension cell cultures are initiated from an explant that has been isolated from plant material (e.g., embryo, needle, bark, stem, leaf). This explant can be placed on a solid growth medium, which must be "personalized" specifically for different species. Growth medium typically consists of a carbon source, minerals, hormones, vitamins, amino acids, and sometimes antioxidants. Under suitable conditions, the explant will grow into a proliferating mass of undifferentiated cells known as a callus culture. This callus can subsequently be transferred from solid to liquid medium, resulting in a suspension cell culture that is incubated under agitation and controlled temperature.

These undifferentiated cells usually show totipotency and, under suitable conditions, these cultures can be used to regenerate fertile plants in most species. However, this undifferentiated state is not constant, and recent evidence showed that the undifferentiated cells lose the ability to regenerate into plants over time and subcultures (Zhang and John, 2005). To this point, little is known about how undifferentiated cells change over time, which is critical for long-term maintenance cell lines used for metabolite production. In fact, the primary challenges impeding a commercial application of plant cell culture technology are low and variable yields of the desired metabolite accumulation.

Metabolite production via plant cell suspension cultures, from a processing point of view, is suitable to strict control, increasing the production and manufacture standards. Technology developed for other cell culture and systems (e.g., mammalian and yeast) can be easily adapted for large-scale plant cells cultures, diminishing the difficulties associated with scale up for commercial production. In fact, there are currently several industrial commercial plant cell culture processes for secondary metabolites production including RA (Table 20.4).

TABLE 20.3
Published Scientific Papers in the Last 10 Years Analyzing RA Production in Different *In Vitro* Plant Cell Culture Systems from Lamiaceae and Boraginaceae Species

Species	Plant Cell Culture Type	References
Melisa officinalis	Cells cultures	Weitzel and Petersen (2011)
Menta longifolia	*In vitro* plants, cell suspension, and callus cultures	Krzyzanowska et al. (2011)
Salvia multiorrhiza	Cell cultures	Dong et al. (2010)
Lavandula vera	Cell suspension cultures	Georgiev et al. (2009)
Origanum vulgare	Shoot cultures	Lattanzio et al. (2009)
Rosmarinus officinalis	Callus cultures	Yesil-Celiktas et al. (2007)
Satureja hortensis	Callus cultures	Tepe and Sokmen (2007)
Ocimum sanctus	Callus cultures	Hakkim et al. (2007)
Lithospermum erythrorhizon	Transformed cell cultures	Bulgakov et al. (2005)
Coleus blumei	Transformed callus cultures	Bauer et al. (2004)
Ocimum basilicum	Suspension cultures	Kintzios et al. (2003)
Salvia officinalis	Callus and suspension cultures	Santos-Gomes et al. (2003)
Agastache rugosa	Callus and suspension cultures	Kim et al. (2001)

TABLE 20.4
Commercial Plant Cell Culture Processes for RA Supply and Examples for Other Metabolites with Their Pharmaceutical Applications

Metabolite	Species	Application	Manufactures
Rosmarinic acid	*Coleus blumei*	Anti-inflammatory	Nattermann (Germany)
Other examples			
Protoberberines	*Coptis japonica* *Halictrum minus* *Taxus* sp.	Antibiotic Anti-inflammatory	Mitsui Petrochem. Industries (Japan)
Paclitaxel	*Taxus* sp.	Antitumor	Phyton Biotech (United States) Samyang Genex (Korea)
Shikonin	*L. erythrorrhizon*	Anti-HIV Antitumor Anti-inflammatory	Mitsui Petrochem. Industries (Japan)

Source: Kolewe, M.E. et al., *Mol. Pharmaceut.*, 5, 243, 2008.

20.7.1 Traditional Strategies to Improve RA Biosynthesis

The commercial success of plant cell cultures relies on the optimization strategies, similar to those used for other cell culture or fermentation processes. This approach needs the manipulation of cell culture operating parameters such as media composition, cell line selection or gas phase composition (Kieran et al., 1997; Roberts and Shuler, 1997). These strategies are starting points, but when yields are what only matters new strategies are needed. So the most notable strategy for improving metabolite yields is elicitation. An elicitor can be defined as any compound that induces the upregulation of genes. Some elicitors target secondary metabolites genes, which are often associated with defense responses. Elicitors are natural hormones, nutrients, and many fungi derived compounds. In particular, jasmonic acid and its methyl ester (methyl jasmonate) are naturally occurring hormones involved in the upregulation of defense genes as part of the signal transduction system. Applied exogenously, they have been shown to induce secondary metabolic activity and promote the accumulation of desired metabolites in numerous plants systems including RA in species of Lamiaceae families (Dong et al., 2010; Xiao et al., 2010).

Product removal *in situ* has received considerable interest over the years, especially when using transgenic plant systems for the expression of foreign proteins (Doran, 2006). Metabolite accumulation in cell cultures may be limited by feedback inhibition and product degradation, so two-phase systems present obvious advantages including simpler metabolite recovery. Many secondary metabolites may also be toxic to cell cultures at high levels when induced by elicitation, making product removal necessary for continued growth and biomass accumulation. In fact, the use of extraction resins and adsorbents has been shown to increase productivity in several culture systems including those of RA production (Pavlov et al., 2001).

In addition, immobilization of plant cell cultures has long been considered for increasing metabolite accumulation due to a number of advantages (Dornenburg and Knorr, 1995). Immobilization can be simply achieved by using a gel matrix such as alginate; however, this becomes costly at a larger scale, especially when the product of interest is not secreted and must be released (Verpoorte et al., 2002). To our knowledge this strategy has not been used for RA production.

Finally, plant cell suspensions in large-scale bioreactors generally exhibited lower biomass and secondary metabolite accumulation due to the hydrodynamic forces resulting from mechanical agitation (Bourgaud et al., 2001). Plant cells are much larger than mammalian cells or microbes, which makes them extremely susceptible to shear forces in the surrounding fluid. Different types of plant cells exhibit different responses regarding to shear forces, and detailed studies have been performed on individual species, evaluating a variety of effects related to shear forces, including reduction in viability, release of intracellular components, changes in metabolism, and changes in morphology (Zhong, 2002). Generally, excessive shear forces can lead to cell death and reduced viability. More recently, efforts have been focused on understanding the processes and underlying mechanisms involved in the cellular responses to shear. Traditionally, research focused on understanding cellular response in order to optimize bioreactor designs. Unfortunately, published data analyzing RA accumulation by using large-scale bioreactors are scarce and only few examples can be found in the bibliography usually with lower RA yields (Su and Humphrey, 1991; Martinez and Park, 1993; Park and Martinez, 1994; Su et al., 1995).

20.7.2 Metabolic Engineering

Metabolic engineering at a plant cell level to enhance secondary metabolite accumulation is today an appealing strategy by which important progress has been made in the past decade. Unfortunately, secondary metabolic pathways are complex and in most cases are partially undefined. In addition, there are few genomes that are sequenced and they lack any medicinal interest. Moreover, many aspects of secondary metabolites transport and regulation, such as transcription factors or other signaling mechanisms, remain unknown. This metabolic engineering approach involves specific targets and techniques for their identification and manipulation. As many secondary metabolic pathways are partially undefined, their genes and regulatory elements receive nowadays much of the researchers' attention.

A variety of tools have been employed to characterize RA unknown pathway regulation including precursor feeding (Xiao et al., 2010), application of metabolic inhibitors (Yang and Shetty et al., 1998), and interspecies transcription factors analysis (Zhang et al., 2010). Additionally, elicitation previously discussed as a powerful tool to improve secondary metabolite yields has also been used to investigate the RA biosynthetic pathway (Petersen et al., 1994, 2009). Plant cell cultures, including hairy root cultures, have proven to be an extremely useful platform for metabolic studies, as a fast growing and renewable source of material (Petersen et al., 1994). Whole plants can also be valuable, particularly as models to study complex control mechanisms related to environmental stimuli and morphogenesis (Yang and Shetty, 1998). In addition, several approaches have been used to identify the enzymes, which are part of the RA biosynthetic pathway (Petersen et al., 1994). Moreover, genes were subsequently identified by using PCR amplification methods, designed to recognize homologous regions from enzymes in other plants whose DNA sequences were known (Huang et al., 2008; Petersen et al., 2009 and references therein).

Finally, overexpressing an identified pathway gene or several genes will complete this metabolic engineering approach, therefore a reliable transformation technology must be available to integrate foreign DNA into plant cells. Although several methods have been proven to work in many plant species, Agrobacterium-mediated transformation is generally used. Agrobacterium contains a Ti plasmid, a portion of which (T-DNA) is integrated into the plant genome after bacterial infection of the host cell. The T-DNA in wildtype bacteria contains genes to promote proliferation (generally hairy roots cultures), which can be replaced with specific genetic sequences allowing their expression in plants cells. Agrobacterium transformation has become the researchers' first choice due to its low cost and easy use; however, hairy root cultures have only been used to produce RA from several species, analyzing their yields and stability over time (Pistelli et al., 2010).

20.8 CONCLUSIONS

This chapter has collected part of current knowledge on rosmarinic acid (RA). Significant progress has been made in understanding RA production dynamics in plant cell cultures on a number of levels. An integrated approach will most likely be necessary in successful engineering efforts to reliably increase RA production. Despite the complexities of plant metabolism and open questions regarding RA gene regulation and RA accumulation, substantial progress has been made in metabolic pathway characterization, gene identification, and culture systems. At the metabolic level, the lack of complete information about the genomes of most RA producing plants makes traditional "omics" approaches difficult to apply. However, inexpensive and more appealing RA producing plants whole genome sequencing may resolve the puzzle in future.

ABBREVIATIONS

C_4H	cinnamic acid 4-hydroxylase
4CL	4-coumaric acid CoA-ligase
DNA	deoxyribonucleic acid
HPPR	hydroxyphenylpyruvate reductase
PAL	phenylalanine ammonia lyase
PCR	polymerase chain reaction
RAS	synthase
RNA	ribonucleic acid
TAT	tyrosine aminotransferase
T-DNA	transfer DNA

REFERENCES

Aung, H.T., Nikai, T., Niwa, M., and Takaya, Y. 2010. Rosmarinic acid in *Argusia argentea* inhibits snake venom-induced haemorrhage. *J. Nat. Med.* 64:482–486.

Awad, R., Muhammad, A., Durst, T., Trudeau, V.L., and Arnason, J.T. 2009. Bioassay-guided fractionation of lemon balm (*Melissa officinalis* L.) using an in vitro measure of GABA transaminase activity. *Phytother. Res.* 23:1075–1081.

Bailly, F. and Cotelle, P. 2005. Anti-HIV activities of natural antioxidant caffeic acid derivatives: Toward an antiviral supplementation diet. *Curr. Med. Chem.* 12:1811–1818.

Bais, H.P., Walker, T.S., Schweizer, H.P., and Vivanco. J.M. 2002a. Root specific elicitation and antimicrobial activity of rosmarinic acid in hairy root cultures of *Ocimum basilicum. Plant Physiol. Bioch.* 40:983–995.

Bais, H.P., Walker, T.S., Stermitz, F.R., Hufbauer, R.A., and Vivanco, J.M. 2002b. Enantiomeric dependent phytotoxic and antimicrobial activity of (±)-catechin. A rhizosecreted racemic mixture from spotted Knapweed. *Plant Physiol.* 128:1173–1179.

Balandrin, M.F., Klocke, J.A., Wurtele, E.S., and Bollinger, W.H. 1985. Natural plant chemicals: Sources of industrial and medicinal materials. *Science* 228:1154–1160.

Bauer, N., Lejak-Levanic, D., and Jelaska, S. 2004. Rosmarinic acid synthesis in transformed callus culture of *Coleus blumei* Benth. *Z. Naturforschung C* 59:554–560.

Bourgaud, F., Gravot, A., Milesi, S., and Gontier, E. 2001. Production of plant secondary metabolites: A historical perspective. *Plant Sci.* 161:839–851.

Bulgakov, V.P., Veselova, M.V., Tchernoded, G.K., Kiselev, K.V., Fedoreyev, S.A., and Zhuravlev, Y.N. 2005. Inhibitory effect of the Agrobacterium rhizogenes rolC gene on rabdosiin and rosmarinic acid production in *Eritrichium sericeum* and *Lithospermum erythrorhizon* transformed cell cultures. *Planta* 221:471–478.

Dong, J., Wan, G., and Liang, Z. 2010. Accumulation of salicylic acid-induced phenolic compounds and raised activities of secondary metabolic and antioxidative enzymes in *Salvia miltiorrhiza* cell cultures. *J. Biotechnol.* 148:99–104.

Doran, P.M. 2006. Foreign protein degradation and instability in plants and plant tissue cultures. *Trends Biotechnol.* 24:426–432.

Dornenburg, H. and Knorr, D. 1995. Strategies for the improvement of secondary metabolite production in plant-cell cultures. *Enzyme Microb. Technol.* 17:674–684.

Gao, L.P., Wei, H.L., Zhao, H.S., Xiao, S.Y., and Zheng, R.L. 2005. Antiapoptotic and antioxidant effects of rosmarinic acid in astrocytes. *Pharmazie* 60:62–65.

Geller, F., Schmidt, C., Gottert, M., Fronza, M., Schattel, V., Heinzmann, B., Werz, O., Flores, E.M.M., Merfort, I., and Laufer, S. 2010. Identification of rosmarinic acid as the major active constituent in *Cordia americana*. *J. Ethnopharmacol.* 128:561–566.

Georgiev, M., Abrashev, R., Krumova, E., Demirevska, K., Ilieva, M., and Angelova, M. 2009. Rosmarinic acid and antioxidant enzyme activities in Lavandula vera MM Cell suspension culture: A Comparative study. *Appl. Biochem. Biotechnol.* 159:415–425.

Grace, S.C. and Logan, B.A. 2000. Energy dissipation and radical scavenging by the plant phenylpropanoid pathway. *Philos. Trans. R. Soc. B* 355:1499–1510.

Hakkim, F.L., Shankar, C.G., and Girua, S. 2007. Chemical composition and antioxidant property of holy basil (*Ocimum sanctum* L.) leaves, stems, and inflorescence and their in vitro callus cultures. *J. Agric. Food Chem.* 55:9109–9117.

Harborne, J.B. 1982. Introduction to the ecological biochemistry. Academic Press, London, U.K.

Huang, B.B., Yi, B., Duan, Y.B., Sun, L.N., Yu, X. J., Guo, J., and Chen, W.S. 2008. Characterization and expression profiling of tyrosine aminotransferase gene from Salvia miltiorrhiza (Dan-shen) in rosmarinic acid biosynthesis pathway. *Mol. Biol. Rep.* 35:601–612.

Kieran, P.M., MacLoughlin, P.F., and Malone, D.M. 1997. Plant cell suspension cultures: Some engineering considerations. *J. Biotechnol.* 59:39–52.

Kim, H.K., Oh, S.R., Lee, H.K., and Huh, H. 2001. Benzothiadiazole enhances the elicitation of rosmarinic acid production in a suspension culture of *Agastache rugosa* O. Kuntze. *Biotechnol. Lett.* 23:55–60.

Kintzios, S., Makri, O., Panagiotopoulos, E., and Scapeti, M. 2003. In vitro rosmarinic acid accumulation in sweet basil (*Ocimum basilicum* L.). *Biotechnol. Lett.* 25:405–408.

Kolewe, M.E., Gaurav, V., and Roberts, S.C. 2008. Pharmaceutically active natural product synthesis and supply via plant cell culture technology. *Mol. Pharmaceut.* 5:243–256.

Krzyzanowska, J., Janda, B., Pecio, L., Stochmal, A., Oleszek, W., Czubacka, A., Przybys, M., and Doroszewska, T. 2011. Determination of polyphenols in *Mentha longifolia* and *M. piperita* field-grown and in vitro plant samples using UPLC-TQ-MS. *J. AOAC Int.* 94:43–50.

Lattanzio, V., Cardinali, A., Ruta, C., Fortunato, I.M., Veronica M.T., Linsalata, V., and Cicco, N. 2009. Relationship of secondary metabolism to growth in oregano (*Origanum vulgare* L.) shoot cultures under nutritional stress. *Environ. Exp. Bot.* 65:54–62.

Li, G.S., Jiang, W.L., Tian, J.W., Qu, G.W., Zhu, H.B., and Fu, F.H. 2010. In vitro and in vivo antifibrotic effects of rosmarinic acid on experimental liver fibrosis. *Phytomedicine* 17:282–288.

Luis, J.C. and Jonson, C.B. 2005. Seasonal variations of rosmarinic and carnosic acids in rosemary extracts. Analysis of their in vitro antiradical activity. *Span. J. Agric. Res.* 3:106–112.

Luis, J.C., Martín-Pérez, R., Frías, I., and Valdés, F. 2007a. Enhanced carnosic acids levels in two rosemary accessions exposed to cold stress conditions. *J. Agric. Food Chem.* 55:8062–8066.

Luis, J.C., Martín-Pérez, R., Frías, I., and Valdés, F. 2007b. UV-B radiation effects on foliar concentration levels of Rosmarinic and Carnosic acids in Rosemary plants. *Food Chem.* 101:1211–1215.

Makino, T., Furata, Y., Wakushima, H., Fujii, H., Saito, K., and Kano, Y. 2003. Anti-allergic effect of *Perilla frutescens* and its active constituents. *Phytother. Res.* 17:240–243.

Martinez, B.C. and Park, C.H. 1993. Characteristics of batch suspension cultures of preconditioned *Coleus blumei* cells: Sucrose effect. *Biotechnol. Prog.* 9:97–100.

Park, C.H. and Martinez, B. 1994. Growth and production characteristics of permeabilized Coleus blumei cells in immobilized fed-batch culture. *Plant Cell Rep.* 13:459–463.

Park, D.H., Park, S.J., Kim, J.M., Jung, W.Y., and Ryu, J.H. 2010. Subchronic administration of rosmarinic acid, a natural prolyl oligopeptidase inhibitor, enhances cognitive performances. *Fitoterapia* 81:644–648.

Pavlov, A., Ilieva, M., and Mincheva, M. 2001. Release of rosmarinic acid by *Lavandula vera* MM cell suspension in two-phase culture systems. *World J. Microbiol. Biotechnol.* 17:417–421.

Petersen, M. 1997. Cytochrome P-450-dependent hydroxylation in the biosynthesis of rosmarinic acid in *Coleus. Phytochemistry* 45:1165–1172.

Petersen, M., Abdullah, Y., Benner, J., Eberle, D., Gehlen, K., Hücherig, S., Janiak, V. et al. 2009. Evolution of rosmarinic acid biosynthesis. *Phytochemistry* 70:1663–1679.

Petersen, M., Häusler, E., Karwatzki, B., and Meinhard, J. 1993. Proposed biosynthetic pathway for rosmarinic acid in cell cultures of *Coleus blumei* Benth. *Planta* 189:10–14.

Petersen, M., Haeusler, E., Meinhard, J., Karwatzki, B., and Gerlowski, C. 1994. The biosynthesis of rosmarinic acid in suspension cultures of Coleus blumei. *Plant Cell Tiss. Org.* 38:171–179.

Pistelli, L., Giovannini, A., Ruffoni, B., Bertoli, A., and Pistelli, L. 2010. Hairy root cultures for secondary metabolites production. Bio-farms for nutraceuticals: Functional food and safety control by biosensors. *Adv. Exp. Med. Biol.* 698:167–184.

Psotova, J., Chlopcikova, S., Miketova, P., and Simanek, V. 2005. Cytoprotectivity of *Prunella vulgaris* on doxorubicin-treated rat cardiomyocytes. *Fitoterapia* 76(6):556–561.

Roberts, S.C. and Shuler, M.L. 1997. Large-scale plant cell culture. *Curr. Opin. Biotechnol.* 8:154–159.

Rosenthal, G.A. 1982. Plant non-protein amino acids and iminoacids. Academic Press, London, U.K.

Sanbongi, C., Takano, H., Osakabe, N., Sasa, N., Natsume, M., Yanagisawa, R., Inoue, K.I., Sadakane, K., Ichinose, T., and Yoshikawa, T. 2004. Rosmarinic acid in perilla extract inhibits allergic inflammation induced by mite allergen, in a mouse model. *Clin. Exp. Allergy* 34:971–977.

Santos-Gomes, P.C., Seabra, R.M., Andrade, P.B., and Fernandes-Ferreira, M. 2003. Determination of phenolic antioxidant compounds produced by calli and cell suspensions of sage (Salvia officinalis L.). *J. Plant Physiol.* 160:1025–1032.

Su, W.W. and Humphrey, A.E. 1991. Production of rosmarinic acid from perfusion culture of *Anchusa officinalis* in a membrane-aerated bioreactor. *Biotechnol. Lett.* 13:889–892.

Su, W.W., Lei, F., and Kao, N.P. 1995. High density cultivation of *Anchusa officinalis* in a stirred-tank bioreactor with in situ filtration. *Appl. Microbiol. Biot.* 44:293–299.

Tepe, B. and Sokmen, A. 2007. Production and optimisation of rosmarinic acid by Satureja hortensis L. callus cultures. *Nat. Prod. Res.* 21:1133–1144.

Verpoorte, R., Contin, A., and Memelink, J. 2002. Biotechnology for the production of plant secondary metabolites. *Phytochem. Rev.* 1:13–25.

Weitzel, C. and Petersen, M. 2011. Cloning and characterisation of rosmarinic acid synthase from *Melissa officinalis* L. *Phytochemistry* 72:572–578.

Wink, M. 1988. Plant breeding: Importance of plant secondary metabolites for protection against pathogens and herbivores. *Theor. Appl. Genet.* 75:225–233.

Xiao, Y., Gao, S.H., Di, P., Chen, J.F., Chen, W.S., and Zhang, L. 2010. Lithospermic acid B is more responsive to silver ions (Ag^+) than rosmarinic acid in *Salvia miltiorrhiza* hairy root cultures. *Biosci. Rep.* 30:33–40.

Xu, Y.C., Jiang, Z.J., Ji, G.A., and Liu, J.W. 2010. Inhibition of bone metastasis from breast carcinoma by Rosmarinic acid. *Planta Med.* 76:956–962.

Yang, R. and Shetty, K. 1998. Stimulation of rosmarinic acid in shoot cultures of oregano (Origanum vulgare) clonal line in response to proline, proline analogue, and proline precursors. *J. Agric. Food Chem.* 46:2888–2893.

Yeoman, M.M. and Yeoman, C.L. 1996. Manipulation of secondary metabolites in cultured plant cells. *New Phytol.* 134:553–569.

Yesil-Celiktas, O., Nartop, P., Gurel, A., Bedir, E., and Vardar-Sukan, F. 2007. Determination of phenolic content and antioxidant activity of extracts obtained from *Rosmarinus officinalis'* calli. *J. Plant Physiol.* 164:1536–1542.

Youn, J., Lee, K.H., Won, J., Huh, S.J., Yun, H.S., Cho, W.G., and Paik, D.J. 2003. Beneficial effects of rosmarinic acid on suppression of collagen induced arthritis. *J. Rheumatol.* 30:1203–1207.

Zhang, K.R. and John, P.C.L. 2005. Raised level of cyclin dependent kinase A after prolonged suspension culture of *Nicotiana plumbaginifolia* is associated with more rapid growth and division, diminished cytoskeleton and lost capacity for regeneration: Implications for instability of cultured plant cells. *Plant Cell Tiss. Organ Cult.* 82:295–308.

Zhang, Y., Yan, Y.P., and Wang, Z.Z. 2010. The arabidopsis PAP1 transcription factor plays an important role in the enrichment of phenolic acids in Salvia miltiorrhiza. *J. Agric. Food Chem.* 58:12168–12175.

Zhong, J. 2002. Biochemical engineering of the production of plant specific secondary metabolites by cell suspension cultures. *Adv. Biochem. Eng. Biot.* 72:2–26.

21 Enhancement of Natural Antioxidants in Plants by Biosynthetic Pathway Modulation

Kanakapura K. Namitha and Pradeep S. Negi

CONTENTS

21.1 INTRODUCTION

Antioxidants play an important role in maintaining human health. Living organisms have evolved several effective defense mechanisms to protect themselves from free radicals, which include enzymes catalase, superoxide dismutase, and glutathione peroxidase. Consumption of plant-derived antioxidants like carotenoids, flavonoids, ascorbic acid, and vitamin E also forms an additional strategy against these degenerative diseases. Carotenoids are widely distributed in nature and dietary sources of carotenoids are primarily derived from crop plants, flowers, fruits, and vegetables. There has been considerable interest in the dietary carotenoids due to their provitamin A activity, high antioxidant potential, and their ability to prevent the onset of certain cancers. Flavonoids represent a family of aromatic molecules with variable phenolic structures naturally occurring in vegetables, fruits, flowers, seeds, grains, bark, stems, roots, and beverages such as tea and wine. Being potential antioxidants, their contribution to human health is mainly due to their estrogenic, antiviral, antibacterial, antiobesity and anticancer properties; and, a diet rich in flavonoids reduces the risk of certain cancers, coronary heart disease, chronic inflammation, and diabetes. L-Ascorbic acid and tocochromanols, commonly known as vitamin C and vitamin E, respectively, are well-known plant secondary metabolites, possess high antioxidant activity and play an important role in human health and nutrition. Increase in public awareness on the health benefits of these compounds have led to consumption of foods rich in these phytonutrients. However, the presence of these secondary metabolites in low levels in plants does not always meet the recommended daily allowance (RDA) requirements. Genetic engineering approaches have been used to modulate the phytonutrient content in plants to the desirable levels.

The biosynthetic pathway of all the four phytonutrients has been elucidated and almost all the genes have been cloned and characterized from various plants. Several strategies have therefore been used at different levels of the pathway to enhance the content of a particular compound or to produce a new compound by altering the flux through the pathway. The following sections will give details of the structure, distribution, function, biosynthesis, and metabolic engineering of these antioxidant compounds in plants.

21.2 CAROTENOIDS

Carotenoids are one of the largest groups of isoprenoid compounds biosynthesized by several organisms, and more than 600 carotenoids have been characterized so far. In plant systems, carotenoids play a major role in light harvesting, photoprotection, photomorphogenesis, nonphotochemical quenching, and in the biosynthesis of abscisic acid (ABA) (DellaPenna and Pogson 2006; Grotewold 2006). Animals rely upon the diet as a source of these compounds as they are not able to synthesize carotenoids. There has been considerable interest in the dietary carotenoids due to their provitamin A activity (DellaPenna 1999; Hirschberg 1999), high antioxidant potential (Sies and Stahl 2003), and their ability to prevent the onset of certain cancers (Gann et al. 1999; Giovannucci 1999).

21.2.1 Structure and Distribution

Carotenoids are a class of hydrocarbons consisting of eight isoprenoid units (ip), joined in a head-to-tail pattern, except at the center to give symmetry to the molecule. The two central methyl groups

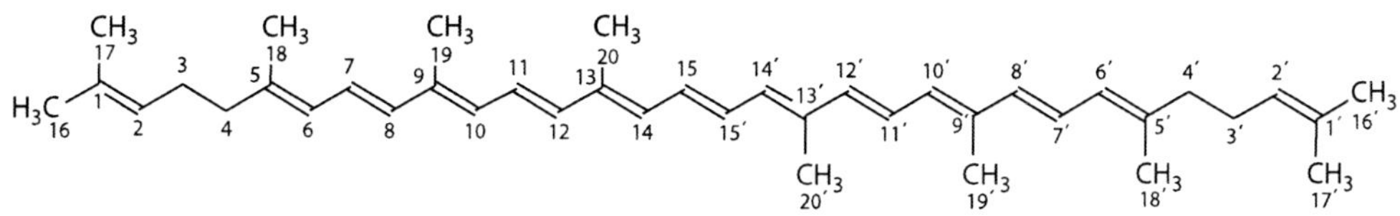

FIGURE 21.1 Basic structure of carotenoid (lycopene) with numbering system.

are in a 1, 6-positional relationship and the remaining nonterminal methyl groups are in a 1, 5-positional relationship (Figure 21.1).

Carotenoids are widely distributed in nature. Different colors produced by this class of pigments including brilliant red, pink, orange, and yellow are found in every form of life. In higher plants, carotenoids are found in plastids, in chloroplasts of photosynthetic tissues, and in chromoplasts in fruits and flowers. Dietary sources of carotenoids are primarily derived from crop plants where they are naturally present in edible leaves, flowers, fruits, and vegetables. The majority of carotenoids are derived from a 40-carbon polyene chain, which could be considered as the backbone of the molecule. On the basis of their chemical structure, carotenoids are classified into two groups, namely, carotenes containing hydrocarbons and xanthophylls, the oxygenated derivatives of these hydrocarbons.

Hydrocarbon carotenes such as β-carotene and lycopene are typically present in free form, which is entrapped within chloroplast and chromoplast bodies. Some of the richest sources of lycopene and β-carotene are tomato, carrot, and watermelon. Xanthophylls, such as lutein and zeaxanthin, are abundant in a number of yellow or orange fruits and vegetables such as peaches, mango, papaya, prunes, squash, and oranges (Namitha and Negi 2010). The uses and sources of a few dietary carotenoids are presented in Table 21.1.

TABLE 21.1
Uses and Sources of Dietary Carotenoids

Use	Carotenoids	Source
Food and feed additive	β-carotene	*Dunaliella* sp.
		Vegetables (carrots, mango, pumpkins, sweet potatoes) and vegetable oils
	Astaxanthin	*Haematococcus* sp.
	Lutein and zeaxanthin	Marigold (*Tagetes erecta*), yellow flowers, and green leafy vegetables
	Capsanthin and capsorubin	Paprika (*Capsicum annum*)
Food colors	Lycopene and β-carotene	Tomato, red grapes, watermelon, pink
	Capsanthin and capsorubin	Grapefruit, papaya, and apricots
		Paprika (*Capsicum annum*)
	Crocetin and crocin	Saffron (*Crocus sativus*)
	Bixin and norbixin	Annatto (*Bixa orellana*)
Nutraceutical	Lycopene and β-carotene	Tomato, red grapes, watermelon, pink
	Astaxanthin	Grapefruit, papaya, and apricots
		Haematococcus sp.
Pharmaceutical	Crocetin and crocin	Saffron (*Crocus sativus*)
Cosmetics and Textiles	Bixin and norbixin	Annatto (*Bixa orellana*)

Sources: Compiled from Rodriguez-Amaya, D.B., *A Guide to Carotenoid Analysis in Foods*, ILSI Press, Washington, DC, 2001; Namitha, K.K. and Negi, P.S., *Crit. Rev. Food Sci. Nutr.*, 50, 728, 2010.

21.2.2 Functions of Carotenoids

In all photosynthetic organisms, carotenoids play a vital and crucial role in photosynthesis. They are involved in photosystem assembly and light harvesting and provide protection from excess light through energy dissipation and free-radical detoxification, which helps in limiting membrane damage. Xanthophylls such as lutein, neoxanthin, and violaxanthin are present in the light-harvesting complex and are involved in the "Xanthophyll cycle" activity induced by high light stress, thus acting as a regulator in the light-harvesting process (Caffarri et al. 2001). β-Carotene is the only carotene found in the core complex of photosystem II (PS II) and plays a dual role at the reaction center either by providing incoming photons to the reaction center (Hanley et al. 1999; Vrettos et al. 1999) or quenching singlet oxygen for the repair of PS II (Anderson and Chow 2002; Tefler 2002). Carotenoids also protect plants against photooxidative damage by dissipating excess energy via nonphotochemical quenching mediated by xanthophylls as well as quenching of triplet chlorophyll by hydrocarbon carotenoids (Niyogi 1999).

Apart from their function in plants, carotenoids have health benefits in animals also. Provitamin A activity is the best-established function of carotenoids, and carotenoids with β-ionone end groups, such as β-carotene, α-carotene, and β-cryptoxanthin have provitamin A activity (Mayne 1996). Carotenoids are considered to be most potent quenchers of singlet oxygen (Boileau et al. 1999; Paiva and Russel 1999) and can react with reactive oxygen species. In majority of these reactions, carotenoids break down to biologically active degradation products (Krinsky and Yeum 2003). The polyene chain length of carotenoids is chemically responsible for quenching the singlet oxygen. Other factors which contribute to antioxidant activity include its isolated double bond, open chain, and lack of oxygen substituents. Lycopene with 11 conjugated and 2 nonconjugated double bonds is the most efficient singlet oxygen quencher of the natural carotenoids (Krinsky 1998).

The carotenoids lutein and zeaxanthin has been shown to be inversely associated with cataracts and age-related macular degeneration, and the intake of vegetables rich in lutein and zeaxanthin gives protection against age-related macular degeneration (Snodderly 1995; Khachik et al. 2002). Dietary intake of tomato and its products, which are rich in lycopene, has been associated with a decrease in the risk of chronic diseases such as cancer and cardiovascular disease (Giovannucci 1999; Rao and Rao 2007). Intake of dietary foods rich in lycopene, such as tomato paste, tomato juice, and lycopene oleoresin capsules, significantly reduced the levels of oxidized low-density lipoproteins (LDL) (Agarwal and Rao 1998). The mechanism responsible for the ability of β-carotene or lycopene to reduce the LDL cholesterol levels may include antioxidant mechanism and feedback mechanism which inhibits HMG-CoA reductase. A small concentration (10 μM) of either β-carotene or lycopene lowers the risk of cardiovascular disease (CVD) by inhibiting cholesterol synthesis in macrophage cell lines (Fuhramn et al. 1997).

Carotenoids also play an important role in boosting the immune response by mechanisms such as increasing lymphocytes' response to mitogens, increasing natural killer cell activity in aging cells, and increasing total white blood cells and CD4/CD8 ratio in HIV-infected persons (Boileau et al. 1999). Various investigations on the role of lycopene in other human diseases such as decreasing hypertension (Paran 2006), neurodegenerative diseases like Alzheimer's disease, Parkinson's disease, and vascular dementia (Foy et al. 1999) have also been reported. Commercially, carotenoids are used as colorants for human food and nutritional supplements, as feed additives to enhance the pigmentation of fish and eggs, as pharmaceutical products, in agriculture, and in the cosmetic industry (Bramley 2003).

21.2.3 Biosynthesis of Carotenoids

Various classical, biochemical, and mutational studies have led to the elucidation of biosynthetic pathways involved in the formation of carotenoids during the last century (Hirschberg 2001). Carotenoids are synthesized and localized in plastids in higher plants and are linked biosynthetically to other isoprenoids such as tocopherol, phylloquinones, gibberellins, chlorophylls, and abscisic acid via the five-carbon compound isopentenyl diphosphate (IPP) (Fraser and Bramley 2004). IPP, the precursor of many isoprenoids, including carotenoids, is produced from mevalonate which is synthesized from acetyl-CoA via the mevalonic acid (MVA) pathway (McGarvey and Croteau 1995), and also from 1-deoxy-D-xylulose-5-phosphate (DXP) via the methyl erythritol 4-phosphate (MEP) pathway (Eisenreich et al. 2001). The MVA pathway is the main pathway of isoprenoid synthesis in eukaryotes, with the exception of the photosynthetic eukaryotes, and occurs in the cytoplasm (Delgado-Vargas and Paredes-Lopez 2003). The MEP pathway is chloroplastidic in nature and is well established in plants as well as in many bacteria (Rodriguez-Concepcion and Boronat 2002). However, plastid isoprenoids at some point of their developmental stage are reported to arise partially from the MVA pathway (Kasahara et al. 2002, Fraser and Bramley 2004).

The first reaction of the MEP pathway is the transketolase condensation of pyruvate with glyceraldehyde-3-Phosphate (G3P) to form DXP which is catalyzed by enzyme 1-deoxy-D-xylulose 5-phosphate synthase (DXPS) (Figure 21.2). The expression of this enzyme is seen during the early stage of leaf development (Araki et al. 2000). Methylerythritol 4-phosphate (MEP) is formed by reduction of DXP (Arigoni et al. 1997; Rohmer 1999), and most carotenoids are produced from MEP-derived precursors in plants grown in light (Lichtenthaler 1999; Eisenreich et al. 2001; Rodriguez-Concepcion and Boronat 2002). MEP is then converted to hydroxymethylbutenyl diphosphate (HMBPP) by the enzyme HMBPP synthase (HDS). HMBPP is finally acted upon by the enzyme HMBPP reductase (HDR) to form a 5:1 mixture of IPP and dimethyl allyl pyrophosphate (DMAPP) (Botella-Pavia et al. 2004). These prenyl–diphosphate units are converted to each other by the enzyme IPP isomerase (IPI) in a reversible reaction (Fraser and Bramley 2004).

Three IPP molecules combine with one DMAPP to form geranyl pyrophosphate (GPP). The addition of one IPP to GPP results in farnesyl pyrophosphate (FPP, C-15 compound) and the addition of one more IPP yields geranyl geranyl pyrophosphate (GGPP, C-20 compound), catalyzed by GGPP synthase. GGPP is a common precursor for several groups of plastid isoprenoids. Two GGPP molecules then condense together to form a colorless symmetrical hydrocarbon compound, phytoene, the backbone of most of the plant carotenoids (Britton 1995), through an intermediate pre-phytoene pyrophosphate (Cunningham and Gantt 1998; Hirschberg 2001). This two-step reaction is catalyzed by the membrane-bound enzyme phytoene synthase (PSY).

Phytoene is converted to lycopene through a series of four desaturation reactions with the formation of intermediate compounds, phytofluene, ζ-carotene, and neurosporene. The colorless phytoene is transformed into the pink-colored lycopene by these desaturation reactions, which serve to lengthen and increase the conjugated series of carbon–carbon double bonds that constitute the chromophore in carotenoid pigments. The four sequential desaturations are catalyzed by two related enzymes in plants: phytoene desaturase (PDS) and ζ-carotene desaturase (ZDS). ZDS catalyzes the conversion of ζ-carotene into lycopene via neurosporene. A carotenoid isomerase (CRTISO) activity is additionally required to transform the poly *cis* lycopene (pro-lycopene), a product of PDS and ZDS activities to the all-*trans* isomer (lycopene) found in plant cells (Fraser and Bramley 2004).

The cyclization of lycopene forms the branching point in the plant carotenoid pathway. β-Carotene is formed in one branch in a two-step reaction wherein one β-ionone ring is produced at each end of the lycopene molecule catalyzed by enzyme lycopene β-cyclase (LCYB/CRTL-B). δ-Carotene is produced in the other branch by the addition of one ε-ring to lycopene in the presence of lycopene ε-cyclase (LCYE/CRTL-E). α-Carotene is produced by the addition of a β-ring to the other end of δ-carotene catalyzed by LCYB, and γ-carotene is formed by the addition of another ε-ring by LCYE. Two types of lycopene β-cyclase enzymes, LCY-B (CRTL-B) and CYC-B (chromoplast-specific lycopene cyclase), have been reported in tomatoes (Pecker et al. 1996; Ronen et al. 2000).

Xanthophylls lutein and zeaxanthin are formed from α-carotene and β-carotene by hydroxylation at the C3 position of each ring via α-cryptoxanthin and β-cryptoxanthin catalyzed by hydroxylases (Fraser and Bramley 2004; Botella-Pavia and Rodriguez-Concepcion 2006). The enzyme zeaxanthin epoxidase (ZEP) catalyzes the conversion of zeaxanthin to violaxanthin via antheraxanthin by

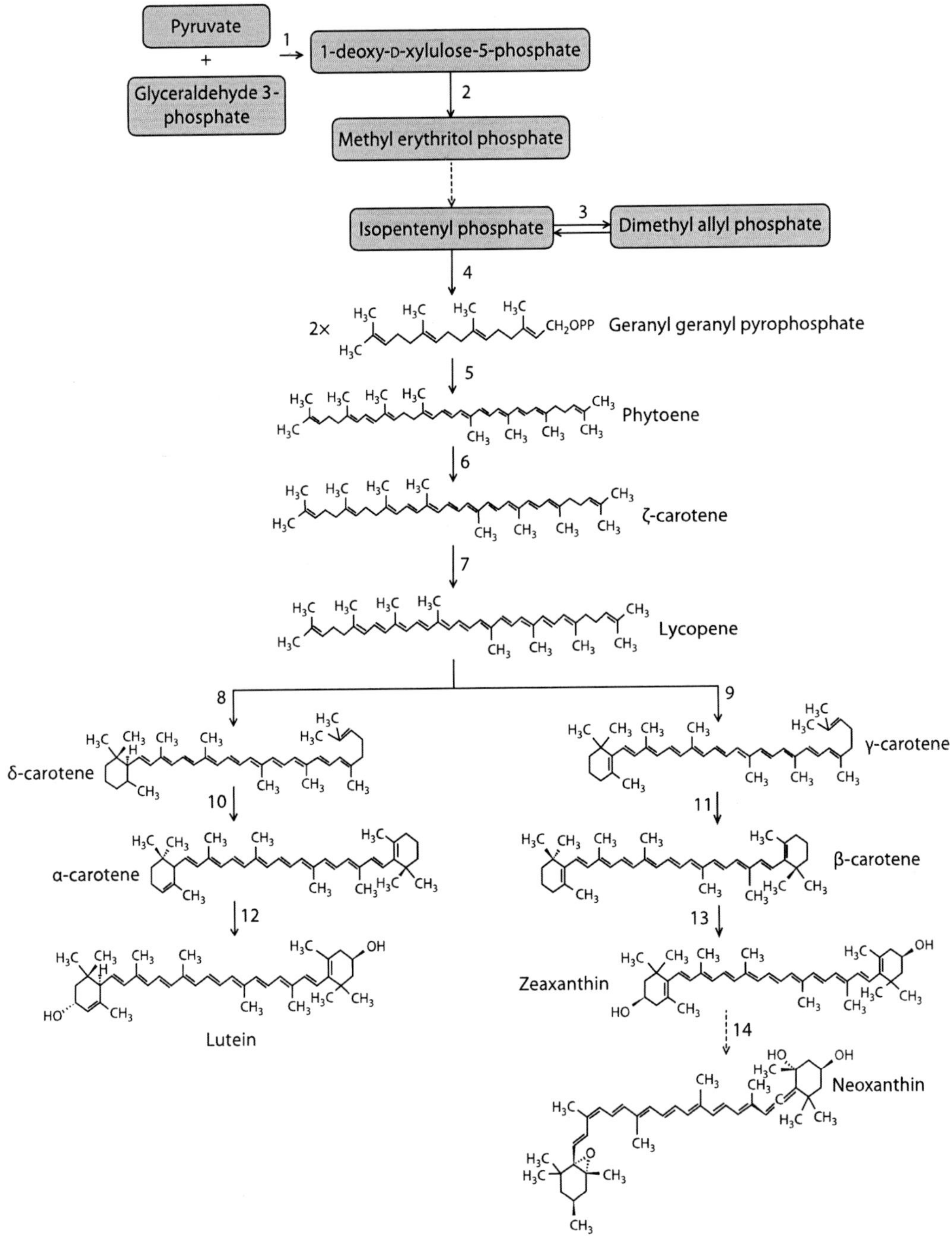

FIGURE 21.2 Generalized pathway of carotenoid biosynthesis in plants. 1—1-deoxy-D-xylulose-5-phosphate synthase; 2—1-deoxy-D-xylulose-5-phosphate reductase; 3—Isopentenyl isomerse; 4—Geranyl geranyl pyrophosphate synthase; 5—Phytoene synthase; 6—Phytoene desaturase; 7—ζ- carotene desaturase; 8—Lycopene ε-cyclase; 9, 10, 11—Lycopene β-cyclase; 12, 13—β-carotene hydroxylase; 14—Neoxanthin synthase. (Compiled from Namitha, K.K. and Negi, P.S., *Crit. Rev. Food Sci. Nutr.*, 50, 728, 2010.)

the introduction of 5, 6- epoxy groups into the 3-hydroxy β-rings (Bouvier et al. 1996). Violaxanthin deepoxidase (VDE) catalyzes the two-step de-epoxidation reaction to transform violaxanthin back to zeaxanthin in leaves exposed to strong light (Pfundel et al. 1994). However, under low-light conditions, transformation of zeaxanthin back to violaxanthin can occur. This interconversion of zeaxanthin and violaxanthin (xanthophyll cycle) is key for plant adaptation to changing environmental conditions (Demmig-Adams et al. 1996). The last step of the carotenoid biosynthesis pathway involves the transformation of violaxanthin into neoxanthin by the activity of neoxanthin synthase (NSY), and genes encoding NSY have been reported in potato (Al-Babili et al. 2000) and tomato (Bouvier et al. 2000).

The ketoxanthophylls, capsanthin, and capsorubin, unique to ripened fruits of pepper (*Capsicum annuum*) are synthesized from antheraxanthin and violaxanthin, respectively by the pepper chromoplast associated enzyme capsanthin–capsorubin synthase (CCS, Bouvier et al. 1994). CCS is similar to tomato CYCB and posses β-cyclase activity (Lefebvre et al. 1998).

21.2.4 Metabolic Engineering of Carotenoids in Plants

With the elucidation of entire carotenoid biosynthetic pathway in plants and almost all the genes being identified, intervention at many points of the pathway has been carried out for enhancement of carotenoids. Metabolite/precursor pool sizes, enzyme activities and location, gene expression profiles, carotenoid catabolism, interaction with other isoprenoid pathways, and regulatory mechanisms influence the choice and combination of genes and promoters necessary to manipulate the pathway (Namitha and Negi 2010). Various strategies have been exploited such as- increasing the flux through the entire pathway by enhancing the production of GGPP; increasing specific steps such as lycopene production, β-carotene production, α-carotene reduction; inhibiting post-target metabolite steps to prevent its conversion to further compounds; to increase the ability of cells to store the compound by providing metabolic sink and preventing feedback inhibition and use of RNA interference technologies (Bai et al. 2011).

21.2.4.1 Carotenoid Enhancement in Cereal Crops

To this date, carotenoid enhancement has been carried out in three cereal crops—rice, maize, and wheat. Rice endosperm lacks provitamin A and other carotenoids, but express several genes of the carotenoid pathway (Burkhardt et al. 1997). Introduction of three heterologous genes, namely, phytoene synthase (*Psy*) and lycopene β-cyclase (*Lcy-B*) from daffodil (*Narcissus pseudonarcissus*) under the control of endosperm specific glutelin promoter and phytoene desaturase (*Crt I*) from bacterium *E. uredovora* under the control of CaMV promoter led to the production of lutein, zeaxanthin, α- and β-carotene in the transformed rice grains (Golden rice 1). The β-carotene reached a maximum level of 1.6 μg/g in the endosperm (Ye at al. 2000). A second generation of golden rice (Golden rice 2) has been produced by using maize phytoene synthase which in conjunction with other two genes elevated the total carotenoid content upto 23-fold (37 μg/g), wherein β-carotene accounted to 31 μg/g in the transgenic rice grains (Paine et al. 2005).

Maize kernels belonging to traditional yellow varieties accumulate lutein, zeaxanthin, and low amounts of β-carotene; while white varieties do not synthesize β-carotene. Aluru et al. (2008) introduced *E. herbicola* phytoene synthase (*Crt B*) and phytoene desaturase (*Crt I*) under the control of γ-*zein* promoter in order to enhance the carotenoid content. Transgenic kernels exhibited a strong endosperm specific expression with 34-fold increase in total carotenoids due to preferential accumulation of β-carotene in endosperm. Zhu et al. (2008) used South African variety M37W to transform multiple enzymes of the carotenoid pathway. Five genes, Maize phytoene synthase (*Psy 1*), *Gentia lutea* lycopene β-cyclase (*Lcy B*), β-carotene hydroxylase (*bch*) and bacterial phytoene desaturase (*Crt I*) and β-carotene ketolase (*Crt W)* were overexpressed under the control of endosperm specific promoter. The transformed plants with random combination of genes had varying carotenoid profile as revealed by the colored endosperms ranging from yellow to scarlet. The kernels had high levels of lycopene, β-carotene, lutein, zeaxanthin, and ketocarotenoids astaxanthin and adinoxanthin. Naqvi et al. (2009) used a different approach wherein multiple genes were transformed to simultaneously enhance three metabolic pathways, thereby increasing three nutrients, namely, β-carotene, ascorbate, and folic acid. To enhance β-carotene content, corn *Psy 1* under the control of wheat LMW glutelin promoter and *Pantoea ananatis Crt I* gene under the control of barley D-hordein promoter were bombarded into 10–14-day-old immature embryos of the South African elite white corn variety M37W. The transgenic kernels were deep orange in color and contained 169-fold more β-carotene (60 μg/g DW), 6 times more ascorbate (110 μg/g DW), and twice the folate (1.94 μg/g DW) than did normal ones.

Wheat cultivars generally have low carotenoid content. Cong et al. (2009) generated transgenic wheat (cultivar EM 12) by expressing maize phytoene synthase (*Psy 1*) under endosperm-specific 1Dx5 promoter *E. uredovora* phytoene desaturase (*Crt I*) gene under constitutive CaMV 35S promoter. The transgenic grains exhibited yellow color with 10.8-fold higher carotenoid levels than the control wild type.

21.2.4.2 Carotenoid Enhancement in Vegetable and Fruit Crops

Among vegetables, potato occupies the first position in being subjected to carotenoid manipulation by various research groups. The carotenoid content in potato is low and it produces only xanthophylls such as lutein and violaxanthin. Carotenoid modulation has been carried out by inhibiting competing enzymes of the pathway wherein endogenous zeaxanthin epoxidase cDNA, under a tuber-specific promoter control, was used both in the antisense and sense direction for the conversion of zeaxanthin to violaxanthin. Downregulation of zeaxanthin epoxidase in potato tubers led to a dramatic increase in zeaxanthin content in some transgenic lines followed by an increase in the tuber carotenoid content upto five- to sevenfold. This also resulted in elevated transcript levels of phytoene synthase and β-*chy* and a two- to threefold increase in α-tocopherol content (Romer et al. 2002). Diretto et al. (2006) carried out silencing of endogenous lycopene epsilon cyclase (*LCY-e*) by using the antisense fragment of this gene under the control of the patatin promoter. Total carotenoids of the antisense tubers were increased by 2.5-fold with maximum increase in β-carotene content (14-fold increase). Higher levels of total carotenoids were achieved by silencing the *bch* gene and thereby preventing the further metabolism of β-carotene. Transgenic plants had a 2.9-fold increase in total carotenoids and a 38-fold increase in β-carotene content (Diretto et al. 2007a). In a separate study, silencing of the β-carotene hydroxylase elevated β-carotene levels to 16.6 μg/g dry weight (Van Eck et al. 2007).

Carotenoid pathway engineering has also been done by introducing transgenes at various stages of the pathway. The overexpression of the *E. uredovora* phytoene synthase (*CrtB*) gene resulted in sevenfold increases in total carotenoids. This also led to increased transcript levels for a protein fibrillin, which functions in carotenoids' storage (Ducreux et al. 2005). To enhance the β-carotene content, potato was transformed with a mini-pathway of bacterial genes, encoding phytoene synthase (*Crt B*), phytoene desaturase (*Crt I*), and lycopene β-cyclase (*Crt Y*) from *Erwinia*, under tuber-specific or constitutive promoter control (Diretto et al. 2007b). Expression of all three genes, under the control of a tuber-specific promoter, resulted in tubers with a deep yellow ("golden") phenotype without any change in leaf carotenoids. These golden tubers accumulated β-carotene to the level of 47 mg/g dry weight (more than 3600-fold increase compared to parental genotype), with simultaneous increase in total carotenoids accounting to 110 mg/g dry weight (approximately 20-fold increase compared to the parental genotype). This is reported to be the highest carotenoid and β-carotene content for bio-fortified potato (Diretto et al. 2007b). To enhance the ketocarotenoid production, a transgenic potato line accumulating zeaxanthin (inactivated zeaxanthin epoxidase) was co-transformed with the *Crt O* (β-carotene ketolase) gene from the cyanobacterium *Synechocystis* under 35 S CaMV promoter. The transformed plants were found to accumulate echinenone, 3′-hydroxyechinenone, and 4-ketozeaxanthin in leaves, as well as 3′-hydroxyechinenone, 4-ketozeaxanthin, together with astaxanthin in the tuber. The amount of ketocarotenoids accounted for 10%–12% of total carotenoids in leaves and tubers (Gerjets and Sandmann 2006). Morris et al. (2006a) produced *Solanum tuberosum* and *Solanum phureja* transgenic lines that expressed an algal *bkt1* gene, encoding a β-ketolase, and accumulated ketocarotenoids. Two major ketocarotenoids, ketolutein and astaxanthin, were accumulated in both the transgenic lines. Potato plants overexpressing *E. coli DXS* gene under the control of a tuber-specific patatin promoter showed an increase in the level of *trans*-zeatin riboside in tubers at the time of harvest. Additionally, in *DXS*-expressing lines, tuber carotenoid content increased by twofold due to a six- to sevenfold increase in phytoene levels when compared with controls (Morris et al. 2006b).

Another approach to enhance carotenoid content involves creating a metabolic sink. Expression of cauliflower *Or* allele in potato under the control of a tuber-specific granule-bound starch synthase promoter produced transgenic potato plants with orange-yellow tubers and carotenoid sequestering organelles similar to those found in mutant cauliflower. A sixfold increase in total carotenoids (24 μg/g DW) over wild-type plants was obtained (Lu et al. 2006). In another study, *Or*-transformed tubers had an increased level of carotenoid intermediates, phytoene, phytofluene, and ζ-carotene suggesting desaturation as the limiting step following *Or* expression. The carotenoid-sequestering structures were also observed in transgenic tubers which were not observed in potato cultivars accumulating a high level of carotenoids (Lopez et al. 2008). The *Or* gene not only enhances carotenoid content but also helps retain carotenoids during storage conditions. Cold-stored *Or* transgenic tubers showed a 10-fold increase in total carotenoids after 5 months than freshly harvested ones. Increased accumulation was associated with lipoprotein-carotenoid sequestering structure formation which was absent in yellow-flesh variety and vector-only control plants (Li et al. 2012).

Most of the cultivated orange carrots have high levels of α- and β-carotene, whereas white-colored roots have low amount of carotenoids. Transgenic strategies have led to two- to fivefold increase in the β-carotene content in transgenic carrot root by the overexpression of *E. herbicola* phytoene synthase (*Crt B*) gene using organ-specific *mas* promoter. The transformed roots exhibited orange/yellow color throughout cross-sections (Ausich et al. 1997; Hauptman et al. 1997). Maass et al. (2009) carried out studies on the effect of overexpression of bacterial *Psy* gene in carrots. Bacterial phytoene synthase (*Crt B*) gene was overexpressed in white root carrot lines under the control of storage-root-specific promoter from yam. The roots of 8 week old transformants exhibited intense yellow color with an increase in carotenoid levels from 180 μg/g DW to 400 μg/g DW at 16 weeks. β-Carotene accounted for 10% of the total carotenoids followed by phytoene, phytofluene, ζ-carotene, and lycopene. Increased Psy protein levels were correlated to the high carotenoid levels as well as crystal formation by sequestration. Engineering ketocarotenoid pathway in carrot tissues has been reported by Jayaraj et al. (2008). β-Carotene ketolase gene from the alga *Haematococcus pluvialis* was introduced separately by three different promoters, double CaMV 35S, *Arabidopsis*-ubiquitin, and RolD from *Agrobacterium rhizogenes* and targeted to plastids in leaf and root tissues. All three promoters provided strong root expression, with double CaMV 35S and ubiquitin promoters exhibiting strong leaf expression. Endogenous expression of carrot β-carotene hydroxylases was upregulated in transgenic leaves and roots, and up to 70% of total carotenoids were converted to ketocarotenoids, which accumulated up to 2.4 mg/g root dry weight level. The most prevalent root carotenoids were astaxanthin, adonirubin, canthaxanthin, echinenone, adonixanthin, and β-cryptoxanthin.

Cassava is an important staple root crop in arid regions such as sub-Saharan Africa. Even though rich in starch, it lacks protein and important micronutrients like iron, zinc, and β-carotene. Gene manipulation studies has been carried out in cassava to increase micronutrient and provitamin A content using root-specific promoters (Arango et al. 2010). The introduction of bacterial phytoene synthase gene (*Crt B*) under the control of cassava CP1 promoter produced deep-orange root transgenics. The total carotenoid content was found to increase to 21.84 μg/g DW as compared to 0.65 μg/g DW in wild roots, while β-carotene increased from 0.41 to 6.67 μg/g DW in transgenic roots. Colorless intermediates phytoene and phytofluene and moderate amounts of xanthophylls such as lutein were also detected (Welsch et al. 2010). In the "biocassava plus program," enhancement of provitamin A content of cassava storage roots were primarily focused by enhancing flux into carotenoid biosynthesis using two-pronged strategies: firstly, expression of the bacterial phytoene synthase gene *crtB* under the control of the storage-root-specific promoter for the potato patatin gene, and secondly, coexpression of *crtB* and the *Arabidopsis* 1-deoxyxylulose-5-phosphate synthase (*DXS*) genes, placed individually under the control of patatin promoters. Transgenic expression of *Crt B* alone increased the carotenoid concentrations 10- to 20-fold higher than nontransformed controls. In engineered roots of 6–8 weeks of age, up to 25 μg/g dry weight (DW) of carotenoids were detected as compared to 1–2.5 μg/g DW in nontransformed roots. Coexpression of phytoene synthase and *DXS* transgenes resulted in enhancing carotenoid concentrations in roots of similar age by 15- to 30-fold higher than those in storage roots from nontransformed plants, reaching concentrations >50 μg/g DW. In the highest carotenoid-producing roots, all-*trans*-β-carotene accounted for 85%–90% of the total carotenoid content of storage roots (Sayre et al. 2011).

Among the fruit crops, tomato is perhaps the most well-studied crop in terms of elucidating the genes responsible for the carotenoid biosynthesis as well as in gene manipulation studies to engineer the carotenoid content in the plant. During ripening of tomato, at breaker stage, the fruit color starts to change from green through pink to red due to tremendous increase in the carotenoid content (400-fold) with lycopene accounting for 90% of the total. This is due to the upregulation of *Psy*, *Pds*, and *Zds*, and the downregulation of *Lcy b* gene (Fraser et al. 1994; Ronen et al. 1999; Namitha et al. 2011). Phytoene synthase exhibits the highest flux control coefficient among the enzymes of the pathway (Fraser et al. 2002); various transgenic strategies have been implemented to induce similar effects. The expression of bacterial *crtB* (from *E. uredovora*) under the control of a tissue-specific tomato polygalacturonase promoter increased total carotenoid levels up to two- to fourfold, with the levels of phytoene, lycopene, β-carotene, and lutein being elevated by 2.4-, 1.8-, 2.2-, and 1.6-fold, respectively (Fraser et al. 2002). D'Ambrosio et al. (2004) reported that the transgenic tomato (HighCaro tomato variety) plants are capable of converting all lycopene into β-carotene under optimal conditions when transformed with tomato lycopene β-cyclase cDNA under the control of CaMV 35S promoter. The engineered plants also showed greater fruit productivity than control under both optimal and reduced water supply conditions. Phytoene desaturation is another step that has been subjected to genetic manipulation in tomato. The constitutive expression of bacterial phytoene desaturase (*crtI*) gene with the CaMV 35S promoter produced orange-colored transgenic fruits with an increase in β-carotene, neoxanthin, antheraxanthin, lutein, zeaxanthin, and tocopherols, and a decrease in carotenoids prior to β-carotene. The amount of β-carotene increased by two- to fourfold, whereas the lycopene levels remain unchanged despite an increase in phytoene desaturase activity but overall carotenoid levels were reduced (Romer et al. 2000). Further investigation showed that endogenous lycopene β-cyclases were upregulated in the transgenic fruits, thus diverting flux toward β-carotene rather than lycopene (Romer et al. 2000). A fivefold increase in β-carotene was achieved in orange-colored ripe fruit of Money maker variety when *Arabidopsis thaliana* lycopene β-cyclase (β-*Lcy*) was expressed under the control of the tomato phytoene desaturase (*Pds*) promoter (Rosati et al. 2000). Attempts were made to engineer both MVA and MEP pathways of carotenoid biosynthesis of tomato by using 3-hydroxymethyl glutaryl CoA (*hmgr-1*) from *A. thaliana* and *DXS* gene from *E. coli* under the control of CaMV 35S promoter. The transgenic tomato plants having additional *hmgr-1*contained elevated phytosterols (up to 2.4-fold), without any change in the IPP-derived isoprenoids. The *DXS*-harboring plants exhibited increased carotenoid content (1.6-fold), with a 2.4- and 2.2-fold increase in the phytoene and β-carotene levels, respectively (Enfissi et al. 2005). Xanthophyll enhancement has also been carried out by various groups using expression of the lycopene β- cyclase (*b-Lcy*) gene from *Arabidopsis thaliana* and β-carotene hydroxylase (β-*Chy*) gene from pepper under the control of fruit-specific *Pds* promoter, which showed a significant increase in β-carotene, β-cryptoxanthin, and zeaxanthin content in the transformed fruits. High levels of hydroxylated β-carotene derivatives coincided with β-chy protein. β-Hydroxylase activity was found only when β-*chy* was expressed in conjugation with *Lcy B*. The expression of endogenous carotenoid biosynthetic genes was unaltered in transformed lines suggesting that the production of xanthophylls was the result of the expression of introduced transgenes rather than the deregulation of endogenous genes (Dharmapuri et al. 2002).

Like potato, RNAi-mediated silencing approach has been used to engineer the carotenoid accumulation in tomato. Silencing of the endogenous photomorphogenesis regulator gene de-etiolated1 (*DET1*) resulted in 8.5-fold higher β-carotene levels in transformed fruits as compared to wild-type plants (Davuluri et al. 2005). Using similar technology, two RNAi expression vectors to silence *Lcy B* and *Lcy E* using gusA introns were constructed and used for the transformation of tomato plants. The transgenic plants had reduced levels of *Lcy B* and *Lcy E* mRNA transcripts and a significant increase in lycopene content was observed up to 13.8 μg/g DW in leaf, which is 4.2-fold higher than in wild plants. The downstream products were also affected by *Lcy B* and *Lcy E* interference. β-Carotene and lutein levels decreased in *Lcy B* RNAi lines. In *Lcy E* RNAi lines, β-carotene levels increased, while lutein levels were decreased. Total carotenoids decreased by 2.9-fold in *Lcy B* lines and increased by 1.7-fold in *Lcy E* lines as compared to that in wild-type controls (Ma et al. 2011).

The overexpression of tomato CRY2, a blue-light photoreceptor, increased carotenoid levels by 1.7-fold, including a 1.3-fold increase in β-carotene (Giliberto et al. 2005). The overexpression of pepper fibrillin, which plays a role in the formation of lipoprotein carotenoid storage structures, resulted in a 95% increase in total carotenoids with a 64% increase in β-carotene and a 118% increase in lycopene levels (Simkin et al. 2007). Chloroplast transformation has also been used to increase carotenoid levels in tomato. Bacterial lycopene β-cyclase (*crtY*) gene driven by the atpl promoter was introduced into tomato plastids by particle bombardment and resulted in the conversion of lycopene to β-carotene. The amount of β-carotene was found to be 28.6 μg/g FW, which amounted to fourfold increase in transplastomic fruits (Wurbs et al. 2007). Using similar technology, the expression of daffodil *lcyb* and bacterial *CrtY* in plastids under the control of the rRNA operon promoter increased β-carotene levels to 95 μg/g fresh weight in tomato fruits, together with a >50% increase in total carotenoids and concomitant decrease in lycopene levels. In leaf tissues, β-ring xanthophylls were elevated with significant reduction in lutein levels, thus suggesting that β-cyclase enhances flux through the β-branch of the carotenoid pathway (Apel and Bock 2009).

Phytoene synthase (*Psy*) gene from Cara cara navel orange (*Citrus sinensis* Osbeck) was transformed into Hongkong kumquat (*Fortunella hindsii* Swingle) fruit under the control of 35S CaMV promoter. The transgenic plants exhibited fruit colors ranging from deep yellow to orange over the wild plants which produced yellow-colored fruits. The transformed fruits were found to contain 171.9 μg/g FW total carotenoids which were twofold higher than the untransformed ones. Transgenic lines showed 1.7- to 3-fold increase in phytoene and 2.2- to 2.9-fold increase in lycopene content over nontransformed controls. The Hongkong kumquat carotenoids such as β-carotene and β-cryptoxanthin were also elevated by 2.0- and 2.3-fold, respectively (Zhang et al. 2009). Recently, Kim et al. (2010) established an efficient genetic transformation procedure for kiwifruit using micro-cross sections of stems. Transgenic leaves constitutively expressing mandarin (*Citrus unshiu*) *GGPPS* or *PSY* accumulated up to 1.3-fold the normal amounts of lutein or β-carotene. Although these two examples show only marginal improvements, these studies may pave the way for additional metabolic engineering studies to modulate carotenoid levels in other fruits.

21.2.4.3 Carotenoid Enhancement in Oilseeds

Canola (*Brassica napus*) seeds contain high carotenoid levels (up to 23 μg/g fresh weight) which includes 0.2 μg/g β-carotene and low levels of lutein, but spectacular increases of 50-fold in the carotenoids content was observed when the bacterial phytoene synthase (*Crt B*) gene was overexpressed in a seed-specific manner (Shewmaker et al. 1999). The transgenic embryos were orange in color and contained β- and α-carotene in the ratio 2:1 along with a significant amount of phytoene. Lutein, the predominant carotenoid in the control seeds, however was not increased. Transgenic canola seeds expressing double constructs of phytoene synthase (*Crt B*) and phytoene desaturase (*Crt I*) and *Crt B* and plant lycopene β-cyclase (*Lcy-b*) showed increase in total carotenoids with minimal changes in the β- to α-carotene ratio. However, the expression of a triple construct consisting of bacterial phytoene synthase, phytoene desaturase, and lycopene β-cyclase in transgenic seeds showed a 50% increase in the β to α ratio (Ravanello et al. 2003). Fujisawa et al. (2009) introduced seven genes—*ipi* from the MEP pathway; bacterial *Crt E, Crt B, Crt I, Crt Y*; marine bacterium *Brevundimonas* SD212 *Crt W*; and *Crt Z* genes—to modulate carotenoid levels in canola. Transgenic seeds were found to accumulate 214 μg/g FW of β-carotene, a 1070-fold increase over control plants. Ketocarotenoids echinenone, canthaxanthin, astaxanthin, and adonixanthin were also synthesized. Studies were carried out to investigate the altered carotenoid accumulation in seeds of *B. napus* as a result of silencing the expression of lycopene ε-cyclase (ε-*CYC*) using RNAi construct. Transgenic seeds expressing this construct had increased levels of β-carotene, zeaxanthin, violaxanthin, and lutein. The β-carotene and lutein concentrations were at least 5.8-fold and 1.9- to 22-fold greater in the ε-*CYC* silenced lines than in control ones. The higher total carotenoid content resulting from the reduction of ε-*CYC* expression in seed suggests that this gene may be a rate-limiting step in the carotenoid biosynthesis pathway (Yu et al. 2008). Recently, Wei et al. (2010) expressed *A. thaliana* micro-RNA gene AtmiR156b by using consitutive as well as napin promoters in *Brassica*. This gene is responsible for the regulation of leaf primordial initiation and transition from the vegetative to the reproductive stage in Arabidopsis. Constitutive expression of AtmiR156b resulted in the enhancement of lutein and β-carotene (upto 4.5-fold) levels in seeds and showed a twofold increase in the number of flowering shoots.

Linseed flax is an important industrial oil crop and linseed oil is an excellent source of α-linolenic acid and lignan. Transgenic studies to increase the carotenoid content in seeds were carried out by Fujisawa et al. (2008). Phytoene synthase (*Crt B*) from *P. ananatis* was expressed under the control of CaMV promoter/*FAE 1* seed-specific promoter. The transformed flax plants produced orange seeds with increased amounts of lutein, phytoene, α-carotene, and β-carotene, while non-transformed flax plants produced light-yellow seeds wherein only lutein was detected. The total carotenoids in the transformed seeds were 65.4–165.3 μg/g FW corresponding to a 7.8- to 18.6-fold increase than in untransformed controls.

21.3 FLAVONOIDS

Flavonoids are an important group of low molecular weight secondary metabolites produced by plants. They consist of a family of aromatic molecules with variable phenolic structures naturally occurring in vegetables, fruits, flowers, seeds, grains, bark, stems, roots, and beverages such as tea and wine. More than 6000 flavonoids are known till date, many of which are responsible for the attractive pigmentation of red, blue, and purple in flowers, fruits, and leaves (Nijveldt et al. 2001; Winkel-Shirley 2001). Basically, flavonoids have a flavan nucleus (Figure 21.3), consisting of two aromatic rings with six carbon atoms (ring A and B) interconnected by a heterocycle with three

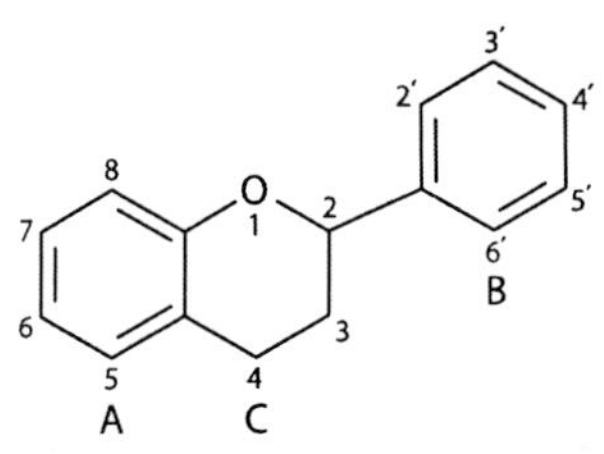

FIGURE 21.3 Basic structure of flavonoid.

TABLE 21.2
Dietary Sources of Flavonoids

General Name	Food Source	Type of Flavonoids
Berries	Blueberries, cranberries	Flavonols (quercetin, myrcetin)
	Blackberries, black grapes	Flavan-3-ols (epicatechin and catechin)
	Raspberries, cherries, red grapes	Anthocyanidins and cyanidins
Vegetables	Peppers, tomatoes, eggplant	Flavonol (quercetin) and flavone (luteolin)
	Onions (red and green)	Flavonol (quercetin)
	Celery, capsicum	flavones (apigenin, luteolin)
	Okra, broccoli	Flavonols (quercetin, kaempferol, myrcetin)
	Lettuce	Flavones
Nuts and beans	Blackbeans, kidneybeans	Anthocyanidins (delphidin, malvidin, petunidin) and flavonol (kaempferol)
	Walnuts	Anthocyanidins
	Pistachios, cashewnuts	Catechins
	Soybean	Catechins and isoflavones (genistein, daidzein)
Fruits	Bananas	Anthocyanidins (cyanidin, delphidin)
	Grapefruit, lemon, lime, orange	Flavanones (hesperetin, naringenin, eriodictoyl)
	Apples, pear, plum, peach, apricot	Catechin and epicatechin
Spices	Dill	Flavonols (quercetin, isohamnetin)
	Parsley	Flavone (apigenin) and flavonol (isohamnetin)
	Thyme	Flavone (luteolin)
	Tea (black, red, green)	(Quercetin, myrcetin)
		Catechins, epigallocatechins and flavonol (thearubigin)
Beverages	Red wine	Anthocyanidins and flavonols
Chocolate	Dark chocolate	Catechins

Source: Compiled from USDA Database for the Flavonoid Content of Selected Foods, March 2003.

carbon atoms (ring C) and are generally represented as C_6-C_3-C_6. Based on the modifications of the central C ring, flavonoids are further divided into different structural classes like flavanones, isoflavones, flavones, flavonols, flavanols, and anthocyanins (Bovy et al. 2007).

21.3.1 Source of Flavonoids

As flavonoids impart color to the fruits, vegetables, nuts, and seeds, they form an integral part of our diet (Parr and Bolwell 2000). The dietary sources rich in flavonoids include soybean isoflavones, onion flavonols, citrus flavanones, celery flavones, apple, tea and cocoa flavanols, and berries anthocyanins (Ross and Kasum 2002). Table 21.2 lists the dietary sources and the flavonoids they contain.

21.3.2 Functions of Flavonoids

Flavonoids are involved in diverse biological functions such as providing eye-catching pigmentation to flowers, fruits, and seeds; facilitating pollination and seed dispersion; protection against UV light; plant defense against pathogenic microbes; plant fertility and germination of pollen; and signal molecules in plant–microbe interactions (Forkmann and Martens 2001; Bovy et al. 2007). Apart from high antioxidant potential, their contribution to human health is mainly due to their estrogenic, antiviral, antibacterial, antiobesity, and anticancer properties (Fowler and Koffas 2009).

The most described property of each group of flavonoids is their ability to act as antioxidants. *In vitro* studies have shown that a majority of flavonoids, especially flavones and catechins are effective and powerful antioxidants in protecting the body against reactive oxygen species (Nijveldt et al. 2001). Flavonol quercitin and flavan-3-ol epicatechin gallate have a fivefold higher total antioxidant activity than vitamins E and C as measured by trolox equivalents (Rice-Evans et al. 1995). Free radicals and reactive oxygen species are produced by body cells and tissues during normal oxygen metabolism or induced by external damage. Living organisms have evolved several effective antioxidant defense mechanisms to protect themselves from these free radicals, which include enzymes catalase, superoxide dismutase, and glutathione peroxidase and nonenzymatic counterparts such as vitamin E and C. Flavonoids contribute to the additive effect of these endogenous scavengers by three different mechanisms. The first mechanism is by direct scavenging of free radicals. Flavonoids are oxidized by free radicals, resulting in a less-reactive radical. Due to the highly reactive hydroxyl group of flavonoids, radicals are made inactive as follows:

$$\underset{\text{Flavonoid-free radical}}{F(OH) + R*} \rightarrow \underset{\text{oxygen-free radical}}{F(O*) + RH}$$

Flavonoids are also capable of directly scavenging superoxides and peroxynitrites (Korkina and Afanasev 1997).

The second mechanism involves interfering with inducible nitric oxide synthase activity. Constitutive nitric oxide synthase activity by macrophages and endothelial cells produces nitric oxide required to maintain the dilation of blood vessels. However, higher concentration of nitric oxide produced due to inducible nitric oxide synthase in macrophages can lead to oxidative damage following a sequence of reactions finally causing irreversible damage to the cell membrane. Flavonoids scavenge free radicals and make them unavailable to react with nitric oxide, thus causing less damage (Shutenko et al. 1999). Flavonoid silibinin has been reported to inhibit nitric oxide in a dose-dependent manner (Dehmlow et al. 1996).

The third mechanism involves the inhibition of xanthine oxidase enzyme. Xanthine dehydrogenase, present in normal physiological conditions, changes to xanthine oxidase during ischemic conditions. This enzyme is a source of oxygen free radicals. Flavonoids such as quercetin, silibinin, and luteolin have been shown to inhibit xanthine oxidase, thereby decreasing oxidative damage (Chang et al. 1993; Cos et al. 1998).

Increased consumption of flavonols quercetin and kaempferol in the form of a balanced diet has shown to protect against cardiovascular diseases (Hertog et al. 1995; Knekt et al. 1996). Isoflavones, predominant in legumes, play an important role in plant–microbe interactions. They are responsible for initiating the root-nodule formation during nitrogen fixation, as they attract the Rhizobium bacteria and induce *nod* gene expression (Van Rhijn and Vanderleyden 1995; Pueppke 1996). They are the precursors of phytoalexins, and some isoflavones also possess antifungal activity (Rivera-Vargas et al. 1993). Isoflavones also contribute to human health by reducing the risk of hormone-related cancers, menopausal symptoms, and heart diseases (Messina 1999; Clarkson 2002; Watanabe et al. 2002). Flavones are involved in various interactions with other organisms, microbes, insects, as well as plants. Siqueira et al. (1991) has shown that flavone chrysin is able to increase mycorrhizal root colonization as well as root growth of *Trifolium repens*. Similarly, certain flavones have been shown to promote *Glomus* hyphal growth and spore germination (Tsai and Phillips 1991). Evidences support the fact that flavones play an important role in the symbiosis between nitrogen-fixing bacteria and legumes. The first flavone identified in such a role was luteolin (Peters et al. 1986). Flavones are exuded by host roots which are recognized by bacterial symbiont, thereby facilitating *nod* gene expression by bacteria, which in turn are recognized by plant hosts (Fisher and Long 1992). Together with anthocyanidins, flavones also help in attraction of pollinators (Harborne and Williams 2000). Flavones also affect insects in various ways: by inhibiting larval feeding and as a feeding deterrent. Flavone 4-hydroxy maysin has been shown to inhibit the development of corn earworm moth *Heliothes zea* (Simmonds 2003). They are also used to control other organisms including plants of family bryophytes and magnoliophytes (Basile et al. 2003); plant parasitic nematodes (Soriano et al. 2004); mollusks (Lahlou 2004); fungi (Weidenborner and Jha 1997; Del Rio et al. 1998); and bacteria (Basile et al. 1999; Xu and Lee 2001).

Apart from their important functions in plant physiology and biochemistry, flavones also contribute to human health and nutrition. Epidemiological and animal studies have revealed that a high dietary intake of flavonoids, especially flavones, flavanones, and isoflavones, has resulted in reduced risk of certain cancers, osteoporosis, coronary heart disease, chronic inflammation, and diabetes (Middleton et al. 2000; Allister et al. 2005; Arts and Hollman 2005; Popiolkiewicz et al. 2005). The flavones baicalein and baicalin from a perennial herb have been used in the treatment of various types of cancer, hepatitis, T-cell leukemia, and inflammation. These compounds also possess strong mutagenic and free-radical scavenging activity (Malikov and Yuldashev 2002; Wozniak et al. 2004). Their mechanism of action includes the reduction of cell-associated matrix metalloproteinase-2 activity, inhibition of migration and proliferation, and *in vitro* capillary formation of vascular endothelial cells (Liu et al. 2003). *In vitro* and *in vivo* activity against prostrate cancer has also been shown, wherein the flavones baicalein and baicalin caused accumulation of cells in G1, induced apoptosis, and the decreased expression of androgen receptor in LNCa P cells (Chen et al. 2001). Another flavone apigenin was also found to be a strong inhibitor of cell proliferation and angiogenesis in human endothelial cells (Osada et al. 2004). Studies have also revealed that apigenin inhibits the growth of human cervical carcinoma cells (HeLa) and neuroblastoma cell lines by inducing p53 expression leading to cell-cycle arrest at G1 phase and apoptosis (Zheng et al. 2005). The glucosylated flavanone hesperidin and the flavanone naringenin were found to be highly effective in improving lipid metabolism by altering hepatic enzyme activities while at the same time lowering blood sugar levels through the downregulation of the hepatic GLUT2 and glucose-6-phosphatase and simultaneously upregulating the hepatic glucokinase and adipocyte GLUT4 in studies carried out using diabetic mice as model systems (Ae Park et al. 2006).

21.3.3 Biosynthesis of Flavonoids

In the last decade, the biosynthetic pathway of flavonoids has been completely elucidated. Many of the structural and some of the regulatory genes have been characterized and cloned from several model plants such as maize, petunia, and arabidopsis (Holton and Cornisch 1995). There have been several attempts to engineer or modify the content of flavonoids in various plants and microorganisms (Schijlen et al. 2004). Two classes of genes are distinguished within the flavonoid pathway: (a) structural genes that encode the enzymes which directly participate in the flavonoids' formation and (b) regulatory genes that control structural genes expression.

The precursors for flavonoid synthesis are malonyl-CoA derived from carbohydrate metabolism and *p*-Coumaroyl-CoA synthesized by the phenylproponoid pathway (Forkmann and Heller 1999). The initial step is the stepwise condensation of three molecules of malonyl-CoA with *p*-Coumaroyl-CoA (Figure 21.4) to yield C15 flavonoid skeleton, yellow-colored chalcone, or isoliquiritigenin catalyzed by the enzyme chalcone synthase (CHS) (Holton and Cornish 1995). This enzyme belongs to type III polyketide synthases (PKS) family. Flavonoid biosynthetic genes including CHS encoding gene have been cloned and characterized from several plant species. Chalcones so formed do not accumulate in plants and are rapidly isomerized into flavanones. The enzyme chalcone isomerase (CHI) brings about cyclization of bicyclic chalcones, isoliquiritigenin, and naringenin into corresponding tricyclic flavanones, liquiritigenin, and flavanone naringenin, respectively (Jez and Noel 2002). Based on substrate specificity, two types of CHI are known: type I that can isomerize 6′-hydroxyl as well as 6-deoxychalcones, found exclusively in leguminous plants, and type II which converts only 6′-hydroxychalcones into flavanones is present in nonlegumes as well. The C3 position in flavanones is subsequently hydroxylated to form dihydroflavonols. The reaction is carried

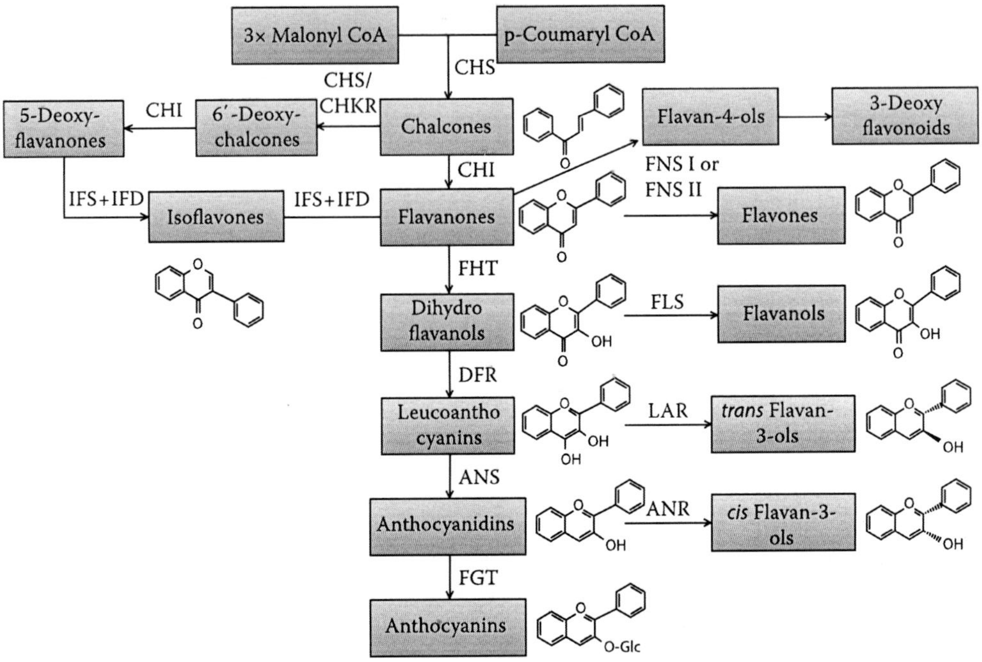

FIGURE 21.4 The generalized pathway of flavonoid biosynthesis in plants. CHS, chalcone synthase; CHI, chalcone isomerase; FHT, flavanone 3-β-hydroxylase; DFR, dihydroflavonol 4-reductase; ANS, anthocyanidin synthase; FGT, flavonoid glycosyltransferase; FNS, flavone synthase; FLS, flavonol synthase; LAR, leucoanthocyanidin reductase; ANR, anthocyanidin reductase; IFS, isoflavone synthase; IFD, isoflavone dehydratase; CHKR, chalcone polyketide reductase. (Compiled from Martens, S. and Mithofer, A., *Phytochemistry,* 66, 2399, 2005.)

out by the enzyme flavanone-3-hydroxylase (F3H). This enzyme is a member of the 2-oxyglutarate-dependent dioxygenase family.

Dihydrokaempferol (DHK), the product of F3H hydroxylation of naringenin, can be further hydroxylated at 3′-position or both at 3′ and 5′ position of the B ring. The former reaction is catalyzed by flavonoid 3′-hydroxylase (F3′H, P450 enzyme) to produce dihydroquercetin (DHQ), which ultimately leads to the production of cyanidin-based pigments. The latter reaction is carried out by flavonoid 3′, 5′-hydroxylase (F3′, 5′H, P450 enzyme) which converts DHK to dihyromyricetin (DHM), finally leading to delphidin-based anthocyanins (Winkel-Shirley 2001; Toda et al. 2002). The next compound in the pathway are leucoanthocyanidins (flavan-3, 4-diol), which are formed due to stereospecific reduction of dihydroflavonols. This reaction is catalyzed by the enzyme dihydroflavanol 4-reductase (DFR), which requires NADPH as cofactor (Kristiansen and Rohde 1991). The leucoanthocyanidins form the immediate precursors for anthocyanins (colored pigments) and also to catechins and proanthocyanidins which are involved in plant resistance and regarded as health-protecting compounds in food and feed.

The leucoanthocyanidins are further converted to anthocyanidins by anthocyanidin synthase (ANS), another member of the 2-oxyglutarate dependent dioxygenase family. The cDNA sequences encoding ANS have been obtained from several plants such as *Arabidopsis*, *Antirrhinum*, Petunia, and maize (Martin et al. 1991; Bradley et al. 1998; Pelletier et al. 1999). Anthocyanidins with a free hydroxyl group at the C3 position (C ring) are highly unstable under physiological conditions and are not found in nature (Forkmann and Heller 1999). The enzyme UDP-glucose flavonoid 3-*O*-glucosyl transferase (3GT or FGT) is responsible for the transfer of glucose moiety from UDP-glucose to hydroxyl group to form anthocyanins. Since this final step is essential to stabilize anthocyanidins to form water-soluble pigments, FGT is considered as an indispensible enzyme in the anthocyanin biosynthetic pathway. Based on the hydroxylation pattern, three types of anthocyanins are distinguished. Each type has a characteristic color depending on the number of hydroxyl groups. Orange, red, or pink colors are imparted by pelargonidin-derived pigments; red or magenta color by cyanidin-derived pigments; and purple or blue color delphinidin-derived pigments (Zuker et al. 2002). Apart from these structural modifications, differences in vacuolar pH, intermolecular stacking, glycosylation, metal complexation, and cell shape also contribute to the innumerable color range of flowers seen in the nature (Tanaka et al. 1998).

Apart from the main pathway leading to anthocyanin production, the modification of flavonoid classes by acylation; additional glycosylation to flavonoid di- or trisaccharides; methylation and hydroxylation within each flavonoid class; and modifications such as prenylation, sulfation, and C-glycosylation to certain flavonoid groups lead to branches in the pathway to form other flavonoid classes such as stilbenes, aurones, flavones, flavonols, isoflavones, catechins, and proanthocyanidins (Schijlen et al. 2004). Stilbene synthase (STS), an enzyme of plant polyketide synthase super family (which also includes CHS), is responsible for the formation of stilbenes. It also catalyzes stepwise condensation of malonyl CoA and P-Coumaryl CoA (aldol condensation) to form stilbene resveratrol. STS has been characterized from several plants such as *Medicago sativa* and *Arachis hypogaea*. Another class of glycosides of flavonoids called aurones (aureusidin and bracteatin) is synthesized from tetra- and penta-hydroxychalcones by the enzyme aureusidin synthase (Nakayama et al. 2000). This enzyme and its cDNA have been isolated and purified from *Antirrhinum* and are responsible for the yellow color of the flower.

Deoxyflavonoids form the branch point in the first step of the flavonoid pathway and are limited to leguminous plants. The presence of only CHS forms 6′-hydroxychalcones from chalcones. The action of two enzymes, CHS and chalcone polyketide reductase (CHKR), results in the formation of isoliquiritigenin (6′-deoxychalcone) (Forkmann and Martens 2001). Formation of isoflavonoids forms another branchpoint in the pathway, wherein the action of CHI on isoliquiritigenin yields 5-deoxyflavonoid liquiritigenin. This is acted upon by the enzyme isoflavone synthase (IFS), a cytochrome P450 enzyme CYP93C which catalyzes the migration of aryl moiety from 2 to 3 position in the B ring to form 2-hydroxyisoflavonone (Steele et al. 1999). This is dehydrated to form corresponding isoflavonoids (daidzein and genistein) by the enzyme 2-hydroxyisoflavone reductase (HID) (Akashi et al. 2005). Another route is the hydroxylation of flavanone at the C2 position, followed by the migration of aryl moiety mediated by IFS and further dehydrated by HID to form isoflavonoids. IFS encoding genes have been identified in many legumes such as licorice, peanut, chickpea, and soyabean which have two IFS genes (*IFS* 1 and *IFS* 2). HID belongs to the family of carboxylesterases and catalyzes the formation of isoflavonoid skeleton; HID encoding genes have been isolated from licorice (Akashi et al. 2005) and soybean.

Formation of three types of 3-deoxy flavonoids—*C*-glycosyl flavones, 3-deoxyflavonoids, and phlobaphenes—form another branch point in the flavonoid pathway. In maize floral organs, this branch is active and is controlled by the MYB-type transcription factor (Grotewald et al. 1998). The 3-deoxyanthocyanins and phlobaphenes are derived from flavan-4-ols, which are produced from flavanones by the action of DHR. *C*-Glycosyl flavones, a special type of flavones with *C*-glycoside attached to the A ring, are synthesized from flavanones by a mechanism that is still unclear (Grotewald et al. 1998). Flavones are synthesized as a branch point from flavanones catalyzed by enzyme flavone synthase (FNS), which introduces a double bond between the C_2 and C_3 positions and abstracts two hydrogen atoms. Two types of FNS are known: NADPH and molecular oxygen-dependent membrane-bound cytochrome P450 monooxygenase FNS II, which is widespread among many plants; and soluble 2-oxoglutarate and Fe^{2+}-dependent dioxygenase FNS I is restricted to Apiaceae members only (Heller and Forkmann 1993).

Flavonols are formed by the introduction of a double bond between C_2 and C_3 in the C ring of dihydroflavonols by the enzyme flavonol synthase (FLS). Catechins are produced as a result of reduction of leucoanthocyanidins by the enzyme leucoanthocyanidin reductase (LAR) and epicatechins (*cis* flavan-3-ols) wherein anthocyanidins are converted to epicatechins by anthocyanidin reductase (ANR). Pro-anthocyanidins (PA, condensed tannins) are polymeric flavonoids, thought to be synthesized by the sequential addition of intermediates derived from leucocyanidins and catechins; however, little is known about the genes and enzymes responsible for the polymerization reaction leading to PA synthesis. Most of the studies on PAs are done on Arabidopsis seed coat mutants (Abrahams et al. 2002).

21.3.4 Metabolic Engineering of Flavonoids in Plants

Cloning and expression of almost all structural genes and identification of transcription factors responsible for the regulation of these structural genes have paved the way to develop strategies to modulate the flavonoid content in various plants. The overexpression or downregulation of structural genes in transgenic plants has been useful in elucidating the flavonoid biosynthetic pathway genes. Further, the overexpression of structural genes have been used in genetic modification to overcome the rate-limiting steps in the pathway so that the flux through the pathway can be increased, leading to an enhanced level of flavonoids or to the production of new flavonoids (Schijlen et al. 2004).

Transcription factors modulate the activity of RNA polymerase II and are important regulators of spatial and temporal expressions of structural genes, and hence are considered as potential tools for manipulating these multienzyme pathways. Molecular and genetic studies of the flavonoid pathway have revealed that the transcription factors are efficient tools in metabolic engineering to increase the production of flavonoids (Du et al. 2010). Boddu et al. (2006) have shown that the MYB gene *y1* has been able to upregulate the expression of *CHS* and *CHI* genes required for the synthesis of 3-deoxyflavonoids in sorghum. Similarly, maize transcription factors C_1 and R have been used to increase the accumulation of anthocyanins and other flavonoids.

21.3.4.1 Enhancement of Isoflavonoids

Isoflavone levels in soybean (*Glycine max*) were increased by the expression of maize C_1 and R transcription factors. Transgenic soyabean seeds had low levels of genistein and high levels of daidzein with small overall increase in total isoflavone levels. However, C_1/R expression in conjunction with co-supression of flavanone 3-hydroxylase (*F3H*) to block anthocyanin biosynthesis produced higher level of isoflavones (Yu et al. 2003). However, RNAi silencing of the IFS gene in soybean resulted in a nearly complete (>95%) reduction of total isoflavonoids in the transgenic roots, leading to a significant reduction in nodule formation on inoculation with *Bradyrhizobium japonicum* and less resistance to *Phytophthora sojae* (Subramanian et al. 2005a). Constitutive expression of *Medicago truncula*ta isoflavone synthase (*Mt IFS1*) was carried out in alfalfa to modulate flavonoid composition and to study the metabolic responses of transgenic plants to biotic/abiotic stress. *Mt IFS1* transgenes showed increased production of genistein glucosides (up to 50 ng/g FW) as well as biochanin and pratensein glucosides to a lesser amount. In response to UV B and *Phoma medicaginis*, the *Mt IFS1* transgenic lines accumulated additional isoflavones such as formononetin and daidzein (Deavours and Dixon 2005).

Soybean *IFS* gene was constitutively expressed in tomato to induce the formation of genistein. Transgenes were found to contain higher levels of genistein and quercetin glycosides in fruit peel (Shih et al. 2008). Attempts were made to produce isoflavones as nodulation signals in nonlegume plants such as rice. Soybean *IFS* gene under the control of CaMV promoter expressed in rice produced genistein glycosides in root tissues of transgenic plants. The root and leaf extracts of these plants were able to stimulate *nod* gene expression in rhizobium (Sreevidya et al. 2006). Canola (*Brassica napus*) produces phenylpropanoids and flavonoids, but due to the absence of isoflavone synthase (*IFS*) gene, it does not naturally accumulate isoflavones. Soybean *IFS* gene (*GmIFS2*) was constitutively expressed in canola plants to check whether exogenous *IFS* is able to use endogenous substrate to produce isoflavone genistein, and the leaves of *GmIFS2* transgenic plants were found to accumulate up to 0.72 mg/g DW genistein derivatives. In addition, expression levels for most of the endogenous phenylpropanoid pathway genes were altered in transgenic canola plants (Li et al. 2011a).

21.3.4.2 Engineering Flavonols

Tomatoes contain only a small amount of flavonoids, much of which is located in the fruit peel. Overexpression of petunia *CHI* gene under the control of strong constitutive double CaMV 35 S promoter in tomato resulted in 78-fold increase in flavonols (mainly rutin) compared to control plants (Muir et al. 2001). Maize transcription factor genes Lc and C_1 were expressed in tomato to enhance flavonoid levels. Transgenic fruits expressing both the genes accumulated high levels of flavonol kaempferol (up to 60-fold) and lesser amounts of flavanone naringenin in flesh. Anthocyanins accumulated in leaves but not in fruits. All structural genes required for the production of kaempferol-type flavonols and pelargonidin-type anthocyanins were strongly induced by transcription factors (Bovy et al. 2002).

Attempts were made to introduce new flavonoids in tomato by introducing several genes from different sources. Grape *STS* under the control of CaMV 35 S promoter, petunia *CHS* and alfalfa *CHR* under the control of CaMV 35 S promoter and petunia *CHI* and gerbera *FNS II* under the control of CaMV 35 S promoter were transformed to tomato plants for the production of stilbene, deoxychalcones, and flavones, respectively. The fruit peels of STS lines accumulated resveratrol aglycon and small amounts of resveratrol glycoside. Significant amounts of stilbenes accumulated in fruit flesh also. The *CHI* and *CHR* lines accumulated deoxychalcone up to 265 mg/kg FW in fruit peels, the main ones being butein and isoliquiritigenin. Tomato plants expressing gerbera *FNS II* construct accumulated small amounts of luteolin and luteolin 7-glucosides in fruit peels. Co-expression of *FNS II* and petunia *CHI* gene resulted in the production of luteolin aglycon (up to 340 mg/kg FW) and luteolin 7-glucoside (up to 150 mg/kg FW); in addition, several flavonols also increased in the peel and flesh of fruits. Total antioxidant activity of tomatoes with high flavones and flavonols increased by threefold as compared to that in control (Schijlen et al. 2006). Similarly, attempts were made to increase health-promoting polyphenols such as kaempferol and chlorogenic acid in potato. Overexpression of modified MYB transcription factor resulted in a 100-fold increase in kaempferol and a fourfold increase in chlorogenic acid content in transgenic plants (Rommens et al. 2008).

Flax (*Linum usitatissimum*) plant is used as a source of oil and fiber and is grown for commercial purposes in many parts of the world. A multigene construct of cDNAs for *CHS*, *CHI*, and *DFR* were used to manipulate flavonoid content in flax plants. The simultaneous expression of genes resulted in a significant increase in the levels of flavanones, flavones, flavonols, and anthocyanins, resulting in enhanced antioxidant capacity of transgenic plants. The increased antioxidative properties of transgenic plants lead to improved resistance to *Fusarium*, the main pathogen of flax. The changes in phenylpropanoids accumulation in transgenic plants were found to affect cell-wall carbohydrate content with significant increase in carbohydrates, constituents of pectin, and hemicellulose. An increase in pectin and hemicellulose content has been proposed to be the reason for enhanced disease resistance of these plants (Lorenc-Kukula et al. 2007). Flax oil is the richest plant source of linoleic and linolenic polyunsaturated fatty acids (PUFA). The overexpression of regulatory genes of the phenylpropanoid pathway was carried out to increase the antioxidant potential of flax for greater accumulation of PUFA and its higher stability against oxidation. Three genes of the flavonoid biosynthetic pathway, namely, *CHS*, *CHI*, and *DFR* from *Petunia hybrida*, were expressed under the control of 35S promoter. Transgenic seed oil extracts exhibited higher levels of quercetin derivatives (46%–90%), kaempferol derivatives (70%–83%), and anthocyanins (198%) than control. The antioxidant capacity of seeds also increased by four- to sixfold than that of the seeds of non-transformed plants, and overproduction of flavonoids also increased the PUFA and total fatty acid content in transgenic seed oil (Zuk et al. 2011).

21.3.4.3 Anthocyanin Enhancement

Constitutive expression of petunia *DFR* gene enhanced anthocyanin and phenolics slightly in potato (Lukaszewicz et al. 2004). Transcription factors from snapdragon (*Antirrhinum majus*); Delila (*Del*), which encodes basic helix-loop-helix; and Rosea I (*Ros I*), encoding an MYB-related transcription factor that interacts to induce anthocyanin biosynthesis in snapdragon flowers were expressed in tomato under the control of fruit-specific E8 promoter. The transgenic fruits developed normally and at the end of the mature green stage showed signs of purple pigmentation in peel, pericarp, as well as pulp. A high amount of anthocyanins were detected both in the peel and pulp of purple fruits with major ones being 3, 5-glucosides acylated with cinnamic acids. The antioxidant activity of the hydrophilic fraction of purple fruits was threefold higher than that observed in control fruits (Butelli et al. 2008).

Rice anthocyanidin synthase (*ANS*) was overexpressed in a rice mutant Nootripathu, which accumulates proanthocyanidins exclusively in pericarp and absolutely no anthocyanins in any tissue. Transgenic plants overexpressing *ANS* channeled the proanthocyanidin precursors to the production of anthocyanins in pericarp. Ten and fourfold increase in the *ANS* transcripts and enzyme activity was observed followed by increased accumulation of a mixture of flavonoids and anthocyanins, with a concomitant decrease in proanthocyanidins, thus increasing the antioxidant potential of transgenic rice (Reddy et al. 2007). The expression of *Arabidopsis* regulatory gene Production of Anthocyanin Pigment 1 (*AtPAP 1*) in canola enhanced the antioxidant capacity in transgenic leaves by fourfold. Plants exhibited intense purple coloration wherein cyanidin and pelargonidin levels were enhanced by 50-fold and sinapic acid and quercitin levels by fivefold. The expression of most of the genes in the flavonoid and phenolic acid pathway were stimulated (Li et al. 2010).

21.3.4.4 Enhancement of Proanthocyanidins

Three anthocyanin regulatory genes of maize (*Zea mays*)—*Lc*, *B-Peru*, and C_1—were introduced into alfalfa (*Medicago sativa*) to stimulate the flavonoid pathway and alter the composition of flavonoids produced in forage. *Lc* transgenic plants showed accumulation of red-purple anthocyanin only under conditions of high light intensity or low temperature. These stress conditions induced chalcone synthase and flavanone 3-hydroxylase expression in *Lc* transgenic alfalfa foliage compared with nontransformed plants. Leucocyanidin reductase activity was also enhanced leading to enhanced proanthocyanidin levels (Ray et al. 2003).

Constitutive expression of maize *Lc* gene in apple (*Malus domestica*) resulted in anthocyanin accumulation in leaves and stems, and higher levels of anthocyanins (mainly idaein by 12-fold), monomeric flavan 3-ols (catechin by 41-fold and epicatechin by 12-fold), and proanthocyanidins were observed. In addition, *Lc* overexpressing *M. domestica* plants showed increased transcription levels of most anthocyanin structural genes, especially *ANS* gene (Li et al. 2007). In a similar study, transgenic apple plants overexpressing the *Leaf Colour* (*Lc*) gene from maize (*Zea mays*) strongly exhibited increased production of anthocyanins and flavan-3-ols (catechins, proanthocyanidins). Further, the transgenic plants showed higher resistance against fire blight (caused by *Erwinia amylovora*) and against scab (caused by *Venturia inaequalis*) diseases (Flachowsky et al. 2010).

21.4 VITAMIN C

Vitamin C (L-ascorbic acid) is one of the best-known plant secondary metabolites, possessing several important biochemical functions such as antioxidant, electron donor and acceptor in electron transport systems and as an enzyme cofactor (Levin 1986). The chemical name of Vitamin C is 2-oxo-L-threo-hexono-1, 4-lactone-2, 3-enediol. It plays an important role in plant growth and metabolism contributing to cell division, expansion, and elongation. Together with vitamin E, polyphenols, and flavonoids, ascorbic acid contributes to the overall intake of free radical scavengers in the human diet (Zhang et al. 2007). Apart from these, various studies suggest that these metabolites either singly or in combination benefit human health acting as anticancer agents and protecting against cardio-vascular diseases (CVD) (Hancock and Viola 2002).

21.4.1 Sources of Ascorbic Acid

L-Ascorbic acid and dehydro ascorbic acid are the major dietary forms of vitamin C. It is mainly found in fruits and vegetables (Table 21.3). Rich sources of fruits include grapefruit, honeydew, kiwi, mango, orange, papaya, strawberries, and watermelon, whereas vegetables include asparagus,

TABLE 21.3
Ascorbic Acid Content in Selected Fruits and Vegetables

Food Source	mg/100 g of Edible Portion
Fruits	
Apple	3–30
Banana	8–16
Cherry	15–30
Grape fruit	30–70
Lemon	40–50
Mango	10–15
Orange	30–50
Papaya	39
Pineapple	15–25
Strawberry	40–70
Vegetables	
Broccoli	80–90
Cabbage	30–70
Cauliflower	50–70
Coriander	90
Onion	10–15
Parsley	200–300
Pepper	150–200
Radish	25
Spinach	35–40
Tomato	10–20

Source: Naidu, K.A., *Nutr. J.*, 2, 7, 2003.

broccoli, brussels sprouts, cabbage, cauliflower, kale, mustard greens, and pepper (red or green). Vitamin C is available as a supplement, in the form of tablets, chewable tablets, capsules, liquid form, and crystalline powder, and also included in many multivitamin formulations. Vitamin C is commonly combined with other selected vitamins to form a complex and is collectively sold as an "antioxidant" supplement (Naidu 2003; Padayatty et al. 2003).

21.4.2 Functions of Ascorbic Acid

Ascorbic acid (AsA) plays an important role in various biological activities such as synthesis of collagen, neurotransmitters, carnitine, and steroid hormones and is also required for the conversion of cholesterol into bile acid. Further, it has been shown to enhance bioavailability of iron, promote calcium absorption, and help in healing of wounds and burns. It is also required to maintain healthy gums and to prevent blood clotting (Patil et al. 2009). Clinical studies have also revealed the role of AsA in the prevention of colorectal carcinoma, ulcers, hypertension, atherosclerosis, and advanced malignancy (Englard and Seifter 1986).

AsA in diet has been shown to protect human sperm against oxidative DNA damage, which may lead to genetic defects following abnormalities of sperm (Fraga et al. 1991). Ascorbic acid has been shown to be useful in controlling diabetes by reducing blood pressure and arterial stiffness in type II diabetic individuals (Mullan et al. 2002). The role played by AsA in the prevention of cancer has been extensively studied. Some of the mechanisms involved in preventing cancer include maintaining intracellular cell integrity by preventing infiltration of malignant cells, restricting nutrition to tumor cells selectively, and also by facilitating collagen encapsulation of healthy tissues (Gey et al. 1987). Enhanced levels of ascorbic acid in plasma has also been shown to protect against gastric cancer through reduction in the number of *Helicobacter pylori* (the major causative agent of gastric cancer [Cameron et al. 1979]), or by radical scavenging activity, and by other nutrition-related mechanisms (Warning et al. 1996). Commercially, L-ascorbic acid and its fatty acid esters are used as food additives, antioxidants, browning inhibitors, reducing agents, flavor stabilizers, dough modifiers, and color stabilizers (Naidu 2003).

21.4.3 Biosynthesis of Vitamin C

Ascorbic acid is synthesized by all higher plants, nearly all higher animals, and a number of yeasts. However, some animals including humans are unable to synthesize the compound due to the presence of nonfunctional L-gulono-1, 4-lactone oxidase gene which acts in the last step of ascorbate synthesis. Hence, the source of vitamin C in our diet mainly comes from plants.

Ascorbate is synthesized in plants by the oxidation of L-galactose (L-Gal) (Wheeler et al. 1998). This pathway reconciles two of the established fundamental principles in ascorbate biosynthesis that were previously thought to be contradictory. *De novo* biosynthesis is believed to be the main reason for its accumulation in plant cells. A combination of radiolabeling, mutant analysis, and transgenic manipulation provides evidence of multiple pathways of ascorbate biosynthesis in plants (Figure 21.5).

The predominant pathway of ascorbic acid biosynthesis is the Smirnoff–Wheeler pathway which is also known as D-Man/L-Gal pathway wherein ascorbic acid is synthesized from D-glucose via a complex 10-step pathway involving phosphorylated sugar intermediates and sugar nucleotides. Briefly, AsA formation involves D-glucose as the initial precursor, with the last step being catalysis by L-galactono-1, 4-lactone oxidase, which oxidizes L-galactono-1, 4-lactone (L-GalL) to produce AsA (Figure 21.5). This is produced from GDP-D- mannose (GDP-D-Man) via GDP-L-Galactose (GDP-L-Gal) (Wheeler et al. 1998; Smirnoff and Gatzek 2004). The first step in directing hexose phosphates into D-mannose metabolism is catalyzed by phosphomannose isomerase (PMI). Two putative genes for this enzyme have been identified in *Arabidopsis* (Fujiki et al. 2001). D-Mannose 6-Phosphate is converted to D-Mannose 1-Phosphate (D-Man 1P) by phosphomannose mutase (PMM). The molecular and functional characterization of PMM from higher plants has been reported (Qian et al. 2007). GDP-D-Mannose (GDP-D-Man) is synthesized from D-Man 1P and GTP which is catalyzed by GDP-D-Mannose pyrophosphorylase (GMP), and GDP-D-Man is converted to GDP-L-Galactose (GDP-L-Gal) by a reversible double epimerization catalyzed by GDP-D-Man-3, 5-epimerase (GME) that was first identified in Chlorella and subsequently in various plant systems (Hebda et al. 1979; Wheeler et al. 1998; Wolucka et al. 2001). Apart from their role in ascorbate synthesis, GDP-D-Man and GDP-L-Gal form the substrates for polysaccharide synthesis and protein glycosylation. L-Gal, in particular, is a component of the pectin rhamnogalacturonan II essential for the proper development of plants (O'Neill et al. 2004).

The steps subsequent to GDP-L-Gal are likely to be dedicated to ascorbate synthesis. Initially, GDP-L-Gal is broken down to L-Galactose 1-Phosphate (L-Gal 1P), which is subsequently hydrolysed to L-galactose (L-Gal) (Smirnoff and Gatzek 2004). GDP-L-Gal is converted to L-Gal 1-P and GDP by a novel and highly specific phosphate- dependent GDP-L-Galactose phosphorylase (GDP-L-GalP) also known as GDP-L-Galactose orthophosphate guanyltransferase. The enzyme L-Galactose 1-Phosphate phosphatase hydrolyses L-Gal 1-P to L-Gal and inorganic phosphate (Laing et al. 2004). The expression of this enzyme has been found to be very high during ripening of tomato and also during abiotic and postharvest stress, including heat, cold, wounding, oxygen supply, and ethylene treatments (Ioannidi et al. 2009). The released L-Gal undergoes oxidation in a two-step reaction, first by a cytosolic NAD-dependent L-Galactose dehydrogenase (L-GalDH) at C_1 position to form

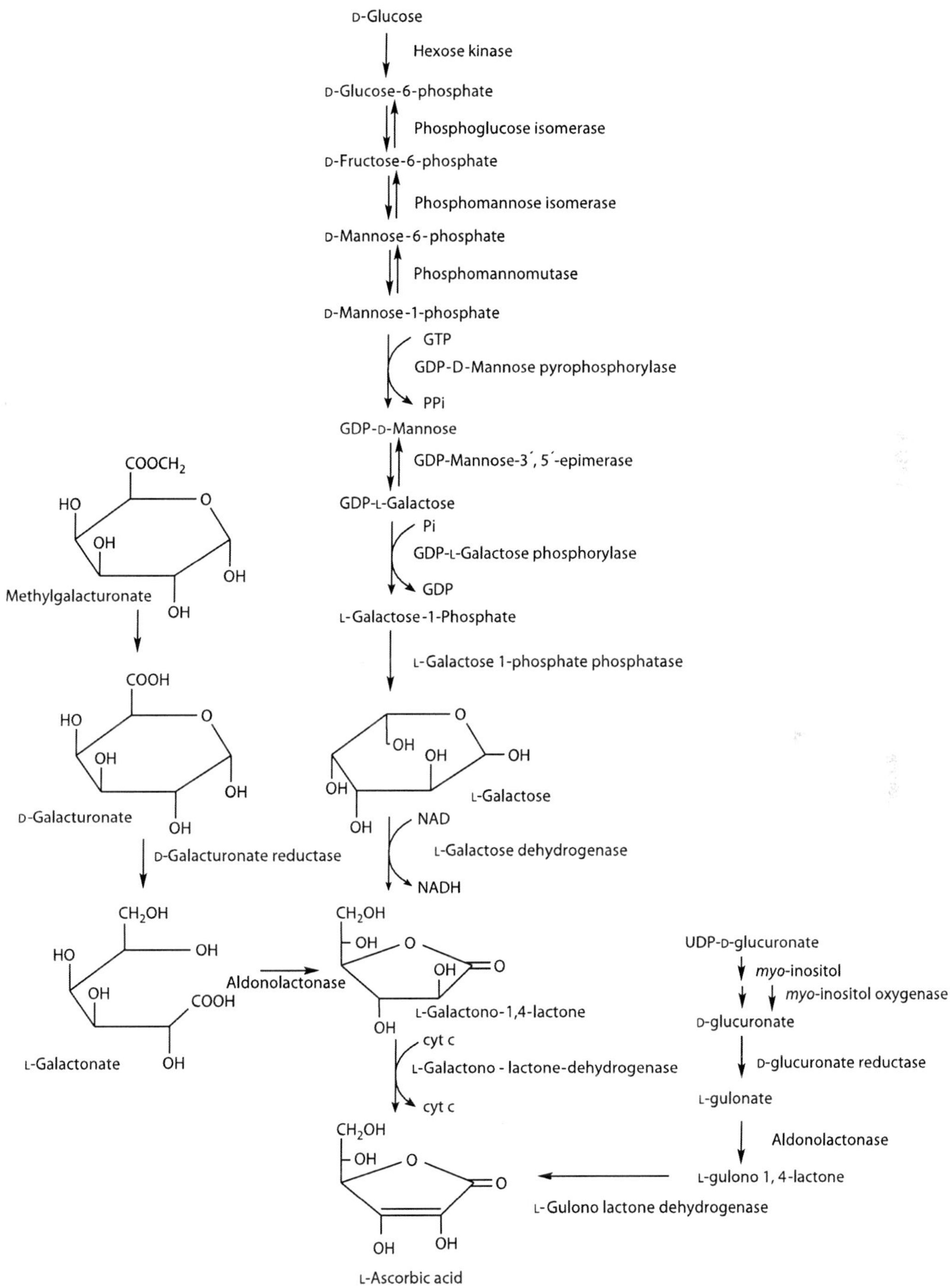

FIGURE 21.5 Generalized scheme of network of ascorbic acid biosynthetic pathway in plants. (Compiled from Valpuesta, V. and Botella, M.A., *Trends Plant Sci.*, 9, 573, 2004. Ishikawa, T. et al., *Physiol. Plant*, 126, 343, 2006.)

L-GalL (Wheeler et al. 1998; Gatzek et al. 2002) and then by L-Galactono-1,4-lactone dehydrogenase (L-GalLDH) at C_2/C_3 position resulting in the production of ascorbate. The final oxidative step occurs on the inner mitochondrial membrane where L-GalLDH uses cytochrome C as an electron acceptor (Bartoli et al. 2000; Millar et al. 2003). L-GalLDH has been characterized from several plant sources (Mutsuda et al. 1995; Oba et al. 1995; Ostergaard et al. 1997; Imai et al. 1998; Yabuta et al. 2000) and it appears to be highly specific for L-GalL.

Although the D-Man/L-Gal pathway appears to be the predominant pathway of AsA biosynthesis in plants, the other biosynthetic pathways via uronic acid intermediates contribute to the ascorbate content of plant tissues and that these may be developmentally regulated (Ishikawa et al. 2006). The cloning and characterization of D-galacturonic acid reductase (D-GalR) from strawberry fruit recently provided molecular evidence for the D-galacturonic acid pathway (Agius et al. 2003). This pathway is also prevalent during fruit ripening and also during photosynthesis of plant cells (Conklin 2001; Fernanda et al. 2003). D-galacturonic acid (Figure 21.5) derived from pectin is reduced to L-galactonic acid by D-GalR, which in turn is readily converted by aldonolactonase to L-GalL and finally oxidized to ascorbate by L-GalLDH. Recently, a myo-inositol oxygenase gene from *A. thaliana* has been cloned which catalyses the oxidation of myo-inositol into D-glucuronate (Lorence et al. 2004). Constitutive expression of this gene resulted in a two- to threefold increase in the ascorbate content of *A. thaliana* leaves compared to controls, suggesting that myo-inositol could be an entry point into ascorbate synthesis. D-glucuronate is then reduced to L-Gulonate by D-glucuronate reductase, which is oxidized by L-Gulono-1, 4-lactone dehydrogenase to ascorbic acid (Ishikawa et al. 2006).

21.4.4 Metabolic Engineering to Modulate Ascorbate Content in Plants

Accumulation of ascorbate, not only due to synthesis but also by recycling as well as existence of multiple biosynthetic pathways makes the task to engineer or predict the strategies to enhance ascorbate content in plants difficult. However, recent progress in understanding the AsA pathway in plants followed by cloning and characterization of several enzymes of the pathway has paved ways to modulate AsA content in plants. Three different strategies, namely, the overexpression of biosynthetic enzymes, inhibition or suppression of AsA degradation, and the overexpression or enhancement of ascorbate recycling have been employed to enhance the ascorbate content in plants.

21.4.4.1 Overexpression of Biosynthetic Enzymes

One general approach in improving the yield of ascorbic acid in plants has been to increase the carbon flux toward AsA by altering the expression levels of the biosynthetic genes of the Smirnoff–Wheeler's (D-Man/L-Gal) pathway and thereby overcoming the rate-limiting steps. Phosphomannomutase catalyzes the interconversion of mannose-6-phosphate to mannose-1-phosphate in the Smirnoff–Wheeler pathway for the biosynthesis of L-ascorbic acid. *PMM* cDNA from acerola (*Malpighia glabra*) overexpressed in transgenic tobacco plants showed about twofold increase in AsA contents as compared with the wild-type, with a corresponding correlation to *PMM* transcript levels and activities (Badejo et al. 2009). GDP-D-mannose pyrophosphorylase (*GMP*) is one of the limiting enzymes in the Smirnoff–Wheeler's pathway for the biosynthesis of AsA (Wheeler et al. 1998). The AsA content of transgenic tobacco plants expressing the Acerola *MgGMP* gene under the control of its own promoter was about twofold higher than that of the wild type plants (Badejo et al. 2008). Over-expression of the kiwifruit GDP-L-galactose guanyltransferase gene in *Arabidopsis* resulted in a fourfold increase in AsA, while up to a sevenfold increase in AsA was observed in transient expression studies where both GDP-L-galactose guanyltransferase and GDP-mannose-3′, 5′-epimerase genes were coexpressed. These studies show the importance of GDP-L-galactose guanyltransferase as a rate-limiting step to AsA, and demonstrate how AsA can be significantly increased in plants (Bulley et al. 2009). Attempts were made to enhance ascorbate concentration in commercially significant edible crops by using GDP-L-galactose phosphorylase (*GGP*) that had previously shown to increase ascorbate concentration in tobacco and *Arabidopsis thaliana*. The *GGP* from *Actinidia chinensis* under the control of the 35S promoter was expressed in tomato and strawberry. Potato was transformed with potato or *Arabidopsis GGP* genes under the control of the 35S promoter or a polyubiquitin promoter. Five lines of tomato, up to nine lines of potato, and eight lines of strawberry were regenerated for each construct. Three lines of tomato had a three- to sixfold increase in fruit ascorbate, and all lines of strawberry showed a twofold increase. All but one line of each potato construct also showed an increase in tuber ascorbate of up to threefold. Interestingly, in tomato fruit, increased ascorbate was associated with loss of seed and the jelly of locular tissue surrounding the seed, which was not seen in strawberry. In both strawberry and tomato, an increase in polyphenolic content was associated with increased ascorbate (Bulley et al. 2012).

L-Galactose dehydrogenase (*L-GalDH*) oxidizes L-Gal to L-galactono-1, 4-lactone. The overexpression of *L-GalDH* in tobacco resulted in a 3.5-fold increase in extractable GalDH activity without any increase in leaf ascorbate concentration. *A. thaliana*, transformed with an antisense *L-GalDH* construct, produced transgenic plants with 30% of wild-type activity. These had lower leaf ascorbate concentration and increased L-Gal pool size when grown under high light conditions showing that L-Gal concentration was negatively correlated with ascorbate. These results suggest that high light intensity modulates L-galactose synthesis, thus increasing the flux in the AsA pathway (Gatzek et al. 2002). The conversion of L-galactono-1, 4-lactone into AsA is a well-characterized step in AsA biosynthesis. Tokunaga et al. (2005) using tobacco suspension cells overexpressing *L-GalLDH* demonstrated an increase in ascorbate content. The cells expressing *L-GalLDH* under the control of the CaMV 35S promoter showed a twofold enhancement in the total ascorbate content. L-Galactono-1, 4-lactone (GalL) dehydrogenase (*GLDH or L-GalLDH*) from sweet potato was introduced into tobacco plants under the control of the CaMV 35S promoter. GLDH protein contents were elevated in three *GLDH*-transformed lines. Further, the transgenic lines showed 6- to 10-fold higher *GLDH* activities in the roots than in the nontransformed plants. Despite the elevated *GLDH* activity, the AsA content in the leaves was 3–7 μmol/g FW, comparable to that in the nontransformed plants. However, the incubation of leaf discs in a GalL solution led to a rapid two- to threefold increase in the AsA content in both *GLDH*-transformed and nontransformed plants in the same manner suggesting that the supply of GalL is a crucial factor in determining the AsA pool size (Imai et al. 2009). Transgenic tobacco and lettuce plants expressing a rat cDNA encoding L-gulono-1, 4-lactone oxidase accumulated up to seven times more ascorbic acid than nontransformed plants (Jain and Nessler 2000).

Attempts to engineer the genes encoding the enzymes of glucuronic acid pathway have also led to moderate increases in the ascorbate contents. The overexpression of D-galacturonate reductase (*GalUR*) from strawberry fruit (*Fragaria ananassa*) in *A. thaliana* resulted in a two- to threefold increase in total ascorbate content (Agius et al. 2003). Transgenic potato plants (*Solanum tuberosum* L. cv. Taedong Valley) were developed with increased ascorbic acid levels by overexpressing strawberry *GalUR* gene under the control of the CaMV 35S promoter. The overexpression of *GalUR* resulted in a 1.6- to 2-fold increase in AsA in transgenic potato and the levels of AsA were positively correlated with increased GalUR activity. The transgenic lines with enhanced vitamin C content showed enhanced tolerance to abiotic stresses induced by methyl viologen (MV), NaCl, or mannitol as compared to nontransformed control plants (Hemavathi et al. 2009). Similarly, the overexpression of myo-inositol oxygenase gene (*miox4*), another enzyme assumed to take part in glucuronic acid pathway, in transgenic *A. thaliana* also produced a moderate enhancement of ascorbate content in the leaves (Lorence et al. 2004).

21.4.4.2 Inhibition or Suppression of AsA Degradation

AsA degradation involves the enzymes ascorbate oxidase (AO) and ascorbate peroxidase (APX). AO functions apoplastically and both AO and APX catalyzes aerobic oxidation of AsA into monodehydro ascorbate (MDHA). Modulation of ascorbate oxidase activity has also led to moderate increases in the ascorbate pool. The sense and antisense reduction of *AO* in transgenic tobacco resulted in a 2.4-fold increase in total AsA content in apoplasts. The proportion of the reduced form in the apoplast was markedly increased (66%) compared with the wild-type plants (40%) in the antisense plants, while it was significantly decreased (3%) in the sense plants (Pignocchi et al. 2003). Similar results were obtained by Yamamoto et al. (2005) using transgenic tobacco expressing *AO* gene in sense and antisense orientation. The proportion of ascorbate/DHA was higher in antisense tobacco plants and lower in sense plants than in wild-type plants in both whole leaf and apoplast.

21.4.4.3 Overexpression or Enhancement of Ascorbate Recycling

AsA is oxidized to monodehydroascorbate, which is short-lived moiety, and gets converted to AsA by monodehydro ascorbate reductase (MDHAR) or nonenzymatically to AsA and dehydroascorbate (DHA). DHA so formed is irreversibly hydrolyzed to 2, 3-diketogulonic acid (Waskho et al. 1992) or is recycled back to AsA by dehydroascorbate reductase (DHAR). Thus, DHAR allows the plant to recycle DHA, and thereby recapturing AsA before it is lost. Because the rate of ascorbate turnover is relatively fast in most of the plants, such as 13% of the pool per h in pea seedlings (Pallanca and Smirnoff 2000) and 40% per 22 h in *A. thaliana* leaves (Conklin et al. 1997), enhancing the ascorbate recycling pathway or downregulating ascorbate oxidation would help in modulating ascorbate accumulation. Encouraging results have been achieved by reducing the rate at which ascorbate is recycled. For example, the overexpression of wheat *DHAR* in tobacco and maize (targeted to chloroplasts) increased the ascorbate pool fourfold in both tobacco leaves and maize kernel. The *DHAR* expression in transgenic tobacco and maize increased up to 32- and 100-fold, respectively (Chen et al. 2003). However, the overexpression of human *DHAR* in the chloroplasts of transgenic tobacco plants showed no difference in total ascorbate content, but an increase in the ascorbate/DHA ratio was observed (Kwon et al. 2003). Transgenic tobacco plants overexpressing *Arabidopsis thaliana MDAR* gene (*AtMDAR)* in the cytosol exhibited up to 2.1-fold higher *MDAR* activity and 2.2-fold higher level of reduced AsA as compared to that in nontransformed control plants. The transgenic plants also showed enhanced stress tolerance in terms of higher net photosynthesis rates under ozone, salt and polyethylene glycol (PEG) stresses (Eltayeb et al. 2007). Transgenic corn kernals overexpressing rice *DHAR* gene showed sixfold increase in the ascorbate content than the control ones (Naqvi et al. 2009). Transgenic potato (*Solanum tuberosum* L.) plants overexpressing the *Arabidopsis thaliana DHAR* gene (*AtDHAR1*) in the cytosol showed up to 4.5 times increase in the DHAR activity and up to 2.8 times the level of reduced ascorbate than that found in the wild-type plants. The transgenic plants exhibited enhanced tolerance in terms of less ion leakage, greater chlorophyll contents, less accumulation of hydrogen peroxide, and less severe visual injury symptoms when subjected to methylviologen treatment. The transgenic plants also exhibited faster growth under drought and salt stress conditions (Eltayeb et al. 2011).

Attempts were made to study the role of *MDAR* and *DHAR* in AsA regeneration during aluminum (Al) stress using transgenic tobacco (*Nicotiana tabacum*) plants overexpressing *Arabidopsis* cytosolic *MDAR* (*MDAR-OX*) or *DHAR (DHAR-OX*). *DHAR-OX* plants showed better root growth than wild-type (SR-1) plants after exposure to Al for 2 weeks, but *MDAR-OX* plants did not. *DHAR-OX* plants also revealed lower hydrogen peroxide content, less lipid peroxidation, and lower level of oxidative DNA damage than did SR-1 plants, whereas *MDAR-OX* plants showed the same extent of damage as SR-1 plants. Compared with SR-1 plants, *DHAR-OX* plants consistently maintained a higher AsA level both with and without Al exposure, while *MDAR-OX* plants maintained a higher AsA level only without Al exposure. *DHAR-OX* plants also maintained higher APX activity under Al stress, suggesting that higher AsA level and APX activity in *DHAR-OX* plants contributed to their higher antioxidant capacity and higher tolerance to Al stress (Yin et al. 2010).

21.4.4.4 Other Approaches

In tomato, an unexpected five- to sixfold increase in ascorbate levels in the leaves was recently achieved through the suppression of mitochondrial malate dehydrogenase (*MDH*) (Nunes-Nesi et al. 2005).

21.5 VITAMIN E

Vitamin E is the collective term given to a family of antioxidants, comprising lipid-soluble, eight structurally related tocochromanol (tocopherols and tocotrienols) compounds synthesized by plants and other photosynthetic organisms (Chen et al. 2006; Hunter and Cahoon 2007), which play an important role in human health and nutrition. The name vitamin E was first given by Evans and Bishop (1922) to the dietary factor important in animal nutrition. Since then their functions in plants, where they are synthesized, in humans, and animals have been extensively studied for health benefits (Falk and Munne-Bosch 2010). Chemically, tocopherols and tocotrienols have a chromanol head group and a prenyl side chain. The α-tocopherols and tocotrienols consist of three methyl groups: the β- and γ-form two methyl groups and δ-form one methyl group in the aromatic ring (Kamal-Eldin and Appelqvist 1996). Tochopherols have fully saturated aliphatic tail while tocotrienols are unsaturated as they contain three double bonds in the side chain (Munne-Bosch and Falk 2004). The structures and names of tocopherol and tocotrienol isomers are shown in Figure 21.6.

21.5.1 Sources of Vitamin E

Although the primary source of vitamin E comes from plants, total tocochromanol composition and content vary with the species and types of plant tissues. Tocopherols occur widely in plants but the form of tocopherol differs in the leaves and seeds of plants. The leaves primarily contain α-tocopherol and the seeds any of the four forms; oilseeds such as soyabean, rapeseed, corn, and perilla (Grusak 1999; Lee et al. 2008) predominantly contain γ-tocopherol and are considered as the richest source of vitamin E with the total content ranging from 300 to 2000 μg/g of oil (Grusak and DellaPenna 1999). Sunflower and olive oils are good sources of α-tocopherol. Occurrence of tocotrienols is however limited only to certain plant species. They form the predominant components of seed endosperm of monocots such as rice, wheat, and barley. In dicots, they are mainly found in seeds of Apiaceae members such as coriander, celery, etc. (Hunter and Cahoon 2007). From the commercial point of view, palm oil is the major source of tocotrienols.

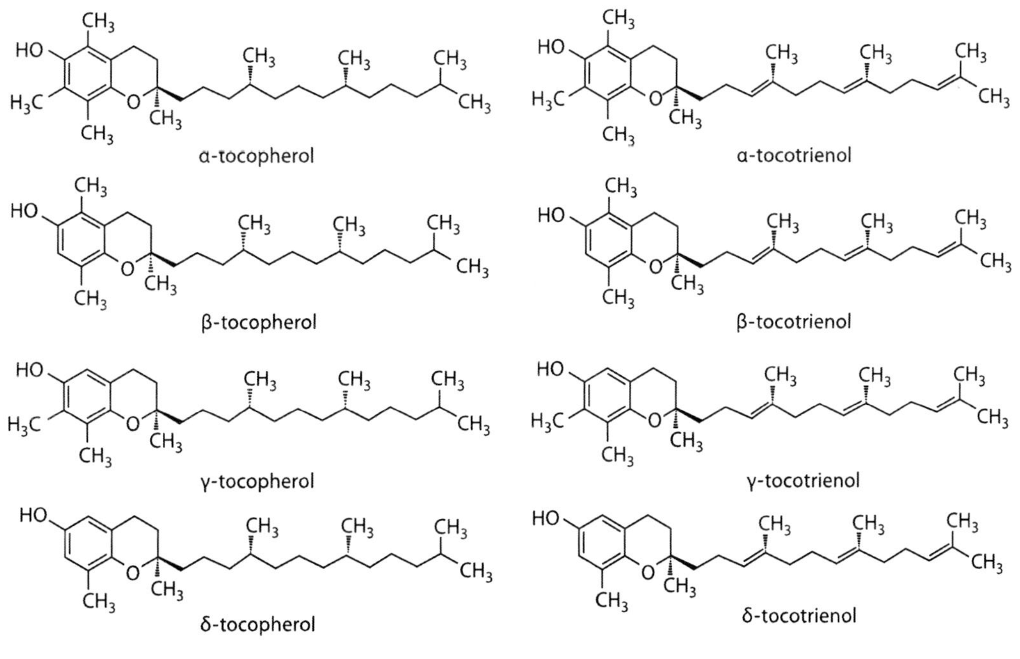

FIGURE 21.6 Structure of tocopherol and tocotrienol isomers.

21.5.2 Functions of Vitamin E

Vitamin E is an antioxidant (Epstein et al. 1966) and its activity has been extensively studied in the last four decades. The antioxidant activity of tocochromanols is due to scavenging of lipid peroxy radicals which are responsible for lipid peroxidation. The factors responsible for their antioxidant potential include degree of methylation in the aromatic ring ($\alpha > \beta = \gamma > \delta$), size of heterocyclic ring, length of the phytyl chain, and stereochemistry at position 2 (Munne-Bosch 2007). The antioxidant potential of tocopherols is generally represented as $\delta > \gamma = \beta > \alpha$ (Kamal-Eldin and Appelqvist 1996).

Peroxidation of lipids starts by generating an alkyl radical from polyunsaturated fatty acid (PUFA) by the action of an initiator (various reactive oxygen species, lipoxygenase, heat, light, and/or trace elements). The radical so formed combines with oxygen to form peroxy radicals. These radicals in turn abstract hydrogen from PUFA to generate lipid hydroperoxides and new alkyl radical and chain reaction continues. The lipid hydroperoxides are further reduced to alcohols or oxidized to jasmonic acid or n-hexanal among other products. Tocopherols scavenge these peroxy radicals before they can abstract hydrogen from lipids and tocopherols lose a hydrogen atom to lipid peroxy radical to form tocopheroxy radical and lipid hydroperoxide. Tocopheroxy radicals are recycled back to tocopherols via the ascorbate-glutathione pathway (Smirnoff and Wheeler 2000). Tocopherols are highly efficient in quenching singlet oxygen and it is estimated that a single α-tocopherol molecule can neutralize up to 120 singlet oxygen molecules *in vitro* before being degraded (Fukuzawa et al. 1982).

α-Tocopherol and α-tocopherol quinone have shown to photoprotect photosynthetic apparatus by dissipating excess energy in thylakoids (Kruk and Strzalka 2001). Recently, some studies have shown that tocopherols may play a more crucial role in the regulation of carbohydrate metabolism (Maeda and DellaPenna 2007). Regulation of membrane fluidity and their role in intracellular transduction have also received much attention (Munne-Bosch and Alegre 2002; Munne-Bosch 2007). Studies on sxd1 mutant of maize (Botha et al. 2000), potato plants (Hofius et al. 2004), and vte1 mutants of *A. thaliana* (Porfiorva et al. 2002) have demonstrated the role of tocopherol in cellular signaling by altering phytohormone levels in plants. Studies using *Arabidopsis* vte1 and vte2 mutants have shown that tocopherols are responsible for reducing nonenzymatic lipid oxidation during seed storage, germination, and early seedling development, thus proving that vitamin E is essential for seed longevity and to prevent lipid peroxidation during germination (Sattler et al. 2004).

Vitamin E has been shown to play an important role in preventing cardiovascular disease and certain cancers, improving immune function, and slowing the progression of a number of degenerative diseases (Traber and Sies 1996), thus benefiting human health. The recommended daily allowance (RDA) for vitamin E for both men and women has been set to 15 mg (Dellapenna 2005a). Like tocopherols, tocotrienols also impart health benefits; they are known to inhibit cholesterol biosynthesis (Qureshi et al. 1986); γ-tocotrienol has been shown to inhibit human hepatoma HepG2 cells (Pearce et al. 1992). Tocotrienols have been shown to reduce the growth of breast cancer cells *in vitro* (Nesaretnam et al. 1998) and studies have revealed that α- tocotrienol imparts greatest protection against oxidative damage to neuronal cells (Osakada et al. 2004).

The oxidative stability offered to vegetable oils by tocochromanols is considered as one of their important functional properties. This property is of much value as it reduces fatty acid oxidation and the formation of off-flavor compounds such as hexanals in fried or processed foods using vegetable oils. δ- and γ-forms of both tocopherols and tocotrienols are effective in reducing oxidative breakdown of vegetable oils during frying applications (Wagner and Elmadfa 2000; Wagner et al. 2001). Commercially, tocochromanols find applications in cosmetics and sunscreens, and are also used as livestock supplements to improve shelf-life and quality of meats (Waylan et al. 2002).

21.5.3 Biosynthesis of Tocochromanols

Tocopherols and tocotrienols are synthesized via two pathways, namely, shikimate pathway and the nonmevolanate or methyl-erythritol pathway (MEP). The precursor of the chromanol aromatic head group, homogentisic acid (HGA) (Figure 21.7), is derived from the cytosolic shikimate pathway and the precursor of the phtyl tail group, phytyl diphosphate (PDP) (Rohmer 2003) or geranylgeranyl diphosphate (GGDP) (Cahoon et al. 2003) from the plastidic MEP/DOXP pathway. The detailed description of the pathway and the enzymes involved therein has been reviewed by several authors (DellaPenna 2005a,b; DellaPenna and Pogson 2006). Several studies involving the mutant and transgenic model and crop plants such as *Arabidopsis thaliana*, tobacco, maize, and potato have led to a better understanding of the vitamin E biosynthetic pathway in the last decade (Falk and Munne-Bosch 2010).

Briefly, the precursor of tocopherol and plastoquinone biosynthesis in plants, the HGA, is synthesized via tyrosine and *p*-hydroxyphenylpyruvate (HPP). HPP is converted to HGA by the enzyme *p*-hydroxyphenylpyruvate dioxygenase (HPPD). The gene encoding the above enzyme has been identified in *Daucus carota*, *Arabidopsis thaliana*, and barley (Garcia et al. 1997, 1999; Falk et al. 2002). Table 21.4 lists the genes encoding the enzymes of the vitamin E biosynthetic pathway and their sources. The first committed step in tocopherol biosynthesis is the prenylation of HGA with PDP which results in the formation of 2-methyl-6-phytylbenzoquinol (2M6PBQ) catalyzed by the homogentisate phytyltransferase (HPT). This is encoded by Vitamin E2 (VTE 2) gene in *Arabidopsis*. Tocotrienol-accumulating plant species such as oil palm, tobacco, and corn have VTE 2 isoform with altered substrate specificity. GGDP is

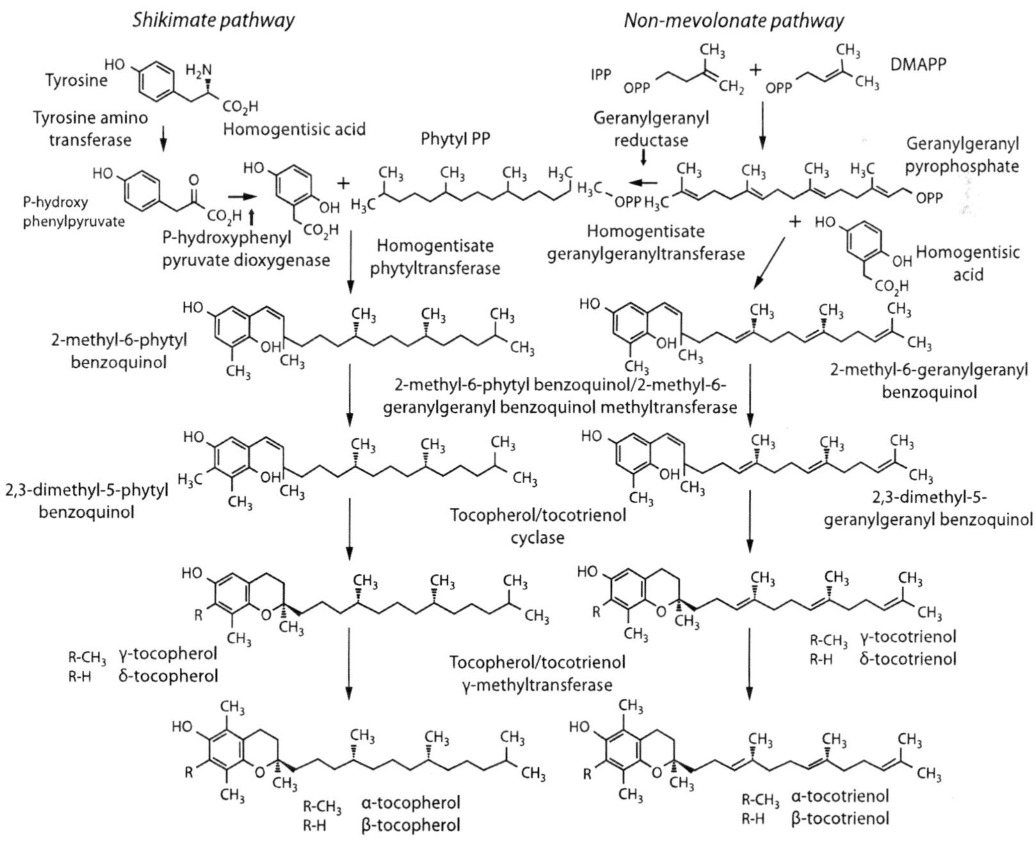

FIGURE 21.7 Generalized scheme of tocochromanol pathway in plants. (Compiled from Valentin, H.E. and Qi, Q., *Appl. Microbiol. Biotechnol.*, 68, 436, 2005; Hunter, S.C. and Cahoon, E.B., *Lipids*, 42, 97, 2007.)

TABLE 21.4
Genes and Enzymes of Vitamin E Biosynthesis in Plants

Gene	Enzyme	Source	Reference
HPT1, VTE2	Homogentisate phytyl transferase	Arabidopsis	Collakova and Dellapenna (2001), Savidge et al. (2002)
HGGT	Homogentisate geranylgeranyl transferase	Arabidopsis	
		Barley, rice, wheat	Cahoon et al. (2003)
VTE3	2-methyl-6 phytylbenzoquinol methyl transferase/2-methyl-6 geranylgeranylbenzoquinol methyl transferase	Arabidopsis	Cheng et al. (2003)
		Spinach	Soll et al. (1985)
		Sunflower	Demurin et al. (1996)
		Maize	Cook and Miles (1992)
VTE1	Tocopherol/tocotrienol cyclase	Arabidopsis	Sattler et al. (2003)
SXD1		Maize	Porfiorva et al. (2002), Kumar et al. (2005)
VTE4	Tocopherol/tocotrienol γ-methyl transferase	Arabidopsis	Shintani and Dellapenna (2003)
		Maize, potato	

preferred over PDP to form 2-methyl-6-geranylgeranylbenzoquinol (2M 6GGBQ), which is a tocotrienol precursor (Cahoon et al. 2003). This reaction is catalyzed by the enzyme homogentisic acid geranylgeranyl transferase (HGGT), a functionally divergent form of HPT. cDNAs encoding HGGT with 40%–50% similarity to Arabidopsis HPT has been identified in seeds of barley, rice, and wheat (Cahoon et al. 2003). The product 2M6PBQ is acted upon by two enzymes: 2M6PBQ-methyltransferase (VTE 3) and tocopherol cyclase (VTE 1). VTE 3 carries out methylation of 2M6PBQ to form 2, 3-dimethyl-5-benzoquinol, which acts as a substrate to VTE 1 to form γ-tocopherol. VTE 1 also utilizes 2M6PBQ to form δ-tocopherol. The final reaction of conversion of γ-tocopherol to α-tocopherol is catalyzed by the γ-methyltransferase (γ-TMT) encoded by VTE4. This gene also methylates δ-tocopherol to β-tocopherol.

For tocotrienol biosynthesis, 2M6GGBQ undergoes the same subsequent methylation and cyclization reactions as 2M6PBQ during synthesis of the corresponding tocopherols to produce γ-, α-, δ-, and β-tocotrienols (Munne-Bosch and Alegre 2002; Cheng et al. 2003; Valentin and Qi 2005).

21.5.4 Metabolic Engineering of Vitamin E in Plants

The purpose of genetic/metabolic engineering in plants is to increase/improve the content of the existing compound, introduce a new compound, or degrade a compound by modulating one or more enzymatic reactions in any biosynthetic pathway.

Several studies on genetic engineering of vitamin E in plants have revealed that the main goal is either to increase the total tocochromanol content by altering the flux through vitamin E biosynthetic pathway or to enhance the tocochromanol content in favor of α-tocopherol. Various strategies have been used at different levels of the pathway to achieve this goal.

21.5.4.1 Enhancing the Total Vitamin E Content

The Arabidopsis *HPPD* was overexpressed in Arabidopsis, wherein the transformed plants showed a 10-fold increase in protein and enzyme activity. The leaf and seed tocopherol content was enhanced by 37% and 28%, respectively, over the control (Tsegaye et al. 2002). Similarly, expression of barley *HPPD* gene resulted in only a modest increase in the total tocopherol content in transgenic tobacco plants (Falk et al. 2003, 2005). The chimeric *HPPD* genes, *HPPD1* and *HPPD2*, when expressed in transgenic canola seeds showed 50% increase in total tocochromanol content (Raclaru et al. 2006). The above studies showed that increased HPPD activity had limited effect on tocopherol flux regardless of the plant species and the constructs used. The gene encoding Arabidopsis HPT (*HPT1*) when constitutively expressed in Arabidopsis resulted in a 1.4- to 1.6-fold and a 4.4-fold increase in total tocopherol content in seeds and leaves, respectively, in comparison to that in wild plants (Savidge et al. 2002; Collakova and DellaPenna 2003a). Abiotic stress such as high light or nutrient stress further enhanced the tocopherol levels by 18- and 8-fold in wild-type and *HPT*-transformed leaves, respectively. This was evidenced by increased HPT mRNA levels and its specific activity in leaves, suggesting that HPT activity increased tocopherol synthesis under abiotic stress (Collakova and DellaPenna 2003b). However, coexpression of *HPT1* along with γ-*TMT* not only increased the total tocopherol levels but conversion of γ- and δ-forms to α- and β-tocopherols was also observed as evidenced by a 12-fold increase in vitamin E activity in transformed plants compared to that found in control plants (Collakova and DellaPenna 2003a). The overexpression of apple *HPT* gene in tomato resulted in higher levels of α- and γ-tocopherols ranging from 1.8- to 3.6-fold and from 1.6- to 2.9-fold, respectively, in transgenic leaves. The levels of α- and γ-tocopherols in transgenic fruits also increased by 1.7- and 3.1-fold, respectively, as compared to that found in control fruits (Seo et al. 2011).

When barley *HGGT* was expressed in *A. thaliana*, the leaves accumulated tocotrienols which were absent in nontransformed plants and a 10- to 15-fold increase in the total tocochromanol content was observed. Over expression of barley *HGGT* in corn seeds also resulted in a sixfold increase in the total tocotrienol and tocopherol content (Cahoon et al. 2003). This increases the antioxidant potential of corn as tocotrienols are known to possess superior *in vitro* antioxidant activity than tocopherols (Kamal-Eldin and Appelqvist 1996).

Constitutive overexpression of two transgenes, *A. thaliana HPPD* and *HPT*, was carried out in potato by Crowell et al. (2008). The transgenic tubers overexpressing *HPPD* and *HPT* genes showed 266% and 106% increase in tocopherol content; however, the content was less when compared to leaves and seeds. These results indicate that other factors may be limiting tocopherol accumulation in tubers. The overexpression of *TyrA* gene alone had little effect on increasing tocochromanol levels, but coexpression of *TyrA* and *HPPD* resulted in a 10-fold increase in total tocochromanol levels in tobacco leaves (Rippert et al. 2004; Matringe et al. 2005). Co-overexpression of bacterial prephenate dehydrogenase encoded by *tyrA* gene, along with *HPPD*, geranylgeranyl reductase (*GGR*), and *HPT* genes, increased the total tocochromanal content in soybean seeds by 10- to 15-fold, which is the highest increase achieved till date in genetically modified seeds. The increase was due to production of tocotrienols which accounted for nearly 90% of the total content rather than tocopherols (Karunanandaa et al. 2005). Recently, studies by Naqvi et al. (2011) have shown that when Arabidopsis cDNA clones corresponding to p-hydroxyphenylpyruvate dioxygenase (*HPPD*) and 2-methyl-6-phytylplastoquinolmethyltransferase (*MPBQ MT*) were constitutively expressed in corn, transgenic kernels exhibited three times more γ-tocopherol content. The coexpression resulted in *HPPD* increasing the overall flux and *MPBQ MT* further redirecting the flux toward the γ–branch to such an extent that γ-tocopherol was the only measurable isomer in the transgenic plants.

Tocopherol cyclase genes from Arabidopsis and maize overexpressed in canola led to an increase of 18%–28% in total tocochromanol content in canola seed oil. The δ-tocopherol content also increased up to 1.6- to 2.7-fold (Kumar et al. 2005). Similarly, transgenic Arabidopsis overexpressing tocopherol cyclase (*ATPT2* sequence) showed an increase of 50% in total tocopherol content and a threefold increase in δ-tocopherol content in seed (Subramanian et al. 2005b). To increase tocopherol content by increasing total flux to the tocopherol biosynthetic pathway, genes encoding *Arabidopsis* homogentisate phytyltransferase (*HPT/VTE2*) and tocopherol cyclase (*TC/VTE1*) were constitutively overexpressed in lettuce (*Lactuca sativa* L.). Total tocopherol content of the transgenic plants overexpressing either of the genes was increased by more than twofold mainly due to an increase in γ-tocopherol. However, chlorophyll content in the *HPT/VTE2* and *TC/VTE1* transgenic lines decreased by up to 20% and increased by up to 35%, respectively (Lee et al. 2007).

21.5.4.2 Modulating the Tocopherol Content

Seeds of *Arabidopsis* and canola harbor predominantly γ-tocopherol. When *Arabidopsis* γ-tocopherol methyl transferase (γ-*TMT*) gene was overexpressed in *Arabidopsis* and canola, transgenic seeds accumulated 95% and 100% α-tocopherol, followed by eight- and fourfold increase in vitamin E activity, respectively (Shintani and Dellapenna 1998, 2003). The overexpression of *Arabidopsis* γ-*TMT* in lettuce resulted in increasing the ratio of α-/γ-tocopherol content up to 0.8–320 from 0.6 to 1.2 in control plants (Cho et al. 2005). Similarly, expression of *Arabidopsis* γ-*TMT* in *Brassica juncea* increased α-tocopherol levels to as high as 62% in transgenes, which amounted to a sixfold increase over the nontransgenic controls (Yusuf and Sarin 2007). Additionally, salt, heavy metal, and osmotic stress induced an increase in the total tocopherol levels. The γ-*TMT* transgenic lines also showed enhanced tolerance to the induced stresses (Yusuf et al. 2010). The above reports suggest the important role of γ-*TMT* in determining the composition of tocopherols.

Soybean seeds contain approximately 20% δ-tocopherol and 60%–70% γ-tocopherol. Seed-specific overexpession of *Arabidopsis VTE3* resulted in nearly complete conversion of δ- and β-tocopherols to γ- and α-tocopherols in transgenic soybean seeds. Coexpression of *VTE3* along with *VTE4* resulted in the conversion of γ-, δ-, and β-forms to α-tocopherol. The transformed seeds accumulated >95% α-tocopherol with a fivefold increase in vitamin E activity (Van Eenennaam et al. 2003). Seed-specific overexpression of *Perilla frutescens* γ-tocopherol methyltransferase gene under the control of vicilin promoter increased the α-tocopherol content from 8.41% in wild-type seed to 81.67% in transgenic soybean seed. This corresponded to a 10.4-fold increase in α-tocopherol content compared to control. In addition, β-tocopherol also increased from 1% in wild-type seed to 18.33% in transgenic seeds. Increase in both α- and β-tocopherol enhanced the vitamin E activity of transformed seeds by 4.8-fold as compared to control seeds (Tavva et al. 2007). A cDNA encoding γ-tocopherol methyltransferase from *Brassica napus* (*BnTMT*) was overexpressed in soybean [*Glycine max* (L.) Merr.] under the control of seed-specific promoter of *Arabidopsis* fatty acid elongase 1 (FAE1) or soybean glycinin G1. Expression of *BnTMT* under the control of FAE1 was higher than that of G1. The seed-specific expression of *BnTMT* resulted in 11.1- and 18.9-fold increase in α- and β-tocopherol content, respectively (Chen et al. 2011). Higher levels of α-tocopherol in soybean was also achieved by a cross between a high α-tocopherol variety and a low α-tocopherol variety. The gene responsible for enhanced accumulation of α-tocopherol has been identified as γ-*TMT3*. The expression analysis of γ-*TMT3* in developing seeds and leaves were in correlation with the higher amount of α-tocopherol in developing seeds. Transgenic Arabidopsis plants with GUS gene driven by γ-*TMT3* promoter also showed higher activity of the promoter as compared to control ones (Dwiyanti et al. 2011).

Genes encoding *Arabidopsis* homogentisate phytyltransferase (*HPT*) and γ-tocopherol methyltransferase (γ-*TMT*) were constitutively overexpressed in lettuce (*Lactuca sativa* L. var. *logifolia*), alone or in combination. Overexpression of *HPT* increased total tocopherol content, while overexpression of γ-*TMT* shifted tocopherol composition in favor of α-tocopherol. Transgenic lettuce lines expressing both *HPT* and γ-*TMT* had higher amount of tocopherol and elevated α-/γ-tocopherol ratio compared to nontransgenic control (Li et al. 2011b). To enhance the tocopherol content, *Perilla frutescens* was engineered by incorporating the γ-*TMT* gene of *A. thaliana* driven by the CaMV 35S promoter. The transgenic plants showed an increase in the α-tocopherol and a decrease in the γ-tocopherol level. The content of α-tocopherol and the α/γ-tocopherol ratio of control plants were 48.60 μg/g tissue and 8.24, respectively, and in transgenic leaves, the α-tocopherol content increased about 1.81-fold to 87.8 μg/g tissue and the α/γ-tocopherol ratio increased 86.61-fold suggesting the conversion of γ-tocopherol into α-tocopherol by the overexpression of the *γ-TMT* gene (Ghimire et al. 2011).

21.6 CONCLUSIONS

Various natural antioxidants like carotenoids, flavonoids, ascorbic acid, and vitamin E not only have nutritional potential but also have the ability to prevent the onset of certain diseases. Several studies have proved that these metabolites either singly or in combination benefit human health. The levels of these phytonutrients in the traditionally bred crop plants are not at par with the RDA requirements. However, advanced genomics and biotechnological approaches have led to the production of the desired level or near-to-the-desired level of these compounds in many crop plants. The production of golden rice, golden potato, tomato, and canola with elevated β-carotene (provitamin A) content on par with RDA has alleviated the problems associated with vitamin A deficiency to a certain extent in developing countries. Similarly, enhanced levels of flavonoids in tomato and potato, α-tocopherol in soybean, and ascorbic acid in corn and lettuce may certainly provide higher levels of antioxidants for the betterment of human health. However, the undesired or unexpected results in the pathway engineering need to be resolved. The activity of the endogenous pathway and its regulation and the behavior of these transgenes in the environment are other important considerations in genetic engineering of crops. To derive the maximum health benefits, the dosage of antioxidants needed for prevention or cure and its bioavailability have to be ascertained. Nevertheless, the emerging technologies in genomics, proteomics, metabolomics, and systems biology may provide strategies to enhance these antioxidants in several crop plants to safe and desirable levels.

ABBREVIATIONS

2M 6GGBQ	2-methyl-6-geranylgeranylbenzoquinol
2M6PBQ	2-methyl-6-phytylbenzoquinol
3GT, FGT	flavonoid 3-O-glucosyl transferase
ABA	abscisic acid
ANR	anthocyanidin reductase
ANS	anthocyanidin synthase
AO	ascorbate oxidase
APX	ascorbate peroxidase
AsA	ascorbic acid
CCS	capsanthin–capsorubin synthase
CHI	chalcone isomerase
CHKR	chalcone polyketide reductase
CHS	chalcone synthase
Crt B	phytoene synthase
Crt I	phytoene desaturase
Crt O	β-carotene ketolase
Crt Y	Lycopene β-cyclase
CVD	cardiovascular disease
DET1	DE-ETIOLATED1
DFR	dihydroflavonaol 4-reductase
D-GalR, GalUR	D-galacturonic acid reductase
DHA	dehydroascorbate
DHAR	dehydroascorbate reductase

DHK	dihydrokaempferol
DHM	dihyromyricetin
DHQ	dihydroquercetin
D-Man 1P	D-Mannose 1-Phosphate
DW	dry weight
DXP	1-deoxy-D-xylulose-5-phosphate
DXPS	1-deoxy-D-xylulose 5-phosphate synthase
F3H	flavanone-3-hydroxylase
FLS	flavonol synthase
FNS	flavone synthase
FW	fresh weight
G3P	glyceraldehyde-3-Phosphate
GDP-D-Man	GDP-D-Mannose
GDP-L-Gal	GDP-L-Galactose
GGDP	geranylgeranyl diphosphate
GGP	GDP-L-galactose phosphorylase
GGPP	geranylgeranyl diphosphate
GMP	GDP-D-mannose pyrophosphorylase
GPP	geranyl pyrophosphate
HGA	homogentisic acid
HGGT	homogentisic acid geranylgeranyl transferase
HID	2-hydroxyisoflavone reductase
HMBPP	hydroxymethylbutenyl diphosphate
HPP	*p*-hydroxyphenylpyruvate
HPPD	*p*-hydroxyphenolpyruvate dioxygenase
HPT	homogentisate phytyltransferase
IFS	isoflavone synthase
IPI	IPP isomerase
IPP	isopentenyl diphosphate
LAR	leucoanthocyanidin reductase
LCYB	Lycopene β-cyclase
LCYE	lycopene ε-cyclase
LDL	low-density lipoproteins
L-Gal	L-Galactose
L-Gal 1P	L-Galactose 1-Phosphate
L-GalDH	L-Galactose dehydrogenase
L-GalL	L-Galactono-1, 4-lactone
L-GalLDH, GLDH	L-Galactono-1,4-lactone dehydrogenase
MDHA	monodehydro ascorbate
MEP	Methyl erythritol 4-phosphate
MPBQ MT	phytylplastoquinolmethyltransferase
MVA	mevalonate
NSY	neoxanthin synthase
PA	Pro-anthocyanidins
PDP	phytyl diphosphate
PDS	phytoene desaturase
PMM	phosphomannose mutase
PS II	photosystem II
PSY	phytoene synthase
PUFA	polyunsaturated fatty acids
RDA	recommended daily allowance

STS	Stilbene synthase
VDE	violaxanthin deepoxidase
VTE 2	vitamin E2
ZDS	ζ-carotene desaturase
ZEP	zeaxanthin epoxidase
β-Chy, bch	β-carotene hydroxylase
γ-TMT	γ-methyltransferase

REFERENCES

Abrahams, S., G.J. Tanner, P.J. Larkin, and A.R. Ashton. 2002. Identification and biochemical characterization of mutants in the proanthocyanidin pathway in Arabidopsis. *Plant Physiol* 130:561–576.

Ae Park, S., M.S. Choi, S.Y. Cho et al. 2006. Genistein and daidzein modulate hepatic glucose and lipid regulating enzyme activities in C57BL/KsJ-db/db mice. *Life Sci* 79:1207–1213.

Agarwal, S. and A.V. Rao. 1998. Tomato lycopene and low density lipoprotein oxidation: A human dietary intervention study. *Lipids* 33:981–984.

Agius, F., R. Gonzalez-Lamothe, J.L. Caballero, J. Munoz-Blanco, M.A. Botella, and V. Valpuesta. 2003. Engineering increased vitamin C levels in plants by overexpression of a D-galacturonic acid reductase. *Nat Biotechnol* 21:77–181.

Akashi, T., T. Aoki, and S. Ayabe. 2005. Molecular and biochemical characterization of 2-hydroxyisoflavanone dehydratase. Involvement of carboxylesterase-like proteins in leguminous isoflavone biosynthesis. *Plant Physiol* 137:882–891.

Al-Babili, S., P. Hugueney, M. Schledz et al. 2000. Identification of a novel gene coding for neoxanthin synthase from *Solanum tuberosum. FEBS Lett* 485:168–172.

Allister, E.M., N.M. Borradaile, J.Y. Edwards, and M.W. Huff. 2005. Inhibition of microsomal triglyceride transfer protein expression and apolipoprotein B100 secretion by the citrus flavonoid naringenin and by insulin involves activation of the mitogen activated protein kinase pathway in hepatocytes. *Diabetes* 54:1676–1683.

Aluru, M., Y. Xu, R. Guo et al. 2008. Generation of transgenic maize with enhanced provitamin A content. *J Exp Bot* 59:3551–3562.

Anderson, J.M. and W.S. Chow. 2002. Structural and functional dynamics of plant photosystem II. *Philos Trans R Soc Lond Ser B Biol Sci* 357:1421–1430.

Apel, W. and R. Bock. 2009. Enhancement of carotenoid biosynthesis in transplastomic tomatoes by induced lycopene-to-provitamin A conversion. *Plant Physiol* 151:59–66.

Araki, N., K. Kusumi, K. Masamoto, Y. Niwa, and K. Iba. 2000. Temperature-sensitive Arabidopsis mutant defective in 1-deoxy-D-xylulose- 5-phosphate synthase within the plastid non-mevolanate pathway of isoprenoid biosynthesis. *Physiol Plant* 108:19–24.

Arango, J., B. Salazar, R. Welsch, F. Sarmiento, P. Beyer, and S. Al-Babili. 2010. Putative storage root specific promoters from cassava and yam: Cloning and evaluation in transgenic carrots as a model system. *Plant Cell Rep* 29:651–659.

Arigoni, D., S. Sagner, C. Latzel, W. Eisenreich, A. Bacher, and M.H. Zenk. 1997. Terpenoid biosynthesis from 1-deoxy-D-xylulose in higher plants by intramolecular skeletal rearrangement. *Proc Natl Acad Sci* 94:10600–10605.

Arts, I.C. and P.C. Hollman. 2005. Polyphenols and disease risk in epidemiologic studies. *Am J Clin Nutr* 81:317–325.

Ausich, R.L., F.L. Brinkhaus, I. Mukharji, J. Yarger, and H.B. Yen. 1997. Beta-carotene biosynthesis in genetically engineered hosts. US Patent 5656472.

Badejo, A.A., H.A. Eltelib, K. Fukunaga, Y. Fujikawa, and M. Esaka. 2009. Increase in ascorbate content of transgenic tobacco plants overexpressing the acerola (*Malpighia glabra*) phosphomannomutase gene. *Plant Cell Physiol* 50:423–428.

Badejo, A.A., N. Tanaka, and M. Esaka. 2008. Analysis of GDP-D-mannose pyrophosphorylase gene promoter from acerola (*Malpighia glabra*) and increase in ascorbate content of transgenic tobacco expressing the acerola gene. *Plant Cell Physiol* 49:126–132.

Bai, C., R.M. Twyman, G. Farre et al. 2011. A golden era-pro-vitamin A enhancement in diverse crops. *In Vitro Cell Dev Biol Plant* 47:205–221.

Bartoli, C.G., G.M. Pastori, and C.H. Foyer. 2000. Ascorbate biosynthesis in mitochondria is linked to the electron transport chain between complexes III and IV. *Plant Physiol* 123:335–343.

Basile, A., S. Giordano, J.A. Lopez-Saez, and R.C. Cobianchi. 1999. Antibacterial activity of pure flavonoids isolated from mosses. *Phytochemistry* 52:1479–1482.

Basile, A., S. Sorbo, J.A. Lopez-Saez, and R.C. Cobianchi. 2003. Effects of seven pure flavonoids from mosses on germination and growth of *Tortula muralis* HEDW. (Bryophyta) and *Raphanus sativus* L. (Magnoliophyta). *Phytochemistry* 62:1145–1151.

Boddu, J., C. Jiang, V. Sangar, T. Olson, T. Peterson, and S. Chopra. 2006. Comparative structural and functional characterization of sorghum and maize duplications containing orthologous myb transcription regulators of 3-deoxyflavonoid biosynthesis. *Plant Mol Biol* 60:185–199.

Boileau, T.W.M., A.C. Moore, and J.W. Erdman Jr. 1999. Carotenoids and Vitamin A. In: *Antioxidant Status, Diet, Nutrition and Health*, ed. A.M. Papas, pp. 133–158. Boca Raton, FL: CRC Press.

Botella-Pavia, P., O. Besumbes, M.A. Phillips, L. Carretero-Paulet, A. Boronat, and M. Rodriguez-Concepcion. 2004. Regulation of carotenoid biosynthesis in plants: Evidence for a key role of hydroxymethylbutenyl diphosphate reductase in controlling the supply of plastidial isoprenoid precursors. *Plant J* 40:188–199.

Botella-Pavia, P. and M. Rodriguez-Concepcion. 2006. Carotenoid biotechnology in plants for nutritionally improved foods. *Physiol Plant* 126:369–381.

Botha, C.E.J., R.H.M. Cross, A.J.E. van Bel, and C.I. Peter. 2000. Phloem loading in the sucrose-export-defective (SXD-1) mutant maize is limited by callose deposition at plasmodesmata in bundle sheath-vascular parenchyma interface. *Protoplasma* 214:65–72.

Bouvier, F., A. D'Harlingue, R.A. Backhaus, M.H. Kumagai, and B. Camara. 2000. Identification of neoxanthin synthase as a carotenoid cyclase paralog. *Eur J Biochem* 267:6346–6352.

Bouvier, F., A. D'Harlingue, P. Hugueney, E. Marin, A. Marion-Poll, and B. Camara. 1996. Xanthophyll biosynthesis—Cloning, expression, functional reconstitution and regulation of beta-cyclohexenyl carotenoid epoxidase from pepper (*Capsicum annum*). *J Biol Chem* 271:28861–28867.

Bouvier, F., P. Hugueney, A. D'Harlingue, M. Kuntz, and B. Camara. 1994. Xanthophyll biosynthesis in chromoplasts: Isolation and molecular cloning of an enzyme catalyzing the conversion of 5, 6-epoxycarotenoid into ketocarotenoid. *Plant J* 6:45–54.

Bovy, A., E. Schijlen, and R.D. Hall. 2007. Metabolic engineering of flavonoids in tomato (*Solanum lycopersicum*): The potential for metabolomics. *Metabolomics* 3:399–412.

Bovy, A., R. de Vos, M. Kemper et al. 2002. High-flavonol tomatoes resulting from the heterologous expression of the maize transcription factor genes LC and C1. *Plant Cell* 14:2509–2526.

Bradley, J.M., K.M. Davies, S.C. Deroles, S.J. Bloor, and D.H. Lewis. 1998. The maize LC regulatory gene up-regulates the flavonoid biosynthetic pathway of Petunia. *Plant J* 13:381–392.

Bramley, P.M. 2003. The genetic enhancement of phytochemicals: The case of carotenoids. In: *Phytochemical Functional Foods*, eds. I. Johnson and G. Williamson, pp. 253–279. Cambridge, U.K.: Woodhead Publishing Ltd.

Britton, G. 1995. Structure and properties of carotenoids in relation to function. *FASEB J* 9:1551–1558.

Bulley, S.M., M. Rassam, D. Hoser et al. 2009. Gene expression studies in kiwifruit and gene over-expression in *Arabidopsis* indicates that GDP-L-galactose guanyltransferase is a major control point of vitamin C biosynthesis. *J Exp Bot* 60:765–778.

Bulley, S., M. Wright, C. Rommens et al. 2012. Enhancing ascorbate in fruits and tubers through over-expression of the L-galactose pathway gene GDP-L-galactose phosphorylase. *Plant Biotechnol J* 10:390–397.

Burkhardt, P.K., P. Beyer, J. Wunn et al. 1997. Transgenic rice *(Oryza sativa)* endosperm expressing daffodil *(Narcissus pseudonarcissus)* phytoene synthase accumulates phytoene, a key intermediate of provitamin A biosynthesis. *Plant J* 11:1071–1078.

Butelli, E., L. Titta, M. Giorgio et al. 2008. Enrichment of tomato fruit with health-promoting anthocyanins by expression of select transcription factors. *Nat Biotechnol* 26:1301–1308.

Caffarri, S., R. Croce, J. Breton, and R. Bassi. 2001. The major antenna complex of photosystem II has a xanthophylls binding site not involved in light harvesting. *J Biol Chem* 276:35924–35933.

Cahoon, E.B., S.E. Hall, K.G. Ripp, T.S. Ganzke, W.D. Hitz, and S.J. Coughlan. 2003. Metabolic redesign of vitamin E biosynthesis in plants for tocotrienol production and increased antioxidant content. *Nat Biotechnol* 21:1082–1087.

Cameron, E.T., L. Pauling, and B. Loibovitz. 1979. Ascorbic acid and cancer: A review. *Cancer Res* 39:687–690.

Chang, W.S., Y.J. Lee, F.J. Lu, and H.C. Chiang. 1993. Inhibitory effects of flavonoids on xanthine oxidase. *Anticancer Res* 13:2165–2170.

Chen, S., H. Li, and G. Liu. 2006. Progress of vitamin E metabolic engineering in plants. *Transgenic Res* 15:655–665.

Chen, S., Q. Ruan, E. Bedner et al. 2001. Effects of flavonoid baicalin and its metabolite baicalein on androgen receptor expression, cell cycle progression and apoptosis of prostate cancer cell lines. *Cell Prolif* 34:293–304.

Chen, Z., T.E. Young, J. Ling, S.C. Chang, and D.R. Gallie. 2003. Increasing vitamin C content of plants through enhanced ascorbate recycling. *Proc Natl Acad Sci USA* 100:3525–3530.

Chen, D.F., M. Zhang, Y.Q. Wang, and X.W. Chen. 2011. Expression of γ-tocopherol methyltransferase gene from *Brassica napus* increased α-tocopherol content in soybean seed. *BiolPlant* doi:10.1007/s10535-011-0192-6.

Cheng, Z., S. Sattler, H. Maeda et al. 2003. Highly divergent methyltransferases catalyze a conserved reaction in tocopherol and plastoquinone synthesis in cyanobacteria and photosynthetic eukaryotes. *Plant Cell* 15:2343–2356.

Cho, E.A., C.A. Lee, Y.S. Kim, S.H. Baek, B.G. de los Reyes, and S.J. Yun. 2005. Expression of gamma-tocopherol methyltransferase transgene improves tocopherol composition in lettuce (*Lactuca sativa* L.). *Mol Cells* 19:16–22.

Clarkson, T.B. 2002. Soy, soy phytoestrogens and cardiovascular disease. *J Nutr* 132:566S–569S.

Collakova, E. and D. DellaPenna. 2001. Isolation and functional analysis of homogentisate phytyltransferase from *Synechocystis* sp. PCC 6803 and *Arabidopsis*. *Plant Physiol* 127:1113–1124.

Collakova, E. and D. DellaPenna. 2003a. Homogentisate phytyltransferase activity is limiting for tocopherol biosynthesis in *Arabidopsis*. *Plant Physiol* 131:632–642.

Collakova, E. and D. DellaPenna. 2003b. The role of homogentisate phytyltransferase and other tocopherol pathway enzymes in regulation of tocopherol biosynthesis during abiotic stress. *Plant Physiol* 133:930–940.

Cong, L., C. Wang, L. Chen, H. Liu, G. Yang, and G. He. 2009. Expression of phytoene synthase1 and carotene desaturase crtI genes result in an increase in the total carotenoids content in transgenic elite wheat (*Triticum aestivum* L.). *J Agric Food Chem* 57:8652–8660.

Conklin, P.L. 2001. Recent advances in the role and biosynthesis of ascorbic acid in plants. *Plant Cell Environ* 24:383–394.

Conklin, P.L., J.E. Pallanca, R.L. Last, and N. Smirnoff. 1997. L-Ascorbic acid metabolism in the ascorbate-deficient *Arabidopsis* mutant vtc1. *Plant Physiol* 115:1277–1285.

Cook, W. and D. Miles. 1992. Nuclear mutations affecting plastoquinone accumulation in maize. *Photosynth Res* 31:99–111.

Cos, P., L. Ying, M. Calomme et al. 1998. Structure-activity relationship and classification of flavonoids as inhibitors of xanthine oxidase and superoxide scavengers. *J Nat Prod* 61:71–76.

Crowell, E.F., J.M. McGrath, and D.S. Douches. 2008. Accumulation of vitamin E in potato (*Solanum tuberosum*) tubers. *Transgenic Res* 17:205–217.

Cunningham, F.X. and E. Gantt. 1998. Genes and enzymes of carotenoid biosynthesis in plants. *Annu Rev Plant Physiol Plant Mol Biol* 49:557–583.

Davuluri, G.R., A. Van Tuinen, P.D. Fraser et al. 2005. Fruit-specific RNAi mediated suppression of DET1 enhances carotenoid and flavonoid content in tomatoes. *Nat Biotechnol* 23:890 895.

D'Ambrosio, C., G. Giorio, I. Marino et al. 2004. Virtually complete conversion of lycopene into β-carotene in fruits of tomato plants transformed with the tomato lycopene b-cyclase (tlcy-b) cDNA. *Plant Sci* 166:207–214.

Deavours, B.E. and R.A. Dixon. 2005. Metabolic engineering of isoflavonoid biosynthesis in alfalfa. *Plant Physiol* 138:2245–2259.

Dehmlow, C., J. Erhard, and H. de Groot. 1996. Inhibition of Kupffer cell functions as an explanation for the hepatoprotective properties of silibinin. *Hepatology* 23:749–754.

Del Rio, J.A., M.C. Arcas, O. Benavente-Garcia, and A. Ortuno. 1998. Citrus polymethoxylated flavones can confer resistance against *Phytophthora citrophthora*, *Penicillium diditatum*, and *Geotrichum* species. *J Agric Food Chem* 46:4423–4428.

Delgado-Vargas, F. and O. Paredes-Lopez. 2003. *Natural Colorants for Food and Nutraceutical Uses*, pp. 113–166. Boca Raton, FL: CRC Press.

DellaPenna, D. 1999. Nutritional genomics: Manipulating plant micronutrients to improve human health. *Science* 285:375–379.

DellaPenna, D. 2005a. Progress in the dissection and manipulation of vitamin E synthesis. *Trends Plant Sci* 10:574–579.

DellaPenna, D. 2005b. A decade of progress in understanding vitamin E synthesis in plants. *J Plant Physiol* 162:29–737.

DellaPenna, D. and B.J. Pogson. 2006. Vitamin synthesis in plants: Tocopherols and carotenoids. *Annu Rev Plant Biol* 57:711–738.

Demmig-Adams, B., A.M. Gilmore, and W.W. Adams. 1996. In vivo functions of carotenoids in higher plants. *FASEB J* 10:403–412.

Demurin, Y., D. Skoric, and D. Karlovic. 1996. Genetic variability of tocopherol composition in sunflower seeds as a basis of breeding for improved oil quality. *Plant Breed* 115:33–36.

Dharmapuri, S., C. Rosati, P. Pallara et al. 2002. Metabolic engineering of xanthophyll content in tomato fruits. *FEBS Lett* 519:30–34.

Diretto, G., S. Al-Babili, R. Tavazza, V. Papacchioli, P. Beyer, and G. Giuliano. 2007b. Metabolic engineering of potato carotenoid content through tuber specific overexpression of a bacterial mini-pathway. *PLoS One* 2:e350.

Diretto, G., R. Tavazza, R. Welsch et al. 2006. Metabolic engineering of potato tuber carotenoids through tuber-specific silencing of lycopene epsilon cyclase. *BMC Plant Biol* 6:13.

Diretto, G., R. Welsch, R. Tavazza et al. 2007a. Silencing of beta-carotene hydroxylase increases total carotenoid and beta-carotene levels in potato tubers. *BMC Plant Biol* 7:11.

Du, H., Y. Huang, and Y. Tang. 2010. Genetic and metabolic engineering of isoflavonoid biosynthesis. *Appl Microbiol Biotechnol* 86:1293–1312.

Ducreux, L.J., W.L. Morris, P.E. Hedley et al. 2005. Metabolic engineering of high carotenoid potato tubers containing enhanced levels of β-carotene and lutein. *J Exp Bot* 56:81–89.

Dwiyanti, M.S., T. Yamada, M. Sato, J. Abe, and K. Kitamura. 2011. Genetic variation of γ-tocopherol methyltransferase gene contributes to elevated α-tocopherol content in soybean seeds. *BMC Plant Biol* 11:152.

Eltayeb, A.E., N. Kawano, G.H. Badawi et al. 2007. Overexpression of monodehydroascorbate reductase in transgenic tobacco confers enhanced tolerance to ozone, salt and polyethylene glycol stresses. *Planta* 225:1255–1264.

Eisenreich, W., F. Rohdich, and A. Bacher. 2001. Deoxylulose phosphate pathway to terpenoids. *Trends Plant Sci* 6:78–84.

Eltayeb, A.E., S. Yamamoto, M.E.E. Habora, L. Yin, H. Tsujimoto, and K. Tanaka. 2011. Transgenic potato overexpressing Arabidopsis cytosolic *AtDHAR1* showed higher tolerance to herbicide, drought and salt stresses. *Breeding Sci* 61:3–10.

Enfissi, E.M.A., P.D. Fraser, L.M. Lois, A. Boronat, W. Schuch, and P.M. Bramley. 2005. Metabolic engineering of the mevalonate and nonmevalonate isopentenyl diphosphate-forming pathways for the production of health-promoting isoprenoids in tomato. *Plant Biotechnol* 3:17–27.

Englard, S. and S. Seifter. 1986. The biochemical functions of ascorbic acid. *Annu Rev Nutr* 6:365–406.

Epstein, S.S., J. Forsyth, I.B. Saporoschetz, and N. Mantel. 1966. An exploratory investigation on the inhibition of selected photosensitizers by agents of varying antioxidant activity. *Radiat Res* 28:322–335.

Evans, H.M. and K.S. Bishop. 1922. Fetal resorption. *Science* 55:650.

Falk, J., G. Andersen, B. Kernebeck, and K. Krupinska. 2003. Constitutive overexpression of barley 4-hydroxyphenylpyruvate dioxygenase in tobacco results in elevation of the vitamin E content in seeds but not in leaves. *FEBS Lett* 540:35–40.

Falk, J., M. Brosch, A. Schafer, S. Braun, and K. Krupinska. 2005. Characterization of transplastomic tobacco plants with a plastid localized barley 4-hydroxyphenylpyruvate dioxygenase. *J Plant Physiol* 162:738–742.

Falk, J., M.N. Kraub, D. Dahnhardt, and K. Krupinska. 2002. The senescence associated gene of barley encoding 4-hydroxyphenylpyruvate dioxygenase is expressed during oxidative stress. *J Plant Physiol* 159:1245–1253.

Falk, J. and S. Munne-Bosch. 2010. Tocochromanol functions in plants: Antioxidation and beyond. *J Exp Bot* 61:1549–1566.

Fernanda, A., G.L. Rocio, L.C. Jose, M.B. Juan, A.B. Miguel, and V. Victoriano. 2003. Engineering increased vitamin C levels in plants by overexpression of a D-galacturonic acid reductase. *Nat Biotechnol* 21:177–181.

Fisher, R.F. and S.R. Long, 1992. Rhizobium–plant signal exchange. *Nature* 357:655–660.

Flachowsky, H., I. Szankowski, T.C. Fischer et al. 2010. Transgenic apple plants overexpressing the *Lc* gene of maize show an altered growth habit and increased resistance to apple scab and fire blight. *Planta* 231:623–635.

Forkmann, G. and W. Heller. 1999. Biosynthesis of flavonoids. In: *Comprehensive Natural Products Chemistry*, pp. 713–748. Amsterdam, the Netherlands: Elsevier.

Forkmann, G. and S. Martens. 2001. Metabolic engineering and applications of flavonoids. *Curr Opin Biotechnol* 12:155–160.

Fowler, Z.L. and M.A.G. Koffas. 2009. Biosynthesis and biotechnological production of flavanones: Current state and perspectives. *Appl Microbiol Biotechnol* 83:799–808.

Foy, C.J., A.P. Passmore, M.D. Vahidassr, I.S. Young, and J.T. Lawson. 1999. Plasma chain-breaking antioxidants in Alzheimer's disease, vascular dementia and Parkinson's disease. *QJM* 92:39–45.

Fraga, C.G., P.A. Motchnik, M.K. Shigenaga, H.J. Helbock, R.A. Jacob, and B.N. Ames. 1991. Ascorbic acid protects against endogenous oxidative DNA damage in human sperm. *Proc Natl Acad Sci USA* 88:11003–11006.

Fraser, P.D. and P.M. Bramley. 2004. The biosynthesis and nutritional uses of carotenoids. *Prog Lipid Res* 43:228–265.

Fraser, P.D., S. Romer, C.A. Shipton et al. 2002. Evaluation of transgenic tomato plants expressing an additional phytoene synthase in a fruit-specific manner. *Proc Natl Acad Sci* 99:1092–1097.

Fraser, P.D., R.M. Truesdale, C.R. Bird, W. Schuch, and P.M. Bramley. 1994. Carotenoid biosynthesis during tomato fruit development: Evidence for tissue-specific gene expression. *Plant Physiol* 105:405–413.

Fuhramn, B., A. Elis, and M. Aviram. 1997. Hypocholesterolemic effect of lycopene and β-carotene is related to suppression of cholesterol synthesis and augmentation of LDL receptor activity in macrophage. *Biochem Biophys Res Commun* 233:658–662.

Fujiki, Y., Y. Yoshikawa, T. Sato et al. 2001. Dark-inducible genes from *Arabidopsis thaliana* are associated with leaf senescence and repressed by sugars. *Physiol Plant* 111:345–352.

Fujisawa, M., E. Takita, H. Harada et al. 2009. Pathway engineering of *Brassica napus* seeds using multiple key enzyme genes involved in ketocarotenoid formation. *J Exp Bot* 60:1319–1332.

Fujisawa, M., M. Watanabe, S.K. Choi, M. Teramota, K. Ohyama, and N. Misawa. 2008. Enrichment of carotenoids in flaxseed (*Linum usitatissimum*) by metabolic engineering with introduction of bacterial phytoene synthase gene *crt B*. *J Biosci Bioeng* 105:636–641.

Fukuzawa, K., A. Tokumura, S. Ouchi, and H. Tsukatani. 1982. Antioxidant activities of tocopherols on iron (2+)-ascorbate-induced lipid peroxidation in lecithin liposomes. *Lipids* 17:511–513.

Gann, P.H., J. Ma, E. Giovannucci et al. 1999. Lower prostate cancer risk in men with elevated plasma lycopene levels: Results of a prospective analysis. *Cancer Res* 59:1225–1230.

Garcia, I., M. Rodgers, C. Lenne, A. Rolland, A. Sailland, and M. Matringe. 1997. Subcellular localization and purification of a p-hydroxyphenylpyruvate dioxygenase from cultured carrot cells and characterization of the corresponding cDNA. *Biochem J* 325:761–769.

Garcia, I., M. Rodgers, R. Pepin, T.F. Hsieh, and M. Matringe. 1999. Characterization and subcellular compartmentation of recombinant 4-hydroxyphenylpyruvate dioxygenase from *Arabidopsis* in transgenic tobacco. *Plant Physiol* 119:1507–1516.

Gatzek, S., G.L. Wheeler, and N. Smirnoff. 2002. Antisense suppression of L-galactose dehydrogenase in *Arabidopsis thaliana* provides evidence for its role in ascorbate synthesis and reveals light modulated L-galactose synthesis. *Plant J* 30:541–553.

Gerjets, T. and G. Sandmann. 2006. Potato tuber as a transgenic production system for keto carotenoids. *Carotenoid Sci* 9:3137.

Gey, K.F., G.B. Brubacher, and H.B. Stahelin. 1987. Plasma levels of antioxidant vitamins in relation to ischemic heart disease and cancer. *Am J Clin Nutr* 45:1368–1372.

Ghimire, B.K., E.S. Seong, C.O. Lee et al. 2011. Enhancement of α-tocopherol content in transgenic *Perilla frutescens* containing the γ-TMT gene. *Afr J Biotechnol* 10:2430–2439.

Giliberto, L., G. Perrotta, P. Pallara et al. 2005. Manipulation of the blue light photoreceptor cryptochrome 2 in tomato affects vegetative development, flowering time, and fruit antioxidant content. *Plant Physiol* 137:199–208.

Giovannucci, E. 1999. Tomatoes, tomato-based products, lycopene, and cancer: Review of the epidemiologic literature. *J Natl Cancer Inst* 91:317–331.

Grotewald, E. 2006. The genetics and biochemistry of floral pigments. *Annu Rev Plant Biol* 57:761–780.

Grotewold, E., M. Chamberlin, M. Snook et al. 1998. Engineering secondary metabolism in maize cells by ectopic expression of transcription factors. *Plant Cell* 10:721–740.

Grusak, M.A. 1999. Genomic-assisted plant improvement to benefit human nutrition and health. *Trends Plant Sci* 4:164–166.

Grusak, M.A. and D. DellaPenna. 1999. Improving the nutrient composition of plants to enhance human nutrition and health. *Annu Rev Plant Physiol Plant Mol Biol* 50:133–161.

Hancock, R.D. and R. Viola. 2002. Biotechnological approaches for L-ascorbic acid production. *Trends Biotechnol* 20:299–305.

Hanley, J., Y. Deligiannakis, A. Pascal, P. Faller, and A.W. Rutherford. 1999. Carotenoid oxidation in photosystem II. *Biochemistry* 38:8189–8195.

Harborne, J.B. and C.A. Williams. 2000. Advances in flavonoid research since 1992. *Phytochemistry* 55:481–504.

Hauptmann, R., W.H. Eschenfeldt, J. English, and F.L. Brinkhaus. 1997. Enhanced carotenoid accumulation in storage organs of genetically engineered plants. US Patent 5618988.

Hebda, P.A., E.J. Behrman, and G.A. Barber. 1979. Guanosine 5′-diphosphate D-mannose—Guanosine 5′-diphosphate L-galactose epimerase of *Chlorella pyrenoidosa*—Chemical synthesis of guanosine 5′-diphosphate L-galactose and further studies of the enzyme and the reaction it catalyzes. *Arch Biochem Biophys* 194:496–502.

Heller, W. and G. Forkmann, 1993. In: *The Flavonoids: Advances in Research Since 1986*, ed. J.B. Harborne, pp. 499–536. London, U.K.: Chapman & Hall.

Hemavathi, C.P. Upadhyaya, K.E. Young et al. 2009. Over-expression of strawberry D-galacturonic acid reductase in potato leads to accumulation of vitamin C with enhanced abiotic stress tolerance. *Plant Sci* 177:659–667.

Hertog, M.G.L., D. Kromhout, C. Aravanis et al. 1995. Flavonoid intake and long-term risk of coronary heart disease and cancer in the Seven Countries Study. *Archs Intern Med* 15:381–386.

Hirschberg, J. 1999. Production of high value compounds: Carotenoids and vitamin E. *Curr Opin Biotechnol* 10:186–192.

Hirschberg, J. 2001. Carotenoid biosynthesis in flowering plants. *Curr Opin Plant Biol* 4:210–218.

Hofius, D., M.R. Hajirezaei, M. Geiger, H. Tschiersch, M. Melzer, and U. Sonnewald. 2004. RNAi-mediated tocopherol deficiency impairs photoassimilate export in transgenic potato plants. *Plant Physiol* 135:1256–1268.

Holton, T.A. and E.C. Cornish. 1995. Genetics and biochemistry of anthocyanin biosynthesis. *Plant Cell* 7:1071–1083.

Hunter, S.C. and E.B. Cahoon. 2007. Enhancing vitamin E in oilseeds: Unraveling tocopherol and tocotrienol biosynthesis. *Lipids* 42:97–108.

Imai, T., S. Karita, G. Shiratori et al. 1998. L-Galactono-g-lactone dehydrogenase from sweet potato: Purification and cDNA sequence analysis. *Plant Cell Physiol* 39:1350–1358.

Imai, T., M. Niwa, Y. Ban, M, Hirai, K. Oba, and T. Moriguchi. 2009. Importance of the L galactonolactone pool for enhancing the ascorbate content revealed by L-galactonolactone dehydrogenase-overexpressing tobacco plants. *Plant Cell Tiss Organ Cult* 96:105–112.

Ioannidi, E., M.. Kalamaki, C. Engineer et al. 2009. Expression profiling of ascorbic acid-related genes during tomato fruit development and ripening and in response to stress conditions. *J Exp Bot* 60:663–678.

Ishikawa, T., J. Dowdle, and N. Smirnoff. 2006. Progress in manipulating ascorbic acid biosynthesis and accumulation in plants. *Physiol Plant* 126:343–355.

Jain, A.K. and C.L. Nessler. 2000. Metabolic engineering of an alternative pathway for ascorbic acid biosynthesis in plants. *Mol Breed* 6:73–78.

Jayaraj, J., R. Delvin, and Z. Punja. 2008. Metabolic engineering of novel ketocarotenoid production in carrot plants. *Transgenic Res* 17:489–501.

Jez, J.M. and J.P. Noel. 2002. Reaction mechanism of chalcone isomerase: pH dependence, diffusion control, and product binding differences. *J Biol Chem* 277:1361–1369.

Kamal-Eldin, A. and L. Appelqvist. 1996. The chemistry and antioxidant properties of tocopherols and tocotrienols. *Lipids* 31:71–701.

Karunanandaa, B., Q. Qi, M. Hao et al. 2005. Metabolically engineered oilseed crops with enhanced seed tocopherol. *Metab Eng* 7:384–400.

Kasahara, H., A. Hanada, T. Kuzuyama, M. Takagi, J. Kamiya, and S. Yamaguchi. 2002. Contribution of the mevalonate and methylerythritol phosphate pathways to the biosynthesis of gibberellins in *Arabidopsis. J Biol Chem* 277:45188–45194.

Khachik, F., L. Carvallo, P.S. Bernstein, G.J. Muir, D.Y. Zhao, and N.B. Katz. 2002. Chemistry, distribution, and metabolism of tomato carotenoids and their impact on human health. *Exp Biol Med* 227:845–851.

Kim, M., S.C. Kim, K.J. Song et al. 2010. Transformation of carotenoid biosynthetic genes using a micro-cross section method in kiwifruit (*Actinidia deliciosa* cv.Hayward). *Plant Cell Rep* 29:1339–1349.

Knekt, P., R. Jarvinen, A. Reunanen, and J. Maatela. 1996. Flavonoid intake and coronary mortality in Finland: A cohort study. *Br Med J* 312:478–481.

Korkina, L.G. and I.B. Afanasev. 1997. Antioxidant and chelating properties of flavonoids. *Adv Pharmacol* 38:151–163.

Krinsky, N.I. 1998. The antioxidant and biological properties of carotenoids. *Ann N Y Acad Sci* 854:443–447.

Krinsky, N. and K. Yeum. 2003. Carotenoid- radical interactions. *Biochem Bioph Res Co* 305:754–760.

Kristiansen, K.N. and W. Rohde. 1991. Structure of the *Hordeum vulgare* gene encoding dihydroflavonol-4-reductase and molecular analysis of ANT18 mutants blocked in flavonoid synthesis. *Mol Gen Genet* 230:49–59.

Kruk, J. and K. Strzalka. 2001. Redox changes of cytochrome b559 in the presence of plastoquinones. *J Biol Chem* 276:86–91.

Kumar, R., M. Raclaru, T. Schusseler et al. 2005. Characterisation of plant tocopherol cyclases and their overexpression intransgenic *Brassica napus* seeds. *FEBS Lett* 579:1357–1364.

Kwon, S.Y., S.M. Choi, Y.O. Ahn et al. 2003. Enhanced stress-tolerance of transgenic tobacco plants expressing a human dehydroascorbate reductase gene. *J Plant Physiol* 160:347–353.

Lahlou, M. 2004. Study of the molluscicidal activity of some phenolic compounds: Structure–activity relationship. *Pharm Biol* 4:258–261.

Laing, W.A., D. Barraclough, S. Bulley et al. 2004. A specific L-galactose-1-phosphate phosphatase on the path to ascorbate biosynthesis. *Proc Natl Acad Sci USA* 101:16976–16981.

Lee, B.K., S.L. Kim, K.H. Kim et al. 2008. Seed specific expression of perilla c-tocopherol methyltransferase gene increases α-tocopherol content in transgenic perilla (*Perilla frutescens*). *Plant Cell Tiss Organ Cult* 92:47–54.

Lee, K., S.M. Lee, S.R. Park et al. 2007. Overexpression of *Arabidopsis* homogenisate phytyltransferase or tocopherol cyclase elevates vitamin E content by increasing γ-tocopherol level in lettuce (*Lactuca sativa* L.). *Mol Cell* 24:301–306.

Lefebvre, V., M. Kuntz, B. Camara, and A. Palloix. 1998. The capsanthin capsorubin synthase gene: A candidate gene for the y locus controlling the red fruit color in pepper. *Plant Mol Biol* 36:785–789.

Levin, M. 1986. New concepts in the biology and biochemistry of ascorbic acid. *N Engl J Med* 314:892–902.

Li, H., H. Flachowsky, T.C. Fische et al. 2007. Maize *Lc* transcription factor enhances biosynthesis of anthocyanins, distinct proanthocyanidins and phenylpropanoids in apple (*Malus domestica* Borkh.). *Planta* 226:1243–1254.

Li, X., M. Gao, H. Pan, D. Cui, and M.Y. Gruber. 2010. Purple Canola: *Arabidopsis PAP1* increases antioxidants and phenolics in *Brassica napus* leaves. *J Agric Food Chem* 58:1639–1645.

Li, X., J.C. Qin, Q.Y. Wang et al. 2011a. Metabolic engineering of isoflavone genistein in *Brassica napus* with soybean isoflavone synthase. *Plant Cell Rep* 30:1435–1442.

Li, L., Y.Q. Xu, K. Owsiany et al. 2012. The *Or* gene enhances carotenoid accumulation and stability during post-harvest storage of potato tubers. *Mol Plant* 5:339–352.

Li, Y., G. Wang, and R. Hou. 2011b. Engineering tocopherol biosynthetic pathway in lettuce. *Biol Plantar* 55:453–460.

Lichtenthaler, H.K. 1999. The 1-deoxy-D-xylulose-5-phosphate pathway of isoprenoid biosynthesis in plants. *Annu Rev Plant Physiol Plant Mol Biol* 50:47–65.

Liu, J.J., T.S. Huang, W.F. Cheng, and F.J. Lu. 2003. Baicalein and Baicalin are potent inhibitors of angiogenesis: Inhibition of endothelial cell proliferation, migration and differentiation. *Int J Cancer* 106:559–565.

Lopez, A.B., J. Van Eck, B.J. Conlin, D.J. Paolillo, J. O'Neill, and L. Li. 2008. Effect of the cauliflower *Or* transgene on carotenoid accumulation and chromoplast formation in transgenic potato tubers. *J Exp Bot* 59:213–223.

Lorenc-Kukula, K., M. Wrobel-Kwiatkowska, M. Starzyckic, and J. Szopa. 2007. Engineering flax with increased flavonoid content and thus *Fusarium* resistance. *Physiol Mol Plant Pathol* 70:38–48.

Lorence, A., B.I. Chevone, P. Mendes, and C.L. Nessler. 2004. Myo-inositol oxygenase offers a possible entry point into plant ascorbate biosynthesis. *Plant Physiol* 134:1200–1205.

Lu, S., J. Van Eck, X. Zhou et al. 2006. The cauliflower *Or* gene encodes a DnaJ cysteine-rich domain-containing protein that mediates high levels of β-carotene accumulation. *Plant Cell* 18:3594–3605.

Lukaszewicz, M., Matysiak-Kata, J. Skala, I. Fecka, W. Cisowski, and J. Szop. 2004. Antioxidant capacity manipulation in transgenic potato tuber by changes in phenolic compounds content. *J Agric Food Chem* 52:1526–1533.

Ma, C., B. Ma, J. He, Q. Hao, X. Lu, and L. Wang. 2011. Regulation of carotenoid content in tomato by silencing of lycopene β/ε-cyclase genes. *Plant Mol Biol Rep* 29:117–124.

Maass, D., J. Arango, F. Wust, P. Beyer, and R. Welsch. 2009. Carotenoid crystal formation in Arabidopsis and carrot roots caused by increased phytoene synthase protein levels. *PLoS One* 4:e6373.

Maeda, H. and D. DellaPenna. 2007. Tocopherol functions in photosynthetic organisms. *Curr Opin Plant Biol* 10:260–265.

Malikov, V.M. and M.P. Yuldashev. 2002. Phenolic compounds of plants of the *Scutellaria* L. genus. Distribution, structure, and properties. *Chem Nat Compd* 38:473–519.

Martens, S. and A. Mithofer. 2005. Flavones and flavone synthases. *Phytochemistry* 66:2399–2407.

Martin, C., A. Prescott, S. Mackay, J. Bartlett, and E. Vrijlandt. 1991. Control of anthocyanin biosynthesis in flowers of *Antirrhinum majus*. *Plant J* 1:37–49.

Matringe, M., A. Sailland, B. Pelissier, A. Rolland, and O. Zink. 2005. p-Hydroxyphenylpyruvate dioxygenase inhibitor- resistant plants. *Pest Manag Sci* 61:269–276.

Mayne, S.T. 1996. β-Carotene, carotenoids and disease prevention in humans. *FASEB J* 10:690–701.

McGarvey, D.J. and R. Croteau. 1995. Terpenoid metabolism. *Plant Cell* 7:1015–1026.

Messina, M. 1999. Soy, soy phytoestrogens (isoflavones), and breast cancer. *Am J Clin Nutr* 70:574–575.

Middleton, E., C. Kandaswami Jr., and T.C. Theoharides. 2000. The effects of plant flavonoids on mammalian cells: Implications for inflammation, heart disease, and cancer. *Pharmacol Rev* 52:673–751.

Millar, A.H., V. Mittova, G. Kiddle et al. 2003. Control of ascorbate synthesis by respiration and its implications for stress responses. *Plant Physiol* 133:443–447.

Morris, W.L., L.J.M. Ducreux, P.D. Fraser, S. Millam, and M.A. Taylor. 2006a. Engineering ketocarotenoid biosynthesis in potato tubers. *Metab Eng* 8:253–263.

Morris, W.L., L.J.M. Ducreux, P. Hedden, S. Millam, and M.A. Taylor. 2006b. Overexpression of a bacterial 1-deoxy-D-xylulose 5-phosphate synthase gene in potato tubers perturbs the isoprenoid metabolic network: Implications for the control of the tuber life cycle. *J Exp Bot* 57:3007–3018.

Muir, S.R., G.J. Collins, S. Robinson et al. 2001. Overexpression of petunia chalcone isomerase in tomato results in fruit containing increased levels of flavonols. *Nat Biotechnol* 19:470–474.

Mullan, B.A., I.S. Young, H. Fee, and D.R. McCance. 2002. Ascorbic acid reduces blood pressure and arterial stiffness in type 2 diabetes. *Hypertension* 40:804–809.

Munne-Bosch, S. 2007. α–Tocopherol: A multifaceted molecule in plants. *Vit Hor* 76:375–392.

Munne-Bosch, S. and L. Alegre. 2002. The function of tocopherols and tocotrienols in plants. *Crit Rev Plant Sci* 21:31–57.

Munne-Bosch, S. and J. Falk. 2004. New insights into the function of tocopherols in plants. *Planta* 218:323–326.

Mutsuda, M., T. Ishikawa, T. Takeda, and S. Shigeoka. 1995. Subcellular localization and properties of L-galactono γ-lactone dehydrogenase in spinach leaves. *Biosci Biotech Biochem* 59:1983–1984.

Naidu, K.A. 2003. Vitamin C in human health and disease is still a mystery? An overview. *Nutr J* 2:7.

Nakayama, T., K. Yonekura-Sakakibara, T. Sato et al. 2000. Aureusidin synthase: A polyphenol oxidase homolog responsible for flower coloration. *Science* 290:1163–1166.

Namitha, K.K., S.N. Archana, and P.S. Negi. 2011. Expression of carotenoid biosynthetic pathway genes and changes in carotenoids during ripening in tomato (*Lycopersicon esculentum*). *Food Funct* 2:168–173.

Namitha, K.K. and P.S. Negi. 2010. Chemistry and biotechnology of carotenoids. *Crit Rev Food Sci Nutr* 50:728–760.

Naqvi, S., G. Farre, C. Zhu, G. Sandmann, T. Capell, and P. Christou. 2011. Simultaneous expression of *Arabidopsis* q-hydroxyphenylpyruvate dioxygenase and MPBQ methyltransferase in transgenic corn kernels triples the tocopherol content. *Transgenic Res* 20:177–181.

Naqvi, S., C. Zhu, G. Farre et al. 2009. Transgenic multivitamin corn through biofortification of endosperm with three vitamins representing three distinct metabolic pathways. *Proc Natl Acad Sci* 106:7762–7767.

Nesaretnam, K., R. Stephen, R. Dils, and P. Darbre. 1998. Tocotrienols inhibit the growth of human breast cancer cells irrespective of estrogen receptor status. *Lipids* 33:461–469.

Nijveldt, R.J., E. van Nood, D.E.C. van Hoorn, P.G. Boelens, K. van Norren, and P.A.M. van Leeuwen. 2001. Flavonoids: A review of probable mechanisms of action and potential applications. *Am J Clin Nutr* 74:418–425.

Niyogi, K.K. 1999. Phoprotection revisited: Genetic and molecular approaches. *Annu Rev Plant Physiol Plant Mol Biol* 50:333–339.

Nunes-Nesi, A., F. Carrari, A. Lytovchenko et al. 2005. Enhanced photosynthetic performance and growth as a consequence of decreasing mitochondrial malate dehydrogenase activity in transgenic tomato plants. *Plant Physiol* 137:611–622.

Oba, K., S. Ishikawa, M. Nishikawa, H. Mizuno, and T. Yamamoto. 1995. Purification and properties of L-galactono-c-lactone dehydrogenase, a key enzyme for ascorbic acid biosynthesis, from sweet potato roots. *J Biochem* 117:120–124.

O'Neill, M.A., T. Ishii, P. Albersheim, and A.G. Darvill. 2004. Rhamnogalacturonan II: Structure and function of a borate cross-linked cell wall pectic polysaccharide. *Ann Rev Plant Biol* 55:109–139.

Osada, M., S. Imaoka, and Y. Funae. 2004. Apigenin suppresses the expression of VEGF, an important factor for angiogenesis, in endothelial cells via degradation of HIF-1alpha protein. *FEBS Lett* 575:59–63.

Osakada, F., A. Hashino, T. Kume, H. Katsuki, S. Kaneko, and A. Akaike. 2004. α-Tocotrienol provides the most potent neuroprotection among vitamin E analogs on cultured striatal neurons. *Neuropharmacol* 47:904–915.

Ostergaard, J., G. Persiau, M.W. Davey, G. Bauw, and M. Van Montagu.1997. Isolation of a cDNA coding for L-galactono-gamma-lactone dehydrogenase, an enzyme involved in the biosynthesis of ascorbic acid in plants. Purification, characterization, cDNA cloning, and expression in yeast. *J Biol Chem* 272:30009–30016.

Padayatty, S.J., A. Katz, Y. Wang et al. 2003. Vitamin C as an antioxidant: Evaluation of its role in disease prevention. *J Am Coll Nutr* 22:18–35.

Paine, J.A., C.A. Shipton, S. Chaggar et al. 2005. Improving the nutritional value of golden rice through increased pro-vitamin A content. *Nat Biotechnol* 23:482–487.

Paiva, S. and R. Russel. 1999. Beta-carotene and other carotenoids as antioxidants. *J Am Coll Nutr* 18:426–433.

Pallanca, J.E. and N. Smirnoff. 2000. The control of ascorbic acid synthesis and turnover in pea seedlings. *J Exp Bot* 51:669–674.

Paran, E. 2006. Reducing hypertension with tomato lycopene. In: *Tomatoes, Lycopene and Human Health*, ed. A.V. Rao, 169–182. Scotland, U.K.: Caledonian Science Press.

Parr, A.J. and G.P. Bowell. 2000. Phenols in the plant and in man. The potential of possible nutritional enhancement of the diet by modifying the phenol content or profile. *J Sci Food Agric* 80:985–1012.

Patil, B.S., G.K. Jayaprakasha, K.N. Chidambara Murthy, and A. Vikram. 2009. Bioactive compounds: Historical perspectives, opportunities, and challenges. *J Agric Food Chem* 57:8142–8160.

Pearce, B.C., R.A. Parker, M.E. Deason, A.A. Qureshi, and J.J. Wright. 1992. Hypocholesterolemic activity of synthetic and natural tocotrienols. *J Med Chem* 35:3595–3606.

Pecker, I., R. Gabbay, F.X. Cunningham, and J. Hirschberg. 1996. Cloning and characterization of the cDNA for lycopene beta-cyclase from tomato reveals decrease in its expression during fruit ripening. *Plant Mol Biol* 30:807–819.

Pelletier, M.K., I.E. Burbulis, and B. Winkel-Shirley. 1999. Disruption of specific flavonoid genes enhances the accumulation of flavonoid enzymes and end-products in Arabidopsis seedlings. *Plant Mol Biol* 40:45–54.

Peters, N.K., J.W. Frost, and S.R. Long. 1986. A plant flavone, luteolin, induces expression of *Rhizobium meliloti* nodulation genes. *Science* 233:977–980.

Pfundel, E.E., M. Renganathan, A.M., Gilmore, H.Y. Yamamoto, and R.A. Dilley. 1994. Intrathylakoid pH in isolated pea chloroplasts as probed by violaxanthin deepoxidation. *Plant Physiol* 106:1647–1658.

Pignocchi, C., J.M. Fletcher, J.E. Wilkinson, J.D. Barnes, and C.H. Foyer. 2003. The function of ascorbate oxidase in tobacco. *Plant Physiol* 132:1631–1641.

Popiolkiewicz, J., K. Polkowski, J.S. Skierski, and A.P. Mazurek. 2005. In vitro toxicity evaluation in the development of new anticancer drugs—Genistein glycosides. *Cancer Lett* 229:67–75.

Porfiorva, S., E. Bergmuller, S. Tropf, R. Lemke, and P. Dormann. 2002. Isolation of an *Arabidopsis* mutant lacking vitamin E and identification of a cyclase essential for all tocopherol biosynthesis. *Proc Natl Acad Sci USA* 99:12495–12500.

Pueppke, J.L. 1996. The genetic and biochemical basis for nodulation of legumes by rhizobia. *Crit Rev Biotechnol* 16:1–51.

Qian, W., C. Yu, H. Qin et al. 2007. Molecular and functional analysis of phosphomannomutase (PMM) from higher plants and genetic evidence for the involvement of PMM in ascorbic acid biosynthesis in Arabidopsis and *Nicotiana benthamiana*. *Plant J* 49:399–413.

Qureshi, A.A., W.C. Burger, D.M. Peterson, and C.E. Elson. 1986. The structure of an inhibitor of cholesterol biosynthesis isolated from barley. *J Biol Chem* 261:10544–10550.

Raclaru, M., J. Gruber, R. Kumar et al. 2006. Increase of the tocochromanol content in transgenic *Brassica napus* seeds by overexpression of key enzymes involved in prenylquinone biosynthesis. *Mol Breed* 18:93–107.

Rao, A.V. and L.G. Rao. 2007. Carotenoids and human health. *Pharmacol Res* 55:207–216.

Ravanello, M.P., D. Ke, J. Alvarez, B. Huang, and C.K. Shewmaker. 2003. Coordinate expression of multiple bacterial carotenoid genes in canola leading to altered carotenoid production. *Metab Eng* 5:255–263.

Ray, H., M. Yu, P. Auser et al. 2003. Expression of anthocyanins and proanthocyanidins after transformation of alfalfa with Maize Lc. *Plant Physiol* 132:1448–1463.

Reddy, A.M., V.S. Reddy, B.E. Schefflerb, U. Wienand, and A.R. Reddy. 2007. Novel transgenic rice overexpressing anthocyanidin synthase accumulates a mixture of flavonoids leading to an increased antioxidant potential. *Metabol Eng* 9:95–111.

Rice-Evans, C.A., N.J. Millier, and G. Paganga. 1995. Structure–antioxidant activity relationships of flavonoids and phenolic acids. *Free Radical Biol Med* 20:933–956.

Rippert, P., C. Scimemi, M. Dubald, and M., Matringe. 2004. Engineering plant shikimate pathway for production of tocotrienol and improving herbicide resistance. *Plant Physiol* 134:92–100.

Rivera-Vargas, L.I., A.F. Schmitthenner, and T.L. Graham. 1993. Soybean flavonoid effects on and metabolism by *Phytophthora sojae*. *Phytochem* 32:851–857.

Rodriguez-Amaya, D.B. 2001. *A Guide to Carotenoid Analysis in Foods*. Washington, DC: ILSI Press.

Rodriguez-Concepcion, M. and A. Boronat. 2002. Elucidation of the methylerythritol phosphate pathway for isoprenoid biosynthesis in bacteria and plastids. A metabolic milestone achieved through genomics. *Plant Physiol* 130:1079–1089.

Rohmer, M. 1999. The discovery of a mevalonate-independent pathway for isoprenoid biosynthesis in bacteria, algae and higher plants. *Nat Prod Rep* 16:565–574.

Rohmer, M. 2003. Mevalonate-independent methylerythritol phosphate pathway for isoprenoid biosynthesis. Elucidation and distribution. *Pure Appl Chem* 75:375–387.

Romer, S., P.D. Fraser, J.W. Kiano et al. 2000. Elevation of the provitamin A content of transgenic tomato plants. *Nat Biotechnol* 18:666–669.

Romer, S., J. Lubeck, F. Kauder, S. Steiger, C. Adomat, and G. Sandmann. 2002. Genetic engineering of a zeaxanthin-rich potato by antisense inactivation and co-suppression of carotenoid epoxidation. *Metab Eng* 4:263–272.

Rommens, C.M., C.M. Richael, H. Yan et al. 2008. Engineered native pathways for high kaempferol and caffeoylquinate production in potato. *Plant Biotechnol J* 6:870–886.

Ronen, G., L. Carmel-Goren, D. Zamir, and J. Hirschberg. 2000. An alternative pathway to β-carotene formation in plant chromoplasts discovered by map-based cloning of Beta and old-gold color mutations in tomato. *Proc Natl Acad Sci* 97:11102–11107.

Ronen, G., M. Cohen, D. Zamir, and J. Hirschberg. 1999. Regulation of carotenoid biosynthesis during tomato fruit development: Expression of the gene for lycopene epsilon cyclase is down regulated during ripening and is elevated in the mutant delta. *Plant J* 17:341–351.

Rosati, C., R. Aquilani, S. Dharmapuri et al. 2000. Metabolic engineering of beta carotene and lycopene content in tomato fruit. *Plant J* 24:413–419.

Ross, J.A. and C.M. Kasum. 2002. Dietary flavonoids: Bioavailability, metabolic effects, and safety. *Annu Rev Nutr* 22:19–34.

Sattler, S.E., E.B. Cahoon, S.J. Coughlan, and D. DellaPenna. 2003. Characterization of tocopherol cyclases from higher plants and cyanobacteria. Evolutionary implications for tocopherol synthesis and function. *Plant Physiol* 132:2184–2195.

Sattler, S.E., Z. Cheng, and D. DellaPenna. 2004. From *Arabidopsis* to agriculture: Engineering improved vitamin E content in soybean. *Trends Plant Sci* 9:365–367.

Savidge, B., J.D. Weiss, Y.H. Wong et al. 2002. Isolation and characterization of homogentisate phytyl- transferase genes from *Synechocystis* sp. PCC 6803 and *Arabidopsis*. *Plant Physiol* 129:321–332.

Sayre, R., J.R. Beeching, E.B. Cahoon et al. 2011. The BioCassava plus program: Biofortification of cassava for sub-Saharan Africa. *Annu Rev Plant Biol* 62:251–257.

Schijlen, E., C.H. Ric de Vos, H. Jonker et al. 2006. Pathway engineering for healthy phytochemicals leading to the production of novel flavonoids in tomato fruit. *Plant Biotechnol J* 4:433–444.

Schijlen, E.G.W.M., C.H. Ric de Vos, A.J. van Tunen, and A.G. Bovy. 2004. Modification of flavonoid biosynthesis in crop plants. *Phytochem* 65:2631–2648.

Seo, Y.S., S.J. Kim, C.H. Harn, and W.T. Kim. 2011. Ectopic expression of apple fruit homogentisate phytyltransferase gene (MdHPT1) increases tocopherol in transgenic tomato (*Solanum lycopersicum* cv. Micro-Tom) leaves and fruits. *Phytochem* 72:321–329.

Shewmaker, C.K., J.A. Sheehy, M. Daley, S. Colburn, and D.Y. Ke. 1999. Seed-specific over-expression of phytoene synthase: Increase in carotenoids and other metabolic effects. *Plant J* 20:401–412.

Shih, C., Y. Chen, M. Wang, I.K. Chu, and C. Lo. 2008. Accumulation of isoflavone genistin in transgenic tomato plants overexpressing a soybean isoflavone synthase gene *J Agric Food Chem* 56:5655–5661.

Shintani, D. and D. DellaPenna. 1998. Elevating the vitamin E content of plants through metabolic engineering. *Science* 282:2098–2100.

Shintani, D.K. and D. DellaPenna. 2003. Transgenic plants with γ-tocopherol methyltransferase. US Patent 118637.

Shutenko, Z., Y. Henry, E. Pinard et al. 1999. Influence of the antioxidant quercetin in vivo on the level of nitric oxide determined by electron paramagnetic resonance in rat brain during global ischemia and reperfusion. *Biochem Pharmacol* 57:199–208.

Sies, H. and W. Stahl. 2003. Non-nutritive bioactive constituents of plants: Lycopene, lutein and zeaxanthin. *Int J Vit Nutr Res* 73:95–100.

Simkin, A.J., J. Gaffe, J.P. Alcaraz et al. 2007. Fibrillin influence on plastid ultrastructure and pigment content in tomato fruit. *Phytochem* 68:1545–1556.

Simmonds, M.S.J. 2003. Flavonoid–insect interactions: Recent advances in our knowledge. *Phytochemistry* 64:21–30.

Siqueira, J.O., G.R. Safir, and M.G. Nair. 1991. Stimulation of vesicular–arbuscular mycorrhiza formation and growth of white clover by flavonoid compounds. *New Phytologist* 118:87–93.

Smirnoff, N. and S. Gatzek. 2004. Ascorbate biosynthesis: A diversity of pathways. In: *Vitamin C. Functions and Biochemistry in Animals and Plants*, eds. H. Asard, J.M. May, and N. Smirnoff, pp. 7–29. Oxford, U.K.: Bios Scientific Publishers.

Smirnoff, N. and G.L. Wheeler. 2000. Ascorbic acid in plants: Biosynthesis and function. *Crit Rev Plant Sci* 19:267–290.

Snodderly, D.M. 1995. Evidence for protection against age-related macular degeneration by carotenoids and antioxidant vitamins. *Am J Clin Nutr* 62S:1448S–1461S.

Soll, J., G. Schultz, J. Joyard, R. Douce, and M.A. Block. 1985. Localization and synthesis of prenylquinones in isolated outer and inner envelop membranes from spinach chloroplasts. *Arch Biochem Biophys* 238:290–299.

Soriano, I.R., R.E. Asenstorfer, O. Schmidt, and I.T. Riley. 2004. Inducible flavone in oats (*Avena sativa*) is a novel defense against plant–parasitic nematodes. *Phytopathol* 94:1207–1214.

Sreevidya, V.S., C. Srinivasa Rao, S.B. Sullia, J.K. Ladha, and P.M. Reddy. 2006. Metabolic engineering of rice with soybean isoflavone synthase for promoting nodulation gene expression in rhizobia. *J Exp Bot* 57:1957–1969.

Steele, C.L., M. Gijzen, D. Qutob, and R.A. Dixon. 1999. Molecular characterization of the enzyme catalyzing the aryl migration reaction of isoflavonoid biosynthesis in soybean. *Arch Biochem Biophys* 367:146–150.

Subramanian, S., M.Y. Graham, O. Yu, and T.L. Graham. 2005a. RNA interference of soybean isoflavone synthase genes leads to silencing in tissues distal to the transformation site and to enhanced susceptibility to *Phytophthora sojae. Plant Physiol* 137:1345–1353.

Subramanian, S.S., S.C. Slater, K. Karberg, R. Chen, H.E. Valentin, and Y.H. Huang Wong. 2005b. Nucleic acid sequences to proteins involved in tocopherol synthesis. US Patent 688069.

Tanaka, Y., S. Tsuda, and T. Kusumi, 1998. Metabolic engineering to modify flower color. *Plant Cell Physiol* 39:1119–1126.

Tavva, V.K., Y.H. Kim, I.A. Kagan, R.D. Dinkins, K.H. Kim, and G.B. Collins. 2007. Increased α-tocopherol content in soybean seed overexpressing the *Perilla frutescens* γ-tocopherol methyltransferase gene. *Plant Cell Rep* 26:61–70.

Tefler, A. 2002. What is beta-carotene doing in photosystem II reaction centre? *Trans R Soc Lond Ser B Biol Sci* 357:1431–1439.

Toda, K., D. Yang, N. Yamanaka, S. Watanabe, K. Harada, and R. Takahashi. 2002. A single-base deletion in soybean flavonoid 30-hydroxylase gene is associated with gray pubescence color. *Plant Mol Biol* 50:187–196.

Tokunaga, T., K. Miyahara, K. Tabata, and M. Esaka. 2005. Generation and properties of ascorbic acid-overproducing transgenic tobacco cells expressing sense RNA for L-galactono-1, 4-lactone dehydrogenase. *Planta* 220:854–863.

Traber, M.G. and H. Sies. 1996. Vitamin E and humans: Demand and delivery. *Annu Rev Nutr* 16:321–347.

Tsai, S.M. and D.A. Phillips. 1991. Flavonoids released naturally from alfalfa promote development of symbiotic *Glomus* spores in vitro. *Appl Environ Microbiol* 57:1485–1488.

Tsegaye, Y., D.K. Shintani, and D. DellaPenna. 2002. Overexpression of the enzyme p-hydroxyphenylpyruvate dioxygenase in *Arabidopsis* and its relation to tocopherol biosynthesis. *Plant Physiol Biochem* 40:913–920.

USDA Database for the Flavonoid Content of Selected Foods. March 2003.

Valentin, H.E. and Q. Qi. 2005. Biotechnological production and application of vitamin E: Current state and prospects. *Appl Microbiol Biotechnol* 68:436–444.

Valpuesta, V. and M.A. Botella. 2004. Biosynthesis of L-ascorbic acid in plants: New pathways for an old antioxidant. *Trends Plant Sci* 9:573–577.

Van Eck, J., B. Conlin, D.F. Garvin, H. Mason, D.A. Navarre, and C.R. Brown. 2007. Enhancing beta-carotene content in potato by RNAi mediated silencing of the beta-carotene hydroxylase gene. *Am J Potato Res* 84:331–342.

Van Eenennaam, A.L., K. Lincoln, T.P. Durrett et al. 2003. Engineering vitamin E content: From *Arabidopsis* mutant to soy oil. *Plant Cell* 15:3007–3019.

Van Rhijn, R. and J. Vanderleyden. 1995. The Rhizobium-plant symbiosis. *Microbiol Rev* 59:124–142.

Vrettos, J.S., D.H. Stewart, J.C. de Paula, and G.W. Brudvig. 1999. Low-temperature optical and resonance Raman spectra of a carotenoid cation radical in photosystem II. *J Phys Chem B* 103:6403–6406.

Wagner, K.H. and I. Elmadfa. 2000. Effects of tocopherols and their mixtures on the oxidative stability of olive oil and linseed oil under heating. *Eur J Lipid Sci Tech* 102:624–629.

Wagner, K.H., F. Wotruba, and I. Elmadfa. 2001. Antioxidative potential of tocotrienols and tocopherols in coconut fat at different oxidation temperatures. *Eur J Lipid Sci Tech* 103:746–751.

Warning, A.J., I.M. Drake, C.J. Schorah et al. 1996. Ascorbic acid and total vitamin C concentrations in plasma, gastric juice, and gastrointestinal mucosa: Effects of gastritis and oral supplementation. *Gut* 38:171–176.

Washko, P.W., R.W. Welch, K.R. Dhariwal, Y. Wang, and M. Levine. 1992. Ascorbic acid and dehydroascorbic acid analyses in biological samples. *Anal Biochem* 204:1–14.

Watanabe, S., S. Uesugi, and Y. Kikuchi. 2002. Isoflavones for prevention of cancer, cardiovascular diseases, gynecological problems and possible immune potentiation. *Biomed Pharmacother* 56:302–312.

Waylan, A.T., P.R. O'Quinn, J.A. Unruh et al. 2002. Effects of modified tall oil and vitamin E on growth performance, carcass characteristics, and meat quality of growing-finishing pigs. *J Anim Sci* 80:1575–1585.

Wei, S., B. Yu, M.Y. Gruber, G.G. Khachatourians, D.D. Hegedus, and A. Hannoufa. 2010. Enhanced seed carotenoid levels and branching in transgenic *Brassica napus* expressing the *Arabidopsis* miR156b Gene. *J Agric Food Chem* 58:9572–9578.

Weidenborner, M. and H.C. Jha. 1997. Antifungal spectrum of flavones and flavanone tested against 34 different fungi. *Mycol Res* 101:733–736.

Welsch, R., J. Arango, C. Bar et al. 2010. Provitamin A accumulation in cassava (*Manihot esculenta*) roots driven by a single nucleotide polymorphism in a phytoene synthase gene. *Plant Cell* 22:3348–3356.

Wheeler, G.L., M.A. Jones, and N. Smirnoff. 1998. The biosynthetic pathway of vitamin C in higher plants. *Nature* 393:365–369.

Winkel-Shirley, B. 2001. Flavonoid biosynthesis. A colorful model for genetics, biochemistry, cell biology, and biotechnology. *Plant Physiol* 126:485–493.

Wolucka, B.A., G. Persiau, J. Van Doorsselaere et al. 2001. Partial purification and identification of GDP-mannose 3′, 5′- epimerase of *Arabidopsis thaliana*, a key enzyme of the plant vitamin C pathway. *Proc Natl Acad Sci* 98:14843–14848.

Wozniak, D., E. Lamer-Zarawska, and A. Matkowski. 2004. Antimutagenic and antiradical properties of flavones from the roots of *Scutellaria baicalensis* Georgi. *Nahrung/Food* 48:9–12.

Wurbs, D., S. Ruf, and R. Bock. 2007. Contained metabolic engineering in tomatoes by expression of carotenoid biosynthesis genes from the plastid genome. *Plant J* 49:276–288.

Xu, H.X. and S.F. Lee. 2001. Activity of plant flavonoids against antibiotic-resistant bacteria. *Phytother Res* 15:39–43.

Yabuta, Y., K. Yoshimura, T. Takeda, and S. Shigeoka. 2000. Molecular characterization of tobacco mitochondrial L-galactono- γ-lactone dehydrogenase and its expression in *Escherichia coli*. *Plant Cell Physiol* 41:666–675.

Yamamoto, A., N.H. Bhuiyan, R. Waditee et al. 2005. Suppressed expression of the apoplastic ascorbate oxidase gene increases salt tolerance in tobacco and Arabidopsis plants. *J Exp Bot* 56:1785–1796.

Ye, X., S. Al Babili, A. Kloti et al. 2000. Engineering the provitamin A (beta-carotene) biosynthetic pathway into (carotenoid-free) rice endosperm. *Science* 287:303–305.

Yin, L., S. Wang, A.E. Eltayeb et al. 2010. Overexpression of dehydroascorbate reductase, but not monodehydroascorbate reductase, confers tolerance to aluminum stress in transgenic tobacco. *Planta* 231:609–621.

Yu, B., D.J. Lydiate, L.W. Young, U.A. Schafer, and A. Hannoufa. 2008. Enhancing the carotenoid content of *Brassica napus* seeds by downregulating lycopene epsilon cyclase. *Transgenic Res* 17:573–585.

Yu, O., J. Shi, A.O. Hession, C.A. Maxwell, B. McGonigle, and J.T. Odell. 2003. Metabolic engineering to increase isoflavone biosynthesis in soybean seed. *Phytochemistry* 63:753–763.

Yusuf, M.A., D. Kumar, R. Rajwanshi, R.J. Strasser, M.T.-M. Govindjee, and N.B. Sarin. 2010. Overexpression of γ-tocopherol methyl transferase gene in transgenic *Brassica juncea* plants alleviates abiotic stress: Physiological and chlorophyll a fluorescence measurements. *Biochim Biophys Acta* 1797:1428–1438.

Yusuf, M.A. and N.B. Sarin. 2007. Antioxidant value addition in human diets: Genetic transformation of *Brassica juncea* with γ-TMT gene form increased α-tocopherol content. *Transgenic Res* 16:109–113.

Zhang, J., N. Tao, Q. Xu et al. 2009. Functional characterization of *Citrus PSY* gene in Hongkong kumquat (*Fortunella hindsii* Swingle). *Plant Cell Rep* 28:1737–1746.

Zhang, L., Z. Wang, Y. Xia et al. 2007. Metabolic engineering of plant L-ascorbic acid biosynthesis: Recent trends and applications. *Crit Rev Biotechnol* 27:173–182.

Zheng, P.W., L.C. Chiang, and C.C. Lin. 2005. Apigenin induced apoptosis through p53-dependent pathway in human cervical carcinoma cells. *Life Sci* 76:1367–1379.

Zhu, C., S. Naqvi, J. Breitenbach, G. Sandmann, P. Christou, and T. Capell. 2008. Combinatorial genetic transformation generates a library of metabolic phenotypes for the carotenoid pathway in maize. *Proc Natl Acad Sci* 105:18232–18237.

Zuk, M., A. Kulma, L. Dyminska et al. 2011. Flavonoid engineering of flax potentiate its biotechnological application. *BMC Biotechnol* 11:10.

Zuker, A., T. Tzfira, H. Ben-Meir et al. 2002. Modification of flower color and fragrance by antisense suppression of the flavanone 3-hydroxylase gene. *Mol Breeding* 9:33–41.

22 Plant as Biofactories of Pharmaceuticals and Nutraceuticals

Dandara R. Muniz, Raquel O. Faria, Vagner A. Benedito, Ângelo de Fátima, and Luzia V. Modolo

CONTENTS

22.1 INTRODUCTION

Secondary metabolism refers to a set of biochemical pathways that are apparently not directly involved in the growth and development of an organism, but is indispensable for its survival. The broad chemical diversity of plant secondary metabolites originates from the hydroxylation, glycosylation, acylation, prenylation, and/or *O*-methylation of a limited number of chemical scaffolds (Modolo et al., 2009). Given the sessile nature of plants, this group evolved secondary metabolite pathways in response to herbivore or pathogen attacks (de Fatima and Modolo, 2008). Besides conferring resistance/tolerance to environment stresses and providing color and fragrance, plant natural products are important health promoters.

During the course of evolution, humans learned early on how to use plant properties for their own benefits to alleviate discomforts and maladies. For instance, during the Middle Stone Age (~77,000 years ago) aromatic leaves of *Cryptocarya woodii* (Lauraceae), which exhibit insecticidal and larvicidal properties, were introduced to bedding (Wadley et al., 2011). Up to our period, many of the drugs currently used in traditional medicine were first discovered in plants (Fabricant and Farnsworth, 2001). Advances in organic chemistry allowed the economically viable synthesis of many plant molecules in the laboratory, making the plant itself no longer required as a supply of valuable metabolites (Faria et al., 2011). This practice minimizes the natural extractivism or the

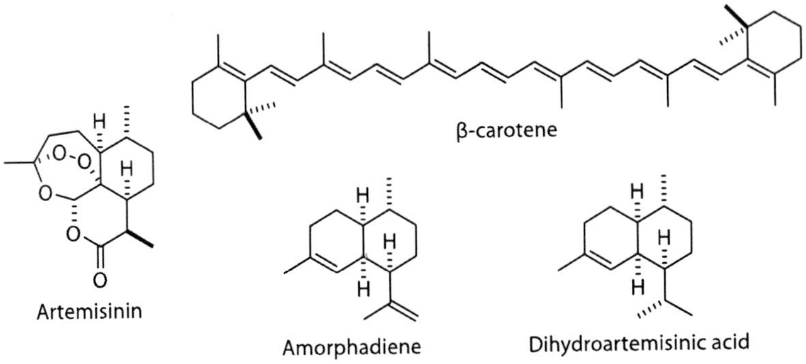

FIGURE 22.1 Example of plant natural products of pharmacological interest.

dependence on crop production. However, the use of organic synthesis approaches can be a challenge when trying to obtain structurally complex plant metabolites in satisfactory yields and a cost-effective fashion. Thus, plant extraction-based systems may comprise alternatives to overcome this drawback. The antimalarial and anticancer agent artemisinin (Figure 22.1) is extracted and purified from *Artemisia annua* (wormwood, Asteraceae) tissues (Ferreira et al., 2010). The worldwide demand for artemisinin to treat malaria reached 130 ton/year in 2010 (Artepal, 2010), costing from US $350.00/kg up to US $1000.00 during shortages. Typically, *Artemisia annua* crops yield about 6–18 kg artemisinin/ha (Kindermans et al., 2007), an amount that has been found insufficient to meet the demand for this drug. Then, the establishment of high-yield *in vitro* culture systems is mandatory to increase artemisinin supplies.

We now have a better understanding on a series of secondary metabolism pathways, since the metabolic intermediates, key regulators, and involved enzymes were disclosed. Some genes encoding for enzymes and key genetic controllers that participate in the biosynthesis of well-studied metabolites are known, especially in model species. In other instances, research is still much required to unveil important pathways in many plants of pharmacological importance, particularly in native species. The evolutionary history of plant species, together with advanced genetic studies in model species related to medicinal herbs, can be a powerful tool to comprehend how plant machinery turns on biochemical pathways that result in the production of small and macromolecules of pharma(nutra) ceutical interest.

The organ where a certain compound is synthesized (leaves, flowers, roots, etc.) and the place of its storage within the cell (vacuole, plastids, special organelles) are also factors to be considered. Accumulation of some compounds in the cell cytosol or in the whole plant is not desired when the target metabolite is highly reactive.

Specialized structures called trichomes evolved on plant epidermis to protect the organism against stresses and threats. In many instances, trichomes function as structural barriers to avoid water loss and minimize the effects of excessive sunlight and insects attack. Also, trichomes evolved as highly specialized glandular appendages that work as biofactories of secondary metabolites that repel insects and prevent pathogen spread in plant cells. The effectiveness of this strategy is easily visualized by the defense of poison ivy (*Toxicodendron radicans*; Anacardiaceae). Leaf trichomes of this species produce the phenolic urushiol, a potent toxin that induces dermatitis in humans and herbivore mammals. Apart from this, there is a plethora of plant metabolites with positive effects on human health.

Understanding where in the plant the molecule of interest is synthesized is a key for choosing the best strategy for massively producing the target compounds *in vivo*. Elicited plant cell suspension cultures are suitable systems for highly producing interesting secondary metabolites that naturally occur in parenchymatic tissues. In contrast, this same system may not provide good yields if the target metabolite (e.g., artemisinin) is specific of a type of glandular trichomes, unless the proper elements are genetically engineered in a particular cell line.

22.2 METABOLIC ENGINEERING

The metabolic engineering approach uses cells, organs, or organisms to genetically manipulate biochemical pathways to diverge metabolic fluxes toward the biosynthesis of compounds of interest. *Genetic engineering* is a key part of the process by enabling introduction or deletion of enzymatic or regulatory genes in the organism's genome. Bioengineering of metabolic pathways is coming to age. The development of the Golden Rice is a notorious example of metabolic engineering. Grains from non-transformed rice plants have no significant amount of β-carotene (Figure 22.1), a provitamin A. Golden Rice can accumulate up to 37 mg β-carotene/g seeds. This is due to the introduction of two key enzymes of carotenoid pathway into rice plants: phytoene synthase from either daffodil plants (*Narcissus pseudonarcissus*; Amaryllidaceae) or maize (in Golden Rice 2) under the control of an endosperm-specific glutein promoter, and a carotene desaturase from the soil bacteria *Erwinia uredovora* under a constitutive promoter (Beyer et al., 2002; Paine et al., 2005). The dealing with intellectual property and transgenic crop regulations has been groundbreaking for functional foods. The Golden Rice is expected to be available in the market in 2012, potentially helping to reduce 6000 deaths daily and avoiding vision problems in several hundred thousand vitamin A-deficient people (Potrykus, 2010). However, the decade-long "bench to belly" road was bumpy, as Golden Rice was ready since 1999 already. This occurred especially because rice is food and a major crop, which is not expected to occur at full extent with medicinal plants. Carotenoid-enriched "golden" potato (Diretto et al., 2007; Llorente et al., 2010) and cassava (Welsch et al., 2010) are now on the way.

Important examples of metabolic engineering of pharma(nutra)ceuticals are the development of production pipelines for omega-3 fatty acids in soybean seeds and the yeast *Yarrowia lipolytica*, paclitaxel in *Escherichia coli* and the artemisinin precursor amorphadiene (Figure 22.1) in baker's yeast, *E. coli* and chicory for further semi-synthesis purposes (Ye and Bhatia, 2012). The latest yeast strains were reported to yield 150 mg of the artemisinin precursor amorphadiene/L culture, denoting commercial potential, although it still requires a somewhat costly semi-synthesis procedure (Ye and Bhatia, 2012). In this regard, chicory (*Cichorium intybus*; Asteraceae) and tobacco species (*Nicotiana* spp.; Solanaceae) have been considered for artemisinin metabolic engineering. The introduction of a few key enzymes of terpene biosynthesis in the roots deviated isoprenoids flux toward dihydroartemisinic acid (Figure 22.1) production (Ye and Bhatia, 2012). This molecule can further furnish artemisinin by using semi-synthesis approaches at a low cost. Artemisinin production in ectopic organs of heterologous species is a challenge, likely because the production of this terpene in plants occurs mostly in flower buds and leaf glandular trichomes (Wetzstein et al., 2009).

22.3 ELICITATION BOOSTS SECONDARY METABOLISM IN PLANT CELL CULTURES

Elicitation is a common technique used to dramatically enhance the biosynthesis of secondary metabolites in plant cell cultures (Goel et al., 2011), often encompassing a cost-effective strategy to increase yield. By mimicking stress conditions, elicitors activate cell defense metabolism of specific secondary compounds (Campos et al., 2009).

Among the most effective treatments to trigger plant secondary metabolism is the addition of defense-related plant hormones (methyl jasmonate, jasmonic acid, salicylic acid, or ethylene) to the culture medium. For example, elicitation of suspension cell cultures of *Salvia miltiorrhiza* (Lamiaceae) with salicylic acid (SA) enhanced up to 10-fold the production of the phenolic compounds salvianolic acid B and caffeic acid (Figure 22.2) (Dong et al., 2010). Production of the stilbene resveratrol (Figure 22.2) in cell suspension of grape (*Vitis vinifera*; Vitaceae) increased significantly upon methyl jasmonate (MeJA) elicitation (Santamaria et al., 2010, 2011). Production of the anticancer paclitaxel (Figure 22.2) was also demonstrated in suspension cell cultures of hazel

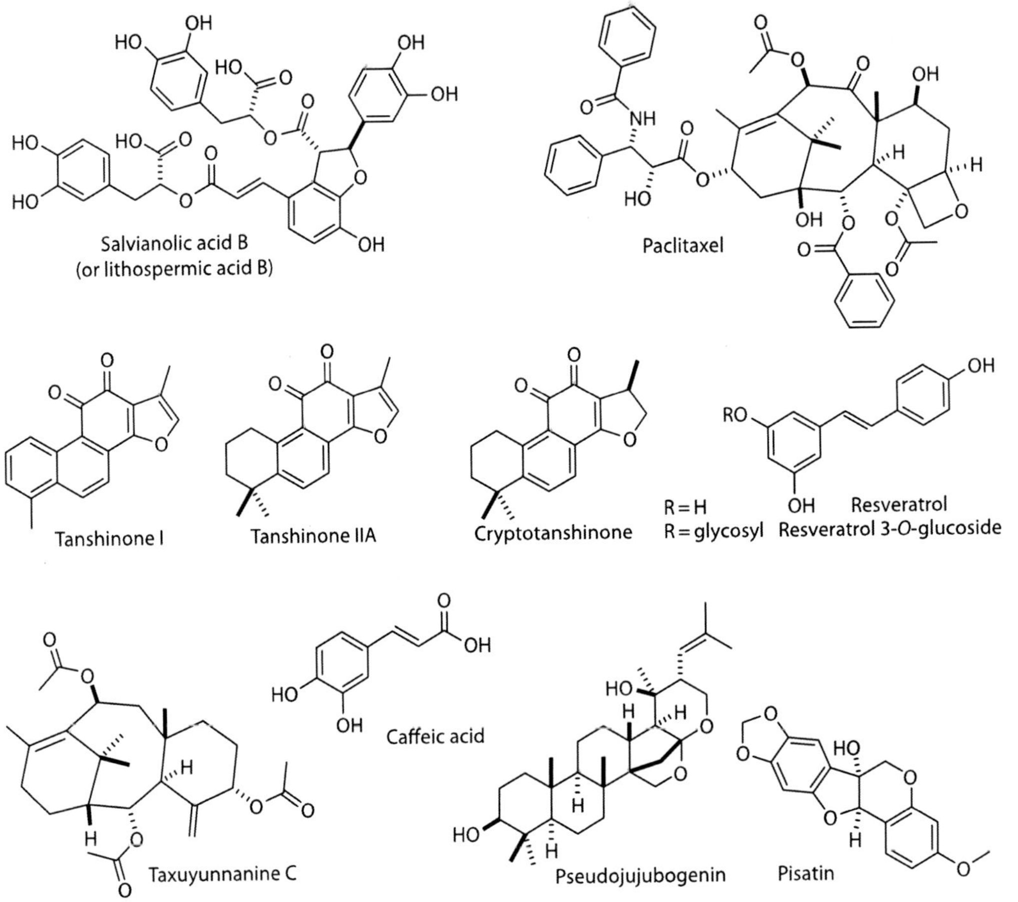

FIGURE 22.2 Example of plant natural products produced from elicited and/or genetically engineered plant systems.

(*Corylus avellana*; Betulaceae). Hazel cell suspension cultures comprise interesting systems for producing paclitaxel since the cells grow much faster than those of *Taxus* species (Taxaceae), plants used to first isolate the referred terpene (Bestoso et al., 2006; Ottaggio et al., 2008). Elicitation of hazel cell cultures with MeJA and the polysaccharide chitosan dramatically increased the production of paclitaxel and taxanes (Bestoso et al., 2006).

Other compounds mimicking biotic attacks are also effective in eliciting secondary metabolism. Yeast extract (YE) treatment induced eightfold accumulation of furanocoumarin in shoot cultures of rue (*Ruta graveolens*; Rutaceae) (Diwan and Maplathak, 2011). Production of tanshinone terpenes (Figure 22.2) was increased by 10-fold when *Salvia miltiorrhiza* cell cultures were challenged with YE (Zhao et al., 2010a).

The synergistic effect of MeJA, SA, and YE has been analyzed on the alkaloid benzophenanthridine production in suspension cultures of California poppy (*Eschscholzia californica*; Papaveraceae). The combined elicitor increased cells secondary metabolism up to 5.5-fold (Cho et al., 2008). Dual elicitation with chitosan (a polysaccharide derived from the fungal cell wall component called chitin) and YE induced a sixfold increase in the biosynthesis of the sapogenin pseudojujubogenin (Figure 22.2) in shoot cultures of *Bacopa monnieri* (Scrophulariaceae) (Kamonwannasit et al., 2008). Combination of pulsed electric field and ethephon (an ethylene-releasing compound) increased the production of resveratrol 3-*O*-glucoside (Figure 22.2) by 20-fold in grape cell culture when compared to non-elicited cells (Cai et al., 2011).

The stimulating effect of Ag^+ and Cd^{2+} on the biosynthesis of tanshinones in *Salvia miltiorrhiza* cell suspension cultures was reported (Zhao et al., 2010a). Cryptotanshinone's biosynthesis was induced by 30-fold and tanshinones I and IIA ones were induced by 5-fold when cell cultures were challenged with Ag^+ and Cd^{2+} at 25 μM.

Physical treatments also act as elicitors. Heat shocks trigger immediate response of cellular systems, inducing reactions that include increased production of specific secondary metabolism. Heat shock in cell suspension cultures of *Taxus yunnanensis* increased the production of paclitaxel by sixfold when compared to non-elicited cells (Zhang and Fevereiro, 2007). Pulsed electric field stimulation increased by 30% the biosynthesis of taxuyunnanine C (Figure 22.2) in *Taxus chinensis* cell suspension culture (Ye et al., 2004) as well as anthocyanin biosynthesis in *Vitis vinifera* cell culture (Cai et al., 2012). Although it remains a largely unexplored subject, electric elicitation is a very promising technique to enhance secondary metabolism in cell and organ cultures.

22.4 PROMOTERS AND MICRORNAs REGULATING GENE EXPRESSION OF SECONDARY METABOLISM

Cell energy expenditure needs to be considered when performing metabolic engineering to optimize the system toward the production of target compounds. The use of promoters (i.e., specific region of the genetic code upstream the protein-coding sequence of the gene that coordinates spatial-temporal gene expression) that trigger gene expression only in cells where the compound is made and at optimal levels is an important factor that is often overlooked. For instance, the use of constitutive plant promoters will induce the production of a certain metabolite in the whole plant while the *in vivo* production of the same metabolite may occur in specific structures such as glandular trichomes. This will lead to unproductive metabolic energy expenditure, with often side effects such as hindered growth rate. The use of promoters that drive gene expression in specific organs at appropriate times is highly recommended.

Molecular and genetic analyses of a series of experiments with organs of plant species that have been extensively studied under particular conditions have allowed the characterization of the gene expression of various promoters. When a closely-related model species is available, plant promoters can be "borrowed" from the model and analyzed in medicinal plants with a marker for gene expression (e.g., promoter: GUS). Glucosinolate biosynthesis in *Brassica rapa* can benefit from studies in *Arabidopsis thaliana* (Wang et al., 2011) since both belong to the Brassicaceae family. Then, it is expected that most Arabidopsis promoters will function in the same fashion when introduced into the *Brassica rapa* genome.

Most herbs, however, do not have closely related model species. In this case, heterologous gene promoters from models may still present similar function in the species of interest. Heterologous gene promoters must be tested first since their activity will largely depend on the biological compatibility of associated gene expression regulators (e.g., transcription factors and regulatory small RNAs) that trigger the genetics of specific metabolic pathways.

The existence of efficient methods of genetic transformation is also an important element of genetic engineering, along with plant tissue culture and plant regeneration protocols that must also be established and optimized to allow testing promoter activities and enhancement of specific secondary metabolism.

The role of gene regulation by small RNAs in secondary metabolism has been recently being explored. *Arabidopsis thaliana* and *Arabidopsis arenosa* are closely-related species with contrasting secondary metabolism regarding plant defense. The microRNA *miR163* targets the degradation of methyltransferase transcripts involved in plant defense. Thus, the production of methyl farnesoate will be suppressed by the action of *miR163* (Ha et al., 2009). It has been recently shown that natural genetic variation in *miR163* (both its promoter and expressed regions) between these two Arabidopsis species account for the more abundant biosynthesis of secondary metabolites in *Arabidopsis arenosa* (Ng et al., 2011). In the same way, the microRNA *miR156* targets the Arabidopsis *SPL* gene transcript, which encodes for a transcription factor that activates flavonoid biosynthesis. Reduction of *miR156* expression indirectly leads to an increase in anthocyanin production (Gou et al., 2011). The expression of the Arabidopsis *miR156*, though, should suffice to boost anthocyanin biosynthesis. Another example is *miR393* that redirects the metabolic flow away from camalexin biosynthesis toward glucosinolate production in Arabidopsis. The *miR393* likely negatively regulates specific auxin response factors (ARF1 and ARF9) that turn off glucosinolate pathway (Robert-Seilaniantz et al., 2011). This intricate genetic regulation of secondary metabolism by specific promoters and small RNAs can be effectively used to trigger the synthesis of specific metabolites without the need of introducing (or even knowing) all genetic components of the metabolic pathway.

22.5 TRANSCRIPTION FACTORS AS HUBS IN GENE REGULATORY NETWORKS OF SECONDARY METABOLISM

An aspect that deserves further exploration in secondary metabolite production in cell cultures is gene regulation by transcription factors (TFs). TFs are key regulators of gene expression in cells, helping coordinate the development, metabolism, and stress responses of the organism (Udvardi et al., 2007; Libault et al., 2009). The comprehension of the basis by which TFs are involved in the activation of regulatory gene networks will potentially facilitate the manipulation of secondary metabolism pathways (Aharoni and Galili, 2011).

The over-expression of the jasmonate-related TF genes *CrORCA3* (AP2/ERF family) and *CrWRKY1* induced the biosynthesis of terpenoid indole alkaloids in the Madagascar periwinkle (*Catharanthus roseus*; Apocynaceae) (van der Fits and Memelink, 2000; Suttipanta et al., 2011). Likewise, glucosinolate biosynthesis was enhanced by the expression of two TF genes (*ART1* and *MYB28*-like) in the model *Arabidopsis thaliana* (Malitsky et al., 2008). Transformation with MYB TF gene *PAP1* from Arabidopsis enhanced phenolic acids production in *Salvia miltiorrhiza* by 10-fold (Zhang et al., 2010). Nicotine synthesis was stimulated in tobacco (*Nicotiana tabacum*) by *NIC2* (an AP2/ERF TF closely related to *ORCA3*) (Shoji et al., 2010; de Boer et al., 2011) with bHLH TFs (*MYC2a,b,c*) being involved in the activation of early steps of this alkaloid biosynthesis (Zhang et al., 2012).

The understanding of gene regulatory networks and TF hubs related to specific processes (including coordination of metabolic pathways) is now facilitated in model species by the availability of genomics tools coupled with specific genetic mutants or knock-down lines (such as via RNAi or VIGS—virus-induced gene silencing) and metabolomics analyses. A significant technical advance was described recently with the identification of key TFs regulating alkaloid biosynthesis in *Nicotiana benthamiana* via analysis of subtractive cDNA libraries contrasting roots treated or not with MeJA, followed by VIGS screening of 69 TFs (amongst the 1898 genes differentially expressed) (Todd et al., 2010). The study has revealed six TFs closely associated with nicotine biosynthesis. Although the work performed by Todd et al. (2010) is not the first report using the referred approach, the mentioned study does set a foundation for functional genetics of secondary metabolite biosynthesis.

As a model species, *Nicotiana benthamiana* has a large repertoire of tools and data available that is not yet accessible in non-model species, such as transcript sequence databanks, well-established gene knock-down protocols (VIGS system) and efficient genetic transformation protocol. However, new technologies promise to benefit "orphan" species by providing open platforms of gene expression analysis, such a RNA-Seq, without the need of first developing a species-specific technology, as it occurs for microarray analyses. Metabolomics approaches are allowing for high-throughput chemical analysis, enabling chemical screening of thousands of samples at a fast pace. Altogether, these technologies facilitate the identification of interesting mutant genotypes in large collections or germplasm banks.

DNA sequencing technologies and bioinformatics tools have been developed to identify specific point mutations by comparing genome sequences between mutants of interest and the reference originator genotype. The steady decreasing costs of these technologies are allowing for an increasing usage of these approaches in less-intensively studied species. This technology revolution in genetic analysis is already becoming available for herbal and medicinal species.

22.6 POTENTIAL OF HAIRY ROOT CULTIVATION SYSTEM FOR PRODUCTION OF SECONDARY METABOLITES

Root hairs are unicellular structures on the epidermis of plants root that contribute to water and nutrient uptake from the soil. A less spoken function of root hairs is the release of substances to the soil (mostly secondary metabolites) to establish a positive or negative below-ground allelochemical communication with microorganisms, insects, and other plants (Dayan et al., 2003; Cook et al., 2010). *Hairy root* cultures are obtained from root transformation with the soil pathogen *Agrobacterium rhizogenes* that inserts T-DNA into the genome of infected root cells, unbalancing their hormone physiology (Pistelli et al., 2011). As a result, high root growth rates can be achieved. Given the metabolic potential of root hairs and the high growth rates of hairy roots, this culture system can be used efficiently for commercial production of bioactive secondary metabolites (Cai et al., 2012). Because roots contain many fully differentiated cells with large vacuoles, high amounts of secondary metabolites may be produced by root cultures. Then, vacuoles can store the produced metabolites to avoid metabolic feedback inhibition in cytosol. Indeed, the Swiss technology company ROOTec (www.rootec.com) performs molecular pharming based on hairy roots cultivated in mist bioreactors. The company announced recently the development of a cost-effective cultivation system for ginseng roots.

Danshen (*Salvia miltiorrhiza*) hairy root systems were found to produce the antioxidant polyphenol rosmarinic acid (Figure 22.3) with the yields largely increased by MeJA elicitation (Chen et al., 2001; Xiao et al., 2009). Besides its nutraceutical property, rosmarinic acid exhibits great potential for Alzheimer's disease treatment (Hamaguchi et al., 2009). Other pharmacological properties of rosmarinic acid are discussed in Chapter 20. A 30-fold increase of resveratrol production in hairy

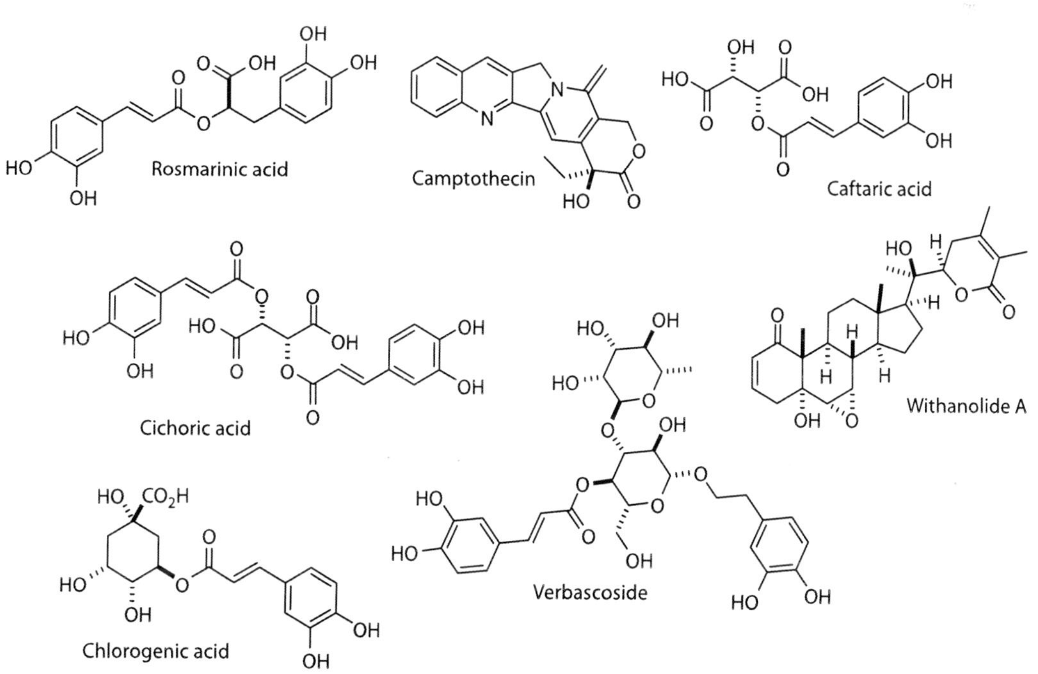

FIGURE 22.3 Example of some plant natural products effectively produced in hairy root culture systems.

root culture of peanut (*Arachis hypogaea*; Fabaceae) was achieved when the system was treated with sodium acetate (Condori et al., 2010). Sub-lethal electrical elicitation (30–100 mA current) of hydroponic hairy root cultures of chickpea (*Cicer arietinum*; Fabaceae) increased by 13-fold the production of pisatin (Figure 22.2) (Kaimoyo et al., 2008), an isoflavonoid recently found to exhibit estrogenic properties (Hegazy, et al., 2011).

Production of antioxidant compounds such as caftaric, chlorogenic, caffeic and cichoric acids (Figure 22.3), and anthocyanins was enhanced in coneflower hairy roots (*Echinacea purpurea*; Asteraceae) by light stimulus (Abbasi et al., 2007). Recombinant human acetylcholinesterase (AChE) was also successfully produced by *Nicotiana benthamiana* hairy root cultures (Woods et al., 2008). This enzyme is under preclinical studies for its great ability to detoxify cells contaminated with organophosphate pesticides.

The production of the anti-inflammatory phenylethanoid verbascoside (Figure 22.3) from hairy root cultures of Devil's claw (*Harpagophytum procumbens*; Pedaliaceae) has been reported (Gyurkovska et al., 2011; Georgiev et al., 2012). Hairy root culture of *Ophiorrhiza alata* (Rubiaceae) produced twice as much camptothecin (Figure 22.3) than soil-grown *Ophiorrhiza alata* plants (Ya-ut et al., 2011). The amount of camptothecin in 250 mL culture medium increased from 151 to 1036 μg when hairy root cultures were supplemented with the polystyrene resin Diaion HP-20 (Ya-ut et al., 2011). Sucrose supplemented at 4% in the medium of hairy root cultures of *Withania somnifera* (Solanaceae) accumulated up to 13.3 mg of withanolide A/g dry weight (Praveen and Murthy, 2012; Figure 22.3). Withanolides were found to be Ca^{2+}-antagonists and potent cholinesterase inhibitors as well (Choudhary et al., 2005).

Although *Artemisia annua* usually produces artemisinin mostly in flower buds and leaf glandular trichomes, its precursors are efficiently produced in hairy root cultures (Souret et al., 2003).

Some other hairy root cultures used for enhancing pharma(nutra)ceuticals production are listed in Table 22.1.

22.7 PLANT TISSUE AND CELL CULTURE SYSTEMS AS PLATFORMS FOR THE PRODUCTION OF HUMAN HEALTH PROMOTERS

Whole-plant system is being used to obtain molecules of pharmacological interests that are either in preclinical or clinical trials. Serum albumin, a protein abundant in blood, functions as a hormone, fatty acid, and steroids carrier besides playing a role as an extracellular fluid volume stabilizer. Human serum albumin (HSA) is clinically used to treat a series of diseases including fetal erythroblastosis, severe burning and hypoproteinemia (Hastings and Wolf, 1992). Commercial HSA is primarily supplied from human plasma purification, which in turn becomes a problem due to the risk of virus transmission from the donor. The induction of recombinant *HSA* gene and protein secretion from transformed *Oryza sativa* (rice; Poaceae) cell cultures was then established (Huang et al., 2005). The *HSA* gene was expressed under the control of the sucrose-depletion inducible rice promoter α *Amy3* and its corresponding signal sequence to drive the secretion of the produced protein from plant cells. The HSA protein was obtained as a stable product in yields as high as 76 mg/L cell culture. This study brought a new perspective for the safe and cost-effective production of commercial HSA. More recently, He et al. (2011) reported the large-scale production of recombinant HSA in rice seeds demonstrating that transgenic plants can comprise feasible systems to attend some worldwide demands. The authors obtained an HSA equivalent to that derived from plasma in an amount of 2.75 g/kg rice seeds (He et al., 2011). Field-grown tobacco plants have been used by Planet Biotechnology Inc. to produce monoclonal IgA antibodies that target *Streptococcus mutans*, the causal agent of tooth decay (Karg and Kallio, 2009). This product, under the name CaroRx™, was approved for human use in the European Union (Kaiser, 2008). SemBioSys Genetics Inc. initiated by the end of 2008 a phase I/II human trial of recombinant human insulin produced by field-grown *Carthamus tinctorius* (safflower, Asteraceae) plants (Karg and Kallio, 2009). Lactoferrin is a human glycoprotein that plays pivotal role in Fe^{2+} absorption by cells. Transgenic *Nicotiana tabacum* cell lines over-expressing human

TABLE 22.1
Hairy Root Systems as Platforms for the Production of Pharma(nutra)ceuticals

Metabolite	Plant Species/Elicitor or Resin	Pharmacological Property	Reference
Alkamides	*Echinacea* spp./JA	Anti-inflammatory, immunostimulatory	Romero et al. (2009)
Anthraquinone	*Rubia tinctoria*/IAA, sucrose	Antifugal	Sato et al. (1991)
Artemisinin	*Artemisia annua*/chitosan, MeJA or YE *Artemisia* spp. *Artemisia annua*/cerebroside or NO	Antimalaria and anticancer	Putalun et al. (2007), Mannan et al. (2008), Wang et al. (2009)
Asiaticoside	*Centella asiatica*/MeJA	Anti-inflammatory	Kim et al. (2007)
Baicalin, wogonin	*Scutellaria baicalensis*	Antioxidants, antiallergic, antitumor	Tiwari et al. (2008)
Betalains and/or rutin	*Beta vulgaris*/Lh or Tween 80 *Beta vulgaris*	Antioxidant, anti-inflammatory, hepatoprotective, etc.	Thimmaraju et al. (2003), Georgiev et al. (2010)
Caffeic acid and derivatives and anthocyanins	*Echinacea purpurea*/light	Antioxidant	Abbasi et al. (2007)
Camptothecin	*Ophiorrhiza pumila* *Ophiorrhiza alata*/resin Diaion HP-20	Anticancer	Sirikantaramas et al. (2007), Ya-ut et al. (2011)
Catharanthine	*Catharanthus roseus*/MeJA	Anticancer, acetylcholinesterase inhibitor	Zhou et al. (2010)
Daidzein, genistein	*Psoralea corylifolia*/SA	Anti-inflammatory, antiangiogenic, antioxidant, estrogen-like effects, etc.	Shinde et al. (2009)
Ferruginol and other diterpenes	*Salvia sclarea*/MeJA	Anticancer, antibiotic, analgesic, anti-inflammatory, etc.	Kuźma et al. (2009)
(Furano)coumarins and/or alkaloids	*Cichorium intybus*/Php *Ammi majus/Es* *Ruta graveolens*/Darkness *Angelica gigas*/Cu^{2+}, MeJA or MeJA	Skin diseases, pigmentation disorders treatment, antiandrogen, anticancer	Bais et al. (2000), Staniszewska et al. (2003), Sidwa-Gorycka et al. (2009), Rhee et al. (2010)
Gluconasturtiin, glucotropaeolin	*Nasturtium officinale*, *Barbarea verna*, and *Arabis caucasica*	Antifungal and antibacterial	Wielanek et al. (2009)
Gossypol and derivatives	*Gossypium barbadense and G. hirsutum* *Gossypium barbadense*/MeJA	Anticancer, antimicrobial, etc.	Triplett et al. (2008), Frankfater et al. (2009)
Harpagoside, verbascoside and isoverbascoside	*Harpagophytum procumbens*	Digestive disorders treatment, antipyretic, analgesic, etc.	Grabkowska et al. (2010)
HBsAg antigen	*Solanum tuberosum*	Hepatitis B vaccine	Sunil-Kumar et al. (2006)
Indigotin	*Polygonum tinctorium*	Chronic leukemia treatment	Park et al. (2008)

(continued)

TABLE 22.1 (continued)
Hairy Root Systems as Platforms for the Production of Pharma(nutra)ceuticals

Metabolite	Plant Species/Elicitor or Resin	Pharmacological Property	Reference
Human tissue-plasminogen activator	*Cucumis melo*	Dissolution of fibrin clots, plasmin formation in blood vessels	Kim et al. (2012)
Indole alkaloids	*Catharanthus roseus* *Catharanthus roseus*/Tv	α_1-Adrenergic receptor antagonist	Ciau-Uitz et al. (1994), Namdeo et al. (2002)
Isoflavonoids	*Pueraria candollei*/Chitosan, MeJA, YE, or SA	Estrogen-like effects, antioxidant	Udomsuk et al. (2011)
Kutkoside and picroside	*Picrorhiza kurroa*	Anti-cholestatic, antiulcerogenic, antiasthmatic, antidiabetic, etc.	Verma et al. (2007)
Licorice flavonoids	*Glycyrrhiza uralensis*/ PEG8000 and YE	Antioxidant, antibacterial, antitumor, etc.	Zhang et al. (2009)
Mouse mAb IgG	*Nicotiana tabacum*	Monoclonal antibody therapy	Sharp and Doran (2001)
Paclitaxel	*Taxus yunnanensis*/OGA	Anticancer	Guo and Wang (2008)
Pisatin	*Cicer arietinum*/Sub-lethal electrical charges	Estrogen-like effects	Kaimoyo et al. (2008)
Plumbagin	*Plumbago indica*	Antimicrobial, cardiotonic, immunosuppressive, etc.	Gangopadhydy et al. (2008)
Resveratrol and derivatives	*Arachis hypogaea*/Sodium acetate	Antioxidant	Medina-Bolivar et al. (2007), Condori et al. (2010)
Riboflavin	*Hyoscyamus albus*/ Fe-deficiency	Vitamin B_2 (enzymes cofactor)	Higa et al. (2008)
Rishitin and lubimin	*Solanum tuberosum*/Rb	Antimicrobial	Komaraiah et al. (2003)
Rosmarinic acid	*Salvia miltiorrhiza*/YE or Ag^+ *Agastache rugosa* *Coleus blumei*/YE or MeJA	Astringent properties, antimutagenic, antimicrobial, etc.	Chen et al. (2001), Yan et al. (2005), Lee et al. (2008), Bauer et al. (2009)
Saponins	*Panax ginseng*/Chitosan or MeJA *Panax ginseng*/HPTS or OCTS *Panax ginseng*/MeJA *Panax quinquefolium*	Immuno-modulatory, adaptogenic, anti-ageing, etc.	Palazon et al. (2003), Zhou et al. (2007), Kim et al. (2009), Mathur et al. (2010)
Sesquiterpene lactones, sonchuside A	*Cichorium intybus*/MeJA or SA	Anticancer, anti-inflammatory, hepatoprotective, etc.	Malarz et al. (2007)
Silymarin	*Silybum marianum*/YE *Silybum marianum*/Ag^+	Hepatoprotective	Hasanloo et al. (2009), Khalili et al. (2010)
Single chain murine IL-12	*Nicotiana tabacum*	Antitumor	Liu et al. (2009)
Solasodine	*Solanum khasianum*/CO_2	Raw material of corticosteroids and antifertility drugs	Jacob and Malpathak (2004)

TABLE 22.1 (continued)
Hairy Root Systems as Platforms for the Production of Pharma(nutra)ceuticals

Metabolite	Plant Species/Elicitor or Resin	Pharmacological Property	Reference
Tanshinones	*Salvia miltiorrhiza*/YE or Ag^+ *Salvia miltiorrhiza*/BcGP *Salvia miltiorrhiza*/YE or HS	Anticancer, antioxidant, etc.	Ge and Wu (2005), Wu et al. (2007), Wu and Shi (2008), Zhao et al. (2010b)
Taxane	*Salvia miltiorrhiza*/BcGP *Taxus* x *media* var. *Hicksii*/ MeJA	Anticancer	Syklowska-Baranek et al. (2009)
Thiophene	*Tagetes patula* *Tagetes patula*/Fc	Biocidal	Croes et al. (1989), Mukundan and Hjortso (1990)
Tropane alkaloids	*Datura innoxia* and *Hyoscyamus niger* *Atropa belladonna* *Brugmansia candida*/$CdCl_2$ or YE *Atropa belladona*/NO_3^- *Brugmansia candida*	Anticholinergic	Shimomura et al. (1991), Bonhomme et al. (2000), Pitta-Alvarez et al. (2000), Chashmi et al. (2008), Cardillo et al. (2010)
Valepotriates	*Valeriana officinalis*	Sedative, spasmolytic effect	Gränicher et al. (1992)
Withanolide A	*Withania somnifera* *Withania somnifera/Sucrose* 4%	Ca^{2+}-Antagonist, cholinesterase inhibitors	Murthy et al. (2008a,b), Praveen and Murthy (2012)

BcGP, glycoproteins from *Bacillus cereus*; Es, *Enterobacter sakazaki* elicitor molecules; HBsAg, Hepatitis B surface antigen; HPTS, heptasaccharide; HS, hyperosmotic stress; IAA, indole 3-acetic acid; IL-12, interleukin-12; JA, jasmonic acid; Lh, *Lactobacillus helveticus* elicitor molecules; MeJA, methyl jasmonate; Mouse mAb IgG, monoclonal antibody IgG from mouse; NO, nitric oxide; OCTS, octasaccharide; OGA, oligogalacturonide elicitor; PEG8000, polyethylene glycol 8000; Php, *Phytopthora parasitica*; Rb, *Rhizoctonia bataticola* elicitor molecules; SA, salicylic acid, Tv, *Trichoderma viridae*; YE, yeast extract or elicitor.

lactoferrin (*hLF*) gene under the control of *SWPA2* (oxidative stress-inducible peroxidase) promoter were also developed (Choi et al., 2003). Transformed cells at the stationary growth phase produced hLF at 4.3% of total soluble protein. The recombinant protein exhibited antibacterial activity similarly to the native human one. Later on, two different constructs harboring either the hLF signal peptide pIG211 or the rice glutein signal peptide pIG200 fused to mature *hLF* gene were introduced into Javanica rice cv. Rojolele (Rachmawati et al., 2005). All tissues of transgenic plants expressed the *hLF* gene under the control of the maize ubiquitin-1 promoter. The recombinant hLF corresponded to 15% of the total soluble protein produced by rice seeds. Likewise, rice-expressed hLF protein was effective against the *Bacillus subtilis* ATCC6633. The company Ventria Bioscience is now producing thousands of kilograms of recombinant hLF from field-grown rice to be used in infant formula and nutrition products to combat acute diarrhea (www.ventria.com).

Lithospermic acid B (also known as salvianolic acid B; Figure 22.2) is usually produced by plants of Labiatae and Boraginaceae families. Besides its nutraceutical value (antioxidant and free radical scavenger effects), lithospermic acid B is known to alleviate renal disease and hypertension. Cell suspension cultures of *Lithospermum erythrorhizon* (Boraginaceae) challenged with Cu^{2+} and blue light produced up to 13 mg lithospermic acid B/g fresh weight (Yamamoto et al., 2002).

Alkaloids exhibit a variety of biological properties including anticancer, antimalarial, and anesthetic activities. Biosynthesis of *N*-β-D-glucopyranosyl vincosamide (Figure 22.4), a novel

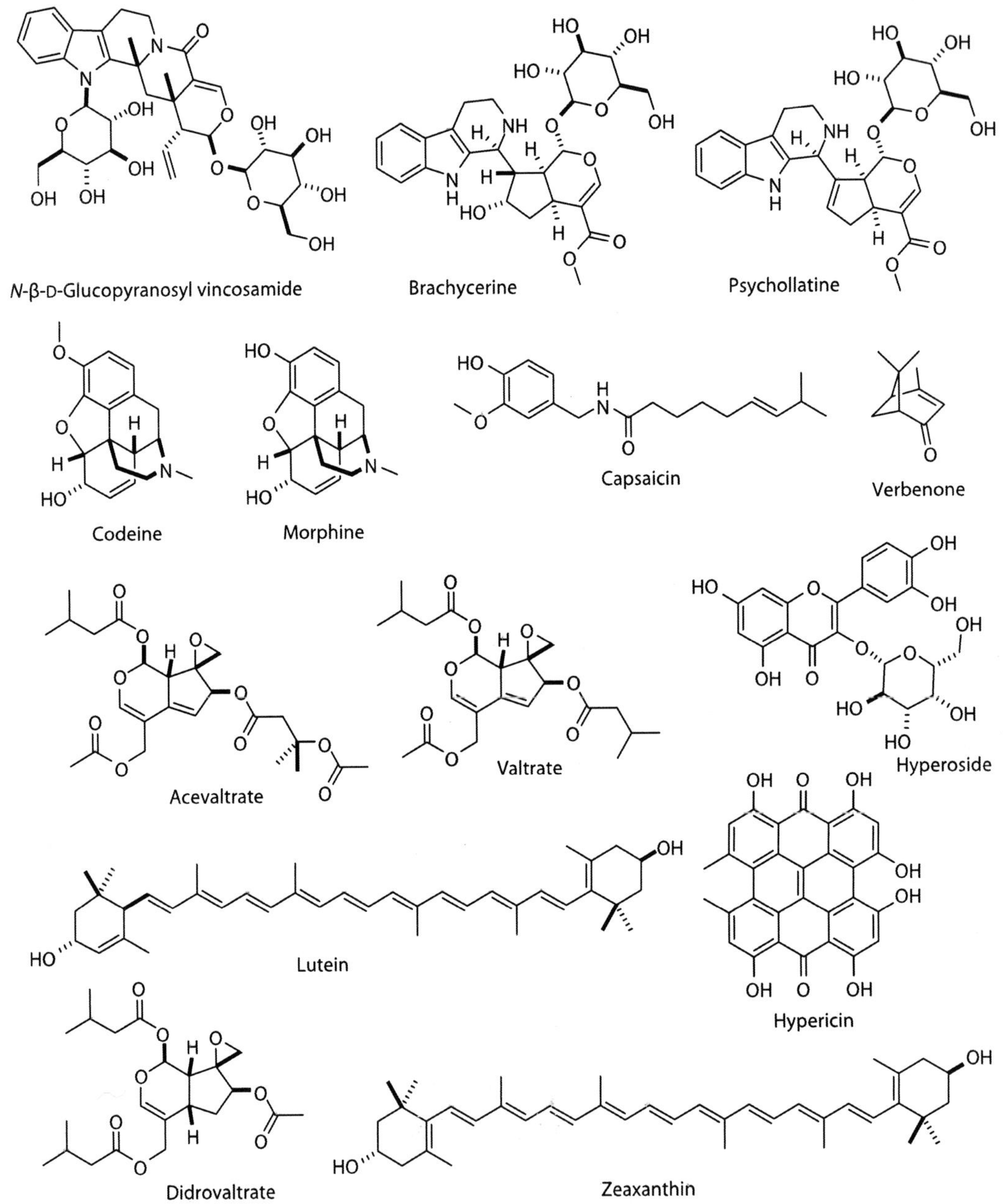

FIGURE 22.4 Example of some plant natural products effectively produced in whole-plant, tissue, or cell cultures.

indole alkaloid glucoside, was enhanced in seedlings cultures of *Psychotria leiocarpa* (Rubiaceae) deprived of sucrose and exposed to light (Henriques et al., 2004). *Psychotria brachyceras* seedlings cultured *in vitro* (14-day old) accumulated the alkaloid brachycerine (Figure 22.4) at the same extent that field-grown trees usually do (Gregianini et al., 2004). Cultured rhizogenic callus of *Psychotria umbellata* accumulated high amounts of the alkaloid psychollatine (Figure 22.4) followed the progression of differentiation of somatic embryo from shoots (Paranhos et al., 2005). Large-scale experiments performed with hormone-free cell cultures of *Papaver somniferum* (Papaveraceae) led to the accumulation of 3.0 mg codeine/g dry weight and 2.5 mg morphine/g dry weight (Figure 22.4), yields threefold higher than those obtained in hormone-supplemented cell cultures (Siah and Doran, 1991). Cell cultures of *Capsicum annuum* (Solanaceae) were shown to biotransform exogenous protocatechuic aldehyde and caffeic acid into capsaicin (Figure 22.4) in both free and immobilized cells (Ramachandra-Rao and Ravishankar, 2000).

Valepotriates, plant natural products belonging to the class of terpenes, are recognized sedative compounds. Non-transformed root cultures of *Valeriana glechomifolia* (Valerianaceae) aging 7–8 weeks were able to produce 2.6 mg acevaltrate/g dry weight, 10.2 mg valtrate/g dry weight and 2.9 mg didrovaltrate/g dry weight (Maurmann et al., 2006; Figure 22.4). Combined treatment of root cultures with 4.5 μM 2,4-dichlorophenoxyacetic acid and 0.9 μM kinetin was essential to guarantee the high yields. Another plant terpene of pharmacological interest is the verbenone (Figure 22.4), an insect attractant compound discovered to possess expectorant activity. Cell suspension cultures of *Psychotria brachyceras* (Rubiaceae) and *Rauwolfia sellowii* (Apocynaceae) were used as biocatalysts to obtain (−)-verbenone and (+)-verbenone through the oxidation of (−)-α-pinene and (+)-α-pinene, respectively (Limberger et al., 2007). *Psychotria brachyceras* cells recognized solely (−)-α-pinene while *Rauwolfia sellowii* ones were able to transform both enantiomers. The constitutive expression of farnesyl diphosphate synthase (FDS/FPS) enzyme in *Artemisia annua* triggered the accumulation of SA that turned on the biosynthesis of artemisinin (Pu et al., 2009).

The nutraceuticals shikonin and ginseng derivatives are commercially supplied via *in vitro* plant cultures of *Lithospermum erythrorhizon* and *Panax ginseng* (Araliaceae), respectively (Weathers et al., 2010). The EU Project Nutra-Snack was created aiming the use of plant systems and biotechnological approaches to produce new high quality ready-to-eat food to promote health (www.nutra-snacks.com). The project conceived 12 *in vitro* plant lines from *Hypericum perforatum* (Hypericaceae) hairy root, some of which accumulate high levels of the antioxidants hyperoside (5 mg/g dry weight; Figure 22.4) and chlorogenic acid (1.1 mg/g dry weight) and the anti-depressant hypericin (0.25 mg/g dry weight; Figure 22.4) (Rea et al., 2011). Researchers also drove their attention to the green algae *Chlamydomonas reinhardtii* (Chlamydomonadaceae). They developed cells that express D1 protein (from photosystem II) fused to valuable peptides (Rea et al., 2011). The D1 protein is highly expressed in chloroplast, and thus only regulatory elements related to the gene that encodes such protein (*psbA*) actually need to be taken into account when expressing a fusing target peptide by this approach. External stimuli (UV-C irradiation, oligogalacturonides, NO donors, etc.) on *Chlamydomonas reinhardtii* cultures also increased the accumulation of lutein and zeaxanthin (Figure 22.4) by up to 20-fold (Rea et al., 2011). These carotenoids are well known by their protective effect on eyes, preventing age-related macular degeneration.

Vitis vinifera cell suspension cultures treated with JA, SA, and the adsorvent HP2MGL produced about 2.7 g resveratrol/L culture (Yue et al., 2011). Resveratrol, a polyphenol that belongs to the stilbene class, is recognized as a health promoter especially because of its antioxidant activity.

Transgenic tomato plants (*Solanum lycopersicum*; Solanaceae) over-expressing the transcription factors *Del* and *Ros1* from snapdragon (*Antirrhinum majus*; Plantaginaceae/Veronicaceae) became purple due to the high content of anthocyanins (2.8 mg/g fresh weight) in both peel and flesh. Life span of cancer-susceptible $Trp53^{-/-}$ mice fed a diet supplemented with anthocyanin-enriched tomatoes increased more than 25%. Over-expression of the MYB transcription factor *LAP1* and the glycosyltransferase *UGT78G1* genes dramatically increased anthocyanins production in *Medicago* species (Peel et al., 2009; Dixon et al., 2011).

Caulifower plants (*Brassica oleracea* var. *botrytis*) may exhibit a deep purple phenotype as a result of abnormal anthocyanin accumulation in curds (3.8 mg/g fresh weight) and other tissues (Chiu et al., 2010). This is due to a natural mutation that implicates a Purple gene (*Pr*), whose identity was disclosed recently as a R2R3 MYB transcription factor involved in anthocyanin biosynthesis (Chiu et al., 2010). Indeed, the mutant phenotype was rescued in transgenic *Arabidopsis thaliana* and cauliflower plants expressing the *Pr*-D allele. The isolation of *Pr* gene is a milestone that will certainly contribute for the development of new food varieties enriched with improved health-promoting properties and visual appeal.

Tocotrienols and tocopherols are forms of vitamin E, an antioxidant found in very low amounts in seeds of most monocot plants (e.g., rice and wheat). Leaves of transgenic *Arabidopsis thaliana* plants over-expressing the barley *HGGT* gene (homogentisic acid geranylgeranyl transferase) accumulated tocotrienols which, in turn, led to an increment in total vitamin E by 15-fold (Cahoon et al., 2003). Total vitamin E levels increased by sixfold when the barley *HGGT* gene was over-expressed in corn seeds (Cahoon et al., 2003).

22.8 NEXT-GENERATION TECHNOLOGIES FOR THE DISCOVERY OF PHARMA(NUTRA)CEUTICALS-RELATED GENES

A better understanding of biochemical upregulation by culture elicitors at the molecular and genetic levels leads to the development of biotechnological tools to breeding plant cell lines with enhanced biosynthesis competence. For example, by comparing gene transcriptional profiles between non-elicited cultures with those at early elicitation stages, the key genes that trigger the biochemical pathway are revealed. While few years ago such studies were expensive, time-consuming, and difficult to carry out with non-model species, the advent of next-generation sequencing technologies (Metzker, 2010) and recent bioinformatics tools (Miller et al., 2010) brought a new perspective to accelerate *de novo* sequencing, assembling, and annotation of full genomes of less-intensively studied species. In the light of these new technologies, comparative transcriptomics and genome sequence assembly were done for *Artemisia annua* (Wang et al., 2009) and *Ginkgo biloba* (Ginkgoaceae; Lin et al., 2011), respectively. Such tools have also been used in the *Medicinal Plant Genome Project* in China (Chen et al., 2011). The project already completed a sequence draft of *Salvia miltiorrhiza* genome, whose roots produce the antioxidant and anti-inflammatory compound tanshinone IIA, a lead compound for the design of new drugs for vascular disorders treatment. *Panax ginseng* and *Panax notoginseng* are other species of interest. Roots of both plant species produce ginsenosides, well-known stimulants and antiallergic compounds. Future efforts of this project involve honeysuckle plants (*Lonicera japonica*; Caprifoliaceae), claimed in Chinese medicine to possess antibacterial and anti-inflammatory properties.

These advances will lead to foundational work to enable discovery of genes involved in the biosynthesis of compounds of pharmacological importance. Then, new breeding programs can be initiated to develop high-yield plant varieties and efficient bioengineered large-scale biosynthetic systems.

Efforts to obtain a more comprehensive knowledge on artemisinin metabolism in *Artemisia annua* are also benefiting from next-generation sequencing. Transcriptomics of glandular trichomes allowed assembling more than 40,000 contigs (Wang et al., 2009). From the functional genetics point of view, such tool constitutes a valuable resource to untangle secondary metabolism in *Artemisia annua*. Transcriptomics, in conjunction with genetic mapping of *Artemisia annua*, led to the identification of chromosomal positions and quantitative trait loci (QTLs) linked to high production of artemisinin. Such discoveries help fast-forward breeding programs obtaining plant varieties that exhibit high artemisinin yields (Graham et al., 2010). Another notable example is the American ginseng *Panax quinquefolius*, in which its root transcriptomics database, with more than 16,500 contigs, is already available. Through bioinformatic analysis, comparison of the dataset for MeJA-elicited plant roots and non-elicited ones revealed five novel candidate genes to encode key enzymes in the ginsenoside pathway (Sun et al., 2010).

22.9 CONCLUDING REMARKS

Plants are undoubtedly sources of an array of metabolites that may contribute to human well-being. Besides this, a set of features exhibited by plants make such organisms suitable for the safe production of pharma(nutra)ceuticals. The advent of metabolic engineering and new-generation technologies has contributed to the achievement of plant systems enriched in valuable substances.

This includes the generation of edible vaccines widely discussed elsewhere (Faria et al., 2011). Thus, many efforts were and are still being driven to standardize protocols for scaling up the production of pharma(nutra)ceuticals *in vitro*, *ex vitro*, and *in vivo* as well. The use of transformed whole-plant or tissue/cell culture systems is an interesting strategy to prevent the shortage of medicinal plants due to indiscriminate extractivism. More importantly, such practices will avoid the extinction of medicinal and native plants species, since they are some of the main sources of lead compounds for the design of new drugs.

ACKNOWLEDGMENTS

This work has been partly supported by Fundação de Amparo à Pesquisa do Estado de Minas Gerais (FAPEMIG) and Conselho Nacional de Desenvolvimento Científico e Tecnológico (CNPq). Luzia V. Modolo is grateful to the Programa de Auxílio à Pesquisa de Doutores Recém-Contratados—Universidade Federal de Minas Gerais (UFMG) for fully supporting this research.

ABBREVIATIONS

AChE	acetylcholinesterase
α *Amy3*	sucrose depletion-inducible promoter from *Oryza sativa*
ARF1 and ARF2	auxin response factors
BcGP	glycoproteins from *Bacillus cereus*
CrORCA3	octadecanoid-responsive *Catharanthus roseus* AP2/ERF domain
CrWRKY1	transcription factor from *Catharanthus roseus*
Es	*Enterobacter sakazakii* elicitor molecules
FDS/FPS	farnesyl diphosphate synthase
GUS	β-glucuronidase gene from *Escherichia coli*
HBsAg	hepatitis B surface antigen
HGGT	homogentisic acid geranylgeranyl transferase from *Hordeum vulgare*
hLF	human lactoferrin
HP2MGL	adsorvent polymer
HPTS	heptasaccharide
HS	hyperosmotic stress
HSA	human serum albumin
IL-12	interleukin-12
JA	jasmonic acid
LAP1	MYB family transcription factor from legumes
Lh	*Lactobacillus helveticus* elicitor molecules
mAb IgG	monoclonal antibody IgG from mouse
MeJA	methyl jasmonate
NO	nitric oxide
OCTS	octasaccharide
OGA	oligogalacturonides elicitor
PAP1	MYB family transcription factor from *Arabidopsis thaliana*
PEG8000	polyethylene glycol 8000
Php	*Phytopthora parasitica*
pIG200 or 211	Human lactoferrin signal peptides
Pr	purple gene
psbA	regulatory element related to the expression of D1 proteins
QTLs	quantitative trait loci
R2R3	MYB family transcription factor involved in anthocyanin biosynthesis
Rb	*Rhizoctonia bataticola* elicitor molecules

SA	salicylic acid
SPL	transcription factor that activates flavonoid biosynthesis
SWPA2	oxidative stress-inducible peroxidase promoter
TF	transcription factor
Tv	*Trichoderma viridae*
UGT78G1	anthocyanin glycosyltransferase from *Medicago truncatula*
VIGS	virus-induced gene silencing
YE	yeast extract/elicitor

REFERENCES

Abbasi, B.H., C.L. Tian, S.L. Murch, P.K. Saxena, and C.Z. Liu. 2007. Light-enhanced caffeic acid derivatives biosynthesis in hairy root cultures of *Echinacea purpurea. Plant Cell Rep* 26:1367–1372.

Aharoni, A. and G. Galili. 2011. Metabolic engineering of the plant primary-secondary metabolism interface. *Curr Opin Biotechnol* 22:239–244.

Artepal, J.P. Artemisinin Market: Quantities and pricing. *Artemisinin Conference 2010*, Antananarivo, Madagascar. http://www.mmv.org/newsroom/events/artemisinin-conference-2010 (accessed December 10, 2011).

Bais, H.P., S. Govindaswamy, and G.A. Ravishankar. 2000. Enhancement of growth and coumarin production in hairy root cultures of Witloof chicory (*Cichorium intybus* L. cv. Lucknow local) under the influence of fungal elicitors. *J Biosci Bioeng* 90:648–653.

Bauer, N., D. Kiseljak, and S. Jelaska. 2009. The effect of yeast extract and methyl jasmonate on rosmarinic acid accumulation in *Coleus blumei* hairy roots. *Biol Plant* 53:650–656.

Bestoso, F., L. Ottaggio, A. Armirotti et al. 2006. In vitro cell cultures obtained from different explants of *Corylus avellana* produce taxol and taxanes. *BMC Biotechnol* 6:45.

Beyer, P., S. Al-Babili, X. Ye et al. 2002. Golden Rice: Introducing the β-carotene biosynthesis pathway into rice endosperm by genetic engineering to defeat vitamin A deficiency. *J Nutr* 132:506S–510S.

de Boer, K., S. Tilleman, L. Pauwels et al. 2011. APETALA2/ETHYLENE RESPONSE FACTOR and basic helix-loop-helix tobacco transcription factors cooperatively mediate jasmonate-elicited nicotine biosynthesis. *Plant J* 66:1053–1065.

Bonhomme, V., D. Laurain-Mattar, J. Lacoux, M. Fliniaux, and A. Jacquin-Dubreuil. 2000. Tropane alkaloid production by hairy roots of *Atropa belladonna* obtained after transformation with *Agrobacterium rhizogenes* 15834 and *Agrobacterium tumefaciens* containing rol A, B, C genes only. *J Biotechnol* 81:151–158.

Cahoon, E.B., S.E. Hall, K.G. Ripp, T.S. Ganzke, W.D. Hitz, and S.J. Coughlan. 2003. Metabolic redesign of vitamin E biosynthesis in plants for tocotrienol production and increased antioxidant content. *Nat Biotechnol* 21:1082–1087.

Cai, Z., A. Kastell, D. Knorr, and I. Smetanska. 2012. Exudation: An expanding technique for continuous production and release of secondary metabolites from plant cell suspension and hairy root cultures. *Plant Cell Rep* 31:461–477.

Cai, Z., H. Riedel, N.M.M. Thaw Saw et al. 2011. Effects of pulsed electric field on secondary metabolism of *Vitis vinifera* L. cv. Gamay Fréaux suspension culture and exudates. *Appl Biochem Biotechnol* 164:443–453.

Campos, M.L., M. de Almeida, M.L. Rossi et al. 2009. Brassinosteroids interact negatively with jasmonates in the formation of anti-herbivory traits in tomato. *J Exp Bot* 60:4347–4361.

Cardillo, A.B., A.M.O. Alvarez, A.C. Lopez, M.E.V. Lozano, J.R. Talou, and A.M. Giulietti. 2010. Anisodamine production from natural sources: Seedlings and hairy root cultures of Argentinean and Colombian *Brugmansia candida* plants. *Planta Med* 76:402–405.

Chashmi, N.A., M. Sharifi, F. Karimi, and H. Rahnama. 2008. Enhanced production of tropane alkaloids by nitrate treatment in hairy root cultures of *Atropa belladonna. J Biotechnol* 136S:S22–S71.

Chen, H., F. Chena, F.C.K. Chiu, and C.M.Y. Lo. 2001. The effect of yeast elicitor on the growth and secondary metabolism of hairy root cultures of *Salvia miltiorrhiza. Enzyme Microb Technol* 28:100–105.

Chen, S., L. Xiang, X. Guo, and Q. Li. 2011. An introduction to the medicinal plant genome project. *Front Med* 5:178–184.

Chiu L.-W., X. Zhou, S. Burke, X. Wu, R.L. Prior, and L. Li. 2010. The purple cauliflower arises from activation of a MYB transcription factor. *Plant Physiol* 154:1470–1480.

Cho, H.-Y., S.Y. Son, H.S. Rhee, S.-Y.H. Yoon, C.W.T. Lee-Parsons, and J.M. Park. 2008. Synergistic effects of sequential treatment with methyl jasmonate, salicylic acid and yeast extract on benzophenanthridine alkaloid accumulation and protein expression in *Eschscholtzia californica* suspension cultures. *J Biotechnol* 135:117–122.

Choi, S.M., O.S. Lee, S.Y. Kwon, S.S. Kwak, D.Y. Yu, and H.S. Lee. 2003. High expression of a human lactoferrin in transgenic tobacco cell cultures. *Biotechnol Lett* 25:213–218.

Choudhary, M.I., S. Yousuf, S.A. Nawaz, S. Ahmed, and A.-U. Rahman. 2005. Cholinesterase inhibiting withanolides from *Withania somnifera. Chem Pharm Bull* 52:1358–1361.

Ciau-Uitz, R., M.L. Miranda-Ham, J. Coello-Coello, B. Chi, L.M. Pacheco, and V.M. Loyola-Vargas. 1994. Indole alkaloid production by transformed and non-transformed root cultures of *Catharanthus roseus. In vitro Cell Dev Biol* 30P:84–88.

Condori, J., G. Sivakumar, J. Hubstenberger, M.C. Dolan, V.S. Sobolev, and F. Medina-Bolivar. 2010. Induced biosynthesis of resveratrol and the prenylated stilbenoids arachidin-1 and arachidin-3 in hairy root cultures of peanut: Effects of culture medium and growth stage. *Plant Physiol Biochem* 48:310–318.

Cook, D., A.M. Rimando, T.E. Clemente et al. 2010. Alkylresorcinol synthases expressed in *Sorghum bicolor* root hairs play an essential role in the biosynthesis of the allelopathic benzoquinone sorgoleone. *Plant Cell* 22:867–887.

Croes, A.F., A.J.R. Vander Berg, M. Bosveld, H. Breteler, and G.J. Wullems. 1989. Thiophene accumulation in relation to morphology in roots of *Tagetes patula*. Effects of auxin and transformation by *Agrobacterium. Planta Med* 179:43–50.

Dayan, F.E., I.A. Kagan, and A.M. Rimando. 2003. Elucidation of the biosynthetic pathway of the allelochemical sorgoleone using retrobiosynthetic NMR analysis. *J Biol Chem* 278:28607–28611.

Diretto, G., S. Al-Babili, R. Tavazza, V. Papacchioli, P. Beyer, and G. Giuliano. 2007. Metabolic engineering of potato carotenoid content through tuber-specific overexpression of a bacterial mini-pathway. *PLoS One* 2(4):e350.

Diwan, R. and N. Malpathak. 2011. Bioprocess optimization of furanocoumarin elicitation by medium renewal and re-elicitation: A perfusion-based approach. *Appl Biochem Biotechnol* 163:756–764.

Dixon, R.A., L.V. Modolo, and G. Peel. 2011. Production of proanthocyanidins to improve forage quality. Patent No. U.S. 7880059 B2.

Dong, J., G. Wan, and Z. Liang. 2010. Accumulation of salicylic acid-induced phenolic compounds and raised activities of secondary metabolic and antioxidative enzymes in *Salvia miltiorrhiza* cell culture. *J Biotechnol* 148:99–104.

Fabricant, D.S. and N.R. Farnsworth. 2001. The value of plants used in traditional medicine for drug discovery. *Environ Health Perspect* 109:69–75.

Faria, A.P., A. deFátima, V.A. Benedito, and L.V. Modolo. 2011. Plant cell culture and transgenic plants: The goldmines for the production of compounds of pharmacological interest. In: *Bioactive Natural Products: Opportunities & Challenges in Medicinal Chemistry*, ed. G. Brahmachari, pp. 631–654. Hackensack, NJ: World Scientific Publishing.

de Fátima, A. and L.V. Modolo. 2008. What chemical weapons do plants use to fight against pathogens? In: *Advances in Plant Biotechnology*, eds. G.P. Rao, Y. Zhao, V.V. Radchuk, and S.K. Bhatnagar, pp. 179–204. Houston, TX: Studium Press, LLC.

Ferreira, J.F.S., D.L. Luthria, T. Sasaki, and A. Heyerick. 2010. Flavonoids from *Artemisia annua* L. as antioxidants and their potential synergism with artemisinin against malaria and cancer. *Molecules* 15:3135–31370.

van der Fits, L. and J. Memelink. 2000. ORCA3, a jasmonate-responsive transcriptional regulator of plant primary and secondary metabolism. *Science* 289:295–297.

Frankfater, C.R., M.K. Dowd, and B.A. Triplett. 2009. Effect of elicitors on the production of gossypol and methylated gossypol in cotton hairy roots. *Plant Cell Tissue Organ Cult* 98:341–349.

Gangopadhydy, M., D. Sircar, A. Mitra, and S. Bhattacharya. 2008. Hairy root culture of *Plumbago indica* as a potential source for plumbagin. *Biol Plant* 52:533–537.

Ge, X.C. and J.Y. Wu. 2005. Tanshinone production and isoprenoid pathways in *Salvia miltiorrhiza* hairy roots induced by Ag^+ and yeast elicitor. *Plant Sci* 168:487–491.

Georgiev, M.I., K. Alipieva, and I.E. Orhan. 2012. Cholinesterases inhibitory and antioxidant activities of *Harpagophytum procumbens* from in vitro systems. *Phytother Res* 26:313–316.

Georgiev, V.G., J. Weber, E.M. Kneschke, P.N. Denev, T. Bley, and A.I. Pavlov. 2010. Antioxidant activity and phenolic content of betalain extracts from intact plants and hairy root cultures of the red beetroot *Beta vulgaris* cv. detroit dark red. *Plant Foods Hum Nutr* 65:105–111.

Goel, M.K., S. Mehrotra, and A.K. Kukreja. 2011. Elicitor-induced cellular and molecular events are responsible for productivity enhancement in hairy root cultures: An insight study. *Appl Biochem Biotechnol* 165:1342–1355.

Gou, J.-Y., F.F. Felippes, C.-J. Liu, D. Weigel, and J.-W. Wang. 2011. Negative regulation of anthocyanin biosynthesis in Arabidopsis by a *miR156*-targeted *SPL* transcription factor. *Plant Cell* 23:1512–1522.

Grabkowska, R., A. Krolicka, W. Mielicki, M. Wielanek, and H. Wysokinska. 2010. Genetic transformation of *Harpagophytum procumbens* by *Agrobacterium rhizogenes*: Iridoid and phenylethanoid glycoside accumulation in hairy root cultures. *Acta Physiol Plant* 32:665–673.

Graham, I.A., K. Besser, S. Blumer et al. 2010. The genetic map of *Artemisia annua* L. identifies loci affecting yield of the antimalarial drug artemisinin. *Science* 327:328–331.

Gränicher, F., P. Christen, and I. Kapetandis. 1992. High yield production of valepotriates by hairy root cultures of *Valeriana officinalis* L. var. *sambucifolia* Mikan. *Plant Cell Rep* 11:339–342.

Gregianini, T.S., D.D. Porto, N.C. Nascimento, J.P. Fett, A.T. Henriques, and A.G. Fett-Neto. 2004. Environmental and ontogenetic control of accumulation of brachycerine, a bioactive indole alkaloid from *Psychotria brachyceras*. *J Chem Ecol* 30:2023–2036.

Guo, Y.T. and J.W. Wang. 2008. Stimulation of taxane production in suspension cultures of *Taxus yunnanensis* by oligogalacturonides. *Afr J Biotechnol* 7:1924–1926.

Gyurkovska, V., K. Alipieva, A. Maciuk et al. 2011. Anti-inflammatory activity of Devil's claw in vitro systems and their active constituents. *Food Chem* 125:171–178.

Ha, M., J. Lu, L. Tian et al. 2009. Small RNAs serve as a genetic buffer against genomic shock in *Arabidopsis* interspecific hybrids and allopolyploids. *Proc Natl Acad Sci USA* 106:17835–17840.

Hamaguchi, T., K. Ono, A. Murase, and M. Yamada. 2009. Phenolic compounds prevent Alzheimer's pathology through different effects on the amyloid-β aggregation pathway. *Am J Pathol* 175:2557–2565.

Hasanloo, T., R. Sepehrifar, H. Rahnama, and M.R. Shams. 2009. Evaluation of the yeast-extract signaling pathway leading to silymarin biosynthesis in milk thistle hairy root culture. *World J Microbiol Biotechnol* 25:1901–1909.

Hastings, G.E. and P.G. Wolf. 1992. The therapeutic use of albumin. *Arch Fam Med* 1:281–287.

He, Y., T. Ning, T. Xie et al. 2011. Large-scale production of functional human serum albumin from transgenic rice seeds. *Proc Natl Acad Sci USA* 108:19078–19083.

Hegazy, M.-E.F., A.E.-H.H. Mohamed, A.M. El-Halawany, P.C. Djemgou, A.A. Shahat, and P.W. Paré. 2011. Estrogenic activity of chemical constituents from *Tephrosia candida*. *J Nat Prod* 74:937–942.

Henriques, A.T., S.O. Lopes, J.T. Paranhos et al. 2004. *N*-β-D-Glucopyranosyl vincosamide, a light regulated indole alkaloid from the shoots of *Psychotria leiocarpa*. *Phytochemistry* 65:449–454.

Higa, A., E. Miyamoto, L. Rahman, and Y. Kitamura. 2008. Root tip-dependent, active riboflavin secretion by *Hyoscyamus albus* hairy roots under iron deficiency. *Plant Physiol Biochem* 46:452–460.

Huang, L.-F., Y.-K. Liu, C.-A. Lu, S.-L. Hsieh, and S.-M. Yu. 2005. Production of human serum albumin by sugar starvation induced promoter and rice cell culture. *Transgenic Res* 14:569–581.

Jacob, A. and N. Malpathak. 2004. Green hairy root cultures of *Solanum khasianum* Clarke—A new route to in vitro solasodine production. *Curr Sci* 87:1442–1447.

Kaiser, J. 2008. Is the drought over for pharming? *Science* 320:473–475.

Kaimoyo, E., M.A. Farag, L.W. Sumner, C. Wasmann, J.L. Cuello, and H. VanEtten. 2008. Sub-lethal levels of electric current elicit the biosynthesis of plant secondary metabolites. *Biotechnol Prog* 24:377–384.

Kamonwannasit, S., W. Phrompittayarat, K. Ingkaninan, H. Tanaka, and W. Putalun. 2008. Improvement of pseudojujubogenin glycosides production from regenerated *Bacopa monnieri* (L.) Wettst. and enhanced yield by elicitors. *Z Naturforsch C* 63c:879–883.

Karg, S.R. and P.T. Kallio. 2009. The production of biopharmaceuticals in plant systems. *Biotechnol Adv* 27:879–894.

Khalili, M., T. Hasanloo, and S.K.K. Tabar. 2010. Ag^+ enhanced silymarin production in hairy root cultures of *Silybum marianum* (L.) Gaertn. *Plant Omics J* 3:109–114.

Kim, O.T., K.H. Bang, Y.C. Kim, D.Y. Hyun, M.Y. Kim, and S.W. Cha. 2009. Upregulation of ginsenoside and gene expression related to triterpene biosynthesis in ginseng hairy root cultures elicited by methyl jasmonate. *Plant Cell Tissue Organ Cult* 98:25–33.

Kim, O.T., K.H. Bang, Y.S. Shin et al. 2007. Enhanced production of asiaticoside from hairy root cultures of *Centella asiatica* (L.) urban elicited by methyl jasmonate. *Plant Cell Rep* 26:1941–1949.

Kim, S.-R., J.-S. Sim, H. Ajjappala, Y.-H. Kim, and B.-S. Hahn. 2012. Expression and large-scale production of the biochemically active human tissue-plasminogen activator in hairy roots of Oriental melon (*Cucumis melo*). *J Biosci Bioeng* 113:106–111.

Kindermans, J.-M., J. Pilloy, P. Olliaro, and M. Gomes. 2007. Ensuring sustained ACT production and reliable artemisinin supply. *Malaria J* 6:125.

Komaraiah, P., G.V. Reddy, P.S. Reddy, A.S. Raghavendra, S.V. Ramakrishna, and P. Reddanna. 2003. Enhanced production of antimicrobial sesquiterpenes and lipoxygenase metabolites in elicitor-treated hairy root cultures of *Solanum tuberosum. Biotechnol Lett* 25:593–597.

Kuźma, L., E. Bruchajzer, and H. Wysokinska. 2009. Methyl jasmonate effect on diterpenoid accumulation in *Salvia sclarea* hairy root culture in shake flasks and sprinkle bioreactor. *Enzyme Microb Technol* 44:406–410.

Lee, S.Y., H. Xu, Y.K. Kim, and S.U. Park. 2008. Rosmarinic acid production in hairy root cultures of *Agastache rugosa* Kuntze. *World J Microbiol Biotechnol* 24:969–972.

Libault, M., T. Joshi, V.A. Benedito, D. Xu, M.K. Udvardi, and G. Stacey. 2009. Legume transcription factor genes: What makes legumes so special? *Plant Physiol* 151:991–1001.

Limberger, R.P., A.M. Aleixo, A.G. Fett-Neto, and A.T. Henriques. 2007. Bioconversion of (+)- and (−)-alpha-pinene to (+)- and (−)-verbenone by plant cell cultures of *Psychotria brachyceras* and *Rauvolfia sellowii. Electr J Biotechnol* 10:500–507.

Lin, X., J. Zhang, Y. Li et al. 2011. Functional genomics of a living fossil tree, Ginkgo, based on next-generation sequencing technology. *Physiol Plant* 143:207–218.

Liu, C., M.J. Towler, G. Medrano, C.L. Cramer, and P.J. Weathers. 2009. Production of mouse interleukin-12 is greater in tobacco hairy roots grown in a mist reactor than in an airlift reactor. *Biotechnol Bioeng* 102:1074–1086.

Llorente, B., V. Rodríguez, G.D. Alonso, H.N. Torres, M.M. Flawiá, and F.F. Bravo-Almonacid. 2010. Improvement of aroma in transgenic potato as a consequence of impairing tuber browning. *PLoS One* 5:e14030.

Malarz, J., A. Stojakowska, and W. Kisiel. 2007. Effect of methyl jasmonate and salicylic acid on sesquiterpene lactone accumulation in hairy roots of *Cichorium intybus. Acta Physiol Plant* 29:127–132.

Malitsky, S., E. Blum, H. Less et al. 2008. The transcript and metabolite networks affected by the two clades of *Arabidopsis* glucosinolate biosynthesis regulators. *Plant Physiol* 148:2021–2049.

Mannan, A., N. Shaheen, W. Arshad, R.A. Qureshi, M. Zia, and B. Mirza. 2008. Hairy roots induction and artemisinin analysis in *Artemisia dubia* and *Artemisia indica. Afr J Biotechnol* 7:3288–3292.

Mathur, A., A. Gangwar, A.K. Mathur, P. Verma, G.C. Uniyal, and R.K. Lal. 2010. Growth kinetics and ginsenosides production in transformed hairy roots of American ginseng—*Panax quinquefolium* L. *Biotechnol Lett* 32:457–461.

Maurmann, N., C.M.B. de Carvalho, A.L. Silva, A.G. Fett-Neto, G.L. Von Poser, and S.B. Rech. 2006. Valepotriates accumulation in callus, suspended cells and untransformed root cultures of *Valeriana glechomifolia. In vitro Cell Dev Biol Plant* 42:50–53.

Medina-Bolivar, F., J. Condori, A.M. Rimando et al. 2007. Production and secretion of resveratrol in hairy root cultures of peanut. *Phytochemistry* 68:1992–2003.

Metzker, M.L. 2010. Sequencing technologies—The next generation. *Nat Rev Genet* 11:31–46.

Miller, J.R., S. Koren, and G. Sutton. 2010. Assembly algorithms for next-generation sequencing data. *Genomics* 95:315–327.

Modolo, L.V., A. Reichert, and R.A. Dixon. 2009. Introduction to the different classes of biosynthetic enzymes. In: *Plant-Derived Natural Products—Synthesis, Function and Application*, eds. A. Osbourn and V. Lanzotti, pp. 143–163. New York: Springer.

Mukundan, U. and M.A. Hjortso. 1990. Thiophene content in normal and transformed root cultures of *Tagetes erecta*: A comparison with thiophene content in roots of intact plants. *J Exp Bot* 41:1497–1501.

Murthy, H.N., C. Dijkstra, P. Anthony, D.A. White, M.R. Davey, J.B. Power, E.J. Hahn, and K.Y. Paek. 2008a. Establishment of *Withania somnifera* hairy root cultures for the production of withanolide. *J Int Plant Biol* 50:975–981.

Murthy, H.N., E.J. Hahn, and K.Y. Paek. 2008b. Adventitious roots and secondary metabolism. *Chin J Biotechnol* 24:711–716.

Namdeo, A., S. Patil, and D.P. Fulzele. 2002. Influence of fungal elicitors on production of ajmalicine by cell cultures of *Catharanthus roseus. Biotechnol Prog* 18:159–162.

Ng, D.W.-K., C. Zhang, M. Miller et al. 2011. *cis*- and *trans*-Regulation of *miR163* and target genes confers natural variation of secondary metabolites in two *Arabidopsis* species and their allopolyploids. *Plant Cell* 23:1729–1740.

Ottaggio, L., F. Bestoso, A. Armirotti et al. 2008. Taxanes from shells and leaves of *Corylus avellana. J Nat Prod* 71:58–60.

Paine, J.A., C.A. Shipton, S. Chaggar et al. 2005. Improving the nutritional value of Golden Rice through increased pro-vitamin A content. *Nat Biotechnol* 23:482–487.

Palazon, J., A. Mallol, R. Eibl, C. Lettenbauer, R.M. Cusido, and M.T. Pinol. 2003. Growth and ginsenside production in hairy root cultures of *Panax ginseng* using a novel bioreactor. *Planta Med* 69:344–349.

Paranhos, J.T., V. Fragoso, A.T. Henriques, A.G. Ferreira, and A.G. Fett-Neto. 2005. Regeneration of *Psychotria umbellata* and production of the analgesic indole alkaloid umbellatine. *Tree Physiol* 25:251–255.

Park, S.U., Y.K. Kim, H.G. Jang, J.N. Kim, and H.W. Ryu. 2008. Auxin treatment improves indigo biosynthesis in hairy root cultures of *Polygonum tinctorium*. *Chem Nat Compd* 44:213–214.

Peel, G.J., L.V. Modolo, Y. Pang, and R.A. Dixon. 2009. The LAP1 MYB transcription factor orchestrates anthocyanidin biosynthesis and glycosylation in Medicago. *Plant J* 59:136–149.

Pistelli, L., A. Giovannini, B. Ruffoni, A. Bertoli, and L. Pistelli. 2011. Hairy root cultures for secondary metabolites production. *Adv Exp Med Biol* 698:167–184.

Pitta-Alvarez, S.I., T.C. Spollansky, and A.M. Guilietti. 2000. The influence of different biotic and abiotic elicitors on the production and profile of tropane alkaloids in hairy root cultures of *Brugmansia candida*. *Enzyme Microb Technol* 26:252–258.

Potrykus, I. 2010. Regulation must be revolutionized. *Nature* 466:561.

Praveen, N. and H.N. Murthy. 2012. Synthesis of withanolide A depends on carbon source and medium pH in hairy root cultures of *Withania somnifera*. *Ind Crops Prod* 35:241–243.

Pu, G.-B., D.-M. Ma, J.-L. Chen et al. 2009. Salicylic acid activates artemisinin biosynthesis in *Artemisia annua* L. *Plant Cell Rep* 28:1127–1135.

Putalun, W., W. Luealon, W. De-Eknamkul, H. Tanaka, and Y. Shoyama. 2007. Improvement of artemisinin production by chitosan in hairy root cultures of *Artemisia annua* L. *Biotechnol Lett* 29:1143–1146.

Rachmawati, D., T. Mori, T. Hosaka, F. Takaiwa, E. Inoue, and H. Anzai. 2005. Production and characterization of recombinant human lactoferrin in transgenic Javanica rice. *Breeding Sci* 55:213–222.

Ramachandra-Rao, S. and G.A. Ravishankar. 2002. Plant cell cultures: Chemical factories of secondary metabolites. *Biotechnol Adv* 20:101–153.

Rea, G., A. Antonacci, M. Lambreva et al. 2011. Integrated plant biotechnologies applied to safer and healthier food production: The Nutra-Snack manufacturing chain. *Trends Food Sci Technol* 22:353–366.

Rhee, H.S., H.W. Cho, S.Y. Son, S.Y.H. Yoon, and J.M. Park. 2010. Enhanced accumulation of decursin and decursinol angelate in root cultures and intact roots of *Angelica gigas* Nakai following elicitation. *Plant Cell Tissue Organ Cult* 101:295–302.

Robert-Seilaniantz, A., D. MacLean, Y. Jikumaru et al. 2011. The microRNA *miR393* re-directs secondary metabolite biosynthesis away from camalexin and towards glucosinolates. *Plant J* 67:218–231.

Romero, F.R., K. Delate, G.A. Kraus, A.K. Solco, P.A. Murphy, and D.J. Hannapel. 2009. Alkamide production from hairy root cultures of *Echinacea*. *In vitro Cell Dev Biol Plant* 45:599–609.

Santamaria, A.R., D. Antonacci, G. Caruso et al. 2010. Stilbene production in cell cultures of *Vitis vinifera* L. cvs Red Globe and Michele Palieri elicited by methyl jasmonate. *Nat Prod Res* 24:1488–1498.

Santamaria, A.R., N. Mulinacci, A. Valletta, M. Innocenti, and G. Pasqua. 2011. Effects of elicitors on the production of resveratrol and viniferins in cell cultures of *Vitis vinifera* L. cv Italia. *J Agric Food Chem* 59:9094–9101.

Sato, K., T. Yamazaki, E. Okuyama, K. Yoshihira, and K. Shimomura. 1991. Anthraquinone production by transformed root cultures of *Rubia ticntorum*: Influence of phytohormone and sucrose concentration. *Phytochemistry* 30:2977–2980.

Sharp, J.M. and P.M. Doran. 2001. Strategies for enhancing monoclonal antibody accumulation in plant cell and organ cultures. *Biotechnol Prog* 17:979–992.

Shimomura, K., M. Sauerwein, and K. Ishimaru. 1991. Tropane alkaloids in the adventitious and hairy root cultures of Solanaceous plants. *Phytochemistry* 30:2275–2278.

Shinde, A.N., N. Malpathak, and D.P. Fulzele. 2009. Enhanced production of phytoestrogenic isoflavones from hairy root cultures of *Psoralea corylifolia* L. using elicitation and precursor feeding. *Biotechnol Bioprocess Eng* 14:288–294.

Shoji, T., M. Kajikawa, and T. Hashimoto. 2010. Clustered transcription factor genes regulate nicotine biosynthesis in tobacco. *Plant Cell* 22:3390–3409.

Siah, C.L. and P.M. Doran. 1991. Enhanced codeine and morphine production in suspended *Papaver somniferum* cultures after removal of exogenous hormones. *Plant Cell Rep* 10:349–353.

Sidwa-Gorycka, M., A. Krolicka, A. Orlita et al. 2009. Genetic transformation of *Ruta graveolens* L. by *Agrobacterium rhizogenes*: Hairy root cultures a promising approach for production of coumarins and furanocoumarins. *Plant Cell Tissue Organ Cult* 97:59–69.

Sirikantaramas, S., H. Sudo, T. Asano, M. Yamazaki, and K. Saito. 2007. Transport of camptothecin in hairy roots of *Ophiorrhiza pumila*. *Phytochemistry* 68:2881–2886.

Souret, F.F., Y. Kim, B.E. Wyslouzil, K.K. Wobbe, and P.J. Weathers. 2003. Scale-up of *Artemisia annua* L. hairy root cultures produces complex patterns of terpenoid gene expression. *Biotechnol Bioeng* 83:653–667.

Staniszewska, I., A. Krolicka, E. Malinski, E. Lojkowska, and J. Szafranek. 2003. Elicitation of secondary metabolites in in vitro cultures of *Ammi majus* L. *Enzyme Microb Technol* 33:565–568.

Sun, C., Y. Li, Q. Wu et al. 2010. *De novo* sequencing and analysis of the American ginseng root transcriptome using a GS FLX Titanium platform to discover putative genes involved in ginsenoside biosynthesis. *BMC Genomics* 11:262.

Sunil-Kumar, G.B., T.R. Ganapathi, L. Srinivas, C.J. Revathi, and V.A. Bapat. 2006. Expression of hepatitis B surface antigen in potato hairy roots. *Plant Sci* 170:918–925.

Suttipanta, N., S. Pattanaik, M. Kulshrestha, B. Patra, S.K. Singh, and L. Yuan. 2011. The transcription factor CrWRKY1 positively regulates the terpenoid indole alkaloid biosynthesis in *Catharanthus roseus*. *Plant Physiol* 157:2081–2093.

Syklowska-Baranek, K., A. Pietrosiuk, A. Kokoszka, and M. Furmanowa. 2009. Enhancement of taxane production in hairy root culture of *Taxus* x *media* var. Hicksii. *J Plant Physiol* 166:1950–1954.

Thimmaraju, R., N. Bhagyalakshmi, M.S. Narayan, and G.A. Ravishankar. 2003. Food-grade chemical and biological agents permeabilize red beet hairy roots, assisting the release of betalaines. *Biotechnol Prog* 19:1274–1282.

Tiwari, R.K., M. Trivedi, Z.C. Guang, G.Q. Guo, and G.C. Zheng. 2008. *Agrobacterium rhizogenes* mediated transformation of *Scutellaria baicalensis* and production of flavonoids in hairy roots. *Biol Plant* 52:26–35.

Todd, A.T., E. Liu, S.L. Polvi, R.T. Pammett, and J.E. Page. 2010. A functional genomics screen identifies diverse transcription factors that regulate alkaloid biosynthesis in *Nicotiana benthamiana*. *Plant J* 62:589–600.

Triplett, B.A., S.C. Moss, J.M. Bland, and M.K. Dowd. 2008. Induction of hairy root cultures from *Gossypium hirsutum* and *Gossypium barbadense* to produce gossypol and related compounds. *In vitro Cell Dev Biol Plant* 44:508–517.

Udomsuk, L., K. Jarukamjorn, H. Tanaka, and W. Putalun. 2011. Improved isoflavonoid production in *Pueraria candollei* hairy root cultures using elicitation. *Biotechnol Lett* 33:369–374.

Udvardi, M.K., K. Kakar, M. Wandrey et al. 2007. Legume transcription factors: Global regulators of plant development and response to the environment. *Plant Physiol* 144:538–549.

Verma, P.C., L. Rahman, A.S. Negi, D.C. Jain, S.P.S. Khanuja, and S. Banerjee. 2007. *Agrobacterium rhizogenes*-mediated transformation of *Picrorhiza kurroa* Royle ex Benth: Establishment and selection of superior hairy root clone. *Plant Biotechnol Rep* 1:169–174.

Wadley, L., C. Sievers, M. Bamford, P. Goldberg, F. Berna, and C. Miller. 2011. Middle Stone Age bedding construction and settlement patterns at Sibudu, South Africa. *Science* 334:1388–1391.

Wang, H., J. Wu, S. Sun et al. 2011. Glucosinolate biosynthetic genes in *Brassica rapa*. *Gene* 487:135–142.

Wang, J.W., L.P. Zheng, B. Zhang, and T. Zou. 2009. Stimulation of artemisinin synthesis by combined cerebroside and nitric oxide elicitation in *Artemisia annua* hairy roots. *Appl Microbiol Biotechnol* 85:285–292.

Weathers, P. J., M.J. Towler, and J. Xu. 2010. Bench to batch: Advances in plant cell culture for producing useful products. *Appl Microbiol Biotechnol* 85:1339–1351.

Welsch, R., J. Arango, C. Bär et al. 2010. Provitamin A accumulation in cassava (*Manihot esculenta*) roots driven by a single nucleotide polymorphism in a phytoene synthase gene. *Plant Cell* 22:3348–3356.

Wetzstein, H., J. Janick, and J.F.S. Ferreira. 2009. Flower morphology and development in *Artemisia annua*, a medicinal plant used as a treatment against malaria. *HortScience* 44:1026.

Wielanek, M., A. Królicka, K. Bergier, E. Gajewska, and M. Skłodowska. 2009. Transformation of *Nasturtium officinale*, *Barbarea verna* and *Arabis caucasica* for hairy roots and glucosinolate-myrosinase system production. *Biotechnol Lett* 31:917–921.

Woods, R.R., B.C. Geyer, and T.S. Mor. 2008. Hairy-root organ cultures for the production of human acetylcholinesterase. *BMC Biotechnol* 8:95–102.

Wu, J.-Y., J. Ng, M. Shi, and S.-J. Wu. 2007. Enhanced secondary metabolite (tanshinone) production of *Salvia miltiorrhiza* hairy roots in a novel root-bacteria coculture process. *Appl Microbiol Biotechnol* 77:543–550.

Wu, J.-Y. and M. Shi. 2008. Ultrahigh diterpenoid tanshinone production through repeated osmotic stress and elicitor stimulation in fedbatch culture of *Salvia miltiorrhiza* hairy roots. *Appl Microbiol Biotechnol* 78:441–448.

Xiao, Y., S. Gao, P. Di, J. Chen, W. Chen, and L. Zhang. 2009. Methyl jasmonate dramatically enhances the accumulation of phenolic acids in *Salvia miltiorrhiza* hairy root cultures. *Physiol Plant* 137:1–9.

Yamamoto, H., P. Zhao, K. Yazaki, and K. Inoue. 2002. Regulation of lithospermic acid B and shikonin production in *Lithospermum erythrorhizon* cell suspension cultures. *Chem Pharm Bull* 50:1086–1090.

Yan, Q., Z. Hu, R.X. Tan, and J. Wu. 2005. Efficient production and recovery of diterpenoid tanshinones in *Salvia miltiorrhiza* hairy root cultures with in situ adsorption, elicitation and semi-continuous operation. *J Biotechnol* 119:416–424.

Ya-ut, P., P. Chareonsap, and S. Sukrong. 2011. Micropropagation and hairy root culture of *Ophiorrhiza alata* Craib for camptothecin production. *Biotechnol Lett* 33:2519–2526.

Ye, V.M. and S.K. Bhatia. 2012. Metabolic engineering for the production of clinically important molecules: Omega-3 fatty acids, artemisinin, and taxol. *Biotechnol J* 7:20–33.

Ye, H., L.-L. Huang, S.-D. Chen, and J.-J. Zhong. 2004. Pulsed electric field stimulates plant secondary metabolism in suspension cultures of *Taxus chinensis*. *Biotechnol Bioeng* 88:788–795.

Yue, X., W. Zhang, and M. Deng. 2011. Hyper-production of ^{13}C-labeled *trans*-resveratrol in *Vitis vinifera* suspension cell culture by elicitation and in situ adsorption. *Biochem Eng J* 53:292–296.

Zhang, H.-B., M.T. Bokowiec, P.J. Rushton, S.-C. Han, and M.P. Timko. 2012. Tobacco transcription factors $NtMYC_2a$ and $NtMYC_2b$ form nuclear complexes with the $NtJAZ_1$ repressor and regulate multiple jasmonate-inducible steps in nicotine biosynthesis. *Mol Plant* 5:73–84.

Zhang, C. and P.S. Fevereiro. 2007. The effect of heat shock on paclitaxel production in *Taxus yunnanensis* cell suspension cultures: Role of abscisic acid pretreatment. *Biotechnol Bioeng* 96:506–514.

Zhang, H.-C., J.-M. Liu, H.-Y. Lu, and S.-L. Gao. 2009. Enhanced flavonoid production in hairy root cultures of *Glycyrrhiza uralensis* Fisch by combining the over-expression of chalcone isomerase gene with the elicitation treatment. *Plant Cell Rep* 28:1205–1213.

Zhang, Y., Y.-P. Yan, and Z.-Z. Wang. 2010. The Arabidopsis PAP1 transcription factor plays an important role in the enrichment of phenolic acids in *Salvia miltiorrhiza*. *J Agric Food Chem* 58:12168–12175.

Zhao, J.-L., L.-G. Zhou, and J.-Y. Wu. 2010a. Effects of biotic and abiotic elicitors on cell growth and tanshinone accumulation in *Salvia miltiorrhiza* cell cultures. *Appl Microbiol Biotechnol* 87:137–144.

Zhao, J.-L., L.-G. Zhou, and J.-Y. Wu. 2010b. Promotion of *Salvia miltiorrhiza* hairy root growth and tanshinone production by polysaccharide protein fractions of plant growth-promoting rhizobacterium *Bacillus cereus*. *Process Biochem* 45:1517–1522.

Zhou, L.G., X.D. Cao, R.F. Zhang, Y.L. Peng, S.J. Zhao, and J.Y. Wu. 2007. Stimulation of saponin production in *Panax ginseng* hairy roots by two oligosaccharides from *Paris polyphylla* var. *yunnanensis*. *Biotechnol Lett* 29:631–634.

Zhou, M.L., X.M. Zhu, J.R. Shao, Y.M. Wu, and Y.X. Tang. 2010. Transcriptional response of the catharanthine biosynthesis pathway to methyl jasmonate/nitric oxide elicitation in *Catharanthus roseus* hairy root culture. *Appl Microbiol Biotechnol* 88:737–750.

Index

A

B

C

D

E

F

G

N

O

P

T

U

V

W

X

Y

Z